**DAVIS' HANDBOOK OF
APPLIED HYDRAULICS**

Other McGraw-Hill Books of Interest

American Water Works Association • WATER QUALITY AND TREATMENT
American Water Works Association • WATER TREATMENT PLANT DESIGN
Brater & King • HANDBOOK OF HYDRAULICS
Brock • FIELD INSPECTION HANDBOOK
Corbitt • STANDARD HANDBOOK OF ENVIRONMENTAL ENGINEERING
Gaylord & Gaylord • STRUCTURAL ENGINEERING HANDBOOK
Gulliver & Arndt • HYDROPOWER ENGINEERING HANDBOOK
Herbich • HANDBOOK OF DREDGING ENGINEERING
Maidment • HANDBOOK OF HYDROLOGY
Merritt • STANDARD HANDBOOK FOR CIVIL ENGINEERS
Parmley • FIELD ENGINEER'S MANUAL
Parmley • HYDRAULICS FIELD MANUAL

DAVIS' HANDBOOK OF APPLIED HYDRAULICS

Vincent J. Zipparro, *Editor in Chief*
Chief Engineer, Harza Engineering Company, Chicago

Hans Hasen, *Coeditor*
Consultant, Harza Engineering Company, Chicago

FOURTH EDITION

McGRAW-HILL, INC.
New York St. Louis San Francisco Auckland Bogotá
Caracas Lisbon London Madrid Mexico Milan
Montreal New Delhi Paris San Juan São Paulo
Singapore Sydney Tokyo Toronto

Library of Congress Cataloging-in-Publication Data

Davis' handbook of applied hydraulics / Vincent J. Zipparro, editor-in-chief,
  Hans Hasen, co-editor.
       p.   cm.
    Rev. ed. of: Handbook of applied hydraulics / Calvin Victor Davis,
  editor-in-chief.
    Includes index and bibliographical references
    ISBN 0-07-073002-4
    1. Hydraulic engineering—Handbooks, manuals, etc. I. Zipparro,
  Vincent J. II. Hasen, Hans. III. Davis, Calvin Victor, date.
  Handbook of applied hydraulics.
  TC145.D3   1993
  627—dc20   92-28842
                                                                  CIP

Copyright © 1952, 1942 by McGraw-Hill, Inc. All rights reserved.
Copyright renewed 1970 by McGraw-Hill, Inc. All rights reserved.

Copyright © 1993, 1969 by McGraw-Hill, Inc. All rights reserved. Printed
in the United States of America. Except as permitted under the United
States Copyright Act of 1976, no part of this publication may be reproduced
or distributed in any form or by any means, or stored in a data base or
retrieval system, without the prior written permission of the publisher.

1 2 3 4 5 6 7 8 9 0   DOC/DOC   9 8 7 6 5 4 3 2

ISBN 0-07-073002-4

*The sponsoring editor for this book was Harold B. Crawford, the editing
supervisor was Peggy Lamb, and the production supervisor was
Pamela A. Pelton. It was set in Times Roman by McGraw-Hill's
Professional Book Group composition unit.*

Printed and bound by R. R. Donnelley & Sons Company.

This book is printed on acid-free paper.

Information contained in this work has been obtained by
McGraw-Hill, Inc., from sources believed to be reliable. However, neither McGraw-Hill nor its authors guarantee the accuracy or completeness of any information published herein and
neither McGraw-Hill nor its authors shall be responsible for any
errors, omissions, or damages arising out of use of this information. This work is published with the understanding that
McGraw-Hill and its authors are supplying information but are
not attempting to render engineering or other professional services. If such services are required, the assistance of an appropriate professional should be sought.

*Dedicated to Leroy F. Harza, Founder
of Harza Engineering Company,
Calvin V. Davis, Chairman, and
Kenneth E. Sorensen, Chairman Emeritus.*

CONTENTS

About the Editors xi
Contributors xiii
Preface xv
Acknowledgments xvii

Section 1. Hydrology 1.1
Precipitation, Evaporation and Evapotranspiration, Storm Rainfall, Runoff, Floods, Computer Programs

Section 2. Basic Hydraulics 2.1
Closed Conduits, Evaluation of Friction Coefficients, Form Losses, Orifices, Open Channels

Section 3. Hydraulic Models 3.1
Definition of a Model, Principles of Hydraulic Modeling, Models of River Channels

Section 4. Reservoir Hydraulics 4.1
Introduction, Sediment Storage Requirements, Analysis of Water Availability, Reservoir Flood Routing, Reservoir Selective Withdrawal

Section 5. Natural Channels 5.1
Hydraulics of Natural Streams, Determination of Discharge, Backwater Curves

Section 6. Regime Canals 6.1
Channels of Alluvium, Theory of Regime Channels, Effects of Sediment and Seepage, Design of Regime Canals

Section 7. Canals and Conduits 7.1
Hydraulic Factors, Conveyance Losses, Design of Canals, Hydraulic Computations, the Lining of Earth Canals

Section 8. River Diversion 8.1
Site Limitations, Diversion in Narrow and Deep Channels, Diversion in Wide Stream Channels, Diversion Through Concrete Dams, Diversion Over and Around Embankment Dams, Diversion Discharge Capacities, Tunnels, Cofferdams

Section 9. Concrete Dams 9.1
General Features, Design Considerations, Uplift Pressure, Seismic Loading

Section 10. Arch Dams 10.1
Arch Dam Types, General Theory of Arch Dams, Loads and Arch Dams, Stress Distribution in Arch Dams, Design of Arch Dams, Analysis of Preliminary Plans, Finite Element Method, Instrumentation of Arch Dams, Model Investigation, Examples of Arch Dams

Section 11. Prestressing/Post-Tensioning and Rehabilitation 11.1
Introduction, Cable and its Anchorage, Examples of Prestresses/Post-Tensioned Dams, Rehabilitation of Dams

Section 12. Barrages and Dams on Permeable Foundations 12.1
Barrage and the River, Barrage Superstructure Design, Substructure Design, Barrage Appurtement Structures, Dams on Permeable Foundations

Section 13. Embankment Dams 13.1
Introduction, Geology, Embankment Types, Foundation Treatment, Subsurface Investigations, Laboratory Tests, Seepage Analysis and Control, Stability Analysis, Earthquake Considerations, Settlement Analysis, Slope Protection and Freeboard, Construction Quality Control, Monitoring and Performance Evaluation

Section 14. Concrete Face Rockfill Dams 14.1
Concrete Face Rockfill Dam (CFRD), Features of the CFRD, Compacted Rockfill for the CFRD, Design, Construction, Performance, Typical Designs

Section 15. Roller Compacted Concrete Dams 15.1
Types of RCC Dams, Physical Properties of RCC, Design of RCC Dams

Section 16. Spillways and Streambed Protection Works 16.1
Discharge Capacity, Spillway Crests, Overfall Spillways, Gates and Orifice Spillways, Chute or Trough Spillways, Design Considering Cavitation and Aeration, Incipient Cavitation and Damage Experiences, Side Channel Spillways, Morning-Glory Shaft and Tunnel Spillways, Siphon Spillways, Scour Protection Below Overfall Dams

Section 17. Gates and Valves 17.1
General, Spillway Gates, High Head Gates, Valves, Hydraulic Design Factors, Equipment Design Considerations

CONTENTS

Section 18. Environmental Aspects and Fish Facilities 18.1
Introduction, General Effects of Impoundment, Biological Criteria, Fish Passage, Fish Ladders, Fish Exclusion Devices, Aeration of Downstream Releases

Section 19. Hydroelectric Plants 19.1
Power from Flowing Water, Water Conductors, Powerhouse Structures

Section 20. Pumped Storage 20.1
Basic Concepts of Pumped Storage, Elements of Pumped Storage, Rating, Performance and Operation, Costs and Economics, Developing Pumped Storage

Section 21. Hydraulic Machinery 21.1
Turbines, Reversible Pump-Turbines, Turbine Speed Governors

Section 22. Hydraulic Transients 22.1
General, Definitions, Basic Waterhammer Equation, Wavespeed in Tunnels and Conduits, Basic Differential Equations for Transient Flow, Characteristic Method of Analysis, Finite Difference Characteristics Equations, Development of Boundary Conditions for Method of Characteristics

Section 23. Navigation Locks 23.1
General, Lock Layouts, Lock Hydraulics, Equipment

Section 24. Irrigation 24.1
Land Classification, Crop Evapotranspiration, Farm-Irrigation Requirements, Conveyance Losses and Waste, Reuse of Drainage Water, Results from Irrigation

Section 25. Drainage 25.1
Drainage Surveys and Investigations, Sources of Water, Soils, Salinity, Water Tables, Surface Drains, Hydraulic Design Drainage Structures, Subsurface Drains, Design Criteria, Drain Size, Materials and Installation, Pumping for Drainage

Section 26. Irrigation Structures 26.1
Diversion Weirs, River Intakes, Distribution System, Canals, Regulating Structures, Protective Structures, Delivery Structures, Miscellaneous Structures, Pumping Installations

Section 27. Water Distribution and Treatment 27.1
Potable Water Requirements, Hydraulics of Water Treatment Systems, Hydraulics of Water Distribution Systems, Water Distribution Pumping Systems, Water Distribution Storage Facilities

Section 28. Wastewater Conveyance and Treatment 28.1
Wastewater Collection/Conveyance, Sewage Quantities, Hydraulics of Sewers, Sewer System Design, Sewage Pumping Stations, Sewage/Wastewater Treatment

Index (follows Sec. 28) I.1

ABOUT THE EDITORS

Vincent J. Zipparro, editor in chief, is vice president and chief engineer of Harza Engineering Company in Chicago, Illinois. He has overall responsibility, under direction of the president, for all technical and administrative activities related to the engineering functions of the company. Mr. Zipparro has served as project manager on more than 10 major Harza projects, and is a Fellow of the American Society of Civil Engineers.

Hans Hasen, coeditor, was a vice president and manager at Harza Engineering Company, where he has worked for more than forty years. Since 1979, he has been manager of the Guri Final Stage Project in Venezuela, and previously served as head of the civil design department. Mr. Hasen is a Fellow of the American Society of Civil Engineers.

CONTRIBUTORS

George C. Antonopoulos *Associate, Harza Engineering Company, Chicago* (SEC. 19)
Rimas J. Banys *Vice President, Harza Engineering Company, Chicago* (SEC. 18)
James E. Borg *Senior Engineer, Harza Engineering Company, Chicago* (SECS. 3, 22)
Henry H. Chen *Vice President, Harza Engineering Company, Chicago* (SEC. 20)
J. Barry Cooke *Consulting Engineer, San Rafael, California* (SEC. 14)
Hans Hasen *Consulting Engineer, Harza Engineering Company, Chicago* (SEC. 19)
Nicholas M. Hernandez *Vice President, Harza Engineering Company, Chicago* (SEC. 26)
Gregory A. Hillebrener *Senior Engineer, Harza Engineering Company, Chicago* (SEC. 15)
Khalid Jawed *Associate, Harza Engineering Company, Chicago* (SEC. 1)
Thomas J. Johnson *Senior Engineer, Harza Environmental Services, Chicago* (SEC. 27)
David E. Kleiner *Vice President, Harza Engineering Company, Chicago* (SEC. 13)
Casey M. Koniarski *Senior Engineer, Harza Engineering Company, Chicago* (SEC. 11)
Istvan T. Laczo *Vice President, Harza Engineering Company, Chicago* (SEC. 23)
James E. Lindell *Senior Associate, Harza Engineering Company, Chicago* (SEC. 1)
R. A. Abdel-Malek *Vice President, Harza Engineering Company International L. P., Caracas, Venezuela* (SEC. 9)
Bernard A. McKiernan *Senior Engineer, Harza Engineering Company, Chicago* (SECS. 2, 4–7, 12)
David B. Palmer *Senior Irrigation Engineer, Harza Engineering Company, Chicago* (SECS. 24, 25)
Richard Persaud *Senior Engineer, Harza Environmental Services, Chicago* (SEC. 28)
John A. Scoville *Chairman, Harza Engineering Company, Chicago* (SEC. 12)
Chander K. Seghal *Associate, Harza Engineering Company, Chicago* (SEC. 17)
James H. T. Sun *Senior Associate, Harza Engineering Company, Chicago* (SEC. 21)
Archivok V. Sundaram *Senior Engineer, Harza Engineering Company, Chicago* (SEC. 14)
Edwin Paul Swatek, Jr. *Consulting Engineer, Lake Bluff, Illinois* (SEC. 8)
James H. Thrall *Vice President, Harza Northwest Inc., Bellevue, Washington* (SEC. 18)
John P. Velon *Vice President, Harza Environmental Services, Chicago* (SECS. 27, 28)
C. Y. Wei *Senior Hydraulic Engineer, Harza Engineering Company, Chicago* (SEC. 16)
Roman P. Wengler *Senior Vice President, Harza Engineering Company, Chicago* (SEC. 10)
Chang-Hua Yeh *Vice President, Harza Engineering Company, Chicago* (SECS. 9, 10)
Vincent J. Zipparro *Vice President and Chief Engineer, Harza Engineering Company, Chicago* (SEC. 11)

PREFACE TO THE FOURTH EDITION

The objectives of *Davis' Handbook of Applied Hydraulics* are to present clearly and concisely the fundamental principles which are basic to each subdivision of hydraulic engineering, and to demonstrate the practical applications of these principles by examples which have been drawn largely from the actual practice of hydraulic engineering. The objective of this fourth edition is to present the most recent developments and current practice consistent with the handbook's objectives.

Since the publication of the third edition in 1969, a considerable amount of new data has been developed and has become available in all aspects of applied hydraulics. The Editors were confronted with the task of incorporating this new information and at the same time condensing the handbook to make it a more convenient reference source. Sections were combined and rewritten and more tables and graphs have been incorporated. Reference has been made to computer programs which are now commonly used to solve hydraulic engineering problems. Also some new sections have been added to the handbook.

All contributors are practicing engineers and have included material based on experience gained in designing and constructing hydraulic projects.

The section on river diversion has been rewritten to incorporate the extensive experience of Edwin Paul Swatek, Jr., who has played a major role in many river diversions and is a world-renowned expert in the field.

Dramatic advances have been made in the design and construction of concrete faced rockfill dams (CFRD) and J. Barry Cooke, renowned authority on this subject, completely rewrote this section, incorporating the latest data and practices.

J. A. Scoville, Chairman of the Board of Directors of Harza Engineering Company, has made substantial contributions to Section 12, Barrages and Dams on Permeable Foundations, incorporating his expertise in this field.

New subject material has been added. Rehabilitation of older hydraulic structures is becoming of increasing interest to hydraulic engineers and has been added to Section 11, Prestressing/Post-tensioning, and Rehabilitation. Environmental protection is a major concern for all hydraulic projects, and a new section on this subject has been written that includes fish facilities at dams.

Since the third edition was published a new technique in the design and construction of hydraulic structures, and dams in particular, has gained prominence, namely, roller compacted concrete (RCC), and a new section has been devoted to it.

Aeration of spillway flows to combat cavitation damage has been another significant advance in hydraulic engineering since the third edition. The subject is covered in detail in the section on spillways and streambed protection works.

The section on waterhammer, surge tanks, speed regulation and governing stability has been combined in a new section on hydraulic transients, emphasizing the new computer methods used in current design practice.

Sections on water supplies, water distribution, and water treatment have been rewritten and combined in a new section on water distribution and treatment. This new section incorporates significant information and focuses on the current practice utilized by engineers.

Similarly, sections on sewage quantities, sewers, and pumping stations have been combined in a new section on wastewater distribution and treatment, expanding the scope and giving information that is valuable for the design engineer.

Vincent J. Zipparro
Hans Hasen

ACKNOWLEDGMENTS

The fourth edition builds on the third edition published under the leadership of the late Calvin V. Davis and the late Kenneth E. Sorensen. Mr. Davis had the initial inspiration to undertake the writing of the first and second edition and incorporated his experiences of a lifetime of successful engineering.

Mr. Sorensen was an outstanding planning engineer, who pioneered the stage construction concept, in which initial developments serve immediate needs, but have the potential to be enlarged in the future. He planned large, successful projects in the United States, Venezuela, and Argentina using this concept.

The Editors' long association with Mr. Sorensen, Chief Planning Engineer and Chairman Emeritus of the Harza Engineering Company, had a definite impact on this book.

The Editors express their deep gratitude to all who have made contributions to the fourth edition and have incorporated their lifelong expertise in this handbook, especially to J. Barry Cooke, who has made outstanding contributions to the design and construction of concrete faced rockfill dams and Edwin Paul Swatek, Jr., who contributed his experience on the diversion of rivers.

Finally, the Editors wish to specially thank Walter J. Bogdovitz, President of the Harza Engineering Companies, without whose support and encouragement the fourth edition would not have been accomplished.

DAVIS' HANDBOOK OF
APPLIED HYDRAULICS

SECTION 1

HYDROLOGY

By Khalid Jawed and James E. Lindell[1]

PRECIPITATION

1. Definition and Measurement. Precipitation consists of rain or snow. Rain is defined as liquid water drops mostly larger than 0.02 in (0.5 mm) in diameter. Rainfall refers to amount of liquid precipitation. Snow is composed of ice crystals, often agglomerated into snowflakes. The density of freshly fallen snow varies greatly. About 5 to 20 in (125 to 500 mm) of snow is equal to 1 in (25 mm) of liquid water. An average density is often assumed to be 0.1.[1]*

The amount of precipitation is measured in inches (to the nearest 0.01 in) or in millimeters (to the nearest 0.2 mm). Nonrecording daily rain gages (precipitation measured at 0800 h daily) and automatic recording rain gages are used. The receiving area of the gage varies from 31 in^2 (200 cm^2) to 78 in^2 (500 cm^2). The standard U.S. Weather Bureau gage[2] has a receiver area of 50.3 in^2 (8 in diameter). For nonrecording gages, the measuring tube provides a magnification of 10 (measurement area 5.03 in^2, 2.53 in diameter). A nonrecording gage is also generally installed at a recording gage site to provide a check on the automatic gage mechanism. A recording gage may be equipped with a 7-day recorder or a strip-chart recorder that serves up to 6 months.

The amount of measured precipitation depends on the exposure of the gage to the wind and also on the nature and height of the surrounding objects. A poor exposure should be avoided and wind barriers such as bushes or fences may be needed at some locations.

The density of precipitation stations (network) is either determined by the project or established for a general assessment of a country's areal and seasonal variations of precipitation. The basic networks in the United States are (1) synoptic network (weather forecasting), (2) special reporting network, (3) recording network, and (4) nonrecording network. At present there are about 11,000 nonrecording gages and about 3500 recording gages.

A comprehensive discussion of exposure, network, sources of error, and types of rain gages is provided by the World Meteorological Organization,[3,4] Chow,[5] and Rodda.[6,7] Techniques for measuring rainfall using radar[8,9] and satellite imagery[10] have been developed in recent years, especially for flood forecasting.

2. Sources of Data. In most countries, meteorological departments and other agencies responsible for water resources developments maintain precipitation records. In the United States, the principal source of precipitation data is the Na-

---

[1]Acknowledgment is made to Phillip Z. Kirpich and Gordon Williams for material in this section which appeared in the third edition (1969).

*Superscripts indicate items in the References at the end of this section.

tional Weather Service (formerly the Weather Bureau). Many other agencies, both public and private, also maintain gages. Published data can be obtained from the National Technical Information Services, Springfield, Va. The National Climatic Data Center, Ashville, N.C., provides both published and unpublished data. The precipitation data publications include state and national summaries, *Hourly Precipitation Data, Weekly Weather and Crop Bulletin,* and *Monthly Weather Review,* among others. However, in a given locality it is always best to obtain data from the appropriate regional office of the National Weather Service, the state climatologist, and other state or private agencies.

3. Estimation of Missing Precipitation Data. Because of malfunctioning of instruments or the absence of the observer, some precipitation data often are missing for short periods. These data gaps should be filled in. The missing data (P_x) can be estimated using a simple arithmetic average of the precipitation of the index stations or by the normal-ratio method.[1,11] The arithmetic-average method is used when the normal annual precipitation at each of three nearby index stations is within 10 percent of that for the station with the missing record. In the normal-ratio method, the amounts at the index stations (A, B, C) are weighted by the ratios of the normal annual precipitation values (NA, NB, NC).

$$Px = \frac{1}{3}\left(\frac{Nx}{NA}P_A + \frac{Nx}{NB}P_B + \frac{Nx}{NC}P_C\right)$$

Multiple linear regression can also be used. The number of index stations depends upon the increase in the correlation coefficient with addition of each nearby station.

$$Px = a + bP_A + cP_B + \cdots$$

The National Weather Service[12] estimates precipitation at a point as the weighted average of that at four stations. Recently stochastic techniques[13,14] have been developed to estimate the missing data.

4. Adjustment of Records. A precipitation record may have been obtained from a poorly exposed gage or from one whose location was changed during the period of record. Generally, the history of exposure conditions and changes in location of a gage is not available to determine the years of changes. All precipitation records should be checked for homogeneity and consistency.[15,16] The homogeneity of a station record is tested by statistical procedures. The consistency of record can be checked by the double-mass curve technique.[17,18]

The adjustment of record can be accomplished by two methods: ratio method and the double-mass curve method. In the ratio method, the adjustment should be made by comparing the ratio between the recorded values of the annual or seasonal precipitation with the corresponding average value for a group of base stations in the vicinity. One computes the ratios for each year or season, and examines them for indication of any sharp changes or trends, which if present would indicate modifications in the regime of the station. In the double-mass curve analysis, a change in the slope of the curve indicates a change in regime of the station. Both methods are based on the fact that seasonal precipitation data at stations in the same general locality are usually consistent with one another. However, this is not true for short-period (hourly or daily) precipitation, for which this type of adjustment is not recommended.

5. Rainfall Frequency. Frequency analysis of point rainfall values is required for the design of many hydraulic structures. The National Weather Service has published the rainfall-frequency data for the United States.[19,20] The rainfall duration varies from 5 min to 24 h for return periods of 2 to 100 years. Several formulas also have been suggested for rainfall-frequency relationships but are rarely used now.

When rainfall-frequency data are required for a particular station, one method of determining extreme rainfall data is the annual series, which involves selecting the maximum value of a given duration (1, 2, 3,... and hours) for each calendar year of record. The other, the partial duration series, involves selecting data so that their magnitudes are greater than a predefined base value (defined to produce values more than the number of years of record and at least one value from each year). Empirical factors of 0.88, 0.96, and 0.99, corresponding to return periods of 2, 5, and 10 years, respectively, can be used to convert the partial duration series to the annual series.

Frequency analysis involves fitting a proper probability distribution to the data. Generally, Gumbel type I, log-Pearson type III, log-normal, or gamma probability distributions are used. Comparisons[21-23] have been made of different distributions. The goodness-of-fit tests[24] should be made for a selected distribution. For most cases Gumbel and log-Pearson type III distributions are applicable.

Rainfall data are published as clock-hour or observational-day amounts. For frequency analysis data based on hourly or daily amounts should be increased by 13 percent to approximate true values for 60 min or 24 h.

6. Geographic Distribution. The following basic factors determine the amount of mean annual precipitation at a station on the earth's surface:[25] (1) latitude, (2) position and size of the continental land mass on which the station is located, (3) distance of the station from the coast or other source of moisture, (4) temperature of ocean and coastwise currents with respect to adjacent land masses, (5) extent and altitude of adjacent mountain ranges, i.e., *orographic effects,* and (6) altitude of the station.

Considering latitude alone, the generalized world pattern is composed of a series of belts resulting from the circulation of the atmosphere. At the equator, there is a belt of relatively low pressure known as the doldrums, where intense solar radiation heats the air and causes it to expand and rise. Warm moisture-laden winds converge on the region and produce high precipitation from frequent thunderstorms. At about 30° north and south latitude, there are high-pressure belts, called the horse latitudes, where warm, dry air descends and precipitation is low. From about latitudes 35° to 65° interaction of the moisture-laden prevailing westerlies with cold, dry polar air generates storms of the frontal type and produces abundant precipitation. Convection-type thunderstorms also occur in this zone in summer but produce less total precipitation than the frontal storms, even though the short-duration intensities at individual points are generally much higher. From latitude 65° to the poles, dry polar air predominates increasingly and causes a decrease in precipitation.

The wide variation in mean annual precipitation in the United States (Fig. 1) is an example of the large departures from the generalized world precipitation pattern caused by factors 2 through 6. These factors produce even greater variations in Asia. High pressure builds up in winter over the cold continental land mass and produces a dry wind (winter monsoon) blowing outward to the Indian Ocean. The summer monsoon is wet and blows in the reverse direction. Thus the normal planetary circulation of the atmosphere is greatly modified by the large size of the Asian continent. The combined effects of the wet summer monsoon and the high

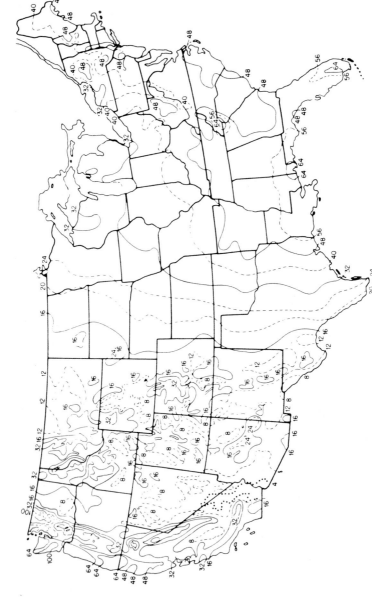

FIGURE 1 Mean annual precipitation in the United States, in inches (1 in = 25.4 mm). (*U.S. Environmental Data Services.*[1])

altitude of the Himalayas in northern India, which it crosses, explain why this region is one of very high precipitation. As an example, the mean annual precipitation at Cherrapunji, India, is 428 in.

7. Seasonal Variation. The belts of different atmospheric pressure described in the preceding article move north away from the equator in summer and south toward the equator in winter (opposite season in the Southern Hemisphere), causing marked changes in the depth and type of precipitation for the various seasons of the year. In the United States, the variation in depth between summer and winter is greatest on the West Coast (Fig. 2) because of two principal factors: (1) the northward movement in summer of the dry high-pressure belt of the horse latitudes and (2) the presence of cool coastal water. The latter lowers the temperature below the dew point and hence reduces the moisture content of air carried landward by the prevailing westerlies. Subsequently, when this air strikes the warm land masses, dehumidification takes place, and almost all opportunity for precipitation is lost.

The East Coast, on the other hand, shows a fairly uniform seasonal depth of precipitation. The origin of precipitation is distinctly different, however, in summer and winter. Summer precipitation is mostly of the convectional thunderstorm type and increases toward the south owing to increased summer convectional activity in that direction. Winter precipitation is almost entirely of the cyclonic or frontal type caused by the interaction of polar and tropical air masses.

In the interior of the country, particularly west of the 95th meridian, still other seasonal patterns exist. In the Great Plains region, summer thunderstorms produce high precipitation as compared with winter, when the region is covered by cold, dry polar air. In the plateau region, mountain ranges seal off incoming moist air during all seasons.

8. Snow. Snowfall is recorded in terms of depth in inches of water using rain gages. In the regions where there is a continuous snow cover, the measurement of the water equivalent and density are made along preselected snow courses. The length of a snow course ranges from about 300 to 1000 ft depending on site conditions and uniformity of snow cover along the snow course. Usually the sampling points are taken at intervals of 25, 50, or 100 ft. The selection of snow courses, method of surveying, snow survey equipment, and method of determining average snow depth and water equivalent are provided in U.S. Department of Agriculture (USDA) publications[26,27] and in many textbooks on hydrology.

The National Weather Service publishes the snowfall totals in the *Monthly Climatological Data Bulletin* and *Climatic Summary of the United States.* For each state (where snowmelt is a major components of spring runoff) the USDA Soil Conservation Service (SCS) publishes[28] the snow pack and water equivalent data for the period January through June for each station. The data from cooperative snow courses also are published.

The distribution of mean annual snowfall in the United States is shown in Fig. 3. This map may be considerably in error for mountainous regions because of limited measurements at high elevations.[1]

Snowmelt is a predominant factor in major floods and in water supply in many areas. Computations and forecasts of snowmelt runoff are required for the operation of existing projects and the design of future projects. Either volumetric or day-to-day runoff forecast is required for the snowmelt period (April through July in some areas). The estimation of design floods requires determination of optimum snow cover and optimum snowmelt rate.

The melting of snow is a physical process, and snowmelt is estimated indirectly from meteorological parameters. Because of variation of snowfall and me-

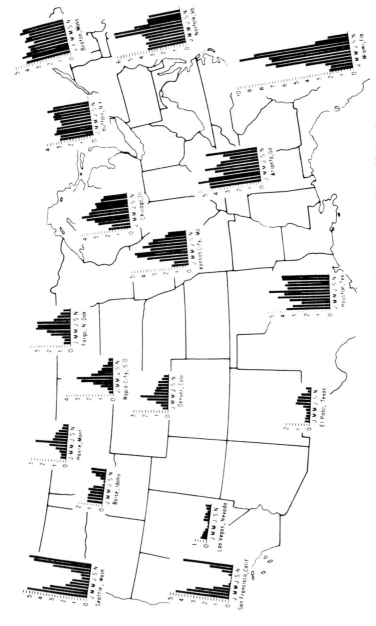

FIGURE 2 Normal monthly distribution of precipitation in the United States, in inches (1 in = 25.4 mm). (*U.S. Environmental Data Service.*[1])

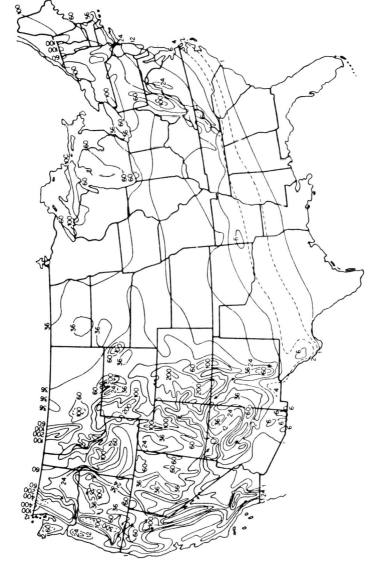

FIGURE 3 Mean annual snowfall in the United States, in inches (1 in = 2.54 cm). *(U.S. Environmental Data Service.[1])*

teorological parameters with elevation, separate snowmelt computations are made for zones of equal elevation within a drainage basin. A number of computer program subroutines[29-31] are available to compute snowmelt rates. U.S. Army Corps of Engineers[32] provides relationships for estimating point snowmelt. The relationships are developed for rainy and rain-free periods and for open and forested areas. These equations are based on the energy-budget concept and require air temperature, dew-point temperature, wind velocity, solar radiation, albedo, and precipitation.

If only temperature data are available, a simple degree-day method can be used:

Snowmelt = coefficient (air temperature − snowmelt temperature)

The snowmelt temperature is usually assumed to be 32°F. The coefficient varies with season, about 0.02 to 0.10 from March to May.[33]

For volumetric forecasts satisfactory empirical relationships often can be derived if there are sufficient data on snow water equivalent, precipitation, temperature, and runoff. Figure 4 shows correlations between (1) a snow-survey index for April 1 and the ensuing April-July runoff, and (2) the basin winter precipitation and the ensuing April-July runoff. The snow-survey index is the estimated water equivalent on the basin in inches, determined as follows:

1. Average water equivalents are measured along 11 snow courses, each at a constant elevation between 5700 and 10,300 ft in the Sierra Nevada (see Ref. 27 for snow-surveying methods).
2. The basin is divided into three zones according to altitude.
3. The average water depth for each zone is computed by averaging the courses in that zone.
4. The average basin depth, or snow-survey index, is the weighted average of the zones, the weights being taken as proportional to the area of the zone.

In recent years, snow-cover satellite data and corresponding runoff records have been used to develop correlation for predicting spring snowmelt runoff.[35] The Soil Conservation Service and U.S. Army Corps of Engineers have pioneered this effort (Fig. 4).

9. Droughts. The U.S. Weather Bureau has defined a "drought" as a "lack of rainfall so great and long-continued as to affect injuriously the plant and animal life of a place and to deplete water supplies both for domestic purposes and for the operation of power plants, especially in those regions where rainfall is normally sufficient for such purposes."[36] Thornthwaite,[37] in a discussion of the definitions of drought from an agronomic point of view, points out that it cannot be defined merely as a shortage of rainfall, because both the crop water demand (which varies according to the time of the year) and the moisture available in the soil must be taken into account. Here, the term "drought" refers to periods of unusually low water supply, irrespective of water demand in a specific area.

Areas with high variation in annual rainfall are most subject to droughts. High variability occurs in areas of low annual rainfall which is generally due to the small number of rainy days. Since the number of rainfall events is low, failure of occurrence of rainfall causes the drought conditions. Droughts may be confined to a small area (a river basin or a state) or may be widespread, covering a climatic region or a part of the world.

Drought conditions may be evaluated using various parameters such as defi-

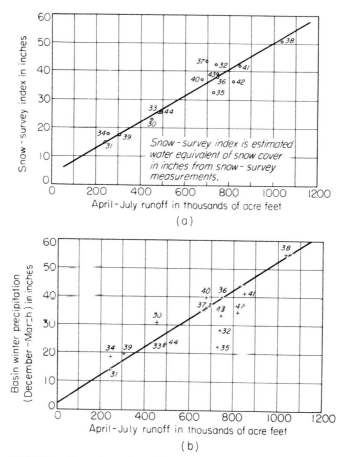

FIGURE 4 Comparison of runoff correlations for the Tuolumne River at Hetchy Hetchy, California.[34] (a) Snow-survey data. (b) Precipitation data.

ciency in rainfall or runoff, and decline in soil moisture contents and groundwater levels. In existing water-supply projects, an inability of a project to meet demand also indicates drought conditions. Hydrologic studies of droughts are based on long-term rainfall or runoff data (daily, monthly, or yearly).

A review of long-term streamflow or rainfall data indicates that dry years do not occur randomly. There is a tendency for dry years to occur together. Therefore, both severity and duration of a drought must be evaluated. The severity may be indicated by expressing observed rainfall (runoff) as percent of normal rainfall[38] (runoff) or as derived from a time series plot of cumulative departure from normal.[18] Figure 5 shows a low flow sequence (plotted as cumulative departure from the mean) on Heath Creek near Rome, Ga., during the 1985–1988 drought period.

A hydrologist is generally concerned with drought studies in predicting magnitude and/or duration and frequency of low flows; this is critical for a water-supply diversion without storage or for water-pollution problems and extended

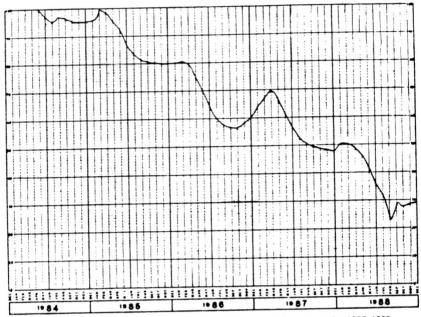

FIGURE 5 Heath Creek near Rome, Ga., cumulative runoff departures for 1985–1988.

periods of low flows that affect reservoir yield. Figure 6 shows examples of low-flow duration-frequency curves.[1,39] The method of developing these curves is discussed by Hudson and Roberts.[40]

EVAPORATION AND EVAPOTRANSPIRATION

10. Definition and Measurement. Evaporation is generally defined as the process by which water is changed from liquid or solid state into vapors through heat energy and carried into the atmosphere. Evapotranspiration is the combination of evaporation from the soil surface and transpiration from vegetation. The factors affecting evaporation are solar radiation, temperature, vapor pressure, humidity of air, and wind. For evaporation from soils and transpiration through vegetation, the availability of moisture is an additional factor. A term "free-water surface" (FWS) evaporation is defined[41] as evaporation from a thin film of water having no appreciable heat storage. The term "evaporation" in this section is confined to the loss of water from a free-water surface (reservoir, lake, etc.). Only evaporation from free-water surface is discussed here.

Direct measurements of evaporation from reservoirs and lakes have not yet been realized. Various methods of computing evaporation rates are given in hydrology textbooks.[1,18,42] Evaporation pans and atmometers are often used to measure evaporation rates which can be converted into FWS evaporation using coefficients. The standard National Weather Service Class A pan is the most

EVAPORATION AND EVAPOTRANSPIRATION

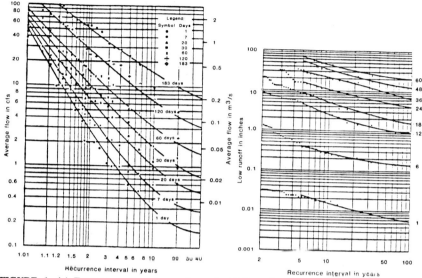

FIGURE 6 (a) Frequency of minimum flows for Yellow Creek near Hammondsville, Ohio, 1915–1935 (*Source: Ref. 1.*); (b) low-flow curves for 1 to 60 months durations for Salt Creek near Rowell. (*Source: Ref. 39.*)

widely used evaporation pan. Attempts have been made to correlate Class A pan evaporation with meteorological factors.[43]

The Class A pan is made of unpainted galvanized iron, 4 ft in diameter and 10 in deep. The bottom, supported on a wooden frame, is raised 6 in above ground to permit air circulation. The water surface in the pan is maintained between 2 and 3 in below the rim of the pan. Daily evaporation is computed as the difference between observed levels (usually at 0800 h each day), corrected for any precipitation measured at the evaporation station. Pan evaporation data are available from National Climatic Center, Asheville, N.C., and from the National Technical Information Service (NTIS), Springfield, Va.

11. Geographic Distribution. Because of variation in climatic conditions at different locations, the evaporation rates vary from one location to another. NOAA Technical Report NWS 33[41] gives maps of Class A pan evaporation (May through October), shallow lake or free-water surface evaporation (May through October and annual), and coefficients to convert Class A pan evaporation to FWS (for the period May through October).

12. Reservoir Evaporation. Reservoir evaporation may be estimated from pan measurements by applying coefficients or using FWS evaporation maps directly. The term "shallow" in the maps refers to water bodies with maximum depths of about 180 ft. Deeper water bodies may have different evaporation characteristics because of the longer time lag required for the water temperature to adjust to seasonal changes in air temperature. However, the difference in evaporation rates between deep and shallow reservoirs is not significant.

Monthly evaporation data often are required for water-resources projects. Pan evaporation data from a nearby station should be used, or the May to October and annual FWS data from the maps can be distributed into monthly evaporation rates based on monthly air temperatures.

Net reservoir evaporation can be computed using the following relationship:

Net evaporation = pan coefficient (pan evaporation)
 − (precipitation − runoff from reservoir area)

The precipitation data from a nearby station are generally used, and runoff from the reservoir drainage area is based on the runoff from the drainage area upstream of the dam site.

The relative importance of evaporation in water-resources planning depends entirely on the climatological characteristics of the region in which a reservoir is located. In humid regions water surface evaporation may be little different from evapotranspiration in the reservoir area, and therefore the creation of a reservoir will not materially change the water yield of the reservoir area. In semiarid regions, where the depth of evaporation may be many times the depth of runoff per unit of area, conversion of only a small percentage of a drainage basin to water area may seriously deplete the water yield available below the reservoir.

On large projects where evaporation is an important consideration, extensive investigations may be justified, including supplemental pan measurements and meteorological measurements at a proposed reservoir site. Such supplemental measurements, even though of short duration, may be compared with similar long-term observations at regular National Weather Service stations, thus enabling a more reliable estimate of evaporation at the reservoir site.

STORM RAINFALL

13. Storm Types. Some features of the various types of storms should be considered in the design of flood control and drainage structures, and spillways for dams. These features include season of occurrence, areal extent, duration, intensity, frequency, and possibility of occurrence of two storms with a short nonrainy period (antecedent and subsequent storms). In a broad sense, all storms result from moisture (water vapor) in the atmosphere being lifted, aided in some situations by topography. The greater the water vapor in the lifted air, the better the chance of greater rainfall. Also, the warmer the lifted air, the more water vapor it can hold.

In plains areas, the rain may be caused by atmospheric processes alone (convergence rainfall). In mountainous regions, convergence rainfall is increased or decreased by terrain effects or orography (orographic rainfall).

Storms are often classified according to the factor mainly responsible for the lifting of air mass. The term cyclone denotes a storm area of low atmospheric pressure, generally circular in shape, in which the winds blow spirally inward counterclockwise in the Northern Hemisphere, clockwise in the Southern. Two types of cyclones are: the tropical cyclone (hurricane or typhoon) and the extratropical cyclone (frontal-type).

Tropical cyclones are comparatively small, violent storms originating in the doldrums belt about 15° of latitude from the equator. Often winds reaching velocities greater than 100 mi/h (160 km/h) accompany the storm. The resulting rapid convergence of warm, moist air causes heavy precipitation. The diameter of a tropical cyclone varies from 100 to 600 miles (160 to 960 km). The data on frequency, paths, and associated precipitation can be obtained from the National Hurricane Center.

The frontal-type storm is associated with the extratropical cyclone. These storms vary greatly in size up to 1000 mi (1600 km) in diameter and move generally eastward across the United States at a speed of 300 to 700 mi (460 to 1130 km) per day.

Thunderstorms are local atmospheric disturbances of short duration characterized by violent vertical air currents, gusty surface winds, torrential rain, lightning, thunder, and sometimes hail. In mountainous areas, orographic lifting tends to increase the frequency and intensity of thunderstorms. A comprehensive discussion of storm types is given in the *Handbook of Applied Hydrology*.[5]

14. Point Rainfall. Rainfall measured at a rain gage station is referred to as a point rainfall in comparison with average depth of rainfall over a river basin area. Generally point rainfall is assumed to be applicable for small areas, up to 10 mi^2 (26 km^2).

Time Distribution. The time distribution of point precipitation obtained from a recording rain gage is generally shown by a mass curve or by a hyetograph, as shown in Fig. 7. The hyetograph shows the depths during a selected time interval. The mass curve represents cumulated depth vs. time; the slope at any point gives the rainfall intensity. In many cases dimensionless mass curves (rainfall as percent of total rainfall and time as percent of total storm duration) are helpful in

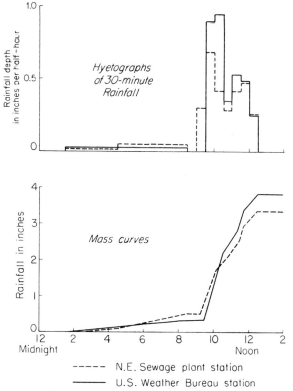

FIGURE 7 Hyetographs and mass curves for the storm of Aug. 3, 1950, at Philadelphia, Pa.

defining the rainfall time distribution characteristics in a meteorologically homogeneous regions.

Terrain Influence. Generally precipitation data are lacking in mountainous areas. A simplified procedure to estimate point rainfall at higher elevations during a storm is to develop a relationship between rainfalls and station elevations and extend the relationship to higher elevations. This technique is generally employed in developing mean annual, season, or individual storm isohyetal patterns. Other methods for determining terrain influence are discussed by Chow[5] and Peck.[44]

15. Basin Average Rainfall. Precipitation data are essentially point observations. The basin average rainfall over a basin on a storm, seasonal, or annual basis is required for hydrologic studies. There are a number of procedures to derive basin average rainfall using point rainfall measurements. The most commonly used methods are arithmetic mean, Thiessen polygon, isohyetal, hyposometric, percent-of-mean annual, abbreviated isopercental, and multiquadric. The details of these methods are given by Chow,[5] Linsley,[1] and Shaw.[18]

16. Hypothetical Design Storms. For the design of small reservoirs (where the failure of the dam would not be disastrous to human life or to important infrastructure), culverts, and storm-water sewers, floods resulting from storms of various return periods are required. The basin average rainfall and its time distribution for a storm of selected return period are determined using the following procedure.

Point rainfall data for various durations and return periods are given in National Weather Service publications.[19,20] The data for other return periods and durations are interpolated using Figs. 8 and 9, respectively. The rainfall depths and corresponding durations are plotted with the return period as the third parameter, and a smooth curve is drawn as shown on Fig. 10.

Basin average rainfall for various durations is derived using Fig. 11. For a given return period, the computed basin average rainfalls and rainfall durations are plotted, and a smooth depth-duration curve is derived. For a selected time increment, the incremental rainfall amounts are determined from the depth-duration curve and arranged in a chronological sequence to produce reasonably credible flood conditions.

A chronological sequence for thunderstorms, which are of short duration (up to 6 h) and small areal extent (maximum about 500 mi^2, or 1300 km^2), recommended by U.S. Army Corps of Engineers[45] is given in Table 1. For storms of longer durations, the alternating block method is a simple way of developing a design hydrograph from depth-duration-frequency data. In this method the maximum increments are placed at about two-thirds of the total storm duration and

TABLE 1 Recommended Chronological Distribution of Rainfall Increments for a 6-Hour Local Storm

1-h increment	Sequence position	15-min sequence	Sequence position
Largest	Fourth	Largest	First
Second largest	Third	Second largest	Second
Third largest	Second	Third largest	Third
Fourth largest	Fifth	Fourth largest	Fourth
Fifth largest	Last		
Least	First		

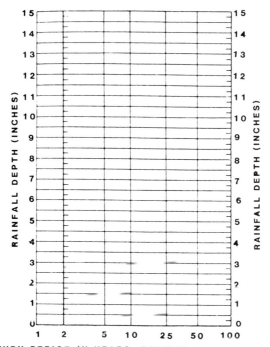

FIGURE 8 Return period—interpolation diagram (NWS).[19]

the remaining increments are arranged in descending order alternately to the left and right of the maximum increment. In addition to the depth-duration-frequency data available from the National Weather Service, some states[46,47] have developed site-specific data. These data also should be reviewed for possible use.

17. Probable Maximum Precipitation. For structures whose failure would cause loss of life and major economic damage, the design rainfall is the probable maximum precipitation (PMP). The U.S. Weather Bureau[48] defines the PMP as the critical depth-area-duration (DAD) rainfall relationship for a particular area during various seasons of the year. In a recent publication by the World Meteorological Organization,[49] the PMP is defined as "the greatest depth of precipitation for a given duration meteorologically possible for a given size storm area at a particular location at a particular time of year, with no allowance made for long-term climatic trends."

Selection of PMP Storms. The PMP for a specific basin is based on "general storms" (frontal cyclone type) or "local storm" (thunderstorm) depending upon the size of the area. A general storm normally lasts more than 6 h and is associated with a major synoptic weather system producing precipitation over a large area. A local storm seldom lasts more than 6 h and covers areas up to 500 mi^2 (1300 km^2).

For a small basin, a PMP duration of 6 h or less may be critical. For a large basin and a large reservoir, a duration of 3 days or more may be critical. For a basin with a critical PMP duration of 6 h or less, a local storm would produce a

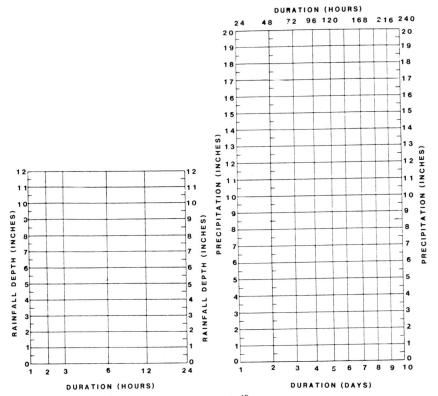

FIGURE 9 Duration-interpolation diagram (NWS).[19]

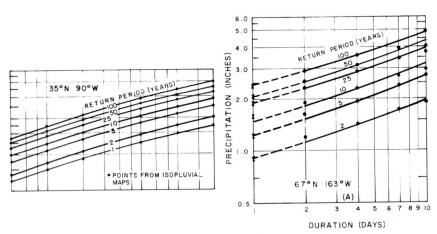

FIGURE 10 Depth-duration frequency curves (NWS).[19]

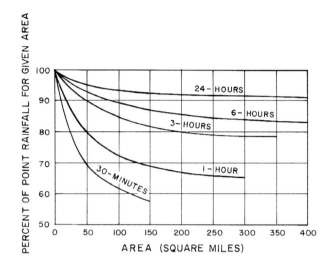

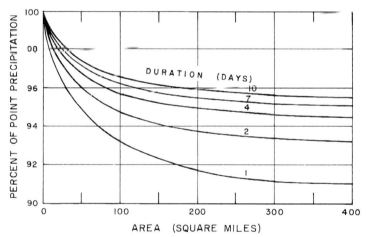

FIGURE 11 Area-reduction curves (NWS).[19]

critical flood. For a critical duration of more than 24 h, a general storm PMP would produce a critical flood. For a basin with critical duration of 6 to 24 h, a 6-h local storm and a 12- or 24-h general storm should be used to determine which storm would produce the critical flood.

Generalized PMP Estimates. Generalized PMPs have been estimated and published by the National Weather Service in collaboration with other agencies such as the Army Corps of Engineers, the Bureau of Reclamation, the Tennessee Valley Authority, and the Soil Conservation Service. These reports include hydrometeorological reports (HMRs),[50] NOAA technical reports (TR), technical memoranda (TM), and technical papers (TP). These publications provide procedures for estimating the magnitude, time distribution, and sequential arrangements of 6-h PMP increments. The procedures also are provided to estimate PMP increments for shorter durations (less than 6 h). Within a 6-h increment, the increments of shorter intervals are arranged in a manner given in Table 1. The procedures for estimating the PMP differ from region to region depending upon the topographic and meteorological characteristics of the regions. In some regions, estimates of both local storm and general storm PMPs are provided. For the region east of the 105th meridian, generalized DAD curves (Fig. 12b) for the specific location are used with synthetic, elliptical isohyetal pattern.[50] The synthetic pattern is superimposed over the projected basin in such a way as to produce the most critical flood at the dam site (Fig. 12a). The values of the isohyets can be computed for periods of 5 min to 72 h.

The magnitude and time distribution of general and local storm PMPs are quite different. For some regions, the National Weather Service has derived the DAD curves for local storms (Fig. 13).

Site-Specific PMP. For basins of unique topographic characteristics or where generalized PMP maps are not available, the site-specific PMP estimate should be made using a hydrometeorological (hydromet) or statistical method. A detailed discussion of these methods is given by WMO.[49] The hydromet method involves the transposition of major historical storms from where they occurred to the project basin and the adjustment of the recorded storm for maximum moisture, terrain effects, and possibly other factors such as wind speed. An example of this method is given below.

Figure 14a shows the Gibson Dam, Montana, storm of June 7–8, 1964, at the place of occurrence. The storm was transposed to a meteorologically homogeneous location in Cheeseman Basin, Colo., placed on a ridge, and oriented on the basis of a detailed comparison of meteorological and topographic conditions of the place of transposition (Fig. 14b) and with those of the location where the storm actually occurred. A moisture maximization factor of 1.70 was used. The terrain effect was assumed to be represented by the ratios between the 100-year, 24-h precipitation in the basin of transposition and in the area where the storm

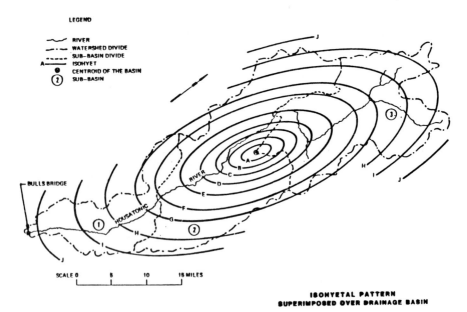

FIGURE 12 (a) Synthetic isohyetal pattern.

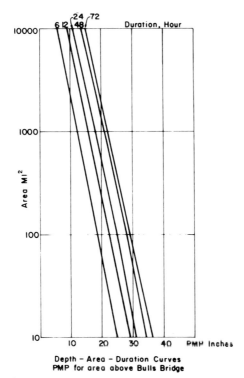

FIGURE 12 (*Continued*) (*b*) Generalized PMP DAD curves.

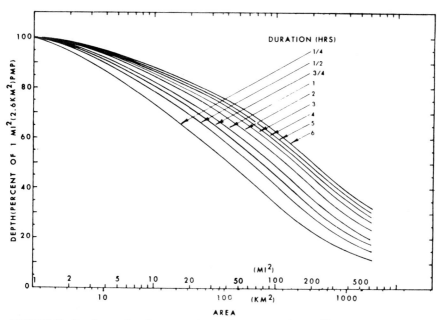

FIGURE 13 Depth-area-duration curves for northwest states (NWS).[50]

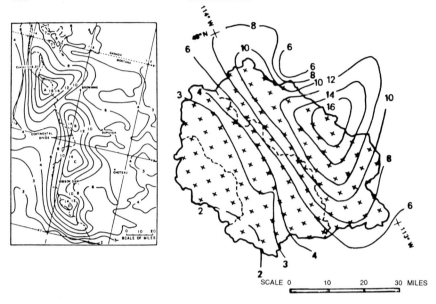

FIGURE 14 Transposition of Montana, June 1964 storm. (*a*) Montana, June 1964 storm at the place of occurrence. (*b*) Montana, June 1964 storm oriented at the place of transposition.

occurred. The ratios were computed at each 5-mi grid joint shown in Fig. 14*b*. The combined adjustment factor was the ratios multiplied by the moisture maximization factor. The transposed and moisture maximized storm is shown in Fig. 15.

The time distribution of the transposed and maximized storm rainfall generally is determined in three ways:

1. Develop envelope DAD curves for various durations for major storms in the region, read rainfall amounts for various durations corresponding to the basin area, and draw depth-duration curve.
2. Construct dimensionless rainfall mass curves of point rainfalls of major storms in the region and draw a near-envelope curve which nearly envelops the low values in the low end of the curves and high values in the high end (Fig. 16).
3. Use actual DAD curves of transposed storm, determine rainfalls for various durations corresponding to the basin size, and develop a depth-duration curve (Fig. 17).

The DAD curves are developed using the procedures outlined by the WMD.[51] Currently computer programs are available to develop the curves. The U.S. Army Corps of Engineers has published[52] the DAD data for major storms up to January 1958. Additional major storms since 1958 are being added by the U.S. Bureau of Reclamation and National Weather Service.

Statistical Procedure. Statistical procedure for estimating PMP is used where long-term maximum precipitation data (daily or shorter durations) at a reasonable number of stations are available within or in the vicinity of a basin, but other meteorological data, such as isohyetal maps and dew-point data and history of major storms, are not available. The method is based on a frequency concept

STORM RAINFALL 1.21

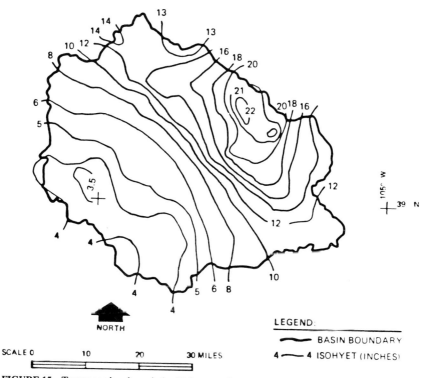

FIGURE 15 Transposed and maximized Montana, June 1967 storm in Cheeseman Basin, Colo.

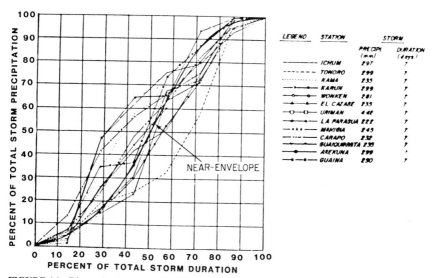

FIGURE 16 Dimensionless storm depth-duration curve based on daily precipitation data.

1.22　　　　　　　　　　　　　HYDROLOGY

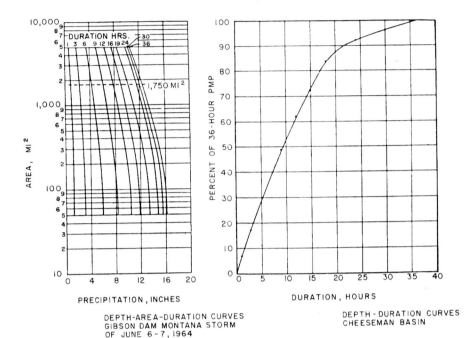

FIGURE 17 DAD and depth-duration curves.

and was proposed by Hershfield.[53,54] The resulting point PMP (1-, 6-, or 24-h) is subject to high uncertainty and should be used with caution. The basin-average PMP is obtained by adjusting the derived PMP by applying depth-area curves.

In some cases a synthetic PMP isohyetal can be constructed using the point PMPs and the depth-area curve, such as that shown in Fig. 18. This isohyetal pattern is superimposed on a basin in such a way as to produce most critical flood at the dam site.

ISOHYET (IN)	AREA ENCLOSED (MI^2)
23.0	10
22.8	25
22.6	50
22.0	100
21.2	175
20.2	300
19.0	450
18.0	700
16.8	1000

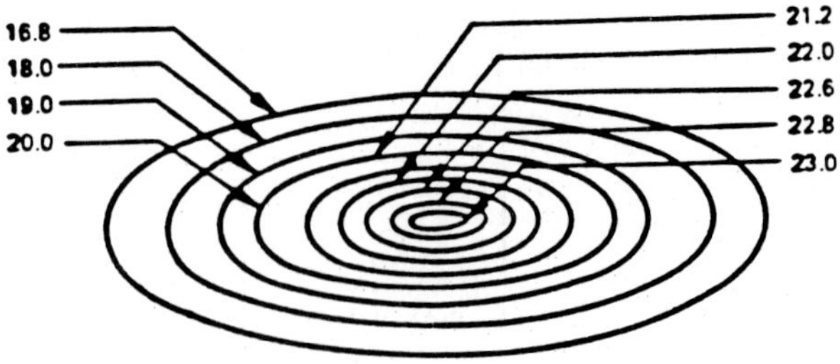

FIGURE 18 Synthetic isohyetal pattern for 24-h PMP.

RUNOFF

18. The Nature of Runoff. Runoff is that part of precipitation and contribution from other sources which appears in surface streams of perennial or intermittent nature. This is measured as the flows collected from a drainage basin (catchment area or watershed) at the basin outlet (point of interest). The runoff units are cubic feet per second, acre-feet, or inches over the drainage basin.

Depending upon the source of runoff, it is defined as surface runoff, subsurface runoff, or groundwater runoff. Also terms like overland flow, streamflow, direct runoff, storm runoff, and base flow are used to define the nature of various components of total runoff measured in a stream.[5]

Runoff is a residual phenomenon which takes place after certain losses have been satisfied. The part of precipitation that forms the runoff is called the precipitation excess or rainfall excess. Figure 19 shows a generalized cross section defining runoff terms. The most important loss is evaporation as defined in Art. 10.

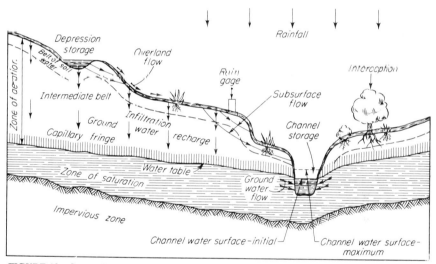

FIGURE 19 Generalized cross section defining runoff terms.

The other losses are interception, infiltration, deep percolation, and depression storage. The volume of water in transit in the overland flow sheet is called surface detention. From total precipitation to total runoff in a stream, the various flow components can be related as indicated below.

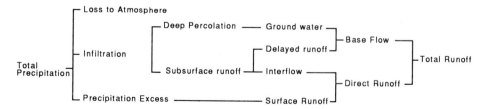

The runoff from a drainage area is influenced by two major factors: climatic factors and physiographic factors. The climatic factors include precipitation, its areal and temporal distribution and frequency of occurrence, and temperature, wind, atmospheric pressure, solar radiation, and humidity affecting the rates of evapotranspiration. The physiographic factors include drainage area characteristics, land use (vegetation, species, composition, age, density, seasonality), soil type (permeability), geological conditions (groundwater aquifers), the presence of lakes and swamps, and channel characteristics.

19. Measurement of Runoff. For hydrology and hydraulic studies of projects, the main concern is the measurement of volume of runoff and its time distribution. Also of interest is changes in runoff values due to human activities. Continuous record of runoff at a point of interest is called streamflow data. The locations at which the streamflows are measured are designated as stream gaging stations. The primary responsibility for operating and maintaining the stream gaging stations and publishing the streamflow data rests with the Water Resources Divisions of the U.S. Geological Survey (USGS). However, a number of state, local, and private agencies also operate and maintain stream gaging stations.

In small streams, irrigation canals, and closed conduit, a continuous measurement of flow rate is feasible. However, on large streams, it is difficult to make a direct, continuous measurement of flow rates. A general procedure for these streams is to make direct continuous measurement of river level (stage), and periodic stage and streamflow (discharge) measurement to develop a stage-discharge relationship. The continuous stage record is converted into discharge data using this relationship. The accuracy of the discharge data depends upon the adequacy of the stage-discharge relationship.

Detailed procedures for measurements of river stages and periodic discharges on large rivers, continuous or periodic measurement of discharges on small streams and conduits, descriptions of instruments and/or facilities required to make stage and discharge measurements, and procedures for interpretation, computation, and publication of data are available in a number of publications.[55-59] Currently, in some cases the stage data are transmitted through telemetric or satellite transmission[60] and processed by computers to derive streamflow records. The stage-discharge relationships are updated periodically if these are expected to change because of aggradation or degradation in the river.

The river stage is generally measured using manual or automatic gages. Peak flood stages are measured using crest stage gages. The discharge of a river is calculated from measurements of velocities and depths at a selected number of subsections across the river, and by adding the discharges in the subsections to obtain total discharge. A number of methods are available to measure velocity and depth.

The USGS collects continuous stage records and periodic discharge measurements (to update or to develop new stage discharge relationships) at about 8000 stream gaging stations, and continuous stage at about 1000 lakes and reservoirs throughout the United States. Discharges at about 8000 additional stream sites are measured periodically or at crest peak stages, at low flows, or in response to other selected hydrologic conditions.[61]

20. Sources of Streamflow Data. Streamflow data are published by the USGS in the form of mean daily flows from midnight to midnight in the water-resources data book for each state. For each gaging station a brief history is provided. Generally the data are reviewed and adjusted for errors resulting from instrumental and observational deficiencies. Data on maximum annual flow peaks and time of occurrence also are given in the data book.

Daily streamflow, hourly gage heights, and flood peak data can be retrieved from USGS WATSTORE System.[62] Earth-Info Inco, formerly U.S. West Optical Publishing, Boulder, Colo., and the National Technical Information Service, Springfield, Va., also provide streamflow data. The USGS district offices also provide data if direct retrieval from WATSTORE system is not available.

Mean daily discharges are computed from hourly or bihourly gage height data and stage-discharge relationships. The gage height data are corrected for silting or scouring of the control defining the stage-discharge relationship and then used with the appropriate rating curve.

In large streams, the mean daily discharges are only slightly less than the instantaneous discharges (Fig. 20b). On small streams, however, the difference between the instantaneous flow hydrograph and mean daily flow hydrograph is quite significant (Fig. 20a). Therefore, depending upon the type of analysis, appropriate flow (hourly or mean daily) data should be collected.

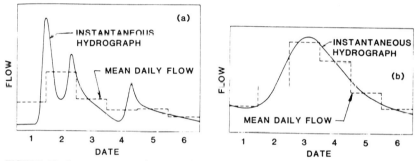

FIGURE 20 Instantaneous and mean daily flow hydrographs. (a) Small stream. (b) Long stream. (*Source: Ref. 1.*)

21. Groundwater Runoff. Although water is present in the ground within all the belts between the impervious zone and the surface (Fig. 19), the term "groundwater" as used in hydraulic engineering refers only to water recoverable by springs and wells. The space between the groundwater surface and the water table is called the *zone of aeration* and includes the *capillary fringe,* where water is held in the soil pores by capillarity; the *intermediate belt,* where suspended water (called "vadose" water) is held by molecular attraction; and the belt of soil water. Plants extend their roots into the intermediate belt to various depths, while trees usually extend their roots into the zone of saturation.

The term *aquifer* refers to a water-bearing geologic formation, i.e., one that is saturated with water. If confined between impervious strata, an aquifer may contain water under pressure, in which case it is called *artesian*.

After the occurrence of direct runoff accompanying a stream rise, streamflow during the descending limb or *recession* side of a flow hydrograph occurs as outflow from channel storage and outflow from groundwater storage (groundwater or base flow). The former outflow occurs relatively rapidly, following which flow is entirely from groundwater storage. From studies of many hydrographs it has been found[63] that groundwater-depletion curves for a given drainage basin are nearly always the same; hence the term *normal groundwater-depletion curve* is used. It has been found further that this curve, or at least segments of it, follows a simple inverse exponential function[64] of elapsed time of the form

$$Q_t = Q_0 K^{-t}$$

where Q_0 is the discharge at any instant, Q_t is the discharge t days later, and K is the "daily depletion factor." As Q_t is the derivative of storage with respect to time, integration of this equation gives

$$S_0 = \frac{Q_0}{\log_e K}$$

where e is the base of natural logarithms and S_0 is the groundwater storage available for runoff at the time of Q_0. From this, it is seen that the discharge at any time is proportional to the water remaining in storage. Then the value of K can be determined by plotting observed recession curves on semilogarithmic paper, taking care to select periods of little or no direct runoff. In Fig. 21, the recession constant K is the average slope of the streamflow hydrographs, plotted on semilogarithmic paper, for 3 years during which typically low summer flows were preceded by periods of relatively high direct runoff.

22. Seasonal and Long-Term Variations. Mean annual runoff in the United States is shown in Fig. 22 as lines of equal runoff. The lines are based on limited data and represent the integrated runoff for the entire basin above a gaging station. Local variation due to nonuniform rainfall, geology, soil cover, etc., could be quite significant.

Mean values of runoff serve an important purpose but do not provide all pertinent information necessary for hydrology and hydraulic analyses. Variation in daily flows throughout a year, seasonal variation, or variation from year to year are important. Figure 23 shows the seasonal variation of various streams in the United States[65] and one in Haiti in terms of average monthly percentages of

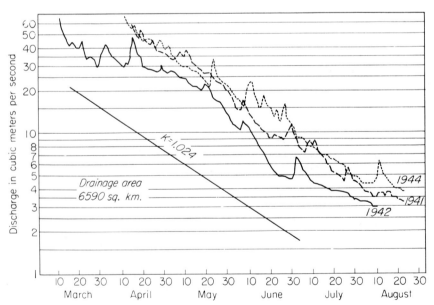

FIGURE 21 Semilogarithmic plotting of streamflow and derivation of the recession constant K. Gediz River at Kizkoprusu, Turkey.

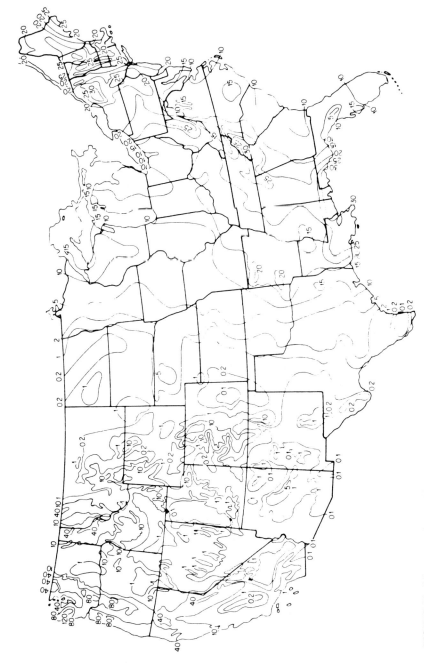

FIGURE 22 Mean annual runoff (inches) in the United States. (*U.S. Geological Survey.*)

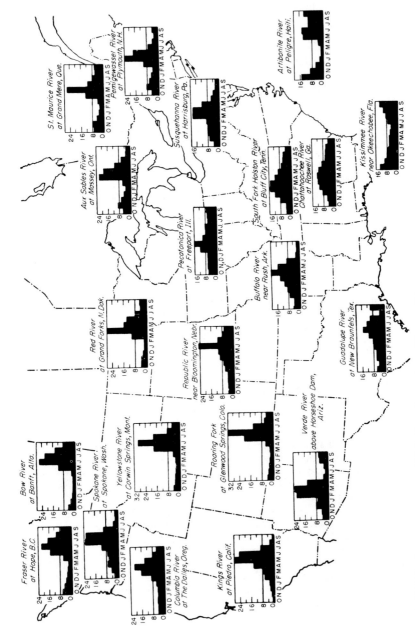

FIGURE 23 Average percentage distribution of annual runoff by months.

the annual runoff. In most of the United States, runoff is generally low in the autumn and winter months. Spring and early summer are periods of high flows.

Long-term variation in streamflow is best studied by a moving-average technique.[5] This technique shows whether a definite trend in the quantity of runoff, either upward or downward, exists in the time series. Short-term variation can be studied by flow-duration analysis.[5]

23. Generation of a Long-Term Flow Sequence. For most hydrologic studies, a long-term daily, monthly, or yearly flow sequence is required. For a specific site the short-term streamflow record overlapping with long-term rainfall or streamflow records at a station in the general vicinity may or may not be available. In any case, the problem requires great caution and judgment and thorough familiarity with the hydrologic characteristics of the region and the two drainage basins being studied.

When concurrent streamflow data exist, monthly or yearly data can be generated using the concurrent data at the project site and at nearby stations with long-term data, using linear or nonlinear regression equations.[5] Daily flows can be generated by this method if a good correlation (correlation coefficient of at least 0.90 and above) is established. In case of a poor correlation, it is better to develop monthly flows and estimate daily flow using daily distribution similar to that observed for a given month at the station with long-term data.

Linear regression technique[5] can also be used for estimating streamflow from long term rainfall data. In most cases, flows correlated with concurrent month and previous month rainfalls indicate better correlation. In some cases, evapotranspiration losses may have to be included to improve the correlation.

Values for missing periods in monthly records can be filled using a simple arithmetic average at the index stations or by the normal-ratio method[1,11] as recommended for precipitation records (Art. 3). Stochastic techniques[13,14] also are used to fill in the missing data and to extend the short-term record.

An estimate of streamflow at a site entirely lacking in streamflow measurement should be made with great caution, particularly in arid and semiarid regions where the quantity of flow, especially during low-flow periods, is greatly affected by climatic and physiographic (particularly geologic) features. In humid regions where fairly uniform precipitation occurs, it is permissible to transpose runoff data from a nearby station using a drainage-area ratio. However, a sufficient number of gaging stations should be used to establish whether or not seasonal runoff area is in fact proportional to the drainage area. In areas with some variability in rainfall, a combined drainage area and seasonal precipitation ratio may be used.

In arid and semiarid areas or humid areas with different geologic conditions, a possible solution is to obtain concurrent spot discharge measurements at the project site and at a station with long-term flow data several times during a year. Using these spot measurements one can develop a correlation[66] to transpose data from the long-term station.

24. Utilization Studies. Long-term streamflow data, preferably including major drought periods, are used for sizing the storage reservoir for irrigation, power, and recreational use. The techniques used include flow-duration curves, sequential and nonsequential mass curves, and sequential computer simulation.

Since future operation of a reservoir and the expected safe yield are based on past records, it is apparent that these records should be of sufficient length. In some cases, it is advisable to derive long synthetic sequences (of 100 years or more) using stochastic procedures.[1] These sequences display the statistical properties of historical sequences but provide a tool to investigate the effects of drought periods on reservoir safe yield.

On run-of-the-river projects, low flows of various durations and recurrence intervals are used to investigate irrigation or power potential of a stream.

FLOODS

25. Flood Hydrograph Components. A hydrograph is a chronological history of flow rates at a selected location. It represents the integrated effect of the physiographic and climatic characteristics that govern the relationships between rainfall and runoff of a particular drainage basin. Detailed analysis of a flood hydrograph is usually important for flood damage studies, estimation of infiltration losses, flood forecasting, and establishing design flows for impoundment structures and structures carrying flood flows.

A typical single-peaked hydrograph (Fig. 24) consists of four parts, approach recession (AB), rising (concentration) segment or limb BC, falling segment or limb CD, and recession DE. At any time during the rising or falling segment, the total runoff rate consists of surface runoff, subsurface runoff, and base flow. For most engineering applications, the hydrograph analyses are confined to the separation of direct runoff (surface plus subsurface) and base flow. A variety of techniques have been suggested for separating base flow and direct runoff.[5,11] Figure 25 indicates three most commonly used methods.[67] The direct runoff volume is equivalent to excess rainfall. A graph of excess rainfall vs. time (hyetograph) is an important component of rainfall-runoff and flood studies. Figure 26 shows a typical hyetograph and flood hydrograph.

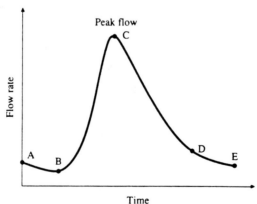

FIGURE 24 Typical hydrograph.[67]

26. Unit Hydrograph. The magnitude of the flood peak, time to the peak, and shape of the flood hydrograph depend upon the characteristics of the storm producing the flood and the physiographic characteristics of the drainage basin. Therefore, the basic requirement to transform rainfall excess (total rainfall minus retention losses) into a flood hydrograph is the determination of a physical response function of the drainage basin. This response was studied by Sherman[68,69]

FLOODS

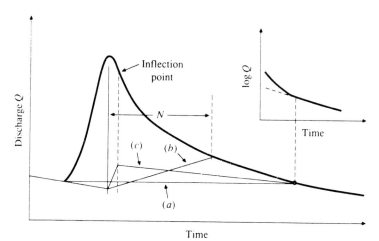

FIGURE 25 Baseflow separation. (*a*) Straight line method. (*b*) Fixed-base method. (*c*) Variable-slope method.[67]

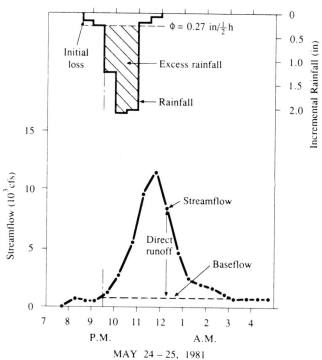

FIGURE 26 Hyetograph and flood hydrograph.[67]

and presented as the concept of a unit hydrograph. The unit hydrograph is defined as a hydrograph at a point of interest from one unit of rainfall excess generated uniformly over the drainage basin upstream of the point of interest at a uniform rate during a specified time period. That definition and the following basic assumptions constitute the unit hydrograph theory.[5]

1. The rainfall excess (or effective rainfall, equal to total rainfall minus retention, surface ponding, and other losses) is uniformly distributed within its duration.
2. The rainfall excess is uniformly distributed over the drainage basin.
3. The base or time duration of the surface runoff hydrograph resulting from the rainfall excess of unit duration is constant.
4. The ordinates of the surface runoff hydrograph of a common base time are directly proportional to the total amount of rainfall excess.
5. The surface runoff hydrograph for a given watershed resulting from a given period of rainfall excess reflects all the combined physiographic and other characteristics of the watershed.

The concept of unit hydrograph application[70] is given in Fig. 27. The hydro-

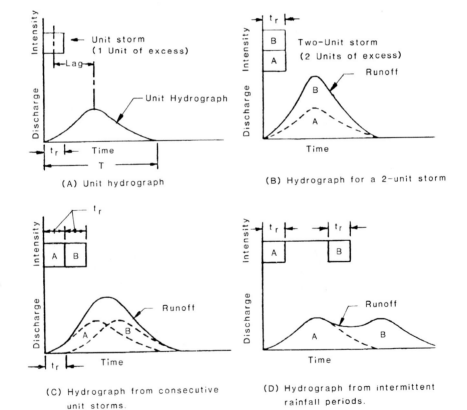

FIGURE 27 Unit hydrograph application.

logic data for unit hydrograph analysis and size of the drainage basin are carefully selected so that they approximately meet the above basic assumptions.

The size of the drainage basin for which a flood hydrograph is required or is to be reconstituted (for deriving unit hydrograph parameters) may vary from a few to thousands of square miles. The large basin should be divided into subbasins depending upon the drainage configuration. The size of each subbasin should be such that uniform rainfall (or snowmelt) can be assumed over the subbasin.

Commonly used methods to derive a unit hydrograph include:

- Direct method
- Dimensionless graph–lag curve method
- Synthetic method

The direct method is used when there is a stream gaging station at or near the point of interest and the drainage area above the point is small enough to treat it as a single basin. Preferably there should be a recording precipitation station in the immediate vicinity or in the drainage basin. The procedure for deriving a unit hydrograph from an observed hydrograph and parameters of the unit hydrograph are shown in Fig. 28. The unit hydrograph of desired unit duration can be obtained from the derived unit hydrograph of specific duration by S-curve technique.[5]

The dimensionless graph–lag curve method requires historic flood hydrographs and corresponding hourly precipitation data. The stream gaging and precipitation stations should be within or near the drainage basin under study. For the gaging stations in the vicinity only the station draining basins having physiographic, physical, and climatic characteristics similar to the basin under study should be selected.

Barnes[71] and the U.S. Bureau of Reclamation[72] have documented the procedures for deriving dimensionless graphs and lag curves. A typical average dimensionless graph (based on a number of gaging station data) and two lag curves developed by Harza Engineering Company for the probable maximum flood studies at three hydroplants in the Housatonic River Basin are shown in Fig. 29.

Synthetic methods are used to synthesize the unit hydrographs for ungaged drainage basins where the estimates of design floods are required. The most commonly used methods are Snyder's method, SCS dimensionless unit graph, and Clark's method. A detailed discussion on the derivation of a unit graph by Snyder's method is given by the U.S. Army Corps of Engineers.[73]

The SCS recommends[74] the use of a triangular or curvilinear unit hydrograph. The time to peak and peak discharge are computed from empirical relationships based on watershed characteristics.

Clark's method[75] has high rationality. The method should be used when relationships linking Clark's parameters to basin characteristics are available or can be developed. A detailed discussion on Clark's method also is provided by the U.S. Army Corps of Engineers.[70]

27. Infiltration Losses. A concept of losses from precipitation falling over a drainage basin was given in Art. 18. For the purpose of deriving a flood hydrograph estimate from a given storm, the most commonly used methods to define infiltration losses (for computing rainfall excess) are:

- Initial retention and a uniform retention rate
- SCS curve number
- Exponential loss rate functions

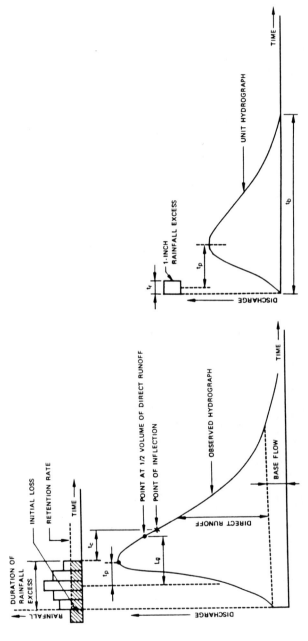

FIGURE 28 Unit hydrograph parameters.

NOTES: t_p = Time from mid-point of rainfall excess duration to time of peak, hours
t_c = Time from end of rainfall excess to point of inflection, hours
L_g = Time from mid-point of rainfall excess duration to time of occurrence of one-half volume, hours
t_r = Duration of rainfall excess, hours
t_b = Time base of unit hydrograph, hours

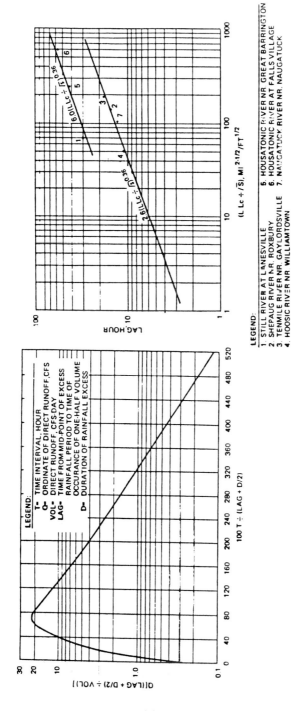

FIGURE 29 Dimensionless graph-lag curves for Housatonic River Basin.

A brief discussion of these methods is given by the U.S. Army Corps of Engineers.[31] These methods assume that the infiltration is lost from the basin and does not contribute to the runoff. Soil moisture accounting and surface storage recovery are not considered. These methods are used when a single flood event is to be estimated. If a continuous simulation of the rainfall-runoff process on an hourly or daily basis is desired, the soil moisture accounting procedures are used.

28. Flood Hydrographs. Flood hydrographs are developed using rainfall-runoff simulation models. A number of deterministic hydrologic simulation models have been developed since 1960. Some of the models simulate single rainfall-runoff events and may be designated as event simulation models. The most commonly used are HEC-1[31] and TR-20.[76] Other models provide a continuous simulation over long time periods. Some of the latter are the Stanford watershed model,[77] the U.S. National Weather Service model,[78] and the BRASS model.[79]

Some models use unit hydrographs to convert rainfall excess into flood hydrographs while others use a linear reservoir concept as discussed by Chow.[5] A schematic diagram of the National Weather Service model[80] is given in Fig. 30.

All models should be calibrated on the basis of historic rainfall and runoff data and other physical characteristics of the drainage basin. The general procedure is to select the flood events to be reconstituted and assemble all known facts regarding the floods including the spatial and temporal distribution of storm precipitation, soil moisture conditions, river discharges, soil types, and related infiltration losses, etc. The calibrated models are then used to predict hydrographs resulting from selected storm patterns.

29. Frequency Analyses. Flood frequency analyses are performed for sizing bridges, culverts, and other facilities, for designing levees, spillways, and other control structures, and delineating flood plains, on the basis of the estimated magnitude of floods of various return periods. The analyses are based on historical flood data, which are available from various federal, state, and private agencies, as discussed in Art. 20.

Many families of probability distributions could be used to describe the distribution of floods.[81] They include both the normal and log-normal distributions, the Pearson and log-Pearson distributions, and the generalized extreme value and Gumbel distributions. Goodness-of-fit tests[24] are generally used to select an appropriate distribution consistent with the available data for a particular site.

The graphic evaluation of a fitted distribution is generally performed by plotting the observed flood peaks on probability paper. Commercially, special probability papers for normal, log-normal, exponential, and Gumbel distributions are available. A number of empirical relationships[82] are available to define the plotting positions of the observed data.

The most commonly used are:

$$Tr = \frac{N}{M} \qquad Tr = \frac{N}{M - 0.5} \qquad Tr = \frac{N + 1}{M}$$

where Tr is the return period in years, N is the length of continuous record in years, and M is the order of magnitude of the flood peak (descending order).

The reliability of a frequency estimate depends on the length of the observed record rather than on the specific method of fitting the probability distributed. Therefore, an estimate of a 10-year flood based on a 50-year record is considered more reliable than an estimate of a 50- or 100-year flood from the same record.

The probability distribution fitted to the flood peaks can be considered to represent the mean value of the data at a given return period. The distribution of the

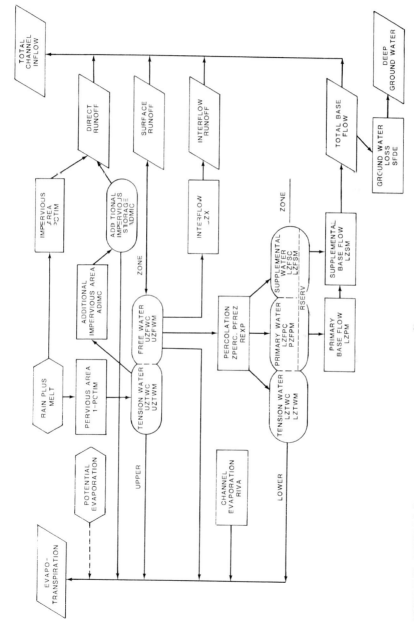

FIGURE 30 NWSRFS (Sacramento) model schematic diagram.[80]

data at a given return period is described by constructing confidence limits on both sides of the fitted curves. Confidence limits, or confidence intervals, define the range within which the estimated value at a given return period could be expected to fall with a specified confidence or level of significance. An upper confidence limit should be considered at a 95 percent confidence level to determine a design flood peak corresponding to larger return periods when the length of observed record is short (less than 30 years).

Uniform procedures for flood frequency analysis are prescribed by the U.S. Water Resources Council[83] Bulletin 17B. This bulletin provides procedures for developing frequency data using log-Pearson type III distribution. Special procedures also are provided for identifying and detecting unusual situations, zero flows, outliers, historic peaks, regional information, confidence intervals, and expected probability.

30. Risk Evaluation. Because of uncertainty in the magnitude of flood for a given return period, the question arises as to the risk that a particular discharge will be exceeded at least once in the selected return period. This can be quantified by the following equation:

$$R = 1 - \left(1 - \frac{1}{Tr}\right)^M$$

in which R is the percent change of a design discharge being exceeded at least once in the design life, Tr is the return period for the design discharge, and M is the design life in years. A brief discussion of risk evaluation is given by the Federal Highway Administration[84] (FHA).

31. Regional Flood Frequency. Reliable flood frequency estimates, as well as estimates of flood frequency for ungaged areas, are made by combining the results of flood frequency analyses of all available flood data in a region. The U.S. Geological Survey (USGS) has published regional regression equations to compute floods of various return periods in nearly all states. These reports can be obtained from the district offices of the USGS.

Other methods generally used are:

- USGS index—flood method
- Regionalization of parameters

A discussion of these methods is provided by the FHA.[84]

32. Floods from Small Areas. Small areas of less than a few square miles may be defined usefully, such as airfields, agricultural drainage units, and natural areas. Urban buildup areas drained by storm sewers may also be classified as "small areas."

Chow[85] has listed a number of empirical and semiempirical formulas to represent the relationship between rainfall and peak discharge. The other methods[86] include the rational method, overland-flow methods, the inlet method, and the unit hydrograph method. Some of the methods provide peak flow only, and the others provide complete flood hydrographs.

INFLOW DESIGN FLOOD

33. Concept. The design flood is the maximum flood against which a structure is protected. If the design flood is larger than those which could possibly occur, a large investment may be wasted in providing the excess capacity which will never be used. On the other hand, if it is smaller than those which could possibly occur, there will be a risk of failure of the structure, with associated consequences. Thus, the determination of a design flood should comprise two steps—first to select the desirable safety standard and second to estimate the flood that meets that standard. The current practices of selecting spillway design floods are discussed by the ASCE.[87]

The Federal Energy Regulatory Commission (FERC)[88] has recommended that the adequacy of a spillway must be evaluated by considering the hazard potential which would result from failure of the project works during flood flows. If failure of the project works would present a hazard to human life or would cause significant property damage, the project works must be designed to withstand either the loading or the overtopping which may occur during a flow up to the probable maximum flood (PMF). A flood level lower than the PMF may be selected if a failure at a level higher than the selected level would not constitute a hazard to downstream life and property higher than what would have occurred if the structure were not present.

34. Probable Maximum Flood (PMF). The derivation of the PMF involves the following tasks:

- Estimation of PMP
- Estimation of snowmelt
- Estimation of infiltration losses
- Transformation of PMP to PMF
- Comparison with flood of record

The procedures for estimations of PMP and infiltration losses are given in Arts. 17 and 27.

In some regions, the PMF could result from a combination of rainfall and snowmelt. The usual practice is to determine the seasonal variation in the PMP and derive PMF for the following two conditions:

1. Rainfall PMF superimposed on a major snowmelt flood
2. Probable maximum snowmelt flood superimposed on a major rainfall flood

For the first condition, all-season PMP (if it occurs during the melt period) or the seasonal PMP is combined with a snowmelt resulting from a snow pack with a 100-year snow water equivalent and a daily temperature equivalent to the maximum 12-h persisting dew-point temperature for the month when the snowmelt would be considered. The flood computations including snowmelt can be made using the HEC-1 program.[31] The temperature and wind data prior to and during the PMP can be derived using the guidelines given in HMR No. 43[50] or in other National Weather Service Publications.

The probable maximum snowmelt flood can be estimated using the guidelines provided by the U.S. Army Corps of Engineers.[32] The rainfall during the melt period could be either a major historical storm or the 100-year rainfall.

Transformation of a PMP to the PMF is accomplished by the unit hydrograph method or the computer-based simulation model discussed in Art. 28.

In most regions, a design flood may be preceded and/or succeeded by floods of lesser magnitude. Therefore, the possibility of antecedent or subsequent floods should be evaluated. As a general guideline, the Bureau of Reclamation uses a

100-year precipitation or flood prior to the PMP with a 3-day interval for areas east of the Sierra Nevada and Cascadas Ranges. West of these ranges an interval between 2 and 3 days is used. West of the 103rd meridian, no antecedent flood is prescribed, although it is expected that initial losses are satisfied by an earlier storm.

The U.S. Weather Bureau recommends an antecedent or subsequent storm of 30 percent of the main storm with a minimum of a 3-day period between the storms.

Before a PMF estimate is adopted for design purposes, its reasonableness should be investigated further by comparing the PMP with greatest storms and the PMF with floods of record. The WMO[49] has provided a reference table to compare the PMP estimated with greatest observed rainfalls.

For a given region, all known maximum discharges (cubic feet per second per square mile) should be plotted against the drainage area in square miles on a logarithmic paper. The envelope curve drawn on these plotted points can be used to compare the estimated PMF. Crippen[89] has provided empirical relationships, based on this procedure, for various regions of the United States. Similar curves are also developed by Costa.[90] However, when comparing the PMF estimates, the similarities in the meteorological and topographical characteristics between the basin where the PMF is estimated and the areas for which the envelope curve is developed should be carefully evaluated.

35. Design Flood. The design flood for a major structure may be a PMF or a flood lower than a PMF, as discussed in Art. 33. The incremental damage study to determine an appropriate design flood is usually performed using dam-break analysis. The inflow flood hydrographs representing selected percentages of the PMF are considered. The flood level at which the incremental damage is insignificant is adopted as the inflow design flood.

36. Standard Project Flood (SPF). The SPF as defined by the U.S. Army Corps of Engineers[45] is a "hydrograph representing runoff from the Standard Project Storm (SPS)." It is a hypothetical flood estimated by transposing the SPS. The SPS is the most severe storm that is considered "reasonably characteristic" of a region. It often undercuts storms of extraordinary severity.

Estimation of SPS can be best accomplished by following the procedures prescribed by the Army Corps of Engineers.[45] Transformation of an SPS to the SPF can follow the procedures for transforming the PMP to the PMF. The use of SPF in the design of major dams is rather uncommon today.

37. Flood Hydrograph of a Selected Return Period. The flood frequency analyses discussed in Art. 29 provide estimates of flood peaks only. In design of many water-bearing structures, flood hydrographs are needed to evaluate the effect of flood volumes. A general procedure is to determine the flood volumes of various durations (1-day, 3-day, 5-day, etc.) for the selected return period through frequency analyses, combine with the corresponding flood peak, and construct the flood hydrograph.

In some cases the flood data (peak and volume) are not available. In that case the flood hydrographs are based on storm rainfalls of selected return periods. The rainfall amounts of various durations and for a selected return period are obtained from depth-duration-frequency curves.[19,20] The transformation of rainfall to the flood hydrograph is made by the unit hydrograph method using appropriate infiltration losses. However, when this procedure is used, the design flood should be designated as "a flood resulting from 50- or 100-year storm." This could be significantly different from a 50- or 100-year flood based on the frequency analyses of flood peaks and volumes.

COMPUTER PROGRAMS

A number of microcomputer programs for hydrologic analyses are available. The programs in the public domain are primarily developed by federal or state agencies. Private agencies and universities also have an extensive libraries of softwares. Some of the most commonly used programs are listed below:

1. HEC-1, Flood Hydrograph Package, Hydrologic Engineering Center, U.S. Army Corps of Engineers, Davis, Calif.
2. HEC-2, Water Surface Profiles, Hydrologic Engineering Center, U.S. Army Corps of Engineers, Davis, Calif.
3. HEC-4, Monthly Streamflow Simulation, Hydrologic Engineering Center, U.S. Army Corps of Engineers, Davis, Calif.
4. HEC-5, Simulation of Flood Control and Conservation Systems, Hydrologic Engineering Center, U.S. Army Corps of Engineers, Davis, Calif.
5. SSARR, Streamflow Synthesis and Reservoir Regulation, North Pacific Division, U.S. Army Corps of Engineers, Portland, Ore.
6. TR20, Computer Program for Project Hydrology Formulation, U.S. Soil Conservation Service.
7. WSP-2, Water Surface Profiles, U.S. Soil Conservation Service.
8. DWOPER, Operational Dynamic Wave Model, U.S. National Weather Service.
9. HECWRC, Flood Flow Frequency Analysis. Hydrologic Engineering Center, U.S. Army Corps of Engineers.
10. MAX, Flood Frequency Analysis for Use with Historic Information, Department of Environmental Engineering, Cornell University, Ithaca, N.Y.
11. DAM BREAK, U.S. National Weather Service.

REFERENCES

1. LINSLEY, RAY K, JR., MAX A. KOHLER, and JOSEPH L. H. PAULHUS, *Hydrology for Engineers*, 3d ed., McGraw-Hill, New York, 1982.
2. U.S. WEATHER BUREAU, "Instructions for Climatological Observers," *Circular* B, 11th ed., 1962.
3. WORLD METEOROLOGICAL ORGANIZATION, "Guide to Meteorological Instrument and Observing Practices," WMO no. 8. TP 3, 4th ed., 1971.
4. WORLD METEOROLOGICAL ORGANIZATION, *Guide to Hydrometeorological Practices*, 1965.
5. CHOW, V. T. (Editor), *Handbook of Applied Hydrology*, McGraw-Hill, New York, 1964.
6. RODDA, J. C., "The Systematic Error in Rainfall Measurement," *Journal of the Institute of Water Engineering*, no. 21, pp 173–177, 1967.
7. RODDA, J. C., "Hydrological Network Design—Needs, Problems and Approaches," WMO/IHD *Report* no. 12, 1969.
8. GREENE, D. R., and A. F. FLANDERS, "Radar Hydrology, The State of the Art," First Conference on Hydrometeorology, Ft. Worth, Tex., April 20–22, 1976, American Meteorological Society, Boston.

9. COLLINGE, V. K., and C. KIRBY (Editors), *Weather Radar and Flood Forecasting,* Wiley, New York, 1987.
10. PARKE, PETER (Editor), *Meteorological Monographs, Satellite Imagery Interpretation for Forecasters,* National Weather Association, Temple Hills, Md., December 1986.
11. PAULHUS, J. L. H., and M. A. KOHLER, "Interpolation of Missing Precipitation Records," *Monthly Weather Review,* vol. 80, no. 8, August 1952.
12. U.S. NATIONAL WEATHER SERVICE, "National Weather Service River Forecast System, Forecast Procedures," *NOAA Technical Memorandum,* NWS HYDRO 14, December 1972.
13. TEXAS WATER DEVELOPMENT BOARD, Systems Engineering Division, "Stochastic Optimization and Simulation Techniques for Management of Regional Water Resources Systems," vol. IIB, Fill-in Program Description, December 1970.
14. U.S. ARMY CORPS OF ENGINEERS, Hydrologic Engineering Center, "Monthly Streamflow Simulation, Computer Program HEC-4," 723-X6-12340, Davis, Calif., 1971.
15. YEVJEVICH, V., and R. I. JENG, "Properties of Non-Homogeneous Hydrologic Series," Colorado State University, Hydrology Paper no. 32, Fort Collins, Colo., April 1969.
16. YEVJEVICH, V., *Probability and Statistics in Hydrology,* Water Resources Publications, Fort Collins, Colo., 1972.
17. SEARCY, J. K., and C. H. HARDISON, "Double-Mass Curves," Manual of Hydrology, Part I, General Surface Water Techniques, U.S. Geological Survey Water Supply Paper 1541-B, 1960.
18. SHAW, ELIZABETH M., *Hydrology in Practice,* 2d ed., Van Nostrand Reinhold (International) London, 1988.
19. U.S. WEATHER BUREAU, "Rainfall Frequency Atlas of the United States for Durations from 30 Minutes to 24 Hours and Return Periods from 1 to 100 Years," Technical Paper 40, 1961; "Two- to 10- Day Precipitation for Return Periods of 2 to 100 years in the Contiguous United States," Technical Paper 49, 1964; Technical Paper 47, 1963, and 52, 1965, for Alaska.
20. MILLER, F. J., R. H. FREDERICK, and R. J. TRACEY, *Precipitation Frequency Atlas of the Western United States,* vols. I–XI, NOAA Atlas 2, National Weather Service, 1973.
21. HERSHFIELD, D. M., "An Empirical Comparison of the Predictive Values of Three Extreme Value Procedures," *Journal of Geophysical Research,* vol. 67, April 1962.
22. HUFF, F. A., and J. C. NEIL, "Comparison of Several Methods for Rainfall Frequency Analysis," *Journal of Geophysical Research,* vol. 64, May 1959.
23. MARKOVIC, R. D., "Probability Functions of Best Fit to Distribution of Annual Precipitation and Runoff," Colorado State University, Hydrology Paper 8, August 1965.
24. HAAN, CHARLES, T., *Statistical Methods in Hydrology,* The Iowa State University Press, Ames, Iowa, 1979.
25. TREWARTHA, GLENN T., *An Introduction to Climate,* 3d. ed., McGraw-Hill, New York, 1959.
26. U.S. DEPARTMENT OF AGRICULTURE, Soil Conservation Service, *National Engineering Handbook, Snow Survey and Water Supply Forecasting,* Sec. 22, April 1972.
27. U.S. DEPARTMENT OF AGRICULTURE, Science and Education Administration, "Field Manual for Research in Agricultural Hydrology," *Agriculture Handbook* 224, Revised February 1979.
28. U.S. DEPARTMENT OF AGRICULTURE, Soil Conservation Service, Washington, "Annual Data Summary of Federal, State and Private Corporation Snow Surveys, Water Year 1986," Spokane, Wash.

REFERENCES

29. ANDERSON, E. A., "A Point Energy and Mass Balance Model of a Snow Cover," *NOAA Technical Report* NWS19, February 1976.
30. ANDERSON, E. A., "National Weather Service River Forecast System—Snow Accumulation and Ablation Model," *NOAA Technical Memorandum* NWS HYDRO-17, November 1973.
31. U.S. ARMY CORPS OF ENGINEERS, Hydrologic Engineering Center, "HEC-1 Flood Hydrograph Package," Users Manual, September 1990.
32. U.S. ARMY CORPS OF ENGINEERS, "Runoff from Snowmelt," Engineering and Design Manual EM 1110-2-1406, January 1960.
33. LINSLEY, RAY K., "A Simple Procedure for the Day-to-Day Forecasting of Runoff from Snowmelt," *Transactions of the American Geophysical Union* 24 (part III), 1943.
34. BOARDMAN, H. P., "Snow Survey versus Winter Precipitation for Forecasting Runoff of the Tuolumne River, California," *Transactions of the American Geophysical Union*, Oct. 28 1947.
35. RANGO, A., V. V. SALOMONSON, and J. L. FOSTER, "Seasonal Streamflow Estimation in the Himalayan Region Employing Meteorological Satellite Snow Cover Observations," *Water Resources Research*, vol. 13, no. 2, 1977.
36. HAVENS, A. V., "Drought and Agriculture," *Weatherwise*, vol. 7, 1954.
37. THORNTHWAITE, C. W., "Climate and Moisture Conservation," *Annals of the Association of American Geographers*, vol. 37, no. 2, June 1947.
38. HUFF, F. A., and S. A. CHANGNON, JR., "Drought Climatology of Illinois," *Illinois State Water Survey Bulletin* 50, 1963.
39. TERSTRIEP, M. L., M. DEMISSIE, D. C. NOEL, and H. V. KNAPP, "Hydrologic Design of Impounding Reservoir in Illinois," *Illinois State Water Survey Bulletin* 67, 1982.
40. HUDSON, H. E., and W. J. ROBERTS, "1952–55, Illinois Drought with Special Reference to Impounding Reservoir Design," *Illinois State Water Survey Bulletin* 43, 1955.
41. FARNSWORTH, RICHARD K., EDWIN S. THOMPSON, and EUGENE L. PECK, "Evaporation Atlas for the Contiguous 48 United States," *NOAA Technical Report* NWS33, Office of Hydrology, National Weather Service, June 1982.
42. VIESSMAN, WARREN, JR., JOHN W. KANPP, GARY L. LEWIS, and TERENCE E. HARBAUGH, *Introduction to Hydrology*, 2d ed., IEP, A Dun-Donnelley Publisher, New York, 1977.
43. KOHLER, M. A., T. J. NORDENSON, and W. E. FOX, "Evaporation from Ponds and Lakes," U.S. Weather Bureau Research Paper 38, 1955.
44. PECK, EUGENE, L., "Relation of Orographic Winter Patterns to Meteorological Parameters, Distribution of Precipitation in Mountainous Areas," WMO/OMM no. 326, Geilo Symposium, Norway, vol. II, no. July 31–Aug. 5, 1972.
45. U.S. ARMY CORPS OF ENGINEERS, "Standard Project Flood Determinations," EM 1110-2-1411, March 1952, Revised March 1965.
46. KNAPP, H. VERNON, and MICHAEL L. TERSTRIEP, "Effects of Basin Rainfall Estimates on Dam Safety Design in Illinois," *State Water Survey Report* 253, May 1981.
47. HUFF, FLOYD A., and JAMES R. ANGEL, Frequency Distribution and Hydroclimatic Characteristics of Heavy Rainstorms in Illinois, Illinois State Water Survey, *ISWS Bulletin* 70/89, 1989.
48. U.S. WEATHER BUREAU, "Seasonal Variation of the Probable Maximum Precipitation East of the 105th Meridian for Areas from 10 to 1,000 Square Miles and Durations of 6, 12, 24 and 48 Hours," *Hydrometeorological Report* 33, prepared by J. T. Riedel, J. F. Appleby, and R. W. Scholeomer, Washington, D.C., April 1956.
49. WORLD METEOROLOGICAL ORGANIZATION, "Manual for Estimation of Probable Maximum Precipitation," *Operational Hydrology Report* 1, WMO 332, 2d ed., Switzerland, 1986.

50. U.S. National Weather Service, "Probable Maximum Precipitation Estimates, United States East of the 105th Meridian," *Hydrometeorological Report* 51, June 1978; "Application of Probable Maximum Precipitation Estimates, United States East of 105th Meridian," *Hydrometeorological Report* 52, August 1982; "Probable Maximum Precipitation—United States between the Continental Divide and the 103rd Meridian," *Hydrometeorological Report* 55A, June 1988; "Probable Maximum Precipitation Northwest States," *Hydrometeorological Report* 43, November 1966; "Probable Maximum Precipitation Estimates, Colorado River and Great Basin Drainages," *Hydrometeorological Report* 49, September 1977.

51. World Meteorological Organization, "Manual for Depth-Area-Duration Analysis of Storm Precipitation," WMO 237 T.P. 129, Geneva, 1969.

52. U.S. Army Corps of Engineers, "Storm Rainfall in the United States, Depth-Area-Duration Data," 1960.

53. Hershfield, D. M., "Method for Estimating Probable Maximum Rainfall," *Journal of American Water Works Association*, vol. 57, no. 8, 1965.

54. Hershfield, D. M. "Estimating the Probable Maximum Precipitation," *ASCE Transactions*, Paper 3431, vol. 128, part I, 1963.

55. U.S. Geological Survey, Office of Data Coordination, *National Handbook of Recommended Methods for Water-Data Acquisition*, 1982.

56. U.S. Geological Survey, "Techniques of Water-Resources Investigations," Book 3, chaps. A1, A2, A3, A4, A5, A6, A7, A8, A11, and A13, published from 1967 through 1983.

57. Grant, D. M., *ISCO Open Channel Flow Measurement Handbook*, 3d ed., 1989.

58. Leopold and Stevens, Inc., *Stevens Water Resources Data Book*, 2d ed., Beaverton, Ore., 1974.

59. U.S. Bureau of Reclamation, *Water Measurement Manual*, 2d ed., Denver, Colo., 1984.

60. Paulson, R. W., and W. G. Shope, Jr., "Development of Earth Satellite Technology for the Telemetry of Hydrologic Data," *Water Resources Bulletin*, vol. 20, no. 4, August 1984.

61. Buchanan, T. J., and B. K. Gilbert, "U.S. Geological Survey's Water Data Program, New Approaches in the 1980's," *Proceedings of the International Symposium on Hydrometeorology*, AWRA, edited by A. I. Johnson and R. A. Clark, Denver, Colo., June 13-17, 1982.

62. U.S. Geology Survey, "National Water Data Storage and Retrieval System (WATSTORE) User's Guide," available from Chief Hydrologist, U.S. Geological Survey, 437 National Center, Reston, Va.

63. Horton, R. E., "Surface Runoff Phenomena." Part I, *Analysis of the Hydrograph*, Edwards Brothers, Inc., Ann Arbor, Mich., 1935.

64. Barnes, B. S., "The Structure of Discharge—Recession Curves," *Transactions of the American Geophysical Union* 20, 1939.

65. American Society of Civil Engineers, *Hydrology Handbook*, ASCE Manual of Engineering Practice 28, reprinted 1957.

66. Searcy, James K., "Flow-Duration Curves," Manual of Hydrology, part 2, Low-Flow Technique, USGS Water Supply Paper 1542-A, 1959.

67. Chow, V. T., D. R. Maidment, and L. W. Mays, *Applied Hydrology*, McGraw-Hill, New York, 1988.

68. Sherman, L. K., "Streamflow from Rainfall by the Unit-Graph Method," *Engineering News-Record*, vol. 108, pp. 501-505, Apr. 7, 1932.

69. Sherman, L. K., "Unit Hydrograph Method," chapter XI of O. E. Meinzer, *Hydrology*, 1942.

70. U.S. Army Corps of Engineers, Hydrologic Engineering Center, "Hydrograph Analysis," vol. 4, *Hydrologic Engineering Methods for Water Resources Development*, Davis, Calif., October 1973.

71. BARNES, B. S., *Unit Graph Procedures,* U.S. Bureau of Reclamation, 1952, revised 1965.
72. U.S. BUREAU OF RECLAMATION, *Design of Small Dams,* A Water Resources Technical Publication, 3d ed., 1989.
73. U.S. ARMY CORPS OF ENGINEERS, "Flood Hydrograph Analysis and Computations," Engineering and Design Manuals, EM 1110-2-1405, Government Printing Office, Washington, D.C., August 1959.
74. SOIL CONSERVATION SERVICE, U.S. Department of Agriculture, *SCS National Engineering Handbook,* sec. 4, *Hydrology,* August 1972.
75. CLARK, C. O., "Storage and the Unit Hydrograph," *Transactions of the American Society of Civil Engineers,* vol. 110, 1945.
76. SOIL CONSERVATION SERVICE, "TR-20, Computer Program for Project Formulation Hydrology," *Technical Release* 20, Northeast NTC and Hydrology Unit, 1982.
77. CRAWFORD, N. H., and R. K. LINSLEY, "Digital Simulation in Hydrology, Stanford Watershed Model IV," Department of Civil Engineering, Stanford University, *Technical Report* 39, July 1966.
78. DAY, G. N., "Extended Streamflow Forecasting Using NWSRFS (National Weather Service River Forecasting System)," *Journal of the Water Resources Planning and Management Division, ASCE,* vol. III, no. 2, 1985.
79. COLON, R., and G. F. MCMAHON, "BRASS (Basin Runoff and Streamflow Simulation) Model Application to Savannah River System Reservoirs," *ASCE Journal of Water Resources Planning and Management,* vol. 113, no. 2, March 1987.
80. PECK, E. L., R. S. MCQUIREY, T. N. KEEFER, E. R. JOHNSON, and J. L. EREKSON, "Review of Hydrologic Models for Evaluating Use of Remote Sensing Capabilities," NASA CR 166674, prepared by Hyden Corporation, Fairfax, Va., 1981.
81. CUNNANE, C., "Statistical Distributions for Flood Frequency Analysis," *Operational Hydrology Report* 33, World Meteorological Organization (WMO) no. 718, Geneva, Switzerland, 1989.
82. CUNNANE, C., "Unbiased Plotting Positions—A Review," *Journal of Hydrology,* vol. 37, no. 3/4, pp. 205–222, 1978.
83. U.S. WATER RESOURCES COUNCIL, Hydrology Committee, "Guidelines for Determining Flood Flow Frequency," *Bulletin* 17B, Interagency Advisory Committee on Water Data, U.S. Geological Survey, Office of Water Data Coordination, Reston, Va., 1982.
84. FEDERAL HIGHWAY ADMINISTRATION, U.S. Department of Transportation, "Hydrology," *Hydraulic Engineering Circular* 19, October 1984.
85. CHOW, VEN TE, "Hydrologic Determination of Waterway Areas for the Design of Drainage Structures in Small Drainage Basins," *University of Illinois Engineering Experimental Station, Bulletin* 462, 1962.
86. AMERICAN SOCIETY OF CIVIL ENGINEERS, "Design and Construction of Sanitary and Storm Sewers," WPCF (Water Pollution Control Federation) Manual of Practice 9, *ASCE Manual of Engineering* 37, 5th printing, 1982.
87. AMERICAN SOCIETY OF CIVIL ENGINEERS, "Evaluation Procedures for Hydrologic Safety of Dams," A report prepared by the Task Committee on Spillway Design Flood, Selection of the Committee on Surface Water Hydrology of the Hydraulic Division, 1988.
88. FEDERAL ENERGY REGULATORY COMMISSION, Office of Hydropower Licensing, "Engineering Guidelines for Evaluation of Hydropower Projects," FERC 0119-1, July 1987.
89. CRIPPEN, JOHN R., "Envelope Curves for Extreme Flood Events," *Journal of Hydraulic Division, ASCE,* vol. 108 no. HY10, October 1982.
90. COSTA, JOHN E., "A Comparison of the Largest Rainfall-Runoff Floods in the United States with Those of the Peoples Republic of China and the World," *Journal of Hydrology,* vol. 96, 1987.

SECTION 2

BASIC HYDRAULICS

By B. A. McKiernan[1]

PART A. CLOSED CONDUITS

FLUID PROPERTIES

1. General. Applied hydraulics is concerned with the physical action and the interaction of particular masses of fluid in either the static or kinetic states. These actions and interactions are analyzed by attributing to the fluid certain physical properties each of which is defined to be the controlling property for a particular type of action. Viscosity plays an important role in the problem of hydraulic friction. Mass density is important in nonuniform flow. Unit weight is of concern in stratified flow. Surface tension is a factor in model experiments. Compressibility is a factor in water hammer. Vapor pressure is a factor in high-velocity flow.

2. Conservation of Mass. The fundamental law of continuity is frequently expressed as

$$Q = AV \tag{1}$$

where Q = discharge in volume per second
A = cross-sectional area of the flow
V = average velocity normal to the flow section

In a more general sense, the temporal mean velocity may be constant at a point, but it usually varies from point to point in a section so that

$$Q = \int_0^{A_0} v \, dA \tag{2}$$

where v = temporal mean velocity
dA = an element of area for which v is the velocity
A_0 = total area of the flow cross section

Because the mean of the values of v is equal to the average velocity, Eq. (1) is exact for all distributions of velocity. Written between two sections 1 and 2 through which the same steady discharge is passing, Eq. (1) also provides the familiar

$$A_1 V_1 = A_2 V_2 \tag{3}$$

3. Conservation of Momentum and Energy. Equations similar to Eq. (2) are written to define the flux of momentum and energy past a flow cross section as follows:

---

[1]Acknowledgment is made to W. J. Bauer, David S. Louie, and W. L. Voorduin for material in this section that appeared in the third edition (1969).

Momentum flux

$$\rho \int_0^{A_0} v^2 \, dA = \beta \rho V^2 A_0 \tag{4}$$

Energy flux

$$\frac{\rho}{2} \int_0^{A_0} v^3 \, dA = \frac{\alpha \rho V^3 A_0}{2} = \alpha Q \frac{V^2}{2} \rho \tag{5}$$

where ρ = mass density
α = kinetic-energy factor
β = momentum factor

If v is constant across the section, both α and β are unity. If v is variable across the section, both α and β are greater than 1.

4. Bernoulli Equation. One of the equations used in the analysis of one-dimensional, steady, incompressible flow is the Bernoulli equation, a special form of the law of conservation of energy. For a constant discharge in a closed conduit the theorem states that the energy head at any cross section must equal that in any other downstream section plus the intervening losses. Figure 1 shows these relationships for a typical closed conduit.

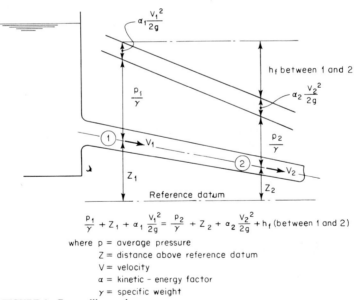

$$\frac{p_1}{\gamma} + Z_1 + \alpha_1 \frac{V_1^2}{2g} = \frac{p_2}{\gamma} + Z_2 + \alpha_2 \frac{V_2^2}{2g} + h_f \text{ (between 1 and 2)}$$

where p = average pressure
Z = distance above reference datum
V = velocity
α = kinetic - energy factor
γ = specific weight

FIGURE 1 Bernoulli equation.

It is common practice to determine the velocity head at a section by means of the average velocity of the flow. In the usual case, the velocity distribution is nonuniform and the energy head so computed is somewhat less than the true value. A kinetic-energy factor, as defined in Eq. (5), should be applied to the velocity head of the mean velocity in order to obtain the correct energy head. That is,

$$h_v = \alpha \frac{V^2}{2g} \qquad (6)$$

Because it usually has a small effect and is tedious to evaluate, α is usually neglected in practical engineering. The Bernoulli equation then becomes

$$\frac{p_1}{\gamma} + Z_1 + \frac{V_1^2}{2g} = \frac{p_2}{\gamma} + Z_1 + \frac{V_2^2}{2g} + h_f \qquad (7)$$

FRICTION LOSSES

5. General. Application of the Bernoulli equation requires a clear understanding of the factors which affect the head loss H_L. Head losses are commonly classified as boundary losses and form losses. Boundary losses are those arising from shear forces between the fluid and the boundary materials. In addition, cross-sectional shapes are significant to boundary losses because they affect the ratio of the flow area to the wetted perimeter. The effects associated with the cross-sectional shape of a uniform conduit are not classified as form losses. Form losses arise from recirculating eddies produced by the geometry of the containing vessel such as bends and either expanding or contracting transitions.

Two types of flow must be considered: laminar flow in which the fluid may be envisioned as flowing in parallel layers and turbulent flow in which the particles are moving in all directions, causing a complete mixing of the fluid. The concept of laminar flow as moving in parallel layers is actually an artificial one which is useful as an aid to theoretical analyses—for engineering practice, conditions of turbulent flow are encountered more frequently than those of laminar flow. As will be shown elsewhere in this section, different laws govern the two types of flow. The principles and equations of fluid resistance, as developed by Chezy, Manning, Kutter, Darcy-Weisbach, and Hazen and Williams, apply to turbulent-flow conditions.

6. The Chezy Formula. Friction losses h_f are usually determined by formulas which have a long background of use. Of fundamental importance is the Chezy formula

$$V = C\sqrt{RS} \qquad (8)$$

where V = mean velocity, fps
R = hydraulic radius, ft
S = slope of hydraulic gradient
C = a coefficient

7. The Manning Formula. The Manning expression

$$C = \frac{1.486}{n} R^{1/6}$$

combined with the Chezy formula yields the widely used Manning formula

$$V = \frac{1.486}{n} R^{2/3} S^{1/2} \qquad (9)$$

where n = a coefficient of frictional resistance which, under many conditions, is common to both the Kutter and Manning formulas.

As will be shown elsewhere (Art. 4, Sec. 3), n is a function of roughness, viscosity, diameter, and velocity. Figure 2 shows a chart for the solution of the Manning formula using average values of n. Figure 2 may be applied to both open and closed conduits.

In computing the flow in pipes, the formulas which result from combining the Manning and Chezy formulas may take several forms:

$$V = \frac{0.590}{n} d^{2/3} S^{1/2} \qquad (10)$$

$$Q = \frac{0.463}{n} d^{8/3} S^{1/2} \qquad (11)$$

$$h = 2.87 n^2 \frac{lV^2}{d^{4/3}} \qquad (12)$$

$$h = 4.66 n^2 \frac{lQ^2}{d^{16/3}} \qquad (13)$$

$$d = \left(\frac{2.159 Qn}{S^{1/2}} \right)^{3/8} \qquad (14)$$

Figure 3 shows a chart which may be used in applying Manning's formula to the computation of flow in pipes flowing full. Table 1 gives average values of n in Manning and Ganguillet and Kutter formulas in common use for various materials.

These values may be used for preliminary investigations.

8. The Kutter Formula. The historic Kutter formula is still used to determine the value of C in the Chezy formula. Ganguillet and Kutter determined empirically in 1869 that

$$C = \frac{41.6 + 0.00281/S + 1.811/n}{1 + \dfrac{(41.6 + 0.00281/S)n}{\sqrt{R}}} \qquad (15)$$

9. The Darcy-Weisbach Formula. Still in use at the present time is the Darcy-Weisbach formula proposed in 1857:

$$h_f = f \frac{l}{d} \frac{V^2}{2g} \qquad (16)$$

where h_f = friction loss, ft
 l = length of pipe, ft
 d = inside diameter, ft
 f = coefficient of frictional resistance

Like n in the Kutter and Manning formulas, f is a function of roughness, viscosity, diameter, and velocity.

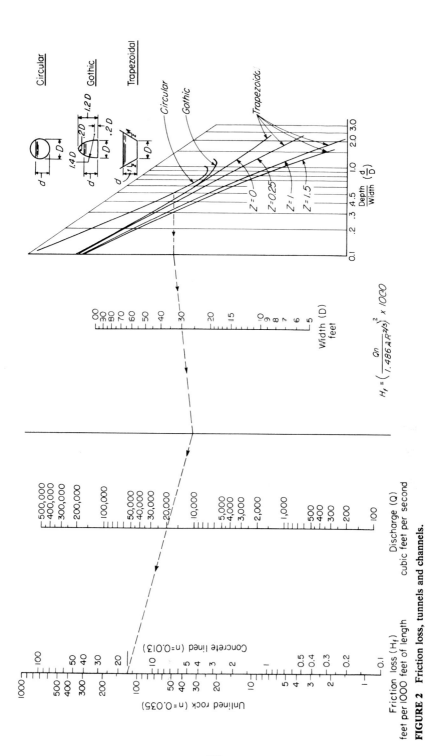

FIGURE 2 Friction loss, tunnels and channels.

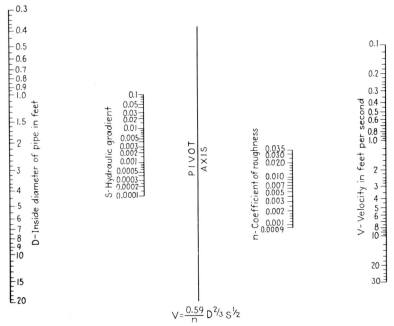

FIGURE 3 Manning's formula for pipes flowing full.

$$V = \frac{0.59}{n} D^{2/3} S^{1/2}$$

TABLE 1 Values* of Coefficient of Roughness n in Manning and Ganguillet and Kutter Formulas

Type of pipe	Condition	Best	Good	Fair	Bad
Cast iron	Clean uncoated	0.012	0.013	0.014	0.015
	Clean, coated	0.011	0.012	0.013	
	Dirty or tuberculated			0.015	0.035
Wrought iron	Commercial, black	0.012	0.013	0.014	0.015
	Commercial, galvanized	0.013	0.014	0.015	0.017
Lock bar or welded	Smooth and clean	0.010	0.011	0.013	
Brass or glass	Smooth	0.009	0.010	0.011	0.013
Riveted steel or spiral steel	Clean	0.013	0.015	0.017	
Vitrified sewer pipe		0.011	0.013	0.015	0.017
Common clay drainage tile		0.011	0.012	0.014	0.017
Concrete	Rough joints		0.016	0.017	
	Dry mix, rough forms		0.015	0.016	
	Wet mix, steel forms		0.012	0.014	
	Very smooth		0.011	0.012	
Wood stave		0.010	0.011	0.012	0.013

*Values are based on tables by Horton and values recommended by King. See H.W. King and E.F. Brater, *Handbook of Hydraulics,* 6th ed., McGraw-Hill, New York, 1976. Values given under good or fair may be used for designing.

10. The Hazen and Williams Formula. The Hazen and Williams formula for the flow in pipes has been used extensively in the design of water-distribution systems. This formula is usually expressed as

$$V = 1.318 C R^{0.63} S^{0.54} \tag{17}$$

where C = a coefficient
R = hydraulic radius
S = slope of hydraulic gradient

EVALUATION OF FRICTION COEFFICIENTS

11. General. It has been pointed out that the friction factors f in the Darcy-Weisbach formula and n in the Manning and Kutter formulas both vary with roughness, viscosity, diameter, and velocity. The evaluation of friction factors under the widely varying conditions usually encountered in practice has been made possible by the contributions of Reynolds, Nikuradse, von Kármán, Prandtl, and Colebrook-White.

12. The Contribution of Reynolds. The Reynolds criterion, developed by Sir Osborne Reynolds late in the nineteenth century, relates the inertial forces per unit of volume to the viscous forces per unit of volume. This relationship provides a rational basis for establishing the dynamic similarity of fluid motion in closed conduits.

As an approach to his analysis, Reynolds considered the Newtonian concept that the tangential stress between contiguous strata of the fluid should be proportional to the rate dv/dy at which the velocity varies across the section of flow. This proportionality has been adopted as a measure of the fluid viscosity. The relationship may be expressed as

$$\tau = \mu \frac{dv}{dy} \tag{18}$$

where τ = intensity of shear
μ = dynamic viscosity

In accordance with the Newtonian equation for shear, a typical viscous force per unit of volume may be written as

$$\mu \frac{V}{L^2} \tag{19}$$

in which L is a length parameter sometimes expressed as D, the diameter of the conduit.

A typical inertial reaction per unit of volume may be designated by

$$\rho \frac{V^2}{L} \tag{20}$$

in which ρ is the mass density. The ratio of the inertial force per unit of volume to the viscous force per unit of volume reduces to the form

$$\rho \frac{VL}{\mu} \tag{21}$$

The factor μ/ρ, known as the kinematic viscosity ν, may be compared with the gravitational factor $\gamma/\rho = g$. The introduction of the kinematic viscosity factor ν into the expression $\rho(VL/\mu)$ yields

$$R = \frac{VL}{\nu} \tag{22}$$

which is known as the Reynolds number. The kinematic viscosity of water relative to temperature is shown by Fig. 4. The basic relationship between boundary shear stress τ_o and friction factor f is

$$v_* = \sqrt{\frac{\tau_o}{\rho}} = V\sqrt{\frac{f}{8}} \tag{23}$$

where v_* = shear velocity or friction velocity
  ρ = density of the fluid (for water, ρ = 1.935 slugs/ft^3)

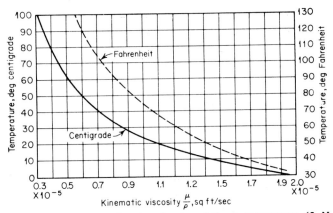

FIGURE 4 Kinematic viscosity of water relative to temperature. (J. N. Bradley and L. R. Thompson, "Friction Factors for Large Conduits Flowing Full," U.S. Department of Interior, Bureau of Reclamation.)

Using the Newtonian definition of laminar flow, Eq. (18) may also be expressed as

$$\tau = -\mu \frac{dv}{dr} \tag{24}$$

For a circular pipe the shearing stress at the boundary, developed by equating shear and pressure forces acting on a cylindrical body of fluid at radius r and length L, would be

$$\tau_o = \frac{\gamma h_f r}{2L} \tag{25}$$

The equation for velocity distribution in laminar flow is

$$v = 2V\left[1 - \left(\frac{r}{R}\right)^2\right] \qquad (26)$$

where R is the radius of the pipe. Using Eqs. (24), (25), and (26), it follows that

$$h_f = \frac{32\mu l V}{\gamma d^2} \qquad (27)$$

From this it can be developed that the Darcy-Weisbach coefficient of friction in the laminar regime is

$$f = \frac{64}{\mathbf{R}} \qquad (28)$$

Equation (28) is valid for values of **R** up to about 2000. The work of Blasius and Nikuradse verified Eq. (28) and showed that head loss in this flow regime is independent of boundary-surface roughness.

13. The Contribution of Nikuradse. During 1932 and 1933, Nikuradse, working under the direction of Prandtl and von Kármán, published the results of his now famous experiments on artificially roughened pipe. Small, smooth pipes of different diameters were coated with sand grains of uniform size and subjected to a wide range of velocities. The resistance to flow represented by the Darcy-Weisbach friction factor f was plotted with respect to the Reynolds number for various values of the relative roughness r_0/k where r_0 represents the radius of the pipe and k the average diameter of sand grains. The Nikuradse k is not the absolute roughness or rugosity ϵ, as the term is defined by more recent writers. The results of these experiments are shown by Fig. 5. The Nikuradse criterion offers a means of grouping pipes having similar roughness characteristics for partially

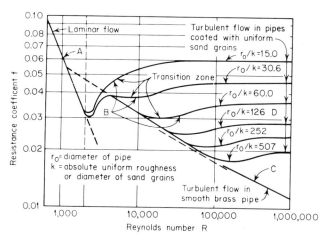

FIGURE 5 Variation of the resistance coefficient with the Reynolds number for artificially roughened pipes. (*Bradley and Thompson, "Friction Factors for Large Conduits Flowing Full," Engineering Monograph no. 7, U.S. Department of Interior, Bureau of Reclamation, March, 1951.*)

and fully developed turbulent flow. The straight line A in Fig. 5 represents laminar flow where $f = 64/\text{R}$ for values of R. Line C represents the results obtained for turbulent flow in smooth brass pipe. The lines denoted as D are for turbulent flow in pipes coated with uniform sand grains. The size of the pipe and diameter of the sand-grain coating were varied in the experiments. The results were then plotted in terms of the relative roughness r_0/k.

14. The Colebrook and White Contributions. Referring to Fig. 5, the lines marked B represent a transition zone between laminar and turbulent flow in smooth brass pipe. The lines marked D are for turbulent flow. The curves of Nikuradse consistently show a sharp drop followed by a reverse curve in the transition zone. The analyses of von Kármán and Prandtl, based on the Nikuradse experiments, showed some disagreement in the transition zone. This disagreement was not explained satisfactorily until 1939, when Colebrook and White developed a practical form of transition to bridge the gap.

Colebrook and White demonstrated that the deviation of von Kármán–Prandtl analyses from those of Nikuradse stemmed from the fact that resistance to flow for uniform sand is different for equivalent, nonuniform roughness, such as that which exists in commercial pipes. The coarser grains disturbed the laminar sublayer before the smaller irregularities became effective. The resulting formula, as proposed by Colebrook and White, follows the trend of experimental results and is asymptotic to both the smooth and rough pipe equations of von Kármán and Prandtl.

The Colebrook and White formula is

$$\frac{1}{\sqrt{f}} - 2 \log_{10} \frac{r_0}{k} = 1.74 - 2 \log_{10}\left(1 + 1.87\frac{r_0/k}{\text{R}\sqrt{f}}\right) \qquad (29)$$

The U.S. Army Engineers, Waterways Experiment Station, after careful study of several large prototype structures concluded that Eq. (29) could not be verified for large Reynolds numbers.

15. The von Kármán and Prandtl Contribution. Concurrently with the Nikuradse experiments, von Kármán and Prandtl developed a theoretical analysis for pipe flow with suitable formulas for smooth and rough pipes. Smooth pipes are defined as those having small irregularities when compared with the thickness of the boundary layer. Rough pipes are defined as having irregularities in the walls which break up the laminar boundary layer, with the result that completely turbulent flow is developed.

The von Kármán–Prandtl resistance equation for turbulent flow in smooth pipes is

$$\frac{1}{\sqrt{f}} = 2 \log_{10}\left(\text{R}\sqrt{f}\right) - 0.8 \qquad (30)$$

which corresponds to line C of Fig. 5. The equation for rough pipes is

$$\frac{1}{\sqrt{f}} = 2 \log_{10} \frac{r_0}{k} + 1.74 \qquad (31)$$

16. The Moody Chart. The Prandtl–von Kármán experiments and the contribution of Colebrook and White were finally brought together by Moody in a comprehensive chart (Fig. 6) which may be superimposed on all the experimental

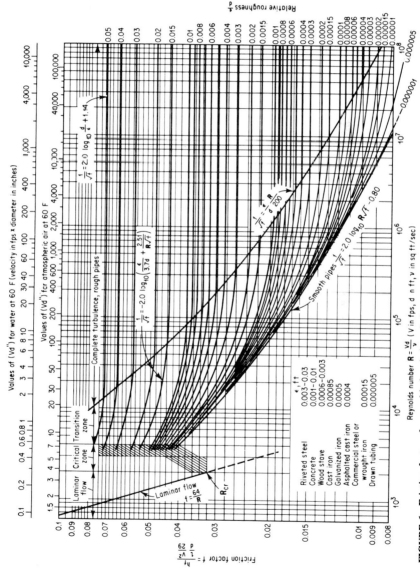

FIGURE 6 Friction factors for any kind and size of pipe.

f vs. R curves for larger pipes. In contrast to the Nikuradse sand roughness, the roughness elements of commercial pipe are not uniform. The protuberances vary not only in size but also in pattern of spacing. As a means of differentiating the Nikuradse uniform sand-grain roughness k, the nonuniform roughness or rugosity of commercial pipes will be designated as ϵ. This is a practical simplification of more complex definitions of roughness. In practice, the measurement of natural roughness has been based upon the use of twice the root-mean-square height of the roughness elements ϵ, which is measured by moving a small probe across a rough surface or a plaster cast of that surface.

17. Deposits and Organic Growth. The discharge capacities of tunnels and other water passages may decrease with aging because of deposits and organic growths on the interior surfaces. These accumulations increase boundary-friction losses with resulting decreases in discharge capacities.

Corrosion on the interior surface of water-conveyance structures reduces carrying capacity in water-supply mains, produces discolored water, and creates taste and odor problems. In metal pipes, corrosion gives rise to tubercules on the interior surface, and these in turn decrease the carrying capacity by roughening of the interior of the pipe surface.

Improper filtration of water or operation of treatment plants beyond their capacities may result in floc passing through filters, depositing aluminum hydroxide on the pipe walls. Even a thin coating will reduce the carrying capacity materially. Other coatings may consist of calcium carbonate and silica slime.

In a number of tunnels in Chicago, an analysis of the gelatinous material showed that it was principally aluminum hydroxide and silica. Measurement of the friction loss in these tunnels indicated that the Manning's coefficient of friction n had increased from 0.014 when new to as high as 0.0196 for some tunnels after 7 years of service.

After 10 years of service, the 18-ft-diameter tunnel leading to the Appalachia hydroelectric plant (TVA) was found to have the following changes in roughness: the average values of Manning's n rose from 0.011 to 0.018 for the concrete section, from 0.010 to 0.018 for the steel section, and from 0.030 to 0.038 for the unlined rock section.

After only 3 years of service, the 10-ft by 10-ft concrete fish-passing conduit at the Priest Rapids Hydroelectric Project, Columbia River, accumulated from ½ to 5/8 in. of algae and sponge sliming, which increased the value of n from 0.0104 to 0.0187. After cleaning, this value dropped to 0.0108.

Figure 7 shows the effect of aging of water-supply mains in several large cities in the United States as measured by increases in the Hazen and Williams coefficient C. The loss in capacity may range between 25 and 30 percent.

FORM LOSSES

18. General. Between intake and exit, flow encounters a variety of shape configurations in the flow passageway such as changes in section from rectangular to circular, partial obstacles, branches, bends, slots, expansions, and contractions. These impose losses in addition to those resulting from frictional resistance. Form losses are the result of fully developed turbulence and thus can be expressed in the general form

FORM LOSSES

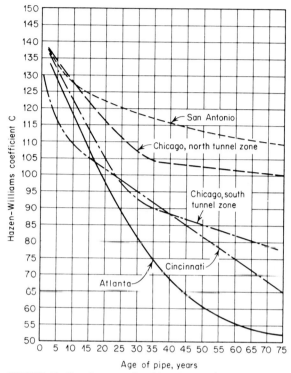

FIGURE 7 Trend curve, head-loss tests. (*W. D. Hudson, "Studies of Distribution System Capacity,"* Journal of the American Water Works Association, *February 1966.*)

$$H_L = K_e \frac{v^2}{2g} \qquad (32)$$

where K_e = coefficient of form loss
H_L = form loss

There follow discussions of a few of the commoner types of form losses encountered in engineering practice.

19. Sudden Enlargements. A sudden enlargement, such as that shown by Fig. 8, results in intense shearing action between the incoming high-velocity jet and the surrounding water. As a result, much of the kinetic energy of the jet is dissipated by eddy action. Most of the turbulence disappears and the velocity becomes practically uniform across the section of the enlarged pipe at a distance of about five diameters from the enlargement. Rapid pressure fluctuations accompany the enlargement.

Using the symbols shown by Fig. 8, the loss of head at a sudden enlargement is

$$H_L = \frac{(V_1 - V_2)^2}{2g} \qquad (33)$$

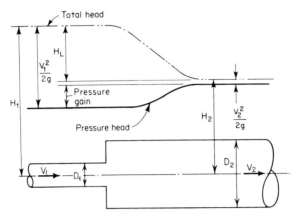

FIGURE 8 Sudden enlargement.

Practical applications of this principle may be found in several high-head outlets which are designed to release discharges at velocities which will not damage the water-passage linings. The sudden-enlargement energy dissipator for Mica Dam, on the Columbia River in British Columbia, offers one example.

One of the two 45-ft diversion tunnels at Mica was selected as the most logical and economic location for a gate-controlled expansion chamber which would serve to reduce the velocities of the discharge into the tunnel during the filling period of the reservoir. With outlets of conventional design, the maximum velocity would have been 170 fps. The most promising arrangement provided for a sudden-enlargement energy dissipator in the tunnel upstream of the main plug which would contain three gated outlets.[1]*

Possible cavitation of the concrete tunnel lining was an important consideration. As shown by the hydraulic gradient (Fig. 8) pressures in the fast-moving eddies are lower in the surrounding water and under certain conditions can fall as low as the vapor pressure. If pressures fall sufficiently low, the collapse of vapor bubbles may occur at the boundary of the tunnel lining, with resulting damage to the surface. From comprehensive model tests, including one made on a concrete-lined steel pipe under approximately prototype discharge conditions, it was determined that the design cavitation number K for the Mica expansion chamber, as expressed by the formula

$$K = \frac{H_2 - H_v}{H_t - H_2} \tag{34}$$

in which H_v = vapor pressure of water, H_t = total head of flow entering enlargement, and H_2 = pressure head downstream from the enlargement, should be approximately 3. The value of K at the point of incipient cavitation damage was less than 1.

20. Sudden Contraction. The head loss resulting from a sudden contraction may be expressed as

---

*Superscripts indicate items in the References at the end of this section.

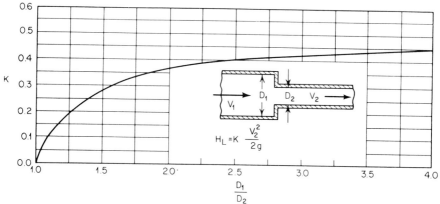

FIGURE 9 Head loss, sudden contraction. (*Pipe Friction Manual, Hydraulic Institute.*)

$$H_L = K \frac{V_2^2}{2g} \tag{35}$$

Values of K for various ratios of D_1 to D_2 are shown by Fig. 9.

21. Gradual Conical Expansion. The losses resulting from gradual expansion, such as shown by Fig. 10, may be expressed in the following form:

$$H_t = K_t \frac{(V_1 - V_2)^2}{2g} \tag{36}$$

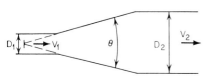

FIGURE 10 Gradual conical expansion.

where V_1 and V_2 are velocities in the upstream and downstream sections, respectively, H_t the transition head loss, and K_t the transition loss coefficient.

Early experiments by Gibson[2] show that the loss coefficient k_t is a function of the central angle θ of the truncated cone. The tests show that the loss coefficient is also a function of the ratio of areas A_1 and A_2. The results of the Gibson tests[3] are shown by Fig. 11.[4]

Figure 11 shows that the loss coefficient increases rapidly as θ increases up to about 60°.

The results of more comprehensive tests published by V. Tatarinov, which appeared in *Product Engineering*, May 1946, are shown by Fig. 12. These tests confirm some of the earlier results obtained by Gibson.

22. Gradual Conical Contraction. Less laboratory information is available on the loss in contraction transitions, such as shown by Fig. 13, than for expansion transitions. Tatarinov also presented the results of tests on gradual conical contractions shown by Fig. 14.

23. Bend Losses. The bend loss, excluding friction loss, for a circular conduit is a function of the bend radius, pipe diameter, and deflection angle of the bend. Bend losses for pipes having 90° bends and varying degrees of rugosity ϵ/D are shown by Fig. 15. Curves showing recommended bend-loss coeffi-

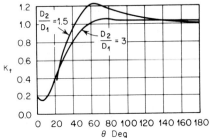

FIGURE 11 Loss coefficients for gradual conical expansion.

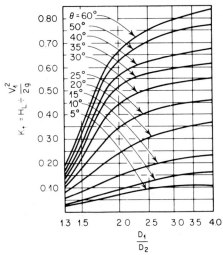

FIGURE 12 Head loss in conical expansion.

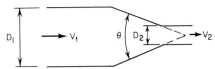

FIGURE 13 Gradual conical contractions.

cients for R/D ratios from 1 to 10 and for deflection angles from 0 to 90° are given in Fig. 16.[5]

24. Minor Losses. Minor losses may be expressed in the equivalent length of pipe that has the same energy loss for the same discharge. The chart in Fig. 17 offers a convenient method of estimating these losses.

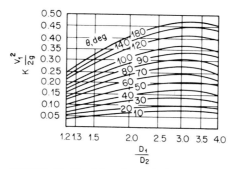

FIGURE 14 Head loss in conical contraction. (*V. Tatarinov*, Product Engineering, *May 1946.*)

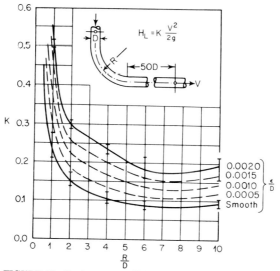

FIGURE 15 Resistance coefficients for 90° bends of uniform diameter. (*Pipe Friction Manual, Hydraulic Institute.*)

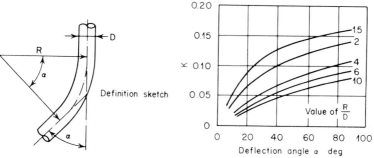

FIGURE 16 Friction losses at bend.

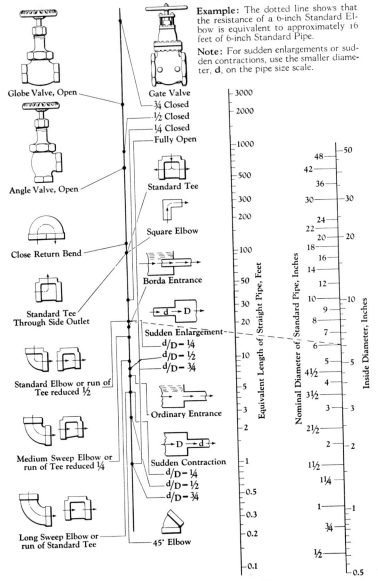

FIGURE 17 Resistance of valves and fittings to flow of fluid.

ORIFICES

25. High Head. When the head is relatively large, as compared with the size of the orifice, the following equation will apply:

$$Q = CA\sqrt{2gH} \tag{37}$$

where Q = discharge, cfs
H = head on centerline of orifice, ft
A = area of orifice, ft^2

Sharp-edged circular orifices, as shown by Fig. 18 and the accompanying table of discharge coefficients, will operate through a wide range of heads with a value of C of approximately 0.60. Rectangular orifices, such as those which are formed by opening a gate having a shape such as that shown by Fig. 19, will have much higher discharge coefficients. Discharge coefficients for submerged orifices of varying forms are shown by Fig. 20.

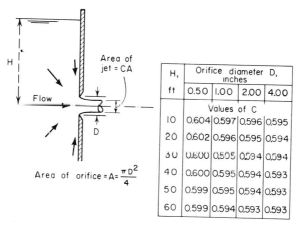

FIGURE 18 Discharge coefficients, circular sharp-edged orifices, varying heads. (*F. W. Medaugh and G. D. Johnson,* Civil Engineering, *p. 424, July 1940.*)

26. Low Head. When the dimensions of the orifice are large, as compared with the head, the discharge formula may be written

$$Q = \frac{2}{3} C\sqrt{2g} L (H_2^{3/2} - H_1^{3/2}) \tag{38}$$

with dimensions and symbols as indicated by Fig. 21. The expression ⅔ $C\sqrt{2g}$ may be designated as the overall coefficient m which appears in the equations and coefficients determined by prototype measurements of the spillway discharge capacity of Wilson Dam. These and other prototype experiments are described more fully in Sec. 20.

PART B. OPEN CHANNELS

27. Critical Flow. Flow in open channels is governed primarily by the action of gravity upon the moving fluid. The acceleration of gravity g or the slope S is expressed or implied in all open-channel flow formulas. In most problems the acceleration of gravity may be considered to be a constant. This makes possible its implicit presence in some flow formulas, even though it is not expressed.

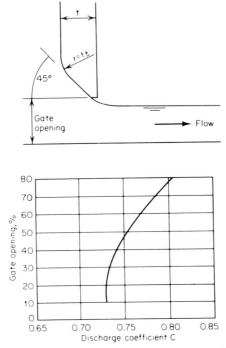

FIGURE 19 Discharge coefficient C. (*Source: Hydraulic Design Criteria, Waterways Experiment Station, Corps of Engineers, U.S. Army, Sheet 320-1.*)

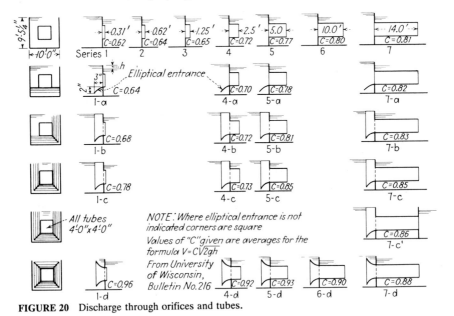

FIGURE 20 Discharge through orifices and tubes.

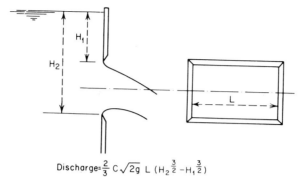

Discharge $= \frac{2}{3} C \sqrt{2g} \, L \, (H_2^{\frac{3}{2}} - H_1^{\frac{3}{2}})$

FIGURE 21 Discharge for large vertical orifice compared with head.

The dimensionless parameter used to describe and classify open-channel flow is the Froude number

$$F = \frac{V}{\sqrt{gy}} \quad (39)$$

where F = Froude number
V = average velocity of the flow
g = acceleration of gravity
y = depth of flow

Flows are classified as follows:

$F < 1$: flow is subcritical
$F = 1$: flow is critical
$F > 1$: flow is supercritical

Critical flow requires the least energy for a given rate of discharge. Assuming a hydrostatic distribution of water pressure within the moving fluid,

$$y_c = \sqrt[3]{\frac{q^2}{g}} \quad (40)$$

where y_c = critical depth
q = unit discharge per lin ft width
g = acceleration of gravity

The critical velocity is the average velocity of flow at the critical condition and is expressed by

$$V_c = \frac{q}{y_c} = \sqrt[3]{gq} = \sqrt{gy_c} \quad (41)$$

In uniform flow in a rectangular channel at critical depth, the velocity head $V^2/2g$ is exactly half the depth of the flow. For other prismoidal sections Eq. (41) may be used without appreciable error by using the average depth of flow. The aver-

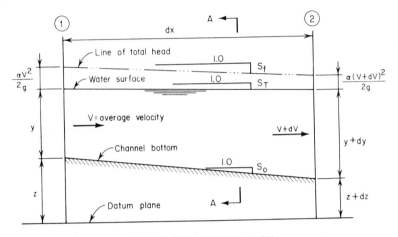

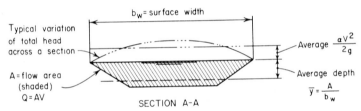

FIGURE 22 Profile of open-channel flow and terms and notations for open-channel flow.

age depth is the area of flow divided by the surface width of the flow. Figure 22 defines terms and notations commonly used in analysis of open-channel flow.

As shown by Fig. 22, the velocity is a variable throughout the section, being typically low near a boundary and high at some distance from a boundary. The square of the average velocity is used in calculating the velocity head, which is considered to be a measure of the total kinetic energy of the moving fluid.

28. Steady Uniform Flow. This is flow in which identical conditions prevail from point to point and from time to time within the flow. The depth, cross section, discharge, velocity distribution, density, viscosity, turbulence, and all other aspects of the flow remain constant from point to point and from instant to instant. Steady uniform flow in open channel is rare, even in the laboratory.

Under usual field conditions, flow is nonuniform and unsteady. Nevertheless, the hypothetical simple condition of steady uniform flow is frequently substituted for the more complex real condition because it is easier to handle mathematically. Although not strictly correct, it is sometimes convenient to analyze nonuniform and unsteady flows by considering the behavior to be comprised of basic steady uniform flow upon which the effects of unsteadiness and nonuniformity are superimposed.

29. Flow Formulas and Frictional Resistance. The most widely used formulas for determining the flow in open channels are those of Chezy, Kutter and Manning, as set forth, respectively, in Eqs. (8), (15), and (9).

For most of the small channels and conduits used in irrigation works, the Kutter-Chezy formula will yield conduit sections comparable with those computed by the Manning formula. Using an n of 0.014, hydraulic radii between 2 and 6, and with slopes between 0.01 and 0.0001, both formulas give approximately the same average velocity. Outside these limits different values of n must be used in one or the other of these formulas to obtain the same results. For example, in a reach of one canal tested with a hydraulic radius of 11.3 and a slope of 0.00061, a value of n of 0.0152 is required in Manning's formula to provide the same average velocity given by Kutter's formula with an n of 0.014.[6]

The design chart for the solution of the Manning formula (Fig. 2) may be used for determining the flow in open channels.

30. Steady Nonuniform Flow. Steady nonuniform flow remains constant with time but varies in velocity from point to point within the fluid. Steady nonuniform flow in an open channel is complicated by the fact that the pressure at the free surface must remain atmospheric. This requirement results in pronounced changes in the configuration of the free surface in zones of rapid convergence or divergence in open channels. The dynamic equation for steady nonuniform flow involves a direct application of the principles of conservation of mass, momentum, and energy.

31. Backwater Curves. The general approach to the solution of the problem of backwater curves is presented in Sec. 5 as it pertains to natural channels. The same approach is applicable to prismatic open channels. However, certain simplifications are possible in prismatic channels which can simplify the calculation of backwater curves. The general equation for nonuniform steady flow in open channels is used as a beginning:

$$S_0 - \frac{\partial y}{\partial x} - S_f = \frac{\alpha V}{g} \frac{\partial V}{\partial x} \tag{42}$$

where S_0 = bottom slope
$\partial y/\partial x$ = rate of change of depth y with distance x
S_f = friction slope
α = kinetic-energy factor
g = acceleration of gravity
V = average velocity
$\partial V/\partial x$ = rate of change of velocity with respect to distance x

Both S_f and $V(\partial V/\partial x)$ can be expressed in terms of a constant q and the depth y, producing an equation in which only x and y are variables. This permits integration of the equation, usually with the aid of some tabulated numerical values of some of the more complex functions. Bresse's backwater functions are among the better known of this type of function.

32. Unsteady Flow. Unsteady, nonuniform flow is a general type of flow commonly analyzed in two-dimensional forms. The fundamental idea is that water-surface slope not only offsets friction but also produces acceleration:

Total slope − friction slope = acceleration slope

Let terms be defined as shown in Fig. 22. The dimensionless form of the terms is

$$\text{Total slope} = S_0 + \frac{\partial y}{\partial x}$$

$$\text{Friction slope} = S_f$$

$$\text{Acceleration slope with distance} = \frac{\alpha V}{g}\frac{\partial V}{\partial x}$$

$$\text{Acceleration slope with time} = \frac{1}{g}\frac{\partial V}{\partial t}$$

The equation becomes

$$S_0 + \frac{\partial y}{\partial x} - S_f = \frac{\alpha V}{g}\frac{\partial V}{\partial x} + \frac{1}{g}\frac{\partial V}{\partial t}$$

Care must be taken with signs. Slopes S_0 and S_f are called positive when sloping down in the direction of flow as in the usual case. The term $\partial y/\partial x$ produces forces in this same direction where y is decreasing. Therefore, if a reduction in y is called negative, the sign of $\partial y/\partial x$ becomes negative and the equation becomes

$$S_0 - \frac{\partial y}{\partial x} - S_f = \frac{\alpha V}{g}\frac{\partial V}{\partial x} + \frac{1}{g}\frac{\partial V}{\partial t} \tag{43}$$

$$A - B - C = D + E$$

Term D contains the kinetic-energy factor α and also the average velocity V. This term is present even in *steady* nonuniform flow.

Term E contains the time t. It is the only term which accounts for changes with time.

All five terms are dimensionless slopes. It is significant to examine the magnitude of each term in any real problem to evaluate its relative significance. An example follows:

Example: A river in flood is found to have the following characteristics:

$$\text{Bottom slope} = S_0 = 0.0002$$

$$\text{Usual friction slope at observed discharge} = S_f = S_0$$

Velocity V increased from 1.5 to 2.0 fps in a period of 1 h while depth y increased from 8 to 10 ft in the same period. Since $S_0 = S_f$, the equation becomes

$$-\frac{\partial y}{\partial x} = \frac{\alpha V}{g}\frac{\partial V}{\partial x} + \frac{1}{g}\frac{\partial V}{\partial t}$$

The time effect is evaluated as follows:

$$\frac{1}{g}\frac{\partial V}{\partial t} \approx \frac{1}{g}\frac{\Delta V}{\Delta t} = \frac{1}{g}\frac{2.0 - 1.5}{3600 \text{ sec}} = 4.3 \times 10^{-6}$$

This is seen to be rather small compared with the friction slope and bottom slope, being only 2.15 percent as large. As a first trial, neglect it so that

$$-\frac{\partial y}{\partial x} = \frac{\alpha V}{g}\frac{\partial V}{\partial x}$$

which when integrated merely says that the change in depth is equal to α times the change in velocity head. As a refinement, the 4.3×10^{-6} slope effect can be added by multiplying this slope by the length of reach involved.

33. Wave Profiles and Velocities. Unsteady flow in open channels involves waves of one sort or another. The shape of these waves is an important factor in the analysis, as are wave velocities which are related to waveshape. The first trace of a wave is transmitted with a celerity $c = \sqrt{gy}$ with respect to the moving fluid. Thus the absolute propagation velocity is the vector sum of the fluid velocity V and the celerity c. As the wave front builds up, it moves at wave velocity V_w with respect to the moving fluid, this V_w being considerably less than the celerity c.

With reference to Fig. 23 the wave velocity for a constant shape of wave front is

$$V_w = \frac{V_1 y_1 - V_2 y_2}{y_1 - y_2} \tag{44}$$

The moving waves of Fig. 23 can be transposed with a steady flow situation by the device of subtracting V_w from all velocities, causing the wave front to remain stationary with respect to the observer. This gives rise to a unit discharge $q = y_1(V_w - V_1) = y_2(V_w - V_2)$ which is useful in comparing wave shapes. Just as in steady flow, the shape of the water surface (wave front) depends upon the Froude number. The critical depth $y_c = \sqrt[3]{q^2/g}$ is used as a criterion. If the depth is greater than y_c, the water surface is a gentle curve. If y_2 is less than y_c, the water surface resembles a hydraulic jump. This then becomes a moving hydraulic jump, or hydraulic bore.

This technique of transposing the moving-wave problem into a steady-flow

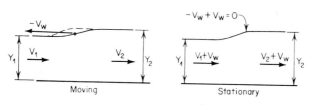

POSITIVE WAVE

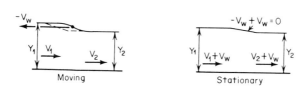

NEGATIVE WAVE

FIGURE 23 Positive and negative waves.

problem is very useful as a first approximation. If the friction slope is parallel to the bottom slope, accurate results are obtained. However, the passage of a wave front changes the discharge and depth and consequently the friction slope. A more elaborate analysis can be used if more precision is required, as it would be in relatively long channels.

34. Typical Wave Velocities. Based on the assumption of a constant shape of wave front, wave velocities V_w in typical channel cross sections have been computed as follows:

Type of cross section	By Manning	By Chezy
Wide rectangular channel	1.67 V	1.50 V
Wide parabolic channel	1.44 V	1.33 V
Triangular channel	1.33 V	1.25 V

The average channel velocity is symbolized by V.

The channel shape is seen to have a significant effect, so that the calculation of V_w in a real channel can be approximate at best. Large-scale model studies and detailed observations along a particular reach of real channel are required to set up working procedures which involve accurate values of V_w. In the absence of such information, the usual practice is to exercise judgment in the selection of V_w with the aid of the typical factors given above.

35. Solitary Waves. Solitary waves of small height in rectangular channels have celerities approximated by

$$c = \sqrt{g(y + h)} \tag{45}$$

where c = celerity, fps, with respect to the fluid beneath the wave
y = depth
h = wave height above the average depth

36. Surges in Power Channels. A long power channel flowing at maximum discharge will produce a positive surge when a sudden interrupting of the electrical load occurs. The governors automatically shut down the turbines, the flow through the plant suddenly stops, and a positive surge develops which travels upstream. Figure 24 illustrates this condition. The principle of conservation of momentum leads to the following:

$$\frac{(V_1 - V_2)^2}{g} = \frac{(y_1 - y_2)^2}{2} \frac{y_1 + y_2}{y_1 y_2} \tag{46}$$

This equation can be solved by trial and error. In the case in which $V_2 = 0$, the equation may be greatly simplified by the following approximation:

$$\frac{y_1 + y_2}{2} = \bar{y} \quad \text{the average depth}$$

$$y_1 y_2 = \bar{y}^2$$

Making these substitutions and letting $V_2 = 0$ and $F_1 = V_1/\sqrt{gy_1}$

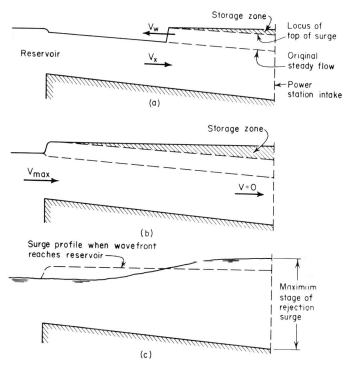

FIGURE 24 Rejection surge in power canal. (*Ven Te Chow*, Open-Channel Hydraulics, *p. 569.*)

$$y_1 - y_2 = \bar{y} F_1 \tag{47}$$

The initial surge height is then approximately equal to the product of the average depth and the initial Froude number. If the average initial depth is assumed equal to y_1 for a first trial, successive trials will estimate the initial surge height.

Example: $V_1 = 6.8\ fps$, $y_1 = 43\ ft$, $V_2 = 0$. *Find the initial surge height.*

$$F_1 = \frac{V_1}{\sqrt{gy_1}} = \frac{6.8}{\sqrt{32.2 \times 43}} = 0.183$$

Assume $\bar{y} = y_1 = 43\ ft$

$$y_2 - y_1 = 0.183 \times 42 = 7.88\ ft$$

Reuse $\bar{y} = 43 + 7.88/2 = 47$ *approximately*

$$y_2 - y_1 = 0.183 \times 47 = 8.6\ ft$$

Reuse $\bar{y} = 43 + 8.6/2 = 47.3$

$$y_2 - y_1 = 0.183 \times 47.3 = 8.6\ ft,$$

which is the required estimate

An important effect of the sloping channel must be taken into account for a more accurate prediction of surge height. The shaded zone labeled "storage zone" in Fig. 24 must be filled with water as the wave progresses upstream. Although this area is small when the surge begins, and V_2 is essentially zero, it becomes increasingly significant as the wave moves farther upstream, giving rise to an appreciable V_2. The trial-and-error stepwise solution of the more exact equation for surge height gives smaller and smaller surge heights as the surge approaches the reservoir. However, at the instant the surge reaches the reservoir, there is now an appreciable V_2 which produces a secondary surge over the level surface calculated on the basis of the initial shutdown. The height of this secondary surge is less than V_2^2/g, which may be sufficiently precise for most design purposes. The velocity V_2 is calculated on the basis of the time rate of filling of the "storage zone," which in turn depends upon wave velocity, which comes from a step-by-step solution of the more exact equation for surge height.

37. Negative Surges. Small negative surges in power channels (or in tail tunnels of underground hydroelectric plants) are treated in much the same manner as are positive surges. Again, the principle of conservation of momentum is used to write the equation relating surge height, depths, and velocities. As before, a trial-and-error solution is necessary, and again one must take into account the effects of slope and the requirements of continuity.

If the surge height is large, however, the fact that the top of the negative surge moves more rapidly than the bottom causes a significant change in wave-front shape.

More elaborate analyses are required to evaluate this effect. An approximate solution which is usually adequate for design purposes is shown in Fig. 25.

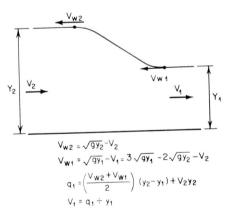

$V_{w2} = \sqrt{gy_2} - V_2$
$V_{w1} = \sqrt{gy_1} - V_1 = 3\sqrt{gy_1} - 2\sqrt{gy_2} - V_2$
$q_1 = \left(\dfrac{V_{w2} + V_{w1}}{2}\right)(y_2 - y_1) + V_2 y_2$
$V_1 = q_1 \div y_1$

Example: $y_2 = 10$, $y_1 = 5$, $V_2 = 0$ are given.
Then $V_{w2} = 18 - 0 = 18$
$V_{w1} = 3\sqrt{161} - 2\sqrt{322} - 0 = 2.2$
$q_1 = \left(\dfrac{18 + 2.2}{2}\right)(10 - 5) + 0 = 50.5$
$V_1 = 50.5 \div 5 = 10.1$ fps

FIGURE 25 Negative surge. (*Ven Te Chow*, Open-Channel Hydraulics, *p. 567.*)

38. Hydraulic Jump on Horizontal Floor. The hydraulic jump on a horizontal floor may be analyzed on the basis of the conservation of momentum. The momentum of the flow entering the jump, plus the hydraulic force of this flow, must equal the sum of the momentum of the flow leaving the jump, plus the hydrostatic

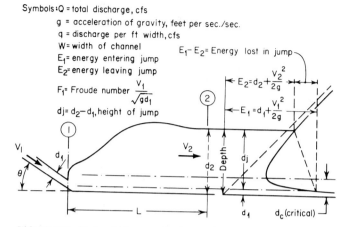

FIGURE 26 Hydraulic jump on horizontal floor.

force of that flow. Such an analysis ignores the friction forces on boundaries of the flow and additional forces introduced by piers, baffles, sills, and other energy-dissipating structures in the channel.

The depth and energy relationships of a jump on an unobstructed horizontal floor are shown by Fig. 26. Equating momentum and hydrostatic relationships for points 1 and 2 in Fig. 26 gives

$$d_2 = -\frac{d_1}{2} + \sqrt{\frac{d_1^2}{4} + \frac{2V_1^2 d_1}{g}} \qquad (48)$$

where d_1 = depth of channel at point 1
d_2 = depth of channel at point 2
V_1 = average velocity at point 1
V_2 = average velocity at point 2
g = acceleration due to gravity

The depths d_1 and d_2 may also be expressed in terms of the Froude number at point 1, designated as F_1, as follows:

$$\frac{d_2}{d_1} = \frac{1}{2}\left(\sqrt{1 + 8F_1^2} - 1\right) \qquad (49)$$

This relationship is shown graphically by Fig. 27. The lengths of the jump L, expressed in terms of d_1, d_2, and F_1, are shown by Fig. 28 and the loss of energy in length L by Fig. 29. The relationships shown by Figs. 28 and 29 have been confirmed experimentally.

39. Hydraulic Jump on Sloping Apron. Several expressions have been developed for the hydraulic jump on a sloping apron as typified by Fig. 30. The following expression presented by Kindsvater has had wide acceptance:

$$\frac{d_2}{d_1} = \frac{1}{2\cos\phi}\left(\sqrt{\frac{8F_1^2 \cos^3\phi}{1 - 2K\tan\phi} + 1} - 1\right) \qquad (50)$$

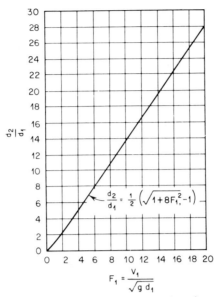

FIGURE 27 Ratio of tailwater depth to d_1.

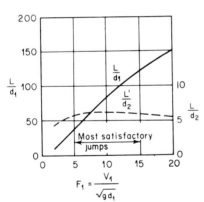

FIGURE 28 Length of hydraulic jump in terms of d_2 and d_1.

All symbols have been referred to previously except K, a dimensionless parameter called the shape factor, which varies with F_1 and the slope of the apron. Figure 31 shows values of K which were determined experimentally.

It was found that the Froude number F_1 affected K only slightly; therefore, K may be related directly to tan ϕ in making preliminary determinations. Figures 32, 33, and 34 were plotted from the results of experiments by Bradley and Peterka. Figure 32 shows the relationships between d_1 and d_2, tailwater depths, apron slopes, and F_1. Figure 33 shows the relationships between length of jump L, F_1, and apron slopes. Figure 34 shows the relationships between length of jump, tailwater depth, apron slope, and F_1.

These charts should be used as aids in the preparation of preliminary designs only. Final designs for important hydraulic structures should be verified by model tests. Further reference to application of the sloping-apron principle will be found in Sec. 17.

A small triangular sill is usually placed near the end of the apron. This serves to lift the flow as it leaves the apron and thus acts to control scour.

40. Sharp-crested Weirs. Figure 35 shows a typical sharp-crested weir with

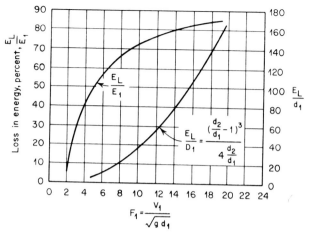

FIGURE 29 Loss of energy in jump on horizontal floor.

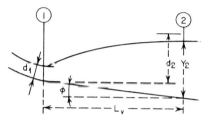

FIGURE 30 Hydraulic jump on sloping apron.

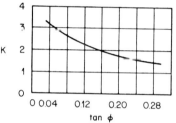

FIGURE 31 K as a function of $\tan \phi$.

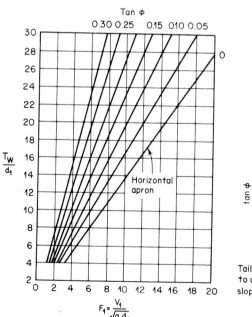

Tailwater depth related to conjugate depth for sloping aprons

FIGURE 32 Tailwater depth for jump on sloping apron.

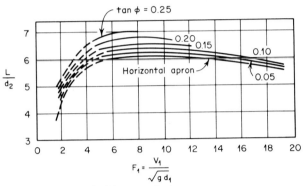

FIGURE 33 Length of jump on sloping apron.

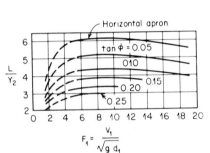

FIGURE 34 Length of jump on sloping apron.

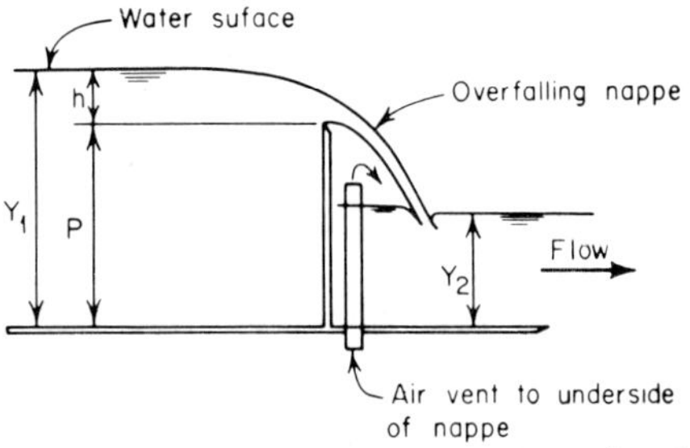

FIGURE 35 Sharp-crested weir with ventilated two-dimensional overflow.

two-dimensional flow. If the underside of the nappe is well-ventilated, the flow over such a weir without end contractions may be analyzed by

$$Q = KLh^{3/2} \tag{51}$$

in which Q is the discharge, cfs, L is the length of the weir crest, ft, and h, ft, is the difference between the elevation of the water surface upstream from the weir and the elevation of the sharp crest. Note that h does not include the velocity head of the approach flow at the section where the elevation of the water surface is taken.

The coefficient K has been evaluated as follows by the experiments indicated:

Francis:

$$K = 3.33\left[\left(1 + \frac{h_v}{h}\right)^{3/2} - \left(\frac{h_v}{h}\right)^{3/2}\right] \tag{52}$$

Bazin:

$$K = \left(3.248 + \frac{0.079}{h}\right)\left(1 + 0.55\frac{h^2}{y_1^2}\right) \tag{53}$$

Rehbock:

$$K = 3.228 + 0.435 \frac{h_v}{P} \tag{54}$$

The submerging of a sharp-crested weir, illustrated in Fig. 36, has the effect of reducing the flow over the weir to Q_1 as compared with the free flow Q determined from Eq. (51). Villemonte has found the following equation will evaluate this effect with useful accuracy, the maximum deviation probably being 5 percent.

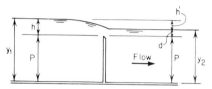

FIGURE 36 Sharp-crested weir with submerged two-dimensional overflow.

$$\frac{Q}{Q_1} = \left[1 - \left(\frac{d}{h}\right)^n\right]^{0.385} \tag{55}$$

The terms d and h are defined in Fig. 36. The exponent n is 1.5 for sharp-crested horizontal weirs without end contractions, and 2.5 for 90° triangular V-notch weirs in which flow may be calculated by means of

$$Q = 2.5 h^{2.5} \tag{56}$$

REFERENCES

1. Russell, Samuel O., and James W. Ball, "Sudden-Enlargement Energy Dissipator for Mica Dam," *Proceedings of the ASCE, Journal of the Hydraulics Division,* July 1967, p. 41.
2. Department of the Army, Office of the Chief of Engineers, Engineer Manual EM 1110-2-1602. p.7, August 1963.
3. Gibson, A. H., "The Conversion of Kinetic to Pressure Energy in the Flow of Water through Passages Having Divergent Boundaries," *Engineering,* vol. 93, no. 205, 1912.
4. Streeter, Victor L., *Fluid Mechanics,* 2d ed., p. 188, McGraw-Hill, New York, 1958.
5. Factors Influencing Flow in Large Conduits, Report of Task Force on Flow in Large Conduits. Committee on Hydraulic Structures, *Proceedings of the ASCE, Journal of the Hydraulic Division,* November 1965, p. 138.
6. Tilp, Paul J., Capacity Tests in Large Concrete-Lined Canals, *Proceedings of the ASCE,* May 1965.

SECTION 3

HYDRAULIC MODELS

By James E. Borg[1]

GENERAL

1. Definition of a Model. A model may be defined as a system by whose operation the characteristics of other similar systems may be predicted. This definition is general and applies not only to hydraulic models. A model is not necessarily smaller than the system to which it is to be compared. Actually, it may even be larger. Hydraulic models, however, are generally smaller than their prototypes; in fact, the chief difficulty is in making them large enough.

This definition of a model implies nothing as to its appearance, although it is generally thought of as a small-scale reproduction of the prototype. Hydraulic models usually bear a recognizable resemblance to their prototypes, although they are frequently considerably distorted. There is undoubtedly a psychological advantage in building models that look like their prototypes.

2. Principles of Hydraulic Modeling. Flow conditions in a physical model can be assumed to be similar to a scaled representation of those in the prototype if the model displays similarity of form (geometric similitude), of motion (kinematic similitude), and of forces (dynamic similitude).

Geometric similitude implies that the ratios of all homologous lengths in the model and the prototype are equal. Length, area, and volume are the parameters involved in geometric similitude, i.e.:

$$L_r = \frac{L_p}{L_m} \quad \text{(length)}$$

$$A_r = \frac{A_p}{A_m} = \frac{L_p^2}{L_m^2} = L_r^2 \quad \text{(area)}$$

$$B_r = \frac{B_p}{B_m} = \frac{L_p^3}{L_m^3} = L_r^3 \quad \text{(volume)}$$

Kinematic similitude implies that the ratios of time for all homologous particles to travel homologous distances in the model and prototype are equal. Time, linear and angular velocity, linear and angular acceleration, and discharge are the parameters involved in kinematic similitude, i.e.:

---

[1] Acknowledgment is made to Frank B. Campbell and Ellis B. Pickett for material in this section that appeared in the third edition (1969).

$$T_r = \frac{T_p}{T_m} \quad \text{(time)}$$

$$V_r = \frac{V_p}{V_m} = \frac{L_p/T_p}{L_m/T_m} = \frac{L_r}{T_r} \quad \text{(linear velocity)}$$

$$a_r = \frac{a_p}{a_m} = \frac{L_p/T_p^2}{L_m/T_m^2} = \frac{L_r}{T_r^2} \quad \text{(linear acceleration)}$$

$$Q_r = \frac{Q_p}{Q_m} = \frac{L_p^3/T_p}{L_m^3/T_m} = \frac{L_r^3}{T_r} \quad \text{(discharge)}$$

$$w_r = \frac{w_p}{w_m} = \frac{V_p/R_p}{V_m/R_m} = \frac{V_r}{L_r} = \frac{L_r/T_r}{L_r} = \frac{1}{T_r} \quad \text{(angular velocity)}$$

$$\alpha_r = \frac{\partial_p}{\partial_m} = \frac{1/T_p^2}{1/T_m^2} = \frac{1}{T_r^2} \quad \text{(angular acceleration)}$$

Dynamic similitude implies that the ratios of all homologous forces in the model and prototype are equal. Force, work, and power are the parameters involved in dynamic similitude, i.e.:

$$F_r = \frac{F_p}{F_m} = \frac{M_p a_p}{M_m a_m} = M_r a_r = \frac{M_r L_r}{T_r^2} \quad \text{(force)}$$

$$W_r = \frac{W_p}{W_m} = \frac{F_p L_p}{F_m L_m} = F_r L_r = \frac{M_r L_r^2}{T_r^2} \quad \text{(work)}$$

$$P_r = \frac{F_p L_p/T_p}{F_m L_m/T_m} = \frac{F_r L_r}{T_r} = \frac{M_r L_r^2}{T_r^3} \quad \text{(power)}$$

Fluid flow around or through a hydraulic structure is caused by an energy differential and is resisted by shear stresses within the fluid and along the flow boundary. The degree of resistance is affected by the shape and roughness of the flow boundary; the fluid density, viscosity, and compressibility; gravity; and surface tension. Hence the function for the flow phenomenon can be defined as

$$f(V, D, \rho, L, \mu, \gamma, \Delta P, \sigma, E, H, \lambda, s, r, \ldots) = 0$$

Further reduction of the function by dimensional analysis yields:

$$f\left(\frac{D}{L}, \frac{VD\rho}{\mu}, \frac{\rho V^2}{\gamma D}, \frac{\rho D V^2}{\Delta P}, \frac{\rho D V^2}{\sigma}, \frac{\rho V^2}{E}, \frac{D}{h}, \frac{D}{\lambda}, \frac{D}{s}, \frac{D}{r}\right) = 0$$

Terms 2 through 6 represent the ratio of inertial forces always found in a flowing fluid to unit forces due to viscosity, gravity, pressure differential, surface tension, and elasticity, respectively. These five terms are commonly encountered in fluid mechanics and have become known as

Reynolds number $\quad N_R = \dfrac{DV\rho}{\mu} = \dfrac{DV}{\nu} \quad \left(\dfrac{\text{inertia}}{\text{viscosity}}\right)$

Froude number $\quad N_F = \dfrac{\rho V^2}{\gamma D} = \dfrac{V^2}{gD} \quad \left(\dfrac{\text{inertia}}{\text{gravity}}\right)$

Euler number $\quad N_E = \dfrac{\rho V^2}{\Delta P} \quad \left(\dfrac{\text{inertia}}{\text{pressure}}\right)$

Weber number $\quad N_W = \dfrac{\rho D V^2}{\sigma} \quad \left(\dfrac{\text{inertia}}{\text{surface tension}}\right)$

Mach number $\quad N_M = \dfrac{\rho V^2}{E} \quad \left(\dfrac{\text{inertia}}{\text{elasticity}}\right)$

All areas of similitude can be achieved only by utilizing a fluid in the model which allows the simultaneous scaling of fluid properties needed to produce identical N_R, N_F, N_E, and N_M in both the model and the prototype. It should be noted, however, that if the fluid in both prototype and model are the same, i.e., water, the ratios of fluid characteristics such as ρ, μ, γ, σ, and E become unity and the ratios of N_W and N_M are unaffected. However, the use of water prohibits simultaneously satisfying the Froude, Reynolds, and Weber number scaling criteria because the

- Froude number requires that V is proportional to $L^{1/2}$
- Reynolds number requires that V is proportional to L^{-1}
- Weber number requires that V is proportional to $L^{-1/2}$

Hence judgment must be practiced by the engineer to determine whether the force of gravity, viscosity, or surface tension predominates in the study and how the selection of model scale can be used to help minimize scale effects resulting from the neglect of the remaining fluid forces.

The Reynolds number can be used to establish modeling laws when the effects of inertia and viscosity predominate the flow path. This is a valid assumption when modeling flow in closed conduits, extensive river models, and around submerged objects.

The Froude number can be used to establish the modeling laws when the effects of inertia and gravity drive the flow path. Froude number modeling is used extensively where friction losses are negligible and flow is highly turbulent, such as with spillways and other high-velocity open-channel hydraulic structures. Froude number scaling is also common for studies involving relatively large waves associated with breakwater or ship models.

Modeling laws for studies where inertia and surface tension forces predominate should be based on the Weber number. Studies involving the entrainment of air in flowing water, the deaeration of flow in hydraulic structures, or the effects of entrained air on scour depth or impact loads should be based on Weber number scaling.

The Euler number is particularly useful in scale models involving air, rather than water, as the test fluid, and is commonly used for models in wind tunnels or models of bifurcations or manifolds. Table 1 shows the various scaling laws required to achieve geometric, kinematic, and dynamic similitude based on Reynolds,

TABLE 1

Characteristics	Dimensions	Scale ratios for the laws of:		
		Froude (gravity)	Reynolds (viscosity)	Weber (surface tension)
Geometric properties				
Length	L	L_r	L_r	L_r
Area	L^2	L_r^2	L_r^2	L_r^2
Volume	L^3	L_r^3	L_r^3	L_r^3
Kinematic properties				
Time	T	$\sqrt{(L\rho/\gamma)_r}$	$(L^2\rho/\mu)_r$	$\sqrt{(L^3\rho/\sigma)_r}$
Velocity	LT^{-1}	$(L\gamma/\rho)_r$	$(\mu/L\rho)_r$	$\sqrt{(\sigma/L\rho)_r}$
Acceleration	LT^{-2}	$(\gamma/\rho)_r$	$(\mu^2/\rho^2 L^3)_r$	$(\sigma/L^2\rho)_r$
Discharge	L^3T^{-1}	$(L^{5/2}\sqrt{\gamma/\rho})_r$	$(L\mu/\rho)_r$	$(L^{3/2}\sqrt{\sigma/\rho})_r$
Kinematic viscosity	L^2T^{-1}	$(L^{3/2}\sqrt{\gamma/\rho})_r$	$(\mu/\rho)_r$	$\sqrt{(L\sigma/\rho)_r}$
Dynamic properties				
Mass	M	$(L^3\rho)_r$	$(L^3\rho)_r$	$(L^3\rho)_r$
Force	MLT^{-2}	$(L^3\gamma)_r$	$(\mu^2/\rho)_r$	$(L\sigma)_r$
Density	ML^{-3}	ρ_r	ρ_r	ρ_r
Specific weight	$ML^{-2}T^{-2}$	γ_r	$(\mu^2/L^3\rho)_r$	$(\sigma/L^2)_r$
Dynamic viscosity	$ML^{-1}T^{-1}$	$(L^{3/2}\sqrt{\gamma\rho})_r$	μ_r	$\sqrt{(L\rho\sigma)_r}$
Surface tension	MT^{-2}	$(L^2\gamma)_r$	$(\mu^2/L\rho)_r$	σ_r
Volume elasticity	$ML^{-1}T^{-2}$	$(L\gamma)_r$	$(\mu^2/L^2\rho)_r$	$(\sigma/L)_r$
Pressure intensity	$ML^{-1}T^{-2}$	$(L\gamma)_r$	$(\mu^2/L^2\rho)_r$	$(\sigma/L)_r$
Momentum impulse	MLT^{-1}	$(L^{7/2}\sqrt{\gamma\rho})_r$	$(L^2\mu)_r$	$(L^{5/2}\sqrt{\rho\sigma})_r$
Energy and work	ML^2T^{-2}	$(L^4\gamma)_r$	$(L\mu^2/\rho)_r$	$(L^2\sigma)_r$
Power	ML^2T^{-3}	$(L^{3/7}\gamma^{3/2}/\rho^{1/2})_r$	$(\mu^3/L\rho^2)_r$	$(\sigma^{3/2}\sqrt{L/\rho})_r$

Soure: "Fluid Property Scales," Table 1 in ASCE paper titled *Hydraulic models*, prepared by the Committee of the Hydraulics Division on Hydraulic Research, July 23, 1942.

Froude, and Weber number modeling. The derivation of these modeling laws is covered in greater detail in publications by Rouse,[1] Stevens,[2] Streeter,[3] Morris,[4] and White.[5]

3. Types of Physical Hydraulic Models. Hydraulic model studies associated with the design of hydroelectric or water resources projects may be divided into three principal categories: models of structures, models of rivers, and models of closed-circuit systems. These categories are distinguished by the behavior of the water surface. In models of hydraulic structures, there is usually a rapid change in the elevation of the water surface and a corresponding dependence on the Froude number for similarity. In models of rivers, the change in water-surface elevation is very gradual, being governed chiefly by friction. In such models, similarity is governed by Reynolds number and the laws of friction in river channels. Both types of flow conditions sometimes occur in one model, such as a sudden constriction in a river model which may be caused by bridge piers or cofferdam structures. In such cases, it is necessary to satisfy both Froude law scaling criteria and a friction law at the same time, or if this is not possible, the model must be designed so that one feature or the other is unimportant.

GENERAL 3.5

Models of Hydraulic Structures. Physical hydraulic model studies are generally performed on hydraulic structures to determine the effects of three-dimensional flow on:

- The discharge capacity of the structure
- The generation and magnitude of cross-waves
- The magnitude and frequency of pressure fluctuations
- The efficiency of energy dissipation devices
- The initiation and degree of downstream scour and determination of required protective measures.

Physical hydraulic model studies should be undertaken only when significant project savings can be realized as a result of the studies, when existing design charts and empirical formulas are inadequate, when accurate flow capacity is crucial for safe project operation, or when the costs associated with numerical modeling are prohibitive.

In the design of models of hydraulic structures, consideration must be given to:

- The results desired
- The space available for construction of the model
- The water supply
- The cost

In general, the model should be built as large as possible in order to obtain the most accurate results. Space limitations frequently dictate the size, and in many cases it is further limited by the amount of water or the funds available. The fundamental relationships for models of hydraulic structures are based on the Froude law.

1. *Three-dimensional comprehensive spillway models*
   a. *Spillway crests and discharge capacity.* The effects of three-dimensional flow on the discharge capacity of a spillway are determined in a comprehensive physical model. The scale of the model should be sufficiently large to reduce the effects of friction in the approach channel and the effects of internal friction and surface tension in the crest area during low flow conditions. Ideally, model scales of 1:40 to 1:100 should be adequate to neglect the influences of friction and surface tension provided that head on the spillway crest is sufficiently large. In some instances, scales as small as 1:200 have been successfully used to model full-head discharges.

The discharge over an uncontrolled crest structure can be defined as

$$Q = CLH^{3/2}$$

where C = discharge coefficient
L = the crest length
H = head in the crest, defined as the vertical difference between crest level and reservoir water level

The relationship between model and prototype parameters, following Froude law scaling, becomes:

$$Q_r = C_r L_r H_r^{3/2} = C_r L_r^{5/2}$$

If the model scale is adequately large, it can be assumed that the flow patterns in model and prototype are similar and that the discharge coefficients for the model and the prototype are the same. Where quantitative comparison of model and prototype discharges are available, they indicate that the flows obtained by a model operating under a head of at least 0.2 ft are reliable within 2 or 3 percent. Friction in very small spillway models having multiple bays is not negligible, and methods of correction similar to those developed by Russel (Sharp[6]) should be used for low-head discharges.

b. *Spillway chute and spillway tunnel hydraulics.* Comprehensive models are also used to determine the effects of asymmetric approach flow or asymmetric gate operations on flow conditions on the spillway crest and in the downstream conveyance chute or tunnel. Spillway crest water surface profiles resulting from asymmetric approach flow conditions must be established in order to determine the location of gate trunnions and gate lip elevations when gates are in their fully opened position. The discharge capacity of an ungated side channel spillway crest is extremely sensitive to unsatisfactory trough hydraulics which can cause submergence of the crest ogee. Asymmetric approach flow conditions can reduce the discharge capacity of morning glory spillway crests by inducing vortices during transition from free overfall to throat or tunnel control.

Chute or tunnel cross-wave patterns resulting from asymmetric approach flow conditions, asymmetric gate operation scenarios, abrupt offsets such as the end of piers, or converging or expanding transitions must be established in order to determine wall heights or tunnel diameters, slopes, and alignments. Chute flow concentrations and cross-wave patterns can also effect the efficiency of energy dissipation devices such as stilling basins, roller buckets, and flip buckets. Downstream scour patterns and return flow currents are also subjected to the degree of flow asymmetry leaving the spillway outlet structure.

c. *Spillway outlet structures.* An important phase of model testing is determining the effect of spillway discharge on the downstream river bed. It is not necessary to simulate approach conditions as accurately as for discharge coefficient measurements, but the model should have a channel that will simulate conditions below the dam. Preliminary tests of erosion tendencies may be made on a large scale with a sectional model. However, where gates are provided on the crest, and it is possible for the distribution of discharge to vary, tests of the entire structure should be made, and the river channel below should be reproduced in detail.

The type of device that will be most applicable in a given project and effective in reducing energy may be determined by considering tailwater levels, foundation materials, project layout, and overall project head and release flow conditions. The effect of the overfalling water on the erodible material at the end of the device in the prototype cannot be determined precisely. In many cases, the river bed material is rock and its erosion characteristics are unknown. If the river bed is alluvium, sand or gravel, it will erode easily. One laboratory expedient is the use of an erodible river bed composed of material that is fine enough to be moved readily, yet that will not be carried away by the normal velocity that exists below the spillway. It is important that the material be carefully screened so that its range of size is limited. If a well-graded material is used, the fines wash away first. In order to reproduce tests, it is necessary that the fines be collected and remixed with the coarse particles. This is practically impossible, and comparative results are very difficult to obtain with such material.

Sometimes it is possible to simulate the rock bed and banks of a stream in a qualitative manner by forming them in concrete so weak that it will be eroded by the appropriate model velocities. This is particularly true when the banks are steep and cannot be made to hold their slopes with sand or gravel. A reasonably satisfactory material has been produced by the use of lumnite cement and sand. Portland cement cannot be used because of its characteristic increase of strength with age. Large gravel or preformed cubes of like density may also be used to simulate erodible rock material. The size of the gravel or cubes can be based on the degree of expected fracturing of the prototype rock material. When energy dissipation occurs in a narrow valley, additional care must be taken to reproduce nonerodible rock outcrops and rock discontinuities which may alter the direction of high-velocity flow and influence the resulting scour pattern.

It is not possible to predict the precise amount of erosion that may be expected below a spillway. This does not restrict the use of models in design, however, because it is not usually desired that erosion shall occur. A properly constructed model is capable of predicting tendencies toward erosion or deposition of material, and it is entirely feasible to construct the dissipation device such that the action of the water leaving it will tend to deposit material at the toe of the spillway or dam rather than to remove it.

A great many models have been tested to determine the best means of dissipating energy and preventing erosion. The conditions are different at each site, and a number of devices have been developed. These devices include stilling basins, submerged roller buckets, and flip buckets.

The hydraulic jump stilling basin is used to dissipate spillway release flow energy where projects are founded on highly erodible materials such as alluvium, overburden, outwash or friable sandstone, and weak rock. Controlled energy dissipation occurs within the deep, high-walled basin structure before flow is returned at a reduced velocity to the downstream river bed. Since the stilling basin is a massive, expensive structure, every effort should be made to reduce its size. The stilling basin costs can be reduced by shortening the basin length, providing chute and basin blocks or downstream riprap protection, lowering or sloping the basin sidewalls, raising the basin invert, altering the end sill height, utilizing a trapezoidal rather than rectangular basin cross section, or altering exit channel geometry and slope. Comprehensive hydraulic model studies are required to evaluate the effects of chute flow distribution and basin design modifications on energy dissipation and basin hydraulics.

The measurement of pressure fluctuation magnitudes and frequencies on the chute and basin blocks and basin floor will indicate potential for cavitation and the need for chute aeration. Pressure fluctuation measurements on the basin floor and the location of the initiation of the jump are used to determine the degree of uplift during design flow conditions. Water surface profile measurements can be used to verify basin wall freeboard requirements. Downstream velocity measurements, wave height and frequency measurements, and exit channel erosion contours can be used to determine the height of endsill, the depth of erosion cutoff required to prevent the undermining of the basin endsill, the lateral extent of exit channel wing walls, and riprap protection for the exit channel and the toe of the adjacent dam.

Flip buckets are generally used on projects which rely on sound, nonerodible rock for structural support and energy dissipation. Unlike the stilling basin, the energy dissipation occurs downstream of the structure where the high-velocity free jet plunges into the river tailwater. The degree of energy dissipation is evidenced by the depth and pattern of the plunge pool erosion, and the strength of the return flow currents near the project structures. Hydraulic model studies should be made to determine the effects of bucket width, radius, and exit angle; plunge pool material; and flow duration on the efficiency of energy dissipation and extent of additional project protection such as bucket and dam cutoffs, riprap blankets, trenches and spur dikes, and bucket apron slabs. The design of flip buckets that use nonconventional lip shapes to divide or spread the impact pattern and reduce erosion should also be confirmed by model studies. In some instances, where preexcavated plunge pools are concrete-lined or expose highly fractured rock formations, hydraulic model studies are performed to measure design impact pressure fluctuations and invert velocity magnitudes and directions.

Submerged roller buckets are used to dissipate spillway release flow energy when the project is founded on highly competent rock. Roller buckets are generally located at the toe of concrete gravity dams and rely on both flipping action and high tailwater for energy dissipation. Although generally located on sound rock, hydraulic model studies are required to measure the erosion contours to verify the bucket radius and the setting of the roller bucket lip elevation. Measurements of pressure fluctuation magnitudes and frequencies are taken to determine the potential for cavitation damage to the submerged roller bucket lip.

d. *Special spillway structures.* Hydraulic model studies are extremely useful in verifying the design and operation of spillways having headworks structures displaying a shift of discharge control with varying head which could be further influenced by unfavorable approach flow conditions or if the geometry of the structure is significantly altered from preceding designs. Side channel spillways, morning glory spillways, siphon spillways, orifice spillways, labyrinth spillways, and fuse plug spillways fall into this category.

Conventional side channel spillway designs employ standard ogee crests parallel to the downstream conveyance chute. The spillway discharge capacity is controlled by the crest until it becomes submerged by the water surface within the collection trough. As long as the flow enters the trough perpendicular to the trough centerline, the trough water surface profile and resulting crest submergence may be estimated by the analysis of spatially varied flow as developed by Hinds.[7] Computational procedures for the estimation of the trough water surface are provided by the Bureau of Reclamation[8] and Chow.[9] Hydraulic model studies are recommended to confirm the spillway capacity and trough hydraulics when the conventional side channel arrangement is modified to allow flow to enter the trough parallel to its centerline, as is the case of L-shaped or U-shaped "duckbill" crest structures.

Morning glory spillways use circular crest structures to achieve the required crest length and desired discharge capacity. The circular control headworks is connected to the spillway tunnel by a vertical transition shaft. The control of the flow shifts from the circular ogee crest to the throat transition tube and finally to the spillway tunnel as the head on the crest increases. Although standard design procedures have been established and are presented by the Bureau of Reclamation,[10] the shift of the flow control is greatly influenced by vorticity induced by asymmetric approach flow patterns and the amount of air entrainment. Hydraulic model studies can be used to identify measures to improve approach flow conditions and reduce vorticity, such as modifications to approach-channel excavations, provisions for crest piers and hood structure, or the insertion of a radiating sidewall connecting the crest structure to adjacent topography to cut off circulating flow patterns. Frequently, a small protrusion is provided in the crown of the tunnel at the vertical curve at the base of the shaft to ensure the controlled formation of open channel flow and allow an opening in the crown of the tunnel for the release of air entrained by the crest structure. The size and location of this flow control feature are often determined by model studies. Hydraulic models of morning glory tunnel spillways will be based on Froude-law scaling until tunnel pipe flow controls and submerges the headworks. At this point, friction forces predominate and the scaling of the tunnel roughness must be considered.

The discharge characteristics of syphon spillways may also be determined by model studies, although it is possible to predict the operation of a syphon spillway with reasonable accuracy by analytical methods. If Froude law scaling is to be used, a scale of sufficient size must be chosen and care in the selection of the model material roughness should be taken to minimize the effects of fluid viscosity on the test results. It must be remembered that test results become invalid if negative model pressures, when transferred to prototype values, fall below vapor pressure. For example, a negative pressure of 3 ft of water as a 1:15 scale model would correspond to a prototype pressure of −45 ft, far below the approximate −33-ft vapor pressure, the point at which the model ceases to represent prototype conditions.

Gated orifice spillway model studies are conducted to determine (1) flow contraction coefficients and discharge capacities during part-gate conditions, (2) the transition point between free and orifice flow conditions, (3) the effects of asymmetric-approach flow and vorticity on full gate flow capacity, (4) the design of antivortex entrance vanes, grills, and hoods, (5) the headworks and gate dynamic pressure fluctuation magnitudes and frequencies, and (6) the effects of downstream submergence on flow capacity. As with the syphon spillway, care must be taken in the selection of model scale and material roughness to minimize the effects of fluid viscosity in the test results, and also in transferring negative model pressures to prototype values.

Overflow spillway discharge capacity can also be improved by providing additional crest length. When space is limited, the crest length can be arranged in a cyclic, accordion shape called a *labyrinth spillway*. Unlike the conventional ogee crest, however, the discharge coefficient of a labyrinth spillway decreases with head because of crest submergence from the intervening troughs formed downstream of each crest cycle. A number of hydraulic model studies have been performed by the Bureau of Reclamation[11] in order to develop standardized design procedures and coefficients. Further model studies are recommended where labyrinth crests (1) are subjected to asymmetric-approach flow conditions, (2) experience head-to-crest design head ratios in excess of standard design graph values, (3) are influenced by downstream submergence, or (4) may result in excessive river bed erosion.

Like the labyrinth spillway, recent model studies have been performed by the Bureau of Reclamation[12] to standardize the design of erodible fuse-plug dikes. The principal reason for model studies of fuse-plug dikes is to determine the effect of material compaction, dike cross section, and dike volume on erosion rate. Once again, additional model studies are recommended when departures from standard design conditions are expected, such as unavailability of suitable material, tailwater submergence effects during activation, or the effects of adjacent hydraulic structures on the fuse-plug erosion rate.

2. *Two-dimensional sectional spillway models.*
   a. Spillways without piers or gates. Uncontrolled crest spillways of this type have long crest lengths, and the effect of end contractions need not be considered. It is customary not to model the entire length of the spillway crest but to make a model of a short section, suppressing the end contractions. The length of crest should be at least equal to the head of the water over it in order to minimize the effect of the channel walls.

   The correct reproduction of approach conditions is important in the determination of discharge coefficients, and they should be simulated as nearly as possible. This will frequently result in a smaller scale model than would otherwise be built, owing to space limitations. If the spillway crest is submerged, care should be taken to reproduce the exit flow conditions, as apron and tailwater submergence influences affect the discharge coefficient.

   b. Spillways with piers. Many spillways with long crests are divided into multiple bays by a number of spillway piers. These piers may be used to support an access bridge over the structure, or gates. When building a model of such structure, it is necessary to use an integral number of sections or bays. The design of the model is similar to the spillway without piers except that the length of the model crest is a multiple of the length of one bay instead of an arbitrary length. The model should have at least one center bay and two half-bays so that the effect of end contractions may be observed, and provisions should be made for eliminating the contractions for a portion of the tests.

3. *Miscellaneous structure models.*
   a. Emergency-gate loads. Gates designed to operate in flowing water under considerable heads are subjected to loads parallel to their direction of motion. Water passing under a downstream sealing gate in a partially opened position may reduce the pressure on its lower face, whereas the upper surfaces within the gate shaft, being in region of low velocity, are under full pressure. The load due to this differential pressure may be much greater than the weight of the gate itself. Both the design of the gate and the capacity of the hoist are directly affected. The load may be reduced by altering the gate lip geometry to increase uplift pressures or providing an upstream sealing gate to eliminate the gate slot water pressure. In addition, gate vibrations have occurred when the gate reaches a position approximately 90 percent open and the net downpull and horizontal load on the gate approach zero.

   Satisfactory determination of the hydraulic load can be made only by means of a model. The model must simulate the prototype in all respects that influence the flow including both approach and downstream conditions. The load may be determined by weighing. Gate vibrations can be measured by accelerometers or transducers. It is important to minimize mechanical friction of the gate in the model.

  b. Intake draft-tube structures. Hydraulic model studies have been made to predict entrance and draft-tube flow conditions, to develop more efficient and cost-effective intake and trashrack designs, and evaluate expected turbine performance. Run-of-river low-head hydroelectric projects must be evaluated to determine the effects of asymmetric approach flow conditions and excessive intake and draft-tube exit losses on project generation. Comprehensive model studies provide point velocity measurements at the powerhouse entrance to document the degree of flow asymmetry which may affect turbine output guarantees. Water level measurements in the approach and exit channels are used to estimate head losses. Piezometric pressure measurements within the powerhouse entrance and draft tube are used to compute intake and draft-tube exit losses and provide the net head on the unit for power output computations, all of which are required to evaluate the unit performance as defined by International Electrotechnical Commission (IEC) standards.[13]

  The design of conventional hydroelectric and water-supply intake structures are generally verified by hydraulic model studies where excessive vorticity may be encountered that may result in operational problems or reduction of capacity by air entrainment. Tendencies toward vorticity can be evaluated by qualitative tests in which the discharge is increased to a value much larger than that anticipated during normal operation. The increase in flow, controlled by the downstream end of the test conduit, results in a reduction of the influence of fluid viscosity and increases asymmetric or vortex-inducing flow conditions. Additional studies are performed on conventional intake structures to determine intake bellmouth, gate slot, and transition head losses and pressure profiles and drop coefficients. Pressure-drop coefficients are used to indicate cavitation potential in high-velocity intake structures.

  Turbine model test results are frequently used to design draft-tube trashracks for pumped-storage hydroelectric projects. The model results generally provide the draft-tube velocity distributions and pressure fluctuation magnitudes and frequencies for various generating gate positions and heads required for the economical design of the draft-tube trashrack.

  c. Aeration devices. In recent years it has been demonstrated that aeration of high-velocity flow can reduce or eliminate the potential for cavitation damage in hydraulic structures. The design of spillway and outlet works aeration devices, such as offsets, wedges, and troughs away from the flow had been based almost exclusively on the results of physical hydraulic model studies until prototype observations became available. The air-entrainment mechanism is initiated by drag from the high-speed spray forming along the flow surface when flow turbulence overcomes surface tension effects. Hence, model scales must be chosen sufficiently large to simulate the air-entrainment mechanism and corresponding air inflow discharge.

  Based on the correlation of model and prototype aerator performance data of the Foz do Aeria spillway, Pinto[14] suggests that in order to model the air entrainment process and reproduce the spray mechanism, the model Weber number N_W must be greater than 1000. The Weber number is

$$N_W = \frac{V}{\sqrt{\dfrac{\sigma}{\rho L}}}$$

where V = flow velocity, m/s
L = ramp trajectory length, m
σ = fluid surface tension, N/m
ρ = fluid density, kg/m^3

Hydraulic model studies are generally made to determine the pressures on the aerator, within the supply air slot, beneath the lower nappe, and in the flow reattachment zone downstream of the aerator. The measurements of the amount of air introduced into the flow and the length of the aerator cavity are also useful in determining the efficiency of the aerator. Assuming that the Reynolds number is sufficiently high in the model to assure the formation of turbulent flow and the Weber number is large enough to adequately reproduce the aeration mechanism, the aerator and impact pressure may be modeled according to Froude law scaling.

The amount of air entering the flow is a function of the demand of the lower nappe of the jet and the efficiency of the air delivery system. The criteria developed by Colorado State University[15] during the Tarbela tunnel and aerator model studies for establishing the scale ratio for modeling the prototype air supply system is much more complex, however.

The mean axial velocity of spiral air flow through the air supply conduit to the aeration is given by the expression

$$V_a = \sqrt{\frac{\dfrac{2\Delta P}{\rho} + gL}{K_e + K_b + \dfrac{fl}{D} + \alpha} - V_C^2}$$

where V_a = mean axial component of the air velocity
V_C = circumferential component of the air velocity
ΔP = pressure drop through the delivery system
ρ = density of air
g = gravitational constant
L = length of the conduit
K_e, K_b = entrance and bend loss coefficients
F = conduit friction factor
D = conduit diameter
α = kinetic energy flux correction coefficient

For convenience, set

$$\Omega = K_e + K_b + \frac{fl}{D} + \alpha$$

and

$$V^2 = V_a^2 + V_C^2$$

so that

$$V = \sqrt{\dfrac{\dfrac{2\,\Delta P}{\rho} + gL}{\Omega}}$$

Generally, in large-scale aeration models,

$$\dfrac{2\,\Delta P}{\rho} \gg gl$$

and $V_c = 0$, so that

$$V_a = V = \sqrt{\dfrac{2\,\Delta P}{\rho\Omega}}$$

may be assumed.

Hence, the air system scaling laws for air velocity and flow rates may be assumed to be

$$V_r = \dfrac{V_m}{V_p} = \sqrt{\dfrac{\Delta P_m}{\Delta P_p}\dfrac{\rho_p}{\rho_m}\dfrac{\Omega_p}{\Omega_m}} \quad \text{(velocity)}$$

and

$$Q_r = \dfrac{Q_m}{Q_p} = \dfrac{V_m A_m}{V_p A_p} \quad \text{(discharge)}$$

In measuring airflow quantities care should be taken not to introduce a measurement device such as a propeller meter or orifice plate which may restrict the airflow path and cause additional losses in the delivery system. Airflow velocities may be measured with hot-wire anemometers, and airflow discharges may be measured with venturi meters or by determining the amount of time required to deflate a known-volume air reservoir which can be quickly attached to the air supply system. Smoke can also be used to observe how the airflow pattern varies within the chute air slot and how the air entrainment varies across the lower nappe.

d. *Navigation locks.* Successful operation of navigation locks requires minimizing wave disturbances to boats moored in the lock chamber and an acceptable level of surges in the lock approach and exit channels. Unacceptably high surge levels resulting from filling and emptying of the lock may cause excessive wave action on adjacent riprap protection and could cause operational problems at adjacent powerhouses in combined hydropower navigation projects. The hydraulic design of the model offers no particular difficulties. The determination of stresses in the hawsers mooring the boat is another matter. Hawser force magnitudes and fluctuations are generally measured with strain gage potentiometers.

If the filling system is composed of a culvert through the lock walls, with ports into the chamber, friction undoubtedly plays at least a minor part in governing the operation of the system, and care must be taken to reduce conduit roughness. However, if the lock is filled from the upper end by short culverts either around or through the upper miter sill or by discharge through the upper gates such as sector or Tainter gates, friction may be neglected entirely.

e. Fish conveyance facilities. The design of fish conveyance systems by means of model tests is a comparatively recent development. Hydraulic model studies have been made to predict slotted weir discharge coefficients and velocities and pool-to-pool energy dissipation in fish ladders for upstream migrants. More recently, model studies have been made to determine the most efficient orientation, depth, and spacing and resulting head losses through louvered fish diversion barriers for downstream migrants.

Structures in Which Friction Is Important. The types of models described in the foregoing section have been those in which friction played a relatively unimportant part. However, in models of closed conduit systems, the scaling relationships should be based on friction formulas such as the Darcy-Weisbach or Manning-Strickler equations. Friction head loss coefficients are constant only when the flow is turbulent and uniform flow has been fully developed. The Darcy-Weisbach head loss coefficient f is a function of the Reynolds number in the transition range between laminar and turbulent flow, and friction loss adjustments must therefore be made when prototype and model test flows fall within this range. Included in the closed conduit systems which fall into this category where friction predominates the flow are lock conduits and tunnels.

1. *Lock flow conveyance systems.* Lock flow conveyance systems are composed of culverts and manifolds in the lock walls and lateral ports leading to the lock basin. These conveyance features are subject to friction losses and may result in losses appreciably higher than expected, particularly in small-scale models.

2. *Spillway and diversion tunnels.* The performance of tunnel spillways and diversion tunnels flowing full is governed by friction. If the conduit roughness and friction slope in the conduit are greater for the model than for the prototype, as is usually the case, the discharge of the system is reduced and the performance of the intake and outlet structures for given headwater and tailwater elevations is not correct. Pressures in the conduit are not correctly represented.

The remedy in such a situation is a careful study of friction losses in the conduit and a comparison with similar friction losses to be expected in the prototype structure. The conduit can then be omitted from quantities consideration except as it affects the back pressure on the intakes or the discharge into the downstream energy dissipators, and these elements may be tested independently. The ratio of model roughness to prototype roughness that will give similarity by the Froude law is $L_r^{1/6}$. If the actual roughness of model and prototype are n_m and n_p, respectively, the discharge that will give similar gradients in the conduit for model and prototype is obtained by multiplying that obtained by the Froude law by

$$n_r = \left(\frac{n_p}{n_m}\right) L_r^{1/6}$$

Models of River Channels

1. *General considerations.* Hydraulic model studies of river channels are performed primarily to determine the effects of alterations in the river channel geometry or the influence of hydraulic structures on flow characteristics and

sediment transport. Channel modifications may include deepening or straightening a reach of the river or placing groins on the river banks to improve navigation conditions during periods of low river flow. Hydraulic structures altering the flow and sediment transport characteristics of the river include lower-head hydropower dams, irrigation diversion weirs, water intake structures, and flood control levees. Physical river hydraulic models are also frequently used to study the thermal and chemical pollutant dispersion downstream of an outfall structure.

The design of models of river channels presents more difficulty than the design of models governed solely by the Froude law. The flow of water in rivers is governed largely by friction. The change in water surface is very gradual so that gravity forces play a relatively minor part, except at marked changes in the cross section where the flow becomes decidedly nonuniform. The scale relationships in river models is therefore usually based on flow friction formulas similar to the Manning-Strickler equation:

$$V_r = \frac{D_r^{7/6}}{L_r^{1/2} n_r} \quad \text{(velocity)}$$

$$Q_r = \frac{L_r^{1/2} D_r^{13/6}}{n_r} \quad \text{(discharge)}$$

$$S_r = \frac{D_r}{L_r} \quad \text{(slope)}$$

$$n_r = L_r^{1/6} \quad \text{(roughness)}$$

The source of many of the troubles with river models is the size of the area to be investigated. Space limitations are usually such that, when a scale has been selected that will permit an undistorted model to be placed in the space available, the depths of flow are so small that laminar flow may occur. The laws governing laminar flow are considerably different from those governing turbulent flow, and since turbulent flow almost always occurs in nature, it should also exist in the model. The use of an undistorted model usually results in water-surface slopes so flat that differences in elevation cannot be measured satisfactorily. These objections can be overcome by distorting the model. Distortion may be accomplished in several ways. The slope may be increased by tilting the model, by arbitrarily changing the discharge scale, by using different horizontal and vertical scales, or by changing the roughness.

2. *Undistorted Models.*
   a. *Fixed-bed models.* Models of river channels are frequently built with fixed beds, i.e., beds in which no erosion can occur. They are useful chiefly in investigating the effect of permanent changes in the channel caused by dredging, by obstructions such as bridge piers, or by levees. The investigation may be concerned with changes in navigation conditions or with backwater during floods. Models of this type have been used to investigate navigation conditions and backwater caused by cofferdamming operations during the construction of dams. The final stages of diversion channel closure with dumped rock methods may be studied according to Froude law scaling of the closure rock because of the high local velocities at the closure section. The principles of design are identical with those for movable-

bed models, except that it is not necessary to consider tractive force. In cases where an obstruction is so great that sudden changes in the water surface occur, it is better to use an undistorted model, if at all possible.

The dimensions of the model can be determined to a first approximation by the use of the foregoing relationships. In order to minimize the effects of viscous forces in the model, the flow must be turbulent. It has been established by measurements made at the U.S. Waterways Experiment Station that flow in the model will be turbulent if the Reynolds number N_r is sufficiently large, that is,

$$N_r = \frac{VR}{\nu} \cong 4000$$

where V = mean velocity, fps
R = hydraulic radius, ft
ν = kinematic viscosity, ft^2/s

For water at 70°F, the kinematic viscosity is approximately 0.00001. For this value of ν, turbulent flow in the model is assured if $VR = 0.04$. A VR product less than 0.04 should be considered sufficient reason for enlarging the model, or if this is not possible, of interpreting the results with extreme caution. Values of $VR < 0.02$ should not be used under any conditions as laminar flow is almost certain to occur. If the criteria of flow turbulence has been satisfied, the relationships for scaling may be based on Froude law.

b. *Movable-bed models.* A type of river model often requiring investigation is one that involves movement of the bed material. This is the most difficult type of model to design and operate, and also the one that gives the least satisfactory results.

The laws governing the movement of bed material in a river channel are subject to many parameters. The initiation of movement of bed load depends on the shear stress exerted on the channel bed by the flowing water. Shield[16] has developed the relationships, shown on Fig. 1, between the dimensionless shear stress τ and the bed boundary Reynolds number R, to determine when the initiation of bed load movement occurs.

Studies of flow resistance in channels lined with cohesionless materials by Anderson[17] has demonstrated the following relationship between channel roughness n and the material d_{50}:

$$n = 0.0395 d_{50}^{1/6}$$

The shear stresses in models are extremely small, and it is difficult to obtain the necessary force to initiate bed material movement without very steep water-surface slopes when sand is used as the bed material. The dimensionless shear stress can be increased to initiate bed load motion by using a smaller material size or using a much lighter bed material. Lighter-weight materials which have been used in movable-bed models for this purpose are lignite, haydite, and cornmeal. The effect of these lighter materials is to lessen the need for extreme distortion of the model and to decrease the time necessary to accomplish completion of the experiment.

The dimensions of the model may be determined from the foregoing equations, with due consideration being given to space available, water supply, shear stress and boundary Reynolds numbers. Frequently, the

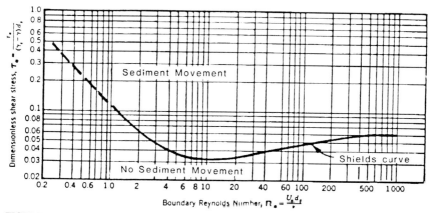

FIGURE 1 Shields diagram for incipient sediment movement, where $\tau_0 = \gamma RS$; τ_0 = shear stress of the flow on the bed material, psf; R = flow hydraulic radius, ft; S = flow energy slope, ft/ft; d_s = diameter of the bed material, ft (usually the 50 percent size estimated from a grain size distribution curve); γ_s = unit weight of the material, pcf; γ_w = unit weight of water; U_o = shear velocity in fps, defined as $(\tau_0/\rho)^{1/2}$, ν = kinematic viscosity, ft²/s.

material used for the movable bed is not of the proper size to satisfy the requirements of friction in obtaining similarity. As a result of this, and of the lack of knowledge of laws governing transportation of material, movable-bed models are built as the result of experience and the model ratios are determined experimentally. The U.S. Waterways Experiment Station has probably developed the technique of movable-bed models further than any other agency. At this laboratory, the model scales are determined by repeated experiments with varying rates of discharge, slope distortion, and times, until the model will reproduce a set of conditions known to exist at two specified times in the prototype. The set of conditions producing satisfactory correspondence is chosen to give the model scales. The process of determining these scales is called *verification*. Subsequent operation of the model is based on the assumption that these model scales will apply equally well after the desired modifications have been made. Since the changes generally consist of cutoffs or dikes and a resulting nonuniformity of flow, whereas the original channel is usually more nearly uniform, the results probably should be used with caution. They should be regarded as qualitative rather than as quantitative.

Comparisons of model tests with actual river operation, made by the U.S. Waterways Experiment Station, have been reported by Vogel[18] and Thompson[19] and the results are encouraging. Even though model tests conducted in such a fashion can be relied on only for qualitative information, they can give positive information as to areas where scour or deposits may be expected to occur.

3. *Distorted models.*

   a. *Fixed-bed models.* In general, an undistorted model is preferable to a distorted model where limitations of space and funds do not preclude its use. It reproduces flow phenomena accurately where gravitational forces predominate. River banks, earth cuts, and dikes have their natural slopes, making it possible to mold them in sand. However, the disadvantages of undistorted river models are that (1) surface slopes are extremely flat, (2) it is difficult to measure depths in small-scale models, and (3) because the turbulent flow criteria severely limit the selection of scale, the model is apt to be relatively large and very expensive.

Scale distortion can be used to reduce the horizontal scale and hence the area and cost of this model. In addition a distorted model has the advantages of:

- Increasing the flow velocities and turbulence in the model
- Reducing the time of the study
- Increasing the model Reynolds number and improving the model to prototype flow similarity
- Exaggerating the slope of the model and allowing a greater accuracy of flow depth measurement
- Reducing the required test discharges and laboratory flow requirements

Among the disadvantages of distorted models are (1) the magnitude and distribution of velocities are incorrectly reproduced, the degree of error depending upon the amount of distortion; (2) river bends, earth cuts, and dikes may become so steep that they cannot be molded satisfactorily in sand or other movable material; (3) they are not adapted to satisfying gravitational and frictional forces simultaneously; and (4) strictly speaking, they are not applicable to any case of nonuniform flow.

The principal concept behind the use of distorted models is that exact geometric similitude may be sacrificed in order to achieve an improvement in the similarity of the flow processes in the model and the prototype. The most common laboratory practice involves a vertical exaggeration in the model by assigning separate models to prototype scale ratios for horizontal and vertical dimensions. These scale factors and their ratio, called the *distortion factor*, can be defined as follows:

$$\frac{L_m}{L_p} = L_{rh} = \text{horizontal scale}$$

$$\frac{h_m}{h_p} = L_{rv} = \text{vertical scale}$$

$$e = \frac{L_{rh}}{L_{rv}} = \text{distortion factor}$$

Hence, for vertically distorted models, e will be greater than unity and will equal one for undistorted models. For simplicity, the model scales will be defined only in terms of the horizontal scale, that is,

$$L_{rh} = L_r \quad \text{and} \quad L_{rv} = \frac{L_r}{e}$$

Table 2 shows a comparison of scale numbers which satisfy Froude number scaling criteria for selected kinematic and dynamic properties for undistorted and distorted models using water as the model fluid.

TABLE 2

		Scale number	
Property	Definition	Distorted model	Undistorted model
Velocity	$V = \sqrt{gh}$	$V_r = L_r^{1/2} e^{-1/2}$	$V_r = L_r^{1/2}$
Time	$T = L/V$	$T_r = L_r^{1/2} e^{1/2}$	$T_r = L_r^{1/2}$
Acceleration	$a = V/T$	$a_r = e^{-1}$	$a_r = 1$
Slopes	$S = h/L$	$S_r = e^{-1}$	$S_r = 1$
Reynolds number	$N_r = Vh/\nu$	$N_{r_r} = L_r^{3/2} e^{-3/2}$	$N_{r_r} = L_r^{3/2}$
Force	$F = Ma$	$F_r = L_r^3 e^{-2}$	$F_r = L_r^3$
Pressure	$P = F/A$	$P_r = L_r e^{-1}$	$P_r = L_r$
Discharge	$Q = VA$	$Q_r = L_r^{5/2} e^{-3/2}$	$Q_r = L_r^{5/2}$
Relative roughness	$R = k/h$	$R_r = e^{-3}$	$R_r = 1$

Knauss[20] has developed guidelines for establishing the distortion factor and modeling scales of a distorted river model:

- In order to use the Manning-Strickler equation to describe the flow processes in the model and prototype, the relative roughness, or the ratio of the roughness height k to flow depth h, should fall in the following range of values:

$$2 \times 10^{-3} < k/h < 2 \times 10^{-1}$$

- The natural river channel width should be at least 10 times the flow depth to allow the simplifying assumption that the hydraulic radius equals the flow depth.
- In order to satisfy the assumption that the flow is uniform, the flow in the model must be fully turbulent, i.e.,

$$N_R \geq (2.35 \times 10^3)(k/h)^{-7/6}$$

- The distortion factor e should be less than approximately one-tenth the ratio of the prototype channel top width at the water surface to the flow depth. Although models with distortion factors as high as 10 to 20 have been successfully used, Press and Schroder have found that good test results can be obtained when the maximum distortion factor is limited to 5.

b. *Movable-bed models.* The scaling relationships for simultaneously modeling hydraulic processes and sediment transport phenomenon in a distorted movable-bed river model is quite complex and involves:

- Froude number scaling to achieve greater similarity of the model and prototype hydraulic process
- Establishing similarity of the transport phenomenon by equating the boundary Reynolds number and dimensionless shear stress of the model and prototype

If the test fluid is water, the boundary Reynolds number and dimensionless shear stress ratios become

$$R_{*r} = \frac{h_r d_r}{L_r^{1/2}} \quad \text{(boundary Reynolds number)}$$

$$\tau_{*r} = \frac{h_r^2}{\rho_r d_r L_r} \quad \text{(dimensionless shear stress)}$$

where

$$d_r = \frac{dp}{dm} \quad \text{(grain size scale)}$$

$$\rho_r = \frac{(\gamma_s - \gamma_w)_p}{(\gamma_s - \gamma_w)_m} \quad \text{(submerged density scale)}$$

The scaling relationship for model roughness as related to the grain diameter derived from the Manning-Strickler equation becomes

$$n_r = \frac{L_r^3 d_r}{h_r^4} \quad \text{(Manning's } n \text{ scale)}$$

Gehrig[20] in his derivation of the time scale from the sediment transport equation has shown that, if the grain size distribution of model and prototype are similar,

$$T_{sr} = \frac{L_r^{5/2} \rho_r}{h_r^2} = L_r^{1/2} e^2 \rho_r$$

Hence, the time scale for modeling the sediment transport phenomenon is quite different from the time scale for modeling the hydraulic process. This discrepancy in sediment transport and flow process time scales shows further the need for model verification and the practice of engineering judgment to achieve meaningful and useful study results.

Additional Hydraulic Models. Under this heading may be grouped those models which are not readily classified otherwise.

1. *Harbor and tidal models.* Tidal models have been studied by the U.S. Waterways Experiment Station and by the Tidal Model Laboratory at the University of California. Some of the methods and equipment used by the U.S. Waterways Experiment Station have been described by Tiffany.[21]

2. *Ship models.* The testing of ship models is a specialized field. Most of the ship testing in the United States has been done by the Navy's Bureau of Ships at the David Taylor Model Basin. A few university and commercial laboratories maintain small towing tanks. The field is of interest to hydraulic engineers because it was during a series of tests on ship models that Froude discovered the relationship that now bears his name. The Taylor Model Basin also tests ship propellers for thrust and cavitation characteristics.

3. *Turbine models.* Hydraulic turbines are designed largely on the basis of model tests conducted by the manufacturers. The principal manufacturers maintain their own laboratories and very little turbine testing is done elsewhere. As a result, the field is specialized, with most of the experimental data in the hands of the manufacturers.

4. Ice models. Hydraulic model studies are frequently made to determine the flow patterns and distribution of surface ice approaching intake structures on run-of-river hydroelectric projects in order to design ice sluicing structures and ice boom arrangements and determine the effectiveness of ice sluicing operations. The movement of river ice after spring breakup, when ice adhesion is low and ice floes are relatively rigid, has been accurately modeled by considering only geometric and dynamic similarity. Such studies have been made by simulating river surface ice with low-density materials such as wax, sawdust, and polyethylene blocks, pellets, and granules.

The modeling of two-phase frazil ice movement, that is, the movement of ice subjected to friction or fracturing, is extremely complex and must take into account the strength and elasticity of the ice. Provisions must be made for the use of natural ice in such models. Discussions and examples of models using natural ice and the development of ice model scaling relationships are described in detail by Sharp[6] and Schwarz.[20]

REFERENCES

1. ROUSE, HUNTER, "Fluid Mechanics for Hydraulic Engineers," McGraw-Hill, New York, 1938.
2. STEVENS, J. C., et al., "Hydraulic Models," ASCE Manual of Engineering Practice No. 25, 1942.
3. STREETER, VICTOR L., "Fluid Mechanics," 4th ed., McGraw-Hill, New York, 1966.
4. MORRIS, HENRY M., "Applied Hydraulics in Engineering," Ronald Press, New York, 1963.
5. WHITE, FRANK M., "Fluid Mechanics," McGraw-Hill, Kogakusha, Tokyo, 1979.
6. SHARP, J. J., "Hydraulic Modelling: Theory and Practice," Buttersworth, New York, 1981.
7. HINDS, JULIAN, "Side Channel Spillway," *Trans. ASCE,* **89,** 1926.
8. U.S. BUREAU OF RECLAMATION, Boulder Canyon Project Report, Part VI, Hydraulic Investigations, Model Studies on Spillway, 1938.
9. CHOW, V. T., "Open Channel-Hydraulics," McGraw-Hill, New York, 1959.
10. U.S. BUREAU OF RECLAMATION, "Design of Small Dams," 2d ed., U.S. Government Printing Office, Washington, 1973.
11. HINCHLIFF, D. L., and K. L. HOUSTON, "Hydraulic Design of Labyrinth Spillways," USBR Publication, Denver, 1984.
12. PUGH, C. A., "Hydraulic Model Studies of Fuse Plug Embankments," Publication REC-ERC-85-7, Bureau of Reclamation, Denver, 1985.
13. INTERNATIONAL ELECTROTECHNICAL COMMISSION, "International Code for Model Acceptance Tests of Hydraulic Turbines," Publication 193, Geneva, 1965.
14. PINTO, N. L., "Prototype and Laboratory Experiments on Aeration at High Velocity Flows," *Water Power and Dam Construction,* March 1982.
15. KARAKI, S., M. A. STEVENS, and T. E. BRISBANE, Air Slot for End of Gate Chambers, Tunnels 3 and 4, Hydraulic Model Study; Colorado State University; January 1977.
16. VANONI, VITO A. (ed.), "Sedimentation Engineering," ASCE Manual of Engineering Practice No. 54, 1977.
17. ANDERSON, A. G., A. S. PAINTEL, and J. T. DAVENPORT, "Tentative Design Procedure for

Riprap-Lined Channels," National Cooperative Highway Research Program Report 108, 1970.
18. VOGEL, H. D., Hydraulic Laboratory Results and Their Verification in Nature, *Trans. ASCE,* **101,** 1936.
19. THOMPSON, P. W., The Use and Trustworthiness of Small-Scale Hydraulic Models, *Civil Eng.,* **8,** 1938.
20. KOBUS, HELMUT, "Hydraulic Modelling," German Association for Water Resources and Land Development Bulletin 7, Pitman Publishing, London, 1980.
21. TIFFANY, J. B., Small-Scale Simulation of Tidal Phenomenon, *Civil Eng.,* B, 1938.

SECTION 4

RESERVOIR HYDRAULICS

By B. A. McKiernan[1]

INTRODUCTION

1. General Discussion. A reservoir site is a natural resource. As such, it is important that it not be wasted by using it to less than its optimum potential. In this sense the word *optimum* may mean the economic limit for a particular purpose or the inclusion of a variety of purposes to be served by a single reservoir.

Reservoir planning involves consideration of hydrology, hydraulics, design, economics, and sociology. This section is concerned with the hydrologic and hydraulic aspects of reservoir planning.

2. Definitions. An understanding of the reservoir-planning process requires a precise definition of terms. The following terms, considered generally acceptable, are used in this section:

Dead storage, sometimes called unusable storage, represents that portion of the reservoir volume below the elevation of the lowest outlet. Water in this portion of the reservoir can be evacuated only by pumping or by evaporation.

Inactive storage represents the portion of the reservoir volume which, by agreement or by legislation, will not be evacuated. Such storage may coincide with dead storage, but not necessarily.

Active storage, sometimes called *usable* or *effective storage,* represents the volume of the reservoir between the top of the inactive storage and the normal maximum operating level of the reservoir.

Conservation storage represents the part of the reservoir volume dedicated to impoundment of water for later release to serve some beneficial purpose, such as municipal supply, power, irrigation, or public health. It frequently coincides with active storage.

Flood-control storage represents the part of the reservoir volume to be utilized for impoundment of floodwaters. Such storage may be *inviolate,* in which case the dedicated volume can be used only for flood-control operations, or it may be *joint-use,* where the storage volume is used to serve both conservation and flood-control operations.

Surcharge storage represents that portion of the reservoir volume above the normal maximum operating level.

Normal maximum operating level is the maximum level at which the reservoir is operated to serve any of its planned purposes. Generally, this corresponds to the elevation of the top of the spillway gates. Operation of the reservoir to protect against a spillway design flood ordinarily will cause the reservoir level to rise above the normal maximum operating level.

---

[1]Acknowledgment is made to Victor A. Koelzer for material in this section that appeared in the third edition (1969).

Freeboard, as used here, represents the vertical distance between the maximum elevation reached in routing of the spillway design flood and the top of the dam. Some authors have used this term to represent the vertical distance between the normal maximum operating level and the top of the dam.

BASIC DATA FOR RESERVOIR PLANNING

3. Reservoir Area-Volume Curves. Sometimes referred to as *area-capacity curves,* these curves are typified by Fig. 1. The curves show total reservoir volume (or storage capacity) and the reservoir area plotted against reservoir elevation. The best basis for such curves is a topographic map of the reservoir area.

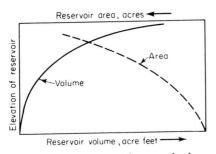

FIGURE 1 Typical reservoir area and volume curves.

If such a map cannot be made available, a preliminary estimate of the available storage volume may be obtained from river profiles and valley cross sections.

4. Streamflow. A reliable estimate of the available streamflow is essential to reservoir planning. Such estimates can best be made from records of streamflow at the reservoir site, if available. In the absence of such records, estimates of streamflow may be made from records of streamflow at another location on the same stream, from records of another stream in the same area, or from records of precipitation. The availability of records and the methods of estimation of streamflow are given in Sec. 1.

5. Climatological Data. The climatological data that may be needed in reservoir planning include data on precipitation, temperature, humidity, wind velocity, and evaporation. The availability of this information is discussed in Sec. 1.

6. Sedimentation. Estimates of the sediment load of a stream are essential in reservoir planning to determine the rate of depletion of reservoir storage that can be anticipated through accumulation of sediment. Sediment records collected by the U.S. Geological survey are published periodically in its "Water Supply Papers." Much unpublished data have been collected by various government agencies, which are inventoried periodically by a Federal Inter-Agency Committee.[1]

Various government agencies periodically survey reservoirs which have been in operation for some time. Frequently, special reports on such surveys are made, which include data on rate of sediment accumulation, distribution of the sediment in the reservoir, trap efficiency of the reservoir, and density of sediment deposits. Summaries of such surveys also are issued periodically by the Inter-Agency Committee.[2]

SEDIMENT-STORAGE REQUIREMENTS

7. Rate of Sedimentation. Estimation of the rate of reservoir depletion by sediment accumulation usually involves three basic steps. The estimates required are (1) sediment inflow, (2) trap efficiency of the reservoir, and (3) density of the sediment deposits.

Sediment Inflow. The inflow of sediment to a reservoir consists of suspended load and bed load. Generally, suspended load is computed from available sediment records or by comparison of the basin in question with watersheds having sediment records. Bed-load estimates usually are made on the basis of judgment and are expressed as a percent of suspended load.

Several methods of estimating suspended load are used:

1. Use of average recorded sediment load
2. Computation by flow-duration–sediment-rating-curve method
3. Estimates by unit yield of watershed
4. Estimates of erosion

Sediment concentration usually increases rapidly with discharge, with the result that the load carried during the highest flood of a year may be as much as the total load carried for the remainder of the year. This highlights the necessity of obtaining as reliable a reflection of the influence of peak flood periods as possible, which is usually accomplished only if the entire period of streamflow record is considered.

The flow-duration–sediment-rating-curve method of analysis is the most desirable method of giving appropriate attention to the entire period of streamflow record. In this method a sediment-rating curve is first prepared from historical records, relating sediment discharge (expressed usually in tons per day) to the water discharge. A flow-duration curve is then prepared for the entire period of streamflow record (or by procedures described later in this section). The computation of average annual load is then made by combination of these curves.[3]

Figure 2 shows a typical sediment-rating curve. Preparation of the rating curve involves plotting all measured values of sediment discharge against the corresponding values of water discharge. The scatter in the plotted points in Fig. 2 is typical of most of such plottings. Some of the scatter frequently can be reduced by developing separate curves for different seasons of the year, when differences in sediment-runoff characteristics can be expected. In extrapolation of rating curves, care should be taken to avoid extending the curves to an unreasonable value of concentration. For this purpose, plotting lines of equal concentration frequently is useful.

The average annual sediment load is computed in the flow-duration–sediment-load method by using an incremental process. The water-discharge values of the midordinate of increments of the flow-duration curve are used to obtain matching values of sediment load from the sediment-rating curve. The values of sediment load are multiplied by the percentage of time represented by the increment, and a summing of these products gives the total load. Table 1 gives an example of the method of computation used by the U.S. Bureau of Reclamation.

Where sediment records are not available (or as a check on other methods of computation), the sediment load may be estimated by estimating the sediment

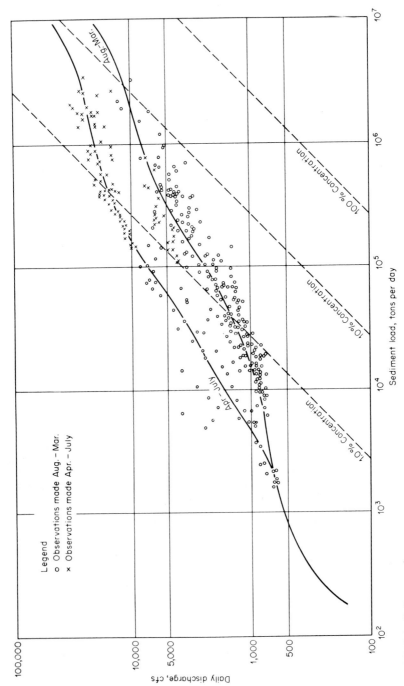

FIGURE 2 Sediment-rating curve.

TABLE 1 Computation of Suspended Sediment Load

Limits of % of time (1)	% Interval (2)	% Mid-ordinate (3)	Instantaneous water discharge, cfs (4)	Instantaneous sediment discharge, tons/day (5)	Daily sediment discharge [(2) × (5)], tons/day (6)
0.00–0.02	0.02	0.01	13,000	2,700,000	540
0.02–0.1	0.08	0.06	8,100	1,600,000	1,280
0.1–0.5	0.4	0.3	4,600	820,000	3,280
0.5–1.5	1.0	1.0	2,700	425,000	4,250
1.5–5.0	3.5	3.25	1,400	185,000	6,470
5–15	10	10	830	93,000	9,300
15–25	10	20	420	39,000	3,900
25–35	10	30	250	19,500	1,950
35–45	10	40	190	13,600	1,360
45–55	10	50	155	10,200	1,020
55–65	10	60	127	7,500	750
65–75	10	70	105	5,700	570
75–85	10	80	81	3,800	380
85–95	10	90	51	1,720	172
95–98.5	3.5	96.75	25	325	11
98.5–99.5	1.0	99.0	13.5	17.5	
99.5–99.9	0.4	99.7	7.0		
99.9–99.98	0.08	99.94	2.75		
99.98–100	0.02	99.99			

Total average daily load [Col. (6)] = 31,953 tons.
Average annual suspended sediment load = 31,953 × 365 = 11,600,000 tons.

yield from a unit of area of the watershed. This may be done by selecting a unit value from a watershed believed to have similar characteristics of sediment runoff. The previously referred to Inter-Agency publication on reservoir surveys gives summaries of sediment yields in different watersheds.

Sediment yield also may be estimated by computing rates of erosion. A method developed by the U.S. Department of Agriculture[4] may be used for this purpose. This procedure uses the so-called universal equation, which involves a number of variables related to localized conditions. The variables include storm energy, rainfall intensity, soil erodibility, land slopes, cropping patterns, and extent of conservation practices.

Only in rare instances is bed load known to exceed 25 percent of suspended load. In such instances, special methods of computation of total load may be used.[5] Bed load is usually estimated as a percentage of suspended load, on the basis of inspection of the load-carrying characteristics of the system.

Trap Efficiency. Trap efficiency represents the percentage of sediment inflow which is deposited in the reservoir. In major reservoirs this is frequently near enough to 100 percent to allow use of such a value. However, there may be instances on major reservoirs in which a portion of the sediment load is passed through the reservoir. This may occur where the reservoir volume is small with respect to the maximum rates of water discharge through the reservoir.

For smaller reservoirs, data developed by Brune[6] are useful. He relates trap

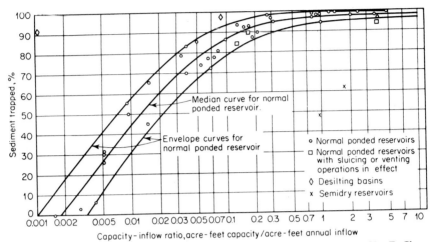

FIGURE 3 Trap efficiency as related to capacity-inflow ratio. (*After Brune. From Ven Te Chow, "Handbook of Applied Hydrology," pp. 17–23 McGraw-Hill, New York, 1964.*)

efficiency to the capacity-inflow ratio (ratio of reservoir capacity to annual inflow), as shown in Fig. 3. The capacity-inflow ratio can be considered to be equivalent to the period of detention in the reservoir. For large reservoirs, data developed by the TVA[7] allow estimates of trap efficiency to be made on the basis of the relationship between the period of detention and the mean velocity of flow through the reservoir. Consideration of velocity of flow is desirable because deposition of sediment cannot take place if the velocity and resulting turbulence are too high. The TVA curve is shown in Fig. 4.

Density of Deposited Sediments. The procedures described for estimating sediment deposited in the reservoir, after applying the trap efficiency to the rate

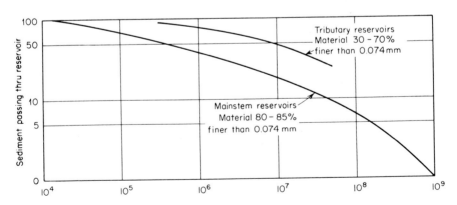

FIGURE 4 Trap efficiency in TVA reservoirs. (*Proceedings of Federal Inter-Agency Sedimentation Conference, May 1947, and correspondence from TVA.*)

of sediment inflow, will result in a value expressed in terms of weight. It then will be necessary to convert the result to a volume basis by estimating the space occupied by a unit of weight of sediment. This conversion factor is referred to as *specific weight, unit weight,* or *density.*

Many published data are available on the many observations that have been made of the density of deposited sediment.[8] It has been found to vary from 18 to 125 lb/ft^3, depending on the sediment size, the depth of deposit, and the degree of submergence or exposure of the deposit. The density also depends on the length of time the material has been deposited, since the rate of consolidation greatly affects the sediment density.

The procedure presented by Lane and Koelzer[9] has gained general acceptance. The equation is

$$W = W_1 + K \log_{10} t \tag{1}$$

where W = density after t years, pcf
W_1 = density after 1 year, pcf
K = constant for each sediment class and operation condition, to reflect consolidation
t = number of years of consolidation

The values of W_1 and K vary with the method of operation and the size of sediment material. The values are

Reservoir operation	Sand (> 0.05 mm)		Silt (0.005 to 0.05 mm)		Clay (< 0.005 mm)	
	W_1	K	W_1	K	W_1	K
Sediment always submerged or nearly submerged	93	0	65	5.7	30	16.0
Normally a moderate reservoir drawdown	93	0	74	2.7	46	10.7
Normally considerable reservoir drawdown	93	0	79	1.0	60	6.0
Reservoir normally empty	93	0	82	0.0	78	0.0

For materials covering a range of particle sizes, it was recommended that the equation be used with appropriate weights, applied in proportion to the percent by weight in each size classification.

Using the above procedure, the average density of deposits W_{av} after t years is

$$W_{av} = W_1 + 0.434 K \left[\frac{t}{t-1} (\log_e t - 1) \right] \tag{2}$$

The U.S. Bureau of Reclamation[10] uses the above procedures but considers that the values of W_1 are too high. It prefers to use the initial densities determined by Trask[11] which are as follows:

Classification	Size range, mm	Initial density, pcf
Sand	0.5–0.25	89
Sand	0.25–0.125	89
Sand	0.125–0.064	86
Silt	0.064–0.016	79
Silt	0.016–0.004	55
Silt	0.004–0.001	23
Clay	0.001–0	3

8. Distribution of Sediment Deposits. The sediment deposited in a reservoir will deposit on a slope, so that the area-volume relationship will be affected throughout by deposition of sediment. Where sediment load is a significant factor, the distribution of deposits should be estimated, to allow determination of the extent to which different allocations of storage are affected by sediment accumulation. The Bureau of Reclamation[12] has developed procedures for estimating the distribution of deposits, based on observations of historical deposition in different types of reservoirs.

RESERVOIR EVAPORATION

Reservoir planning requires considerations of the losses to be expected in evaporation from the reservoir surface. This can be quite significant in some instances. In Lake Mead, for instance, the annual evaporation loss at full reservoir approaches 1 million acre-ft, compared with an average annual water supply of the order of 10 million acre-ft. Methods of estimating reservoir evaporation are given in Sec. 1.

ANALYSIS OF WATER REQUIREMENTS

The design of a reservoir must be based on considerations of water availability as related to water requirements, since both generally vary throughout the year. The development of water requirements for various uses is discussed individually in Secs. 18, 19, 20, 24, 27, and 28.

ANALYSIS OF WATER AVAILABILITY

General. Various techniques have been used to analyze water availability. Among the well-established methods are flow-duration analyses, mass-curve analyses, and period-by-period simulated reservoir operation through the historical period of streamflow record. The latter two methods are the commonest traditional approaches used in reservoir planning, and involve examination of the most critical combinations of historical streamflow in relation to the water demands expected to be placed on the reservoir. Stochastic procedures, which are basically probability approaches, have gained increasing acceptance for reservoir planning.

9. Flow-Duration Analyses. The procedures for flow-duration analyses of the records for a given stream are presented in Sec. 1. Duration curves are useful in computing sediment load, in establishing the degree of flood control to be provided, in establishing urban, highway, and agricultural drainage requirements, and in evaluating the power capabilities of a river.

10. Mass-Curve Analysis. The mass-curve method of water-availability analysis is well-established and is extremely useful. It can be utilized to great advantage for the complete analysis of a single reservoir having a simple pattern of water requirements. On a more complicated system, it can be used for preliminary analytical purposes, to identify periods that are apt to be most critical and in need of more detailed study.

Mass curves show cumulative flow and can be used to indicate cumulative utilization and storage requirements. Adjustments for evaporation and other losses must be made in determining the net volume available from accumulated streamflow. This may be done by subtracting from the natural flow the losses anticipated in order to ensure the required net yield. To illustrate, Fig. 5 shows an accumulation of streamflow OA and an accumulation of demand OB. The curve OA has been adjusted by subtracting the expected losses before the plot was made. If a line parallel to OB is drawn tangential to point a, the beginning of the longest dry period of record, the ordinate dc will represent the volume of storage required to maintain a rate of flow not less than that represented by the slope of the line OB.

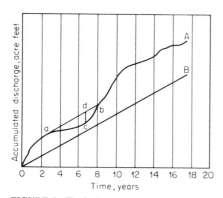

FIGURE 5 Typical mass curve.

The mass curve of water utilization need not be a straight line. Figure 6 shows a curve of irregular demand plotted with a curve of reservoir inflow for a typical dry period. The upper curve shows the cumulative requirements for water use from Oct. 1. The lower curve shows the total natural flow in acre-feet from Oct. 1 to May 7 to be 110,000. On Dec. 15 the total was 15,000 acre-ft; on April 1 it was 40,000 acre-ft; likewise the total flow from Oct. 1 to other dates may be read from the curve. The maximum vertical distance between the curves is the storage required to meet the needs of the project. If the worst period of record is selected for the study, the storage requirements are obtained from the mass curve. In the illustration, it was assumed that the reservoir was full on Oct. 1, the beginning of the period. The greatest amount of storage was used on Jan. 31, 74,000 acre-ft. The reservoir was full again on May 4, when the two curves intersected.

11. Tabular Reservoir-Operation Analysis. Where the relation between supply and demand is complicated, reservoir planning usually is accomplished by a tabular accounting of inflow, reservoir evaporation, release requirements, and reservoir storage. In a complex system where several reservoirs, water uses, and points of water diversion are involved, this becomes a complicated process.

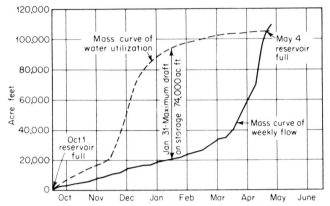

FIGURE 6 Mass curves of water utilization and weekly flow.

Table 2 shows a reservoir-operation study of the critical period for a large reservoir. This is a multiple-purpose reservoir, planned for irrigation, power, and such incidental flood control as may be possible. The study was on a monthly basis and represented the final of several trial-and-error studies to develop the required reservoir size to meet specified water-delivery requirements.

The studies made for the mammoth Indus River Basin investigations in Pakistan represent a vast expansion of the type of study described above. The studies involved four major tributaries, two major storage reservoirs, pumped supply of groundwater into numerous locations of an irrigation distribution system, five points of diversion, and many interconnected canals. The conjunctive operation of all these facilities required a method of operation that could only be carried out by computer, because of the sheer magnitude of work involved in manual studies and the impossibility of eliminating errors in the manual process.

RESERVOIR WAVE ACTION

12. Freeboard Allowances. The term *freeboard* is frequently used in different ways. As defined previously, freeboard must include consideration of the following:

1. Height of wind tide (referred to also as *setup*)
2. Height of waves in deep water generated by winds
3. Effect of wave run-up on sloping embankments on height of waves
4. Any additional margin of safety considered necessary

Final design decisions on freeboard allowances usually involve considerations of the type of dam, the situation governing the spillway design flood, and the effect of waves. This section is concerned only with wave action. An excellent article by Sabille, McClendon, and Cochran[13] and a manual by the U.S. Corps of Engineers[14] form the basis of procedures given in this section for computing waves.

TABLE 2 Reservoir-Operating Study for Critical Period

All values in 1000 acre-ft

Year	Month	Inflow to reservoir	Total irrigation water demands at diversion dam	Usable inflow between reservoir and diversion dam	Required reservoir release to meet irrigation demand	Evaporation	Required reservoir release for power demands	Change in reservoir storage	Reservoir storage at end of month	Reservoir spill
1950	May	106.7	41.2	11.5	29.7	0.3		+30.1	190.0	46.6
	June	76.9	37.7	1.5	36.2	1.2		0	190.0	39.5
	July	33.7	34.3	2.9	31.4	1.2		0	190.0	1.1
	August	19.0	32.9	1.0	31.9	1.2		−14.1	175.9	
	September	13.6	31.1	1.6	29.5	0.9		−16.8	159.1	
	October	13.1	24.6	1.0	23.6	0.6		−11.1	148.0	
	November	12.9	22.3	0	22.3	0		−9.4	138.6	
	December	11.9	22.0	0	22.0	0		−10.1	128.5	
1951	January	12.0	22.5	0	22.5	0		−10.5	118.0	
	February	13.2	21.1	0	21.1	0		−7.9	110.1	
	March	21.5	22.3	6.4	16.9	0	19.2	+2.3	112.4	
	April	40.5	36.0	3.7	32.3	0.3		+7.9	120.3	
	May	52.4	41.2	5.1	36.1	0.3		+16.0	136.3	
	June	35.9	37.7	0.8	36.9	1.2		−2.2	134.1	
	July	19.1	34.3	1.0	33.3	1.2		−15.4	118.7	
	August	15.6	32.9	0.6	32.3	1.2		−17.9	100.8	
	September	12.4	31.1	0	31.1	0.9		−19.6	81.2	
	October	16.3	24.6	3.8	20.8	0.6	21.5	−5.8	75.4	
	November	22.6	22.3	6.8	15.5	0	20.9	+1.7	77.1	
	December	15.3	22.0	3.4	18.6	0	21.8	−6.5	70.6	
1952	January	14.3	22.5	2.5	20.0	0	21.6	−7.3	63.3	
	February	20.2	21.1	5.9	15.2	0	21.0	−0.8	62.5	
	March	34.5	23.3	10.4	12.9	0	22.1	+12.4	74.9	
	April	129.5	36.0	14.3	21.7	0.3		+107.5	182.4	
	May	191.1	41.2	21.8	19.4	0.3		+7.6	190.0	163.8
	June	113.0	37.7	2.1	35.6	1.2		0	190.0	76.2

13. Basic Assumptions. A number of formulas have been developed for computation of heights of wind tide, waves, and run-up. Most of these involve use of wind velocity and fetch as basic parameters. While the different formulas will yield different results, the variation between formulas is frequently not so great as the variation possible in results that are due to assumptions of wind velocity and fetch. Thus, the development of reasonable assumptions is of prime importance.

Wind. The magnitude of wind tide, wave height, and run-up will vary with the magnitude and wind velocity and the duration of that velocity. Thus, it is desirable to develop wind data on a duration basis, if possible. The proper combination of velocity and duration is not always subject to precise determination, although procedures are available in the computation of wave heights for determining the minimum required time to reach maximum wave heights. Frequently, maximum wave conditions will not result unless the duration is of the order of 1 h. Therefore, in wind-tide calculations the maximum observed 60-min average wind is frequently taken as the first trial for design. This assumption then can be checked against the value of wind tides derived as described in Art 14.

Care should be taken to utilize only those wind conditions possible at the same time as or immediately following the meteorological conditions causing the pool level under consideration. For example, if the spillway design storm is not of the hurricane type, the winds used to compute freeboard allowances should not be of the hurricane type.

Fetch. Fetch length is the horizontal distance of open water surface over which the wind blows. The use of the greatest straight-line distance over open water in wave computations will result in computed wave heights that are too high, since the amount of adjoining open water having shorter but significant fetches influences the waves. Observations on artificial reservoirs have indicated that use of an *effective fetch* is more reliable. The effective fetch is computed by dividing the 45° angle on either side of the maximum fetch line into about 15 equal segments, multiplying the fetch length for each segment by the cosine of the angle of deviation from the maximum fetch line, and dividing the sum of the products by the sum of the cosines.

Wind velocities over water are generally higher than over land under comparable meteorological conditions, because of lesser roughness. The following values represent averages observed on artificial reservoirs:

Fetch, miles	0.5	1.0	2.0	3.0	4.0	5.0 (or over)
Wind ratio, over water/over land	1.08	1.31	1.21	1.26	1.28	1.30

The maximum potential wind velocity may not always coincide in direction with the direction of maximum fetch. If observations of maximum winds of given directions are available, the use of the effective fetch length can be carried one step further, utilizing the appropriate design wind velocities with the fetches indicated.

14. Wind Tide. Wind tide, or *setup,* is the piling up of water at the leeward end of an enclosed body of water, as a result of the horizontal stress on water exerted by the wind. The magnitude of wind tide can be expressed by the following modification of the Zuider Zee formula:

$$S = \frac{U^2 F}{1400 D} \qquad (3)$$

in which S is the wind tide (in feet) above still water, U is average wind velocity (in statute miles per hour) over the fetch distance F (in miles), and D is the average depth of water along the fetch line (in feet).

15. Wave Height and Other Characteristics. Wind-generated waves in a large body of water are not uniform in height. Successive waves will not be identical—each wave will be preceded and succeeded by a higher or lower wave. Data obtained from recordings of 45 storm periods at Fort Peck and Denison reservoirs have shown a very close comparison between the observed frequency distribution of wave heights on inland reservoirs with observed data on oceans. The following characteristics have been observed for the spectrum of waves observed at a given time and place:

% of total number of waves averaged to compute specific wave height H	Ratio of H to average wave height H_{av}	Ratio of H to significant wave height H_s	% of waves exceeding H
1	2.66	1.67	0.4
5	2.24	1.40	2
10	2.03	1.27	4
33½	1.60	1.00	13
50	1.42	0.89	20
100	1.00	0.62	46

The significant wave height H_s is defined as the average height of the highest one-third of all waves in a spectrum. As will be seen from the above tabulation, 13 percent of all waves can be expected to exceed H_s. These values would be reached at the end of a buildup period and give measures of the variations that can be expected in wave-height distributions. H_s may be computed by the set of curves in Fig. 7. Knowing the effective fetch and the wind velocity, the curve can be entered with these values to give the minimum time duration and value of H_s.

Once the value of H_s is computed, the occurrence frequency of a wave of any height can be computed from the preceding tabulation. The design height for

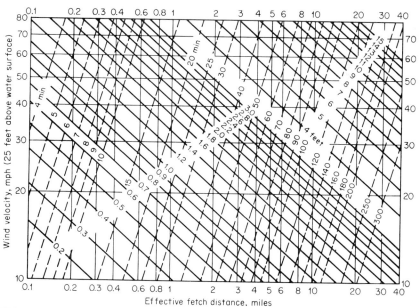

FIGURE 7 Generalized correlations of significant wave height H_s with related factors (deepwater conditions). Solid lines represent significant wave heights, in feet; dashed lines represent minimum wind duration, in minutes, required for generation of wave heights indicated for corresponding wind velocities and fetch distance. (*Beach Erosion Board, U.S. Corps of Engineers, Tech. Mem. 132.*)

waves H can be selected on the basis of consideration of frequency of winds of a given magnitude, duration of winds, and frequency of waves of a given size. The finally selected design height must be a judgment value, involving consideration of the type of dam involved, as well.

16. Wave Run-up on Slopes. A wind-generated wave will be influenced when it runs up the slope of an embankment. The effect may be either to increase or to decrease the height of the wave in relation to the still-water surface, depending on wave characteristics and the slope, roughness, and permeability of the embankment. Therefore, the effect of run-up is usually combined with the actual wave height in computing allowances for a wave action, into a single item designated as wave run-up height R.

In this sense R is the vertical distance between the maximum elevation obtained by a wave running up an embankment and the water elevation at the toe of the slope. The water elevation at the toe of the slope is the still-water elevation plus wind tide. Because of the relationship between wave height and run-up, it usually is convenient to compute run-up as a function of wave height.

The wave characteristics are represented by the wave steepness ratio H_0/L_0, where H_0 = specific weave height and L_0 = wave length, measured from crest to crest, in deep water. H_0 may, for practical purposes in deep reservoirs, be taken as equal to H. L_0 may be computed from the following formula:

$$L_0 = 5.12T^2 \tag{4}$$

where T is the wave period, which may be determined from Fig. 8. This wave period is approximately the same for waves ranging between the signficant wave H and the maximum wave H_{max}. Thus, in deep water the L_0 determined for the

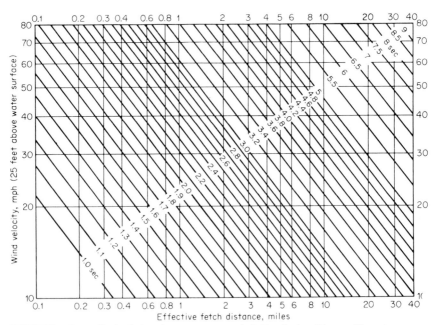

FIGURE 8 Generalized relations between wave periods T and related factors (deep-water conditions). (*Beach Erosion Board, U.S. Corps of Engineers, Tech. Mem. 132.*)

wave steepness ratio can be used for any value of H between H_{max} and H_s. Deepwater conditions can be considered to be present when the depth at the toe of the slope is more than one-third of the calculated wave length.

Using the values of H_0 and L_0, the effect of run-up on wave height may be computed from Fig. 9. Curves are shown for smooth slopes and for rubble mounds. Smooth slopes include surfaces such as well-graded earth embankments covered with sod and asphalt or concrete facing. Run-up on hand-placed riprap slopes approaches that computed for smooth slopes. Run-up on dumped riprap slopes can be considered to be about 50 percent of computed run-up on smooth slopes.

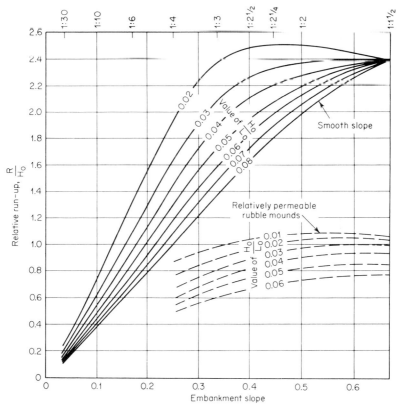

FIGURE 9 Wave run-up ratios versus wave steepness and embankment slopes. (*Trans. ASCE*, **128**, 217, 1963.)

17. Total Allowance for Wave Action. The total allowance for wave action is the sum of the following:

1. Wind tide, computed as in Art. 14.
2. The combination of wave height and wave run-up, utilizing wave height as computed in Art. 15 and the ratios shown in Fig. 9.

RESERVOIR FLOOD ROUTING

General. Routing of floods through reservoirs is far simpler than through river channels, because of the unique relationship existing between reservoir stage and discharge. This relationship is established from the discharge characteristics of the reservoir outlets and the spillway, as they relate to reservoir elevation.

A number of methods have been developed for flood routing. The arithmetic method presented here has the advantage of easy checking and filing. Also, because it utilizes curves developed from small differences as described below, it possesses greater accuracy in reading of curves than most arithmetic methods.

18. Arithmetic Method. For any time interval, the storage equation can be expressed as

$$I = (S_2 - S_1) + \left(\frac{Q_1}{2} + \frac{Q_2}{2}\right) \tag{5}$$

where I = average inflow during period
S_1, S_2 = storage at beginning and end of period
Q_1, Q_2 = outflow at beginning and end of period

This may be rewritten as

$$\left(S_1 - \frac{Q_1}{2}\right) + I = S_2 + \frac{Q_2}{2} \tag{6}$$

If we let $S - Q/2 = (S + Q/2) - Q$, then substituting in the first step gives

$$\left(S_1 + \frac{Q_1}{2}\right) - Q + I_1 = S_2 + \frac{Q_2}{2} \tag{7}$$

Also, in the second step,

$$\left(S_2 + \frac{Q_2}{2}\right) - Q + I_2 = S_3 + \frac{Q_3}{2} \tag{8}$$

If S and Q for the first step are known, and I for all steps also is known, the equations can be solved by a curve relating $(S + Q/2)$ to Q (for a given time interval). A typical curve for this purpose is shown in Fig. 10. This curve is constructed from data on reservoir volume and spillway discharge, as related to reservoir elevation. Since Q is small with respect to $(S + Q/2)$, the routing can be carried out with a much smaller sheet of graph paper than would be required to obtain similar accuracy with the usual graphs relating $(S + Q/2)$ to $(S - Q/2)$.

By using Fig. 10, the routing can be carried out in the following manner:

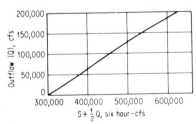

FIGURE 10 Working curve for reservoir flood routing. Routing interval, 6 h.

6-h period	Item in routing	Volume, 6-h, cfs	Origin of value
1	$S+Q/2$	300,000	Known
	$-Q$	0	Known
	$+I$	40,000	Given
2	$S+Q/2$	340,000	Algebraic sum
	$-Q$	23,000	From Fig. 10
	$+I$	80,000	Given
3	$S+Q/2$	397,000	Algebraic sum
	$-Q$	60,000	From Fig. 10
	$+I$	150,000	Given
4	$S+Q/2$	487,000	Algebraic sum
	$-Q$	118,000	From Fig. 10
	$+I$	90,000	Given
	$S+Q/2$	459,000	Algebraic sum

RESERVOIR SELECTIVE WITHDRAWAL

Selective withdrawal of water from a reservoir takes advantage of density stratification to withdraw water by means of a multilevel intake from a depth where temperature and/or quality of water are preferred. Typical temperature profiles of a stratified reservoir are shown in Fig. 11.

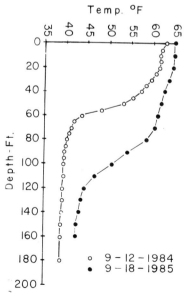

FIGURE 11 Typical temperature profiles of a stratified reservoir.

19. Need for Selective Withdrawal. Increasing demands of impounded water for hydroelectric, municipal, industrial, and fishery purposes and the expanding development of the nation's rivers and streams have generated awareness and legislation to develop and operate projects that maintain high water quality. If the water quality concerns are not addressed during the planning and design of a project, many undesirable problems may develop during subsequent operation. Glacier-fed reservoirs may require intakes to minimize sand and silt entrainment that causes premature wear on pumps and turbines. Water supply reservoirs may require intakes that minimize entrainment of decaying algae which create undesirable taste and odors. For an impoundment project where sport and/or commercial fisheries are important, intakes capable of releasing water with a temperature range suitable for normal maturation and reproduction of fish may be required. Selective withdrawal of water from a reservoir though multilevel intake structures is a practical and efficient way to improve both reservoir and river water quality. Many reservoirs in the United States are currently operated to control the quality of releases using multilevel intakes for selective withdrawal.

20. Reservoir Thermal Hydraulic and Water Quality Modeling. The requirements for number, location, capacity, and operating arrangement of multilevel intake openings for selective withdrawal vary widely among different reservoirs. They depend mainly on the hydrothermal condition and water quality objectives of the individual reservoir. The thermal hydraulics and the temperature structure of a reservoir are affected by the meteorological and hydrologic conditions and operation of the reservoir. These phenomena are difficult to evaluate analytically. Hydraulic models are expensive and their applications are limited. However, computer model simulation provides a more economical means of investigating reservoir dynamics. Using the dynamic reservoir simulation technique DYRESM, the Harza Engineering Co. has developed a reservoir simulation model[15] called HARZA/DYRESM to simulate the thermal hydrodynamics and water quality of a reservoir and the operation of a multilevel intake for selective withdrawal. A schematic of the DYRESM model is shown in Fig. 12.

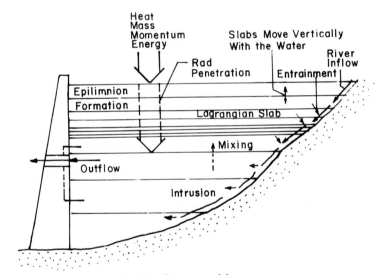

FIGURE 12 A schematic of the Dyresm model.

The DYRESM approach is internationally recognized as one of the most advanced modeling techniques for simulating reservoir thermal hydrodynamics and has been applied to many lakes and reservoirs in the world. The modeling technique is based on the hydrodynamic mixed-layer approach, parameterizations of the principal physical processes, and a variable grid system to obtain a solution economically and accurately. The model requires less calibration and is more generally applicable than other empirically based models. The HARZA/DYRESM model is capable of simulating not only the temperature structure of a reservoir, including ice cover in cold regions, but also the salinity and suspended sediment concentration. The model has been extensively tested with data collected from Eklutna Lake near Anchorage, Alaska. Excellent test results were obtained. The model can be easily coupled with other existing water quality models such as CEQUAL, which is an updated version of the U.S. Corps of Engineers water quality model WQRRS, to perform simulations of other water quality variables. The HARZA/DYRESM model has been applied to locate selective withdrawal intake ports required to meet downstream water quality objectives and to determine the effects of intake structural modifications and project operation changes on the water quality of a reservoir and its releases.

REFERENCES

1. "Inventory of Published and Unpublished Sediment Load Data in the U.S.," April 1949, and supplements of later dates, prepared for Inter-Agency Committee on Water Resources by U.S. Department of the Interior.
2. "Summary of Reservoir Sediment Deposition Surveys Made in the U.S. through 1975," U.S. Dept. of Agr. Misc. Publ. 1362, February, 1978.
3. MILLER, CARL R., "Analysis of Flow Duration, Sediment-rating Curve Method of Computing Sediment Load," U.S. Bureau of Reclamation, April 1951.
4. CHOW, VEN TE, "Handbook of Applied Hydrology," pp. 17-6 to 17-9, McGraw-Hill, New York, 1964. "A Universal Equation for Predicting Rainfall-erosion Losses," U.S. Agr. Res. Ser. Spec. Rept., March 1961, pp. 22–26.
5. SCHROEDER, K. B., and D. B. RAITT, "Total Suspended Load from Vertical Transport Distribution," U.S. Bureau of Reclamation, November 1952. KOELZER, V. A., and M. BITOUN, "A Review of Sediment Problems and Possible Solutions," West Pakistan Engineering Congress, Lahore, April 1962.
6. BRUNE, GUNNAR M., Trap Efficiency of Reservoirs, *Trans. Am. Geophys. Union,* June 1953, U.S. Dept. Agr. Misc. Publ. 970, p. 884.
7. "Proceedings of the Federal Inter-Agency Sedimentation Conference," May 1947, U.S. Bureau of Reclamation, Denver, Colorado.
8. LANE, E. W., and V. A. KOELZER, "Density of Sediments Deposited in Reservoirs," U.S. Interdepartmental Comm. Rept. 9, St. Paul District, Corps of Engineers, St. Paul, Minn., November 1943. HEMBREE, C. H., B. R. COLBY, H. A. SWEBSIB, and J. R. DAVIS, "Sedimentation and Chemical Quality of Water in the Powder River Drainage Basin, Wyoming and Montana," U.S. Geol. Survey Circ. 170, Washington, D.C., 1952.
9. KOELZER, V. A., and JOE M. LARA, Density and Compaction Rates of Deposited Sediments, *Proc. ASCE, J. Hydraulics Div.,* Paper 1603, April, 1958.
10. MILLER, CARL R., "Determination of the Unit Weight of Sediment for Use in Sediment Volume Computation," U.S. Bur. Reclamation Mem., Feb. 17, 1963.
11. TRASK, PARKER, Compaction of Sediments, *Bull. Am. Assoc. Petrol. Geologists,* **15,** 271–276.
12. "Interim Report—Distribution of Sediment in Reservoirs," Project Investigations Division, U.S. Bureau of Reclamation, June 1954.
13. SAVILLE, THORNDIKE J., ELMO W. MCCLENDON, and ALBERT L. COCHRAN, Freeboard Allowances for Waves in Inland Reservoirs, *Trans. ASCE,* 1963, pt. IV, pp. 195, 226.
14. "Waves in Inland Reservoirs," Tech. Mem. 132, Beach Erosion Board, U.S. Corps of Engineers, November 1962.
15. WEI, C. Y., and P. F. HAMBLIN, "Reservoir Water Quality Simulation in Cold Regions," Cold Regions Hydrology Symposium, American Water Resources Association, July 1986.

SECTION 5

NATURAL CHANNELS

By B. A. McKiernan[1]

INTRODUCTION

1. Flow Characteristics. Of the several types of open-channel flow, only those commonly found in natural channels are discussed here.

Flow in natural channels is varied because the depth changes along the channel. It is gradually varied if the change in depth is gradual and rapidly varied if the change is abrupt. Flow in natural channels is almost always turbulent flow with water particles moving in an irregular pattern. Viscous forces are small compared with inertial forces. The Reynolds number **R** is less significant than the Froude number **F** (Sec. 2), which expresses the relationship between inertial forces and gravity forces. In the case of natural streams, the length characteristic L is the hydraulic mean depth.[1]

The commonest flow in natural channels is subcritical flow, in which $V < \sqrt{gD}$ and **F** is less than unity. The flow is termed *tranquil* and the velocity is less than the velocity of gravity waves so that downstream conditions affect the depth of flow. In critical flow, $V = \sqrt{gD}$ and the water-surface profile is usually marked by standing waves. In supercritical flow, $V > \sqrt{gD}$ and the flow is described as *rapid* or *shooting*. Downstream conditions cannot affect the depth of flow. This section is concerned with subcritical flow.

HYDRAULICS OF NATURAL STREAMS

2. Discharge Formulas. Bernoulli's theorem applied to open channels states that the sum of the water-surface elevation and velocity head at any section is the same as at any upstream section minus the intervening loss in head. In Fig. 1,

$$Z_2 + h_{v2} = Z_1 + h_{v1} + h_f + \text{other losses} \qquad (1)$$

where Z = elevation of the water surface above some datum
h_v = velocity head at the indicated cross section
h_f = head loss in reach due to friction
l = length of reach

---

[1]Acknowledgment is made to Russell W. Revell for material in this section that appeared in the third edition (1969).

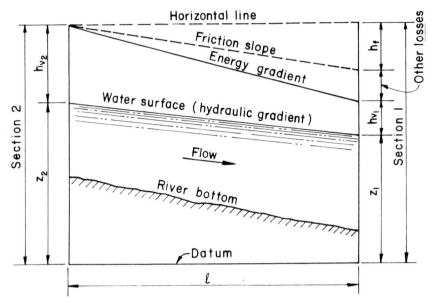

FIGURE 1 Energy profile in open-channel flow.

Although the velocity head is usually assumed to be $V^2/2g$, where V is the mean velocity and g is the acceleration of gravity, this is not strictly true where velocity distribution is nonuniform. The true velocity head, in energy computations, is $\alpha V^2/2g$ where α is the energy coefficient. Kolupaila[2] proposed values of α for natural channels ranging from 1.15 to 1.50 with an average of 1.30. This refinement is frequently omitted where velocities are low.

Water flowing in a channel continually loses energy to its surroundings. This loss is expressed as the slope of the energy gradient. In Fig. 1 the energy gradient is $(h_f + \text{other losses})/l$. Discharge formulas define the relationship between the friction slope h_f/l, the discharge, channel roughness, and the geometry of the channel cross section.

The other losses shown in Fig. 1 may result from stream curvature, form factors, contractions at bridges, and other channel obstructions. If these losses are fairly uniformly distributed along the channel, they are included in the friction head h_f. If not, they must be computed separately, as explained in Art. 5.

The Manning formula (Sec. 2) is by far the most commonly used in the United States for determining flow in natural channels. The Chezy formula, the Ganguillet and Kutter formula, and the Darcy-Weisbach formula (all in Sec. 2) are also in use. The values represented by the terms of these discharge formulas all vary from place to place in natural channels and also vary with discharge. Therefore, application of discharge formulas to natural channels is usually less accurate than for artificial channels. For this reason, their use for natural channels usually is restricted to relatively short reaches where reliable average values of the terms can be determined.

3. Roughness Factor. In the discharge formulas referred to in Art. 2, the resistance to flow is evaluated by roughness factors: n in Manning's and Kutter's formulas, C in the Chezy formula, and f in the Darcy-Weisbach formula.

Energy losses in turbulent flow are due primarily to skin friction, internal distortion, and impact. Roughness factors cover all these losses. Discharge formulas are empirically derived. The roughness factors are adjusted to make the formulas fit observed data. The formulas then will give reliable results for flow conditions having a similar roughness factor. For most flow conditions, impact will be small, and it is not too difficult to select a roughness coefficient that will account for the skin friction and the internal distortion. In natural streams having steep slopes and high velocities, large impact losses result from channel irregularities or obstructions that would cause only negligible impact losses for lower slopes and velocities. In such cases, the n values must be substantially increased unless the additional impact losses are otherwise accounted for.

The effective roughness of a channel depends as much on the concentration of the roughness elements as it does on the size of the elements. The greatest effective roughness is found when the roughness elements cover from 15 to 25 percent of the channel.[3] Greater concentrations reduce rather than increase the distortion of the flow caused by the roughness elements.

A comprehensive discussion of roughness factors is given in a report of the American Society of Civil Engineers.[4]

4. Values of n for Natural Streams. Manning's formula is convenient to use with natural streams. Its n value can be computed if all the other terms are known. An n determined for one stage might not be usable at a significantly different stage because of differences in impact losses and changes in configuration, and because n varies slightly with hydraulic radius. When the bed is moving, sand dunes may form on its bottom which significantly increase the n value over what it would be at lower velocities when the bed is not moving. There are no adequate criteria as to when bed movement starts.

Where n cannot be computed directly, it must be estimated. This can be done by visually comparing the roughness of the channel with that for other channels of known roughness, or from descriptive tables of n values. Errors in estimating n values result in equal and opposite percentage errors in computing velocity and discharge. Therefore, care in selecting n values should be consistent with desired accuracy. Values of n for natural channels are given in Tables 1 and 2.

Table 1, which lists n values for natural streams for a broad range of conditions, is excerpted from Table 5-6 of "Open-Channel Hydraulics" by Chow.[1] The table is useful as a guide in the quick selection of an appropriate n value.

Table 2, excerpted form Table 5-5 of "Open-Channel Hydraulics"[1] lists values enabling the estimation of the value of n by a procedure developed by Cowan.[5] In this procedure, n is estimated by

$$n = (n_0 + n_1 + n_2 + n_3 + n_4)m_5 \qquad (2)$$

where n_0 = basic value for straight, uniform, smooth conditions
n_1 = value to n_0 to adjust for degree of irregularity
n_2 = value to adjust for variation in cross section
n_3 = value to adjust for the effect of obstructions
n_4 = value to adjust for degree of vegetation
m_5 = correction factor for degree of meandering

As an example, an estimated n value for a channel in rock with minor irregularity, occasionally alternating cross section, negligible obstructions (no debris deposits, tree stumps, exposed roots, boulders or lodged logs), no vegetation,

TABLE 1 Values of the Roughness Coefficient n for Natural Streams

Type of channel and description	Min.	Normal	Max.
Minor streams (top width at flood stage < 100 ft)			
a. Streams on plain			
1. Clean, straight, full stage, no rifts or deep pools	0.025	0.030	0.033
2. Same as above, but more stones and weeds	0.030	0.035	0.040
3. Clean, winding, some pools and shoals	0.033	0.040	0.045
4. Same as above, but some weeds and stones	0.035	0.045	0.050
5. Same as above, lower stages, more ineffective slopes and sections	0.040	0.048	0.055
6. Same as 4, but more stones	0.045	0.050	0.060
7. Sluggish reaches, weedy, deep pools	0.050	0.070	0.080
8. Very weedy reaches, deep pools, or floodways with heavy stand of timber and underbrush	0.075	0.100	0.150
b. Mountain streams, no vegetation in channel, banks usually steep, trees and brush along banks submerged at high stages			
1. Bottom: gravels, cobbles, and few boulders	0.030	0.040	0.050
2. Bottom: cobbles with large boulders	0.040	0.050	0.070
Flood Plains			
a. Pasture, no brush			
1. Short grass	0.025	0.030	0.035
2. High grass	0.030	0.035	0.050
b. Cultivated areas			
1. No crop	0.020	0.030	0.040
2. Mature row crops	0.025	0.035	0.045
3. Mature field crops	0.030	0.040	0.050
c. Brush			
1. Scattered brush, heavy weeds	0.035	0.050	0.070
2. Light brush and trees, in winter	0.035	0.050	0.060
3. Light brush and trees, in summer	0.040	0.060	0.080
4. Medium to dense brush, in winter	0.045	0.070	0.110
5. Medium to dense brush, in summer	0.070	0.100	0.160
d. Trees			
1. Dense willows, summer, straight	0.110	0.150	0.200
2. Cleared land with tree stumps, no sprouts	0.030	0.040	0.050
3. Same as above, but with heavy growth of sprouts	0.050	0.060	0.080
4. Heavy stand of timber, a few down trees, little undergrowth, flood stage below branches	0.080	0.100	0.120
5. Same as above, but with flood stage reaching branches	0.100	0.120	0.160
Major streams (top width at flood stage > 100 ft). The n value is less than that for minor streams of similar description, because banks offer less effective resistance.			
a. Regular section with no boulders or brush	0.025	...	0.060
b. Irregular and rough section	0.035	...	0.100

and a minor degree of meander would be = (0.025 + 0.005 + 0.005 + 0 + 0) 1.0 = 0.035.

5. Curvature, Form Factors, and Other Losses. The n in Manning's formula normally accounts for the friction from the roughness of the channel boundary and for normal internal distortion. Other losses in natural channels result from

TABLE 2 Values for the Estimation of n by Eq. (2)

Channel conditions			Values
Material involved	Earth Rock cut Fine gravel Coarse gravel	n_0	0.020 0.025 0.024 0.028
Degree of irregularity	Smooth Minor Moderate Severe	n_1	0.000 0.005 0.010 0.020
Variations of channel cross section	Gradual Alternating occasionally Alternating frequently	n_2	0.000 0.005 0.010–0.015
Relative effect of obstructions	Negligible Minor Appreciable Severe	n_3	0.000 0.010–0.015 0.020–0.030 0.040–0.060
Vegetation	Low Medium High Very high	n_4	0.005–0.010 0.010–0.025 0.025–0.050 0.050–0.100
Degree of meandering	Minor (ratio 1.0 to 1.2) Appreciable (ratio 1.2 to 1.5) Severe (ratio 1.5 and greater)	m_5	1.000 1.150 1.300

velocity disturbances caused by curvature of the channel bank irregularities, expansion or contraction of the channel, and obstructions in the channel. These cause abnormal internal distortion and impact losses. Before applying additional corrections for such losses, the selected n value should be carefully reviewed to determine whether it already accounts for these losses.

Losses not covered by the n value usually are expressed as percentage of velocity head or percentage of changes in velocity head. Bends in natural channels induce energy loss in addition to that from an equal length of straight channels. Spiral flow is induced in and below the bend, with the water near the surface being deflected toward the outside of the bend and water near the bottom being forced toward the inside of the bend. Also, the water-surface elevation on the outside of the bend is somewhat higher than on the inside, because of the centrifugal force.[6]

Few quantitative data are available on the head loss in bends in natural channels. Experimental observations on losses in bends are largely restricted to artificial channels where variables can be controlled. Adapting head losses for artificial channels to natural channels requires considerable judgment.

Yen and Howe[7] list three sources of additional head loss due to bends: internal friction from secondary currents in and below the bend, reduction of effective cross-sectional area due to eddies accompanying separation of flow from the bank, and loss of head due to repeated velocity changes. These combined effects can exceed those of boundary friction. They found the energy loss due to a 90° bend with uniform width of 11 in and 5-ft radius of curvature to be

$$H_b = 0.380 \frac{V^2}{2g} \tag{3}$$

where H_b = head lost in the bend, ft
V = mean velocity in the approach section, fps
g = acceleration of gravity

Experiments by Scobey[8] on the flow of water around bends in flumes indicated that Manning's n value should be increased by 0.001 for each 20° of curvature per 100 ft.

Yarnell and Woodward[9] found the loss in 180° bends in flumes to be

$$H_b = 0.21 \frac{\text{channel width}}{\text{inner radius}} \frac{V^2}{2g} \tag{4}$$

A rough empirical guide for energy loss due to expanding or contracting sections follows. In a contracting section the energy loss due to contraction is considered to vary from 10 percent of the difference in velocity head where the contraction is gradual and smooth to 50 percent where the contraction is abrupt. For expanding sections, the energy loss is considered to vary from 20 percent of the difference in velocity head where the expansion is gradual and smooth to 50 percent where the expansion is abrupt.

Other sources of energy losses are channel obstructions, such as bridge piers, which have been covered in technical literature.[10] Such losses are discussed in Art. 11.

DETERMINATION OF DISCHARGE

6. Methods. Discharge in natural channels usually is determined from current-meter measurements, from evaluating the terms in formulas for turbulent flow, as in the slope-area method, by special indirect methods utilizing known hydraulic properties, or, for small streams, from weir measurements.

7. Weir Measurements. Weir formulas are derived from experiments, so the geometry and approach conditions of the weir must be similar to those of the experimental weir from which the formula is derived. In addition, the underside of the nappe must be well-ventilated so that the pressure will be atmospheric.

The use of weirs for determining discharge of natural channels is limited by the difficulty of obtaining free overflow conditions for the ranges of discharge usually experienced and, for larger streams, by installation cost. In addition, variable approach conditions often cause the discharge to vary appreciably from that indicated by weir formulas. Therefore, where accuracy is required, weirs in natural channels should be rated frequently by current-meter measurements instead of depending on a weir formula.

An outgrowth of the weir for natural channels has been the artificial control whose purpose is to stabilize the stage-discharge relationship rather than to fit a standard weir formula. The artificial control creates supercritical velocity up to the stage at which it becomes drowned out, thereby eliminating any effects of changing channel conditions downstream for such stages. The control must be rated by discharge measurements or model studies. It is subject to the same inaccuracies due to changing upstream pool conditions as are weirs.

8. Current-Meter Measurements. Current-meter measurements offer a relatively accurate method of determining flow in natural channels. Current meters have a rotating element actuated by the current, a device for indicating rotations of this element, and usually fins for orienting the meter with the current. The rate of rotation of the rotor indicates the velocity of the water. There also must be a device for lowering the current meter to the desired depth below the water surface. The commonest current meter in the United States is the Price-type meter which gives reliable results for velocities ranging from about 0.1 to more than 15 fps. Appropriate procedures in obtaining representative average velocities are described in U.S. Geological Water Supply Paper 888.[11]

9. Area-Velocity Method. Computation of discharge from a current-meter measurement is most conveniently performed by the area-velocity method, as shown in Fig. 2. The cross-sectional area is divided into many small areas of measured size, and the mean velocity is determined for each area. The sum of the incremental products of area and velocity is the total discharge.

An incremental area is found by multiplying the depth at each vertical by the sum of half the distances to each adjacent vertical. The average velocity in the vertical is taken to represent the average for the entire incremental area.

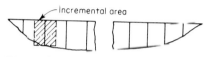

FIGURE 2 Subdividing the cross section.

10. Slope-Area Method. Practical considerations sometimes prevent making discharge measurements during floods. The slope-area method is useful for determining such flood discharges from field data. It is essentially the reverse of computing n from field data as explained in Art. 4. It makes use of an open-channel formula, usually Manning's formula. The following conditions are desirable for a slope-area reach:

1. Sufficient high-water marks to determine the flood-crest profile along each bank
2. A single, fairly uniform channel with no significant gain or loss in the reach
3. No significant erosion or deposition in the reach during or following the flood

Cross sections, at least two and preferably three, are surveyed. They are so spaced that the fall between them can be measured accurately, but close enough together so that uniformity of conditions in the reach is not sacrificed unduly. Elevations in the cross sections should be to the nearest tenth of a foot, and in the profiles to the nearest hundredth of a foot. Where there are pronounced breaks in the slope or roughness of a cross section, such as between main channel and overbank areas, the conveyances are computed separately for each subsection.

The discharge will not be reliable without an accurate determination of n; so considerable care should be given to its selection. A contracting section is preferable to an expanding section because the effects of errors in determining n are reduced.

Since the submerged cross sections almost certainly will have different areas, velocity-head corrections must be made to determine the energy gradient. If there are large variations of velocity in a cross section, the velocity head must be determined from weighted velocity head for different parts of the cross section.

In making the computations, approximations of discharge will have to be tried until the proper discharge is found that will give compatible values of energy gradient and surface gradient.

11. Indirect Determinations. A contracted section in a channel causes an abrupt drop in water surface because of conversion of static head to velocity head. A contracted bridge opening is a typical example. This drop can be used to determine discharge, using Bernoulli's theorem of continuity of head. The equation for discharge through the contracted opening is

$$Q = kA \sqrt{2g\left(H + \frac{V^2}{2g} - h_f\right)} \qquad (5)$$

where Q = discharge
k = a coefficient for sharp-edged or square entrances, usually ranging from 0.90 to 0.95
A = effective area of the most contracted section (eliminating parts of the cross section not contributing to the flow)
g = acceleration due to gravity
H = drop in water surface caused by the contraction
$V^2/2g$ = velocity head in the approach section
h_f = loss in head caused by friction

A sketch of the site will aid materially in making the computations. Longitudinal profiles upstream and downstream from the contraction, when extended to the point of contraction, will give the value of H. If the determination is made later from flood marks, care must be exercised to eliminate the effects of standing waves in the downstream profile. Velocity head in the approach section can be determined from the cross-sectional area by trial and error. The friction loss h_f can be determined from Manning's formula using for the average velocity the square root of the average of the squares of the velocities in the approach section and in the contracted section. The value of h_f has a minor effect in short, abrupt contractions.

STREAMFLOW RECORDS

12. General Requirements. The accumulation of streamflow records requires three steps: (1) the collection of stage or gage-height records, (2) the determination of the relationship between stage and discharge, and (3) the computation of discharge from the stage using this relationship.

13. Gage-Height Records and Recorders. The record of river stage is derived from periodic readings of a nonrecording gage, usually a graduated vertical staff gage, or obtained continuously from a water-stage recorder.

Nonrecording gages are adequate for rivers where the stage changes gradually, provided a reliable gage reader is available. On the other hand, flashy streams would require very frequent gage readings to provide a good record of stage. Therefore, most river gages are provided with an automatic water-stage recorder.

All gages should be checked periodically by levels to assure that they are at correct datum. Therefore, there should be at least one, and preferably more, bench marks at known elevations and completely separated from the gage structure so that they will remain unaffected by any casualty to the gage.

14. Stage-Discharge Relationships. The second step in determining streamflow is to establish the relationship between river stage and discharge. This requires the simultaneous observation of river stage and measurements of river discharge. Except for the unusual case of a completely stable channel, the relationship between stage and discharge will not long remain constant. Therefore, the determination of the stage-discharge relationship is a continuing process.

Stage is plotted as ordinate against discharge as abscissa on either rectangular or logarithmic coordinates, and a representative curve is drawn through the plotted points. Where a rating curve is used for a considerable period of time, it is convenient also to express the relationship in tabular form. The curve and/or table are referred to as the *station rating*.

The daily discharge is computed directly from daily average gage heights and the rating. In cases of rapidly changing stage, the day should be subdivided into several shorter periods and discharge computed for each period and averaged.

BACKWATER CURVES

15. Problem and Purpose. A common problem in river hydraulics is to determine what the water-surface profile of the river would be under specific river-discharge and channel conditions. The term *backwater curves* is applied to such profiles.

Backwater curves are used to determine grade lines for flood-protection works, highways, and bridges. They are used to determine tailwater ratings for hydroelectric power plants and elevations for pumping plants, canal headworks, and energy dissipaters. They are often used to determine areas subject to flooding.

16. Theory and Principles. It has been shown that there are 12 possible types of backwater curves,[12] but as used herein, the term applies to low-gradient profiles in rivers or other open channels with subcritical velocities. Such conditions are found above dams or other obstructions. River backwater curves usually are derived by one of many step methods, in which the reach of river in question is subdivided into many short subreaches and the change in water-surface elevation is determined for each subreach, either analytically or graphically.

Figure 1 shows the energy profile for open-channel flow. Equation (1), based on that figure, can be rearranged to read

$$Z_2 = Z_1 + h_f + \text{other losses} + h_{v1} - h_{v2} \qquad (6)$$

Determination of a backwater curve for a specific discharge consists of successively computing Z_2 for the upstream end of a reach from known values of Z_1 at the downstream end and the physical characteristics of the reach. The value Z_2 then becomes Z_1 for the next upstream reach. This continues until the desired backwater curve has been computed.

The key to the computations is the determination of the friction loss h_f and other losses, if significant, usually with Manning's formula. Manning's formula, when used for backwater curves may be written in the form

$$h_f = Sl = \frac{V_m^2 n^2 l}{2.21 R_m^{4/3}} \qquad \text{ft} \qquad (7)$$

where S = friction slope
$V_m = Q/A_m$
Q = discharge, cfs

$$A_m = \frac{A_1 + A_2}{2} \quad \text{ft}$$

$$R_m = \frac{A_m}{P_m} \quad \text{ft}$$

$$P_m = \frac{P_1 + P_2}{2} \quad \text{ft}$$

and A_1, A_2 are flow areas at sections 1 and 2 and P_1, P_2 are wetted perimeters at sections 1 and 2. The difference between downstream and upstream water levels may be computed as

$$\Sigma \Delta = \Delta_1 + \Delta_2 + \Delta_3$$

where friction loss $\Delta_1 = h_f$ as computed in Eq. (7).

$$\Delta_2 = K \frac{V_1^2 - V_2^2}{2g} \tag{8}$$

where $V_1 = Q/A_1$, $V_2 = Q/A_2$, and $K = 1$ when $V_1 > V_2$ and $K = \frac{1}{2}$ when $V_2 > V_1$. Δ_3 represents other losses such as bend losses and bridge pier losses, which are estimated separately.

The step-by-step method used in the computations is illustrated in Art. 18.

17. Basic Data. All step methods of deriving backwater curves require a knowledge of the shape and elevations of cross sections at each end of the various subreaches and also of the channel roughness and the water-surface elevation at the starting point.

Backwater curves start at a point where the water-surface elevation is known for the particular discharge and proceed upstream for flow at subcritical velocities. If there are no such points of known water-surface elevation, then a starting elevation must be assumed for a point well downstream from the reach for which the backwater curve is desired. A good procedure for such cases is to assume two different starting elevations, one higher and one lower than the elevation normally expected for such a discharge. Backwater curves are computed from each starting point. If the two curves merge downstream from the reach for which the backwater curve is desired, then the procedure is acceptable. If not, the computations should be started at a point farther downstream.

Field surveys are required unless accurate large-scale topographic maps showing water depths are available. Cross sections should extend up both banks to above the expected backwater elevation and should be approximately at right angles to the current. The field survey determines the values in the Manning formula.

Where several backwater curves are to be computed, it is convenient to plot the wetted perimeter and the cross-sectional area at each section against water-surface elevation as shown in Fig. 3.

The distance between cross sections depends on the uniformity of the channel and on the desired accuracy. Cross sections should be located at transitions between expanding, contracting, or uniform sections. Preferably, they should be spaced so that average velocities will not vary by more than 20 percent between adjacent cross sections. On large rivers, cross sections spaced at an average interval of a mile or more may be adequate, while on small rivers the spacing might be a few hundred feet. Distances between cross sections should be measured along the main part of the current. In some cases, such as overbank flow at a

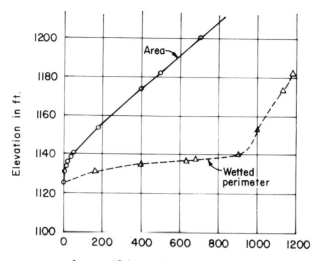

FIGURE 3 Elements of cross section 40 + 62 listed in Table 3.

river bend, the distance between cross sections, and hence the slope, will be different for the main channel and the overbank flow. If pronounced, this will require special procedures.[13,14]

18. Mathematical Solutions. The standard step method is particularly suited to natural channels and is illustrated here.

The actual steps in determining the backwater curve are carried out in Table 3.

FLOW ROUTING

19. Definition and Methods. Flow routing is the determination of the timing and shape of a flow wave in an open channel or reservoir. Since flow routing is frequently used for predicting or reconstructing the progress of a flood, it is often referred to as *flood routing* regardless of the magnitude of the flow involved.

Flow-routing methods include mathematical, graphical,[15] and computer[16] methods. A common mathematical method is described in Art. 23. Most flow-routing methods will work to a degree for either a river or a reservoir. The choice depends on convenience and required precision. This article is concerned with flow routing through open channels, such as natural rivers, in which there is no significant backwater from tributaries.

20. Practical Applications. Routed streamflows are used to provide flood warnings, to determine benefits from existing or proposed flood-control works, to determine the effects of upstream power-plant operation on downstream plants, for scheduling irrigation deliveries, etc. They also can be used to reconstruct natural flood hydrographs on regulated rivers to extend the period of flood-frequency data and to check the consistency of streamflow records at different points on a river.

21. Theory and Principles. Flow routing has two basic aspects, time displacement of the flood wave, and the reduction of the peak and spreading out of

TABLE 3 Determination of Backwater Curve

Section (1)	Reach l, ft (2)	A, ft² (3)	P, ft (4)	V, fps (5)	Friction loss Δ₁, ft (6)	Velocity head loss Δ₂, ft (7)	Other losses, Δ₃, ft (8)	Total losses ΣΔ, ft (9)	W.S. el., ft (10)	Remarks (11)
74 + 25		4,300	220	23.26					1155.0	Flow at critical depth assumed
	155				0.19	7.44		7.63		Q = 100,000 cfs, n = 0.035
72 + 70		12,700	540	7.87					1162.63	
	270				0.12	0.16		0.28		
70 + 00		13,900	582	7.19					1162.91	
	845				0.21	0.41		0.62		
61 + 55		20,000	694	5.00					1163.53	
	1067				0.22	− 0.07		0.15		K = 0.5
50 + 88		17,000	684	5.88					1163.68	
	1026				0.13	0.36		0.49		
40 + 62		29,800	1068	3.36					1164.17	
	1052				0.15	− 0.16		− 0.01		K = 0.5
30 + 10		17,700	954	5.65					1164.16	

Column 1 identifies and shows the distance of the cross section from some reference.
Column 2 is the length of the reach in feet.
Column 3 shows the cross-sectional area of the cross section.
Column 4 shows the wetted perimeter of the cross section.
Column 5 shows the velocity in the cross section.
Column 6 is the friction loss in the reach. It is equal to h_f, computed by Eq. (7) of Art. 16.
Column 7 shows the difference in velocity head, modified as appropriate for negative values as computed by Eq. (8) of Art. 16.
Column 8 lists other losses, as applicable. Where energy losses not covered by the roughness factor n are significant, as discussed in Art. 5, they must be taken into account in computing backwater curves.
Column 9 shows the total difference in head between the cross section and the previous (downstream) cross section.
Column 10 shows the elevation at the cross section. It is the previous value plus Column 9. If it differs materially from the value assumed in deriving the quantities in Columns 3 and 4, a new value must be assumed and the computations for the entire line repeated. The elevation in Column 10 is a point on the desired backwater curve. The function of the space in Column 11 is self-explanatory.

the wave base. Complex procedures utilizing all the known factors affecting open-channel flow[17] are used where there are tidal influences or where the quality of the basic data and the required accuracy of the results warrant the considerable additional work involved. For most problems, approximate methods that keep the computations within manageable limits are indicated. These methods ascribe the change in shape of the flood wave to channel storage.[18] A common method is described in Art. 23.

The solution of flow-routing problems must conform to the law of continuity. In any time interval and using comparable units,

$$O = I - \frac{\Delta S}{\Delta t} \tag{9}$$

where O = mean outflow during period Δt
I = mean inflow during period Δt
ΔS = change in storage during period Δt
Δt = length of period

The above equation can also be written

$$S_2 - S_1 = \frac{1}{2}(I_1 + I_2)\Delta t - \frac{1}{2}(O_1 + O_2)\Delta t \qquad (10)$$

where subscripts 1 and 2 denote values at the beginning and end of a period, respectively.

Figure 4 shows a typical drainage area for which the flow is to be routed from A to B. A tributary is gaged at C, and the remaining inflow between A and B is ungaged.

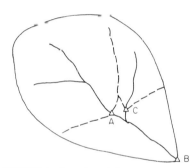

FIGURE 4 Location of flow-routing reach.

Tributaries present a special complication. There is no exact direct solution for the case in which the flow of one stream induces significant backwater in the other stream. Gilcrest[19] presents a trial-and-error procedure for such a case. Where the backwater induced by the confluence is small, a simple procedure is adequate. If the tributary enters the reach near the upstream end as in Fig. 4, its flow can be combined with the inflow from the main channel. If it enters near the downstream end of the reach, its flow can be added to the outflow after completing the routing. For tributaries entering midway in the reach and for ungaged inflow, the inflow can be distributed between the two ends of the reach or, for greater accuracy, treated as an independent variable.[20] Ungaged inflow can be determined from rainfall data using a synthetic unit hydrograph or by assuming it to be a percentage of the gaged flow at a nearby gaging station.

Figure 5 shows typical hydrographs at the upstream and downstream end of a reach of channel. The inflow to the reach is denoted by *abcfg* and the outflow from the reach by *acdeg*. The downstream or outflow hydrograph has a lower peak and is more spread out than the upstream hydrograph. These changes result from the volume *abca* going into channel storage during the period *ac* and an equal volume *cdegfc* being released from channel storage during the period *cg*.

22. Basic Data Requirements. All methods of flow routing require knowledge or assumptions regarding the river channel. The most basic requirement is the relationship between discharge and channel storage.

Surveys, including cross sections for various discharges, will give this storage. Such surveys usually are not available. An alternative, and more common, method is to compute channel storage from inflow and outflow records of an observed flood. Figure 5 will serve as an example of effective inflow and outflow for the river reach AB shown in Fig. 4.

The first step in determining channel storage from inflow and outflow records is to determine the average inflow and outflow by periods for the entire flood

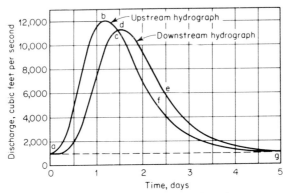

FIGURE 5 Change in flood wave due to channel storage.

event. The computations are shown in Table 4. In this example, the flood event is divided into 6-h periods. The average inflow and outflow for each period, $\frac{1}{2}(I_1 + I_2)$ and $\frac{1}{2}(O_1 + O_2)$, respectively, in Eq. (10), are shown in Columns 3 and 4, respectively, of the table. Column 5 shows the change in channel storage, expressed in $\frac{1}{4}$ (6 of 24 hr) second-feet-days ($\frac{1}{4}$ SFD), during the period, and Column 6 shows the total channel storage at the end of the period above an arbitrary base. The computations should cover a period such that the beginning and ending discharges are about the same. If the resulting beginning and ending storage val-

TABLE 4 Determination of Channel Storage

Day (1)	Hours (2)	Average inflow, cfs (3)	Average outflow, cfs (4)	Change in channel storage ¼ SFD (5)	Total channel storage (end of period) ¼ SFD (6)
1	0–6	1,300	1,000	300	300
	6–12	2,600	1,400	1200	1,500
	12–18	6,000	2,600	3400	4,900
	18–24	9,800	5,300	4500	9,400
2	0–6	11,700	8,500	3200	12,600
	6–12	11,700	10,600	1100	13,700
	12–18	10,400	11,200	−800	12,900
	18–24	8,300	10,400	−2100	10,800
3	0–6	6,300	8,800	−2500	8,300
	6–12	4,900	7,000	−2100	6,200
	12–18	3,700	5,400	−1700	4,500
	18–24	2,900	4,100	−1200	3,300
4	0–6	2,300	3,200	−900	2,400
	6–12	1,900	2,600	−700	1,700
	12–18	1,600	2,100	−500	1,200
	18–24	1,400	1,800	−400	800
5	0–6	1,200	1,500	−300	500
	6–12	1,100	1,300	−200	300
	12–18	1,000	1,200	−200	100
	18–24	1,000	1,100	−100	0

ues are not approximately equal, then either the inflow or the outflow, or both, should be adjusted.

Another basic data requirement is the time of travel of the flood wave through the reach. The time of travel of the flood wave will be the time between the center of mass of the flood hydrographs for A and B. This time will be somewhat greater than the time interval between the peaks at the two points.

Gilcrest[19] has shown by the Manning formula that the velocity of the flood wave is 1.67 times the average velocity of flow for a wide rectangular channel, 1.44 times the average velocity for a wide parabolic channel, and 1.33 times the average velocity for a triangular channel. Another method of determining time of travel of the flood wave through the reach is given in Art. 23.

23. Mathematical Solution. One of the most satisfactory mathematical methods of flow routing is the Muskingum method, developed by G. T. McCarthy and others in 1934–35 during studies of the Muskingum Conservancy District Flood Control Project of the Corps of Engineers in Ohio.

In the Muskingum method, storage is related by a weighting process to both inflow and outflow. It is convenient to consider the storage in the reach as being composed of both prism storage and wedge storage as shown in Fig. 6. A reservoir has no wedge storage; so the discharge for a given gate opening is a function only of the reservoir stage. In a river channel the outflow is a function of both stage at outflow point and slope at the outflow point. The slope at the outflow is related to difference in flow within the reach. During the rising phase of a flood wave, inflow exceeds outflow, creating the wedge storage and increasing the slope of the water surface. During the falling phase, the wedge storage is negative. The total storage is expressed mathematically by the Muskingum equation as follows:

$$S = K[xI + (1 - x)O] = KO + Kx(I - O) \tag{11}$$

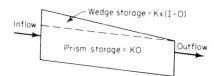

FIGURE 6 Flood-wave storage.

where S is storage volume and I and O are simultaneous instantaneous values of inflow and outflow, respectively. K is a coefficient with the dimensions of time and is equal to the time of passage of the centroid of the flood wave through the reach. The parameter x weighs the relative effects of inflow and outflow on storage within the reach.

The Muskingum method assumes that K and x are constant throughout the flood and that there is a unique relationship between storage and weighted discharge at the two ends of the reach. This is not strictly correct in most cases, but the method usually gives adequate results and the simplifying assumptions greatly reduce the computations.

The actual flow routing is performed for reaches and time intervals of predetermined length. The objective is to determine outflow from the reach at the end of successive time intervals. The general equation for outflow is

$$O_2 = C_1 I_2 + C_2 I_1 + C_3 O_1 \tag{12}$$

where subscripts 1 and 2 to the O and I terms denote values at the beginning and end, respectively, of the routing time intervals, and where

$$C_{1'} = \frac{-(Kx - 0.5\,\Delta t)}{0.5\,\Delta t + K(1 - x)} \tag{13}$$

$$C_{2'} = \frac{Kx + 0.5\,\Delta t}{0.5\,\Delta t + K(1 - x)} \tag{14}$$

$$C_{3'} = \frac{-0.5\,\Delta t + K(1 - x)}{0.5\,\Delta t + K(1 - x)} \tag{15}$$

Adding these equations shows that

$$C_{1'} + C_{2'} + C_{3'} = 1 \tag{16}$$

An alternative and somewhat simpler form of the equation follows:

$$O_2 + O_1 + C_1(I_1 - O_1) + C_2(I_2 + I_1) \tag{17}$$

where

$$C_1 = \frac{\Delta t}{K(1 - x) + 0.5\Delta t} \tag{18}$$

$$C_2 = \frac{0.5\Delta t - Kx}{K(1 - x) + 0.5\Delta t} \tag{19}$$

It will be noted that all the C values in the above equations are dimensionless. K and Δt must be in the same units. Equation (17) is used in the routing example in this section.

Before computing the C values, it is necessary to determine Δt, K, and x. The first step in flow routing is to select the routing period Δt. Flow-routing procedures give the ordinate of the outflow hydrograph at the end of each routing period. The periods should be sufficiently short to define that hydrograph adequately. Also the period should be not longer than the time of travel of the flow through the reach and not less than $2Kx$.

If the length of the reach is such that the flood is at an appreciably different phase at the two ends of the reach, the reach should be subdivided into two or more subreaches. The value of K must be reduced accordingly. The flow is routed successively through each subreach, the outflow for the upstream subreach becoming the inflow for the next downstream subreach.

A value of x is chosen so that the storage will be the same for a given weighted discharge for rising and falling stages [see Eq. (11)]. Its value will be between zero and 0.5. It will be zero for a reservoir and 0.5 for uniformly progressive flow. Its determination for a natural river requires inflow and outflow records. In the absence of such records, a value of 0.25 is often used. A lower value should be used where the reach is constricted at the lower end. A higher value should be used for steep, fairly uniform channels where the flow is confined within well-defined banks or levees at all stages.

The values of K and x can be found simultaneously where dependable records of storage vs. discharge are available. In the example, inflow and outflow are taken from Fig. 5. Weighted discharge, $xI + (1 - x)O$, is plotted as ordinate against storage for that instant (Column 6 of Table 5) in Fig. 7 using several values of x. It will be noted that curves b and c for x values of 0.2 and 0.3, respectively, have the smallest loops; that is, the relationship for rising and falling stages is most nearly the same. Therefore, 0.25 would be an acceptable

FLOW ROUTING

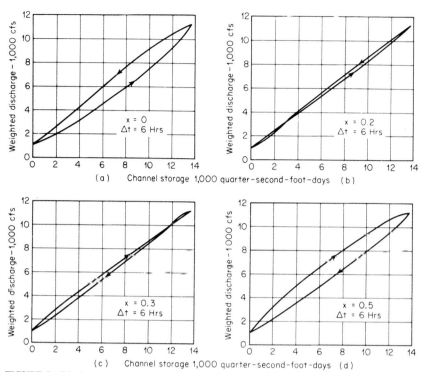

FIGURE 7 Discharge-storage loops.

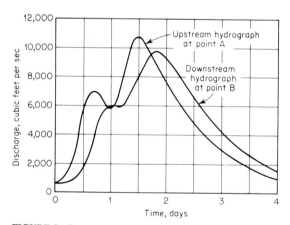

FIGURE 8 Routed hydrograph.

TABLE 5 Routing Computations

$O_2 = O_1 + C_1(I_1 - O_1) + C_2(I_2 - I_1)$

Units: 1000 cfs; $C_1 = 2/3$, $C_2 = 1/9$

Day and hour (1)	I (2)	$I_2 - I_1$ (3)	$C_2(I_2 - I_1)$ (4)	$I_1 - O_1$, Col. 2–Col. 8 (5)	$C_1(I_1 - O_1)$ (6)	$C_1(I_1 - O_1) +$ $C_2(I_2 - I_1)$ Col. 4 + Col. 6 (7)	O (8)
1–0	0.6	1.0	0.11	0	0	0.11	0.6
6	1.6	3.9	0.43	0.89	0.59	1.02	0.71
12	5.5	1.4	0.16	3.77	2.51	2.67	1.73
18	6.9	−1.1	−0.12	2.50	1.67	1.55	4.40
24	5.8	2.6	0.29	−0.15	−0.10	0.19	5.95
2–6	8.4	2.3	0.26	2.26	1.51	1.77	6.14
12	10.7	−1.2	−0.13	2.79	1.86	1.73	7.91
18	9.5	−1.7	−0.19	−0.14	−0.09	−0.28	9.64
24	7.8	−1.6	−0.18	−1.56	−1.04	−1.22	9.36
3–6	6.2	−1.4	−.016	−1.94	−1.29	−1.45	8.14
12	4.8	−1.1	−0.12	−1.89	−1.26	−1.38	6.69
18	3.7	−0.7	−0.08	−1.61	−1.07	−1.15	5.31
24	3.0	−0.6	−0.07	−1.16	−0.77	−0.84	4.16
4–6	2.4	−0.6	−0.07	−0.92	−0.61	−0.68	3.32
12	1.8	−0.4	−0.04	−0.84	−0.56	−0.60	2.64
18	1.4	−0.4	−0.04	−0.64	−0.43	−0.47	2.04
24	1.0						1.51

value for x. The corresponding value of K is the reciprocal of the slope of the curve, or 8 h.

With the values of Δt, K, and x determined, the coefficients C_1', C_2', and C_3' in Eq. (12) or the coefficients C_1 and C_2 in Eq. (17) are computed. The latter equation is used for illustration.

Figure 8 shows a new hydrograph for the upstream end of the reach AB in Fig. 4. This hydrograph is routed in Table 5, and the routed downstream hydrograph is also shown in Fig. 8.

In Table 5, Columns 1 through 4 can be computed first. Since the flow wave starts at zero hour on the first day, the inflow (Column 1) and the outflow (Column 8) are the same at that instant. Column 5 is found by subtracting Column 8 from Column 2. Column 7 is found by adding Columns 4 and 6. Column 8 for the next line (O_2 for the first period or O_1 for the second period) is computed next. It is found by adding Columns 7 and 8 for the line above. Column 5 for that line is then computed. This routine is then followed until the routing is completed.

REFERENCES

1. Chow, Ven Te, "Open-Channel Hydraulics," McGraw-Hill, New York, 1959.
2. Kolupaila, Steponas, Methods of Determination of the Kinetic Energy Factor, *The Port Engineer*, Calcutta, India, **5** (1), 12–18, January 1956.

REFERENCES

3. ROUSE, HUNTER, Critical Analysis of Open Channel Resistance, *Proc. ASCE, J. Hydraulics Div.*, July 1965.
4. Task Force on Friction Factors in Open Channels of the Committee on Hydromechanics of the Hydraulics Division of the American Society of Civil Engineers, Friction Factors in Open Channels, Progress Report, *Proc. ASCE, J. Hydraulics Div.*, **89** (HY2), pt. 1, March 1963.
5. COWAN, WOODY I., Estimating Hydraulic Roughness Coefficients, *Agricultural Engineering*, **37** (7), 473–475, July 1956.
6. BLUE, F. L., Jr., J. K. HERBERT, and R. L. LANCEFIELD, Flow around a River Bend Investigated, *Civil Eng.*, **4** (5), May 1934.
7. YEN, C. H., and J. W. HOWE, Effects of Channel Shapes on Losses in a Canal Bend, *Civil Eng.*, **12** (1), January 1942.
8. SCOBEY, FRED C., "The Flow of Water in Flumes," U.S. Dept. Agr. Tech. Bull. 393.
9. YARNELL, DAVID L., and SHERMAN M. WOODWARD, "Flow of Water around Bends," U.S. Dept. Agr. Tech. Bull. 526.
10. YARNELL, DAVID L., "Bridge Piers as Channel Obstructions," U.S. Dept. Agr. Tech. Bull. 442, 1934. LIU, H. K., J. N. BRADLEY, and E. J. PLATE, "Backwater Effects of Piers and Abutments," prepared by the Civil Engineering Section, Colorado State University, Fort Collins, Colo., in cooperation with the U.S. Bureau of Public Roads, October 1957. BRADLEY, J. N., "Hydraulics of Bridge Waterways," 2d ed., rev., March 1978, U.S. Dept. of Transportation/Federal Highway Administration, Washington, D.C.
11. CORBETT, DON M., et al., "Stream-Gaging Procedure," U.S. Geol. Survey Water Supply Paper 888.
12. WOODWARD, SHERMAN M., and CHESLEY J. POSEY, "Hydraulics of Steady Flow in Open Channels," Wiley, New York, 1941.
13. U.S. CORPS OF ENGINEERS, "Engineering Manual," Civil Works Construction, Part 114, Chap. 9, Hydrologic and Hydraulic Analyses, Computation of Backwater Curves in River Channels, p. 9, May 1952.
14. LARA, JOE M., and KENNETH B. SCHRODER, Two Methods to Compute Water Surface Profiles, *Proc. ASCE*, **85** (HY4), April 1959.
15. LAWLER, EWARD A., Hydrology of Flood Control, Part II, Flood Routing, in "Handbook of Applied Hydrology" by Ven Te Chow, McGraw-Hill, New York, 1964. LINSLEY, RAY K., Jr., MAX A. KOHLER, and JOSEPH L. H. PAULHUS, "Applied Hydrology," McGraw-Hill, New York, 1949.
16. KOHLER, MAX A., Electrical Analogies and Electronic Computers: A Symposium, Application to Streamflow Routing, *Trans. ASCE*, **118**, 1953. ROCKWOOD, DAVID M., Columbia Basin Streamflow Routing by Computer, *Proc. ASCE, J. Waterways, Harbors Div.*, Paper 1874, WW 5, December 1958.
17. U.S. CORPS OF ENGINEERS, "Engineering Manual, Civil Works Construction," Part 114, Chap. 8, Hydrologic and Hydraulic Analysis, Routing of Floods through River Channels, p. 2, September 1953. THOMAS, H. A., "Hydraulics of Flood Movement in Rivers," Carnegie Institute of Technology, Pittsburgh, 1934.
18. CARTER, R. W., and R. G. GODFREY, "Storage and Flood Routing," U.S. Geol. Survey Water Supply Paper 1543-B, 1960.
19. GILCREST, B. R., Flood Routing, Chap. 10 in "Engineering Hydraulics," Hunter House (ed.), Wiley, New York, 1950.
20. U.S. CORPS OF ENGINEERS, "Engineering Manual," Civil Works Construction Hydrologic and Hydraulic Analysis, Routing of Floods through River Channels, Part 114, Chap. 8, September 1953.

SECTION 6

REGIME CANALS

By B. A. McKiernan[1]

CHANNELS IN ALLUVIUM

1. Stable and Regime Channels. Channels formed in alluvial soils and other erodible materials may be lined with nonerodible materials such as concrete or with materials which increase resistance to erosion such as gravel, or they may be designed so that their unprotected banks and beds will not be eroded by the flow. The relative economy of lined and unlined canals depends on the particulars of each case. Channels lined with nonerodible materials are dealt with in Sec. 7. Channels lined with gravel are similar to unlined channels excavated in gravel and may be designed by the tractive-force method described in Art. 3. Channels with unprotected beds and banks formed of alluvial materials in the silt-sand range may be designed by the regime-channel formulas given in Art. 4.

As is generally the case with canals drawing supplies from rivers, the flow carries sediment in suspension and as bed load. Suspended load is sediment which is supported and distributed by the turbulence of the flow. It is subdivided into bed-material load, comprising particles of a size range found also in the bed, and wash load, comprising smaller particles generally absent from the bed. Bed load is sediment which is transported by rolling and sliding along the bed or by intermittent entrainment in the turbulent flow near the bed. It is important to prevent deposits of sediment which would reduce the discharge capacity of a canal. It is equally important to avoid erosive velocities which would attack banks and the foundations of structures. The regime formulas have been derived for this case. Also they provide guidance for the design of lined canals which are to carry sediment-laden flow.

2. Channels in Alluvium. Channels formed in alluvial or other granular material are said to be *stable* if their geometry remains substantially unchanged by scour or sediment deposit. Such stability will be achieved if (1) the materials forming the boundaries are sufficiently coarse to resist scour from flow in the channel and (2) the sediment carried by the flow does not exceed in size or quantity the transport capability of the channel.

Channels are said to be *in regime* when scour and deposition occur, but the balance of these is such that the boundaries remain essentially in equilibrium over a period of time. In this condition the materials transported by the flow and forming the boundaries are of similar origin and accordingly have generally the same physical characteristics.

Formulas for the design of stable and regime channels have been derived by two different approaches: one, based on the effect of shear stress (or tractive force) at the boundary (applying primarily to stable but not necessarily regime

---

[1]Acknowledgment is made to Franklyn C. Rogers and A. Rylands Thomas for material in this section that appeared in the third edition (1969).

channels) and the other, a more empirical approach, based on observations of channels seen to be in regime.

3. Tractive Force. The tractive-force formulas are useful for the design of channels with erodible boundaries carrying clear water and also channels in which the material forming the boundaries is appreciably coarser than the transported sediment. Typical of the latter are channels with gravel boundaries carrying fine sand as suspended sediment and channels with rigid boundaries conveying flow with sediment of any size.

Tractive-Force Formula. The mean shear stress or tractive force per unit area may be obtained by resolution of forces along the wetted boundary of the channel, assuming that the component of weight of water acting in the direction of the channel slope is balanced by the total shear resistance of water acting on the wetted perimeter.

$$\tau_0 P = \gamma A S_0 \tag{1}$$

where τ_0 is the mean shear stress on the bed, P the wetted perimeter, γ the unit weight of water, A the cross-sectional area, and S_0 the longitudinal slope of the bed. The mean shear stress is expressed as

$$\tau_0 = \frac{\gamma A S_0}{P} = \gamma R S_0 \tag{1a}$$

where R is the hydraulic radius. In relatively wide channels the mean shear stress on the bed is given by

$$\tau_0 = \gamma d_m S_0 \tag{1b}$$

where d_m is the mean depth.

Maximum Shear Stress. Shear stress generally varies over the wetted perimeter. In natural channels, it tends to be maximum on the bed and minimum on the slopes. Correspondingly, the boundary material is often coarser on the bed and finer near the banks. Figures 1 and 2 show the maximum shear stress acting on

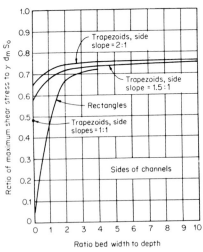

FIGURE 1 Maximum shear stress on sides of channel.

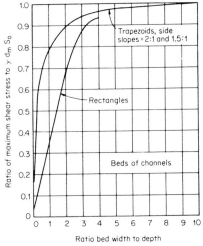

FIGURE 2 Maximum shear stress on channel bed.

the sides and beds of rectangular and trapezoidal channels.[1]

Limiting Tractive Forces. To avoid erosion of the material forming the boundary, the shear stress should not exceed certain limiting values which are related to the grain size of the material and to the angle of side slope. The theory is directly applicable to the side slopes of noncohesive granular material and to the beds of channels consisting of smooth (plane) surfaces formed in such material where there is no bed load. Such conditions may prevail in channels formed in coarse gravelly material.

The tractive-force formula is also applied to channels in fine material transporting sediment as bed load. In such cases, accumulation of transported sediment may occur unless the shear stress at the bed exceeds the limiting value for the material transported (but, to avoid erosion of the sides, the shear stress should be within the given limits). Where there is bed load, however, it is preferable to use the regime-channel formulas given in Art. 4.

Figure 3 shows the limiting shear stress or tractive force recommended for the design of stable channels by the U.S. Bureau of Reclamation. The values in Fig. 3 apply to horizontal or near-horizontal beds. For sloping banks or sides of channels, a reduction of factor K should be used, depending on the angle of repose of the material. Values of K may be determined from Fig. 4.

Slope Formulas. The formula in most general use is that of Manning,

$$V = \frac{1.486}{n} R^{2/3} S_0^{1/2} \qquad (2)$$

where V is the mean velocity and n a roughness factor.

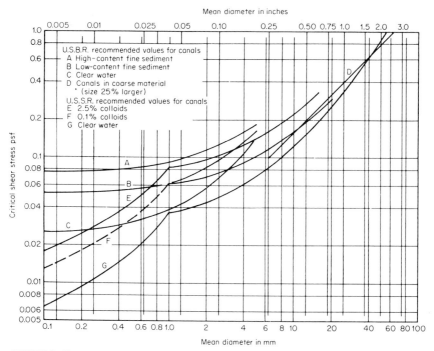

FIGURE 3 Recommended values of limiting shear stress. (*From American Society of Civil Engineers.*)

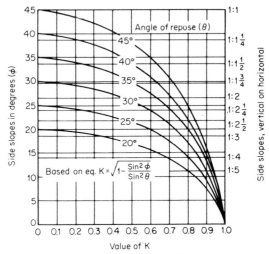

FIGURE 4 Critical shear stress on inclined slopes.

There are no precise means of determining n, and an estimate has to be based on judgment and records of similar channels; n depends on the particle or grain size of the boundary material and on larger-scale form roughness. For example, n is affected by ripples in the bed, by channel geometry such as bends and changes in cross sections, and also in some cases by the sediment load. Plant growth increases the value considerably. A wide range of n values appears in Tables 1 and 2 in Sec. 5.

Channels with Gravel Beds. Channels with beds of gravel, shingle, or boulders may be considered in two categories according to whether or not there is bed movement.

If there is no bed movement and the banks are stable (which may be verified by Figs. 1 to 4), the relation of slope to grain size is indicated approximately by the following formula:

$$V = 8.2 \left(\frac{R}{k_s}\right)^{1/6} (gRS_0)^{1/2} \quad (3)$$

where k_s is the equivalent grain roughness, which is taken as the mean size of the material forming the boundary, and g is the gravitational constant; g, R, and k_s are in the same units. Equation (3) approximates the Nikuradse rough-pipe formula over a range of R/k_s from 10 to 150.[2]

In terms of the Manning formula, this is equivalent to

$$n_g = 0.021 k_s^{1/6} \quad (4)$$

where n_g is the coefficient for grain roughness when k_s is in inches. In nearly all cases, however, because of irregularities in bed and banks, grain roughness is superimposed on form roughness and the value of n is higher than indicated by Eq. (4). The Manning roughness factor is expressed as

$$n = \left(n_g^2 + n_f^2\right)^{1/2} \quad (5)$$

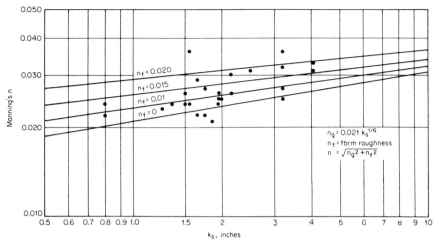
FIGURE 5 Manning's n for gravel channels.

where n_f is the coefficient for form roughness based only on the irregularities. Values of n_f are given in Fig. 5, which shows curves based on Eqs. (4) and (5) and data of the San Luis Valley canals, Colorado.[3] It will be seen that, in gravel of representative size ¾ to 4 in, n_f ranged, with few exceptions, from 0 to 0.02 and n from 0.022 to 0.033. It is stated that there was bed movement in some of these canals, which may explain the wide range of n_f. Elsewhere n values of 0.023 (Upper Jhelum Canal, Pakistan, 10,000 cfs) and 0.025 (Dun Canal, India, 80 cfs) have been observed in canals with gravel beds.

If there is general movement, the bed becomes irregular and n is consequently increased. This is very marked in the case of natural streams with a wide range of discharge and considerable bed movement at high stages. The value of n may vary considerably with stage, often being maximum at low stages when the large formations resulting from flood flows create discontinuities and surface waves. The value of n in the River Beas, India, where the mean size of bed material was between 3 and 6 in, ranged from 0.023 with 56,000 cfs to 0.037 with 3300 cfs.[4] When designing a canal with a low value of n it is therefore important to check that the bed and side slopes will be stable under all possible conditions.

Channels with Sand Beds. Although there is a tendency for Manning's n to be higher with coarse sand than with finer materials, the predominant factor is the degree of roughness resulting from formation of ripples or dunes on the bed. In the case of small channels, the condition of the side banks has some influence. The following values of Manning's or Kutter's n have been used for many years as a guide in India and Pakistan.[5]

Condition of channel	n
Above average	0.0225
Tolerably good	0.025
Below average	0.0275
Bad	0.030

Actual values observed have ranged from 0.011 with nearly plane bed to 0.039 with bed of large dunes of coarse sand.

THEORY OF REGIME CHANNELS

4. Regime-Channel Formulas. The regime formulas of Kennedy, Lacey, and others apply to channels with erodible boundaries in alluvial soils carrying small sediment loads. They were derived mainly from data of irrigation channels in the alluvial plains of India and Pakistan where the canals draw supplies directly from large rivers and in the flood season carry sediment consisting of silt and fine and medium sand, the load generally not exceeding 2000 ppm.

Unless the dimensions and slope of a channel are constructed to regime requirements, the geometry will suffer gradual change and performance will deteriorate. Inadequate slope leads to siltation and loss of discharge capacity and excessive slope, to scour, endangering structures. Inadequate width leads to bank scour, excessive width, to meandering.

Kennedy's Formula. Now mainly of historic interest, Kennedy's formula[6]

$$V_0 = 0.84 d_b^{0.64} \tag{6}$$

where V_0 = nonsilting, nonscouring, mean velocity and d_b = average depth over the bed (excluding the side slopes), was widely used for many years. The coefficient and exponent were developed for the Bari-Doab canal system, Punjab, where the channel bed material was medium sand of approximately 0.32-mm mean diameter.

Lindley's Regime Concept. It was later recognized that, for stability, a channel must also have the correct width and slope. Lindley[7] advanced a *regime* concept embracing all channel dimensions and slope, which he believed were fixed by nature by given conditions. Lindley stated, "When an artificial channel is used to convey silty water, both bed and banks scour or fill, changing depth, gradient and width, until a state of balance is attained at which the channel is said to be regime."

Lacey's Formulas. Lacey[8–12] analyzed data from irrigation channels and rivers, mainly in India and Pakistan, but also from some in Egypt, Europe, and America. His results supported Lindley's regime concept and permitted quantitative evaluation. They were obtained by the correlation of dimensions and slopes of apparently stable channels with discharges and size of bed materials. The Lacey formulas express three primary relationships.

Velocity-to-depth relationship

$$V = 1.15 f^{1/2} R^{1/2} \tag{7}$$

Velocity-to-slope relationship

$$V = 16 R^{2/3} S_0^{1/3} \tag{8}$$

Width-to-discharge relationship

$$P = 2.67 Q^{1/2} \tag{9}$$

where V = mean velocity
R = hydraulic radius
S_0 = longitudinal slope

P = wetted perimeter
Q = discharge
f = silt or sediment factor

All units are in foot-seconds. Lacey proposed an approximate relationship of f to the grade of bed materials:

$$f = 1.76 D_g^{1/2} \qquad (10)$$

where D_g = mean diameter, mm.

The Harza Method.[13] An extensive program of hydraulic and sediment measurements was initiated in Pakistan in 1961 by the Water and Power Development Authority of Pakistan (WAPDA) and was named the Canal and Headworks Data Observation Program (CHOP).[14] The comprehensive CHOP continued from August 1961 through October 1964 in order to provide a data base for design of the canals of the Indus Basin Project. Analyses of the data confirmed the correctness of the forms of the Lacey equations, (8) and (9), above but, however, indicated that the constant in these equations varied from canal system to canal system. Accordingly, the Harza Engineering Co. in reviewing the designs of Indus Basin Project Canals adopted the following three basic regime relationships:

$$f_{VR} = 0.75 \frac{V^2}{R} \qquad (11)$$

$$P = K_1 Q^{1/2} \qquad (12)$$

$$V = K_2 R^{2/3} S^{1/2} \qquad (13)$$

where f_{VR} is the Lacey silt factor, in terms of velocity and hydraulic radius.

From Eqs. (11), (12), and (13), the following regime relationships may be derived:

$$A = \frac{0.909 K_1^{1/3} Q^{5/6}}{f_{VR}^{1/3}} \qquad (14)$$

$$S = \frac{1.615 K_1^{1/3} f_{VR}^{5/3}}{K_2^3 Q^{1/6}} \qquad (15)$$

and

$$\text{Manning's } n = \frac{1.61 K_1^{1/18} f_{VR}^{5/18}}{K_2^{3/2} Q^{1/36}} \qquad (16)$$

Ideally, a designer should base values of f_{VR}, K_1, and K_2 on analyses of data from regime channels (Art. 2) in the same area as a proposed channel. If this is not possible, values from channels in a similar environment with respect to anticipated sediment characteristics should be used.

Plots illustrating Eqs. (12) and (13) are shown on Figs. 6 and 7, respectively. From Fig. 8, where values of K_2 from several canals in Pakistan have been plotted against median diameter of bed materials, an envelope to the plotted points indicates a relationship for maximum values of K_2 in the form

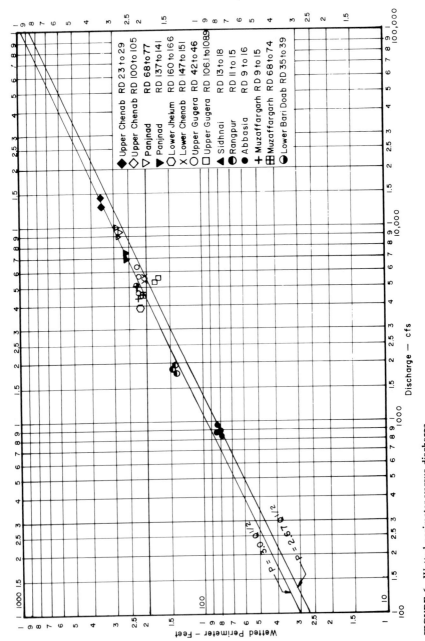

FIGURE 6 Wetted perimeter versus discharge.

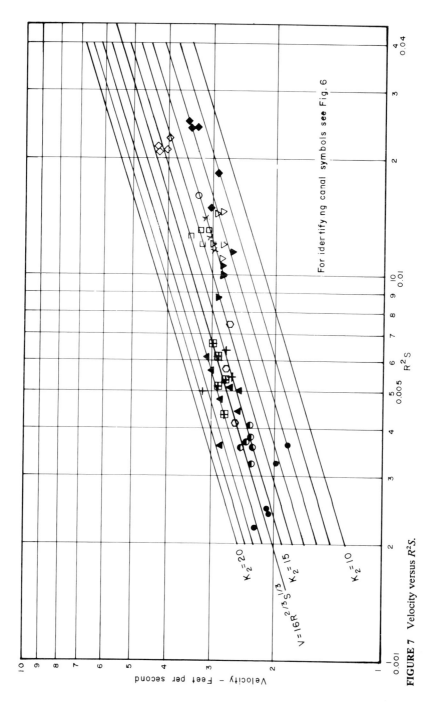

FIGURE 7 Velocity versus R^2S.

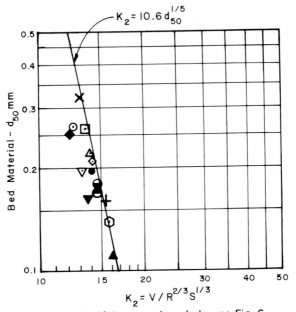

For identifying canal symbols see Fig. 6

FIGURE 8 Regime channels—bed material size versus K_2.

$$K_2 = \frac{10.6}{d_{50}^{1/5}} \qquad (17)$$

where d_{50} is the median diameter, millimeters, of material from the canal bed.

EFFECTS OF SEDIMENT AND SEEPAGE

5. Sediment Transport. The balance of erosion and deposition which characterizes regime channels is closely related to the sediment-transport capabilities of the flow. The sediment forming the bed and banks of the regime channel originates from the rivers and streams from which the supplies are drawn and to a lesser degree from the beds and banks.

Sediment is classified according to size. The classification proposed by the Subcommittee on Sediment Terminology of the American Geophysical Union[15] is given in Table 1.

Particle size is determined by sieving (coarser particles) or sedimentation (finer particles). For the study of the mechanics of sediment transport and computation of sediment load both size and shape are conveniently defined by the fall velocity, i.e., the terminal velocity of fall of a particle in water. The relation of size to fall velocity for quartz spheres at various temperatures is shown in Fig. 9.[16] Flat (e.g., micaceous) particles have lower fall velocities than shown, the ef-

TABLE 1 Sediment Grade Scale

Class name	Size range		Approx. sieve mesh openings per inch	
	Millimeters	Inches	Tyler	United States Standard
Very large boulders	4096–2048	160–80		
Large boulders	2048–1024	80–40		
Medium boulders	1024–512	40–20		
Small boulders	512–256	20–10		
Large cobbles	256–128	10–5		
Small cobbles	128–64	5.0–2.5		
Very coarse gravel	64–32	2.5–1.3		
Coarse gravel	32–16	1.3–0.6		
Medium gravel	16–8	0.6–0.3	2½	
Fine gravel	8–4	0.3–0.16	5	5
Very fine gravel	4–2	0.16–0.08	9	10
Very coarse sand	2.000–1.000		16	18
Coarse sand	1.000–0.500		32	35
Medium sand	0.500–0.250		60	60
Fine sand	0.250–0.125		115	120
Very fine sand	0.125–0.062		250	230
Coarse silt	0.062–0.031			
Medium silt	0.031–0.016			
Fine silt	0.016–0.008			
Very fine silt	0.008–0.004			
Coarse clay	0.004–0.0020			
Medium clay	0.0020–0.0010			
Fine clay	0.0010–0.0005			
Very fine clay	0.0005–0.00024			

fect being small for particles up to 0.1 mm, but it may be 25 percent for 1-mm particles.

The size distribution of a sediment mixture can be represented by a frequency diagram as in Fig. 10, which shows characteristic grain size for typical sandy suspended sediment and bed materials. The mean or effective diameter of a mixed sediment is often described by its median or 50 percent size. For further information on the properties of sediment see Ref. 17.

Sediment is transported partly in suspension and partly as bed load. Transport results from the action of turbulent flow on the bed. Basically this consists of drag on individual particles due to relative velocity. Transport action is complicated by variations in velocity, bed forms, presence of particles of a wide size range, action of the flow in sorting and lifting particles of various sizes into sus-

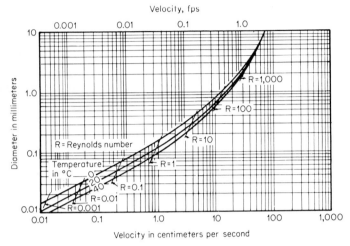

FIGURE 9 Fall velocity of quartz spheres in water.

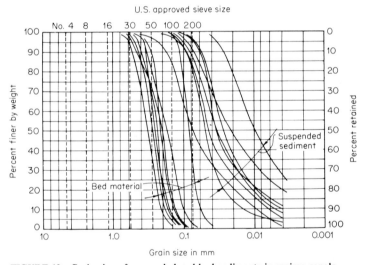

FIGURE 10 Grain size of suspended and bed sediments in regime canals.

pension, and variation of fluid properties with varying concentration of suspended sediment. The mechanics of entrainment are very complex and are not yet adequately understood. Continuous interchange of sediment occurs between bed and suspended load, but because sand and finer particles are more easily entrained than coarser material, there is a marked difference in behavior between active channels in which the bed material is sand or finer and those in which it is gravel or coarser. In the former, the suspended load is composed of particles representing the whole range of bed-material size, though the fine grades generally predominate. In the latter, with normal velocities, the suspended load consists only of sand and finer particles without any of the dominant sizes of bed material. In both cases, the volume of suspended load may greatly exceed the bed load.

Typical grain sizes of suspended and bed load, sampled in eight regime canals in Pakistan, are shown in Fig. 10. The effect of alluvial sorting is indicated by the narrow range of grain sizes of the bed material. In most regime canals, the sizes of the bed material and suspended sediments usually show a marked seasonal variation.

Although the finer suspended sediments (less than about 0.06 mm) have a measurable effect on the performance of a canal, the impairment of canal performance usually results from an excess of coarser materials (greater than about 0.06 mm). Accordingly, it is the concentrations of the coarser sizes and the variation in such concentrations that are significant in canal operation. A typical annual variation, as measured near the head of the Upper Chenab Canal, Pakistan, is shown in Fig. 11.

The transport capability of a channel is a function of capacity, measured as the quantity of sediment which will be moved, and competence, measured by the maximum size of bed particles which will be moved. Capacity increases with decrease in particle size, and transport capability in any given size range can be far greater than the volume of sediment available.

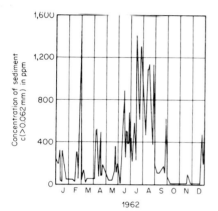

FIGURE 11 Annual variation of sediment concentration—Upper Chenab Canal, Pakistan.

The size of transportable grain is indicated by the tractive force [Eq. (1) and Fig. 3]. In a regime channel, the bed material is usually composed of sediment which deposited during periods of decreased competency or insufficient transport capacity.

A simple empirical formula which has been shown to conform moderately well with sediment-load data of flumes and canals in medium-fine sand is that of Ahmad and Rehman:[18]

$$\frac{1000 q^{2/3} S_0}{\omega^{1/2}} = 1 + 7C^{2/3} \qquad (18)$$

where q is discharge per unit width, S_0 is energy slope, ω is mean fall velocity of the sediment, fps, and C is the concentration of sediment by weight in parts per thousand, excluding sediment finer than 0.06 mm.

6. Sediment Control. The introduction of excessive sediment into a canal frequently results in deposition in the canal with loss of capacity. If available head is sufficient to increase velocity and carry the sediment through the canal, the problem must be dealt with at the point where the water is discharged from the canal. Usually either or both aspects are objectionable and provisions are made for excluding the sediment from the canal.

A canal drawing supplies at an angle from a river will tend to draw bed water and therefore a heavy load of coarse sediment. The most effective way of excluding coarse material is conversely to locate the offtake on the outer bank of a bend in the river so that it draws mainly surface water. This method, which has been successfully used by Inglis[19] requires smooth approach-flow conditions. Divide walls, which are provided at most canal headworks in India and Pakistan, can ensure proportional distribution of sediment where the approach is straight. Sediment excluders with tunnels discharging downstream[20,21] are sometimes provided between the divide wall and the canal-head structure; these also require smooth approach conditions to avoid mixing bed and surface flow. The canal-head structure should be given a crest high above the river bed and the canal should be designed with energy level below that of the river, thus providing a margin of head so that a slight steepening of slope in the canal will not drown the head control.

A form of sediment excluder in operation on a number of large canals in Pakistan employs a depressed area or pocket to collect the coarser sediments before the flow enters the canal-head regulator. Sediments are flushed intermittently from the pocket through low-level culvert-type tunnels which extend past the face of the regulator to a point downstream from the canal intake.

The flow approaching the pocket and head regulator should be guided so as to minimize turbulence through the entire range of canal operation. Guide walls are usually provided to assure a short length of straight approach channel leading to the regulators. Observations have established that sediment excluders utilizing pockets and tunnels can keep out 30 to 70 percent of the coarse sediment.[21]

The desilting works built in 1938 at the headworks of the All-American Canal on the California bank of the Colorado River uses six stilling basins 269 by 769 ft by 12.5 ft deep to precipitate sediment down to 0.075-mm diameter.[22] Sediment is collected from the basin floors by seventy-two 125-ft-diameter electrically driven rotary scrapers and flushed to a sluiceway through sludge discharge pipes varying from 15 to 36 in in diameter. The works are designed to remove 80 percent of the incoming sediment load, which is estimated as 60,000 tons/day for a flow of 12,000 cfs. The sediment control is sufficient to achieve successful operation of the unlined canal on a slope of 1 to 19,000 at a flow of 15,000 cfs.

Another type of basin is used to desilt the flow entering the 34-mile-long unlined channel which supplies 3000 to 3500 cfs to the hydroelectric plants of the Loup River Public Power District, Nebraska.[23] The basin is a widened canal section 10,000 ft in length, 16 ft deep, 200 ft in bed width, and 264 ft wide at the water surface. The flow, at velocities of 0.8 to 0.9 fps, deposits all material including medium sand down to 0.25-mm diameter and about half of the fine sand (0.25 to 0.075 mm). A dredge with a 24-in pump powered by a 1200-hp electric motor (nominal dredge capacity 62.5 cfs) removes about 1.8 million tons of silt per year, which is equivalent to an average of 1100 ppm for the flow passing through the basin. The canal has operated for over 50 years without appreciable deposition or scour on a slope of 1 to 20,000.

When desilting works are provided, a means of control is required to allow some fine sediment to enter the canal to avoid serious erosion downstream.

7. Seepage Losses. The loss of water by seepage from unlined canals can be a significant factor in selecting a design and appraising the economic value of a canal. Seepage losses have been measured in Pakistan as ranging from 2 cfs per million square feet of wetted area in canals cut in clayey soils to as high as 15 cfs per million square feet in coarse sand. Values of 6 to 8 cfs per million square feet are common for fine to medium sand.

In the United States seepage losses are usually estimated by the Moritz formula:[24]

$$S = 0.2C\left(\frac{Q}{V}\right)^{1/2} \qquad (19)$$

in which S is the loss in cfs per mile of canal, Q is the discharge of the canal in cfs, V is the mean velocity in fps, and C is the rate of water loss in cubic feet per 24 h per square foot of wetted area. Average observed values of C are listed in Table 2. The values are suitable for preliminary estimates.

TABLE 2 Average Observed Values of Losses in Earth Canals

Type of material	C, ft^3 in 24 h for 1-ft^2 wetted area
Cemented gravel and hardpan with sandy loam	0.34
Clay and clayey loam	0.41
Sandy loam	0.66
Volcanic ash	0.68
Volcanic ash with sand	0.98
Sand and volcanic ash or clay	1.20
Sandy soil with rock	1.68
Sandy and gravelly soil	2.20

DESIGN OF REGIME CANALS

8. Design Procedure. The design of a regime canal proceeds from the proposed discharge Q and desired shape of cross-section. Figure 12 shows a canal section having compound side slopes, for which a regime design is developed as follows:

1. Values of f_{VR}, K_1, and K_2 are selected as outlined in Art. 4.
2. The cross-sectional area A is computed from Eq. (14)
3. V is computed from $V = A/Q$
4. S is calculated from Eq. (15)
5. D is computed from the relationship (see Fig. 12):

$$D = \frac{P - \sqrt{P^2 - 4A\{2[F\sqrt{1 + S_1^2} + (1 - F)\sqrt{1 + S_2^2}] - [F^2(S_1 - S_2) + S_2]\}}}{2\{2[F\sqrt{1 + S_1^2} + (1 - F)\sqrt{1 + S_2^2}] - [F^2(S_1 - S_2) + S_2]\}} \quad (20)$$

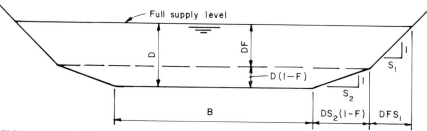

FIGURE 12 Typical regime canal cross section.

where F is the depth factor at which the lower slope, 1 on S_2, changes to the upper slope, 1 on S_1, and P is obtained from Eq. (12).

6. The bed width B is given by

$$B = P - 2D\left[F\sqrt{1 + S_1^2} + (1 - F)\sqrt{1 + S_2^2}\right] \quad (21)$$

7. The water surface width W_s is given by

$$W_s = B + 2[DS_2(1 - F) + DFS_1] \quad (22)$$

8. Manning's n may be computed from Eq. (16) as a check on expected friction factors for channels in the area under consideration.
9. Approximate sediment-carrying capacity may be checked by Eq. (18).

Example 1. A canal to carry a maximum discharge of 4000 cfs is to be designed for an area in which values of $f_{VR} = 0.80$, $K_1 = 3.0$, and $K_2 = 14.5$ have been selected as appropriate and a side slope of 1 on 1.5 has been considered adequate, resulting in $S_1 = S_2 = 1.5$ and $F = 1.0$ (Fig. 12). Note that, from Eqs. (10) and (17), a bed-material d_{50} of about 0.2 mm would be compatible with $f_{VR} = 0.8$ and $K_2 = 14.5$. From the design procedure outlined above, the following canal properties would result from Eqs. (14), (15), (20), (21), (22), and (16):

$A = 1418$ ft^2 ($V = 2.82$ fps), $S = 1/7563$
$D = 8.22$ ft, $B = 160.1$ ft, $W_s = 184.8$ ft
Manning's $n = 0.023$

Example 2. A canal with a maximum discharge of 20,000 cfs is required for an area having the same values of f_{VR}, K_1, and K_2 as in Example 1, above. However, side slopes of 1 on 3 to half depth and 1 on 2 above half depth are considered appropriate for the materials to be excavated in the construction of the canal.

Accordingly, $S_1 = 2$, $S_2 = 3$, and $F = 0.5$.
The resulting canal properties are

$A = 5421$ ft^2 ($V = 3.69$ fps), $S = 1/9891$
$D = 14.00$ ft, $B = 352.2$ ft, $W_s = 422.2$ ft
Manning's $n = 0.022$

REFERENCES

1. LANE, E. W., "Progress Report on the Results of Studies on Design of Stable Channels," U.S. Bur. Reclamation Hydraulic Lab. Rep. Hyd-353, Denver, 1952.
2. KEULEGAN, G. H., Laws of Turbulent Flow in Open Channels, *J. Res. Natl. Bur. Std.,* **21,** Washington, 1938.
3. LANE, E. W., and E. J. CARLSON, Some Factors Affecting the Stability of Canals Constructed in Coarse Granular Materials, *Proc. IAHR,* Minneapolis, 1953.
4. MALHOTRA, J. K., "Hydraulic Data for a Boulder River (Beas at Sujanpur Tira)," Central Board of Irrigation, *Ann. Rep. (Tech.),* Simla, 1943.

REFERENCES

5. BUCKLEY, R. B., "Irrigation Pocket Book." Spon, London, 1928, p. 178.
6. KENNEDY, R. G., The Prevention of Silting in Irrigation Canals, *Proc. Inst. Civil Engrs.*, **119**, 281, 1895.
7. LINDLEY, E. S., Regime Channels, *Proc. Punjab Eng. Congr.*, Lahore, **7**, 1919.
8. LACEY, G., Stable Channels in Alluvium, *Proc. Inst. Civil Engrs.*, **229**, 259, 1929–1930.
9. LACEY, G., Uniform Flow in Alluvial Rivers and Canals, *Proc. Inst. Civil Engrs.*, **237**, 421, 1933–1934.
10. LACEY, G., "Regime Flow in Incoherent Alluvium," Central Board of Irrigation, India, Publication 20, Simla, 1939.
11. LACEY, G., A General Theory of Flow in Alluvium, *J. Inst. Civil Engrs.*, **27**, 16, 1946.
12. LACEY, G., Flow in Alluvial Channels with Mobile Sandy Beds, *Proc. Inst. Civil Engrs.*, **9**, 145, 1958.
13. MAHMOOD, KHALID, and H. W. SHEN, The Regime Concept of Sediment—Transporting Canals and Rivers, "River Mechanics," vol. II, H. W. Shen (ed.), Colorado State University, Fort Collins, 1971.
14. WATER AND POWER DEVELOPMENT AUTHORITY OF PAKISTAN (WAPDA), "Canal and Headworks Data Observation Program," 1962 Data Tabulation in April 1963, 1962–63 Data Tabulations Parts I and II, and 1963–64 Data Tabulation.
15. LANE, E. W., Reports on the Subcommittee on Sediment Terminology, *Trans. Am. Geophys. Union*, **26** (6), 936, 1947.
16. ROUSE, H., "Nomogram for the Settling Velocity of Spheres," Division of Geology and Geography Exhibit D of the Report of the Committee on Sedimentation, 1936–37, National Research Council, Washington, D.C., October, 1937, pp. 57–64.
17. Sediment Transportation Mechanics: Introduction and Properties of Sediment, Progress Report, Task Committee on Preparation of Sedimentation Manual, *Proc. ASCE*, paper 3194, July, 1962.
18. AHMAD, MUSHTAQ, and ABDUL REHMAN, Appraisal and Analysis of New Data from Alluvial Canals of West Pakistan in Relation to Regime Concepts and Formulae, *Proc. West Pakistan Eng. Congr.*, 1963, p 69.
19. INGLIS, SIR C. C., "The Behavior and Control of Rivers and Canals, Central Waterpower Irrigation and Navigation Research Station, Poona, India," Res. Publ. 13, pp. 217–279, 1949.
20. International Association for Hydraulic Research, *Proc. 4th Meeting*, Bombay, 1951, several papers on Q.2.
21. HAIGH, F. F., The Emerson Barrage, *J. Inst. Civil Engrs.*, December 1941, p. 107.
22. FORESTER, D. M., Desilting Works for All-American Canal, *Civil Engrg.*, October 1938, p. 649.
23. Information supplied by Loup River Public Power District, Columbus, Neb.
24. U.S. BUREAU OF RECLAMATION, "Canals and Related Structures," Design Standard No. 3, December 1967.

SECTION 7

CANALS AND CONDUITS

By B. A. McKiernan[1]

HYDRAULIC FACTORS

1. Introductory Statement. Artificial channels for the conveyance of fluids fall into two primary divisions: those which merely guide the fluid as it flows down a sloping surface, and those which confine and guide its movement under pressure, commonly referred to as *free-flow* and *pressure conduits*. Free-flow conduits may be simple open channels or ditches, or they may be pipes or other enclosed structures, flowing partly full. Pressure conduits may be of wood, metal, glass, steel, concrete, ceramic, plastic, or other suitable substance and, although usually circular in cross section, may be of any form. Consideration is to be given here only to those types of conduits commonly utilized in the civil works of hydraulic projects.

2. Effect of Slope. The velocity at which water flows depends on the steepness of the slope, the size and shape of the channel, the roughness of its walls, and the viscosity and density of the water. Reference should be made to Sec. 2 for flow formulas in common use. For the problems of this section, and within the usual temperature range, the effect of viscosity is negligible.

3. Mean Velocity. The velocity used in most flow formulas is equivalent to the quantity of flow divided by the area of the water prism and is called the *mean velocity*. Only turbulent flow will be considered in this section. Reference is made to Sec. 2 for the principles governing turbulent flow and for tables and diagrams useful in the design of both free-flow and pressure conduits.

4. Best Hydraulic Shape. The wetted perimeter offers resistance to flow and generally, for "hard" conduits, contributes to cost. Hence, it should be held to the minimum value consistent with the conditions. The degree to which a conduit conforms to this criterion is a measure of its hydraulic efficiency.

The most hydraulically efficient section on this basis is a semicircle, open at the top and flowing full. The best closed section, flowing full, is a circle.

Although minimum area and perimeter contribute to economy, they may be overshadowed by practical considerations. Figures 1 and 2 show, respectively, typical sections of aqueducts and tunnels which are now in service. A square closed conduit is obviously hydraulically inefficient, and its flat sides are structurally undesirable, for either internal or external load. Yet such conduits are occasionally used to meet practical conditions.

Both unlined and lined canals have traditionally been trapezoidal, for construction reasons. With development of mechanical trimming and lining machines, these reasons are no longer always valid, and curved bottoms are appear-

---

[1]Acknowledgment is made to Calvin V. Davis and Kenneth E. Sorensen for material in this section that appeared in the third edition (1969).

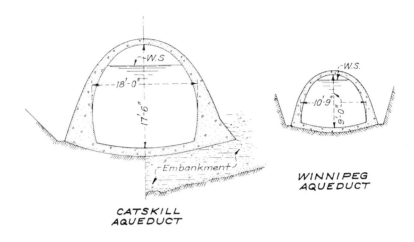

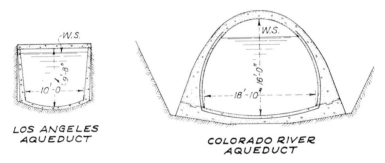

FIGURE 1 Typical aqueduct sections.

ing. The best shape for a lined tunnel usually is a circle, but the section is frequently modified to provide increased floor space for hauling equipment, and for other purposes. Figure 3 shows the proportions of a standard horseshoe tunnel. Tables 1 and 2 summarize the principal geometric hydraulic design parameters for partially filled circular and horseshoe conduit sections.

5. Allowance for Critical Flow. As explained in Sec. 2, flow in an open channel, at or near the critical depth (Froude's number $F = 1.0$), is in indifferent equilibrium, and slight boundary irregularities may produce marked irregularities in flow. The relation of designed depth to critical depth should always be known, and where a safe margin above or below critical cannot be maintained by changing the shape of the channel, or otherwise, ample freeboard to care for possible disturbances should be provided.

6. Permissible Velocities. The cost of a conduit varies with its size. Therefore, if available slope permits, the cost of the initial construction may be reduced by using the highest safe velocity. However, if the velocity is made too high, the conduit or open channel may be damaged or destroyed by erosion. This must be avoided by limiting velocities according to the boundary materials.

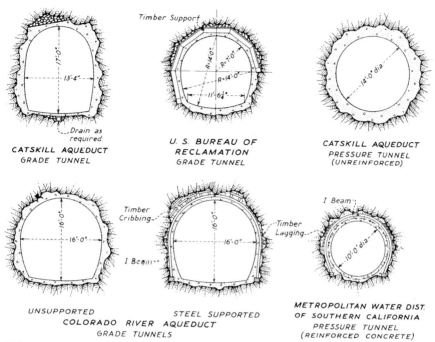

FIGURE 2 Typical tunnel sections.

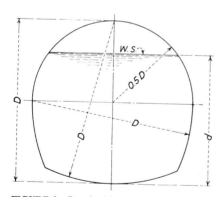

FIGURE 3 Standard horseshoe tunnel.

For clear water in smooth concrete or other hard-surfaced water conductors, the limiting velocity is beyond practical requirements, except perhaps in very steep chutes. Velocities above 40 fps for clear water in concrete channels have been found to do no harm. If the water carries abrasive materials, damage may occur with lower velocities. Unless the abrasive material is particularly erosive, velocities up to 10 or 12 fps should not prove injurious to wood or concrete. Thin metal flumes may be damaged by coarse sand or gravel at 6 to 8 fps, and the galvanizing may be injured by lower velocities. The abrasive materials may be in the water at its source or may enter along the channel. No definite relation has been established between the nature of abrasive materials, material of channel bank, and permissible velocity. Abrasive materials should be excluded as far as practicable.

In unlined earthen channels, the limiting velocity involves many factors. Generally, a fine soil is more easily eroded than a coarse one, but the effect of grain size may be obscured by the presence or absence of a cementing or binding ma-

TABLE 1 Area, Wetted Perimeter (WP), and Hydraulic Radius (HR) of Partially Filled Circular Conduit Sections

$\dfrac{d}{D}$	$\dfrac{\text{Area}}{D^2}$	$\dfrac{\text{WP}}{D}$	$\dfrac{\text{HR}}{D}$	$\dfrac{d}{D}$	$\dfrac{\text{Area}}{D^2}$	$\dfrac{\text{WP}}{D}$	$\dfrac{\text{HR}}{D}$
0.01	0.0013	0.2003	0.0066	0.51	0.4027	1.5908	0.2531
0.02	0.0037	0.2838	0.0132	0.52	0.4127	1.6108	0.2562
0.03	0.0069	0.3482	0.0197	0.53	0.4227	1.6308	0.2592
0.04	0.0105	0.4027	0.0262	0.54	0.4327	1.6509	0.2621
0.05	0.0147	0.4510	0.0326	0.55	0.4426	1.6710	0.2649
0.06	0.0192	0.4949	0.0389	0.56	0.4526	1.6911	0.2676
0.07	0.0242	0.5355	0.0451	0.57	0.4625	1.7113	0.2703
0.08	0.0294	0.5735	0.0513	0.58	0.4724	1.7315	0.2728
0.09	0.0350	0.6094	0.0575	0.59	0.4822	1.7518	0.2753
0.10	0.0409	0.6435	0.0635	0.60	0.4920	1.7722	0.2776
0.11	0.0470	0.6761	0.0695	0.61	0.5018	1.7926	0.2799
0.12	0.0534	0.7075	0.0755	0.62	0.5115	1.8132	0.2821
0.13	0.0600	0.7377	0.0813	0.63	0.5212	1.8338	0.2842
0.14	0.0668	0.7670	0.0871	0.64	0.5308	1.8546	0.2862
0.15	0.0739	0.7954	0.0929	0.65	0.5404	1.8755	0.2881
0.16	0.0811	0.8230	0.0968	0.66	0.5499	1.8965	0.2900
0.17	0.0885	0.8500	0.1042	0.67	0.5594	1.9177	0.2917
0.18	0.0961	0.8763	0.1097	0.68	0.5687	1.9391	0.2933
0.19	0.1039	0.9021	0.1152	0.69	0.5780	1.9606	0.2948
0.20	0.1118	0.9273	0.1206	0.70	0.5872	1.9823	0.2962
0.21	0.1199	0.9521	0.1259	0.71	0.5964	2.0042	0.2975
0.22	0.1281	0.9764	0.1312	0.72	0.6054	2.0264	0.2987
0.23	0.1365	1.0004	0.1364	0.73	0.6143	2.0488	0.2998
0.24	0.1449	1.0239	0.1416	0.74	0.6231	2.0715	0.3008
0.25	0.1535	1.0472	0.1466	0.75	0.6319	2.0944	0.3017
0.26	0.1623	1.0701	0.1516	0.76	0.6405	2.1176	0.3024
0.27	0.1711	1.0928	0.1566	0.77	0.6489	2.1412	0.3031
0.28	0.1800	1.1152	0.1614	0.78	0.6573	2.1652	0.3036
0.29	0.1890	1.1374	0.1662	0.79	0.6655	2.1895	0.3039
0.30	0.1982	1.1593	0.1709	0.80	0.6736	2.2143	0.3042
0.31	0.2074	1.1810	0.1756	0.81	0.6815	2.2395	0.3043
0.32	0.2167	1.2025	0.1802	0.82	0.6893	2.2653	0.3043
0.33	0.2260	1.2239	0.1847	0.83	0.6969	2.2916	0.3041
0.34	0.2355	1.2451	0.1891	0.84	0.7043	2.3186	0.3038
0.35	0.2450	1.2661	0.1935	0.85	0.7115	2.3462	0.3033
0.36	0.2546	1.2870	0.1978	0.86	0.7186	2.3746	0.3026
0.37	0.2642	1.3078	0.2020	0.87	0.7254	2.4039	0.3018
0.38	0.2739	1.3284	0.2062	0.88	0.7320	2.4341	0.3007
0.39	0.2836	1.3490	0.2102	0.89	0.7384	2.4655	0.2995
0.40	0.2934	1.3694	0.2142	0.90	0.7445	2.4981	0.2980
0.41	0.3032	1.3898	0.2182	0.91	0.7504	2.5322	0.2963
0.42	0.3130	1.4101	0.2220	0.92	0.7560	2.5681	0.2944
0.43	0.3229	1.4303	0.2258	0.93	0.7612	2.6061	0.2921
0.44	0.3328	1.4505	0.2295	0.94	0.7662	2.6467	0.2895
0.45	0.3428	1.4706	0.2331	0.95	0.7707	2.6906	0.2865
0.46	0.3527	1.4907	0.2366	0.96	0.7749	2.7389	0.2829
0.47	0.3627	1.5108	0.2401	0.97	0.7785	2.7934	0.2787
0.48	0.3727	1.5308	0.2435	0.98	0.7816	2.8578	0.2735
0.49	0.3827	1.5508	0.2468	0.99	0.7841	2.9413	0.2666
0.50	0.3927	1.5708	0.2500	1.00	0.7854	3.1416	0.2500

TABLE 2 Area, Wetted Perimeter (WP), and Hydraulic Radius (HR) of Partially Filled Horseshoe Conduit Sections

$\dfrac{d}{D}$	$\dfrac{\text{Area}}{D^2}$	$\dfrac{\text{WP}}{D}$	$\dfrac{\text{HR}}{D}$	$\dfrac{d}{D}$	$\dfrac{\text{Area}}{D^2}$	$\dfrac{\text{WP}}{D}$	$\dfrac{\text{HR}}{D}$
0.01	0.0019	0.2831	0.0067	0.51	0.4466	1.7161	0.2603
0.02	0.0053	0.4007	0.0133	0.52	0.4566	1.7361	0.2630
0.03	0.0098	0.4911	0.0199	0.53	0.4666	1.7562	0.2657
0.04	0.0150	0.5676	0.0264	0.54	0.4766	1.7762	0.2683
0.05	0.0209	0.6351	0.0329	0.55	0.4865	1.7963	0.2709
0.06	0.0275	0.6963	0.0394	0.56	0.4965	1.8164	0.2733
0.07	0.0346	0.7528	0.0439	0.57	0.5064	1.8366	0.2757
0.08	0.0422	0.8054	0.0523	0.58	0.5163	1.8568	0.2780
0.0886	0.0491	0.8482	0.0578	0.59	0.5261	1.8771	0.2803
0.09	0.0502	0.8512	0.0590	0.60	0.5360	1.8975	0.2825
0.10	0.0585	0.8731	0.0670				
0.11	0.0669	0.8949	0.0747	0.61	0.5457	1.9179	0.2845
0.12	0.0753	0.9165	0.0822	0.62	0.5555	1.9385	0.2865
0.13	0.0839	0.9381	0.0894	0.63	0.5651	1.9591	0.2885
0.14	0.0925	0.9596	0.0964	0.64	0.5748	1.9799	0.2903
0.15	0.1012	0.9810	0.1032	0.65	0.5843	2.0008	0.2921
0.16	0.1100	1.0023	0.1097	0.66	0.5939	2.0219	0.2937
0.17	0.1188	1.0235	0.1161	0.67	0.6033	2.0430	0.2953
0.18	0.1277	1.0447	0.1223	0.68	0.6127	2.0644	0.2968
0.19	0.1367	1.0657	0.1283	0.69	0.6219	2.0859	0.2982
0.20	0.1458	1.0867	0.1341	0.70	0.6312	2.1076	0.2995
0.21	0.1549	1.1077	0.1398	0.71	0.6403	2.1296	0.3007
0.22	0.1640	1.1285	0.1453	0.72	0.6493	2.1517	0.3018
0.23	0.1733	1.1493	0.1507	0.73	0.6582	2.1741	0.3028
0.24	0.1825	1.1701	0.1560	0.74	0.6671	2.1968	0.3037
0.25	0.1919	1.1908	0.1611	0.75	0.6758	2.2197	0.3044
0.26	0.2013	1.2114	0.1662	0.76	0.6844	2.2430	0.3051
0.27	0.2107	1.2320	0.1710	0.77	0.6929	2.2666	0.3057
0.28	0.2202	1.2525	0.1758	0.78	0.7012	2.2905	0.3061
0.29	0.2297	1.2730	0.1805	0.79	0.7094	2.3149	0.3065
0.30	0.2393	1.2934	0.1850	0.80	0.7175	2.3396	0.3067
0.31	0.2489	1.3138	0.1895	0.81	0.7254	2.3649	0.3068
0.32	0.2586	1.3341	0.1938	0.82	0.7332	2.3906	0.3067
0.33	0.2683	1.3545	0.1981	0.83	0.7408	2.4169	0.3065
0.34	0.2780	1.3747	0.2022	0.84	0.7482	2.4439	0.3062
0.35	0.2878	1.3950	0.2063	0.85	0.7554	2.4715	0.3057
0.36	0.2975	1.4152	0.2102	0.86	0.7625	2.4999	0.3050
0.37	0.3074	1.4354	0.2141	0.87	0.7693	2.5292	0.3042
0.38	0.3172	1.4555	0.2179	0.88	0.7759	2.5594	0.3032
0.39	0.3271	1.4757	0.2216	0.89	0.7823	2.5908	0.3020
0.40	0.3370	1.4958	0.2253	0.90	0.7884	2.6234	0.3005
0.41	0.3469	1.5159	0.2288	0.91	0.7943	2.6575	0.2989
0.42	0.3568	1.5360	0.2323	0.92	0.7999	2.6934	0.2970
0.43	0.3667	1.5560	0.2357	0.93	0.8052	2.7314	0.2948
0.44	0.3767	1.5761	0.2390	0.94	0.8101	2.7720	0.2922
0.45	0.3867	1.5961	0.2423	0.95	0.8146	2.8159	0.2893
0.46	0.3966	1.6161	0.2454	0.96	0.8188	2.8642	0.2859
0.47	0.4066	1.6361	0.2485	0.97	0.8225	2.9188	0.2818
0.48	0.4166	1.6561	0.2516	0.98	0.8256	2.9831	0.2767
0.49	0.4266	1.6761	0.2545	0.99	0.8280	3.0666	0.2700
0.50	0.4366	1.6961	0.2574	1.00	0.8293	3.2669	0.2539

terial. The tendency to erode is reduced by seasoning. Groundwater conditions exert an important influence. Seepage out of the channel, particularly if the water is turbid, tends to toughen the banks, infiltration reduces resistance to erosion. Erosion can be reduced or avoided by designing for low velocities. If carried to an extreme, this results in large and costly canals, encourages the growth of aquatic plants, and increases seepage and evaporation.

If the water carries an appreciable amount of silt in suspension, too low a velocity will cause the canal to fill up until the capacity is impaired. It is necessary to choose a velocity that will keep the silt in motion but that will not erode the banks of the canal. The margin of permissible velocities depends on the amount and nature of the silt in the water, the nature of the bank material, the size and shape of the canal, and many other factors. The silt content of most turbid water varies with the season, as does also the demand for water and the resultant velocity of the flow. Thus, as demonstrated in Sec. 6, a canal that will scour at one season may silt at another.

The determination of nonscouring, nonsilting velocities for earth canals has attracted the attention of many investigators over a long period of time, and a considerable mass of data and formulas have been accumulated. Because of its complexities and importance, this problem is discussed separately in Sec. 6. Reference should be made to this section for design information on all important canal projects in alluvial soils.

However, for preliminary purposes, and for design in many cases, use may be made of the approximate values proposed by Fortier and Scobey, in 1926, as shown in Table 3.[1] Where the silt burden is important, it is better to make the slope a little too steep rather than a little too flat. A gradient that proves to be too

TABLE 3 Permissible Canal Velocities

Original material excavated for canal (1)	Velocity, fps, after aging, of canals carrying:		
	Clear water, no detritus (2)	Water transporting colloidal silts (3)	Water transporting noncolloidal silts, sands, gravels, or rock fragments (4)
Fine sand (noncolloidal)	1.50	2.50	1.50
Sandy loam (noncolloidal)	1.75	2.50	2.00
Silt loan (noncolloidal)	2.00	3.00	2.00
Alluvial silts when noncolloidal	2.00	3.50	2.00
Ordinary firm loam	2.50	3.50	2.25
Volcanic ash	2.50	3.50	2.00
Fine gravel	2.50	5.00	3.75
Stiff clay (very colloidal)	3.75	5.00	3.00
Graded, loam to cobbles, when noncolloidal	3.75	5.00	5.00
Alluvial silts when colloidal	3.75	5.00	3.00
Graded, silt to cobbles, when colloidal	4.00	5.50	5.00
Coarse gravel (noncolloidal)	4.00	6.00	6.50
Cobbles and shingles	5.00	5.50	6.50
Shales and hardpans	6.00	6.00	5.00

steep can be controlled by checks. In hard-surfaced channels, silting is easily controlled if fall for scouring velocity is available.

The values in Columns 3 and 4 apply only to particles in suspension and not to coarser particles, usually referred to as the *bed load,* which are rolled along the bottom. Every precaution should be taken to exclude these coarser particles, and where this cannot be fully accomplished, means for removal should be provided.

CONVEYANCE LOSSES

7. Losses from Concrete, Metal, and Wood Conduits. Losses are caused by leakage, absorption, and evaporation. The leakage from well-constructed and well-maintained concrete, metal, and wood conduits is relatively small. Where such conduits occur in short lengths in systems composed chiefly of earthen channels, losses from them are negligible by comparison. However, no conduit is completely tight, and in long lined systems, the accumulation of even small leakage may be important. An example is the Colorado River Aqueduct, bringing water from the Colorado River to Los Angeles. The main line of this aqueduct is 242 miles in length, made up as follows: tunnel (16 ft) 92.0 miles, cut-and-cover conduit (18 ft) 55.0 miles, lined canal (20 ft) 62.4 miles, pressure conduits 29.7 miles, unlined canal 1.0 mile, reservoir passage 2.2 miles. The specifications required that all visible leaks be closed, yet an appreciable loss was allowed in economic studies. Using data available at the time, the loss for final full operation, including reservoir seepage and evaporation for some 3000 acres of canal and reservoir surfaces, was set at about 80,000 acre-ft/year, or 7.5 percent of the designed annual capacity. This allowance has been found adequate.

Data on losses from large aqueduct conduits are scarce. Such leakage is frequently expressed in terms of gallons per inch diameter of conduit per mile per day. Tests on 17 ft by 17 ft 6 in concrete horseshoe conduits in the New York City water system showed leakage of 167 to 463 gal per inch diameter, with an average of 283. An average leakage allowance of 300 to 400 gal per inch diameter is liberal for any well-constructed concrete, steel, or timber free-flow conduit. Leakage from new concrete conduits may exceed this allowance, but if visible leaks are repaired, the loss reduces as the concrete swells and as small openings fill with silt or algae.

8. Losses from Canals. The losses discussed in Art. 7 are applicable primarily to substantially built channels of structural materials, or excavated in rock or firm earth. The boundary materials of canals, both lined and unlined, are subject to wide variations in permeability, as discussed in Arts. 34 and 35.

FLOW RESISTANCE

9. Basic Data. The subjects of flow resistance and flow formulas for computing it are discussed in Sec. 2. Flow formulas are to varying degrees empirical. Their application depends on experimental coefficients or constants, of which many published lists are available. Such data are usually derived from tests on newly constructed or well-maintained channels, the designer being left to use judgment for any allowance needed to provide for deterioration with use.

10. Unlined Earth Channels. An unlined earth channel is generally more subject to progressive deterioration than other waterways. It is easy to compute, with reasonable accuracy, the hydraulic performance of a newly constructed channel. However, once put into service, resistance begins to change because of erosion, silting, aquatic growths such as weeds, grass, tules, willows, moss, Crustacea, and other reasons. Aquatic growths increase flow resistance and reduce the effective waterway area. If left unattended, the usefulness of the channel may be seriously impaired. The speed with which choking growth may appear is difficult to predict. Some channels require only occasional weeding. Others, more exposed to seeding opportunities, may deteriorate rapidly.[2]

In the case of very light annual growth, a liberal design friction factor may largely eliminate the trouble. For rapid, or choking, growth, a liberal allowance for friction is not a complete remedy. It is not practicable to tabulate a friction value for a mossy or otherwise fouled canal. The ultimate remedy is cleaning, which in serious cases must start early and proceed continuously and may be expensive. A more liberal design of freeboard may extend the time between cleaning and reduce costs. Cleaning may be accomplished mechanically, by fallowing, or chemically. Table 4, which lists Manning's n values for unlined channels, is excerpted from Tables 5 and 6 of "Open-Channel Hydraulics" by Chow.[3] The extent to which aquatic growths affect flow resistance may be assessed from the values listed.

11. Hard-Surface Conduits. The effects of fouling in hard-surface conduits usually are less pronounced and more easily controlled than in earth channels. Several examples of the effects of sliming and other capacity-reducing factors

TABLE 4 Values of the Roughness Coefficient n for Unlined Channels

Type of channel and description	Minimum	Normal	Maximum
Excavated or dredged			
a. Earth, straight and uniform			
1. Clean, recently completed	0.016	0.018	0.020
2. Clean, after weathering	0.018	0.022	0.025
3. Gravel, uniform section, clean	0.022	0.025	0.030
4. With short grass, few weeds	0.022	0.027	0.033
b. Earth, winding and sluggish			
1. No vegetation	0.023	0.025	0.030
2. Grass, some weeds	0.025	0.030	0.033
3. Dense weeds or aquatic plants in deep channels	0.030	0.035	0.040
4. Earth bottom and rubble sides	0.028	0.030	0.035
5. Stony bottom and weedy banks	0.025	0.035	0.040
6. Cobble bottom and clean sides	0.030	0.040	0.050
c. Dragline-excavated or dredged			
1. No vegetation	0.025	0.028	0.033
2. Light brush on banks	0.035	0.050	0.060
d. Rock cuts			
1. Smooth and uniform	0.025	0.035	0.040
2. Jagged and irregular	0.035	0.040	0.050
e. Channels not maintained, weeds and brush uncut			
1. Dense weeds, high as flow depth	0.050	0.080	0.120
2. Clean bottom, brush on sides	0.040	0.050	0.080
3. Same, highest stage of flow	0.045	0.070	0.110
4. Dense brush, high stage	0.080	0.100	0.140

will be found in Sec. 2. In many cases flow capacity has been tested and found to be less than the designed capacity.

At one time, flow formulas and tables for cast-iron and steel pipes were generally accompanied by tables of aging factors, to allow for the gradual accumulation of tubercules on the interior surface. With modern coatings, this difficulty has been mitigated. As a result, the aging factors are frequently ignored. Severe tuberculation is unlikely, unless the coating fails or spalls off. Such failures are relatively rare, making it cheaper to clean and repair occasionally than to provide excessive capacity initially. However, modern linings do not completely inhibit deterioration by organic slimes and other aquatic growth which may affect any hard-surfaced conduit.

Table 5, excerpted from Table 5-6 of Ref. 3, lists Manning's n values for hard-surface conduits for a range of surfaces and flow conditions.

12. Lined Channels. Between 1957 and 1962 the U.S. Bureau of Reclamation carried out hydraulic tests[4,5] in nine trapezoidal concrete-lined canals in the western United States. Linings of all but one of the canals had been placed by traveling rail-mounted slipforms.

A principal purpose of the tests was to measure flow resistance, including that due to canal surfaces; aquatic growths, adhering to surfaces; structures such as bridge piers; and horizontal curves in canal alignment. Design discharges in the test canals varied from 700 to 13,200 cfs and test discharges ranged from 555 to 6820 cfs. Measured Manning's n values ranged from about 0.013 to 0.018. Lower n values were generally obtained in canals which had been treated with copper sulfate to control weed growth. Freshwater clam deposits were found in one canal in which n values up to about 0.018 were measured. The range of measured n values is of the order of magnitude listed for concrete surfaces C.2 (float finish) and C.3 (finished, with gravel on bottom) listed in Table 6. Table 6, excerpted from Table 5-6 of "Open-Channel Hydraulics" by Chow[3] lists n values for a broad range of lined channel surfaces.

13. Pier Losses. The expression

$$h_p = K_p h_v \tag{1}$$

was used by the U.S. Bureau of Reclamation in analyses of flow resistance due to bridge piers, in the tests referred to in Art. 12 above,[4] in which h_p is the incremental head loss caused by piers, K_p is the pier loss coefficient, and h_v is the velocity head in the unobstructed canal section.

Figure 4 (reproduced from Fig. 7 on p. 15 of Ref. 6) is a chart from which a backwater coefficient K may be selected to compute the rise in water surface across piers of various shapes and configurations. Figure 4 shows K_p values (ΔK) to be a function of the amount of channel contraction at the piers.

The contraction ratio J is defined as the ratio of the total projected area of piers normal to the flow to the gross flow area as illustrated in Fig. 4. In the Ref. 4 canals, values of J ranged from 0.018 to 0.055.

The bridge opening ratio M (equal to 1.0 when bridge abutments do not project into the flow) defines the degree of flow constriction and is expressed as the ratio of the flow which can pass between bridge abutments without causing a backwater effect to the total flow.

The term σ is a multiplication factor for the influence of M on the backwater coefficient for piers. As shown on Fig. 4, $\Delta K_p = \Delta K \times \sigma$.

The backwater effect from piers can be estimated from Fig. 4 as shown in the following example:

For a canal in which $V = 5.0$ fps, $J = 0.04$, $M = 1.0$, with a single roundnosed

TABLE 5 Values of the Roughness Coefficient n for Hard-Surface Conduits

Closed conduits flowing partly full

Type of channel and description	Minimum	Normal	Maximum
1. Metal			
a. Brass, smooth	0.009	0.010	0.013
b. Steel			
1. Lockbar and welded	0.010	0.012	0.014
2. Riveted and spiral	0.013	0.016	0.017
c. Cast iron			
1. Coated	0.010	0.013	0.014
2. Uncoated	0.011	0.014	0.016
d. Wrought iron			
1. Black	0.012	0.014	0.015
2. Galvanized	0.013	0.016	0.017
e. Corrugated metal			
1. Subdrain	0.017	0.019	0.021
2. Storm drain	0.021	0.024	0.030
2. Nonmetal			
a. Lucite	0.008	0.009	0.010
b. Glass	0.009	0.010	0.013
c. Cement			
1. Neat, surface	0.010	0.011	0.013
2. Mortar	0.011	0.013	0.015
d. Concrete			
1. Culvert, straight and free of debris	0.010	0.011	0.013
2. Culvert with bends, connections and some debris	0.011	0.013	0.014
3. Finished	0.011	0.012	0.014
4. Sewer with manholes, inlet, etc., straight	0.013	0.015	0.017
5. Unfinished, steel form	0.012	0.013	0.014
6. Unfinished, smooth wood form	0.012	0.014	0.016
7. Unfinished, rough wood form	0.015	0.017	0.020
e. Wood			
1. Stave	0.010	0.012	0.014
2. Laminated, treated	0.015	0.017	0.020
f. Clay			
1. Common drainage tile	0.011	0.013	0.017
2. Vitrified sewer	0.011	0.014	0.017
3. Vitrified sewer with manholes, inlet, etc.	0.013	0.015	0.017
4. Vitrified subdrain with open joint	0.014	0.016	0.018
g. Brickwork			
1. Glazed	0.011	0.013	0.015
2. Lined with cement mortar	0.012	0.015	0.017
h. Sanitary sewers coated with sewage slimes, with bends and connections	0.012	0.013	0.016
i. Paved insert, sewer, smooth bottom	0.016	0.019	0.20
j. Rubble masonry, cemented	0.018	0.025	0.030

pier, from (A) in Fig. 4, for $J = 0.04$, $\Delta = 0.07$. From (B) in Fig. 4, for $M = 1.0$, $\sigma = 1.0$ and $\Delta K_p = \Delta K \times \sigma = 0.07$. Pier losses $h_p = \Delta K_p h_v$ are then $0.07 \times V^2/2g = 0.03$ ft.

It was noted during test measurements[4] that pier losses at structures located in or immediately downstream of bends were about double those obtained across similar bridges in straight reaches of canal.

TABLE 6 Values of the Roughness Coefficient n for Lined Channels

Type of channel and description	Minimum	Normal	Maximum
1. Metal			
a. Smooth steel surface			
1. Unpainted	0.011	0.012	0.014
2. Painted	0.012	0.013	0.017
b. Corrugated	0.021	0.025	0.030
2. Nonmetal			
a. Cement			
1. Neat, surface	0.010	0.011	0.013
2. Uncoated	0.011	0.013	0.015
b. Wood			
1. Planed, untreated	0.010	0.012	0.014
2. Planed, creosoted	0.011	0.012	0.015
3. Unplaned	0.011	0.013	0.015
4. Plank with battens	0.012	0.015	0.018
5. Lined with roofing paper	0.010	0.014	0.017
c. Concrete			
1. Trowel finish	0.011	0.013	0.015
2. Float finish	0.013	0.015	0.016
3. Finished, with gravel on bottom	0.015	0.017	0.020
4. Unfinished	0.014	0.017	0.020
5. Gunite, good section	0.016	0.019	0.023
6. Gunite, wavy section	0.018	0.022	0.025
7. On good excavated rock	0.017	0.020	
8. On irregular excavated rock	0.022	0.027	
d. Concrete bottom float finished with sides of			
1. Dressed stone in mortar	0.015	0.017	0.020
2. Random stone in mortar	0.017	0.020	0.024
3. Cement rubble masonry, plastered	0.016	0.020	0.024
4. Cement rubble masonry	0.020	0.025	0.030
5. Dry rubble or riprap	0.020	0.030	0.035
e. Gravel bottom with sides of			
1. Formed concrete	0.017	0.020	0.025
2. Random stone in mortar	0.020	0.023	0.026
3. Dry rubble or riprap	0.023	0.033	0.036
f. Brick			
1. Glazed	0.011	0.013	0.015
2. In cement mortar	0.012	0.015	0.018
g. Masonry			
1. Cemented rubble	0.017	0.025	0.030
2. Dry rubble	0.023	0.032	0.035
h. Dressed ashlar	0.013	0.015	0.017
i. Asphalt			
1. Smooth	0.013	0.013	
2. Rough	0.016	0.016	
j. Vegetal lining	0.030		0.500

14. Bend Losses. Analyses made by the U.S. Bureau of Reclamation[4,5] from the tests described in Art. 12 showed that head losses caused by curves, h_c, could be expressed as a coefficient K_c times the summation of deflection angles in a reach $\Sigma\Delta°$ times the velocity head, $h_c = K_c (\Sigma\Delta°) h_v$.   (2)

It was found that bend losses were approximated when $K_c = 0.001$. The ratio

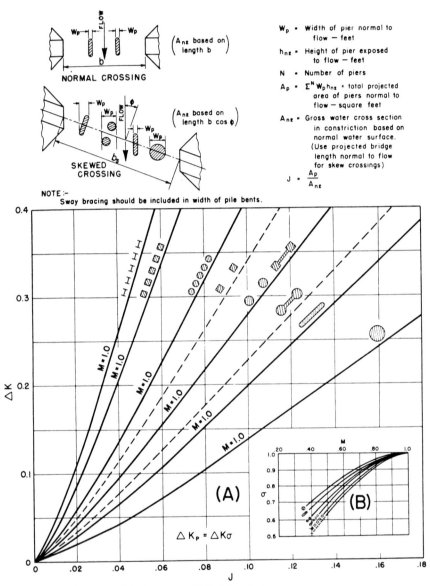

FIGURE 4 Incremental backwater coefficient for piers.

R_1/T of curve radius R_1 to design water surface width T was generally greater than 5 in the canals measured and varied from 4.1 to 22.1.

Using the formula $h_c = 0.001 \ \Sigma\Delta° \ h_v$ for $\Sigma\Delta° = 45°$ and $V = 5$ fps gives $h_c = 0.017$ ft. For four successive deflection angles of 22.5° ($\Sigma\Delta° = 4 \times 22.5° = 90°$) and $V = 5$ fps, $h_c = 0.035$ ft.

A similar order of magnitude for h_c is obtained by using Fig. 5 (Fig. 8 of the paper by Rouse,[7] figure attributed to Hayat[8]). For example, from data for the

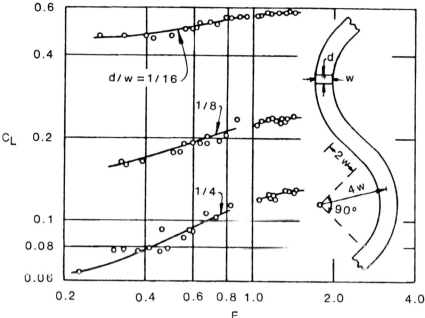

FIGURE 5 Loss at one of a series of channel bends.[7,8]

Gateway Canal from Ref. 4, the depth-to-width ratio d/w is 7/30 (about 1/4) and the Froude number is about 0.33. From Fig. 5 the loss coefficient C_L in one 90° bend is about 0.07 and $C_L h_v = 0.027$, which compares well with the value of 0.035 ft obtained previously for $\Sigma \Delta° = 90°$ by using Eq. (2).

DESIGN OF CANALS

15. General Approach. Earth canals have traditionally been trapezoidal in form, but with modern materials and construction facilities, curved bottoms are possible for specific jobs. Side slopes are determined by stability of the bank materials, on the basis of experience. The dimensions of the channel and its setting in the ground are governed by costs and practical considerations. Hydraulic efficiency is not usually a determining factor. The heights and widths of banks are determined by freeboard and stability requirements. Typical unlined trapezoidal canal sections are shown in Figs. 6 to 9, inclusively.

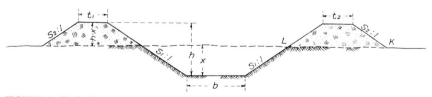

FIGURE 6 Typical canal section.

7.14 CANALS AND CONDUITS

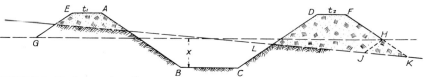

FIGURE 7 Typical canal section.

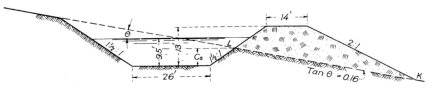

FIGURE 8 Typical canal section.

FIGURE 9 Typical deep-cut canal section.

16. Bank Slopes. The side slopes of cuts and fills not exposed to the action of water must conform to the angle of repose of the materials, with allowance for possible saturation by seepage. The steepest safe slopes are usually most economical. If the slopes within the waterway of an unlined canal are made too steep, the banks will erode or slough. Etcheverry and Harding[9] suggest slopes within unlined waterways as shown in Table 7.

These suggested slopes, based on judgment and experience, have been proved by many years of use and are generally adequate for normal canal work.

TABLE 7 Canal Bank Slopes

For cuts in firm rock	¼:1
For cuts in fissured rock, more or less disintegrated rock, tough hardpan	½:1
For cuts in cemented gravel, stiff clay soils, ordinary hardpan	¾:1
For cuts in firm, gravelly, clay soil, or for side-hill cross section in average loam	1:1
For cuts or fills in average loam or gravelly loam	1½:1
For cuts or fills in loose sandy loam	2:1
For cuts or fills in very sandy soil	3:1
For banks not within the waterway:	
Rock and gravel fills	1¼:1
Fills of average loam, gravelly loam	1½:1
Sandy loam and sandy soil	2:1

Lining protects the banks of the waterway from weathering and from the action of the water in the canal, but for bank slopes steeper than 1:1 or 1.5:1, the lining may have to act as partial retaining wall and, unless specially designed, may fail. Almost any coherent, free-draining material can be maintained on a 1:1 slope if substantially lined.

17. Freeboard. An inadequate friction coefficient, accumulation of sand or silt, the growth of moss or other vegetation, centrifugal force on curves, wave action, increase in flow resulting from error at diversion, or the inflow of storm waters may raise the water level above that computed for a normal flow. Therefore, canal banks must extend above the designed water level to provide a factor of safety. The lower limit for freeboard is usually 1 ft for small canals, and 4 ft is a usual upper limit for a canal having a capacity of 2000 to 3000 cfs. Between these limits, the freeboard may be made about 1 ft plus 25 percent of the depth, with special allowance for unusual conditions.

The top of the lining, in lined canals, is not usually extended for the full height of the bank freeboard. Figure 10 shows the U.S. Bureau of Reclamation recommendations for bank height for canals and freeboard for hard-surface, buried-membrane, and earth linings.[10]

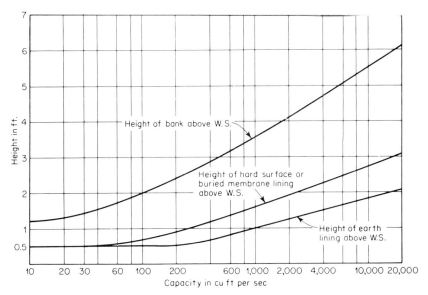

FIGURE 10 Bank height for canals and freeboard for hard-surface, buried membrane, and earth linings. (*Reproduction of Fig. 7, p. 36, "Linings for Irrigation Canals," U.S. Department of Interior, Bureau of Reclamation.*)

18. Top Width and Thickness of Banks. A canal bank must have sufficient thickness and strength to withstand the water pressure against it and to prevent too free an escape of water by seepage. The top width is usually made about equal to the depth of the water with a minimum of 4 ft, or 12 ft if a patrol road is required. If the embankment is to be exposed to the water pressure for a considerable height, it should be widened as required and carefully compacted. For first-class gravelly soil with just sufficient clay to ensure cohesion, the horizontal distance from L to K (Figs. 6, 7, and 8) may be as little as 4 or 5 times the depth of water above L. For less stable soils, this distance should be increased to a ratio of 8 or 10. The bank thickness must be sufficient to avoid piping along the outer toe.

19. Spoil Banks and Berms. Deep cuts may yield more materials than needed for banks. If the excess materials are deposited adjacent to the canal, a level space, or berm, should be provided to protect the waterway from sloughing materials. Berms are also desirable in excavated slopes above the freeboard level. Spoil banks should be regular in form and roughly level. Waste materials may be used to construct an embankment along the uphill side of the canal to exclude cross drainage or, if not needed for this purpose, may be used to reinforce the downhill bank. A typical deep-cut canal section is shown in Fig. 9. Berms usually vary from 5 to 10 ft in width, depending on the height of fills or cuts. Where practicable, the tops of banks should slope away from the canal.

20. Shape and Size of Waterway. In an unlined canal, the area of the waterway is determined by the permissible velocity or by the available slope if the maximum permissible velocity is not to be attained. In a lined canal, the area and velocity are usually determined by the available slope, unless some other condition controls. A steep hydraulic slope and high velocity reduce the size and cost of the canal but may use up head needed for other purposes. Where such factors can be evaluated, the canal velocity may be made such that the cost of the canal, plus the loss due to steepened slopes, is a minimum. In large canals, it may be necessary to limit the depth to avoid the expense of making high banks safe against water pressure or to minimize the danger of a bank failure. For moderate-sized canals, depths in excess of 10 ft are usually avoided, but appreciably greater depths are permissible where required. In rock cut or other firm material, the danger resulting from increased depth is negligible. In unlined canals where topographic conditions make the use of flat hydraulic slopes difficult, wide shallow sections may be used to reduce velocities, but this device should not be carried to an extreme.

If the canal is to be constructed across a nearly level terrain, without restriction as to the relation of water level to ground surface, the banks become lower and hence of smaller volume as the base is widened, decreasing the volume of earthwork. If the width is made extreme, the cost per cubic yard of excavation and the cost for foundation trimming increase, and the cost of the whole project may be greater. A bottom width in excess of 6 times the water depth is seldom justified.

A usual procedure is to choose arbitrarily a ratio of base to depth b/d and find the value of d from the equation

$$d = \sqrt{\frac{A}{b/d + S_s{:}1}} \tag{3}$$

where d = depth of water
  b = bottom width
  A = area of water prism found by dividing the flow by the permissible mean velocity
  $S_s{:}1$ = side slope

The bottom width is usually chosen to the nearest foot or 2 ft, and the depth is adjusted as required.

Equation (3) applies only to trapezoidal sections. Similar equations may be derived for other fixed shapes.

HYDRAULIC COMPUTATIONS

21. Basic Procedures. It will be necessary, as the discussion proceeds, to illustrate hydraulic computations for various types of conduits. Appropriate flow formulas and basic constants are chosen from Sec. 2. Once the required formula is selected, the computations are conceptionally simple.

22. Computation Aids. Tabular aids for fixed-shape conduits, such as circles and horsehoes, appear in Tables 1 and 2, respectively. For canals it is not customary to hold the water prism to an exact fixed shape; hence tabular or diagrammatic aids are sometimes complicated. In any event, their intelligent use is helped by an understanding of the basic procedures. For this reason the following examples will be solved with minimum use of such aids.

23. Example 1. The simplest and most basic problem is the computation of the flow capacity in an existing canal. Assume such a problem with the following information:

1. *Data*
   Trapezoidal, unlined, firm loam, good repair, clear water
   Bottom width b = 12.00 ft
   Flow depth d = 5.60 ft
   Side slopes S_s:1 − 1.5:1
   Hydraulic (friction) slope S = 0.000144
   Manning's n = 0.0225
   Wanted: velocity V and discharge Q

2. *Symbols.* Symbols used in this and in subsequent examples are as follows:
   A = Area of waterway, ft^2
   b = bottom width of canal, ft
   d = water depth, ft
   P = wetted perimeter, ft
   R = hydraulic radius = A/P, ft
   S = hydraulic slope
   V = mean velocity, fps
   n = roughness coefficient in Manning's formula
   Q = discharge, cfs

3. *Canal Properties.* From the specified data, find the physical canal properties as follows:
   $A = (b + 1.5d)d = (12.00 + 8.4)5.6 = 114.24$ ft^2
   $P = b + 2d\sqrt{3.25}$ (for S_s:1 = 1.5:1) = 32.19 ft
   $R = A/P = 3.55$ ft
   $R^{2/3} = 2.33$
   $S^{1/2} = \sqrt{0.000144} = 0.012$

4. *Computing flow from Manning's formula, Sec. 2*
   $V = (1.486/n)R^{2/3}S^{1/2} = 1.85$ fps
   $Q = AV = 114.24 \times 1.85 = 211$ cfs

The flow Q was not specified; hence 211 merely indicates what to expect.

24. Example 2. Assume that the actual flow in the canal of Example 1 is measured and found to be only 200 cfs, indicating n = 0.0225 to be low. From the given data, find the correct value of n. This is a problem of frequent occurrence in investigational work.

7.18 CANALS AND CONDUITS

1. *Procedure.* Find A, R, $R^{2/3}$, and $S^{1/2}$, as in Example 1. Then find the actual value of V by dividing the known flow by A, thus:

$$V = \frac{200}{114.24} = 1.75 \text{ fps}$$

2. *Computation of n.*
   Insert this value of V into the Manning formula, Example 1, leaving n as an unknown, and solve as follows:

$$1.75 = \frac{1.486}{n} R^{2/3} S^{1/2}$$

$$n = \frac{1.486}{1.75} \times 2.33 \times 0.012 = 0.024$$

This is the answer desired.

25. Example 3. Find dimensions and hydraulic slope for a new canal in firm loam, for a flow of 1020 cfs, at maximum permissible velocity.

1. *Data and Assumptions.* From Table 3, Art. 6, the maximum allowable velocity in firm loam is $V = 2.50$ fps. From Sec. 2, assume Manning's formula, with $n = 0.0225$. From Art. 16, assume side slopes of 1.5:1. Assume a desirable bottom width-to-depth ratio of $b/d = 4$ with b restricted to the nearest foot.

2. *Canal Properties.* Compute the physical properties of the canal as follows:

$$A = \frac{Q}{V} = \frac{1020}{2.50} = 408 \text{ ft}^2$$

From Eq. (3) with $b/d = 4$,

$$d = \sqrt{\frac{A}{b/d + S:1}} = \sqrt{\frac{408}{5.5}} = 8.61 \text{ ft}$$

$$b = 4d = 34.44 \text{ ft}$$

3. *Computation of d.* Rounding b out to the nearest foot (34 ft), the resulting flow area = $1.5d^2 + 34d = 408$, giving $d = 8.68$ ft.

4. *Computation of S.* The computations for S are as follows:

$$P = 34 + 2 \times 8.68\sqrt{3.25} = 65.30 \text{ ft}$$

$$R = \frac{A}{P} = \frac{408}{65.3} = 6.25$$

$$R^{2/3} = 3.39$$

With these values and the specified value of n, Manning's formula yields

$$2.5 = \frac{1.486}{0.0225} \times 3.39 \times S^{1/2}$$

from which
$$S = 0.000125$$

26. Example 4. As a further illustration, assume that in Example 3 the terrain makes it impossible or impracticable to use a slope flatter than 0.00015 without relocation or costly checks. Investigate the possibility of velocity control by a change in canal shape.

Enter Manning's equation with the specified values of n and S, leaving R unknown, thus:

$$2.50 = \frac{1.486}{0.0225} R^{2/3} \sqrt{0.00015}$$

from which
$$R^{2/3} = 3.09$$
$$R = 5.43$$

The wetted perimeter is
$$P = \frac{A}{R} = \frac{408}{5.43} = 75.14$$

From the equations area $= bd + 1.5d^2 = 408$ and wetted perimeter $= b + 2d\sqrt{3.25} = 75.14$, it is found that $b = 51.05$ and $d = 6.68$ satisfy the values of $V = 2.50$ fps and $S = 0.00015$. This width is excessive. Whether it should be used depends on factors not specified in this example.

27. Example 5. A more usual condition, particularly in lined canals, is to have only the slope and flow prescribed, the velocity and canal dimensions being optional. Assume a flow of 1000 cfs in a trapezoidal lined canal with Manning's $n = 0.014$, side slopes 1.5:1 and hydraulic slope 0.0004.

This condition may be analyzed using the Manning equation in a form convenient for use with a programmable calculator as follows:

$$\frac{1.486}{n} \times S^{1/2}(db + S_s d^2)^{5/3} - Q\left(b + 2d\sqrt{1 + S_s^2}\right)^{2/3} = 0$$

For values of $d = 7.07$ and $b = 14.14$, $Q = 1000$; for values of $d = 7.10$ and $b = 14.00$, $Q = 1001$ and for values of $d = 6.14$ and $b = 20.00$, $Q = 1000$. If the value of n increased to $n = 0.016$, for values of $d = 6.14$ and $b = 20.00$, $Q = 875$ or for values of $d = 6.60$ and $b = 20.00$, $Q = 1000$. These examples do not cover all eventualities but are illustrative of procedures.

THE LINING OF EARTH CANALS

28. Purposes of Lining. A partial list of purposes is

1. To avoid excessive loss of water by seepage
2. To avoid piping through or under banks
3. To provide needed stability

4. To avoid erosion
5. To promote the continued movement of sediments
6. To facilitate cleaning
7. To help in the control of weeds and aquatic growths
8. To reduce flow resistance
9. To avoid waterlogging of adjacent lands
10. To promote economy by a reduction in excavation

29. History and Progress of Lining. The U.S. Bureau of Reclamation and other water-transporting agencies have been working on the problem of effective canal lining at minimum cost since the beginning of the present century, and about 1945 began an intensive study, giving particular attention to newly developed materials and to the cost reductions which would be made possible by improvement in the design and construction of conventional types.

The accomplishment of this study up to 1963 is effectively analyzed in a report entitled "Linings for Irrigation Canals."[10] This report is drawn on freely for essential information in the following discussion of alternative lining materials and procedures.

PORTLAND-CEMENT CONCRETE LINING

30. General. Because of its many desirable qualities, portland-cement concrete has long held a dominant place in the canal-lining field. It meets more of the purposes of lining than any other material so far developed, and meets them more effectively. However, as generally used in the past, it has been relatively expensive, and in this respect it fails to meet the ever-increasing need for low-cost linings. The high cost of concrete is due in part to the inherent cost of the concrete itself, but there are controllable factors, especially applicable to canals, which can be improved. Among these are thickness, reinforcement, placing procedures, and design and construction tolerances.

31. Thickness Requirements. The required thickness of a concrete lining is determined by the purpose to be served and by operating conditions. If seepage control were the only purpose, a relatively thin lining (if uncracked, or if cracks are closed) would suffice.

A brief discussion of thicknesses for use in cement concrete, asphaltic concrete, and shotcrete is given on p. 33 of "Linings for Irrigation Canals,"[10] followed by a diagram of the relation of thickness to canal capacity, derived from Bureau practice. The data on this diagram are transposed into Table 8. The values shown are somewhat arbitrary but are based on long experience under a wide variety of conditions and are valuable as guides.

If surface deterioration from alkali, freezing and thawing, or other cause, or if heavy icing or other extraneous influence is possible, the indicated thicknesses may need to be increased.

32. Minimum Reinforcement. The Bureau of Reclamation has found that, for usual irrigation requirements, nominal reinforcement or none at all provides adequate strength and seepage control and recommends that reinforcement on irrigation canals be used only where required for structural reasons. For thin linings,

TABLE 8 Suggested Thicknesses of Portland-Cement and Asphaltic Concrete Linings[10]

Thickness, in	Flow, cfs
Unreinforced concrete	
2.00	0–200
2.50	200–500
3.00	500–1500
3.50	1500–3500
4.00	Above 3500
Asphaltic concrete	
2.00	0–200
3.25	200–1500
4.00	Above 1500
Reinforced concrete	
3.50	0–500
4.00	500–2000
4.50	Above 2000
Gunite	
1.25	0–100
1.50	100–200
1.75	200–400
2.00	400 up

nominal reinforcement may defer complete collapse by holding cracked segments together.

33. Joints and Grooves in Concrete and Mortar Lining. A concrete or mortar canal lining is subject to swelling and shrinkage under varying temperature and moisture conditions. In concrete lining of normal thickness, swelling is unimportant. For uncracked thin concrete or mortar, swelling can cause buckling. Buckling is generally not a serious problem, and expansion joints usually can be omitted except at junctions with rigid structures. Short of very heavy and costly reinforcement, concrete lining cannot be designed to overcome cracking. Partial control can be secured by contraction joints at proper intervals, with or without light reinforcement.

The use of reinforcement steel for this purpose, except in special cases, such as for high-velocity channels, has been practically abandoned. For continuously placed lining, a weakened-plane-type joint, or "sidewalk" groove, formed in the concrete to a depth of about one-third of the lining thickness, is effective in controlling the spacing and consequently the width of cracks. For a canal perimeter of more than 30 ft, longitudinal as well as transverse joints are advisable. Recommended spacing of transverse grooves in unreinforced concrete varies from 10 to 15 ft, depending on the size of canal and the thickness of lining. Figure 11 and the accompanying table show recommended spacing and groove dimensions. A more detailed discussion of the spacing of grooves and methods of forming them is contained in the Bureau's "Concrete Manual." Similar grooves may be provided in gunite linings. Expansion joints also may be required for gunite linings if placed in cold (less than 50°F) weather. One-inch-wide joints at 100-ft centers have been found to be effective.

For thick concrete linings, particularly in high-velocity channels such as spillway chutes, thicker linings and more substantial joints, designed to permit

CANALS AND CONDUITS

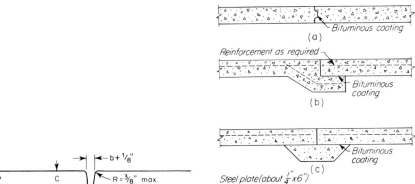

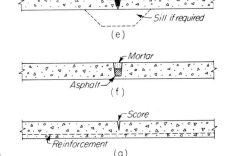

t, inches	b, inches	c, inches	Approximate groove spacing, center to center (feet-inches)
2	1/4 to 3/8	5/8 to 3/4	10-0
2 1/2	1/4 to 3/8	3/4 to 7/8	10-0
3	3/8 to 1/2	1 to 1 1/8	12-0 to 15-0
3 1/2	3/8 to 1/2	1 1/8 to 1 1/4	12-0 to 15-0
4	3/8 to 1/2	1 1/4 to 1 3/8	12-0 to 15-0

Dimensions b and c show allowable tolerance

FIGURE 11 Recommended groove dimensions for unreinforced-concrete canal linings and accompanying table. (*Reproduced from Table 6, p. 35, "Linings for Irrigation Canals," U.S. Department of Interior, Bureau of Reclamation.*)

FIGURE 12 Typical canal-lining joints.

TABLE 9 Statistical Losses for Canals Not Affected by the Rise of Groundwater

Character of material	Loss, ft^3/ft^2 in 24 h
Impervious clay loam	0.25–0.35
Medium clay loam underlain with hardpan at depth of not over 2 to 3 ft below bed	0.35–0.50
Ordinary clay loam, silt soil, or lava-ash loam	0.50–0.75
Gravelly clay loam or sandy clay loam, cemented gravel, sand, and clay	0.75–1.00
Sandy loam	1.00–1.50
Loose sandy soils	1.50–1.75
Gravelly sand soils	2.00–2.50
Porous gravelly soils	2.50–3.00
Very gravelly soils	3.00–6.00

TABLE 10 Seepage from Lined Canals*

Item no.	No. tests	Thickness, in	Range of C_s†	Ave. water depth, ft	Ave. C_s cu. ft/ft.²/day
Unreinforced concrete:					
1	1	3.5		17.2	0.07
2	3	4.0			0.53
Concrete blocks:					
3	1			2.55	0.20
Gunite mortar:					
4	2	1.5		14.0	0.30
Exposed prefabricate membrane:					
5	11	½	0.01–0.53	0.82	0.12
6	3	½	0.05–0.48	0.75	0.20
Buried hot applied asphalt:					
7	1	³⁄₁₆–¼		2.10	0.16
Buried prefabricated organic fiber:					
8	5	⅛	0.03–0.39	0.69	0.16
9	4	⅛–⁵⁄₁₆	0.09–0.54	0.91	0.35
Buried prefabricated asphalt asbestos fiber:					
10	13	³⁄₃₂	0.02–0.13	0.86	0.07
Buried prefabricated asphalt glass fiber:					
11	22	¹⁄₁₆	0.01–1.57	0.76	0.23
Exposed prefabricated asphaltic membrane					
12	17	¼–½	0.01–0.53	0.77	0.16
Thick compacted earth:					
13	1	24–32		3.76	0.08
14	1	24–36		17.2	0.07
15	2	12–24		7.0	0.10
16	1	18		2.79	0.05
Loose earth:					
17	4	12	0.54–1.47	0.68	0.83
Soil cement:					
18	7	3	0.03–0.10	2.20	0.09
Sedimentation					
19	7		0.54–1.06	1.89	0.76

*Condensed and rearranged from Table 4, "Linings for Irrigation Canals," U.S. Bureau of Reclamation.
†C_s = cubic feet per square foot per day.

limited slippage without leakage, may be required. Typical examples are shown in Fig. 12.

SEEPAGE FROM CANALS

34. Need for Information. A canal may require lining for any of many purposes including those listed earlier in Art. 18. The reduction of seepage is usually a prime item, calling for advanced knowledge of seepage losses, before and after lining. The value of water at point of loss is needed, also relative construction costs. The determination of seepage losses may be approached statistically, theoretically, or by direct measurement.

35. Unlined Canals. Seepage loss from an unlined canal depends on the canal's dimensions, the gradation of the materials of which its perimeter is composed, groundwater conditions, and other geological factors.

36. Historical Unlined-Canal Data. A very simple and reasonably reliable procedure for unlined canals is based on historical statistics. An example is shown in Table 9, which was compiled and published by Etcheverry and Harding in 1933.[9] The values shown were carefully compiled from many field measurements and are still usable. However, these simple visual graduations should be used as a guide only in preliminary investigations.

37. Statistical Leakage Data, Lined Canals. In connection with its canal-lining studies, the Bureau of Reclamation has performed many seepage tests on existing and for proposed canals. Selected data from these tests are tabulated in Table 10.

REFERENCES

1. FORTIER and SCOBEY, Permissible Canal Velocities, *Trans. ASCE,* **89**, 940, 1926.
2. STEPHENS, J. C., R. D. BLACKBURN, D. E. SEAMAN, and L. W. WELDON, Flow Retardance by Weeds and Their Control, *Proc. ASCE, J. Irrigation Drainage Div.,* June 1963.
3. CHOW, VEN TE, "Open Channel Hydraulics," McGraw-Hill, New York, 1959.
4. TILP, PAUL J., and MANSIL W. SCRIVNER, "Analyses and Descriptions of Capacity Tests in Large Concrete-lined Canals," Technical Memorandum No. 661, U.S. Bureau of Reclamation, Denver, April 1964.
5. TILP, PAUL J., Capacity Tests in Large Concrete Lined Canals, *Proc. ASCE J. Hydraulic Div.,* May 1965, p. 189.
6. BRADLEY, J. N., Hydraulics of Bridge Waterways, U.S. Department of Transportation/Federal Highway Administration, 2d ed., rev. March 1978, Washington, D.C.
7. ROUSE, HUNTER, Critical Analysis of Open-Channel Resistance, *J. Hydraulics Div., ASCE,* **91** (HY4), July 1965.
8. HAYAT, S., "The Variation of Loss Coefficient with Froude Number in an Open-Channel Bend," thesis presented to the University of Iowa, at Iowa City, in January, 1965, in partial fulfillment of the requirements for the degree of Master of Science.
9. ETCHEVERRY and HARDING, "Irrigation Practice and Engineering, vol. 2, p. 124, McGraw-Hill, New York, 1933.
10. U.S. BUREAU OF RECLAMATION, "Linings for Irrigation Canals," 1963.

SECTION 8

RIVER DIVERSION

By Edwin Paul Swatek, Jr.[1]

GENERAL

1. **Site Limitations.** Construction of a dam in a river channel generally requires that the site be unwatered. The scheme used for the diversion and cutting off of the flow often will rank in importance with other project features and may be a major cost item. In fact, the layout of the principal features of the project may be adjusted to aid in the diversion scheme.

Each site will have some limitations. A stream channel may be narrow and deep, wide and shallow, or some combination of these. Depths of alluvium vary widely, and these variations must be known. Topography and local construction materials are important factors in the selection of a scheme.

If navigation is to be maintained on the river, staging must provide for this, and current velocities in diversion channels will have to be lowered. Upstream fish migration will also affect channel current velocities. Whether the river is a virgin stream or is developed with upstream control will impact the scheme. Concentrations of population and property downstream will be considered in evaluating the risk from damage due to cofferdam overtopping and failure.

2. **Diversion in Narrow and Deep Channels.** *Boulder Dam* is located in deep, narrow Black Canyon. This dictated the use of tunnels driven through the canyon walls on each side to carry the large diversion flows. The upstream tunnel portals were located to provide enough room for the upstream earth cofferdam and slopes for the earth excavation. Another earth cofferdam downstream permitted the complete dewatering of the site to bedrock. By the addition of sloped risers these tunnels also carried the discharge from the permanent spillways.

Mossyrock Dam, on the Cowlitz River in the state of Washington, is also located in a narrow canyon. This is a beautiful site for an arch dam. Overburden in the river bed over rock was more than 200 ft deep. To provide for the slopes for this excavation, which was done without sheeting or cutoff, the diversion tunnels had to be planned long enough for clearance for the cofferdams and the earth slopes for this deep excavation. An extra allowance for room for dewatering sumps and/or wells, and access roads down into the excavation, was also needed. The site had been extensively explored with borings to reveal this depth to rock.

San Lorenzo Dam on the Rio Lempa in El Salvador has a very deep hidden rock channel at the dam site, 80 m depth of overburden below the main river channel. The valley is wide enough to locate the powerhouse and spillway on the right bank terrace out of the main river. It was deemed feasible to construct a

---

[1]Acknowledgment is made to Arthur P. Geuss for material in this section that appeared in the third edition (1969).

30-m-deep slurry trench cutoff through the overburden while work was proceeding on the powerhouse and spillway.

During the first two dry seasons the slurry trench was constructed in two parts through fills placed successively in each half of the river. After this, the river flow was restored to the original channel until the spillway and powerhouse were ready for river diversion and permanent closure. In this way the main channel of the river was never excavated. The slurry trench was not disturbed or eroded by the river flowing over it. During these operations there was some control of the river by the dam upstream.

Wynoochee Dam was built in a small, deep, and narrow canyon in western Washington state. This is a stream of modest flow, so the diversion could be carried through the dewatered canyon in a 13-ft-diameter steel pipe supported partly on the canyon walls (see Fig. 1).

Upstream concrete and sheetpile headworks provided for the cutoff and pipe

FIGURE 1 Wynoochee Dam. Flow at approximately 4000 cfs.

conduit closure gate. The pipe was then embedded in one of the concrete dam monoliths and exited over the downstream earthen cofferdam. The discharge end was hung from overhead beams spanning between the canyon walls. This diversion pipe eliminated the driving of a costly tunnel, thereby saving construction time.

Maximum annual flows are in the order of 5000 cfs. A 4-month shutdown during December, January, February, and March was built into the construction schedule. The 13-ft-diameter bypass pipe could accommodate 3500 cfs, the remaining 8 months' diversion flow.

3. Diversion in Wide Stream Channels. *Caruachi Dam and powerhouse,* owned by CVG-Electrificacion del Caroni (EDELCA), on the Caroni River in Venezuela, was originally sited in a canyon of moderate but restricted width. A great deal of study was made to locate a 12-unit powerhouse, a nine-gated concrete spillway, and a diversion channel occupying the original deep central river gorge. However, to provide adequately for construction clearance and cofferdam layouts the site eventually moved to a location some 6 km farther downstream.

Here the river disgorges onto a flatter landscape. The powerhouse and spillway could be comfortably built on the right bank in a large cofferdam that left 500 m of shallow riverbed on rock on the left bank for first stage river diversion. Closure is to be eventually made against moderate velocity heads by dumping two rockfill dikes to form the toes of the upstream second-stage cofferdam. An embankment dam will be constructed over this left bank diversion channel. Figure 2 shows this layout.

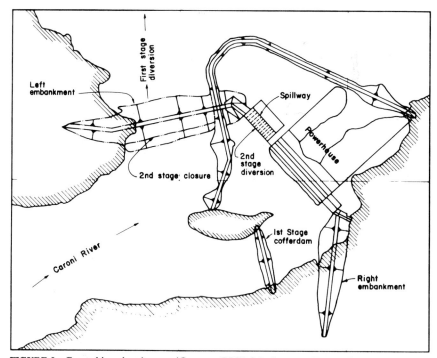

FIGURE 2 Caruachi project layout. (*Courtesy EDELCA, Caracas, Venezuela.*)

A large hydraulic model, built near the site, provided valuable information to establish optimum cofferdam layouts, and heights, closure velocities, spillway and powerhouse tailrace velocity studies, and hydraulic capacity rating curves.

Melvin Price locks and dam located at Alton, Ill., on the Mississippi River have been built principally for an improved navigation system. The original locks and dam, built in the 1930s, have been replaced with a modern high-lift dam 2 mi downstream. In the process of this reconstruction a typical cofferdam layout and staging is shown in Fig. 3.

The first-stage cofferdam, shown in Fig. 3, is on the right Missouri bank. In it six spillway gate bays were constructed, and tied into the Missouri bank. Meanwhile, river traffic and flood flows continued unimpeded through the left channel. The width of this channel was such that high-water velocities did not exceed the capabilities of tows proceeding upstream.

In the second-stage cofferdam the midchannel navigation lock was constructed. Meanwhile, the river was diverted through the five completed gate bays and a reduced-width left-bank navigation channel. The flow through the five gate bays kept velocities reasonably low in the left-bank navigation channel for river traffic.

The third and final stage cofferdam, after completion of the midriver lock, tied the completed lock structure into the left Illinois bank. During this third stage all river traffic was confined to the completed lock. The second lock was built within this third-stage cofferdam. During construction within the third-stage cofferdam all river flow was diverted through the seven completed spillway gates.

Yacyreta Dam project, on the Paraná River in Argentina, is another diversion greatly affected by the local topography. Rio Paraná is one of the largest rivers in South America. High water flows are large, exceeding 30,000 m^3/s; the 50-year flood is 44,000 m^3/s. At Yacyreta dam site, on the lower reaches of the river, the landscape is very flat and the river divides into several channels around islands. Managing such a river on flat topography requires large diversion structures and wide channels to avoid flooding the landscape.

Yacyreta Island stretches some 60 km above the dam site. The right descending channel, the Brazo Ana Cua, is the smaller of the two, carrying one-third of the total flow. Where the 50-km right embankment crosses this channel a gated spillway, closure dike, and bridge features in the first-stage diversion and access to the island (see Fig. 4). Neither channel has an effect on the other, hydraulically.

The main project, consisting of the main spillway, the lock, 2760-MW powerhouse, and embankments, is located across the main, left channel near the foot of Yacyreta Island. The entire complex stretches in a linear arrangement across almost 6 km of river and island. During the first stage the spillway, powerhouse, and lock are located out of the main channel. Thus, navigation was maintained for up to 3 years until closure. Floods did not bother construction, which was protected on each bank by low earthen cofferdams. Cofferdam heights were set to protect against a flood of 37,500 m^3/s in the principal channel; a 50-year event.

Since the spillway is located on the island, out of the main river channel, a 350-m-wide diversion channel had to be excavated upstream and downstream to access the spillway for second-stage diversion (see Fig. 5). After completion of the lock, and with a temporary lower crest spillway ready for diversion, closure of the main river channel was made in an unusual fashion.

The permanent embankment, across the closure portion of the main channel, consists of two dumped rockfill cofferdams forming the upstream and downstream toes of the embankment (see Fig. 6). These had to be placed in deep and

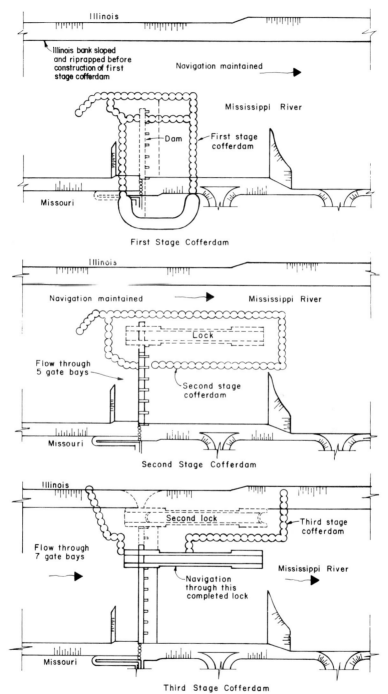

FIGURE 3 Melvin Price locks and dam cofferdam layout and staging.

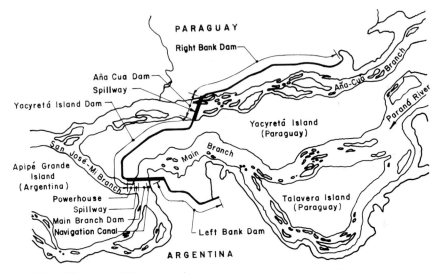

FIGURE 4 Plan view of Yacyreta project.

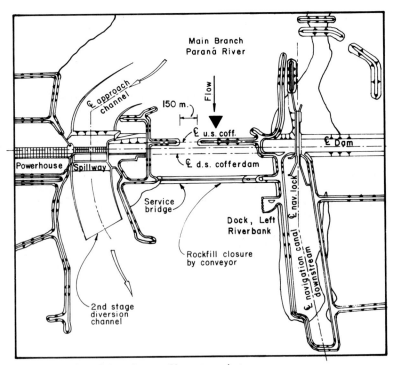

FIGURE 5 Plan of river closure—Yacyreta project.

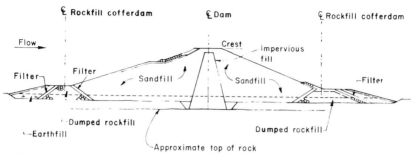

FIGURE 6 Yacyreta project. Section of Brazo principal closure dam.

fast-flowing water against a 2.5-m diversion head. If placed by the end-dump method much of the 20 ft of overburden in the riverbed would have been eroded down to rock. This would greatly increase the volume of rock needed, and the time to complete the diversion. It would also have required large rock in quantities much greater than available at the site.

Extensive model studies were run for different flows up to 30,000 m³/s. The superior scheme proved to be a combination of a submerged level weir placed with a conveyor from a bridge spanning the main channel, 300 m downstream of the main embankment, and an end-dumped fill placed afterward from both sides to form the upstream rock cofferdam.

The trick proved to be to place the submerged level weir first with the conveyor. Maximum rock size needed for this was 0.6 m. This diverted considerable flow to the spillway, reducing the diversion head to 2 m. While this was being topped out, preliminary headway was made advancing the upstream dumped-rockfill cofferdam incrementally from each side.

Finally, with the submerged weir brought to grade, the upstream end-dump rock dike was pinched off in water of lower velocity. The size of rock placed in the upstream cofferdam ranged from 2 to 5 ft in the final closure. This large rock had been stockpiled on each side for the round-the-clock final closure.

The planned total river flow to divert would be 20,400 m³/s, 14,200 m³/s in the principal channel. The actual was less, 8500 m³/s, because of fortunate lower than expected flows.

The access bridge had been constructed early in the project across the main channel. Using this, the contractor employed a 900-meter-long 72-in-wide conveyor belt with a traveling spiller to place rock from the left bank back and forth across the channel. This built up the submerged level weir and maintained it level during the closure of the upstream dike.

The rate of placement achieved was 1500 tons/h. A total 270,000 m³ of rock was placed by the conveyor in about 6 weeks.

Rock Material for Diversion Dikes. The absence of a sound rock quarry, or working with a highly jointed rock, can compromise the planning for a rock dike river closure. Such was the case at Yacyreta and will be in the planning of the *Corpus Dam* upstream. Here, the maximum size rock that can be produced from a local basalt quarry is from 1 m on down. Though perhaps not decisive, this small-size rock will rule out an end-dump river closure in lieu of a conveyor or cableway placed level weir type of closure.

On the other hand, at the *LG2 Dam,* a rock-fill dam in Canada on the *James*

FIGURE 7 LG2 Dam project upstream rock dike river closure.

Bay Project, there was an excellent massive granite quarry on top of the left bank just adjacent to the dam. This provided not only an economical source for the dam fill but also a supply of large-size rock for the upstream closure dike.[1] With an extreme closure head of 23 ft across two 1500 ft dikes a large quantity of 2-m-size rock was needed. In the final closure (see Fig. 7) the rock size could be tailored, up to 40 tons, to the swift-moving currents. Haul trucks of 85 ton capacity helped. This was a case of overpowering the river without adding cost. The total river flow at closure was 122,000 cfs.

Steel sheetpile cellular cofferdams for diversion are an effective tool for building cofferdams and diverting rivers. But large quantities of inexpensive select, clean sand and gravel fill are needed for the cell fill. These materials are not always available locally.

4. Diversion through Concrete Dams—Spillways. Figure 8 shows river diversion through low monoliths at *Chief Joseph Dam* on the Columbia River. Here, five low monoliths in the dam structure built in the first-stage right-bank cofferdam will carry the full flow during the second-stage construction. Note that closure of the left channel is being completed while the final cells of the first-stage upstream are being pulled.

Two rock dikes are used to make the closure against a head of 7 ft, with a river flow of 77,000 cfs. The cells of the upper arm of the second-stage cofferdam on the left will be constructed in the quiet water between the two dikes to tie in to the central common river arm. Fifteen- to 20-ton rocks were needed to make the final 50 ft of closure. These were cabled to an overhead wire line to reduce the loss of rocks swept downstream by the high currents. The closure scheme was modeled in the Bonneville Hydraulic Laboratory.

The sills of the low concrete monoliths should be as low as possible, compatible with the elevation of the riverbed upstream.

Bluestone Dam is a concrete gravity dam on the New River in West Virginia. As with the Chief Joseph Dam, the second-stage diversion was carried through

FIGURE 8 Chief Joseph Dam. Second-stage river closure.

four low 38-ft-wide spillway monoliths. Note in Fig. 9 the manner of closing these low monoliths with steel arch beams and Z-sheetpiles. This could now be improved with fewer beams and stronger sheetpiles. While the closure head was conveniently low, although designed for 50-ft head, it is conceivable to design for a much higher head. The height of the sheetpiles on the final closure is determined by the elevation of the permanent sluices.

The beams and starter sheetpiles connected to the upstream face of the dam were placed before the first-stage cofferdam was breached. Upon the closure of the low monoliths the total river flow was carried through the permanent 5- by 10-in dam sluices.

The permanent spillway sluices and tainter gates at *Caruachi Dam* on the Caroni River (see Fig. 10) are well planned for second-stage diversion and final closure. The low sill of the 5.5-m-wide by 9.0-m-high sluiceways, close to riverbed, aids by reducing the second-stage closure head. Final closure is already incorporated in the permanent spillway sluice and tainter gate arrangement.

5. Diversion through Concrete Dams—Powerhouses. At *Wanapum Dam,* in the Columbia River, second-stage diversion was provided in part through six skeleton powerhouse intake units. These were the upstream portion of an integrated powerhouse unit for future power additions. Stability for the 68-ft-wide by 145-ft-high structure was provided by post-tensioned anchor cables. These intakes were bulkheaded with temporary concrete stop logs.

Figure 11 shows TVA practice for diversion through powerhouse units as used at *Fort Loudoun Dam.*

Figure 12 shows powerhouse diversion cross sections for the *Kainji* project.

FIGURE 9 Bluestone Dam closure of spillway monoliths.

6. Diversion over and around Embankment Dams. *Ord River Dam,* in Australia, was designed so that seasonal flood flows overtopped the partially completed fill. With proper design the partially completed fill can stand a certain degree of overtopping without severe damage or failure. The overtopping flow can be maximized by ensuring uniform depth of flow and by the use of cabled mats or fencing anchored to the downstream face and top of the fill. This may require flattening of the downstream face. In addition, consideration must be given to the differential height of fill and river stage downstream of the fill dam. This type of operation may be feasible during the first season but may entail too much risk of major damage or loss of the completed fill if used during the second or subsequent high-flow seasons.

LG2 Dam is a 525-ft-high rock-fill dam on the *Le Grande River* in Quebec, part of the *James Bay Project.* Many typical river diversion features were incorporated in its construction (Fig. 13).

The 33,000,000 yd^3 dam occupies the entire river canyon. The underground powerhouse, intakes, and penstock tunnels are located off the main dam site in another arm of the reservoir. Two 2540-ft-long unlined diversion tunnels are located on the left bank (see Fig. 14). There horseshoe-shape tunnels are 48.5 ft wide and 59 ft high. They can carry a combined flow of 264,000 cfs, which is the protection offered by the 170-ft-high upstream cofferdam. This is equivalent to a return flood of 64 years. Design maximum velocity in the tunnels is 50 fps. Actual flows in the tunnels were about 10 percent above design due to overbreak.

Figure 15 shows a cross section of the dam. It is a zoned rock-fill dam with an impervious till core. There is sand in the river bottom: about 15 to 25 ft deep

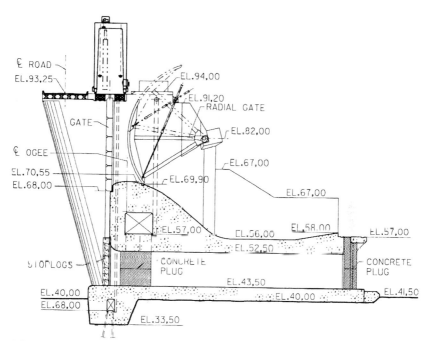

FIGURE 10 Caruachi project. Spillway sluice. (*Courtesy EDELCA, Caracas, Venezuela.*)

across the channel upstream, and as much as 60 ft deep at the downstream cofferdam. The main upstream cofferdam 2 is incorporated in the upstream heel of the main embankment. This cofferdam is founded on rock with an upstream impervious blanket placed in the dry. To place this blanket, a turning dike, cofferdam 1, was used to divert the river through the two tunnels. This permits dewatering the area just downstream of the turning dike, cleaning out to top of rock, and sealing the blanket to rock.

Scheduling the placing of cofferdam 1, diverting the river into the tunnels, is a high-risk operation. Peak floods occur in June. Preferably, advantage is taken of the low-flow fall season, although secondary flood peaks can occur during the autumn.

This turning dike must be completed, the area immediately downstream dewatered, the foundation for the main upstream cofferdam cleaned to rock, and cofferdam 2 built to a secure height before the following spring flood season. At LG2, the closure was planned for November, for a 20-year flood of 80,000 cfs. This would provide comfortable time to raise cofferdam 2 to sufficient height to protect the site during the following spring flood. The closure was started in late June and completed in 1 week against a flow of 122,000 cfs.

Figure 16 shows the scheme of pushing cofferdam 1 from the left bank all the way across the 1500 ft of river to the right bank. Cofferdam 2 would also be advanced, somewhat behind, so as to take advantage of two drops in the total closure head of 23. The only rock quarry available was on top of the left bank adjacent. The depth of water at final closure was 50 ft.

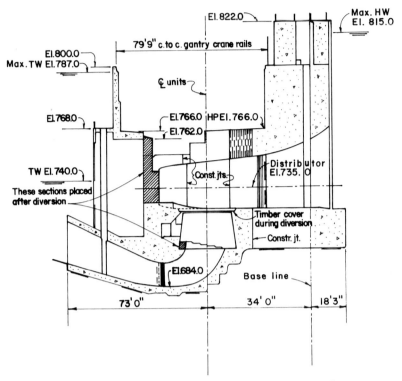

FIGURE 11 Fort Loudoun Dam. Powerhouse unit used for river diversion.

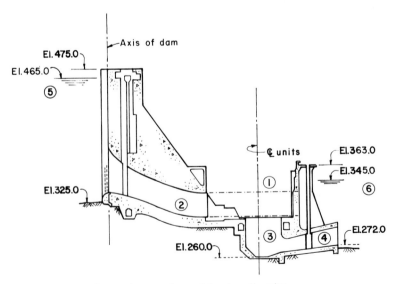

FIGURE 12 Kainji powerhouse units used for river diversion.

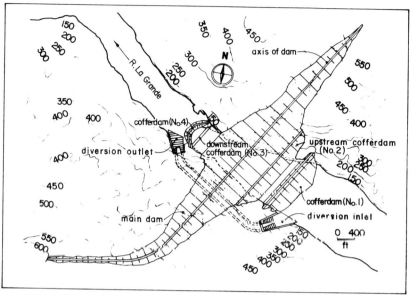

FIGURE 13 LG2 Dam layout.

The head drop just before closure was 23 ft, 17 ft across cofferdam 1 and 6 ft across cofferdam 2. Rock placed in the last 500 ft in cofferdam 1 was of 2 m maximum size; for the last 50 ft up to 3 m size. A few pieces of 100-ton single rocks were finally used. Maximum velocities were up to 30 fps. The maximum height of cofferdam 1 was 60 ft.

The south tunnel had one gate, 48.5 ft wide. This was closed well in advance of the north tunnel, and the tunnel was plugged. The north tunnel had two gates, each 23 ft wide. These were closed 6 months later. They were designed for full head, and the tunnel was later plugged.

DIVERSION DISCHARGE CAPACITIES

Expected river flow pattern, seasonal fluctuations, timing and duration of the low-water season, length of construction period, limitations of the site, and type of diversion schemes are the principal factors which will influence the selection of diversion capacity. Requirements for passage of upstream fish migrants may limit velocities through diversion facilities and require intermediate resting areas. Maintenance of minimum downstream releases for power, navigation, water supply, recreation, or fish may be required after diversion facilities are closed.

Where concrete structures are endangered from overtopping, a return flood of 5 to 10 years may be acceptable. Where cofferdams protect partially completed

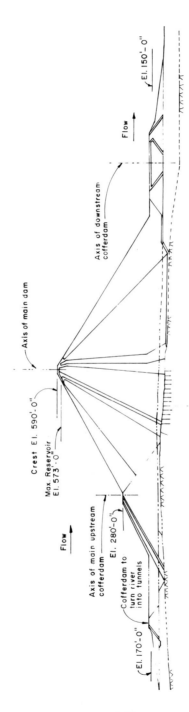

FIGURE 14 LG2 Dam main embankment and cofferdams.

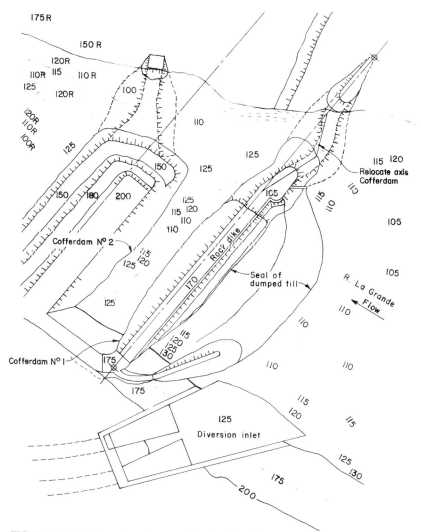

FIGURE 15 LG2 Dam river closure, cofferdams 1 and 2.

major embankment dams, a return flood of 50 years or better is likely indicated. Downstream property damage and loss of life may be involved.

TUNNELS

The *Dworshak Dam* on the North Fork of the Clearwater River in Idaho, is a concrete gravity-type dam having a height of approximately 700 ft. The design of

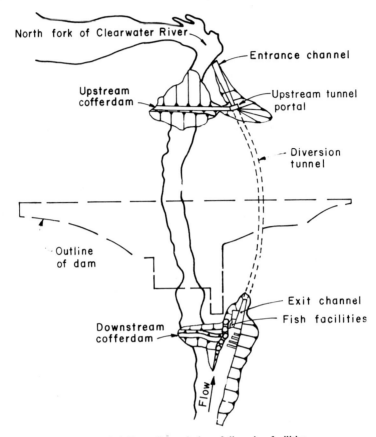

FIGURE 16 Dworshak Dam. General plan of diversion facilities.

the diversion facilities had three basic requirements: (1) to pass a river flow of 68,000 cfs, which is a flood with a 25-year recurrence level; (2) to accommodate the annual log drive, which occurs during the high-water period each spring; and (3) to pass upstream migratory fish during construction.

The characteristics of the valley walls indicated that diversion should be accomplished by the construction of a single tunnel through the left abutment. The general plan is shown by Fig. 16.[2]

It was assumed that logs with a maximum length of 40 ft would be passed between 10,000 and 44,000 cfs with approximately 6 ft of minimum clearance between the water surface and the roof along the centerline of the tunnel. The smooth concrete lining was expected to reduce any erosion resulting from high-velocity flow or abrasion from passing logs. Mean velocities as high as 52 fps were anticipated for the conditions of design flow. In computing the head loss due to friction in the tunnel a value for Manning's n of 0.011 was used for concrete. The pool elevation was computed for the design discharge of 68,000 cfs with an n value of 0.013. It is recognized that a period of service could increase the value of n. Generally, energy dissipation works should be designed for the discharge conditions arising from a low friction coefficient resulting from the most favorable conditions, whereas discharge capacity should be designed for the most unfavorable conditions that will exist after the tunnel has been in service for a time and perhaps roughened by the passage of logs or gravel.

Prototype experience in passing large floods through both concrete-lined and unlined diversion tunnels has revealed that average discharge velocities ranging from 50 to 75 fps have resulted in little or no damage to the contact surfaces.

The *Dworshak Dam diversion tunnel* had a portal width of 35 ft. The gate slots were 7 ft wide; the sill was flat with the invert of the tunnel and approach floor. The tunnel closure gate consisted of structural steel boxes stacked one on top of another. As each was placed in the gate slots it was filled with concrete. The flat bottom of the lowest box had a full neoprene gasket on the bottom in full contact with the tunnel floor.

Because of the bottom neoprene gasket seal the engineers directed that the sill surface under the gate closure be dewatered before closure and the abraded surface repaired to a tolerance of 1/16 in. It was also required that the gate slots be widened slightly by chipping the upstream face of the slot.

Figure 17 shows how the bottom surface and side walls of the slots were dewatered. The river flow of 4000 cfs was passed over and through a 40-ton structural steel cofferdam that was lowered by two cranes to the floor. There could be no diversion of the river flow. The U-shape box was guided down in the slots. The edges of the box in contact with the tunnel walls were sealed. The soft rubber glued to the conveyor belting seal was squeezed by the water pressure against the rough concrete wall, and a tight seal was secured. Very little water leaked into the cofferdam. When ready to remove, a strain was taken by the cranes, water was pumped into the cofferdam, and the frame was pulled out.

Quality of workmanship was an important factor in designing the diversion tunnels for the 400-ft-high *Mangla Dam,* an embankment-type structure in the Jhelum River in West Pakistan.

A study of recorded discharge of the River Jhelum disclosed that flood peaks in excess of 1 million cfs could occur during the months of July and August. It was determined that a flood of 1,240,000 cfs, having a return period of approximately 73 years, could be passed by five 30-ft-diameter diversion tunnels, each about 1900 ft long, located in the left abutment, with a closure dam level of 1080 (approximately 65 ft above the stream bed) before the cofferdam would be overtopped.[3,4]

Surprisingly high velocities have been passed without damage by unlined tunnels in rock. For example, the 25-ft-diameter horseshoe, 530-ft-long diversion tunnel for the *Mayfield Dam,* a 200-ft-high concrete dam constructed on the Cowlitz River in the state of Washington, passed during the flood of April 7, 1961, a flood of 30,000 cfs. The average area of the tunnel, including overbreak, was about 600 ft^2. The average velocity was approximately 50 fps. The rock through which the tunnel was driven was a sound basalt. No damage was observed.

COFFERDAMS

A cofferdam must be designed as a temporary dam. The type used is governed by the materials available and by foundation and placement problems.

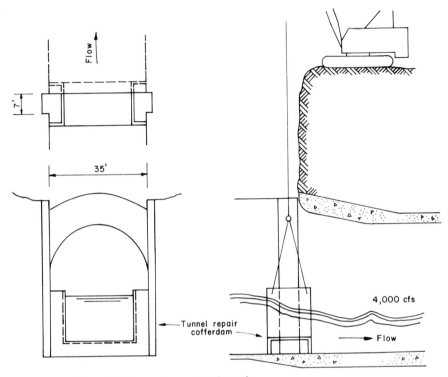

FIGURE 17 Dworshak Dam upstream tunnel portal.

In the past many cofferdams were constructed in the form of timber cribs. Timber cribs are usually floated into place and then filled with rock or cobbles. Watertightness is achieved by the use of wood plank and canvas. Alluvium overlying the top of a rock foundation should be removed prior to placement of the cribs, which are constructed with a bottom that approximately fits the rock surface. Fill or blanketing material is then placed at the upstream heel to reduce leakage through this contact. Earth- or rock-fill cofferdams or filled sheetpile cells are most frequently used at present.

Flat-web steel sheetpiling is most commonly used to construct cellular circular cells which are connected to one another. High interlock strength is required. Filling material is usually gravel, sand, or a combination which has sufficient shear strength. When filled, the pressure of the fill material will develop hoop tension in the interlocks sufficient to provide watertightness and structural integrity. Proportions of the cells are determined by strength of the interlocks, shear resistance of the fill material, the head of the cofferdam, design of the fabricated connectors, and the weight to resist sliding.[5,6] Cellular cofferdams without berms have been built to an 80-ft height. With berms, cellular cofferdams can be built to a height of 120 ft. Figure 18 shows one such cofferdam with a height of 115 ft.

A cellular cofferdam, placed on a hard rock foundation that will not permit driving, can be sealed at the rock contact to prevent piping of fill material. This can be accomplished by having divers place sand cement bags around the inside perimeter at the bottom of the sheets, or by grouting a fill of gravel placed on the

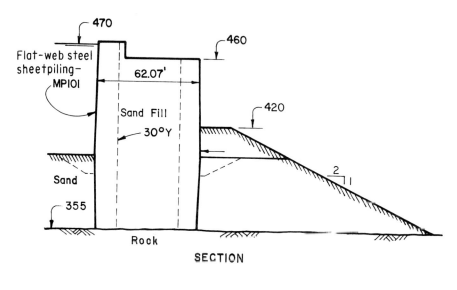

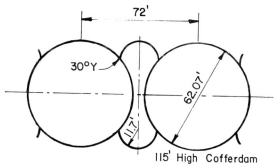

FIGURE 18 Markland powerhouse cellular cofferdam.

rock around the inside of the cell, or by placing a tremie concrete curb, 3 by 3 ft, around the outside of the cell.[7]

Seepage through or under the cofferdam and through or around the abutment will determine the rate and extent of pumping. Good contact with rock or to an impervious stratum along the outside (wet side) sheetpiles is essential to good water control. The large cofferdam for Melvin Price locks and dam on the Mississippi River required a pumping capacity of 90,000 gpm. This cofferdam was founded on a deep sand alluvium.

Crib and cellular cofferdam are usually protected against overtopping by placement of large stone or rock on top of the fill. A concrete cap on cells offers good protection. In fact, if adequately tied to the sheetpiles, it can accommodate a considerable overtopping. This gives superior survival for an upstream cellular cofferdam to that of an embankment, and permits the risk of a lower cofferdam in some cases.

Cells can be constructed in currents up to 4 fps with ordinary templates and

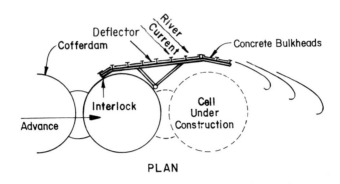

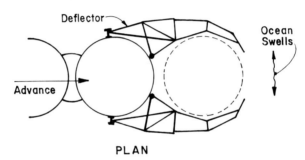

FIGURE 19 Arrangement of deflectors.

spuds. Above this speed special current deflectors have been developed to allow the construction of cells in currents of 15 fps (see Fig. 19).[8] Full-circle cells are generally used for constructing in fast water. Straight-wall diaphragm cells can be used in the river arms where construction is parallel to the current. Figure 20 shows a straight-wall diaphragm cell, with inside berm, of 76-ft height.

Cells have been used in conjunction with embankment cofferdams, especially at the river ends of the upstream arm where fast water is encountered and a current force must be turned. They are also used to parallel a fast watercourse such as a diversion channel. Cells require much less width for a certain height than an embankment cofferdam. This sometimes helps in a river of restricted width.[9]

A straight single row of sheetpiling will serve as a cofferdam, if adequately braced. Figure 21 shows an arrangement used on the first-stage cofferdam at *Lock and Dam 6* on the Arkansas River at Little Rock. This was used on the upstream arm to turn the main river channel into a right-bank diversion channel. First, a dike of quarry-run stone was end-dumped on top of a 5-ft blanket of stone. The stone blanket was placed to protect the riverbed from scour. Using the top of this stone dike, a crane then set and drove the sheetpile to grade. After backfilling with sand, anchors and tie rods were attached to the sheetpiles, with additional fill and stone protection placed. This wall could then withstand overtopping.

Debris should be controlled to minimize the damage of plugging of diversion

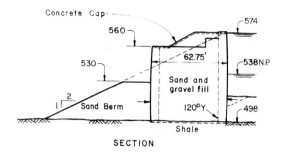
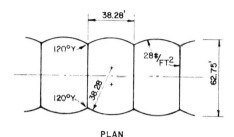

FIGURE 20 Racine locks cofferdam.

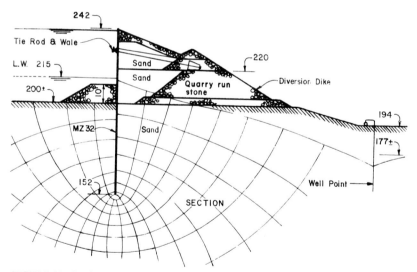

FIGURE 21 Lock and dam 6, Arkansas River, cofferdam and overflow section.

tunnels or conduits. The upstream watershed may produce considerable quantities of debris during high-flow periods, particularly if clearing of the reservoir is in progress. Floating drift and debris can be controlled to some extent by log booms across the river channel. If placed in a reach of the river where velocities are lower, a log boom can be very effective. A log boom will not control debris that is moving along the river bottom.

Figure 22 shows a method of closure in the downstream arm of the second-stage cofferdam at *Millers Ferry lock and dam* across the Alabama River. All but the last cell had been placed, leaving a 70-ft-wide by 20-ft-deep channel to close. With water running at 12 fps, and a rise of 15 ft required to divert the river through the spillway, a heavy box girder was framed to brackets welded on the cells (Fig. 22a). Vertical double W24 soldier beams were driven hard into the chalk rock (Fig. 22b). Two horizontal wales were slid down the upstream face of the W24s. Z-piles were then driven across these wales to make the closure.

Cellular cofferdamming provides flexibility in scheming for diversion of river flows. Refer to Chap. 7 in *Advanced Dam Engineering,* in the References.

(a)

(b)

FIGURE 22 Millers Ferry lock and dam across Alabama River.

REFERENCES

1. Amyot, P., C. Laliberte, and G. S. Larocque, "Meeting Quebec's Power Needs for the Eighties," part 1, Water Power and Dam Construction, July 1976; part 2, August 1976.
2. Pearce, Robert O., "Hydraulic Design of Dworshak Dam Diversion Facilities," *Proceedings of the ASCE, Journal of the Hydraulics Division,* January 1968.
3. Thomas, A. R., and J. R. Gwyther, Diversion of the River Jhelum during Construction of Mangla Dams, Ninth International Congress on Large Dams, Istanbul, 1967, Q.33, R9.
4. Binnie, F. M., et al., "Mangla Dam," Paper 7063, *Proceedings of the Institute of Civil Engineers,* November 1967.
5. Fetzer, Claude A., and E. P. Swatek, Jr., Chap. 7, "Cofferdams," in *Advanced Dam Engineering, for Design, Construction, and Rehabilitation,* edited by Robert B. Jansen, Van Nostrand Reinhold, New York, 1988.
6. Swatek, Edwin Paul, Jr., "Cellular Cofferdam Design and Practice," *Journal of the Waterways and Harbors Division, ASCE,* vol. 93, no. WW3, Proceedings Paper 5398, pp. 109–132, August 1967.
7. Patterson, J. H., "Installation Techniques for Cellular Structures," ASCE Conference on Design and Installation of Pile Foundations and Cellular Structures, Lehigh University, April 1970.
8. "Big Show Now Only Days Away: Stage Set for Big St. Lawrence Diversion," *Engineering News-Record,* Oct. 18, 1956; May 2, 1957.
9. Swatek, Edwin Paul, Jr., Summary of the Day Remarks, "Cellular Structure Design and Installations," Lehigh University Conference on Design and Installation of Pile Foundations and Cellular Structures, pp. 413–423, April 1970.

SECTION 9

CONCRETE DAMS

By Chang-Hua Yeh and Refaat Abdel-Malek[1]

GENERAL FEATURES

1. Types of Dams. There are two types of concrete dams, based on different design philosophies. The first type is the arch dam in which the geometry is carefully designed for each site such that most of the dam is under compression when subjected to water load. A portion of the water load is transferred to the abutments by arch action. This design utilizes the superior compressive strength of the concrete and depends on the geometry of the structure to resist water load. Arch dams are discussed separately in Sec. 11. The second type is the gravity dam and its predecessor, the masonry dam. In this design the water load is resisted principally by the weight of the dam. Water load is transferred entirely to the foundation below, without counting on the abutments to resist it. In gravity dams, most of the dam is also subjected to compression. The magnitude of the stress, however, is usually lower than an arch dam of the same height and is not a governing factor in most designs.

The highest gravity dam in the world is the 935-ft-high (285-m) Grand Dixence Dam in Switzerland (Fig. 1), constructed in 1961. It was also the highest dam of any type in the world at the time of construction. Grand Dixence held this title until 1980 when the 984-ft-high (300-m) earthfill Nurek Dam was constructed in the USSR.

The highest gravity dam in the United States, the Dworshak Dam [719 ft. (219 m)], was completed in 1973 on the North Fork of the Clearwater River in Idaho.

The buttress dam is a variation of the gravity dam. In addition to its own weight, a buttress dam also utilizes the vertical component of the water load as part of the stabilizing force. This principal is demonstrated in Fig. 2. There are several variations, including cored gravity dams, round-head buttress dams, and Ambursen dams. Figures 3, 4, and 5 show typical horizontal sections of these designs.

The design of the buttress dam was motivated by the desire to reduce construction cost by using less concrete while at the same time reducing uplift pressure. The uplift pressure is a major factor in gravity dam design and is discussed in more detail later in this section. The reduction in concrete volume is obvious for buttress dams as compared to a solid gravity dam of the same height, but higher internal stresses are caused, especially in the buttresses. However, the savings due to reduction in concrete are offset by the increasing cost of more formwork and higher strength requirements. In addition, reinforcing steel is needed in a thin buttress dam, further increasing the cost. The relative costs of

---

[1] Acknowledgment is made to Calvin V. Davis, Claudio Marcello, and Edgar H. Burroughs for material of Arts. 11, 12, and 13 that appeared in the third edition (1969) and has been used in this section.

FIGURE 1 Grand Dixence Dam in Switzerland.

concrete, formwork, steel, and labor vary with location. Therefore, the economy of a buttress dam versus a solid gravity dam must be evaluated for each project individually.

There are two other factors to be considered in designing buttress dams. First, if the buttresses are not braced laterally, the dam is susceptible to earthquake load in the direction transverse to the flow. The other factor is that, in severe climates, the solid gravity dam is more resistant to damage because of temperature load than the thin-sectioned buttress dam.

Like buttress dams, hollow gravity dams eliminate redundant concrete, provide massive unreinforced-concrete sections, and have adequate lateral bracing. Form costs are approximately comparable with those of the gravity dams. Uplift pressures are largely eliminated, stress patterns are improved, and, as compared with dams of equal height, safety factors are increased. Construction costs compare favorably with those of the buttress dams.

Arch-gravity dams are a cross between arch and gravity dams, and multiple-

GENERAL FEATURES 9.3

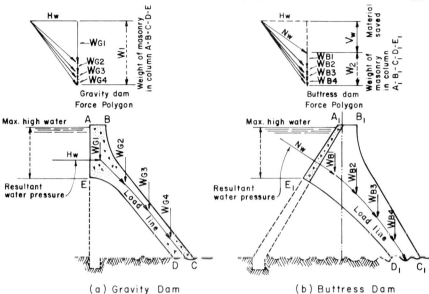

FIGURE 2 Comparison of loading action and quantity of materials in gravity and buttress dams.

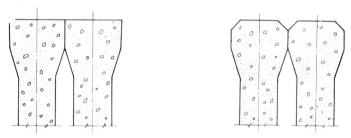

FIGURE 3 Typical cored gravity-type buttresses.

arch dams are a cross between the arch and buttress dams. The arch-gravity dam is usually designed as an arch dam. A landmark arch-gravity dam is probably the 725-ft (221-m) Hoover Dam on the Colorado River (Fig. 6). This dam was constructed in 1936. In a multiple-arch dam, the arch barrels are designed as arch dams, while the stability of the buttresses is checked the same way as a gravity dam. The highest multiple-arch dam in the world is the 702-ft-high (214-m) Daniel Johnson Dam on the Maniconagan River in Quebec, Canada, constructed in 1965. Figure 7 shows a downstream view of this magnificent structure.

2. Features of Gravity Dams. Gravity dams are constructed in blocks called *monoliths*. The width of the monolith varies generally between 50 and 90 ft (15 and 28 m). Each monolith is designed to be independently stable without transferring load to the adjacent monoliths. Because of this feature, there is a great flexibility in dam alignment and project layout. The dam alignment can be straight, curved, bent, or even crooked. The flexibility in layout of a gravity dam facilitates the arrangement of a large number of generating units and multiple

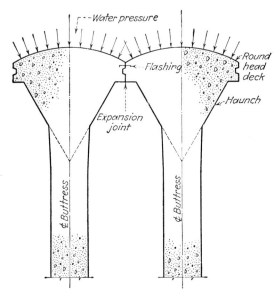

FIGURE 4 Typical round-head buttress type.

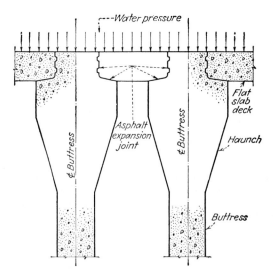

FIGURE 5 Typical Ambursen massive-buttress type.

bays of spillway. It is of interest to note that the three largest hydroelectric projects in the world have either a solid gravity dam or a hollow gravity dam. The Guri Dam on the Caroni River in Venezuela (Fig. 8) and the Grand Coulee Dam on the Columbia River in the United States are solid gravity dams. The Itaipu Dam on the Parana River between Paraguay and Brazil is a hollow gravity dam.

FIGURE 6 Hoover Dam on the Colorado River.

An even larger and still to be constructed project, the Three Gorges Project of China, would also employ a solid gravity dam.

Waterstops are installed in the contraction joints between monoliths. Practice varies with regard to these joints. They can be constructed with or without keys. They can be grouted or left ungrouted. Some older designs such as Fontana Dam in Tennessee, shown in Fig. 9, also have longitudinal joints in the monoliths.

FIGURE 7 Daniel Johnson Dam on the Maniconagan River in Quebec, Canada.

FIGURE 8 Guri Dam on the Caroni River in Venezuela. (*Courtesy CVG-EDELCA, Caracas.*)

These joints were placed for crack control. Advances in temperature control in modern construction have largely eliminated the need for longitudinal joints.

Conventional stability analysis of gravity dams is simple and straightforward. It provides an approximation of the overall stress distribution in the dam and its foundation. Adequate safety factors are selected for the dam, when simple analysis methods are used, in order to ensure stability requirements.

The stresses in the dam are generally low. For a 500-ft (150-m) solid gravity dam under normal reservoir loading, the maximum compressive stress in the dam is about 500 psi (3.4 MPa). Therefore, in most cases, concrete strength is not a governing factor. In modern construction, when large aggregate, up to 6 in (150 mm), is used, the cement content of the dam concrete is usually governed by workability, not by strength requirements. Lean mix also has the advantage of reducing heat of hydration with consequent savings in the cost of temperature and crack control. Usually a rich mix is needed on the surface of the dam: on the upstream face for water tightness, on the downstream face for weather endurance.

DESIGN CONSIDERATIONS

3. Loads. Loads to be considered in the design of the gravity dam consist of the following:

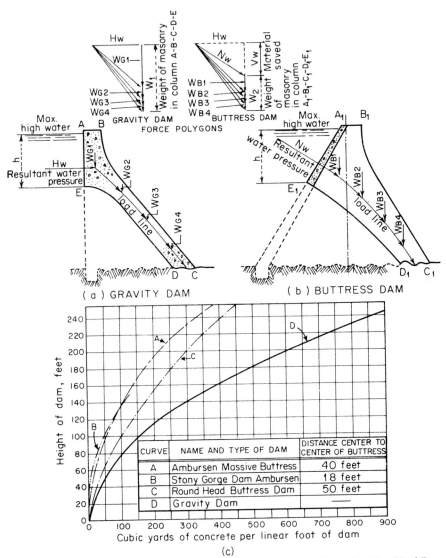

FIGURE 9 Fontana Dam and powerhouse, Little Tennessee River, North Carolina. (*Civil Eng.*, July 1943, p. 306.)

Dead load. This load is primarily the weight of the concrete. Sometimes the weight of major appurtenant structures, such as gates, is also considered for small dams.

Water load. Hydrostatic water pressure is the principal load on the dam. For dams with inclined upstream face, the vertical component of the water load is a stabilizing force.

Silt load. Silt load varies from project to project. It is usually included as an equivalent hydrostatic load with a density of 90 lb/ft^3 (1440 kg/m^3).

Ice load. Ice load is usually important only for small dams. Ice pressure of 5 kip/ft^2 is normally used. In North America, ice load is usually assumed to be between 5 to 10 kip/ft, depending upon the thickness of the ice.

Thermal load. Temperature load is a very important factor in crack control. Usually, it does not affect the stability of dams. Details will be discussed later in this section.

Uplift. Uplift is a very important load on the dam, widely discussed and yet not fully understood. It is commonly assumed to vary linearly from full headwater pressure at the upstream heel to full tailwater pressure at the downstream toe, if there is no drainage curtain. With a drainage curtain, a reduction of pressure between 25 and 50 percent at the drain is usually assumed. A full discussion of this subject is presented later in this section.

Earthquake. Like uplift, earthquake load is important and not fully understood. Traditionally, earthquake load is considered an equivalent horizontal force on the dam. Its magnitude depends on the intensity of the expected earthquake at the project site. The most difficult part of evaluating the earthquake load is determining its effect on the reservoir and the interaction of the reservoir and the dam during earthquake. A separate article is devoted to this subject.

Loading Combinations. The following four loading combinations are normally considered for gravity dam design:

I.	Normal case	Normal high headwater and tailwater with appropriate uplift plus dead, ice, and silt loads
II.	Construction case	Dead load with and without earthquake
III.	Flood case	Headwater and tailwater during flood with appropriate uplift plus dead and silt loads
IV.	Earthquake case	Case I plus earthquake load

4. Method of Analysis. By the late 1960s, the finite-element method was well-developed. General-purpose computer programs have been readily available for analyzing various types of structures including gravity and buttress dams. However, it has not been a general practice to use finite-element analysis in the design of gravity dams. The conventional stability analysis is simple and its results are satisfactory for design purposes in most cases. The finite-element analysis has been used for special studies in the design. Some examples include heterogeneous or difficult foundations, stress concentrations around openings in the dam, stresses due to temperature load, and dynamic analysis for seismic conditions.

The conventional stability analysis of the gravity dam is simple statics. When judiciously applied, this analysis can be extended to include buttress dams. Referring to Fig. 10, the dam is considered to be a rigid body with different loads acting on it. The loads include dead load (F_1), water load (F_2, F_3), silt load (F_4), ice load (F_5), earthquake (F_6, F_7), and uplift (F_8). The resultant of these loads (R) and its location are computed and the dam is checked for overturning and sliding stability.

It should be noted that this analysis is also made for selected horizontal sec-

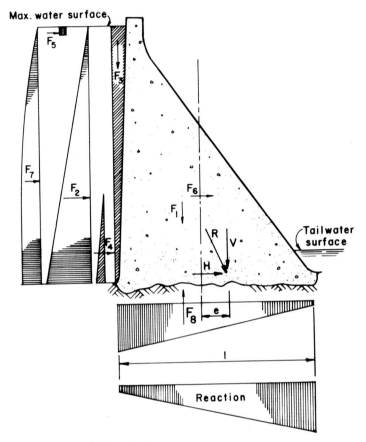

FIGURE 10 Typical gravity dam.

tions through the dam. However, the section at the dam-foundation contact is usually the most critical one.

Overturning Stability. The overturning stability is evaluated by calculating the vertical stresses at the section using the familiar beam-column equation

$$\sigma = \frac{V}{A} \pm \frac{Mc}{I}$$

where $M = Ve$. For a solid gravity dam section of unit width and length l, this equation becomes

$$\sigma = \frac{V}{l} \pm \frac{6M}{l^2}$$

The resulting stress diagram is trapezoid-shaped, as shown in Fig. 10. Thus, this method is sometimes referred to as the *trapezoidal law*. The criteria for evaluating these stresses are discussed below.

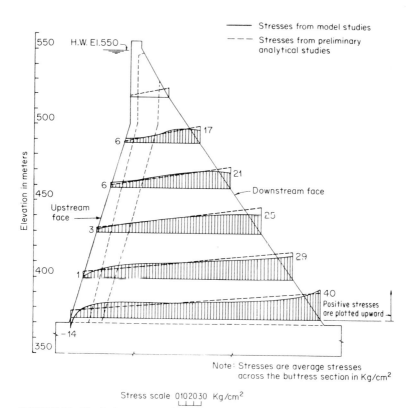

FIGURE 11 Vertical stresses on horizontal planes. (Numbers refer to stresses from model studies at up- and downstream faces, in kg/cm².)

The calculated stresses are approximately correct for sections in the dam away from the foundation, as shown in Fig. 11. In the figure, stresses from model studies are compared with calculated stresses. For a section near the foundation, substantial difference appears between the measured and calculated stress on the upstream side. While the calculations indicate the heel of the dam to be in compression, substantial tension was measured in the model studies. This point is further demonstrated in Fig. 12, which compares the stresses computed by the conventional stability analysis and by a linear elastic finite-element analysis. The maximum compressions for these two cases are reasonably close, 640 psi versus 500 psi. This difference has very little effect on the design, but the difference at the upstream heel invites further discussion.

The tensile stress at the heel, calculated by the finite-element method, is a stress concentration caused by the discontinuity of geometry at that location. It is a localized stress existing in an elastic body such as the model study shown in Fig. 11. In reality the rock beneath the dam is jointed. If the detailed joint formation is known, finite-element analysis can model these joints and calculate a more realistic stress distribution in this area. If this information is not available, a simplified approach can be employed.

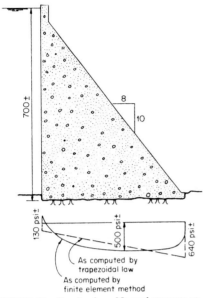

FIGURE 12 Gravity dam. Normal stresses at foundation line σ_x. Load conditions: dead, live, horizontal, and vertical earthquake loads.

A study was made for the stage-constructed Guri Dam.[1] In this analysis, the rock was assumed to be homogeneous and has the same tensile strength as the concrete, 150 psi (1 MPa). It was found that the maximum principal tension occurred in a rock element immediately upstream of the heel. High tension was released by introducing cracks in that element and an iterative analysis was performed until all tensile stresses in the finite-element model were lower than the assumed tensile strength. The final cracked zone and pattern are shown in Fig. 13, and the elastic and final stress distributions are presented in Fig. 14. Also shown in this figure is the stress distribution computed by the conventional stability analysis. It is interesting to note that the overall stress distribution from the simple stability analysis is very close to that calculated by a nonlinear finite-element analysis. However, the con-

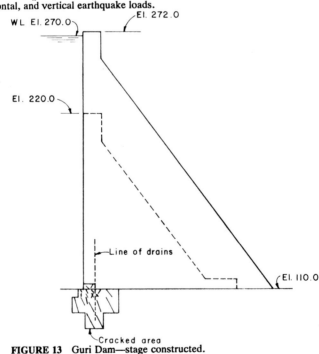

FIGURE 13 Guri Dam—stage constructed.

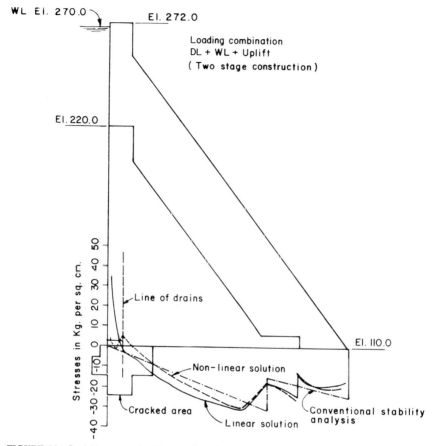

FIGURE 14 Guri Dam—cracked base and analysis.

ventional stability analysis indicated that the heel of the dam was in compression while the finite-element study showed a substantial cracked zone under the dam. More discussions of this subject is provided later in Art 5.

Sliding Stability. The sliding stability is checked by calculating the shear friction factor of safety as

$$Q = \frac{Vf + l'c}{H}$$

where V, H = vertical and horizontal components of resultant force (see Fig. 10)

f, c = friction coefficient and cohesion of section

l' = length of the section under compression as calculated by the stability analysis

Different values of Q are required for different loading combinations to ensure the sliding stability of the dam. The calculation of Q is independent of the shear stress distribution along the section. This approximation is covered by the ample factor of safety required for the design.

5. Criteria. The following criteria are applicable to the results of the conventional stability analysis. For loading combination I, normal case, no tension is allowed on the section. This requirement is sometimes stated differently: the resultant shall fall within the middle third of the section. The safety factor for maximum compression shall be 3, but this condition rarely governs. The sliding factor of safety shall also be 3. These criteria are virtually universally accepted for the normal case.

For other loading combinations, the practice varies. For loading combinations II and III, a safety factor of 2 usually is required for compression and sliding. The safety factor shall be greater than 1 for loading case IV. A certain amount of tension may be allowed for the construction, flood, and earthquake cases.[2] Cracked-base analysis may be sometimes required.[3] In this analysis, horizontal cracking is assumed to occur in a section when vertical stress exceeds the allowable tension. In this case, if all external loads remain unchanged, the magnitude and location of the resultant also remain unchanged. The cracked-base analysis is simply a recalculation of the stress distribution, as shown in Fig. 14. The dam is considered stable if the recalculated compression stays within allowable limits, which is usually the case. The sliding factor of safety has to be based on the uncracked portion of the section ($3l_1$).

The following criteria were published by the Federal Energy Regulatory Commission (FERC) in the late 1980s.[4] FERC requires that (1) no tension is allowed along the dam-foundation contact, usually the most critical section (a crack must be assumed to develop in the area under tension); (2) when a crack is developed, full uplift pressure is assumed in the crack; and (3) when the crack extends beyond the drainline, the drain is assumed to be inoperative. The second and the third requirements change the external loads as the crack develops. The problem becomes a nonlinear one and an iterative procedure has to be used for the analysis. A dam designed prior to the publication of FERC guidelines could become "unstable" under these new criteria.

A well-designed gravity dam is sized such that, under normal reservoir loading, vertical stress at the upstream heel of the dam is, or is close to, zero, as computed by the conventional stability analysis. An example is shown in Fig. 15. Since there is no tension, no crack would develop. Now, as the reservoir level starts to increase gradually during a flood, the stability analysis will show that tensile stress starts to appear at the heel of the dam. A horizontal crack will develop, together with an increase of uplift pressure in the crack, according to the FERC guidelines. An iterative procedure has to be used for the stability analysis. At first the crack will stabilize after a few iterations. As the water level continues to rise, the crack gradually extends downstream. Eventually, the water will rise to a critical level at which a portion of the base remains uncracked and the dam is still theoretically stable. Then, with only a slight increase in reservoir level, the iterations fail to converge, the crack propagates through the entire base, and the dam becomes theoretically unstable. In other words, the stability evaluation can be reversed completely by a slight change of water level in either direction. Careful analysis must be performed to determine the correct application of the criteria to each particular project. Recent research[5] has addressed the question of uplift pressure and cracks. This will be discussed in the next article.

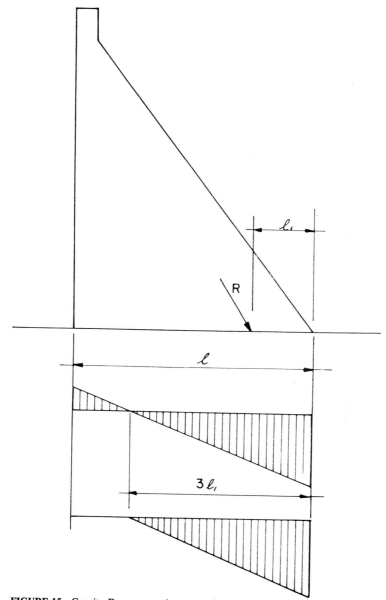

FIGURE 15 Gravity Dam—normal stress at foundation contact.

UPLIFT PRESSURE

6. Conventional Assumption. Uplift pressure is a very important factor in evaluating the stability of the gravity dam. The characteristics of this load are different from other loads acting on the dam. Unlike dead load or reservoir water load, uplift pressure in most cases cannot be determined precisely, particularly during the design phase. Unlike earthquake load, which acts for only a few short durations during the life of the project, if at all, uplift pressure is constantly acting on the dam. Uplift may be reduced by the grout curtain and the drain curtain. In design, common practice is to rely primarily on the drain curtain to reduce uplift pressure. Drainage curtains, when appropriately maintained, can effectively reduce uplift pressures. On the other hand, the effectiveness of grout curtain is very difficult to quantify.

Uplift pressure on existing dams can be measured. For the design of new dams, certain assumptions have to be adopted. Without a drain curtain, uplift pressure is usually assumed to vary linearly from the headwater pressure at the heel to the tailwater pressure at the toe. This assumption is pretty accurate if the foundation is reasonably homogeneous. With a drain curtain, the uplift pressure is reduced as shown in Fig. 16. The amount of reduction is based primarily on experience and the degree of conservatism employed. It varies among different design institutes.

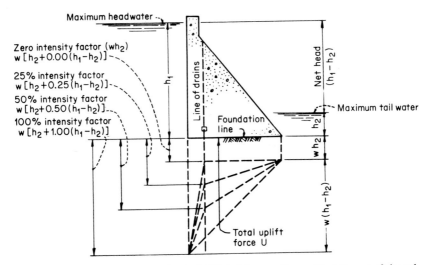

FIGURE 16 Uplift assumptions. (*From Uplift in Masonry Dams, Final Report of the subcommittee on Uplift in Masonry Dams of the Committee Dams of the Power Division 1951, Trans. ASCE, Paper 2531, vol. 117, 1218, 1952.*)

Field measurements have shown that, for well-designed and constructed dams, the actual uplift pressures are lower than those assumed during the design stage. Two examples are shown in Figs. 17 and 18 for Hiwassee Dam and Guri Dam, respectively. The figures show that substantial reduction in the uplift pressure can be achieved by providing drains.

With the development of the finite-element method, it is a simple matter to calculate uplift pressure beneath the dam. The resulting uplift pressures for the Guri Dam calculated by finite-element analysis are presented in Figs. 19 and 20. The calculation assumed a homogeneous permeability coefficient in the foundation rock. If the variation of this coefficient in the foundation is known, it can be easily incorporated in the analysis. If changes in the permeability coefficient in

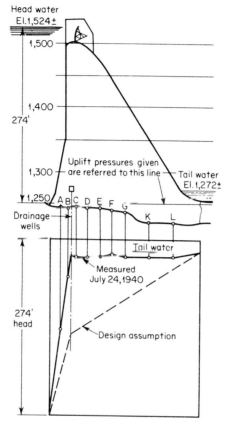

FIGURE 17 Uplift pressures at the base of Hiwassee Dam. (*C.E. Pearce, Design of Hiwassee Dam, Basic Considerations, Civil Eng., June 1940, p. 340.*)

the grouted area are known, then the effectiveness of the grout curtain can be evaluated numerically by using the finite-element method.

7. Uplift Pressure and Cracked-Base Analysis. The most restrictive cracked-base analysis, horizontal crack is assumed to develop if the vertical tension exceeds the tensile strength. Along the base of the dam, the tensile strength between the concrete and the rock is assumed to be zero. Therefore, the crack is assumed to develop in the portion of the base not in compression. Further, it is assumed that full uplift pressure will develop in the crack and an iterative analysis is required. If the crack extends beyond the drain curtain, the drain is assumed to be ineffective (Fig. 21). This is an extremely conservative approach. The additional uplift pressure in the assumed crack could tip the scale between a theoretically stable dam and a theoretically unstable one.[6]

Extensive research has been conducted at the University of Colorado to study the relationship between uplift in postulated cracks and drain holes. This work is still in progress as of 1990. Reference 5 provides a summary of the results to date.

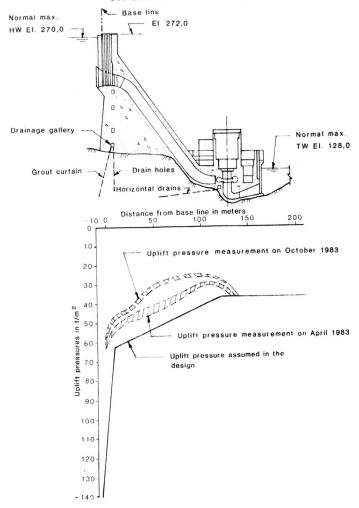

FIGURE 18 Guri Dam, uplift-pressure diagram.

This research uses a combination of laboratory and field tests and mathematical models to investigate the uplift pressure in cracks and the effectiveness of drain holes in relieving the pressure. The results showed that drain holes remain effective in reducing uplift pressure even if a crack that communicates with the headwater extends beyond the drain curtain. The assumption that full uplift would develop in the crack is too conservative and cannot be justified. The uplift pressure is governed by the location and spacing of the drain holes as long as the drain holes are properly cleaned to prevent clogging. This conclusion is intuitively correct, since a drain hole of the size of 150 mm should be effective in intercepting the flow in cracks with thickness in the order of millimeters or a fraction of a millimeter.

8. Uplift Reduction and Buttress Dams. When the concept of buttress dams was developed in the late nineteenth century, the reduction of uplift pressure was

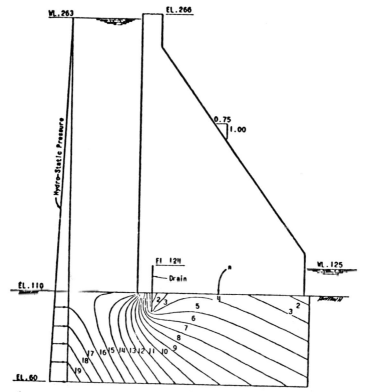

FIGURE 19 Guri Dam raising—equal pressure lines.

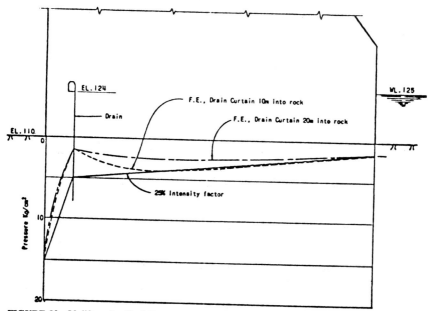

FIGURE 20 Uplift under Guri Dam.

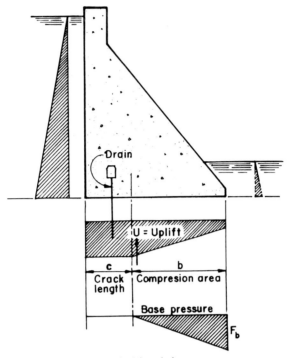

FIGURE 21 Uplift-cracked foundation.

probably not considered. It was recognized later that the spaces between buttresses provide positive relief of uplift pressure beneath the dam. This principle applies to buttress dams of all types and to hollow gravity dams. An interesting example is the 110-m-high Albinga Dam[7] in Switzerland, shown in Fig. 22. The construction joints between the 20-m monoliths were built as openings 5 m wide. Drain holes were drilled in the foundation in the openings and from the galleries and uplift pressures were monitored. Most measurements have shown little or no uplift pressure, demonstrating that the openings are effective in relieving the pressure. The need to reduce pressure, however, does not dictate the selection of dam type; it is determined by the construction cost when all factors are considered.

SEISMIC LOADING

9. Pseudo-Static Method. Up to the 1950s, earthquake design of most gravity dams in the United States consisted of applying an equivalent horizontal load ranging between 0.05 and 0.10 g.[8] This approach was later refined by the development of the seismic zone map. The one commonly used in the United States was developed by the U.S. Army Corps of Engineers.[9] An equivalent horizontal force of up to 0.20 g was specified for high seismic activity areas such as California.

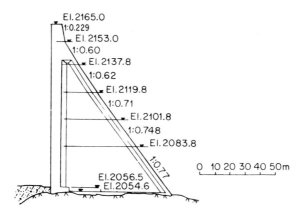

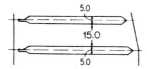

FIGURE 22 Albinga Dam vertical and horizontal cross sections. (*From "Behavior of Large Swiss Dams," p. 116, Swiss National Committee on Large Dams, 1964.*)

In the pseudo-static analysis, the hydrodynamic effect of the reservoir water is represented by added mass based on the Westergaard formulation, which assumes that the dam is rigid and the water is incompressible.

This simple and approximate method appears to be adequate, since no gravity dam has been reported to have failed due to earthquake, although some have been severely damaged. This is at least partly due to the ample factor of safety associated with the method. On the other hand, one can always argue that very few dams have experienced the maximum credible earthquake prescribed by modern seismicity studies. In any case, it is important to have a good understanding of the behavior of such major structures during severe earthquake, utilizing up-to-date techniques.

10. Finite-Element Method. On Dec. 11, 1967, a strong earthquake hit the 103-m-high Koyna Gravity Dam in India, causing severe damage.[10] Horizontal cracks were developed in the upper part of the dam on both upstream and downstream face. The maximum ground acceleration at the site was about 0.5 g. Shortly afterward, extensive research was initiated at the University of California to investigate the seismic behavior of the gravity dam by finite-element analysis. The work was led by Professor A. K. Chopra.

Initial results showed that maximum seismically induced stresses occurred in the upper part of the dam. This was contradictory to the results of conventional stability analysis but was confirmed by field-observed damage. The research continued for more than 10 years and turned out to be the most comprehensive study on the subject of linear elastic earthquake behavior of gravity dams to date. This study considered all factors affecting the linear elastic response of the dam, i.e., flexibility of the dam, compressibility of the water, dam-water-foundation interaction, and energy absorption by the silt on the reservoir bottom. A comprehensive summary of this research can be found in Ref. 11. A computer program was developed[12] to assist designers in evaluating the seismic response of gravity dams.

For preliminary investigation, a simplified analysis procedure was also developed.[13] This hand-calculation procedure computes maximum response due to the fundamental mode of vibration using equivalent lateral forces estimated from the earthquake design spectrum. The effects of dam-water-foundation interaction, compressibility of water, and reservoir bottom absorption are considered together with a procedure to estimate the contribution of higher vibration modes. This provides a simple and practical method for including the most important factors in calculating earthquake response of gravity dams. It produces results accurate enough for preliminary design.

Under a severe earthquake, the response of the dam inevitably lands in the nonlinear region. While the compressive strength of the concrete is rarely exceeded, its tensile strength almost always is. Research has been done to evaluate the nonlinear seismic response of gravity dams.[14,15] As of 1990, its use by the designers is still not common. It is conceivable, however, that nonlinear analysis of gravity dams under seismic load will be common practice at design offices in the near future.

TEMPERATURE AND CRACK CONTROL

11. Cause of Cracking. Cracking in mass concrete structures is not unusual, but it is definitely undesirable. It affects water tightness, durability, and appearance and may affect the ability of the structure to function as designed. Cracking occurs when tension in the concrete exceeds its tensile strength. Tension in a gravity dam may be caused by external load, but, more often, it is the result of restraint against volumetric change in the mass concrete. The main cause of this volumetric change is temperature variation.

There are two kinds of restraints against volumetric change, external and internal. When fresh concrete is placed on the foundation rock, temperature rises due to the heat of hydration of the cement, causing expansion. Since the concrete is still fresh, very little stress is developed. The temperature in the mass concrete usually reaches its peak within a few days. It then starts to drop back to the ambient temperature and the concrete begins to shrink. At this time, the concrete is already hardened. The friction and bond with the rock below prevents free shrinking, causing tensile stresses to develop in the concrete. This is called *external constraint*.

When the ambient temperature falls, the temperature on the surface of the concrete also drops. The relatively warm interior concrete prevents free shrinkage of the surface, causing tension. This is *internal constraint*. Cracks caused by internal restraint can be prevented in large measure by using more durable concrete and by air entrainment. In severe climates, thermal protection, such as insulation, may be needed. The rest of this discussion will be concentrated on problems related to external restraint.

Tensile stress in a concrete block, caused by a temperature drop, can be calculated by the simple equation

$$\sigma = E_{ef}\alpha TR$$

where σ = thermally induced stress
  E_{ef} = effective elastic modulus, a function of the elastic modulus of concrete E_c and the modulus of rock E_r
  $E_{ef} = E_c/(1 + 0.4E_c/E_r)$
  α = coefficient of thermal expansion
  T = temperature drop
  R = restraint factor, a function of block geometry and height above the foundation

R values derived by Carlson and Reading from test data[16] are presented in Fig. 23.
The most effective way of reducing the thermally induced tension and preventing crack development is to limit the amount of temperature drop. Table 1,

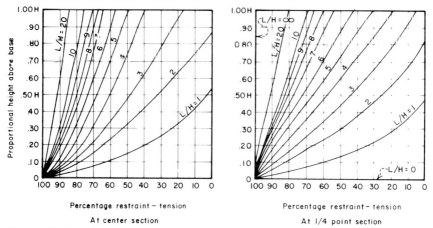

FIGURE 23 Foundation restraint factors.

TABLE 1

Block length	Treatment		
Over 200 ft	Use longitudinal joint. Stagger longitudinal joints in adjoining blocks by minimum of 30 ft		
Shorter blocks:	Maintain temperature drop, maximum concrete temperature to grouting temperature, °F, no greater than:		
	Foundation to $H = 0.2L^*$	$H = 0.2L$ to $0.5L$	Over $H = 0.5L$
150 to 200 ft	25	35	40
120 to 150 ft	30	40	45
90 to 120 ft	35	45	No restrictions
60 to 90 ft	40	No restrictions	No restrictions
Up to 60 ft	45	No restrictions	No restrictions

*H = height above foundation; L = block length.

developed by the U.S. Bureau of Reclamation (USBR) provides a guide for limiting the temperature drop in a concrete block as a function of the block size, based on experience. This table has been used successfully in the design of many dams. However, one phenomenon was not explained by the empirical data. The equation above clearly shows that the thermal stress σ should be independent of the block size, yet experience has demonstrated that more stringent temperature control is needed for larger blocks.

The explanation for this apparent contradiction was found by examining the method used for computing the temperature change. The average peak temperature in a concrete block can be calculated by considering the initial placing temperature, construction sequence, heat of hydration based on the type and amount of cement, ambient temperature, and heat loss through the exposed top and side surfaces. The difference between this average peak temperature and the expected final temperature of the block is the temperature drop to be checked against allowable values in Table 1. This calculation assumes that the foundation rock is adiabatic. In reality, a small amount of heat will be absorbed by the rock, raising its temperature slightly and lowering the temperature in the concrete immediately above the rock as shown in Fig. 24. The amount of heat exchange through a unit surface area of the rock is virtually independent of the size of the concrete block. However, the influence on tensile stress in the concrete due to this drop in concrete temperature at the interface is a function of the block size.

An investigation has been made using finite-element analysis to evaluate the influence of the block size on the induced stresses. First, the temperature distribution in the concrete and the rock was computed using a heat-transfer program.[17] This temperature field was then used as load in a stress calculation employing a plane strain analysis program. The results showed that the maximum

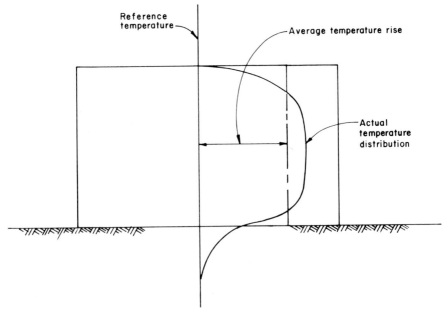

FIGURE 24 Temperature distribution in a concrete block.

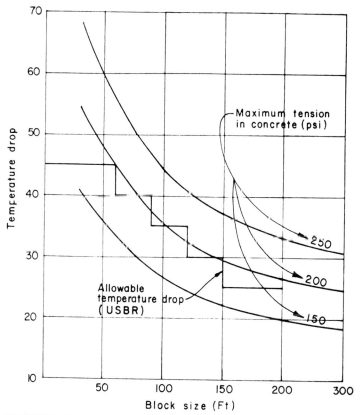

FIGURE 25 Temperature stress versus block size.

tension in a concrete block is indeed a function of its size. These results are presented in Fig. 25, which shows the allowable temperature drop as a function of block size and allowable tension. The allowable temperature drop of Table 1 is also plotted on the figure. This comparison showed that the USBR empirical guideline corresponds to an allowable tension of about 200 psi (1.4 MPa). The closeness of the empirical guideline and the analytical results is striking and very reassuring.

12. Method of Temperature Control. The most effective way to control cracks in mass concrete is to limit the temperature drop, or in other words, to limit the peak temperature reached by the concrete. This can be achieved by reducing the heat-generating cement, reducing the initial placing temperature, and accelerating heat dissipation after the concrete is placed.

Since compressive strength is not a governing factor in most gravity dam designs, the amount of cement may be minimized and a very lean mix used for the interior mass concrete. The mix design is governed by requirements other than strength, such as workability. Using Type II low-heat cement is a common practice for temperature control. In addition, a portion of the cement can be replaced by fly ash or pozzolan to further reduce the heat of hydration.

Reducing placing temperature is an effective method of reducing peak temperature. This can be achieved by precooling. The most commonly used method is to replace water in the mix with ice. Further cooling can be achieved by cooling the coarse aggregate before mixing. The injection of liquid nitrogen in the concrete mix has been tested but it is not a common practice.

Heat dissipation can be accelerated by placing concrete during cool season, if possible, or at night and by placing smaller lifts. Usually, small lifts are placed near the foundation where the external restraint is the greatest. Postcooling by circulating cold water in embedded metal pipe is not an effective way of reducing peak temperature of the concrete. It is used primarily in the construction of arch dams and is discussed in Sec. 10.

ADDITIONAL FEATURES

Simple geometry, individually stable monoliths, and flexibility in project layout are some of the special characteristics of gravity dams. With these characteristics, gravity dams can be designed to fit project needs in ways that are difficult to achieve with other types of dams.

13. Dam Raising. It is sometimes desirable to raise an existing dam to increase its storage capacity. In some other cases, dams are designed to be constructed in stages, that is, to be raised in the future. Because of its simple geometry, a gravity dam is much easier to raise than any other type of dam.

The most prominent achievement in dam raising to date is the Guri Project in Venezuela[1]. The highest monolith of the gravity dam was raised from a height of 110 m to 162 m (Fig. 26). The dam was designed to be constructed in stages. The powerhouse was set away from the first stage dam so that the dam can be raised by adding concrete to its top and downstream face.

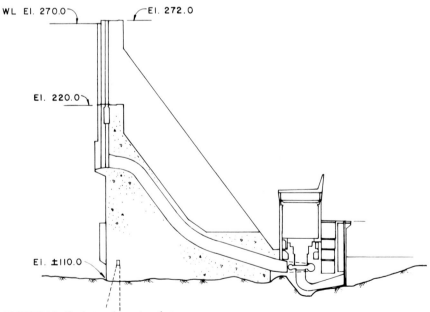

FIGURE 26 Guri staged construction.

ADDITIONAL FEATURES

During the design phase of the project, detailed studies were made to investigate the possibility of raising the dam in one or more stages to its final height. Finite-element analysis were made to study the stress distribution and heat transfer among different parts of the dam as constructed in stages. The dam was finally constructed in two stages and the project was completed in 1987.

Many other dams have been raised to lesser amounts, worldwide. Most of them are gravity-type dams.

14. Adding Powerhouses. The Grand Coulee Dam, constructed in the 1940s, is a straight gravity dam with a spillway in the middle section and two powerhouses, one on each side of the spillway (Fig. 27a). The construction of a third

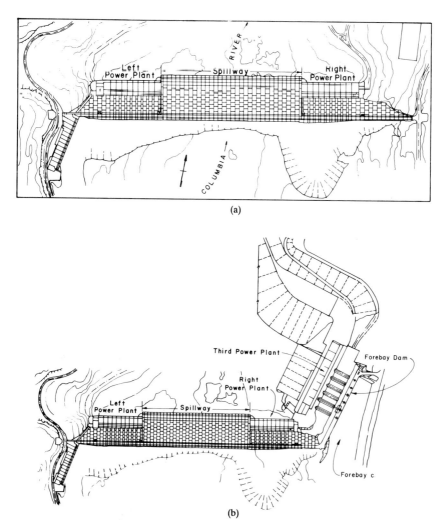

FIGURE 27 Grand Coulee Dam. (a) Original two powerhouses. (b) Construction of third powerhouse.

power plant started in 1967. This included removal of a section of the dam on the right side, excavation of a forebay channel, and construction of a forebay dam and power plant on the right bank (Fig. 27b). The axis of the forebay dam intercepts the axis of the main dam at an angle of about 64°. All construction work was done in the dry. The forebay channel was flooded after the forebay dam was completed.

The Guri powerhouse no. 2 was constructed in a similar manner, with provision for extending to powerhouse no. 3 in the future. This is another example of the flexibility of gravity dam layout. Such arrangements of features and additions are difficult, if not impossible, with other types of dams.

15. Posttensioned Anchorages. In evaluating the stability of existing dams, it is not unusual to find that some of the old designs do not meet current stability requirements. This is because many existing dams were designed with the more liberal criteria of the past or because modern flood loads and earthquake loads are higher than those of the past, or both. In raising existing dams, it is often impossible to provide a corresponding increase in base width, unless the dam was designed to be raised in the first place. In such cases, the stability of the dam may be in question and needs to be strengthened. A commonly used remedial measure to enhance stability of gravity dams is to provide posttensioned anchors. This is a relatively easy task because of the simple geometry of the gravity dams. More detail on this subject can be found in Sec. 11.

16. Roller-Compacted Concrete. In the 1980s, roller-compacted concrete (RCC) gained increasing acceptance worldwide as a rapid and economical method of concrete dam construction. The problems encountered in the early construction of the 1970s, such as leakage, have been largely resolved. RCC can

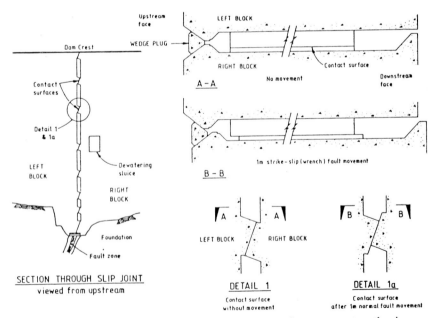

FIGURE 28 Diagram of the slip joint designed to accommodate movement on the river channel fault.[18]

be best adopted in the construction of gravity dams because of simple geometry and large volume. With only a few exceptions, all roller-compacted concrete dams built to date are gravity dams. More details are presented in Sec. 15.

17. Dams Designed for Foundation Movement. The Clyde Dam on the Clutha River, in the South Island of New Zealand, is a concrete gravity dam, 102 m high. The site is close to the active Dunstan Fault. Seismicity studies showed that a major rupture of that fault could induce a 200-mm sympathic movement on the River Channel Fault passing through the dam site.[18] To accommodate this possible movement, the dam was constructed with a slip joint (Fig. 28). The slip face has been aligned parallel to the fault and the joint was designed to move up to 2 m horizontally and 1 m vertically. To keep the joint watertight during normal operation and to limit leakage in the event of fault movement, a wedge plug was constructed on the upstream side of the joint. The plug is held in place by the water pressure of the reservoir. If the joint moves, the plug will move to close the joint. This is a unique and innovative design. Again it is very difficult to achieve for other types of dams.

REFERENCES

1. CHAVARRI, G., A. DEFRIES, W. Y. J. SHIEH, and C. H. YEH, Raising Guri Gravity Dam—Stability and Stress Investigations, 13th International Congress on Large Dams, New Delhi, India, October 1979.
2. "Current United States Practice in the Design and Construction of Arch Dams, Embankment Dams and Concrete Gravity Dams," Joint ASCE-USCOLD Committee, 1967.
3. "Design Criteria for Concrete Arch and Gravity Dams," USBR Engineering Monograph No. 19, 1974.
4. "Guidelines for the Evaluation of Hydropower Projects," Chap. III, Gravity Dams, Federal Energy Regulatory Commission, July 1987.
5. AMADAI, B., T. ILLANGASEKARE, C. CHINNASWAMY, and D. I. MORRIS, Reducing Uplift Pressures in Concrete Gravity Dams, *Water Power and Dam Construction,* February 1990.
6. STELLE, W. W., D. I. RUBIN, and N. J. BUHAC, Stability of Concrete Dam—Case History, *ASCE J. Energy Engineering,* 109 (3), September 1983.
7. "Behavior of Large Swiss Dams," Swiss National Committee on Large Dams, 1964.
8. "Gravity Dam Design," U.S. Army Corps of Engineers, EM 1110-2-2200, Washington, D.C., September 1958.
9. "Earthquake Design and Analysis for Corps of Engineers Projects," U.S. Army Corps of Engineers, ER 1110-2-1806, Washington, D.C., May 1983.
10. CHOPRA, A. K., and P. CHAKRABARTI, The Earthquake Experience at Koyna Dam and Stresses in Concrete Gravity Dam, *Earthquake Engineering and Structural Dynamics,* 1 (2), 1972.
11. FENVES, G., and A. K. CHOPRA, Earthquake Analysis of Concrete Gravity Dams Including Reservoir Bottom Absorption and Dam-Water-Foundation Rock Interaction, *Earthquake Engineering and Structural Dynamics,* 12 (5), September–October 1984.
12. FENVES, G., and A. K. CHOPRA, "EAGD-84: A Computer Program for Earthquake Analysis of Concrete Gravity Dams," report no. UCB/EERC-84/11, Earthquake Engineering Research Center, University of California, Berkeley, August 1984.
13. FENVES, G., and A. K. CHOPRA, Simplified Earthquake Analysis of Concrete Gravity Dams, *ASCE J. of Structural Engineers,* 113 (8), August 1987.

14. FELTRIN, G., D. WEPF, and H. BACHMANN, Seismic Cracking of Concrete Gravity Dams, *Dam Engineering* **1** (4), October 1990.
15. PEKAU, O. A., C. ZHANG, and L. FENG, Seismic Fracture Analysis of Concrete Gravity Dams, *Earthquake Engineering and Structural Dynamics,* **20** (4), April 1991.
16. "Control of Cracking in Mass Concrete," United States Bureau of Reclamation, EM34, October 1965.
17. POLIVKA R. M., and E. L. WILSON, "Finite-Element Analysis of Nonlinear Heat Transfer Problems," report no. SESM 76-2, University of California, June 1976.
18. HATTON, J. W., J. C. BLACK, and P. F. FOSTER, New Zealand's Clyde Power Station, *Water Power and Dam Construction,* December 1987.

SECTION 10

ARCH DAMS

By Roman P. Wengler and Chang-Hua Yeh[1]

An arch dam is a curved dam that carries a major part of its water load horizontally to the abutments by arch action, the part so carried being primarily dependent on the amount of curvature. Massive masonry dams, slightly curved, are usually considered as gravity dams, although some parts of the loads may be carried by arch action. Many early arch dams were built of rubble, ashlar, or cyclopean masonry. However, practically all arch dams constructed during the last 60 years have been built of concrete.

Arch principles have been used in bridges and buildings since about 200 B.C. Apparently, Pontalto Dam, built in Austria in 1611 A.D., was the first arch dam recorded in engineering history.[1] The 64-ft Bear Valley Dam, built in the San Bernardino Mountains of southern California in 1883, was the first arch dam constructed in America. It was followed by the 95-ft Sweetwater Dam, in 1888, and the 88-ft Upper Otay Dam, in 1900, both built near San Diego, California. Lake Chessman Dam, a 236-ft curved gravity dam constructed near Denver in 1904, was the first high dam for which a careful attempt was made to analyze arch action. Since 1904, many arch dams have been built throughout the world.

ARCH-DAM TYPES

Arch dams are sometimes classified on the basis of thickness. Thick arch dams have a base-thickness-to-height ratio of 0.3 or larger, while medium-thickness dams have ratios between 0.3 and 0.2, and thin arch dams, ratios less than 0.2. The ratio is governed primarily by the shape of the valley. Very thin arch dams can be designed only for sites in very narrow valleys. Another way of classification is based on design considerations as explained below.

1. Design Philosophy. Early arch dams were designed as cylindrical shells with a constant radius and a vertical upstream face. These are single-curvature arch dams. The design assumed that water load is carried entirely by individual arches without considering cantilever action. This type of design is actually quite satisfactory for small dams. Since the differential deflections between arches are not considered in the design, tensile stresses may develop in the cantilever direction, causing cracks. However, because the arches are designed to resist all water load, the safety of the dams is generally not affected.

Jorgensen's constant-angle arch dam[2] was a major improvement over the cylindrical arch. This design has gradually decreasing radii from the crest to base such that the central angles are kept nearly constant to maximize arch action at

---

[1]Acknowledgment is made to Ivan E. Houk for material in this section that appeared in the third edition (1969).

all elevations. The cantilever section is slightly curved, resembling modern double-curvature arch dams.

The concept of the double-curvature arch dam was developed after World War II. The arch dam is designed to be curved in both the horizontal and the vertical planes. The vertical sections are designed to bulge upstream so that cantilever compressions due to dead load compensate for the cantilever tensions caused by water load. This type of design is very efficient and has been adopted for most modern arch dams.

2. Evolution of Arch Shapes. The geometry of the arch evolved gradually over the years due to better understanding of its behavior under load. The simplest form of arch is defined by two concentric circular arcs forming a uniform thickness arch. Stresses near the crown are relatively uniform through the thickness. Near the abutments, stress distributions become nonuniform because of the presence of bending moments. In order to reduce the bending stress, the thickness of the arch near the abutments must be increased. This can be achieved by reducing the radius of the intrados. The two nonconcentric arcs form an arch with gradually increasing thickness toward the abutments. This type of design is very satisfactory for a narrow valley.

For a wide valley, it is desirable to have an arch with larger curvature (smaller radius) near the crown to fully develop the arch action and smaller curvature (larger radius) near the abutments to reduce the bending moments. This leads to the development of the three-centered arch. In this design, the intrados and the extrados are each defined by three circular arcs with smaller radius at the center and larger radii on each side.

Further development has lead to the parabolic arch and the elliptical arch. The advantage of these last two designs over the three-centered arch is that the curvature changes gradually along the arch instead of a sudden change at the junction of two circular arcs of different radii. There appears relatively little advantage between these two designs.

GENERAL THEORY OF ARCH DAMS

The general theory of arch dams now used in design, whether it be by trial-load or finite-element analysis, constitutes a comparatively recent development in engineering science. However, the mathematical principles, laws of mechanics, and theories of elasticity involved in an arch dam analysis have been known for many years. Summarized arch dam formulas are given later. This section is confined to the general theory of arch dam action and relates primarily to the trial load method.

3. Arch Action Only. Many arch dams have been designed on the theory that all horizontal water loads are carried horizontally to the abutments by arch action and that only the dead load plus the vertical water loads, in the case of a sloping upstream face, are carried vertically to the foundations by cantilever action. In some of the earlier designs, arch thicknesses were determined by the unreliable thin-cylinder formula $t = RP/S$, where t is the thickness of the arch, R is the radius of the upstream face, P is the water load, and S is the allowable concrete stress. In other cases, thicknesses were determined by analyses of elastic arches, formulas being used such as those developed by the late William Cain.[3]

Designs that ignore cantilever action can seldom be considered wholly satisfactory. Actually, with the ready availability of computers in modern times, there is no longer any good reason for ignoring cantilever action. The vertical cantilevers that make up the dam are restrained at the foundation. They must bend until their deflected positions coincide with the deflected positions of the arch elements. Since the cantilever bending can be produced only by the transfer of water load through the cantilever elements to the foundation, the theory that the entire water load is carried horizontally to the abutments by arch action is obviously less accurate than the theory which assumes water load distribution between arches and cantilevers.

4. Cantilever and Arch Action. For many years, the most commonly accepted method of analyzing arch dams assumed that the horizontal water load is divided between the arches and cantilevers so that the calculated arch and cantilever deflections are equal at all conjugate points in all parts of the structure. During the development of this method, and for many years thereafter, the load distribution required to satisfy this criterion was determined by trial. Consequently, the method was called the *trial-load method*. At the present time, the load distribution is often obtained directly through the use of flexibility matrices and computers. However, the analysis is still generally called the trial-load method. After the load distribution between arches and cantilevers is found, whether by trial or directly, the stresses in the arches and cantilevers are calculated and are considered to be the true stresses in the dam.

With the trial-load method, the first analysis is usually made on the assumption that any element can move in a radial direction without being restrained by adjacent elements and without being subjected to tangential or twisting deformations. Since this assumption is inaccurate, the discrepancies must be corrected by subsequent trial-load adjustments which make adequate allowances for tangential shear and twist effects.

The cantilever elements are assumed to be fixed at the foundation and the arch elements fixed at the abutments. However, the rock formations may be moved by loads transferred through the dam and by direct reservoir pressures. Although foundation and abutment materials are probably never uniformly elastic, owing to the presence of cracks, fissures, faults, and bedding planes, their movement may be roughly calculated by elastic formulas and included in the analysis of arch and cantilever deflections. Since the dam is curved, the cantilever elements are vertical slices bounded by vertical radial planes. Arch elements are horizontal slices, with constant vertical thicknesses from abutment to abutment.

5. Basic Assumptions. The basic assumptions usually made in designing arch dams, when using the trail-load method, may be briefly listed as follows:[4]

1. The foundation and abutment rock is homogeneous, isotropic, and uniformly elastic.
2. The concrete is homogeneous, isotropic, and uniformly elastic.
3. The stresses are well within the elastic limit, and Hooke's law applies.
4. Stresses vary as a straight line between the upstream and downstream faces of the dam in both arch and cantilever elements.
5. Plane surfaces in the unloaded structure remain plane after the load is applied.
6. Temperature changes in the arches vary with the horizontal thickness.
7. Temperature strains and stresses are proportional to temperature changes.
8. Effects of flow of concrete and rock materials may be neglected.

9. Tension stresses are relieved by cracking, so that all loads are carried by compressive and shearing stresses in the uncracked portions of the dam.
10. Radial construction joints are grouted or open slots filled, so that the dam acts as a monolith.
11. Vertical shrinkage is completed before the joints are grouted or the slots filled, so that no loads are transferred laterally by vertical arching.

Some of the above simplifying assumptions can be made more general when the finite-element method is used.

LOADS AND ARCH DAMS

Loads on arch dams are essentially the same as loads on gravity dams, except that temperature changes, which usually are not important considerations in straight dams, cause important deflections and stresses in curved dams. The principal dead load is the concrete weight. The principal live load is the reservoir water pressure. Additional loads may be imposed by tailwater pressure, uplift pressure, upward water pressure under overhanging sections, deposition of silt on sloping faces, presence of silt in flood flows, and formation of ice surfaces. Earthquake accelerations cause momentary changes in water pressure and an additional live load due to the inertia of the concrete.

The general subject of forces on dams is treated in Sec. 9. Discussions presented here are confined to additional considerations required in designing arch dams.

6. Uplift Pressure. Uplift pressure seldom has an important bearing on the safety of an arch dam. If no cracking occurs, it can be neglected. If cracking occurs, uplift pressure in the cracks causes increases in downstream deflections, a change in load distribution, and increases in maximum compressive stresses in both arch and cantilever elements. Uplift in horizontal cantilever cracks usually has a greater effect on stress conditions than uplift pressure in vertical arch cracks.[5]

7. Ice Pressure. Ice pressure causes a continuous concentrated load along the arch element at the elevation of the ice. This load is carried partly by arch action and partly by cantilever action. The actual distribution can be determined by a trial-load analysis.

8. Temperature Loads. Temperature changes cause internal forces that move the dam upstream during the summer and downstream during the winter, the former condition working against the reservoir load and the latter with it. Consequently the winter condition is usually the more important in the stress analysis.

Since zero temperature stresses occur at the time of closing the arches, the closures should be made after the setting heat has been developed and dissipated. Most high arch dams are cooled artificially with refrigerated-pipe cooling. This generally allows the designer to make closure by grouting the construction joints at the opportune time. Unless the concrete is artificially cooled,[6] it may be necessary to include some of the setting-heat effects in analyzing temperature stresses.[7]

If closure can be deferred until the setting heat has been fully developed and completely dissipated, the designer may assume that the temperature changes to be considered in the arch analyses will be the reductions from mean annual to minimum concrete temperature expected during full reservoir load. Figure 1 shows the maximum drop in average concrete temperature, below mean annual, which may occur in arches of different thickness. This curve is based on actual observations, but was drawn so as to be well above the average of all actual measurements.

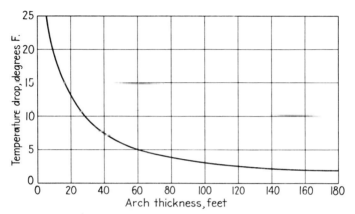

FIGURE 1 Maximum drop in average concrete temperature, below mean annual.

STRESS DISTRIBUTION IN ARCH DAMS

The distribution of stress in an arch dam varies with the horizontal curvature, shape of vertical cross sections, general dimensions of the structure, and uniformity of canyon profile. Pronounced humps in the rock surface cause stress concentrations in adjoining concrete, sometimes resulting in the formation of diagonal cracks. Maximum cantilever stresses often occur at such humps, even at elevations appreciably higher than the base of the maximum cross section.

9. Cantilever Stresses. Maximum cantilever stresses in arch dams, built at sites free from pronounced irregularities, usually occur at the base of the highest cantilever. During full reservoir load, maximum compressive stresses usually occur at the downstream edge of the base, but may occur at the upstream edge in comparatively high and thick dams provided with an upstream batter. Tension often occurs at the upstream edge of the base in relatively thin arch dams and at the downstream face in the upper central portion of the dam.

10. Arch Stresses. Arch stresses in the central and upper portions of arch dams are commonly higher than in the lower portions. Maximum arch stresses usually occur at the crown and abutment sections. At the crown sections, relatively high compressive stresses usually occur at the upstream face of the dam and relatively low compressive or tension stresses at the downstream face. At the abutment sections, stress conditions are usually reversed in accordance with the change in moment sign which generally takes place near the quarter points. Stress conditions at the abutments may be somewhat different in the top arches of long thin dams, owing to the upstream deflections that sometimes occur near such locations. Shearing stresses at the crown section are zero in symmetrical arches symmetrically loaded.

11. Principal Stresses. Major principal stresses along the contact between concrete and rock usually act in planes approximately horizontal at the top of the dam, practically vertical at the base of the maximum cross section, and at gradually varying inclinations along the intervening parts of the profile. The principal stresses for water load only are shown in Fig. 2 for Mossyrock Dam. Mossyrock Dam is a 606-ft-high variable-radius dam located on the Cowlitz River in Washington. These stresses were obtained by trial-load analysis including tangential and twist adjustments.

DESIGN OF ARCH DAMS

The design of an arch dam is a cut-and-try problem. Preliminary plans must be prepared, stresses analyzed, and costs compiled. The best design will have the stresses as uniformly distributed as possible, tension stresses as low as possible, maximum compressive and shear stresses kept within allowable limits, and the total cost of the structure held to a minimum. The following sections briefly discuss technical problems involved in determining the best design.[8] Details of structural features and construction methods are not considered.

12. Allowable Stresses. Stresses in arch dams, analyzed on the assumption of a straight-line distribution of stress, should normally not exceed one-fourth of the mass concrete strength at 1 year as determined on 18- by 36-in cylinders or from correlation with 6- by 12-in cylinders. In recent years, design criteria have allowed higher stresses because of improvements in concrete quality control. Increases up to 33 percent may be permissible, momentarily, during intense earthquake shocks. However, decreases of 25 to 35 percent should be made if the dam is analyzed by approximate methods, such as placing the full water load on the arch elements or bringing the arch and cantilever deflections into agreement at the crown section only.

Ordinarily, vertical tension at the upstream face may be as high as 100 psi without analyzing secondary cantilevers, when the corresponding compression at the downstream face does not exceed 500 psi. Horizontal tension at the upstream face may be as much as one-third the corresponding compression at the downstream face without analyzing secondary arches, when the sum of the tension and compression does not exceed 600 psi.

13. Maximum Stresses. Table 1 gives maximum arch, cantilever, and principal stresses in some arch dams analyzed by trial-load methods. Effects of tangential shear and twist action were included in all cases except Gibson Dam. Effects of rock deformation were considered in all cases.

14. Constants Needed in Analyses. Table 2 gives general values of constants needed in analyzing arch dams. These values may be used in preliminary studies where more accurate information is not available. They should be replaced by data based on field and laboratory measurements before adopting final designs. Tabulated values of modulus of elasticity are for sustained load conditions. Great accuracy in determining elastic properties of canyon rock is not necessary, since effects of foundation and abutment movements are of a secondary nature. The modulus of elasticity for direct stress may be assumed to be the same for tension and compression, for both rock and concrete materials. The modulus for shear

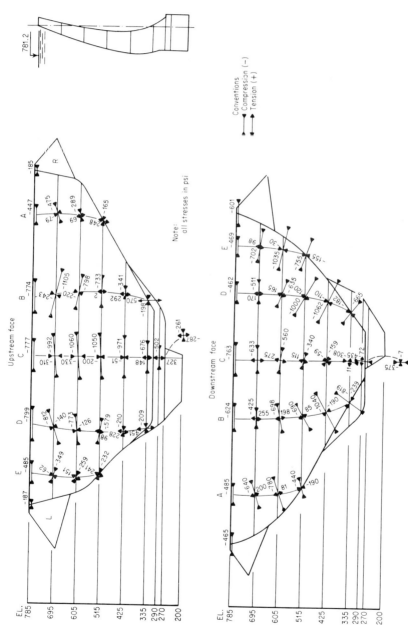

FIGURE 2 Principal stresses in Mossyrock Dam.

TABLE 1 Maximum Stresses in Arch Dams Determined by Trial-Load Analyses, psi

Name of dam	Type[c]	Max height, ft	Cantilever stresses			Arch stresses			Principal stresses[a]			Loading condition
			Comp.	Tens.	Shear	Comp.	Tens.	Shear	Comp.	Tens.	Shear	
Hoover	C.R.	731	565	None	154	231	31	120	565	16	160	Full reservoir, 5 deg subcooling
Owyhee	C.R.	421	358	6	86	294	242	175	413	344	143	W.S. at top of dam[f]
Arrowrock[a]	C.R.	356	466	39	164	305	314	128	496	314	207	Full reservoir with earthquake
Parker	C.R.	335	289	C[d]	65	542	1	76	451	C	95	Full reservoir with earthquake 7 deg subcooling
Ariel	V.R.	313	808	C		560	107	—	—	—	—	W.S. at top of parapet
Horse Mesa	V.R.	305	956	C	250	1061	C	273	1074	C	297	W.S. at top of parapet
Seminoe	C.R.	261	303	8	100	429	193	103	485	193	134	Full reservoir with earthquake, 5 deg subcooling
Mormon Flat	V.R.	229	633	C	74	893	C	181	1095	C	322	W.S. at top of parapet
Stewart Mountain	V.R.	212	862	C	56	625	C	182	990	C	331	W.S. at top of parapet
Gibson[b]	C.R.	199	605	C		364	66	—	—	—	—	W.S. at top of parapet
Deadwood	C.R.	168	472	C		360	184	—	—	—	—	W.S. at top of parapet
Cat Creek	V.R.	118	413	C		277	114	—	—	—	—	W.S. 1.5' above top of dam
Mossyrock	V.R.	606	643	−83		1178	57	—	1062[g]	570[g]	—	Full reservoir
Mayfield	C.R.	245	647	0		726	149	—	—	—	—	Full reservoir
Karadj	V.R.	590	845	131		1003	13	—	1220[g]	356[g]	—	Full reservoir

[a] Analysis for 5-ft increases in height.
[b] Tangential shear and twist not included.
[c] C.R., constant radius, V.R., variable radius.
[d] C, cracked, no load carried by tension.
[e] Along abutment planes.
[f] W.S., water surface.
[g] Water load only.

10.8

TABLE 2 Constants Needed in Analyzing Arch Dams

Constant	Material	Values	Units
Weight, saturated	Concrete	150	pcf
Weight, saturated	Silt	110–120	pcf
Weight, saturated	Sand	110–120	pcf
Temperature coefficient	Concrete	0.0000040–0.0000060	ft/ft · °F
Poisson's ratio	Concrete	0.15–0.22	
Poisson's ratio	Rock	0.10–0.30	
Modulus of elasticity	Concrete	2–3.0 million	psi
Modulus of elasticity	Limestone	1–2 million	psi
Modulus of elasticity	Granite	2–4 million	psi
Modulus of elasticity	Sandstone	1–1.5 million	psi

can be computed by the formula $E_s = E/2(1 + \mu)$, where μ is Poisson's ratio, E is the modulus for direct stress, and E_s is the modulus for shear.

15. Preliminary Plans. In preparing preliminary plans for an arch dam, the engineer should study designs adopted for similar sites, where dimensions and curvature were accurately determined. Published descriptions of constructed dams and data are helpful in preliminary investigations. In order to avoid high-tension stresses at the reservoir face and to secure maximum arch efficiency, central angles should be as large as possible. Theoretical considerations, based on the thin-cylinder formula, show that a central angle of 133° 34′ is most advantageous from the viewpoint of economy.[9] However, practical considerations, together with topographical conditions, usually prevent the adoption of such angles for the lower arch elements.

The extrados and intrados curves should be located so that the ends of the arches converge in a downstream direction. Otherwise radial buttresses at the abutments may be needed to carry the loads transferred horizontally by radial shear. Such buttresses are often necessary where arch elements abut against gravity tangents. Radial arch ends are most desirable; but smaller amounts of convergence usually suffice where radial construction requires excessive excavation, as in large thick dams.

Top widths of arch dams are usually made constant from abutment to abutment. Arch thicknesses at lower elevations may be constant or may increase toward the abutments, depending on stress conditions. Abutment thickening should be warped between adjunct arch elements as the elevation varies, so as to avoid undesirable appearances at the downstream face.

The ratio of length to thickness at the top of the dam should not exceed about 60. Usually the ratio will be smaller, owing to the desirability of stiffening the upper part of the structure or the necessity for providing a roadway along the top. Considerations of slenderness ratio are not important at the lower arches. The additional thicknesses needed from the stress viewpoint, together with the reduced widths of the canyon, will reduce the ratio to satisfactory values. Furthermore, the restraining effect of the cantilevers on the bending of the arch elements increases as the depth below the top increases.

16. Foundations and Abutments. Depths of required excavations must be estimated in determining dimensions for preliminary analyses. Sometimes humps in rock profiles, which may cause stress concentrations, can be removed in preparing rock surfaces. Sometimes deep holes, or relatively narrow gorges, can be plugged with concrete and treated as parts of the foundation instead of parts of the dam. Excavated surfaces should be gradually warped between adjacent elevations, avoiding pronounced stepping along abutment planes. Adequate grouting and draining should always be specified. Geological conditions at the dam site should be approved by competent foundation experts before proceeding with detailed designs.

ANALYSES OF PRELIMINARY PLANS

Analyses of preliminary plans for arch dams are usually made for full reservoir load plus maximum temperature drop. Analyses for other loads, which seldom require major changes in dimensions, can be made after general designs are tentatively adopted. Analyses of preliminary plans may be made by the following simplified methods.

1. Assigning full horizontal loads to arch elements.
2. Dividing horizontal loads between arch and cantilever elements on the basis of a radial adjustment of deflections at the crown section.
3. Dividing horizontal loads between arch and cantilever elements on the basis of radial adjustments at several vertical sections.

The method to be used in a particular case depends on the shape of the canyon and the type, height, and importance of the structure. If the rock profile contains pronounced irregularities, or the shape of the canyon is not symmetrical, the analyses should be made by the trial-load method, listed as 3 above, regardless of the size of the dam. If the canyon is V-shaped, with comparatively uniform sides, and if the dam is of nominal size and importance, the second method may suffice. If the canyon is relatively regular and narrow, and the dam of low height, so that a symmetrical thin arch structure with large central angles can be adopted, the first method may be sufficient. However, with high-speed computers the time required to use method 2 is negligible, and the costs have been reduced sufficiently so that the first method is seldom used.

17. Full Load on Arches. Formulas for analyzing circular arches of constant thickness, under uniform radial loads, have been developed by various engineers. The studies made by William Cain were especially noteworthy.[10] Slightly modified forms of Cain's equations for thrust and moment at the crown and abutment sections, due to uniform water loads, are as follows:

Thrust at crown:

$$H_0 = pr - \frac{pr}{D} 2\varphi \sin \varphi \frac{t^2}{12r^2} \tag{1}$$

Moment at crown:

$$M_0 = -(pr - H_0)r\left(1 - \frac{\sin \varphi}{\varphi}\right) \tag{2}$$

Thrust at abutments:

$$H_a = pr - (pr - H_0) \cos \phi \tag{3}$$

Moment at abutments:

$$M_a = r(pr - H_0)\left(\frac{\sin \varphi}{\varphi} - \cos \varphi\right) \tag{4}$$

In the preceding formulas, r is the radius to the centerline of the arch, p is the normal radial pressure at the centerline, t is the horizontal arch thickness, and φ is the angle between the crown and abutment radii. The centerline pressure p is the extrados pressure times the ratio of the upstream radius to the centerline radius (see Fig. 3).

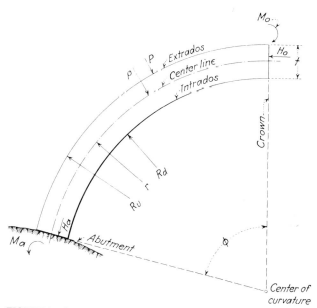

FIGURE 3 Constant-thickness circular arch, fixed at abutments.

If shear is neglected, values of D_S are given by the equation

$$D = \left(1 + \frac{t^2}{12r^2}\right)\varphi\left(\varphi + \frac{\sin 2\varphi}{2}\right) - 2\sin^2\varphi \tag{5}$$

In order to simplify the formulas for crown thrust, D has been used in Eqs. (1) and (8), in lieu of the lengthy right-hand part of Eq. (5), which appears in the original formulas.

When shear is included, D is replaced by D_S, the value of which is given by

$$D_S = \left(1 + \frac{t^2}{12r^2}\right)\varphi\left(\varphi + \frac{\sin 2\varphi}{2}\right) - 2\sin^2\varphi + 3.00\frac{t^2}{12r^2}\varphi\left(\varphi - \frac{\sin 2\varphi}{2}\right) \tag{6}$$

When thrust and moments have been calculated, intrados and extrados stresses may be found by the usual formula

$$S = \frac{H}{t} \pm \frac{6M}{t^2} \tag{7}$$

More complicated formulas, referred to the neutral axis, with water pressures referred to the extrados, were later developed by Cain.[11] Frederick Hall Fowler, using Cain's formulas as a basis, worked out diagrams from which intrados and extrados stresses at the crown and abutment sections may be easily obtained for different values of central angle and ratio of thickness to radius.[12] Philip Cravitz later prepared similar diagrams which included effects of abutment deformations.[13]

For temperature loads, Cain's equations, shear being neglected, are as follows:

$$H_0 = \frac{2\varphi \sin \varphi}{D} \times \frac{Et^3 cT}{12r^2} \tag{8}$$

$$M_0 = H_0 r \left(1 - \frac{\sin \varphi}{\varphi}\right) \tag{9}$$

$$H_a = H_0 \cos \varphi \tag{10}$$

$$M_a = H_0 r \text{ vers } \varphi - M_0 \tag{11}$$

In the preceding formulas, E is the modulus of elasticity, c is the coefficient of thermal expansion, and T is the change in concrete temperature. Other quantities are the same as before. In the preceding equations, the moment of inertia I has been replaced by the quantity $t^3/12$ which applies to rectangular sections. Formulas for temperature thrusts and moments, including shear, are given in the subsequent section on arch analyses.

18. Radial Adjustment at Crown. In analyzing an arch dam by dividing horizontal loads between arch and cantilever elements on the basis of a radial deflection adjustment at the crown section, formulas for cantilever and arch deflections are needed. Since such methods assume the partial water loads on the arch elements to be constant from abutment to abutment, Cain's arch equations may be used. His crown deflection equations for constant thickness, circular arches, slightly modified, are as follows:

Water-load deflection:

$$\Delta = \frac{PR_u r C}{ET} \tag{12}$$

Temperature deflection:

$$\Delta = crTC \tag{13}$$

In these equations, P is the normal radial pressure at the extrados, R is the radius of the extrados, and C is a coefficient depending on r, t, and φ, previously defined. If shear is neglected, C is given by the formula

$$C = \frac{(\varphi - \sin \varphi)(1 - \cos \varphi)}{\left(\varphi + \dfrac{\sin 2\varphi}{2}\right) - \left[\dfrac{1 - \cos 2\varphi}{\varphi\left(1 + \dfrac{t^2}{12r^2}\right)}\right]} \tag{14}$$

When shear is included, C is replaced by C_S, the value of which is given by the formula

$$C_S = \frac{(1 - \cos\varphi)\left[\left(1 + \frac{t^2}{12r^2}\right)(\varphi - \sin\varphi) + \frac{t^2}{4r^2}(\varphi + \sin\varphi)\right]}{\left(1 + \frac{t^2}{12r^2}\right)\left(\varphi + \frac{\sin 2\varphi}{2}\right) - \left(\frac{1 - \cos 2\varphi}{\varphi}\right) + \frac{t^2}{4r^2}\left(\varphi - \frac{\sin 2\varphi}{2}\right)} \quad (15)$$

Figure 4 shows values of C_S for various values of φ and the ratio t/r.[14]

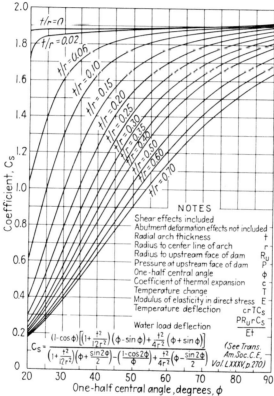

FIGURE 4 Values of C_S, Eq. (15).

Cantilever forces, moments, deflections, and stresses may be calculated by methods described in "Cantilever Analysis," below. If the dam is relatively thin, cantilevers may be considered as vertical slices with parallel sides 1 ft apart. However, they generally should be considered as vertical slices with radial sides 1 ft apart at the upstream face or at a circular vertical plane passing through the upstream edge of the top, herein referred to as the *axis of the dam*. Analyses of cracked cantilevers seldom are necessary in preliminary studies.

In determining the water-load distribution, temperature deflections must be added to water-load deflections in the case of the arch elements, but not in the case of the cantilever elements. The load distribution having been determined, arch stresses may be obtained from the Fowler or Cravitz diagrams. Arch stresses due to temperature changes may be calculated by Eq. (7) after thrusts and moments have been computed by Eqs. (8) to (11).

19. Radial Adjustment at Several Sections. In analysis of an arch dam by adjusting radial deflections at several vertical sections, the division of the water load between the different horizontal and vertical elements can be done by trial and error or directly by use of simultaneous equations. Arch and cantilever stresses are then computed for the final load distribution. The analysis of five or six arch elements and an equal number of cantilever elements usually is sufficient in preliminary studies.

Cantilever elements may be analyzed by methods given later. Arch elements must be analyzed by more complicated methods than those given above, for the loads are not constant along the extrados curves. In preliminary trial-load calculations, arch elements may be analyzed by the voussoir summation process[15] or by theoretical formulas given in the subsequent section on arch analyses. In relatively thick dams, arch analyses should include radial shear effects. Effects of tangential shear, twist, rock deformations, and other secondary influences usually may be omitted in preliminary trial-load computations.

ANALYSES OF ADOPTED PLANS

General plans of arch dams, adopted on the basis of preliminary investigations, should be reanalyzed by the finite-element or detailed trial-load methods, including all important secondary effects. Necessary alterations in dimensions or curvature can then be made before beginning construction. Final analyses should consider all possible load conditions, including earthquake shocks, ice forces, silt pressures, maximum flood stages, and maximum temperature increases, as well as normal full reservoir loads plus maximum temperature reductions. However, special load conditions generally may be analyzed on the basis of radial adjustments of deflections if the trial-load method is used. One complete analysis, including effects of tangential shear and twist action, usually is adequate. Repeated trial-load studies have shown that such effects for the same dam under different conditions of loading are of the same sign and very similar magnitude, unless the change in applied load is sufficient to change the direction of the deflections. Such a change may sometimes occur when maximum temperature increases are used instead of maximum temperature reductions. The consideration of eight or ten arch elements and an equal number of cantilever elements usually is sufficient in the final analyses.

Radial shear buttresses at the ends of the arched section, if needed, are analyzed by methods used for gravity dams. Radial shear forces are added to direct water pressures and are assumed to decrease uniformly from maximum values at the edge of the buttresses, adjoining the arched section, to zero at the opposite edge.

THE TRIAL-LOAD METHOD

The development of the trial-load method was begun by the U.S. Bureau of Reclamation in 1923,[16] about the time that a similar method was being investigated in

Europe.[17] The Bureau's first use of the method included effects of thrust, moment, and temperature in the arch analyses and thrust, moment, and horizontal radial shear in the cantilever elements. The first analyses brought the deflection into adjustment in the radial direction only. The next step in the development was the inclusion of radial shear effects in the arch calculations. Since that time, the method has been gradually amplified so that now the effects of rock deformations, tangential shear, twist action, and other secondary considerations may be included whenever necessary. The introduction of tangential shear and twist effects requires adjustments of deflections in circumferential and angular directions as well as in radial directions.[18]

20. Rock Movements. Considerations of rock movements and their effects on the action of arch dams may be based on approximate formulas.[19] If the ends of the arch elements are vertical and the bases of the cantilever elements are horizontal, rock rotations and deflections of elements with parallel sides 1 ft apart may be calculated by the following equations:

Rotation due to moment:

$$\alpha' = \frac{MK_1}{E_r t^2} \quad (16)$$

Deflection due to thrust:

$$\beta' = \frac{HK_2}{E_r} \quad (17)$$

Deflection due to shear:

$$\gamma' = \frac{VK_3}{E_r} \quad (18)$$

Rotation due to twist:

$$\delta' = \frac{M_t K_4}{E_r t^2} \quad (19)$$

Rotation due to shear:

$$\alpha'' = \frac{VK_5}{E_r t} \quad (20)$$

Deflection due to moment:

$$\gamma'' = \frac{MK_5}{E_r t} \quad (21)$$

where M, V = arch and cantilever moments and shears
H = arch thrust
M_t = cantilever twisting moment
E_r = elastic modulus of the rock
t = radial thickness of the element

$K_1, K_2, K_3, K_4,$ and K_5 are constants depending on Poisson's ratio and the ratio of the average length of the dam b to the average width a. Table 3 gives values of K constants for a Poisson's ratio of 0.20 and different values of b/a.

TABLE 3 Values of K Constants in Eqs. (16) to (21), for Poisson's Ratio = 0.20

Values of b/a	Values of K				
	K_1	K_2	K_3	K_4	K_5
1.0	4.32	0.62	1.02	4.65	0.345
1.5	4.65	0.78	1.23	4.86	0.413
2.0	4.83	0.91	1.39	5.18	0.458
3.0	5.04	1.0	1.60	5.04	0.515
4.0	5.15	1.25	1.77	5.90	0.550
5.0	5.22	1.36	1.89	6.08	0.574
6.0	5.27	1.47	2.00	6.20	0.592
8.0	5.32	1.63	2.17	6.37	0.614
10.0	5.36	1.75	2.31	6.46	0.630
15.0	5.41	1.98	2.55	6.59	0.653
20.0	5.43	2.16	2.72	6.66	0.668

Equations (16), (18), (20), and (21) give movements at the ends of the arch and cantilever elements. Equation (17) gives horizontal movements caused by arch thrusts. Vertical movements at cantilever bases and twist movement at arch abutments are not needed. Equation (19) gives twist movements at cantilever bases. Rotations and deflections given by Eqs. (20) and (21) are of a secondary nature and relatively unimportant.

If pounds, feet, and radians are used as dimensional units, calculated deflections and rotations are feet and radians, respectively. Further discussions of rock movements in trial-load analyses are given in subsequent sections on cantilever and arch analyses.

Cantilever Analysis

In cantilever analyses, the vertical elements are divided into sections by horizontal planes at small increments of height, as shown in Fig. 5. Total loads, shears, and moments, acting on the horizontal planes, are then summed from the top downward to the foundation; and slopes of neutral axis, moment deflections, and shear deflections are summed from the foundation upward to the top, rock deformations being inserted as initial movements in beginning the upward summations. Radial deflections at assumed horizontal planes are then found by adding moment and shear deflections.

In the following discussions, cantilevers with parallel sides are treated from the viewpoint of radial loads. Effects of tangential shear and twist loads on uncracked elements with radial sides are then considered separately.

21. Cantilever with Parallel Sides. For an uncracked cantilever with parallel vertical sides, 1 ft apart, increments of concrete weight, vertical water loads on

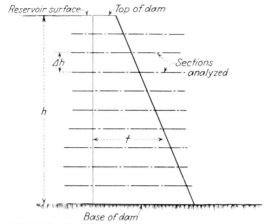

FIGURE 5 Vertical element of arch dam.

the upstream face, where sloped, horizontal water pressure, centers of gravity, shears, moments, and moments of inertia are easily calculated by usual methods. Slopes of the neutral axis, moment deflections, and shear deflections are then obtained by the following summation formulas:

Slope of neutral axis:

$$\frac{dy}{dh} = \alpha' + \alpha'' + \sum \frac{12M}{Et^3} \Delta h \qquad (22)$$

Moment deflection:

$$\Delta_m = \sum \left(\alpha' + \alpha'' + \sum \frac{12M}{Et^3} \Delta h \right) \Delta h \qquad (23)$$

Shear deflection:

$$\Delta_S = \left(\gamma' + \gamma'' + \sum \frac{KV}{tE_S} \Delta h \right) \qquad (24)$$

where M = resultant bending moment
V = total horizontal shear in the radial direction
Δh = increment of height
dy = differential movement in the horizontal radial direction
E = shearing modulus of elasticity
K = constant allowing for nonuniform distribution of shear, usually taken as 1.25

Other quantities are the same as in preceding sections. The moment of inertia I has been replaced by its equivalent $t^3/12$.

Vertical stresses at the faces of the dam may be computed by Eq. (7), the vertical force W being used instead of the horizontal thrust H. Inclined stresses at the edges of the cantilever, acting parallel to the slopes, and unit shearing stresses on horizontal and vertical planes at the edges of the cantilever may be calculated by the formulas

Inclined stresses:

$$S_i = S \sec^2 \alpha - p \tan^2 \alpha \qquad (25)$$

Shearing stress:

$$N = \pm (S - p) \tan \alpha \quad \text{(negative at upstream face)} \qquad (26)$$

where S = vertical stress
 p = water pressure
 α = angle between the face of the dam and the vertical direction

If the face of the dam is vertical, the shearing stress at the edge of the cantilever is zero.

For cantilever with radial sides or for cracked cantilevers and for tangential shear and twist loads, the equations become more complex. For the development of these formulas, see Ref. 20.

Arch Analyses

Moments, forces, and movements of arch elements caused by radial, tangential, twist, and temperature loads, may be analyzed by flexure formulas for curved cantilever beams, amplified to allow for rib-shortening and transverse-shear effects. The method consists of cutting the loaded arch at the crown, introducing initial moments, thrusts, and shears to compensate for crown displacements, developing equations for crown movements for both parts of the arch, equating the two sets of formulas, and solving for crown forces. Equations for moments, thrusts, and shears may then be written in terms of crown forces, and moments, thrusts, and shears due to external loads. The moments, thrusts, and shears having been determined, stresses may be calculated by the usual formulas.

The basic theory of analyzing the arch is identical whether done manually or by computer. However, since there can be a considerable difference in technique, the following treatment is divided into two parts: manual method and computer method.

Abutment movements, determined by Eqs. (16) to (21), may be inserted in the general deflection formulas. However, for the sake of simplicity, such movements are neglected in the following treatment.

22. Notation. The following notation is used, all quantities being measured in horizontal planes:

R_u = radius to upstream face

R_d = radius to downstream face

r = radius to centerline

t = radial arch thickness

A = area of radial cross section

I = moment of inertia of radial cross section about axis along arch centerline

s = length along centerline

φ = angle from arch point under consideration to any point on arch

$\chi = r \sin \varphi$

$y = r \text{ vers } \varphi$

φ_a = angle from arch point where deflections are desired to abutment

φ_0 = angle from arch point where deflections are desired to beginning of external load

φ_1 = angle from beginning of external load to abutment

M = moment

H = thrust

V = shear

P = intensity of external load

E = modulus of elasticity of concrete in tension and compression

E_S = modulus of elasticity of concrete in shear

μ = Poisson's ratio

K = constant to allow for nonuniform distribution of shear

c = coefficient of thermal expansion of concrete

θ = angular movement of arch centerline

Δr = radial deflection of arch centerline

Δs = tangential deflection along centerline

T = temperature change, positive when rising

The subscript 0 means at the crown, a at the abutment, L at the left of the crown, and R at the right of the crown. In the case of M, H, and V, subscripts L or R mean that the moment, thrust, and shear are due to external loads on the left or right portions of the arch, respectively.

If μ equals 0.20, and K 1.25, the ratio K/E_s in some of the subsequent equations may be replaced by $3/E$.

23. Signs. The convention of signs shown in Fig. 6 is as follows:

Positive moments cause compression at the extrados.

Positive thrusts cause compression.

Positive shears produce positive moments on the section of the arch at the left in the case of the left part of the arch and positive moments at the right in the case of the right part, except V_0, which acts as shown in Fig. 6.

Radial loads are positive when acting toward the arch center. Uniform tangential loads are positive when acting from left to right, in both parts of the arch. Triangular tangential loads are positive when acting from the abutments toward the crown. Uniform twist loads are positive when acting clockwise in the left part and counterclockwise in the right part.

Positive moments, thrusts, and shears (M_L, H_L, V_L, or M_R, H_R, V_R) due to external loads are in the same direction as the moments, thrusts, and shears of

ARCH DAMS

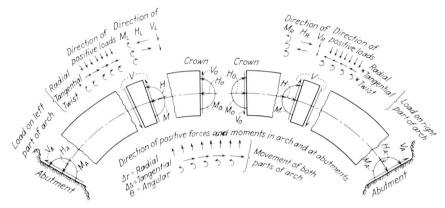

*Uniform tangential and twist loads are continuous along the arch and their directions are those for load on left part of arch

FIGURE 6 Direction of positive loads, forces, moments, and movements.

positive radial loads. Following this convention, moments, thrusts, and shears of all positive triangular loads are positive except thrusts of tangential loads, which are negative. Since the portion of the uniform tangential or twist load on the right part of the arch is applied in the same direction as the load on the left part, the M_R, H_R, and V_R of these loads will change sign.

Positive radial deflections are upstream.

Positive tangential deflections are toward the right.

Positive angular movements are counterclockwise.

Trial-Load Manual Method

General formulas are given for a circular arch subjected to symmetrical or nonsymmetrical loads. Special formulas for constant-thickness circular arches are then given for the terms that are functions of the arch properties, called *arch constants*; for the moments, thrusts, and shears due to external loads, called *load formulas*; and for the terms that are functions of both arch and load properties, called *load constants*.

Consider a differential element of length ds in the left part of an arch cut at the crown, as shown in Fig. 7. From mechanics, the equations for the arch movements at the crown due to a moment, thrust, and shear acting on the element, are

$$d\theta_0 = \frac{M\,ds}{EI} \tag{27}$$

$$d(\Delta r)_0 = \frac{M\chi\,ds}{EI} - \frac{H \sin \alpha\,ds}{EA} + \frac{KV \cos \varphi\,ds}{E_sA} \tag{28}$$

$$d(\Delta s)_0 = -\frac{My\,ds}{EI} - \frac{H \cos \varphi\,ds}{EA} - \frac{KV \sin \varphi\,ds}{E_sA} \tag{29}$$

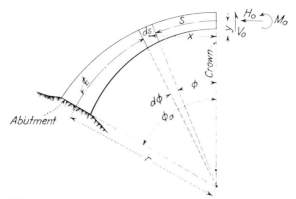

FIGURE 7 Left part of arch cut at crown.

The first term in Eqs. (28) and (29) gives the movement caused by bending; the second term, the movement caused by rib shortening; and the third term, the movement caused by shear.

By integrating the preceding equations and using s to designate the total length along the centerline from crown to abutment, the following equations are obtained for the left part of the arch:

$$\theta_0 = \int_0^s \frac{M\,ds}{EI} \tag{30}$$

$$\Delta r_0 = \int_0^s \frac{M\chi\,ds}{EI} - \int_0^s \frac{H \sin \varphi\,ds}{EA} + \int_0^s \frac{KV \cos \varphi\,ds}{E_s A} \tag{31}$$

$$\Delta s_0 = -\int_0^s \frac{My\,ds}{EI} - \int_0^s \frac{H \cos \varphi\,ds}{EA} - \int_0^s \frac{KV \sin \varphi\,ds}{E_s A} \tag{32}$$

Quantities M, H, and V may be replaced by their equivalents in terms of moment, thrust, and shear at the crown (M_0, H_0, and V_0) and moment, thrust, and shear due to external loads between the differential element and the crown (M_L, H_L, and V_L).

$$M = M_0 + H_0 y + V_0 x - M_L \tag{33}$$

$$H = H_0 \cos \varphi - V_0 \sin \varphi + H_L \tag{34}$$

$$V = H_0 \sin \varphi + V_0 \cos \varphi - V_L \tag{35}$$

If these substitutions are made and the ratio K/E_s is replaced by $3/E$, the following formulas are obtained:

$$\theta_0 = M_0 \int_0^s \frac{ds}{EI} + H_0 \int_0^s \frac{y\,ds}{EI} + V_0 \int_0^s \frac{x\,ds}{EI} - \int_0^s \frac{M_L\,ds}{EI} \tag{36}$$

$$\Delta r_0 = M_0 \int_0^s \frac{\chi\,ds}{EI} + H_0 \left(\int_0^s \frac{xy\,ds}{EI} - \int_0^s \frac{\sin \varphi \cos \varphi\,ds}{EA} + 3 \int_0^s \frac{\sin \varphi \cos \varphi\,ds}{EA} \right)$$

$$+ V_0\left(\int_0^s \frac{x^2\, ds}{EI} + \int_0^s \frac{\sin^2 \varphi\, ds}{EA} + 3\int_0^s \frac{\cos^2 \varphi\, ds}{EA}\right)$$

$$- \left(\int_0^s \frac{M_{Lx}\, ds}{EI} + \int_0^s \frac{H_L \sin \varphi\, ds}{EA} + 3\int_0^s \frac{V_L \cos \varphi\, ds}{EA}\right) \quad (37)$$

$$\Delta s_0 = - M_0 \int_0^s \frac{y\, ds}{EI} - H_0\left(\int_0^s \frac{y^2\, ds}{EI} + \int_0^s \frac{\cos^2 \varphi\, ds}{EA} + 3\int_0^s \frac{\sin^2 \varphi\, ds}{EA}\right)$$

$$- V_0\left(\int_0^s \frac{x y\, ds}{EI} - \int_0^s \frac{\sin \varphi \cos \varphi\, ds}{EA} + 3\int_0^s \frac{\sin \varphi \cos \varphi\, ds}{EA}\right)$$

$$+ \left(\int_0^s \frac{M_L y\, ds}{EI} - \int_0^s \frac{H_L \cos \varphi\, ds}{EA} + 3\int_0^s \frac{V_L \sin \varphi\, ds}{EA}\right) \quad (38)$$

If symbols are substituted for the multipliers of M_0, H_0, and V_0, and for the terms depending on load, the preceding equations may be written

$$\theta_0 = A_1 M_0 + B_1 H_0 + C_1 V_0 - D_1 \quad (39)$$

$$\Delta r_0 = C_1 M_0 + B_2 H_0 + C_2 V_0 - D_2 \quad (40)$$

$$\Delta s_0 = - B_1 M_0 - B_3 H_0 - B_2 V_0 + D_3 \quad (41)$$

Similar equations for the right part of the arch may be developed in the same manner. In this case, the values of M, H, and V are

$$M = M_0 + H_0 y - V_0 x - M_R \quad (42)$$

$$H = H_0 \cos \varphi + V_0 \sin \varphi + H_R \quad (43)$$

$$V = H_0 \sin \varphi - V_0 \cos \varphi - V_R \quad (44)$$

The resulting equations for the right part of the arch are

$$\theta_0 = - A_1' M_0 - B_1' H_0 + C_1' V_0 + D_1' \quad (45)$$

$$\Delta r_0 = C_1' M_0 + B_2' H_0 - C_2' V_0 - D_2' \quad (46)$$

$$\Delta s_0 = B_1' M_0 + B_3' H_0 - B_2' V_0 + D_3' \quad (47)$$

24. Crown Forces. The moment, thrust, and shear at the crown, M_0, H_0, and V_0, may be obtained by equating the values of θ_0, Δr_0, and Δs_0 for the two parts of the arch as given by Eqs. (39), (40), (41), (45), (46), and (47). The equations so derived are

$$(A_1 + A_1') M_0 + (B_1 + B_1') H_0 + (C_1 - C_1') V_0 = (D_1 + D_1') \quad (48)$$

$$(C_1 - C_1') M_0 + (B_2 - B_2') H_0 + (C_2 + C_2') V_0 = (D_2 - D_2') \quad (49)$$

$$(B_1 + B_1') M_0 + (B_3 + B_3') H_0 + (B_2 - B_2') V_0 = (D_3 + D_3') \quad (50)$$

If the quantities in parentheses are replaced by a, b, c, and d, the equations may be written

$$a_1 M_0 + b_1 H_0 + c_1 V_0 = d_1 \quad (51)$$

$$c_1 M_0 + b_2 H_0 + c_2 V_0 = d_2 \quad (52)$$

$$b_1M_0 + b_3H_0 + b_2V_0 = d_3 \tag{53}$$

By solving Eqs. (51), (52), and (53) simultaneously and introducing an additional symbol K', the following equations for M_0, H_0, and V_0 are obtained:

$$M_0 = \frac{1}{K'}[d_1(b_3c_2 - b_2^2) - d_3(b_1c_2 - c_1b_2) - d_2(b_3c_1 - b_1b_2)] \tag{54}$$

$$H_0 = \frac{1}{K'}[-d_1(b_1c_2 - b_2c_1) + d_3(a_1c_2 - c_1^2) + d_2(b_1c_1 - a_1b_2)] \tag{55}$$

$$V_0 = \frac{1}{K'}[-d_1(b_3c_1 - b_1b_2) + d_3(b_1c_1 - a_1b_2) + d_2(a_1b_3 - b_1^2)] \tag{56}$$

The value of K' is given by the equation

$$K' = a_1(b_3c_2 - b_2^2) - b_1(b_1c_2 - c_1b_2) - c_1(b_3c_1 - b_1b_2) \tag{57}$$

In the case of a symmetrical arch, the preceding equations reduce to

$$M_0 = \frac{1}{K'}(d_1b_3 - d_3b_1) \tag{58}$$

$$H_0 = \frac{1}{K'}(-d_1b_1 + d_3a_1) \tag{59}$$

$$V_0 = \frac{d_2}{c_2} \tag{60}$$

$$K' = a_1b_3 - b_1^2 \tag{61}$$

The functions included in the a, b, c, and d terms of Eqs. (51), (52), and (53) are given in Table 4. These are the quantities needed in determining the moment, thrust, and shear at the crown. In the case of the b_2, c_1, and d_2 terms, the signs of the quantities for the right part of the arch are negative, in accordance with the signs in Eqs. (48), (49), and (50). Consequently, the algebraic sums of the a, b, c, and d terms in Table 4 may be substituted directly in Eqs. (51) to (61).

25. Deflections. The deflections at any point on an arch may be obtained by considering the portion of the arch between the given point and the abutment as a curved cantilever beam. The desired movements are the sum of the movements due to the moment, thrust, and shear at the point and the movements due to the external load between the point and the abutment. Consequently, the movement may be calculated by the general formulas given in the preceding section.

In calculating deflections at a point in the left part of the arch, the moment, thrust, and shear at the crown are first determined as previously explained. The moment, thrust, and shear at the point are then obtained from Eqs. (33), (34), and (35), by using formulas for M_L, H_L, and V_L given in the subsequent section on load formulas. The deflections at the point are then obtained from Eqs. (36), (37), and (38), the radial section through the point being considered as a new crown section and the moment, thrust, and shear at the point being used as new values of M_0, H_0, and V_0. Deflections at points in the right part of the arch may be determined by similar methods.

TABLE 4 Functions Needed in Determining Crown Forces

Term	Functions		Term	Functions	
	Left part	Right part		Left part	Right part
a_1	$\int_0^s \dfrac{ds}{EI}$	$\int_0^s \dfrac{ds}{EI}$	b_3	$\int_0^s \dfrac{y^2\,ds}{EI}$	$\int_0^s \dfrac{y^2\,ds}{EI}$
				$\int_0^s \dfrac{\cos^2\phi\,ds}{EA}$	$\int_0^s \dfrac{\cos^2\phi\,ds}{EA}$
				$3\int_0^s \dfrac{\sin^2\phi\,ds}{EA}$	$3\int_0^s \dfrac{\sin^2\phi\,ds}{EA}$
b_1	$\int_0^s \dfrac{y\,ds}{EI}$	$\int_0^s \dfrac{y\,ds}{EI}$	d_1	$\int_0^s \dfrac{M_L\,ds}{EI}$	$\int_0^s \dfrac{M_R\,ds}{EI}$
c_1	$\int_0^s \dfrac{x\,ds}{EI}$	$-\int_0^s \dfrac{x\,ds}{EI}$	d_2	$\int_0^s \dfrac{M_L x\,ds}{EI}$	$-\int_0^s \dfrac{M_R x\,ds}{EI}$
				$\int_0^s \dfrac{H_L \sin\phi\,ds}{EA}$	$-\int_0^s \dfrac{H_R \sin\phi\,ds}{EA}$
				$3\int_0^s \dfrac{V_L \cos\phi\,ds}{EA}$	$-3\int_0^s \dfrac{V_R \cos\phi\,ds}{EA}$
b_2	$\int_0^s \dfrac{xy\,ds}{EI}$	$-\int_0^s \dfrac{xy\,ds}{EI}$	d_3	$\int_0^s \dfrac{M_L y\,ds}{EI}$	$\int_0^s \dfrac{M_R y\,ds}{EI}$
	$-\int_0^s \dfrac{\sin\phi\cos\phi\,ds}{EA}$	$\int_0^s \dfrac{\sin\phi\cos\phi\,ds}{EA}$		$-\int_0^s \dfrac{H_L \cos\phi\,ds}{EA}$	$-\int_0^s \dfrac{H_R \cos\phi\,ds}{EA}$
	$3\int_0^s \dfrac{\sin\phi\cos\phi\,ds}{EA}$	$-3\int_0^s \dfrac{\sin\phi\cos\phi\,ds}{EA}$		$3\int_0^s \dfrac{V_L \sin\phi\,ds}{EA}$	$3\int_0^s \dfrac{V_R \sin\phi\,ds}{EA}$
c_2	$\int_0^s \dfrac{x^2\,ds}{EI}$	$\int_0^s \dfrac{x^2\,ds}{EI}$			
	$\int_0^s \dfrac{\sin^2\phi\,ds}{EA}$	$\int_0^s \dfrac{\sin^2\phi\,ds}{EA}$			
	$3\int_0^s \dfrac{\cos^2\phi\,ds}{EA}$	$3\int_0^s \dfrac{\cos^2\phi\,ds}{EA}$			

26. Arch Constants. The quantities A_1, B_1, B_2, B_3, C_1, and C_2 in Eqs. (39), (40), and (41) and the similar quantities in Eqs. (45), (46), and (47) consist of integrals or groups of integrals which are functions of the arch properties. Consequently, they are designated *arch constants*. These constants are really deflections at a point due to a unit force or moment at the point. Their meanings may be briefly stated as follows:

A_1 = angular movement due to a unit moment

B_1 = angular movement due to a unit thrust, or the tangential deflection due to a unit moment

C_1 = angular movement due to a unit shear, or the radial deflection due to a unit moment

B_2 = radial deflection due to a unit thrust, or the tangential deflection due to a unit shear

C_2 = radial deflection due to a unit shear

B_3 = tangential deflection due to a unit thrust

If the arch has a constant thickness and the centerline is used instead of the neutral axis, quantities I, s, ds, and A in Eqs. (36), (37), and (38) may be replaced by $t^3/12$, $r\varphi$, $r\,d\varphi$, and t. Since E is a constant, the integrals of the arch constants, for either side of the arch, may then be evaluated and the constants determined by the following equations in which φ_a is the angle from the point where the deflections are desired to the abutment:

$$A_1 = \frac{12r}{Et^3}(\varphi_a) \tag{62}$$

$$B_1 = \frac{12r^2}{Et^3}(\varphi_a - \sin \varphi_a) \tag{63}$$

$$C_1 = \frac{12r^2}{Et^3}(\text{vers } \varphi_a) \tag{64}$$

$$B_2 = \frac{12r^3}{Et^3}\left(\text{vers } \varphi_a - \frac{\sin^2 \varphi_a}{2}\right) + \frac{r}{Et}(\sin^2 \varphi_a) \tag{65}$$

$$C_2 = \frac{12r^3}{Et}\left(\frac{\varphi_a - \sin \varphi_a \cos \varphi_a}{2}\right) + \frac{r}{Et}\left[\frac{(\varphi_a - \sin \varphi_a \cos \varphi_a)}{2} + \frac{3(\varphi_a + \sin \varphi_a \cos \varphi_a)}{2}\right] \tag{66}$$

$$B_3 = \frac{12r^3}{Et^3}\left[\varphi_a - 2\sin \varphi_a + \frac{(\varphi_a + \sin \varphi_a \cos \varphi_a)}{2}\right] \tag{67}$$

$$+ \frac{r}{Et}\left[\frac{(\varphi_a + \sin \varphi_a \cos \varphi_a)}{2} + \frac{3(\varphi_a \sin \varphi_a \cos \varphi_a)}{2}\right]$$

Since the quantities contained in the brackets of Eqs. (62) to (67) depend only on the arch angle, suitable tables may be prepared for use in calculating arch constants.

27. Load Formulas. Formulas for moment, thrust, and shear due to external loads, M_L, H_L, and V_L or M_R, H_R, and V_R, must be obtained before the D terms in the preceding equations can be evaluated. Such formulas may be written in terms of the external load P, the upstream radius R_u, the centerline radius r, the total central angle subtended by the load φ_1, and different functions of the central angle φ from the beginning of the load to any point on the loaded section of the arch. Equations for the uniform and triangular radial, tangential, and twist loads are developed in Ref. 20. Equations for M_R, H_R, and V_R due to similar loads on the right side of the arch are the same as the equations for M_L, H_L, and V_L, except as noted in the section on signs.

P is usually expressed in pounds per square foot in the case of radial and tangential loads, and in foot-pound per square foot in the case of twist loads. Since the twist loads are couples applied along the arch centerline, they do not produce thrusts or shears. Consequently no formulas for H_L or V_L due to twist appear in the table. The load formulas may be used to calculate moment, thrust, and shear due to external loads at the right of any point between the beginning of the load and the abutment.

28. Load Constants. The quantities $D_1, D_2,$ and D_3 in Eqs. (39), (40), and (41) and the similar quantities in Eqs. (45), (46), and (47) consist of integrals or groups of integrals that depend on both arch properties and external loads. These quantities are designated *load constants*. They are really deflections at a point due to the loads applied between the point and the abutment. D_1 is the angular movement, D_2 the radial movement, and D_3 the tangential movement. See Ref. 20 for the development of these formulas.

Trial-Load Computer Method

The arch constants and load constants developed in the previous section are for constant-thickness single-center circular arches. The integrations for these constants become considerably more complicated for variable-thickness, multicenter circular arches and are not included here. When using a computer, it becomes feasible to use a voussoir summation procedure. The voussoirs can be made small enough to be considered constant thickness with uniform load. The use of the voussoir method eliminates the need for integrating the rather complex formulas for the load and arch constants. Variable-thickness, multicentered arches can therefore be analyzed almost as readily as the constant-thickness arches. The voussoir method does require, however, considerable geometrical manipulation in order to define the various voussoirs and their loading geometrically.

During the early development of the trial-load method, it was found convenient to use triangular loads which varied from maximum at the abutment to zero at some distance away from the abutment. By varying the sign, magnitude, and distance from the abutment, it was possible to approximate, reasonably, any varying distributed load. This load scheme made it possible to set up procedures and tables which facilitated calculation of the load constants by hand. With computers, these tables assume a position of lesser importance. Once the equations are programmed, it becomes easier in some cases to use the computer than the tables. The computer programmer is therefore free to choose any shape of load. Shapes shown in Figs. 8, 9, and 10 are often used. The points of maximum load are located so as to coincide with the intersections of the cantilevers. The load then varies linearly to zero at the adjacent cantilever.

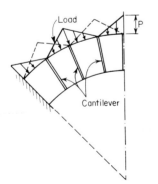
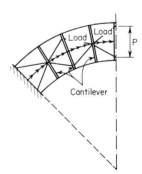

FIGURE 8 Triangular radial loads. **FIGURE 9** Triangular tangential loads.

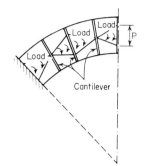

FIGURE 10 Triangular twist loads.

The geometry of a nonsymmetrical multicentered arch is defined in Fig. 11. The geometrical properties of the individual voussoirs can be derived as follows:

When $\phi < \phi_1$,

$$t_m = R_1 - (R_1 - r_1 - T_0) \cos \frac{m}{N_S} \phi_1$$

$$- \left\{ r_1^2 - \left[(R_1 - r_1 - T_0) \sin \frac{m}{N_S} \phi_1 \right]^2 \right\}^{1/2}$$

where m is the voussoir number for which t is being computed.

$$X_m = \left(R_1 - \frac{t_m}{2} \right) \sin \frac{m}{N_S} \phi_1$$

$$Y_m = R_1 - \frac{t_0}{2} - \left(R - \frac{t_m}{2} \right) \cos \frac{m}{N_S} \phi_1$$

$$\delta_S = \left(R_1 - \frac{t_m}{2} \right) \frac{\phi}{N_S}$$

Similarly, values can be derived for t_m, N_m, and Y_m for $\phi > \phi_1$.

Having determined the required geometrical properties of the voussoirs, one can now proceed with the arch analysis. The basic equations (27), (28), and (29) as developed for manual calculations still apply. However, the integral signs in Eqs. (30), (31), and (32) are replaced by summation signs. Equations (33), (34), and (35) remain the same. The integral signs in (36), (37), and (38) are replaced by summation signs. The development of Eqs. (39) through (61) remains unchanged. The functions needed in determining the crown forces shown in Table 4 remain identical except that the integral sign changes to a summation sign.

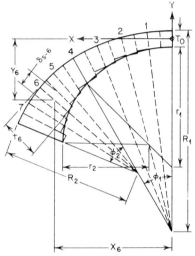

FIGURE 11 Geometry of multicentered arch.

The values for the arch constants can now be obtained directly through summation. Formulas still need to be derived for obtaining moments, thrust, and shear due to external loads. Assume a typical triangular load as shown in Fig. 12. Assume the distributed load over any voussoir is a concentrated load p at the center of the voussoir. For a load with intensity of unity at cantilever 2, the value of P to the right of cantilever 2 is

$$P_m = \frac{m - N_1}{N_2 - N_1} \delta_S$$

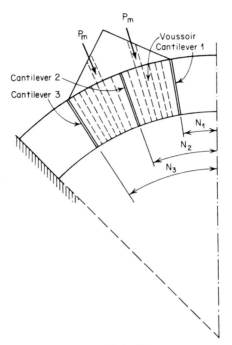

FIGURE 12 Voussoir loading.

and the left of cantilever 2,

$$P_m = \left(1 - \frac{m - N_2}{N_3 - N_2}\right)\delta s$$

where m is the voussoir number for which P is being computed and $N_{1,2,3}$ are the numbers of voussoirs from the crown to the first, second, and third cantilevers, respectively. The formulas for moment, thrust, and shear due to external loads become:

Radial loads:

$$M_L = \Sigma P_m(\Delta x) \cos \phi + \Sigma P_m(\Delta y) \sin \phi$$
$$H_L + \Sigma P_m \sin \Delta\phi$$
$$V_L = \Sigma P_m \cos \Delta\phi$$

Tangential loads:

$$M_L = \Sigma P_m(\Delta y) \sin \phi + \Sigma P_m(\Delta x) \cos \phi$$
$$H_L + \Sigma P_m \cos \Delta\phi$$
$$V_L = \Sigma P_m \sin \Delta\phi$$

Twist loads:

$$M_L + \Sigma P_m$$

where Δx, Δy, and $\Delta \phi$ are incremental values for x, y, and ϕ, respectively, between P_m and the point where M_L is being computed. All the load constants can now be evaluated and the crown reactions determined with Eqs. (36) through (38).

Adjustments

29. Adjustments. Having determined the deflections due to unit loads of the various arches and cantilevers, it is now possible to divide the external load by trial and error so that the deflections in the radial directions are equal. In the present use of the method, adjustments of deflections are first made in radial directions, including effects of radial shear and rock deformations in both arch and cantilever elements. If considerations of tangential shear and twist effects are necessary, adjustments are next made in circumferential directions, then in angular directions. In circumferential adjustments, equal and opposite tangential shear loads are introduced, by trial, to compensate for differences in tangential movements caused by radial loads, one set of loads being applied to the arch elements and the balancing set to the cantilever elements. In angular adjustments, equal and opposite twist loads are applied to arch and cantilever elements, to compensate for discrepancies in rotation caused by radial loads. Radial movements, caused by tangential shear and twist loads, are then considered in a radial readjustment and their effects considered in circumferential and angular readjustments, until resultant deflections are in agreement in all three directions.[1]

This procedure of determining the load distribution by trial and error was developed prior to the development of the electronic computer. It can also be adapted to a computer solution. It is probably more expedient, however, to write a set of equations which equate the deflections of the arches and cantilevers and to solve these equations for the unknown load distribution. Using the matrix notation, let $[A]$ be the deflection matrix of the independent arches and $[C]$ that of the cantilevers. Let $[P_A]$ be a column matrix of the external load applied to the arches and $[P_C]$ that of the external load applied to the cantilevers. Let $[I_A]$ be an unknown column matrix representing internal arch loads and $[I_C]$ be a matrix representing unknown cantilever loads required to bring the arches and cantilevers in agreement. Then

$$[A]\begin{bmatrix}P_A\\P_C\end{bmatrix} + [A]\begin{bmatrix}I_A\\I_C\end{bmatrix} = [C]\begin{bmatrix}I_A\\I_C\end{bmatrix} + [C]\begin{bmatrix}P_A\\P_C\end{bmatrix}$$

Because the internal loads have to be self-balancing,

$$[I_A] = [-I_C]$$

Then

$$[A]\begin{bmatrix}-I_C\\I_C\end{bmatrix} - [C]\begin{bmatrix}-I_C\\I_C\end{bmatrix} = -[A]\begin{bmatrix}P_A\\P_C\end{bmatrix} + [C]\begin{bmatrix}P_A\\P_C\end{bmatrix}$$

It is now possible to solve for the adjusting loads $[I_C]$.

30. Arch Stresses. When the arch deflections have been brought into satisfactory agreement with the cantilever deflections in all parts of the dam, arch thrusts, moments, and shears caused by the combined radial, tangential, twist, and temperature loads are calculated for the locations where stresses are desired, usually the crown and abutment sections. Direct stresses at the extrados and intrados curves are then computed by Eq. (7). Stresses at different depths in the concrete may be computed on the assumption of a straight-line variation between the faces of the dam.

Average shearing stresses are the shearing forces divided by the areas on which they act. Shearing forces at crown sections of symmetrical arches, symmetrically loaded, are zero. Horizontal shearing stresses at the upstream and downstream edges of arch elements, acting in vertical tangential planes, may be computed by the formula

$$N = \pm (S - p) \tan \beta \quad \text{(negative at extrados)}$$

where S = direct stress acting normal to the plane
p = water pressure at the face of the dam
β = angle between the normal to the plane and a line tangent to the edge of the arch element

In circular constant-thickness arches, shearing stresses at the extrados and intrados are zero; and maximum shearing stresses in the interior may be estimated at three-halves the average shearing stress, a parabolic distribution being assumed.

FINITE-ELEMENT METHOD

31. General. The development of the finite-element method since 1956 and the rapid advance in computer hardware have had profound influence on the method of structural and stress analysis. The development of the three-dimensional solid elements needed for the analysis of the arch dam started in the middle 1960s. Shortly afterward, design firms began to experiment with this new method in arch dam analysis. By the middle of the 1970s, finite-element analysis of arch dams had become a common practice worldwide. The method is completely flexible with respect to complex geometries, arbitrary boundary conditions, variable material properties, and different loading conditions. As of 1990, most analyses performed in design firms are linear elastic. A common procedure is to model the dam by three-dimensional solid elements or by thick shell elements. The dam-foundation interaction is usually accounted for by including a portion of the foundation rock in the finite-element model (Fig. 13) or by providing elastic support to the dam such as the Vogt coefficient type of formulation.[21] A double-curvature arch dam can be easily discretized as an assemblage of three-dimensional solid elements regardless of the complexity of its geometry. Openings in the dam, appurtenant structures, heterogeneous foundations, and joints in the foundation rock can be readily included in the finite-element model. Dead weight, water load, temperature change, ice and silt load, and even foundation movement can be analyzed and resulting deformations and stresses in the arch dam computed for evaluation. Codes for dynamic analysis have also been devel-

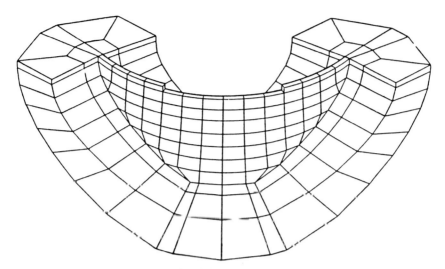

FIGURE 13 Finite-element model of arch dam.

oped to evaluate the response of arch dams under seismic loadings. This is an important subject. An article will be devoted to it later.

Most general-purpose, three-dimensional structural analysis programs can be used for the analysis of arch dams. In the United States, the most commonly used program is SAPIV,[22] developed at the University of California. Most users developed their own pre- and postprocessing programs with different degrees of sophistication. The preprocessors are used for data preparation and verification, and the postprocessors are for presenting results in graphic or tabulated forms. There are also special-purpose programs developed exclusively for arch dam analysis. These programs usually have some built-in pre- and postprocessing capabilities. One example is ADAP,[23] also developed at the University of California.

Simplified nonlinear analyses have been made for design purposes. However, most nonlinear studies are performed in universities or research institutes as of 1990. The finite-element method has been developed to such a stage that the analytical capability has far exceeded the knowledge of input data and the ability to evaluate the analytical results. Particularly, research is needed to establish the material properties of the foundation rock and the material behavior beyond elastic limit.

32. Criteria. While finite-element analysis has become the primary method—sometimes the only method—in the design and analysis of many types of structures, it has not yet replaced the traditional trial-load analysis in the design of arch dams. The reason is that there are generally accepted and well-tested criteria to evaluate the results of the trial-load analysis, such as those published by the U.S. Bureau of Reclamation (USBR),[20] whereas, for the finite-element analysis, the criteria are still in the developing stage.

The above USBR criteria were developed to be compatible with the theory and assumptions associated with the trial-load method. Although the finite-element method is theoretically more complete than the trial-load method, not all of its results should be evaluated by these criteria. Comparative studies show that compressive stresses calculated by these two methods are similar. Therefore, it is reasonable to use USBR criteria to evaluate compressions calculated by the finite-element analysis. On the other hand, the finite-element method often yields higher tensile stresses near the dam foundation contact than the trial-load analysis. This is caused by localized effects due to discontinuities in geometry which are not reflected in the trial-load results. These localized tensions should not be evaluated by criteria developed for the trial-load analysis.

There are two issues involved in developing criteria to evaluate tensions: the tensile strength and the safety factor. First, test results of the tensile strength of the concrete vary over a wide range, depending upon the condition of the samples and methods of testing. Even under ideal test conditions, results are usually spread over a wider range than the compressive tests. Therefore, it is difficult to define the tensile strength of the concrete. Detailed discussions of this subject can be found in Refs. 24 and 25. Second, arch dam concrete usually has very high compressive strength, and a relatively generous safety factor is used to evaluate the calculated compressions. This same safety factor may not be appropriate for evaluating tension because tensile strength is substantially smaller and because compression is the primary load-carrying mechanism of the arch dam, while minor tensile cracks usually do not affect the safety of the dam. The problem is further complicated by the presence of vertical contraction joints in the arch dam. The tensile strength across these joints is usually very small, making it more difficult to evaluate the calculated tensile stresses.

Because of these problems in evaluating results of the finite-element analysis of arch dams, it has not replaced the trial-load method in arch dam design. The finite-element method is most useful in special studies such as stresses around openings in the dam, inhomogeneous foundations, and dynamic analysis.

33. Dynamic Analysis. In the United States, practically all dynamic analysis of arch dams is performed by the finite-element method. This analysis is used primarily to evaluate the seismic response of arch dams. Sometimes it is also used to investigate other dynamic loadings such as machine-induced vibration. Usually the same finite-element model is used for both the static and dynamic analyses. The dynamic response is computed either by the time history analysis or by a response spectrum analysis. These can be performed by most general-purpose three-dimensional finite-element analysis programs. The dynamic analysis of the arch dam is much more involved than the static one. Many phenomena are not yet fully understood by engineers and researchers alike, mainly because of uncertainties in the following areas: the earthquake mechanism, dam foundation-reservoir interaction, and the material behavior beyond elastic limit.

A China-U.S. Workshop on Earthquake Behavior of Arch Dams was held in June 1987 in Beijing. Twenty-three invited participants, including researchers and engineers from China, the United States, and other countries, attended this workshop. The principal findings and comments of this workshop provide a comprehensive summary of what the finite-element method can do (the current practice) and cannot do (future research needs) in earthquake analysis of arch dams as of 1987. They are as follows:

On Current Practice

Mathematical model. A finite-element model of the dam plus a massless block of foundation rock is developed as the subject of the seismic analysis.

Seismic input. The earthquake motions are specified as three components of free-field surface rock motions. These motions are applied directly to the rigid base of the foundation block. Foundation interaction is represented by the flexibility of the massless rock.

Reservoir interaction. (1) In the least-refined analyses presently done, the reservoir is represented as added mass applied to the face of the dam in accordance with some modified version of the Westergaard procedure for gravity dams. In more-refined analyses, a finite-element model of the incompressible reservoir is used to evaluate the added mass to be added to the dam face. (2) An alternative, more powerful capability now is provided by the computer program EACD-3D,[26] which recognizes the reservoir compressibility and a bottom reflection (absorption) coefficient. However, because experience with this program is still limited, a range of absorption coefficients should be used in this program to determine the variation of results and to adopt the appropriate result with judgment according to the bottom geological conditions.

Nonlinearity effects. It is recognized that the earthquake behavior of arch dams may be subject to various types of system nonlinearities, especially in response to a maximum credible earthquake (MCE). However, current practice considers only linear behavior; hence the linear response must be evaluated carefully to determine the extent to which nonlinear behavior may be developed during the MCE, and to estimate its effects on the safety of the dam. An actual nonlinear analysis is not considered feasible in standard practice at the present time because of its complexity and the inaccurate earthquake input.

Abutment behavior. Deformation and stability of abutment rock may be evaluated by a pseudostatic procedure that accounts for the static and dynamic forces applied to the rock by the dam, plus the pseudostatic inertial forces of the abutment rock mass.

Evaluation of performance. In general, the adequacy of the dam design is judged by comparison of the calculated static plus dynamic stresses in the concrete with the estimated strength of the concrete. Usually both the tensile and the compressive strengths of the concrete are expressed as fractions of the nominal concrete cylinder strength.

On Future Research Needs

Seismic input. Procedures for applying the earthquake motions to the dam-foundation model should be greatly improved. The spatial distribution of the ground motions about the canyon walls should be represented for the situation without the dam in place, including possible traveling-wave effects. However, much more data on actual canyon wall seismic motions are needed to establish suitable earthquake input. Hence, measurement of earthquake motions about the canyon walls and in the nearby regions is also an urgent need.

Foundation-dam interaction. It is essential that foundation rock in contact with the dam should serve as an energy radiation boundary, so substructure techniques that treat the foundation as a continuum should be adapted to arch dam analysis. These substructure techniques should be organized for use in time-domain analysis as well as in the frequency-domain methods.

Reservoir-dam interaction. It is now recognized that radiation of energy from the reservoir into the bottom rock or silt is an important factor in the dam behavior, hence it is clear that compressibility of the reservoir water must be considered in order to model this radiation effect. However, experimental research is needed to determine appropriate bottom reflection coefficients for typical reservoirs. Also the effect of bottom topography on this energy loss mechanism must be studied to determine the sensitivity of seismic response to this factor.

Nonlinear dam response. (1) Nonlinear analyses such as those reported by Hall[27] must be continued and extended in scope to develop understanding of the influence of nonlinearity on the dam response, always keeping in mind that damage to the dam in an MCE is acceptable so long as the reservoir is not released. The ultimate analytical objective would be to predict the final static stability of an arch dam that has been severely fractured by an earthquake. The opening of transverse joints of arch dam during low reservoir water level and strong earthquake will greatly influence the response of the dam and therefore it is important that it be further studied. (2) Shaking-table tests of physical models, carried to the point of collapse, will be needed to validate the nonlinear analysis procedures, as well as to demonstrate the cracking and postcracking behavior of arch dams.

Field observations. Seismograph networks are needed to measure the spatial distribution of earthquake motions in canyons without dams, as well as canyons with dams, to determine the influence of dam interaction. Also, selected dams located in highly seismic regions should be instrumented to record the response of the dam, the reservoir, and the foundation rock during actual earthquakes. Such actual earthquake measurements are the only means to provide a comprehensive validation of analysis procedures.

Concrete material tests. Comprehensive testing programs are needed to determine the biaxial and triaxial strength and stiffness of concrete materials. Static and dynamic strengths of large concrete specimens in states of tension, compression, and combined stress must be evaluated in order to assess the safety of existing or proposed dams.

Abutment stability. Extensive studies are needed to develop and improve procedures for assessing the strength, deformability, and stability of the dam abutment, including the spatial variation effects of earthquake motion along the abutment rock. The abutment stability is very much a concern in arch dam design.

Earthquake resistance measures. Design provisions and construction procedures that lead to increases of the inherent seismic safety of arch dams should be studied and catalogued systematically to improve the general practice of arch dam engineering in seismic regions.

Simplified analysis procedures. Efforts should be made to develop simplified procedures for earthquake response analysis of arch dams. Possibly, methods that account for reservoir and foundation interaction by empirical modifications of simplified response spectrum analyses could be used effectively in the early stages of dam design or for preliminary assessment of the safety of existing structures.

INSTRUMENTATION OF ARCH DAMS

Most high arch dams being built today have an instrumentation program. The purpose of these programs is to give continuing assurance of the structural integrity of the dam and to furnish data leading to further refinements in design. The Stevenson Creek Dam was an early example of measuring stresses in dams. It was a constant-radius structure 60 ft high, and built solely for research measurements.[28] The measurements obtained were so accurate and comprehensive that research engineers were able to analyze satisfactorily the action of the structure. It gave an early verification that trial-load analysis furnished a satisfactory basis for the design of arch dams.

The placement of instruments in the 600-ft-high Karadj Dam designed by Harza Engineering Co. is shown in Fig. 14. Strain meters are embedded in locations of cantilever and arch intersection to simplify subsequent comparison with theoretical analysis. Joint meters are used to check on the extent of joint openings prior to grouting of the contraction joints. Foundation-deformation meters give an indication of stresses created in the foundation by changes in load.[30]

To obtain principal stresses, strain meters are generally placed in groups of five (Fig. 15) at the faces of the dam and groups of nine (Fig. 16) in the interior. The interpretation of the strain-meter readings into stresses is a rather lengthy calculation. See Ref. 29.

MODEL INVESTIGATIONS

Structural model tests are used extensively in arch-dam design. Certain designers prefer to carry out their design primarily with the aid of analytical tools, using models only for verification of the analysis of the final shape. On the other hand, model testing techniques have been refined and simplified to the point where some designers now prefer to carry out the design primarily by models, using analytical tools only occasionally for checking purposes. Occasionally problems arise which are difficult if not impossible to analyze with existing mathematical tools. In these cases models are extremely valuable.

Mossyrock Dam, designed by Harza Engineering Co., was model-tested in the LNEC Laboratory in Lisbon. The results of tests for water load are shown in Fig. 17. These can be compared directly with the analytical value shown in Fig. 2. The model generally shows somewhat higher stresses along the abutments. This is attributed to stress concentrations which can be expected where the intersection of the arch with the massive abutments forms a physical discontinuity. The modulus of elasticity of the concrete was assumed to be 3×10^6 psi, whereas the rock abutment was generally assumed to have a modulus of 1.5×10^6 psi and a small zone of 2.5×10^6 psi in the upper left abutment. As the foundation and arch were formed from the same plaster material, the variation in modulus of elasticity in the rock was obtained by drilling holes in the foundation according to a pattern determined experimentally that would reflect the difference. A picture of the model is shown in Fig. 18.

Figure 19 shows an ultimate-load test of the model after failure. As can be seen, failure occurred in the upper central portion of the model. Failure was caused by crushing of the model material.

Model tests are often used to investigate arch dams for earthquake. Either shaking tables or electromagnetic vibrators are used.[31] With the aid of models, one can determine the natural frequency of the prototype. The models can be subjected to specified accelerations. Considerable work has been done recently in analyzing available accelerograms of strong-motion earthquakes. The various spectra such as those of velocity, acceleration, and power are obtained from these studies. Assuming an earthquake to be a random vibration, the strains can

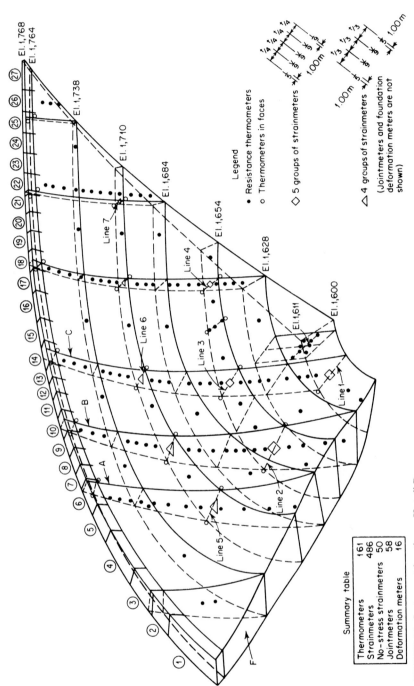

FIGURE 14 Instrument locations—Karadj Dam.

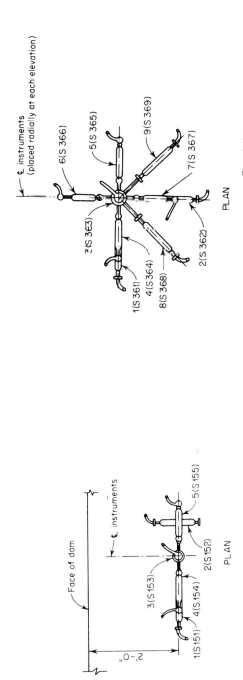

FIGURE 15 Group of five strain meters.

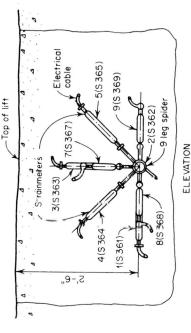

FIGURE 16 Group of nine strain meters.

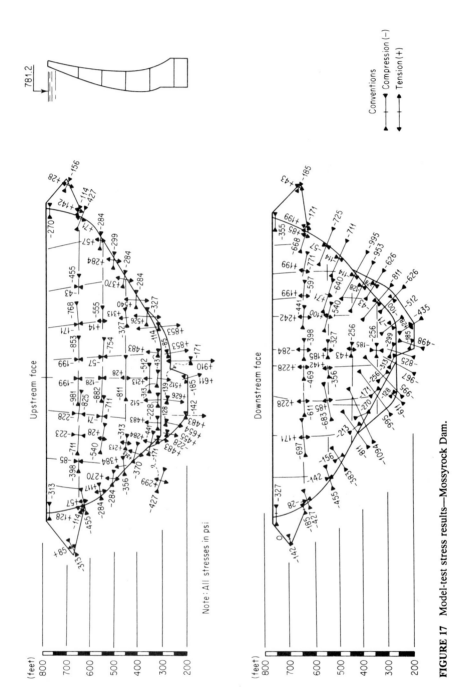

FIGURE 17 Model-test stress results—Mossyrock Dam.

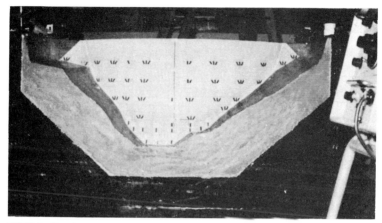

FIGURE 18 Model for hydrostatic load, Mossyrock Dam.

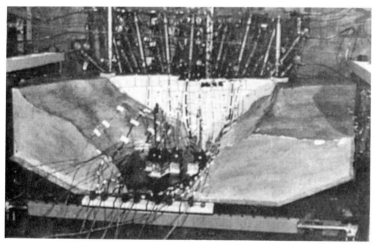

FIGURE 19 Ultimate-load test, Mossyrock Dam.

be measured in a model when subjected to the power spectrum of any historical earthquake.

EXAMPLES OF ARCH DAMS

The following examples illustrate practical design of arch dams that have been built and are operating satisfactorily. Hoover Dam represents an unusually high and massive arch-gravity type of constant-radius dam, whereas Karadj and Mossyrock are both high, variable-radius, double-curvature dams.

34. Hoover Dam. Hoover Dam, built by the Bureau of Reclamation, was completed on the Colorado River near Las Vegas, Nev., in 1936. Figure 20 shows a general plan of the structure, a vertical cross section along the line of centers,

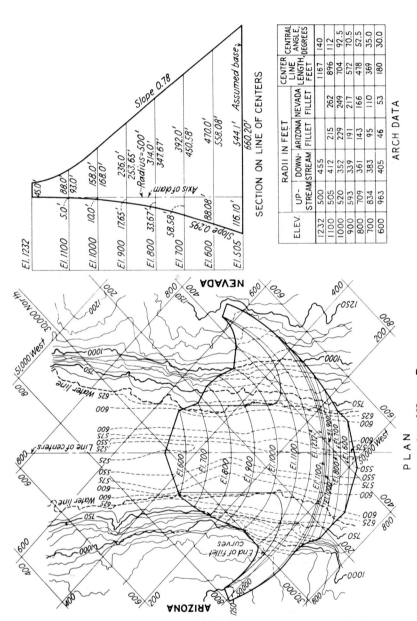

FIGURE 20 Plan and maximum section of Hoover Dam.

EXAMPLES OF ARCH DAMS

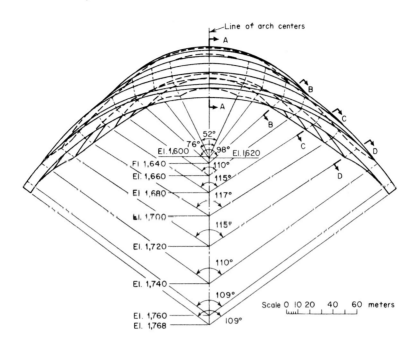

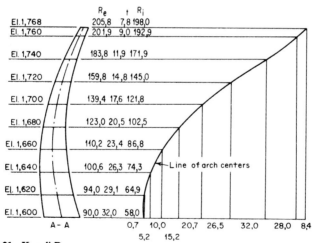

FIGURE 21 Karadj Dam.

and a tabulation of arch data at 100-ft intervals of elevation. The dam was designed on the basis of trial-load analyses. It is curved on a radius of 500 ft to the upstream edge of the crest, has extrados curves of gradually increasing radii as the depth below the crest increases, and is provided with long-radius fillets at the abutment ends of the intrados curves in the regions of pronounced arch stress.

FIGURE 22 Mossyrock Development, Washington, Harza Engineering Co., Owner, City of Tacoma.

Short-radius fillets connect the dam with the abutment and foundation rock along the entire profile at the upstream face. Several articles describing its design and construction have been published.[32]

35. Karadj Dam. Karadj Dam, built by the Plan Organization of the Government of Iran, was completed in 1962. It is located on the Karadj River approximately 25 mi northwest of Tehran. The dam is a multipurpose development that serves the joint interests of water supply, irrigation, and power. Figure 21 shows a general plan of the dam and a cross section of the crown cantilever. A description of the project features and its history, as well as significant data, has been published.[33]

36. Mossyrock Dam. Mossyrock Dam (Fig. 22) was built by the City of Tacoma. The project was completed in 1968. It is located on the Cowlitz River approximately 60 mi southwest of Tacoma, Washington. The dam is 605 ft high with thrust blocks 140 and 125 ft high on the left and right sides, respectively. An overflow spillway is provided with four 30- × 30- × 30-ft radial Taintor gates.

37. Strontia Springs Dam. Strontia Springs Dam (Fig. 23), built by the Denver Board of Water Commissioning, was completed in 1981. It is located on the South Platte River approximately 20 mi upstream (southwest) of the city of Denver. The dam serves primarily as a reregulation reservoir for diversion to a water treatment plant. The dam is 300 ft high and has a thickness of 10 ft at the crest and 30 ft at the base. An article describing its design has been published.[34]

FIGURE 23 Strontia Springs Diversion Dam, South Platte River, Colo., Harza Engineering Co., Owner, Denver Board of Water Commissioning.

REFERENCES

1. NOETZLI, FRED A., Pontalto and Madruzza Arch Dams, *Western Construction News,* April 10, 1932, pp. 451–452.
2. JORGENSEN, LARS R., The Constant-angle Arch Dam, *Trans. ASCE,* **78,** 685–733, 1915.
3. CAIN, WILLIAM, The Circular Arch under Normal Loads, *Trans. ASCE,* **85,** 233–283, 1922.
4. HOUK, IVAN E., and KENNETH B. KEENER, Masonry Dams—Basic Design Assumptions, *Trans. ASCE,* **106,** 1115–1130, 1941.
5. HOUK, IVAN, E., Uplift Pressure in Gravity Dams, *Western Construction News,* July 25, 1930, pp. 344–349.
6. STEELE, BYRAM W., Cooling Boulder Dam Concrete, *Eng. News-Record,* **133,** 451–455, 1934.

7. HOUK, IVAN E., Setting Heat and Concrete Temperature, *Western Construction News,* Aug. 10, 1931, pp. 411–415.
8. HOUK, IVAN E., Technical Design of High Masonry Dams, *Engineer,* Aug. 4, 1933, pp. 105, 106, Aug. 11, 1933, pp. 128–130.
9. JORGENSEN, LARS R., The Constant-angle Arch Dam, *Trans. ASCE,* **78,** 689, 1915.
10. CAIN, WILLIAM, The Circular Arch under Normal Loads, *Trans. ASCE,* **85,** 233–283, 1922.
11. CAIN, WILLIAM, Discussion of Stresses in Thick Arches of Dams by B. F. Jakobsen, *Trans. ASCE,* **90,** 522–547, 1927.
12. FOWLER, F. H., A Graphic Method for Determining the Stresses in Circular Arches under Normal Loads by the Cain Formulas, *Trans. ASCE,* **92,** 1512–1560, 1928.
13. CRAVITZ, PHILIP, Analyses of *Thick Arch Dams,* including Abutment Yield, *Trans. ASCE,* **101,** 501–523, 1936.
14. HOUK, IVAN E., Arch Deflections and Temperature Stresses in Curved Dams, No. II, *Engineer,* Apr. 9, 1937, pp. 414–415.
15. HOWELL, C. H., and A. C. JAQUITH, Analysis of Arch Dams by the Trial Load Method, *Trans. ASCE,* **93,** 1191–1316, 1929.
16. HOWELL, C. H., and A. C. JAQUITH, Analysis of Arch Dams by the Trial Load Method, *Trans. ASCE,* **93,** 1191–1316, 1929.
17. STUCKY, ALFRED, Study of Arch Dams, *Bull. Tech. Suisse Romande,* Lausanne, 1922.
18. HOUK, IVAN E., Trial Load Analyses of Curved Concrete Dams, *Engineer,* July 5, 1935, pp. 2–5.
19. VOGT, FREDRIK, Ueber die Berechnung der Fundamentodeformation, *Det Norske Videnskaps Akademi,* 1925.
20. U.S. BUREAU OF RECLAMATION, *Design of Arch Dams,* Denver, 1977.
21. HARTLEY, G. A., G. M. MCNEICE, and W. STENSCH, Vogt Boundary for Finite Element Arch Dam Analysis, *ASCE Structural Division Journal,* January 1974.
22. BATHE, K. J., E. L. WILSON, and F. E. PETERSON, "SAPIV: A Structural Analysis Program for Static and Dynamic Response of Linear Systems," UCB/EERC Report 73-11, Berkeley, Calif.
23. CLOUGH, R. W., J. M. RAPHAEL, and S. MAJTAHEDI, "ADAP: A Computation Program for Static and Dynamic Analysis of Arch Dams," UCB/EERC Report 73-14, Berkeley, Calif.
24. RAPHAEL, J. M., Tensile Strength of Concrete, *ACI Journal,* March–April 1984.
25. YEH, C. H. "Tensile Stresses in Arch Dams," China-U.S. Workshop on Earthquake Behavior of Arch Dams, Beijing, June 1987.
26. FOK, K. L., J. F. HALL, and A. K. CHOPRA, "EACD-3D: A Computer Program for Three-Dimensional Earthquake Analysis of Concrete Dams," UCB/EERC Report 86-09, Berkeley, Calif.
27. HALL, J. F., "Analysis of the Nonlinear Seismic Response of Arch Dams," China-U.S. Workshop on Earthquake Behavior of Arch Dams, Beijing, June 1987.
28. *Report on Arch Dam Investigation,* Vols. I and III, The Engineering Foundation, 1927 and 1933.
29. JONES, KEITH, "Calculation of Stress from Strain in Concrete," U.S. Bureau of Reclamation Tech. Memo 653.
30. VELTROP, J. A., R. P. WENGLER, and S. AZRI, "Structural Behaviour of Karadj Dam," 8th Congress on Large Dams, Edinburgh, 1964.
31. BORGES, J. F., J. PEREIVA, A. RAVARA, and J. PEDRO, "Seismic Studies on Concrete Dam Models," Symposium on Concrete Dam Models, Lisbon, 1964.

32. HOUK, IVAN E., Technical Design Studies for Hoover Dam, *Western Construction News,* Apr. 10, 1932, pp. 187–193. Also see *Eng. News-Record,* Feb. 6, 1930, p. 109; Dec. 15, 1932, p. 111; Dec. 21, 1933, p. 115.
33. HARZA, R. D., and R. EDBROOKE, Design of Karadj Hydroelectric Project, *Proc. ASCE, J. Power Div.,* **86**, 4, Proc. Paper 2579. VELTROP, J. A., and R. P. WENGLER, Design of Karadj Arch Dam, *Proc. ASCE, J. Power Div.,* p. D1, Proc. Paper 3827.
34. PARSONS, J., R. P. WENGLER, and C. H. YEH, "Design of Strontia Springs Arch Dam," *Proc. ASCE Convention,* Denver, May 1, 1985.

SECTION 11

PRESTRESSING/POST-TENSIONING AND REHABILITATION

By Vincent J. Zipparro and Casey M. Koniarski[1]

1. Introduction. Prestressing and post-tensioning techniques are applied to hydraulic structures for a variety of reasons, generally to improve the load-carrying capability of the structure. These techniques consist of imposing a defined external load which results in a state of stress in the structure which, in turn, results in an improved (or desired) state of stress when the structure is subjected to the working loads.

In general, if the imposed load is applied prior to the working load, the technique is termed prestressing. If the imposed load is applied while the structure is subjected to a working load, the technique is termed post-tensioning.

There are two basic mechanical methods of prestressing and post-tensioning. The method most commonly used is to apply a force to steel cables which are anchored at one end during the application of the force, and subsequently anchored at the end where the force is applied, thereby resulting in a tensioned cable. The other method is to apply a jacking force at some internal location of the structure which is subsequently filled with concrete, creating a "wedge" which maintains the imposed state of stress of the structure. Figure 1 illustrates these two methods.

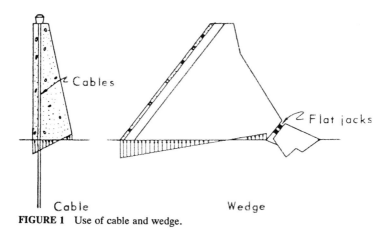

FIGURE 1 Use of cable and wedge.

---

[1]Acknowledgment is made to O.C. Zienkiewics for material in this section that appeared in the third edition (1969).

While there are several examples of dams that have been prestressed by the use of wedges and jacking systems, developments in steel cable anchoring technology have greatly exceeded those of wedge and jacking systems, particularly for high dams and for rehabilitation projects.

Discussions of prestressing and post-tensioning methods have been combined since the vast majority of current applications employ the same methods and techniques for accomplishing the desired stress distributions. Whether the application is a new dam or hydraulic structure, or rehabilitation of an existing dam or structure, most applications consist of drilling cable holes into the structure, or forming holes as the structure is constructed, installing steel cables, anchoring the cables, and applying tension to the cables to achieve the desired stress condition in the structure or its foundation.[1] For discussion purposes, the term post-tensioned will be used for both prestressing and post-tensioning methods.

Design methodology for the anchoring system is consistent whether the application is for a new or an existing structure. New structures offer the designer greater flexibility in approach, existing structures greater challenges in installation conditions.

2. Design Criteria for Dams and Gravity Structures. Current dam safety requirements for existing dams often result in the need for stability improvement at many older projects, especially concrete gravity dams. Post-tensioned anchors are a common means of achieving the required safety factors.

Design of a post-tensioned dam, whether solid or hollow in section, must account for the same forces as comparable gravity or buttress dams. The principal design criteria for stability are:

1. No tensile stresses at the dam base or in the foundation.
2. Tensile stress is permitted in concrete, to the material's allowable value.
3. Maximum compressive stress in the analyzed plane should not exceed one-quarter of the ultimate compressive strength of the concrete (concrete-concrete plane) or the allowable bearing capacity of the foundation rock (concrete–foundation rock plane).
4. The fixed anchor zone should be deep enough to ensure safety against pullout and possible overturning.
5. Adequate safety factor against sliding or shear failure should be provided.

In addition to these principal stability criteria, additional considerations such as foundation integrity, seismic environment, and seepage must be taken into account in providing an adequate factor of safety against failure.

3. Analysis Procedures and Calculating Anchor Load. Since the early 1900s dam designers have understood the importance of designing dam geometry so that the resultant of all forces on the dam falls within the middle third of the base.[2] When this condition exists, the factors of safety against overturning and sliding are usually acceptable, and the foundation pressure diagram is such that the entire dam base is in compression. This condition is highly desirable, since it is commonly assumed that a dam-rock interface cannot withstand tensile stresses. This is a prudent approach, and the assumption of the interface's intolerance to tension is widely accepted and currently used in the design of new gravity dams.[3]

The principles governing the analysis procedures and determination of the required anchor load are similar to those used in the design of other post-tensioned-concrete structures as described in standard texts on this subject. Sufficient evidence from solutions based on the theory of linear elasticity is available to justify the usual assumptions of linear vertical stress distribution in the case of stresses caused by water pressures or gravity in the usual gravity dam cross sections. When considering the stress distribution due to post-stressing forces, similar reasoning can be applied.

The required anchor force for a dam can be derived so as to satisfy design criteria 1 through 3 with due consideration to criterion 5, which is usually expressed in terms of a permissible "friction coefficient" known as the shear-friction factor of safety against horizontal sliding S_{s-f}, such that

$$S_{s-f} = \frac{f(\Sigma V + P_v) + A_c \times \tau}{\Sigma H + P_H} \geq \text{required}$$

where f = coefficient of friction
ΣV = sum of vertical forces
P_V = vertical component of anchor force
A_C = area of base in compression
τ = shear strength of foundation interface
ΣH = sum of horizontal forces
P_H = horizontal component of anchor force acting on cross section under consideration.

The problem of complying with criteria 1, 2, and 3 is illustrated in Fig. 2. In the cross section shown, the forces P, W, U, and H represent, respectively, the post-stressing force, weight of the dam, uplift force, and external water thrust. If no post-stressing force were present, calculations identical to those used to determine stresses in a gravity dam would result in the vertical stresses shown in Figs. 3 and 4 for the reservoir full and design conditions, respectively.

The analysis procedure described results in the determination of the minimum post-tensioning force required to meet stability requirements. A two-dimensional linear stability analysis is performed and determination is made of the anchor force per linear foot that will maintain the resultant at the desired location. There are several choices for design criteria as far as location of the resultant force is concerned, i.e., resultant within the middle one-third, one-half, or two-thirds of the base. The criteria adopted for location of the resultant force in the anchored condition for the design case can have a significant effect on the required anchor force and thus the cost of the anchoring system.

Having determined the force per unit of length, one must then consider the spacing of the monolith joints and select a reasonable spacing for the anchors. This will determine the number of anchors and their force. If the base elevation changes significantly across the length of the dam, several stability analyses may be required, and the anchor force, and possibly spacing, may be adjusted accordingly.

It is recommended that the spacing be not greater than 15 ft on centers and that there be no less than two anchors per monolith if the dam is divided into discrete monoliths. As a general rule most of the cost of a post-tensioned anchor is in the drilling. Thus it is most cost-effective to use as large a hole spacing as possible and an anchor hole diameter which is reasonably large yet common to contractors involved in installing post-tensioned rock anchors.

4. Bond-Length Determination. While the determination of the required anchor force presents no difficulties and can be dealt with rationally on the basis of

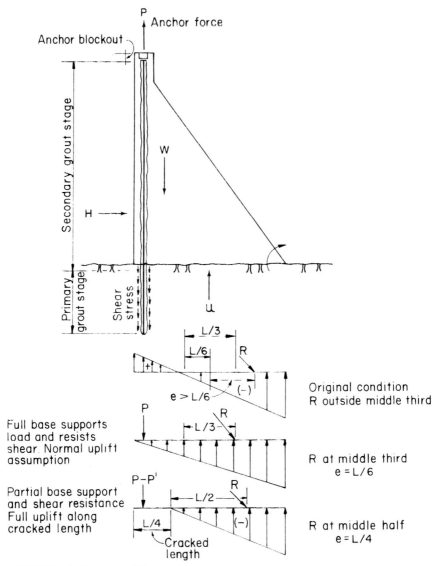

FIGURE 2 Anchorage condition.

conventional beam theory, the question of the depth and the anchorage zone at which the cables are to be anchored in the foundation is more complex.

The anchor transmits a tensile force (pullout) to the foundation. The shear strength of the foundation is used to resist this tensile force. The major part of the transfer of the force from the anchor to the foundation occurs in the initial anchorage (bond) zone. The depth at which to place the anchorage zone has to be determined to ensure that failure does not occur in the rock mass. A common

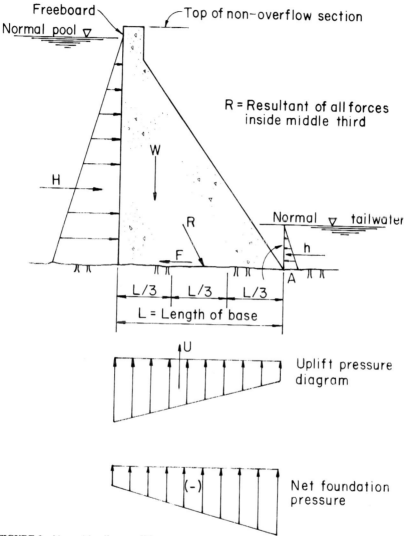

FIGURE 3 Normal loading condition.

approach for anchor-depth determination is to consider an inverted cone of rock to be pulled out by the anchor load. The location of the base of the cone has been the subject of some discussion. The three locations most commonly used are the base of the anchor, the middle of the fixed anchor length, and the top of the fixed anchor length.

Foundation rock is seldom intact and homogeneous. It is most often jointed and fractured, requiring experience in rock mechanics to use the cone method properly. Hobst (1965)[4] developed simple empirical expressions for the estimation of the anchor depth, allowing for rock strength in homogeneous and irregularly fissured rock. Empirical expressions were developed for closely spaced or grouped anchors.

Standard practice is to size the anchor so that the design load does not exceed 60 percent of the guaranteed ultimate tensile strength of the anchor.[5] The load from the anchor to the foundation is primarily transmitted by the bond between the steel tendon and the surrounding grout column. The bond length into the foundation required to provide the necessary resistance force is determined from the following equation:

$$L_b = \frac{P}{\pi \times d \times \tau_w}$$

where L_b = bond length required, at a minimum of 15 ft
  P = design load for anchor
  d = diameter of drilled hole
  τ_w = working bond stress along annular interface between rock and grout

The working bond stress used to determine the bond length is normally 25 to 50 percent of the ultimate bond stress. This equation assumes that there is no local debonding, that a shear failure occurs at the grout-rock interface, and that the bond stress is uniformly distributed through the entire bond length. A common practice is to determine the required bond length into the rock and then to add a few feet to account for any possible damage or deterioration that may have resulted to the rock near the concrete-foundation interface during construction.

CABLE AND ITS ANCHORAGE

5. Types of Anchors. A number of anchor types are available, usually made of one of the following steel materials: bar, wire, or strand. Cost considerations usually dictate which type to use. Cost considerations are generally related to allowable stress levels, manufacture, transportation, and corrosion protection.

Solid steel bars are more suitable for small loads, where a bar anchor is easier and less expensive to install. It is a generally preferred practice to use wire or strand anchors in appropriately greater numbers. Wire or strand anchors have the following advantages over bar-type anchors:

1. Greater ultimate strengths
2. Greater flexibility of the tendon, permitting complete lengths with adjustments easily completed in the field
3. Decreased risk of complete failure should corrosion attack on a single strand occur

Main considerations or features of anchor steel are:

1. Strength characteristics
2. Elasticity characteristics
3. Creep behavior
4. Relaxation behavior

Steel bars, plain or deformed, range in diameter from 12 to 40 mm, conforming to ASTM A722. Bar anchors are generally used as low-capacity anchors, mainly in single-bar situations. However, clusters of up to four bars have been used.[6]

Wire ranges in diameter from 2 to 8 mm, and is manufactured from cold-drawn plain carbon steel that is subjected to a stress-relieving heat treatment conforming to ASTM A421. The size of the anchor depends on the capacity required. In general, anchors range between 10 and 100 wires. A typical installation was performed in 1963 on Wanapum Dam on the Columbia River. At Wanapum, each anchor consisted of four cables, containing 90 ¼-in-diameter wires and was post-tensioned to a load of 2400 kips. These are believed to be the largest-capacity wire-type anchors ever installed in a hydraulic structure. In service since 1963 in the Wanapum intake structure, they have performed satisfactorily. (See Fig. 5.)

Strands range in diameter from 12.7 to 15.2 mm, consisting of a group of wires spun in helical form around a straight wire. Seven-wire strand is the most commonly used, comprised of six wires wound around a longitudinal axis of a single straight wire. The strand is stress relieved after winding, conforming to ASTM A416.

6. Corrosion and Corrosion Protection. The corrosion of post-tensioning steel is mainly electrolytic, with the electrolyte having contact with an anode and a cathode, both of which have a metallic connection. The electrolyte is usually aqueous. Corrosion protection should be taken into account in the design of an anchor system to ensure that unacceptable corrosion does not occur. The problem facing the designer is quantifying the design life and the amount of corrosion protection required.

There are three major mechanisms of corrosion of post-tensioning steel: corrosion by pitting, corrosion through tension (stressing) or hydrogen embrittlement, and corrosion involving oxygen. Corrosion by pitting occurs under chemical or physical variations causing electrical potential or ionization over the metal surface forming a bimetallic cell. This causes the protective coating or oxide film on the metal to break locally, causing pitting and crevices which promote corrosion. This can be very severe especially in the presence of aggressive ions such as chlorides.

Corrosion involving hydrogen embrittlement is the result of the steel's being subjected to stressing. The mechanism is such that the effect of local corrosion produces a pit in the surface of the steel, causing a highly concentrated stress to develop when subjected to tension or stressing. This promotes crack propagation through the metal until a reduction in cross-sectional area leads to metal failure.

Corrosion involving oxygen leads to oxygen concentrations at an anode that leads to the formation of rust. This is a chemical reaction which, in the presence of water, results in metal loss and the conversion of water and oxygen to hydroxyl (OH) ions. In addition, electrons are transferred from the metal to form the hydroxyl ion and metal ions migrate into the aqueous electrolyte. The region of the metal loss is the anode and the location of the hydroxyl ions is the cathode. The force which causes this reaction is the potential between the anode and the cathode and the chemistry of the environment. Corrosion of this type is clearly very complex, and provision of corrosion protection is warranted.

Corrosion protection systems generally come in two types, preprotection and post-protection systems. The former system is provided prior to installation and the latter after installation. The most common preprotection system currently used is a PVC sheath and greased strands in the free length of the anchor. Epoxy-coated steel tendons are also gaining acceptance.

Post-protection systems consist of basically filling the annulus around the ten-

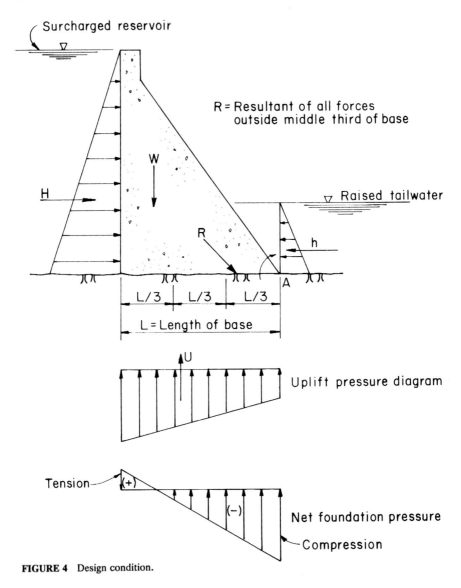

FIGURE 4 Design condition.

don in the free length of the anchor after stressing and coating the strands with a corrosion inhibitor such as grease, bitumens, and cement. The choice of material is dependent on the aggressiveness of the ground and the design life of the system.

7. Anchor Stressing Operations. The stressing of the anchor is generally one of the last construction operations to be carried out. The upper end of the anchor is anchored in a suitably sized anchor head and is stressed by a direct pull from a hydraulic jack. During stressing a special manufactured piece or "jack chair" is used to separate the jack from the anchor head and allow the anchor, as it elon-

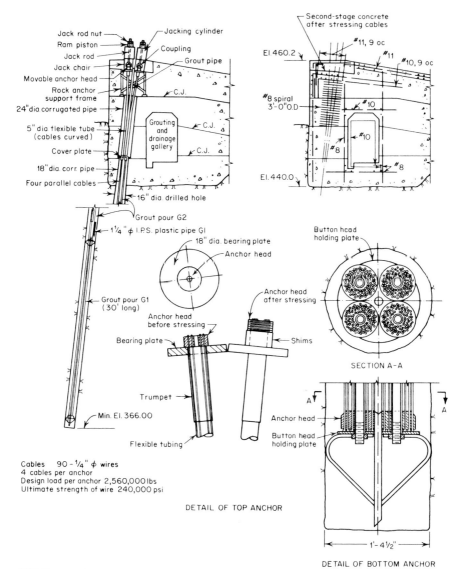

FIGURE 5 Wanapum intake structure. Details of the cable and its anchorage. Installation procedure: *a.* Drill hole at foundation level. *b.* Form hole in concrete. *c.* Lower complete cable unit into hole and support from rock anchor support frame. Make grout, pour G1, and remove plastic pipe G1. *d.* Concrete 5-in flexible conduits in formed hole. Enclose support frame. *e.* Stress cables. Place split ring shims. Remove jacks. Make grout; pour G2.

gates, to be fixed by steel wedges as the jack load is removed. When the load is locked off, the steel wedges in the anchor head bite into the strand and, remaining in place, wedged into the anchor head. In Figs. 6 and 7, details of a typical stressing head and a multistrand anchor are shown. Prior to the start of the stressing operations, it is important to check that the individual wires or strands of an anchor are not twisted around each other. This can be avoided with the proper use

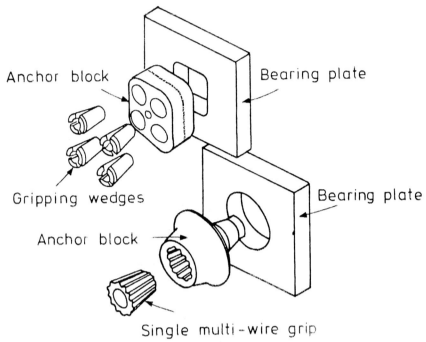

FIGURE 6 Method of gripping multistrand tendon at anchor head. (*After Hanna, Ref. 1.*)

of centralizers and spacers through the free length of the anchor. Figure 7 is an example of a multistrand anchor.

The jacking system should load all the strands or wires simultaneously to provide the same loading in each strand or wire. Anchor stressing is performed in a controlled manner to load each anchor with a known load level and to record the respective elongation. Comparison of the data is made with test anchors or with established criteria. Stressing methods and testing procedures are given in the Post-Tensioning Institute's *Post-Tensioning Manual.*[5]

As in all post-tensioned construction, the possibility of relaxation of the initial stress has to be taken into account. The relaxation is due primarily to the creep of concrete and to the relaxation of the steel. Allowance for these effects must be accounted for in the design of the anchor. Information on relaxation effects is available from steel suppliers for calculating long-term losses with reasonable accuracy.

During stressing operations, strand wires tend to unwind, and therefore the relaxation values also depend to a great degree on the restraint provided by the jacking system. It is preferable to have jacks which have a mechanism to prevent rotation during jacking.

Permissible stress levels are usually identified as a certain percentage of the guaranteed ultimate strength (GUTS) of the steel. For determination of working stress levels, the *PTI Manual* requires that the maximum test load for an anchor should not exceed 80 percent of GUTS of the tendon and the magnitude of the lock-off load should not exceed 70 percent of GUTS of the tendon. Design load is usually taken as 60 percent of GUTS.

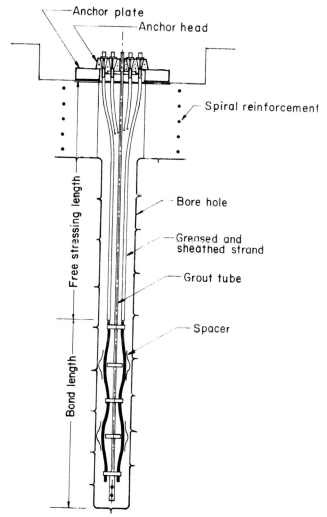

FIGURE 7 Multistrand anchor with simple corrosion protection. (*Courtesy of Dywidag Systems International.*)

8. Anchor Heads. The load from the anchor has to be transferred to the structure through a stressing head assembly. For anchors consisting of multiwire strands, the top of the tendon is fastened by wedges, which are pressed against the strands and are forced into tapered holes of a steel anchor head, under which is a bearing plate, which transfers the load to the structure. The detail of the anchor head depends on the anchor system being used. The anchor head and bearing plates must be adequately designed to accommodate the maximum loads during stressing and testing operations. Details of the anchor components are given in the various catalogs of the manufacturers of the equipment.

EXAMPLES OF PRESTRESSED/POST-TENSIONED DAMS

9. Existing Projects. Since the original application of raising Cheurfas Dam in Algeria in 1934, the use of post-tensioned cables in dams has grown significantly. The technique has been applied in dam construction worldwide. Applications of this technique have ranged from strengthening and raising of numerous old gravity dams to the construction of new dams.[7] Table 1 shows a range of types and sizes of anchors used in a few projects over the past 30 years.

There are also advantages to using post-tensioned anchors for structures such as spillway piers. An example is the gate trunnion piers for Cushman no. 1 Dam which were reinforced using anchor cables to support large radial gates. The layout of the anchors for the center pier is shown in Fig. 8.

Figure 9 shows the spillway profile of Boyd's Corner Dam in New York which has recently undergone major rehabilitation. Anchors were installed to increase stability of the structure for the design flood case. Figure 10 shows the anchorage detail adopted for Boyd's Corner anchors.

Shepaug Dam in Connecticut incorporated the first known application of a permanent anchor utilizing the second-stage grout as a permanent anchor head. The anchors were the largest-capacity strand tendons ever installed in the United States for a gravity dam stability improvement at the time of project completion in 1988. The project involved nearly 100 high-capacity tendons.

The elimination of a permanent anchor head was necessary because embedded steel at the top of the dam precluded the use of large bearing plates. In lieu of the anchor head, the anchors were second-stage grouted in the top portion of the dam. Shepaug Dam was high enough to permit adequate stressing length and bonding length through the body of the dam.

TABLE 1

Name and location	Year of installation	Anchors		Force, kips per anchor (max)
		Hole diameter, in	Type	
Wanapum, Wash.	1961	16	270 wires at 0.25 in ϕ	2560
Oneida, Idaho	1968	7	98 wires at 0.25 in ϕ	695
Cushman no. 1, Washington	1976	5½	66 wires at 0.25 in ϕ	468
Lock and Dam no. 1, Minnesota	1979	6	16 strands at 0.6 in ϕ	396
Piney, Pa.	1982	6	21 strands at 0.5 in ϕ	520
Hinckley, N.Y.	1984	9	32 strands at 0.6 in ϕ	1125
Elwha, Wash.	1987	6½	26 strands at 0.5 in ϕ	644
Stevenson, Conn.	1987	10	43 strands at 0.5 in ϕ	1062
Shepaug, Conn.	1988	10	52 strands at 0.6 in ϕ	1825
Long Lake, Wash.	1989	10	54 strands at 0.6 in ϕ	1898
Boyd's Corner, N.Y.	1989	4½	9 strands at 0.6 in ϕ	316
Glines Canyon, Wash.	1989	6	26 strands at 0.6 in ϕ	910
Iron Bridge, Tex.	1990	8	22 strands at 0.6 in ϕ	772
Upper Occoquan, Va.	1992	10	53 strands at 0.6 in ϕ	1855

Source: Harza Engineering Company.

EXAMPLES OF PRESTRESSED/POST-TENSIONED DAMS 11.13

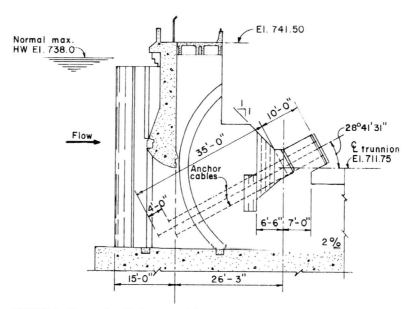

FIGURE 8 Center pier, Cushman no. 1 Dam.

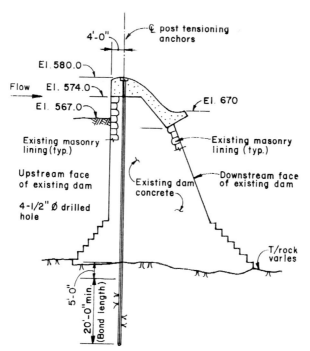

FIGURE 9 Spillway profile of Boyd's Corner Dam.

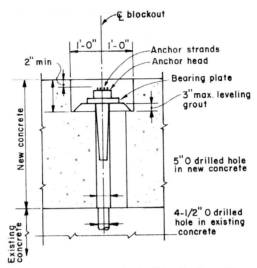

FIGURE 10 Anchorage detail of Boyd's Corner Dam.

The longest anchor at Shepaug was 192 ft long and consisted of 52 0.6-in-diameter strands with a design capacity of 1825 kips. The primary bond length of the anchor was 49 ft into the rock foundation. The secondary bond length, in lieu of the permanent anchor head, was 33 ft. The bond length was based on allowable bond stresses of 150 psi in the concrete and 100 psi in the rock foundation. Extensometers were added to the secondary bond portion of the anchor to monitor the performance of the anchor. Measurements taken since the project completion indicate movements in the tendon are due to temperature effects only.

REHABILITATION OF DAMS

10. Introduction. Just as concrete and embankment dams behave differently and are subject to differing operating requirements, they also react differently to the effects of time. Aging, deterioration, and increasingly stringent safety standards necessitate a wide variety of repair approaches that must often go beyond restoring a dam to its original condition. Increased public sensitivity to the potential effects of dam failure often necessitate general structural stability improvements. Advances in analytical capabilities, increased understanding of material properties, and improvements in construction materials and techniques contribute substantially to the types of repairs that can be performed.

11. Concrete Dams. Concrete dams often require superficial concrete repairs to correct the usual effects of aging, operation, and weather exposure. More than 30 percent of the existing dams in the United States were built before 1950, and significant changes in construction technology have occurred in the intervening decades. A significant number of dams built before or near the turn of the century require special consideration when evaluating the current quality of the construction materials and techniques in light of present practice and standards.[8]

Probably more significant than the question of the performance of the concrete material in these dams are questions related to dam stability. Geotechnical evaluations are necessary components for design of remedial measures, and they also provide data that are essential to accurate stability analyses.[9] In the absence of accurate estimates of material properties for the dam and its foundation, computed factors of safety can lead to uncertainty regarding safety. Through careful drilling and testing, parameters such as cohesion, friction angle, and unconfined compressive strength can be determined. The increased confidence with which these parameters can be used allows computation of factors of safety in stability analyses with greater degrees of certainty.

Detailed laboratory testing of geotechnical samples can also indicate whether the foundation interface can withstand tensile stress. Table 2 provides selected testing standards currently being used.[10] If it can be shown through laboratory testing that a dam foundation can withstand tensile stresses equal to about 1.5 times those theoretically calculated in stability analyses, anchors may not be needed, depending on a detailed examination of and accounting for the jointing and stratification of the rock, downstream conditions, and the possible consequences of a dam failure.

The rehabilitation of Piney Hydroelectric Project on the Clarion River, Pennsylvania, is an example of efforts to prolong the life of an economical source of power while assuring the safety of the project. Piney Dam, originally constructed in 1923, is a concrete gravity arch with a crest length of 813 ft and a height of 125 ft.

Detailed investigations of the conditions of the concrete in Piney Dam included a visual condition survey and mapping of concrete distress, nondes-

TABLE 2 Selected Testing Standards Available for Concrete Dam Investigations

ASTM D2936	Direct Tensile Strength of Intact Rock Core Specimens
ASTM D3148	Elastic Moduli of Intact Rock Core Specimens in Uniaxial Compression
ASTM D3967	Spitting Tensile Strength of Intact Rock Core Specimens
ASTM D2938	Unconfined Compressive Strength of Intact Rock Core Specimen
ASTM D4543	Preparing Rock Core Specimens and Determining Dimensional and Shape Tolerances
ASTM C39	Test Method for Compressive Strength of Cylindrical Concrete Specimens
ASTM C42	Method of Obtaining and Testing Drilled Cores and Sawed Beams of Concrete
ASTM C138	Test Method for Unit Weight, Yield, and Air Content (Gravimetric) of Concrete
ASTM C227	Test Method for Potential Alkali Reactivity of Cement-Aggregate Combinations
ASTM C295	Practice for Petrographic Examination of Aggregates for Concrete
ASTM C496	Test Methods for Splitting Tensile Strength of Cylindrical Concrete Specimens
ASTM C856	Practice for Petrographic Examination of Hardened Concrete

11.16 PRESTRESSING/POST-TENSIONING AND REHABILITATION

tructive pulse velocity (sonic) tests of internal concrete, and dynamic modulus, compressive strength, and rapid freeze-thaw tests on concrete core specimens.

It was determined that damage to the concrete was principally the result of freeze-thaw deterioration, consisting of visible cracking, laminar cracking beneath and parallel to the concrete surface, surface spalling and erosion, and localized disintegration.

Recommended concrete repairs included:

- Removal and replacement of unsound concrete to a minimum depth of 4 in using reinforcement and anchor bars where necessary
- Repair of cracks in otherwise sound concrete, including laminar cracks, by epoxy injection methods
- Cleaning of all surface and coating the surfaces with waterproofing material

In addition to the concrete work, rehabilitation measures included installation of a new trash gate, new gate seals, addition of a 10-ft-high parapet wall to prevent overtopping by the PMF, and installation of post-tensioned anchors to improve stability of the right nonoverflow wall. Typical details used for the concrete repair are shown in Fig. 11.

The rehabilitation of the Stewart Mountain Dam on the Salt River, Arizona (October 1991) is the first application of post-tensioning of an existing thin-arch dam. The dam is a 212-ft-high double-curvature thin-arch dam constructed in 1930. Crest thickness is 8 ft and base thickness 33 ft. The crest is 583 ft long. The dam was constructed in sections with vertical keyed contraction joints. It has been subjected to alkali aggregate reaction causing surface cracking and upstream displacement.

Dynamic analysis of the dam, performed by the Bureau of Reclamation in the 1980s, indicated that during the maximum credible earthquake a portion of the top 40 ft of the arch might pull apart horizontally. The adopted solution was to install 62 post-tensioned anchors through the arch from the crest into the foundation. In addition 22 post-tensioned anchors were installed in the left thrust block to protect against sliding failure. Anchor spacing varied between 8.5 and 10 ft. Anchors placed in the arch section varied in design load from 545 to 740 kips. Anchors in the thrust block had a design capacity of 985 kips. Anchor inclination varied from vertical to a maximum inclination of about 8° (Fig. 12).

12. Embankment Dams. Embankment dam safety and rehabilitation is a complex issue because of the many types of embankment dams that have been constructed over the years. Design of embankment dams varies because of availability of material and performance requirements. Low-permeability materials are used in the dam to reduce seepage.

Adequate performance of an embankment dam is dependent on various factors. Items to monitor during the life of the dam include erosion of outer slopes; seepage appearing on the downstream slopes, abutments, or foundation; changes in pore pressure within the dam; and increased settlement with time. Information on embankment dams and their design and monitoring performance evaluation are given in Sec. 13.

13. Types of Repairs. For many dams, current criteria require that the dam be able to withstand and safely pass the probable maximum flood (PMF), as estimated from the probable maximum precipitation (PMP). Under these criteria many dams do not possess the required factor of safety for structural stability.

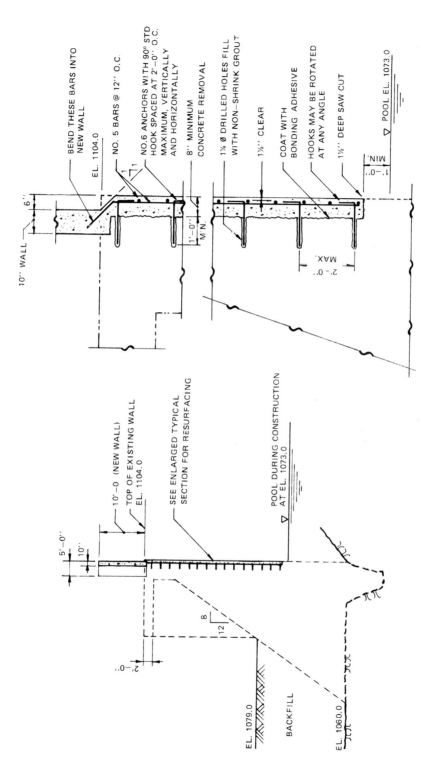

FIGURE 11 Typical details of concrete repair at Piney Dam. (*Owner Pennsylvania Electric Company, Engineer Harza Engineering Co.*)

11.18 PRESTRESSING/POST-TENSIONING AND REHABILITATION

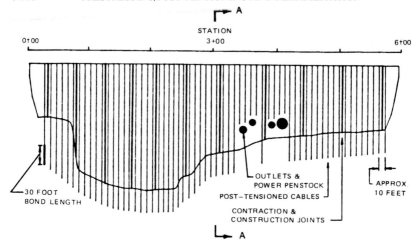

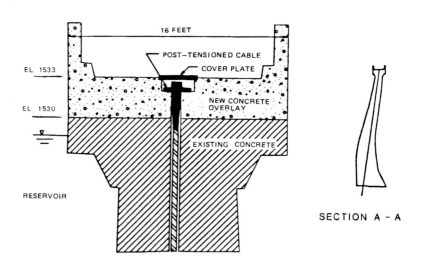

FIGURE 12 Stewart Mountain thin-arch dam rehabilitation. (*USBR.*)

Dams are also expected to meet more stringent criteria in resisting sliding. In these instances, post-tensioned anchoring is frequently the repair of choice.

More restrictive flood-protection requirements also necessitate modifications to spillways and mechanical equipment such as gates. Spillway expansion design is discussed in detail in subsequent sections, but it is worthwhile to note here that flood-protection repairs may include reshaping the existing spillway and/or pier geometry, adjusting gate positions to accommodate higher water surface profiles, upgrading the electrical and mechanical systems, and designing plunge pool systems to avoid downstream scouring.

Grouting is a frequently used component in concrete dam repairs. Typically when deteriorated concrete is removed and replaced, grouted anchors are used to improve the bonding of old and new concrete. Chemical or cement grouting is an additional and sometimes alternative measure which has been used to seal leakage, arrest deterioration, and improve stability. This method can be used to eliminate leakage through construction or lift joints, or through the mass of a masonry structure, without undertaking removal and replacement of any of the concrete mass of the structure.

Cracking and spalling of concrete are among the most common forms of deterioration at aging dams. Temperature stress, chemical reaction, reinforcing-steel corrosion, and differential movement are among the causes of such deficiencies. After initial deterioration of the concrete surface, subsequent cycles of freezing and thawing lead to continued deterioration. This can be aggravated by non-air-entrained concrete that was placed in most older structures. In 1984, Committee 224 of the American Concrete Institute released a report detailing the causes, evaluation, and repair of cracks in concrete structures, and identified 12 techniques most commonly used for the repair of cracks in concrete (Webster and Kukacka, 1987).[11]

Techniques in concrete crack repair include pressure injection, routing and sealing; stitching; addition of reinforcement; drilling and grouting; flexible sealing; grouting; drypack motor; crack arrest; impregnation; overlays and surface treatment; and autogenous healing. Table 3 provides information on the typical methods used in crack repair.

The types of repairs available for embankment dams are as various as the number of dams constructed. The type of repair method available is contingent on the desired results. For example, inadequate slope protection can be repaired by the addition of riprap, reinforced concrete, or roller-compacted concrete. The amount of slope protection required is determined by the method shown in Sec. 13.

Adequate freeboard is an important safety issue for embankment dams. Freeboard is the vertical distance between the water surface and the crest of the dam. It is required for design flood events and for wave action and reservoir runup. Providing additional freeboard can be accomplished in various ways depending on the increase in freeboard required. Measures include raising the dam with additional material embankment (0 to 2 ft) and adding highway median barriers (2 to 4 ft) or sheetpile wall (5 ft and higher).

REHABILITATION OF SPILLWAYS

14. Introduction. Methods for hydrologic analysis and flood forecasting have changed radically since hydrology's infancy at the turn of the century. Improvements in the understanding of meteorology and the duration and effects of floods, data collection, numerical methods, and computer analysis have greatly refined the hydrologist's ability to conduct flood routing studies. Probable maximum precipitation (PMP) and resulting probable maximum flood (PMF) are now determined with greater detail and accuracy. One outgrowth of this development

TABLE 3 Typical Methods Used in Concrete Crack Repair

Repair technique	Comments
Pressure injection	Little surface preparation is needed; scar marks may be left on surface where crack was injected. Limited to areas where concrete has not yet spalled. Structural-quality bond is established but if large structural movements are still occurring, new cracks may open. Process can be used against a hydraulic head
Routing and sealing	Simplest method available for repair of cracks with no structural significance. Process not applicable to repair of cracks subjected to hydraulic head
Stitching	Process will not close or seal cracks but can be used to prevent them from progressing. Generally used when it is necessary to reestablish tensile strength across crack
Addition of reinforcement	Primarily used to restore or upgrade structural properties of cracked members
Drilling and grouting	Technique applicable only when cracks run in straight line and are accessible at one end
Flexible sealing	Technique is applicable where appearance is not important and in areas where cracks are not subjected to traffic or mechanical abuse
Grouting	Wide cracks may be filled with portland-cement grout. Narrow cracks may be filled with chemical grouts
Drypack mortar	For use in cavities that are deeper than they are wide. Convenient for repair of vertical members
Crack arrest	Commonly used to prevent propagation of cracks into new concrete during construction
Impregnation	Technique can be used to restore structural integrity of highly deteriorated or low-quality concrete. Can be used to seal small crack networks
Overlays and surface treatment	Slabs containing fine dormant cracks can be repaired using bonded overlays. Unbonded overlays should be used to cover active cracks
Autogenous healing	A natural process of crack repair has practical applications for closing dormant cracks in moist environments

Source: Adapted from Ref. 11.

is that currently predicted PMPs and PMFs can differ drastically from those originally incorporated in the design of older dams.

Of particular concern is the situation when current flood forecasts require that dams pass far greater volumes of floodwaters than those for which the dams were originally designed. Larger modern floods generally mean higher reservoir levels than were assumed in the original designs. This in turn means larger horizontal and uplift loads which have the effect of shifting the resultant force outside of the middle one-third of the base of the dam. By its definition this condition results in a foundation bearing pressure diagram with tension at the upstream face.

If the dam-foundation interface cannot withstand tensile loading, cracking can develop in the tensile zone. If it cracks, it is logical to assume full uplift pressure can enter along the crack, corresponding to the headwater pressure. This increases the total uplift force, further increasing the total of the destabilizing forces and moments, leading to a shifting of the resultant force in the downstream direction. Since a smaller portion of the base is now in compression, the peak compressive load also increases, introducing the possibility of a crushing failure at the dam toe or the rock foundation along the downstream edge of the dam.

Under the National Dam Inspection Act of 1972, approximately 8800 dams were inspected in a program administered by the U.S. Army Corps of Engineers. This program revealed that roughly one-third of the dams were unsafe by modern design criteria and that 81 percent of the unsafe dams were considered to have inadequate spillway capacity.

15. Inflow Design Flood. Evaluation of spillway-capacity adequacy at existing projects involves knowledge of the magnitude of the inflow design flood (IDF). The design flood is the maximum flood against which a structure is protected. The studies involve work based on National Weather Service (NWS) generalized probable maximum precipitation (PMP) data from NWS hydrometeorological reports and site-specific PMP studies.

Determination of the design flood consists of two steps: (1) select the desired safety standard and (2) estimate the flood that meets that standard. The current practices of selecting spillway design floods are discussed by ASCE (1988).[12] Information on hydrology is presented in Sec. 1.

16. Spillway Expansion Methods. Options for spillway capacity expansion commonly involve consideration of new gated structures, design of fuse plug spillways, armoring dams with concrete or rollcrete to permit safe overtopping, or raising dams to increase head on existing spillways or contain floodwaters.

Many earth and concrete dams require raising to contain PMF headwaters, increase head and capacity of existing spillways, or provide adequate freeboard under extreme flood conditions. Evaluation of structural alternatives, evaluation of freeboard requirements, embankment stability, or seepage analyses, and flood routing studies are some of the studies required to determine the manner in which to raise a dam for PMF containment.

Where requirements for additional capacity are small relative to existing service spillway capacity, minor modifications or additions can sometimes be made to meet the requirements. These include the construction of additional bays, reshaping the spillway crest and/or pier nose geometry to improve the coefficient of discharge, or the raising of gates to permit higher flows. If increased spillway discharges have the potential to develop scour problems, energy-dissipation measures must be developed as well, such as plunge pools or flip-bucket-type spillways.

The rehabilitation of Boyd's Corner Dam (Fig. 13) is an example of a spillway capacity expansion that includes several of the rehabilitation techniques discussed in this section. Located on the West Branch of the Croton River in Putman County, New York, and constructed in 1870, Boyd's Corner Dam is approximately 78 ft high, 625 ft long, and founded on rock. The original spillway was a rock cut approximately 30 ft wide through the left abutment rock. Boyd's Corner Dam is a cyclopean masonry structure with a granite-stone veneer.* The

---

*Cyclopean masonry is a stone construction method employed in the late 1800s in which large, irregular blocks of stone are embedded in concrete. At Boyd's Corner Dam, a stone masonry veneer on either side of the dam confines the cyclopean feature.

FIGURE 13 Downstream elevation of rehabilitated Boyd's Corner Dam. (*Photograph by C. Koniarski.*)

dam's stone veneer masonry construction is an excellent example of the type of construction predominant at and before the turn of the century. During the ensuing years, the dam experienced extensive structural deterioration due to seepage through the cyclopean masonry concrete.

The reconstruction of Boyd's Corner Dam required a careful and creative balance between the modification demanded by current safety and operating criteria, maintenance of the structure's historic significance, and economical considerations. The reconstruction of the dam required an innovative combination of materials and construction techniques, including the use of reinforced concrete, post-tensioned anchors, and grouting of the masonry structure.

Roller-compacted concrete (RCC) may also be used in the construction of auxiliary spillways. RCC has properties similar to those of mass concrete but is proportioned in a manner to facilitate placement by conventional earth-moving equipment and compacted with vibratory rollers. RCC also provides resistance to erosion from seepage and overtopping. RCC has been used to form stair-step-type cascading spillways on dams where space limitations require the use of the entire crest width of the dam to pass the design flood.

Roller-compacted concrete was recently used on Boney Falls Dam in the upper peninsula of Michigan to remedy an inadequate spillway problem. The dam was built in the early 1920s.

Additional spillway capacity was provided by removing a portion of the existing left embankment behind an existing concrete core wall and constructing a new free-standing uncontrolled-overflow roller-compacted concrete (RCC) emergency spillway in its place. The RCC section was just over 500 ft long, 16 ft high, 21 ft wide at the base, and 11 ft wide at the top. An earth berm was placed over the RCC section and fuse plugs that would wash out the earth berm to provide the necessary spillway capacity.

REFERENCES

1. HANNA, T. H., *Foundations in Tension—Ground Anchors,* Trans Tech Publications, McGraw-Hill, New York, 1982.
2. U.S. Department of the Interior, Bureau of Reclamation, *Design of Gravity Dams,* 1976.
3. U.S. Department of the Interior, Bureau of Reclamation, *Gravity Dam Design Stability,* ETL-1110-2-184, 1974.
4. Hobst, L., "Vizepitmenyek Kihorgonyzasa," Vizugi Kozlemenyek, Vol. 4, 1965, 475–515.
5. Post-Tensioning Institute, *Post-Tensioning Manual,* 5th ed., 1990.
6. Berardi, G., "Richerche teoriche e sperimentali sugli ancoraggi in roccia," *Geotecnia,* Vol. 6, 1960, p. 6.
7. LITTLEJOHN, G. S., and D. A. BRUCE, *Rock Anchors: State of the Art,* Foundation Publications Ltd., 1977.
8. U.S. Department of the Interior, Bureau of Reclamation, *Concrete Manual,* 8th ed., 1975.
9. SWIGER, W. F., *Foundation Treatment, Advanced Dam Engineering for Design, Construction, and Rehabilitation,* R. B. Janson, editor, Van Nostrand Reinhold, New York, 1988.
10. American Society for Testing and Materials, *Annual Book of ASTM Standards,* vols. 04.02 and 04.08, 1991.
11. WEBSTER, R. P., and L. E. KUKACKA, "In Situ Repair of Deteriorated Concrete in Hydraulic Structures: Feasibility Studies," *Technical Report* REMR C 5-6, U.S. Army Engineer Waterways Experiment Station, May 1987.
12. American Society of Civil Engineers, "Evaluation Procedures for Hydrologic Safety of Dams," A Report prepared by the Task Committee on Spillway Design Flood, Selection of the Committee on Surface Water Hydrology of the Hydraulics Division, 1988.

SECTION 12

BARRAGES AND DAMS ON PERMEABLE FOUNDATIONS

By J. A. Scoville and B. A. McKiernan[1]

BARRAGE AND THE RIVER

1. Introduction. The name *barrage* has been given to the relatively low head, diversion-type dam constructed chiefly in Pakistan, India, Egypt, Iraq, and other countries in the Middle East.

Such structures consist of a number of gated, flood-passing bays of masonry or concrete. The function of these structures is to raise the river level sufficiently to divert the river flow or a part of it into the main supply canal of an irrigation system.

Barrages include canal regulators and low-level sluices to maintain proper flow approach to the regulator and may include silt-excluder tunnels to control entrance of silt into the canal, fish ladders, navigation locks, and low-head hydro plants.

2. Nomenclature. Figures 1 and 2 show the nomenclature usually applied to the various parts of a barrage and which is used in this text.

3. Siting. In selection of a barrage site, the location and elevation of the irrigation canal or canals which the barrage will command play an important part. Adjustments in location of the canal to suit the barrage might be made to some extent, but the topography will generally limit this. The rivers on which barrages are constructed are generally unstable, but if it is possible to find a relatively stable site, where the floodplain is locally comparatively narrow, it is desirable to adopt this for the barrage location. Other factors which will influence location are

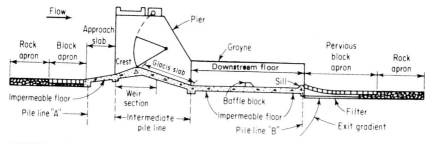

FIGURE 1 Section through barrage.

---

[1]Acknowledgment is made to Sir Thomas Foy and H. Spencer Green for material in this section that appeared in the third edition (1969).

12.2 BARRAGES AND DAMS ON PERMEABLE FOUNDATIONS

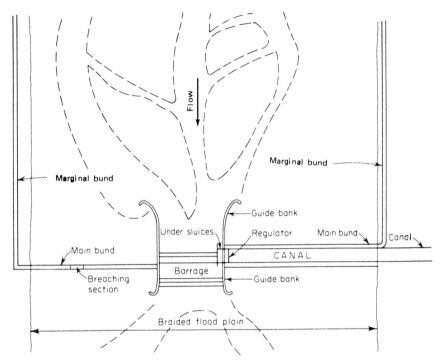

FIGURE 2 Plan of barrage and bunds on a braided river.

slope and curvature of the river, volume of pondage, environmental considerations such as relocation of riverain population, and interference with existing features such as bridges, roads, railroads, towns, and valuable farmland.

4. Orientation. If the river flows in a braided floodplain, the orientation and exact lateral location of the barrage must be carefully considered before a final decision is made. The requirements are that operation must be satisfactory after completion and construction must be easy.

To assist in the decision, an aerial photomosaic of the river for several miles upstream and downstream, taken at low stage, showing the pattern of river channels, is invaluable. Old maps of the river reaches under consideration may indicate regions, if any, of comparative channel stability.

In general, because the channels are unstable and a major channel which is on the right one year may in one flood shift to the left, and because of the general downstream progression of loops and meanders, the requirements are best met by selecting a tentative site for the barrage centrally in the floodplain and normal to the overall river axis in that reach.

In addition to satisfactory operation after completion, an important consideration is the facility of diversion of river flows during construction. In a number of cases where the floodplain is wide enough, the barrage is constructed on high ground toward the canal side surrounded by earth cofferdams, while the river continues to flow in the original deep channel. After completion of barrage construction, the river is diverted through leading cuts over the barrage, with simultaneous closure of the deep channel.

A large comprehensive model is a useful tool that can be used as a final check and confirmation of the orientation selected. The model can be constructed by using data from recent topographic maps and aerial photographs, care being taken to reproduce in the model important vegetative cover, such as forest growth, and differences in the nature of the soil, such as a local occurrence of clay.

5. Accretion and Retrogression. After a barrage has been constructed, the pond formed by it will act as a settling basin for the sediments carried by the river. The process of settling out of solids is called *accretion*. Accretion starts upstream of the barrage just beyond the limit of the scouring effects induced by the operation of the barrage gates and gradually works its way upstream, eventually raising the river bottom parallel to the prebarrage slope.

For a number of years after construction of a barrage, because of accretion, the water discharging from the barrage will be freer of silt and sediment than before construction. This clearer water, not having its normal load, will pick up silt and sediment from the river bed downstream of the barrage, causing a lowering or *retrogression* of the bed. The amount of water surface retrogression in some existing barrages has been between 4 and 14 ft for low river flows but has been only 1 to 1½ ft for maximum flood flows.

After accretion upstream has ceased, the prebarrage amount of silt will pass through the barrage. The downstream flows, which will carry higher concentrations after upstream accretion has stabilized, will no longer pick up bed material, but the reverse will occur. Retrogression will gradually change to accretion, with the river bed rising again to its original grade or perhaps even higher. This fact of accretion and retrogression is of great significance in design of barrages or dams on alluvial streams, and must be taken into consideration in the hydraulic design of the downstream floor of a barrage. Some reasonable estimate must be made of retrogression before the final level of the downstream floor slab of a barrage has been established. The slab must be low enough so that the hydraulic jump stays on it for all stages of flow with the river in a retrogressed condition.

6. Afflux. *Afflux* is a term applied to the increase in the maximum flood level upstream of a barrage as a result of its construction. It is also the difference in elevation between the headwater and tailwater at the time of maximum flood. The effect of afflux will extend upstream many miles, gradually tapering off to the prebarrage floodwater surface profile. In the rivers of the Punjab, where the sand is quite fine, designers have generally limited afflux to between 3 and 4 ft to obtain a discharge of 250 to 300 cfs per foot run of barrage width. A greater amount of afflux would result in a narrower barrage with higher discharges per foot run. This would necessitate lowering of the downstream floor, the pervious block apron, and the rock apron. The result could be uneconomic and could increase construction difficulties. In river beds of coarser-grained sand or consolidated or cohesive material, where discharge intensities greater than 300 cfs per foot can be accepted, more than 3 or 4 ft of afflux could be permitted.

It must be remembered that the figures quoted, i.e., 250 to 300 cfs per foot run, are average figures over the entire width of barrage. Because of the inequalities of intensity of flow, they will be greatly exceeded in certain bays during high floods, and this must be taken into consideration in the detailed design.

Top levels of guide banks will be determined by the afflux, and levels of flood-protection bunds will be determined by the backwater curve caused by the afflux. Over a number of years, accretion may increase the amount of afflux. The raised levels may occur for many miles upstream, but the effect diminishes as the distance increases. Additional freeboard on the guide banks and bunds may be constructed at the time of original construction. However, since the full effect will not take place for many years, the extra freeboard can be obtained when needed by adding a layer of compacted fill to the tops of these earth structures.

BARRAGE-SUPERSTRUCTURE DESIGN

7. Barrage Width. Three considerations govern the width of a barrage. They are the design flood, the Lacey design width, and the looseness factor.

The *design flood* must be established from flood records; methods used are fully covered in Sec. 1. In Pakistan and India, barrages were formerly designed to pass the maximum flood previously experienced, with some margin of safety. In terms of modern hydrology, the designs in practice were capable of accommodating floods of about 40- or 50-year frequency. Larger floods were experienced in the course of time, and these breached the main earth bunds, which in effect became fuse plugs. Such a principle has been found to be more economical than providing a greater spillway capacity in the barrage, and can be permitted because no great amount of stored water is lost, as the barrage is primarily a diversion structure. To have the main bund breach at a predetermined location, a section of this bund is constructed with a slightly lower crest than the rest of the bunds. This section is located at a safe distance from the barrage so that there will be little chance of barrage washout when the bund breaches.

An important consideration in design of a breaching section arises from the fact that the fuse plug cannot become operative until shortly before the flood peak so that erosion of the bund is limited. This, combined with the fact that the channel approach to the fuse plug soon gets silted up and overgrown with brushwood, reduces the capacity of the escapeway so that the relief afforded by the fuse plug is only a small percentage of the maximum flood.

The second consideration governing the barrage width is the combined widths of the channels approaching the barrage. These channels flow in erodible material and tend to adopt the Lacey regime given by the formula

$$W = 2.67\sqrt{Q}$$

where W = minimum stable width, ft, or *Lacey width*. The width W would be developed by the discharge Q flowing continuously for a length of time sufficient to develop regime conditions. The bed and banks of the channel, though erodible, have been formed by discharges less than the maximum flood, which never flows long enough to establish regime conditions. While the bed adjusts fairly rapidly to increasing discharge, the banks are often protected by a clay cover reinforced by vegetation. Away from the barrage, the flood is carried partly by the existing channels and partly over the floodplain. The barrage is accordingly designed for a width exceeding W, partly to accommodate this floodplain discharge, and partly to take advantage of the dispersion of the channel flow induced by the obstruction caused by the barrage itself.

This ratio of actual width to regime width is the *looseness factor,* the third factor affecting barrage width. It has been used in the design of barrages generally to increase the minimum width indicated by the Lacey formula. Values used have varied from 1.9 to 0.9, the larger factor being applied in the earlier designs. The extrawide barrage was advocated because of the convenience of construction resulting from not-too-low downstream floors and protections. At many of the "loose" barrages, sand islands developed immediately upstream, no doubt a direct result of too much width. The slower velocities permitted settling out of sediment and gradual accumulation until islands formed. Islands became anchored and fixed by growth of vegetation. They disturbed the flow patterns during floods, both upstream and downstream, resulting in undesirable cross currents and concentrations of flow. Very deep scour holes have been a consequence, sometimes causing failure or partial failure of the barrage. More recent designs have recognized this fault of too much width by using smaller looseness factors.

An additional reason for selecting a low value for the looseness factor is the increasing utilization of supplies in the upper reaches of a river for storage and diversion. These withdrawals reduce the low and medium supplies which shape and form the approach channels. The withdrawals have little, if any, effect on the size of the major floods. A barrage designed with a low looseness factor should have better control of the approach under the changed conditions.

8. Crest Level. Four factors determine the crest elevation of a barrage: the afflux, the pond level, the discharge per foot, and the coefficient of discharge. The afflux is usually set at 3 or 4 ft, and the average discharge per foot q is determined by dividing the design flood by the sum of the clear span distances. The coefficient of discharge may be taken as 3.3 in preliminary design for unsubmerged crests. For submerged crests, the coefficient varies between 2.5 and 2.8, depending on the degree of submergence. A value of 2.6 is usually satisfactory for first trials and can be refined as the design advances, using submergence factors given elsewhere in this book. By substituting the known values of q and C in the spillway discharge formula $q = Ch^{3/2}$, a value of head on crest h is obtained. Pond level is determined by adding the afflux to the natural flood stage at the location selected, and crest elevation is determined by subtracting h from the pond level.

9. Width of Bays. Clear distance between piers is governed by the magnitude of the structure, type of structure, bridge requirements, and kind and size of flotsam. The larger barrages generally have clear distances of 60 ft and the smaller ones about 40 ft. If the barrage will have to pass large logs or trees, the larger spans are in order. A design with reinforced-concrete floor slabs may be more economical than mass concrete design, but extensions of the piers downstream to the sill are required to provide the slab with weight against uplift.

10. Piers. Pier thickness has varied from 7 to 15 ft in different designs. However, 7 ft has been usually adopted and is considered to be adequate even for 60-ft spans with balanced vertical lift gates. For Tainter gates, thicker piers may be necessary. Forces that must be considered in pier design are the thrust coming from the Tainter gates or wheeled gates, particularly when an adjacent bay is unwatered by stop logs; bridge forces; and earthquake forces.

11. Glacis. *Glacis* is the name given to the surface which slopes down from the crest to the downstream floor. It is on the glacis that the upstream end of the hydraulic jump should occur. The glacis controls the location of the jump, keeping it in a well-defined zone for the full range of discharges. A jump forming on a horizontal surface is unstable and may move downstream with slight changes of discharge or tailwater. If the jump moves off the downstream floor, the structure is in jeopardy. Glacis slopes designed to follow the theoretical trajectory from a 1-ft gate opening under a normal pool ending in a 1 vertical on 3 horizontal slope have been frequently adopted in recent practice.

12. Downstream Floor. In the design of downstream floor levels, the possibility of nonuniform approach flow to the barrage is taken into account by increasing flow intensities by an appropriate concentration factor, generally 20 percent of the average approach flow intensity.

With a properly designed glacis and appropriate downstream floor elevation, the upstream end of the hydraulic jump will always occur on the glacis for all discharges. With the upstream end of the jump so confined, the downstream end of the downstream floor can be established. The length of downstream floor is made equal to the length of jump. Although the length of jump cannot be precisely determined, experience shows that most of the turbulence has ended in a length which is equal to $5(D_2 - D_1)$. In design of barrages of the Indus Basin Project in Pakistan, a length equal to $4.5Ef_2$ was adopted, where Ef_2 (see Fig. 3) is the energy head of flow in the stilling basin.

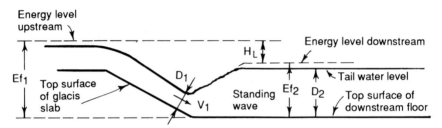

FIGURE 3 Hydraulic jump.

The top surface of the floor must be set deep enough so that the upstream end of the jump will always occur on the glacis. This depth is equal to D_2 ft below the surface of the tailwater. As shown in Fig. 3, D_2 is the subcritical flow depth occurring downstream of the standing wave of the jump, and can be computed from the hydraulic-jump formula

$$D_2 = -\frac{D_1}{2} + \sqrt{\frac{2V_1^2 D_1}{g} + \frac{D_1^2}{4}}$$

in which D_1 = supercritical flow depth and V_1 = supercritical flow velocity. V_1 is computed by using the potential head h in the formula $V = 2gh$, where h equals the difference between the energy level upstream and the level of the glacis at the point where the standing wave forms. Potential head should include the effect of nonuniform approach flow referred to above.

For making preliminary studies, the curves in Fig. 4 are useful in quickly obtaining the elevation of the downstream floor. Elevations of floor for several values of discharge and corresponding tailwater should be obtained. The lowest elevation so obtained would be the correct level. In determining the tailwater level to use in these computations, possible retrogression of the downstream river bed should be considered. As noted in Art. 5, it has been found from past experience on alluvial-sand rivers that retrogression over several years can lower tailwater 1 to 1½ ft during maximum flood, but 4 to 14 ft during normal or low flow stages.

The thickness of downstream floor and glacis must be sufficient to resist uplift and bending moments. Conditions of uplift with gates closed with low tailwater

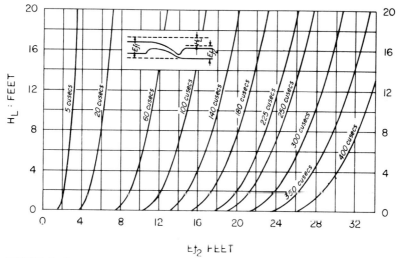

FIGURE 4 Curves for setting the level of the downstream floor.

and open with high tailwater must be considered. In the latter condition, a large unbalanced upward load occurs under the trough in the hydraulic jump, requiring the thickest portion of the floor. Some reduction of thickness can be made by bridging this large load over an upstream-downstream length of slab which is longer than the length of the trough.

13. Baffle Blocks. Baffle blocks of proper design will assist in concentrating the center of turbulence of the jump close to the glacis, and so will permit reduction in the length of the downstream floor. Size, location, and spacing of blocks can best be determined during hydraulic-model studies.

14. End Sill. A concrete end sill at the downstream end of the floor is essential to proper design. A well-designed sill will deflect the bottom currents of water off the floor at an upward angle. The angle of this current will cause a horizontal vortex or ground roller immediately downstream of the sill. Desirable and undesirable patterns for the ground roller are shown in Figs. 5 and 6.

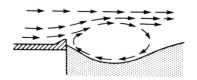 

FIGURE 5 Desirable ground-roller pattern—with end sill.

FIGURE 6 Undesirable ground-roller pattern—without end sill.

Sheetpiling under the sill is necessary, as will be discussed in subsequent paragraphs. The final shape of sill should be obtained from observing results on a movable bed model for several shapes and sizes of sills in a series of model runs.

15. Approach Slab. A concrete slab upstream of the weir section is a necessary feature. It will increase the seepage path and so reduce uplift pressures. It will provide a smooth erosion-resistant transition in an area where the velocity increases considerably.

The upstream end of the approach slab must be securely joined to the upstream line of sheetpiling or to the vertical concrete cutoff, if that is used. The downstream end is joined to the weir section in many designs with a flexible watertight joint. With this type of design, the approach slab is provided with temperature reinforcement only.

Alternatively, if it is desired to make the approach slab monolithic with the weir section in order to utilize its potential capacity to provide the structure with additional resistance against sliding, it will be necessary to determine the uplift pressures, to establish a suitable coefficient of friction between the slab and the foundation sand, and to design the reinforcement to suit these conditions.

Determination of uplift can be obtained by the method of independent variables, as discussed in Art. 20. Maximum values of coefficients of friction are 0.45 for clay and 0.55 for sand and gravel, but values 15 or 20 percent higher than this should be used when computing the reinforcing that resists sliding and 15 to 20 percent lower when computing the sliding resistance contributed by the slab to the whole structure. The slab will slope down to the upstream end gradually from the 1 vertical on 3 horizontal slope of the crest section. The elevation of the upstream end should be made low enough, if economically feasible, so that the average maximum approach velocity on the upstream blocks and stone apron will not exceed 10 fps, which is considered to be the upper limit for water flow on loose-stone aprons. The level is also set below the crest level a minimum of 0.2 times the maximum head on the crest to achieve a good coefficient of discharge.

16. Design of Weir Section, Glacis, and Downstream Floor. The weir section, glacis, and downstream floor are required to withstand the uplift pressure, and this objective can be achieved in two ways:

1. By making the section at all points of sufficient thickness and weight equal to the maximum uplift pressure at the point under consideration. This design is known as a *gravity section*.
2. By designing the barrage slab as a reinforced-concrete raft and utilizing the weight of the piers and groins to assist the glacis and floor slabs in resisting uplift. This design is known as a *raft design*; where it is adopted, care should be taken to provide some positive pressure or weight in excess of the total uplift pressure.

Where a raft design is adopted, the reinforcement will be designed in accordance with reinforced-concrete practice. Where the gravity design is used, the slab is often reinforced to prevent temperature and shrinkage cracks.

The selection of one type or the other is a matter of judgment, taking into consideration the comparative costs, facilities for construction, and available materials and skills.

17. Concrete-Block Aprons and Scour. Aprons of unreinforced-concrete blocks are usually placed immediately upstream of the approach slab and downstream of the downstream floor. The blocks are cast in place, usually measuring about 3 by 5 ft in plan and 2 ft deep for those upstream, and 3 to 4 ft deep for those downstream. The 4-ft-deep blocks are preferable as being less likely to shift under swirling scour. Downstream blocks are cast with a 3-in space between each and all adjoining blocks; upstream blocks are contiguous. The 3-in space is to permit relief of uplift pressure and is packed with gravel. Between the upstream blocks, the direction of flow is downward. It is therefore better to have the blocks as close as possible to reduce water flow and uplift.

It has been the practice in the Punjab to provide total lengths of block aprons equal to 1.5D to 2.0D downstream and about 1.0D upstream, where D is the design depth of scour below the top of a slab. Upstream the blocks are placed on a 2-ft thickness of loose stone, making a total thickness of 4 ft equal to the stone apron which continues farther upstream. The following definitions apply:

D = XR minus depth of water above top of slab, ft

X = 1.24 to 1.75 for upstream aprons, 1.75 to 2.0 for downstream aprons (the factor X provides for the additional scour due to concentration and any local eddy formation)

R = $0.9(q^2/f)^{1/3}$ = Lacey scour depth, ft below floodwater surface, based on unit flow without allowance for concentration

f = 1.0 for fine sand, about 2.5 for medium sand, and 4 for coarse sand

q = average flow, cfs per foot

Immediately downstream of the downstream floor slab, an inverted filter is provided for a length of 1 to 1.5 times the design scour depth without allowance for flow concentration in order to have an adequate cover for the downstream pile line to safeguard against the steepening of the exit gradient in case of scour downstream. The inverted filter consists of two or three layers designed for protection of the base soil. Downstream of the inverted filter blocks, a 2-ft layer of rock is provided.

The rules for designing block and stone aprons set out above and in Art. 18 were evolved by A. N. Khosla and are still widely followed.

18. Stone Aprons. In addition to the concrete-block aprons, additional protection against scour is provided by means of stone aprons upstream and downstream. Where there is the combination of concrete-block and stone aprons, the total quantity per foot run required in either upstream or downstream pervious aprons is 10D minimum, which includes blocks, filter, and riprap.

Stone protection is also placed on the compacted-earth guide-bank slopes and toes. According to Spring,[1] the quantity of stone in aprons should be sufficient to provide about 3 ft of cover over a slope of 1 on 2 to the level of deepest probable scour as shown on Fig. 7. The depth of cover should also be 1.24T, where T is the thickness of stone on the slope. Based on these criteria, the required stone volume in an apron width of 1.5D is about 7D per foot run.

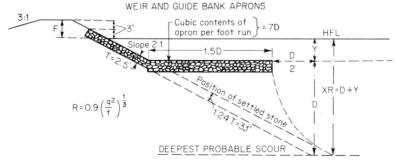

FIGURE 7 Criteria for stone protective blankets and aprons on barrage and guide banks.

Table for X in $XR = D + Y$ in Fig. 7*

Upstream of barrage and concrete blocks	1.25–1.75
Downstream of barrage and concrete blocks	1.75–2.25
Noses of guide banks	2.0–2.5
Transition from nose to straight guide banks	1.25–1.75
Straight reaches of guide banks	1.0–1.5

*These factors include allowance for concentration and are to be used when q does not include concentration allowances. Experience has shown that, for the guide-bank aprons, the upper limits of the suggested factors are preferable.

BARRAGE-SUBSTRUCTURE DESIGN

19. Steel Sheetpiling. The principal function of sheetpiling under a barrage or dam on a permeable-sand or other soft foundation is to confine and hold secure the foundation material under the structure during all conditions of river flow and turbulence. Other functions are to increase the seepage path to reduce uplift pressure and the exit gradient. A well-designed sheetpile system will make a continuous and closed circuit around the periphery of the entire structure. The lines under the flanking walls will guard against high subsoil water level on the flanks blowing up the floor slab and will ensure confinement of the foundation if an adjacent bund is washed out when the barrage is outflanked by a superflood (Fig. 8).

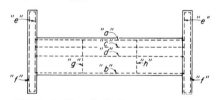

FIGURE 8 Plan of pile lines.

The two lines of piling given most consideration during design are under the upstream end of the approach slab a and under the downstream end of the floor slab b. These lines are the ones most vulnerable to attack by the river.

Other lines include intermediate piles parallel to the centerline of the structure c and d, which help to reduce uplift pressures, transverse lines under the toe and heel of the flanking structures e and f, and transverse lines between the two flanks g and h.

The upstream line of piling a has two functions:

1. To prevent the loss of foundation material in case a scour hole should occur near the structure
2. To increase the seepage path and thus reduce the uplift force under the structure

The downstream line b also has two functions:

1. Same as 1 above
2. To limit the exit gradient to a safe value

Function 1 governs the depth of both lines a and b. Intermediate lines c and d are installed in some designs to act as a second line of defense against loss of the weir, piers, gates, and bridge in case of failure of the approach or the floor slab. The intermediate lines also effectively increase the seepage path to reduce uplift under the floor slab.

Where the barrage is very long, transverse lines g and h are used to divide the foundation into sections. This is another precaution against progressive and total failure by undermining. Complete enclosure of the flanking-structure foundation material is recommended, as shown by pile lines e and f.

Perimeter piles are driven to a depth to prevent failure by scour. The Lacey formula, defined previously for scour depth, has been used as a guide in many barrage designs. In this application, q, the maximum discharge for foot run, is increased by some concentration factor, usually of the order of 20 percent.

Scour holes are usually the result of concentrated flows that develop because gates are opened unevenly, because sand islands create disturbance in laminar flow, or because lack of river training or changed conditions result in cross flows upstream, producing a higher water level and discharge at one flank than at the other. To have the design recognize these possibilities, the concentration factor is applied to the discharge before the Lacey formula is solved. The upstream line a can be driven deeper than indicated by the formula to be extra safe. However, the downstream line b should not be driven any deeper than necessary for proper exit gradient, since increased depth would increase uplift under the downstream floor. Depth of downstream piles is primarily determined by the exit-gradient considerations. Protection against scour is provided by the inverted filter and blocks and by the downstream flexible protection.

20. Uplift. Hydraulic pressure acting in an upward direction under a barrage or dam is termed *uplift*. Uplift varies under a structure, decreasing in magnitude from upstream to downstream in a manner depending on the nature of the foundation and the configuration of the structure. Determination of the magnitude and distribution of uplift is necessary in order to provide a structure which will safely resist uplift.

Several procedures are available for the determination of uplift, and the methods used depend on the complexity of the foundation and the analytical means available to the designer. Procedures include the graphical determination of flow nets,[2–10] preparation of flow nets by electric analogy,[3–5,8,9,11,12] the use of physical models,[5,8,9] analytical methods,[2,3,5,8,9,11] and the use of finite-element techniques.[13] Among the analytical methods, the method of fragments[9] is useful in obtaining approximate results quickly, and the method of independent variables[5] has been used extensively in India and Pakistan where foundation conditions permit the use of this method, in which it is assumed that foundation materials are homogeneous, are equally permeable in all directions, and are without provisions for drainage.

The method of independent variables evolved from a mathematical solution initiated by Weaver,[11] of the University of Wisconsin, which was developed and refined to an accurate method by Khosla et al.,[5] of the Punjab Irrigation Research Institute. The method, which is semiempirical, is applicable to any shape of structure having one or more lines of cutoffs. Harza[4] obtained agreement between theoretical values of uplift pressures and values from electric models for some of the simple cases dealt with by Weaver.

Khosla[5] developed a set of curves to give pressures for particular elements to be used in the solution by the method of independent variables. These curves are given in Fig. 9. As an example of use of the curves, the uplift under a typical type of barrage as shown in Fig. 10 has been solved. Each foundation element is considered independently with appropriate corrections as required by this method.

12.12 BARRAGES AND DAMS ON PERMEABLE FOUNDATIONS

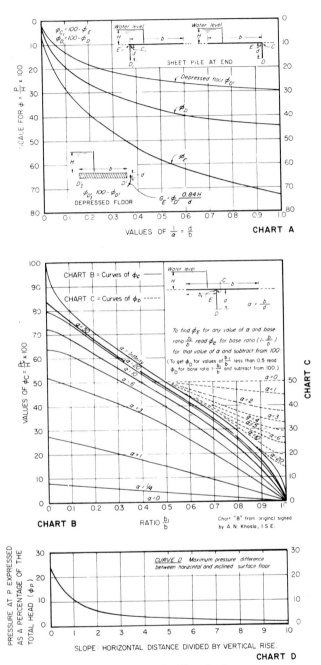

FIGURE 9 Curves for solutions of uplift under barrages.

BARRAGE-SUBSTRUCTURE DESIGN 12.13

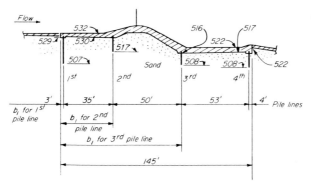

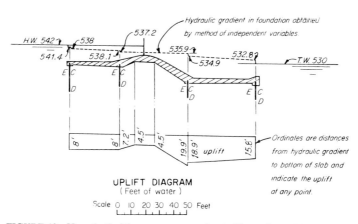

FIGURE 10 Hypothetical barrage cross section in illustrative uplift problem.

Using Khosla's notation, the symbol ϕ denotes the proportion of head H at any point; $\phi = 100P/H$. The suffix D applies to the bottom of a pile line; E and C apply to the junction of the pile and floor upstream and downstream, respectively; α is the ratio of the length of floor b to the depth of pile d; and b_1 is the distance of the pile from the upstream end of the floor. It is recommended that Ref. 5 be studied by anyone applying this method.

First pile line

$b = 145$ ft total length of monolithic slabs

$d = 532 - 507 = 25$-ft depth of pile below top of slab

$b_1 = 3$ ft

$$\alpha = \frac{145}{25} = 5.8$$

12.14   BARRAGES AND DAMS ON PERMEABLE FOUNDATIONS

$$\frac{b_1}{b} = \frac{3}{145} = 0.021$$

$\phi_E = 98\%$, computed as follows:
On chart B, read $\phi_C = 2\%$ for $1 - b_1/b = 1.0 - 0.021 = 0.98$ and for $\alpha = 5.8$. Then $\phi_E = 100.0 - 2 = 98\%$.

$\phi_D = 75\%$, computed as follows:
On chart C, since b_1/b is less than 0.5, subtract it from 1.0 and read 25% on right scale for $\alpha = 5.8$; $\phi_D = 100 - 25 = 75\%$.

$\phi_C = 63.0\%$, computed as follows:
On chart B, read 63.0% directly for $\alpha = 5.8$ and $b_1/b = 0.021$.

Instead of using the curves of Fig. 9 as outlined above, values of ϕ_E, ϕ_C, and ϕ_D may be computed from equations given in Ref. 5, as follows:

$$\phi_E = \frac{1}{\pi} \cos^{-1}\left(\frac{\lambda_1 - 1}{\lambda}\right)$$

$$\phi_C = \frac{1}{\pi} \cos^{-1}\left(\frac{\lambda + 1}{\lambda}\right)$$

$$\phi_D = \frac{1}{\pi} \cos^{-1}\left(\frac{\lambda_1}{\lambda}\right)$$

where angles are in radians and

$$\lambda = \frac{\sqrt{1 + \alpha_1^2} + \sqrt{1 + \alpha_2^2}}{2}$$

$$\lambda_1 = \frac{\sqrt{1 + \alpha_1^2} - \sqrt{1 + \alpha_2^2}}{2}$$

$$\alpha_1 = \frac{b_1}{d} \qquad \alpha_2 = \frac{b_2}{d} \qquad b = b_1 + b_2$$

In the example above,

$$b = 145 \qquad b_1 = 3 \qquad b_2 = 142 \qquad d = 25$$

$$\alpha_1 = 3/25 = 0.12$$

$$\alpha_2 = 142/25 = 5.68$$

$$\lambda = 3.387 \qquad \lambda_1 = -2.380$$

From the above,

$$\phi_E = 0.980 \quad \text{or} \quad 98.0\%$$

$$\phi_C = 0.634 \quad \text{or} \quad 63.4\%$$

$$\phi_D = 0.748 \quad \text{or} \quad 74.8\%$$

Corrections to be applied to ϕ_E and ϕ_C:

$$\phi_E \text{ correction for depth} = \frac{532 - 529}{532 - 507} \times (\phi_E - \phi_D) = 2.8\%$$

$$\phi_C \text{ correction for depth} = \frac{532 - 530}{532 - 507} \times (\phi_D - \phi_C) = 1.0\%$$

Influence of second pile line on ϕ_C:

$$D_2 = 530 - 517 = 13 \text{ ft}$$
$$D_1 = 530 - 507 = 23 \text{ ft}$$

$$\text{Percentage correction to } \phi_C = +19 \sqrt{\frac{D_2}{b'}} \times \frac{D_1 + D_2}{b}$$

$$= +19 \sqrt{\frac{13}{35}} \times \frac{23 + 13}{145} = +2.9\%$$

where b' = distance between piles.

This correction is added when the influencing pile is downstream and subtracted when it is upstream of the pile line in question. Then:

$$\phi_E \text{ corrected} = 98 - 2.8 = 95.2\%$$
$$\phi_C \text{ corrected} = 63.0 + 1.0 + 2.9 = 66.9\%$$

Second pile line

$$d = 532 - 517 = 15 \text{ ft}$$
$$b_1 = 38 \text{ ft}$$
$$\alpha = \frac{145}{15} = 9.7$$
$$\frac{b_1}{b} = \frac{38}{145} = 0.26$$

$\phi_E = 72.0\%$, computed as follows:
  On chart B, read $\phi_C = 28.0$ for $b_1/b = 1.0 - 0.26 = 0.74$ and $\alpha = 9.7$. Then $\phi_E = 100 - 28 = 72.0\%$.

$\phi_D = 64.0\%$, computed as follows:
  On chart C, since b_1/b is less than 0.5, subtract from 1.0 and read 36.0% on right scale for $\alpha = 9.7$, $\phi_D = 100 - 36.0 = 64.0\%$.

$\phi_C = 58.0\%$, computed as follows:
  On chart B, read 58.0% directly for $b_1/b = 0.26$ and $\phi = 9.7$.
  Corrections to be applied to ϕ_E and ϕ_C:

$$\phi_E \text{ correction for depth} = -\frac{532 - 530}{532 - 517}(\phi_E - \phi_D) = -1.1\%$$

$$\phi_C \text{ correction for depth} = +\frac{532 - 530}{532 - 517}(\phi_D - \phi_C) = +0.8\%$$

Influence of first pile line on ϕ_E:

$$\text{Correction to } \phi_E = -19\sqrt{\frac{D_2}{b'}} \times \frac{D_1 + D_2}{b}$$

$$= -19\sqrt{\frac{23}{35}} \times \frac{13 + 23}{145} = -3.8\%$$

$$D_2 = 530 - 507 = 23 \text{ ft}$$
$$D_1 = 530 - 517 = 13 \text{ ft}$$
$$b' = 35 \text{ ft}$$

Influence of third pile line on ϕ_C:

$$\text{Correction to } \phi_C = +19\sqrt{\frac{D_2}{b'}} \times \frac{D_1 + D_2}{b}$$

$$\text{Correction to } \phi_C = +19\sqrt{\frac{22}{50}} \times \frac{13 + 11}{145} = +2.1\%$$

$$D_2 = 530 - 508 = 22 \text{ ft}$$
$$D_1 = 530 - 517 = 13 \text{ ft}$$
$$b' = 50 \text{ ft}$$

Correction to ϕ_C for slope:
Slope is 1 vertical on 2 horizontal upward; from chart D, correction is -6.5% (minus for slope upward with flow). Proportional length is 10/50; therefore correction = $6.5 \times 1/5 = -1.3\%$. Then

$$\phi_E \text{ corrected} = 72.0 - 1.1 - 3.8 = 67.1\%$$

$$\phi_C \text{ corrected} = 58.0 + 0.8 + 2.1 - 1.3 = 59.6\%$$

Third pile line

$$d = 522 - 508 = 14 \text{ ft}$$
$$b_1 = 88 \text{ ft}$$
$$\alpha = \frac{145}{14} = 10.4$$
$$\frac{b_1}{b} = \frac{88}{145} = 0.61$$

$\phi_E = 48.5\%$, computed as follows:
On chart B, read $\phi_C = 51.5$ for $b_1/b = 1.0 - 0.61 = 0.39$ and $\alpha = 10.4$; then $\phi_E = 100 - 51.5 = 48.5\%$.

$\phi_C = 38.0\%$, computed as follows:
On chart B, read 38 percent directly for $b_1/b = 0.61$ and $\alpha = 10.4$.

ϕ_D = 43.0%, computed as follows:
On chart C, since b_1/b is greater than 0.5, read 43 percent directly on right scale.

Corrections to be applied to ϕ_E and ϕ_C:

$$\phi_E \text{ correction for depth} = -\frac{522-516}{522-508}(\phi_E - \phi_D) = -2.4\%$$

$$\phi_C \text{ correction for depth} = \frac{522-516}{522-508}(\phi_D - \phi_C) = +2.2\%$$

The influence of the second pile on ϕ_E is nil, since the bottom of second pile line is at elevation 517 and ϕ_E relates to elevation 516.
Influence of fourth pile line on ϕ_C:

$$\text{Correction to } \phi_C = +19\sqrt{\frac{D_2}{b'}} \times \frac{D_1 + D_2}{b}$$

$$= +19\sqrt{\frac{8}{53}} \times \frac{8+8}{145} = 0.8\%$$

$$D_2 = 516 - 508 = 8 \text{ ft}$$
$$D_1 = 516 - 508 = 8 \text{ ft}$$
$$b' = 53 \text{ ft}$$

Correction to ϕ_E for slope:

Slope is 1 vertical on 2 horizontal upward; from chart D, correction is +6.5 percent (plus for slope upward against flow). Proportional length is 25/50; therefore, correction = +6.5 × ½ = +3.25%. Then:

$$\phi_E \text{ corrected} = 48.5 - 2.4 + 3.25 = 49.35\%$$
$$\phi_C \text{ corrected} = 38.0 + 2.2 + 0.8 = 41.0\%$$

Fourth pile line

$$d = 522 - 508 = 14 \text{ ft}$$
$$\frac{1}{\alpha} = \frac{d}{b} = \frac{14}{145} = 0.097$$
$$b = 145 \text{ ft}$$
$$\phi_E = 27.5 \quad \text{by chart A, reading directly}$$
$$\phi_D = 18.8 \quad \text{by chart A, reading directly}$$

Corrections to be applied to ϕ_E:

$$\phi_E \text{ correction for depth} = -\frac{522-516}{14}(\phi_E - \phi_D) = -3.7\%$$

ϕ_E correction for influence of third pile line $C = -19\sqrt{\dfrac{8}{53}} \times \dfrac{8+8}{145} = -0.8\%$

ϕ_E corrected = 27.5 − 3.7 − 0.8 = 23.0%

Values of head loss at locations E and C for each pile line location may be computed from the corrected ϕ values calculated above.

$$\text{Head loss} = HW_1 - TW(1-\phi)$$

$$= (542 - 530)(1 - \phi) \text{ in the Fig. 10 example.}$$

The value of the hydraulic gradient equals

$$542 - (542 - 530)(1 - \phi) \text{ in the example.}$$

The uplift value is obtained at a location by subtracting the elevation of the bottom of the slab at that location from the hydraulic gradient elevation at that point. Values of the hydraulic gradient and uplift are listed in the table of hydraulic gradient and uplift and shown in Fig. 10.

Exit gradient (see Art. 21):

$$G_E = \dfrac{H}{d} \dfrac{1}{\pi\sqrt{\lambda}}$$

where

$$\lambda = \dfrac{1 + \sqrt{1 + \left(\dfrac{b}{d}\right)^2}}{2}$$

$$G_E = \dfrac{12}{14} \dfrac{1}{\pi\sqrt{5.7}} = 0.114$$

Uplift pressures and exit gradients for barrages of the Indus Basin Project[14,15] were generally evaluated in accordance with the Khosla[5] method outlined above. However, in the case of the Marala barrage, where a clay layer was known to exist, flow nets were used to determine uplift.

Nearly all authorities on the subject of uplift have concluded that uplift pressures should be applied to 100 percent of the base area of the structure. This holds particularly true for structures on permeable foundations where there is no adhesion between foundation and concrete.

In the past, barrages were designed without underdrainage to reduce uplift pressures. It was considered that underdrains might become clogged with fines as time passed, rendering them inoperative and resulting in greater uplift pressures than the design assumed. However, a number of American structures, such as the Petenwell and Castle Rock spillways in Wisconsin, and the Santee spillway in South Carolina, were built on deep fine-to-medium sand foundations and have well-designed underdrains. These structures have performed well for 40 years or more. Perhaps as a result, some modern barrage designs provide and rely on underdrainage systems such as drainage wells and drainage blankets, protected with filters as necessary.

Underdrainage systems are usually designed with the aid of electric analog studies. Redundancy is provided and great care must be taken with the design of filter and drainage layers.

Table for Hydraulic Gradient and Uplift in Fig. 10

Location	Values at E					Values at C			
	φ, %	Head loss, ft.	Hydr. grade elev., ft.	Slab bottom elev., ft.	Uplift, ft.	φ, %	Head loss, ft.	Hydr. grade elev., ft.	Slab bottom elev., ft.	Uplift, ft.
First pile line	95.2	0.6	541.4	530	11.4					
Second pile line	67.1	3.9	538.1	530	8.1	66.9	4.0	538.0	530	8.0
Third pile line	49.4	6.1	535.9	516	19.9	59.6	6.2	537.2	530	7.2
Fourth pile line	23.0	9.2	532.8	517	15.8	41.0	7.1	534.9	516	18.9

21. Exit Gradient. Loss of material from the riverbed at the end of the structure by piping may occur if the weight of the sand is overcome by the upward flow of water through it. Good design requires that the pressure gradient in the foundation material at the end of the structure, causing this upward flow, be as low as possible consistent with other requirements of design. The gradient is an inverse function of the depth of the downstream line of sheetpiling and the length of the structure, and a direct function of the head. Deep downstream piling will reduce the gradient; however, it will also increase uplift and therefore increase concrete and excavation quantities. A depth should be used that is just sufficient for reducing the gradient to a proper value. Experiments with sands of various densities have shown that flotation can occur with gradients from $G_E = 0.44$ to 1.44, or with an average value of G_E of about 1.0. A factor of safety of between 4.0 and 6.0 may be applied for coarse to very fine sand, respectively. Khosla has derived the following formula for determination of the exit gradient at point C for a single row of downstream piling, as illustrated in the upper right corner of chart A.

$$G_E = \frac{H}{d} \frac{1}{\pi\sqrt{\lambda}} \quad \text{with} \quad \lambda = \frac{1 + \sqrt{1 + (b/d)^2}}{2}$$

where G_E = exit gradient
H = head between headwater and tailwater
d = depth of downstream sheetpiling below the underside of the downstream inverted filter
b = length of concrete structure from upstream to downstream ends

22. Sliding Factor. In barrage design, as in any other dam design, a basic objective is to maintain a proper ratio of the total horizontal forces to the total net downward vertical forces. This ratio, known as the sliding factor, must be limited so that the frictional resistance of the foundation will not be exceeded. Since structures constructed on sand have no adhesion with the foundation, the shear-friction safety factor criterion is not applicable; only the friction or sliding factor can be used. In the case of fine sand, medium sand, coarse sand, or gravel foundations, a value of $\Sigma H/\Sigma V = f = 0.4$ is considered safe, and for clay the value should never exceed 0.3. Some English authorities recommend a more conservative 0.3 for sand and 0.2 for clay.

The condition of loading which causes the greatest tendency to slide is the normal condition with gates closed in which tailwater is low and net uplift is high. In computing resistance to sliding of the whole structure, no credit is usually given to the anchoring effect of sheetpiling. However, reinforcement should be included in the approach slab and around the top of the piling, assuming an anchoring effect.

Article 15 covers resistance to sliding of the approach slab.

BARRAGE APPURTENANT STRUCTURES

23. Gates and Hoists. Details of gate design can be found in Sec. 17. The varieties of gates used on modern barrages are the vertical lift fixed-wheel gate and the radial gate. They require less maintenance compared to the counterbalanced Stoney-type vertical lift gate, installed at many barrages in the first half of this century. Stoney gates, however, are still operating in several projects around the world, notably in India and Pakistan. To save capital expenditures that would be incurred in their replacement, a systematic program is under way in Pakistan to rehabilitate worn and corroded Stoney gates, tracks, and hoists.

Selection of the gate type, whether radial or vertical lift, is based on the same criteria as described for spillway crest gates in Sec. 17. Because of ready availability of power at most places, counterbalanced weights are not usually used to reduce the hoisting capacity. Hoist selection criteria are also the same as described in Sec. 17.

24. Flank Walls. The barrage flank walls are primarily retaining walls which support the abutting earth bunds. Therefore, retaining-wall principles of design should be followed, but with these precautions:

1. Design should provide for good positive contact between the earth fill and the backs of the flank walls to prevent seepage of water along the plane of contact. Backs of walls should therefore have a definite batter to create better contact as the fill settles. Vertical cutoff walls on the back of the flank wall into the abutting earth structure are usually provided along the upstream and downstream sheetpile lines. Depths of cutoffs are determined by horizontal flow net at the back of the wall. A check is made for safe exit gradient at the downstream end along the back side of the wall.

2. Water levels in the abutting earth should be considered for various loading conditions. In the case of fast lowering of tailwater, there can be a residual water level above tailwater.

3. Scour possibilities along the upstream and downstream ends of the flank walls beyond the limits of the monolithic concrete aprons should be carefully investigated. This can best be determined by model studies. Adequate depth of steel sheetpiling should be provided to protect the structure from possible scour.

25. Regulators. Barrages are constructed to raise the level of a river sufficiently to permit diversion into a supply canal. The control structures at the point of diversion are referred to as regulators and are located at one or both ends of a barrage. Model tests have indicated that a slight inclination upstream gives better approach conditions than a normal takeoff.

Design of the regulators follows the same principles used in design of barrages. Regulators are smaller versions of barrages designed to pass several thousand cubic feet per second, whereas barrages are designed to pass up to several hundred thousand cubic feet per second. Openings and crest elevations must be set to permit passage of the maximum canal flow at various pond levels. Silt exclusion from the canals is a very important consideration in connection with regulator design. Quite often silt-excluder tunnels are constructed in the barrage bay or bays adjacent to the regulator. These tunnels discharge onto the downstream side of the barrage and are designed to bypass the heavier silt-laden bottom layers of water which would otherwise go through the regulator into the canal.

26. Guide Banks. Guide banks are necessary for barrages on rivers with erodible beds and variable discharges. The ideal aimed at is to direct the main current of the river normally onto the barrage as centrally as possible, and so to center the channel that the embankments upstream and downstream of the barrage are kept away from river attack and erosion. They may also be designed to induce a favorable curvature for silt exclusion from the canals.

The upstream length of the guide banks should be at least equal to the width between the abutments of the barrage. In plan they may be normal to the barrage, may converge to form a funnel, or, where silt exclusion is an important consideration, may be given a curve concave to the river axis. At their upstream end they must be given a curve which is generally of 1000-ft radius and is carried round through at least 135° to the river axis.

The downstream guide banks should not be less than 500 ft long and curved outward in plan. They are usually from 0.1 to 0.2 times the barrage width.

The guide banks must be protected by stone placed on a base course on the side slope adjacent to the river and must be provided with an adequate self-launching stone apron in advance of the stone slope protection. The apron should be laid at the lowest practicable level.

Model tests are generally carried out to determine the optimum layout of the guide banks.

27. Marginal Embankments or Bunds. Barrages normally raise the pond levels above the level of the floodplains; flood levels are higher still, with the backwater curves extending many miles upstream. It is necessary to provide marginal embankments, or bunds as indicated on Fig. 2.

The function of these embankments is to prevent outflanking of the barrage by the pond or during floods. Where there has been encroachment on the floodplain, the embankments are aligned to limit the flooding of valuable agricultural land or village sites.

The banks are aligned to connect from the abutments with land at elevations above the afflux level and backwater curve. They generally cross the floodplain more or less at right angles to the river axis in the vicinity of the barrage and are then turned parallel to the river axis to protect adjoining land. In this portion of their length, which may extend for more than 10 mi, they should be set back far enough from the central river axis to avoid embayments, since it is usually not feasible from an economic aspect to protect the banks themselves against river scour. Setting back, however, in many cases is not practically feasible, so that it is often necessary to provide stone or brushwood armored spurs to protect the embankments, more particularly in the area immediately upstream of the barrage. Sometimes it is necessary to construct these spurs simultaneously with the barrage, but it is often possible to defer their construction until river attack develops. In the latter case it is desirable, if possible, to envisage the final line to which it is desired to train the river and to make financial provision for the spurs.

The design of the embankments follows the rules and practice for any water-containing bank as to geometry, freeboard, wave action, and resistance to seepage both through the bank and under it. By reason of their extent, local soil must be used and the design fitted to the available soil.

The tops of all banks should be made wide enough and hard enough to provide a roadway for vehicles for inspection and maintenance purposes.

DAMS ON PERMEABLE FOUNDATIONS

Many of the principles and criteria given above for barrages are equally applicable to dams on permeable foundations. Usually, in the case of a spillway for a dam, the afflux will be greater than for a barrage. This results in greater discharge per foot, requiring a deeper and possibly longer downstream floor. Dams generally resist greater heads than barrages, with attendant greater uplift, overturning, and sliding problems. Examples of dams on soft foundations are well-illustrated in the transverse sections given in Fig. 11. Spillway capacities are proportioned for design floods more of the order of a 1000-year recurrence rather than for a 100-year recurrence, as used for barrages. Reference 16 has useful information on a concrete dam constructed on a deep sand foundation.

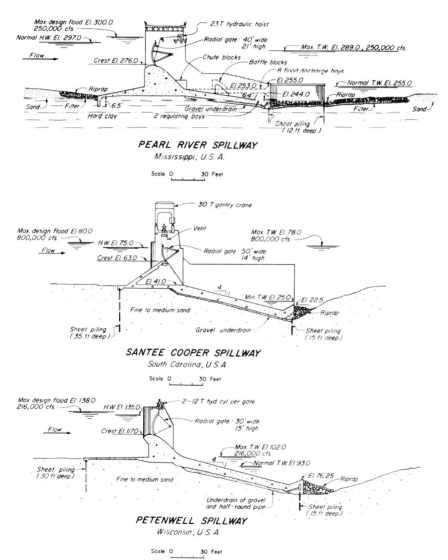

FIGURE 11 Examples of spillways constructed in the United States on permeable foundations.

REFERENCES

1. Spring, F. J. E., "River Training and Control on the Guide Bank System," Technical Paper No. 153, Government of India, 1903.
2. Forchheimer, P., *Hydraulik*, 3d ed., Teubner, Leipzig, 1930.
3. Pavlovsky, N. N., "Motion of Water Under Dams," report no. 36, vol. 4. First Congress of Large Dams, Stockholm, 1933.
4. Harza, L. F., "Uplift and Seepage Under Dams on Sand," paper no. 1920, American Society of Civil Engineers, 1935.
5. Khosla, A. N., N. K. Bose, and E. McKenzie-Taylor, "Design of Weirs on Permeable Foundations," Central Board of Irrigation, India, Publication 12, 1936.
6. Casagrande, A., Seepage Through Dams, *J. New England Water Works Association*, June 1937.
7. Taylor, D. W., *Fundamentals of Soil Mechanics*, Wiley, New York, 1948.
8. Leliavsky, S., *Irrigation and Hydraulic Design*, vol. 1, Chapman and Hall, London, 1959.
9. Harr, M. E., *Groundwater and Seepage*, McGraw-Hill, New York, 1962.
10. Cedergren, H. R., *Seepage, Drainage, and Flow Nets*, 2d ed., Wiley, New York, 1977.
11. Weaver, Warren, Uplift Pressure on Dams, *J. Mathematics and Physics*, Massachusetts Institute of Technology Press, vol. XI, no. 2, June 1932.
12. Zangar, C. N., "Theory and Problems of Water Percolation," U.S. Bureau of Reclamation Engineering Monograph No. 8, April 1953.
13. Finn, W. D. Liam, "Finite Element Analysis of Seepage," Soil Mechanics Series No. 2, University of British Columbia, March 1966.
14. Wilkinson, A. R., The Design and Construction of the Mailsi Siphon, paper no. 6889, *Proc. Inst. Civ. Engrs.*, December 1965.
15. Wilson, P. F., A. R. Wilkinson, and O. Andronov, "Barrages of the Indus Basin Project," paper no. 7483S, *Proc. Inst. Civ. Engrs.*, 1972.
16. Fucik, E. Montford, Petenwell Hydroelectric Project, *Trans. ASCE*, 1952.

SECTION 13

EMBANKMENT DAMS

By David E. Kleiner[1]

INTRODUCTION

Basic Requirements. The following criteria must be met to assure the satisfactory performance of earth and rockfill dams:

1. The embankment, foundation, and abutments must be stable during construction and reservoir operation and during and following unusual events such as earthquake and flood.

2. Seepage through the embankment, foundation, and abutments must be collected and controlled to prevent excessive uplift pressures, piping, sloughing, removal of material by solution, and erosion of material into cracks, joints, and cavities. In addition, some projects require a limitation on the rate of seepage.
Include seepage control measures such as foundation cutoffs, select core material, upstream impervious blankets, chimney filter and drain systems, blanket drains, finger drains, toe drains, multiple transition filters between core and rockfill shell material, drainage adits and tunnels, drain holes, and relief wells. Redundancy and multiple defenses are necessary.

3. Freeboard must be sufficient to prevent overtopping by wave action. Include an allowance for postconstruction settlement of the dam and its foundation and deformation caused by earthquake. In addition, freeboard must be sufficient to pass the maximum design flood, often chosen as the probable maximum flood. Spillways and outlets must be designed with sufficient capacity such that overtopping of the dam does not occur.

4. Outer slope protection on both the upstream and downstream slopes must prevent erosion by wave action, rainfall, and wind. Materials must be durable and resistant to wetting and drying and to freeze/thaw. Materials which will weather or erode cannot be used.

5. The foundations must be properly prepared and treated, and the dam must be constructed according to appropriate quality control and quality assurance procedures. Appropriate changes to the design must be made during construction should site conditions show the need to modifiy the design. The ultimate performance of the dam depends on proper construction practices, especially regarding foundation treatment, moisture and density control of the fill, and filters and drains.

---

[1]Acknowledgment is made to John Lowe III for material in this section that appeared in the third edition (1969). Acknowledgment is also made to Art Stukey and Kea-Ling Wong, Harza Engineering Co., for their contributions.

6. During reservoir filling and project operation, visually observe dam performance and evaluate instrumented measurements to identify problems and the necessity for remedial treatment. Maintain the dam to assure long-term acceptable performance. Look for the following and respond, if necessary: erosion of outer slopes, seepage emerging on the downstream slope or from abutments and foundations, changes in seepage rate or in the pore pressure distribution within the dam, clogged drains, seepage carrying fines, cracks, and increased settlement with time.

GEOLOGY

1. Introduction. The geology of a project site strongly influences design and construction. Factors such as foundation rock type, rock strength, structural geology, distribution and types of soil, groundwater occurrence, and regional seismicity often govern the dam and major structure location, dam type and size, construction methods and sequence, and reservoir operation.

Geologic investigations progress in stages from reconnaissance, through feasibility, to design, construction, and project operation. Similarly, investigation methods progress in stages ranging from general techniques to increasingly more complex or mechanized methods.

The geologic framework of the dam site, the reservoir, and the entire watershed must be investigated and understood. Information is added progressively to establish technical and economic feasibility and project design. Critical elements include structural and bedrock geology, geomorphology, groundwater geology, and earthquake evaluation.

2. Early Stage/Prefeasibility Investigations. These investigations include review of existing literature and site reconnaissance by experienced engineering geologists. The site visit provides the opportunity to assess the surface geologic and soil conditions and to evaluate site access.

Geologic mapping of the reservoir and potential damsites is conducted at the outset of project investigations. Mapping detects the presence of major faults, the orientation of regional joint sets, structure and sequence of bedrock units, distribution of soil units, and geomorphic evidence of slope instability in the project area.

Map data are interpolated and projected to the subsurface and used to suggest the extent of specific features or deposits. Mapping is most useful in planning the extent and types of subsequent investigations.

3. Intermediate/Feasibility Stage Investigations. Feasibility investigations confirm the absence of "fatal flaws," provide input to conceptual design, and allow estimates of project costs to within an accuracy level of approximately 20 percent.

The range of available tools and methods for geologic investigations is listed in Table 1 (Hunt, 1984). Each site is reviewed individually by geologists and engineers to evaluate the method or combination of methods that are most effective.

The most common exploratory methods include refraction geophysics, auger holes and test pits to find material sources and obtain samples, and bedrock coring to define the structural sequence and integrity of foundation rock and to conduct borehole permeability tests. The data are summarized, interpreted, and used to select the type, size, and location of major project features and to estimate project costs.

TABLE 1 Tools and Methods for Subsurface Investigations

Category	Applications	Limitations
Geophysical methods		
Surface seismic refraction	Determine stratum depths and characteristic velocities.	May be unreliable unless velocities increase with depth and bedrock surface is regular. Data are indirect and represent averages.
Uphole, downhole, and cross hole surveys (seismic direct methods)	Obtain velocities for particular strata; dynamic properties and rock-mass quality.	Data are indirect and represent averages, and may be affected by mass characteristics.
Seismic reflection	Used for deep engineering studies on land. Useful offshore for continuous profiling.	Does not provide velocities. Computations of depths to stratum changes require velocity data obtained by other means.
Electrical resistivity	Locate saltwater and freshwater boundaries, clean granular and clay strata; rock depth.	Difficult to interpret and subject to wide variations. Does not provide engineering properties.
Radar subsurface profiling	Provides subsurface profile; used to locate buried pipe, bedrock, boulders.	In development stages. Does not provide depths or engineering properties. Shallow penetration.
Video-pulse radar	Used to locate faults, caverns, voids, buried pipe, general rock structure.	Same for radar subsurface profiling.
Reconnaissance methods		
Test pits and trenches	Provides visual examination of soil stratigraphy, groundwater and rock depth, fault features.	Limited depth when machine-excavated. Deep excavation below groundwater level is costly when sheeting and pumping are required.
Adits and tunnels	Examination of rock quality in situ and access for in situ testing.	Costly for small projects. Normally not used in soil.
Continuous cone penetrometer	Continuous penetration resistance for all but strong soils. Fast and efficient.	Samples not recovered. Cannot penetrate strong soils or rock.
Hand augers	Continuous profiling in granular soils above groundwater level and in clayey soils of firm or greater consistency. Location of groundwater.	Samples disturbed. Cannot penetrate below groundwater level in granular soils. Penetration in strong soils very difficult.
Bucket auger	Similar to hand auger but greater penetration in strong soils.	Not used in unstable soils or below groundwater level.
Test borings		
Wash boring	Obtain soil samples primarily for identification and index testing, and to perform the Standard Penetration Test.	Slow procedure. Cannot penetrate strong soils or rock.
Rotary drilling	Obtain samples of all types in soil or rock for identification and laboratory testing of index and engineering properties, and in situ testing.	Requires large and costly equipment. Soil samples and rock cores normally limited to 6-in dia.
Continuous flight auger	Rapid drilling and disturbed sampling in soils with cohesion and greater than soft consistency. Normal sampling possible if hole remains open. Can penetrate soft rock.	Hole collapses in soft soils, dry granular soils without cohesion, and many soils below groundwater level.

TABLE 1 Tools and Methods for Subsurface Investigations (*Continued*)

Category	Applications	Limitations
Test borings		
Hollow-stem flight auger	Similar to continuous flight but hollow-stem serves as casing, permitting normal soil sampling.	Cannot penetrate strong soils, boulders, or rock.
Wireline drilling	Fast and efficient for deep core drilling on land and offshore.	Equipment costly and no more efficient than normal rotary drilling for most land investigations.
Borehole sensing and logging		
Borehole and TV cameras	Obtain continuous image of borehole in rock showing orientation of faults and joints. Small cavern examination possible when equipped with telephoto lens and spotlights.	Requires open hole. Images are affected by water quality.
Rock detector (acoustical sounding)	Differentiates boulders from bedrock and locates bedrock.	Results are qualitative.
Electric well logger	Provides continuous record of resistivity from which material types can be deduced when correlated with test borings.	Generally provides qualitative information. Best used with test boring information. Limited to uncased hole.
Scintillometer	Measures gamma rays. Used to locate shale and clay beds and in mineral prospecting.	Data generally of limited engineering use.
Gamma-gamma probe	Provides continuous measure of material density.	Value limited to density measurements.
Neutron probe	Provides continuous measure of natural moisture content. Has been used with the density probe to locate failure zones in slopes.	Value limited to in situ moisture content measurements.
Ultrasonic acoustical devices	Provides continuous image of borehole wall showing fractures and other discontinuities. Can be used to compute dip.	Images are much less clear than those obtained with borehole cameras.
Three-dimensional velocity logger	Provides an image of shear and compression waves for a short distance beyond borehole, and reveals fracture patterns. Used with acoustical device for rock quality and with gamma-gamma probe to obtain dynamic elastic properties.	Penetration depth beyond hole wall of a meter or so. Hole diameter must be known accurately and is measured with borehole calipers.
Caliper logging	Used to continuously measure and record borehole diameter.	Maximum range about 32 in.
Temperature logging	Continuous measurement of borehole temperature after fluid has stabilized.	Normally used in deep-hole drilling for petroleum exploration and well development.
Borehole surveying	Measures borehole inclination and direction in rock.	Relatively deep boreholes in good-quality rock.

Source: Adapted from Hunt, 1984.

4. Design Stage Investigations. Exploration at the design level concentrates on obtaining data for final design. The data collected at this and previous investigations are presented in contract bid documents so that prospective contractors can estimate the impact of geologic conditions on construction and cost.

At the early design stage, geologic questions are related to sources and volumes of construction materials, the extent and nature of permeable features, the stability of slopes at the site or in the reservoir, and the competence of foundations for the dam, spillways, and outlets. As the design stage progresses, data are obtained to evaluate the necessity and extent of foundation surface preparation, grout curtain, and drainage, the depth of excavation to reach acceptable or treatable foundation rock, and foundation conditions for structures. Material sources are identified and the characteristics and volumes of available material are estimated.

Specific exploration tools used in design investigations do not vary significantly from those used at earlier stages, but the number of specific boreholes or test pits, and the nature and quantity of materials tests are larger. *Detailed knowledge is required to complete the design, to prepare better cost estimates, and to avoid construction claims relating to changed foundation conditions.*

5. Field Supervision of Investigations. Geologic field investigation involves careful selection of techniques and close coordination between the design staff and field investigation team. The scope of investigations is flexible to permit rapid evaluation of data so that modifications to the remaining program may be made. Other considerations include:

Contracts. The scope of work must be clearly stated so that contractors are aware of the expected results, necessary equipment, and working procedures. Contract language is reviewed by experienced field geologists, preferably those who will be responsible for administering the contract in the field.

Staffing. Safety and logistic support of field personnel are foremost in planning and conducting field investigations. Logistic support includes vehicles, communication, realistic working schedules, and prompt decisions during the conduct of the field work.

Field activities must be led by experienced personnel. These individuals understand the objectives of the investigation and can make independent, responsible decisions.

Interpretation of field data throughout the investigations involves the active participation of the principal field investigators. The field staff contributes a first-hand understanding of the materials and site conditions, which often cannot be conveyed in normal paper documentation such as logs and daily reports. Useful information is obtained by observing the action of a drill rig, such as changes in drill water color and in penetration rates.

6. Construction Supervision. Perhaps the most critical period of geologic involvement is the construction phase when excavations reveal actual geologic conditions and materials. The condition of exposed foundations is studied and mapped to assure that the foundation treatment is consistent with the design intent.

Geologic activities during construction include mapping foundations and recommending the extent of grouting, drainage, overexcavation, dental excavation, and concrete, shotcrete, and rock support. The geologist interfaces with the construction personnel responsible for concrete and fill placement and provides a link between the design staff and the construction inspection staff.

EMBANKMENT TYPES

7. General Considerations. The zoning of a dam and the method of construction depend on the materials available from required excavations and from borrow. Low-permeability materials are used in the dam to reduce seepage to a permissible limit; other materials, often more pervious, are incorporated to assure stability with economical outer slopes.

Seepage rates and the control of seepage are studied to estimate flow and gradient through and around the dam and to design the necessary filter and drainage systems. The uppermost line of seepage must be kept well within the dam to protect the downstream slope against softening and sloughing. Modern dam design uses chimney and blanket filters and drains to capture seepage within the body of the dam and to convey these flows safety to the downstream toe. The chimney filter also serves to protect the core, if, for any reason, core cracking or other defects allow water to move freely through the core. Filters and drains are key elements of embankment dams.

Shell materials vary from the higher-quality, hard, durable rock and well-graded sands, gravels, and cobbles to the lower-quality, weathered, low-strength rock and mixtures of rock and soil. Core materials consist of broadly graded glacial till; broadly graded mixtures of silts, sands, and weathered rock; moderately plastic silts and clays with some sand sizes; and nonplastic silts and sands. Most materials encountered at the dam site can be used in the embankment, provided the characteristics of the materials and their potential use are clearly understood.

8. Earth Dams. An earth dam is composed of suitable soils obtained from borrow areas or required excavation. Following preparation of the foundation, earth fill is transported to the site, dumped, and spread in layers of a specified thickness. The soil layers are then compacted by heavy pad-foot compactors or, in some cases, by vibratory pad-foot or smooth drum rollers.

A major advantage of the earth dam is that it can be adapted to a weak foundation by flattening the outer slopes or constructing weighting berms at the upstream and downstream toes. Figure 1 illustrates various types of earth dams.

9. Rockfill Dams. A rockfill dam consists largely of fragmented rock with an impervious earth core, Fig. 2. The core is separated from the rock shells by a series of transition filters built of properly graded material. The rockfill zones are compacted using smooth drum vibratory rollers in layers 12 to 48 in thick, depending on the characteristics of the rock.

Materials for rockfill dams range from hard, durable, and free-draining rock to the more friable materials such as weakly cemented sandstones, siltstones, and shales that break down under handling and compacting. Because these latter materials are weak or weathered, flatter outer slopes must be used. Often, the best methods of construction and compaction are determined on the basis of test excavations in the proposed quarry and test fills using various compactors. These tests provide data on the amount of breakdown, the appropriate lift thickness, and provide insight regarding the shear strength of the compacted material.

Free-draining, well-compacted rockfill is placed with steep slopes if the dam is on a competent rock foundation. Slopes as steep as 1.3H:1V are used for concrete-faced rockfill dams (H = horizontal, V = vertical). Commonly, slopes varying from 1.6 to 1.75H:1V are used for central core rockfill dams. Slopes must be flattened if the rockfill is placed on a weak foundation such as decomposed or highly weathered rock or a foundation with low-angle bedding shears.

FOUNDATION TREATMENT

10. General Considerations. The choice of type of dam and the design of the dam are often dictated by the foundation conditions at the site. The foundation is treated to control seepage, to provide additional bearing capacity or stability, or

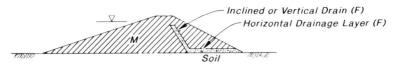
(a) Homogeneous dam with internal drainage on impervious foundation

(b) Central core dam on impervious foundation

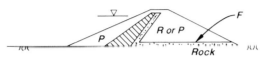
(c) Inclined core dam on impervious foundation

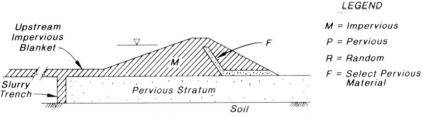
(d) Homogeneous dam with internal drainage on pervious foundation

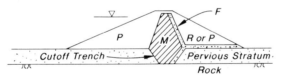
(e) Central core dam on pervious foundation

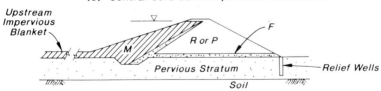
(f) Dam with upstream impervious zone on pervious foundation

FIGURE 1 Typical cross sections of earth-fill dams with filter/drain systems. (*U.S. Army Office of the Chief of Engineers, 1982.*)

to reduce differential settlement. Treatment consists of excavation and replacement, cutoffs, impervious upstream blankets, foundation drainage, and combinations of these techniques.

The dam must accommodate the given foundation conditions and the foundation treatment must be compatible with the dam and with the foundation characteristics. The outer slopes selected for the dam must be compatible with the strength of the foundation.

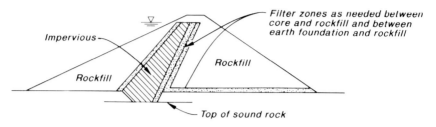
(a) Dam with inclined impervious zone

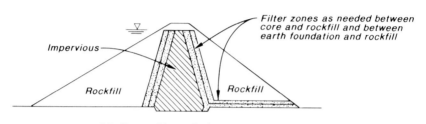
(b) Dam with central core

FIGURE 2 Typical rockfill dams with filter/drain systems. (*U.S. Army Office of the Chief of Engineers, 1982.*)

Often, foundation conditions are erratic and difficult to define. Success depends on careful subsurface investigations to disclose strata or lenses critical to stability or seepage, design of appropriate foundation treatment, and careful execution of foundation excavation and treatment during construction.

Construction often reveals conditions that were not observed during the site investigations. When this occurs, foundation treatment concepts must be reviewed and appropriate changes made.

11. Foundation Treatment. Tables 2 and 3 provide outlines of foundation treatment considerations to fulfill the following fundamental purposes of the treatment:

1. Obtain positive control of underseepage.
2. Remove unstable or unsuitable foundation material.
3. Prepare soil and rock surfaces to achieve satisfactory contact with overlying compacted fill.
4. Minimize differential settlements and thereby prevent cracking in the fill.

Tables 2 and 3 provide a brief overview of common foundation treatment techniques. Foundation conditions must be evaluated with care; often, a combination of treatments is required.

References such as U.S. Army Office of the Chief of Engineers, 1977, and Swiger, 1988, provide detailed discussions of the various methods to prepare and treat foundations, including unfavorable foundation conditions such as:

1. Highly compressible and low-strength soils
2. Clay shales

TABLE 2 Surface Foundation Treatment

A. Clearing and grubbing. Remove trees, brush, stumps, buried debris, and other objectionable material.
B. Stripping. Remove organic topsoil to a firm overburden surface to receive earth fill or to a sound rock or weathered rock surface to receive earth or rock fill. The foundation must posses strength equal to or greater than the overlying fill.
C. Preparation of earth surfaces
   1. Remove all debris
   2. Drain or pump standing or running water
   3. Disk to depth of 9 in and compact in accordance with specifications for earth fill
D. Cleaning and preparation of rock surfaces
   1. Remove all debris
   2. Drain or pump standing or running water
   3. Remove unsuitable material from beneath the outer shells of the dam
   4. In addition, beneath core and filters:
      a. Excavate rock irregularities and protrusions to minimize hand compaction.
      b. Excavate soft material from cracks, crevices, joints, fractures, and cavities.
      c. Backfill cracks, crevices, joints, fractures, and cavities with concrete. If necessary, use concrete to backfill irregularities in the rock surface to minimize hand compaction to fill material.
      d. Clean surfaces with air and high-pressure water, unless the rock surface can be damaged by water.
      e. Immediately prior to placement of the core material, broom a minimum thickness of ½ in of slush grout onto the surface. The grout consists of water, sand, and portland cement and has the consistency of a thick cream. On steep slopes, the grout may be placed pneumatically.
      f. Place the first lift of core material prior to the set of the slush grout. The grout seals the rock surface, fills small cracks, and provides a thin, nonerodible, soil-cement layer which bonds to the rock surface.

3. Collapsible soils
4. Loose granular soils
5. Old river channels
6. Weathered rock
7. Open joints and fractures
8. Cavities and solution features
9. Overhangs and surface depressions
10. Springs and seeps from the foundation

SUBSURFACE INVESTIGATIONS

12. General Considerations. A variety of field investigation and testing methods are available to evaluate the nature of subsurface conditions at a dam site. Many of these procedures are described by ASTM, 1991, and by various government agencies [U.S. Bureau of Reclamation (USBR), 1974]. Tables 4 and 5 list some of the most pertinent subsurface investigation methods.

TABLE 3 Control of Underseepage

A. Types of cutoffs
  1. Open trenches. Backfilled with earth fill and filter material.
  2. Slurry trenches
     a. Soil/bentonite. A bentonite slurry is used to maintain the stability of the vertical walls of the trench. Backfill consists of a mixture of well-graded soil and the bentonite slurry.
     b. Cement/bentonite. The slurry to maintain wall stability consists of a mixture of portland cement, bentonite, and water; this same mixture also forms the backfill. Permeability of the cement/bentonite backfill commonly is about 10^{-6} cm/s, whereas the soil/bentonite backfill commonly has a permeability of about 10^{-7} cm/s.
  3. Grout curtains. The foundation is grouted with cement or chemicals. Most rock foundations and open coarse granular materials can be effectively grouted with cement grout. Use of chemical grouts is beyond the scope of this handbook.
  4. Concrete diaphragm walls. Concrete backfilled walls are constructed from the ground surface by specialized drilling and backfilling method or by mining methods to excavate adits, drifts, and shafts which are then filled with concrete to form a continuous wall.
B. Upstream impervious blankets. An upstream impervious blanket controls underseepage by lengthening the path of underseepage. The effectiveness of the blanket depends on its length, thickness, continuity, and the relative permeabilities of the blanket and the foundation.
C. Pressure-relief wells. Large-diameter wells are placed at the downstream toe of an embankment dam to intercept underseepage and relieve excess uplift pressures.
D. Toe drains and horizontal blanket drains. These collect and intercept underseepage below the downstream shell of the dam at locations where the weight of the dam far exceeds the uplift pressures. Water is carried to the downstream toe by means of pipe drains or coarse, uniform granular material. The pipes and coarse drain material require protection with properly graded filters.
E. Drainage galleries and tunnels. These are constructed within the abutments or foundation to intercept and control seepage at locations downstream of diaphragm walls or grout curtains. Holes are drilled from the galleries and tunnels to provide drainage and pressure relief within the rock mass above or below the galleries.

---

13. Logging and Testing of Soil. The USBR procedures E-1, E-2, and E-3 provide a summary of several techniques to sample and classify soil materials. Materials are obtained from borings, pits, and trenches to evaluate potential borrow sources and foundation conditions. Options include continuously sampling, sampling every 5 ft or at a change in soil character, or not sampling, as during the cone penetration or vane shear test. Combinations of sampling and in-situ testing for specific purposes are often conducted.

Several characteristics of the soil column are most important:

Classification of material with depth. This requires the preparation of a detailed log of the boring or pit. Materials are classified in the field and in the laboratory in accordance with the Unified Soil Classification System (ASTM, 1991, D2487-85; USBR, 1974), Fig. 3. Locations of samples and the degree of disturbance are noted. The position of the water level during drilling and general information on location, method of drilling, significant dates, logger, and unusual occurrences are recorded.

Strength of material with depth. In-situ tests such as the Standard Penetration

TABLE 4 American Society for Testing and Materials (ASTM), Book of ASTM Standards, Volume 04.08, Selected Field Tests

D4428M-84	Crosshole Seismic Testing
D2922-81	Density of Soil and Soil-Aggregate In Place by Nuclear Methods (Shallow Depth)
D2937-83	Density of Soil In Place by the Drive-Cylinder Method
D2167-84	Density and Unit Weight of Soil In Place by the Rubber Balloon Method
D1556-82	Density of Soil In Place by the Sand-Cone Method
D2573-72(1978)	Field Vane Shear Test in Cohesive Soil
D3385-88	Infiltration Rate of Soils in Field Using Double-Ring Infiltrometers
D4719-87	Pressuremeter Testing in Soils
D4750-87	Determining Subsurface Liquid Levels in a Borehole or Monitoring Well (Observation Well)
D3017-88	Water Content of Soil and Rock in Place by Nuclear Methods (Shallow Depth)
D3441-86	Deep, Quasi-Static, Cone and Friction-Cone Penetration Tests of Soil
D1586-84	Penetration Test and Split-Barrel Sampling of Soils
D2113-83(1987)	Diamond Core Drilling for Site Investigation
D3550-84	Ring-Lined Barrel Sampling of Soils
D1452-80	Soil Investigation and Sampling by Auger Borings
D1587-83	Thin-Walled Tube Sampling of Soils
D4553-85	Determining In Situ Creep Characteristics of Rock
D4554-85	In Situ Determination of Direct Shear Strength of Rock Discontinuities
D4395-84	Determining In Situ Modulus of Deformation of Rock Mass Using the Flexible Plate Loading Method
D4506-85	Determining In Situ Modulus of Deformation of Rock Mass Using a Radial Jacking Test
D4394-84	Determining In Situ Modulus of Deformation of Rock Mass Using the Rigid Plate Loading Method
D4623-86	Determination of In-Situ Stress in Rock Mass by Overcoring Method—USBM Borehole Deformation Gage
D4729-87	In Situ Stress and Modulus of Deformation Determination Using the Flatjack Method
D4645-87	Determination of In Situ Stress in Rock Using the Hydraulic Fracturing Method
D4525-85	Permeability of Rocks by Flowing Air
D4622-86	Rock Mass Monitoring Using Inclinometers
D4630-86	Determining Transmissivity and Storativity of Low Permeability Rocks by in-Situ Measurements Using the Constant Head Injection Test
D4631-86	Determining Transmissivity and Storativity of Low Permeability Rocks by in-Situ Measurements Using the Pressure Pulse Technique
D4435-84	Rock Bolt Anchor Pull Test
D4436-84	Rock Bolt Long-Term Load Retention Test
D4403-84	Extensometers Used in Rock

Test (ASTM, 1991, D1586-84), the cone penetration test (ASTM, 1991, D3441-86), and the vane shear test (ASTM, 1991, D2573-78) are commonly used to estimate the consistency, shear strength, and susceptibility of foundation or embankment materials to liquefy during an earthquake. Table 6 presents a summary of consistency and strength based on the Standard Penetration Test.

The cone penetration test provides data on strength, density, and classification of material by correlating the tip penetration resistance with frictional

TABLE 5 U.S. Bureau of Reclamation Earth Manual, 1974, Field Sampling and Testing Procedures

Designation	Description
E-1	Disturbed Sampling of Soils
E-2	Undisturbed Sampling of Soils
E-3	Visual and Laboratory Methods for Identification and Classification of Soils
E-18	Field Permeability Tests in Boreholes
E-19	Field Permeability Test (well permeameter method)
E-20	Inplace Vane Shear Test
E-21	Field Penetration Test with Split-tube Sampler
E-23	Field Density of Dry, Gravel-Free Soils
E-24	Field Density Test Procedure
E-36	Field Permeability Test (shallow-well permeameter method)
E-39	Investigations for Rock Sources for Riprap

resistance. A device known as the piezocone measures piezometric pressure at the tip of the cone. The time response to obtain a stable pressure allows a qualitative evaluation of permeability.

Permeability of material with depth. Several techniques to evaluate the in-situ permeability of soil are available. These include the procedures presented by the USBR in tests E-18, E-19, and E-36, Table 5. Tests in cased borings are performed at various depths as drilling proceeds. Figure 4 illustrates the test setup (from E-18, Table 5).

Permeability is estimated by using the following relationship:

$$k = \frac{Q}{5.5rH}$$

where k = permeability
Q = constant rate of flow into the hole
r = internal radius of casing
H = differential head of water

Any consistent set of units may be used. For convenience, if k is measured in cm/s, Q in gal/min, r in in, and H in ft, the above equation can be written:

$$k = \frac{CQ}{H}$$

where values of C vary with the size of the hole as shown below:

Radius r, in	Coefficient C
1	0.148
2	0.074
3	0.049
4	0.037
5	0.030
6	0.025

Criteria for Assigning Group Symbols and Group Names Using Laboratory Tests[A]				Soil Classification	
				Group Symbol	Group Name[B]
Coarse-grained soils more than 50% retained on No. 200 sieve	Gravels More than 50% of coarse fraction retained on No. 4 sieve	Clean gravels Less than 5% fines[C]	$Cu \geq 4$ and $1 \leq Cc \leq 3$[E]	GW	Well-graded gravel[F]
			$Cu < 4$ and/or $1 > Cc > 3$[E]	GP	Poorly graded gravel[F]
		Gravels with fines more than 12% fines[C]	Fines classify as ML or MH	GM	Silty gravel[F,G,H]
			Fines classify as CL or CH	GC	Clayey gravel[F,G,I]
	Sands 50% or more of coarse fraction passes No. 4 sieve	Clean sands Less than 5% fines[D]	$Cu \geq 6$ and $1 \leq Cc \leq 3$[E]	SW	Well-graded sand
			$CU < 6$ and/or $1 > Cc > 3$[E]	SP	Poorly graded sand[I]
		Sands with fines More than 12% fines[D]	Fines classify as ML or MH	SM	Silty sand[G,H,J]
			Fines classify as CL or CH	SC	Clayey sand[G,H,I]
Fine-grained soils 50% or more passes the No. 200 sieve	Silts and clays Liquid limit less than 50	Inorganic	$PI > 7$ and plots on or above "A" line[J]	CL	Lean clay[K,L,M]
			$PI < 4$ or plots below "A" line[J]	ML	Silt[K,L,M]
		Organic	Liquid limit-oven dried < 0.75	OL	Organic clay[K,L,M,N]
			Liquid limit-not dried		Organic silt[K,L,M,O]
	Silts and clays Liquid limit 50 or more	Inorganic	PI plots on or above "A" line	CH	Fat clay[K,L,M]
			PI plots below "A" line	MH	Elastic silt[K,L,M]
		Organic	Liquid limit-oven dried < 0.75	OH	Organic clay[K,L,M,P]
			Liquid limit-not dried		Organic silt[K,L,M,Q]
Highly organic soils	Primarily organic matter, dark in color, and organic odor			PT	Peat

FIGURE 3 Soil Classification Chart.

Notes to Fig. 3 Soil Classification Chart

A	Based on the material passing the 3-in (75-mm) sieve.
B	If field sample contained cobbles or boulders, or both, add "with boulders, or both" to group name.
C	Gravels with 5 to 12% fines require dual symbols: GW-GM, well-graded gravel with silt GW-GC, well-grade gravel with clay GP-GM, poorly graded gravel with silt GP-GC, poorly graded gravel with clay
D	Sands with 5 to 12% fines require dual symbols: SW-SM, well-graded sand with silt SW-SC, well-graded sand with clay SP-SM, poorly graded sand with silt SP-SC, poorly graded sand with clay
E	$Cu = \dfrac{D_{60}}{D_{10}} \qquad C_c = \dfrac{(D_{30})^2}{D_{10} \times D_{60}}$
F	If soil contains $\geq 15\%$ sand, add "with sand" to group name.
G	If fines classify as CL-ML, use dual symbol GC-GM or SC-SM.
H	If fines are organic, add "with organic fines" to group name.
I	If soil contains $\geq 15\%$ gravel, add "with gravel" to group name.
J	If Atterberg limits plot in hatched area, soil is a CL-ML, silty clay.
K	If soil contains 15 to 29% plus no. 200, add "with sand" or "with gravel," whichever is predominant.
L	If soil contains $\geq 30\%$ plus no. 200, predominantly sand, add "sandy" to group name.
M	If soil contains $\geq 30\%$ plus no. 200, predominantly gravel, add "gravelly" to group name.
N	PI ≥ 4 and plots on or above A line.
O	PI < 4 plots below A line.
P	PI plot on or above A line.
Q	PI plots below A line.

TABLE 6 Relationship between Penetration Resistance and Angle of Internal Friction of Cohesionless Soils

Type of soil	Penetration resistance N	Angle of internal friction ϕ
Very loose sand	< 4	< 29
Loose sand	4–10	29–30
Medium sand	10–30	30–36
Dense sand	30–50	36–41
Very dense sand	> 50	> 41

Source: Adapted from Peck, Hanson, and Thornburn, 1974.

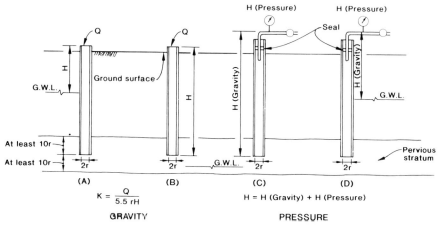

FIGURE 4 Open-end pipe test to evaluate field permeability. (*USBR, 1974.*)

In addition, permeability may be estimated by using a test similar to the packer test described in the next section. This test is performed through the sand section of an open standpipe piezometer constructed with perforated pipe.

14. Logging and Testing of Rock. In subsurface investigations for a dam, it is essential to estimate the depth to bedrock and the nature of the bedrock. The log of the hole in bedrock must include the lithologic description and the physical characteristics of the rock, emphasizing the occurrence of fractures and joints, the degree of weathering and alteration, and the occurrence of water.

Usually 2⅛-in (NX) cores obtained with a diamond bit are satisfactory, although 4- and 6-in-diameter cores are secured when the rock is fractured or when the holes are deep and telescoping is required. Commonly, double-tube core barrels with core catcher and check valve are used. Occasionally, triple-tube core barrels are required for ground conditions that are easily disturbed.

Use of polymer drilling fluids allows the retrieval of deeply weathered and decomposed rock and dense glacial tills. These are materials that are too hard to sample with ordinary soil sampling tools and too weak or friable for normal coring techniques.

The occurrence of core loss is noted on the log and the percentage of core recovery for each run is recorded. In addition, the Rock Quality Designation (RQD) (Deere, 1967) is computed for each core run. *The RQD is estimated by dividing the sum of the total length of core pieces 4 in or longer in each run by the length of the run, expressed as a percentage.* If the core is broken by handling or by the drilling process, the fresh broken pieces are fitted together, marked with Magic Marker, counted as one piece, and included in the RQD, provided that the pieces form the required length of 4 in. The 4-in length is valid only for N-size core and must be modified for other core sizes. The relationship between RQD and rock quality is shown below:

RQD, %	Description of rock quality
0–25	Very poor
25–50	Poor
50–75	Fair
75–90	Good
90–100	Excellent

Observation of the color of the drilling fluid, the loss or gain of drilling fluid, and the behavior of the drill provide considerable insight into the nature of the bedrock. Use of a borehole camera or television helps to establish jointing characteristics and the condition of the walls of the boring. The core is photographed, placed in sturdy wooden boxes, and stored in a secure, weather-protected location.

15. Pressure Tests in Rock. Data on the permeability of rock in situ and an indication of the soundness of the rock are obtained from pressure tests in drill holes. Figure 5 shows schematically the test setup. The water pressure in the sealed-off section is raised above the piezometric level within the rock at this section. The leakage into rock joints and fractures under various pressures such as 15, 30, and 45 psi above natural piezometric pressure is observed. A linear variation of leakage with pressure indicates stable joint openings for flow of water in the rock. If the leakage increase is less than the linear increase, the passageways are plugging; if greater, then the joints or fractures are unplugging or perhaps enlarging by hydraulic jacking.

The test data are evaluated using methods developed by the USBR (USBR, 1974) to estimate the mass permeability of rock:

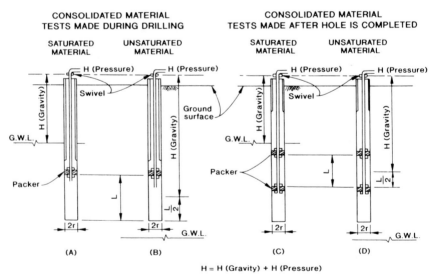

FIGURE 5 Packer test for rock mass permeability. (*USBR, 1974.*)

$$k = \frac{Q}{2\pi LH} \log_e \left(\frac{L}{r}\right) \quad \text{if } L \geq 10r$$

where k = permeability
 Q = constant rate of flow into the hole
 L = length of the portion of the hole tested
 H = differential head of water
 r = radius of hole
 $\log_e$ = natural logarithm

The above yields an approximate value of k because the effects of flow to the borehole around the packer are not considered and several simplifying assumptions are included. This does not change the order of magnitude of the permeability obtained or lessen the usefulness of the test.

LABORATORY TESTS

16. General Considerations. There are three general types of laboratory tests which are performed on materials for embankment dams: classification and identification tests, compaction tests, and physical property tests.

The classification and identification tests are performed on both foundation and borrow materials. The tests amplify and confirm the visual classification of materials and allow an estimate of the physical properties of the soil. On the basis of the identification tests, samples are chosen for the compaction and physical property tests. Figure 6 presents fundamental relationships and definitions within a moist or saturated soil.

Compact tests are utilized to estimate the optimum and allowable moisture contents for fill placement and to predict the unit weight of fill after compaction.

Physical property tests include the various shear strength, permeability, and compressibility tests and are performed on undisturbed samples of foundation materials and on compacted samples of borrow materials. The test procedures and the composition and condition of the test specimens must duplicate field conditions as closely as possible.

Standard test procedures are available for the classification and identification tests and for compaction tests. The physical property tests have been standardized by various agencies and test organizations. It is difficult to standardize the property tests because research provides improved techniques for testing and because projects frequently require modifications for the particular soil or conditions encountered at the site. The engineer must understand the test procedure and apply the results with judgment.

17. Test Procedures. Test procedures applicable to the design and construction of embankment dams are available through the American Society for Testing and Materials (ASTM, 1991), the U.S. Corps of Engineers (U.S. Army, Office of the Chief of Engineers, 1970*b*), and the U.S. Bureau of Reclamation (USBR, 1974). Tables 7, 8, and 9 list pertinent laboratory test procedures for soil and rock.

The design and construction of embankment dams is founded on a fundamental understanding of the nature of the materials within the dam and its foundation. The compaction characteristics, permeability, shear strength, and compressibility of foundation and fill must be understood. The use of the several types of shear strength tests is presented in the section on stability analysis. The section

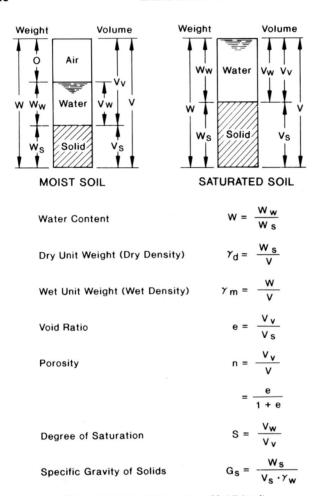

FIGURE 6 Fundamental relationships and definitions. (*U.S. Army Office of the Chief of Engineers, 1970b.*)

on seepage control presents ample evidence of the need to understand the permeability of the foundation and of the materials within the body of the dam.

SEEPAGE ANALYSIS AND CONTROL

The discussion which follows relies on several key references. These include:

1. Casagrande, 1937.
2. Harr, 1962.

TABLE 7 American Society for Testing and Materials, Book of ASTM Standards, Volume 04.08, Soil and Rock—Selected Laboratory Tests

D4753-87	Evaluating, Selecting, and Specifying Balances and Scales for Use in Soil and Rock Testing
D2487-85	Classification of Soils for Engineering Purposes
D2166-85	Unconfined Compressive Strength of Cohesive Soil
D4767-88	Consolidated-Undrained Triaxial Compression Test on Cohesive Soils
D4221-83	Dispersive Characteristics of Clay Soil by Double Hydrometer
D4647-87	Identification and Classification of Dispersive Clay Soils by the Pinhole Test
D4318-84	Liquid Limit, Plastic Limit, and Plasticity Index of Soils
D4253-83	Maximum Index Density of Soils Using a Vibratory Table
D4254-83	Minimum Index Density of Soils and Calculation of Relative Density
D698-78	Moisture-Density Relations of Soils and Soil-Aggregate Mixtures Using 5.5-lb (2.49-kg) Rammer and 12-in. (305-mm) Drop
D1557-78	Moisture-Density Relations of Soils and Soil-Aggregate Mixtures Using 10-lb (4.54-kg) Rammer and 18-in. (457-mm) Drop
D2435-80	One-Dimensional Consolidation Properties of Soils
D4186-82	One-Dimensional Consolidation Properties of Soils Using Controlled-Strain Loading
D4546-85	One-Dimensional Swell or Settlement Potential of Cohesive Soils
D2434-68(1974)	Permeability of Granular Soils (Constant Head)
D854-83	Specific Gravity of Soils
D2850-87	Unconsolidated, Undrained Compressive Strength of Cohesive Soils in Triaxial Compression
D4643-87	Determining Water (Moisture) Content of Soil by the Microwave Oven Method
D653-88	Soil, Rock, and Contained Fluids
D3080-72(1979)	Direct Shear Test of Soils Under Consolidated Drained Conditions
D422-63(1972)	Particle-Size Analysis of Soils
D2216-80	Laboratory Determination of Water (Moisture) Content of Soil, Rock, and Soil-Aggregate Mixtures
D4718-87	Correction of Unit Weight and Water Content for Soils Containing Oversize Particles
D4083-83	Description of Frozen Soils (Visual-Manual Procedure)
D2488-84	Description and Identification of Soils (Visual-Manual Procedure)
D421-85	Dry Preparation of Soil Samples for Particle-Size Analysis and Determination of Soil Constants
D3740-88	Evaluation of Agencies Engaged in the Testing and/or Inspection of Soil and Rock as Used in Engineering Design and Construction
D3584-83(1988)	Indexing Papers and Reports on Soil and Rock for Engineering Purposes
D420-87	Investigating and Sampling Soil and Rock for Engineering Purposes
D4220-83	Preserving and Transporting Soil Samples
D2217-85	Wet Preparation of Soil Samples for Particle-Size Analysis and Determination of Soil Constants
D2936-84	Direct Tensile Strength of Intact Rock Core Specimens
D3148-86	Elastic Moduli of Intact Rock Core Specimens in Uniaxial Compression
D4644-87	Slake Durability of Shales and Similar Weak Rocks
D3967-86	Splitting Tensile Strength of Intact Rock Core Specimens
D4535-85	Measurement of Thermal Expansion of Rock Using a Dilatometer
D2664-86	Triaxial Compressive Strength of Undrained Rock Core Specimens Without Pore Pressure Measurements
D2938-86	Unconfined Compressive Strength of Intact Rock Core Specimens
D4543-85	Preparing Core Specimens and Determining Dimensional and Shape Tolerances

TABLE 8 U.S. Department of the Army, Office of the Chief of Engineers, Laboratory Soils Testing EM 1110-2-1906, November 1970

Water Content—General
Unit Weights, Void Ratio, Porosity, and Degree of Saturation
Liquid and Plastic Limits
One-Point Liquid Limit Test
Shrinkage Limit Test
Grain-Size Analysis
Compaction Tests
Compaction Test for Earth-Rock Mixtures
Permeability Tests
   Constant Head Permeability Test with Permeameter Cylinder
   Falling Head Permeability Test with Permeameter Cylinder
   Permeability Tests with Sampling Tubes
   Permeability Test with Pressure Chamber
   Permeability Tests with Back Pressure
   Permeability Tests with Consolidometer
Consolidation Test
Drained (S) Direct Shear Test
Drained (S) Repeated Direct Shear Test
Triaxial Compression Tests
   Q Test (Unconsolidated, Sheared Undrained)
   R Test (Consolidated, Sheared Undrained)
   S Test (Consolidated, Sheared Drained)

TABLE 9 U.S. Department of the Interior, Bureau of Reclamation, Earth Manual, 1974, Laboratory Tests

E-3	Visual and Laboratory Methods for Identification and Classification of Soils
E-4	Lists of Laboratory Equipment
E-5	Preparation of Soil Samples for Testing
E-6	Gradation Analysis of Soils
E-7	Soil Consistency Tests
E-8	Soluble Salts Determination of Soils
E-9	Moisture Determination of Soils
E-10	Specific Gravity of Soils, Aggregate, and Density of Irregular Blocks of Soil
E-11	Proctor Compaction Test (Moisture-Density Relations of Soil)
E-12	Relative Density of Cohesionless Soils
E-13	Permeability and Settlement of Soils
E-14	Permeability and Settlement of Soil Containing Gravel
E-15	One-Dimensional Consolidation of Soils
E-16	Measurement of Capillary Pressures in Soils
E-17	Triaxial Shear of Soils
E-38	Compaction Test for Soil Containing Gravel (Moisture-Density Relations)

3. U.S. Army Corps of Engineers, 1986.
4. Cedergren, 1989.

   The complete reference for each of the above is presented in the bibliography.
   18. Permeability. Henry Darcy, a French engineer, conducted a laboratory experiment to study the flow of water through filters. The results of his studies indicated that

$$v = ki$$

Using $v = q/A$ gives

$$q = kiA$$

Using $q = Q/t$ gives

$$Q = kiAt$$

where v = discharge velocity, defined as the quantity of fluid that flows through a unit cross-sectional area of the soil oriented at a right angle to the direction of flow in a unit time (discharge velocity is used in determining the quantity of flow or rate of discharge through the soil)
k = coefficient of permeability
i = hydraulic gradient (head loss divided by the length over which the head loss occurs)
Q = total quantity of discharge
A = total cross-sectional area of flow
t = time over which flow occurs
q = rate of flow or discharge

Darcy's law is illustrated in Fig. 7.

The discharge velocity must not be confused with the seepage velocity, which is the average rate of movement of water through the soil, as might be measured by dye tracers, for instance. Seepage velocity always exceeds the discharge velocity and is related to the discharge velocity as follows:

$$\bar{v} = \frac{v}{n}$$

where n = porosity of the soil, defined as the ratio of the volume of voids within the soil to the total volume
$\bar{v}$ = seepage velocity

Darcy's law is valid for laminar flow only. Experiments indicate that when the Reynolds number exceeds about 1, flow becomes turbulent. Figure 8 relates the discharge velocity v, average diameter D_{50} of the soil, and flow conditions. Thus, for materials exceeding 10 mm in diameter and for discharge velocities above 1 cm/s the Darcy relationship is no longer valid. Leps, 1973, presents the following relationship to estimate the seepage velocity $\bar{v}$ in in/s:

$$\bar{v} = Wm^{0.5}i^{0.54}$$

where W = an empirical constant, which depends primarily on the shape and roughness of the particles and the viscosity of water (according to Wilkins, 1956, and others, W varies from about 33 for crushed gravel to about 46 for polished marbles), in units of in$^{0.5}$/s
m = hydraulic mean radius of the rock voids (for a given volume of particles, m is equal to the volume of voids divided by the total surface area of the particles, or the void ratio divided by the surface area per unit volume of solids) in in
i = hydraulic gradient

Leps, 1973, states:

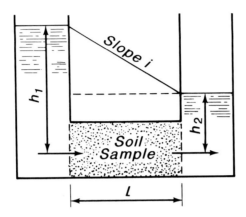

Quantity of Seepage $\quad Q = k \cdot i \cdot A \cdot t$

Discharge Velocity $\quad v = k \cdot i$

Hydraulic Gradient $\quad i = \dfrac{h_1 - h_2}{L}$

Area of Sample $\quad A$

Time $\quad t$

Coefficient of Permeability $\quad k$

FIGURE 7 Darcy's law for flow through soils. (*Casagrande, 1937.*)

A practical determination of m is fairly reliable for clean, monosized rock but is very uncertain for a well-graded or nonhomogeneous rockfill. For monosized rock, assuming that the specific gravity of the rock is 2.87, the void ratio is 1.0, and $W = 33$, data presented by Wilkins, 1956, for specific surface (based on physical measurement of the surface area of monosized crushed rock particles) would give the values listed below:

Rock size, inches	m	$m^{0.5}$	$Wm^{0.5}$
¾	0.09	0.30	10
2	0.24	0.49	16
8	0.75	0.87	28
24	3.11	1.76	58
48	6.43	2.54	84

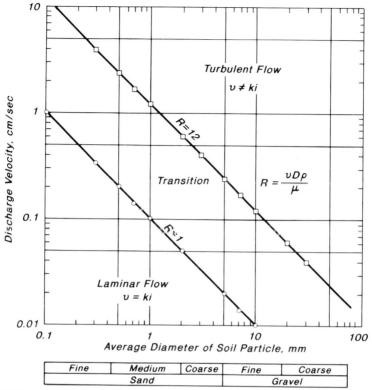

FIGURE 8 Boundary between laminar and turbulent flow determined by using Reynolds number for temperature of 20°C. (*U.S. Army Corps of Engineers, 1986.*)

For well-graded rockfills, Leps, 1973, suggests that the D_{50} size can be used to calculate the hydraulic mean radius provided that the minus 1-in size material is less than about 30 percent by weight. If the minus 1-in material exceeds 30 percent, the equivalent permeability should be determined experimentally.

Except possibly for flow through a gravel underdrain or through rockfill, seepage flow through an embankment dam is laminar. Thus, the determination of the Darcy coefficient of permeability is important in estimating the total flow through or around the dam.

The coefficient of permeability may be determined in the laboratory by tests on undisturbed samples or compacted specimens and by field permeability tests in overburden or rock. Various means are available to estimate the coefficient of permeability using the grain-size distribution of the soil. The range in permeability of various soil types on the basis of grain size is shown in Fig. 9 together with embankment characteristics and methods for determining permeability. An estimate of the coefficient of permeability can be made by using Hazen's expression (Hazen, 1892):

$$k = CD_{10}^2$$

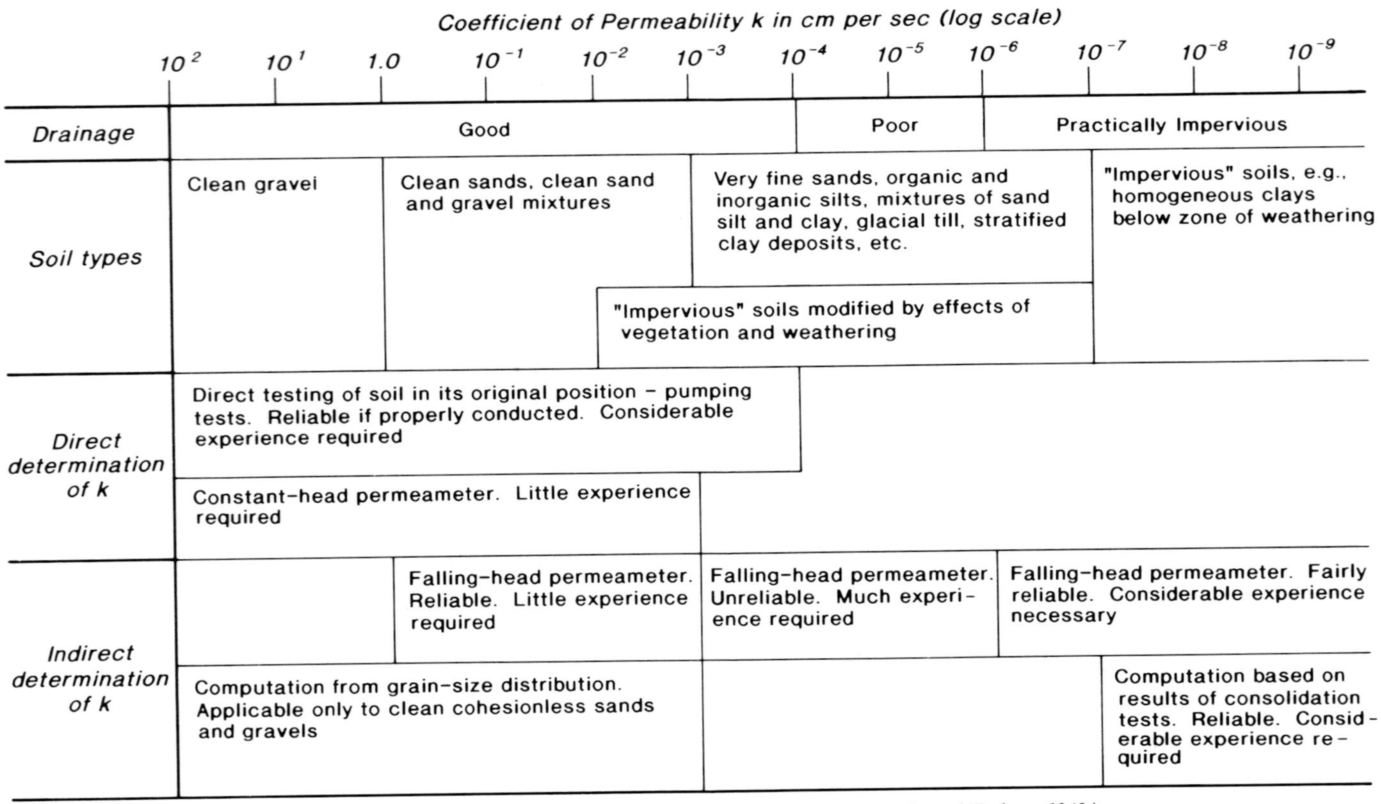

Coefficient of Permeability k in cm per sec (log scale)												
10^2	10^1	1.0	10^{-1}	10^{-2}	10^{-3}	10^{-4}	10^{-5}	10^{-6}	10^{-7}	10^{-8}	10^{-9}
Drainage	Good					Poor			Practically Impervious			
Soil types	Clean gravel	Clean sands, clean sand and gravel mixtures				Very fine sands, organic and inorganic silts, mixtures of sand silt and clay, glacial till, stratified clay deposits, etc.			"Impervious" soils, e.g., homogeneous clays below zone of weathering		
						"Impervious" soils modified by effects of vegetation and weathering					
Direct determination of k	Direct testing of soil in its original position – pumping tests. Reliable if properly conducted. Considerable experience required										
	Constant-head permeameter. Little experience required										
		Falling-head permeameter. Reliable. Little experience required			Falling-head permeameter. Unreliable. Much experience required			Falling-head permeameter. Fairly reliable. Considerable experience necessary			
Indirect determination of k	Computation from grain-size distribution. Applicable only to clean cohesionless sands and gravels								Computation based on results of consolidation tests. Reliable. Considerable experience required			

FIGURE 9 Permeability and drainage characteristics of soils. (*Terzaghi and Peck, 1948, Casagrande and Fadum, 1940.*)

where k = coefficient of permeability, cm/s
C = constant for which Hazen reported values ranging from 41 to 146 for fine sands as used in water-treatment filters; a value of 100 is commonly used in the Hazen expression
D_{10} = the grain-size diameter for which 10 percent of the material is finer by weight, in cm

On the basis of laboratory testing of filters, Sherard et al., 1984a, suggested the following relationship:

$$k = 0.35 D_{15}^2$$

where k = coefficient of permeability, cm/s
D_{15} = the grain-size diameter for which 15 percent of material is finer by weight, in mm

Natural soils almost invariably occur in strata of varying thickness and composition, with the result that the permeability of the soil varies from a maximum parallel to the strata to a minimum perpendicular to the strata. Compacted earth and rockfill stratify because of the inevitable variation in density and composition of material from layer to layer and within each layer. The seepage analysis must include these variations.

Casagrande suggests that the ratio of k_{max} to k_{min} for rolled earth fill is probably at least 9. The maximum and minimum permeabilities for thinly stratified soil may be estimated by using:

$$k_{max} = \frac{k_1 d_1 + k_2 d_2 + \cdots + k_n d_n}{d_1 + d_2 + \cdots + d_n}$$

$$k_{min} = \frac{d_1 + d_2 + \cdots + d_n}{d_1/k_1 + d_2/k_2 + \cdots + d_n/k_n}$$

where k_1, d_1, d_2, etc., are the corresponding permeabilities and thicknesses of the strata or layers making up the soil. The methods described in Art. 20 should be used to evaluate seepage through the thicker strata found in alluvial foundations.

19. Flow Nets. The pattern of flow for steady seepage of an incompressible fluid through a porous soil of constant pore space is expressed mathematically by a laplacian equation. Direct solution of the laplacian equation is difficult and is available for a limited number of cases. Examples of direct solutions as well as examples of solutions obtained by using modern mathematical tools for close approximations are given by Harr, 1962.

For two-dimensional flow, a graphical solution known as the *flow net* was proposed by Forchheimer and is widely used. This method is described in detail by Casagrande, 1937, and in various textbooks on soil mechanics.

The simplest procedure is to assume that the dam rests on an impermeable foundation, then estimate the flow pattern through the embankment cross section. The flow through the foundation is then estimated by assuming the embankment is an impermeable boundary. Each of these problems are solved independently, then compared at the common boundaries for compatibility. Unless a poor comparison results, the extra effort to develop a flow net for the dam and foundation together is not required.

For the embankment dam resting on an impervious foundation, the free surface is estimated using methods described by Casagrande, 1937; Harr, 1962; and U.S. Army Corps of Engineers, 1986. These techniques involve plotting the basic parabola then adjusting the construction to describe the free surface. Solutions for slopes of the discharge face that vary from nearly flat through vertical and overhanging to horizontal, i.e., from 10 to 180°, are presented.

A solution for the angle of the discharge face from 10° to vertical is shown in Fig. 10. The solution was first obtained by Gilboy, 1933. This chart can be used to estimate flow through central core dams, slightly sloping upstream core dams, and homogeneous or zoned dams with vertical or downstream sloping chimney drains.

For the earth dam, with horizontal permeability exceeding the vertical perme-

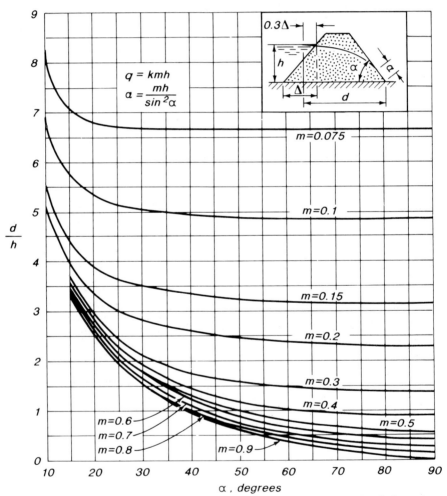

FIGURE 10 Determination of rate of discharge and position of free surface for discharge face angles from 10 to 90°. (*Harr, 1962.*)

ability by up to 10 times, the flow net of squares is drawn on a section transformed such that all dimensions in the direction of maximum permeability are reduced by dividing them by the square root of the ratio of k_{max} to k_{min} (or all the dimensions in the direction of minimum permeability are increased by multiplying them by the square root of the ratio k_{max} to k_{min}). After the flow net has been drawn on the transformed section, the flow lines and equipotential lines are transposed back to the true scale section. The process of constructing the free surface using two assumptions of horizontal vs. vertical permeability is illustrated in Fig. 11. This demonstrates the desirability of inserting a chimney filter and drain within the dam.

A relatively simple technique for flow net construction, using an electronic spreadsheet, was presented by Kleiner, 1985. The method, based on finite-difference techniques, is described below.

20. Seepage by Relaxation Methods. In the calculus of finite differences, the governing differential equation is replaced by a finite-difference approximation. Southwell, 1946, in his classic text, "Relaxation Methods in Theoretical Physics," presents a variety of problems and solutions using the finite-difference approximation. Nearly all approximation can be eliminated by making the finite differences as small as desirable by successive iterations and by decreasing the mesh size in local regions of interest. One of the applications presented by Southwell is the analysis of seepage through porous media. Southwell and his co-workers solved problems with irregular boundaries, layered materials with dissimilar permeabilities, and determination of the free surface for unconfined flow problems.

The electronic spreadsheet solves problems using finite-difference approximations because of its ability to recalculate as many times as necessary to reduce differences to insignificant quantities and because cell relationships within the spreadsheet can be specified. The spreadsheet is programmed so that the value computed at any given coordinate (cell) location is dependent on the values of the surrounding cells. For example, the numerical value of a cell can be set equal to the average of the cells above, below, and to the sides of the cell. This relationship between cells can be replicated to all cells that have the same relationship to their neighbors by activation of the COPY command in the spreadsheet software. Specific relationships are specified for cells at or near boundaries and at or near the interface between materials of differing permeabilities.

Figure 12 illustrates a typical seepage problem, that of seepage through a two-layer system beneath a dam. The upper layer, a silty sandy material, is 10 times less permeable than the lower layer, a more sandy material. In this example, the spreadsheet was recalculated 150 times in a few seconds. After this many recalculations, changes in values at any cell occur only beyond several decimal places. The problem is formatted on the screen so that the vertical and horizontal scales are only slightly distorted. Line spacing on modern printers or plotters is adjusted to within a small fraction of an inch to eliminate scale distortion when printing a hard copy.

Figure 13 includes both inclined boundaries and a free surface in the example of seepage through the central core of an embankment dam. Values of cells adjacent to boundaries are based on the distance of the cell to the boundary and the distance to neighboring cells. Figure 14 is a summary of typical cell relationships for interior cells, for cells adjacent to or on a boundary, and for cells on an interface between materials of differing permeability.

Since the free surface is not initially known, a best guess is made using techniques suggested above. The precision of the assumed free surface can be tested

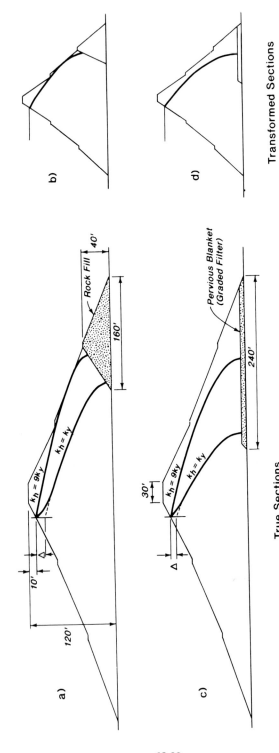

FIGURE 11 Effect of anisotropy for homogeneous embankment on impervious foundation. (*Casagrande, 1937.*)

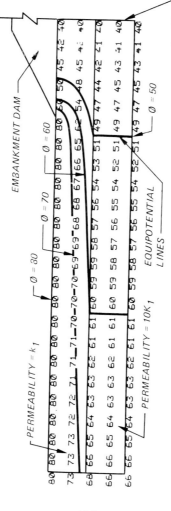

FIGURE 12 Equipotential lines for foundation layers with differing permeability determined from use of an electronic spreadsheet. (*Kleiner, 1985.*)

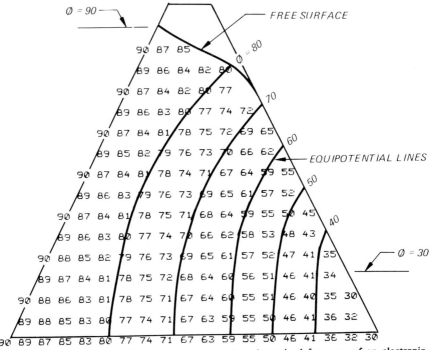

FIGURE 13 Equipotential lines for central core dam determined from use of an electronic spreadsheet. (*Kleiner, 1985.*)

by using the relationship shown in Fig. 14*f*. An adjustment to the free surface can then be made, new relationships input for those cells adjacent to the free surface, and the spreadsheet recalculated. For most practical problems, it is not necessary to make an adjustment to the free surface after a careful hand approximation of the free surface has been made.

21. Discharge Rate of Seepage. The discharge rate of seepage is computed directly from a flow net or, in certain instances, from charts or equations without the construction of a flow net. The discharge rate of seepage based on the flow net is computed as follows:

$$q = k \frac{n_f}{n_d} h_t L$$

where q = discharge rate of seepage through the length of dam under consideration
k = effective coefficient of permeability: $\sqrt{k_{max} k_{min}}$
n_f = number of flow channels within the flow net
n_d = number of equipotential drops within the flow net
h_t = head difference between headwater and tailwater
L = length of dam to which the flow net applies

The ratio of n_f/n_d is termed the *shape factor* of the flow net.

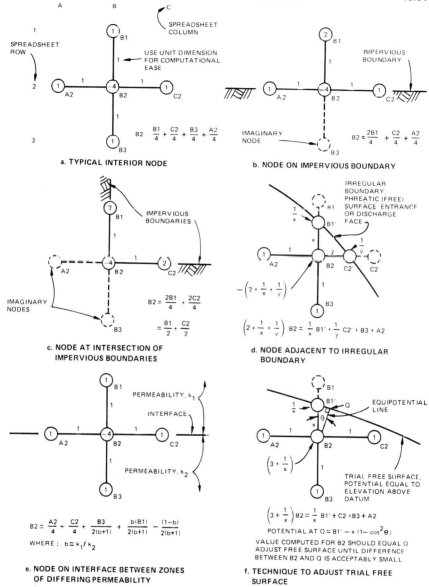

FIGURE 14 Seepage by the relaxation method for use with an electronic spreadsheet. (*Kleiner, 1985.*)

Both graphical and analytical methods are available for computation of the quantity of seepage through an earth dam on an impervious foundation. The results of such computations are presented in Fig. 10 as a family of curves.

Approximate solutions for the quantity of seepage passing beneath an impervious embankment founded on a pervious foundation are available in chart form.

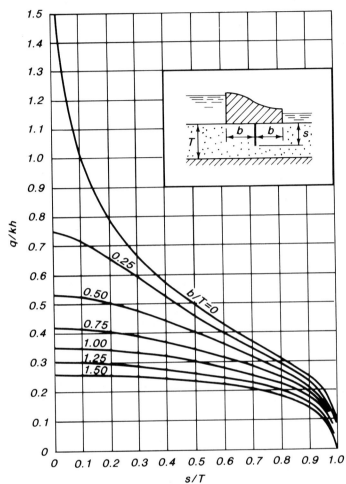

FIGURE 15 Seepage through pervious foundation below an impervious dam. (*Harr, 1962.*)

Figure 15 is taken from Harr, 1962. Note that the discharge rates can be computed for cases with no cutoff and partial cutoff.

22. Piping. Piping, internal erosion, and/or heave occur when the force exerted on the soil by seeping water exceeds the resistive force offered by the soil. Filters and drains are necessary at critical locations where seeping water emerges from fine-grained to coarse-grained soils within or below the dam and at the downstream toe or slope of the dam. The potential energy represented by the differential head across an embankment dam must be dissipated as frictional loss as the water flows through the soil. The seepage force J exerted on the soil by the water is equal to the unit weight of water γ_w times the hydraulic gradient i and is always in a direction perpendicular to equipotential lines. For either an isotropic or stratified soil, the hydraulic gradient is determined by dividing the head loss between two equipotential lines by the perpendicular distance between the lines measured on a true scale drawing.

When water flows into the upstream face of an embankment, the seepage force has a stabilizing effect, but when seepage flows out of an embankment the seepage force has a destabilizing effect. *When upward flow under a hydraulic gradient of 1 exists in a cohesionless soil which has a total weight of about 125 pcf the material will be in a condition of unstable equilibrium, termed a quick condition.*

Whenever water flows from a less pervious material to a more pervious material, or emerges to the surface, the possibility of migration of fines or piping must be considered. Even a minor washing away of fines at the downstream side of a dam is serious. As soon as some fines are washed away, the resistance to flow along the path of seepage is reduced and increased flow results. Because of the increased flow, the rate of washing away of fines is increased.

In order to prevent piping or internal erosion, movement of soil particles under the action of seepage forces must be prevented. Piping is prevented either by reducing the seepage gradient at the exit so that the seepage force is too small to cause movement of particles or by holding the soil at the exit in place by a filter or a series of filters.

23. Filters. There can be no doubt in the mind of the dam designer concerning the importance of filters and drains within the body of an embankment dam. Indeed, many incidents of failure or near failure can be attributed to the absence of filters and/or drains or to filter protection which was not appropriate to the application. The professional literature provides ample case histories of accidents related to the lack of proper filter protection.

Thus, the structural safety of the embankment dam depends on the proper design and construction of filter and filter/drain systems.

Two fundamental functions are required of filters in earth and earth-rock dams:

Retention Criterion. The filter must prevent internal erosion by blocking the migration of soil particles from adjacent foundation or fill materials. Thus, a fine filter must prevent internal erosion of finer-grained impervious fill or foundation material; a coarse filter must prevent any tendency for movement of the fine filter. This first requirement is often referred to as the *piping* or *stability criterion.*

The classic Terzaghi criterion $D_{15f}/d_{85b} < 4$ addresses this requirement. In this expression the following symbols are used: D_{15f} = particle size in filter for which 15 percent by weight of particles are smaller, and d_{85b} = particle size in base for which 85 percent by weight of particles are smaller.

Permeability Criterion. The filter must accept seepage flows from adjacent foundation or fill materials without the buildup of excess hydrostatic pressure. Thus, a fine filter must readily accept seepage flows from a finer-grained impervious fill or foundation material; a coarse filter must readily accept flow from an adjacent fine filter. Permeability ratios between adjacent materials of at least 25 are necessary.

The classic Terzaghi criterion $D_{15f}/d_{85b} > 4$ addresses this requirement.

To achieve the above functions, the filter:

1. Must not segregate during processing, handling, placing, spreading, or compaction. The filter gradation must be sufficiently uniform such that segregation is avoided, especially at the interface between adjacent materials.
2. Must not change in gradation (degrade or break down) during processing, handling, placing, spreading, and compaction; or degrade with time as might be

caused by freeze-thaw or seepage flow. The filter must consist of hard, durable particles not susceptible to degradation.
3. Must not have apparent or real cohesion or the ability to cement as a result of chemical, physical, or biological action. The filter must remain cohesionless and free of cracks even though an adjacent core zone may have been damaged by cracking.
4. Must be internally stable; that is, the coarser fraction of the filter with respect to its own finer fraction must meet the retention (piping) criterion.
5. Must have sufficient discharge capacity such that seepage entering the system is conveyed safely and readily with little head loss. Thus, chimney and blanket filter/drain systems must be designed with ample discharge capacity.

24. Flow Conditions and Forces Acting on Filters. Figure 16 illustrates the two basic flow conditions that can occur between a filter and base material. Typical examples of these two flow conditions are

1. Flow perpendicular to the interface:
    a. At the downstream contact between the core and chimney filter in an earth, earth-rock, or rockfill dam.
    b. At the contact between foundation soils and the bottom filter layer in a downstream blanket filter/drain or finger drain system.
    c. At the contact between earth fill and the top filter layer in a downstream blanket filter/drain or finger drain system.
    d. At the contact between the fine filter and coarse filter (drain) in downstream chimney, blanket, and finger drains.
    e. At the contact between sand/gravel layers and silt/clay layers within alluvial foundations near the upstream and downstream toes of embankment dams. These are locations where seepage enters or discharges from the foundation; flow is vertical or nearly so.
2. Flow parallel to the interface:
    a. At the upstream contact between the core and fine filter in earth, earth-rock or rockfill dam, locations subject to a fluctuating reservoir such as in pumped-storage projects.
    b. At the contact between bedding filters and base material and between bedding filter and riprap or revetment on the upstream slopes of embankment dams.
    c. At the contact between gravel/cobble slope protection and base material on the downstream slopes of embankment dams.
    d. At the contact between sand/gravel layers and silt/clay layers within allu-

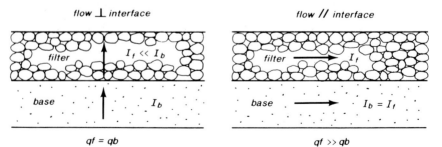

FIGURE 16 Flow perpendicular and parallel to filter/base interface. (*Bakker, 1990.*)

vial foundations below embankment dams, locations where seepage is horizontal or nearly so.

Dr. James Sherard's great interest in cracking and piping in embankment dams is amply described in his writings (Sherard, 1973, 1979, 1985). This interest led to the extensive research conducted in the Lincoln, Nebraska, soil mechanics laboratory of the U.S. Soil Conservation Service during the early 1980s. This research is widely reported (Sherard, 1984a, 1984b, 1985, 1989). The findings and conclusions are included in the design criteria for filters adopted by the U.S. Soil Conservation Service and by the U.S. Bureau of Reclamation (U.S. Soil Conservation Service, 1986; USBR, 1987). The basic conclusions from this research are presented in Tables 10 and 11. Segregation criteria are presented in Table 12.

TABLE 10 Categories of Base Soil Materials (U.S. Soil Conservation Service, 1986)

Category	Percent finer than the no. 200 (0.075 mm) sieve
1	> 85
2	40–85
3	15–39
4	< 15

TABLE 11 Criteria for Filters (U.S. Soil Conservation Service, 1986)

Base soil category	Base soil description, and percent finer than no. 200 (0.075 mm) sieve[1]	Filter criteria[2]
1	Fine silts and clays; more than 85% finer	$D_{15} \leq 9 \times d_{85}$[3]
2	Sands, silts, clays, and silty and clayey sands; 40 to 85% finer	$D_{15} \leq 0.7$ mm
3	Silty and clayey sands and gravels; 15 to 39% finer	$D_{15} \leq \dfrac{40 - A}{40 - 15} (4 \times d_{85} - 0.7 \text{ mm}) + 0.7 \text{ mm}$[4,5]
4	Sands and gravels; less than 15% finer	$D_{15} \leq 4 \times d_{85}$[6]

[1]Category designation for soil containing particles larger than 4.75 mm is determined from a gradation curve of the base soil which has been adjusted to 100% passing the no. 4 (4.75 mm) sieve.
[2]Filters are to have a maximum particle size of 3 in (75-mm) and a maximum of 5% passing the no. 200 (0.075 mm) sieve with the plasticity index (PI) of the fines equal to zero. PI is determined on the material passing the no. 40 (0.425 mm) sieve in accordance with ASTM-D-4318. To ensure sufficient permeability, filters are to have a D_{15} size equal to or greater than $4 \times d_{15}$ but no smaller than 0.1 mm.
[3]When $9 \times d_{85}$ is less than 0.2 mm, use 0.2 mm.
[4]A = percent passing the no. 200 (0.075 mm) sieve after any regrading.
[5]When $4 \times d_{85}$ is less than 0.7 mm, use 0.7 mm.
[6]In category 4, the d_{85} may be determined from the original gradation curve of the base soil without adjustments for particles larger than 4.75 mm.

TABLE 12 D_{10f} and D_{90f} Limits to Prevent Segregation (U.S. Soil Conservation Service, 1986)

Minimum D_{10}, mm	Maximum D_{90}, mm
< 0.5	20
0.5–1.0	25
1.0–2.0	30
2.0–5.0	40
5.0–10	50
10–50	60

Other conclusions include:

1. For typical coarse glacial moraines, graded from cobbles to fines, and other similarly graded impervious soils, the Soil Conservation Service research demonstrated that a sand or gravelly sand with $D_{15f} \leq 0.7$ mm is needed for a conservative downstream filter.
2. The no-erosion filter test (Sherard, 1985, 1989), is superior to the slot and slurry tests used during the earlier research. "It was found to be the best test for routine laboratory evaluation of filters for specific projects. It is applicable for tests on coarse impervious soils as well as fine clays and silts." (Sherard, 1985).

Before a filter can be designed, an analysis of the grain size distribution curve of the base soil must be made. This analysis determines whether the material is broadly graded and potentially internally unstable. Materials which are not internally unstable in the controlled environment of the laboratory, may be internally unstable in the field if segregation occurs such that pockets or lenses of differing gradations exist within the fill. Substantial differences in gradation often occur during dumping and spreading of broadly graded materials, at the interface between adjacent zones, and when compaction is poor such as adjacent to instrument locations or in trenches.

The internal stability of the base material is checked by separating the grain size curve into two parts at any arbitrary point of separation, as indicated in Fig. 17. For an internally stable material, the D_{15} size of coarser fraction should be no more than 5 times the d_{85} of the finer fraction. As is evident from the figure, whenever the slope of the grain size curve is flatter than 15 percent for a 5 times change in grain size, the material is not self-filtering and is, therefore, internally unstable.

Whenever the base material is not self-filtering, then the D_{15} of the filter must be based on the d_{85} of the finer fraction of the base rather than on the d_{85} of the total material.

25. Seepage Control within the Dam. Drainage measures are required within the body of the dam, at the downstream toes of dams, and within the foundation to reduce uplift pressure and seepage gradients. Within the dam, these measures consist of inclined or vertical chimney filter/drain systems, blanket drains, toe drains, drainage trenches, finger drains with or without internal pipe drains, and/or various combinations of drain types.

Within an earth foundation, seepage control provisions consist of various combinations of horizontal blanket drains, partial or complete cutoffs, upstream impervious blankets, downstream seepage berms, toe drains and relief wells. Cutoffs are constructed by using slurry trenches, concrete walls, or open

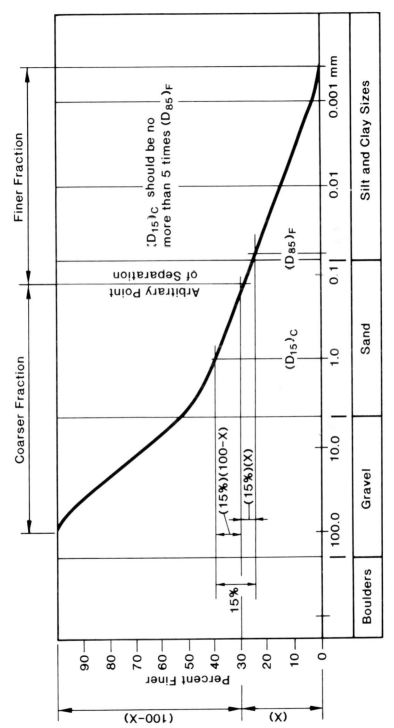

FIGURE 17 Criteria for self-filtering material. (*Lowe, 1988a.*)

backfilled trenches. Open trenches are preferred because the trench side walls and invert are accessible for inspection and treatment.

Within a rock foundation, seepage control measures consist of rock surface cleaning, application of slush grout, shallow grouting to treat geologic defects, curtain grouting, drainage adits and tunnels, and drain holes. The specific uses of these various techniques are dependent on the ground conditions as revealed during site investigation and during construction when foundations are excavated.

Filter and drainage systems within embankment dams must conduct seepage safely and without the buildup of excess pressure. While there are a number of ways of analyzing flow in filters and drains, one of the simplest to use (after potential flow rates have been estimated) is Darcy's law in the form:

$$\frac{q}{i} = kA$$

In this equation, q is the estimated rate of flow which must be handled by the filter or drain (per unit length of dam), i is the allowable (available) hydraulic gradient in the filter or drain, k is the required coefficient of permeability of the filter or drain having an area A normal to the direction of flow in the filter or drain. Any practical combination of k and A that ensures the required discharge capacity, with an adequate factor of safety, is used. Relatively thin layers of highly permeable materials are more economical than thicker layers of low permeability.

Figure 18, taken from an appendix to the USBR design standard (USBR, 1987), emphasizes the need to evaluate the discharge capacity of filter and drainage systems.

Internal drainage capacity can be increased dramatically with the use of perforated collector pipes, properly protected with fine and coarse filter material. The following retention criteria relate the hole or slot size to the gradation of the coarse filter surrounding the pipe:

$$\frac{D_{85f}}{\text{Max. pipe opening}} > 2 \quad \text{(USBR, 1987)}$$

$$\frac{D_{50f}}{\text{Hole diameter}} > 1.0 \quad \text{(U.S. Army Corps of Engineers, 1986)}$$

$$\frac{D_{50f}}{\text{Slot width}} > 1.2 \quad \text{(U.S. Army Corps of Engineers, 1986)}$$

The above criteria are valid for relatively uniform coarse filters, i.e., uniformity coefficient $C_u = D_{60}/D_{10}$, ≈ 5 or less.

STABILITY ANALYSIS

26. Basic Concepts. Whenever the shearing force along any continuous surface through the embankment or through the embankment and its foundation exceeds the shearing resistance along that surface, a stability failure occurs. The trace of the surface of sliding on a cross-sectional view is most often approximated by either a series of straight lines or the arc of a circle. The stability of the dam is evaluated by considering various possible surfaces of sliding and computing the factor of safety against a stability failure for each surface. *The factor of*

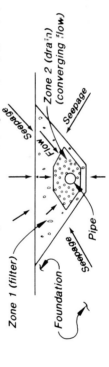

FIGURE 18 Filter and transmissibility needs of drains in embankment dams. (*USBR, 1987.*)

TABLE 13 Minimum Required Factor of Safety

Condition	Upstream slope	Downstream slope
End of construction	1.3	1.3
With earthquake	1.0	1.0
Steady seepage, partial or full pool	1.5	1.5
With earthquake	1.1	1.1
Rapid drawdown	1.3	—
Rapid drawdown, pumped storage reservoir	1.5	—
With earthquake	1.1	—

safety against slope failure is defined as the available shear strength on the postulated failure surface divided by the mobilized shear strength on that same surface. The sliding surface with the least factor of safety is considered critical.

The various cases of loading for which embankment slopes are analyzed and the minimum factors of safety recommended for these cases are listed in Table 13.

The factor of safety provides a margin for inaccuracies in the method of stability computation, in the location of the critical surface for sliding, and in the estimate of the shear strength of the embankment and foundation materials.

The shear strength of the embankment and foundation materials are different for the various cases listed above. Test procedures to estimate the shear strengths for the various conditions of consolidation and drainage are described in manuals and test standards (ASTM, 1991; U.S. Army Office of the Chief of Engineers, 1970; USBR, 1974). Choice of the test method for analysis of a particular case is described in the literature (U.S. Army, Office of the Chief of Engineers, 1970a; Lowe, 1967; Lowe and Karafiath, 1959; Lowe, 1988b) and in Art. 27.

For each condition of consolidation and drainage during shear, the shear strength may be expressed in terms of the parameters c and ϕ according to the following expression:

$$s = c + \sigma \tan \phi$$

where s = available shear strength
  c = cohesion
  ϕ = angle of internal friction
  σ = normal stress on the failure surface

The linear relationship indicated by the expression is an approximation of actual shear strength and applies to a limited range of normal stresses.

27. Shear Strength Parameters. The shear strength parameters to be used for specific loading conditions are summarized in Table 14.

For the *end-of-construction* case, the analysis usually assumes instantaneous construction. Some consolidation of materials in impervious zones in the dam occurs during construction, but the means to predict this consolidation is empirical. The shear strength parameters are estimated from tests performed on specimens compacted to the same water content and to the same unit weight as anticipated in the prototype embankment. The tests are performed with drainage lines closed both during application of the chamber pressure as well as during shear, i.e., the

TABLE 14 Shear Strength Parameters

Design condition	Test to determine shear strength	Normal stress on failure surface*
End of construction	Q or S†	Total Effective
Steady-state seepage	S	Effective
Rapid drawdown	R or S‡	Effective Effective
Earthquake	R or S‡	Effective Effective

*When water is present, either within the foundation or within the embankment as might be caused by groundwater, a partial pool at the upstream toe, the presence of a reservoir, or the drawdown of a reservoir, proper consideration of uplift and, therefore, reduction in normal stress on the failure plane must be included. The effective normal stress must be used such that the frictional component of the shear strength is properly reduced.
†For materials with no excess pore pressure as a result of construction.
‡For materials with no excess pore pressure as a result of shear.

Q test (U.S. Army, Office of the Chief of Engineers, 1970b). Where no excess pore pressures are generated during placement and compaction of the fill, the consolidated drained shear strength, as obtained by the S test (U.S. Army, Office of the Chief of Engineers, 1970b), is appropriate.

For the foundation consisting of a saturated compressible soil, the in-situ strength of the foundation before placement of the fill may be used. This means that each point in the foundation has a particular strength in undrained shear which is independent of the loads applied by the fill during construction. The *in-situ shear strength* is the consolidated undrained shear strength for the condition of consolidation under the weight of overburden before placement of fill. Q tests on undisturbed specimens of the in-situ material are used to estimate this shear strength. If significant consolidation of the foundation occurs during construction, the consolidated undrained strength of the foundation material, estimated from the R test (U.S. Army, Office of the Chief of Engineers, 1970b), is used for the estimated amount of consolidation. If complete consolidation occurs and no excess pore pressure is generated during construction, the consolidated drained strength, estimated from the S test, is the appropriate choice.

For *steady seepage* through the dam, the stability of the upstream slope is estimated by varying the reservoir level from partial to full pool. Under the condition of partial pool, the weight of the material in the upper part of the slope is the moist weight, which results in the maximum driving force, whereas the effective weight of the material in the lower part of the slope is the saturated weight minus the uplift water force. This results in the development of minimum resistance at the toe. Stability analyses are performed for levels of the reservoir from partial to full pool to derive the critical level of pool and the corresponding minimum factor of safety.

It is assumed that the embankment and foundation materials have reached equilibrium under the loading conditions; therefore, the consolidated drained strength, derived from the S test, is used. Alternatively, an effective stress parameter is derived from the consolidated undrained R test and used in the analysis.

Pore water pressures within the embankment and foundation are derived from the piezometric levels as indicated by the equipotential lines of the flow net or by measurements of the pore water pressure in an existing dam. A less rigorous but usually more conservative analysis simply assumes hydrostatic water pressure to the location of the reservoir or the phreatic surface within the embankment.

The stability analysis of the downstream slope is made in a manner similar to that for the upstream slope. Partial pool conditions are not evaluated. Pore water pressure is estimated from a flow net or from an assumption of hydrostatic pressure to the estimated phreatic surface or the measured surface in an existing dam.

In the *rapid drawdown* case it is assumed that the reservoir has been at full pool for a sufficiently long time such that steady-state seepage forces exist within the embankment. Instantaneous drawdown from full pool to some minimum pool is then assumed. This means that there is no time for drainage to occur to relieve the excess pore pressures which develop in low-permeability embankment materials.

The method of rapid drawdown analysis presented herein includes an estimate of the pore pressure retained within the embankment materials during the drawdown process. For free-draining granular materials, no excess pore pressures are retained, i.e., the pore pressure reduces as the level of the reservoir is lowered. For materials of low permeability and compressibility, such as a jointed rock foundation or a dirty rock fill with rock-to-rock contact, the pore pressure reduces as the level of the reservoir is lowered. But when the pool level drops below the level of the material, drainage must occur to reduce pore pressure. The amount of drainage and, therefore, the amount of pore pressure reduction will depend on the drawdown rate with respect to the permeability of the material.

For materials of low permeability which consolidate over long periods of time, excess pore pressure above the hydrostatic level after drawdown will be retained. The pore pressure after drawdown may be estimated by assuming that the change in pore pressure from the full reservoir pore pressure is equal to the change in total stress above the point in question.

Thus:

$$\Delta_u = \Delta\sigma_{\text{total}}$$

$$u_{\text{drawdown}} = u_{\text{fullpool}} - \Delta_u$$

The above is simple consolidation theory; the instantaneous change in pore water pressure equals the change in total stress. The pore pressure changes that occur under drawdown are illustrated in Fig. 19.

Since shearing occurs in the undrained state, the consolidated undrained shear strength, derived from the R test, is used. Consolidated drained tests (S test) are appropriate to select the shear strength parameters for free-draining materials and for those materials below the drawdown pool in which the pore pressure responds rapidly to loading or unloading. Materials such as jointed rock, dirty rock fill with rock-to-rock contact, and dense, incompressible granular fills fall into this category.

In the pseudo-static method of analysis for *earthquake loading,* the consolidated undrained shear strength, as estimated from the R test, should be used for those materials which will not drain readily during the earthquake. The consolidated drained shear strength, estimated from the S test, should be used for those materials which will drain readily during the earthquake. The use and validity of

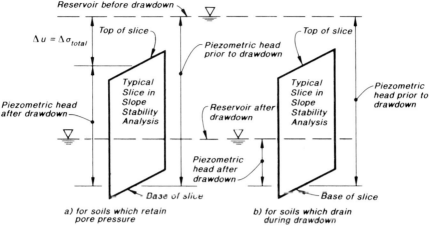

FIGURE 19 Pore pressure retention under rapid drawdown.

the pseudo-static method are discussed below under "Earthquake Considerations."

28. Stability Analysis Procedures. Several methods are available for performing the stability analysis, including the ϕ circle method, the log-spiral method, the slices method, and the sliding-block method. The sliding-block method can be considered a variation of the method of slices with specific shapes for the slices. The slices method and sliding-block method are commonly used and are described herein. The ϕ circle and log-spiral methods are described by Taylor, 1949.

29. Infinite-Slope Analysis. Simplified methods such as the infinite-slope analysis are useful for cohesionless materials. This analysis considers a typical vertical slice of a long shallow sliding mass. End effects on the sliding mass are negligible because the length of the sliding mass is large compared with the depth. Thus, each slice is identical to adjacent slices and, therefore, the side earth and water forces on each side of the slice are equal in magnitude and direction. An illustration of a typical slice is shown in Fig. 20.

The equilibrium equations, $\Sigma H = 0$ and $\Sigma V = 0$ are solved simultaneously to find an expression for the factor of safety. The following equation for the factor of safety considers uplift on the base of the slice and a horizontal seismic coefficient:

$$FS = \frac{m \tan \phi - fm^2 \tan \phi \sin i - a \tan \phi - f \tan \phi \sin i}{1 + ma}$$

where FS = factor of safety
  i = angle of the face of the slope
  m = cotangent of the angle i
  a = horizontal seismic coefficient
  ϕ = angle of internal friction
  f = uplift coefficient

The uplift coefficient f is defined as follows:

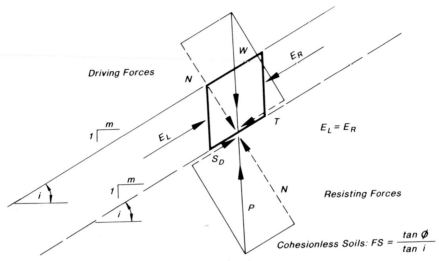

FIGURE 20 Infinite slope, forces acting on a unit volume.

1. Dry, submerged, or free-draining slopes under drawdown: $f = 0$. Then, for $a = 0$:

$$FS = m \tan \phi$$

2. Seepage parallel to the slope:

$$f = \frac{\gamma_w}{\gamma_{sat}}$$

Then, for $a = 0$ and $\gamma_w/\gamma_{sat} = 0.5$:

$$FS = (m - 0.5m \cos i - 0.5 \sin i) \tan \phi$$

3. Seepage horizontal or retained pore water pressure equal to the height of the slice:

$$f = \frac{\gamma_w}{\gamma_{sat} \cos i}$$

Then, for $a = 0$ and $\gamma_w/\gamma_{sat} = 0.5$:

$$FS = \left(0.5m - \frac{0.5}{m}\right) \tan \phi$$

The above general equation is solved using an electronic spreadsheet and inserting various values of the several parameters. The above cases are plotted in Fig. 21 for various values of m and the angle of internal friction ϕ.

30. Slices Method of Stability Analysis. The slices method of stability analysis was introduced by Fellenius, 1936. Modifications have been made to the details of the method, for example, by May, 1936, and Bishop, 1955. May's method does not consider earth forces on the vertical sides of the slices. Bishop's method

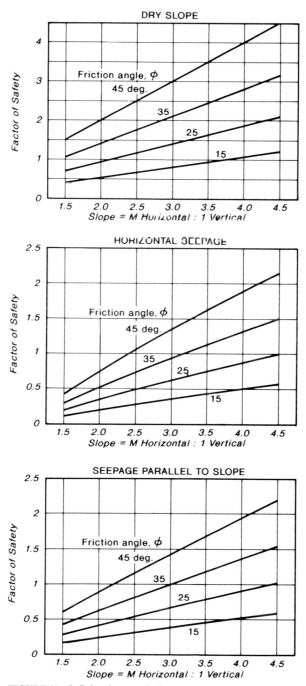

FIGURE 21 Infinite slope analysis.

l;.assumes that the earth forces on the sides of the slices act in an horizontal direction. Spencer, 1967, introduced a moment equation in an attempt to derive the appropriate angle on the sides of the slice. Spencer lumps both the side water and side earth forces together and translates these forces to the base of the slice.

The variation presented herein is general in that the earth forces are applied at an angle with the vertical sides of the slices, and the water forces acting on the sides, top, and bottom of the slices are considered separately in the analysis. Water forces are applied perpendicular to the face on which they act. Both vertical and horizontal seismic coefficients may be included in the analysis, if needed. With the aid of the computer, the basic method of slices is easily solved.

The first step is to divide the sliding mass into a number of vertical slices, Fig. 22. The sliding surface is a series of straight lines which bound a series of sliding blocks or which approximate a circular arc. The number of slices chosen usually is about 10. Widths of slices are adjusted so that the entire base of each slice is located within a single material.

The forces acting on a typical slice are shown in Fig. 23 and consist of the following:

n = nth slice
i = angle of the top of the slice
α = angle of the base of the slice
W_n = total weight of the nth slice
E_n, E_{n-1} = earth forces on the far side and near side of the vertical faces of the slice; the near and far sides of the slice are defined by the potential direction of movement of the sliding mass
θ_n, θ_{n-1} = angle between the horizontal and the earth forces on the sides of the slice; the angle may vary from one side of the slice to the other
$U_n, U_{n-1}, U_{pn}, U_{tn}$ = water forces on the far side, near side, base, and top of the slice; the water forces are applied perpendicular to the sides, base, and top of the slice; when slices are submerged below the reservoir surface, a top water force must be introduced

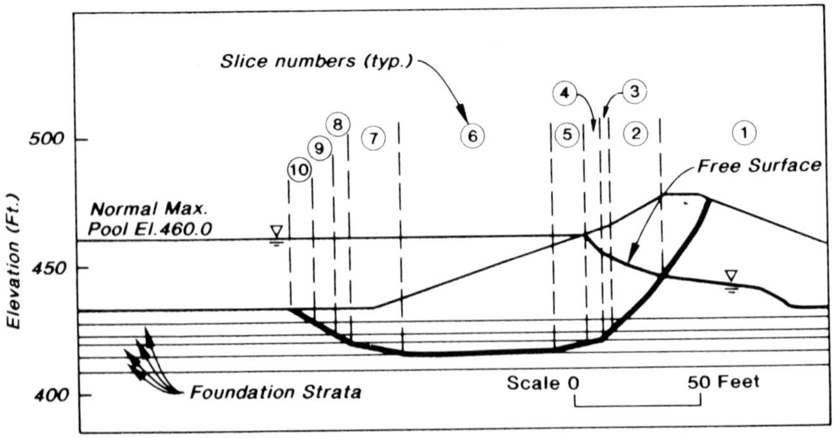

FIGURE 22 Potential sliding mass subdivided into slices, Beech dam, full pool.

STABILITY ANALYSIS

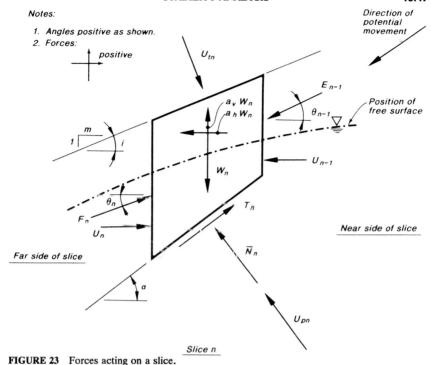

FIGURE 23 Forces acting on a slice.

a_h, a_v	=	horizontal and vertical seismic coefficients. Seismic forces are computed from the total weight of the slice, i.e., $a_h W_n$ and $a_v W_n$
$\overline{N}_n$	=	effective normal force on the base of the slice
T_n	=	shear force on the base of the slice
	=	$\overline{N}_n \tan \phi / FS + cL / FS$
ϕ	=	angle of internal friction
FS	=	factor of safety
c	=	cohesion
L	=	length of slice base

The basic equations of equilibrium, $\Sigma H = 0$ and $\Sigma V = 0$ are used to find the factor of safety for various conditions of shear strength, pore water, and shape of the assumed failure mass:

$$\sum V = 0 = a_v W_n - W_n + E_n \sin \theta_n - E_{n-1} \sin \theta_{n-1}$$
$$+ T_n \sin \alpha + \overline{N}_n \cos \alpha + U_{pn} \cos \alpha - U_{tn} \cos i$$

$$\sum H = 0 = -a_h W_n + E_n \cos \theta_n - E_{n-1} \cos \theta_{n-1} + U_n - U_{n-1}$$
$$+ T_n \cos \alpha - \overline{N}_n \sin \alpha - U_{pn} \sin \alpha + U_{tn} \sin i$$

The water forces are determined from the water pressure acting on the sides, base, and top of the slice estimated from static water conditions if no seepage occurs or from flow nets if seepage occurs. Often, sufficient accuracy is obtained by using the assumption of hydrostatic pressure below the phreatic surface.

In the above equations, all forces are known except for the side earth forces and the normal effective earth force on the base of the slice. For slice 1 at the top of the slope, the near-side earth force is equal to zero, and, for the last slice at the bottom of the slope, the far-side earth force is equal to zero. The above equations are used to eliminate the normal effective force on the base of the slice so that a solution for the value of the far-side earth force of the first slice is found for any given value of θ. Trial factors of safety for given values of θ are input until a solution that yields a value of zero for the far-side earth force of the last slice is obtained. Since the value of the angle θ is unknown, an assumption must be made. It is obvious that should shearing strain occur, some shearing force would act between the several slices. This implies that the value of θ is greater than zero and less than the value derived by using the shear strength of the material.

The U.S. Army, Office of the Chief of Engineers, 1970a, recommends a constant value of θ equal to the average of the outer slope of the failure mass. Lowe and Karafiath, 1959, suggest that θ should be equal to the average of the outer slope and the failure surface for each slice. Since the method of slices is itself an approximation of the real problem, it seems reasonable to use a value of θ that approximates the average of the outer slope of the failure mass, unless the failure mass consists of a few sliding blocks, in which case the value of θ might be chosen to vary in accordance with the average of the outer slope and the failure surface.

The U.S. Army, Office of the Chief of Engineers, 1970a, presents detailed examples of the above method using graphical procedures with stacked polygons of forces for each slice. The graphical procedure requires considerable effort which can be avoided with the use of an electronic spreadsheet running on a personal computer (PC). Input values include the geometry of slices, unit weights, shear strength parameters, water conditions, and the seismic coefficients. Output values include the total weight of the slices, the various water forces, earth forces, and the normal effective and shear forces on the base of the slices. The trial factor of safety is adjusted until the value of the earth force on the far side of the last slice is equal to zero. Various values of θ can be input to determine the variation of the factor of safety with θ.

A more rigorous estimate of θ can be introduced by including a moment equation for the entire sliding mass. All internal forces on the sides of the slices are balanced, because the far-side earth and water forces of one slice are the same as the near-side earth and water forces of the adjacent downslope slice. These internal forces are excluded from the moment equation. The balance of external moments is dependent on the value of mobilized shear resistance on the base of the slide, which in turn is dependent on the assumed values of θ and factor of safety. The method requires inputting trial values of θ and factor of safety until the summation of forces equals zero and the summation of moment equals zero. The basic spreadsheet is modified to accommodate the moment equation.

Several computer solutions are commercially available which incorporate a search routine to find that failure surface which yields the minimum factor of safety for various given conditions. Graphics are included to present the output. Procedure checks such as that outlined above are used to understand the method more fully and to analyze the internal manipulation of forces used by the specific computer program.

31. Sliding-Block Method. This method is similar to the slices method of stability analysis except that only two or three slices, called *blocks*, are used and the surface of sliding is composed of two or three planar surfaces. Examples of sliding surfaces are shown in Figs. 24 and 25. Again, the analysis is performed by using the general method of slices, but with the angle θ between slices or blocks chosen by the investigator. The U.S. Army, Office of the Chief of Engineers, 1970a, presents a discussion of the sliding-block stability analysis and the angle of the earth force to be used between the blocks.

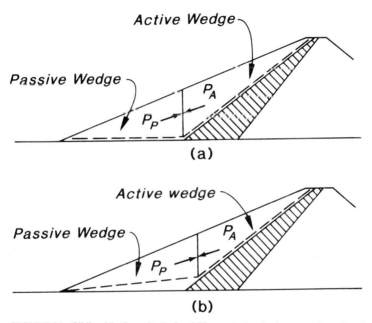

FIGURE 24 Sliding-block method of stability analysis, sloping-core dam. (*Davis, 1969.*)

As shown in Figs. 24 and 25, the angle of the earth force between the active wedge and the central block or between the active wedge and the passive wedge should be approximately the average of the outer slope and the failure surfaces. The earth force angle between the central block and the passive is often taken as zero. The variation of factor of safety with the angle θ is easily investigated if the analysis is performed using an electronic spreadsheet.

32. Stability Charts. The stability chart presented in Fig. 26 provides a conservative and simple method for determining the factor of safety against stability failure for a homogeneous dam and foundation in:

1. The dry condition
2. The completely submerged condition
3. The condition of complete rapid drawdown from crown to toe for an embankment with negligible drainage

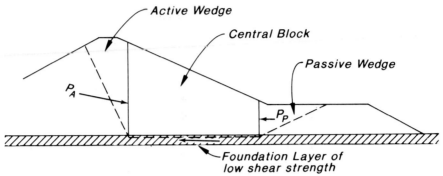

FIGURE 25 Sliding-block method of stability analysis, dam with counterweight berm situated on foundation of low shear strength. (*Davis, 1969.*)

The chart gives the relationship between the factor of safety FS, the height of slope H, the angle of slope α, the friction angle ϕ, and the cohesion c. Note that the factor of safety is with respect to the cohesion. The frictional resistance must be adjusted by using the expression

$$\frac{\tan \phi}{\text{FS}}$$

such that FS values with respect to friction and cohesion are equal. This results in a trial-and-error procedure. The effect of a firm stratum at shallow depth in the foundation and the effect of counterweights a short distance away from the toe of slope are also included.

Shear strength parameters are selected based on the Q, R, or S test in accordance with Table 14.

No adjustment is required to estimate the factor of safety for the dry or completely submerged condition. The factor of safety must be reduced by approximately one-half for the rapid drawdown case for materials with negligible drainage.

EARTHQUAKE CONSIDERATIONS

33. General. The following summary of earthquake considerations is drawn from the 19th Rankine Lecture, presented by Professor H. Bolton Seed, March 6, 1979 (Seed, 1979). In his lecture, Dr. Seed lists the many ways in which an earthquake may cause damage or failure of an earth dam:

1. Disruption of dam by major fault movement in the foundation
2. Loss of freeboard because of differential tectonic ground movements
3. Slope failures induced by ground motions
4. Loss of freeboard because of slope failures or soil compaction
5. Sliding of dam on weak foundation materials
6. Piping failure through cracks induced by ground motions

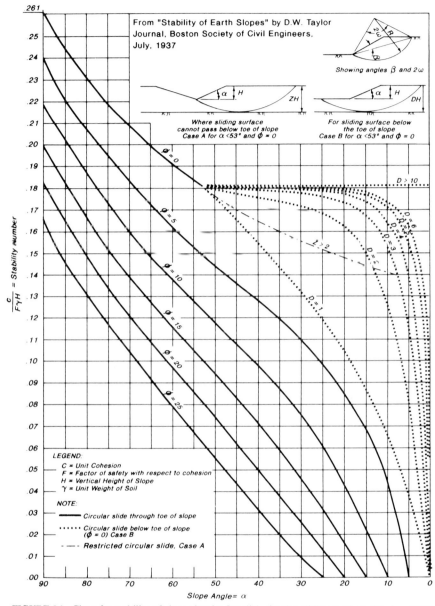

FIGURE 26 Chart for stability of slopes by the Swedish circular sliding method. (*Taylor, 1937.*)

7. Overtopping of dam because of seiches in reservoir
8. Overtopping of dam because of slides or rock falls into reservoir
9. Failure of spillway or outlet works

34. Defensive Measures. Dr. Seed lists a series of defensive measures to include in the design to resist the damaging effects of an earthquake:

1. Provide ample freeboard to allow for settlement, slumping, or fault movements.
2. Use wide transition zones of material not vulnerable to cracking (and which meet the criteria for filters).
3. Use chimney drains near the central portion of the embankment.
4. Provide ample drainage zones to allow for possible flow of water through cracks.
5. Use wide core zones of plastic materials not vulnerable to cracking.
6. Use a well-graded filter zone upstream of the core to serve as a crack stopper.
7. Provide crest details which will prevent erosion in the event of overtopping.
8. Flare the embankment core at abutment contacts.
9. Locate the core to minimize the degree of saturation of materials.
10. Stabilize slopes around the reservoir rim to prevent slides into the reservoir.
11. Provide special details if danger of fault movement in the foundation exists.

Defensive measures, especially the use of wide filters and transition zones provide a major contribution to earthquake-resistant design and should be the first consideration by the prudent engineer in arriving at a solution to problems posed by the possibility of earthquake effects.

35. Embankment Dam Performance. Based on studies of embankment dams during earthquakes, Seed concludes (Seed, 1978):

1. Hydraulic fill dams have been found to be vulnerable to failures under unfavorable conditions. One particularly unfavorable condition is the shaking produced by strong earthquakes. However, many hydraulic fill dams have performed well for many years, and, when they are built with reasonable slopes on good foundations, they can apparently survive moderately strong shaking, i.e., with accelerations up to about $0.2g$ from magnitude $6\frac{1}{2}$ earthquakes with no harmful effects.
2. Virtually any well-built dam on a firm foundation can withstand moderate earthquake shaking, say with peak accelerations of about $0.2g$, with no detrimental effects.
3. Dams constructed of clay soils on clay or rock foundations have withstood extremely strong shaking ranging from 0.35 to $0.8g$ from a magnitude $8\frac{1}{4}$ earthquake with no apparent damage.
4. Two rockfill dams have withstood moderately strong shaking with no significant damage, and, if the rockfill is kept dry by means of a concrete facing, such dams should be able to withstand extremely strong shaking with only small deformations.
5. Dams which have suffered complete failure or slope failures as a result of earthquake shaking seem to have been constructed primarily with saturated sand shells or on saturated sand foundations.
6. Since there is ample field evidence that well-built dams can withstand moderate shaking with peak accelerations up to at least $0.2g$ with no harmful effects, we should not waste our time and money analyzing this type of problem—

rather we should concentrate our efforts on those dams likely to present problems, either because of strong shaking involving accelerations well in excess of 0.2g or because they incorporate large bodies of cohesionless materials (usually sands) which, if saturated, may lose most of their strength during earthquake shaking and thereby lead to undesirable movements.

7. For dams consisting of saturated cohesionless soils and subjected to strong shaking, a primary cause of damage or failure is the buildup of pore water pressures in the embankment and the possible loss of strength which may accrue as a result of these pore pressures. It is not possible to predict this type of failure by pseudo-static analyses, and other types of analysis techniques are required to provide a more reliable basis for evaluating field performance.

36. Design Criteria. Seed suggests (Seed, 1979) the use of the pseudo-static method of analysis "for embankments constructed of soils which do not build up large pore pressures due to earthquake shaking nor show more than 15% strength loss (usually cohesive soils such as clays, silty clays, sandy clays or very dense cohesionless soils), based on acceptable deformations due to earthquake shaking and crest acceleration less than 0.75g." Under these conditions the following is suggested:

Earthquake magnitude	Minimum factor of safety	Seismic coefficient
6½	1.15	0.10
8¼	1.15	0.15

Seed considers that up to 3 ft of deformation is acceptable. Also, foundation materials must exhibit at least the same strength characteristics as the embankment.

37. Liquefaction Analysis. For those materials which do exhibit large pore pressure buildup due to earthquake shaking or which show more than 15 percent loss in strength, the following analysis procedure is recommended (Seed, 1979):

1. Determine the cross section of the dam to be used for analysis.
2. Determine, with the cooperation of geologists and seismologists, the maximum time history of base excitation to which the dam and its foundation might be subjected.
3. Determine, as accurately as possible, the stresses existing in the embankment before the earthquake; this is probably done most effectively at the present time using finite-element analysis procedures.
4. Determine the dynamic properties of the soils comprising the dam, such as shear modulus, damping characteristics, bulk modulus, or Poisson's ratio, which determine their response to dynamic excitation. Since the material characteristics are nonlinear, it is also necessary to determine how the properties vary with strain.
5. Compute, using an appropriate dynamic finite-element analysis procedure, the stresses induced in the embankment by the selected base excitation.
6. Subject representative samples of the embankment materials to the combined effects of the initial static stresses and the superimposed dynamic stresses and determine their effects in terms of the generation of pore water pressures and

the development of strains. Perform a sufficient number of these tests to permit similar evaluations to be made, by interpolation, for all elements composing the embankment.

7. From the knowledge of the pore pressures generated by the earthquake, the soil deformation characteristics, and strength characteristics, evaluate the factor of safety against failure of the embankment either during or following the earthquake.
8. If the embankment is found to be safe against failure, use the strains induced by the combined effects of static and dynamic loads to assess the overall deformations of the embankment.
9. Be sure to incorporate the requisite amount of judgment in each of steps 1 to 8, as well as in the final assessment of probable performance, being guided by a thorough knowledge of typical soil characteristics, the essential details of finite-element analysis procedures, and a detailed knowledge of the past performance of embankments in other earthquakes.

The analysis involves considerable time and cost and must be conducted by knowledgeable investigators. Various simplified versions of the analysis have been used to reduce the effort. Seed states (Seed, 1979), "The ultimate simplification is, of course, the total elimination of all analysis procedures and a simple evaluation, based on a knowledge of the materials comprising the dam and the judgment resulting from conducting many previous analyses and observing the performance of existing dams."

The references listed by Seed, 1979, present in-depth discussions, analyses, and procedures which are beyond the scope of this handbook.

SETTLEMENT ANALYSIS

38. General Considerations. Estimates of the magnitude and sources of settlement within the dam and its foundation provide design data for the selection of freeboard and for conduit design. At the end of construction, adequate camber, i.e., allowance for settlement, must be provided so that, after settlement, the crest of the dam provides sufficient freeboard. After postconstruction settlement has occurred, some camber should remain for appearance. Where a conduit passes underneath or through an embankment dam on a soil foundation, the design of the conduit requires knowledge of the pattern of settlement along its length.

The possible sources of settlement as listed by Seed, 1965, for cohesive materials apply to embankment materials and include:

1. Immediate or elastic settlement
2. Volumetric compression resulting from consolidation
3. Axial and lateral strains induced by effective stress changes during consolidation
4. Secondary compression
5. Plastic flow or creep at constant water content

Deformation as caused by earthquake loading must be added to this list.

Often, the settlement due to volumetric compression resulting from consolidation is most significant. The rate of reduction in excess pore pressures which results from consolidation is significant since the shear strength of a consolidating layer increases directly with decrease in the pore pressure.

39. Settlement from Consolidation of Foundation. When an increment of stress is applied to a saturated soil, the entire increment of stress is at first carried by an increase in pore water pressure. Because of the high gradients that exist near drainage faces, water is forced out of the soil, the soil structure consolidates, the effective stresses in the soil structure increase, and the pore water pressures decrease. When the entire increment of stress is carried as intergranular pressure, the soil is considered to be 100 percent consolidated under the particular increment of load.

Often, the volume change that takes place as a result of consolidation occurs during the construction period. Higher-permeability materials such as sands, silty sands, and silts consolidate rapidly. Secondary features in fine-grained soils such as dessication cracks, slickensides, open void structure, or relict rock structure in residual soils allow consolidation to occur during construction.

If low-mass-permeability clays are present within the foundation, an estimate of settlement is made using classic consolidation theory. Terzaghi and Peck, 1967, provide a good discussion of consolidation theory and the procedures to estimate magnitude and time rate of consolidation.

40. Settlement from Compression within the Embankment. Wilson, 1977, presented procedures to estimate the compression *during construction* based on field measurements at several dams. The following conditions are assumed:

1. Construction proceeds upward from the lowest point in the foundation
2. Materials are placed in horizontal layers and compacted
3. Compression of any element of compacted fill is linearly proportional to the applied vertical load above the soil element

The settlement of the surface of each layer is then proportional to the product of the number of layers below that elevation and the number of layers of fill added above. If the embankment is divided into 10 layers, then the settlement of each layer is approximately proportional to the following:

Layer no.	Layer below	Layer above	Wilson distribution factor
1	0	9	0
2	1	8	8
3	2	7	14
4	3	6	18
5	4	5	20
6	5	4	20
7	6	3	18
8	7	2	14
9	8	1	8
10	9	0	0

Thus, the distribution of vertical compression within the embankment is roughly parabolic, with the maximum settlement occurring at midheight. The settlement of the crest is zero on the day the embankment is completed. Wilson suggests that the method is approximately valid for dams up to about 300 ft in height.

One-dimensional consolidation data is often used in conjunction with the above to estimate settlement within the embankment. Experience with actual settlements indicates that these data will overestimate settlement by up to a factor of 3.

Consolidation tests performed in the triaxial chamber with control on lateral stress or strain provide better data for use in estimating vertical settlement.

41. Postconstruction Settlement. Postconstruction settlement has been studied by various investigators (Dascal, 1987; Clements, 1984; Soydemir and Kjaernsli, 1975; Sowers et al., 1965). From a study of 15 rockfill structures built by Hydro-Quebec after 1950, Dascal concludes the following:

1. The postconstruction crest settlement of compacted rockfill dams (central or slightly inclined cores) remains less than 0.35 percent of the height of the dam.
2. The settlement of the rock fill on the downstream shoulder can reach 0.7 to 0.8 percent of the height.

At the Guri hydroelectric project in Venezuela, postconstruction crest settlement of central core rockfill and earthfill dams ranging in height from about 80 to 260 ft varied from about 0.2 to 0.3 percent of the height after about 6 years. At the Bath County Pumped Storage project, the 480-ft-high upper dam settled 20 in in 6 years, or 0.35 percent of the height; the 150-ft-high lower dam settled 5.6 in, or 0.31 percent of the height. Both dams are central core rockfill structures. The data support the findings of Dascal.

The U.S. Bureau of Reclamation (USBR, 1977) recommends a camber of 1 percent of the height of the dam plus anticipated foundation settlement to accommodate postconstruction settlement. Their recommendations apply to the design of small dams, i.e., structures less than about 50 ft high.

SLOPE PROTECTION AND FREEBOARD

42. General Considerations. The slopes of embankment dams must be protected against erosion by waves, rainfall runoff, frost, and wind. Freeboard and spillway capacity must be conservatively chosen because protection cannot economically be provided against the erosion which would occur if the dam were overtopped.

43. Freeboard. Freeboard is defined for two extreme events:

For the probable maximum flood, freeboard is the vertical distance between the maximum water surface of the probable maximum flood (PMF) and the crest of the dam. Because the estimated water surface caused by the PMF is the result of an extreme event, freeboard is small, on the order of 1 ft. This freeboard remains after the estimated postconstruction settlement and deformation from the design earthquake have occurred. Thus, camber, included in the design to account for settlement and deformation, is added to the freeboard to obtain the crest elevation of the dam.

For maximum reservoir setup and wave action, freeboard is the vertical distance between the maximum normal water surface (crest of spillway ogee or top of gates) and the crest of the dam. Thus, freeboard consists of reservoir setup, wave runup, plus an additional amount depending on the judgment of the designer. USBR, 1977, suggests the following freeboard:

Fetch, mi	Freeboard, ft
Less than 1	4
1.0	5
2.5	6
5.0	8
10.0	10

Fetch is the average over-water distance from the dam to the opposite shore. Fetch is estimated by drawing nine radials at 3° increments centered on the maximum fetch and arithmetically averaging the radial lengths (U.S. Army Corps of Engineers, 1984). For large reservoirs, the more rigorous techniques for estimating wave heights and runup should be followed as outlined by the U.S. Army Corps of Engineers, 1973.

44. Slope Protection. Materials used for the protection of upstream and downstream slopes vary greatly. The following is only a partial list:

Upstream slope:

1. Grass (low dams, small fetch, farm ponds)
2. Rockfill (quarry run rock, substantial width, placed in horizontal layers)
3. Dumped riprap (processed to obtain required sizes, placed in single layer)
4. Reinforced concrete
5. Asphaltic concrete
6. Soil cement
7. Roller-compacted concrete.

Downstream slope:

1. Grass
2. Gravel and cobbles (processed to obtain required sizes, placed in single layer)
3. Downstream shell material, if nonerosive, such as rockfill.

If riprap is used, the following procedure is suggested:

1. *Estimate wave height* using the following relationship suggested by USBR, 1977, or the more rigorous technique presented by the U.S. Corps of Engineers, 1984, 1973:

Fetch, mi	Wind speed, mi/h	Wave height, ft
1.0	50	2.7
1.0	75	3.0
2.5	50	3.2

Fetch, mi	Wind speed, mi/h	Wave height, ft
2.5	75	3.6
2.5	100	3.9
5.0	50	3.7
5.0	75	4.3
5.0	100	4.8
10.0	50	4.5
10.0	75	5.4
10.0	100	6.1

2. *Estimate riprap weights* using the following relationships, as suggested by the U.S. Army Corps of Engineers, 1975, and the U.S. Army, Office of the Chief of Engineers, 1978:

$$W_{50} = \frac{\gamma H^a}{K(G-1)^3(\cot\theta)^b}$$

$$W_{max} = 4W_{50}$$

$$W_{min} = \frac{W_{50}}{8}$$

where W_{max} = maximum weight of stone, lb
W_{min} = minimum weight of stone, lb
W_{50} = weight of average-size stone, lb
H = wave height, ft
γ = unit weight of stone material, lb/ft^3
G = specific gravity of stone material
θ = angle of embankment slope from horizontal, degrees
a, b = empirical coefficients
K = stability coefficient

The coefficients for tolerable damage are suggested, i.e., $a = 3$, $b = 1$, and $K = 4.37$ (U.S. Army Corps of Engineers, 1975, and U.S. Army, Office of the Chief of Engineers, 1978).

3. *Estimate riprap sizes* using the following relationship, which assumes that the shape of individual rock particles is the average between a sphere and a cube:

$$D = \left[\frac{7W}{5\gamma}\right]^{1/3}$$

where W = weight of stone, lb
γ = unit weight of stone material, lb/ft^3

4. *Estimate the riprap layer thickness.* The thickness of the layer must be sufficiently large to accommodate the maximum size stone. In addition, the thickness should be no less than 1.5 times the D_{50} size or 24 in, whichever is greater.

5. *Design the riprap bedding.* If coarse granular material underlies the riprap, no bedding is required. If a bedding layer is required, it should be designed in accordance with the principles outlined in Art. 23, Filters. A single bedding layer is commonly used for small reservoirs, whereas two layers are required for large reservoirs. Bedding thickness approximates 0.5 times the riprap layer thickness, but need not exceed 18 in measured perpendicular to the slope.

An alternative to processed riprap and bedding layers consists of a quarry-run rockfill with maximum stone sizes consistent with an equivalent riprap placed in horizontal layers and compacted. The minimum width is 15 ft for ease in placement and the thickness, measured perpendicular to the slope, is at least twice the required riprap layer thickness. The rock source must yield hard, durable material similar in quality and size to a riprap source.

CONSTRUCTION QUALITY CONTROL

45. General. Foundation conditions and embankment materials encountered during construction may be different from those anticipated earlier because design concepts are based on limited site investigations. Many or few field modifications may be required to satisfy the design intent. Some changes are significant and, in some instances, require substantial changes to the construction contract. But, most often, the changes are minor and require no substantial change to the construction contract.

The constructed embankment dam must satisfy design concepts. For this reason, the design engineer is represented on site during construction. This assures that the design intent is achieved during construction and that appropriate changes to the design are made as foundation conditions and material characteristics are revealed. Selection of the required foundation grade, interpretation of foundation conditions, and selection of foundation treatment must be performed by on-site personnel with a clear understanding of the design intent.

46. Quality Control. Routine quality control of the various embankment materials is achieved by sampling, laboratory testing, and continuous on-site inspection. Standard test procedures are available, as listed below:

Test	Designation (USBR, 1974; ASTM, 1991)
Field sand cone density	USBR Earth Manual, E-24
Field water balloon density	ASTM D2167 (USBR 7221-86?)
Rapid compaction control	USBR Earth Manual, E-25
Moisture content	ASTM D2216 or D4643
Standard Proctor compaction	ASTM D698
Modified Proctor compaction	ASTM D1557
Gradation	ASTM C136, D422
Maximum index density	ASTM D4253

Testing frequency varies with the type of the material. In general, earthfill, filter, and drain materials require frequent testing. Rockfill in the shells requires less frequent testing or no testing. The following are minimum suggested testing frequencies.

Test Frequency, Cubic Yard per Test

Material	Gradation	Field density	Moisture	Laboratory density
Earthfill	5000*	5000*	5000*	5000*
Filter/drain	1000*	1000*		5000
Rockfill	As needed			

*Minimum one test per shift per work area.

Testing frequency is increased or decreased depending on field conditions, weather, and variation of material characteristics. In addition, test frequency is increased when new borrow areas are opened or when questions arise concerning material quality.

For control of earthfill compaction, the rapid compaction method (USBR, 1974) is a practical method that yields results in hours. Weekly Proctor tests are performed to verify selected results.

47. Records. Records are maintained of tests compliant or noncompliant with the specifications and tests on reworked, replaced, or repaired materials. Geologic maps are prepared to record foundation surface conditions as encountered in core trench invert and walls and in foundations of concrete structures. Undisturbed samples of earthfill are taken at specific locations within the constructed fill to verify the strength, density, and permeability parameters used in design.

MONITORING AND PERFORMANCE EVALUATION

48. General. As discussed previously, the design engineer participates in quality control during construction to assure that the constructed embankment dam meets the design intent. In addition, the design engineer monitors the behavior of the partially completed structure during construction and the completed structure during reservoir filling and during operation. The performance of the embankment dam is evaluated qualitatively and quantitatively and compared with the expected behavior. Unexpected performance is investigated promptly and remedial treatment undertaken, as needed.

49. Performance Evaluation. Performance evaluation is accomplished by evaluating two general types of information. The first is derived from visual observations of the general appearance of the structure and of noticeable changes with time. Questions such as the following are answered during visual observations:

1. Are cracks visible on the crest or slopes of the dam? Have preexisting cracks or repaired cracks changed in appearance or size since the last inspection?
2. Has slumping or settlement occurred since the last inspection?
3. Is seepage emerging on the slopes, at the toe, or from the abutments?
4. If seepage is emerging, has the flow increased? Is the flow carrying solids?
5. Is the upstream and downstream slope protection in good repair?
6. Has erosion of slopes, crest, or abutments occurred since the last inspection?

50. Instrumentation. The second type of information is obtained from instrumentation. Some are direct measurements that can be made from the surface of the dam, abutments, downstream valley floor, and from within galleries or adits. Other measurements are made within the body of the dam or its foundation.

Common parameters measured and instruments used are listed below:

Parameter	Instrument
1. Water pressures in soil and rock	Open-end standpipe piezometers, pneumatic piezometers, vibrating-wire piezometers
2. Total pressure	Total pressure cells
3. Seepage flow	Container/stopwatch, weirs, flumes
4. Internal settlement	Cross-arm device, inclinometers, piezometer-type settlement device
5. Surface deformation	Survey monuments
6. Internal deformation	Single-point extensometers, multiple-point extensometers, inclinometers
7. Ground vibration	Strong motion recorder

The selection of specific types and number of instruments varies from project to project. Table 15 lists a series of questions, features, and potential parameters to be monitored. The table is a useful guide during the instrument selection process.

TABLE 15 Monitoring Embankment Dams (Modified from Dunnicliff, 1990)

Question	Feature	Potential parameters to be monitored
What are the initial site conditions?	Foundation and abutments	Pore water pressure (soil) Joint water pressure (rock)
Is performance satisfactory during construction?	Zone in foundation with low shear strength or high compressibility	Pore water pressure Horizontal deformation Vertical deformation
	Zone in embankment dam with high compressibility	Pore water pressure Horizontal deformation Vertical deformation
	Zone in embankment dam where transfer of stress is caused by differential settlements, e.g., near irregular foundations or steep abutments, or near contact between zones of different compressibility	Strain within embankment Total stress Pore water pressure Horizontal deformation Vertical deformation
	Deformation pattern of entire embankment dam, for comparison with predictions and with other dams	Vertical deformation Horizontal deformation Strain within embankment

TABLE 15 Monitoring Embankment Dams (Modified from Dunnicliff, 1990) *(Continued)*

Question	Feature	Potential parameters to be monitored
	Design without precedent	Any of the parameters indicated in this table
Is performance satisfactory during first filling?	Entire dam and foundation	Leakage quantities Solids content in leakage water Pore/joint water pressure Crest settlement Crest cracking (visual observations)
Is performance satisfactory during drawdown?	Upstream shell (especially important for pumped storage projects)	Pore water pressure

Is long-term performance satisfactory?	Entire dam and foundation	Leakage quantities
		Solids content in leakage water
		Pore/joint water pressure
		  Within downstream shell
		  At base of chimney drain
		  Within foundation
		  Within core
		Vibration (seismic events)
		Vertical deformation
		Horizontal deformation
		Strain within embankment
		Total stress at contact between embankment and structure
		Crest settlement
		Crest cracking (visual observations)

	Entire dam: no special problem, but designer fears criticism if no instrumentation	Leakage quantities Solids content in leakage water Pore/joint water pressure

BIBLIOGRAPHY

American Society of Civil Engineers Proceedings, Review of Slope Protection Methods, Subcommittee on Slope Protection, Soil Mechanics and Foundations Division, vol. 74, June 1948.

AMERICAN SOCIETY FOR TESTING AND MATERIALS, Annual Book of ASTM Standards, vol. 04.08, Natural Building Stones, Soil and Rock, Geotextiles, 1991.

BAKKER, K. J., M. K. BRETELER, H. DEN ADEL, New Criteria for Filters and Geotextile Filters under Revetments, International Conference on Coast Engineering, Delft, 1990.

BISHOP, A. W., The Use of the Slip Circle in the Stability Analysis of Slopes, *Geotechnique,* 5 (1), 11, 7–17, 1955.

CASAGRANDE, ARTHUR, "Seepage Through Dams," New England Water Works Association, June 1937.

CASAGRANDE, A., and R. E. FADUM, "Notes on Soil Testing for Engineering Purposes," Harvard Univ. Grad. School of Engineering Publ. 268, 1940.

CEDERGREN, HARRY, *Seepage, Drainage, and Flow Nets*, Wiley, New York, 1989.

CLEMENTS, RONALD P., Post-Construction Deformation of Rockfill Dams, *J. Geotechnical Engineering*, ASCE, **110**, July 1984.

DASCAL, OSCAR, Postconstruction Deformations of Rockfill Dams, *J. Geotechnical Engineering*, ASCE, **113**, January 1987.

DAVIS, C. V., and K. E. SORENSEN, *Handbook of Applied Hydraulics*, 3d ed., McGraw-Hill, New York, 1969.

DEERE, D. U., A. J. HENDRON, F. D. PATTON, and E. J. CORDING in *Failure & Breakage of Rock*, Proceedings of the Eighth Symposium on Rock Mechanics, American Institute of Mining and Metallurgical Engineers, Minneapolis, Minn., 1967, pp. 237–303.

DUNNICLIFF, JOHN, Twenty-Five Steps to Successful Performance Monitoring of Dams, *Hydro Review*, HCI Publications, Kansas City, Mo., August 1990.

FELLENIUS, W., Calculation of the Stability of Earth Dams, *Trans. 2d Congress on Large Dams*, Washington, D.C., 1936.

GILBOY, G., Hydraulic-Fill Dams, *Proc. International Commission on Large Dams*, Stockholm, 1933.

HARR, MILTON, *Groundwater and Seepage*, McGraw-Hill, New York, 1962.

HAZEN, A., "Physical Properties of Sands and Gravels with Reference to Their Use in Filtration," Report Mass. State Board of Health, 1892.

HUNT, ROY, E., *Geotechnical Engineering Investigation Manual*, McGraw-Hill, New York, 1984.

KLEINER, DAVID, Electronic Spreadsheets and Civil Engineering, *Civil Engineering*, 1985.

LEPS, THOMAS, Flow Through Rockfill, *Embankment Dam Engineering*, Hirschfeld and Poulos (eds.), Wiley, New York, 1973.

LOWE, JOHN, III, Stability Analysis of Embankments, *J. Soil Mechanics and Foundations Div.*, ASCE, **93**, July 1967.

LOWE, JOHN, III, Seepage Analysis, *Advanced Dam Engineering for Design, Construction, and Rehabilitation*, R. B. Jansen (ed.), Van Nostrand Reinhold, New York, 1988a.

LOWE, JOHN, III, Stability Analysis, *Advanced Dam Engineering for Design, Construction, and Rehabilitation*, R. B. Jansen (ed.), Van Nostrand Reinhold, New York, 1988b.

LOWE, J., III, and L. KARAFIATH, Stability of Earth Dams upon Drawdown, *Proc. First Panamerican Conference on Soil Mechanics and Foundation Engineering*, Mexico City, 1959.

MAY, D. R., Application of the Planimeter to the Swedish Method of Analyzing the Stability of Earth Slopes, *Trans. 2d Congress on Large Dams*, Washington, D.C., 1936.

MILLIGAN, V., Field Measurement of Permeability in Soil and Rock, *In-situ Measurement of Soil Properties*, ASCE, **2**, 1976.

PECK, R. B., W. E. HANSON, and T. H. THORNBURN, *Foundation Engineering*, Wiley, New York, 1974.

SEED, H. BOLTON, Settlement Analysis, A Review of Theory and Testing Procedures, *Proc. ASCE*, SM2, March 1965.

SEED, H. BOLTON, 19th Rankine Lecture, Considerations in the Earthquake-Resistant Design of Earth and Rockfill Dams, *Geotechnique*, September 1979.

SEED, H. BOLTON, MAKDISI, FAIZ, I., and DEALBA, PEDRO, Performance on Earth Dams During Earthquakes, *J. Geotechnical Engineering Div.*, ASCE, **104** (GT7), July 1978.

SHERARD, J. L. "Embankment Dam Cracking," *Embankment Dam Engineering*, A. Casagrande (ed.), Wiley, New York, 1973.

SHERARD, J. L., Sinkholes in Dams of Coarse, Broadly Graded Soils, ICOLD, 13th Congress on Large Dams, Q.49, R2, New Delhi, 1979.

SHERARD, J. L., Hydraulic Fracturing in Embankment Dams, *Proc. Symposium on Seepage and Leakage from Dams and Impoundments,* ASCE, May 1985.

SHERARD, J. L., and L. P. DUNNIGAN, Filters for Silts and Clays, *J. of Geotechnical Engineering,* ASCE, June 1984*b*.

SHERARD, J. L., and L. P. DUNNIGAN, Filters and Leakage Control in Embankment Dams, *Proc. Symposium on Seepage and Leakage from Dams and Impoundments,* ASCE, May 1985.

SHERARD, J. L., and L. P. DUNNIGAN, Critical Filters for Impervious Soils, *J. Geotechnical Engineering,* ASCE, July 1989.

SHERARD, J. L., L. P. DUNNIGAN, and J. R. TALBOT, Basic Properties of Sand and Gravel Filters, *J. Geotechnical Engineering,* ASCE, June 1984*a*.

SOUTHWELL, R. V., *Relaxation Methods in Theoretical Physics,* Oxford University Press, 1946.

SOWERS, G. F., R. C. WILLIAMS, and T. S. WALLACE, Compressibility of Broken Rock and the Settlement of Rockfill, *Proc. Sixth International Conference on Soil Mechanics and Foundation Engineering,* vol. II, Montreal, 1965.

SOYDEMIR, C, and B. KJAERNSLI, "Deformation of Membrane-faced Rockfill Dams with Unyielding Foundations in Relation to the Design of Storvass Dams," Report 53203, Norwegian Geotechnical Institute, Oslo, November 1975.

SPENCER, E., A Method of Analysis of the Stability of Embankments Assuming Parallel Inter-slice Forces, *Geotechnique,* **17,** 11–26, 1967.

SWIGER, WILLIAM F., Foundation Treatment, *Advanced Dam Engineering for Design, Construction, and Rehabilitation,* R. B. Jansen (ed.), Van Nostrand Reinhold, New York, 1988.

TAYLOR, D. W., *Fundamentals of Soil Mechanics,* Wiley, New York, 1949.

TAYLOR, D. W., Stability of Earth Slopes, *Contributions to Soil Mechanics, 1925–1940,* Boston Society of Civil Engineers, 1937.

TECHNICA, LTD., Report on Beech Dam, 1989.

TERZAGHI, K., and R. B. PECK, *Soil Mechanics in Engineering Practice,* Wiley, New York, 1948.

U.S. ARMY CORPS OF ENGINEERS, Coastal Engineering Research Center, Shore Protection Manual, 3 vol., 1973.

U.S. ARMY CORPS OF ENGINEERS, Coastal Engineering Research Center, "Large Wave Tank Tests of Riprap Stability," Technical Memorandum No. 51, May 1975.

U.S. ARMY CORPS OF ENGINEERS, "Determining Sheltered Water Wave Characteristics," Engineer Technical Letter, ETL 1110-2-305, February 1984.

U.S. ARMY CORPS OF ENGINEERS, "Seepage Analysis and Control for Dams," Engineer Manual EM 1110-2-1902, September 1986.

U.S. ARMY OFFICE OF THE CHIEF OF ENGINEERS, Engineer Manual EM 1110-2-1902, "Stability of Earth and Rock-fill Dams," April 1970*a*.

U.S. ARMY OFFICE OF THE CHIEF OF ENGINEERS, Engineer Manual EM 1110-2-1906, Laboratory Soils Testing, November 1970*b*.

U.S. ARMY OFFICE OF THE CHIEF OF ENGINEERS, "Construction Control for Earth and Rock-Fill Dams," EM 1110-2-1911, January 1977.

U.S. ARMY OFFICE OF THE CHIEF OF ENGINEERS, "Slope Protection Design for Embankments in Reservoirs," Engineer Technical Letter, ETL 1110-2-222, July 1978.

U.S. ARMY OFFICE OF THE CHIEF OF ENGINEERS, "Earth and Rock Fill Dams, General Design and Construction Considerations," EM 1110-2-2300, May 1982.

U.S. BUREAU OF RECLAMATION, "Earth Manual," 2d ed., 1974.

U.S. BUREAU OF RECLAMATION, "Design of Small Dams," 2d ed., 1977.

U.S. BUREAU OF RECLAMATION, Design Standards No. 13, "Embankment Dams," Chap. 5, Protective Filters, May 1987.

U.S. SOIL CONSERVATION SERVICE, Soil Mechanics Note No. 1, Guide for Determining the Gradation of Sand and Gravel Filters, January 1986.

WILKINS, J. K., Flow of Water Through Rockfill and its Application to the Design of Dams, *Proc. 2d Australian/New Zealand Soils Conference,* 1956.

WILSON, S. D., Influence of Field Measurements on the Design of Embankment Dams, *Proc. Ninth International Conference on Soil Mechanics and Foundation Engineering,* Tokyo, 1977.

SECTION 14

CONCRETE FACE ROCKFILL DAMS

By J. Barry Cooke and A. V. Sundaram

INTRODUCTION

1. Evolution. Evolution of the rockfill dam began with the concrete face rockfill dam and then two main types were developed: the concrete face rockfill dam (CFRD), and the earth-core rockfill dam (ECRD). Figure 1 shows the transition from dumped to compacted rockfill and the continued increase in heights of both types.

The early rockfill dams are seen to be CFRD dams of dumped rock. Up to a height of 75 m they were satisfactory, but higher dams developed face cracks and excessive leakage, because of the high compressibility of the dumped rockfill. These experiences resulted in fewer CFRD dams being adopted. However, the steep slopes of the CFRD dams demonstrated the high shear strength of dumped rockfill and its usefulness as a dam building material with earth core and filters of the ECRD. Consequently, the ECRD of dumped rock was developed and became the typical rockfill dam.

Vibratory rollers, developed earlier particularly for use on road construction, became standard machines for rockfill dam compaction in about 1960 (see Fig. 1). There was no experience to indicate that compacted rockfill was necessary for the ECRD of sound rock. The ECRDs of dumped rockfill, including a number higher than 120 m and one at 150 m, had all performed well. Measurements of the modulus of compressibility of the compacted rockfill indicated values several times greater than that for dumped rockfill. However, for the ECRD, the high-modulus rockfill was not required.

In contrast, the high CFRD of dumped rockfill had experienced concrete face difficulties. As a result of the high modulus of compressibility of compacted rockfill, observed in the high ECRD dams, the adoption of the CFRD and its design development were resumed. Compacted rockfill, in addition to reviving the CFRD, enabled small size rocks and rocks with low compressive strength to be used.

The development of the CFRD moved slowly between 1960 and 1970, and progressed rapidly after Cethana in Australia (100 m), Alto Anchicaya in Colombia (140 m), and Foz do Areia in Brazil (160 m) in the 1970–1980 decade. Today selection of the type of rockfill dam, ECRD or CFRD, is based principally on cost. At reasonably equal estimated costs, there are inherent advantages of the CFRD.

THE CONCRETE FACE ROCKFILL DAM (CFRD)

2. General. Important improvements in the design principles of concrete face rockfill dams adopted during the past 25 to 30 years have resulted in in-

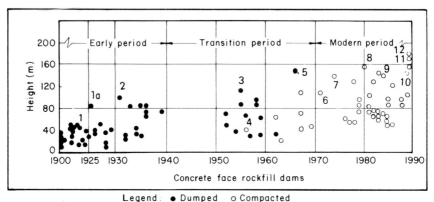

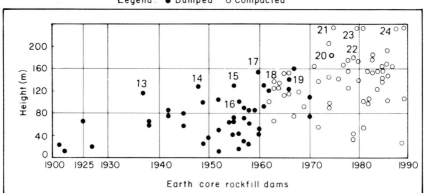

Concrete Face Rockfill Dams:
1. Strawberry
1a. Dix River
2. Salt Springs
3. Paradela
4. Quioch
5. Exchequer
6. Cethana
7. Anchicaya
8. Areia
9. Khao Laem
10. Segredo
11. Aguamilpa
12. Tianshenqiao

Earth Core Rockfill Dams:
13. San Gabriel
14. Mud Mountain
15. Ambuklao
16. Browniee
17. Goeschenersip
18. Congar
19. Akosombo
20. New Melones
21. Chivor
22. Dertmouth
23. Nurek
24. Guavio

FIGURE 1 Trends in the type and height of rockfill dams.[1]

creased use of this type of rockfill dam and its adoption for higher dams. The advent of construction technology, through the use of properly zoned compacted rockfill and/or gravel, results in a dam of reliable performance in terms of safety and leakage. The development of concrete toe slabs with grouted cutoffs and face slab improvements, notably abandoning the highly articulated pattern of slabs and compressible joints, are the principal factors in current design trends and a resulting higher frequency of acceptance of CFRD.

3. Typical Section. No type of dam is actually of standard design. Adaptation to the foundation and available materials, and consideration of each design ele-

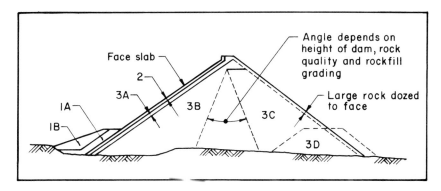

Rockfill zones for Concrete Face Rockfill Dams of sound rock:

1A = Impervious soil
1B = Random 0.5 m layers
2 = Processed small rock transition
3A = Selected small rock placed in same layer thickness as zone 2
3B = Quarry-run rockfill, approximately 1 m layers
3C = Quarry-run rockfill, approximately 2 m layers
3D = Dumped rockfill. Compaction for rock zones: four passes of 10 t vibratory roller.

FIGURE 2 Typical section and material zones for CFRD of sound rock.[1]

ment, are necessary for each dam. However, the CFRD has evolved to a stage where the main elements are common. Figure 2 shows a typical cross section and zone designations for the CFRD of sound rockfill on bedrock.

The concrete elements are similar in all dams: a toe slab (plinth), a monolithic face slab with joints only as necessary for construction, and a parapet wall. However, significant variations in fill, filter zone 2, and rockfill zone 3 have been allowed, depending on the available construction materials.

Zone 2 is usually subdivided into two processed zones:

- Zone 2A: a fine filter zone within a 3-m radius in plan from the perimeter joint
- Zone 2B: a minus 7.5-cm crusher-run rock.

The several leakage events in CFRDs of compacted rockfill have been in or near the perimeter joint, and the filter allows convenient sealing by dirty fine sand (minus no. 10 or no. 20 mesh). There have been no crack or leakage events elsewhere in the face slab, and the more economical crusher-run zone provides a workable and dense zone on which to place the concrete face.

Zone 3 rockfill is further divided into several subzones. The increased layer thickness from 3A to 3B to 3C and possibly to dumped rockfill, 3D, is to provide a high modulus where it is needed, and to provide an increase in permeability. The thicker downstream layer, 3C, gives cost savings in terms of lower construction costs and reduced tonnage of rockfill. The density of the 2-m layer of rockfill is about 7 percent less than for the 1-m layer of rockfill, with consequent savings in rock volume. Dumped rockfill has been used in the downstream toe of several recent CFRDs. It could be used more often, particularly if no future raising of the dam is planned.

Dumped rockfill can be used in the downstream toe without affecting the face slab movement. At Segredo, a 20-m-high underwater dumped rockfill cofferdam has been used in the downstream toe. At Xingó, underwater dumped rockfill is used in depths of up to 20 m, except within 20 m of the toe slab, to permit rockfill placement on each abutment prior to diversion.

For dams of up to 75 m high, use of an upstream core feature consisting of zone 1A, impervious fill, and zone 1B, random fill, shown in Fig. 2, is not required. However, it is the general practice to use the upstream core feature for dams higher than 75 m. The combination of upstream core feature and concrete face rockfill dam is used only in the lower part of the valley or canyon, while a simple concrete face rockfill dam is adopted in the upper part of the valley. The purpose of the upstream core feature is to seal any cracks or openings that might develop in the perimeter joint in the lower elevations.

4. Adaptability to Topography. The dam shape is not critical, since the water load principally causes only perimeter joint opening, and the slab just floats on the rockfill. For steep abutments, however, special attention needs to be paid to perimeter joint design, to accommodate greater offset movements. Typical longitudinal profiles of selected high rockfill dams are shown in Fig. 3.

5. Adaptability to Cold Sites. Four concrete face rockfill dams have demonstrated the feasibility of rockfill dams in cold climates and under severe freeze-thaw conditions. These dams were Outardes 2 in northern Canada, Cabin Creek at elevation 12,000 ft (3660 m) in the Rockies, Courtright at elevation 8200 ft (2500 m) in the California Sierras, and Golillas at elevation 10,000 ft (3000 m) in the Andes, where temperatures reach minus 20°C. The coldest temperatures are at Outardes 2, where the winter range is minus 25 to minus 35°C, and the summer day range is 20 to 25°C. The 180-ft-high (55-m) Outardes 2 Dam, constructed in part in cold weather, has settled only 0.015 ft (0.5 cm) in 3 years. No cracks have occurred, and consequently no seepage was observed. The concrete faces at

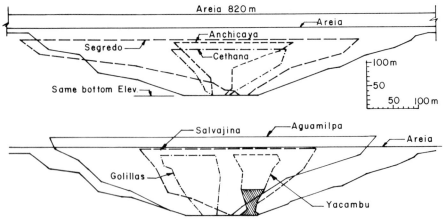

FIGURE 3 Longitudinal profiles of selected compacted concrete face rockfill dams.[14]

Courtright and at Cabin Creek have shown no temperature cracks or freeze-thaw damage after their respective service lives of about 30 and 22 years. No ice plucking or ice thrust damage has been reported. The 300-ft (98-m) Courtright Dam has had face damage and leakage problems because of settlement related to dumped rockfill.

From this data base on the performance of compacted CFRD rockfill dams in areas subject to extremes of weather, it may be concluded that rockfill dam design criteria do not need any major modifications to accommodate these extreme weather conditions. The measures of using air entrainment and pozzolan in the concrete have proved successful in preventing freeze-thaw damage.

6. Adaptability to Pumped Storage. The loading cycle of the upper reservoir of pumped storage projects has provided some interesting data on cyclic resistance of rockfills. The 100- to 140-ft-high (31- to 43-m) Taum Sauk Dam constructed of dumped rhyolite rockfill, completed in 1963, with daily operation range of 80 ft (24 m) for the 140-ft (43-m) section, settled 1.0, 0.31, and 0.13 ft (0.3, 0.1, and 0.04 m) in the first, second, and third 5-year periods, respectively. While this is more than the normal annual cyclic settlement performance for dumped rockfill, it is nevertheless within tolerable limits, since it has not adversely affected performance. Cabin Creek, a 250-ft-high (76-m) high compacted rockfill completed in 1967, has a daily operating cycle of 90 ft (27 m). Settlement after 10 years of operation was 0.4 ft (0.12 m), 0.02 percent of its height, while leakage has never exceeded 0.9 cfs (28 L/s). These data suggest that daily cyclic loading does not affect the performance of concrete face rockfill dams, and thus CFRD is readily adapted to pumped storage reservoirs.

7. Future Raising. Among all the dam types, the CFRD is the best-suited for stage development and future raising. The added rockfill causes only moderate settlement of the first-stage dam. The face thickness should be designed for the ultimate height, or, in the event of unexpected raising, a higher ratio of head to thickness will have to be accepted.

8. Spillways over CFRDs. The compacted rockfill forming the essential load-bearing feature of a CFRD provides a spillway foundation which is as safe and secure as many of the steep and weathered rock valley walls where spillways are commonly constructed. It is now well established that the total settlements of the modern CFRDs are relatively small and such small movements mostly occur within the first few years of construction. Gated spillways cannot tolerate any movements; however, ungated spillways can tolerate small movements without significant damage. For a relatively common situation, where river flood flows occur for a few weeks or less with a peak discharge of about 300 to 350 cfs per foot of chute width, ungated spillway over a CFRD offers large potential cost savings. In addition to a separate service spillway located beyond the dam, ungated spillway over a CFRD can take the place of an emergency spillway to pass the excess flood.

When placing spillway over a CFRD, the downstream portion of the rockfill forming the spillway foundation must be of small size rock compacted in the same layer thickness as the upstream rockfill supporting the face slab.

9. Seismic Considerations. Since the entire CFRD embankment is dry, earthquake shaking cannot cause pore pressure in the rockfill voids. The CFRD foundation is generally rock, which does not magnify the incoming acceleration forces. The embankment is heavily compacted in thin layers to a dense state with vibratory rollers. Earthquakes can cause only small deformations during the short period of strong shaking. After the earthquake is over, the CFRD is as stable as before.

The concrete face may be cracked by strong earthquakes, resulting in increased leakage. The potential cracking and leakage cannot threaten the overall safety of the dam, because the amount of leakage which can get through the cracks and the zone of small rock under the face slab can easily be passed safely through the main rockfill embankment. For these reasons, the CFRD is considered to have the highest fundamental conservatism against earthquakes, and the same basic design has been used in regions of high seismicity and in nonseismic areas.

Measured settlement of rockfill dams caused by earthquakes confirms that very few dams have been strongly shaken, and the measured settlements to date are small.[17] For six among seven reliable records, the crest settlements during the earthquakes are less than 0.1 percent of the dam height. For the seventh—the 84-m-high Cogoti Dam in Chile, judged to have been shaken considerably more strongly than any of the others—the measured crest settlement was about 0.4 m, or about 0.45 percent of its height, during a 1943 earthquake with magnitude variously reported in the range of 7.9 to 8.3. The epicenter of the earthquake was located about 89 km away. Cogoti Dam is of dumped rockfill and its performance demonstrated that even with the high settlement of dumped rockfill, the concrete face was not damaged.

For the great majority of sites which may be very strongly shaken, such as near the epicenter of a magnitude 7.5 earthquake, the same CFRD design can be used as in nonseismic areas. For these sites, all present experience with dam behavior and overall results of current dynamic calculations give confidence that the worst earthquake-induced crest settlement will be substantially less than 1 percent of the dam height. A sudden crest settlement of 1 percent of the dam height will not threaten the safety of a modern CFRD.

For the smaller number of sites where the world's strongest ground shaking could occur, within a short distance of major faults capable of generating earthquakes of magnitude 8 or greater, there is very little evidence from any source to guide judgment about the maximum probable earthquake-induced crest settlement of a modern CFRD. At these sites, it is desirable to include additional freeboard to the dam height to account for the maximum possible earthquake-induced crest settlements of a compacted CFRD. The provision of conservative extra freeboard is probably the appropriate and economical seismic design provision. Since the worst damage to the dam during strong earthquake shaking is a general slumping of the upper part of the embankment, there can be no threat to the basic dam safety unless the crest settlement exceeds the freeboard. In the extreme case, some provision to prevent individual rocks near the top of the dam on the downstream sloping surface from being dislodged may be justified.

FEATURES OF THE CFRD

10. General Considerations. Selection of dam type includes initial consideration of all types; some can be eliminated early as not being adaptable to the site conditions. For the dams that may be suitable, each type has its particular features to be taken into consideration. For the rockfill option, the choice is between the ECRD and the CFRD. Some features of the CFRD, to be taken into account in the selection of the type of rockfill dam, and in the comparison with the alternative concrete dam, are design, construction, and schedule features.

11. Design Features. The design features to be considered are

- The zoned rockfill is semi-pervious and an effective energy dissipator. It is zoned to be safe against internal erosion and instability for full reservoir loading without the concrete face. The sliding safety factor exceeds 7.
- The dam is of zoned rockfill, all downstream from the reservoir water.
- With all the rockfill downstream from the reservoir water, drainage galleries in the abutment rock are not required.
- Water load is transmitted into the foundation upstream from the dam axis, an inherently safe feature.
- Uplift is not involved. The pressure on the foundation exceeds reservoir pressure over three-quarters of the base width.
- Postconstruction movements are small and cease within 5 years of construction.
- With the now known small and quickly stabilized movements of compacted rockfill, it is safe and reasonable to build moderate-size spillways over CFRDs.
- Since all the rockfill is dry, earthquake shaking cannot cause pore pressure.
- The conditions of high shear strength, no pore pressure, and small settlement under seismic loading make zoned rockfill inherently resistant to seismic loading.
- Instrumentation for the CFRD is to contribute to a data base for future dams and is not required for safety monitoring.
- The only credible mechanism of failure of a CFRD is erosion by sustained overtopping flow. Spillway and freeboard design to handle the probable-maximum flood (PMF) is the response to this risk.
- The CFRD of competent rockfill on sound rock is inherently safe. It is essential for this type to be well-engineered and well-constructed when rock and foundation properties are poor.

12. Construction and Schedule Features. Major construction and schedule features are

- Ramps are used with 1.3H:1V slopes in any direction. This minimizes haulage roads to the dam and facilitates construction. There are no traffic or schedule restrictions caused by the presence of the core and filters as in the case of ECRD.
- Rockfill may be placed on abutments at 1.3H:1V slope so that rock from required excavation can be used directly, in advance of diversion, and to minimize the volume of rockfill in the closure section.
- The relatively unrestricted placement of rockfill results in its lower unit cost.
- The short base width of the dam results in auxiliary structure economies.
- Rainfall does not interfere with rockfill or concrete face placement. A plastic sheet has been pulled by the slipform to protect newly placed concrete during rainfall.
- There only needs to be minimal contingency in foundation, materials, construction, and schedule.
- Both the rockfill and concrete face can be placed at scheduled rates, and a fast schedule can be met reliably.

- The concrete face can be placed in stages which are convenient for the contractor.
- The toe slab (plinth) and its grouting are outside the dam, and are thus not on the critical path of the schedule.

COMPACTED ROCKFILL FOR THE CFRD

13. General. Compacted rockfill properties that are particularly useful to the CFRD are high shear strength and a high modulus of compressibility. The high shear strength has been demonstrated by steep slopes of some existing CFRDs and by triaxial tests. The modulus of compressibility is 5 to 8 times higher than that of the lower zone of dumped rockfill. The modulus of gravels, in general, is several times greater than that of compacted rockfill. Therefore, gravels may be used exclusively or in combination with rockfill for CFRD.

14. Placing. For sound rock, the dumping and spreading of rockfill is intentionally done to obtain segregation (see Fig. 4). End dumping is on the edge of the layer being placed, and several passes of the dozer spread the rock. There is inherent segregation both in the dumping and in the spreading. The smooth surface of the previous layer is desirable. The large rocks tend to contact the smooth surface with a flat face. Scarifying or removal of fines is not required. The smooth surface of fines is desirable to reduce truck tire and dozer track costs. Horizontal permeability is much higher than vertical permeability.

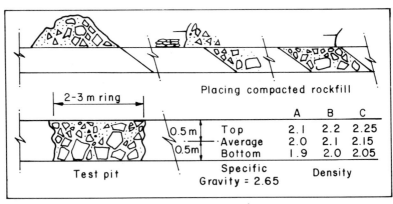

FIGURE 4 Placing and density of compacted rockfill.[1]

15. Compaction. The traffic of loaded trucks and the movements of the spreading dozer provide effective compaction which is supplemented by passes of the smooth-drum vibratory roller. For sound rock, the fines and small rocks in the upper zone of a layer are well-compacted. In the lower zone, energy is effectively transmitted through large rocks, causing wedging and crushing of contact edges and points. For sound rock, four passes of a 10-t vibratory roller has become a standard practice.

Inadequate compaction near the perimeter joint has resulted in more face offset than desired. It is not possible to compact effectively the rockfill near the joint with the vibratory roller. Erratic and sometimes large offsets have been measured. Future offsets will be less, with the current specified use of a backhoe-mounted plate vibrator on the face and on the plan surface within 3 m of the perimeter joint. For weak rock, the factors of number of passes, layer thickness, and use of water must all be considered.

16. Density. Density is of interest, but does not have the significance that it has in soil compaction. The influence of soil mechanics thinking and practice on rockfill placement has resulted in excessive waste and higher density, and in more restrictive specifications than required. Density tests have been carried out on many dams, not as a control, but to learn about compacted rockfill and to have a record of the composition of the dam. For grading tests of sound rock, where gradings and densities have been taken in the upper and lower half of a 1-m layer, the upper, finer-graded rockfill is about 8 percent more dense than the lower, coarse-graded rockfill.

17. Grading. Since all rockfill of sound rock is highly segregated in each layer, grading of the quarry-run rock is not important. Well-graded quarry-run rock will give the highest density and modulus, but all quarry-run rock, even when poorly graded, has been satisfactory. Strength in a layer of rockfill of sound rock comes from the density of the upper zone of fine rockfill and from the wedged and interlocked rocks in the lower zone. Foz do Areia is an example of an excellent CFRD of poorly graded quarry-run basalt.

For weak rock, grading is meaningless, and procedures (layer thickness, compaction, use of water) are selected to ensure breakdown and high density. Strength comes from the high density of the pieces of the rock in a dense matrix of fines from crushed rock.

18. Water. The use of water always improves rockfill properties, particularly in reducing compressibility. For rock having low water absorption (less than about 2 percent), the benefit is small, and the use of water may be justified only in selected zones of high dams. For low compressive strength rock with high water absorption, the loss of strength on saturation can be 40 to 60 percent and the use of water requires serious consideration even for low dams.

A rockfill condition for hard rock that may require water during placement is the excessive presence of fines (minus 200 mesh). Enforcing a specified maximum percentage is not practicable. If there is an appearance of excessive fines, a check can be made by saturating the area with a water wagon and then observing whether a loaded dump truck and the vibratory roller are supported. For rock from a source known to contain a high percentage of fines, the application of water during placement may be specified.

19. Shear Strength. Early Californian dumped rockfill dams of 80 to 130 ft (25 to 40 m) height had steep slopes and demonstrated shear strengths of the order of 60° even with high field sample void ratios. A ϕ value of 45° can be assumed for compacted rockfill of sound rock. However, since there have been no slope stability failures of the CFRDs of dumped or compacted rockfill, slopes have generally been based on precedent rather than stability analyses.

20. Modulus. The vertical modulus of compressibility during construction is an indication of rockfill quality. It is obtained by water level devices or crossarms. The modulus determined from face deflection measurements is 1.5 to 3 times higher. Values of vertical moduli obtained in ECRD and CFRD dams have ranged from 4500 to 19,500 psi (30 to 130 MPa), depending on the rock, rockfill grading, layer thickness, and other factors. The usefulness of moduli data is principally for the highest dams and future higher CFRD dams.

DESIGN

21. General. The design of CFRD has been mainly empirical and based on experience and judgment. The CFRD dams incorporating the design and construction changes leading to current practice have all performed well. Changes have been principally to effect economies in design and construction. The criteria and guidelines presented herein are proven features of design and of requirements in specifications that provide for a safe and satisfactory dam at low cost.

22. Foundation Excavation and Treatment

Under Toe Slab. The toe slab is usually on hard, nonerodible fresh rock which is groutable. For less favorable foundation rock, after a trench is made to an estimated acceptable foundation, many methods are available to treat local imperfections.[2] The criterion is to eliminate the possibility of erosion or piping in the foundation. Careful excavation is used to minimize fracturing of the rock surface on which the toe slab is placed. Air or air-water cleanup, just prior to placing concrete, is required to obtain a bonded contact of the concrete to the foundation.

While most CFRDs have erosion-resistant hard rock formations, many have been built on rocks with faults or badly weathered seams, or on fairly large areas of softer rock of questionable erosion resistance.[3,4] For these, the length of the foundation seepage path has been increased to one-fourth, or more, of the water head by placing a layer of shotcrete (or concrete) on the final foundation surface downstream of the toe slab, with a filter placed for a distance downstream of the shotcrete, and sometimes a filter placed also on top of the shotcrete to anticipate possible cracking. Reinforced (fibrous) shotcrete has sometimes been used[5] in an effort to distribute any settlement cracking into a series of small cracks rather than a few large cracks.

At the 485-ft (148-m) high Salvajina Dam, where the upper right abutment of the dam foundation contained a wide intrusion dike of severely weathered to soillike properties, the rockfill foundation was covered with a two-layer filter, to prevent the foundation material from being eroded into voids in the dam embankment. This is considered good practice at locations where there is any possibility that soillike foundation material (weathered rock, faults, and soil-filled joints) could be caused to migrate by erosion into the embankment voids.

Under Rockfill Embankment. For the CFRD, all the rockfill is downstream from the water thrust. The base width is more than 2.6 times the height and essentially all the water load is taken into the rock foundation upstream from the dam axis. Measurements on many dams have confirmed that the reservoir filling causes very little movement in the downstream shell.[6] The overall sliding factor (weight divided by horizontal reservoir thrust) is about 7.5. The criterion for foundation excavation and treatment under the downstream half of the dam is therefore lower than for the ECRD.

The foundation excavation criterion employed at Khao Laem is an example of current practice for a CFRD on a deeply weathered and highly compressible foundation.[4] At the dam axis, the foundation settled 4 ft (1.25 m) during construction, and only 1 inch (0.025 m) during and since reservoir filling.

Typical current accepted practice requires trimming overhangs and vertical faces higher than about 6.5 ft (2 m) for a horizontal distance downstream of the toe slab equal to about 30 percent of the dam height, or about 35 ft (10 m) minimum. In the rest of the foundation area, cliffs and overhangs are left in place. In this main foundation area, abutment voids or zones of lower density under overhangs are arched over by the highly frictional rockfill embankment and have no influence on the face-slab movements or CFRD performance.

Over most of the foundation area, the excavation is made only with earthmoving equipment, removing soillike surface deposits and exposing the points of in situ hard rock. Dozer removal is adequate; ripping is seldom required. Under an upstream portion of high-dam foundations, most of the soil and soft weathered rock between hard rock points is excavated, with a backhoe or similar equipment, but final cleanup by hand labor is not required. Under the downstream portion of the foundation, and most of the upstream portion for dams of low to moderate height, soil and surface material between the hard rock points is left in place. Alluvial gravel deposits, not sand, in the riverbed are commonly left in place except for a short distance downstream from the toe slab. Such gravel deposits usually have a high modulus of compressibility, far exceeding that of well-compacted rockfill. Settlement is not a problem, but alluvium that is judged to be possibly subject to liquefaction is removed.

23. Toe Slab

Dimensions. Toe slab widths for hard and groutable foundation rock are made of the order of $\frac{1}{20}$ to $\frac{1}{25}$ the water depth.[2] At a given dam, the width is changed in several steps and is not tapered, mainly for construction convenience. The minimum width has generally been 10 ft (3 m). For a poor local rock condition, or for a site in poor rock, greater widths are used.

The design thickness has been frequently made about equal to the face slab thickness. Excavation overbreak and irregular topography usually provide significantly greater thickness, so that a minimum design thickness of 1.0 to 1.3 ft (0.3 to 0.4 m) is generally reasonable for most toe slabs. Since the toe slab is uniformly supported on sound, hard, thoroughly grouted rock, it is difficult to justify a thickness much greater than 1.0 ft (0.3 m). A design thickness of 1.3 to 1.6 ft (0.4 to 0.5 m) may be considered for the lower toe slab of high dams.

Stability. The toe slab must resist the high horizontal water thrust without support from the rockfill. For the toe slab of normal thickness, there is ample frictional resistance to resist the water thrust, unless the foundation has unfavorably oriented planes of low shearing resistance just below the toe slab. For high toe slabs, and usually in local areas, stability analyses are required. Uplift pressure under the slab is assumed to vary linearly from full reservoir pressure to zero over the width. Water loading on the face slab opens the perimeter joint so there is no interaction between toe slab and face slab. The rockfill may exert a resistance force on the downstream side of the toe slab, but this cannot be relied upon, and is neglected in stability computations.

In the toe slab alignment, it is attempted to minimize the height of the toe slab, not only from the standpoint of stability, but to limit the thickness of rockfill that must resist water loading in the starter slab area, and thus minimize perimeter joint offset.

Layout. Various excavation methods and toe slab orientations are used, depending primarily on the abutment steepness and the thickness and nature of soil and weathered rock overlying a satisfactory foundation. Rock excavation is made by blasting. Occasionally, only earth moving is needed to expose satisfactory foundation along parts, or all, of the length. Sometimes a continuous trench is made along the toe slab, digging to refusal with earthmoving equipment, before a decision is made to locate the toe slab rock excavation reference line. In steep canyons with suitable hard rock near the surface, the slab is sometimes placed on or very near the existing surface in order to minimize difficult excavations.[7]

The toe slab is laid out as a series of straight lines with joints determined by the contractor. The angle points are selected to suit the foundation conditions and topography, and have no required relation to the vertical joints of the face slab. The reference line for the excavation is at the bottom of the toe slab, and for forming the toe slab, at the contact of the rockfill face and toe slab.

In each straight segment of length, the toe slab is usually made horizontal in a direction at right angles to the reference line. This puts the upstream edge at the deepest line of excavation, and gives the most convenient surface for toe slab construction and grouting. However, the toe slab plane can be at any angle in plan to the reference line. For very steep abutments, the toe slab may be nearly normal to the dam axis.

Reinforcing, Joints, and Anchors. The main purpose of the reinforcing is to function as temperature steel, and to spread out and minimize the widths of any cracks. Reinforcing and dowels are useful as a grout cap. A single layer should be used.[3,4] The steel is put 4 to 6 in (10 to 15 cm) clear of the upper surface as temperature steel, where it is hooked by the anchors; 0.3 percent of the slab thickness each way is adequate. A double row of longitudinal steel has a slight theoretical disadvantage because it makes the toe slab stiff and less able to adjust to any small differential settlements of the underlying rock, which has no tensile properties.

In the past, construction joints with waterstops located at predetermined distances were commonly used. Current practice is to continue longitudinal reinforcing through the joints without waterstops. This is considered good practice, is more economical, and has been adopted for recent dams.[8]

Anchors in the toe slab are used simply to pin the concrete to the rock. The anchors are not to resist any given uplift loads. Lengths, spacing, and bar diameters should be chosen on the basis of precedent and the characteristics of the rock foundation. The anchors and temperature reinforcing do improve the slab as a grout cap. Anchors used in common practice have generally been no. 8 to no. 11 (25- to 35-mm-diameter) bars spaced about 4 to 5 ft (1.2 to 1.5 m) each direction, with lengths usually 10 to 15 ft (3.0 to 4.6 m). The anchors are simple dowels of reinforcing steel, grouted full length in the rock, and hooked (90°) on the one layer of reinforcing.

24. Face Slab

Concrete. For the concrete, durability and impermeability are more important than strength, a 28-day compressive strength of about 3000 psi (20 MPa) being adequate.[3,6] Maximum-size aggregate of 1.5 in (38 mm), air entrainment, and use of pozzolan[6,9] are common features in current practice. However, a maximum size of 2.5 in (64 mm) has sometimes been used and is satisfactory with special care being taken at construction joints and waterstops.

There has been no record of alkali aggregate reaction in face slabs for CFRDs, and use of pozzolan in many dams may be a factor. However, it should generally be considered good practice to use pozzolan, even with apparently nonreactive aggregates, for providing a more impervious and durable concrete, and for reducing the risk, albeit rare, of concrete deterioration from some as yet little-understood and possibly slowly occurring cement-aggregate reaction.

Thickness. The thickness of slab on the early, dumped rockfill dams was traditionally 1 ft (0.3 m) + $0.0067H$, where H is height. The face was underlain by a layer of large, derrick-placed rocks. For the CFRD with compacted rockfill and a compacted upstream face, the thickness increment was decreased to $0.003H$,[3,4,10,11] and even $0.002H$ or less.[5,12,13] These slabs have given satisfactory performance, and there is a current general trend toward thinner slabs.

For four dams, 130 to 246 ft (40 to 75 m) high, the Hydro-Electric Commission in Tasmania has used a uniform design thickness of 10 in (0.25 m), with thickening at the perimeter joint. For the 308-ft- (94-m-) high Murchison Dam, completed in 1982, the slab has a constant design thickness 1 ft (0.30 m) + 0.001H.

Another recent example is the long, low dam for the major 2600-MW Macagua hydro plant on the Caroni River in Venezuela, where the face slab has a constant design thickness of 10 in (0.25 m).[18]

On the basis of presently available experience and current practice, it is reasonable to use slabs with constant thickness of 10 to 12 in (0.25 to 0.30 m) for dams of low to moderate height of 250 to 300 ft (75 to 100 m) and to use an incremental thickness of about 0.002H for the very important and high dams.

Reinforcing. The use of reinforcing of 0.4 percent of slab thickness in each direction, for compacted rockfill, has been an economical and successful change from the traditional 0.5 percent used with dumped rockfill dams.[14] Current practice is to use 0.3 percent steel in the large central area of known compression, and 0.4 percent near the perimeter and in the starter slabs.[2] The steel area should be calculated on the basis of the design concrete thickness.

A trend which appears to be desirable and economical is to carry horizontal reinforcing through the vertical joints, reverting to a practice used successfully on some older CRFDs.[14] The trend is based on the fact that the major area of the face is under compression. Several vertical joints near abutments are contraction joints to minimize perimeter joint opening. Where reinforcing passes through vertical joints, a bottom waterstop has sometimes also been used[4] as a carryover from the earlier practice. With the steel passing through, there is little or no more tendency for a crack to open at the construction joint than at other locations in the slab. Bonded joints are assumed in reinforced concrete design. Waterstops are not used in CFRD face-slab horizontal construction joints.

Reinforcing steel is always placed in one mat in the center of the slab, or a little above the centerline. The purpose is to make a slab of given thickness as flexible as possible, allowing it to follow small differential settlements without developing high bending stresses and to provide equal bending resistance in both direction.

Slab Width and Joints. The selection of face slab width and thus location of vertical joints are mostly dependent on the capability of the available slip-forming equipment and the convenience of construction. Traditionally a slab width of 30 to 60 ft (10 to 18 m) has been used, but the current trend is to leave it to the option of the contractor.

25. Parapet Wall and Camber. The early dams had a parapet wall of 4 ft (1.2 m). A higher 10- to 16-ft (3- to 5-m) wall has become an economical and desirable practice.[6] The economy is in saving a slice of upstream rockfill when the saving exceeds the cost of the wall. The rockfill volume saved in this slice increases with the height of dam. The wide surface of rockfill at the base elevation of the wall provides desirable surface width for the slip form operation and parapet placement.[5]

The freeboard for the CFRD is calculated from the top of the parapet wall, rather than from the top of the dam, with the walls being extended into the abutments.

The small postconstruction and reservoir-filling settlements require only nominal camber.[4,6] Camber is conveniently obtained by variable height of the parapet wall with oversteepening of the upper zone of the downstream face.

26. Joints. The perimeter joint is the most sensitive part of the concrete facing because of its potential opening due to reservoir loading and causing leakage. Several types of designs have been used and further improvements (easier construction) are being developed with new projects. However, vertical and horizontal joints have been similar in design over the years with minor difference in details. Typical details of vertical, horizontal, and perimeter joints which have been successfully used in large projects are illustrated under "Typical Designs."

27. Embankment Zoning

Zoning Designations. It is useful to use standard zoning designations and to adopt those common for the ECRD materials (Fig. 2); i.e., zone 1 for impervious fill, zone 2 for the filter or transition zone directly under the concrete slab, and zone 3 for the main rockfill. Figure 2 does not show the zoning used in special cases of impermeable or weak rock, but the zone designations would apply to the properties of the material.

Zone 1. A blanket of compacted impervious soil was placed on the lower part of the concrete face at Alto Anchicaya Dam,[10] since the dam height was breaking precedent. This detail has since been repeated on the Areia, Khao Laem, and Golillas Dams and on several other high dams. The purpose is to cover the perimeter joint and slab in the lower elevations with impervious soil, preferably silt, which would seal any cracks or joint openings. A minimum practical construction thickness of impervious soil can be used directly adjacent to the concrete slab and rock foundation, covered with less costly waste material for stability (zone 1B, Fig. 2).

Zone 1 is useful only if a problem develops. The current practice is to place zone 1 to a level only several meters above the original riverbed for dams less than about 250 ft (75 m) high. For higher dams, zone 1 is placed to a level of about one-fourth the height of the dam.

Zone 2. The early and primary use of the face zone of finer rock was to provide uniform and firm support of the concrete face and to minimize excess concrete beyond minimum design thickness. Crusher-run -6 to -3 in (-15 to -7.5 cm) rockfill has been used. Current practice is to use a pocket of fine filter, zone 2A, within about 3 m of the perimeter joint and a zone 2B of a crusher-run -6 to -3 in (-15 to -7.5 cm) rockfill downstream of zone 2A and across the face.

For zone 2 face compaction, the roller is first pulled up the slope without vibration for several passes, and then given four upward passes with vibration.[5,6,14] A new and promising development is the use of a backhoe-mounted plate vibrator.[15] On several new dams, the plate vibrator is being specified for zone 2 compaction adjacent to the toe slab, and optional for face compaction.

Zone 3. The rockfill embankment is in three zones of increasing layer thickness to give a desirable transition of compressibility and permeability from upstream to downstream. Lowest compressibility is desirable in the portion of the upstream shell, which transmits water load to the foundation. The increasing permeability from zone 2 progressively through zones 3A, 3B, and 3C (Fig. 2) is desirable during construction, in the event of a flood before the concrete face is placed. After the concrete face is placed, there is no credible face problem that could cause more leakage than the rockfill could handle without damage.

Zone 3A is a transition between zone 2 and the main rockfill, placed and compacted at the same time and in the same layer thickness 16 to 20 in (0.4 to 0.5 m) as zone 2. The main purpose is to limit the size of voids and ensure that zone 2 material could not be washed into large voids in the main rockfill. Similar zones of small rockfill placed in thin layers are commonly used in ECRDs at the filter–rockfill interfaces for the same purpose. Usually zone 3A rock is obtained by selected loads; however, on some jobs, the zone 3A is obtained by grizzling.

Because most of the water load passes into the foundation through the upstream shell,[3,6] it is desirable that the compressibility of zone 3B be made as low as practical to minimize slab settlement. Much experience has shown that embankments placed in layers about 1 m thick and compacted with four passes of a 10 t (static) smooth steel-drum vibratory roller give satisfactory performance.

The downstream zone 3C takes negligible water load, and its compressibility has little influence on the settlement of the face slab. Consequently, zone 3C is commonly placed satisfactorily in thicker layers, usually about 70 to 80 in (1.5 to 2 m) and also compacted with four roller passes.[15] It is desirable to specify the thicker layers for zone 3C because it will be highly permeable and will result in a substantial cost saving due to less equipment wear. Also, the thicker zone 3C layers provide a location into which maximum size rocks are placed, thickening the layer if necessary.

CONSTRUCTION

28. Quarry. Project investigations and studies include an evaluation of rock and rockfill from required excavations and potential quarries. The specifications and zoning in the dam are developed to make use of the required rock excavation and quarry-run rock with little wastage.

All compacted rockfills of quarry-run rock have performed satisfactorily. Test rockfills are certainly not necessary for rock of medium to high compressive strength. Geological appraisal, drill cores, and saturated unconfined compressive test results are adequate. For weak rock with significant loss of strength for saturated specimens, conservative specifications for placement can be established from precedent, but a saturated test fill is desirable for a major or high dam.

29. Specifications. For sound rock, the specifications can be general and concise. Grading is not specified, except to give a fines limit that provides a basis for rejecting loads of predominantly earth. Limits of 50 percent −1 in (−2.5 cm) and 12 percent minus no. 200 mesh assure the rockfill properties, the quarry source being known. The maximum size of rock is specified to be the layer thickness, when incorporated into the layer, and to provide a relatively smooth surface for compaction. Rockfill adjacent to a rock of full layer thickness is not fully compacted, and does not need to be. The large rock will attract the load in the area. For placement and compaction, a method specification is used. The layer thickness and number of passes of the 10-t vibratory roller are specified.

30. Haul Roads. The dam being constructed from one material, rockfill, construction traffic is unrestricted and haul roads may be in any direction with 1.3H:1V slopes. This allows rockfill to be placed on abutments prior to diversion. It also greatly facilitates construction. Such haul roads in the dam, with no restriction on differential levels, result in lower unit cost of rockfill for the CFRD in comparison to the ECRD.

31. Protection of Zone 2. After placing, trimming, and compacting the zone 2 material, the compacted surface needs to be protected against damage from erosion and construction activities until the face slab concrete is poured. Such protection is often obtained by applying a layer of quick curing asphaltic emulsion at the rate of 2 to 4 L/m^2 or about 4- to 5-cm thick shotcrete on the compacted zone 2 surface.

PERFORMANCE

32. General. The performance of the CFRDs of compacted rockfill has been excellent. Large leakage and settlement (Paradela, Exchequer) were associated with the earlier practice of using dumped (with no compaction) rockfill for high CFRD dams. The advent of using compacted zoned rockfill and improved joint details have virtually eliminated these earlier problems of large settlements and leakage.

33. Settlement. Compaction of rockfill embankments has been shown to be very effective in reducing settlements. Figure 5 shows representative crest settlement measurements for five compacted CFRDs and three older dams of the dumped-rock type.[16]

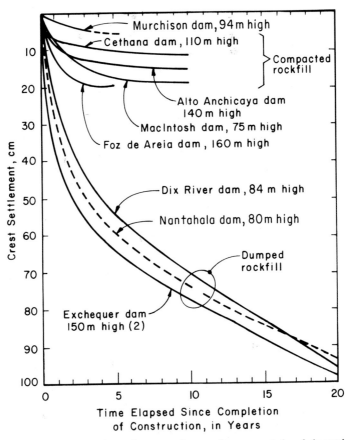

FIGURE 5 Comparison of crest settlements for compacted and dumped rockfill dams.[16]

The rate of magnitude of crest settlement is influenced by both the rockfill properties and the shape of the valley. In narrow canyons, the compacted rockfill arches between the abutments, reducing the crest settlement in the early years. Both Cethana and Alto Anchicaya Dams are built in relatively narrow canyons and the measured crest settlements apparently continue for many years at a decreasing rate, probably due to slow progressive release of the arch action by creep. Areia Dam is constructed in a broad valley where no significant arch action between abutments would be expected, and the measured crest settlement practically stopped after 3 years. Also, the Areia Dam rockfill had larger average rock size and less rock fines than the average quarried hard rock used for CFRDs. Measurements during construction show that the Murchison and Mackintosh rockfill embankments are probably the most incompressible and compressible, respectively, of the modern CFRDs for which good records are available.[5] Therefore, the range of crest settlement-time records shown in Fig. 5 should be near the total range expected for CFRDs of different rock types, in valleys of different width, with dam heights from 250 to 525 ft (75 to 160 m).

The crest settlements for the compacted CFRD is a small fraction of that measured on the dumped (uncompacted) rockfill CFRDs of older practice. For a modern compacted CFRD of 330-ft (100-m) height, the expected crest settlement will generally be about 4 to 6 in (10 to 15 cm) in 15 years and 6 to 10 in (15 to 25 cm) in 100 years. For the purpose of design, 0.30 to 0.35 percent of the dam height could be used to predict the estimated postconstruction settlement.

34. Leakage. The several leakage events in CFRDs of compacted rockfill have been in or near the perimeter joint, and the filter allows convenient sealing by dirty fine sand (minus no. 10 or minus no. 20 mesh). There have been no crack or leakage events elsewhere in the face slab, and the more economical crusher-run zone provides a workable and dense zone on which to place the concrete face.

Weak rock is used in CFRDs (Kangaroo Creek, Little Para, Mangrove Creek) with special consideration of the rock properties, placement procedures, zoning, and drainage provisions. For foundation rock with permeable and possibly erodible features, the cutoff can be extended downstream from the toe slab by shotcrete covered by filter material (Reece, Salvajina, Winneke). This is in contrast to the CFRD of sound rock on nonerodible rock, where designs are similar.

Where leakage has occurred in compacted rockfill dams, it has been at the perimeter joint, and nominal in amount. The current practice of using a fine filter zone, for several meters from the perimeter joint, limits leakage in the event of local malfunction at the joint, and allows sealing by underwater application of dirty fine sand.

TYPICAL DESIGNS

35. Cethana. Cethana Dam, located in Tasmania, Australia, is 360 ft (110 m) high and was completed in 1972. Design and construction was by the Hydro-Electric Commission (HEC). The rockfill is sound, clean quartzite obtained from quarry and required excavation. The typical dam section and zoning is shown in Fig. 6.

Cethana is noteworthy in the development of the CRFD of compacted rockfill, not only for being the highest in 1972, but also for pioneering the development of the following important design and construction features:

1. A face zone, zone 2, of well-graded small-size rockfill which included fines
2. Sealing the face rockfill with a bitumen-rock chip seal
3. A toe slab width of $1/20$ the head
4. Two types of waterstop in perimeter joints, one being a copper waterstop at base of slab and the other a PVC waterstop at the middle of the slab
5. Rockfill zones of three layer thicknesses: 18 in (45 m), 3.0 ft (0.9 m), and 4.5 ft (1.35 m)

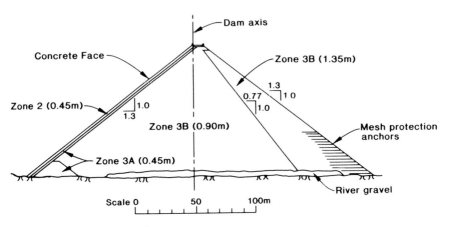

FIGURE 6 Cethana dam section.

6. Use of water with rockfill placement, 0.15 percent of rockfill volume
7. Concrete face slab thickness of 1 ft + 0.002H (0.3 m + 0.002H)
8. Use of ramping haul roads at 1.3H:1V side slopes within the dam
9. Continuous slip-form placement of concrete face panels for the dam of unprecedented height
10. Use of steeper outer slopes of 1.3H:1V for the dam

These pioneering features resulted in excellent performance and became a guide for subsequent CFRDs.

36. Foz do Areia Dam. The 525-ft (160-m) high Foz do Areia Dam[6] in the Iguaçu River, Paraná, Brazil, built by Companhia Paranaense de Energia (COPEL) between 1975 and 1980, is part of the 2500-MW Governador Bento Munhoz da Rocha Netto powerplant. The main characteristics of the dam are summarized below:

Height	160 m
Crest length	828 m
Rockfill volume	14 × 10^6 m^3 of basalt and basaltic breccia
Toe slab (plinth)	Width ≥ 1/20 the hydraulic load, doweled to the rock foundation
Face slab	Thickness = 1 ft + 0.00357H (0.30 + 0.0357H m)
	Maximum thickness: 32 in (0.80 m)
	Thickness at the top: 12 in (0.30 m)
	Reinforcement: 0.4% of design concrete thickness in each direction
	Horizontal construction joint at midheight, el. 680 m
	Joint spacing: 52 ft (16 m)
	Concrete: pozzolanic cement: 310 kg/m^3; 4.5% entrained air; water cement ratio = 0.53; slump 8 cm; strength = 20.6 MPa

The region is made up of subhorizontal basalt flows in which massive basalts predominate over basaltic breccias. The properties of the rock are shown in Table 1.

The dam section and rockfill zoning are shown in Fig. 7. The basalt was known to have few fines and to have relatively low modulus of compressibility. Therefore, water was used during placement of rockfill.

TABLE 1 Properties of Rocks

	Compressive strength, MPa	Modulus of elasticity, MPa	Apparent density, g/cm³	Apparent porosity, %	Soundness test sodium sulfate, % of losses	Unconfined compression strength ratio, soaked/dry, %
Massive basalts	235	68,000	2.8	1.3	2–5	80
Basaltic breccias	37	25,500	2.3	12.0	35–60	67

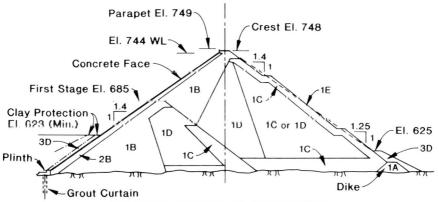

MATERIAL ZONING AND COMPACTION

Material	Classification	Zone	Method of Placement	Compaction
Rockfill I	Massive basalt (up to 25% basaltic breccia)	1A	Dumped	
		1B	Compacted in 0.80-m layers	Four passes of vibratory roller (10 t), 25% water
		1C	Compacted in 1.60-m layers	Four passes of vibratory roller (10 t), 25% water
	Intercalation of massive basalt and basaltic breccia	1D	Compacted in 0.80-m layers	Four passes of vibratory roller (10 t), 25% water
	Massive basalt (selected rock—min. 0.80 m)	1E	Placed rock (downstream face)	
Transition II	Crushed sound basalt	2B	Well-graded, max. size 6-in, compacted in 0.40-m layers	Layers: 4 passes of vibratory roller; face: 6 passes of vibratory roller (upslope)
Earthfill III	Impervious soil	3D	Maximum size ¾ in, compacted in 0.30-m layers	Pneumatic roller or construction equipment

FIGURE 7 Areia Dam section and rockfill zoning.[7]

Perimeter Joint. In the design of Foz do Areia Dam, the greatest attention was given to the perimeter joint, where, because of the discontinuity between the toe slab founded on rock and the slab supported on the fairly deformable rockfill, large movements were expected. A single waterstop system was disregarded in view of the experience of Alto Anchicaya Dam, in which relatively high leakage had resulted at first filling because of localized weak points in the single waterstop barrier and poor compaction near the toe slab. A multiple defense line system was selected, in order to guarantee a satisfactory overall performance even in case of local defects in any of the joint water tight elements.

Main features of the joints are shown in Figure 8. The copper waterstop at the bottom of the slab was placed over a neoprene strip resting on an asphalt-sand pad. The asphalt used was an asphalt cement, penetration grade 50 to 60, and the sand was the same as that used for the concrete.

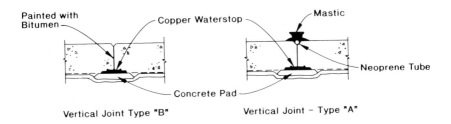

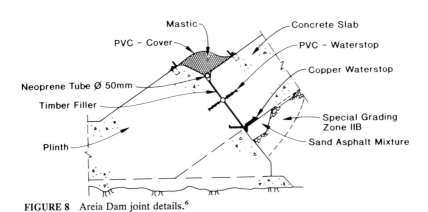

FIGURE 8 Areia Dam joint details.[6]

During construction, the transition material in the vicinity of the joint was improved by extending the grading and adding a small percentage of cement. The modified hand compacted material, used above el. 665 m, ensured better transition properties at this critical area, and because of its slight cohesion, made construction of the asphalt pad much easier.

A clay fill was spread between the cofferdam and the toe of the dam covering the toe slab up to el. 623 m at the concrete face. The earth cover was extended along the abutment up to el. 662 m, providing an additional protection to the perimeter joint.

Construction. Construction and completed dam photos are shown in Fig. 9.

FIGURE 9 Areia Dam construction photos.[14] (*a*) Right abutment rockfill placed before diversion upstream rockfill being placed to divert 1:200-year flood, November 1977. (*b*) First-stage upstream rockfill, August 1978.

(d)

FIGURE 9 (*Continued*) Areia Dam construction photos.[14] (*c*) Completed first stage, September 1979. (*d*) Completed project, September 1980.

37. Macagua. The Macagua project[18] developed in two stages is located near Puerto Ordaz, Venezuela, about 6 mi (10 km) upstream from the mouth of the Caroni River. The second stage development presently (1991) being built by CVG Electrificacion del Caroni, C.A., of Caracas, Venezuela, includes the construction of a 22-m-high and 2800-m-long concrete face rockfill dam, and an intake, 2400-MW powerhouse, and 14-bay spillway capable of passing 1.06 Mcfs (30,000 m^3/s). A typical section of the dam, toe and face slab is shown in Fig. 10.

Reason for Selecting a Concrete Face Rockfill Dam. Since the required excavation for the powerhouse and tailrace channel would provide a large supply of rock at relatively short hauling distances, the alternatives of ECRD and CFRD were evaluated.

This evaluation showed that a CFRD would be less costly, and would have the following advantages at this site:

1. The foundation will be generally a rough surface of hard rock. Since the dam is long and low, the cost of foundation treatment relative to embankment will be greater than for a higher dam. For an alternative dam with earth core and filters the cost of smoothing up the foundation under those zones would be considerably greater than the cost of the foundation seal under the narrow toe slab of a concrete face rockfill dam.
2. For a low rockfill dam with earth core, the relative cost of the filter is much greater than for a higher dam, giving substantial advantage to a concrete face rockfill dam.
3. The concrete face rockfill dam with steeper slopes than other rockfill alternatives minimizes the extent of concrete structures that interface with the main dam.

Rockfill Specifications. The rockfill zones are zone 2, face zone of minus no. 1½ in material in 20 in (0.5 m) thick layers; zone 3A, transition zone of rockfill in 20-in (0.5-m) layers; and zone 3B, main rockfill, in 5-ft (1.5-m) layers. Compaction for all zones is by five passes of a 10-t vibrator roller. The maximum size of rocks in zones 3A and 3B is the layer thickness, subject to providing a relatively smooth surface for compaction.

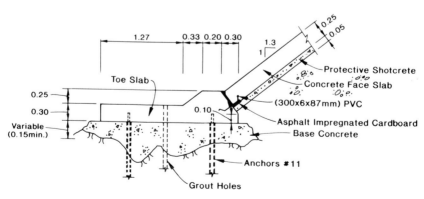

b) Toe and Face Slab

FIGURE 10 Macagua Dam section.[18]

Toe Slab, Face Slab, and Joints. The width of the toe slab was limited to 6.5 ft (2.0 m) only because of the low dam height and the hard, nonerodible nature of the foundation rock. The rock under the slab was cleaned and smoothed by using high density lean concrete. Cracks were cleaned, filled with concrete, and covered with zone 2 material downstream of the toe slab. The thickness of the toe slab is 10 in (0.25 m).

Reinforcement was placed in a single layer in the upper third point of the slab and was carried through the vertical joints. The toe slab was doweled to the rock before grouting.

The face slab is of constant 10-in (25-cm) thickness with 0.4 percent reinforcement both horizontally and vertically. The horizontal steel crosses the vertical joints, which are spaced at 50-ft (15-m) intervals. The concrete is of 3000- to 3500-psi (20-MPa to 25-MPa) 28-day strength with 1½-in maximum size aggregate and air entrainment.

Perimeter joints are provided with PVC bottom waterstops, and grooves are filled with plastic asphalt. Special attention was given to the seal between the slab joint and connecting concrete structures. These are provided with double waterstops, and special details. Photos of construction details are shown in Fig. 11.

38. Khao Laem Dam. The Khao Laem Multipurpose Project[4] is in Central West Thailand on the Quae Noi River, 188 mi (300 km) from Bangkok, and was completed in December 1982. The main feature is a concrete face rockfill dam, 426 ft (130 m) high and 3280 ft (1 km) long at the crest, on a very poor and karstic foundation. The volume of rockfill was 10.5×10^6 yd^3 (8×10^6 m^3).

The bedrock under the dam consists of shale, sandstone, and siltstone, both calcareous and noncalcareous, locally interbedded with limestone beds. Extensive karst commonly occurs as solution cavities up to several meters across, partially infilled with clay or river alluvium consisting of cobbles, gravel, sand, and silty clay. In some areas, karstic weathering was found during construction up to 492 ft (150 m) below the toe slab. Therefore, foundation treatment included the provision of a concrete diaphragm wall up to 180 ft (55 m) deep under the upstream toe of the dam and an extensive system of grouting galleries, totaling 13.7 mi (22 km) in the right abutment of the dam.

Selection of Dam Type. The poor foundations eliminated a concrete gravity dam and an earth core rockfill dam on the basis of technical, schedule, and cost considerations. The CFRD was the only type of dam considered to be adaptable to the site. Foundation treatment could continue throughout the construction period and foundation excavation would be minimized. A typical dam section is shown in Fig. 12.

The tender design provided for removal of all weathered and cavernous material from the dam foundation. However, wide variations of foundation, with pinnacles and ridges of rock, deep and extensive pockets of weathered material, and blocks of fresh rock surrounded by wide clay seams, forced changes in the foundation stripping criteria in the valley as follows (Fig. 12):

Zone A. From toe slab to $\frac{1}{6}W$ (where W = dam bottom width), this zone carries some rock load and the highest water load. The foundation was required to be moderately weathered rock or better and highly weathered pockets were removed.

Zone B. From $\frac{1}{6}W$ to dam axis, this zone carries moderate water load and is farther away from the concrete face. The foundation was stripped generally down to rock for 50 percent exposure over the area, but pockets of highly to completely weathered rock were allowed to remain. Highly weathered material was removed from a large area.

FIGURE 11 Photos of Macagua Dam construction. (*a*) Slip form. (*b*) Reinforcing, timber slide form, and toe slab.

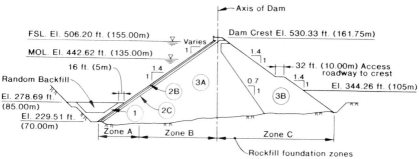

FIGURE 12 Khao Laem Dam section.[4]

Zone C. From centerline to downstream toe, this zone carries shell load only. The foundation was stripped to reveal 30 percent of evenly distributed sound rock.

The embankment is divided into five zones: 2A, 2B, 2C, 3A, and 3B. Zones 2B and 2C form a transition layer between the concrete face and the rockfill.

Zone 2A consists of minus 2-in (minus 50-mm) material. It is located adjacent to the toe slab and in other areas where mechanical compaction equipment cannot be used. Its purpose is to limit the differential movement between the face slab and toe slab and to reduce leakage should it occur through the perimeter joint. Zone 2B provides a firm, smooth base for the concrete face and a surface that can be trimmed to a close tolerance so as to minimize excess concrete over design thickness. Zone 2C is a graded transition zone to prevent fine material from zone 2B being washed into zone 3A and leaving cavities behind the face, in the event of a leakage developing in the concrete face.

To maximize compaction during construction in order to limit deflection of the face slab by water load, the layer thickness was reduced and compaction requirements increased from downstream to upstream as follows:

Zone	Maximum rock size		Layer thickness	
	Inches	Meters	Feet	Meters
2A	5	0.125	0.4	0.125
2B	20	0.500	1.6	0.5
2C	20	0.500	1.6	0.5
3A	36	0.9	3.3	1.0
3B	60	1.5	6.6	2.0

Rockfill was compacted with four passes of a 10-t vibratory roller. The upstream face of 2A was compacted with 12 passes (4 without vibration, 4 with half vibration, 4 with full vibration) of a 10-t vibratory roller. Vibration was applied only on the upward travel of the roller. Immediately after compaction, the face was sprayed with a nominal 1.0-in layer of shotcrete to protect against erosion and to provide a working surface for face slab construction.

Face Slab. Face slab thickness t was 1 ft (0.3 m) at the top. It increased linearly with dam height H according to the formula $t = 1 + 0.003H$. Reinforcement used was 0.5 percent of the design slab thickness in both the horizontal and vertical directions.

Toe Slab. The contract documents included a conventional reinforced and doweled toe slab. Foundation excavation disclosed even worse conditions than anticipated, and consequently a design was developed which included a miniplinth and a gallery. The miniplinth was to permit face rockfill to be placed on schedule while the diaphragm wall and grouting proceeded. The gallery was added to gain additional time for grouting and for possible future use. The perimeter joint connecting toe slab to face slab is identical to that used at Areia. Typical joint details used at Khao Laem are shown in Fig. 13.

Performance. The general performance of the dam has been excellent. Though the foundation settled as much as 3.7 ft (1.25 m) at the axis during construction, maximum crest settlement after 5 years of operation was only 4 in (10 cm). Maximum deflection normal to the face slab was 8 in (20 cm). Surface leak-

TYPICAL DESIGNS

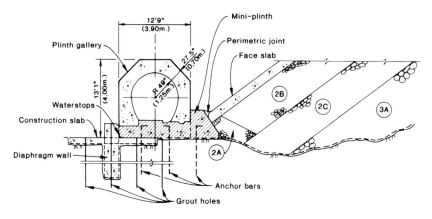
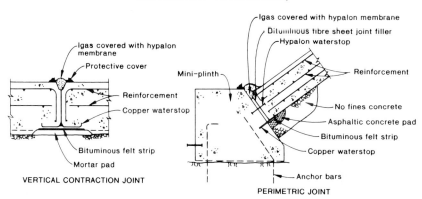
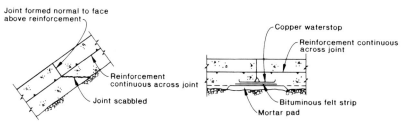

FIGURE 13 Khao Laem Dam joint details.[4]

age at the toe of the dam is negligible. Total leakage, including inflow from abutments, measured in the river 2 km downstream, is about 18 to 35 cfs (0.5 to 1.0 m³/s).

39. Segredo. The Segredo Project[8] is the second hydroelectric plant comprising a 480-ft-high (145-m) concrete face rockfill dam with a rockfill volume of 7.2×10^6 m³ presently (1991) being built by COPEL in southern Brazil. The excellent performance of the Foz do Areia concrete face rockfill dam, after 10 years of continuous operation, was a major factor in the selection of this type of dam. A typical dam section is shown in Fig. 14.

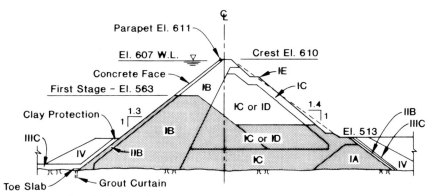

FIGURE 14 Segredo Dam section.

The rockfill material was mostly obtained from the required excavation for the hydraulic structures. Sound basalt is the predominant rock, with some occurrences of amygdaloid basalt and basaltic breccia. In spite of sound rockfill, a somewhat high deformability of the rockfill embankment is expected, as was observed at Foz do Areia, because of the lack of sufficient fines. The void ratio of the fill after compaction was higher than 0.35.

For the upstream portion of the dam, where sound basalt with less than 25 percent of basaltic breccia is required, compaction was by six passes of a 9-t vibratory roller over 0.8-m layers with the addition of 25 percent water by volume.

The transition material between the concrete slab and the main rockfill (2B) was placed in 0.4-m layers and compacted horizontally by six passes of a 9-t vibratory roller, and that along the slope by six passes of a 6-t roller.

A vibratory plate was used in restricted corners near the toe slab. A special transition zone (2B-B) was specified for the region immediately below the perimeter joint, to guarantee a better compaction condition in this confined area and to provide filter properties in case of localized leakage.

The rockfill embankment and the concrete slab were built in two stages, following the completion of each rockfill construction stage. The 52-ft- (16-m-) high slabs, separated by longitudinal joints, were made by slip-forming. The thickness of the concrete slab is 23 in (0.7 m) at the maximum section, tapering to 12 in (0.3 m) at the dam crest.

Great attention has been given to the perimeter joint details. Leakage is to be controlled by a copper waterstop at the bottom of the slab, supported on a cement mortar pad, and by mastic filler placed over the joint and protected by a watertight fabric-reinforced rubber membrane (Hypalon). A strip of silty sand over the joint will act as an additional seal against eventual leakage, along the most critical deeper zone (see Fig. 15).

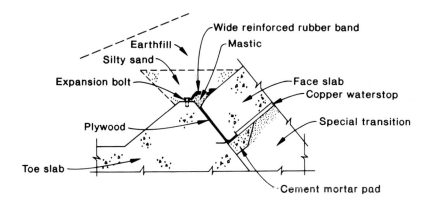

(a) PERIMETRIC JOINT

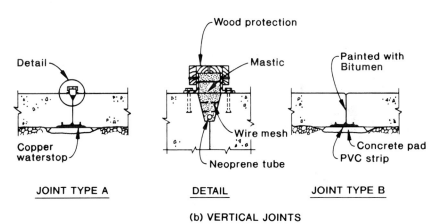

(b) VERTICAL JOINTS

FIGURE 15 Segredo Dam joint details.[8]

The main differences in Segredo Dam compared to Foz do Areia Dam are as follows:

- The upstream slope was steepened to 1.3H:1V. The downstream nominal slope was maintained at 1.4H:1V, but actual slopes between access road berms are 1.2H:1V
- Compaction of the upstream third of the rockfill embankment, including the transition zone, was by six passes of a 9-t roller, with 25 percent water. For the remaining fill, compaction was by four passes, with no water.
- The grading of the transition material was improved by limiting the maximum size to 4 in (102 mm) and processing for a better distribution of fines. A special transition zone, under the toe slab, is a well-graded mixture of 1½-in (38-mm)

(a)

(a)

FIGURE 16 Segredo Dam construction photos. (*a*) First-stage rockfill (*top*) and slab construction (*bottom*), April 1990.

(b)

(b)

FIGURE 16 (*Continued*) Segredo Dam construction photos. (*b*) Simultaneous construction of first-stage concrete face (*top*) and downstream rockfill placement (*bottom*), November 1990.

maximum size, for better compaction and filter properties in the event of leakage.
- Details of the perimeter joint have been altered: the bottom copper waterstop is supported by a concrete pad, instead of a sand-asphalt pad as used at Foz do Areia; the central waterstop was eliminated because of difficult concrete placing conditions and weakening of the concrete slab; the mastic filler was covered by a fabric-reinforced rubber membrane, tightly tied to the concrete; and the strip of silty sand was placed over the joint, as a selected material of the earthfill which covers the toe slab below el. 532 m.

TABLE 2 The World's Highest Concrete Face Rockfill Dams (>100 m) Completed and under Construction

Name	Country	Year of completion*	Height, m
Aguamilpa	Mexico	(1993)	190
Tianshengqiao	China	(1995)	180
Foz do Areia	Brazil	1980	160
Salvajina	Colombia	1985	148
Yacambu	Venezuela	(1995)	146
New Exchequer†	U.S.A.	1966	146
Segredo	Brazil	(1992)	145
Alto Anchicaya	Colombia	1974	140
Xingo	Brazil	(1994)	140
Al Wehdah	Jordan	(1996)	135
Massochora	Greece	(1994)	135
Koman	Albania	1986	133
Khao Laem	Thailand	1984	130
Siah Bishe Lower	Iran	(1994)	130
Golillas	Colombia	1978	130
Cirata	Indonesia	1987	125
Shiroro	Nigeria	1984	125
Reece	Australia	1986	122
Neveri	Venezuela	1981	115
Paradela	Portugal	1955	112
Cethana	Australia	1971	110
Rama	Yugoslavia	1967	110
Paradela†	Portugal	1958	110
Fortuna	Panama	(1993)	105
Salt Springs†	U.S.A.	1931	100
Siah Bishe Upper	Iran	(1994)	100
Courtright†	U.S.A.	1958	100

*Parentheses indicate projected completion year.
†Dumped rockfill.

- The face-slab horizontal reinforcing steel has been reduced from 0.4 to 0.3 percent of the nominal concrete section for the central zone of the face, where compression stresses are expected.
- The toe slab was cast as a continuous slab, doweled to the rock, without joints, except for construction joints with reinforcement passing through.

- Rockfill instrumentation includes monitoring of horizontal movements in three lines across the main dam section, to obtain a better understanding of the rockfill deformation under load.

Construction. Stages of construction seven months apart are shown in Fig. 16.

The design details employed at Segredo are the results of wide experience of CFRD not only in Brazil but also in other parts of the world. Thus Segredo Dam can be considered to be representative of current practice for the CFRD.

40. High CRFD Dams. In the development of the CFRD of compacted rockfill, there has been caution in adopting the CFRD of unprecedented height. Cautious steps in height have been taken: Cethana, 110 m; Alto Anchicaya, 140 m; and Foz do Areia, 160 m. In 1991, Tianshengqiao, 180 m, and Aguamilpa, 190 m, were under construction, and projects with CFRD dams of 210 to 220 m high were under consideration. The CFRD dams under construction and under consideration include the basic design features of Segredo Dam. A list of the world's highest CFRD dams completed and under construction in 1991 is presented in Table 2.

REFERENCES

1. COOKE, J. BARRY, The Concrete-faced Rockfill Dam, *Water Power and Dam Construction,* January 1991, pp. 11–15.
2. COOKE, J. BARRY, and J. L. SHERARD, Concrete-Face Rockfill Dam: II—Design, ASCE, *Journal of Geotechnical Engineering,* 113(10), October 1987, pp. 1113–1132.
3. SIERRA, J. M., C. A. RAMIREZ, and S. J. E. HACELAS, Design Features of Salvajina Dam, Concrete Face Rockfill Dams, Design Construction and Performance, *Symposium Proceedings,* ASCE, 1985, pp. 266–285.
4. WATAKEEKUL, S., G. J. ROBERTS, and A. J. COLES, Khao-Laem—A Concrete Face Rockfill Dam on Karst, Concrete Face Rockfill Dams, Design, Construction and Performance, *Symposium Proceedings,* ASCE, 1985, pp. 336–361.
5. FITZPATRICK, M. D., B. A. COLE, F. L. KINSTLER, and B. P. KNOOP, Design of Concrete-Faced Rockfill Dams, Concrete Face Rockfill Dams, Design, Construction and Performance, *Symposium Proceedings,* ASCE, 1985, pp. 410–434.
6. PINTO, N. L., F. P. L. MARQUES, and E. MAURER, Foz do Areia Dam Design, Construction and Behavior, Concrete Face Rockfill Dams, Design, Construction and Performance, *Symposium Proceedings,* ASCE, 1985, pp. 173–191.
7. AMAYA, F., and A. MARULANDA, Golillas Dam—Design, Construction and Performance, Concrete Face Rockfill Dams, Design, Construction and Performance, *Symposium Proceedings,* ASCE, 1985, pp. 98–120.
8. PINTO, N. L., F. P. L. MARQUES, and E. MAURER, Segredo Dam—Basic Design Aspects, Concrete Face Rockfill Dams, Design, Construction and Performance, *Symposium Proceedings,* ASCE, 1985, pp. 587–593.
9. DASCAL, O., The Outardes 2 Concrete Face Rockfill Dam, Concrete Face Rockfill Dams, Design, Construction and Performance, *Symposium Proceedings,* ASCE, 1985, pp. 121–139.
10. MATERON, B., Alto Anchicaya Dam—Ten Years' Performance, Concrete Face Rockfill Dams, Design, Construction and Performance, *Symposium Proceedings,* ASCE, 1985, pp. 73–87.
11. PINKERTON, I. L., S. SISWOWIDJONO, and Y. MATSUI, Design of Cirata Concrete Face

Rockfill Dam, Concrete Face Rockfill Dams, Design, Construction and Performance, *Symposium Proceedings,* ASCE, 1985, pp. 642–656.

12. CASINADER, R., and R. E. WATT, Concrete-Face Rockfill Dams of the Winneke Project, Concrete Face Rockfill Dams, Design, Construction and Performance, *Symposium Proceedings,* ASCE, 1985, pp. 140–162.

13. GOSSCHALK, E. M., and A. N. S. KULASINGHE, Kotmale Dam and Observations on CFRD, Concrete Face Rockfill Dams, Design, Construction and Performance, *Symposium Proceedings,* ASCE, 1985, pp. 379–395.

14. COOKE, J. BARRY, Progress on Rockfill Dams, 18th Terzaghi Lecture, ASCE, *Journal of Geotechnical Engineering,* 110(10), October 1984, pp. 1381–1414.

15. PHILLIPS, P. R., Batang Ali—Transition Zone, Concrete Face Rockfill Dams, Design, Construction and Performance, *Symposium Proceedings,* ASCE, 1985, pp. 396–409.

16. SHERARD, J. L., and J. B. COOKE, Concrete-Face Rockfill Dam: I—Assessment, ASCE, *Journal of Geotechnical Engineering,* 113(10), October 1987, pp. 1096–1112.

17. BUREAU, G., R. L. VOLPE, W. H. ROTH, and T. UDAKA, Seismic Analysis of Concrete Face Rockfill Dams, Concrete Face Rockfill Dams, Design, Construction and Performance, *Symposium Proceedings,* ASCE, 1985, pp. 479–508.

18. PRUSZA, Z., K. DE FRIES, and F. LUQUE, "The Design of Macagua Concrete Face Rockfill Dam," Concrete Face Rockfill Dams, Design, Construction and Performance, Symposium Proceedings, ASCE, 1985, pp. 608–617.

SECTION 15

ROLLER-COMPACTED CONCRETE DAMS

By Gregory Hillebrenner

TYPES OF RCC DAMS

1. General. Roller-compacted concrete (RCC) is a relatively new method of dam construction which has gained wide acceptance since first being used in the early 1980s. RCC construction is a proven technology which is both economical and time-saving and is used throughout the world for dam construction and rehabilitation of existing dams. It combines the benefits and safety of a concrete gravity dam with the rapid continuous-placement methods normally associated with earth-embankment dam construction. Typical RCC nonoverflow and stepped-spillway sections are shown in Figs. 1 and 2, respectively.

The RCC mixture consists of portland cement, water, aggregate, and pozzolan (if desired) and is proportioned in a manner that produces a zero-slump mix which can support the weight of a vibratory roller. The RCC mix is normally placed and compacted in lifts 1 to 2 ft thick (0.3 to 0.6 m) in an around-the-clock continuous operation to produce a monolithic structure free of the many construction joints normally associated with conventional concrete dam construction. Physical properties of hardened RCC can vary over a wide range, depending on the quality of the aggregates used, content of cementitious materials (cement plus pozzolan), degree of compaction, and the quality control exercised during construction. Material selection and proportions of the RCC mix are controlled by design requirements, availability of materials, and planned placement procedures. The RCC mix should be designed to produce a compactable and stable mass which, when hardened, meets the strength, durability, and permeability requirements for the specific dam being designed.

Development of RCC dam design has essentially taken three paths, which are based on two different philosophies: (1) the soils approach, which includes the lean RCC dam method and (2) the concrete approach, which includes the high-paste RCC dam and Japanese roller-compacted dam (RCD) methods. Both concrete approach methods have a more fluid consistency and are more workable because of their higher paste content than the lean RCC mixes which use the soils approach, yet both approaches produce an RCC mix with zero slump which can support the weight of the vibratory compaction equipment.

Initial RCC dam design and construction followed the lean RCC soils approach, first used in 1982 by the United States Army Corps of Engineers for construction of Willow Creek Dam in Oregon.[25,26] During the 1980s, the trends shifted toward the high-paste-content concrete approach, which was developed by Malcolm R. Dunstan and modified by the United States Bureau of Reclamation for construction of Upper Stillwater Dam, completed in 1988 in Utah.[24] Current designs seem to favor an approach somewhere between the soils and concrete approaches, referred to as *medium-paste-content RCC*, although dams

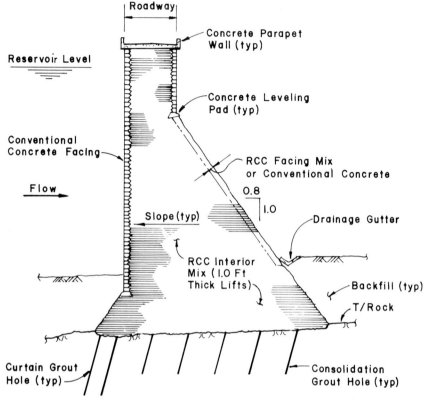

FIGURE 1 Typical RCC nonoverflow section (shown without drainage gallery).

using all approaches are currently being designed and constructed. The approach to be used depends on the strength, durability, and permeability requirements desired for the finished dam structure.

2. Lean RCC Dam. Lean RCC mix design incorporates the maximum density philosophies commonly associated with the principles of soil compaction developed by Proctor. For a particular aggregate and cementitious material content, an optimum moisture content is determined which will produce the maximum density of the RCC mixture for a given compactive effort. While the cement content is sufficient to blend the aggregate particles together and produce a hardened RCC with the required strength properties, it does not produce a paste content sufficient to fill all the voids after compaction. This results in a dam structure that may not have adequate durability against freeze-thaw damage and is likely to be more permeable and require additional measures to reduce seepage. It is common to overbuild the downstream unformed face of lean RCC dams to provide at least 1 ft of sacrificial RCC to account for freeze-thaw damage.

To reduce seepage, additional measures such as membrane liners, bedding mixes at lift surfaces, or conventional concrete facings are often used at the upstream face. In addition to higher permeability and lower durability, a lean RCC dam will also have reduced bond strength at the lift surfaces and, because of its

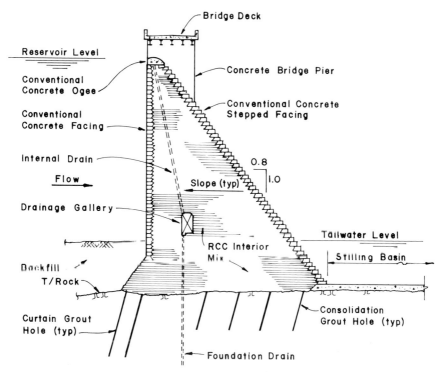

FIGURE 2 Typical RCC spillway section (shown with drainage gallery).

lower density, may require more mass for stability than a conventional concrete dam or an RCC dam designed using the concrete approach.

Typical cement content for a lean RCC dam will be 80 to 170 lb/yd^3 (45 to 100 kg/m^3). Some cement may be replaced with a pozzolan, such as fly ash, although adiabatic heat generation is usually not a problem with lean RCC mixes. Sufficient sand and other fines should be included in the mix to prevent direct contact between the coarse aggregate particles. The amount of nonplastic fines passing the no. 200 sieve are generally limited to 5 to 10 percent. Coarse aggregate sizes are normally limited to 3-in maximum size aggregate (MSA). For a typical RCC aggregate gradation curve, refer to Fig. 3.

Advantages of using the lean RCC approach are that no pozzolans are required to reduce heat generation due to hydration, and the finished dam normally has a low elastic modulus and a high creep ratio, which reduces the potential for cracking and may eliminate the need for vertical joints.

3. High-Paste-Content RCC Dam. High-paste-content RCC mixes were initially studied by the Tennessee Valley Authority during the early 1970s and later developed more fully by Malcolm R. Dunstan in the United Kingdom.[23] Dunstan found that a relatively high paste/mortar ratio was required to obtain a density having a low percentage of air voids, which leads to increased strength properties. The paste/mortar ratio is the ratio of the volume of paste [i.e., cementitious material, water, and entrained air (if used)] to the volume of mortar (i.e., paste and fine aggregate). The high-paste-content RCC mix has a much more fluid con-

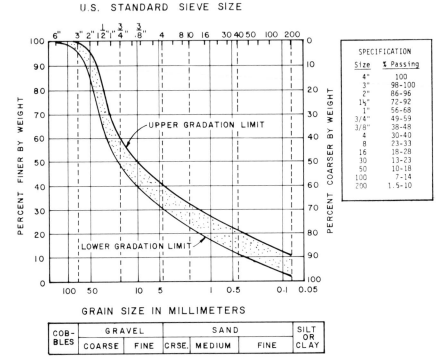

FIGURE 3 RCC aggregate, typical gradation requirements.

sistency than that produced with the soils approach, yet results in a zero-slump mix capable of supporting the weight of a vibratory roller compactor. The increased workability is a result of sufficient paste being present to fill all the voids between the aggregates. Mixture proportioning for high-paste-content mixes is based on designing a mix which will produce hardened RCC that is dense and impermeable, and has excellent bond strength (cohesion) at lift surfaces to prevent seepage. Bond strengths are generally much higher than for lean RCC mixes because of the greater moldability of the high paste mix, and bedding mixes are generally not required except for treatment of cold joints. Unlike the lean RCC approach, the high-paste-content RCC is considered to be the seepage barrier for the dam with the facing mix serving only as a durable skin. The interior of the dam is constructed using the highly impermeable RCC mix and the facings are normally constructed of air-entrained conventional concrete to provide increased durability against freeze-thaw damage.

Workability of the mix is tested using the Vebe method, which establishes a correlation between water content and desired consistency as determined by placement of a test section. The Vebe test uses a vibratory table in conjunction with a weighted surcharge and measures the time required for excess cement paste to rise to the surface of the RCC mix.

Typical cementitious paste contents for a high-paste-content RCC mix generally exceed 250 lb/yd^3 (150 kg/m^3). To reduce costs and control internal heat generation due to hydration of the cement, 60 to 80 percent of the cementitious material content may consist of a pozzolan such as fly ash. A typical mix for a 160-

ft-high (50-m) dam having acceptable permeability levels would consist of 85 lb/yd³ (50 kg/m³) of cement and 165 lb/yd³ (100 kg/m³) of fly ash. For higher dams, the cementitious material content must be increased to maintain an acceptable level of permeability against increased water pressure. While the high percentage of fly ash will slow the strength gain of the RCC, compressive strength requirements for hardened RCC are normally based on 90-day or 180-day requirements, since the dam will not be operational and, therefore, not subject to loading within that time period.

The higher cement content used for this approach will result in hardened RCC, which has a higher elastic modulus and lower creep rate than for lean RCC. This will cause a higher potential for cracking through the dam because of thermal stresses and may require the use of contraction joints at locations determined by an in-depth thermal analysis.

4. Roller-Compacted Dam (Japanese Method). The roller-compacted dam method was developed in Japan during the 1970s to increase speed of concrete placement and lower construction costs for concrete dams. Most dams in Japan are constructed of concrete because of the high seismic risk and because most sites have topographical and hydrological problems associated with them which favor a concrete dam.

Concrete dams constructed using the RCD method have some very significant differences from those constructed with the high-paste content RCC mix. Lift thicknesses for the RCD method generally range from 20 to 40 in (0.5 to 1 m), while those for the high-paste-content method are usually 12 in (0.3 m). The Japanese also restrict placement of successive lifts for up to 4 days, to avoid heat buildup from hydration of the cement, and require all lift surfaces to be treated as cold joints and bedded with a mortar mix before placement of subsequent lifts. Dams constructed in Japan by the RCD method are required to have an inspection gallery.

To increase watertightness, this approach uses upstream and downstream facings constructed of conventional concrete, which is placed between forms and the freshly placed RCD mix. It is then consolidated into the RCD mix by immersion vibrators. These conventional concrete facings are generally 10 ft (3 m) or more in thickness. In addition, an elaborate system of joints, which includes transverse contraction joints, is placed throughout the length of the dam at 50-ft (15-m) spacings.

Temperature control is an important detail in RCD design. Particular attention is given to restricting the type and amount of cement, precooling the mix constituents, and adjusting the lift thicknesses and placement schedule to match ambient temperature conditions.

An RCD mix is a relatively wet, high-fines-content material that follows the concrete approach. Consistency is measured by vibratory compactor (VC) time, which is similar to the Vebe time used for the high-paste-content mix approach. cementitious material content is approximately 210 lb/yd³ (125 kg/m³), with 20 to 30 percent normally consisting of fly ash. The aggregates used are well-graded, processed rock with a sand content of approximately 30 percent. Normal maximum size of the coarse aggregate is 3 in, although aggregates as large as 6 in have been used.

PHYSICAL PROPERTIES OF RCC

5. General. The significant physical properties of hardened RCC include strength, elasticity, permeability, durability, volume change, and unit weight.

Differences between the hardened properties of RCC and conventional concrete are primarily a result of mixture proportions, aggregate quality and grading, and void content. In general, RCC will have lower cement, paste, and water contents than conventional concrete and have little or no entrained air.

RCC may also use silt or nonplastic fines to fill aggregate voids. Aggregate quality, grading, and physical properties influence the properties of hardened RCC to a greater degree than for conventional concrete. Use of aggregates not meeting the requirements of ASTM C33, "Standard Specifications for Concrete Aggregates," is acceptable but will result in a wider range of properties for the RCC.

6. Strength. Strength requirements of interest for RCC include compressive, tensile, and shear (bond) strength.

When RCC mixtures are proportioned with a paste volume exceeding the voids between the aggregates, known as *high-paste-content RCC,* the RCC exhibits compressive strengths similar to those for conventional concrete having the same water-cement relationships and aggregate properties. Where sufficient paste is not present to fill the voids between aggregate particles, such as for lean RCC, workability will be lower and the strength relationships will be significantly different than for conventional concrete. For this condition, compressive strength will be a function of the crushing strength of the aggregates resulting from particle-to-particle contact.

Tensile strength for RCC is generally lower than for conventional concrete, especially for lean RCC mixes. This also applies to RCC mixes utilizing marginal or poor-quality aggregates. RCC mixes made using high-quality aggregates and having a high paste content result in hardened RCC having tensile strengths comparable to conventional concrete. The ratio of tensile strength to compressive strength for RCC typically ranges from 7 to 13 percent, depending on aggregate quality, age, cement content, and strength. However, except as determined by laboratory testing, tensile strengths should not be considered in the design of RCC. The shear or bond strength of RCC is dependent on tensile bond properties (or cohesion) and the angle of internal friction at the lift surface.

The factors which control bonding of one lift of RCC to another are (1) the conditions of the surface to be bonded, (2) the moldability of the fresh covering concrete, and (3) the compaction effort applied in consolidating the covering concrete.

Because minimum strengths occur at construction joints and lift surfaces, preparation of the lift surface prior to placement of fresh RCC is extremely important. The surface must be clean, damp, and free of loose materials. When a dry, damaged, or contaminated surface has developed, or the specified maturity has been exceeded, the result is a cold joint, where special treatment of the joint is required prior to placing the next layer of RCC. This may include cleaning the surface by water blasting and spreading conventional concrete bedding mix over the surface immediately prior to placing the fresh RCC. For high-paste-content RCC mixes, a bedding mix would normally not be required to improve bonding properties, except where a cold joint has occurred.

For a lean RCC mix, because of its lower paste content, a bedding mix can be very beneficial in increasing the bond strength at lift surfaces and, subsequently, results in a more watertight joint. The amount of compactive effort required for maximum consolidation of the fresh RCC should be determined during construction of the test fill section. The number of passes required by the vibratory roller will depend on the actual equipment being used and the characteristics of the RCC mix design. Typical values to be used for preliminary design are 100 psi for shear strength (cohesion) and 1.0 for coefficient of friction. Actual values used for the final design should be determined by testing drilled cores taken from the test placement, constructed with actual materials and placement methods the contractor will use during construction of the dam.

7. Elasticity. The modulus of elasticity E, also known as *Young's modulus*, is defined as the ratio of change in stress with respect to change in strain (deformation), below the elastic limit for a material. It is a measure of a material's ability to resist deformation when subject to load. For conventional concrete, the modulus of elasticity is proportional to unit weight and compressive strength, and the average value of E can be determined from the formula

$$E = 33W^{1.5}\sqrt{f'_c} \quad \text{psi}$$

where W is the unit weight of concrete in lb/ft^3.

Principal factors affecting the elastic properties of RCC are compressive strength, aggregate type, and content of cementitious materials. An RCC mixture made with a high-quality aggregate and a high-cementitious-materials content (i.e., cement plus pozzolan) can develop a modulus of elasticity similar to that obtained for conventional concrete.

Consequently, a mass concrete dam constructed with a lean RCC mix and lesser-quality aggregates would have a lower modulus and would have less potential for cracking than a dam constructed of high-paste-content RCC.

8. Poisson's Ratio. Poisson's ratio v is defined as the ratio of lateral (transverse) strain to axial (longitudinal) strain resulting from a uniform axial stress. Values for RCC are similar to those for conventional concrete and have been found to range from 0.17 to 0.22, with lower values occurring at earlier ages and lower strengths. A typical value to be used for design would be 0.20.

9. Permeability. The permeability of a mass concrete dam is related to the air void content and the porosity of the hydrated cement matrix, which provide a channel for the water to penetrate the dam. For a fully compacted RCC mix containing sufficient fines to fill the voids, the permeability of the RCC mass will be similar to that for conventional mass concrete. According to previous studies of existing RCC dams and their permeabilities, a cementitious materials content of approximately 250 lb/yd^3 (150 kg/m^3), which is representative of a medium- to high-paste-content RCC mix, is necessary to produce permeabilities consistent with conventional mass concrete. Lean RCC mixes having sufficient fines contents will also have acceptable permeabilities throughout the parent material. However, because of reduced bond strength at lift surfaces, the overall permeability for the dam structure may be unsatisfactory and require special measures such as an upstream membrane, conventional concrete facing, or bedding mix between lifts to eliminate seepage at lift surfaces or reduce it to an acceptable level. Existing dams which experience high levels of seepage during initial filling of the reservoir have noticed large reductions in seepage with time because of siltation of voids, autogenous healing, and calcification.

10. Durability. Durability of hardened RCC is an important property if the dam will be subject to severe weather variations and/or hydraulic forces. Freeze-thaw resistance of existing RCC structures has been very good when mix proportioning has taken this into account. High-cement-content mixes that result in RCC with high strength and density with low permeability have demonstrated greater resistance to freeze-thaw cycles than the lean RCC mixes which have lower strengths and higher permeability. In practice, RCC has shown better durability than that determined by laboratory testing using the severe ASTM C566 rapid freezing and thawing test. Except for the high-paste-content mixes, it has generally not been possible to entrain air into an RCC mix. However, high-paste-content mixes, usually with a larger percentage of fly ash, tend to neutralize the effects of the air-entrainment admixture. Therefore, it is usually necessary to rely on the higher strength and density of high-paste-content mixes for freeze-thaw resistance. For lean concrete mixes, sacrificial material or an air-entrained conventional concrete facing should be provided.

11. Volume Changes
Drying Shrinkage. As for conventional concrete, drying shrinkage in RCC is related to water content of the mix. Since the water content for RCC is generally lower than for conventional concrete because of the reduced slump, drying shrinkage is greatly reduced.

Creep. Creep is a function of the properties and gradation of the aggregates and the volume of cement used. Generally, aggregates with a lower modulus of elasticity will provide a higher creep rate and vice versa. High-strength mixes usually have a lower creep rate, which results in higher internal thermal stresses. However, since compressive stresses are usually not large in a concrete gravity dam, creep rates are not a major concern.

12. Thermal Properties. Thermal properties of concrete, which include specific heat, conductivity, and coefficient of expansion, are largely dependent on the aggregate properties. Typical values for RCC are very similar to those for conventional concrete when aggregates come from the same source.

After the aggregate source of an RCC mix has been determined, testing should be performed to determine actual thermal properties. A thermal analysis should then be performed to determine thermal stresses and the potential occurrence of transverse cracking in the dam. Further discussion of thermal studies can be found in Art. 23.

13. Unit Weight. The unit weight of RCC generally exceeds that of conventional concrete as a result of the low initial water content, low amount of entrained air, and compaction during placement. It is common for the unit weight to exceed 150 lb/ft^3 (2400 kg/m^3) by several percent.

DESIGN OF RCC DAMS

14. General. RCC dams are primarily designed as gravity-type structures, although in recent years dams of the arch-gravity type have been designed for RCC. Arch-gravity dams, if designed properly, will have the entire upstream face in compression under the normal loading condition, which will result in a more impermeable structure. Since most RCC dams are designed as two-dimensional gravity structures, this section will concentrate on the design of that type of structure, which is very similar to the design of conventional concrete dam structures. High dams or those having unusual foundation conditions may benefit from a three-dimensional finite-element analysis.

Important aspects of RCC dam design which must be considered include design philosophy, foundation properties, structural stability, thermal studies, seepage control, and appurtenant structures.

15. Design Philosophy. One of the primary tasks for design of an RCC dam is to decide whether to use the design philosophy which follows the lean RCC mix (i.e., soils approach) or the high-paste-content RCC mix (i.e., concrete approach). Many factors must be considered before choosing a particular approach for RCC dam construction. Lean RCC dams normally incorporate an upstream face membrane when seepage through the dam is a concern and tend to be more economical for RCC dams less than 100 ft high (30 m) than for those following the concrete approach. This concept recognizes that, for lean RCC dams, complete bonding of the lift joint is not likely to have occurred, as previously discussed in Art. 2. Those dams designed using the concrete approach do not rely on an upstream membrane to reduce seepage and instead consider the entire body of the dam to be the primary water barrier, the same as for a conventional concrete dam, as discussed in Art. 3. To make a proper decision as to which approach to use, the designer must have a clear understanding of the project goals as related to cost, construction schedule, purpose of the finished structures, required watertightness, final appearance, and maintenance requirements. Availability of materials such as cement, fly ash, and aggregates at the construction site will also be a major factor when considering the type of RCC dam to be designed and constructed.

Whichever design approach is selected, the designer must have a thorough understanding of the RCC construction methods utilized by contractors and should design the details to be as simple as possible to allow the contractor to take full advantage of the rapid construction possible by using RCC.

16. Foundation Properties. In general, foundations for RCC dams should be of similar quality to those that would be acceptable for a conventional concrete dam. Ideally, they should be founded on fresh sound rock that has no potential for differential settlement and is of a quality that can be treated with grout to form a waterproof barrier under the dam. For most dams, it is common practice to provide a grout curtain near the upstream edge to reduce seepage under the dam.

Irregularities in the foundation shape may cause uneven transfer of loads from the RCC dam to the foundation which cannot be computed from a conventional two-dimensional analysis. Should this situation be present, a finite-element analysis which correctly models the actual foundation conditions may be appropriate. Where possible, irregularities in foundations for RCC dams should be avoided or eliminated by using dental excavation and/or conventional concrete fill to avoid stress concentrations in either the foundation rock or the RCC dam.

As is the case for conventional concrete dams, a thorough foundation investigation program is extremely important to determine critical properties of the foundation rock. Those properties of most importance are compressive strength, shear strength, deformation modulus, Poisson's ratio, and permeability.

Compressive strength of the foundation will be important in determining the base width of the RCC dam section. Base pressures produced by loadings in the dam should not exceed the allowable bearing stresses in the rock as determined by a laboratory testing program.

Shear strength of the rock will depend on whether the foundation rock is intact or contains discontinuities such as joints, shears, and faults. Shear strength of an intact rock can be determined using Coulomb's equation for the linear relationship between normal load and shear resistance (Fig. 4a). Shear strength for a discontinuous foundation rock can be determined only by physical tests on the material. From these tests, a plot of shear strength versus normal applied load can be developed (Fig. 4b). Appropriate factors of safety should be applied to the ultimate values to determine an allowable shear strength for a given loading combination.

The deformation modulus for a dam foundation will indicate what degree of settlement can be expected to occur. The actual value of the modulus is not as important as how it varies over the extent of the foundation. A rock type having

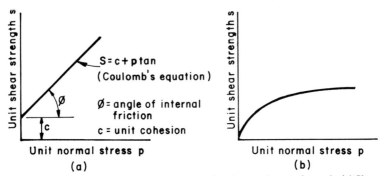

FIGURE 4 Relationship between shear strength and normal stress for rock. (*a*) Shear strength of intact rock. (*b*) Shear strength of an existing joint in rock.

a uniform low-deformation modulus is not nearly as critical as one which has large variations in the modulus. Abrupt changes in deformation modulus can result in large differential settlements that can cause cracking in the RCC dam. Those zones identified as having a weaker material should be grouted or be excavated and replaced with conventional concrete or RCC.

As with most dams, permeability of the foundation rock for an RCC dam is extremely important. Curtain grouting is usually specified to control seepage below dams regardless of the rock quality. This usually consists of drilling and grouting a single line of 1½- to 3-in-diameter (38- to 75-mm) holes spaced at 10 ft (3 m) on centers near the upstream face of the dam. Depth of the grout curtain depends on the hydrostatic loads and foundation rock conditions. The depth of the grout curtain normally ranges from 40 percent of the water head for dense foundations to 70 percent for poor-quality rock foundations.

Poisson's ratio, the ratio of transverse strain to axial strain, is normally not required unless a complex foundation situation exists which requires a three-dimensional finite-element analysis to determine deformation patterns.

17. Structural Stability. Gravity structures such as RCC dams are designed for stability against overturning and sliding failures. Design of concrete gravity dams is controlled mainly by foundation considerations. No concrete gravity dam has been known to fail due to initial failure within the concrete above the foundation. Failures of concrete dams in the past have been the result of sliding failures at planes of weakness within the foundation or of erosion of the foundation material.

The basic criteria to be checked for the design of a gravity RCC dam are

- The dam's cross section should be established such that the allowable stresses for both the RCC and the foundation are not exceeded.
- The dam must have a satisfactory factor of safety against the occurrence of a sliding failure at any plane within the dam, at the foundation interface, or at any potential plane of failure within the foundation.
- In general, no tensile stresses should exist within the dam, specifically at the upstream face. If tensile capacity of the RCC is required and the RCC mix has been designed specifically for this purpose, careful quality control should be maintained during placement of the RCC. In addition, laboratory and field testing should be performed to verify that the actual tensile capacity exceeds that required.

Most RCC dams are of a triangular shape and have a vertical or near-vertical upstream face, much the same as a conventional concrete gravity dam. This shape concentrates a significant portion of the structure dead load near the upstream face where it can most effectively resist uplift forces and eliminate tensile stresses. Downstream face slopes normally range between 0.6 and 1.0, horizontal to vertical, with 0.8 being the most common. A crest width of 15 to 30 ft (5 to 10 m) is normally required to accommodate construction equipment during RCC placement and also permit future traffic across the dam. If an overflow spillway is to be incorporated into the dam design, it should have a section similar to that used for the nonoverflow section to simplify RCC construction. It has become very common to utilize stepped spillways for RCC dams because of their horizontal-layered construction and effectiveness as an energy dissipator. Stepped spillways can dissipate as much as 70 percent of the total energy of the overflowing water, thereby reducing stilling basin or other energy dissipator requirements. Drainage galleries are usually required on higher dams, but should be eliminated on dams lower than 100 ft (30 m) where reduction of uplift pressures is not required for stability purposes. Construction of galleries in RCC dams causes major interruption to RCC placement methods.

    18. Design Loads. Loads that should be considered for the design of a gravity RCC dam are shown in Fig. 5 and include the following:

Vertical Loads

V_1 Gravity dead load of the RCC dam mass and any permanent appurtenant structures. Unit weight of the RCC and any conventional concrete should be taken as 150 lb/ft^3 (2400 kg/m^3) for design.

V_2, V_3 Weight of water and/or silt acting on inclined portions of upstream and downstream faces, respectively. Unit weights of water and saturated silt should be taken as 62.5 lb/ft^3 (1000 kg/m^3) and 120 lb/ft^3 (1925 kg/m^3), respectively.

V_4 Uplift pressure at the base of the dam or at any horizontal plane within the dam. If no drains are present, its value is taken as full headwater at the upstream face, decreasing linearly to full tailwater at the downstream face. If drains are used, the uplift pressure may be reduced between the upstream and downstream faces to account for the effectiveness of the drains. Uplift pressures are assumed to act over 100 percent of the base area.

V_5 Inertial force of the dam mass during an earthquake.

Horizontal Loads

H_1 Horizontal silt pressure equal to an equivalent fluid pressure of 85 lb/ft^3 (1360 kg/m^3).

H_2, H_3 Hydrostatic pressure of headwater on upstream face and tailwater on downstream face, respectively. Unit weight of water should be taken as 62.5 lb/ft^3 (1000 kg/m^3).

H_4 Ice load on upstream face. Where applicable, an ice load of 5 to 10 kips per linear foot (7450 to 14,900 kg per linear meter) should be applied 1 ft (30 cm) below the water surface.

H_5, H_6 Inertial force of the reservoir water acting on the upstream and downstream faces of the dam, respectively. Magnitudes of forces should be determined in accordance with the Westergard theory.

H_7 Inertial force of the dam mass during an earthquake.

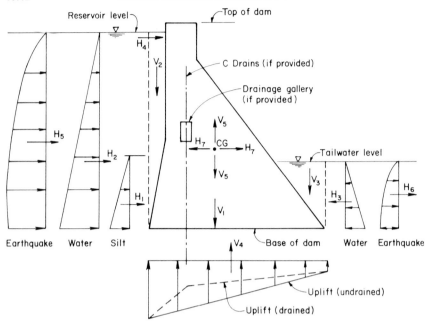

FIGURE 5 Design loads acting on dam.

19. Loading Combinations. In designing a gravity concrete dam, stability for a minimum of three loading combinations, consisting of the appropriate vertical and horizontal loads, should be investigated:

1. *Usual (normal).* The usual loading combination consists of gravity dead loads, hydrostatic loads from maximum reservoir and tailwater levels during normal operation, uplift, and silt and ice loads.
2. *Unusual (PMF).* The unusual loading combination consists of gravity dead loads, hydrostatic loads from reservoir and tailwater levels during a probable maximum flood (PMF) occurrence, uplift, and silt loads. Ice loads would normally not occur in conjunction with a flood condition.
3. *Extreme (earthquake).* The extreme loading combination consists of gravity dead loads, hydrostatic loads from maximum reservoir and tailwater levels during normal operation, uplift, silt and ice loads, and the effects of the maximum credible earthquake for the seismic zone in which the dam is located.

Stability for the RCC dam should be checked not only at the base of the dam, but also at horizontal planes throughout the height of the dam. Any loading combinations, other than the three mentioned, which the designer identifies which might govern the design should also be investigated.

20. Factors of Safety. The required factors of safety for the specified loading combinations vary and should be determined in accordance with the appropriate agency responsible for regulating the safety of the dam. Two of the more common guidelines are based on the United States Bureau of Reclamation (USBR)[5] and the Federal Energy Regulatory Commission (FERC).[6] The USBR requires factors of safety for foundation stresses of 4.0, 2.7, and 1.3 for the usual, unusual, and extreme loading combinations, respectively. FERC requires values of

3.0, 2.0, and 1.0 for the same combinations. It would benefit the designer to review the most recent editions of USBR's *Design of Gravity Dams*[5] and FERC's guidelines[6] for additional information regarding required factors of safety, reduction of uplift pressure due to drain effectiveness, and cracked-base theory.

21. Sliding Stability. The stability of the RCC dam against a sliding or shearing failure is determined by the shear-friction factor (SFF) of safety method.

The shear friction factor of safety is defined as

$$SFF = \frac{CA + (W - U) \tan \phi}{H}$$

where C = unit cohesion
A = area of section in compression
W = vertical weight above section
U = uplift force acting on section
ϕ = angle of internal friction
H = horizontal driving force

Representative values of C and ϕ for RCC mixes can be found in Ref. 13. Assumptions used for C and ϕ during design should normally be verified by testing of hardened RCC. For design purposes, C is normally taken as 100 psi. Friction factor $f = \tan \phi$ is normally assumed as 1.0 within the mass of the dam and 0.75 at the contact between the RCC and sound, relatively smooth foundation rock.

22. Overturning Stability (Stress Analysis). Factor of safety against overturning is a misnomer, since no gravity dam has actually ever failed by overturning. The stability analysis is used to determine maximum compressive stresses at the downstream face and minimum compressive stresses or tensile stresses at the upstream face due to the overturning effects of the design loads. Most agencies will not permit tensile loads to occur in new dams except for the extreme loading condition. After maximum stresses have been determined, the RCC mix should be designed to satisfy the appropriate factors of safety for allowable compressive stresses (and tensile stresses, if required).

23. Thermal Studies. Cracking of concrete which results from volume changes induced by temperature and moisture variations has always been a major concern for the design of mass concrete-dam structures. Designers of RCC dams must also be concerned with the cracking potential of RCC; however, dams constructed of RCC have advantages over conventional concrete dams which reduce the potential for thermal cracking. Lower cementitious contents and the use of fly ash replacement for cement greatly reduce the adiabatic heat rise which occurs. Also, the rapid placement of RCC in horizontal layers 1 to 2 ft (0.3 to 0.6 m) in thickness can greatly reduce the time the dam lift surfaces are exposed to higher ambient temperatures and results in a more uniform temperature distribution throughout the dam. If it can be determined by a thermal analysis that the likelihood of thermal cracking is low, it may be possible to construct the dam without the vertical contraction joints commonly used for conventional concrete dam monolith construction. Placement of contraction joints in RCC dams complicates the RCC placement procedure and offsets some of the economic advantages.

With one of several available computer programs, a thermal analysis should be performed to determine transient temperature distribution throughout the dam cross section, assuming continuous placement of RCC. One such program which has been used for the thermal analysis of a number of RCC dams is THERM, a finite-element program initially developed at the University of California at Berkeley and subsequently modified by the U.S. Army Corps of Engineers to make it compatible with RCC construction techniques. A thermal study of an RCC dam requires the designer to make a number of assumptions regarding ambient temperatures during placement, placement rate, RCC mix temperature, and thermal properties of the RCC mixture such as specific heat, coefficient of thermal expansion, diffusivity, conductivity, adiabatic heat rise, tensile strength, modulus of elasticity, and creep coefficients. It becomes apparent that, because of the many assumptions which must be made, the correctness of the results will be greatly dependent on the accuracy of the original assumptions. For this reason, once the aggregate source is known, the designer should verify as many of the RCC mix properties as possible through laboratory testing prior to performing the thermal analysis. Changes in the construction schedule which significantly affect ambient temperatures or result in a slower placement rate than originally assumed may require the designer to perform a new thermal analysis to evaluate the effect that these changes might have on the cracking potential of the RCC.

A number of measures can be introduced to reduce the potential for cracking in an RCC dam. Among these are

- Night placement of RCC, which reduces the radiant effects of the sun
- Precooling of aggregates by stockpiling during the winter or spraying them with water
- Precooling of the RCC mix with ice chips as replacement for mixing water or the injection of liquid nitrogen into the mix
- Construction during the cooler months of the year

The results of the thermal analysis indicate the magnitude and locations of tensile stresses throughout the dam. This information can then be compared to the tensile strengths of the RCC at various ages to determine the likelihood that cracking will occur. These results will provide the basis for establishing the need for and location of vertical contraction joints in the RCC dam.

24. Seepage Control. Seepage through a dam is one characteristic by which a dam is commonly judged. While visible seepage at the downstream face of a dam does not mean structural stability problems exist, seepage is unsightly, and every effort should be made to either eliminate or control it.

Various methods have been utilized for RCC dams to control seepage. For dams designed using the high-paste-concrete approach, the body of the dam is considered to act as the seepage barrier. This approach depends on a dense, hardened RCC which will have a permeability similar to that found in a conventional concrete dam. It also depends on good bonding at the lift surface between horizontal layers of RCC to eliminate seepage paths for the water. Normally precast concrete panels or conventional concrete facing is used to provide a durable skin to resist freeze-thaw damage to the RCC at the upstream face of the dam. Dams constructed by using the lean RCC soils approach must rely on a separate seepage barrier at the upstream face, since complete bonding between lifts is much less certain than for the concrete approach. Typical types of seepage barriers used on previous RCC dams are

Precast Panel with Membrane Liner (Fig. 6). This barrier consists of precast concrete panels with a polyvinyl chloride (PVC) membrane bonded to them during manufacture. A common size for the 4-in-thick (10-cm) concrete panels is 4 ft (1.3 m) high and 16 ft (5.3 m) long. At all horizontal and vertical joints between panels, a strip of PVC material is heat-welded to the panel liner to form a continuous membrane. Anchor rods extend from the panel into the hardened RCC for support of the panel. This system was first used for Winchester Dam in

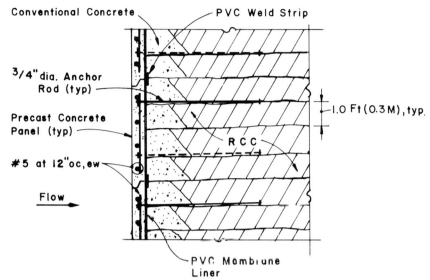

FIGURE 6 Precast-concrete panel facing (typical sections).

Kentucky.[27] In addition to being the seepage barrier, the panels also serve as forms for placement of the RCC. Because this system is patented, payment of royalty fees is required when this method is used.

Reinforced Concrete (Fig. 7). A reinforced concrete facing is placed against the upstream RCC face after the temperature within the RCC has reached its peak and begins to decrease, approximately 30 days after RCC replacement. Anchors embedded in the hardened RCC support the reinforced-concrete facing. Reinforcing is placed in both the horizontal and vertical directions to control crack widths, and waterstops are provided at all construction joints. The spillway at Stacy Dam in Texas utilizes this facing system.[28]

Unreinforced Concrete (Fig. 8). An unreinforced conventional concrete facing is placed for a specified width at the upstream face between the forms and the RCC. The conventional concrete is vibrated into the RCC by using an immersion vibrator. There have been many variations of this type of facing with regard to width, which is most commonly 1 to 3 ft (0.3 to 1.0 m), and whether it should be placed before or after the RCC lift. Because of the difficulty involved with coordinating two different mixes, quality control is extremely important when using this system to prevent the formation of a cold joint between lifts. This scheme has been the most common used for RCC dams.

Extruded Concrete Curb (Fig. 9). A concrete curb, constructed with a modified, laser-guided slip-form curb machine is used to place a continuously extruded concrete curb along the upstream face of the dam. The curb is constructed approximately 3 ft (1.0 m) high. After the curb hardens, it serves as a form for placement of RCC, in addition to being a durable skin for the upstream face of the dam. This system was developed by Dunstan[23] and was used for construction of Upper Stillwater Dam in Utah.[24] For small dams, with a relatively small cross section and short crest length, this system is not practical, since construction of the curbs will continually interfere with placement of RCC.

Because of the large number of lift joints in an RCC dam, bonding at lift joints

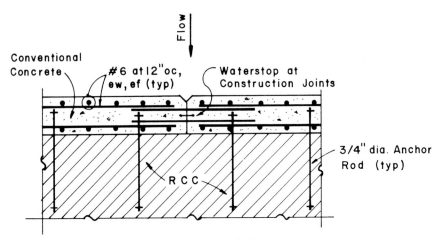

FIGURE 7 Reinforced-concrete facing (sectional plan).

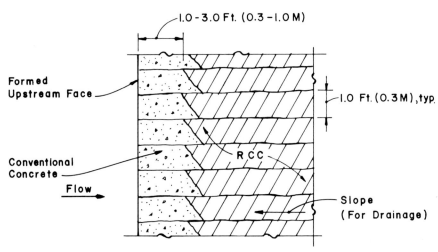

FIGURE 8 Unreinforced-concrete facing (typical section).

is extremely important in reducing seepage through the dam. There are many factors affecting bonding at the lift surfaces, but the most important are

- The condition of the previous lift surface
- Time delay between placement of lifts
- Moldability of the covering RCC
- Degree of compaction of the covering RCC

Before placement of a subsequent RCC lift, the previous lift surface must be clean and free of all loose material. The previous lift surface should be kept moist

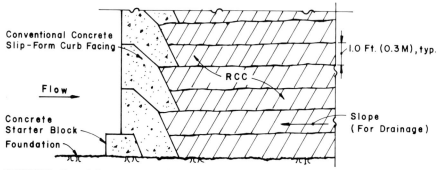

FIGURE 9 Extruded concrete curb facing (typical section).

at all times, but all standing water should be removed. If the next RCC lift is placed before a cold joint develops at the surface of the previous lift, satisfactory bond strength should develop. Joint maturity, the measure of a cold joint, is commonly defined in one of two ways. One way defines joint maturity in terms of degree-hours, which is the product of the time of exposure (hours) and air temperature (°F) at the lift surface. While designers have been unable to agree on a common value for maturity, values normally range from 500 to 2000 degree-hours. A second way defines joint maturity strictly in terms of hours since the previous placement was made. Laboratory or field tests should be used to confirm the correct joint maturity value. For the lean RCC dam, treatment of cold joints consists of the application of a bedding mix prior to placement of the next RCC lift. Dams designed using the concrete approach commonly use the joint maturity in terms of hours. If more than 24 h but less than 72 h has elapsed since the previous placement, a conventional concrete bedding mix is specified. If more than 72 h has elapsed, water-blasting of the lift surface is normally specified in addition to applying a bedding mix.

In addition to following proper treatment of cold joints, compaction of the new RCC lift using a vibratory roller should be performed in a manner which ensures complete consolidation. The required compactive effort should be determined during placement of the test section prior to placement of RCC for the dam.

Most RCC dams also incorporate some type of internal drainage system to collect seepage which penetrates the dam to prevent it from reaching the downstream face where it would be unsightly and may cause freeze-thaw damage to the exposed face. Some of the types of internal drainage systems which have been utilized to date are vertically drilled drain holes through the dam which connect to a gallery, half-round drain pipes installed on lift surfaces, open-gravel drains, drainage tubes, and filter fabric wick drains. Where a drainage gallery has not been provided, these drains are routed to a common collection point or discharged into toe drains along the downstream face of the dam. References 11 and 16 provide additional information with regards to types of drainage systems, their limitations, and their effectiveness in controlling seepage.

25. Galleries and Adits. The construction of galleries and adits in an RCC dam continues to have the most significant negative impact on rapid placement of RCC. Construction progress is reduced by the special labor-intensive measures which are required to form the gallery while attempting to continue placement of RCC without the formation of cold joints at lift surfaces. The designer should attempt to eliminate the need for galleries in all dams less than 100 ft (30 m) in height. This will require that the dam be designed for stability against full uplift pressure and may require more RCC mass. For those dams which are higher, elimination of the galleries will most likely not be possible because of the need to provide for internal and foundation drainage to satisfy stability requirements.

Several methods have been used to construct galleries in RCC dams. Among the more common methods are

- Forming the sides with conventional forming systems.
- Placing precast concrete stay-in-place panels for walls and ceiling.
- Placing a noncementitious fill (such as gravel or RCC aggregate) simultaneously with lifts of RCC and later mining the fill out using hand methods or high-strength industrial vacuums. The fill is often separated from the RCC with sideboards or polyethylene to prevent bonding and to form a relatively smooth surface.
- Extruded concrete curbs formed by a modified slip-form paver machine.

Galleries should be located a distance of approximately 16 ft (5 m) away from the upstream face of the dam to provide enough space for equipment placing and compacting the RCC. The height of the gallery should be an even multiple of the lift thickness and should be located at a single level to simplify construction as much as possible. Where galleries or adits are located transverse to the dam, they should be constructed against or into the rock foundation to minimize interference with RCC placement.

26. Appurtenant Structures. Appurtenant structures for an RCC dam commonly include an intake, outlet conduit, control gates, and stilling basin. These structures are normally constructed of conventional reinforced concrete and are located so as to have the least interference with placement of RCC. The intake usually consists of a tower which can be constructed against the upstream face of the dam after the RCC placement is complete. It is common practice to extend anchors from the RCC which can be embedded into the intake concrete. The outlet conduit should be located in rock either below the base of the dam or in one of the abutments. Conventional concrete fill should be used to cover the conduit and provide a smooth surface over which to place RCC. If the control gate is placed at the downstream toe of the dam, the outlet conduit will be subject to full hydraulic pressures, and this should be considered for design of the conduit. Locating the control for the outlet at the downstream face of the dam normally facilitates access for maintaining the required mechanical and electrical equipment. As for conventional concrete dams, a stilling basin or other type of energy dissipator will be required at the discharge end of the outlet conduit. This can normally be constructed independent of the dam after all RCC operations have been completed. Sometimes it may be economical to place RCC fill between the rock surface and the bottom of the stilling basin to prevent differential foundation settlement, which might damage and shorten the life of the structures.

27. Summary. In summary, there are many choices to be made in designing a roller-compacted concrete dam. RCC has many economical benefits, and it is vitally important that the designer of an RCC dam have a thorough understanding of the materials, details, and construction methods used. There is no standard design which can be specified, and many of the design decisions are site-dependent. Each RCC dam is unique, and for the designer to determine the most economical design, complete consideration must be given to which design philosophy, facing scheme, seepage control method, galleries, etc. to use. There are still many changes occurring with this relatively new dam construction method and the designer should spend time reviewing past and present articles and other information on this subject to learn about the actual performance of dams constructed by the different methods. References 1 and 2 are two of the most complete sources of information available on RCC design and construction. Other references listed at the end of this section provide an excellent source of information for roller-compacted concrete design and construction. Only with a complete understanding of the RCC construction process and its limitations can the most economical design be developed.

REFERENCES

1. "Roller-Compacted Mass Concrete," American Concrete Institute, Report 207.5R, 1988.
2. HANSEN, K. D., and W. G. REINHARDT, *Roller-Compacted Concrete Dams,* McGraw-Hill, New York, 1991.
3. "Roller-Compacted Concrete for Dams," Electrical Power Research Institute Report AP-4715, prepared by Morrison-Knudsen Engineers, Inc., 1986.
4. "Design of Gravity Dams," U.S. Bureau of Reclamation, 1976.
5. "Design Criteria for Concrete Arch and Gravity Dams," U.S. Bureau of Reclamation, monograph no. 19, 1977.
6. "Engineering Guidelines for the Evaluation of Hydropower Projects," Federal Energy Regulatory Commission, July 1987.
7. DUNSTAN, M. R., "Whither Roller Compacted Concrete for Dam Construction?," *Roller-Compacted Concrete II,* ASCE, 1988, pp. 294–308.
8. DUNSTAN, M. R., A Review of Roller Compacted Concrete Dams in the 1980's, *Water Power and Dam Construction,* May 1990, pp. 43–45.
9. DUNSTAN, M. R., Recent Developments in Roller Compacted Concrete Dam Construction, *Water Power and Dam Construction Handbook,* 1989, pp. 39–47.
10. HOPMAN, D. R., O. KEIFER, and F. ANDERSON, Current Corps of Engineers Concepts for Roller Compacted Concrete in Dams, "Roller Compacted Concrete for Dams," EPRI, 1986, Appendix B8, pp. 4–13.
11. KOLLGAARD, E. B., and H. E. JACKSON, Design Innovations for Roller Compacted Concrete Dams, 15th ICOLD Congress, Q. 57, June 1985.
12. DOLEN, T. P., and S. D. TAYABJI, Bond Strength of Roller Compacted Concrete, *Roller Compacted Concrete II,* ASCE, 1988, pp. 170–186.
13. MCLEAN, F. G., and J. S. PIERCE, Comparison of Joint Shear Strengths for Conventional and Roller Compacted Concrete, *Roller-Compacted Concrete II,* ASCE, 1988, pp. 151–169.
14. "Bonding Roller-Compacted Concrete Layers," Portland Cement Association, 1987.
15. MOLER, W. A., and J. F. MOORE, Design of Seepage Control Systems for RCC Dams, *Roller Compacted Concrete II,* ASCE, 1988, pp. 61–75.
16. SCHRADER, E. K., Watertightness and Seepage Control in Roller Compacted Concrete Dams, *Roller Compacted Concrete,* ASCE, 1985, pp. 11–30.
17. CANNON, R. W., Compaction of Mass Concrete with Vibratory Roller, *ACI Journal,* October 1974, pp. 506–513.
18. TATRO, S. B., and E. K. SCHRADER, Thermal Considerations for Roller-Compacted Concrete, *ACI Journal,* March-April 1985, pp. 119–128.
19. LOWE, J., Roller Compacted Concrete Dams—An Overview, *Roller Compacted Concrete II,* ASCE, 1988, pp. 1–16.

20. CANNON, R. W., Proportioning Non-Slump Concrete for Expanded Applications, *Concrete International*, ACI, August 1982, pp. 43–47.
21. SCHRADER, E. K., Behavior of Completed RCC Dams, *Roller-Compacted Concrete II*, ASCE, 1988, pp. 76–91.
22. HOLLINGWORTH, F., D. J. HOOPER, and J. J. GERINGER, Roller-Compacted Concrete Arch Dams, *Water Power and Dam Construction*, November 1989, pp. 29–34.
23. DUNSTAN, M. R., "Rolled Concrete for Dams—Construction Trials Using High Flyash-Content Concrete," CIRIA Technical Note 106, 1981.
24. OLIVERSON, J. E., and A. T. RICHARDSON, Upper Stillwater Dam—Design and Construction, *Concrete International*, ACI, May 1984, pp. 20–28.
25. SCHRADER, E. K., and R. E. MCKINNON, Construction of Willow Creek Dam, *Concrete International*, ACI, May 1984, pp. 38–45.
26. SCHRADER, E. K., First Concrete Gravity Dam Designed and Built for Roller Compacted Construction Methods, *Concrete International*, ACI, October 1982, pp. 15–24.
27. Membrane-Lined Panels Face New RCC Dam, *Highway and Heavy Construction*, February 1985, pp. 64–65.
28. LEMONS, R. M., A Combined RCC and Reinforced Concrete Spillway, *Roller Compacted Concrete II*, ASCE, 1988, pp. 51–60.
29. PARENT, W. F., W. A. MOLER, and R. W. SOUTHARD, Construction of Middle Fork Dam, *Roller Compacted Concrete for Dams*, EPRI, 1986 Appendix B2, pp. 20–38.
30. KOLLGAARD, E. B., H. E. JACKSON, and J. J. COLLINS, Design of Galesville Dam, *Roller Compacted Concrete for Dams*, EPRI, 1986, Appendix B4, pp. 10–44.
31. SNIDER, S. H., and E. K. SCHRADER, Monksville Dam: Design Evolution and Construction, *Roller Compacted Concrete II*, ASCE, 1988, pp. 220–235.

SECTION 16

SPILLWAYS AND STREAMBED PROTECTION WORKS

By C. Y. Wei[1]

DISCHARGE CAPACITY

1. Design Flood. A spillway must have the capacity to discharge major floods without damage to the dam and the associated structures and to maintain the reservoir at a safe level. The hydrological determination of the flood for which a spillway is to be designed is treated in Sec. 1. The operation of spillways and the method of routing are treated in Sec. 4.

Emergency capacity, in excess of that required to discharge the spillway design flood, may be obtained by (1) encroaching on the freeboard, which under design-flood conditions would make adequate provisions for wave action, reservoir tilt, and run-up; (2) providing emergency spillways; and (3) providing a total freeboard which would permit a sizable flood, say the flood of record, to pass over the top of the gates, if they become inoperative, without overtopping the dam. In evaluating the possible effects of emergency capacity, it must be remembered that the concurrence of maximum peak inflow, maximum wave heights, and reservoir tilt and run-up is an extremely remote possibility and that this additional freeboard may be brought into service under such conditions.

2. Approach Flow Conditions. During the design of a spillway, it must be ascertained that good smooth approach flow conditions can be obtained for safe and efficient passage of flows under various operating conditions.

Analyses of approach conditions are of two general types—those performed analytically and those performed in the hydraulic laboratory. Both are required for a large and important project, as the benefits in the form of improved performance and reductions in construction costs, which may be achieved through both analytical and model studies, far outweigh the cost of these studies. On the other hand, for small structures it may be more economical to provide generous low-velocity approach conditions rather than invest in a model study.

The analytical method, in most cases, implies application of numerical methods using computers. These may include a relatively simple flownet analysis, a two-dimensional depth-integrated surface water flow analysis, or a three-dimensional flow analysis. Since computers are becoming more accessible, with improved capabilities, the cost of utilizing computer models has become relatively competitive against the cost of physical models. However, there are cases where computer models may not be applicable because of assumptions made during the development of the model. Therefore the users of computer models must

---

[1]Acknowledgment is made to William J. Bauer and Earl J. Beck for material in this section that appeared in the third edition (1969).

be fully aware of the assumptions and exercise judgment in applying analytical results obtained by using computer models.

The following basic principles should be considered when the approach to a spillway control structure is designed:

1. The maximum velocity of approach under the most critical combination of reservoir elevation and discharge must not exceed the scouring velocity of the material of which the approach channel is constructed.
2. Curvature of flow in a horizontal plane should be gradual such that an excessive differential in the water-surface elevations would not be produced on opposite sides of the spillway.
3. End walls which guide flow to the control structure should extend upstream from the crest to a point where the velocity is low enough to avoid the development of strong eddies which would be carried over the crest. Walls can also be flared or curved at this upstream end to form, in plan, a streamlined entrance to the channel, thus avoiding the formation of flow separation and circulating eddies.
4. The depth of an approach channel should be established in combination with the design of the control structures so as to develop the most economical combination for the design capacity.

SPILLWAY CRESTS

3. Types. Nearly all spillways fall into one of six types or are made up of a combination of them: (1) overfall, (2) gates and orifice, (3) trough or chute, (4) side channel, (5) morning-glory shaft or tunnel, and (6) siphon. Streambed protection works, more or less common to all these types, are described under a separate heading.

The overfall type is by far the most common and is usually the most economical for passing large floods.

Overfall spillways provided with crest gates will act as orifices under partial gate openings and as open-crest weirs under full gate openings.

Trough or chute spillways are commonly used for earth dams. Side-channel and shaft-spillway types are most frequently found in narrow canyons. The siphon spillway is usually used to provide, automatically, a nearly constant headwater level under varying flow.

OVERFALL SPILLWAYS

4. Shape of Crest. The crest of an overflow dam is generally formed to fit the shape that the overflowing water would take for the selected design head h, as shown by Fig. 1a. Table 1 gives the ordinates of the upper and lower nappe for various slopes of the upstream face. Figure 2 shows the shape of the upper and lower nappes which correspond to the values given in Table 1. Table 1 and Fig. 2 are derived from an analysis based on studies of the U.S. Bureau of Reclamation of the results of experiments by Bazin and Scimemi. The curve shown by Fig. 2 is expressed in terms of the design head measured from the highest point of the

OVERFALL SPILLWAYS

16.3

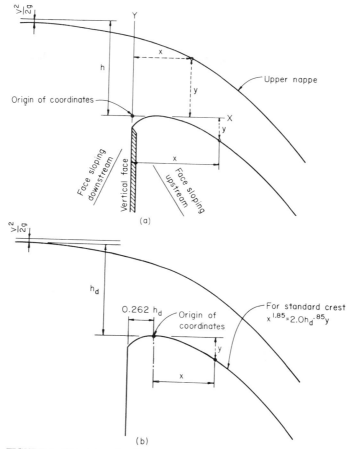

FIGURE 1 Geometry of spillway crests.

crest. When the actual head on the crest is equal to the head for which the crest is designed, the pressure along the crest would be atmospheric for a frictionless fluid. However, with a real fluid, slightly greater than atmospheric pressure results because of the developing boundary layer.

An actual head smaller than the design head will result in greater pressure along the curved surfaces and in lower discharge capacities. Conversely, an actual head greater than the design head will result in subatmospheric pressures over some portion of the crest surface, and an increased discharge capacity. Excessive subatmospheric pressures can result in pulsating, inefficient spillway operation, and possibly damage to the structure as a result of cavitation. A certain amount of subatmospheric pressure can be attained without undesirable effects. Figure 3 provides a guide for determining the minimum pressures on the crest for various ratios of design head and actual head on the crest.

Designing the crest shape to fit the nappe for a head less than the maximum head expected often results in economy in construction. The resulting increase in

TABLE 1 Nappe Coordinates for Spillway Design

Horizontal coordinates x/h	Lower nappe — Vertical coordinates y/h								
	Vertical	Downstream					Upstream		
		1H:3V	2H:3V	1H:1V	2H:1V	4H:1V	1H:3V	2H:3V	1H:1V
0.0	−0.125	−0.098	−0.066	−0.045					
0.05	−0.066	−0.051	−0.032	−0.021					
0.10	−0.033	−0.026	−0.015	−0.008					
0.15	−0.014	−0.011	−0.005	−0.001					
0.20	−0.004	−0.003	−0.001	0.000					
0.25	0.000	0.000	−0.001	−0.003					
0.30	0.000	−0.001	−0.004	−0.009					
0.35	−0.004	−0.005	−0.010	−0.018					
0.40	−0.011	−0.012	−0.019	−0.030					
0.45	−0.021	−0.022	−0.031	−0.045					
0.50	−0.034	−0.035	−0.046	−0.062	−0.093	−0.143	−0.033	−0.033	−0.033
0.55	−0.049	−0.051	−0.064	−0.082	−0.117	−0.173	−0.048	−0.048	−0.048
0.60	−0.066	−0.069	−0.084	−0.104	−0.141	−0.206	−0.065	−0.065	−0.065
0.65	−0.085	−0.090	−0.106	−0.128	−0.168	−0.240	−0.084	−0.084	−0.084
0.70	−0.106	−0.113	−0.131	−0.154	−0.197	−0.277	−0.105	−0.105	−0.105
0.75	−0.129	−0.138	−0.158	−0.182	−0.228	−0.315	−0.128	−0.128	−0.128
0.80	−0.157	−0.165	−0.187	−0.212	−0.261	−0.356	−0.153	−0.153	−0.153
0.85	−0.185	−0.195	−0.218	−0.244	−0.297	−0.400	−0.180	−0.180	−0.180
0.90	−0.216	−0.227	−0.251	−0.278	−0.333	−0.443	−0.210	−0.210	−0.210
0.95	−0.249	−0.261	−0.286	−0.314	−0.373	−0.490	−0.242	−0.242	−0.242
1.0	−0.283	−0.297	−0.323	−0.352	−0.413	−0.538	−0.277	−0.277	−0.277
1.1	−0.358	−0.375	−0.403	−0.434	−0.501	−0.641	−0.351	−0.351	−0.351
1.2	−0.441	−0.461	−0.491	0.524	−0.597	−0.752	−0.433	−0.433	−0.433
1.3	−0.532	−0.555	−0.587	−0.622	−0.701	−0.871	−0.523	−0.523	−0.523
1.4	−0.631	−0.657	−0.691	−0.728	−0.813	−0.998	−0.621	−0.621	−0.621
1.5	−0.738	−0.767	−0.803	−0.842	−0.933	−1.133	−0.727	−0.727	−0.727
1.6	−0.853	−0.885	−0.923	−0.964	−1.061	−1.276	−0.841	−0.841	−0.841
1.7	−0.976	−1.011	−1.051	−1.094	−1.197	−1.427	−0.963	−0.963	−0.963
1.8	−1.107	−1.145	−1.187	−1.232	−1.341	−1.586	−1.093	−1.093	−1.093
1.9	−1.246	−1.287	−1.331	−1.378	−1.493	−1.753	−1.321	−1.321	−1.321
2.0	−1.393	−1.437	−1.483	−1.532	−1.653	−1.928	−1.377	−1.377	−1.377
2.2	−1.711	−1.761	−1.811	−1.864	−1.997	−2.302	−1.693	−1.693	−1.693
2.4	−2.061	−2.117	−2.171	−2.228	−2.373	−2.708	−2.041	−2.041	−2.041
2.6	−2.443	−2.505	−2.563	−2.624	−2.781	−3.146	−2.421	−2.421	−2.421
2.8	−2.857	−2.925	−2.987	−3.052	−3.221	−3.616	−2.833	−2.833	−2.833
3.0	−3.303	−3.377	−3.443	−3.512	−3.693	−4.118	−3.277	−3.277	−3.277
3.2	−3.781	−3.861	−3.931	−4.004	−4.197	−4.652	−3.753	−3.753	−3.753
3.4	−4.291	−4.377	−4.451	−4.528	−4.733	−5.218	−4.261	−4.261	−4.261
3.6	−4.833	−4.925	−5.003	−5.084	−5.301	−5.816	−4.801	−4.801	−4.801
3.8	−5.407	−5.505	−5.587	−5.672	−5.901	−6.446	−5.373	−5.373	−5.373

TABLE 1 Nappe Coordinates for Spillway Design (*Continued*)

Horizontal coordinates x/h	Lower nappe (*Continued*)								
	Vertical coordinates y/h								
	Vertical	Downstream					Upstream		
		1H:3V	2H:3V	1H:1V	2H:1V	4H:1V	1H:3V	2H:3V	1H:1V
4.0	−6.013	−6.117	−6.203	−6.292	−6.533	−7.108	−5.977	−5.977	−5.977
4.2	−6.651	−6.761	−6.851	−6.944	−7.197	−7.802	−6.613	−6.613	−6.613
4.4	−7.321	−7.437	−7.531	−7.628	−7.893	−8.528	−7.281	−7.281	−7.281
4.6	−8.023	−8.145	−8.243	−8.344	−8.621	−9.286	−7.981	−7.981	−7.981
Overall spillways									
4.8	−8.767	−8.885	−8.987	−9.092	−9.381	−10.076	−8.713	−8.713	−8.713
5.0	−9.523	−9.657	−9.763	−9.872	−10.173	−10.898	−9.477	−9.477	−9.477
5.2	−10.321	−10.461	−10.571	−10.684	−10.997	−11.752	−10.273	−10.273	−10.273
5.4	−11.151	−11.297	−11.411	−11.528	−11.853	−12.638	−11.101	−11.101	−11.101
−2.4	0.980	0.988	0.985	0.983	0.980	0.973	0.990	0.990	0.990
−2.2	0.987	0.980	0.981	0.977	0.972	0.957	0.988	0.988	0.988
−2.0	0.984	0.983	0.977	0.971	0.964	0.940	0.985	0.985	0.985
−1.8	0.980	0.979	0.971	0.964	0.955	0.922	0.981	0.981	0.981
−1.6	0.975	0.974	0.965	0.957	0.945	0.904	0.976	0.976	0.976
−1.4	0.969	0.968	0.958	0.949	0.934	0.885	0.970	0.970	0.970
−1.2	0.961	0.959	0.950	0.941	0.921	0.865	0.962	0.962	0.962
−1.0	0.951	0.948	0.939	0.930	0.904	0.842	0.953	0.953	0.953
−0.8	0.938	0.935	0.926	0.917	0.883	0.817	0.940	0.940	0.940
−0.6	0.921	0.918	0.908	0.899	0.858	0.788	0.923	0.923	0.923
−0.4	0.898	0.895	0.885	0.875	0.826	0.754	0.900	0.900	0.900
−0.2	0.870	0.865	0.853	9.841	0.786	0.712	0.872	0.872	0.872
0.0	0.831	0.826	0.811	0.796	0.737	0.659	0.833	0.833	0.833
0.05	0.819	0.814	0.798	0.783	0.723	0.643	0.822	0.822	0.822
0.10	0.807	0.802	0.785	0.768	0.708	0.627	0.810	0.810	0.810
0.15	0.793	0.788	0.770	0.752	0.692	0.610	0.796	0.796	0.796
0.20	0.779	0.774	0.755	0.736	0.675	0.591	0.782	0.782	0.782
0.25	0.763	0.758	0.739	0.719	0.657	0.572	0.766	0.766	0.766
0.30	0.747	0.742	0.721	0.700	0.638	0.550	0.750	0.750	0.750
0.35	0.730	0.724	0.702	0.680	0.617	0.528	0.733	0.733	0.733
0.40	0.710	0.704	0.681	0.659	0.596	0.504	0.713	0.713	0.713
0.45	0.690	0.683	0.659	0.626	0.572	0.480	0.693	0.693	0.693
0.50	0.668	0.661	0.637	0.613	0.549	0.452	0.671	0.671	0.671
0.55	0.646	0.638	0.613	0.588	0.523	0.424	0.650	0.650	0.650
0.60	0.621	0.612	0.587	0.562	0.497	0.394	0.625	0.625	0.625
0.65	0.596	0.586	0.560	0.535	0.470	0.363	0.600	0.600	0.600
0.70	0.568	0.558	0.531	0.505	0.439	0.330	0.572	0.572	0.572
0.71	0.539	0.529	0.501	0.475	0.408	0.298	0.543	0.543	0.543
0.80	0.509	0.498	0.470	0.442	0.375	0.261	0.513	0.513	0.513

TABLE 1 Nappe Coordinates for Spillway Design (*Continued*)

Horizontal coordinates x/h	Overfall spillways (*Continued*)									
	Vertical coordinates y/h									
	Vertical	Downstream					Upstream		
		1H:3V	2H:3V	1H:1V	2H:1V	4H:1V	1H:3V	2H:3V	1H:1V
0.85	0.478	0.466	0.438	0.409	0.340	0.223	0.482	0.482	0.482
0.90	0.444	0.431	0.402	0.373	0.303	0.183	0.449	0.449	0.449
0.95	0.410	0.395	0.366	0.337	0.264	0.141	0.415	0.415	0.415
1.0	0.373	0.358	0.327	0.297	0.223	0.098	0.379	0.379	0.379
1.1	0.295	0.278	0.245	0.214	0.135	0.005	0.302	0.302	0.302
1.2	0.210	0.191	0.156	0.124	0.039	−0.096	0.218	0.218	0.218
1.3	0.118	0.097	0.060	0.026	−0.065	−0.205	0.127	0.127	0.127
1.4	0.019	−0.005	−0.044	−0.080	−0.177	−0.322	0.029	0.029	0.029
1.5	−0.088	−0.115	−0.156	−0.194	−0.297	−0.447	−0.077	−0.077	−0.077
1.6	−0.203	−0.233	−0.276	−0.316	−0.425	−0.580	−0.191	−0.191	−0.191
1.7	−0.326	−0.359	−0.404	−0.446	−0.561	−0.721	−0.313	−0.313	−0.313
1.8	−0.457	−0.493	−0.540	−0.584	−0.705	−0.870	−0.443	−0.443	−0.443
1.9	−0.596	−0.635	−0.684	−0.730	−0.857	−1.027	−0.581	−0.581	−0.581
2.0	−0.743	−0.785	−0.836	−0.884	−1.017	−1.192	−0.727	−0.727	−0.727
2.2	−1.601	−1.109	−1.164	−1.216	−1.361	−1.546	−1.043	−1.043	−1.043
2.4	−1.411	−1.465	−1.524	−1.580	−1.737	−1.932	−1.391	−1.391	−1.391
2.6	−1.793	−1.853	−1.916	−1.976	−2.145	−2.350	−1.771	−1.771	−1.771
2.8	−2.207	−2.273	−2.340	−2.404	−2.585	−2.800	−2.183	−2.183	−2.183
3.0	−2.653	−2.725	−2.796	−2.864	−3.057	−3.282	−2.627	−2.627	−2.627
3.2	−3.131	−3.209	−3.284	−3.356	−3.561	−3.796	−3.103	−3.103	−3.103
3.4	−3.641	−3.725	−3.804	−3.880	−4.097	−4.342	−3.611	−3.611	−3.611
3.6	−4.183	−4.273	−4.356	−4.436	−4.665	−4.920	−4.151	−4.151	−4.151
3.8	−4.757	−4.853	−4.940	−5.024	−5.265	−5.530	−4.723	−4.723	−4.723
4.0	−5.363	−5.465	−5.556	−5.644	−5.897	6.172	−5.327	−5.327	−5.327
4.2	−6.001	−6.109	−6.204	−6.296	−6.561	−6.846	−5.963	−5.963	−5.963
4.4	−6.671	−6.785	−6.884	−6.980	−7.257	−7.552	−6.631	−6.631	−6.631
4.6	−7.373	−7.493	−7.596	−7.696	−7.985	−8.290	−7.331	−7.331	−7.331
4.8	−8.107	−8.233	−8.340	−8.444	−8.745	−9.060	−8.063	−8.063	−8.063
5.0	−8.873	−9.005	−9.116	−9.224	−9.537	−9.862	−8.827	−8.827	−8.827
5.2	−9.671	−9.809	−9.924	−10.036	−10.361	−10.696	−9.623	−9.623	−9.623
5.4	10.501	−10.645	−10.764	−10.880	−11.217	−11.562	−10.451	−10.451	−10.451

unit discharge may make possible a shortening of the crest length, or a reduction in freeboard allowance for reservoir surcharge under extreme flood conditions.

Because the occurrence of design floods is usually so infrequent, several water-control agencies design a spillway crest for a head which is 75 percent of that resulting from the actual discharge capacity. Tests have shown that the subatmospheric pressures on a nappe-shaped crest do not exceed about one-half

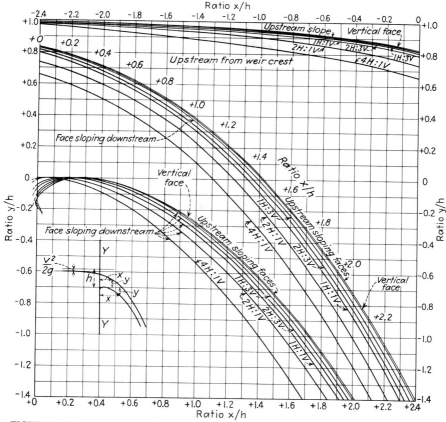

FIGURE 2 Shape of upper and lower nappe for weirs with faces of various slopes.

of the design head when the design head is not less than 75 percent of the maximum head.[1] An approximate diagram of the subatmospheric pressures, as determined from model tests, is shown by Fig. 4. Figure 3 shows the manner in which the minimum pressure varies with the particular actual and design heads adopted for a given spillway. The minimum crest pressure must be greater than cavitation pressure. It is suggested that the minimum pressure allowable for design purposes be 20 ft of water below sea-level atmospheric pressure and that the altitude of the project site be taken into account in making the calculation. For example, assume a site where the atmospheric pressure is 5 ft of water less than sea-level pressure, and in which the maximum head contemplated is 60 ft; then, only 15 additional feet of subatmospheric pressure is allowable. Entering Fig. 3 with an actual head of 60 ft and an allowance of 15 ft of subatmospheric pressure, read the design head to be 50 ft. This is seen to be greater than the 45 ft which would be used under the $0.75 = h_d/h_a$ rule; so consideration of cavitation potential governs in this instance.

An overfall spillway crest which approximates closely the lower portion of a

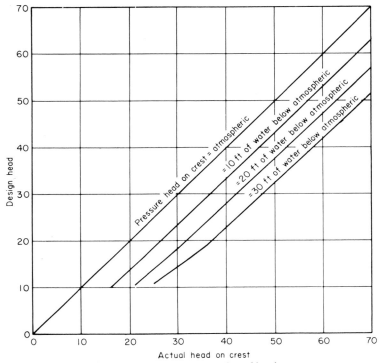

FIGURE 3 Minimum pressure on crest as function of head.

jet issuing from a sharp-crested weir is designated as a *standard crest*. A curve plotted from the formula

$$x^{1.85} = 2.0 h_d^{0.85} y \tag{1}$$

as shown by Fig. 1*b* fits closely to the curve of the standard crest.[2]

5. Overfall Spillway Discharge Coefficients. The discharge for an overfall spillway of nappe shape varies as the following equation:

$$Q = CLh^{3/2} \tag{2}$$

where Q = total discharge over the spillway, cfs
L = length of spillway crest, ft
h = total head on crest including velocity head, $V^2/2g$
C = coefficient of discharge

The discharge coefficients for vertical-faced ogee crests which are fitted to the lower nappe of the design head are given in Fig. 5*a*. Figure 5*b* presents the correction factors for the discharge coefficients for other than the design head. End contractions at spillway piers reduce the net length between the pier faces. Equation (2) would then be modified as follows:[2]

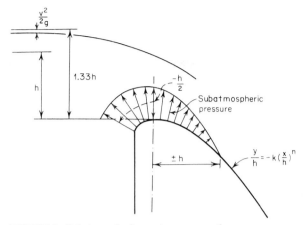

FIGURE 4 Subatmospheric crest pressure ratio.

$$Q = C[L - 2(Nk_p + k_a)h]h^{3/2} \tag{3}$$

where k_p = pier-contraction coefficient
k_a = abutment contraction coefficient
$h = H_d + V^2/2g$
L = net length between piers
V = velocity of approach
N = number of piers

Figure 6 shows the effect of three commonly used pier shapes on the coefficient of pier contraction k_p for various ratios of head on crest to design head.

As a basis for design, spillway discharge coefficients are usually determined by model tests. Figure 7 shows the discharge coefficient obtained by model tests of the spillways of the Tennessee Valley Authority dams, and Fig. 8 shows the principal features of these spillways.

Flow conditions both upstream and downstream from the controls are also factors which may influence the discharge capacity. The depth of approach, friction losses in approach channels, slopes of upstream faces, downstream convergence, and downstream submergence are examples of factors which must be taken into consideration. Figure 9 shows the effect of the sloping upstream face on discharge coefficient for ogee-shaped crests. Figure 10 shows the effect of discharge coefficients for various depths of approach and for downstream submergence.

GATED AND ORIFICE SPILLWAYS

6. Gate-Controlled Ogee Crests. Releases from partially opened gates controlling the discharge over ogee crests will occur as orifice flow. With the gate open a small amount under full head, the path of the jet can be expressed by the parabolic equation

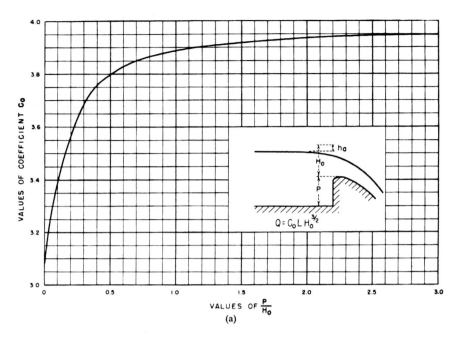

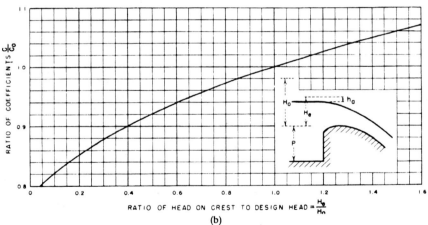

FIGURE 5 (a) Discharge coefficients for vertical-faced ogee crest. (b) Correction factors for discharge coefficients for other than design head.

$$-y = \frac{x^2}{4h} \qquad (4)$$

where h is the head on the center of the opening.

For an orifice inclined at an angle θ from the vertical, the equation will be

$$-y = x \tan \theta + \frac{x^2}{4h \cos^2 \theta} \qquad (5)$$

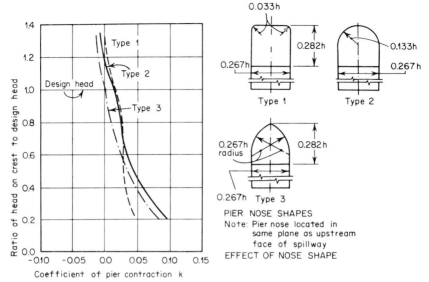

FIGURE 6 Effect of pier nose shape.[2]

The adoption of a jet-trajectory profile rather than a nappe profile will result in a wider ogee and in reduced discharge efficiency for full gate opening. Where the ogee is shaped to the ideal nappe profile, subatmospheric pressures may be reduced by placing the gate sill a short distance downstream from the crest.

The discharge for a gated ogee crest at partial gate openings, in cases where the openings are large, may be computed by the equation

$$Q = \frac{2}{3}\sqrt{2g}cL(h_1^{3/2} - h_2^{3/2}) \tag{6}$$

where h_1 and h_2 are defined in Fig. 11. In some cases it may be convenient to express $(\frac{2}{3}\sqrt{2g}c)$ as an overall coefficient m as in Eq. (7).

The orifice discharge coefficient c will differ with gate and crest arrangements. This coefficient is also affected by the approach and downstream conditions as they influence the jet contractions. Thus, the contractions for a vertical leaf gate will differ from those for a curved inclined radial gate. Figure 11 shows the approximate coefficients of orifice discharge c for various ratios of gate openings d_1 to total head h_1.

The rating curves for gate-controlled crests with partial gate openings are usually determined from model tests. Figure 12 shows typical discharge rating curves from a hydraulic model study for a 40-ft-wide by 38-ft-high Tainter gate.

The basic formula for spillway discharge, used in analyzing prototype measurements for a spillway, with a vertical leaf gate operating at partial opening (Fig. 11), is

$$Q = mL(h_1^{3/2} - h_2^{3/2}) \tag{7}$$

where h_1 = head on gate lips, ft

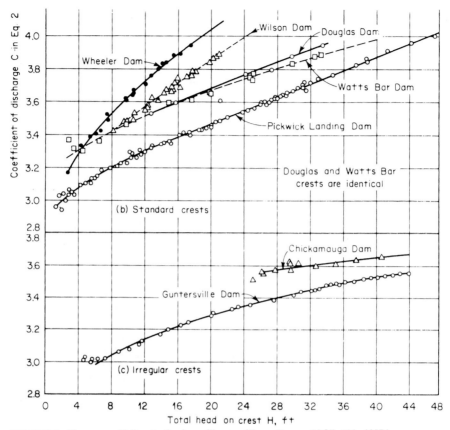

FIGURE 7 Tennessee Valley Authority spillways. (*From Trans. ASCE,* **122,** *1957.*)

h_2 = head on ogee crest, ft
L = net length, ft
m = coefficient of discharge = $\tfrac{2}{3}\sqrt{2g}c$

Coefficient m will not be identical with C in Eqs. (2) and (3) except where there is free overfall.

Interesting prototype experiments were conducted at Wilson Dam, Alabama, to determine the values of m.[3] These measurements were made at the dam with various heights of gate openings and with various combinations of gates in operation.

It is sometimes desirable to taper the downstream ends of spillway piers to provide for the gradual spreading of the discharge. When this is done, care must be taken to avoid negative pressures which could result in cavitation along the pier sides.

The possibility that cavitation may occur under high-velocity conditions at pier slots should be given consideration. Comprehensive studies of pier-slot de-

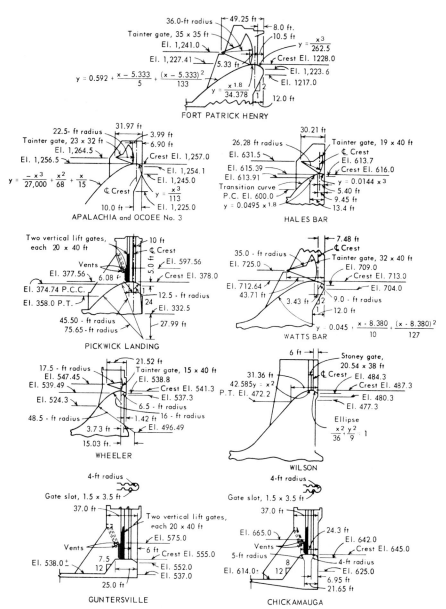

FIGURE 8 Tennessee Valley Authority spillway crests (*From Trans. ASCE,* **122,** *1957.*)

sign have been made by the U.S. Army Corps of Engineers and the Bureau of Reclamation.

7. Orifice Spillway. One means of providing a submerged control is with an orifice-type spillway. This type may be designed to discharge as an orifice under partial gate openings, and as an overflow spillway with an ogee crest under full

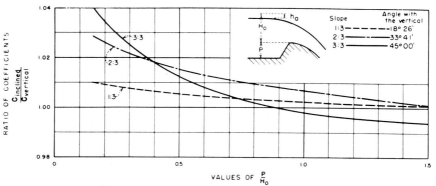

FIGURE 9 Discharge coefficient for ogee-shaped crest with sloping upstream face.

gate opening. By moving the gate downstream from the axis and providing an adequate headwall, the spillway gate may be designed to control a head much higher than the gate height. By way of illustration, a 50-ft-high gate could be designed to control the discharge from a 100-ft head on the crest of an ogee-type spillway. The discharge resulting from this orifice-overflow combination would be approximately 4000 cfs per foot of opening. In comparison with 50-ft gates mounted on the crest, assuming 10-ft surcharge, the discharge would be approximately half that of the orifice-overflow combination. The length of the spillway could be reduced proportionately. At sites such as that at Mangla Dam on the Indus river (described elsewhere), where only a limited area and relatively short length of suitable foundation material are available for the spillway structure, the orifice-type spillway offers the most economic means of passing the design flood. It should be noted, however, that the orifice type results in high flow concentration, which will increase the size and cost of energy-dissipation works below.

The gates controlling orifice discharge would be opened fully only under the extremely rare occurrence of larger than design-flood flow conditions. Under all other conditions, the partially opened gates would discharge as orifices. The orifice-type gate offers the advantages of permitting a deep reservoir drawdown in advance of floods. In some situations, these gates may be designed to serve both as spillway gates and as reservoir outlets. Either radial or vertical lift gates are usually used to control the discharge from the orifice-type spillway. Because of higher heads the gates are much heavier than a crest gate of equal size.

The spillway of the Roseires Dam on the Blue Nile River, constructed for the Republic of Sudan, is provided with seven 10-m-wide by 13-m-high spillway radial control gates operating under a design head of 28.5 m, as shown in Fig. 14, and five 6.0-m-wide by 11.1-m-high sluice radial control gates which operate under a design head of 55.1 m, as shown by Fig. 15. The high-level spillway will operate as an orifice during partial gate openings and as an ogee spillway when the gates are opened fully. The concrete headwall reduces the height of the radial gates by 4.5 m and increases the range of flows discharged by orifice action.

The spillway section of the Wilson Dam consists of a gravity overflow dam with the permanent crest about 80 ft above the stream bed. Flow is controlled by 58 gates 18 ft high by 38 ft wide. The piers are 8 ft thick. The pier noses are circular in form and flush with the upstream face of the dam.

Figure 13 shows the rating curve and discharge coefficients for two operating

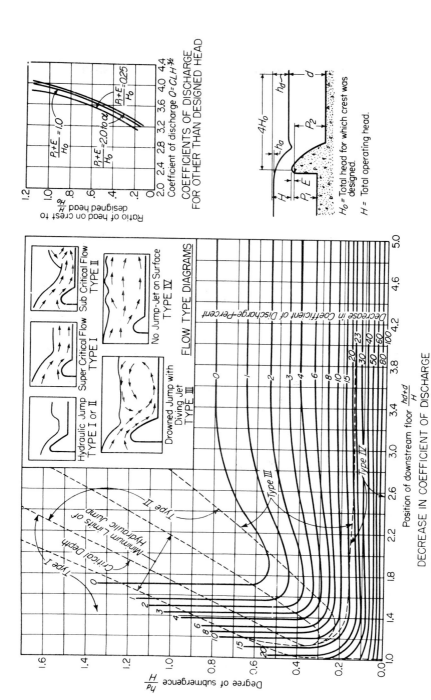

FIGURE 10 Discharge coefficients for various approach depths and degrees of submergence. *(From "Boulder Canyon Projects, Final Reports," Part VI, Hydraulic Investigations, Bulletin 3, Studies of Crests for Overfall Dams, Fig. 44, p. 150, U.S. Bureau of Reclamation.)*

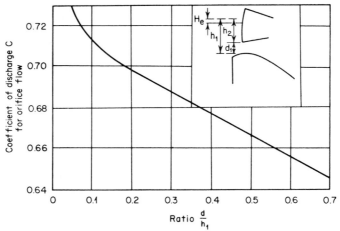

FIGURE 11 Approximate coefficients of orifice discharge. (*From "Design of Small Dams," U.S. Bureau of Reclamation.*)

conditions: (1) with adjacent gates completely closed and (2) with adjacent gates completely open. Both sets of tests were made with the pool surface held constant at 18 ft above the permanent crest. The value of h_2 should not be taken as the distance from the pool level down to the point where the gate clears the water, but rather it should be equal to the head on the crest minus the depth of the discharging jet.

8. Design of Piers. The hydrostatic water pressures acting on the pier sides resulting from one gate discharging under maximum head, and with one or more adjacent gates closed, will result in lateral loadings which must be transferred to other parts of the structure. The thickness and reinforcement of the piers must be adequate to withstand, at acceptable factors of safety, the resulting bending and shearing stresses. Pier thicknesses usually range between 9 and 15 ft. The anchorages which transfer water loads to the piers from the crest gates will also be a factor in determining pier widths. Piers may also contain aeration ducts and serve other purposes such as supporting highway bridges and hoist or crane structures. Piers with vertical downstream face may be used to vent the flow for preventing cavitation damages on the concrete surface.

The water passages leading to the orifice outlet must be formed so as to minimize entrance vortices and negative pressures. Since the design of an orifice outlet varies from project to project, model tests should be conducted to determine the required submergence. For preliminary design, the data compiled by Gulliver[29] as shown in Fig. 16 can be used. The water passages for the Mangla orifice-type spillway (Fig. 17) were shaped after extensive model tests had been made on the headworks structure.

CHUTE (OR TROUGH) SPILLWAYS

9. Features. The chute is the commonest type of water conductor used for conveying flow between control structures and energy dissipators. Chutes can be

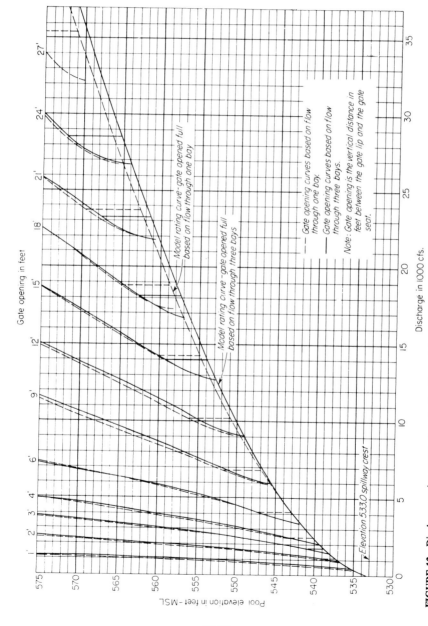

FIGURE 12 Discharge rating curves from hydraulic-model study of one 40- by 38-ft Tainter gate.

16.18   SPILLWAYS AND STREAMBED PROTECTION WORKS

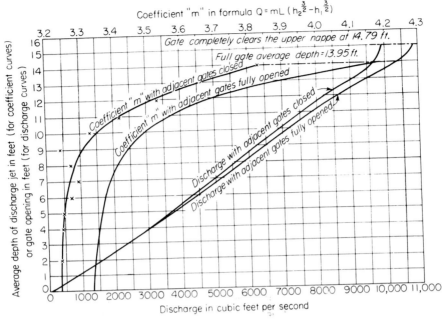

FIGURE 13   Spillway rating curve, Wilson Dam.

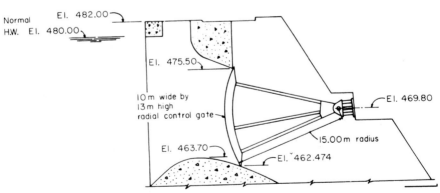

FIGURE 14   Spillway, Roseires Dam, Blue Nile River, Republic of Sudan.

formed on the downstream face of gravity dams, cut into rock abutments and either concrete-lined or left unlined and built as free-standing structures on foundations of rock or soil.

One of the commonest causes of spillway failures has been the improper design of chutes. The flow in a chute is usually supercritical; in many cases the velocity is greater than 100 fps (30.5 m/s). As a result of the high-velocity flow and large dynamic forces, the hydraulics of the chute is a critical factor. For spillway

CHUTE (OR TROUGH) SPILLWAYS   16.19

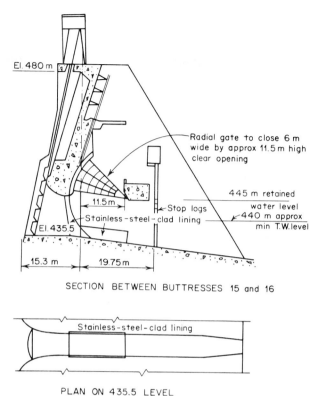

FIGURE 15 Low-level outlet, Roseires Dam, Republic of Sudan. (*Courtesy of Sir Alexander Gibb and Partners, London.*)

chutes which convey high-velocity flows, the occurrence of cavitation is often unavoidable, and implementation of aeration devices to reduce potential damage due to cavitation must be considered.

10. Principles of Design. Water flowing down a steep chute with an improperly aligned guide wall and/or floor slab will produce standing waves, piling up of water against walls, overtopping, excessive dynamic impact, cavitation, erosion, poor bucket action, and other adverse effects.

In general, the following principles should govern design: (1) nearly all guiding or direction changing of water should be limited to locations upstream from the control structure where velocities are comparatively low and (2) the alignment of the flow should usually be as straight and symmetrical as possible once the water is accelerated.

The velocity of water increases rapidly as it passes over the control structure. Downstream of the ogee crest, the velocities become supercritical and increase with drop in elevation. Therefore, any changes in alignment of the walls and chute floor downstream of the crest must be handled with extreme caution. Chute alignment for high-velocity flow can be curved in the lower reaches only if the chute floor and walls are shaped adequately to force water into a turn without

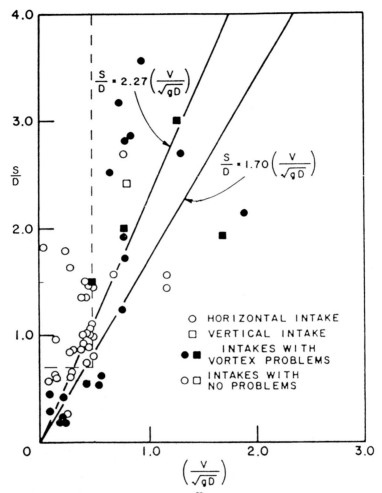

FIGURE 16 Submergence and vortices.[29]

overtopping the walls. Aeration devices should be incorporated when cavitation is expected.

Since the upstream width of a spillway is established by the required discharge at the controlling section near the crest, quite often the control section will be larger than the remaining chute, requiring a transition section between the two. In other cases, it is possible to limit the width of the control section by providing deep gates or submerged orifice gates with headwall or a labyrinth weir. The width of the chute is limited by topography, construction cost, and width of the river channel into which it discharges. The transition between the control section and the chute should be governed by the following considerations: (1) the convergence should be symmetrical to balance hydraulic forces; (2) the transition

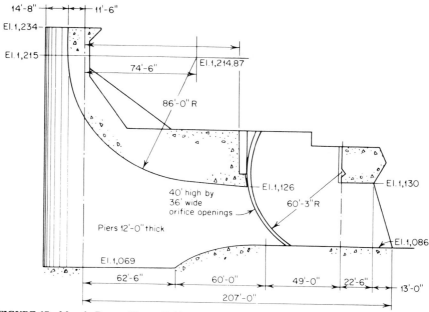

FIGURE 17 Mangla Dam spillway, Pakistan.

should be as far up the chute as practical, since velocities will be least at the upstream end; (3) the transition should be smooth and gradual and should be so proportioned that supercritical flow will be maintained throughout; and (4) where the chute is narrower than the ogee, spillway openings should be arranged radially to eliminate a reverse curve in the guide walls.

These principles are illustrated by Fig. 18, which shows the spillway chute at the Derbendi Khan Dam on the Diyala River in northern Iraq. The main dam is a rock-fill structure about 430 ft high. Spillway discharges are controlled by three 49- by 49-ft radial gates. Under design flood conditions, both the control structure and chute are designed to pass a flood of 400,000 cfs. The maximum concentration of flow at the crest is about 2700 cfs/ft. The maximum concentration of flow in the chute is close to 4000 cfs/ft.

At maximum capacity the discharge coefficient C in the formula $Q = CLh^{3/2}$ is 4.09, indicating low negative pressures at the crest.

To attain satisfactory individual operation of each of the three spillway gates, internal training walls were extended down the chute to the bucket. The bucket type which showed the best characteristics was the V-channel type shown in Fig. 18.

In testing the models of the Derbendi Khan spillway, measurements of pressure and water-surface profiles were made for various flows. The increased pressures in the bucket locations must be given consideration in designing both the sidewall and floor structures. Centrifugal forces in these locations actually have an effect equivalent to that of increasing the density of the water. At the downstream bucket the actual pressure on both the walls and floor for a discharge of 400,000 cfs is approximately 3 times the hydrostatic pressure resulting from the depth of water on the bucket.

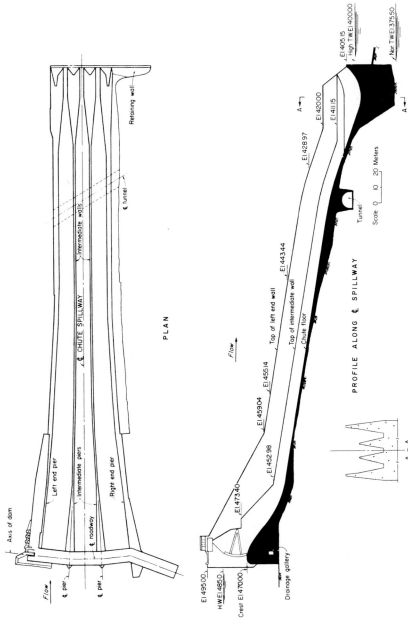

FIGURE 18 Derbendi Khan Dam.

11. Sidewalls. In addition to these centrifugal forces, the following variables are considered in the design of the sidewalls:

1. The average depth of flow.
2. The effect of changes in direction of flow which result in the formation of cross waves. This can occur where flow impinges on a sidewall, or it can occur upstream at a pier or some other obstruction. These waves are then reflected back and forth between the sidewalls of the chute.
3. The effect of the development of the boundary layer on the sidewalls and on the bottom of the spillway chute, which requires flow area in addition to that which would be called for by the application of the laws of conservation of mass and energy.
4. The effect of the entrainment of air by the turbulent high-velocity flow which produces a general bulking, particularly along the sidewalls.
5. An arbitrary allowance of additional freeboard over and above that required by the preceding four considerations.

Hydraulic-model studies are useful in addition to the analytical studies which could be made to evaluate all the foregoing criteria except the one pertaining to air entrainment. Observations obtained from operating spillways provide the most reliable basis for estimation of the bulking effect due to air entrainment. Hydraulic models may be applied, provided that the scales selected are greater than about 1:10 with Weber number exceeding about 1000. Small models do not provide satisfactory simulation of the prototype air-entrainment behavior.

The air entrainment is highly dependent on the level of turbulence in the flow. In deeper portions of the flow at considerable distance from the sidewalls, turbulence is confined to the boundary layer along the bottom of the chute. At some point the thickness of the growing boundary layer becomes equal to the depth of flow. This point of inception marks the location where the surface air entrainment (aeration) begins forming turbulent white water downstream. Along the sidewalls, white water appears much farther up in the spillway because the edge of the boundary layer development along the side walls is continuously exposed to the atmosphere. Air entrainment along the walls then begins at the point where the intensity of turbulence is sufficient to project small masses of water required to entrain the air.

The development of the boundary layer on the floor of a spillway chute can be approximated by the following equation given in the U.S. Army Corps of Engineers *Hydraulic Design Criteria,* Sheets 111-18 to 111-18/5.

$$\frac{\delta}{L} = 0.08 \left(\frac{L}{k}\right)^{-0.233} \tag{8}$$

where δ = boundary layer thickness, ft
L = surface distance from upstream end of the ogee crest, ft
k = absolute concrete-surface roughness (ranges from 0.002 to 0.007 ft)

Some guidelines which are useful in preliminary design are

1. The typical rate of growth of boundary layer thickness in concrete-lined chutes, computed from Eq. (8).
2. The amount of energy lost in flow down spillway chutes may be taken to be

the potential head multiplied by the boundary layer thickness divided by 5 times the depth of flow at the point in question.

3. The allowance to be made for wave action and surface bulking produced by piers on the spillway crest should be approximately 25 percent of the estimated thickness of the flow at the downstream edge of the piers at maximum discharge. Actual model tests might produce better estimates of the amount of the wave action. Additional bulking produced by air entrainment should also be added. It is believed that the provision of 25 percent of the thickness of the flow at the downstream edge of the piers would be sufficient to accommodate all bulking along the sidewalls of the spillway, assuming that the alignment was carefully made.

4. The amount of additional freeboard above the sum of the potential thickness of the flow, 10 percent of the boundary layer thickness and 25 percent of the thickness of the flow at the downstream edge of the piers, should be an additional 2 to 5 ft depending on the size of the spillway involved.

12. Floor Slabs. Floor offset must be small and within allowable limits so that cavitation on the downstream slab surface will not occur. More discussions of this subject will be given in the following sections. The possibility of the development of stagnation pressures beneath the slab as a result of impingement of high-velocity jets on offsets at joints must be considered. Should such large forces develop, the slab would in all probability be torn from its place as would all the spillway floor downstream from this point. Such failures have occurred in a large number of chute spillways. This type of failure can be guarded against by the following procedures:

1. Joints in floor slabs should be designed to accommodate possible differential movements without the formation of an abrupt surface protrusion on which the high-velocity flow may impinge and cavitation may occur.
2. Joints between sections of floor slab should be keyed or sloped to minimize differential motion.
3. Forces under floor slabs which could develop by underseepage from the headwater or from high tailwater must be resisted structurally, or reliable underdrainage must be provided to eliminate the development of such high underpressures.

SPILLWAY DESIGN CONSIDERING CAVITATION AND AERATION

13. Introduction. With more and more dams of medium to large size built in the world, many spillways have experienced flow velocity exceeding 10 to 40 m/s. As the velocity approaches 10 to 15 m/s, cavitation is likely to occur. A slight offset at a construction joint or a surface irregularity may be sufficient to induce cavitation and cause damage on the downstream surface. A good construction specification or tolerance is required to protect the concrete floor from cavitation damage. When the velocity exceeds 20 to 30 m/s, the occurrence of cavitation will be highly likely and additional protective measures must be provided to guard the concrete surface against potential cavitation damage. The most effective method of eliminating or reducing cavitation damage known to date has been the aeration of the flow through incorporation of aeration devices.

14. Cavitation. Cavitation in water is the formation of a water vapor cavity resulting from reduction of local pressures to the vapor pressure of the water. On a spillway surface, when the streamlines of high-velocity flow curve significantly at a surface irregularity and converge locally, the pressure will drop in the direction along the converging streamlines. The curvature of the converging streamlines is generally abrupt and the pressures are often dropped to the vapor pressures, causing the formation of cavities. Examples of cavitation at surface irregularities are shown in Fig. 19.

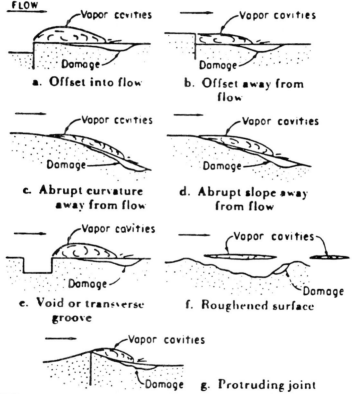

FIGURE 19 Typical isolated roughness elements found in hydraulic structures.[7,14] (*Reproduced by permission of U.S. Bureau of Reclamation and ASCE.*)

Cavitation damage on a spillway occurs when the small cavitation bubbles are swept downstream from the low-pressure area into an adjacent higher pressure zone which will not support cavitation, and the bubbles collapse suddenly. The surrounding water will rush in to fill the void. At the point of disappearance of the cavity, the inrushing water comes together, causing an implosion, momentarily raising the local pressure within the water body to a very high value. If a bubble collapses near the concrete surface, the implosion is asymmetric and a microsize jet will be developed and expelled toward the surface. The maximum speed of this jet has been estimated to exceed roughly 400 km/h. The concrete surface will receive blows from such tiny jets. With additional high-pressure waves emanating from other imploding cavities, the surface material will be stressed locally and eventually destroyed by fatigue and failure. The impingement of the high-speed microjet against the concrete surface may also produce a locally high dynamic pressure which is transmitted through microcracks or voids into the interior of the concrete mass. A significant pressure differential may occur around a small concrete mass and result in its detachment from the main concrete mass. In Fig. 20a, an imploding cavity in a liquid irradiated with ultrasound is captured in a high-speed flash photomicrograph. The formation of a microjet is clearly shown.[8] Computer simulation of the microjet formations for vapor cavities situated at the concrete surface and a short distance above it is shown in Fig. 20b. The onset of cavitation depends on the pressure and velocity of the water in the flow and can be expressed as a function of the cavitation number σ, which is defined in the form of the Euler number:

$$\sigma = \frac{P - P_v}{\rho V^2/2}$$

in which P is the pressure, P_v is the vapor pressure of water, and V is the flow velocity. Cavitation occurs if the cavitation number drops below a critical value. This critical value is a function of the spillway geometry and the flow condition.

INCIPIENT CAVITATION AND DAMAGE EXPERIENCE

Incipient cavitation characteristics of vertical, chamfered, and elliptically rounded offsets are shown in Fig. 21. Some cavitation damages experienced by several large dam projects are shown in Fig. 22. The figure shows that cavitation is a major concern when the velocity in a spillway reaches or exceeds approximately 25 m/s or the local cavitation number is less than 0.20. Also indicated is a relatively high probability of cavitation damage at the end of a vertical concave section of a spillway chute. As the high-velocity flow enters the curved section, the pressure distribution is readjusted and the velocity in the boundary layer becomes more uniformly distributed.[24] This in turn leads to a substantial increase in velocity at the concrete surface. A substantial reduction in the value of the cavitation number is realized, and damage to the concrete surface is often initiated in this area.

15. Cavitation Potential. To assess the cavitation potential over a spillway chute, the cavitation number is computed for the entire chute length. The areas where the values of the cavitation number are near or less than that of the corresponding critical cavitation numbers are identified. The cavitation potential of the Glen Canyon Dam spillway analyzed for a discharge of 475 m³/s is shown in Fig. 23. This discharge produced the lowest value of the cavitation number which occurred upstream of the point of curvature and downstream of the point of tangency of the vertical bend. To reduce the cavitation potential at problem areas, the chute profile (geometry) may be designed to increase the local pressure and to raise the value of the cavitation number. The chute profile can be improved by introducing constant cavitation number profile to the vertical bend

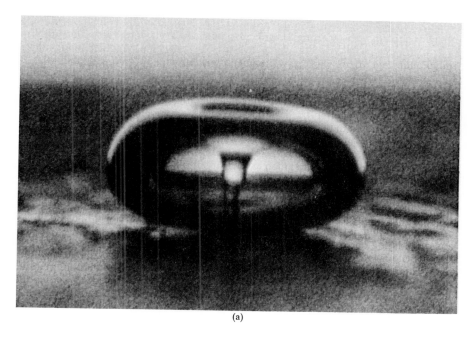

(a)

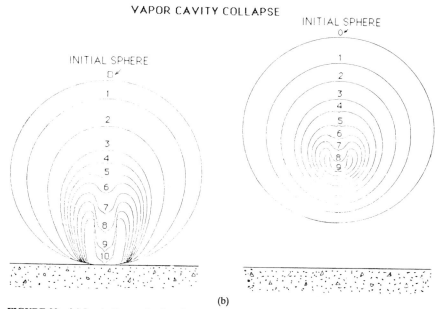

(b)

FIGURE 20 (a) Imploding cavity in a liquid irradiated with ultrasound. The diameter of the cavity is about 150 μm.[11] (b) Numerical simulation of cavity implosions.[12]

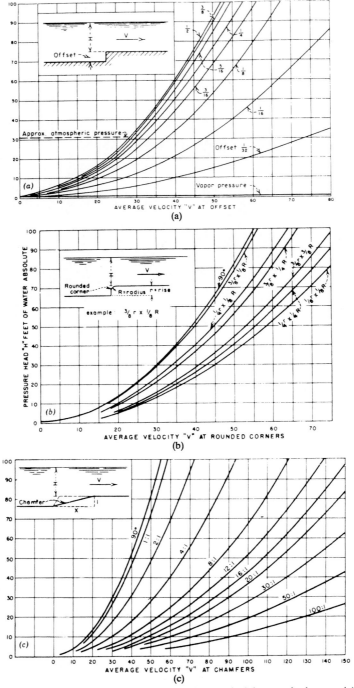

FIGURE 21 Pressures and flow velocities for incipient cavitation at: (*a*) square-edged offset into flow,[7] (*b*) round-edged offsets into flow, and (*c*) slopes into and away from flow.

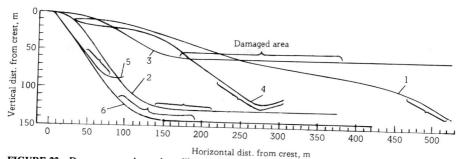

FIGURE 22 Damage experience in spillways.

	Dam	Head, m	Max. V, m/s
1.	Mica	186	50
2.	Yellowtail	140	42
3.	Liujiaxia	120	40–45
4.	Karun	118	35–41
5.	Bratsk	95	24–33
6.	Glen Canyon	175	>40

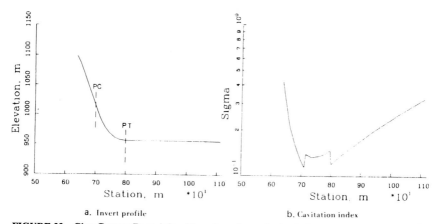

FIGURE 23 Glen Canyon Dam, left spillway tunnel—cavitation index for flow of $Q = 475$ m^3/s.[7]

section.[19] A procedure for obtaining constant cavitation number profile has been given by Falvey.[7] If modification of the chute geometry is ineffective in reducing the cavitation potential to a safe level, other types of protective measures must be implemented. Provision of aeration devices is the most economical and effective way of reducing or eliminating potential cavitation damage.

16. Finish Requirement and Surface Treatment. The maximum irregularities allowed in the concrete surface can be specified on the basis of the expected flow velocities. The surface irregularities present in a hydraulic structure consist of three basic types: offset, slope, and uniformly distributed roughness. An into-the-flow offset greater than 3 mm high should be ground by using the bevels given in Table 2 as guidance.[14]

TABLE 2 Grinding Tolerances for High-Velocity Flow

Max. level height to length	Velocity range	
	ft/s	m/s
1:20	40–90	12–27
1:50	90–120	27–37
1:100	>120	>37

Source: Ref. 14. Reproduced by permission of ASCE.

Falvey[7] classifies the surface tolerances for offset and slope variations into three categories, shown in Table 3. He also relates these tolerance classes to the cavitation number of the flow. The effect of aerated flow is included in the specifications of the required tolerance, as shown in Table 4.

TABLE 3 Surface Tolerance

Tolerance	Offset, mm	Slope
T1	25	1:4
T2	12	1:8
T3	6	1:16

Source: Ref. 7. U.S. Bureau of Reclamation.

TABLE 4 Specification of Surface Tolerance

Cavitation number	Tolerance without aeration	Tolerance with aeration
>0.6	T1	T1
0.40–0.60	T2	T1
0.20–0.40	T3	T1
0.10–0.20	Design changes required	T2
<0.10	Design changes required	Design changes required

Surce: Ref. 7. U.S. Bureau of Reclamation.

Data on tolerances for uniformly distributed roughnesses are too limited for useful application.

Field experience indicates that finishing the surface to limit the gradual irregularities to a bevel of 1:50 or smoother is practically difficult. It is often ineffective, and the incorporation of aeration devices is desirable.

17. Need for Aeration. Damage experience, for flows in spillway chutes and tunnels, indicates that damage due to cavitation becomes significant when water velocities exceed 25 m/s. However, a small quantity of air introduced into the flow near the concrete surface will significantly reduce the potential for the cavitation to damage the surface. Laboratory and field tests have shown that surface irregularities will not cause cavitation damage if the air introduced into the flow near the surface exceeds about 8 percent by volume.[24] Other studies indicate that use of more expensive higher-strength concrete will not totally eliminate the requirement for aeration.[5] The indications are that aeration devices can be designed to furnish air to the flow near concrete surfaces at critical locations to protect the spillway from cavitation damage.

18. Types of Aerators. The principal types of aeration devices consist of ramps (or deflectors), steps, troughs, and combinations of these as shown in Fig. 24. The ramps and steps are used to create separation zones behind the aerator for the air to be entrained underneath the water nappe lifted from the concrete surface. A step, when combined with a ramp, is useful in preventing the returning flow, caused by the impact of the free jet on the downstream surface, from reducing the air entrainment capacity of the free jet. Aeration troughs are used to distribute air across the entire width of the ramp. In some cases, aeration troughs are replaced by air galleries (ducts) with openings or ports for distribution of air.

19. Location of Aeration Devices. As indicated previously, significant damage on spillways can be expected when the cavitation number of the flow is less than 0.20 or the velocity exceeds about 25 m/s. Careful consideration must be given to the placement of the aerators when these criteria are applied. The maximum discharge is not necessarily the flow rate that produces the lowest value of the cavitation number. The cavitation numbers computed at the aerators and the critical discharges for a variety of spillways constructed by the U.S. Bureau of Reclamation are given in Table 5.[7] Placement of aerators in vertical bends which are concave upward should be avoided or considered with caution. The range of flow rates over which aerators will function satisfactorily is limited. The air trough may be filled with water and the effectiveness of the free jet to entrain air may be greatly reduced.

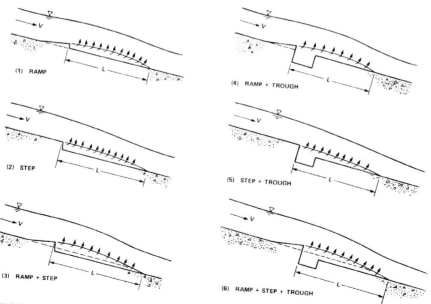

FIGURE 24 Aeration devices.

TABLE 5 Location of United States Bureau of Reclamation Aerators and Critical Discharge

Spillway	Spillway type	Cavitation number at aerator	Critical discharge (% of maximum)
Blue Mesa	Tunnel	0.22	30
Flaming Gorge	Tunnel	0.19	52
Glen Canyon	Tunnel	0.14	14
Hoover	Tunnel	0.18	19
McPhee	Chute	0.19	100
Yellowtail	Tunnel	0.13	16

20. Ramp Design. The purpose of a ramp is to cause the flow to separate from the spillway surface. The separation at the lip of a ramp creates a free jet that allows the lower surface of the jet to entrain air. The amount of air entrained by the turbulent flow is proportional to the size of the cavity formed by the separation and the average flow velocity. The size of the cavity can be determined from the trajectory of the free jet and the spillway geometry. The free-jet trajectory is affected by: (1) ramp height; (2) ramp length; (3) ramp slope (or ramp angle); (4) pressure in the cavity, termed *underpressure*; and (5) the flow condition at the ramp. Wei and DeFazio[5] solved the governing equations for the free-jet potential flow problem by the finite-element method. A constant underpressure was assigned to the lower boundary of the jet to simulate the effect of the air entrainment. An iterative procedure was introduced to determine the profile of the free jet. This method produced excellent results for both ramps and free overfalls. The results of a series of finite-element trajectory analyses for the Guri chute No. 3 for a 17° test ramp are shown on Fig. 25 together with the prototype and the 1:50 hydraulic model test results.[23] In the prototype tests, the cavity lengths were more difficult to determine under higher discharge conditions because of higher degrees of turbulence and sprays. A cavity underpressure of −0.03 bar recorded during the field tests was used in the analyses. The computed cavity lengths agree very well with the field test data. Because of the scale effect, the hydraulic model did not simulate the effect of the air entrainment satisfactorily and resulted in significant overestimation of the cavity lengths. Based on the finite-element analyses, the model produced only about −0.01 bar of underpressure. To reasonably simulate the surface breakup and the air entrainment action, a model scale of about 1:15 to 1:10 is needed to reduce the effects of the surface tension to a reasonable level.[4] The nappe profile (flow net) computed by the finite-element method for the King Talal spillway (Jordan) upper ramp and the design of the lower ramp are shown in Fig. 26. In addition to solving for the free jet profile, the pressure distribution around the ramp is determined. Knowledge of the pressure distribution is useful for the design of sidewalls in the vicinity of the ramp. An example of the computed pressure distribution at a ramp is given in Fig. 27. The finite-element free-jet model can be used to analyze various ramp geometries over a range of flow conditions and underpressures, resulting from a finite-element analysis. The effects of the ramp angle, ramp length, and the chute curvature of the approach section on the cavity length are demonstrated in Fig. 28. Such analyses are useful for optimization of the ramp configuration.

For the preliminary design of a ramp, a simplified approach can be taken. For a ramp on a straight chute, Pan et al.[6] ignored the underpressure of the free jet and

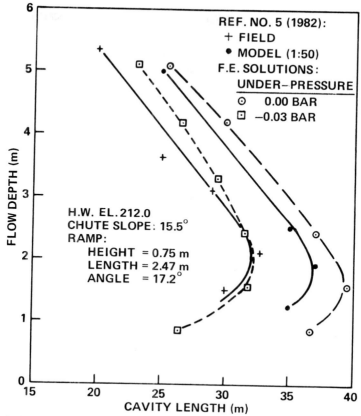

FIGURE 25 Guri chute 3 operation (left) and test results (right).[23]

developed a one-dimensional trajectory equation referenced to the jet centerline. Corrections were then made to account for the effects of the ramp and the energy loss on the trajectory of the jet. Glazov[17] ignored the effect of the relative height of the ramp to the flow depth but included the effects of the underpressure and the offset height in his equation. From the studies of Glazov and Pan, Falvey[7] developed a useful procedure for determining the free-jet trajectory. A computer program was developed to solve simultaneously the equation for the airflow through an air duct and the equation for air entrainment into the jet produced by an aerator. The estimate of airflow into the jet is based on the computed jet trajectory.

An alternative of solving the free-jet problem numerically by the finite-difference method was introduced by Xu and Gu.[9] The air-entraining free-jet flow was solved by an air-water two-phase unsteady turbulent-flow model. Good results were obtained for simulation of a hydraulic model experiment. The simulated free-jet profile and velocity vectors for flow over a 11.6° ramp are shown in Fig. 29. The computed cavity length was within 10 percent of the observed value. To analyze flow over a curved section, a coordinate transformation scheme is applied to create an orthogonal boundary-fitted grid system.

21. Air-Supply System. Several methods have been developed to supply air to the lower surface of the air-entraining jet. Major types of air-supply systems

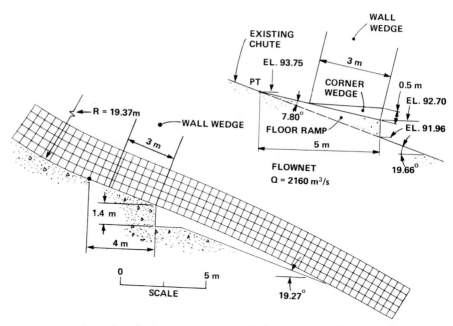

FIGURE 26 King Talal spillway aeration ramps: (left) upper ramp with computed nappe profile and (right) lower ramp.[5]

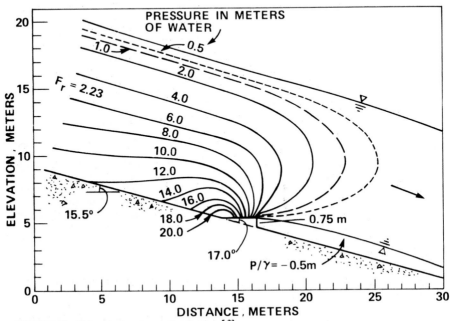

FIGURE 27 Computed pressure distribution.[5,23]

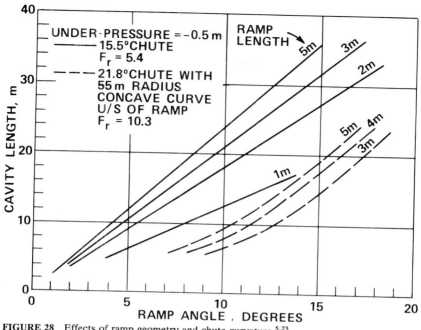

FIGURE 28 Effects of ramp geometry and chute curvature.[5,23]

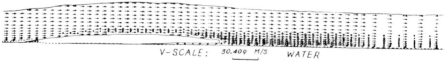

FIGURE 29 Free-surface shape and velocity vector field with cavity pressure $P_c = 0.005$ bar; $u_c = 5.22$ m/s.[9]

used by existing projects are shown in Fig. 30. These include: (1) piers, (2) wall slots, (3) wall offsets, (4) wall wedges (deflectors), (5) wall chimneys (ducts), and (6) air galleries (ducts) underneath the floor.

Wall wedges or deflectors, wall offsets, and piers are frequently used for flow aeration downstream of control gates. Since it is impractical to incorporate a large wedge or offset on the sidewalls or to use large piers, capability of these devices to supply air is generally limited and thus they are not applied to wide chutes. Piers, in addition to their normal functions, can be used to supply air to the aeration ramp. Flow separations at the downstream face of a pier produce a vertical cavity; the air is pulled in to vent the free jet through this vertical cavity, but venting capability is limited and the method is not applied to wide chutes. Wall slots are often used in conjunction with existing control gate structures. The downstream end of the slot may be offset and chamfered to keep water from entering the slot. Wall chimneys or ducts are used on wide chutes when the required slot size or the offsets are excessive.

22. Estimation of Air Entrainment Capacity of the Free Jet. The most commonly used empirical equation for the estimation of an aerator's air entraining capacity is

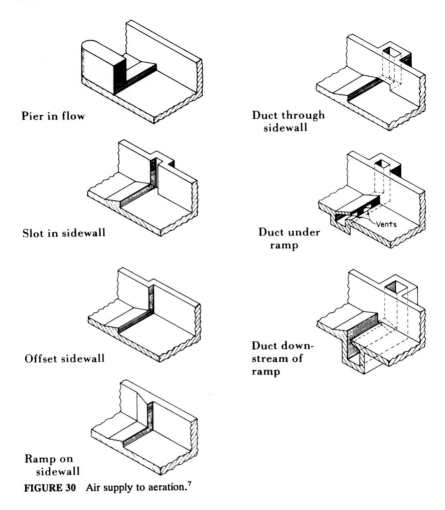

FIGURE 30 Air supply to aeration.[7]

$$q_a = C_a VL \qquad (9)$$

in which q_a is the unit discharge of air, C_a is a constant, V is the average jet velocity over the ramp, and L is the cavity length. This empirical relationship was recognized by Hamilton in an analysis of the Tarbela dam stilling basin test data.[4] Model and prototype measurements indicate that the value of the coefficient C_a lies between 0.01 and 0.04. In an analysis of the air concentration data obtained from experiments, Pan et al.[6] obtained a value of 0.022 for the coefficient C_a.

23. Aerator Spacing. An aerator device provides protection to the downstream surface against cavitation damage by introducing air into the flow. To furnish a minimum level of protection, an air concentration of at least 8 percent by volume must be introduced at the concrete surface. Because of buoyancy of the air bubbles, the air concentration along the surface decreases rapidly in the downstream direction from a maximum value near the point where the free jet impacts the surface. The area effectively protected by the aerator against cavitation erosion is limited by 8 percent concentration. For the area farther downstream, another aeration device is needed. The distribution of air concentration in the flow depends on the buoyancy of the bubbles and the diffusive effects of turbulence in the flow and flow instability near the concrete surface. The distribution of the air entrained in the flow due to an aerator can be described by the following transport equation:[13]

$$\frac{\partial}{\partial y}\left(\epsilon_c \frac{\delta C}{\partial y}\right) - V_b \frac{\partial C}{\partial y} = U_x \frac{\partial C}{\partial y} \qquad (10)$$

in which C is the air concentration, x is the direction of the flow, y is the distance in the transverse direction, ϵ_c is the eddy diffusivity, V_b is the rising velocity of the air bubble, and U_x is the flow velocity. The solution of this equation is an expression of the distribution of the air concentration in the flow:[13]

$$C(x, y) = e^{\lambda x}\left(\frac{y}{y_o}\right)^m \qquad (11)$$

in which $m = V_b/Au_*$, u_* is the wall shear velocity, y_o is the y value at the impact point, and λ is a constant. It was assumed that $C = 1$ when $x = 0$. The value of $C(x, y)$ computed by the above equation is a relative concentration with respect to the concentration at the impact point. The value of λ is obtained experimentally and has been found to range from 0.017 m^{-1} to 0.04 m^{-1}.[7,13] From an analysis of Semenkov's study,[21] a value of 0.02 can be obtained. In Liang and Wang's analysis,[13] the air concentration was computed at a distance one-eighth turbulent boundary layer thickness (i.e., a distance equal to the displacement thickness) above the spillway surface. The above analyses do not consider the surface air entrainment process or self-aeration of flow in a chute or spillway. The effect of the self-aeration on the air concentration distribution of the flow has been studied by several researchers.[7,18,26] A method for computing mean air concentration of developed aeration with the effect of the surface tension taken into account was proposed by Falvey.[7]

24. Design Procedure. The following steps[7,10,27] can be applied to design a spillway where cavitation is a major concern and implementation of aeration devices must be considered:

1. Develop a preliminary profile (geometry) of the spillway.
2. Compute the cavitation numbers along the spillway surface for the flow rates considered.
3. Determine the required surface tolerances based on the computed cavitation numbers.
4. For areas of high cavitation potential (i.e., cavitation numbers less than 0.2), try to modify the spillway profile (geometry) to reduce the cavitation potential.
5. Review steps 2 and 4 and the general spillway layout. If the cavitation potential cannot be reduced to the desired level, consider the following steps.
6. Determine the locations of the aerators and methods of introducing air to vent the jet. For the most upstream aerator, the downstream end of the di-

vider piers near the spillway crest may be used to vent the flow. Other possible locations for aerators may include the areas upstream of a convex curve and upstream of a flip bucket.

7. For each aerator, select trial ramp geometry (ramp height, length, angle, etc.)
8. For the selected ramp geometry, determine the air demand curve of the jet (i.e., the air entrainment rate of the jet versus the assumed underpressure).
9. Select a design for the air venting, and determine the air-supply curve (i.e., air-supply flow rate versus underpressure) for the venting system. Try to limit the air velocity to less than 50 m/s.
10. Determine the air discharge and the underpressure in the cavity or the separation zone based on steps 8 and 9.
11. Determine the air concentration at the impact point considering the boundary layer thickness at the ramp and the surface self-aeration effect.
12. Determine the air concentration distribution along the spillway surface and the area protected by the aerator.
13. Review steps 2 and 12 and determine if a second aerator is needed to protect the area farther downstream.

25. Examples of Aeration Device Design. The Guri Dam owned by CVG-Electrificacion del Caroni (EDELCA) in Venezuela was raised 55 m in three stages to its final height in 1986. The spillway was designed to pass 9000 m^3/s of flow and consists of three chutes, each with an average width of 50 m. These chutes have a drop of about 110 m in elevation over a length of 140, 145, and 150 m, respectively. During the raising of the spillway, a total of 12 aerators of different types were incorporated in the design to protect the concrete surface from cavitation damage.[28] They include simple ramps and ramps with offsets. The air was introduced to aerate the flow with divider piers, wall wedges, and air galleries. Corner wedges were also introduced at the intersections of the air ramps and the vertical walls to reduce the lateral sprays of the free jets. The profiles of the chutes during the initial and the second stages exhibited both concaving and convexing vertical curvatures. These curvatures made the flow curvilinear and the designs of the aerators more complicated. A finite-element free-jet trajectory program,[5] as described in Sec. 20, was applied during the design of these aerators. Figure 31 shows the profiles of the Guri spillway chute no. 1 during the raising and the implementation of aerators, as shown on details A, B, and C.

SIDE-CHANNEL SPILLWAYS

26. General. The side-channel spillway is commonly used in sites where the sides are steep and rise to a considerable height above the dam. In this form, the water falls over the spillway crest into a channel in which the flow is parallel to the crest. This channel leads eventually to the stream below the dam. A complete explanation of the hydraulics is given by Hinds.[22] This analysis is based on the assumption that all the energy of the overfalling water is dissipated in turbulence and that the slope in the side channel must be sufficient to accelerate the overfalling water in the direction of flow down the channel. Observations on many spillway models have confirmed the essential accuracy of this analysis. In the spillways of Hoover Dam, a cross weir was constructed at the downstream

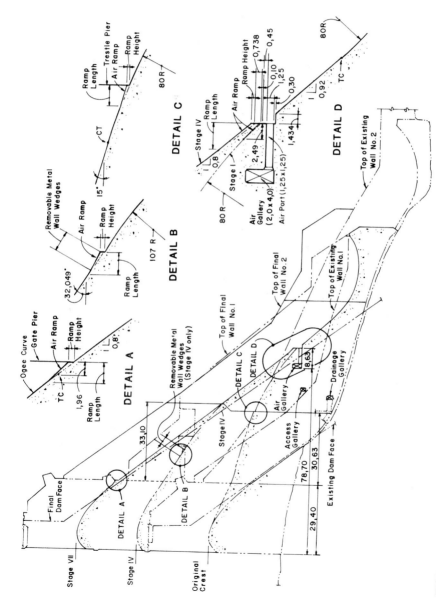

FIGURE 31 Aerators considered for the Guri chute No. 1 spillway.[5,28]

end of the channel section. At all flows, the weir provided sufficient depth of water for the energy of the overflowing water to be dissipated without causing excessive turbulence in the tunnel that carried the water back to the river. In this case, it was very desirable to avoid turbulence in order that the water might flow through the long tunnels without undesirable effects due to entrained air. The increase in the size of channel required with the weir was negligible. Extensive model experiments on the Hoover Dam spillways indicated that turbulence in the side channel can probably be reduced at less expense by the weir device than by any form of baffles that might be used. Attempts to drive the overfalling water in a downstream direction, increasing the velocity of flow in the channel and decreasing the necessary channel size, were unsuccessful in evolving any practical method. The tests demonstrated that the flow conditions in the channel downstream from the overflow section will be improved if the channel is narrowed, downstream from the overflow section, by offsetting the side toward the dam inward by an amount equal to the thickness of the stream of water falling over the weir.

In connection with the design of the spillways of the Hoover Dam, extensive studies were made of crest shapes and discharge coefficients. Detailed information can be found in the Bureau of Reclamation reports.

27. Flow Characteristics. Basic side-channel flow characteristics are illustrated by Fig. 32. The theory of flow is based on the law of conservation of linear momentum. For any short reach of the side channel, the momentum at the beginning of the reach plus any increase due to external forces must equal the momentum at the end of the reach.

Consider a short reach Δx in length, with a velocity and discharge at the upstream section of v and Q, respectively. At the downstream section the velocity and discharge will be $v + \Delta v$ and $Q + q\,\Delta x$, where q is the inflow per foot of length of the weir crest.

By applying the law of conservation of linear momentum, it can be demonstrated that change in the water elevation Δy in the reach Δx can be expressed in the following formulas:

$$\Delta y = \frac{Q}{g}\left(\frac{v + \tfrac{1}{2}\Delta v}{Q + \tfrac{1}{2}\Delta Q}\right)\left[\Delta v + \frac{g(\Delta x)}{Q}(v + \Delta v)\right] \tag{12}$$

If Q_1 and v_1 are values at the beginning of the reach and Q_2 and v_2 are the values at the end of the reach, Eq. (12) can be written

$$\Delta y = \frac{Q_1}{g}\left(\frac{v_1 + v_2}{Q_1 + Q_2}\right)\left[(v_2 - v_1) + \frac{v_2(Q_2 - Q_1)}{Q_1}\right] \tag{13}$$

This derivation can also be developed so that

$$\Delta y = \frac{Q_2}{g}\left(\frac{v_1 + v_2}{Q_1 + Q_2}\right)\left[(v_2 - v_1) + \frac{v_1(Q_2 - Q_1)}{Q_2}\right] \tag{14}$$

By the use of Eqs. (13) and (14), the water-surface profile can be determined for any particular side channel by assuming successive short reaches of a channel once a starting point is found. Commonly, a control section where critical flow occurs is the starting point. The solution of Eqs. (13) and (14) is obtained by a trial-and-error procedure.[1]

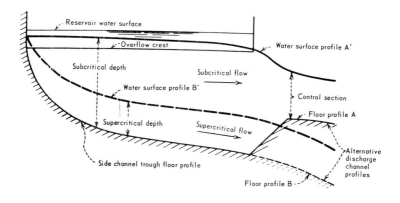

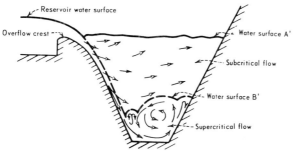

FIGURE 32 Schematic definition of flow in a side-channel spillway.

MORNING-GLORY SHAFT AND TUNNEL SPILLWAYS

28. General. In morning-glory shaft or tunnel spillways, the water flows over the lip of a funnel-shaped spillway and discharges down a shaft or tunnel. This form of spillway is adapted to narrow canyons where room for a spillway is restricted. A disadvantage of this type is that the discharge beyond a certain point increases only slightly with increased depth of overflow and therefore does not give so great a factor of safety against underestimation of flood discharge as do most other forms.

The morning-glory type has been tested extensively in models, and some limited observations have been made of prototype performance.

29. Typical Morning-Glory Spillway. The Hungry Horse Dam was completed in 1953 by the Bureau of Reclamation and is located on the South Fork of the Flathead River near Columbia Falls, Montana. The principal features of the spillway are shown by Fig. 33. Discharges are controlled by an adjustable 64-ft-diameter by 12-ft-high ring-gate structure which releases flow into a tapering and sloping tunnel. The throat of the converging section, as shown by Fig. 33, is 37 ft

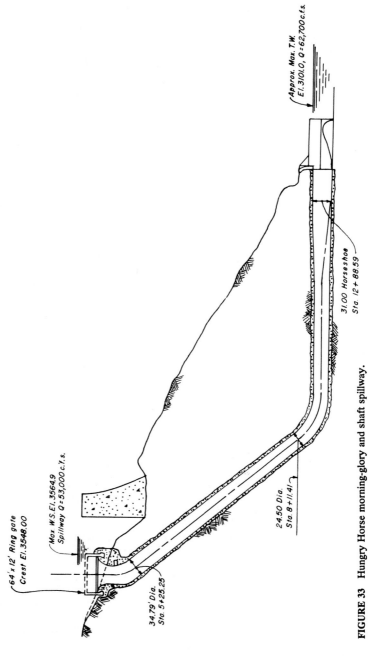

FIGURE 33 Hungry Horse morning-glory and shaft spillway.

in diameter. This tapers to a diameter of 34.79 ft at the upstream end of the inclined tunnel.

The incline has a vertical drop of 341.3 ft and tapers to a diameter of 24.5 ft at the downstream end. A vertical bend connects the inclined tunnel to a nearly horizontal tunnel which continues to the outlet portal at a slope of 0.0019. The tunnel is 24.5 ft in diameter throughout the lower bend and for a distance of 219 ft downstream, then is transformed through a 166-ft-long transition section to a 31-ft-diameter horseshoe tunnel.

The spillway is designed to pass a flow of 53,000 cfs. In discharge tests, made in July 1954, the reservoir release through the spillway was 30,000 cfs.

30. Hydraulics. The morning-glory shaft spillway may operate under three conditions:

1. *With crest control.* Under this condition, there is accelerating flow in the vertical and transition sections of the shaft and decelerating open-channel flow in the outlet leg of the conduit.
2. *With tube or orifice flow.* Under this condition, the top of the morning glory is partially submerged with orifice control at the throat of the transition.
3. *With full pipe flow.* Under this condition, the morning-glory intake is completely submerged, the entire conduit runs full, and the control moves to the downstream portal of the outlet leg of the conduit.

For the nomenclature shown by Fig. 34, the basic equation for the discharge of a nappe-shaped circular weir is

$$Q = C_0(2\pi R_s)H_0^{3/2} \tag{15}$$

It is apparent that the coefficient of discharge for a circular crest differs from that of a straight crest because of the effects of submergence and back pressure incident to the joining of convergent flows. Thus, C_0 must be related to H_0 and R_s

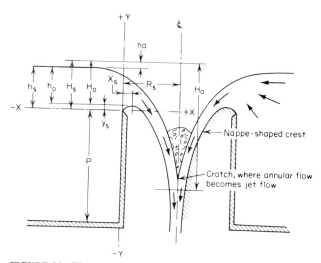

FIGURE 34 Elements of nappe-shaped profile for circular weir.

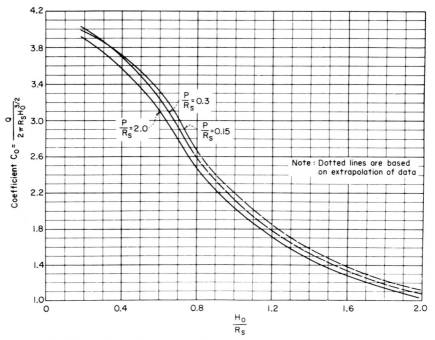

FIGURE 35 Hoover Dam side-channel spillway.

and, expressed in terms of H_0/R_s, is shown in Fig. 35 for three conditions of approach depth.

31. Typical Tunnel Spillway. Glen Canyon Dam, an important feature of the Colorado River storage project is located on the Colorado River in Arizona, 13 mi south of the Utah border. The main dam consists of a concrete arch which rises over 700 ft above the bedrock foundation.

Identical spillways in each rock abutment discharge into inclined tunnels connecting with each diversion tunnel downstream from their plugs. The principal features of the spillways are shown by Fig. 36. The total spillway capacity is 276,000 cfs. The discharge is controlled by two 40-ft-wide by 52.5-ft-high radial gates in each spillway. The spillway had suffered cavitation damages[7] near the downstream end of the vertical bend after passing floods. Aeration devices have been added to mitigate the problem.

SIPHON SPILLWAYS

32. General. Where the available space is limited and where the discharge is not extremely large, siphon spillways are often superior to other forms. They are also useful in providing automatic surface-level regulation within narrow limits. Because the siphon spillways prime rapidly and bring into action their full capacity, they are especially useful as an escape feature at the powerhouse end of long power canals with limited forebay capacity. In case of plant trip-out and rapid

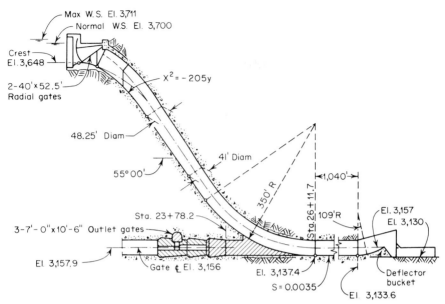

FIGURE 36 Glen Canyon Dam—profile of left spillway and tunnel plug.

turbine shutdown, the siphon provides a large discharge capacity within a very short time and can prevent overflow of the canal banks.

Such an arrangement has been incorporated in the Ghazi-Barotha Hydropower Project being developed in Pakistan. The siphon spillway, located at the downstream end of powerhouse 1, has a discharge capacity of 2000 m^3/s. In this case, a relatively large forebay has been incorporated to increase the control of the surges in the channel.

Figure 37 shows a cross section of one of 18 siphon spillways in the O'Shaughnessy Dam as initially constructed, which have a combined capacity of about 20,000 cfs. Siphon spillways are often built with a basin at the lower end so that the discharge end will be submerged, to facilitate priming. Alternatively, an ejector action can be introduced by placing a bend or lip in the downstream leg, which deflects the water when flowing over the crest at a slight depth to the opposite side of the siphon barrel. The lower end of the siphon barrel will thus be sealed, and the flowing steam, by carrying along bubbles of air from the inside of the siphon, will reduce the pressure inside enough to cause the siphon to start.

After the siphon action is started, unless the siphon is vented, the upstream water level will be drawn down to the level of the entrance before flow ceases. The magnitude of the draw-

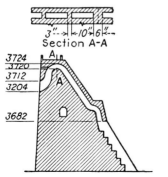

FIGURE 37 Siphon spillway in O'Shaughnessy Dam.

down can be controlled by means of a vent, through which the air enters the siphon and destroys the siphon action as soon as the water level has fallen below the vent.

The siphons should be made with gradually contracting entrances and curves of as large radius as possible. The capacity can sometimes be increased by gradually expanding the downstream section of the tube. It is not possible to accurately compute the action of siphon spillways. Much better results can be obtained by model tests.

SCOUR PROTECTION BELOW OVERFALL DAMS

33. Plunge Pools. A plunge pool frequently offers a simple and effective means of dissipating the energy of falling water. Natural waterfalls usually erode a pool having a depth of approximately one-third the head above the pool. Thus, nature offers a rough guide in determining pool depths. Figure 38 shows a plunge pool below the overfall spillway of the 600-ft-high double-curvature arch dam constructed by the City of Tacoma in 1968 on the Cowlitz River, Washington.

The spillway discharge for the design flood is 276,000 cfs. This discharge is controlled by four 42-ft, 6-in wide by 50-ft high radial crest gates. During the period of the design flood the maximum depth of the pool below the overfall spillway would be approximately 260 ft.

In the design of plunge pools, model tests are usually required to determine that the energy of the spillway discharge is dissipated before it reaches the foundation. For example, the discharge of an ogee-type spillway with a radial bucket

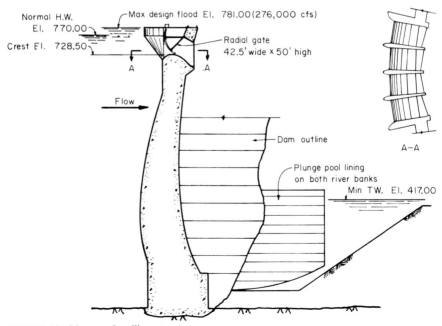

FIGURE 38 Mossyrock spillway.

into a relatively deep pool may follow the submerged face of the dam and travel horizontally along the foundation as a submerged jet. In such cases, the energy of the submerged jet will be diminished only slightly by frictional losses, and undercutting of the streambed protection works may result.

34. Deflector Buckets. Where the spillway discharge may be delivered directly to the river without additional streambed protection works, the jet may be projected beyond the structure by a deflector bucket. Flow from these deflectors leave the structure as a free-discharging upturned jet.

The trajectory of the jet depends on the energy of the flow at the lip and the angle at which the jet leaves the bucket. With the origin of the coordinates taken at the end of the lip, the path of the trajectory may be expressed by the equation

$$y = x \tan \theta \frac{x^2}{K[4(D + h_v) \cos^2 \theta]} \quad (16)$$

where θ = angle of edge of lip with horizontal
K = factor usually assumed as 0.9 to compensate for loss of energy
d = depth of water on bucket
h_v = velocity head of discharging jet

Ordinarily, the exit angle should not exceed 30° and the minimum radius of curvature should not be less than 5 times the depth of the water on the bucket.

In cases where the defector bucket discharges under submerged or partially submerged conditions, model tests are usually required to finalize the design.

Under some conditions of submergence, large eddy currents may circulate around the guide walls and possibly undercut the bucket and terminal structures. Model tests show that the omission of the guide walls near the end of the bucket will introduce eddies which will counteract this effect. These eddies result from the centrifugal force acting on the bucket, which spreads laterally some part of the spillway discharge.

35. Stilling Basins. Unless proper precautions are taken, the velocity of the spillway discharge may erode the streambed material and undermine the dam until failure occurs. The hydraulic jump, in many cases, is the most effective way of preventing this erosion, since it quickly dissipates excessive energy and reduces the water velocity to a point where it is incapable of scouring the streambed.

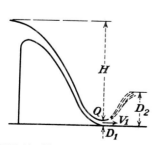

FIGURE 39 Elements of a hydraulic jump.

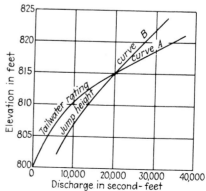

FIGURE 40 Tailwater rating curve versus jump height.

The simplest kind of protection could be used if a jump would form at all stages on a horizontal floor, at the streambed level, extending from the dam to the downstream end of the jump. The formula for the hydraulic jump in a horizontal channel of rectangular section is

$$D_2 = \frac{D_1}{2} \pm \sqrt{\frac{2V_1^2 D_1}{g} + \frac{D_1^2}{4}} \tag{17}$$

where, as shown by Fig. 39, D_1 and D_2 are the depths upstream and downstream from the jump and V_1 and V_2 are the corresponding velocities. A more complete discussion of the hydraulic jump on both horizontal and sloping aprons will be found in Sec. 2.

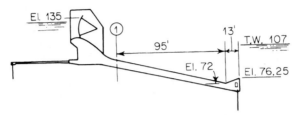

FIGURE 41 Sloping apron of Petenwell.

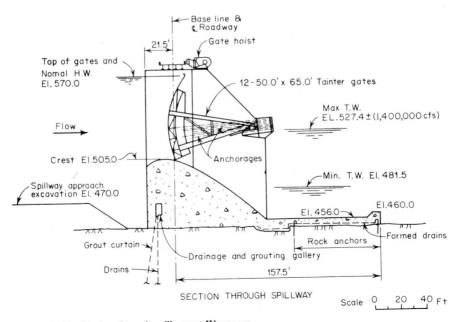

FIGURE 42 Section through spillway at Wanapum.

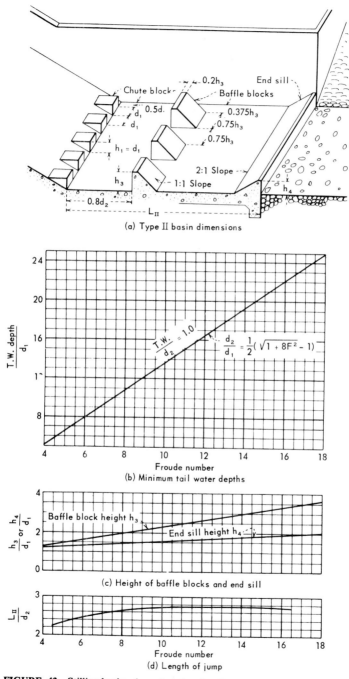

FIGURE 43 Stilling-basin characteristics for Froude numbers above 4.5 where incoming velocity does not exceed 50 fps.

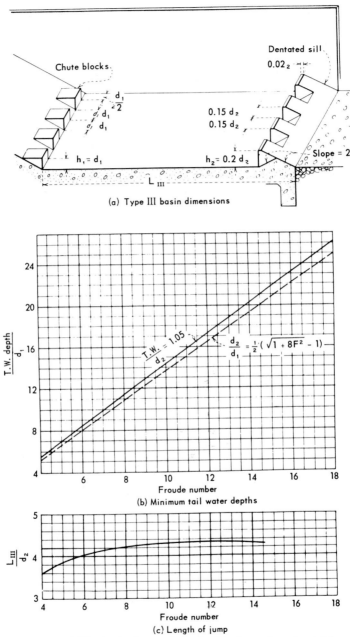

FIGURE 44 Stilling-basin characteristics for Froude numbers above 4.5.

The height of the tailwater for each discharge seldom corresponds to the height of a perfect jump. Spillway curves of jump height and of the tailwater rating are as shown by Fig. 40. The relations between the positions of these curves fall into four classes: (1) jump height curve above tailwater rating curve, (2) jump height curve below tailwater rating curve, (3) jump height curve above tailwater rating curve at low discharges and below at high discharges, (4) jump height curve below tailwater rating curve at low discharges and above at high discharges. The best form of protection depends largely on which of these four conditions exists.

In some cases, a sloping apron of the general type shown by Fig. 41 will permit a hydraulic jump to form at the proper depth within the limits of the apron throughout the entire range of spillway discharges and corresponding tailwater depths. It will be noted that the sloping apron terminates in a deflector sill which directs the discharge upward toward the surface of the tailwater where the residual energy is dissipated.

The Wanapum spillway shown in Fig. 42 is another type of streambed protection structure which has operated successfully under a wide range of discharge and tailwater conditions. Features include stilling basins formed by plane surfaces with deflector sills at the end of the hearth.

Model tests for the Wanapum project, completed in 1964 on the Columbia River by the Grant County Public Utility District No. 2, revealed that a relatively short hearth and low terminal sill would be adequate to deflect a large part of the discharge upward and dissipate the energy on the surface of the tailwater. These tests showed a high localized positive pressure at the junction of the ogee and the horizontal apron slab. Model tests indicated that the introduction of a curved bucket at this intersection required a substantial increase in the length of the hearth. Apparently a substantial amount of energy is dissipated at this point and part of the discharge is reflected upward to the surface. The residual flow is directed upward to the surface by the deflector sill.

As a guide to the design of stilling basins for small dams, the Bureau of Reclamation presents two charts in "Design of Small Dams,"[1] reproduced in Figs. 43 and 44. Figure 43 applies to Froude numbers above 4.5 where the incoming velocity does not exceed 50 fps. Where velocities exceed 50 fps, or where impact baffle blocks are not employed, the type of basin shown by Fig. 44 can be used.

It must be emphasized that these charts should be used only in the formulation of preliminary designs. The whole subject of streambed protection is a complex one, and in most cases, no single chart can offer a complete solution.

REFERENCES

1. "Design of Small Dams," U.S. Bureau of Reclamation, 3d ed., U.S. Government Printing Office, Washington, D.C., 1987.
2. Hydraulic Design Criteria, U.S. Army Corps of Engineers.
3. PULS, LOUIS G., Spillway Discharge Coefficients of Wilson Dam, *Trans ASCE,* **91** (316), 1931.
4. HAMILTON, W. S., "Aeration of Flow down Spillway" (unpublished memorandum), Harza Engineering Co., Chicago, February 1978.
5. WEI, C. Y., and F. G. DEFAZIO, Simulation of Free Jet Trajectories for the Design of

Aeration Devices on Hydraulic Structure, *Proceedings,* 4th International Conference on Finite Elements in Water Resources, Hanover, Germany, June 1982.

6. PAN, S. B., Y. Y. SHAO, Q. S. SHI, and X. L. DONG, The Self-Aeration Capacity of the Water Jet over the Aeration Ramp, *Journal of Hydraulic Engineering* (Shuili Xuebao, People's Republic of China), no. 5, 1980.
7. FALVEY, H. T., "Cavitation in Chutes and Spillways," U.S. Bureau of Reclamation, Engineering Monograph No. 42, April 1990.
8. Erosion of Concrete in Hydraulic Structures, *ACI Materials Journal,* American Concrete Institute, March-April 1987.
9. XU, W. X., and Z. X. GU, Two-Field Model: A Method to Simulate the Unsteady Free Surface Turbulent Flow with Irregular Obstacles, The Fourth Asian Congress of Fluid Mechanics, Hong Kong, August 1989.
10. DEFAZIO, F. G., and C. Y. WEI, Design of Aeration Devices on Hydraulic Structures, *Frontiers in Hydraulic Engineering,* American Society of Civil Engineers, 1983.
11. SUSLICK, K. S., The Chemical Effects of Ultrasound, *Scientific American,* February 1989.
12. PLESSET, M. S., and R. B. CHAPMAN, Collapse of an Initially Spherical Vapor Cavity in the Neighborhood of a Solid Boundary, *Journal of Fluid Mechanics,* **47**, part 2, pp. 283–290, 1070.
13. LIANG, Z. C., and D. Z. WANG, On the Carry Capacity in Turbulent Boundary Layer, *Journal of Hydraulic Engineering* (Shuili Xuebao, People's Republic of China), no. 2, February 1982.
14. BALL, J. W., Cavitation from Surface Irregularities in High Velocity Flow, *Journal of the Hydraulics Division,* ASCE, September 1976.
15. "Fibre Reinforced Concrete," ICOLD Bulletin 40, 1982.
16. "Spillways for Dams," ICOLD Bulletin 58, 1987.
17. GLAZOV, A. I., Calculation of the Air-Capturing Ability of a Flow Behind an Aerator Ledge, *Hydrotechnical Construction,* **18** (11), Plenum Publishing Corp., 1985, pp. 554–558.
18. FALVEY, H. T., "Air-Water Flow in Hydraulic Structures," U.S. Bureau of Reclamation, Engineering Monograph No. 41, 1980.
19. LIN, B. N., Z. GONG, and D. PAN, A Rational Profile for Flip Buckets of High Head Dams, *Scientia Sinica,* series A, **25** (12), China, December 1982, pp. 1343–1352.
20. Spillways, Shockwaves and Air Entrainment: Review and Recommendations," ICOLD Bulletin 81, 1992.
21. SEMENKOV, V. M., and L. D. LENTYAEV, "Spillway with Nappe Aeration," *Gidrotekhnicheskoe Stroitel'stvo,* **5**, May 1973.
22. HINDS, J., Side Channel Spillways, *Transactions ASCE,* **89**, 1926, p. 881. (Also appears in "Design of Small Dams," U.S. Bureau of Reclamation, 3d ed., U.S. Government Printing Office, Washington, D.C., 1987, p. 376.)
23. ZAGUSTIN, K., T. MANTELLINI, and N. CASTILLEJO, "Some Experience on the Relationship between a Model and Prototype for Flow Aeration in Spillways." *Proceedings,* International Conference on the Hydraulic Modelling of Civil Engineering Structures, Coventry, England, September 1982.
24. JIANG, SHUHAI, "Turbulent Boundary Layers in Vertical Curves," *Journal of Hydraulic Engineering,* ASCE, July 1988.
25. QUINTELA, A. C., "Flow Aeration to Prevent Cavitation Erosion," *Water Power & Dam Construction,* January 1980.
26. VOLKART, P., Transition from Aerated Supercritical to Subcritical Flow and Associated Bubble De-Aeration, International Association for Hydraulic Research, Proceedings of the XXIst Congress, Melbourne, Australia, August 1985.

REFERENCES

27. COLEMAN, H. W., A. R. SIMPSON, and L. M. DE GARCIA, Aeration for Cavitation Protection of Uribante Spillway, *Frontiers in Hydraulic Engineering,* American Society of Civil Engineers, 1983.
28. CASTILLEJO, N., and A. MARCANO, "Operation of Guri Dam Spillway during Raising of the Dam," Sixteenth International Congress on Large Dams, San Francisco, Calif., June 1988.
29. GULLIVER, J. S., A. RINDEL, and K. C. LINDBLOM, "Guidelines for Intake Design without Free Surface Vortices," *Waterpower*, Knoxville, September, 1983.

SECTION 17

GATES AND VALVES

By Chander K. Sehgal[1]

GENERAL

1. **Scope.** This section provides general guidelines pertaining to the selection, application, and design of gates and large discharge and shutoff valves. Usually more than one basic arrangement or type of gate or valve can be used for a given spillway or outlet. (The term *outlet* used in this section refers to the outlets from a dam and includes all outlets as well as penstock intakes and other intakes.) The selection must consider the overall arrangement, including the arrangement of the hoist, handling of the gate or valve for installation and removal for major maintenance, structural factors, operating requirements, and cost to determine the best possible technical and economic solution.

2. **Spillway Crest Gates.** Spillway crest gates represent the movable crest, which permits control of the flood stage above the dam. Many types of movable crests are in successful operation, including flashboards, radial gates, vertical lift wheel gates, flap gates including drum and bear trap gates, rolling gates (circular cross section), and stoney-type roller gates (with a roller chain sandwiched between the gate and the embedded track). The most popular modern types include radial gates, wheel gates, and flap gates. Each has a specific purpose as described in the related parts of this section. The gate type and size selection depends on several hydraulic, hydrological, mechanical, and control factors including the type of foundation, flood expectancy, the amount of debris in the flow, winter conditions, and operation requirements including the frequency of operation and the need for automatic control.

A spillway concentrates its discharge in a narrow width compared to the natural flood flow; consequently there is a large amount of energy to be dissipated. This provides a valid argument for a long spillway, with corresponding reduction in depth of overflow to reduce the energy to be dissipated per unit length, which is helpful especially where foundation erosion is a problem. Spillways that are too small may cause surface turbulence downstream, resulting in navigation problems or bank erosion. Cost of spillways including gates should be compared with the cost of a stilling basin or other types of protection downstream in finalizing the arrangement. Wide gates with fewer piers are preferable where heavy runs of drift, logs, or ice are expected, thereby avoiding jams caused by obstruction by the piers. Gates in freezing climates should ensure minimal leakage, and heating and/or ice removal should be provided for. Where pool levels must be closely maintained, automatic operation of the gates is usually necessary.

---

[1]Acknowledgment is made to Warren H. Kohler, James W. Ball, P. R. Mayer, and Joseph R. Bowman for material in this section that appeared in the third edition (1969).

3. Gates for Orifice Spillways. In many projects, economics will dictate whether the spillway is combined with another structure such as an irrigation release structure, low level outlet structure, diversion structure, or sediment passage structure. In such cases an orifice spillway is provided instead of a crest spillway. The gates for an orifice spillway are usually high-head gates, designed with the same considerations as outlet gates.

4. Outlet Gates and Valves. Outlet gates and valves serve to control flow from the outlets from a reservoir created by a dam so that the water stored can be used economically for irrigation, domestic use, flood control, navigation, and power. Gates designed for operation at heads greater than 50 ft are generally characterized as *high-head gates*. The high-head gates require special design considerations such as gate bottom geometry and aeration downstream to avoid cavitation damage. Gates used for flow regulation are usually slide gates, which, because of the continuous contact between the gate end plates and the embedded bearing plates, provide stability against vibrations.

The 5- by 10-ft slide gates (220-ft head) installed to regulate flows at Roosevelt Dam in Arizona in 1908 and at Pathfinder Dam in Wyoming in 1909 were designed without the special consideration required for high heads. These gates were seriously damaged by cavitation even though used for regulation at considerably less than the design head. In contrast, the 7- by 10.5-ft slide gates of modern design used for free discharge release of over 2 million acre-ft of water at nearly 340 ft of head at Glen Canyon Dam in 1965 have experienced relatively minor cavitation damage. Properly designed slide gates are also giving satisfactory performance in other installations at heads once considered too high for slide gates.

In addition to slide gates, valves are also used to control and regulate flow. The U.S. Bureau of Reclamation (USBR) Ensign-type needle valve, first used in 1908, proved to be better than the slide gates of that time for precision of regulation. However, the Ensign design did not eliminate cavitation problems; also, since the valve was designed for mounting on the face of a dam, valve repairs required drawing the reservoir level below the outlets, which led to the abandonment of most of the Ensign-type needle-valve installations.

A noncavitating needle shape was developed in laboratory testing, but at the cost of a considerable reduction in the discharge capacity as compared with the original divergent-cone type of needle valve. Pursuit of a cavitation-free valve led to the development of the tube valve, which is essentially a needle valve with the downstream conical needle portion of the closure member omitted. However, in some cases, eliminating the conical needle resulted in instability. The 102-in-diameter tube valves installed in the outlet at Shasta Dam, designed to regulate within a conduit, required very large air-admission ducts to contain instability. The high-cost problems of the Shasta tube valves resulted in studies of alternatives for the remaining 14 middle- and upper-tier conduits. The jet-flow gate was developed as a result, and the 96-in size has given satisfactory service at Shasta Dam.

Outlet gates include bonnetted and unbonnetted slide gates, jet-flow gates, radial gates, and wheel and roller (caterpillar) gates. The last two types, because of their relative instability at partial openings, are generally used for nonregulating service only. The most common outlet valves include needle valves, tube valves, butterfly valves, sphere valves, hollow-jet valves, and fixed-cone valves. Because of vibration and cavitation problems associated with outlet gates and valves, their selection must be based on past experience. Model tests should be performed for installations which exceed past experience.

5. Definitions. The terminology used in this section to describe the various types of closure devices is as follows:

Gate. A gate is a closure device in which a leaf (closure member) is moved across the fluidway from an external position to control the flow of water.

Valve. A valve is a closure device in which the closure member remains fixed axially with respect to a fluidway and is either rotated or moved longitudinally to control the flow of water.

Guard gates or valves. Guard gates or valves operate fully open or closed and usually function as a secondary device for shutting off the flow of water in case the primary closure device becomes inoperable. Guard gates are usually operated under balanced-pressure, no-flow conditions, except for closure in emergencies.

Regulating gates and valves. Regulating gates and valves operate under full pressure and flow conditions to throttle and vary the rate of discharge.

Bulkhead gates. Bulkhead gates are usually installed at the entrance and are used to unwater fluidways for inspection or maintenance; they are nearly always opened or closed under balanced pressures.

Stoplogs. The name *stoplogs* is often used interchangeably with the name *bulkhead.*

SPILLWAY GATES

6. Flashboards. Flashboards, the simplest of the movable-crest devices, are essentially flap gates which, at times of flood, may be removed by deliberate failure or by a tripping mechanism. They are very economical and are used only during the extreme floods. Flashboards provide an unobstructed crest when lowered. The wear and corrosion of moving parts, fouling with trash, and similar troubles could, however, cause the failure of a given scheme. It is therefore preferable that positively operated gates such as hydraulic-cylinder-operated flap gates be used in their place.

7. Radial Gates. A conventional radial gate consists of a curved skinplate supported by a framework of horizontal and/or vertical girders and stiffeners and by two or more radial struts (arms) which converge downstream to trunnions about which the skinplate and arms rotate while the gate is operated. The trunnions are supported by girders anchored in the piers and transfer the entire thrust of the water load to the anchorage. The skinplate is usually concentric to the trunnion pin; the resultant of the water pressure on the skinplate thus passes through the pin, and its moment about the pin centerline, about which all moments acting on the gate are calculated to determine the required hoisting force, is zero. Figure 1 shows the general arrangement of a radial gate. Typically the gate sill is located a short distance downstream from the crest; because of the slope downstream of the crest, this arrangement reduces the possibility of subatmospheric pressure on the crest. Often a wave deflector is provided (1- to 2-ft vertical height) on top of the gate to prevent waves from overtopping the gate. Flow splitters may be provided instead of wave deflectors to break the overtopping flow to disintegrate the flow energy and to aerate the flow nappe, but wave deflectors, although more expensive, are preferable, because overtopping may cause gate vibrations unless flow splitters are precisely placed.

Radial gates are the simplest, most reliable, and usually the least expensive type of crest gates, and thus have become the most widely used type. They require no slots in the piers and have good discharge characteristics.

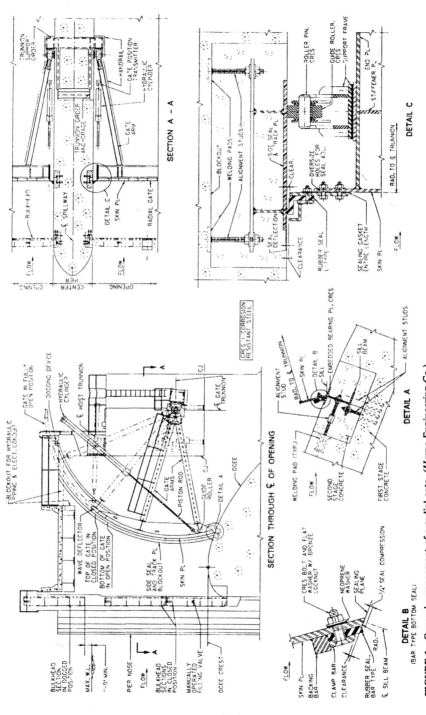

FIGURE 1 General arrangement of a radial gate. (*Harza Engineering Co.*)

The conventional radial gate may be altered for the passage of floating material by adding a flap to the top of the gate (Figs. 2 and 3b) or by making the gate submersible (Fig. 3c).

Radial gates may be operated by chains or wire ropes or by hydraulic cylinders. In modern design, chains are rarely used; wire ropes, seldom (because of maintenance costs); and versatile hydraulic cylinders, most frequently.

8. Flap Gates. This type of gate is a curved leaf hinged along its lower edge, usually quite wide and primarily used for passing floating material and/or for close regulation of reservoir elevation. Submersible radial and vertical gates may be used for the same purpose, but they have width limitations and require more maintenance. Flap gate hoists are usually arranged with hydraulic cylinders either on the sides or underneath the leaf, depending on the length of the gate. Cylinders arranged on the sides are preferable for ease of maintenance, but for the sake of economical structural design with long gates, the cylinders may be arranged underneath. Figure 4 shows the typical arrangements for flap gates. Several cylinders may be provided underneath a flap gate. Because fluid always flows to the line of least resistance, coordination of their motion will be automatic as long as the gate is structurally rigid. At the City of Waco's Lake Brazos Project in Texas, Harza Engineering Company designed the provision of 12 cylinders for each of the two existing drum gates (which were rehabilitated into automatically operating flap gates); automatic hydraulic coordination of these cylinders presented no difficulty.

In applications requiring close regulation of reservoir levels, operation of the flap gate is automated by actuation of the hydraulic system, usually by a reservoir-level-sensing device or a timing device. Flap gates, to be operated in partially opened positions, must be designed for the hydrodynamic effects of the overflowing water. If not properly designed and vented, severe vibration forces may occur. Gates provided for flow regulation require embedded sealing plates on the sides throughout the full travel of the gate (similar to the side sealing plate shown in Fig. 2). Because of their size, these plates add substantially to the cost of the flap gates. Gates provided only for passing of floating materials may not have extensive side-sealing plates.

Drum gates and bear-trap gates are different versions of flap gates. Both types are hoistless and use headwater pressure to automatically operate the gate. In hoistless gates, the balance of forces needed for automatic operation is difficult to maintain because of the effects of corrosion and silt accumulation. Both of these types are being increasingly eliminated on new projects in favor of regular flap gates which can be positively operated by hydraulic cylinders.

9. Vertical-Lift Gates. Vertical-lift gates are rectangular gates supported by vertical guides. Vertical gates for spillways normally have the skinplate and seals on the upstream side of the framework to minimize the effect of the gate slot on the flow. The gate hoist may be hydraulic, screw-stem, or wire-rope type. For spillway applications, an overhead structure above the deck level is usually necessary so that a vertical-lift gate can be raised above the maximum level. Figure 5 shows the typical arrangement of a vertical-lift spillway gate.

Vertical-lift gates may be slide type or fixed-wheel type. Slide gates have their frames directly bearing against the plates (usually stainless steel) embedded in concrete, and therefore the sliding friction to be overcome in operating the gate is very high. The high operating force limits the maximum water size of the gate. Slide gates in spillway applications are limited to small spillways, wasteways, log flume inlets, and similar applications, and in these cases slide gates are particularly suited for passing the discharge over the top.

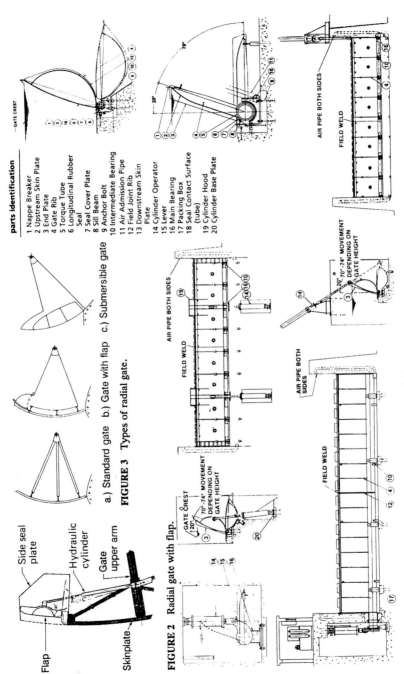

FIGURE 2 Radial gate with flap.

FIGURE 3 Types of radial gates.
a.) Standard gate b.) Gate with flap c.) Submersible gate

FIGURE 4 Typical arrangement for flap gates. (*Rodney Hunt Co.*)

parts identification

1 Nappe Breaker
2 Upstream Skin Plate
3 End Plate
4 Gate Rib
5 Torque Tube
6 Longitudinal Rubber Seal
7 Seal Cover Plate
8 Sill Beam
9 Anchor Bolt
10 Intermediate Bearing
11 Air Admission Pipe
12 Field Joint Rib
13 Downstream Skin Plate
14 Cylinder Operator
15 Lever
16 Main Bearing
17 Packing Box
18 Seal Contact Surface (tube)
19 Cylinder Hood
20 Cylinder Base Plate

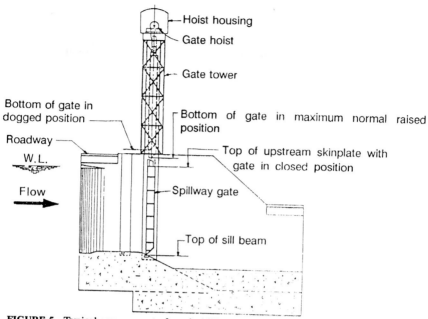

FIGURE 5 Typical arrangement of vertical-lift spillway gate. (*Manitoba Hydro.*)

Fixed-wheel gates have a series of wheels mounted along the height of the gate on each side. The wheels carry the water load to a vertical track supported by embedded bearing plates on the downstream side of the gate slot. The wheels substitute the sliding friction of the slide gates with rolling friction and thus help reduce the hoist capacity required to raise the gate. The reduced friction also helps the gate to self-close under its own weight. For passage of floating material and to reduce the headroom for hoist installation, a fixed-wheel gate may be designed as follows:

1. Split the gate horizontally into two or more sections that travel in the same guides. The topmost section is of a height suitable for surface spilling. The sections are progressively raised or removed from the guides, depending on the upstream water level. For example, in a gate with four sections, the two top sections may be removed from the guides and the third section raised to allow water to discharge under it and over the bottom section. This method also considerably reduces the hoisting load.

2. Divide the gate into two leaves arranged so that the upper leaf may be lowered alongside the lower leaf to allow flow over the top of the gate. Both leaves are equipped with separate guides and hoists and may be operated as necessary for flow regulation.

Although vertical-lift gates are simpler to construct and install and, unlike radial gates, do not require trunnion support girder anchorages embedded in long piers, radial gates are more economical for most spillway applications, especially with regard to the hoisting arrangement. Radial gates are also more suited to cold regions because there are no slots, which, in vertical gates, can become lodged with ice, requiring considerable heat to free the gate and the guides. Modern spillway gates are of the radial type. Vertical gates are usually more suitable for spillways with high tailwater level; raising the trunnions of a radial gate above the tailwater level to protect them from debris and freezing could make a radial gate more uneconomical. Vertical-lift gates are mostly used in high-head applications.

10. Bulkheads. Bulkheads provided for maintenance of spillway gates are essentially similar to the bulkheads for outlets, described in Art. 19. The same description applies except as follows:

1. Storage of spillway bulkheads when they are not in use is an important issue to be considered in the design of a spillway. Unlike the bulkheads of an outlet structure, the spillway bulkhead cannot usually be stored (dogged) at the top of the bulkhead slot. This is so because the bottom of the bulkhead (or a bulkhead section) would interfere with the flow (the spillway deck usually being only a few feet above the maximum water level), unless there is a right combination of the number of spillway gate openings and the size and number of bulkhead sections. Figure 1 shows a case where it is possible to store the bulkhead sections at the top of the slot.

2. If a filling valve needed to fill the space between the spillway bulkhead and the spillway gate is to be mounted on a bulkhead section (a separate fill line could be provided instead), it must be mounted on the lowermost bulkhead section to cover the full range of water-level variation above the spillway crest. For outlet bulkheads, the filling valve must be provided on the topmost section so that it can be actuated by a lifting beam. The valve on the spillway bulkhead bottom section is usually opened manually (see A-A, Fig. 1); therefore, there must be adequate space between the bulkhead and the gate for access by the operator.

11. Inflatable Gates. Inflatable gates are usually made of synthetic fiber, single- or multilayer fabric, rubberized on one or both sides or coated with plastic film. The rubber or plastic encasement is pressurized with fluid, air, or both. Materials have vastly improved since the first inflatable gates were used in 1950s, including rubber with better resistance to air, sunlight, and abrasion and better rubber-to-fabric bonding materials. When deflated, the gate lies flat on the side. The inflatable gates can be made in very large widths (400 to 500 ft); are very convenient for passage of silt, debris, and ice; and are very economical. The disadvantages include limited design head and susceptibility to damage, especially by vandalism.

Figure 6 shows typical arrangements of inflatable gates. The gate encasing is anchored to the sill with one (Fig. 6a) or two (Fig. 6b) lines of anchorage, and the gate is supported in the raised position by the inflation pressure. The inflation pressure may be applied by gravity or by a pump or compressor. Automatic deflation is provided in many cases using a counterweight to open an exhaust valve to deflate the gate when the reservoir water level reaches a predetermined deflation point. Figure 6c shows an inflatable gate with an outer casing in addition to the inner casing to provide protection to the inner casing. Figure 6d shows an anchoring arrangement which includes metallic wickets.

HIGH-HEAD GATES

12. General. Figure 7a shows schematically some of the typical installations of high-head gates. Figure 7b and 7c show typical gating arrangements. Service

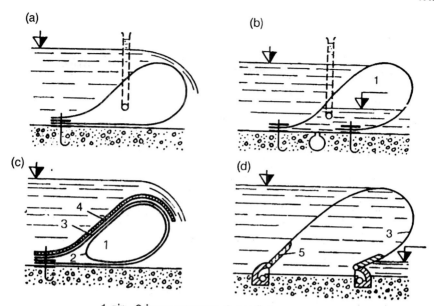

1 air; 2 inner encase; 3 confining encase;
4 protection shell; 5 metallic wicket

FIGURE 6 Schematic examples of fabric gate designs. (*Hydraulic structures* by M. M. Grishin.)

gates must be capable of being raised and lowered under the maximum flow conditions (full unbalanced head). Guard gates also must be designed to close under the full head and maximum flow, but are normally operated under balanced-pressure no-flow conditions; a fill line is usually provided for balancing the head across the guard gate. The fill line is sized for the desired time for filling the conduit and is usually between 3 and 6 in in diameter. To shorten the time to balance the pressure across the gate, the guard gate may be designed for crack opening (usually 3 to 4 in) against full head. Bulkheads are provided at the upstream end of an outlet to permit inspection or repair of the gate and conduit; they are designed to be placed and removed under balanced head. Bulkheads are usually in sections to reduce the hoisting capacity and are normally handled by a mobile crane. One set of bulkhead sections can be used for several outlet openings. For balancing the pressure across the bulkhead, either a fill line or a filling valve is provided on the top bulkhead section. The filling valve is operated by the weight of the lifting beam or by the hoist pull.

Over the years, many types of high-head gates have been designed and built, but only a relatively few types have survived; those surviving are rugged, easy to maintain, and economical to build. Slide gates, radial gates, and jet-flow gates are the only ones which are specifically used for throttling. Wheel, roller-mounted, and cylinder gates are normally used only as fully opened or closed guard gates because, unlike slide, radial, and jet-flow gates, they are not supported continuously and therefore are subject to vibrations. They may, however, be used for regulation under low-head conditions with due consideration given to gate bottom

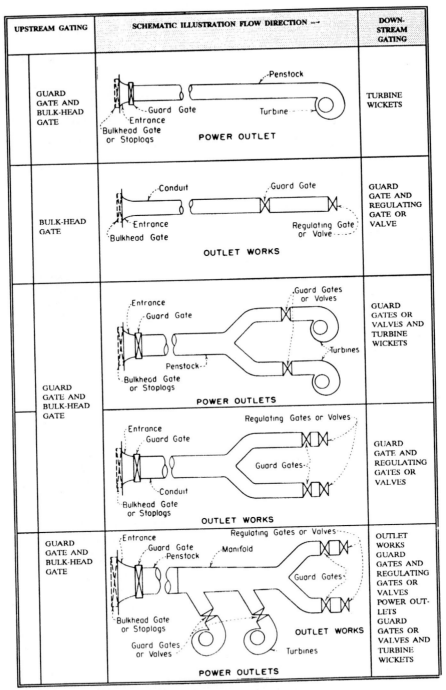

FIGURE 7a Schematic of typical outlet arrangement.

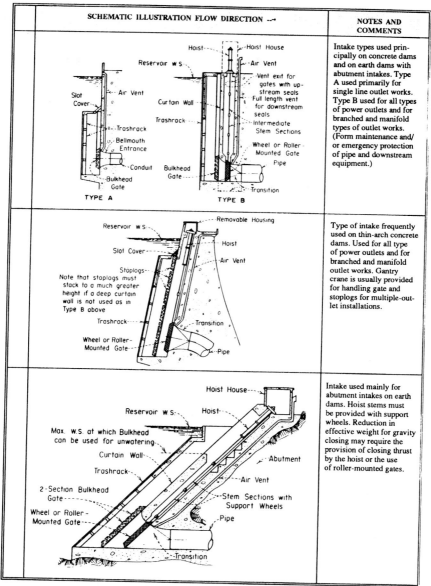

FIGURE 7b Typical intake gating arrangement.

shape and slot size as related to the hydrodynamics of flow. Bulkhead gates (or stoplogs) are never used for regulation.

13. Slide Gates. Slide gates are normally used for regulating the flow through low-level outlet works. Because of high friction forces associated with their operation, they are not used for power outlets, which need wheel- or roller-mounted

17.12 GATES AND VALVES

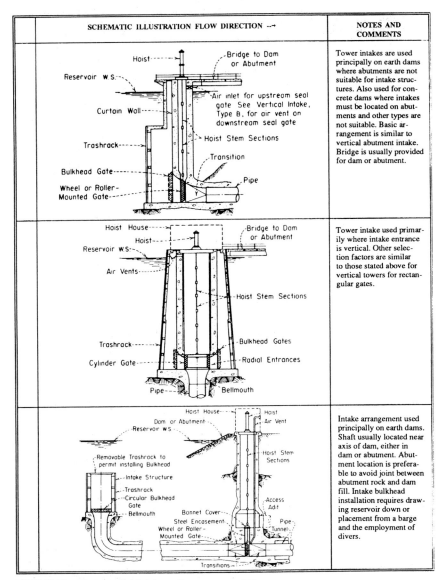

FIGURE 7c More typical intake gating arrangements.

gates for emergency closure by gravity (without power) to safeguard the generating unit against runaway. Slide gates may be installed either near the entrance of the outlet conduit (using a gate shaft for access) or within the conduit (using galleries for access and a high-pressure bonnet). Selection of the type of installation depends on cost studies including the cost of the gate, accessories, and civil structure. Structures with bonnetted gates are generally more economical for heads above 75 ft. Figures 8 and 9 show typical installations. The gate of Fig.

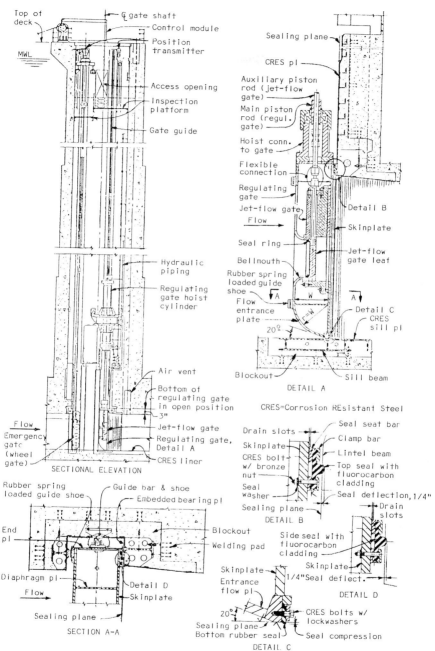

FIGURE 8 Unbonnetted slide gate. (*Virginia Power/Harza Engineering Co.*)

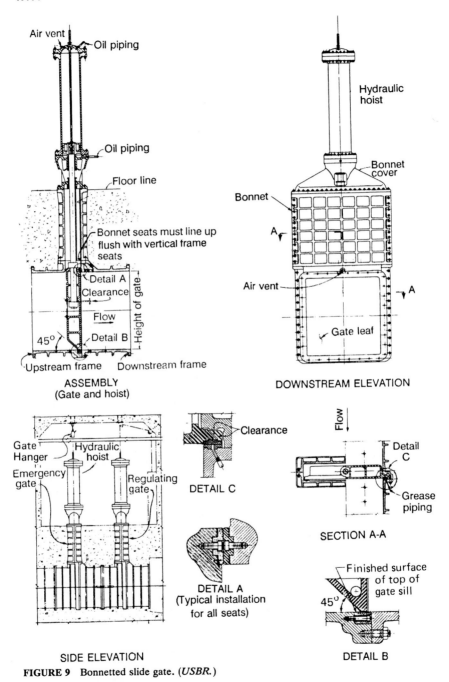

FIGURE 9 Bonnetted slide gate. (*USBR.*)

8 includes a jet-flow gate within its body so that the main gate does not have to be opened for small flows.

Frequently, for slide gates located in conduits, identical bonnetted gates are bolted together in tandem (the upstream gate functions as the guard gate for the downstream regulating gate) as shown in Fig. 9; this arrangement is generally more economical than using a different type of guard gate (a gravity-closure wheel gate would be preferable, but difficulty of access for maintenance is a deterrent). For slide gates located in gate shafts, the guard gate is usually a gravity-closure wheel gate. Some gate installations are made on a slope so that the discharge is downward into the stilling basin. This arrangement reduces the required length and cost of the stilling basin.

Bonnetted slide gates have been successfully used with only minor cavitation damage at heads up to 500 ft. At heads above 200 ft, fluidway surfaces and the bottom seating and sloping surfaces of the gate leaf should be protected with stainless-steel linings for improved cavitation resistance.

The slide gate arranged in a gate shaft is usually a conventional gate leaf with a hydraulic hoist located just above the gate's fully open position; the hydrostatic load is transferred by the sliding surfaces of the gate end plates to the bearing plates embedded in concrete (Fig. 8). For downstream sealing gates, an air vent is provided a short distance downstream of the gate to aerate the conduit and the gate; for upstream sealing gates, the gate shaft acts as the air vent. The gate leaf is made of structural steel (ASTM A-36 or A-572) and is equipped with rubber seals. Frequently, to reduce sliding friction, fluorocarbon pads are attached to the gate's endplates. The use of pads, however, depends on the design head and the resulting bearing pressure on the pads. For heads up to 50 ft, standard AWWA (Specification C501) cast-iron gates with bronze seals and either screw-stem or hydraulic hoists can also be used.

The bonnetted slide gate includes a gate leaf, a body consisting of upstream and downstream halves, and a bonnet made in halves. The leaf typically has bronze seals and opens by being withdrawn into the bonnet by a hoist mounted on the bonnet cover. The mating seats on the gate leaf, body and bonnet serve as the sliding surfaces as well as the sealing surfaces. The body and bonnet (except for bonnet top flange and cover) are not designed to withstand the internal pressure; the load is assumed to be transferred to the surrounding concrete. The body and the bonnet are heavily ribbed to minimize distortion when they are embedded in concrete. The bonnet cover and flange connections must resist the full internal pressure as well as the full load of the maximum hoisting effort as the gate leaf contacts the bottom seat in the closed position.

Cast steel has been commonly used for bonnetted gate leaves; however, structural steel weldments are becoming increasingly popular. Similarly, bonnets and bonnet covers made of weldments are becoming more popular because of economy and material quality. Air-inlet manifolds are usually provided on the downstream side of the gates when the gates are located at some distance upstream from the outlet end of the conduit. For guard gates which are directly coupled or are located very close to the regulating gate or valve, air manifolds can usually be omitted; however, a small manually operated vent is always provided to permit release of air trapped during filling of the conduit between the guard gate and the regulating gate or valve. The air vent will avoid vacuum and reduce the noise and vibration during closure against flow. Gate seats (which also act as gate seals) of bonnetted gates should be provided with a lubrication system to reduce sliding friction and wear. Bearing pads are not preferred because of difficulty of access and maintenance.

For cavitation-free operation of the slide gates, the fluidway should be smooth (250 μm or better), straight, and without offsets. Gate slot and gate bottom geometry should be carefully selected to provide a converging fluidway and definite spring point for flow discharge and to help reduce the hydraulic downpull caused by the separation of flow at the gate bottom. The gate slot should be as narrow as possible and steel liners provided upstream and downstream of the gates to protect the concrete from erosion due to high-velocity turbulent flow. The gate leaf should be installed very accurately with respect to the conduit and the body. Slide gates can be used either for free discharge into the atmosphere or for submerged discharge in water. The latter case requires more care to be sure adequate water can circulate and flow readily to the critical regions around the gate orifice. Laboratory model tests are very helpful.

Slide gates should not be operated at openings so small that flow beneath the gate does not spring clear of the lip on the bottom of the gate. At small openings, the short-tube effect of the issuing flow may result in flow contact at the downstream edge of the gate lip and produce cavitation damage on the bottom sealing surfaces of the gate. The minimum opening for regulation is usually limited to not less than one-half the width of the seating lip on the bottom of the gate.

The bottom of downstream sealing gates usually has a slope of about 45° (Fig. 9) to provide a convergence of the flow passage when the gate is partially open for regulating flows. The sloped surface has positive pressures which will reduce downpull and ensure positive control at the spring point at the bottom of the leaf. A curved surface formed at the bottom with radius of curvature equal to gate thickness and a slope of about 20° at the bottom (Fig. 8) has also been used. Figure 8's arrangement will also improve the coefficient of discharge for partially open regulating gates. Upstream sealing gates are usually sloped 35° on the downstream side so that the flow jet clears the gate body on the downstream side, ensuring aeration of flow. All values of slope and radius are empirical data based on experience and model studies.

14. Ring-Follower Gates and Ring-Seal Gates. Prior to 1930, bonnetted slide gates were commonly used as guard gates for needle valves which regulated the flow from outlet works. This required transition of the fluidway from circular to rectangular for slide gate installation and again to circular for the needle valve, adding expense and hydraulic losses to outlet works. The ring-follower gate corrects this problem.

A ring-follower gate is similar to the bonnetted slide gate except that it includes two bonnets, one below and one above the conduit; the leaf is composed of a bulkhead portion, which blocks the conduit when the gate is closed, and a follower portion, which has a circular opening of the same size as the conduit and aligns concentrically with the conduit when the gate is open. The follower sits inside the lower bonnet when the gate is closed.

A ring-seal gate is a ring-follower gate which has roller trains and wheels to reduce friction and hoisting capacity, can be opened and closed against unbalanced head, and has a movable, hydraulically actuated rubber seal ring.

Ring-follower gates have been primarily used as guard gates for needle valves, and ring-seal gates, as guard gates for turbines. There are practically no hydraulic losses with this type of gate because the fluidway closely matches the conduit diameter and avoids pronounced boundary discontinuities. Because of their shape, these gates are not suitable for throttling.

15. Jet-Flow Gates. The jet-flow gate was developed in the mid-1940s by the U.S. Bureau of Reclamation to provide a better and less costly alternative to the tube valves originally scheduled for installation at Shasta Dam. The fundamental features of the jet-flow gate (Fig. 10) are the truncated conical nozzle, a floating seal ring which forms a circular discharge orifice at the downstream end of the nozzle, and a leaf which contacts and is moved across the seal-ring orifice to regulate flow discharges. A contracted, jet-type discharge is produced which is responsible for the name of the gate.

Some of the notable jet-flow gate installations around the world include 96-in gates (225-ft head) at Shasta Dam, 96-in gates at Bhakra Dam in India, 90-in gates (430-ft head) at Roosevelt Dam, an 84-in gate at Tumut Pond Dam in Australia, an 84-in gate (400-ft head) at Trinity Dam in California, 77-in gates (150-ft head) at Canyon Ferry Dam, 68-in gates (610-ft head) at Hoover Dam, 60-in gates (230-ft head) at Echo Dam in Utah, a 54-in gate (250-ft head) at Kangneung Project in Korea, and two 22-in gates (150-ft head) at Bath County Pumped Storage Project in Virginia. The design at Bath County (Fig. 8) is unique because the jet-flow gate is built inside the 6- × 9-ft regulating slide gate for economic reasons. As with most small-size jet-flow gates, this gate is used to meet minimum stream-flow requirements and avoids the damage that would occur if the larger 6- × 9-ft gate was crack-opened for small discharges.

The simplicity and excellent flow-regulation characteristics of jet-flow gates have resulted in the development of standard designs in 10-, 12-, and 14-in sizes by the Bureau of Reclamation. Bureau Reports MECH-3 and HYD-569 provide the construction details and the calibration of a 10-in jet-flow gate.[7] Jet-flow gates operate smoothly without vibration or serious cavitation damage at any opening for most applications; under free-discharge conditions, there is considerable air demand at partial openings.

A common version of the jet-flow gate is shown in Fig. 10. It consists of a leaf with a sloping bottom (to provide for better aeration at partial openings), the fundamental features stated above, a body consisting of upstream and downstream halves, and a body cover on which the operator is mounted. The fluidway just upstream of the jet spring point is the frustum of a 45° cone which forms a nozzle and causes the discharging jet to contract and spring free of the gate leaf slots. The smaller diameter of the nozzle forms the jet orifice. The larger diameter of the nozzle is made at least 1.2 times the smaller diameter so that proper contraction of the jet will occur. If the upstream conduit diameter is smaller than the larger diameter of the jet nozzle, a conical diffuser which will expand the size to the required diameter must be provided. The 45° conical nozzle should be machined to provide a hydraulically smooth and accurate orifice boundary to assure a definite spring point on both the conical nozzle and the bottom of the leaf. The leaf, body, and body cover are usually made of structural steel, except that the upstream face of the gate leaf which is in sliding contact with the seal seat is made of corrosion-resistant material. The sealing ring is made of bronze.

As only the circular upstream fluidway is under reservoir pressure, heavy reinforcement is usually not required around the body. The pressure in the interior spaces of the body is usually atmospheric. The body cover, however, will be subject to the gate hoisting forces. An adequate number of ribs must be provided on the body if the body is not embedded in concrete (for jet-flow gates located on the end of a conduit) to enable transfer of the load on the fully closed gate to the conduit without deflection of the upstream portion of the body.

Jet-flow gates may be installed within a conduit or at the downstream end of a conduit (Fig. 10). An opening 25 to 30 percent larger than the jet orifice is provided in the downstream body for adequate aeration of the jet. For gates located

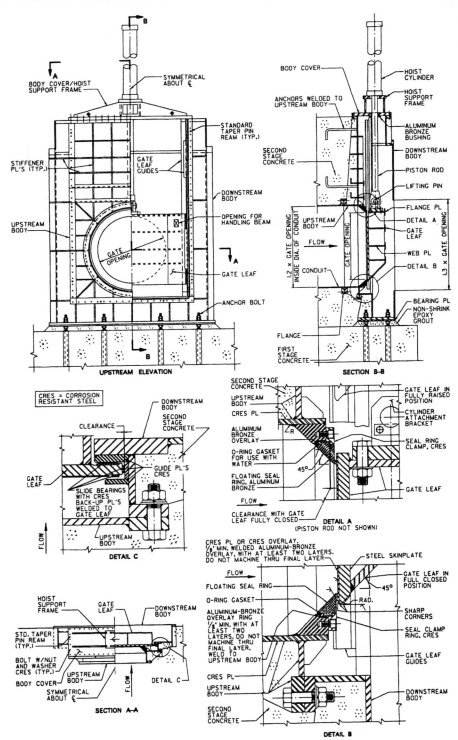

FIGURE 10 Jet-flow gate mounted at the downstream end of a conduit.

within a conduit, the diameter of the downstream conduit follows the same criteria. Shaping and sizing the downstream conduit to be certain that adequate air can be delivered around the entire periphery of the jet are mandatory.

Usually, the upstream body and the connection points between the upstream and downstream bodies are embedded in concrete. The bodies, therefore, are not separable once embedded; the opening in the downstream body should be sufficiently large so that maintenance work on the floating seal ring, including its removal for replacement when necessary, is possible through this opening.

16. Wheel- and Roller-mounted Gates. Wheel-mounted gates (fixed-wheel gates) and roller-mounted gates (also called *tractor, coaster,* or *caterpillar* gates) normally serve as primary shutoff (guard) gates for a conduit or penstock and are usually designed to close by gravity. Compared to the roller-mounted gates, wheel-mounted gates are simpler and more economical to build. As the gate size and head increase, structural constraints (the maximum wheel load and maximum number and size of wheels that can be accommodated on the gate) favor roller-mounted gates.

Because of the lower friction of roller trains and their high load-carrying capacity, the gravity closure criterion also favors roller-mounted gates for high heads. However, because of the complexity of their construction and associated maintenance problems, roller-mounted gates are rarely used.

Wheel- or roller-mounted gates are usually installed at the entrance of conduits or penstocks, at the face of a dam, or in intake towers located in a reservoir. Typical wheel gates are shown in Figs. 11 and 12 and a typical roller gate in Fig. 13a. Dam face installations are usually feasible only for a concrete dam (too costly for earth-fill dams, in which case gates are installed in concrete-lined gate shafts).

Wheel- or roller-mounted gates are usually used only in the fully open or closed positions except that gates for penstocks are usually also opened a crack (3- to 4-in opening) to fill the penstock. They are suspended just above the intake so that closure can be made rapidly in case of an emergency. Hydraulic hoists are preferable (see Art. 38) for operating wheel or roller gates, but mechanically operated wire-rope hoists can also be used. The maintenance of wire rope and other mechanical components and lack of protection against uplift forces due to hydrodynamic effects of emergency closure against flow are the major disadvantages of wire-rope hoists. The closure speed for wheel- and roller-mounted gates depends on the protection requirement of the downstream equipment; usually a maximum closure time of 3 to 5 min is provided for protection against turbine runaway. Opening speed is not critical. In some cases, gantry cranes have been used to handle the emergency closure of the gates. This arrangement can be very economical, especially for a project with large number of gate openings. For gates used as guard gates for generating units, the gate closure time must be carefully compared to the turbine runaway situation before gantry cranes are chosen.

Large wheel or roller gates are made in sections because of shipping limitations. The sections are connected such that the skinplate splice acts as a hinge to permit the upper and lower sections to rotate slightly and accommodate minor deviations in the straightness of the tracks. The articulation avoids the possibility of high bending stresses in the vertical plane at the joint and also reduces the possibility of excessive load on the wheels or rollers by better wheel or roller contact with the track. In some instances, rubber pads are provided under the tracks to make the tracks flexible and help equalize the load on the wheels or rollers, but maintenance of these pads is costly. In wheel gates on which each section can be designed with a maximum of four wheels, two on each side, free rotation at the

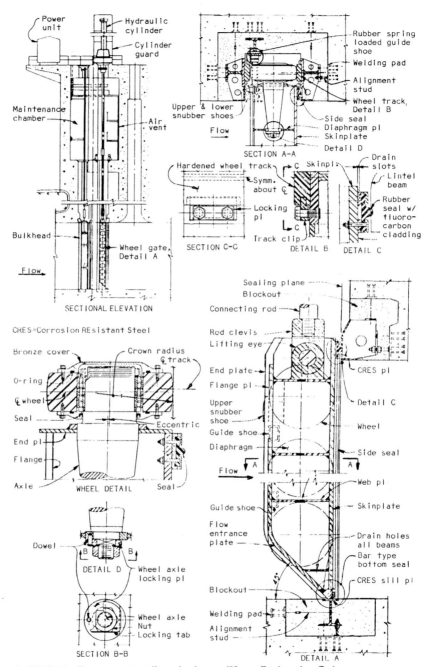

FIGURE 11 Downstream sealing wheel gate. (*Harza Engineering Co.*)

HIGH-HEAD GATES

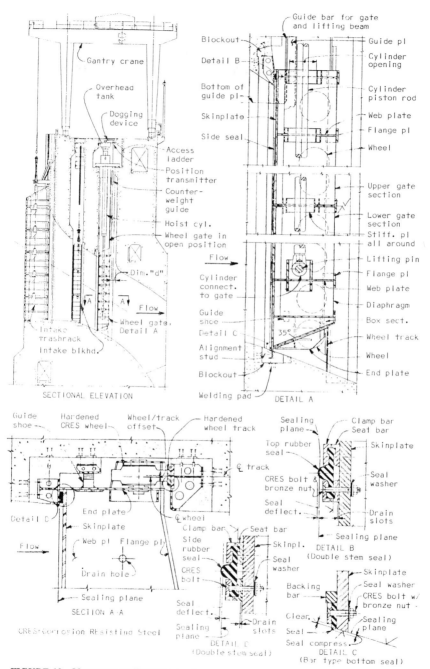

FIGURE 12 Upstream sealing wheel gate.

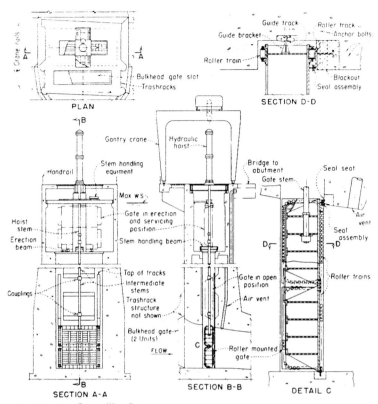

FIGURE 13a Caterpillar Gate.

hinges between the sections is usually provided because, unlike sections with more than four wheels, free rotation of the sections does not hinder contact of all wheels with the track.

As with slide gates, the wheel- or roller-gate skinplate and seals can be provided on the upstream or downstream side. Unless the downstream side of the gate is submerged, hydrodynamic downpull on a gate with an upstream skin plate and seals is not significant, because of the aeration provided by the gate shaft; lower downpull means lower hoisting capacity and hoist cost. Upstream sealing gates submerged on the downstream side and downstream sealing gates can have substantial downpull unless the gate leaf bottom is carefully designed to avoid flow separation (gate bottom shapes for wheel and roller gates generally follows the same criteria as the slide gates) or air is forced into the area below the gate bottom beam. However, though the downpull increases the hoisting capacity, it helps gravity closure, which can be critical for the emergency lowering of the gate. In making a decision regarding upstream or downstream sealing, it should also be noted that the wheels of a downstream sealing gate, being exposed to silt and debris, may get easily clogged. On upstream sealing gates, uplift forces acting on the upstream projection of gate top seals and on downstream projection of the gate members because of the hydraulic transient conditions (caused either by the

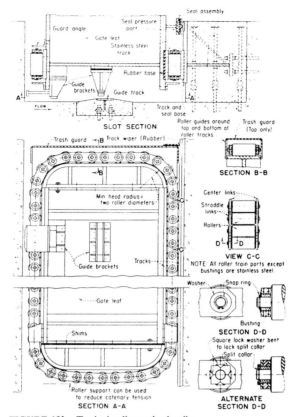

FIGURE 13b Typical roller-train details.

gate overshooting the crack-opened position or by sudden closure of the turbine wicket gates once the gate has started to lower against full flow) can prevent gate closure and must be carefully evaluated. Dimension d (Fig. 12) can be increased to minimize the uplift forces, but that would result in increased cost of the overall structure. Because there is minimal downstream projection on the downstream sealing gates, the effect of the downstream transient conditions on those gates is minimal. Among the advantages of the upstream sealing gates is the use of gate shaft as air vent (downstream sealing gates require separate air vents).

Wheel and roller gates are made of structural steel (ASTM A-36 or A-572). Gate wheels, rollers, tracks, and roller gate links are usually made of ASTM A-276, type 410 martensitic stainless steel which can be hardened above 300 Brinell hardness number (BHN) to provide safe contact stresses within limited size of the wheels or rollers. Solid stainless-steel wheels or cast-steel wheels with a stainless-steel shrunk-fit rim may be used. The wheels or roller axles are usually made of high-strength ASTM A-564, type 630 (17-4 PH) steel to keep their size minimal. The use of roller bearings (low friction) instead of sleeve bearings increases the feasible heads for wheel-mounted gates, but self-lubricating sleeve bearings are preferable because of the negligible maintenance required for them.

Also in projects where silt is present, roller bearings could become clogged, creating gate operating problems. Wheels are usually arranged eccentric to axles to facilitate adjustment at the site for contact with the track, and are crowned so that proper contact between the wheel and the track is maintained when the gate bends under water load. Wheel tracks are made at least 20 BHN harder than the wheels; wear will then occur on the wheels which are more accessible for replacement. The embedded frames on which the seal seats and tracks are mounted must be made and installed to close tolerances, especially for upstream sealing gates which have seals on the upstream side and wheels or rollers on the downstream side.

Gate slots for wheel- and roller-mounted gates do not usually pose any hydraulic design problems because the gates are not ordinarily used for throttling except as required occasionally for filling a conduit or penstock.

17. Cylinder Gates. Cylinder gates provide a relatively simple and effective installation for vertical intakes primarily as shutoff (gravity closure) gates for conduit and penstock intakes and sometimes for regulating discharges.

A cylinder gate is composed of a cylindrical shell which is raised and lowered to control flow through radial openings in the gate (Fig. 14a). Cylinder gates may be located on the inside or the outside of a circular intake structure, but an inside location is preferable for maintenance reasons. Seals at the top and bottom of the cylinder make contact with mating seats when the gate is closed. Up to three equally spaced screw stems have been used for hoisting; with proper gate guidance, however, a single hoisting stem is usually satisfactory. Hydraulic hoists have also been used.

There is very little operating friction for a cylinder gate except for minor rubbing on the guides. Because of the absence of the damping effect of friction, sometimes there can be strong vertical vibrations when flow is throttled at small openings. The tendency to vibrate is intensified when the gate is nearly closed, where the downpull forces can become unstable with rapid changes in magnitude. To alleviate this condition, the spring point on the bottom of the gate must be well-defined (Fig. 14a, detail D) with assurance of adequate fluid circulation to the spring point. Enlargement of air vents to reduce vibrations is not always successful. Large amounts of air entrained in the discharge may result in unstable surging when the air is released in the downstream tunnel. Figure 14b shows the clamping cylinders used at Tolt River to control vibrations.

18. Radial Gates. High-head radial gates are similar in design to the radial gates for spillway crests except for the top seal, which is not required for crest gates. Figure 15 shows the arrangement of 40-ft-wide × 20-ft-high radial gates (40-ft head) used at Cushman No. 2 Dam. Note two top seals, one fixed to the gate skinplate and the second fixed to the embedded lintel beam to provide sealing at the top when the gate is partially open. The top seal fixed to the gate is essential so that the gate can properly seal when it is fully closed, because it is very difficult to obtain perfect sealing between the side seals, which must be installed on the gate, and the top seal, fixed to the embedded lintel. The flat bottom seal fixed to the sill is expected to extend seal life; it also permits a sharp point at gate bottom for high-velocity throttling flow at high heads. Radial gates can be designed for very high heads, limited only by the capacity of the piers to withstand loading caused by trunnion supports and by the increased cost of the structure because of the large space needed for installation and operation.

19. Bulkheads. Bulkheads or stoplogs are essentially slide gates except that, because they are placed and removed under no-flow conditions, the bottom shape is not critical. Figure 16 shows the typical arrangement of a two-section

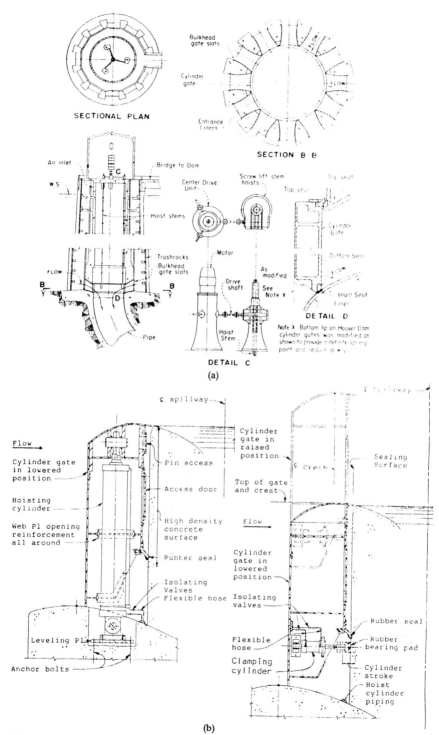

FIGURE 14 (*a*) Typical cylinder gate. (*b*) Tolt River cylinder gate.

17.26 GATES AND VALVES

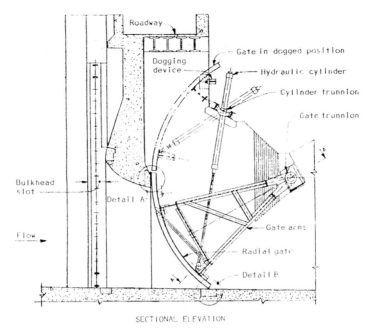

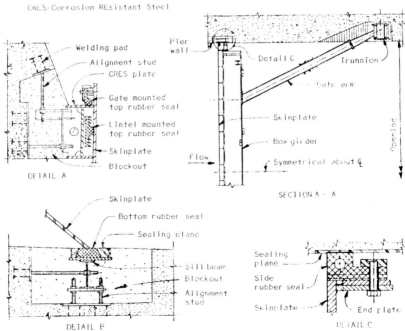

FIGURE 15 Cushman Dam top sealing radial gate. (*City of Tacoma/Noell/Harza Engineering Co.*)

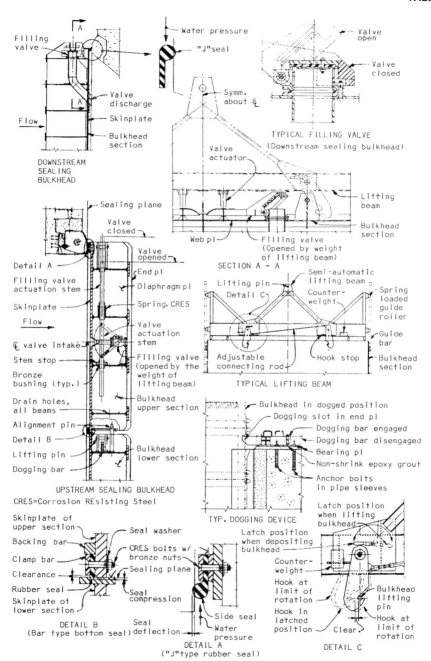

FIGURE 16 Typical bulkhead with lifting beam.

bulkhead including the seal detail between the sections and the essential features of the lifting beam and the bulkhead-installed filling valve. The lifting beam is equipped with an automatic latching device to permit connecting and disconnecting the bulkhead sections under water. In some cases, for economy, a common slot is used for the intake trashracks and the maintenance bulkheads; the trashracks are removed from the slot whenever the bulkhead is to be placed in the slot. The lifting beam is then designed for use for both the bulkhead and trashrack. In some installations, the bulkhead slot upper end may be submerged (Fig. 7b); in such cases, a cover is usually provided over the bulkhead slot. The covers are either removed with the lifting beam or with lifting ropes attached to the covers by divers.

Bulkheads for high-head installations should be designed with the same integrity as the service gates. Bulkheads are handled either by a mobile crane or a gantry crane. Gantry cranes are usually more economical if there are several bulkhead openings involved; they often have to be provided when a mobile crane of proper capacity is not available in the vicinity of the project site. If a mobile crane is to be used, handling studies must be performed to verify space requirement on the deck for stationing and maneuvering a mobile crane. Bulkhead design studies should include design of their storage facility. Bulkheads for high-head installations are usually stored by dogging at the top of their slots.

VALVES

20. General. Large discharge and shutoff valves serve the same purpose as the gates, but, because of their flow path configurations, can be designed for much higher heads (up to 1000 ft for throttling valves and up to 2000 ft for guard valves) with respect to flow efficiency, cavitation, and vibration potential. Their maximum size is limited by the structural constraints as well as by their in-conduit installation. Because of their relative inaccessibility, they can be more difficult to maintain than gates; for that reason, in cases where it is possible to use a gate or a valve for a given application, a gate is preferred unless a cost comparison heavily favors a valve. Also the bonnetted slide gate and the jet-flow gate are being used successfully for regulation at heads up to 500 ft where needle valves were once considered the only suitable type. Figures 17 to 19 show comparative data for various types of high-head regulating gates and valves. Regulating valves may have guard gates or guard valves upstream for protection. The comparative data were published by Warren H. Kohler in 1957 but is still valid except that it is understood that USBR is close to a new design for a sleeve valve suitable for heads up to 1300 ft; also guard valves suitable up to 2000 ft are currently being designed.

As with gates, the modern design of valves involves only a few simple and rugged valve types. The needle valve has been largely replaced by fixed-cone (Howell-Bunger) valves or by Staats-Hornsby hollow-jet-type valves for regulating high-head releases. The newer types of discharge valves are less costly, require less maintenance, and have a higher discharge capacity than a comparable needle valve. The common type of shutoff valves in use are the butterfly- and sphere-type valves. Butterfly valves are most common because they are less costly. The older-style butterfly valve with a lenticular disk has the disadvantage of incurring relatively high head losses; where head losses are critical, the trend is to use an open-frame, flow-through type disk. There is, however, an increasing

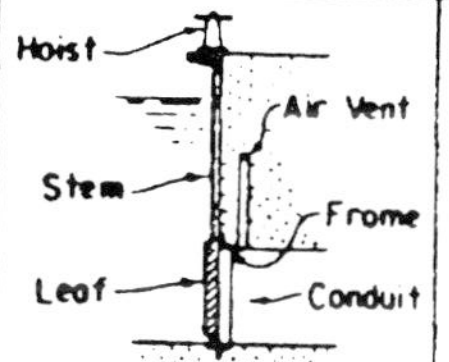

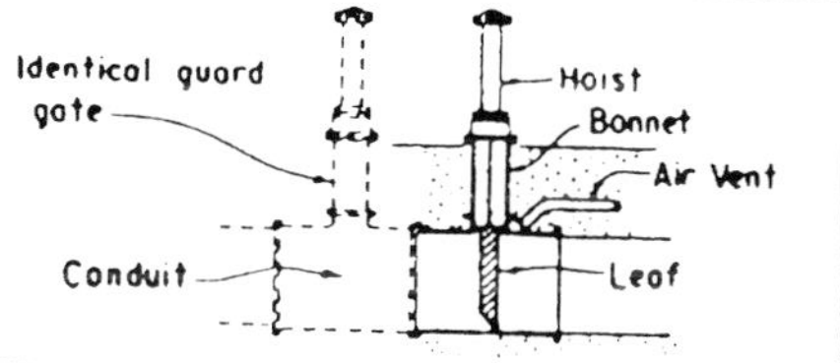

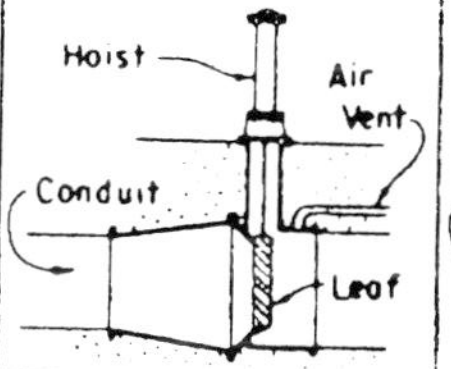

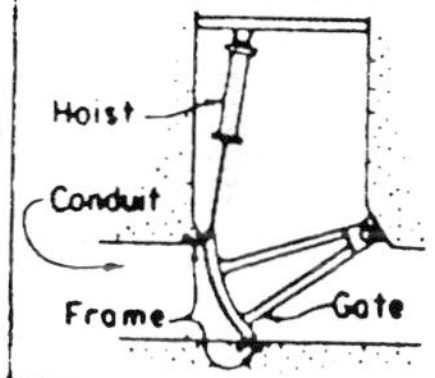

THROTTLING GATES

SERVICE CLASSIFICATION	UNBONNETED SLIDE GATE	BONNETED SLIDE GATES		JET-FLOW GATE	TOP-SEAL RADIAL GATE
		HIGH PRESSURE TYPE	STREAMLINED TYPE		
SCHEMATIC DIAGRAM / FLOW DIRECTION	Hoist, Stem, Leaf, Air Vent, Frame, Conduit	Identical guard gate, Conduit	Hoist, Bonnet, Air Vent, Leaf	Hoist, Conduit, Air Vent, Leaf	Hoist, Conduit, Frame, Gate
NAME	UNBONNETED SLIDE GATE	HIGH PRESSURE TYPE	STREAMLINED TYPE	JET-FLOW GATE	TOP-SEAL RADIAL GATE
MAX. HEAD (APPROX.)	75'	200'	500'+	500'+	200'-250'
DISCH. COEFFICIENT (a)	0.6 TO 0.8	0.95	0.97	0.80 TO 0.84	0.95
SUBMERGED OPERATION	NO	NO	YES (1)	YES (1)	NO
THROTTLING LIMITATIONS	AVOID VERY SMALL DISCH.	AVOID VERY SMALL DISCH.	AVOID VERY SMALL DISCH.	NONE	NONE
SPRAY	MINIMUM	MINIMUM	MINIMUM	SMALL	MINIMUM
LEAKAGE	SMALL	SMALL	SMALL	NONE	SMALL TO MODERATE
NOMINAL SIZE RANGE (b)	TO 12' WIDE & 12' HIGH	TO 6' WIDE & 9' HIGH	TO 10' WIDE & 20' HIGH	10" TO 120" DIA.	TO 15' WIDE & 30' HIGH
AVAILABILITY	COMMERCIAL STD. (1)	SPECIAL DESIGN	SPECIAL DESIGN	SPECIAL DESIGN	SPECIAL DESIGN
MAINTENANCE REQUIRED	PAINT	PAINT	PAINT (2)	PAINT	PAINT - SEALS (1)
COMMENTS AND NOTES:	(1) Gates are readily available from several commercial sources. They are not on "off-the-shelf" item, however.		(1) Air vents required (2) Use of stainless steel surfaced fluidways, will reduce painting requirements and cavitation damage hazard.	(1) Air vents required	(1) Seal replacement in 5-15 years is probable depending on design and use
(a) Coefficients are approximate and may vary somewhat with specific designs.					
(b) Size ranges shown are representative, and are not limiting.					

FIGURE 17 Throttling gate data.

THROTTLING VALVES

SERVICE CLASSIFICATION	FIXED-CONE VALVE	HOLLOW-JET VALVE	NEEDLE VALVE	TUBE VALVE	SLEEVE VALVE
SCHEMATIC DIAGRAM / FLOW DIRECTION →	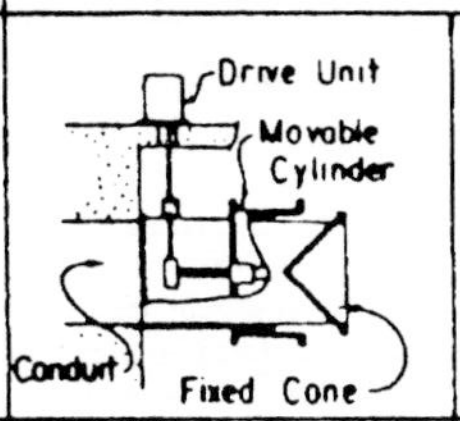 Drive Unit, Movable Cylinder, Conduit, Fixed Cone	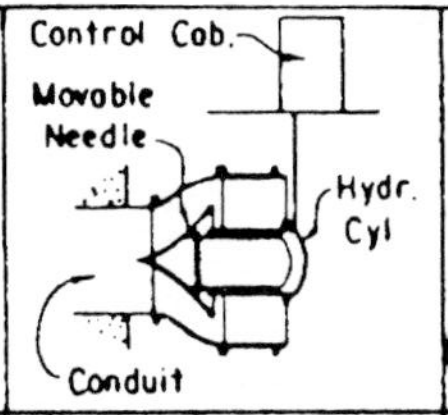 Control Cab., Hydr. Cyl, Movable Needle, Conduit	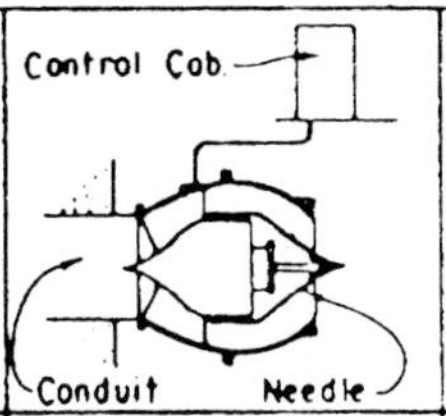 Control Cab., Conduit, Needle	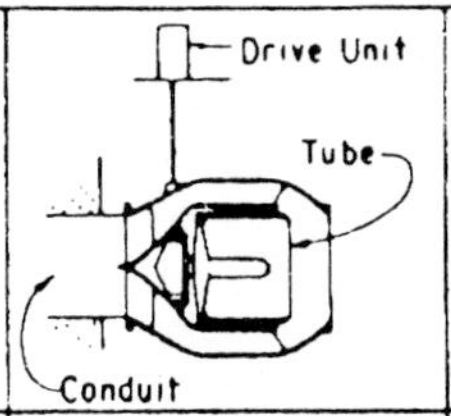 Drive Unit, Tube, Conduit	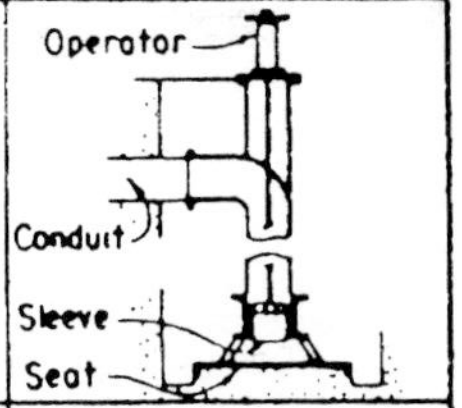 Operator, Conduit, Sleeve, Seat
NAME	FIXED-CONE VALVE	HOLLOW-JET VALVE	NEEDLE VALVE	TUBE VALVE	SLEEVE VALVE
MAX. HEAD (APPROX.)	1000'	1000'	1000'+	300'	250'+ (1)
DISCH. COEFFICIENT (a)	0.85	0.70	0.45 TO 0.60	0.50 TO 0.55	0.80
SUBMERGED OPERATION	YES (1)	NO (1)	NO	YES	YES (2)
THROTTLING LIMITATIONS	NONE	AVOID VERY SMALL DISCH.	NONE	NONE	NONE
SPRAY	VERY HEAVY (2)	MODERATE	SMALL	MODERATE (1)	NONE
LEAKAGE	NONE	NONE	NONE	NONE	NONE
NOMINAL SIZE RANGE (b)	6" TO 108" DIA	30" TO 108" DIA	10" TO 96" DIA	36" TO 96" DIA	12" TO 24" + DIA (2)
AVAILABILITY	COMMERCIAL STD. (3)	SPECIAL DESIGN	SPECIAL DESIGN	SPECIAL DESIGN	SPECIAL DESIGN
MAINTENANCE REQUIRED	PAINT	PAINT	PAINT (1)	PAINT	PAINT
COMMENTS AND NOTES: (a) Coefficients are approximate and may vary somewhat with specific designs. (b) Size ranges shown are representative, and are not limiting.	(1) Air venting required (2) Spray rating will change to moderate if a downstream hood is added (3) Valves are not "stock" items but standard commercial designs are available.	(1) Submergence to ℄ of valve is permissible	(1) If water operation is used, disassembly at 3 to 5 year intervals for removing scale deposits is usually necessary	(1) Spray is heaviest at openings of less than 35%. At the larger openings the rating would be better than moderate	(1) Sleeve valve suitable for up to 1300 ft. head are presently being developed by USBR (2) Valve is designed for use only in fully submerged conditions

FIGURE 18 Throttling valve data.

SERVICE CLASSIFICATION	GUARD GATES					GUARD VALVES	
SCHEMATIC DIAGRAM FLOW DIRECTION →	SEE FIG. 1 FOR DIAGRAMS			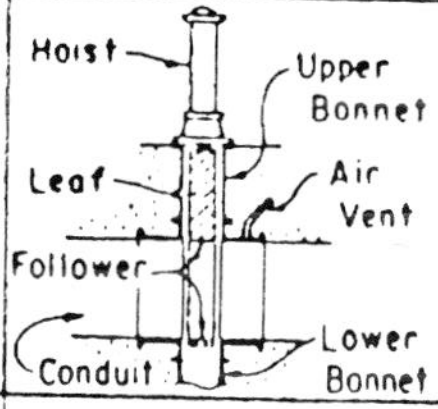	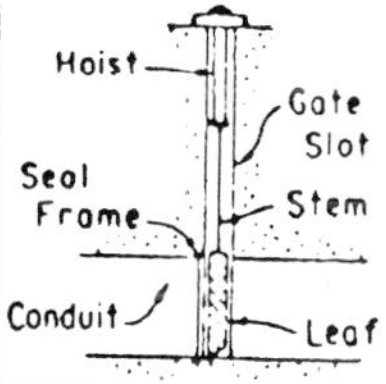	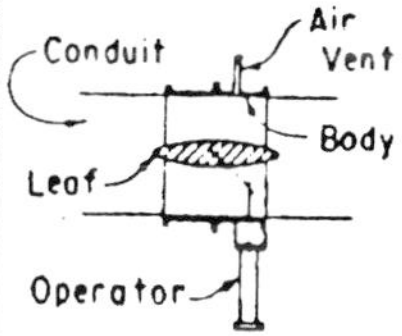	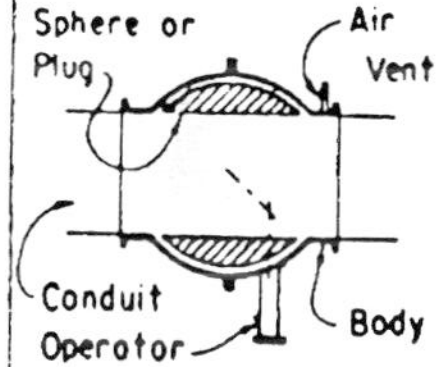
NAME	SLIDE GATES			RING-FOLLOWER GATE	WHEEL-MOUNTED OR ROLLER-MOUNTED GATES	BUTTERFLY VALVE	SPHERICAL AND PLUG VALVES
	UNBON.	HI-PRES.	STRML.				
MAX. HEAD (APPROX.)	100'	250'	500'	500' +	500'	750' +	1500'+
HEAD LOSS, H_L (a)	(1)	$.005V^2/2G$	$.003V^2/2G$	NEGLIGIBLE	$0.1\ V^2/2G$	0.2 TO $0.3\ V^2/2G$	NEGLIGIBLE
LEAKAGE	*	*	*	NONE	SMALL TO MODERATE	NONE TO SMALL (1)	NONE
NOMINAL SIZE RANGE (b)	*	*	*	36" TO 120" DIA.	TO 10' WIDE & 30' HIGH	12" TO OVER 12' DIA.	12" TO OVER 10' DIA.
USED AS GUARD UNIT FOR	(2)	(2)	(3)	ALL TYPES OF CIRCULAR CONDUIT THROTTLING GATES AND VALVES	TOP SEAL RADIAL GATES AND OTHER SQUARE OR RECTANGULAR UNITS	ALL TYPES OF CIRCULAR CONDUIT THROTTLING GATES AND VALVES	
AVAILABILITY	*	*	*	SPECIAL DESIGN	SPECIAL DESIGN	STD. AND SPECIAL (2)	STD. AND SPECIAL (1)
MAINTENANCE REQUIRED	*	*	*	PAINT	PAINT - RUBBER SEALS	PAINT - SEALS (3)	PAINT
COMMENTS AND NOTES: (a) Head losses are approximate and may vary somewhat with specific designs. (b) Size ranges shown are representative, and are not limits.	* See data on Fig. 1 (1) Head loss coefficients will vary from about 0.2 to 0.4 depending on entrance. (2) Usually used with a similar type throttling gate. Sometimes used for other types (3) Used close coupled with similar throttling gate. See Fig. 17				Normally wheel-mounted gates are used except for high heads.	(1) Rubber seated valves have no leakage when new. Metal seats will have some leakage. (2) Sizes to 36" or 48" are fairly standard. Larger sizes and high pressures are usually special. (3) Metal seals may require periodic adjustment.	(1) Sizes to about 24" are fairly standard. Larger sizes and high pressures are special.

FIGURE 19 Guard gate and valve data.

use of sphere- or analogous-type valves, such as plug valves, which also have straight-through fluidways because of their very low losses and less difficult sealing problems.

Valve operators are usually mechanical (screw stem) or oil-pressure hydraulic cylinder systems. In some cases reservoir water pressure is used to avoid external power, but corrosion is a big problem with this type of system.

21. Needle Valves. The needle valve is an excellent device for regulating high-velocity flows which is amply demonstrated by the almost universal use of such valves for controlling high-head flow for impulse turbines. However, for outlet works, needle valves are generally less economical and less hydraulically efficient than gates such as jet-flow gates and valves such as fixed-cone and hollow-jet valves. Needle valves should not be used for regulation within a pipe unless the operating chambers are designed to prevent a water-hammer condition resulting from the needle slamming shut when nearly closed.

Figure 20 shows a typical installation arrangement and cross sections through commonly used interior differential needle valve. This particular valve is closed by admitting water pressure to chambers C and C_1 and connecting chamber O to drain. To open the valve, water pressure is admitted to chamber O and chambers C and C_1 are opened to drain.

Valve operation by reservoir water normally requires valves to be disassembled every 2 to 6 years (depending on the quality of water) to remove accumulated scale. In some installations, such as Hoover Dam, where the water has extremely high scaling characteristics, it is necessary to drain the water completely from the valve controls when not in use and to fill the control chambers with an oil, similar to kerosene, to avoid frequent disassembly. Maintenance, obviously, is very expensive.

Fundamental fluidway geometry to produce a noncavitating needle valve, and discharge curves for such a valve, are shown in Fig. 21. It is essential that the downstream cone angle of the needle be less than the downstream cone angle on the body to avoid cavitation at high heads. Also a sharp, well-defined orifice and spring point is necessary for the needle-valve jet (Fig. 21, detail A). The rounded seat as shown in the phantom outline will produce cavitation and should not be used.

Venting of air from the interior pressure chambers and the high point of the fluidway is essential to avoid severe hydraulic shocks caused by rapid expansion of air compressed in the valve chamber and by the sudden needle movement which may occur when a chamber under high pressure is opened to low drain pressure for operating the valve. A gooseneck overflow having the high point above the maximum chamber elevation is usually provided to ensure that the operating chambers remain full of water.

22. Tube Valves. The tests conducted by the Bureau of Reclamation to develop a cavitation-free needle valve resulted in the development of the tube valve. A tube valve is a needle valve minus the conical needle tip which was eliminated to avoid the cavitation occurring on the conical surface of the needle. Although this elimination resulted in considerable instability of the jet at openings of less than about 35 percent (as characterized by the jet no longer remaining concentric with the valve orifice and by having the direction of discharge of the jet change position in an unpredictable pattern around the valve orifice), the jet instability is not objectionable. This instability does not pose any vibration or hydraulic problems except for creating some minor problems of spray from the discharge.

In free-discharge installations, tube valves have given excellent service. How-

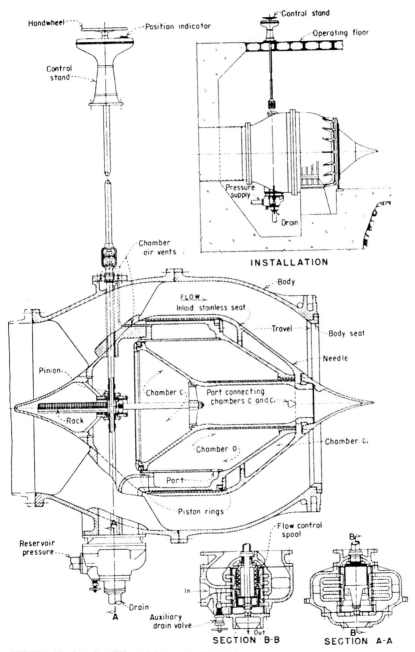

FIGURE 20 Interior differential-type needle valve.

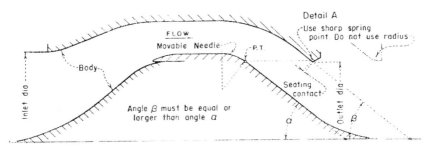

FIGURE 21 Fluidway geometry for noncavitating needle valve.

ever, four 102-in special long-body tube valves which were installed in the lower tier of outlets at Shasta Dam have had very limited use because the amount of spray in the outflow discharge is objectionable at the adjacent switchyard, one valve had a tendency to vibrate severely at openings above 96 percent, and some noise and rumble were evident in all valves toward the closed position of travel. Studies of these problems (and the fact that the tube valves proved quite costly to build) resulted in the development of the jet-flow gate, which was installed in the middle and upper tiers of outlets instead of tube valves as originally planned.

Like needle valves, tube valves can be used to discharge under submerged conditions. In one such installation, made to bypass the pump turbine at the Flatiron power plant, the Bureau of Reclamation found no evidence of cavitation on a 42-in tube valve after considerable operation.

As shown in Fig. 22, the fluidway configuration of a tube valve is quite similar to that of a needle valve without a downstream needle tip. Only relatively small unbalanced hydraulic thrust loads exist on the cylindrical tube as compared with the large thrust loads on a needle valve. It is therefore feasible to operate a tube valve by a relatively simple mechanical drive (or a small oil cylinder) to move the cylindrical tube longitudinally for controlling flow.

In the design of tube valves, an effective seal must be provided for the tube in the body bore. Both simple packings and water-pressure-actuated seals, as shown on detail A, have been used for the seal between the body and tube. Care must also be used to provide a well-defined control orifice between the tube and body seat, as shown on detail B.

23. Fixed-Cone (Howell-Bunger) Valves. The fixed-cone valve, which is also known by the surnames of the inventors, C. H. Howell and Howard Bunger, is an excellent and widely used regulating valve. Although the primary use of the valve has been for free discharge into the atmosphere, the valves have also been used for submerged-discharge operation. Unless spray is objectionable, free discharge into the atmosphere poses no particular problems; but for submerged discharges, severe vibration and cracking of the valve members has occurred, and, as a result, very careful design usually involving model testing is required. The basic arrangement and typical features of a fixed-cone (Howell-Bunger) valve installation are shown in Fig. 23, which includes a set of discharge curves. The discharge curves are based on a wide-open discharge coefficient of $C = 0.85$, in accordance with the following formula:

$$Q = CA\sqrt{2gH}$$

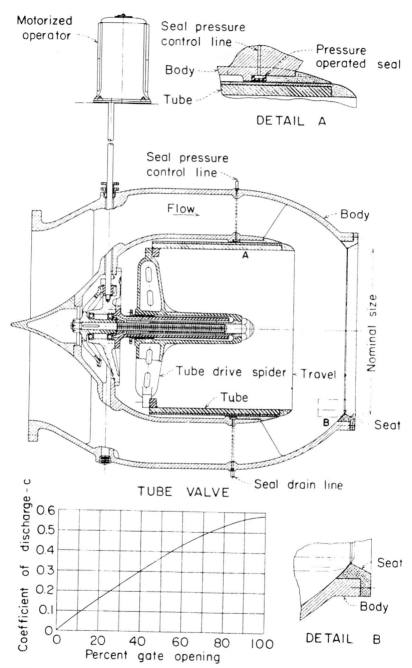

FIGURE 22 General arrangement and discharge-coefficient curve for tube valves.

where Q = discharge, cfs
A = area of valve, ft², based on nominal inside diameter
g = gravitational acceleration of 32.2 ft/s²
H = net head, ft

By combining the constants, the formula can be simplified and rewritten as follows:

$$Q = 5.357 D^2 \sqrt{H} \quad \text{(as shown on discharge curves)}$$

where D = nominal valve diameter, ft.

The expanding cone-shaped discharge pattern of fixed-cone (Howell-Bunger) valves does an excellent job of aerating the water and dispersing the energy. Where the spray from a widely dispersed jet is not acceptable, a downstream hood is used to confine and redirect the discharge. Because of the wide dispersion of the jet, stilling basins are not normally used for these valves. In all installations, adequate air must be supplied to the region upstream from the jet. This requires large ducts on hooded-type installations. Because of the large air demands, the entrances to the ducts should be located and protected to avoid the danger of persons being sucked into the duct.

The basic fixed-cone valve is composed of four essential elements: (1) the body, (2) a cylindrical gate member, (3) the seals, and (4) the operating mechanisms. The inside diameter of the cylindrical portion of the body is the same as of the upstream pipe or conduit to which it is attached by a bolted flanged connection. Radial ribs, which are attached to the cylindrical portion of the body and which extend downstream from the end of the cylindrical shell, support the concentric conical head on the end of the valve. The apex of the cone is pointed upstream, and the cone angle with respect to the centerline of the fluidway is about 45°. Flow from the radial discharge ports, formed by the arrangement of the body cylinder, ribs, and cone, is controlled by a cylindrical gate member on the outside of the valve body. The mating surfaces on the body and cylindrical gate member are made of corrosion-resistant material. An operating mechanism moves the cylindrical gate downstream to cover the radial discharge ports in the body and close the valve. In opening the valve, the amount of discharge is governed by the varying area of the orifice between the cylindrical gate and the fixed cone as the cylinder is moved longitudinally.

The upstream end of the cylindrical gate is provided with a seal which slides and seals on the exterior cylindrical portion of the body. The sealing surface on the downstream end of the cylindrical gate is made of corrosion-resistant material and contacts the corrosion-resistant seat ring on the periphery of the conical portion of the valve body. The mating metal seats on the cylinder and cone are frequently made of stainless steel to resist corrosion and cavitation damage.

Although hydraulic cylinders, bell cranks, and other types of operators are used to operate these valves, the twin-screw type of operator shown in Fig. 23a is most commonly used.

The vanes in a fixed-cone valve have to be sized to resist resonant vibrations involving the vanes and valve body. A. G. Mercer, in the paper "Vane Failures of Hollow Cone Valves," concludes that vane failures are closely associated with a kinematic condition of flow related to the vibration frequencies of the vanes.[6] He develops a parameter based on the valves which have not failed:

VALVES

$$\text{Parameter} = \frac{Q}{CDT_V} + \sqrt{\frac{E}{\rho}}$$

where Q = flow
C = dimensionless coefficient (see table below)
D = valve diameter
T_V = vane thickness
E = Young's modulus of elasticity
ρ = density

in any consistent units.

N^*	4	5	6	6	6	6	6
T_S/T_V†	1.00	1.00	0.50	0.90	1.00	1.20	2.00
C	2.22	2.35	1.98	2.40	2.48	2.53	2.75

*N = number of vanes.
†T_S/T_V = ratio of thickness of body shell and vanes.

To avoid vane failure, the parameter must be less than 0.115.

Fixed-cone valves are available in sizes from 8 to 108 in. Large-size valves have been installed for heads up to 420 ft and smaller sizes for heads up to 900 ft. Special designs can be made to meet almost any size and head requirements. The relatively small force required to move the cylindrical sleeve permits manual operation up to about the 42-in size; however, all sizes from 18 in upward are usually provided with motor operators.

24. Hollow-Jet (Staats-Hornsby) Valves. The hollow-jet valve, developed at the Bureau of Reclamation, is also known by the names of the inventors, B. H. Staats and G. J. Hornsby. Like fixed-cone valves, the hollow-jet valve is an excellent and widely used valve for regulating high-pressure outlets. The valve is designed to be used for free discharge into the atmosphere, and should not be installed where discharge under fully submerged conditions can occur. It is permissible, however, to operate the valve partially submerged provided the tailwater is not higher than the centerline of the valve.

A discharge-coefficient curve, as shown in Fig. 23b, may be used to compute valve discharges in accordance with the following formula:

$$Q = CA\sqrt{2gH}$$

where Q = discharge, cfs
A = area of valve inlet, ft^2
g = gravitational acceleration of 32.2 ft/s^2
H = total head (static plus velocity) at inlet flange of valve

As shown in Fig. 24, a hollow-jet valve consists basically of a body, a needle, and the operating means for moving the needle upstream or downstream to vary the area of an annular orifice between the needle and body for controlling discharges. There is no converging fluidway on the downstream end of the valve body; consequently the flow emerges in the form of an annular cylinder (hence

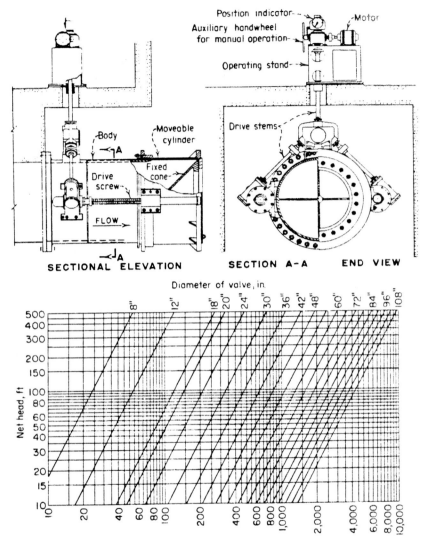

FIGURE 23a General arrangement and discharge curves for fixed-cone valves. (*Rodney Hunt.*)

the name *hollow-jet*) which is segmented by the splitter ribs, as shown in Fig. 24. The operator may be mechanical (screw stem) or hydraulic.

The splitter ribs support the central structure which contains the needle and provide openings for admitting air into the jet interior to avoid excessive subatmospheric pressures and jet instability. Air supply must be carefully checked if the valve is discharging into a tunnel.

The discharge of a hollow-jet valve is dispersed and aerated considerably

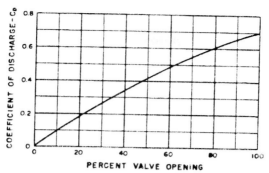

FIGURE 23b Coefficient curve for hollow-jet valves.

more than that of a needle valve but not so much as that of a fixed-cone valve. The discharging jet has a fairly clean, well-defined cylindrical shape and does not require a hood to confine the discharge and avoid spray dispersion. However, unless there is sound rock or a deep pool into which the valve can discharge, a stilling basin is usually required.

In the mechanically operated valve (Fig. 24), fluidway water pressure is admitted through ports in the needle face to the balancing chamber in the valve to minimize the unbalanced hydraulic load on the needle. By properly locating the balancing ports on the face of the needle, it is possible to limit the unbalanced water load on the needle to approximately ±12 percent of the total water load on the needle at any opening. The resulting operating forces are low, and manual operation can be used satisfactorily for up to 36-in valves. Most valves, however, are supplied with motor-driven operators of a type which limits the maximum driving torque.

The hydraulically (oil) operated valve (Fig. 25), developed in 1958, has been used as the standard design since then, because it is simpler and more economical to build and requires less maintenance after installation. It is necessary to periodically disassemble the mechanically operated valve to remove water-deposited scale from the sliding surfaces in the balancing chamber to prevent damage to the valve seals. Inaccessible sliding surfaces which could collect scale have been entirely eliminated from the hydraulically operated valve, and all sliding surfaces are lubricated by the hydraulic oil.

The hydraulic-type valve also does not require a balancing chamber to reduce the hydrostatic thrust on the needle. Oil pressure is directed to chamber C to close the valve. When the conduit upstream from the valve is filled with water under pressure, the valve can be opened by releasing oil from chamber C. If no conduit water pressure is available, the needle can be moved to the open position by admitting oil pressure to the annular chamber O.

Hollow-jet valves at openings of less than 5 percent may result in cavitation damage downstream from the seat region. No other cavitation, except that due to obvious surface imperfections in the valve fluidways, has been noted in these valves. As with all valves, it is essential that the surfaces be smooth (no pronounced offsets).

Hollow-jet valves have been built in sizes from 24 to 96 in. The highest-head valves are the 96-in hydraulic-type valves which are designed for discharging under a 535-ft head at Glen Canyon Dam. The 96-in mechanical-type valves at Hungry Horse Dam are designed for a head of 460 ft.

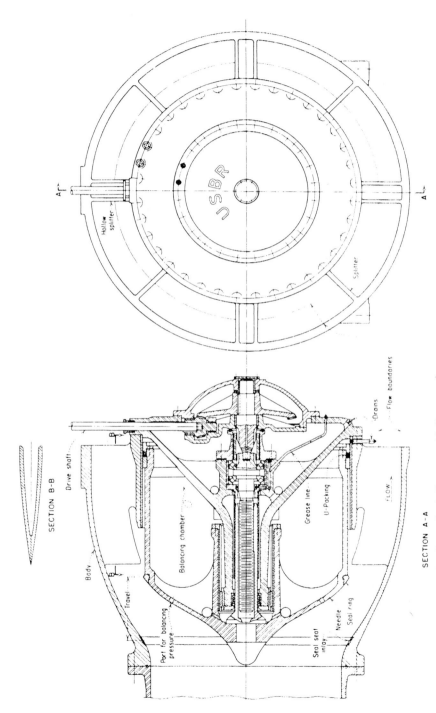

FIGURE 24 General assembly of a mechanically operated hollow-jet valve.

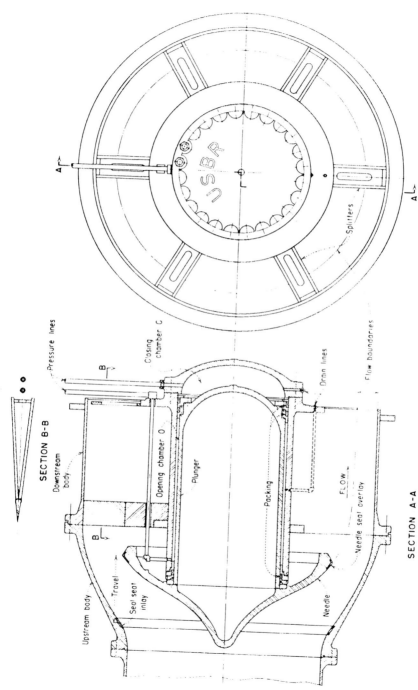

FIGURE 25 General assembly of a hydraulically operated hollow-jet valve.

25. Sleeve-Type Valves. A sleeve valve (Fig. 26) is capable of regulating under fully submerged conditions and is extensively used in vertical stilling wells. The basic regulating member is a cylindrical sleeve. The valve is similar in some respects to the cylinder gate and the fixed-cone valve. It differs in operation from the usual cylinder gate in that the flow is from, rather than into, a central pipe. It differs from the fixed-cone valve in that the sleeve is internal rather than external and the issuing jet discharges at 90° with respect to the axial centerline of the sleeve rather than at about 45°. The sleeve valve embodies the basic principle of the sudden enlargement and is designed to release the high-velocity discharge into an enlarged, water-filled chamber for dissipating the energy.

The valve shown in Fig. 26 was developed by the Bureau of Reclamation. The outflow from the square stilling well is not under a large back pressure, and either a conduit or an open flume may be used. No air supply is required. Manual operation is possible for up to about 36-in valves. Sleeve valves have been made both with and without a central conical member mounted on the base plate. Various shapes and angles of central members have also been used to vary the discharge-rate characteristics. The bureau is presently in the process of improving the valve shown in Fig. 26 and is developing new designs of sleeve valves.

The valve discharge is controlled by moving the sleeve axially with the control-stand operator to vary the orifice between the sleeve and cone (or the base plate if no cone is used). The base plate, cone, and sleeve are usually made of stainless steel to resist corrosion and cavitation damage. The tip of the sleeve must be designed to provide a definite spring point for the issuing flow to avoid cavitation damage (see detail A in Fig. 26).

26. Butterfly Valves. Butterfly valves are commonly used as the primary guard valve for turbines in power outlets, and frequently as the guard valve for regulating gates or valves in outlet works. Butterfly valves have also been used to some extent for regulating free discharges in outlet works. Standard valves in the small to medium sizes, built according to American Water Works specifications, are also widely used in water-distribution systems and outlet works.

The principal problems in the design and use of butterfly valves has been the seals. The seal must be continuously effective around the trunnions and the periphery of the leaf and also be able to accommodate leaf deflection. The junction between the circumferential and trunnion sealing points is critical. In some designs, the seal has been offset from the trunnions so that a single circular seal ring can be used to avoid the problem of sealing around the trunnions. Most butterfly-valve seals, however, are designed to seal around both the leaf and trunnions. For a number of years, butterfly valves generally used only metal-to-metal seals. The development and standardization of rubber-seated butterfly valves in the 1950s resulted in widespread use of rubber seats and a large reduction in the use of metal-seated valves. However, for large or specially designed valves for high-pressure outlet works and penstocks, a basic metal seal is still widely used. Rubber inserts are sometimes added to metal sealing faces to reduce the seal leakage.

A valve configuration commonly used by the Bureau of Reclamation for large turbine guard valves is shown in Fig. 27. The valve illustrated has 156-in seal diameter and is designed to operate under a maximum head of slightly over 400 ft. This design utilizes the conventional piston and crank mechanism for rotating the leaf, and is arranged so that all operating forces for the leaf are internal with respect to the valve unit. This arrangement avoids the problems of anchorage and alignment which are encountered when the operating unit and valve are mounted on the floor separately.

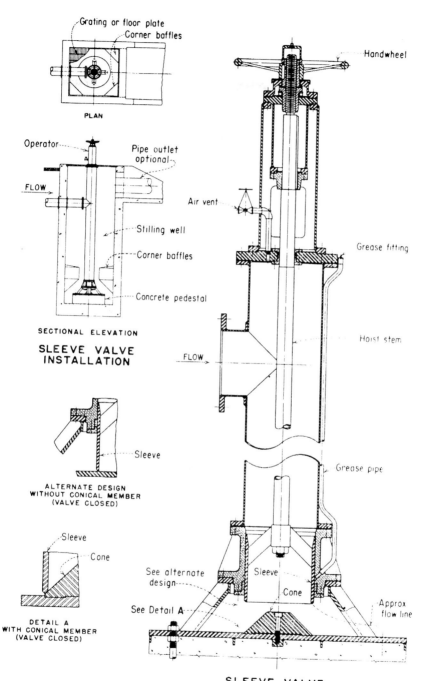

FIGURE 26 Bureau of Reclamation–type sleeve-valve installation and details.

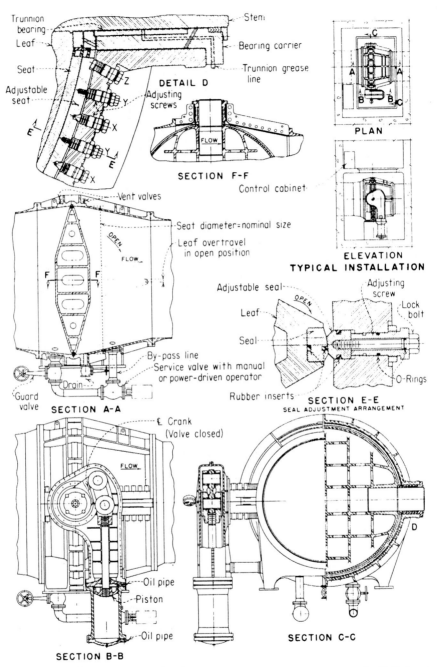

FIGURE 27 General arrangement and details of butterfly valve, based on the 13-ft-diameter valves designed by the Bureau of Reclamation for the Central Valley project power plants.

The valve may be held open or closed by shutting off the oil-line valves to the top and bottom of the operating cylinder, which provides an effective lock. Another method is to rotate the leaf slightly beyond the open position (see section A-A in Fig. 27) so that flow-induced torque will tend to hold the leaf open rather than tend to close it. In high-velocity outlet-works installations, a leaf-holding method to ensure that unbalanced turbulent flow past the leaf does not induce closure may need to be considered. In the closed position, the trunnion friction from the water load on the leaf provides an effective brake on leaf movement. Arranging the leaf rotation so the bottom portion seats in the direction of flow results in a positive sealing torque on the leaf. The torque is developed by the differential pressure on the top and bottom halves of the valve leaf. A manual lock, usually a bolt arrangement which physically locks the operating lever, is sometimes provided to hold the valve in the closed position. The manual lock is designed to withstand the full operating force of the oil servomotor so as to prevent opening of the valve when it is being serviced.

Operator design must consider that the butterfly-valve leaves tend to self-close under the dynamic closing torque produced by the flowing water. For valves which have relatively low velocities and high pressures, such as turbine guard valves, the dynamic closure torque is usually less than the torque required to overcome the seal and trunnion friction at the valve seats. For valves which must be either opened or closed under conditions approaching free discharge, the dynamic torque becomes very large and nearly always governs the torque capacity of the operator.

The following formula governs the operation of the butterfly valve:

$$T = KV^2D^3$$

where T = operating torque, in·lb or ft·lb, depending on the units of the K factor
K = constant determined by the leaf position and the units of torque as defined for T
V = velocity of flow through valve, fps
D = diameter of leaf, ft

The constant K varies with the geometry of the valve and is usually determined by test of a specific shape by the valve manufacturer. The maximum value of K for free discharge usually occurs with the valve leaf about 20° from the fully open position. The variation of torque as the square of the velocity requires accurate determination of the maximum velocity to ensure adequate operator capacity. For valves which are relatively close-coupled to turbines, the maximum turbine discharge is usually assumed for determining the velocity. For valves which are used to guard considerable lengths of exposed conduits or penstocks, the assumption of free-discharge conditions which could occur with a line break is usually used for determining the operator torque capacity. Free-discharge valves require the most operating torque.

Although butterfly guard valves are normally opened and closed under no-flow pressure-balanced conditions, they are always designed for closure under the maximum flow condition. When used as a turbine guard valve, the butterfly valve must be self-closing without power. This feature is provided by counterweights, which close the valve unless held open by the column of oil trapped in the operating servo motors. One or two bypass lines are usually provided on guard valves for balancing upstream and downstream pressures. Such bypasses frequently have a power-operated service valve and a manually operated guard valve. A drain line to remove the maximum expected seal leakage is provided just downstream from the leaf. A drain diameter about 5 percent of the leaf diameter will normally ensure ample drain capacity.

The externally adjustable seal design shown in Fig. 27 has been used on a number of 13-ft valves for heads from 300 to 500 ft. Shop test leakages ranged from 3 gpm to a maximum of 15 gpm. Leakage should preferably be less than 25 gpm to avoid excessive spray and paint-application problems downstream. Durable metal seals (with or without rubber inserts) are preferable to rubber seals even though the former have greater leakage.

Section E-E in Fig. 27 shows the typical seal and seal-adjusting arrangement. The seal shown is a metal-to-metal seal with rubber inserts. The seal-adjusting arrangement provides both circumferential and radial adjustment of the seals which is essential. The conical-nosed screws X shown on detail D provide circumferential adjustment for butting the seal tightly against the trunnion. The push-pull adjusting screws Y as well as the unidirectional push screw Z provide radial adjustment of the seal. Except for the initial circumferential adjustment of the seal bearing on the trunnion, all final adjusting of the seal may be made with the seal under pressure.

Figure 28 shows a cross-section of the 10-ft and 14-ft butterfly valves used at Hoover Dam.

27. Sphere Valves. Sphere valves have a circular fluidway through the valve body which is approximately spherical in shape. Two of the commonly used arrangements are shown in Fig. 29. In the type A valve, the cylindrical fluidway through the valve is an integral part of the rotating member. In type B, the circular fluidway is made an integral part of the body. In both cases, closure is effected by turning the rotating member through 90°. Both types of valves in the open position have a fluidway which is essentially a straight cylinder having practically no flow disturbance or hydraulic losses. They are well-suited as guard valves for installations where the guard valve is located close to a turbine, requiring minimal turbulence, and for high-velocity outlet works.

Sphere valves can be readily sealed to provide practically droptight closures, even at very high heads. The sealing contact surfaces are usually made of corrosion-resistant metals such as bronze or stainless steel. Metal-to-metal sealing is the most common, particularly for high-head valves, but a rubber-to-stainless-steel combination is also used. Most seals are hydraulically actuated so that they can be retracted during rotation of the valve; water pressure in the pipe or an independent oil-pressure source may be used for that purpose.

Valve operators may be mechanical or hydraulic, the latter being the preferred type, especially for the larger valves. Hydraulic cylinders may be powered by oil pressure or by reservoir water pressure. The operation of sphere valves type A in which the central fluidway rotates is usually subject to dynamic torques which tend to close the valve. The operator torque requirement must consider the dynamic effects in addition to the trunnion friction. Type B valves having spherical shell-type rotating members do not have a pronounced dynamic closure torque characteristics, and the torque required for their operation is governed principally by the trunnion friction. Sphere valves have been made in sizes over 10 ft in diameter and for heads in excess of 1500 ft.

The use of sphere valves and other analogous types, such as plug valves, which have straight-through fluidways is increasing because of their high hydraulic efficiency and minimal flow turbulence compared to butterfly valves. How-

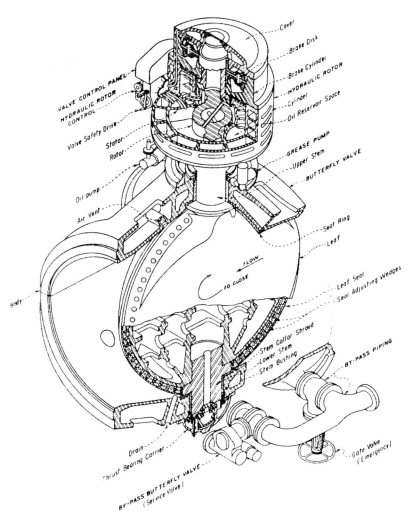

FIGURE 28 General assembly of the 10- and 14-ft-diameter butterfly-valve design used at Hoover Dam.

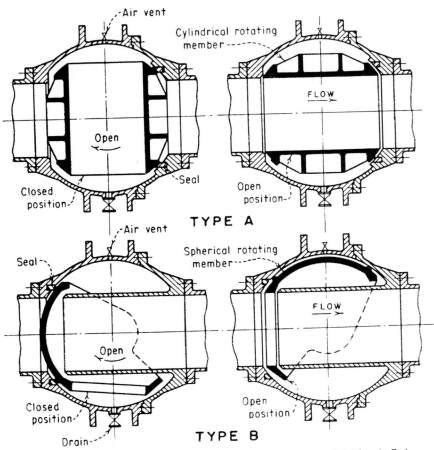

FIGURE 29 Sphere-valve types (*type B shown by courtesy of the English Electric Co.*)

ever, for a given installation, the overall economics of employing a sphere valve must be evaluated against the lower weight and cost, as well as the shorter length and smaller installation space, required for a butterfly valve.

HYDRAULIC DESIGN FACTORS

28. General. To limit cavitation and attain a high discharge coefficient, proper choices of gate slot configuration, gate bottom shape, alignment and smoothness of fluidways, and inlet shape are critical.

While the basic quantities and overall flow characteristics can be readily calculated, the effect of various changes in boundary surfaces cannot be predicted or determined mathematically with reliable accuracy except in cases which are geometrically similar to shapes verified by repeated tests. Model testing of new types of gates and valves and of outlet-works arrangements should always be performed to provide a basis for predicting full-scale performance. Model tests are especially useful in determining and studying flow patterns, estimating air demand, determining hydraulic forces such as downpull and uplift forces on gates, avoiding fluidway geometry which could cause cavitation, and providing discharge-coefficient and calibration information. Location and sizing of air vents is critical for minimizing cavitation and vibration problems as well as for preventing subatmospheric pressures in the conduits which could result in their collapse.

29. Gate Slots. Gate slots disrupt the smooth boundary lines for fluidways. Figure 30 shows various methods of conditioning the gate slots for high-head gates, including the use of fillers for ring-follower gates. To bridge the slots and smooth the path, slot fillers, where feasible, are also sometimes used for spillway bulkhead slots upstream of the spillway gates. No part of the filler should form a sharp offset into the flow.

For high-velocity outlets, one method is to force a contraction in the flow so that the discharge can jump across the slots; provision of adequate air downstream from the point of jet impingement on the downstream wall would usually prevent cavitation damage. At some partial gate openings, the lateral spread of the jet may impinge and enter the gate slots. This may be corrected by increasing the degree of forced contraction. However, an increase in contraction would reduce the discharge capacity.

Another method is to use a narrow slot. The upstream slot face has a sharp (not rounded) corner at the fluidway. The downstream corner may be sharp or only slightly rounded; however, there is an outward offset of the fluidway at the downstream side of the slot which is gradually sloped inward and faired smoothly with the fluidway surfaces. The amount of slot offset is closely related to the slot width, but variations in the depth of the slot have not been found to be critical. In general, a downstream offset of about 0.075 to 0.1 of the slot width is recommended. The slope of the converging surface from the slot offset to the nominal fluidway size should follow the criteria given in Fig. 30a.

30. Gate Bottom Shapes. Gate bottom shapes are critical with regard to the cavitation as well as the downpull and uplift forces acting on high-head gates. Bottom shapes of various high-head gates, which have been successfully used in the past are discussed in Arts. 13 and 16. Bottom shape also affects the vibration potential of the gates. The calculations for the downpull and uplift forces are complex. References for performing the calculations include "Computation of the Dynamic Forces on High Head Gates" by E. Naudascher and "Hydrodynamic Analysis for High-head Leaf Gates" by E. Naudascher, H. E. Kobus, and R. P. R. Rao.[1,2]

31. Other Hydraulic Design Factors. In addition to the geometry of gate slots and gate bottom shapes, the coefficient of discharge, cavitation, and gate vibration are governed by the shape of flow entrance way (which, for high-pressure conduits, should be bellmouthed), the smoothness of fluidway surface, use of cavitation-damage-resistant materials such as stainless steel or epoxy grouts, and the size and location of air vents. It is nearly impossible to predict the exact behavior of a given installation in view of the several variables involved. In many cases it has been found that the best solution for a bad situation relating to cavitation or vibration at a given gate opening is simply to alter the operation of the gate so that it is not kept open for extended periods of time at that opening.

The free-discharge coefficients C_D based on the total head immediately upstream from the various types of gates and valves are about as follows:

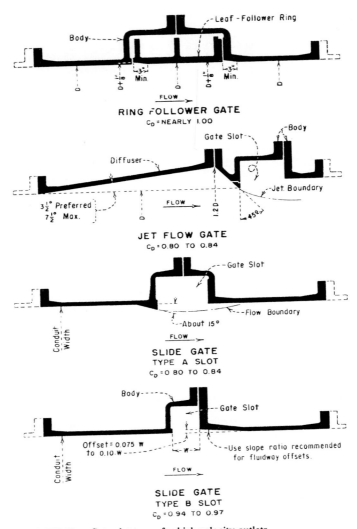

FIGURE 30a Gate-slot types for high-velocity outlets.

Slide gates	0.95–0.97
Ring-follower and ring-seal gates	Nearly 1.00
Jet-flow gates	0.80–0.84
Cylinder gates	0.80–0.90
Needle valves	0.45–0.60
Tube valves	0.50–0.55
Fixed-cone valves (Howell-Bunger)	0.85
Hollow-jet valves (Staats-Hornsby)	0.70
Sleeve valves (submerged discharge)	0.85
Butterfly valves	0.60–0.80
Sphere valves (full-diameter fluidway)	Nearly 1.00

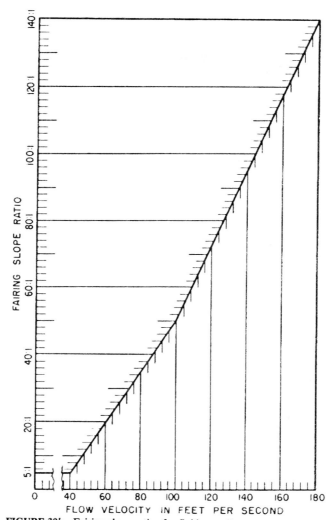

FIGURE 30b Fairing slope ratios for fluidway offsets.

Head loss H_L through gates and valves at the end of conduits may be determined by using the C_D of the gates or valves in the following formula:

$$H_L = \left(\frac{1}{C_D^2} - 1\right) H_V$$

where H_V = velocity head. The value of the exit loss is usually taken as unity, based on the velocity-head loss with free discharge into atmosphere.

EQUIPMENT DESIGN CONSIDERATIONS

32. General. For all gates and valves, safety and reliability are of paramount importance. Any failure is usually a major disaster, and it is far better to err on the side of overdesign than underdesign. Minimizing maintenance work is critical because reservoir releases can be interrupted only for limited periods. The emphasis should always be on simple, rugged, and minimal-maintenance design. Also the useful life of the equipment is critical. The equipment, with proper maintenance, should have the same useful life as the dam (100 years in most cases). The small additional incremental first cost to ensure a long life for equipment is basically a sound expenditure. Designing for replacement, even in 50 to 75 years, is economically unsound.

Corrosion-resistant materials, such as brass, bronze, stainless steels, and monel metal, must be used for all critical areas which cannot be adequately protected with paint. In addition, insulators must be used to isolate submerged materials with other than minor potential difference to avoid serious galvanic corrosion.

Gate and valve design must include provision of proper installation and maintenance equipment (including spare parts) and procedures. Failure to make adequate provisions can result in inefficient procedures in the field and may result in damage and deterioration to critical equipment. Definite maintenance schedules must be established and strictly adhered to.

33. Safety Factors and Friction Coefficients. Conservative safety factors and friction coefficients must be adopted because of the indeterminate structural and variable operational characteristics of gates and valves. Continuity in service is far more important that the slightly greater cost which may result from using conservative safety factors and friction coefficients. Maximum values of friction coefficients should be used while determining the hoisting capacity or the safety of closure and minimum values used while determining the force required to control the lowering. Calculated values of forces including weight, downpull, and uplift should be reduced or increased by 5 to 10 percent to obtain conservative values of lowering and raising forces. The following general rules are recommended:

1. For structural members, the stresses shall not exceed 90 percent of those stated in the AISC (American Institute of Steel Construction) *Manual of Steel Construction.*
2. The tensile and compressive stresses in mechanical components (including threaded fasteners, wire ropes, axles, gears, hydraulic cylinders, and so on) of carbon steels (including cast steels) and stainless steels shall not exceed one-third the specified minimum yield strength or one-fifth the specified minimum ultimate tensile strength, whichever is the lower; and the shear stresses shall not exceed two-thirds of the corresponding tensile stresses.
3. For gray and ductile iron, the tensile stress shall not exceed one-tenth the specified minimum ultimate tensile strength, the compression stress shall not exceed 2 times the tensile stress, and the shear stress shall not exceed one-half the tensile stress.
4. The bearing stress in bronze (including self-lubricating bronze) shall not exceed 10 percent of the yield point. The bending stresses in structural bronze shall be based on the same criteria as the structural steel.

5. The following seal friction coefficients are recommended:

	Maximum	Minimum
Rubber on steel	1.0	0.3
Rubber on corrosion-resistant steel	0.8	0.2
Fluorocarbon on corrosion-resistant steel	0.15	0.05
Bronze on corrosion-resistant steel	0.5	0.15
Bronze on bronze or cast iron	0.4	0.10

6. The following sliding/rolling friction coefficients are recommended:

	Maximum	Minimum
Corrosion-resistant steel on carbon steel, nonlubricated	0.5	0.1
Corrosion-resistant steel on carbon steel, lubricated	0.18	0.08
Corrosion-resistant steel on corrosion-resistant steel	Not acceptable	
Bronze on corrosion-resistant steel, nonlubricated	0.4	0.1
Bronze on corrosion-resistant steel, lubricated	0.2	0.07
Self-lubricating bearing on corrosion-resistant steel	0.2	0.06
Roller bearing at bearing bore	0.010	0.00

34. Seals. As stated above for various gates and valves, seals may be metal-metal or rubber-metal. Rubber-metal seals are the rule for most gates because of rubber's flexibility, which provides better leakage control. For high-head bonnetted gates and valves, where access to the sealing areas is difficult for maintenance, metal-to-metal seals are frequently used.

Rubber seals usually have a hardness of 60 to 70 durometer Shore type A, to provide them adequate strength. Single-stem J-type seals (Fig. 16) are usually used for gates which are operated at lower heads (up to 50 ft) and for bulkheads which are operated under balanced heads. Double-stem seals (Fig. 12) are used for higher heads and where rolling of a single-stem seal may be a problem (such as the lintel-attached top seal for top-sealing radial gates). Hollow bulb is recommended for heads less than 20 ft to effect a better seal. L-type seals are mostly used as side seals for the radial gates because of their greater flexibility (Fig. 1). Bottom seals are usually bar-type (Figs. 1, 12, and 16); they provide better sealing and stability because they do not have to be extended much below the skinplate bottom. Fluorocarbon cladding is recommended for the side and top seals of high-head gates to reduce the friction coefficient and thus the hoisting loads. It is typical to provide a nominal seal interference (precompression) of ¼ in for top and side seals to ensure that seals make proper contact while the gate is in its slot. A minimum side and top seals precompression of ¹⁄₁₆ in should be ensured after considering the most unfavorable fabrication and erection tolerances; adjustability of items such as guide shoes should be provided for to get a precompression of ¼ in after adjustment. The bottom seal projects ¼ in below the skinplate so that it can compress ¼ in while the gate is closed. Bottom seals may be provided upstream or downstream of the skinplate; upstream seals provide a better spring point for the flow jet but also result in greater hydrostatic downpull. Approximately 200 lb/ft compressive load on the bottom seal is usually provided by adjusting gate weight to assure proper sealing at the bottom; the weight of the gate minus the friction and uplift forces must at least be equal to 200 lb multiplied by the width of the gate in feet. Single-stem side and top seals are arranged so that the water pressure helps the sealing action (Fig. 16). Double-stem side and top seal assemblies have slots for the same purpose (Figs. 8, 11, and 12). Permissible seal leakage is usually 0.1 gpm per foot of seal perimeter for seating heads (downstream seals) and 0.2 gpm for unseating heads (upstream seals). Generally, a newly installed gate has more leakage. As various parts of the gate set in and seals wear off somewhat, the leakage improves.

35. Gate Wheels and Embedded Parts. The most common gate-wheel-mounting arrangements are cantilevered mounting (Fig. 11) and straddle mounting (Fig. 12).

The straddle-mounted type minimizes the required wheel-pin size, permitting the use of smaller wheels, which means lower rolling friction and accommodation of more wheels on the gate, resulting in smaller contact stresses and hence the use of readily achievable wheel hardness. This also permits the use of less costly sleeve bearings instead of roller bearings. The straddle-mounted arrangement, however, requires more complex gate-leaf framing and machining and the need to remove the wheels for painting the wheel recesses makes maintenance more difficult.

The cantilever-mounted wheels provide a maximum of accessibility for inspection and maintenance. The units can be subassembled before being installed on the gates. An eccentric wheel pin is required for both the cantilevered and straddle-mounted wheels to obtain proper alignment of the wheel tread faces.

Both sleeve and antifriction bearings are used for the wheels on both types of mountings. Sleeve bearings are usually of the self-lubricating type, although leaded-bronze bearings are also sometimes used. It is usual practice to provide seals on the ends of the bearings to minimize the entrance of water and silt.

When roller bearings are used on wheels, it is important that the bearing seals be effective, especially when gates are submerged for long periods of time between inspection and service, including clearing of deposits and relubrication. The use of stainless-steel antifriction bearings, even at considerably higher costs, is recommended to avoid rusting damage which could prevent closure of critical gates. It is preferable to use upstream sealing gates and/or sleeve bearings instead of roller bearings where too much silt is present in the water. If bearing seals are not maintained, the silt may clog the roller bearings.

In all cases the provision of grease fittings is recommended. Although not essential for lubrication of self-lubricating bearings, the grease helps prevent the entrance of water and silt.

Wheel tracks should be of the same material as the wheels (to prevent galvanic corrosion) but of greater hardness so that any wear takes place on the more accessible wheels instead of tracks. In any case, tracks should be bolted to the embedded parts, as shown in Fig. 11, so that they can be easily replaced if necessary. Tracks welded to the embedded parts would be very difficult and expensive to replace. Harza Engineering Company has been involved since 1990 in building up worn embedded cast-iron tracks of nearly 250 stoney-type roller gates (60 ft × 20 ft) for several barrages in Pakistan. A special machine had to be developed to mill the worn tracks in place. Replaceable nodular cast-iron plates are being mounted into the milled tracks by countersunk screws.

For the wheel design, it is usual to provide a ratio of 10 to 15 between the diameter of the wheel crown and the diameter of the wheel. The maximum allowable contact pressure (lb/in^2) between the crowned wheels and flat track is usually assumed as 778 times the Brinnel hardness number of the wheel crown. The wheel track is frequently designed by using data from the following sources:

University of Illinois Bulletin 212, "Beams on an Elastic Foundation," and "Bending of an Infinite Beam on an Elastic Foundation." The foundation modulus k is calculated by the following formula in Biot's paper:[5]

$$k = 1.29 \left[\frac{1}{C(1 - v^2)} \frac{Eb^4}{E_b I} \right]^{0.11} \frac{E}{C(1 - v^2)}$$

The shear Q, bending moment M, and deflection y are calculated in accordance with the following formulas in M. Hetényi's book:[22]

$$Q = -\tfrac{1}{2} P D_{\lambda x}$$

$$M = \frac{P C_{\lambda x}}{4\lambda}$$

$$y = \frac{P \lambda A_{\lambda x}}{2k}$$

The cited references should be consulted for the meanings of the nomenclature in the formulas and for the calculation procedures. The allowable shear stress in the track would be the same as for the wheels if the Brinell hardness were the same. If cast-steel wheels with stainless-steel rings are provided, the rim thickness should be able to contain the heavy shear forces under the Hertz area.

The embedded bearing plates of the slide gates are also designed according to the above formulas. As far as possible, all embedded parts must be placed in second-stage concrete. They must be adjusted to correct tolerances and secured tightly by alignment studs (and other means) field-welded to first-stage embedded welding pads, before embedment. It is important to check the bearing strength of the concrete supporting the track and bearing plates. Many a failure has occurred because the second-stage concrete was not properly reinforced. Rebars from first-stage concrete should be extended into second-stage concrete to reinforce the latter. The design of the gate guides is also important; they should be checked for loading conditions resulting from the rotation of the gate when the gate hits the sill (because of the eccentricity of the gate's center of gravity with respect to the contact point of the gate with the sill.) For wire-rope hoists, where single-point holding (due to wire rope snapping on one side) is a consideration, the load on the guide can be minimized by locating it close to the gate's center of gravity.

36. Roller Trains. Separate roller trains are usually mounted on each gate section as shown in Fig. 13b to provide the flexibility to distribute the loading to the embedded track more uniformly than would be possible with roller trains extending the full height of a gate.

Experience and the forces involved indicate that the track radius for the top and bottom of the trains should not be less than two roller diameters. The track radius may be a full semicircle or may be made in quadrants as shown in Fig. 13b. The quadrant arrangements is used for large fairly thick gate leaves so that the maximum length for vertical distribution of the leaf load to the embedded track can be obtained. Semicircular tracks at the ends of roller trains are commonly used on the smaller gates.

Roller-train link arrangement is also critical. The straddle-mounted link design (Fig. 13b, view C-C) is recommended because it provides balanced loading on the chain bushings and thus avoids binding of the bushing on the roller trunnions. Fabrication methods (use jigs) must ensure uniformity of roller trains. For roller diameter, a tolerance of about 0.0002 times the diameter in inches is recommended.

Both self-lubricating bronze bushings and leaded-bearing bronze bushings have been used with success on roller trains (self-lubricating bushings for more heavily loaded chains). With straddle link-chain construction, the bushing load is primarily the load of the chain and the principal friction is a result of the knee action of the links as the chain travels around the curved tracks. The roller trunnions theoretically carry only the weight of the roller and serve to position the roller accurately in the chain bushings. Bushing tolerances must be closely held to ensure satisfactory chain operation.

Two types of locking devices for holding the link and bushing assemblies on rollers are illustrated. The stainless-steel snap ring is the simplest. The split collar which is held in place by bending over the four corners on a square washer is a more rugged and positive type of locking device, but also more expensive.

The rollers, tracks, and track bases for roller-mounted gates are designed on the same basic principles and formulas used for wheel-mounted gates.

The track on the gate is sometimes mounted on a rubber base to allow movement of the track for adjusting to the slope at the sides of the leaf; this eliminates the hazard of getting high edge loadings on the rollers, but the rubber should be confined so that it cannot flow from under the track.

Tests indicate that the force required to move an unloaded chain around a track is quite small, provided the catenary sag at the bottom end is not too tight. If the bottom rollers are supported on a guide track, the tension in the links due to catenary action can be eliminated. Arbitrarily assuming that 20 percent of the roller train weight is the force required to move a roller chain should provide an ample operating-force allowance.

It is desirable that a trash guard be placed over the top of roller trains as shown in Fig. 13b to prevent damage to both the roller and track surfaces by debris caught in between.

37. Radial-Gate Trunnion Support and Anchorage. Radial-gate trunnion support and anchorage are the critical items in radial-gate design because all the load acting on the gate is transferred to the trunnion support and anchorage. Figure 31 shows three types of systems generally in use. Types 2 and 3 are the posttensioned type and are generally more economical for larger gates (no exact comparative studies are known, but it is estimated that a posttensioned system would be more economical for gates larger than 35 ft × 35 ft for crest-type radial gates; the size limitation for high-head radial gates would depend on the head on the gate). Because of larger hydrostatic loads associated with larger gates, the anchor girder to be embedded in concrete in type 1 design would have to be very long, complicating the pier design. Type 2 has a concrete girder similar to the steel girder of type 1 but utilizes a posttensioned anchoring system. Type 3 also utilizes a posttensioned anchoring system but a massive reinforced-concrete block integral with the pier is provided instead of a separate steel or concrete girder and the pier is narrowed to a thin neck to enable the gate load to be located closer to the centerline of the pier. While the narrowing of the pier helps reduce the load moment (which is critical with one of the two adjacent gates open), the shape of the pier becomes complicated and must be carefully analyzed.

The load transferred to the support and the anchorage must consider the load contributed by the hoisting forces (in the direction of the hydrostatic load) in addition to the hydrostatic and hydrodynamic loads. As a rule of thumb, the hydrostatic loading is increased 15 percent to account for the hydrodynamic loading, which is difficult to compute. The contribution of the hoisting force can be

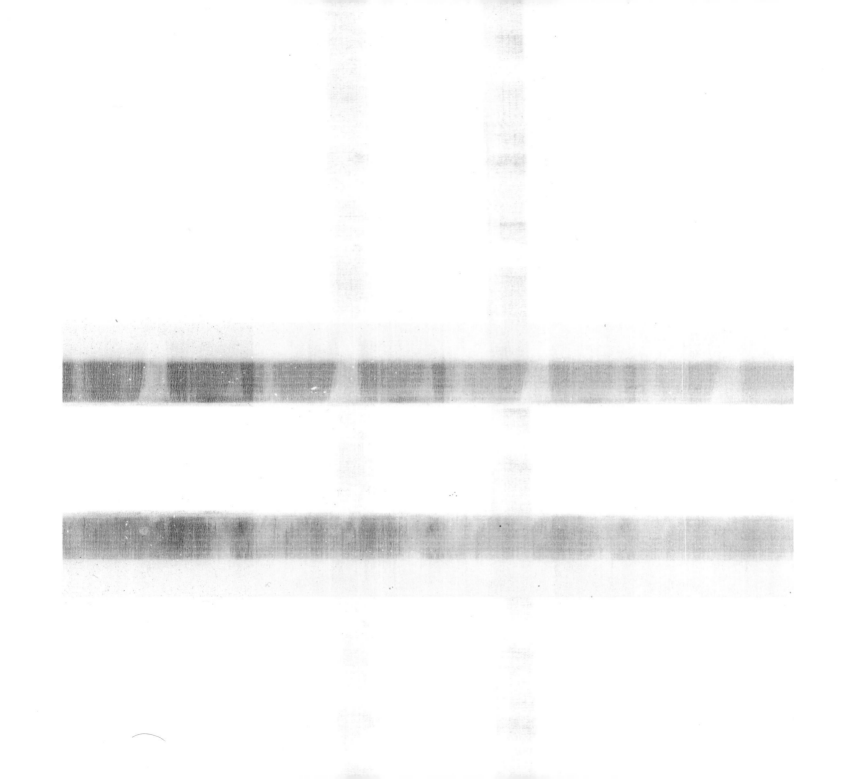

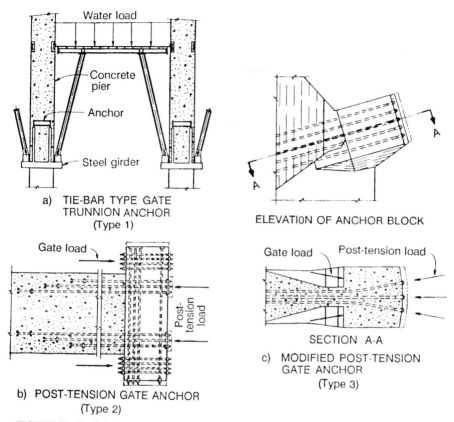

FIGURE 31 Radial-gate trunnion girder anchorage system.

readily computed; however, as an initial estimate, it may be assumed to be 15 percent of the hydrostatic load. The worst loading condition usually is with the gate slightly above the fully closed position.

38. Hoists. For high hoisting capacities (say above 50 t), hydraulic operation has widely superseded the screw lift and geared drive units for most types of gates and valves because of the ease with which such capacities can be attained with a hydraulic operator and because of design simplicity, ease and flexibility of control, and operating reliability of the hydraulic operator. The economics also usually favors hydraulic hoists, specially for capacities above 30 t. Wire-rope hoists, however, are still used for many new gates because of the inexperience some project personnel have with the hydraulic hoists. Chain hoists, used extensively before, are not used much anymore because of the problems associated with their maintenance.

Hydraulic and screw stem hoists can be used for all types of gates including slide gates which usually require a push force to close. Wire rope hoists can be used only on gates which can close by gravity.

Because of their compactness, screw-stem hoists are usually used for smaller gates with pull-push requirements (say, up to 10-t hoisting capacity), especially where hand operation is possible and where standard manufactured units can be used. For higher capacities, it is preferable to use a hydraulic hoist instead of a motorized screw hoist. All screw-stem hoists need to be aligned as perfectly as possible with respect to the gate guides to avoid excessive wear on the stem and stem nut, but the alignment is more critical for a motorized hoist because the motor power could force even a much misaligned stem to move and damage it. It is recommended that all stems be of stainless steel and all nuts be of bronze to eliminate problems caused by corrosion.

A wire-rope hoist is the only choice if the gate must travel more than the opening height to be stored above the water level (such as draft tube gates which must be protected from the turbulence in the tailwater downstream of a powerhouse), because hydraulic hoists cannot perform that function. Advantages of wire-rope hoists include a simple, easily understandable system, less headroom requirement for vertical gates, lower hoisting capacity for radial gates (if attached on the upstream face of the gate), possibility of providing a reliable mechanical position indicator, and no problems associated with gate drifting and oil spill. Wire-rope hoists provided at the upstream face of a radial gate result in lower hoisting capacity because of the larger moment arm available about the trunnion (hydraulic cylinders cannot be conveniently installed in that position); however, the rope in that location would be subject to damage by debris. Besides being uneconomical at higher capacities, the reason wire-rope hoists have given way to hydraulic hoists is the large amount of inspection and maintenance related to the wire ropes. They must be kept clean and greased regularly and also inspected for wire breaks. If neglected, serious safety problems could occur. In any case, all wire ropes used for permanent gate hoists must be of stainless steel for prolonged life.

Advantages of hydraulic hoists may be summed up as versatility in meeting changes in loading and speed, easy provision of end-of-stroke cushioning, possibility of operating several gates using only one power unit, ease of manual lowering without power, and low maintenance costs due to very few moving parts. Also, for gates requiring gravity closure without power, the hydraulic hoists need no special equipment (devices to retard the free fall must be provided for wire-rope hoists). Gate drift problems can be readily taken care of by providing an electrically operated gate repositioning circuit. Oil spill is usually not a problem because modern hydraulic systems are very reliable.

Figure 32 shows the cross-section of a typical screw hoist; Fig. 33 of a wire rope hoist for a radial gate, and Fig. 34, of a hydraulic cylinder. Figures 1, 4, 8, 9, 10, 11, 12, 13a, 14b, and 15 show a few typical installations for hydraulic hoists. Figure 35 shows typical layouts of the hydraulic power unit and control cabinet.

Location of the hoist must be fully studied before proceeding with the hoist design. For vertical-lift spillway gates, an overhead structure is needed to locate the hoist, as stated in Art. 9 and as shown in Fig. 5. A wire-rope hoist can be located on the top of the overhead structure without problems, but a hydraulic cylinder over and above the overhead structure would be very unsightly especially for tall gates. If a hydraulic hoist is used for operating a vertical-lift spillway gate, the cylinder should preferably be installed with its top near the top of the overhead structure and the gate should be provided with openings to accommodate the cylinder as the gate moves up. It should be noted that it is desirable to have a single cylinder for vertical-lift gates; if two cylinders are provided, they must be electronically coordinated by the use of proportional valves in the hydraulic system.

For high-head vertical-lift gates, it is preferable to locate the hydraulic cylin-

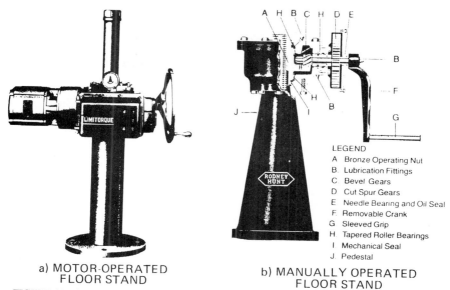

LEGEND
A Bronze Operating Nut
B Lubrication Fittings
C Bevel Gears
D Cut Spur Gears
E Needle Bearing and Oil Seal
F Removable Crank
G Sleeved Grip
H Tapered Roller Bearings
I Mechanical Seal
J Pedestal

a) MOTOR-OPERATED FLOOR STAND

b) MANUALLY OPERATED FLOOR STAND

FIGURE 32 Screw hoists. (*Courtesy of Rodney Hunt.*)

der directly above the gate's fully raised portion so that it can better withstand the buckling loads; it also helps quick closure of the gates where the gates are used as emergency guard gates. Submersion of hydraulic cylinders in water (for gate-shaft-located vertical gates) is not a problem as long as they are accessible for maintenance by provision of bulkheads upstream. Location of the cylinder above the top of the structure would be advantageous from a ready-access point of view. In some designs, cylinders are erected with the rod above the cylinder, in which case the rod is drilled to route the oil to the cylinder bottom. This is a cumbersome arrangement and should be avoided except in cold climates where an upside cylinder connected to the gate bottom can be kept warm by the heaters provided to heat the skinplate.

For radial gates, wire rope may be attached to the upstream face of the gate skinplate to reduce hoisting force requirements or to the downstream side to protect the wire ropes from debris. Hydraulic cylinders cannot be conveniently located upstream and therefore they are always connected to the gate on the downstream side. Provision of two hydraulic cylinders for a radial gate does not present a problem because the gate is guided at the trunnions and cylinders can be hydraulically coordinated automatically by providing a structurally rigid gate. The cylinders are usually connected to the lower main horizontal girders as shown in Fig. 1, which appears to be the best from a structural viewpoint as well as from the viewpoint of orienting of the cylinders for minimal hoisting capacity. In many cases, especially in Europe, the cylinders are attached to the upper gate arms closer to the trunnions to reduce the cylinder stroke (in some cases that would make the cylinder economical), but that arrangement requires high hoisting capacity requirements and heavier gate arms. For the Corps of Engineers Winfield Project, Harza Engineering Company performed an optimization study for a 110-ft-wide by 28-ft-high radial gate and concluded, on the basis of the economics of the hoisting system and the trunnion

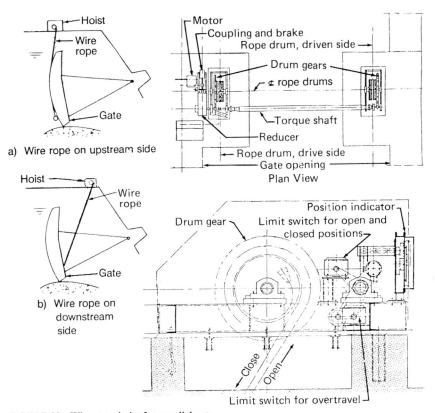

FIGURE 33 Wire-rope hoist for a radial gate.

reactions for trunnion anchorage and pier design, that the best location would be as shown in Fig. 36. The constraints of the study included a restriction on the maximum elevation of the top of the cylinder.

For flap gates, the hoist location is discussed in Art. 8.

A typical wire-rope hoist includes two wire-rope drums, one at each end of the gate width, an electric drive motor, a solenoid brake, and the necessary reduction gearing. The drive motor and the solenoid brake (which is coupled to the drive motor) may be located at one end only (as shown in Fig. 33) or equidistant from each drum on the top of a hoisting bridge. The two drums are usually connected (through necessary reduction gearing) by a line shaft spanning the distance between the drums or between the drive motor and each drum. In some cases, the line shaft is not used; a separate drive motor is provided for each side, and the two sides are synchronized electrically. The drums may be either regular or spiral type. Regular drums permit the wire rope to wind in a single layer, whereas, in the case of spiral type, the wire rope winds on the drum in multiple layers. Regular drums are provided with up to two wire ropes per drum, whereas the spiral types can have several ropes, each rope winding separated from the next by plates bolted or welded to the drum.

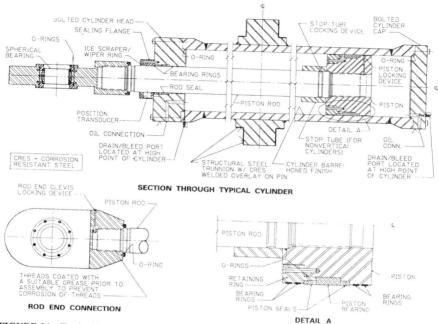

FIGURE 34 Typical hydraulic cylinders.

FIGURE 35 Hydraulic power unit and electrical control cabinet. (*Cushman Project, orifice spillway radial gate.*)

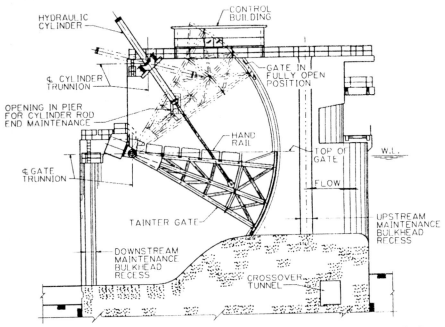

FIGURE 36 110-ft-wide × 28-ft-high Winfield radial gate. (*Corps of Engineers/Harza Engineering Co.*)

A hydraulic hoist includes one or more hydraulic cylinders, a hydraulic power unit (including an electric motor, one or more pumps, a fluid reservoir, filters, necessary valves, and piping) and piping interconnecting the hydraulic cylinders with the power unit. The cylinders can be conveniently trunnion-mounted to suit the angular movement of the radial or flap gates. Practically any hoisting capacity can be achieved by simply selecting a suitable cylinder bore size and/or pump pressure output. Accumulators are sometimes provided to operate the system through a limited number of cycles in case of power failure.

To protect against corrosion, the piston rod of a hydraulic cylinder is usually of stainless steel with hard chrome plating (0.003-in-thick plating applied in a minimum of two coats). Recently, ceramic-coated carbon-steel rods were developed in Europe and have been successfully used at a few projects. In addition to being slightly more economical than chrome-plated stainless-steel rods, they appear to have longer life. Experience in their use, however, is still limited.

Typically a hydraulic system is designed to operate at pressure between 2000 and 3000 psi. Higher pressures would result in more economical size of the equipment, but more leakage problems are also likely, although the recent advances in the fluid-power industry have produced fairly reliable seals. Pressure and return filters must be used in a system to keep it clean; it is usually preferable that suction strainers or filters not be used because the large mesh size (100 μm) normally provided to prevent pump suction from starving does not serve a very useful purpose, but if the mesh does get clogged, pump damage could occur. Figure 37 shows a typical hydraulic schematic. This schematic will serve the gravity clo-

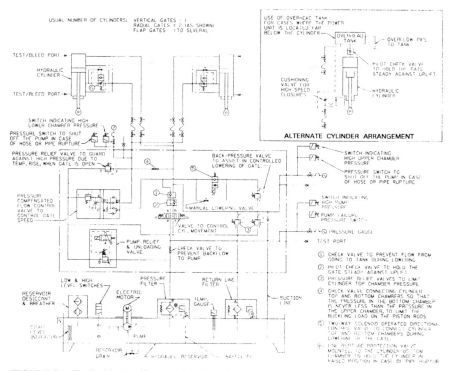

FIGURE 37 Typical hydraulic control schematic.

sure gates as well as the push-type gates except that, for push-type gates, if closure without power is required, hydraulic accumulators should be provided in the system. For gravity-closure gates, it is preferable to locate the cylinder below the hydraulic power unit (or at a height above the power unit not exceeding the height of the oil column equivalent to atmospheric pressure) so that oil can flow to the top of the cylinder while the gate is lowered without power. Many times, that is not possible, in which case it is desirable to equip the cylinders with an overhead tank (Figs. 12 and 37) which would provide sufficient oil volume for one gate closure. For radial- and flap-gate cylinders which rotate, an overhead tank is not convenient, and therefore, in these cases, location of power unit above the cylinders is critical, unless accumulators are provided in the system. It is important that even for gravity-closure gates the pump must run during lowering of the gate so that the cylinder upper chamber always remains full; dependence on a suction line is not desirable. The pump will also keep the overhead tank or the accumulators, where provided, full. The pressure on the top of the piston can be limited by provision of relief valves connected to the cylinder upper chamber; the piston rod buckling check (for the gate's fully closed position) then would have to be made only for the upper chamber relief valve setting.

Each hydraulic hoist is usually provided with a drift control system which would restore the gate to its set position if the gate drifts from that position because of leakage in the system. The system locks in the set position and constantly compares the gate position with the set position, raising the gate back up if the drift exceeds a premeasured amount (2 to 3 in). When the gate is operated by the use of raise or lower commands, the set position lock is automatically canceled.

Gates subject to partial openings must be equipped with a position indication system. Wire rope or screw hoists can be more readily fitted with a reliable mechanical indication system through reduction gearing, or, in the case of screw hoists, by directly measuring the vertical movement of the stem. Hydraulic cylinders (for which mechanical indication systems are more cumbersome), are usually provided with an electronic magnetic linear displacement transducer (MLDT) fitted into a cylinder head. An MLDT combines ultrasonic and magnetic effects to measure movement of a piston rod.

Most gates and valves are operated manually from a local control station provided on the deck of the structure containing the gate. The local control units may be self-standing units in weatherproof cabinets or within a control building. Remote control (generally in the powerhouse for hydroelectric projects) is provided for emergency operation and for remotely monitoring the system faults. Automatic operation is also provided, especially for the gates required to precisely control reservoir levels. Spillway crest gates are usually automatically stopped every 2 to 3 ft of travel during raising, requiring repeated manual actuation of the raise button, to prevent accidental spill. Electrical controls may be the conventional relay type, but an electronic programmable logic controller (PLC) is being used more and more because the PLC uses less parts and permits faster troubleshooting and corrective measures.

39. Corrosion Protection. Although some installations have used cathodic protection in addition to painting of the gate equipment for corrosion protection, a good painting system alone, maintained properly (every 5 to 6 years), is usually the best way from an economic as well as a convenience point of view. It is very difficult to estimate the number, dimensions, and locations of the sacrificial anodes required for proper cathodic protection, and, if the spent anodes are not replaced promptly, corrosion could rapidly progress specially if cathodic protection is depended on to a significant extent. There are many projects in existence for several years maintained in good shape by regular painting.

Coal-tar epoxy paint (*SSPC-PS11.01) is usually the best for the structural steel parts of the gate equipment. A zinc-based primer (SSPC-Paint 20) could be an additional help. Vinyl paint has been used in some installations and, according to the Steel Structures Painting Council (SSPC), is superior to coal-tar epoxy, but application of vinyl paint is very cumbersome and it is very difficult to get rid of pinholes. Several modern paints, polymer-based and zinc/liquid glass-based are currently available in the market; they appear to be economical substitutes for the coal-tar epoxy and vinyl paint, especially considering practically no maintenance requirements. These paints should be looked into before deciding on the final paint selection for a particular gate application.

The exposed surfaces of the embedded items for all permanent installations should normally be of stainless steel, because paint is difficult to maintain on these items. All gate and hoist items, including pins, wheels, hoist piston rods, and piping, and other similar items, on which paint cannot be maintained and which are critical for proper gate operation should be of stainless steel. Self-lubricating bronze bearings (working on stainless-steel pins) should be used throughout the equipment as far as possible. On all submerged parts, insulators should be provided to separate the surfaces of different materials to prevent anodic corrosion.

40. Ice Problems. Spillway gates, because of their proximity to the water surface, present operational difficulties under icy conditions, including possible inability to operate, because of increased hydrostatic loading and increased operating friction caused by the ice. The hoist is normally not designed to carry the full extent of increased loading, for reasons of economy. In extreme cases, the gate could be overwhelmed by the ice and frozen in place. Frazil ice (fine spicules of ice suspended in water, formed in supercooled, turbulent waters) can affect even intake gates and trashracks 30 to 40 ft below the water surface.

To keep a gate operable in the winter, the following provisions are essential:

1. Provide means to prevent ice formation on the gate-embedded parts and, if necessary, on the gate skinplate.
2. Design the hoist with generous capacity, sufficient to overcome, to a reasonable extent, the increased ice loading.
3. Select material to suit site conditions.
4. Provide adequate drainage to prevent ice buildup on the downstream face of the gate.
5. Do not open the gate at too small an opening (say less than 3 in).
6. At locations subject to extreme ice formation, provide ice booms upstream of the gate (in addition to the other provisions listed above) to prevent ice buildup closer to the gate.

Gate-embedded parts should be maintained ice-free to minimize frictional forces and to prevent damage to the gate seals and gate bearing areas by rubbing against ice buildup. Where heavy ice buildup is expected next to the gate skinplate, heating should be provided for the skinplate to keep the gate operable; this is most likely on projects with relatively small reservoirs.

Gate design and gate hoisting capacity usually consider a hydrostatic ice load of 5000 lb per foot run of gate width acting approximately for 2 ft below the maximum water level. This value is somewhat conservative but will go a long way toward assuring adequate design, especially considering that it is not simple to calculate the dynamic effects of ice buildup, expansion, and contraction.

Proper selection of the hoist and of the materials used in gate and hoist design are equally important. Hydraulic hoists are preferable because of their push capability. Rope or chain hoists, especially those connected to the gate on the upstream side, can cause difficulty if the wire rope or chain gets frozen in ice. Structural materials should resist embrittlement at cold temperatures and grease or hydraulic fluid should not freeze. Gate seals should preferably be fluorocarbon-clad to minimize damage. Heating piping should be insulated on the exposed areas to minimize heat loss.

Adequate drainage should be provided to minimize addition of ice weight to the gate. During operation, the gate should not be opened at too small an opening; besides being hydraulically unacceptable, a small opening will have the same effect as a leaking gate and accelerate ice formation.

The most common methods of combating ice on gates are

1. Use of heaters for gates and embedded parts
2. Use of bubbler system
3. Installation of flow mixers

Methods 2 and 3 are less common than method 1 because they are relatively slow and less efficient. Direct heating presents the most effective means of combating ice. A direct-heating system may be operated on a cycled basis, controlled by timers and/or thermostats. The heat may be applied either by electric heaters (nichrome strip heaters, mineral-insulated heating cables, radiant heaters, tubular heaters, etc.) or by circulating heated fluid (using a 50-50 water-glycol solution) or hot air through piping or ducts attached to the parts. The fluid or hot-air system requires greater initial investment than the electric heater system. Most types of electric heaters, however, have to be frequently replaced because their life is limited by high wattage concentrated over a narrow band. In addition, the heaters are affected by the humid conditions prevalent and by gate vibrations. Frequent replacement of heating elements can be a nuisance and would offset their initial cost superiority.

A typical value normally chosen for heat requirements is 0.5 kW/m (0.15 kW per linear ft) for the embedded parts and 1 kW/m^2 (100 W/ft^2) for the gate skinplate and other parts. These values are based on experience at many installations and may be reduced or increased, depending on the project variables such as ambient conditions, the amount of ice expected to be formed, heating cost consideration, and the time in which the gate must be deiced, the last depending on the urgency of gate operation. A larger gate at one project may need less overall heating capacity than a smaller gate at another project.

The height of the skinplate to be heated may vary depending on the expected long-term winter fluctuation of the upstream water level. A minimum of 2 ft is usually heated if no long-term winter fluctuation is expected. The embedded parts should normally be heated throughout their full height; the heating should include the sill beam.

The ice problems relating to intake trashracks are more complex because frazil ice can clog the racks, resulting in a shutdown of the powerhouse. The methods of combating ice on trashracks include suppression of frazil production by creation of ice covers, use of heat, mechanical removal of ice with trashrakes, back-flushing techniques, use of special coatings and materials, creation of vibrations, blasting of ice, and actual removal of trashracks during winter. Removal of trashracks should be resorted to only after carefully weighing the damage that may be caused by not having trashracks in place.[9]

41. Operation and Maintenance. The keystone of proper operation and long service life of gates and valves is adequate inspection and maintenance. Gates and valves are critical to the safety of the powerhouses and dams; their serviceability must be maintained at all times. Because of the rugged nature of the equipment, the need for periodic testing, inspections and maintenance is sometimes overlooked. With a proper schedule of testing, inspection, and maintenance the useful life of such equipment is practically unlimited. Without such a program, gates and valves can be junk in 30 to 50 years.

When gates and valves are installed, their operation should be carefully checked to be sure they work as intended. A detailed checklist should be prepared to include all desired characteristics including speed and smoothness of operation; stops at fully open, fully closed, and other desired positions; position indication; and leakage control. Corrective measures should be taken, as necessary, to ensure that the equipment, as installed, functions satisfactorily. Many times, because of project delays and flow conditions, gates or valves may have to be closed before they can be fully checked. In that case, checks should be completed as soon as flow conditions permit and gates or valves can be opened.

In addition to performing the complete check at the time of installation, a schedule of testing should be established for checking the equipment periodically to be sure it continues to operate properly. The time interval for making such operational tests will vary with the nature, importance, and frequency of use of the equipment. Where no specific factors indicate otherwise, an annual operating check is recommended. It is also recommended that all gates and valves be given a full operating test after every major maintenance overhaul. Gates that normally remain in a specific position (fully open or fully closed) should be exercised every three months to prevent jamming.

A schedule of routine maintenance must be established and adhered to. This schedule will depend on the type of gate and valve and the frequency of operation. Generally an annual inspection of moving parts including wheels, rollers, and seals; inspection of damage due to corrosion, cavitation, or abrasion; lubrication of bearing surfaces including screw threads; lubrication of wire ropes and gears; inspection of hoses and pipes; and inspection of electric motors and pumps is recommended. Sliding surfaces on gate stems and seal seats should be inspected for deposits of scale from the water. Seals of stoplogs and bulkheads which are usually exposed to sunlight should be protected from deterioration by use of a water-based latex paint. The normally inaccessible regions around regulating gates and valves should be specifically checked for cavitation damage. Repair of corrosion and cavitation damage should be promptly taken care of. Cavitated surfaces must be built up by the use of a suitable rebuilding compound; also cause of cavitation must be investigated and removed. All corroded surfaces must be cleaned to bare metal and touched up with the original paint or with a paint compatible with the original. In order to ensure structural integrity, the thickness of structural members should be ultrasonically measured and recorded every 5 years. Gates should be completely sandblasted (to remove subsurface corrosion) and repainted every 15 to 20 years; this will, of course, vary depending upon the operating condition and the type of protective coating. It is essential that bulkheads or stoplogs or other means are provided to isolate service gates and valves for inspection and repairs. The author has come across several projects where this provision was not made; the gates could not be inspected and maintained, and therefore had to be replaced at considerable expense in relatively short time (30 to 35 years).

A careful record should be kept of all repairs for future reference and for establishing future maintenance schedules.

42. Weight Estimates. To estimate the weight of a gate of a given size and head at the feasibility stage of a project, formulas for approximate gate weights have been developed, based on statistical samples of manufactured gates. As gate opening width and height ($w \times h$) and hydrostatic head on sill (H) are the only data generally known at the feasibility stage, the weight-estimating formulas have been developed in terms of these variables.

It should be recognized that the weights of gates would vary with the materials used and the design criteria adopted. Refer to Tables 1 to 7 (pp. 17.71 to 17.76) for the actual weights of the various types of large gates in operation (or designed) around the world. This information was taken from a paper by P. C. Erbiste[3] and supplemented by several of the large gates recently designed by Harza Engineering Company. Statistical formulas developed by Erbiste for various gates are included in Table 8. Formulas developed by Harza Engineering Company and successfully used for the past decade are indicated below.

The results in some cases are expected to be different between Erbiste's formulas and Harza's formulas because they are based on different statistical data. Also formulas may give faulty results when the size of the gate being estimated is outside the range of gate sizes which was used to determine the weight formula. It is best to compare the results based on various formulas and then make a judgment based on experience.

Harza Formulas. The terminology used is as follows:

w = opening width, m
h = opening height, m
h_1 = height of individual section, m
H = head on sill, m
H_0 = sill-to-deck height, m
W = gate weight, kg
W_0 = weight of embedded parts, kg (except trunnion support girder and anchorage)
Q = hoist capacity, tonnes

Spillway Tainter Gates

$$W = 360w^{1.25}h(0.01h + 0.5)$$

$$W_0 = 70w + 300h$$

$$Q = 2.08\left(\frac{W}{1000} + 20\right) \quad \text{for hydraulic hoist}$$

$$Q = 1.27W \quad \text{for mechanical hoist located downstream of the skinplate}$$

$$Q = W \quad \text{for mechanical hoist located upstream of the skinplate}$$

Spillway Bulkheads

$$W = 255w^{1.3}h(0.014h + 0.5) \quad \text{for complete bulkhead}$$

$$W = 255w^{1.3}h_1\left[0.028\left(H - \frac{h_1}{2}\right) + 0.5\right]$$

for individual bulkhead sections

$$W_0 = H_0[210 + 0.026(H_0 w)^{1.5}] + 85L$$

Vertical-Lift Spillway Gates

$$W = 270w^{1.3}h(0.014h + 0.5) + (wh)^{1.5}\left(0.27h + \frac{12h^{0.5}}{w}\right)$$

$$W_0 = h[150(wh)^{0.4} + 190] + 90w$$

$$Q = 0.0014[W + 24w^2 + h^2(5.33 + 2.9w) + h(16w + 22)]$$

Vertical-Lift Fixed-Wheel Gates

$$W = 33w^{1.78}h\left(H - \frac{h}{2}\right)^{0.55} \quad \text{for}\left(H - \frac{h}{2}\right) \geq 22$$

$$W = 162w^{1.78}h\left[0.028\left(H - \frac{h}{2}\right) + 0.5\right] \quad \text{for}\left(H - \frac{h}{2}\right) \leq 22$$

$$W_0 = h(3.3wH + 860) + 240w$$

$$Q = \frac{w}{1000}\left(H - \frac{h}{2}\right)\left[4.5h + 30w + \frac{50w^{0.78}h}{\left(H - \frac{h}{2}\right)^{0.45}}\right]$$

Submerged Bulkhead or Stoplogs (for Intakes and Draft Tubes, etc.)

$$W = 43w^{1.35}h\left(H - \frac{h}{2}\right)^{0.55} \qquad \text{for}\left(H - \frac{h}{2}\right) \geq 22$$

$$W = 210w^{1.35}h\left[0.028\left(H - \frac{h}{2}\right) + 0.5\right] \qquad \text{for}\left(H - \frac{h}{2}\right) \leq 22$$

$$W_0 = 650w + 100H$$

43. Cost Estimates. Based on international quotations analyzed by Harza Engineering Company over the last 10 years, the following are the estimated 1991 rates [cost including freight (CIF)] for budget estimates for gate and hoist crane equipment (add approximately 30 percent if the fabrication is to be done in the United States). The prices would be slightly lower if a large quantity (say greater than five) is involved. For future years, these prices must, of course, be indexed for inflation.

Bulkheads and stoplogs	$7.50/kg
Gates and lifting beams	$8.00/kg
Embedded parts and trashracks	$8.50/kg

For hoists the price varies from approximately $0.12 per kilogram capacity per meter stroke for capacities over 400 to $0.40/kg · m for capacities less than 50 t.

44. Cranes. Overhead bridge cranes, gantry cranes, and monorail cranes are provided throughout a hydro project (including powerhouse and spillways) to facilitate the maintenance, and in some cases installation as well, of various equipment, including gates. Figure 38 shows a cross section through a typical powerhouse which includes an overhead bridge crane (called *powerhouse crane*) and two gantry cranes.

Powerhouse cranes are used for the installation and maintenance of the turbine-generating unit. Crane capacity is based on the weight of the heaviest component to be lifted, which is usually the generator rotor. These cranes have been designed for capacities up to 1000 tons, and can certainly be designed for yet higher capacities. In order to reduce the crane capacity required for the installation and maintenance of large size generators, a generator rotor design that permits assembly of the generator rotor in the generator pit is adopted. A typical powerhouse crane has a moving bridge, one or two trolleys for transverse motion, and a main and an auxiliary hoist mounted on the trolleys. Geared drives are provided for each motion. A twin-trolley arrangement results in reduced height of the powerhouse, because less headroom is required for smaller trolleys, and in improved load distribution on the bridge girders. Also, in many cases, two cranes (each crane with half the total capacity required for rotor/stator handling) are provided for the same reasons. The two-crane arrangement also maximizes the utility of the cranes. For the very occasional generator rotor/stator lifts, the two cranes are coupled together to combine the lifting capacity of each crane and/or each trolley.

17.70 GATES AND VALVES

TABLE 1 Spillway Tainter Gates

No.	Year	Project	Span w, m	Height h, m	Head H, m	Gate weight, kN	Manufacturer
1	1982	Itaipu	20	21.34	20.84	2756	Badoni-ATB and others
2	1978	Salto Grande	15.3	22.33	22.33	1579	ATB
3	1974	Salto Osorio	15.3	20.77	20.77	1667	Confab/Kurimoto
4	1979	Salto Santiago	15.3	20.77	20	1815	Ishibrás
5	1977	Itaúba	15	20.3	20	1471	BSI
6	1974	Agua Vermelha	15	19.6	18.6	1805	Mec. Pesada
7	1977	São Simão	15	18.78	18.98	1446	Badoni-ATB
8	1980	Foz de Areia	14.5	19.45	19.45	1620	Badoni-ATB
9	1979	Itumbiara	15	18.86	18.36	1275	Badoni-ATB
10	1975	Marimbondo	15	18.85	18.35	1216	Badoni-ATB
11	1972	Mascarenhas	13.8	16.87	16.87	963	Ishibrás
12	1976	Capivara	15	15.6	15.2	938	Badoni-ATB
13	1969	Ilha Solteira	15	15.46	15.13	1177	Mec. Pesada
14	1979	Paulo Afonso IV	11.5	19.65	19.65	1049	Badoni-ATB
15	1978	Emborcação	12	18.7	18.7	1315	Mec. Pesada
16	1970	Porto Colômbia	15	14.9	14.9	814	Badoni-ATB
17	1971	Cedillo	14	14.16	14.16	975	ATB
18	1968	Jupiá	15	13.7	12.7	945	Ishibrás
19	1974	Sobradinho	13.75	13.4	15.14	702	Voith
20	1964	Funil	11.5	16.5	16.5	942	Bardella
21	1964	Funil	13	14.22	14.22	706	Mec. Pesada
22	1967	Estreito	11.5	16.2	15.8	711	Badoni-ATB
23	1961	Furnas	13	15.86	15.4	611	IHI
24	1974	Coaracy Nunes	12.5	14.1	13.6	650	Badoni-ATB
25	1974	Volta Grande	15	11.41	10.9	506	Badoni-ATB
26	1968	Boa Esperança	13	12.5	12.5	642	Ishibrás
27	—	Kafue Gorge	14	11.53	11.53	677	R. Calzoni
28	1960	Três Marias	11	14	14	647	Mec. Pesada
29	1971	Passo Real	11.5	11.77	11.27	483	Ishibrás
30	1977	Banabuiu	15	8.5	8	441	Ishibrás
31	—	Zarga	10	12.1	12.1	477	R. Calzoni
32	—	El Chocon	14	8.41	8.41	405	ATB
33	—	Tavera	9.2	12	12	393	ATB
34	1968	Salto Mimoso	13	8.48	8.18	304	Ishibrás
35	1972	Curuá-Una	10	10.26	10	344	Ishibrás
36	1968	Salto Mimoso	11	8.48	8.18	253	Ishibrás
37	1977	Eng. Avidos	10	9	9	343	Ishibrás
38	—	Cachoeira do Funil	10	8.5	8.62	373	SFAC
39	1972	Paulo Afonso III	7.67	10.5	10.5	277	Voith
40	1967	Foz do Chopim	12.5	6	6	271	Badoni-ATB
41	1990	Rocky Mountain	6.86	7.70	6.70	160	Voest-Alpine
42	1991	Winfield	33.53	8.53	8.53	2402	Design Weight
43	1991	Macagua II	22.00	15.30	15.00	2060	Imosa
44	1988	Yacyreta (Ana-Cua)	15.00	17.20	17.20	1215	ATB and others

TABLE 1 Spillway Tainter Gates (*Continued*)

No.	Year	Project	Span w, m	Height h, m	Head H, m	Gate weight, kN	Manufacturer
45	1987	Yacyreta (main spillway)	15.00	20.20	20.20	1668	ATB and others
46	1985	King Talal	10.00	10.62	10.50	524	Riva-Calzoni
47	1980	Yacyreta	14.94	20.73	20.73	1765	Metanac
48	1980	Ana-Cua	14.94	16.31	16.31	1152	Metanac
49	1979	Bath County	9.75	9.60	9.60	343	MAN
50	1978	San Lorenzo	12.19	16.46	16.46	919	Bynsa
51	1977	Kajakai	12.01	11.46	11.46	569	Hitachi
52	1974	Cerron Grande	12.19	15.64	15.64	827	Voest
53	1969	Reza Shah Kabir	15.00	21.28	21.28	17.16	BVS

TABLE 2 Submerged Tainter Gates

No.	Year	Project	Span w, m	Height h, m	Head H, m	Gate weight, kN	Manufacturer
1	1979	Jebba	12	9.5	36	908	Mitsubishi
2	1974	Sobradinho	9.8	7.5	33.87	734	Voith
3	1962	Roseires	10	13	17.53	954	Sorefame
4	1973	Moxoto	10	8	28	675	Bardella/Sorefame
5	1958	Macagua	10	11	17	571	Voest-Alpine
6	1969	Dez	12.8	5.6	16.3	419	Krupp
7	—	Aguieira	9.5	8.86	15.56	476	Sorefame
8	—	Kariba	4.9	6.17	51.89	305	Sorefame
9	1970	Midorikawa	6.2	6.3	31.4	566	Mitsubishi
10	—	Picote	4.3	5	57.2	265	Sorefame
11	—	Régua	5	5.8	35.15	373	Sorefame
12	1969	Matsubara	4.4	4.4	51.6	279	H. Zosen
13	1970	Lower Tachien	5	7	23.3	333	Mitsubishi
14	1964	Tsuruta	4.3	4.15	45.6	237	H. Zosen
15	1977	Oishi	3.63	5.43	45.63	383	H. Zosen
16	1975	Palagnedra	5.1	4.3	26	181	Z. Wartmann
17	—	Valeira	4.2	3.8	36.57	282	Sorefame
18	1971	Rio Prado	3.5	3.5	54.7	176	Voest-Alpine
19	—	Oguchigawa	4.62	3.4	20.8	171	H. Zosen
20	1966	Tinajones	3.6	2.83	40	147	MAN
21	1969	Capivari-Cachoeira	2.5	4.45	45	107	Ishibrás
22	—	Yado	2.82	2.82	49.7	178	H. Zosen
23	1990	Cushman	12.19	6.20	14.10	4.00	Noell
24	1971	Cerron Grande	7.00	7.00	27.50	389	Voest-Alpine

TABLE 3 Large Fixed-Wheel Gates ($w^2hH > 2000$ m^4)

No.	Year	Project	Span w, m	Height h, m	Head H, m	Gate weight, kN	Manufacturer
1	1982	Itaipu	7.31	22.37	137	3433	Mec. Pesada and others
2	1979	Sobradinho	12.3	19.15	43.28	1913	Badoni
3	1975	Capivara	8.5	12	58	1481	Bardella/Sorefame
4	1978	Agua Vermelha	10.4	9.53	42	1398	Bardella
5	1969	Ilha Solteira	8.5	9	58.5	1128	Mec. Pesada and others
6	1979	Itumbiara	7.3	11.84	58.1	834	Krupp
7	1964	Jupiá	11.65	7.5	31	1256	Bardella and others
8	1981	Nova Avanhandava	6.54	18	37.8	831	Mec. Pesada
9	1980	Foz do Areia	7.4	7.4	63.7	883	BSI
10	1976	Moxotó	7.1	14.3	34.9	785	Mec. Pesada
11	1973	Porto Colômbia	10.85	5.43	35.6	1079	Mec. Pesada
12	1971	Volta Grande	6.33	14.8	36	915	Bardella
13	1975	São Simão	6.5	11.3	43.64	834	Voith
14	1979	Salto Santiago	6.6	11.1	41.6	559	Badoni
15	1968	Jaguara	7.2	9.7	36.3	554	Coemsa/GIE
16	1980	Paulo Afonso IV	8.7	9.35	24.25	716	Ishibrás
17	1955	Euclides da Cunha	6.5	6.6	57.25	530	Neyrpic
18	1978	Emborcação	5.44	8.74	60.25	549	Mec. Pesada
19	—	Malpaso	4.8	8.2	62.1	438	ATB
20	1972	Promissão	9.1	6.2	20.8	589	Voith
21	1968	Boa Esperança	6.75	6.68	33	471	Ishibrás
22	—	Plover Cove (Pos. I)	9.14	8.53	13.72	329	R. Calzoni
23	—	Bandama	5.7	10	29.2	620	R. Calzoni
24	1959	Três Marias	4.5	8.15	57	471	IHI
25	1970	Estreito	6.55	10.63	19	319	R. Calzoni and others
26	1973	Passo Real	5.1	9	37	392	Bardella
27	—	La Angostura	3.45	8.76	81	527	ATB
28	1973	Mascarenhas	4.56	12.17	32.67	513	Ishibrás
29	—	La Villita	3.75	10.5	55.54	444	ATB
30	1974	Salto Osorio	7.4	7.4	19.7	425	Ishibrás
31	1960	Furnas	4.6	9.65	33.5	314	Krupp and others
32	—	Macchu Picchu	12	6.5	7.2	272	ATB
33	1966	Peixoto	6	6	22	246	Ishibrás
34	—	Tachfine	3.5	5.25	71	434	R. Calzoni
35	—	El Novillo	3.6	6.6	51.75	238	ATB
36	—	Inga	5.5	5.5	25.1	221	ATB
37	—	Santa Rosa	4	5	50.12	205	ATB
38	—	Khasm el Girba	5.05	4.7	24	180	ATB
39	—	Cedillo	3.4	4.5	48	182	ATB
40	—	Peligre	2.84	6.35	45.3	206	R. Calzoni
41	1984	Mirorós	4	3.5	40	108	Muller
42	1977	Itaúba	3.8	6.26	22.5	108	Coemsa
43	1984	TARP	1.90	4.65	113.35	222	Mitsubishi
44	1983	TARP	3.12	3.94	74.83	136	Armco
45	1983	TARP	3.55	4.54	63.72	159	Armco
46	1982	Bath County	3.90	8.70	70	431	Mitsubishi
50	1981	San Lorenzo	6.00	13.65	43.6	793	Bynsa

TABLE 4 Small Fixed-Wheel Gates ($w^2hH < 2000$ m^4)

No.	Year	Project	Span w, m	Height h, m	Head H, m	Gate weight, kN	Manufacturer
1	1965	Guandu	8	5.3	5.8	114	Ishibrás
2	—	Fadalto	3.25	4.8	32.22	108	R. Calzoni
3	1965	Guandu	8	4	4.5	89	Ishibrás
4	—	La Spezia (pos. 3-4)	4.4	7.4	6.9	89	R. Calzoni
5	1978	Serraria	3.5	3.5	19.5	82	Badoni-ATB
6	1965	Guandu	8	3	3.5	66	Ishibrás
7	1973	Alecrim	3	3.8	18.1	65	Badoni-ATB
8	1961	Bariri	2.95	2.1	32.4	47	CKD-Blansko
9	—	Diga di Ganda	3	2.5	23.2	51	ATB
10	—	S. Fiorano	4.5	4.6	4.6	53	R. Calzoni
11	—	Pisayambo	2.1	2.6	35.66	48	ATB
12	—	La Spezia (pos. 1)	2.45	7.2	7	45	R. Calzoni
13	1968	Taipu	2	2	36.4	44	Ishibrás
14	—	Valdarda	1.7	1.7	47.35	37	R. Calzoni
15	—	Ponte Liscione	2.5	2.5	11	26	R. Calzoni
16	—	Macchu Picchu	1.9	3.5	4.5	11	ATB
17	1965	Guandu	2.6	2.1	3.92	12	Ishibrás
18	1965	Guandu	2.7	2.15	2.65	10	Ishibrás
19	1982	Bath County	1.97	2.95	41.15	110	Mitsubishi

TABLE 5 Double-Leaf Fixed-Wheel Crest Gates ($w^2hH > 2000$ m^4)

No.	Year	Project	Span w, m	Height h, m	Head H, m	Gate weight, kN	Manufacturer
1	1955	Ybbs-Persenbeug	30	13.5	13.5	2845	W. Biro
2	1980	Crestuma	28	13.8	13.8	2707	Sorefame
3	1961	Aschach	24	15.5	15.5	2344	W. Biro
4	1960	Beauchastel	26	13.5	13.5	2403	BVS
5	1963	Passau-Ingling	23	14.3	14	1723	Voest-Alpine
6	1967	Wallsee-Mitterkirchen	24	13.2	13.2	1913	W. Biro
7	1959	Schärding	23	13.8	13.5	1634	Voest-Alpine
8	1953	Braunau	23	13.5	13.5	2178	T. Klönne
9	1972	Ottensheim	24	12.5	12	1876	Voest-Alpine
10	1954	Jochenstein	24	12.1	11.8	2050	Voest-Alpine
11	1965	Komori	14	17.8	17.76	1226	Mitsubishi
12	—	Belver	17	14.15	14.15	1938	Sorefame
13	1951	Rosenau	16	13.5	14.5	1783	Voest-Alpine
14	1959	Losenstein	13.5	16.2	16.2	1138	W. Biro
15	—	Donzère-Mondragon	24	6.5	6.5	958	—
16	1968	Yamanoi	12.7	9.5	9.5	358	Mitsubishi
17	1864	Miyagoochi	9	12.5	13	600	Mitsubishi
18	1952	Kniepass	11.5	9.3	9	469	Voest-Alpine
19	1960	Ichibusa	7.1	14.6	14.8	404	Mitsubishi
20	1966	Shimbashi	10	7.5	7.5	248	Mitsubishi
21	1952	Rauris Kitzloch	8	6	5.5	185	Voest-Alpine

TABLE 6 Bulkheads or Stoplogs

No.	Year	Project	Span w, m	Height h, m	Head H, m	Gate weight, kN	Manufacturer
1	1982	Itaipu	20	21.6	20.42	3461	Badoni-ATB and others
2	1979	Salto Santiago	15.3	21	20	2188	Ishibrás
3	1977	Itaúba	15	20.58	20.23	2354	Vogg
4	1980	Foz do Areia	15.1	20	19.2	1834	Badoni-ATB
5	1977	Itumbiara	15	18.75	18.26	1748	Badoni-ATB
6	1977	São Simão	15	18.45	18	1727	Badoni-ATB
7	1974	Marimbondo	15	18.3	18	1682	Ishibrás
8	1972	Porto Colômbia	15	15.25	18	1403	Ishibrás
9	1975	Capivara	15	15.4	15	1177	Bardella/Sorefame
10	1977	Itumbiara	7.3	13.14	57.59	750	Badoni/ATB
11	1982	Itaipú	10	9.9	40	1040	BSI and Mec. Pesada
12	1972	Promissão	10	12	30.3	1057	Bardella
13	1974	Marimbondo	8.16	13.67	37.3	811	Ishibrás
14	1969	Estreito	11.5	16	16	806	MAN
15	1972	Promissão	11.2	16.5	16.2	665	Bardella
16	1960	Furnas	11.5	15.75	15.75	844	Krupp and others
17	1972	Porto Colômbia	6.84	21	30.2	755	Ishibrás
18	1976	Moxotó	11.95	8.96	23.16	745	Mec. Pesada
19	1973	Moxotó	10	10.6	27	1089	Mec. Pesada
20	1982	Nova Avanhandava	6.3	19.6	34.5	589	Mec. Pesada
21	1979	Salto Santiago	7	12.35	42	490	Badoni-ATB
22	1977	São Simão	6.5	12.36	42.23	516	Badoni-ATB
23	1972	Porto Colômbia	10.85	6.47	26.3	444	Ishibrás
24	1980	Paulo Afonso IV	8.9	10.2	23.66	649	Ishibrás
25	1980	Paulo Afonso IV	5.8	14.3	38.5	423	Mec. Pesada
26	1977	Banabuiu	15	8.5	8	589	Ishibrás
27	1981	Nova Avanhandava	10.1	7.71	18.79	495	BSI
28	1977	Eng. Avidos	10	12	11.62	603	Ishibrás
29	1968	Boa Esperança	13	9	9	324	Ishibrás
30	1960	Furnas	5.66	12.14	33.5	319	Krupp and others

Powerhouse cranes are generally used for several months during the construction phase of a project for the installation of the generating equipment. During this phase, the powerhouse walls and roof may not have been completed, and the crane will serve as an outdoor crane. It is preferable that the crane design takes that into consideration.

Gantry cranes are primarily used for the handling of bulkheads or stoplogs provided for maintenance of service gates or other equipment, and for the handling of trashracks. They are also sometimes used for the removal of service gates for major maintenance. They are similar to powerhouse cranes except that they have gantry legs to suit the height of the bulkhead, stoplog or trashrack (or other equipment) to be handled. In most instances the trolley support girder is cantilevered with respect to the gantry legs to facilitate handling of the bulkheads/stoplogs and trashracks located on the upstream face of an intake or spillway. The hoisting speeds (usually 10 to 20 ft/min) for gantry cranes are generally

TABLE 6 Bulkheads or Stoplogs (*Continued*)

No.	Year	Project	Span w, m	Height h, m	Head H, m	Gate weight, kN	Manufacturer
31	1980	Foz do Areia	9.78	4.76	27.3	369	BSI
32	1977	São Simão	10.37	4.78	23.98	349	Badoni-ATB
33	1977	Itumbiara	8.77	5.9	25.46	354	Badoni-ATB
34	1969	Estreito	6.91	12.3	19	549	Sta. Matilde
35	1974	Marimbondo	7.4	5.06	39.65	477	Ishibrás
36	1972	Curuá-Una	10	10.5	10.17	451	Ishibrás
37	1968	Jupiá	7.56	8.24	20.77	354	Ishibrás
38	1974	Cachoeira Dourada	6.8	9.25	20.3	388	Ishibrás
39	1968	Bao Esperança	7.75	3.57	40	320	Ishibrás
40	1968	Salto Mimoso	13	7	7	325	Ishibrás
41	—	Cachoeira do Funil	10	8.85	8.85	451	SFAC
42	1961	Bariri	12.1	7	7	238	CKD Blankso
43	1973	Salto Osorio	8.5	4.25	23.1	257	Ishibrás
44	1968	Salto Mimoso	11	7	7	288	Ishibrás
45	1969	Estreito	6.16	6.46	23.9	314	Sta. Matilde
46	1973	Mascarenhas	5.15	7.65	27.31	201	Ishibrás
47	1973	Moxotó	10	6	6	230	Mec. Pesada
48	1974	Coaracy Nunes	5.55	6.54	17	219	Ishibrás
49	1976	Curuá-Una	4.94	6.25	21	193	Ishibrás
50	1977	Itaúba	3.8	6.4	22.5	98	Vogg
51	1991	Rocky Mountain	5.5	5.1	40.4	175	Noell
52	1986	Macagua II					
		PH2 Draft tube	9.5	9.05	33.1	680	Commetasa
		PH3 Draft tube	9.58	8.25	27.9	609	Commetasa
		Spillway	22	16.7	15.90	2741	Imosa
53	1982	Bath County					
		Draft tube	5.94	4.32	54.44	197	MAN
		Intake	3.89	8.69	70	156	MAN
		Spillway	9.75	9.68	9.68	216	MAN
54	1981	San Lorenzo					
		Intake	6	14.62	42	477	Bynsa
		Spillway	13.9	18.74	18.74	1752	Bynsa

higher than for powerhouse cranes (usually 4 to 6 ft/min). Gantry cranes have been designed to hoist loads up to 300 tons. A gantry crane handling an underwater load must be provided with a fail-safe feature (overload sensor) to prevent damage to the gantry structure if an attempt is made to hoist a jammed load. For gantry cranes which hoist gates up an inclined plane, bumpers are usually provided on the crane legs closest to the gate to reduce impact problems should the gate swing against the gantry legs when the gate comes out of its slot.

Monorail cranes are usually standard manufactured units with capacities up to 20 tons for handling smaller loads.

Handling studies must be performed for the powerhouse cranes to check the clearances available in the powerhouse for hoisting, moving, setting down, and assembly of the components. The total height of the crane including the distance from the top of the trolley to the high hook position plus the height of the tallest

TABLE 7 Flap Gates

No.	Year	Project	Span w, m	Height h, m	Head H, m	Gate weight, kN	Manufacturer
1	—	Promissão	8	6	6	169	Sorefame
2	—	Cabora Bassa	12	5.8	5.8	223	Sorefame
3	1958	Ottendorf	30	5.5	5.5	689	MAN
4	1960	Lohmühle	8	5.17	5.17	221	MAN
5	1960	Altrusried	10	5.15	5.15	265	MAN
6	1959	Sihl-Höfe	8.5	4.55	4.55	102	Z. Wartmann
7	1979	Riedenburg	15	4.3	4.3	343	MAN
8	—	Biopio	15	4.25	4.25	288	Sorefame
9	1959	Innerferrera	8	3.7	4.8	129	Z. Wartmann
10	1960	Thun	12	4.15	4.25	142	Z. Wartmann
11	1974	Takase Zeki	6	3.8	3.8	82	Mitsubishi
12	1978	Kemnader See	25	3.6	3.6	436	Krupp and others
13	1962	Hausen	24.1	3.25	3.25	216	MAN
14	—	Limpopo	13.3	3.2	3.2	125	Sorefame
15	1955	Ottenstein	27.7	3	3	185	Voest-Alpine
16	—	Matala	17.5	2.95	2.95	137	Sorefame
17	1962	Hale	6.1	2.59	2.59	42	Voest-Alpine
18	1967	Muro-Matsubara	40	2.5	2.5	329	Mitsubishi
19	1968	Burfell	20	2.5	2.5	116	Krupp
20	1967	Gmunden	22.5	2.37	2.37	127	Voest-Alpine
21	—	Toobetsu A	25	2	2.5	194	H. Zosen
22	1974	Nukui	6	2	2.3	40	Mitsubishi
23	—	Toobetsu B	25	1.7	2.2	161	H. Zosen
24	—	Kamigawara	14	1.7	2.2	95	H. Zosen
25	1970	Kuritsubo	27	1.5	1.8	132	Mitsubishi
26	1958	Altdorf	19.74	1.6	1.6	88	MAN
27	1956	Geinsfurt	55	1.55	1.55	235	Voest-Alpine
28	1978	Bou Heurtma	8.6	1.1	1.1	25	MAN
29	1984	Lake Brazos	38.0	4.26	4.26	596	N.A.

TABLE 8 List of Formulas Derived by Erbiste

Type of gate	Weight, kN
Spillway Tainter	$W = 0.698(w^2 hH)^{0.673}$
Submerged Tainter	$W = 3.688(w^2 hH)^{0.521}$
Large fixed-wheel ($w^2 hH > 2000$ m^4)	$W = 0.706(w^2 hH)^{0.7}$
Small fixed-wheel ($w^2 hH < 2000$ m^4)	$W = 0.888(w^2 hH)^{0.659}$
Double-leaf fixed-wheel	$W = 0.913(w^2 hH)^{0.669}$
Stoplogs	$W = 0.503(w^2 hH)^{0.716}$
Flap	$W = 2.389w(hH)^{0.643}$

assembly to be handled is an important factor in determining the height of the powerhouse; an allowance should be made for the lifting beam, where provided. A load test at 125 percent of the crane capacity is usually performed to verify that the crane design is safe, before the crane is allowed to handle permanent equipment; the dimensions of the test load are critical with respect to distribution of

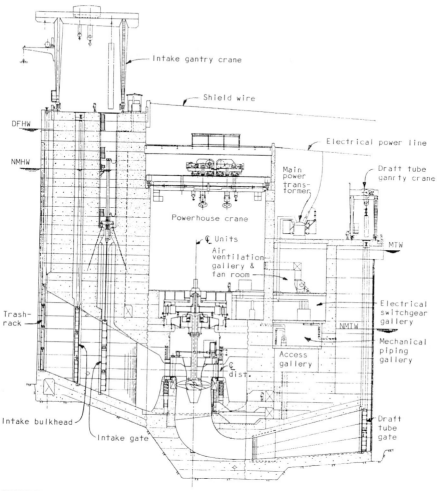

FIGURE 38 Typical crane arrangement.

the load on the floor where the test load will be located, to reduce the intensity of the loading on the floor.

Similar handling studies are also necessary for gantry cranes and monorail cranes. It may be necessary to increase the number of sections on a gate, bulkhead, or trashrack to reduce the height of the gantry structure to improve its stability. The stability must be checked for a gantry crane handling loads with the trollery located on the cantilever.

In the United States the design of all cranes is governed by the Crane Manufacturers Association of America (CMAA) specification no. 70.

REFERENCES

1. Naubascher, E., "Computation of the Dynamic Forces on High Head," *Der Stahlbau,* Institute for Fluid Mechanics, Barrages and Water Supply of the Technical University Karlsruhe/Germany, March 1959.
2. Naubascher, E., H. E. Kobus, and R. P. R. Rao, "Hydrodynamic Analysis for High-Head Leaf Gates," *Journal of the Hydraulics Division,* May 1964.
3. Erbiste, P. C., Estimating Gate Weights, *Waterpower and Dam Construction,* May 1984.
4. "Beams on an Elastic Foundation," University of Illinois bulletin 212.
5. Biot, M. A., Bending of an Infinite Beam on an Elastic Foundation, *Journal of Applied Mechanics,* March 1937.
6. Mercer, A. G., Vane Failures of Hollow-Cone Valves, IAHR Symposium, Stockholm, 1970.
7. Reports MECH-3 and HYD-569, U.S. Bureau of Reclamation.
8. Grishin, M. M., "Hydraulic Structures," vol. 2, Mir Publishers, Moscow.
9. Daly, Steven F., Frazil Ice Blockage of Intake Trashracks, *Cold Regions Technical Digest,* no. 91-1, March 1991.
10. Sehgal, C. K., and D. T. Wafle, Operation of Jet Flow Gate at Bath County Pumped Storage Project, ASME Fourth International Hydropower Symposium, December 1986.
11. Sehgal, C. K., Selection of Spillway Gates for Cold Regions, CEA/EPRI/Hydro-Quebec Ice Problems Workshop, August 1987.
12. Sehgal, C. K., and F. G. Ala, Operation and Maintenance of Hoisting Equipment for Flood Gates for Locks and Dams, ASME Second Biennial Movable Bridge Symposium, November 1987.
13. Sehgal, C. K., Handling of Intake Bulkhead at King Talal Dam, ASCE National Hydraulics Conference, August 1989.
14. Sehgal, C. K., Critical Features of Custom Design Hydro Gates, Water Power '89 Conference.
15. Sehgal, C. K., and F. A. Padilla, Intake Wheel Gates Sealing Arrangements, ASCE National Hydraulics Conference, August 1990.
16. Sehgal, C. K., and Erick Wirzberger, Position Indicator Arrangements for Gates, Waterpower '91 Conference.
17. Kohler, Warren H., Selection of Outlet Works Gates and Valves, ASCE Annual and Environmental Meeting, 1957.
18. DeFazio, F. G., and H. W. Coleman, Analytical and Field Studies of the Intake Gates at Mossyrock Dam, ASCE Hydraulic Specialty Conference, University of Iowa, August 1971.
19. "Hydraulic Downpull Forces on Large Gates," research report no. 4, USBR.
20. Sagar, B. T. A., Downpull in High-Head Gate Installations, Parts 1, 2 and 3, *Water Power and Dam Construction,* March, April, and May 1977.
21. Yeh, C. H., Warren, J. Paul, and James A. Witnik, Design of Prestressed Anchorage for Large Tainter Gates Using the Finite Element Method, International Conference on Finite Element Methods, China, August 1982.
22. Hetényi, Miklos, *Beams on Elastic Foundation: Theory with Applications in the Fields of Civil and Mechanical Engineering,* Univ. of Michigan Press, Ann Arbor, 1946.
23. Sehgal, C. K., S. H. Fischer, and R. G. Sabri, Cushman Spillway Modification, USCOLD Annual Meeting, 1992.

SECTION 18

ENVIRONMENTAL ASPECTS AND FISH FACILITIES

By James H. Thrall and Rimas J. Banys

INTRODUCTION

Society's development and use of water has a wide range of environmental effects incidental to those intended. These can be either adverse or beneficial and are not limited to the aquatic environment (water quality, fish). Rather, they commonly include effects on the surrounding land (soils, vegetation, wildlife) as well as on local and distant socioeconomic systems (recreation resources, public health, industry, agriculture, transportation, etc.).

At each stage of project development, planning, design, construction, and operation, there are decisions or actions that can be taken to ensure that beneficial environmental effects are optimized while negative effects are either avoided or minimized. Experience has shown that altering project plans early to avoid serious environmental problems is almost always less costly than retrofit solutions imposed later. In addition to being cost-effective, however, early incorporation of environmental planning into the project development process is no longer a matter of choice. Beginning with the National Environmental Policy Act of 1969, federal, state, and local governments in the United States have passed legislation and implemented regulations that mandate environmental study of all development projects, and regulatory agencies routinely impose conditions on projects to protect nonproject natural values. The Federal Energy Regulatory Commission (FERC) regulates all hydropower developments in the United States, except those developed by the federal government, and requires a license for those projects.

To obtain the license, extensive environmental studies are necessary to identify potential impacts; plans are required to mitigate impacts imposed in the license to ensure the plans are carried out. As a part of this process, the FERC requires early consultation with all local, state, federal, and tribal agencies having a legitimate interest or regulatory role in the project to ensure that their concerns are included in the studies. Then, environmental studies are necessary to identify potential impacts; plans are required to mitigate impacts identified along with monitoring programs to ensure the plans are carried out. Usually, the regulatory agencies involved in this process include the U.S. Fish and Wildlife Service (Fish and Wildlife Coordination Act, Endangered Species Coordination Act), National Marine Fisheries Service (if anadromous species are present), state fish and game agencies, state Environmental Protection Agency (EPA) and/or U.S. EPA (Clean Water Act, Sec. 401 permit), and various other state and local agencies charged with recreation, land use, coastal zone management, and cultural resource management regulations.

Most other governments around the world have followed suit over the past 20 years, as have all major international financing agencies (World Bank, U.S.

Agency for International Development, etc.). With the added oversight of the many nongovernmental environmental interest groups, who make it their business to evaluate the environmental acceptability of almost any type of development, the successful implementation of a water resource development project today is highly dependent on careful environmental planning.

This section discusses the types of environmental issues commonly faced in the development of a major water resource project and presents the basics of engineering solutions commonly used to environmentally optimize projects. It is limited, however, to developments which involve water resource development for power production and agriculture; impoundment of water in reservoirs; and the diversion of water, either from reservoirs or from free-flowing streams, both for consumptive and nonconsumptive use. Thus, it does not treat the important and complex set of issues surrounding industrial use of water, including pollution by hazardous or toxic chemicals. It is also limited in that it focuses mainly on engineering design and operational solutions to various environmental problems. Therefore, it does not provide detailed discussion of the many types of studies necessary for early alternative analysis and decision making for major projects.

For the benefit of those not familiar with the general issues surrounding dam and reservoir projects, a simplified summary of impact issues is presented. For more extensive discussions of these subjects one should consult the following sources:

Hynes, H. B. N., *The Ecology of Running Waters.* University of Toronto, Canada, 1972.

Ward, J. V., and J. A. Hanford, Editors, *The Ecology of Regulated Streams.* Plenum Press, New York, 1979.

Gore, J. A., and G. E. Petts, Editors, *Alternatives in Regulated River Management,* CRC Press, Boca Raton, Florida, 1989.

GENERAL EFFECTS OF IMPOUNDMENT

When a free-flowing stream or river is impounded by a dam, there are numerous environmental effects, both above and below the dam site. These include the following general categories.

1. Physical Barriers. When a dam is placed in a stream system, an obvious effect is to block the free upstream and downstream movement of fish and other aquatic organisms past the dam. When these movements are a critical part of the life cycle of these animals, as is the case with the various species of salmon in Pacific coastal rivers in the U.S. northwest, as well as the Atlantic salmon, herrings, and shad of the Atlantic coastal rivers of the east and southeast, the effects of blockage are immediate and devastating. These species are examples of *anadromous fish,* that is, species which migrate from the sea upstream into rivers where they spawn. Obviously, preventing them from doing so will result in their extinction in the affected river. This problem has been recognized for many years, and systems to assist the fish past dams, including fish ladders, locks, elevators, and trap and haul facilities, using tank trucks, have been used with varying degrees of success at a large number of dams around the world. Where the results of these fish passage systems have been less than desired, hatchery facilities or artificial spawning channels have been used to supplement the natural reproduction of these species.

There is another group of organisms that migrate into and out of rivers to spawn that is not as well-recognized as the anadromous species mentioned above. These are species which spawn in the sea and have young which enter rivers to feed and grow. Known as *catadromous species,* they include the American eel and several species of freshwater North American shrimp. Partially because they are economically less important than the salmon and herrings, the effects of blockage on these animals is not well-understood and few facilities have been built to provide for their survival. Young eels (elvers) are remarkably adept at overcoming dams, even where no provision for upstream passage has been made, and thus have been able to maintain themselves on river systems with dams. Freshwater shrimp have likely been extirpated from some drainages because the young are blocked from migrating upstream from the estuary into their adult habitat in the river, although a few species apparently have been able to survive by moving their reproductive and early growth life stages to the reservoir from the estuary or sea.

On the Occoquan River in Virginia, for example, it has been reported by the project operators that, on the furthest downstream dam of a two-dam water-supply and hydropower project, outmigrating eels have clogged the water-supply pumps.

Only in the past few years has attention been paid to the fate of the young of the anadromous species as they move back downstream to the ocean. Although it has long been known that the outmigrating juvenile salmon were subject to significant mortality as they passed through the turbines or over spillways, it was generally concluded that the volumes of water involved and the value of that water for power production precluded the possibility of screening or diverting the fish around these hazards. This is no longer the case, however, as it has been demonstrated that the cumulative losses suffered over a number of dams, combined with other losses due to difficulty in negotiating the reservoirs, is having significant effects on salmon population levels. Currently significant time and money is being invested in development of screens, louvers, behavioral barriers, and other guidance devices to divert fish away from the turbines, thereby increasing the survival of young fish as they make their way downstream to the ocean.

Some of this technology has been borrowed from earlier work done to screen downstream migrants out of irrigation and cooling water diversions. Effective fixed screens, traveling screens, and drum screens have long been in existence for this purpose. However, because of the volumes of water involved, these systems are usually viewed as being too expensive for large hydroelectric or pumped-storage projects.

2. Altered Flow Regime. Dams and reservoirs are created to store water for some specific use. Often this involves holding back part of the flow for use in power production, irrigation or municipal water supply at a time of year when natural flow in the stream is low. In other cases, water is released for power production on a daily cycle, increasing during times when energy demand is high and dropping off to some minimum flow at off-peak times. Finally, in the case of agricultural and municipal use of stored water, some portion of the flow is diverted out of the stream for consumptive use, thus not only affecting the timing of the flows but reducing the total flow available for the downstream aquatic ecosystem.

All free-flowing streams have a naturally variable discharge pattern, dependent on seasonal patterns of precipitation (and snow melt) in the drainage basin. Although year-to-year variations in this pattern can be great, depending on short-term weather patterns, over longer periods there is a definite seasonal pattern, to which the fish, other aquatic animals, and riparian vegetation are adapted. Thus, changes in the seasonal or daily discharge pattern in a stream due to dams and reservoirs will have effects on the biological communities living there. For example, some species of fish depend on spring flood flows to carry the young rapidly downstream to the sea or to lower river floodplain lakes for feeding and growth. A large power project which stores a significant part of this spring flood for late generation may negatively affect the movement of fish at this critical time.

More serious, and usually more definitely seen as a detrimental effect, are those cases where water is stored and/or consumptively diverted out of the stream at times when flows are naturally low. Especially in late summer, when water temperatures are likely to be high enough to stress the aquatic fauna even with natural flow, further reduction of the discharge can have severe biological impacts. Lethal temperatures may be reached, killing cold- or cool-water–adapted species or life stages. Further, reduced flow will eliminate habitat as the width and depth of the stream decreases. The issue of minimum flow releases in streams is a major one which, because of the competing resource aspect, is often difficult to address. Very commonly, increasing flows during low-flow periods to accommodate fish has severe economic consequences to the project's primary purpose of power production or water supply. When one factors in the possible need for an entirely different flow regime to meet the needs of recreational rafters and the desire of reservoir recreationists to have a stable shoreline elevation, the entire issue of project operation and downstream flow can become complex very quickly. A number of methods, of varying degrees of complexity, have been developed to address the issue of instream flow needs for fish. The most commonly employed method at this time is the incremental flow instream methodology (IFIM), developed by the U.S. Fish and Wildlife Service (USFWS). Hydraulic models are used to compare stream depth, velocity, wetted perimeter, and substrate conditions at various discharge conditions to the needs of the fish, to determine what flow is required to protect or enhance survival of the different life stages.

For projects where daily flow releases vary in response to power needs (load-following or -peaking power projects) the minimum flow release question also is important. However, an additional set of issues to be considered for these projects is those related to the rate of change of the flow as discharge is increased and decreased in response to the power needs of the system. Too rapid an increase in discharge can wash small fish out of their protected habitat areas. In addition, problems of human safety can be associated with a rapid increase in discharge over a short period. Possibly more important to the fish is the question of rate of change when peak discharges are decreased. Where flows drop rapidly, fish, particularly the young, weaker-swimming life stages (as well as aquatic insects and other invertebrates) can be stranded in off-channel pools which overheat or become stagnant or in areas which become completely dewatered. Obviously, this has a detrimental effect on these species. In these cases, two types of solutions have been implemented. In some cases a small dam is built just below the peaking project to act as a reregulating structure, utilizing the limited storage provided to smooth out the releases downstream. These reregulating dams may or may not include power production. In other cases it has been found that simply controlling the rate at which the discharge is increased and decreased will provide adequate protection to fish and other aquatic life. Often, ramping up or down in a stepwise fashion will provide adequate protection to the fish.

It should be noted that, in some cases, this shaving of the peak flood discharge may protect some fish habitats from damage by erosion, thus providing a benefit for that particular year. However, over longer periods, the lack of these temporarily damaging scouring floods may result in changes in the stream channel morphology from sediment buildup or the growth of riparian vegetation, which has a negative impact on certain species. In any case, the effects are rarely simple and seldom solely detrimental or beneficial.

3. Flow Diversion. Where water is diverted out of the stream for municipal or industrial use, irrigation, or cooling water at thermal plants, fish, especially the younger, weaker swimming states and passive stages such as eggs or larvae, can also be diverted. These organisms are subject to mortality due to various factors including mechanical injury and thermal stress, or may simply become lost and eventually die in off-channel areas. Proper sitting of the diversion intake structure (in areas where few fish are likely to occur and so as to have favorable currents to carry fish past), along with screening, will reduce the problems associated with flow diversion.

4. Water-Quality Effects. The effects of impoundment on the water quality of a river system are highly dependent on the size, depth, and operational rules used to release flows from the reservoir, as well as on the basin elevation, regional geology, and local soil, vegetation, and land-use conditions. Narrow, deep reservoirs with a relatively high inflow in relation to the total available storage (usually operated as run-of-river) will tend to have minimal effects on the water quality. These types of reservoirs do not develop lakelike conditions of stratification with the resultant changes in temperature, dissolved gas, and nutrient content.

Large reservoirs with greater storage capacity in relation to inflow and with a larger surface area can have substantial water-quality effects. Slowing of the flow of water through the reservoir, exposure of a large surface area to solar energy, and allowing development of seasonal stratification (isolation of certain strata or layers of water by formation of thermal gradients) will result in substantial water-quality changes including temperature, dissolved oxygen, and nutrients. Often, a large storage reservoir located in a temperate climate will stratify over the summer, developing a warmer, relatively well-oxygenated water layer on the surface (the epilimnion) and cooler, possibly (but not necessarily) deoxygenated water on the bottom (the hypolimnion). These two layers are separated by a zone where temperatures change rapidly over depth (the thermocline) and will persist until fall cooling is sufficient to cause remixing (overturn). The location of the outlet works in such a system will determine the water-quality conditions in the downstream reaches of the river. In some cases, cooler water released from the bottom of a stratified reservoir, if the dissolved oxygen content is high enough, has been beneficial, in that a high-quality cold-water (trout and salmon) habitat is created. In other cases, however, existing warm-water fisheries have been extirpated below dams with very cold summer releases without the compensation of a cold-water fishery developing.

Frequently, in reservoirs located in warmer climates, usually where high concentrations of nutrients exist in the water, the lower strata of the reservoir can have very low dissolved oxygen levels (even becoming totally anoxic) during late summer and early fall. This is because, in the deeper portions of the water, not enough light penetrates to allow photosynthesis (which produces oxygen) to take place. Respiration and decay do continue to occur, however, and the dissolved oxygen can quickly become depleted, particularly in reservoirs where large amounts of organic material remain on the bottom or in reservoirs receiving nutrient-laden water from the upstream drainage basin. In such cases, the surface water layers, where sunlight does penetrate, can be saturated or even sometimes supersaturated with oxygen during the day, due to very high photosynthesis rates, but, at night, as the algae stop photosynthesizing but continue to respire, even the upper layers of the water can become deoxygenated.

Solutions to these types of problems include: use of multilevel intake structures to control the depth and thus either temperature and/or dissolved oxygen level of the water being released, aeration of the water in the reservoir by mechanical means, injection of air or pure oxygen by means of pumps and aeration devices, injection of air into the turbines at hydro projects, or use of spillways, Howell-Bunger valves, or other outlet works to entrain air into the downstream releases. Many of these solutions are very costly, both in terms of equipment and energy consumption, and may be only partially successful. For projects just in the development stage, careful reservoir clearing, although initially very costly, may eliminate some of the low-dissolved-oxygen problems, because nutrients in the vegetation are not left to cycle into the reservoir water. On a more regional development scale, land management plans to prevent excessive nutrient runoff in the drainage basin are also good solutions to the problem of high nutrients. The last two approaches have the added advantage of decreasing the probability of problems with nuisance growth of algae and rooted aquatic vegetation, which can be a problem in reservoirs with excessive nutrients.

Finally, reservoir clearing will also prevent or ameliorate mercury problems in some reservoirs. In many newly formed reservoirs with high levels of organic matter present in the bottom, bacterial action on these organics methylates mercury, liberating it into the water where it becomes incorporated into biological organisms. By the process of biomagnification (as one type of organism, say an insect, feeds on other organisms with low tissue concentrations of mercury and the total mercury ingested over time builds up to a higher level; a third organism feeds on the insect and the concentrations build up even more; etc.), top predators such as game fish can develop levels of mercury high enough to be a health concern for humans catching and consuming the fish on a regular basis.

A second type of dissolved-gas problem occurs at dams with spillways designed with deep plunge pools. The spilled water, with air bubbles mixed in, plunges to depths where the pressures can significantly exceed 1 atmosphere. In simplified terms, this results in supersaturated conditions, not only in the water, but in any fish which are carried into the plunge pool area. These fish, with supersaturated blood, if they are rapidly returned to the surface, suffer the same fate as deep-sea divers who surface too rapidly. Under reduced atmospheric pressure, the gases begin to boil off, much like bubbles in a carbonated drink when first opened, causing what is known as *the bends* in humans and *gas bubble disease* in fish. Restricting spills by routing flows through the turbines or using energy-dissipating devices such as Howell-Bunger valves, which dissipate the energy in a way that reduces plunging, are effective means of dealing with this problem.

Many other water-quality problems exist in rivers and reservoirs. However, they are mostly related to industrial, agricultural, or municipal waste disposal practices and are beyond the scope of this discussion.

BIOLOGICAL CRITERIA

As can be deduced from the above discussion, fish and other aquatic organisms are adapted to specific ranges of physical and chemical conditions in the water in which they live. These can include optimum conditions, preferred conditions, or tolerance limits, which vary with the species and life stage. The degree to which a project can meet these conditions on biological criteria is important in determining, first of all, how a project development will impact the biota of the river, and, equally as important, whether a proposed engineering solution in fact will be effective.

Identification and establishment of the appropriate biological criteria requires knowledge of the species to be affected. When and where the different life stages of important species of fish occur in the river and how the project will affect these habitat areas on both daily and annual bases needs to be established. This requires field programs to collect baseline data on the existing fauna, review of previous studies on the area fauna, and general studies of the biology of the species of interest, as well as consultation with the resource agencies to determine what their management goals are for the area. From these sources, information on the temporal and spatial occurrence of the different life stages of the important species can be developed, along with information on existing and desired population size and exploitation rates. Then, postproject environmental conditions can be compared to the biological criteria identified as limiting to the various species of interest to determine, first, what effects the project may have and, second, how the project needs to be modified to avoid excessive negative effects.

The most common environmental conditions which can be effectively controlled by design or operational modifications to a water development project are water velocity, temperature, and dissolved gases, especially dissolved oxygen. Basic knowledge of these three sets of conditions, as they pertain to the kinds of game fish most commonly of concern in the United States, provides valuable guidance to the designers attempting to provide suitable environmental conditions for the affected fauna. The following tables summarize the tolerance limits of several species and life stages of important game fish for these parameters. These are not meant to be exhaustive, and it is always necessary to develop project-specific criteria, using experienced, trained specialists. Rather they serve to indicate the general range of values which can be expected when dealing with this type of problem.

In examining these data, also keep clearly in mind that few of the values presented can be taken as absolute. Living organisms have the ability to adapt to adverse conditions under many circumstances, so that occasional, short-term excedance of an upper temperature limit, for example, might not be fatal to a fish population. However, in a case where temperatures are increased to near the upper limit, over the long term, but never exceed it, that same population might not survive because of chronic stress effects on reproduction and growth of the young. Again, trained, experienced biologists must be involved in determining project-specific criteria to avoid serious problems.

5. Temperature. Table 1 presents both critical upper temperature limits for short-term survival and general ranges of preferred temperatures for fish, including some data on eggs, young, and larval life stages, which often are more sensitive. The designations of warm-water, cool-water, and cold-water species, while not always a clearcut distinction, is one generally accepted by most fisheries biologists. These data are not meant to be either exhaustive or absolute. Thermal limits vary within species depending on the geographic region. Moreover, fish can acclimate themselves to different temperature regimes so as to increase their ability to withstand extremes. It is essential, therefore, that data specific to the species in the project area be utilized in setting any criteria for a project. For more information on temperature see the bibliography.

TABLE 1 Thermal Limits and Preferred Temperatures for Selected North American Game Fish

	Spawning	Juvenile		Adult	
		Lethal	Preferred	Lethal	Preferred
Cold-water species					
Chinook salmon	2–14°C	25°C	12–14°C	22°C	12–16°C
Coho salmon	4–8°C	25°C	12–14°C	—	10–13°C
Atlantic salmon	2–8°C	29°C	13–15°C	30°C	12–16°C
Brown trout	2–12°C	26°C	18–23°C	30°C	18–24°C
Cool-water species					
Smallmouth bass	13–20°C	—	26–29°C	35°C	22°C
Wallage	6–14°C	—	—	—	13–20°C
Northern pike	4–11°C	33°C	33°C	29°C	8–24°C
Muskellunge	9–15°C	—	—	33°C	26°C
Warm-water species					
Largemouth bass	17–20°C	36°C	30–32°C	36°C	27–30°C
Bluegill	17–32°C	36°C	32°C	36°C	24–27°C
Channel catfish	21–27°C	38°C	28–32°C	—	30–32°C
White crappie	14–23°C	33°C	25°C	—	27–31°C

Source: BROWN, H. W., "Handbook of the Effects of Temperature on Some North American Fishes," American Electric Power Service Corporation, Environmental Engineering Division. Canton, Ohio, 1974. SCOTT, W. B., and E. J. CROSSMAN, "Freshwater Fishes of Canada," Bulletin 184, Fisheries Research Board of Canada, Ottawa, 1973.

6. Velocity. Fish encountering dams, intake structures, turbines, and outlet works, as well as fish being induced to enter fishways or other guidance structures, rely on current velocity to a large extent in their movements. Further, fish which must pass other artificial obstructions in streams (culverts) must be capable of overcoming maximum velocities. The ability of fish to handle various current velocities is a function of their size as well as their body shape and general design for swimming. As would be expected, larger fish are stronger swimmers. Fish such as salmon and trout, with their streamlined, torpedo-shaped bodies, are better adapted to flowing water and higher velocities than are slack water species with "slab-sided" bodies such as crappies or blue gill.

Fish also have the ability to maintain higher speeds over short distances than they can over long distances. Generally, three swimming speeds are recognized for fish: burst speed, which is a maximum velocity that can be maintained only over a very short period (15 s maximum); prolonged speed, which can be maintained up to as much as 3 h, but which eventually will produce fatigue; and sustained speed, which can be maintained indefinitely without fatigue. Table 2 presents data on swimming performance for several migratory species of North American game fish. Because absolute speeds are a function of size, speeds are presented in terms of body lengths per second.

7. Dissolved Oxygen. Dissolved oxygen (DO) levels that are acceptably high for supporting healthy fish populations have proven to be a difficult subject to

TABLE 2 Swimming Speeds of Some Adult Game Fish, ft/s

Species	Burst	Prolonged	Sustained
1. Chinook salmon	22.4	10.6	3.4
2. Coho salmon	21.5	10.6	3.4
3. Sockeye salmon	20.6	10.2	3.2
4. Steelhead	26.5	13.7	4.6
5. Brown trout	12.7	6.2	2.2
6. Atlantic salmon	23.2	12.0	4.0
7. Shad	15.0	8.0	3.0
8. Herring	7.0	4.5	3.0
9. Carp	14.5	4.0	2.0

Source: BELL, M. C., "Fisheries Handbook of Engineering Requirements and Biological Criteria," Fisheries Engineering Research Program, North Pacific Division, Corps of Engineers, Portland, Oregon, 1973.

address. As is the case for temperature, long-term chronic effects can be subtle and show up only as reduced reproductive success or slowed growth rates. As a result, many regulatory agencies have adapted arbitrary limits of around 5.0 mg/L and applied them to all species and all waters. In truth, however, different species and life stages have different dissolved oxygen needs, as is the case for temperature. In general, cold-water fish such as salmon or trout are much less tolerant of low DO than are warm-water species. Eggs and young life stages are also more sensitive than are adults. Thus, recent trends have been to use different standards, based on existing use and quality of the water bodies in question.

Table 3 presents U.S. EPA transfer for dissolved oxygen in both salmonid and nonsalmonid waters.

FISH PASSAGE

Fish ladders, also referred to as *fishways,* are the most common method of passing adult fish over an obstacle to their upstream migration. Provided that site conditions, topography, hydrology, hydraulics, and fish species present make a fish ladder practical at a certain site, a fish ladder is preferable to trap-and-haul, fish elevator, or other means of passing fish. The primary benefit of a fish ladder, compared to alternative methods of passing adult fish, is that a fish ladder operates continuously and allows fish to swim upstream at their own pace. This minimizes human interference with fish during upstream migrations. A typical layout and terminology for a fish ladder is shown in Fig. 1.

Important factors in the design of fish ladders include swimming performance and behavior of the fish species to be passed, hydraulic characteristics of the type of fishway selected, site conditions, and fish passage experience with the target species. Major steps in the design process include the following:

8. Determination of Biological Criteria. Biological criteria necessary for a fish ladder design include species, age and size of fish, swimming capabilities, run size, seasonal and diurnal timing of migrations, allowable density of fish in holding areas, and knowledge of each species' behavior at barriers and response to flow patterns (Bates 1990). State and federal agency personnel provide the best source of these data in most instances, and there are also several written guides

TABLE 3 Dissolved Oxygen Requirements, mg/L, for Adults, Juveniles, and Eggs of Selected North American Game Fish

Salmonid waters	
Embryo and larval stages*	
No production impairment	11 (8)
Slight production impairment	9 (6)
Moderate production impairment	8 (5)
Severe production impairment	7 (4)
Limit to avoid acute mortality	6 (3)
Other life stages	
No production impairment	8
Slight production impairment	6
Moderate production impairment	5
Severe production impairment	4
Limit to avoid acute mortality	3
Nonsalmonid waters	
Early life stages	
No production impairment	6.5
Slight production impairment	5.5
Moderate production impairment	5.0
Severe production impairment	4.5
Limit to avoid acute mortality	4.0
Other life stages	
No production impairment	8
Slight production impairment	5
Moderate production impairment	4
Severe production impairment	3.5
Limit to avoid acute mortality	3
Invertebrates	
No production impairment	8
Some production impairment	5
Acute mortality limit	4

Source: Copyright © 1990. Electric Power Research Institute. EPRI GS-7001. *Assessment and Guide for Meeting Dissolved Oxygen Water Quality Standards for Hydroelectric Plant Discharges.* Reprinted with Permission.

*Values are water-column concentrations recommended to achieve the required intergravel DO concentrations shown in parentheses. The 3-mg/L difference is discussed in the standards documents.

(U.S. Fish and Wildlife Service 1990). Fish behavior at barriers is complicated, and it is important to involve someone knowledgeable in this field early in the fish-ladder design process.

9. Quantification of Hydrology. Hydrologic data needed for design include flow duration curves for each month during which fish migrations occur; headwater and tailwater rating curves for the fish exit and entrance to the ladder, respectively; an analysis of flood flows; and low flow frequency and duration data (Bates 1990). Fish ladders should operate when fish are migrating; generally this is across a broad range of flows and is negotiated with agency personnel before fish ladder design.

FISH PASSAGE

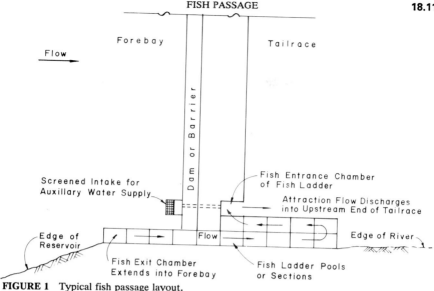

FIGURE 1 Typical fish passage layout.

In a typical situation, fish ladder operation will be required at all river flows up to the 5 or 10 percent excedance flow. The upper limit of river flow that must be accommodated by the fish ladders strongly influences the overall size and cost of the facility, and selection of this flow should be a balance between biological necessity and practical limitations. In many instances, fish ladder operation is not necessary when river flows are extremely low and fish are not migrating; however, most fish ladders designed to operate during high flows also function well (hydraulically) during low flows.

Flood flow calculations are necessary to estimate water levels, hydrostatic forces, and velocity forces for structural design. Fish ladders are designed to withstand flood flows and forces, but not to pass fish during extreme events.

10. Fish Ladder Hydraulics. The most important hydraulic parameters that must be evaluated prior to design are the pattern, velocity, and depth of flow at the downstream end of the fish ladder. Depending on the experience of the designer, these variables can be used to optimize fish attraction to the fish ladder, or can greatly impede fish passage. Placement of the downstream end of the fish ladder (fish entrance) is one of the most important considerations for a successful fish ladder, and a task that requires experience as well as technical knowledge.

Wide rivers and/or irregular barriers commonly require more than one fish ladder at each obstruction to migration, or multiple fish entrances to a single ladder.

The amount and velocity of flow discharging from the fish ladder is another major factor affecting the ability of fish to find the fish ladder, having a direct relationship to fish passage success. In all instances, more water discharged from a fish ladder will result in more (or faster) fish attraction to the ladder, and the amount of attraction flow needs to be balanced between biological needs and economic consideration. When fish ladder entrance conditions are advantageous for directing fish, the amount of fish ladder flow required may be 5 percent or less of river flow. Where conditions are not as favorable, 10 percent or more of the river flow may be needed to attract fish.

Requirements for large volumes of attraction flow at the fish-ladder discharge mandate that an auxiliary water supply system be designed to convey most of the fish attraction water directly from the project forebay to tailrace. Otherwise, the size of the fish ladder becomes enormous. The flow capacity of a fish ladder is determined by the fish ladder type and size.

11. Preliminary Design Report. A preliminary design report is necessary for all fish ladders, to document the biological, hydrologic, and hydraulic criteria (above) that will be used for design. In addition to a clear statement of the design criteria, the report should include conceptual design drawings to show all major elements of construction. Drawings should include fish-ladder plan and layout, configuration of the fish entrance and exit locations, fish-ladder profile, and typical dimensions of fish-ladder section.

Completion of the design report requires that the type and overall size of fish ladder be determined. Fish ladder types include vertical-slot, pool-and-weir/orifice, Denil and steep-pass, and many combinations and variations of these (Fig. 2 presents the three most common types). Vertical-slot or pool-and-weir/orifice fishways are usually employed at large projects; Denil and steep-pass fish ladders are mostly used at small projects.

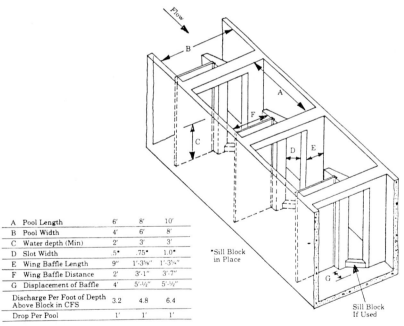

A	Pool Length	6'	8'	10'
B	Pool Width	4'	6'	8'
C	Water depth (Min)	2'	3'	3'
D	Slot Width	.5*	.75*	1.0*
E	Wing Baffle Length	9"	1'-3⅜"	1'-3¾"
F	Wing Baffle Distance	2'	3'-1"	3'-7"
G	Displacement of Baffle	4'	5'-½"	5'-½"
	Discharge Per Foot of Depth Above Block in CFS	3.2	4.8	6.4
	Drop Per Pool	1'	1'	1'

FIGURE 2 (*a*) Vertical slot fishway.

A specific fish ladder type may work for some species, and not for other species. For example, sturgeon will not ascend pool-and-weir ladders, but salmon will. Shad will avoid orifice passageways, and will instead swim through vertical slots or overflow weirs (Bell, 1991). Experience documented by Bell (1991) or

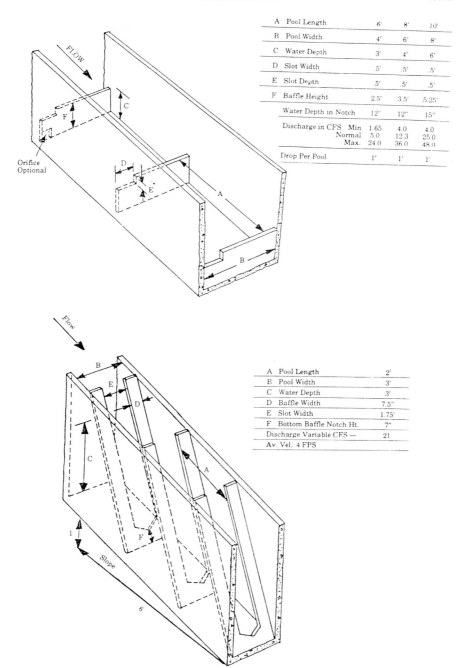

FIGURE 2 *(Continued)* (*b*) Pool and weir fishway. (*c*) Denil fishway.

available from knowledgeable personnel should be used for guidance in selection of a fish-ladder type.

The preliminary report should also list requirements, if applicable, for fish counting stations, trashracks, and periodic maintenance.

12. Fish Ladder Design. Fish ladder design follows completion and approval of the design report, and is usually a straightforward application of engineering principles. Prior to this stage of design, the importance of fish passage expertise for design guidance cannot be underestimated.

13. Fish Ladder Types. The basic types of fish ladders, design principles, and hydraulic characteristics are well described by Bell (1991), Clay (1961), and the U.S. Fish and Wildlife Service (1990). A thorough review and understanding of these references is necessary prior to fish-ladder design. The following narratives and illustration briefly describe each fish-ladder type.

a. Vertical-Slot Fish Ladder. A vertical slot fish ladder consists of a series of pools separated by a wall containing a vertical slot (Fig. 2); some ladders have two slots in each wall separating pools. Water flowing down the fishway accelerates as it passes through each slot, then rapidly decelerates in the pool downstream. Velocity changes, and associated turbulence, are designed to dissipate about 1 ft of head at each slot. Head loss and flow data are best summarized by Katopodis (1990), with added data presented in Bates (1990) and Bell (1991).

Vertical-slot fishways are normally built on a 10 percent slope. Major advantages of vertical-slot fishways surface on relatively large fish passage projects are: (1) vertical-slot fishway flow is higher than flow in other fishway types and (2) vertical-slot fish ladders can accommodate larger variations in forebay and tailrace water levels than other fishway types.

b. Pool-and-Weir/Orifice Fish Ladders. Pool-and-weir fish ladders may or may not have orifices between pools (Fig. 2), depending primarily on which fish species is targeted for passage. The gradient of pool-and-weir, orifice, or combinations of these types is usually 10 percent. Hydraulic design data for these types of fish ladders are presented by Katopodis (1990) and Bell (1991).

c. Denil and Steep-Pass Fish Ladders. Denil and steep-pass fish ladders both consist of baffled rectangular chutes, usually constructed of metal (Fig. 2). Katopodis (1990) gives a good description of each type and presents equations for hydraulic design.

There are three advantages of Denil and steep-pass fishways which frequently apply to relatively small projects: (1) The 20- to 30-cfs capacity of Denil fishways and 5- to 10-cfs capacity of steep-pass ladders is appropriate for many small projects; (2) both fishway types are installed at slopes of 20 to 30 percent, substantially reducing fish-ladder length, size, and cost; and (3) some fish species (e.g., salmonids) move faster through and appear to prefer Denil-type ladders, compared to pool-and-weir and vertical-slot types.

14. Fish Ladder Components. The major components of a fish ladder, with respect to its function, are (1) fish entrance, (2) fish-ladder section (pools, chambers, etc.), (3) fish exit, and (4) auxiliary water system. Each component is critical to the overall performance of the fish ladder, as described below.

Fish which cannot find the fish ladder entrance have no chance of migrating upstream; therefore, the fish entrance is probably the most critical design element in a fish ladder. The most important design objectives are

1. Locate the fish entrances at the most upstream area possible, preferably in an area where fish have been observed to school. Proper location depends on fish

behavior and the cumulative experience of individuals consulted during preliminary design.

2. Design the fish ladder discharge so a distinct, fast-moving current of water enters the tailrace area in sufficient volume to attract fish. Water velocities should be in the range of 4 to 8 fps for adequate attraction of most species, and volume should be in the range of 5 to 10 percent of flow in most instances. The discharge should be aligned so upstream migrating fish encounter the attraction flow without having to change direction.

Fish-ladder sections (pools, chambers, etc.) can readily be designed with existing guidelines (Bell, 1991; Katopodis, 1990). For vertical-slot, pool-and-weir, and orifice fish ladders, each pool functions as a resting pool for fish and additional resting areas are not required. In Denil and steep-pass fish ladders, resting pools are conventionally spaced at intervals of 20 to 60 ft to provide holding areas for upstream-migrating fish.

The upstream end of a fish ladder functions as a water intake and fish exit; it must be designed with both these objectives in mind. For water intake, considerations include maintaining the required volume of flow; ensuring that fish-ladder flow has similar temperature and water chemistry as water downstream of the migration barrier, fluctuating forebay, or river water levels; and providing a trashrack to minimize the amount of debris entering the ladder.

Considerations related to the fish exit from a fish ladder include the following (Bates 1990):

1. Do not locate the fish exit next to a spillway, penstock intake, or other area where fish could be swept back downstream.
2. The fish exit should be placed in an area where current moves downstream, so fish will sense the proper direction (upstream immediately after exiting the ladder).
3. Water exit depth should be 3 to 6 ft in most cases, and preferably near a shoreline to assist with fish orientation.

Requirements for many fish ladders include the need for auxiliary attraction water, because the amount of water needed to attract fish to the ladder greatly exceeds what is needed in the rest of the ladder for fish passage. Auxiliary water is pumped or supplied by gravity to the lowermost sections of fish ladders to fulfill this need.

A large volume of auxiliary water entering the lower pools of a fish ladder must be introduced into the pools through grating or narrow slots, so fish remain in the fish ladder and cannot enter the auxiliary water system. To prevent fish from being falsely attracted to the auxiliary water, grating panels and baffles are used to diffuse the incoming water so velocities do not exceed 1 fps. Design considerations for auxiliary water systems are summarized by Bates (1990).

15. Fish Locks and Fish Lifts. The main use of fish locks is in Scotland and Ireland for passage of Atlantic salmon. A few fish locks have been constructed within some of the Columbia River dams; however, their use has been discontinued. Apparently the fish ladders also constructed at these dams handle fish passage more capably. This is probably because of the enormous fish run in the Columbia River.

The construction of fish locks in small rivers under certain circumstances may be more economical than fish ladders. There is not complete agreement, however, that the two would be of equal efficiency in fish passage. The success of locks with Atlantic salmon does not imply an equal success with Pacific salmon.

The method of operation of fish locks is somewhat similar to navigation locks, with the exception that flow is maintained to induce fish into the chamber and out of it. Water is discharged through the chamber into tail water in order to attract fish into the lock. Then, the lock chamber is filled and the fish are induced to swim out into the reservoir by flow through the exit. The fish lift substitutes a mechanical lift for the locking operation. Typical layouts for fish lift and fish lock systems are shown in Fig. 3.

FISH PROTECTION

Fish protection at hydropower, water resource, and water intake projects is intended to prevent fish from entering project intakes, because facilities downstream of intakes are commonly harmful or fatal to fish. In addition to physical injury, intakes may divert migrating fish from their intended course, or remove resident fish from their native habitat. In almost all cases, entrainment of fish into project intakes is viewed as detrimental to aquatic resources, and methods should be explored for fish protection.

Fish protection methods have been extensively studied for at least 40 years, and a large amount of information is available in published and unpublished form. The comprehensive review published by the Electric Power Research Institute (1986) is an excellent starting point for consideration of available fish protection technologies for any particular location. The many fish protection systems available can be divided into three general categories most applicable to water resource projects: behavioral systems, physical barriers, and diversion systems.

16. Behavioral Systems. Behavioral systems for fish protection take advantage of fish behavior to divert fish from an intake. This category includes air bubble curtains, hanging chains, strobe lights, sound-producing devices, and a host of other methods that may produce an avoidance response in certain fish species. Behavioral systems also include mechanisms such as mercury lights, which have shown some promise to attract fish away from turbine or water supply intakes.

Behavioral systems are usually much less expensive to construct and operate than other fish protection systems, but their biological effectiveness is generally lower than physical barriers such as screens. At present, a large number of behavioral systems and techniques are being researched throughout the United States (for example, Loeffelman, 1990; McKinley et al., 1988; and Taft, 1988).

A major drawback of behavioral systems is that a particular device may work well for one species or life stage during a specific time period or season, but when conditions change, fish response to the system may change. In addition, fish may become habituated to light or sound, eventually ignoring what was originally a repellent.

Behavioral systems appear to be a cost-effective alternative where fish populations are generally stationary (resident) rather than migratory, or where a significant loss of migratory fish is not a concern. Overall, behavioral systems have been demonstrated to be inadequate for protecting migratory species of economic or biological significant, such as Pacific and Atlantic salmon.

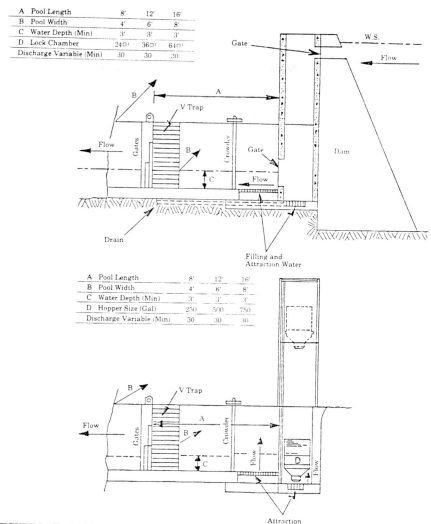

FIGURE 3 Typical layouts for fish lift and lock systems.

17. Physical Barriers. These barriers result in a complete physical blockage to fish entrainment for specific species and/or life stages of fish. Physical barriers include closely spaced bar racks, traveling screens, fixed screens, Eicher screens, rotary drum screens, permeable dikes, infiltration intakes, and barrier nets. This category of fish protection alternatives is usually the most effective from a biological standpoint, but carries with it the largest construction and operation costs, compared to other types of fish protection.

Design of physical barrier systems requires a combination of biological and engineering expertise. In most instances, sufficient experience and information can be collected to design an effective fish protection system. The design process for physical barriers is similar to what was previously described for fish ladder design, namely the following:

- Determination of biological and engineering criteria
- Quantification of hydrology for stream and river intakes
- Evaluation of hydraulic patterns at project intake
- Preliminary design and agency review
- Final design including structural, mechanical, and electrical components

Technical information related to screen design is extensive, including excellent overviews provided by EPRI (1986) and Dorratcague et al. (1985). Biological requirements and criteria for screens are commonly available from state and federal agencies (for example, NMFS 1989, WDF 1986) and engineering information is also widely accessible.

Vertical traveling screens have been installed at thousands of locations for fish protection and interception of debris and design information is available from several manufacturers of complete screen systems. Some vertical traveling screen systems have a separate backflush system for removing fish from screen collector troughs and returning them to the lake or river. Horizontal traveling screens have also been used for fish protection, but their use is less common than vertical traveling screens. A primary disadvantage of horizontal traveling screens is that collected debris is not brought to the water surface, and must be handled using submerged facilities.

Fixed screens have a stationary surface and include inclined plane screens, Coanda screens, and vertical screens installed in front of or over an intake (Fig. 4). Inclined and vertical screens are constructed of slotted or punched aluminum plate, or stainless steel profile bar. These screens require a mechanical, air-burst, or hydraulic method for cleaning debris and also usually have a trashrack upstream to intercept large debris. Coanda screens are a recent innovation for fish protection at small intakes where sufficient water head is available for this type of screen, which is somewhat self-cleaning (Strong and Ott, 1988).

Eicher screens (Fig. 5) have been developed during the last ten years and are specifically intended for diverting fish from penstocks or pipelines, rather than open-channel sites. The surface of an Eicher screen is made with stainless steel profile bars oriented at 15–20 degrees from pipeline or penstock flow, and preliminary results are encouraging for this new technology (Adam et al., 1991).

Rotary drum screens are installed at an angle to the diverted flow or perpendicular to the flow. Angled installations greatly help guide fish to bypass facilities and are the preferred layout where migratory fish are encountered (Johnson, 1991). For exclusion of resident fish or at small installations where the bypass is readily accessible across the screen surface, an installation perpendicular to the flow is probably adequate.

18. Diversion Systems. Diversion systems cover a portion of the turbine intake on hydropower projects, or operate in the project forebay upstream of the intake. These systems do not completely protect fish because diversion systems do not screen 100 percent of the intake flow. However, for large projects, diversion systems using partial screens are the only practical alternative because of the excessive cost of complete screening.

A common diversion system in the western United States is installation of

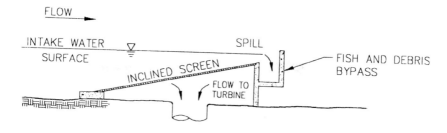

PROFILE – INCLINED PLANE SCREEN

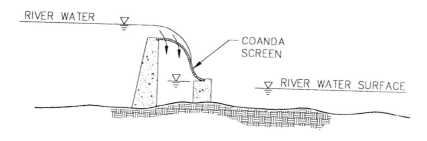

PROFILE – COANDA SCREEN

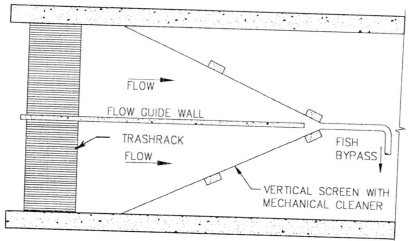

PLAN – FIXED VERTICAL SCREEN

FIGURE 4 Type and arrangement of fixed screens.

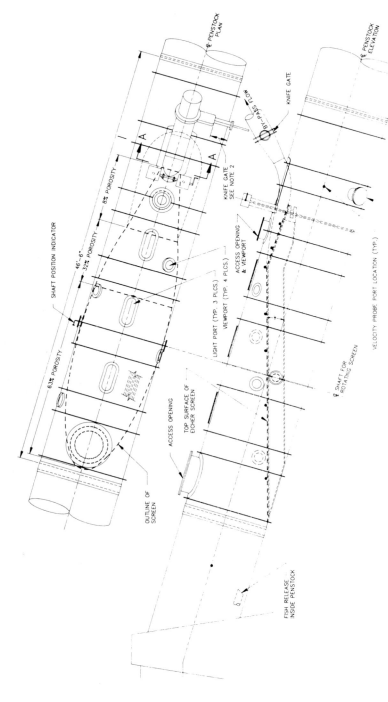

FIGURE 5 Eicher Screen for downstream exclusion of fish from turbines.

traveling or fixed screens over the upper part of submerged turbine intakes, with the screens oriented 50–60 degrees from vertical. Because most migratory fish concentrate near the surface, a relatively high percentage of fish can be diverted from a portion of the flow. For instance, on most Columbia River dams in Washington and Oregon, submerged screens covering less than a quarter of the intake area divert at least half and up to 80 percent of downstream migrating salmon and steelhead. Methods for improving fish guidance and collection are continually being developed for submerged screens (Gessel et al., 1991).

Louvers, skimmers, gulpers, and fish horns are examples of the wide variety of diversion systems that have been constructed in project forebays or reservoirs to try to economically divert a large proportion of migratory fish from water resource or hydropower intakes. EPRI (1986) provides biological and engineering data on these systems, which have mostly been one-of-a-kind applications to meet specific project needs.

AERATION OF DOWNSTREAM RELEASES

Many reservoirs, especially those located in warm climates, which thermally stratify and have moderate high nutrient content in the water seasonally develop low dissolved oxygen conditions in the lower, hypolimnetic strata. This usually occurs in late summer when the hypolimnion has been isolated for some time, water temperatures are high, and inflow to the reservoir is low. Other reservoirs with very high nutrient levels and low inflow to storage ratios, even very shallow, nonstratified systems, can become totally deoxygenated during late summer and fall. This occurs at night when the phytoplankton ceases to produce oxygen via photosynthesis but continues to use oxygen in respiration. This nightly deoxygenation often can occur where daytime oxygen levels reach conditions of supersaturation due to the large phytoplankton blooms. This type of system with a daily change is characterized as exhibiting a diel (daily) oxygen sag pattern. Smaller reservoirs with relatively high inflow to storage ratios and low nutrient content, especially those located in high altitudes or cooler climates are much less likely to have dissolved oxygen problems.

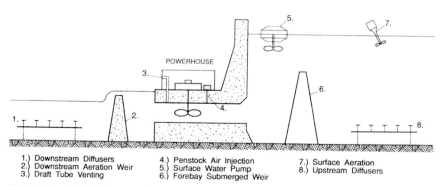

1.) Downstream Diffusers
2.) Downstream Aeration Weir
3.) Draft Tube Venting
4.) Penstock Air Injection
5.) Surface Water Pump
6.) Forebay Submerged Weir
7.) Surface Aeration
8.) Upstream Diffusers

FIGURE 6 Methods for aeration of reservoir release flows.

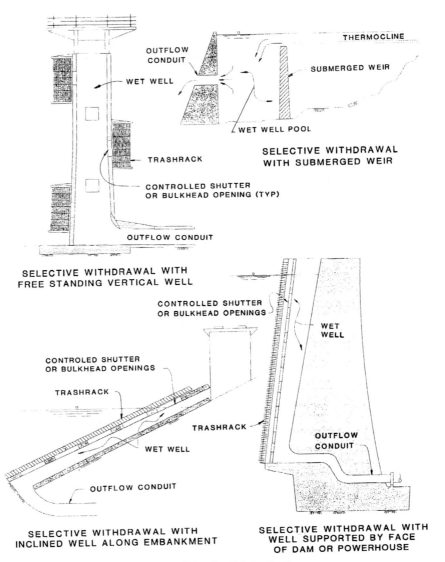

FIGURE 7 Selective withdrawal on multilevel outlet structures.

When a reservoir has unacceptably low dissolved oxygen levels, aeration may be required in order to protect downstream habitat quality. This can be accomplished by a variety of methods including:

- Partial destratification of the reservoir using pumps
- Skimming of epilimnetic water by use of submerged dike upstream of the dam

- Mechanical aeration in the reservoir with paddlewheels or similar mechanical devices
- Air injection in the reservoir by either surface or subsurface diffusers
- Oxygen injection in the reservoir by diffusers
- Turbine aeration/draft tube venting
- Use of epilimnetic release from a multilevel outlet structure
- Aeration with spillway or Howell-Bunger releases
- Aeration with a downstream weir
- Downstream mechanical aeration
- Downstream air injection
- Downstream oxygen injection

A number of these methods are schematically presented in Fig. 6.

EPRI (1990) has completed a detailed assessment of these methods and the reader is referred to that report for more extensive detail as well as to the ASCE Civil Engineering Guidelines for Planning and Designing Hydroelectric Developments (ASCE, 1989) and the work by Bohac et al. (1986) and Sheppard and Miller (1982).

MULTILEVEL OUTLET STRUCTURES FOR TEMPERATURE CONTROL

For reservoirs which stratify, with warmer (but possibly better-oxygenated water) on the surface (epilimnion) and cooler (but possibly deoxygenated water in the lower strata or hypolimnion), use of a multilevel outlet structure, to allow withdrawal and even mixing of water having different temperatures can be very effective in preserving downstream habitat conditions.

Figure 7 shows a multilevel outlet tower. It should be noted that operating experience with such structures have shown that as few as two outlet depths may be sufficient to provide adequate temperature control, as mixing in varying proportions usually allows a wide range of downstream temperatures.

REFERENCES

1. ADAM, P., D. P. JARRETT, A. C. SOLONSKY, and L. SWENSON, "Development of an Eicher Screen at the Elwha Dam Hydroelectric Project," in *Proceedings of Waterpower 1991 Conference*, American Society of Civil Engineers, July 1991.
2. ASCE, "Guidelines for Planning and Designing Hydroelectric Developments," ASCE, 1989.
3. BATES, K., "Fishway Design Guidelines for Pacific Salmon," in *Fish Passageways and Diversion Structures*, U.S. Fish and Wildlife Service, Fisheries Academy, October 1990.
4. BELL, M. C., "Fisheries Handbook of Engineering Requirements and Biological Crite-

ria," Fisheries Engineering Research Program, North Pacific Division, U.S. Army Corps of Engineers, Portland, Oreg., 1991.

5. Bohac, C. E., R. M. Shane, E. O. Harshbarger and H. M. Goranflo, "Recent Progress on Improving Reservoir Releases." *Proceedings of the International Symposium on Applied Lake and Watershed Management Lake Geneva, Wisconsin, North American Lake Management Society, 1986*

6. Brown, H. W., "Handbook of the Effects of Some North American Fishes," American Electric Power Service Co., Environmental Engineering Division, Canton, Ohio, 1974.

7. Clay, C. H., "Design of Fishways and Other Fish Facilities," Canada Department of Fisheries, Ottawa, Canada, 1961.

8. Dorratcague, D. E., G. R. Leidy, and R. F. Ott, "Fish Screens for Hydropower Developments," in *Proceedings of Waterpower 1985 Conference*, American Society of Civil Engineers, July 1985.

9. Electric Power Research Institute, "Assessment of Downstream Migrant Fish Protection Technologies for Hydroelectric Application," EPRI Report AP-4711, Palo Alto, Calif., September 1986.

10. EPRI, "Assessment and Guide for Meeting Dissolved Oxygen Water Quality Standards for Hydroelectric Plant Discharges," Electric Power Research Institute Research Project 2694-8, Palo Alto, Calif., 1990.

11. Gessel, M. H., J. G. Williams, D. A. Brege, and R. C. Krcma, "Juvenile Salmonid Guidance at the Benneville Dam Second Powerhouse, Columbia River, 1983–1989," *North American Journal of Fisheries Management*, 11:400–412, 1991.

12. Johnson, P. L., "Hydraulic Design of Angled Drum Fish Screens," unpublished manuscript, U.S. Bureau of Reclamation, Denver, Colo., 1991.

13. Katopodis, C., "Introduction to Fishway Design," in *Fish Passageways and Diversion Structures*, U.S. Fish and Wildlife Service, Fisheries Academy, October 1990.

14. Loeffelman, P. H., "Aquatic Animal Guidance Using a new Tuning Process and Sound System," unpublished report by the Environmental and Technical Assessment Division, American Electric Power Service Corporation, Columbus, Ohio, 1990.

15. McKinley, R. S., P. H. Patrick, and Y. Mussali, "Controlling Fish Movement with Sonic Devices," *Water Power and Dam Construction*, March 1988.

16. National Marine Fisheries Service (NMFS), "Fish Screening Criteria," unpublished manuscript developed by the Environmental and Technical Services Division, NMFS, Portland, Ore., August 1989.

17. Scott, W. B., and E. J. Crossman, "Freshwater Fishes of Canada," Bulletin 184, Fisheries Research Board of Canada, Ottawa, 1973.

18. Sheppard, A. R., and D. E. Miller, "Dissolved Oxygen in Hydroelectric Discharge Increased by Aeration," *Power Engineer*, Vol. 62, October 1982.

19. Strong, J. J., and R. F. Ott, "Intake Screens for Small Hydro Plants," *Hydro Review magazine, Kansas City, Mo., October 1988.*

20. Taft, N., "Evaluations of Fish Protection Systems for Use at Hydroelectric Plants," *Hydro Review*, Vol. VII, No. IV, 1988.

21. U.S. Fish and Wildlife Service, "Fish Passageways and Diversion Structures," Short course publication for USFWS Fisheries Academy, October 1990.

SECTION 19

HYDROELECTRIC PLANTS

By Hans Hasen and George C. Antonopoulos[1]

1. General. Sections 1 through 28 of this handbook are all more or less common to the subject of hydroelectric development. Multipurpose river-development projects that may involve all water uses are not discussed in this handbook.

POWER FROM FLOWING WATER

2. Energy and Work. Energy is the capacity to perform work. It is expressed in terms of the product of weight and length. The unit of energy is the product of a unit weight multiplied by a unit length, i.e., the foot-pound, the gram-centimeter, or the kilogram-meter.

Work is utilized energy and is measured in the same units as energy. The element of time is not involved.

The energy of water exists in two forms: (1) potential energy, that due to position or elevation, and (2) kinetic energy, that due to velocity of motion. These two forms are theoretically convertible one to the other. Energy may be measured with reference to any datum.

The potential energy of a given volume of stored water with reference to any datum is the product of the weight of that volume and the vertical distance of its center of gravity above that datum. For example, a rectangular tank of water with a 100-ft^2 surface and 20 ft deep, whose water surface is 100 ft above sea level, has a potential energy of $100 \times 20 \times 62.5 \times 90 = 1125 \times 10^4$ ft · lb. This potential energy cannot perform work until it is set in motion. If a stream flows out of that tank and connects with a pipe supplying water to a perfect turbine, 1125×10^4 ft · lb of work may be performed by the turbine as the tank empties and the potential energy is converted to kinetic energy.

3. Power. Power is utilized energy per unit of time, or the rate of performing work, and is expressed in horsepower, 550 ft · lb/s, or kilowatts, 737 ft · lb/s. The power from the tank of the preceding example will be at a decreasing rate because the head and flow diminish as the tank empties. Assume the outflow for the first second is 100 ft^3. The surface of the tank would be lowered 1.0 ft, and the center of gravity (head) of that 100 ft^3 is 99.5 ft. The energy utilized in the first second, therefore, is 621,000 ft · lb, or 1130 hp.

Now assume that a stream flows into the tank as fast as it is drawn off. A

---

[1]Acknowledgment is made to J. C. Stevens and Calvin V. Davis for material in this section that appeared in the third edition (1969).

constant discharge of 100 cfs may then be passed through the turbine under a constant head of 100 ft, for the surface is not lowered and a constant output of 625,000 ft · lb/s may be realized from our perfect turbine, equivalent to 1136 hp, or 848 kW.

The potential energy of a stream of water at any cross section must be measured in terms of power, in which time is an indispensable element. It is the weight of water passing per second times the elevation of its water surface (not center of gravity) above the datum considered. The kinetic energy of a unit weight of the stream is measured by its velocity. It must also be measured in terms of power, since velocity involves time. It is the weight per second times the velocity head, i.e., the height the water would have to fall to produce that velocity.

If the water of the preceding example were drawn off at a velocity of 10 fps, the surface elevation of the outlet channel would have to be $V^2/2g$ = 1.55 ft lower than that in the tank in order to produce that velocity and the kinetic energy would be 6250 × 1.55 = 9650 ft · lb/s. The total energy of a stream is the sum of its potential and kinetic energies. Thus the outlet stream has a total energy of 6250 lb/s × 98.45 = 615,350 potential plus 9650 kinetic, or a total of 625,000 ft · lb/s of total energy. At the perfect turbine, all the potential energy can be converted into kinetic energy where the velocity head is 100 ft.

Of course the perfect turbine does not exist. Some of the potential energy is converted into heat by friction and turbulence so that the useful part is less than the theoretical potential.

4. Energy Line. The energy head is a convenient measure of the total energy of a stream of constant discharge at any particular section. It is the elevation of the water surface, potential energy, plus the velocity head, kinetic energy, of a unit weight of the stream. Although every unit of the stream has a different velocity, what is usually considered is the velocity head corresponding to the mean velocity of the stream. If the stream is flowing in a pipe, the energy head is the elevation of the pressure line, or the height to which water would stand in risers, plus the velocity head of the mean velocity in the pipe.

A line joining the energy heads at all points is the energy line. The energy lines would be horizontal if the energy converted to heat were included. Energy converted to heat, however, is considered lost. Hence the energy line always slopes in the direction of flow, and its fall in any length represents losses by friction, eddies, or impact in that length. Where sudden losses occur, the energy line drops more rapidly. Where only channel friction is involved, the slope of the energy line is the friction slope.

Figure 1 illustrates the principles of the foregoing example. The potential energy head of the water in the forebay without inflow or outflow is that of the center of gravity Z. With inflow and outflow equal, however, the potential energy head is H. As the water passes into the canal, a drop of the water surface equal to the velocity head in the canal $V_1^2/2g$ must occur. At the entrance to the pipeline, an entrance loss h_1 is encountered as well as an additional drop for the higher velocity in the pipe. At any point in the pipeline, the pressure head h_p will be shown in a riser. The energy head at any point is the pressure head plus the velocity head, and the line joining the energy heads is the energy line. The energy lost (converted to heat) is the sum of friction, entrance, bend, and other losses in all the conduits, including the turbine and draft tube. The useful energy is that of the power developed by the turbine. The sum of the useful energy and the lost energy must equal the original potential energy.

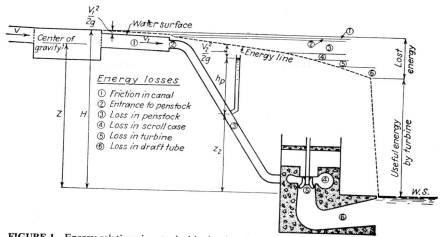

FIGURE 1 Energy relations in a typical hydroelectric plant.

5. The Bernoulli Theorem. Bernoulli's theorem expresses the law of flow in conduits. For a constant discharge in a closed or open conduit, the theorem states that the energy head at any cross section must equal that at any other downstream section plus the intervening losses above any datum:

$$Z_1 + \frac{V_1^2}{2g} = Z_2 + \frac{V_2^2}{2g} + h_c \tag{1}$$

In Fig. 2, Z is the elevation of a free water surface above datum, whether it be in a piezometer tube or a quiescent or moving surface of a stream; V is the mean velocity; h_c is the conduit losses between the two sections considered; and e is the energy head above the chosen datum.

Obviously Z may be made up of a number of elements, such as elevation of stream bed or pipe invert above datum k, pipe diameter D, depth in open channel y, or pressure head above crown of pipe h. Frequently k and h are measured to the centerline of the pipe, but, if the pipe is large, a distinction is necessary.

6. Head. There are several heads involved in a hydroelectric plant, defined as follows:

Gross head, simultaneous difference in elevation of the stream surfaces between points of diversion and return.

Operating head on the plant, simultaneous difference of elevations between the water surfaces of the forebay and tailrace with allowances for velocity head.

Net or *effective head* on the turbine has different meanings for different types of development as follows:

1. For an open-flume turbine, the difference between (1) headwater in the flume at a section immediately in advance of the turbine plus velocity head and (2) the tailwater plus velocity head.

2. For an encased turbine, the difference between (1) elevation corresponding to the pressure head measured at the entrance to the turbine casing plus velocity

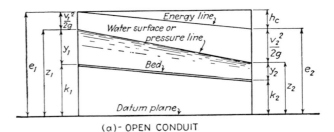

FIGURE 2 Energy relations in open and closed conduits. (*a*) Open conduit. (*b*) Closed conduit.

head at the same point of measurement and (2) the elevation of the tailwater plus velocity head at a section beyond the disturbances of exit from draft tube.

3. For an impulse turbine, the difference between (1) elevation corresponding to the pressure head at an entrance to the nozzle plus velocity head at that point and (2) the elevation of the lowest point of the pitch circle of the runner buckets (to which the jet is tangent).

7. Efficiency. *Efficiencies* of elements composing a hydroelectric system are all measured as the ratio of energy output to input or of useful to total energy.

No element is perfect; its functioning involves lost energy (conversion to heat). The efficiency of a plant or system is the product of the efficiencies of its several elements; thus

$$E_s = E_c E_t E_g E_u E_l E_d \qquad (2)$$

where E_s is the overall system efficiency made up of the product of the several efficiencies of the conduits: canal, penstocks, tailrace, E_c; turbines, including spiral case and draft tube, E_t; generators, including exciters, E_g; step-up transformers, E_u; transmission lines, E_l; stepdown transformers, E_d. Equation (2) expresses the overall efficiency from the river intake to the distribution switches at the substation. To this could be added the efficiency of the distribution system, even to the customer's meters, lights, water heaters, ranges, motors, etc.

The term *efficiency* is not often used for plant elements other than the generating equipment. It has been given here merely to illustrate the relationship of each element to the whole in this regard and to show the effect of the datum of reference on indicated efficiencies. In practice, the lost head in each such element is found and deducted from the gross head to obtain the net power head.

The efficiency of generators is generally greater the larger the unit, but it too depends on the load carried. The efficiency of transformers increases rapidly with capacity and load within certain limits, whereas that of transmission lines increases with capacity but decreases with load.

8. Power Formulas. A cubic foot per second of water at 62.5 lb/ft^3 falling 8.8 ft is equivalent to 1 hp and falling 11.8 ft is equivalent to 1 kW; therefore

$$\text{Theoretical hp} = \frac{Qh}{8.8} \qquad (3)$$

$$\text{kW} = \frac{Qh}{11.8} \qquad (4)$$

where Q is cubic feet of water and h is height in feet.

If E is the efficiency of the plant, the power that can be realized is given by

$$\text{hp} = \frac{Qh}{8.8} E \qquad (5)$$

$$\text{kW} = \frac{Qh}{11.8} E \qquad (6)$$

In the expression, E is the plant efficiency and h is the head on the turbine defined by Eqs. (2), (3), or (4) as may be appropriate.

Useful energy is generally measured in terms of kilowatthours, occasionally in terms of horsepower-hours. Where the discharge and head are constant,

$$\text{hp} \cdot \text{h} = \frac{Qh}{8.8} Et \qquad (7)$$

$$\text{kw} \cdot \text{h} = \frac{Qh}{11.8} Et \qquad (8)$$

where t is the time in hours for which the flow and head are constant or for which Q and h are average values. When the flow and head vary materially, the period considered is divided into smaller time intervals for which they are sensibly constant.

The horsepower-year and the kilowatt-year are terms sometimes used for power sales. On a 100 percent load factor the relationships are

$$1 \text{ hp} \cdot \text{year} = 0.746 \text{ kW} \cdot \text{year} = 8760 \text{ hp} \cdot \text{h} = 6540 \text{ kW} \cdot \text{h}$$

9. Classifications of Power and Energy. Power from any particular plant may be limited by the capacity of the installed equipment, available water supply, head, and storage. *Dependable capacity* may be defined as the load-carrying ability for the time interval and period specified when related to the load to be supplied.

Other definitions are as follows:

Firm power. Power intended to have assured availability to the customer to meet load requirements.

Primary energy. Hydroelectric energy which is available from continuous power.

Secondary energy. All hydroelectric energy other than primary energy.

Surplus system capacity. The difference between assured capacity and the system peak load for a specified period.

Dump energy. Energy generated that cannot be stored or conserved and is beyond the immediate needs of the electrical system producing energy.

The *capacity* of a power plant is not easily defined. Nameplate capacity or rated capacity of a turbine is usually given in horsepower for a given head, discharge, and speed at which the best efficiency obtains. Obviously each of these quantities may vary within definite limits. The rated capacity of ac generators is usually stated in terms of a definite speed, power factor, and temperature rise and is usually given in kilovolt-amperes. Each of these quantities may also vary within definite limits.

The IEEE definition of generating station capacity is the "maximum net power output that a generating station can produce without exceeding the operating limit of its component parts." The station or plant capacity can therefore be determined for a given station. It may be stated for a peak load over a given period such as 15 min or 1 h or for a continuous load. It would be higher for short periods than for continuous service if storage regulation exists but is limited by the temperature rise of generators. Until the station capacity has been fixed, the various factors having to do with capacity cannot acquire definite meanings. Where the capacity of a plant has not been fixed, it is customary to take nameplate capacity of generators as the plant capacity, which is often called *installed capacity.*

The *average load* of a plant or system during a given period of time is a hypothetical constant load over the same period that would produce the same energy output as the actual loading produced (IEEE).

The *peak load* is a maximum load consumed or produced by a unit or a group of units in a stated period of time. It may be the maximum instantaneous load or a maximum average load over a designated interval of time.

The maximum average load is generally used. In commercial transactions involving peak load, it is taken as the average load during a time interval of specified duration occurring within a given period of time, that time interval being selected during which the average power is greatest (IEEE).

The *load factor* is an index of the load characteristics. It is the ratio of the average load over a designated period to the peak load occurring in that period. It may apply to a generating or a consuming station and is usually determined from recording power meters. We may thus have a daily, weekly, monthly, or yearly load factor; it may apply to a single plant or to a system. Some plants of a system may be run continuously at a high load factor, whereas variations in load are taken by other plants of the system, either hydro or steam. Hydro plants designed to take such variations must have sufficient regulating storage to enable them to operate on a low load factor. They are often called *peak-load plants.* Operating on a 50 percent load factor, such a plant must have sufficient storage that it can, in effect, utilize twice the inflow for half the time; on a 25 percent load factor, the plant should be able to utilize 4 times the inflow for a quarter of the time, etc. The lower the load factor, the greater the storage required.

The *capacity factor* is a measure of plant use. It is the ratio of the average load to the plant capacity. It may be computed for a day, month, year, or any other period of time. When the peak load just equals the plant capacity, the capacity factor and load factor are obviously the same. If the maximum demand is less than the plant capacity, the capacity factor may be either greater or less than the load factor, depending largely on the load factor itself.

The *utilization factor* is a measure of plant use as affected by water supply. It is the ratio of energy output to available energy within the capacity and characteristics of the plant. Where there is always sufficient water to run the plant at capacity, the utilization factor is the same as the capacity factor.

WATER CONDUCTORS

10. General. A hydroelectric development includes in some form a water-diverting structure, conduit to carry water to the turbines and governors, generators, control and switching apparatus, housing for the equipment, transformers, and transmission lines to distribution centers. In most cases, a forebay or a surge tank is provided in which head regulation is effected. Trashracks and gates are placed at the head of penstocks. Connected to the waterwheel cases are the draft tubes which utilize the head below the wheels, recovering the kinetic energy of the water. The draft tube delivers the water to the tailrace, through which it flows to the stream.

11. The Forebay. The purpose of a forebay is to store water rejected when the load on the plant is reduced and to supply water for initial increments of an increasing load while the water in the canal or pipeline is being accelerated. Therefore, a forebay is essentially a storage reservoir at the head of the penstocks. It may be a canal, as used at the Box Canyon project on the Pend Oreille River in the state of Washington.

A canal-type forebay should be sized carefully to minimize losses and to equalize the flows into the turbines. Hydraulic-model tests are usually required.

12. The Intake. Intake structures vary widely, depending on the type of plant, river flows, reservoir operations, and site conditions.

The Guri intakes (Fig. 3) lead to 10.5-m-diameter penstocks serving the 700-MW units. A single intake gate could not be optimally designed because of the large spans and high gate-wheel contact pressures involved. Therefore, an intake with a central intake pier was used, dividing the flow in half and allowing the use of two intake gates to regulate the flow. This resulted in an additional benefit during maintenance operations of one intake gate, allowing the intake to pass one half of the turbine flow through the remaining intake passage. The intake gate slots are also used as the air-supply ducts as upstream sealing gates are used.[1]

The Fontana intake (Fig. 4) leads to water passages and structures which are built to accommodate three 67.5-MW generating units. These operate under a rated head of 330 ft. A short concrete transition leads from the 19-ft 2½-in–high by 11-ft 0-in–wide rectangular openings at the gate to a 14-ft-diameter steel-lined penstock. Each intake gate is of the tractor type. These gates are approximately 27 ft 10 in high by 16 ft 6 in wide and close openings which are 20 ft 10 in high by 11 ft 6 in wide. The reservoir serves as the forebay for a plant of this type.

13. Trashracks. Trashracks used in hydroelectric generating stations consist essentially of vertical or slightly inclined steel bars placed parallel to each other and spaced uniformly to permit the use of rakes. Bars are supported in water passages by horizontal supports which transmit loads developed by the flow, especially when partly clogged by trash, into side members or horizontal supporting beams.

By way of example, Fig. 4 also shows the arrangement of the trashracks and gates for the Fontana hydroelectric plant. These precede three 14-ft-diameter

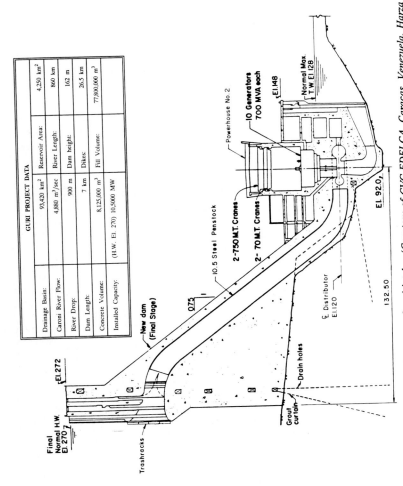

FIGURE 3 Guri powerhouse no. 2 and intake. (*Courtesy of CVG-EDELCA, Caracas, Venezuela, Harza Engineering Co., consulting engineer.*)

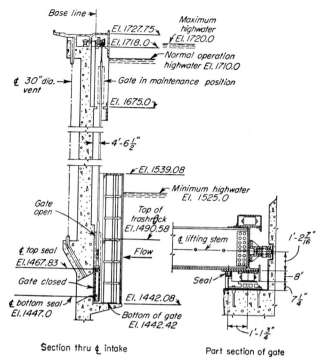

FIGURE 4 Intake, Fontana project.

steel-lined intake conduits leading to the units in the powerhouse at the foot of the dam.

Each of the three trashrack structures is supported on a concrete cantilever slab which projects 22 ft 2 in from the face of the dam. In plan, the trashrack structure is of semicircular shape with a 13 ft 0 in outside radius. The circumference of the semicircle is subdivided into four straight panels, framed by concrete columns with concrete-beam bracing. The column and beam structure has a total height of 97 ft, of which the lower 48 ft has the racks in the panel openings and the upper 49 ft is an extension for withdrawal and maintenance of the racks with the top 14 ft above minimum drawdown.

In the corner columns of the four panels, 7-in guide channels are embedded. Four rack sections fit into each panel, each 12 ft 1½ in high by 8 ft 4 in wide. The vertical rack bars are 4 by ⅝ in and the horizontal supporting members are 6 by ¾ in, spaced to give openings 6 in wide by 2 ft 4 in high. The concrete structure is designed for a differential head of 10 ft, and the steel rack sections are based on a differential head of 5 ft. The maximum velocity through the net trashrack area is 2.4 fps.

The head loss through trashracks is mainly due to the flow contraction at the entry and the sudden enlargement of the area at the exist from bar spaces. Among the many formulas for calculating head loss, the one developed by O. Kirchmer on the basis of experiments at the Munich Hydraulic Laboratory is in widespread use:

$$hr = K\left(\frac{t}{b}\right)^{4/3} \frac{V_0^2}{2g} \sin \alpha \qquad (9)$$

where hr = loss of head through racks, ft
       t = thickness of bars, in
      b = clear spacing between bars, in
     V_0 = velocity of approach, fps
      g = acceleration due to gravity
     α = angle of bar inclination to horizontal, degrees
     K = a factor depending on bar shape in accordance with Fig. 5.

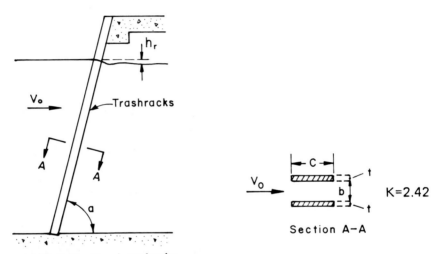

FIGURE 5 Head loss in trashracks.

The losses determined from this formula apply to clean racks.

The spacing of the vertical or inclined rack bars depends primarily on the size and type of turbine to be protected and is also influenced by the size of the trash.

Figure 6 shows the maximum recommended length of rack bars between lateral supports or stiffeners, as limited by vibration characteristics related to bar thickness and velocity through the net area. To avoid objectionable vibration, the length limits shown should not be exceeded.

The intake and trashrack arrangement for the 400-ft-high Mangla Dam (Fig. 7) on the Jhelum River, West Pakistan, completed in 1967 by the Water and Power Development Authority, offers an interesting example of cage-type racks operating on the face on an embankment-type dam.[2]

The 450-ft-wide intake structure is divided by contraction joints into five reinforced-concrete monoliths, each containing a power intake, as shown by Fig. 7. The gate opening of each power intake is 18 ft wide and 36 ft long on a slope of 1.0 on 2.5. The gate openings are followed by radius bends which showed a loss from intake to tunnel of 0.20 velocity head in the tunnel or approximately 0.62 ft at the design discharge of 10,000 cfs.

Each intake screen provided 3000 ft² net area and consists of ½-in galvanized

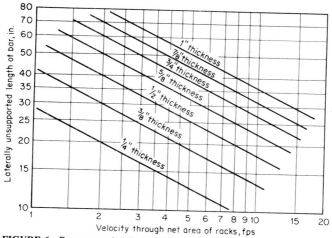

FIGURE 6 Recommended limits of laterally unsupported length of steel rack bars to avoid vibration.

mild steel bars at 6-in centers designed for 60 percent of the yield at a differential head of 10 ft. The supporting members of the bars are designed for a head of 20 ft.

14. Penstocks. Penstocks and tunnels between the intakes and powerhouse are designed to carry water to the turbines with the least possible loss of head consistent with the overall economy of the project. Some dimensional limitations may be imposed if the units need to contribute to the system's transient stability and the quality of frequency regulation. The various losses which occur between the reservoir and the turbine—trashrack, entrance, pipes, valves, and fittings—are discussed elsewhere (Secs. 2, 3, and 22). Where there is an elbow in the penstock just ahead of the turbine, a reducer is required, minimum losses will usually result from shaping the elbow to serve also as a reducer.

The most economical penstock will be the one in which the annual value of the power lost in friction plus annual charges such as interest, depreciation, and maintenance will be a minimum. The variables entering the problem are

1. Daily variation of flow through penstock
2. Estimated load factor over a term of years
3. Profile of penstock
4. Number of penstocks
5. Materials used in construction
6. Diameter and thickness
7. Value of power lost in friction
8. Cost of penstock installed
9. Cost of piers and anchors
10. Total annual charges of penstock in place
11. Maximum permissible velocity

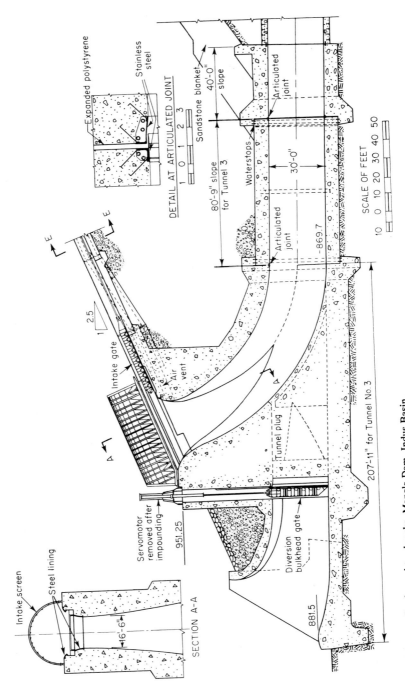

FIGURE 7 Intake and trashracks, Mangla Dam, Indus Basin.

It is extremely difficult to express these variables in a comprehensive formula, although several attempts have been made to do so. An interesting study by Sarkaria[3] resulted in a single empirical formula, which applied to penstocks embedded in gravity intakes:

$$D = 4.44 \frac{P^{0.43}}{H^{0.65}} \tag{10}$$

where D = economical diameter of penstock, ft
P = rated horsepower of turbine, hp
H = rated head of turbine, ft

This formula is applicable primarily to power plants with Francis and propeller-type turbines and gives fairly reliable results for penstocks 5 ft or more in diameter.

Figure 8 shows penstock velocities in selected existing plants having a wide range of heads.

Similar attention should be devoted to the effect of the water conduit dimensions in the regulating characteristics of the unit. Specifically, one should evaluate the need for the unit to contribute to the system's transient stability and the quality of frequency regulation.

Water conduits should be initially dimensioned to give a reasonably good water starting time for unit start-up. This assures that the minimum stability index, as measured by the ratio of mechanical starting time to water starting time (T_m/T_w), will be economically obtained.

Water starting time is measured in seconds by the following equation:

$$T_w = \frac{\Sigma LV}{gh}$$

where $\Sigma LV = L_1V_1 + L_2V_2 + \cdots L_nV_n$
L_n = length of a constant-diameter section of the water conductor

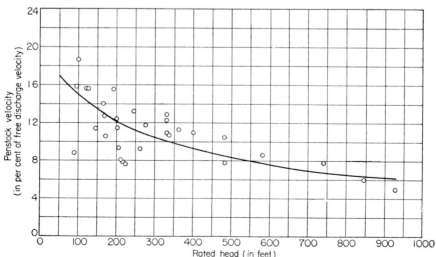

FIGURE 8 Penstock velocities at full turbine discharge, existing plants.

V_n = related average flow velocity in the L_n section of the water conductor
g = gravitational constant
h = related net head at which the turbines are operating

The sum of LV includes all the water conductor elements from the intake to the end of the draft tube (including the spiral case extension and one-half the spiral case for a Francis turbine, in which the velocity is essentially constant), or from an upstream free water surface to a downstream free water surface when surge tanks are used. T_w is the time required to accelerate the water in the water conductors from rest, or zero velocity, to the steady-state velocity at full gate discharge with all units on a given penstock or power conduit in operation.

Realistic water starting times T_w are in the order of 1 to 2 s. A water starting time exceeding 2.5 s will usually require an adjustment in hydraulic conduit dimensions.

When water starting times are less than 2.5 s, the required stability index ratio can usually be obtained by providing more WR^2 [see Eq. (11)] than the amount normally provided by the generator manufacturers.

Mechanical starting time, or *flywheel effect,* is the time in seconds to accelerate the rotating parts of the connected generator rotor and turbine runner of the hydroelectric unit from rest, or zero speed, to operating rotation speed. Mechanical starting time is measured by

$$T_m = \frac{WR^2 \times n^2}{1{,}620{,}000 \times P} \tag{11}$$

where WR^2 = product of weight of revolving parts (shaft, turbine runner, and generator rotor) and the square of their radius of gyration
n = rotational speed of the turbine and the generator for a direct-connected synchronous generator
P = turbine full gate capacity, hp

The ratio of the mechanical inertia (the stabilizing influence) to the hydraulic inertia (the destabilizing influence)—in other words the T_m/T_w ratio—is widely recognized as an important measure of the unit stability. The ratio is a measure of the inertial regulating characteristics of a unit and its ability to follow load changes and contribute to system transient speed regulation.

It is only in the case of a unit operating in parallel at constant load and discharge on a large system that a low index is acceptable. A system is considered large if the WR^2 of the unit is less than or equal to 25 percent of the total WR^2 or rotating inertia of the system. The lowest desirable stability index for the start-up and synchronizing of a unit has been determined to be 2. The desirable minimum for off-line stability is 2.5.

When a unit is expected to follow load and provide frequency regulation, and it is operating isolated, or is expected to operate as an isolated plant during emergency conditions, or will operate as a small system where the WR^2 of the unit is greater than or equal to 45 percent of the total WR^2 of the system, the index of the unit should be equal to 5 or more.

National Electrical Manufacturers Association standards require that hydro generators be provided with a WR^2 during manufacturing such that:

$$\text{Gen. } WR^2 = \frac{4.33\text{MVA}[0.54(\log_e \text{MVA}) \div 0.30] \times 10^9}{(\text{rpm})^2} \quad (12)$$

Losses in bifurcations, which are usually located a short distance upstream from the control valves, may best be determined from model tests. Few prototype performance data are available. Hydraulic-model tests revealed that the maximum head loss for the bifurcation ahead of the Mangla turbines was approximately $0.40(V^2/2g)$ (Fig. 9). The practice of the Corps of Engineers, U.S. Army, is to estimate bifurcation losses at $0.15(V^2/2g)$ and trifurcation losses at $0.50(V^2/2g)$.

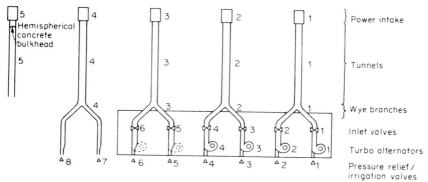

FIGURE 9 Mangla Dam, Pakistan—diagrammatic hydroelectric scheme.

Prototype tests were made to determine the hydraulic losses in the wye branch which divided the flow of the Chelan station power tunnel.[4] A 2-mi-long, 14-ft-diameter concrete-lined pressure tunnel converges below the surge tank to a steel-lined 14-ft-diameter penstock which divides through a bifurcation into two 12-ft 6-in–diameter water passages leading to the turbines. The head loss at this bifurcation, under full-flow conditions, was approximately $0.50 (V^2/2g)$, where V is the average velocity in the penstock before the flow is divided.

15. Turbines and Pumps. Reference should be made to Sec. 21 for a comprehensive presentation of the hydraulics of hydraulic machinery.

16. Draft Tubes. The draft tube is an essential part of a turbine installation. It supplements the action of the runner by utilizing most of the energy remaining in the water at the discharge from the runner.

The draft tube is designed as a diverging passage connecting the runner with the tailrace. It is shaped to decelerate the flow with a minimum of losses so that the kinetic energy remaining in the flow at discharge from the runner may be efficiently regained by conversion into the suction head, thereby increasing the total pressure difference on the runner. This regain of kinetic energy is usually the primary function of a draft tube.

Draft tubes are not used with impulse wheels, and the head from the nozzle to tailwater is necessarily lost.

Energy Relations. Using the tailwater level as datum, we can write the Bernoulli equation for the draft tube thus:

$$Z_1 + 2.3p_1 + \frac{V_1^2}{2g} = \frac{V_3^2}{2g} + h_f + h_i \tag{13}$$

where Z_1 = elevation of turbine exit above tailwater equivalent to the draft head
p_1 = gage pressure, psi, at turbine exit
V_1 = velocity at turbine exit
V_2 = velocity at draft-tube exit
V_3 = velocity in tailrace beyond disturbance from draft-tube exit
h_f = friction loss in draft tube
h_i = eddy loss at draft-tube exit

For the vertical-tube type, the exit velocity head is lost, i.e., $h_i = V_2^2/2g$. For the elbow or symmetrical type, some of the exit velocity is preserved in the direction of flow in the tailrace; hence it may be considered as a sudden enlargement in a conduit for which the loss is

$$h_i = \frac{(V_2 - V_3)^2}{2g} \tag{14}$$

The right-hand member of Eq. (13) is a relatively small quantity, usually not over 2 ft. The exit velocity from the runner will depend on the specific speed of the turbine and may be anywhere between 20 and 40 fps; hence the greater V_1, the less the draft head, for the absolute pressure at the inlet of the draft tube cannot be less than the vapor pressure of water and should be substantially more.

The design of the draft tube is usually supplied by the turbine manufacturer and is an integral part of the turbine in determining turbine performance. Cavitation and pitting on the underside of runner blades may result with low-head high-specific-speed units if the draft head is too great or the rate of expansion too rapid.

The negative pressures at the head of the draft tube may be computed from Eq. (13).

17. The Tailrace. A careful study of tailrace conditions should be made. The first step is to obtain records showing the discharges at different elevations of tailwater. Usually, downstream gages should be established to show the effects of projecting ridges which may influence tailwater elevations. The removal of such ridges may increase the operating head.

POWERHOUSE STRUCTURES

18. Classification. Powerhouse structures may be classified under five general types:

1. Integral intake
2. Separate powerhouse constructed at the toe of the dam
3. Separate powerhouse structure connected to the intake by either penstocks or tunnels
4. Underground powerhouse
5. Low-head powerhouse

Falling within these are plants which may be classified as indoor, outdoor, and semioutdoor. Typical examples of each are described.

19. The Integral Powerhouse. Representative of this type are the Main Canal headworks power plant, Pasco, Washington (Fig. 10), and the Wanapum powerhouse (Fig. 11), located on the Columbia River near the center of the state of Washington. The Main Canal headworks power plant contains a bulb-type unit rated at 27.37 kVA and operated under normal head of 43 ft. The power plant was constructed in 1986 and is operated by the South Columbia Basin Irrigation District.[5]

The Wanapum substructure is of mass concrete and the intake and semispiral case are of reinforced concrete. The superstructure is formed by the intake concrete wall, a downstream reinforced-concrete wall, and a roof deck. There are three wheeled gates, each 42 ft 6 in high by 20 ft wide, one for each intake opening of one unit. Trashracks are designed for a maximum flow velocity of 4 fps through the gross area. Gates are provided for emergency or service closure of one unit. Draft-tube gates are provided for the emergency closure of two units. The powerhouse contains 16 generating units and an erection bay. Generators are rated 87,500 kVa at 0.95 power factor, 60 Hz, 85.7 rpm. The turbines are of the vertical-shaft, adjustable-blade-propeller (Kaplan) type rated 120,000 hp at an 80-ft net head.

The Moses-Saunders project (Fig. 12) on the St. Lawrence River is a notable integral-intake structure with a modified outdoor-type powerhouse. The Moses-Saunders plant is located a short distance above Cornwall, N.Y., and connects the Canadian mainland with Barnhart Island.[6] The powerhouse is approximately 3200 ft long and contains 32 generating units. The turbines are of the fixed-blade-propeller type and are rated at 59 MW each, at 81-ft head and a speed of 94.7 rpm. The three-phase 60-Hz generators are rated at 60,000 kVA.

The intake and substructure form an integral dam and powerhouse which provide waterways for the 32 turbines. The powerhouse is divided into 32 blocks, each of which is 80 ft long. Each intake passage is subdivided by two intermediate piers into three water passages.

Representative of semioutdoor design is the Kentucky Dam power station structure [Tennessee Valley Authority (TVA)] shown by Fig. 13.[7] This type combines some of the features of both the outdoor and indoor types. The generator room at Kentucky is 30 ft high, which is ample to clear the Kaplan head on the unit. The station houses four 32-MW units. The generators are driven by Kaplan adjustable-blade units which operate under a rated head of 48 ft. The head varies from a maximum of 58.5 ft to a minimum, during maximum flood conditions, of 6 ft.

20. Separate Powerhouse Constructed at Toe of Dam. Brief descriptions of two plants will illustrate the wide range of designs which fall under this classification: (1) the Karadj project and (2) the Hartwell project.

The 600-ft-high Karadj Dam is located on the Karadj River about 25 mi from the city of Teheran, Iran. It was completed in 1961 by the Karadj Water and Power Authority, a division of the Plan Organization. Figure 14 shows a section of the arch dam and powerhouse.

The powerhouse, an indoor-type structure, houses two Francis turbines rated at 55,230 hp at a head of 482.3 ft and a speed of 333 rpm. The generator rating is 40 MW. The units are spaced 36.1 ft center to center. Provision is made for the later installation of a third unit. Steel-lined, 8-ft 6-in–diameter penstocks connect the intake, as shown by Fig. 14, with reducers which lead to 86.6-in spherical-type turbine inlet valves.

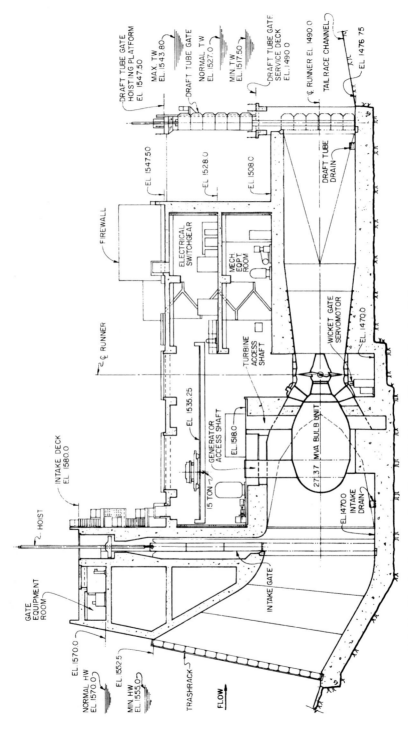

FIGURE 10 Main Canal headworks power plant, Pasco, Washington. *(Harza Engineering Co.)*

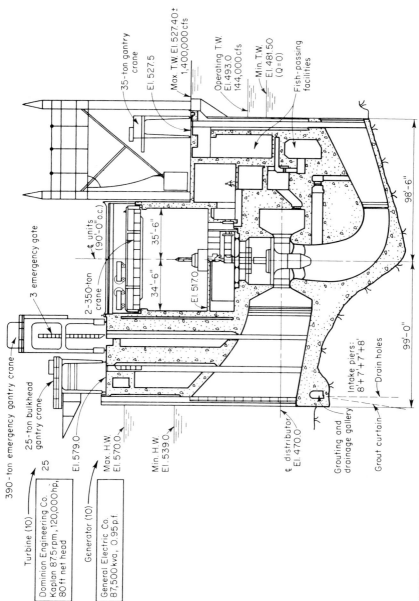

FIGURE 11 Wanapum Powerhouse (1963), 830-MW development on Columbia River, Washington.

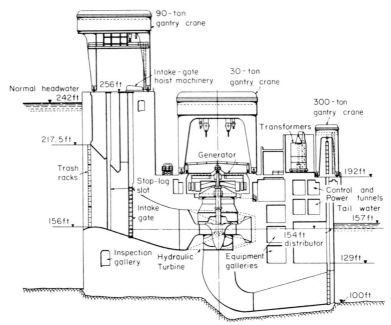

FIGURE 12 Moses-Saunders hydroelectric plant, St. Lawrence River—section, American side.

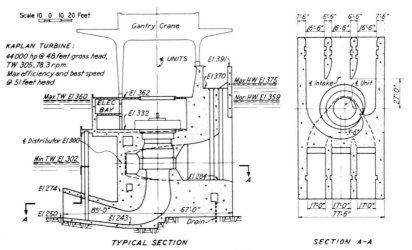

FIGURE 13 Semioutdoor plant, Kentucky Dam, TVA.

The 200-ft-high Hartwell project, located on the Savannah River, Georgia, was completed in 1962 by the Corps of Engineers, U.S. Army. Figure 15 shows a section through the dam and powerhouse. The dam and intake are of the conventional gravity type. An outdoor-type powerhouse contains four generating sets. The Francis-type turbines are rated 91,500 hp each at 170-ft head. The generators are rated at 66 MW each. The speed is 100 rpm. The units are 68 ft on centers.

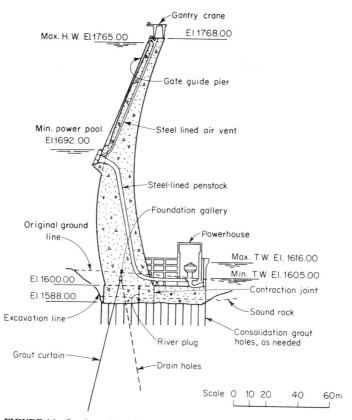

FIGURE 14 Section of arch dam and powerhouse, Karadj.

21. Powerhouse Connected to Intake by Tunnels and Penstocks. The Appalachia powerhouse (Fig. 16) offers an interesting example of this type. A concrete dam 150 ft high, a tunnel nearly 8 mi long, and a powerhouse containing two 37.5-MW units operating under a 360-ft rated head are principal elements of the project. The units are spaced 44 ft center to center.

The substructure and outside walls of the superstructure are of concrete to levels above maximum tailwater elevation. The remainder of the superstructure consists of structural-steel framework with concrete floors and tile or concrete walls. The layout and design of the powerhouse were substantially influenced by the provision for tailwater at elevation 882 ft.[8]

The draft tube is of the plain elbow type. It has two outlet openings separated by a 4-ft-thick center pier. At the bottom of the runner, the draft tube is 8 ft 7 in in diameter; at the discharge end it has a total clear width of 21 ft 9 in and a height of 9 ft 8 in. The velocity head at the draft-tube exit at full gate discharge is 1.02 ft.

Cork-tar mastic is placed over the spiral case in the unit block and over the upper half of the penstock where it goes through the east wall of the powerhouse. Between the unit block and the east wall, only the lower part of the penstock is embedded in concrete. These arrangements allow the penstock and the spiral case to expand and contract freely without causing undue stress in the concrete.

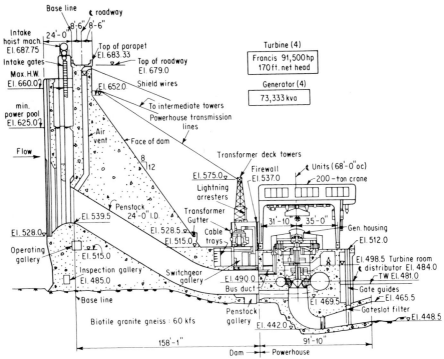

FIGURE 15 Hartwell powerhouse 260-MW development on Savannah River. (*Courtesy of U.S. Army Engineers.*)

Where the station is located at the toe of the dam, it is usual practice to provide a contraction joint between the dam and the powerhouse substructure. A flexible joint should be placed in the penstock where it crosses this contraction joint.

The Robert Moses Niagara power plant, on the Niagara River, operated by the New York Power Authority, is a medium-head plant of this type. Figure 17 shows a section through the plant. The structure is of the semioutdoor type. The plant is about 1840 ft long and contains 13 units. The total installed capacity is 1950 MW. The Francis-type turbines are rated at 210,000 hp each at a net head of 300 ft. The normal operating head is 305 ft. The throat diameter of the runner is 205 in.

Welded-steel penstocks, approximately 462 ft long, connect the intake with the turbines, as shown by Fig. 17. An upper elbow reduces the penstock from 28 ft 6 in to 24 ft in diameter. A lower elbow reduces it from 24 to 21 ft in diameter at the spiral case extension.

The Uribante-Doradas power plant, Venezuela (Fig. 18), contains two impulse-type units (Pelton) rated at 153 MW each and operates under a normal head of 392 m. The project was completed in 1985 and is operated by CADAFE, Empresa de Energia Electrica del Estado, Caracas, Venezuela.

22. The Underground Powerhouse. Underground power plants are usually built in locations where the cost of excavating the caverns required to house the

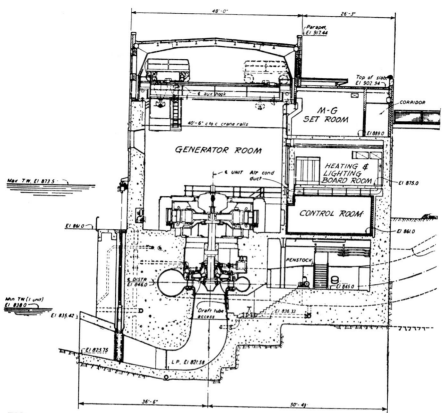

FIGURE 16 Section through the Appalachia powerhouse.

generating and electrical equipment will be less that of constructing a powerhouse of another type. Where foundation conditions permit, the underground plant has many advantages at certain sites. For example, where space is lacking to accommodate the structures, as in a narrow gorge, this type may be used to advantage. In some situations an additional advantage is obtained by cutting across one or more river bends with the tailrace tunnels and thus increasing the head on the plant and removing the tailrace from the influence of the spillway discharge.

A brief description of Churchill Falls powerhouse, Labrador, Canada, Fig. 19, will illustrate the features of this type, which is operated by Churchill Falls (Labrador) Corporation. Churchill Falls has a total head of only 250 ft. By building a control dike some distance above the falls, the water of the Churchill River is directed along the top of the bluff for a distance of 20 mi, where a total head of 1025 ft is being utilized. Constructing an underground powerhouse in the rock, below the bluff, 972 ft long, 81 ft wide, and 154 ft high, 11 sloped penstocks feed 11 turbine generators, rated at 475 MW each. The 11 draft tubes discharge into one large surge chamber 783 ft long, 40 to 64 ft wide, and 140 ft high. Two tailrace

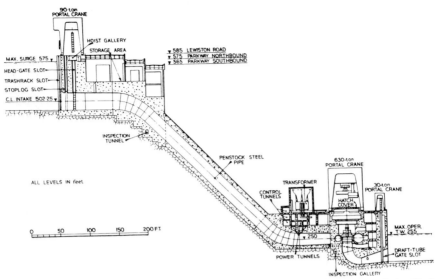

FIGURE 17 Section through the Robert Moses Niagara power plant. (*Courtesy of New York Power Authority.*)

tunnels 45 ft wide, 60 ft high, and 5560 ft long convey the water from the surge chamber to the Churchill River.

A large reservoir, with a surface area of 27,000 mi^2, feeds the Churchill Falls underground powerhouse, which was created by building dikes and control works on the existing lakes above the Churchill River. This reservoir assures an adequate year-round water supply for generating 5428 MW of electricity.

See Table 1 for a tabulation of underground powerhouses with caverns 21 m or wider.

23. Low-Head Plants. Many run-of-river stations such as Kentucky Dam (Fig. 13) fall under the low-head classification. Improvements in the tubular-type and bulb-type generating units have resulted in the utilization of low heads which heretofore have not been economically feasible to develop.

The low-head hydroelectric stations on the Mosel River, Germany, illustrate this type of river development.[9] Figure 20 shows a map and profile of the Mosel River projects and water levels. The accompanying table in Fig. 20 shows some of the principal data for the 14 plants in Germany. It will be noted that the heads vary from 4 to 9 m.

The lowermost station at Coblenz (16 MW) is of conventional design with vertical-shaft Kaplan sets as opposed to the horizontal-shaft tubular sets used in the more recent stations constructed farther upstream.

The Trier station no. 10, Mosel River (Fig. 21), illustrates the advantages of the tubular type. In addition to being a very low structure, with its roof only 2.4 m above the maximum flood level of 132.31 m, the Trier station is much more compact longitudinally than a plant incorporating conventional Kaplan sets. At Coblenz station no. 1, Mosel River, for example, the four vertical Kaplan units occupy a total length of 68.8 m, whereas the tubular set at Trier is accommodated

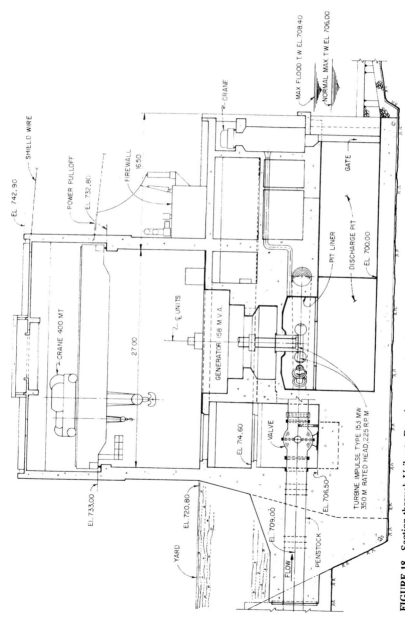

FIGURE 18 Section through Uribante Doradas power plant. (*Courtesy of CADAFE, Caracas, Venezuela, Harza Engineering Co., consulting engineer.*)

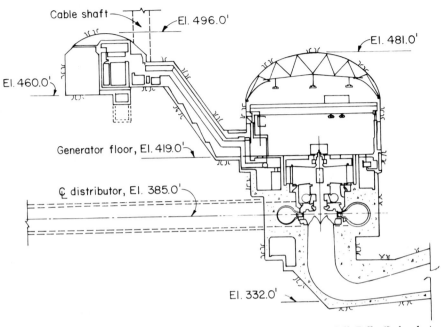

FIGURE 19 Churchill Falls powerhouse, Canada. [*Courtesy Churchill Falls (Labrador) Corp.*]

in only 45 m. The amounts of concrete used in the power stations were 25,100 m³ at Coblenz and 17,200 m³ at Trier.

The Gregory B. Jarvis power plant, Hinckley, New York (Fig. 22), contains two tubular-type units rated at 4.5 MW each and operates under a normal head of 65 ft. The project was completed in 1986 and is operated by the New York Power Authority.[12]

Project development plans called for the installation of an intake structure and a 15-ft-diameter penstock through the existing dam, and a two-unit, 9-MW powerhouse, located 200 ft downstream of the dam. Tunneling through the dam for the installation of the penstock and construction of the intake structure had to be performed while maintaining normal operations of the Hinckley Reservoir. A large permanent bulkhead, frame with guides, and bulkhead panels were installed by underwater construction techniques at the upstream face of the dam to facilitate the tunneling and concrete placement operations.

The King Talal powerhouse, Jordan (Fig. 23), contains two horizontal Francis-type units rated at 3.2 MW each and operates under a normal head of 107 m. The project was completed in 1988 and is operated by Jordan Valley Authority, Jordan.[13]

The project included raising of King Talal Dam and spillway headworks by 15 m, construction of a new 45-m-high concrete gravity dam, modifications to the original chute spillway, low-level outlets, drainage gallery and cutoff wall sys-

TABLE 1 Underground Powerhouses with Caverns 21 m or Wider (Ref. 11)

Project	Country	Units	MW	Cavern size, m W	H	L	Ground cover, m	Rock type	Remarks
1. Cirata	Indonesia	4	125	35	49.5	253	100*	Pyroclastic, shale	
2. Imiachi	Japan	3	350	33.5	51.0	160	400	Sandstone, slate	PS
3. Waldek II	Germany	2	239	33.5	40	106	260	Graywacke, sandstone	PS
4. Veytaux-Hongrin	Switzerland	4	60	29.6	26.8	137.5	330	Fractured limestone	
5. Fadalto	Italy	2	120	29.5	57.3	69.6		Sound limestone	
6. El Cajon	Honduras	4	80	29	49	110	200	Karstic limestone	
7. Grimsel II	Switzerland	4	75	28.9	19	140	200*	Hard gneiss	PS
8. Cabora Bassa	Mozambique	5	415	27.5	56	220	140	Unknown	
9. Ferrara	Switzerland	3	62	27	25	143	150	Hard gneiss	PS
10. Coo-Trois Ponta	Belgium	6	125	27	40	130	75	Thin beds of phylites	PS
11. Robiei	Switzerland	4	41	26.5	28.7	75		Conglomerate, gneiss	PS
12. Shintakasegawa	Japan	4	336	25.3	54.5	163	250	Good-quality granites	PS
13. LaGrande 2	Canada	16	333	25	47.3	483	100	Granitic gneiss	
14. Helms	U.S.A.	3	350	25	38.1	102.4	370	Grandiorite	PS
15. Kemano	Canada	16	100	24.7	42.7	213.4	430	Granite, diorite	
16. Motezic	France	4	228	24.5	42.5	145	380	High-quality granite	PS
17. Churchill Falls	Canada	11	475	24.4	44.8	296.3	300	Gneiss, diorite	
18. Dinorwic	U.K.	6	300	24.4	52.2	180.3	300	Slate	PS
19. Mica	Canada	6	435	24.4	44.2	237.2	220	Schist	
20. El Toro	Chile	4	100	24	40	103		Grandiorite	Pelton
21. Bear Swamp	U.S.A.	2	200	23.5	45.7	68.5	100	Chlorite micaschist	PS
22. Lake Delio	Italy	8	130	23.2	60.5	195.5	150	Gneiss	PS
23. Tusut 1	Australia	4	80	23	33.5	91.5	340	Granite gneiss	
24. Boundary	U.S.A.	6	200	23	53.4	145	150	Limestone	
25. Sackingen	Germany	4	83	22.9	30.6	102	400	Solid gneiss	PS
26. Turlough Hill	Ireland	4	73	22.5	32	82	200	Grained granite	PS
27. Kariba	Zimbabwe	9	100	22	40.2	148	60	Biotate gneiss	
28. Bad Creek	U.S.A.	4	250	22.5	50.0	132	160	Gneiss	PS
29. Racoon Mountain	U.S.A.	4	382	22	50	150	200	Limestone	PS
30. Northfield Mountain	U.S.A.	4	250	21.3	44.2	100	170	Gneiss, quartzite	PS
31. Chute-des-Passes	Canada	5	150	21.2	38	140	120	Granite, gneisses	
32. Edward Hyatt	U.S.A.	6	351	21	36.6	67.7	90	Amphoblite	C/PS

*Approximate value
PS = pumped storage
C/PS = conventional and pumped storage

tems, underwater modifications to the original irrigation intake, installation of a steel penstock inside the original outlet tunnel, and construction of a power tunnel, a powerhouse, and a discharge structure.

The Moose River power plant, Lyonsdale, New York (Fig. 24), contains one Kaplan-type unit rated at 12 MW and operates under a normal head of 135 ft. The

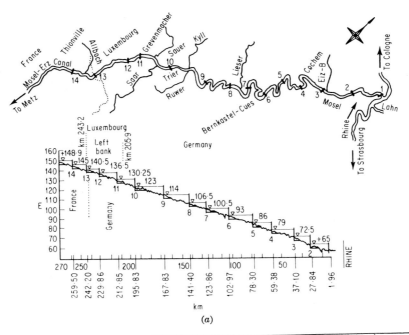

Station	No.	Pond elevation, m	Rated discharge, m³/sec	Head, m	Capacity, mw	Gross output, gwh	Construction dates
Coblenz	1	65.00	380	5.3	16.0	65.0	1941–1951
Lehmen	2	72.50	380	7.5	16.2	81.6	1960–1962
Müden	3	79.00	380	6.5	12.8	66.0	1963–1964
Fankel	4	86.00	380	7.0	14.72	71.6	1963–1964
St. Aldegund (Neef)	5	93.00	380	7.0	14.4	72.3	1962–1964
Enkirch	6	100.50	380	7.5	16.5	81.7	1964–1965
Zeltingen	7	106.50	380	6.0	12.2	61.7	1962–1964
Wintrich	8	114.00	380	7.5	17.7	87.0	1964–1965
Detzem	9	123.00	380	9.0	23.0	111.3	1960–1962
Trier	10	130.25	380	7.2	16.5	79.5	1959–1961
Grevenmacher	11	136.50	165	6.3	7.5	38.8	1963–1964
Palzem	12	140.50	150	4.0	4.1	19.6	1963–1964

(b)

FIGURE 20 (a) Map and profile of Mosel River. (b) Principal features, Mosel plants. (*Water Power, July 1965.*)

project was completed in 1987.[14] The project's main features are a 240-ft-long concrete overflow weir, a bell-mouth entrance intake structure with trashrack, a 15.25-ft-diameter unlined power tunnel approximately 4500 ft long which was excavated by a tunnel boring machine, a 30-ft-diameter by 74-ft-high below-ground concrete-lined air-pressurized surge chamber located at the downstream end of

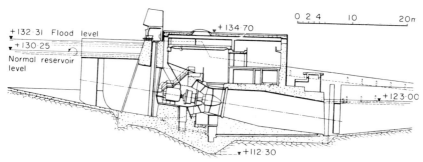

FIGURE 21 Cross section through Trier power station showing arrangement of tubular generating set.

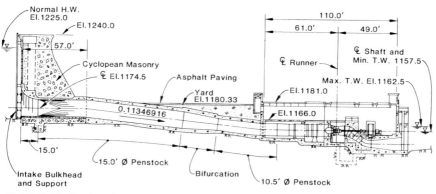

FIGURE 22 Section through Gregory B. Jarvis power plant, New York Power Authority. (*Harza Engineering Co.*)

the power tunnel, a 9.5-ft-diameter steel penstock, and a reinforced-concrete powerhouse structure enclosing one vertical-shaft Kaplan turbine.

The Philadelphia powerplant, Philadelphia, New York (Fig. 25), contains one tubular-type unit rated at 3.6 MW and operated under a normal head of 57 ft. The project was completed in 1986.[15]

24. Economic Design. Overall station economy must be the objective of the designer in selecting the size and spacing of the units, the elevation of the runner in relation to tailwater, and the specific speed. In order to achieve minimum station cost, the turbine setting must be tailored to the particular site and cannot be established by statistical or empirical methods. This is illustrated by Fig. 26,[10] which shows a somewhat exaggerated comparison of alternate deep and shallow settings for an integral intake and powerhouse substructure. In other words, the whole structure acts as a dam and both the overall proportions and the stability are affected materially by the choice of setting. For the same head, the same power, and the same margin of safety against cavitation, a relatively deep setting, with respect to tailwater, means both deeper excavation and more expensive structures. This additional cost, however, is offset in part by a higher allowable

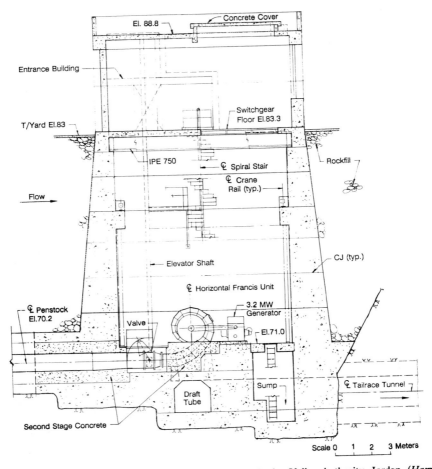

FIGURE 23 Section through King Talal powerhouse, Jordan Valley Authority, Jordan. (*Harza Engineering Co.*)

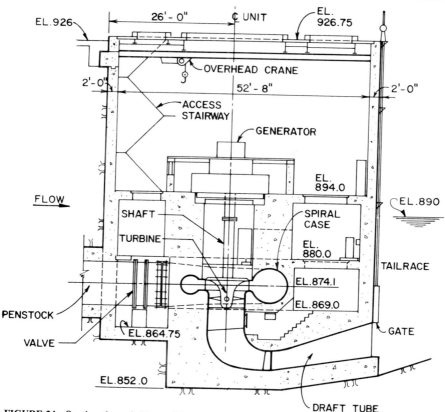

FIGURE 24 Section through Moose River power plant, New York. (*Harza Engineering Co.*)

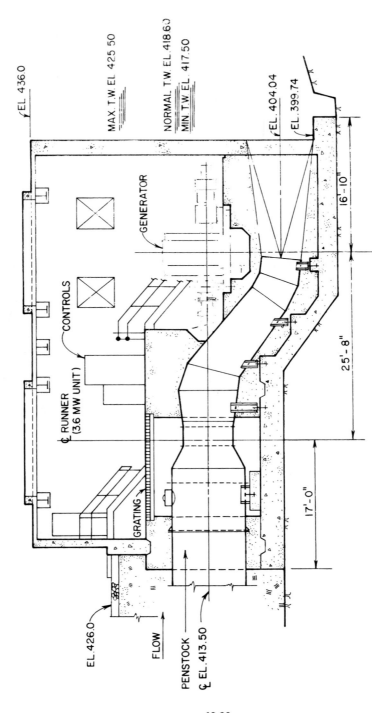

FIGURE 25 Section through Philadelphia power plant, New York. (*Harza Engineering Co.*)

19.32

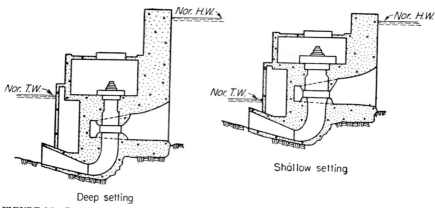

FIGURE 26 Comparison of deep and shallow settings.

operating speed which results in smaller physical dimensions and lower costs for the turbines and generators, and relatively less WR^2 will be required for a given degree of speed regulation.

The shallow setting, on the other hand, minimizes excavation and the structures. A slower operating speed, however, results in larger physical dimensions and costs for the generating units. The choice, therefore, will lie somewhere between the deep setting shown on the left of Fig. 26 and the shallow setting shown on the right. Where the rock is at low depth, the natural selection would be the smaller size, high-speed turbine, and conversely where the rock occurs at higher levels. For most sites, comparative estimates of several settings will be required to achieve the optimum.

It must be emphasized that the selection of the most economical turbine-generator set may not result in minimum overall powerhouse cost.

REFERENCES

1. CARREA, E., R. RODDY, and H. HASEN, "Raising Guri Dam: Stage Construction to Meet Growth in Venezuela," 13th International Congress on Large Dams, New Delhi, India, October 1979.

2. BINNIE, G., et al., Engineering of Mangla, *Proc. Inst. Civil Engrs.*, November 1967, p. 449.

3. SARKARIA, G. S., Economical Diameter for Penstocks, *Water Power*, September 1958, p. 352.

4. FOSDICK, ELLERY R., Tunnel and Penstock Tests at Chelan Station, Washington, *Trans. ASCE*, **101**, 1936, p. 1409.

5. WILLEY, C. KEITH, Wanapum Hydroelectric Development, *Civil Eng.*, September 1960, p. 65.

6. COCHRANE, H. G., The St. Lawrence Seaway and Power Project, *Water Power*, April 1956.

7. PALO, G. P., and C. R,. MARKS, C. R., The Design of Hydroelectric Stations, *Trans. ASCE*, Vol. 111, Pages 1175, Paper No. 2291.

8. "The Appalachia, Ocoee No. 3, Nottley and Chatuge Projects," Technical Report 5, vol. 2, Tennessee Valley Authority.
9. The Mosel Hydroelectric Stations, *Water Power,* July 1965, p. 259.
10. RICH, G. R., Basic Hydraulics of Water Storage Projects, *Civil Eng.,* 1944, p. 351.
11. *Civil Engineering Guidelines for Planning and Designing Hydroelectric Developments,* Volume 3, *Powerhouses and Related Topics,* Table 1-4 Page 1-158. Published by the American Society of Civil Engineers, 1989.
12. WEAVER, R. M., and W. SAYED, "Application of Small Hydro to Hinckley Dam," Pennsylvania Electrical Association, April 1984.
13. ANTONOPOULOS, G. C., V. J. ZIPPARRO, and Z. ALEM, "Raising of King Talal Dam, *Waterpower '91,* Denver, Co., July 1991.
14. ZIPPARRO, V. J., and C. KONSTANTELLOS, "Design and Construction of the Moose River Hydroelectric Project," *Waterpower '89,* August 1989.
15. CLEMEN, D. M., and S. J. HAYES, Design and Commissioning of the Philadelphia Hydroelectric Project, *Water Power,* 1987.

SECTION 20

PUMPED STORAGE

By Henry H. Chen

1. Contents of this Section. This section is a comprehensive treatment on pumped storage. Part A is a description of how pumped storage has been developed and used over the years, and how it is being planned for the future.

In Part B the physical features making up a pumped storage project are described and discussed. The discussions focus on those aspects that pertain specifically to pumped storage, making reference to the earlier sections where there are extensive discussions on conventional hydroelectric power plants.

Part C provides a treatment on pumped storage operation. The discussions cover the performance of the plant itself and how the plant performs in an electric system.

Part D is devoted to costs and economics, which decides whether a pumped storage facility will be built. Part E discusses how to plan a pumped storage project. In planning a modern pumped storage project, the technical issues are closely related to regulatory and environmental issues. The engineer must be knowledgeable about these considerations. The section closes with a bibliography of pumped storage references.

PART A. BASIC CONCEPTS OF PUMPED STORAGE

2. Pumped Storage Concept. A pumped storage plant involves pumping water from a lower reservoir to an upper reservoir, and later returning water from the upper reservoir to the lower reservoir or to a different location and reservoir. Electricity is consumed in the pumping cycle and is produced in the generating cycle when water is returned from the upper reservoir.

Pumped storage may be developed for electric power system use or for water-supply improvements or for both purposes. In defining pumped storage, the following classifications are often given; they are illustrated in Fig. 1.[1]

Pure pumped storage—pumped storage built for electric power system operation, consisting of two offstream reservoirs, where electric power generation is primarily dependent on the pumping operation (rather than from natural inflow into the upper reservoir).

Pump-back—on a river in a hydroelectric power plant where several or all of the units are reversible, so that electric power generation relies on both natural inflow as well as the water that is pumped back.

Seasonal pumped storage—pumped storage operation balances the variability of natural fluctuation of river flow, which is generally seasonal, with demand which may be more uniform and may possibly vary in a direction opposite to sup-

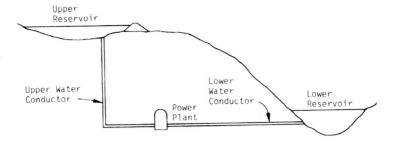

Water is cycled between two reservoirs at different elevations, alternatively pumping water and producing electrical energy.

PUMPED-STORAGE CONCEPT

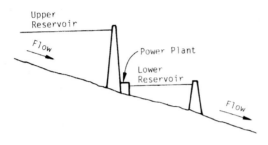

Providing pumped-storage features in a conventional hydroelectric plant permits additional generation in peak hours. The extra release over natural inflow is retained in the lower reservoir and pumped back later.

PUMPED-BACK PUMPED STORAGE

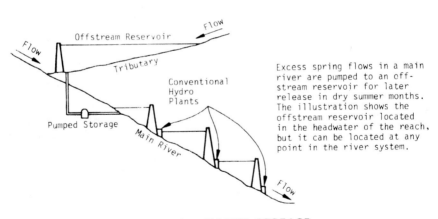

Excess spring flows in a main river are pumped to an offstream reservoir for later release in dry summer months. The illustration shows the offstream reservoir located in the headwater of the reach, but it can be located at any point in the river system.

SEASONAL PUMPED STORAGE

FIGURE 1 Types of pumped storage.

ply. In months of plentiful supply, water is pumped from the river to a reservoir offstream or to a tributary to the main river for later release in months of high water demand and low stream flow. Pumped storage may also be utilized where water is transferred from one river basin to another.

In application to an electric power system, pumped storage balances demand, which tends to vary during the day, with supply, which tends to be constant. Pumping takes place when there is surplus capacity in coal, nuclear, and renewable energy plants; and generation takes place when needed. Pumped storage provides (1) dependable capacity, (2) peaking and intermediate energy, and (3) supply-side management (dynamic benefits) by using its superior operational capability to enable other plants to operate more efficiently.

3. Historical Background. Pumped storage is a century-old technology. It began with building a pumping plant to capture and store water in reservoirs in high-flow seasons for later hydroelectric generation at plants downstream, in either the same or a different basin. These projects initially were built in Europe. Later, projects involved installing both pumping and generating facilities in the same plant. Still later, the machines for pumping and generating were combined, using first reversible generator-motors with separate pumps and turbines, followed by reversible pump-turbines. Most modern pumped storage plants employ reversible generator-motors and pump-turbines.

In the United States, the first pumped storage facility was the Rocky River project, placed in service in 1929. The first reversible pump-turbine and generator-motor units were at Flatiron in Colorado (Bureau of Reclamation) in 1954. The first true pumped storage plant for electric system operation employing reversible pumping and generating machinery was the Taum Sauk plant in Missouri (Union Electric Company) in 1963.

The predominant developments and technological advances in pumped storage are those for electric power production. In the last 30 years, pumped storage is finding wider application in electric systems, with larger units and power plants operating at increasingly higher heads being built. The largest pumped storage plant in the world today is the 2100-MW Bath County project in Virginia commissioned in 1985 (see Fig. 2).[2]

On the other hand, imaginative application of pumped storage can benefit other purposes. The United States and Canada built pumped storage plants on each side of their Niagara River border as a part of the Niagara Falls development. The storage plants pump water into the upper reservoirs at night, reducing the river flow over the Niagara Falls at night. During the daytime water released improves the visual effect of the falls while increasing generation. (See Fig. 3.)

4. Status of Current Developments. Pumped storage is in wide use today. According to a survey by Water Power & Dam Construction,[3,4] 42 countries have pumped storage installations totaling nearly 81,000 MW in capacity. (See Fig. 4.) The developments are led by Japan and the United States, each with about 17,000 MW as of 1991, followed by Italy, Germany, France, Spain, and the United Kingdom. Japan will have 23,000 MW installed before the end of this century and will have more capacity in pumped storage plants than in conventional hydroelectric power. Its present installed pumped storage capacity is about 12 percent of the total generating capacity.

5. Significant Pumped Storage Achievements. Here are some significant pumped storage achievements as of 1991:

Largest plant—2100 MW (generating), Bath County, United States

Highest head for reversible multistage units—4151 ft at Edolo in Italy

Highest head for separate turbines-pumps—5815 ft at Reiszeck in Austria

Highest head for reversible single-stage units—2200 ft at Chaira in Bulgaria

20.4 PUMPED STORAGE

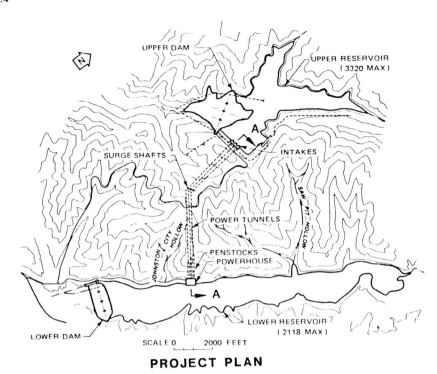

PROJECT PLAN

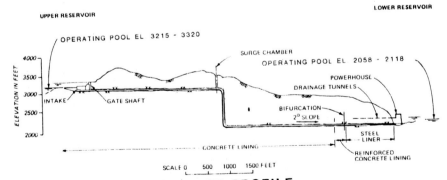

PROJECT PROFILE

FIGURE 2 Bath County project, Virginia.

BASIC CONCEPTS OF PUMPED STORAGE

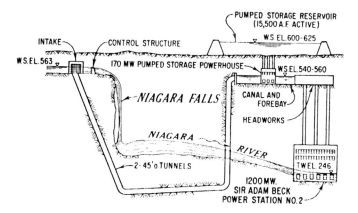

PROFILE THROUGH CANADIAN NIAGARA FALLS PUMPED STORAGE PLANT

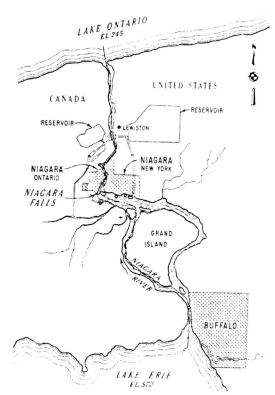

GENERAL PLAN PUMPED STORAGE PLANT ON EACH SIDE OF THE BORDER
FIGURE 3 Niagara Falls pumped storage plants.

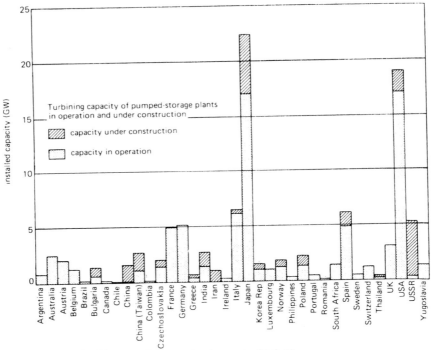

FIGURE 4 Worldwide pumped storage developments. (*Ref. 4.*)

PART B. ELEMENTS OF PUMPED STORAGE

6. Introduction. This part contains a general description of the physical features of a pumped storage project. Emphasis is placed on pure pumped storage projects. In the case of seasonal pumped storage and pump-back, the projects would be much more dependent on the multiple-purpose nature, with a wide range of site conditions.

7. Power and Energy Equations. The reader is directed to Sec. 19 for details on the mathematics of power and energy. For ease of reference, the formulas governing power and energy productions for pumped storage are:

$$\text{Output} = \text{constant} \times \text{discharge} \times \text{head} \times \text{efficiency}$$

This can be expressed in English units as follows:

$$\text{megawatts} = 0.0845 \times \text{cubic feet per second} \times \text{feet} \times \text{efficiency}$$

$$\text{megawatthours} = 1.02 \times \text{acre-feet} \times \text{feet} \times \text{efficiency}$$

In the pumping mode, the efficiency term will need to be placed in the denominator.

8. Operating Heads and Cycling Storage. Pumped storage plants normally have operating heads between 300 and 1500 ft. Heads lower than 300 ft require facilities that have excessive dimensions, resulting in high costs. As is often the case when a limit is placed, one finds exceptions where operating heads fall beyond that range. Heads greater than 2500 ft result in the need for two-stage reversible pump-turbines. Again, as is the case in most technologies, advances keep pushing the upper limit of practicality. One would expect that the upper limit of single-stage reversible units will increase with time.

Cycling storage is a function of installed capacity, operating head, and the amount of energy storage, i.e., operating hours, required for effective operation. Normally, a 10-h storage is reasonable for daily and weekly operation. For purposes of relating energy storage to volume of water, it is useful to remember that each acre-foot of water can produce 1 MWh under 1200 ft of head.

Cycling storage is normally achieved by reservoir drawdown, which in turn affects the operating capability of the pump-turbine. In general, the range of reservoir operating levels should be limited to 100 ft in order that facilities such as intakes can be easily built. In any case, the drawdown must not cause excessive reduction in the operating head on the pump-turbine. These relationships provide a good framework for planning the sizes and surface areas of the reservoirs.

9. Principal Physical Elements. A pumped storage project consists of several essential features required for its operation:

Upper reservoir—created by building an enclosed dike or damming a small stream, or by use of an existing reservoir, or conversion of a natural depression, or a quarry.

Lower reservoir—formed by damming a small stream, use of an existing reservoir or lake, or an underground cavern from mining operation, or one specifically excavated for the project.

Water-conductor system—linking the reservoirs with the power plant. The portion of the water-conductor system between the upper reservoir and the power plant is generally referred to as the high-head system, and the portion between the power plant and the lower reservoir is usually referred to as the low-head water-conductor system.

Power plant—housing the pumping and generating units, equipment for pump-start, electrical and mechanical equipment for control, protection, operation, and maintenance of the power plant.

In addition to the above, there are switchyard and transmission lines to connect the plant to the electric system, access facilities, and a water-supply system for initial filling of the reservoirs and required makeup, if required.

10. Physical Setting. A successful pumped storage project is dependent on topographic and geologic site conditions which permit proper sizing and arranging of the principal features. In addition to high heads, there are other physical conditions under which pumped storage may be practical. A review of the overall dimensions of the facilities is useful.

A 500-MW project operating under a head of 1000 ft requires an active storage of 6000 acre-ft in each reservoir to provide 10 h of generation or 5000 MWh. A higher operating head would permit the use of smaller reservoirs. A 6000-acre-ft reservoir can be provided with an average reservoir area of 100 acres and a fluctuation of 60 ft, or other combinations of reservoir area and drawdown. Ideally, reservoirs could be constructed by damming a stream where the dam does not have to be excessively high or long. If the reservoir is to be formed with a ring dike, the reservoir area should be flat or gently rolling, with no more than 100 ft of topographic relief.

Most pumped storage projects tend to be built with heads in the 700- to 1500-ft range. Sites with head differential less than 300 ft would not be economical unless there are exceptional factors favoring the site. There are few areas in the United States for potential pumped storage where heads would exceed 1500 ft.

In addition to operating head, the other key parameter is the length of water conductors connecting the upper and lower reservoirs. The parameter is usually expressed as a function of the operating head. The shorter the length the better the project, for both result in low cost and low hydraulic head losses. A representative range of the distance between the reservoirs would be 2000 to 15,000 ft. The ratio of length to head falls between 3 and 10 for most pumped storage projects.

The operating head of the pump-turbine is controlled by the head differential between the upper and lower reservoirs as well as reservoir fluctuation. There are upper limits in the operating heads of the reversible pump-turbines. Over the history of the reversible pump-turbine, this limit has been increased from about 1000 ft 25 years ago to about 2500 ft today, for single-stage regulable reversible pump-turbines. While the current technical upper limit may be 2500 ft, the optimum head may be no more than 1200 ft, from the cost standpoint. The reason is that as the size of a reservoir with a certain required energy content goes down with an increased design head, the cost of both the water conductors and the pump-turbines increases, causing an overall increase in cost.

11. Upper Reservoir. Upper reservoirs tend to be located on mountaintops or bluffs or at the headwaters of small streams. In addition to topographic characteristics favorable for the construction of dams and dikes to create the reservoir, the site must possess favorable geotechnical conditions. The foundation must be suitable to support the dams and dikes. The reservoir floor and the ridges must be stable under the increased water load imposed when the reservoir is full, as well as during cycles or filling and emptying. While there will always be some seepage, it must be controlled to a minimal amount by cutoffs of the seepage path, with grout curtains, consolidation grouting, or a reservoir membrane.

Upper reservoirs of pumped storage plants are subject to large fluctuations resulting from daily or weekly cycling. A number of projects have fluctuations approaching 100 ft. When considered in terms of stability for natural slopes or fill-type structures, such fluctuation is practically instantaneous, meaning almost 100 percent of the hydrostatic load is imposed. This requires careful consideration of stability and proper drainage for both the reservoir rims and the reservoir floor.

Upper reservoirs are often sited close to the edge of a bluff or approaching the top of ridges forming the reservoir rim. Consequently, the overall stability of the ridges should be addressed in siting such reservoirs. If necessary, measures to strengthen the ridges should be considered.

Enclosed Ring Dikes. An upper reservoir of the enclosed-ring-dike type can be sited (1) at the edge of a bluff, (2) on a mountaintop, or (3) on a saddle flanked by two or more ridge tops. The challenge in building the dikes for the upper reservoir usually is related to having enough space to site the dikes while retaining enough area for the reservoir. Another challenge is to make use of available material to build the dikes. If the underlying rock is suitable, a zoned rockfill or a concrete-face rockfill would be preferred. The concrete-face rockfill is preferred in the sense that it is inherently a very stable structure.

If a dam is to be created by shaving the top off a mountain, a usual consideration is balancing cut and fill and making the best use of the excavated material as fill, with a minimum of waste to spoil and minimum of borrow from off site. Also, the sequence of construction should be worked out to recognize the limited available space.

Figure 5 shows the Kinzua project with the upper reservoir created by the

FIGURE 5 Kinzua pumped storage development.

construction of a circular ring dike with the dike and reservoir floor provided with an asphalt liner.

Spillways. Spillways are required, both to pass the floods from the basin and to pass overpumped water. Because spillway capacity can be substantially reduced by surcharge, it is always a good strategy to provide some surcharge. An overflow spillway of the fuse-plug type may be advantageous to meet requirements of providing an "emergency action plan." The spillway should be sited to direct water along a path with the least detrimental consequences.

Low-Level Outlets. These are needed to maintain a continuous minimum flow when the upper reservoir is on a stream. In an enclosed reservoir, such outlets would not be required.

12. Lower Reservoir. Lower reservoirs are created by damming a small tributary or by using an existing lake which can be manmade or natural. More than half of the existing U.S. utility pumped storage projects utilize existing reservoirs as lower reservoirs. Examples include Muddy Run, Kinzua, Northfield,

Blenheim Gilboa, Joccassee, Bear Swamp, Raccoon Mountain, Fairfield, Helms, Bad Creek, and Balsom Meadow.

Because lower reservoirs are onstream, the inflow regimes tend to observe the same situations as those of a conventional hydroelectric project. However, because of the daily withdrawal and supply of active storage from pumped storage operation, special considerations must be given. As with the upper reservoir, dams, dikes, and slopes must remain stable, and in particular, the power-plant setting must account for a wide change in reservoir elevation as well as the necessary pump submergence. This may require a very deep structure.

Finally, the spillway provided with the lower reservoir must be capable of passing the natural flood plus any flood intensification due to the pumping and generating cycling operation. A large-capacity low-level outlet may be needed to handle such conditions.

Underground Pumped Storage. Invariably, the environmental acceptability of a pumped storage project revolves around the acceptability of constructing the reservoirs and damming the rivers. The use of underground reservoirs would be a way of overcoming this problem, as well as siting projects in population centers. Such construction would also make pump storage possible in geographical areas that do not have great reliefs, such as the U.S. Middle West. Underground reservoirs imply three concepts: (1) the reservoir would be excavated, (2) the reservoir would be converted from an existing underground quarry, or (3) an existing quarry facility would be used, with the reservoir being excavated.

Underground pumped storage projects requiring the excavation of the entire lower reservoir were studied extensively in the late 1970s and early 1980s.[5] If such a concept could be demonstrated to be viable, many such projects could be built. (See Fig. 6.) The construction cost of such a scheme is relatively high. Economics dictates the size to be at least 2000 MW, with an operating head of at least 3000 ft and as much as 5000 ft. At 5000 ft, the project would employ either (1) two power stations in series or (2) multistage pump-turbines.[6] Ultra-high-head pump-turbines have been studied and tested for such applications.[7]

An essential element in the feasibility of underground reservoirs is the integrity of the rock caverns deep underground. The cavern has to be in very massive and strong homogeneous rock so that supports will not be required, as the in situ stress in the rock at depth is very high. Thus the rock mass suitable for pumped storage is limited to Cambrian or Precambrian rocks of the granitic or sedimentary type subject to minimal structural or tectonic changes.[8]

Underground pumped storage might be developed and be attractive with the use of an existing underground mine. For the worked-out space to be suitable as an underground reservoir it has to be at an appropriate depth from the surface and have a nearly level surface, and the excavated cavern must be structurally stable. Several underground limestone mines seem to meet these requirements; among them is the Summit project near Akron, Ohio.

Seawater Pumped Storage. Because of the environmental difficulties involved in reservoir siting, the use of the sea as a lower reservoir has been considered in many countries, and a demonstration plant is being built in Okinawa. (See Fig. 7.)[9]

13. Water-Conductor System. The water-conductor system connects the upper and lower reservoirs through the pump-turbine. The size of the water-conductor system is designed to pass the required discharge of the pump-turbines in acceptable velocity and head loss. The water-conductor system is usually designated as high-head and low-head portions, referring to the portion between the

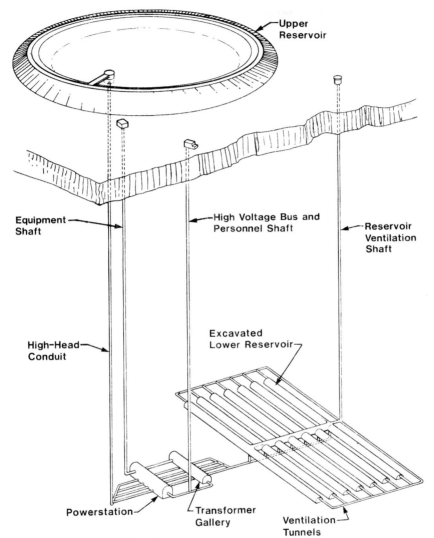

FIGURE 6 3000 MW, 5000 ft underground pumped storage development concept, northern Illinois, Harza Engineering Company.

upper reservoir and the power plant, and the portion between the power plant and the lower reservoir.

The water-conductor system consists of approach channels, large-diameter conduits (tunnels and penstocks) merging into individual penstocks to connect with the inlet valves on the high-head side.

The water conductor should be designed to have good transition with small

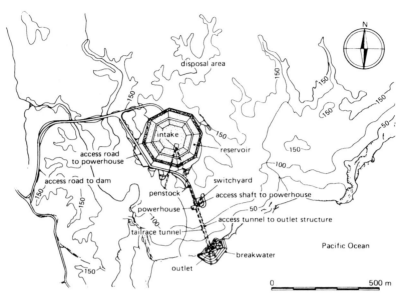

FIGURE 7 30-MW Okinawa seawater pumped storage project, Japan. (*Ref. 9.*)

velocity changes. It is easier to design for flow acceleration than for flow deceleration when flow concentration is likely to occur under less than perfect design conditions.

Water conductors are concrete or steel conduits. This is mostly due to the relatively high head and water velocity in the conduits. Unlined tunnels are rarely used, one exception being the tailrace tunnel for the Northfield Mountain project.

Water velocity in concrete or steel conduits should be between 12 and 25 fps except in the portion near the inlet valve, where it may be much higher. Water velocity is controlled by economics and hydraulic stability requirements. Optimum economic velocity is in the 15- to 20-fps range for concrete-lined tunnels and 20- to 25-fps range for steel penstocks.

Tunnel diameters tend to range between 12 and 35 ft. Penstock diameters may range between 6 and 30 ft, where the smaller diameters may be at the lower end of the water-conductor system and the large diameter may be at the upper end. Where the rock mass is favorable, concrete-lined tunnels with adequate rock support may be used so that the bulk of the internal stress is transferred to the rock mass. This consideration often influences the selection of the project general arrangement and water-conductor route selection.

Intakes. Intakes for pumped storage plants require special attention because of the need to permit water flow in both directions. The intakes are sometimes referred to as intake-outlets. In addition, since the operating range of the water surface is usually large, up to about 100 ft in some reservoirs, the intake design has to be flexible.

The morning glory intake with antivortex horizontal plates is a good design associated with ring-dike-type upper reservoirs. The intake is highly efficient hydraulically and of relatively low cost. It is not suitable for use where water withdrawals are to be made at selective levels.

Vertical intakes, built against a steeply inclined rock face, or tower intakes are often utilized in reservoirs with steep canyons or side slopes. These intakes can be provided with trashracks, intake gates, and structures that permit selective withdrawal of water from any level in the reservoir.

In some cases, fish screens are required in the intakes. Providing fish screens in pumped storage intakes is very difficult because regulatory requirements may specify a very low velocity across the screen. The physical size of the intake would then be extremely large, resulting in a very high cost.

In passing water from reservoirs to water conductors, intakes must include transition sections for water to accelerate or decelerate depending on direction of operation, pumping, or generating. It is often the deceleration that controls the shape of the transition, which should be sufficiently gradual to prevent flow concentration or jet formation. Figures 8 and 9 show typical vertical and horizontal intakes.

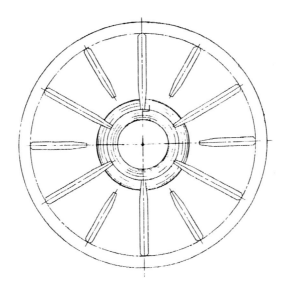

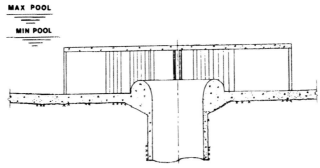

FIGURE 8 Typical vertical intake (Kinzua).

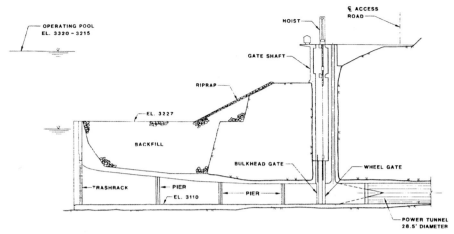

FIGURE 9 Typical horizontal intake (Bath County).

Control of Water-Conductor System. Intake gates are required to protect and control the water-conductor system. The general philosophy is that at least two alternative means are needed to be capable of shutting the water-conductor system under flow conditions. Sometimes, a single water conductor serves two or more units. To assure that any one unit can be isolated for servicing without interfering with the operation of the other units, turbine inlet valves are needed. Except where the head is low (less than 400 ft) and the water-conductor length is short, turbine inlet valves are also needed for pump start and for reversal of operation.

An important consideration is the intake gate control mechanism. An intake gate is used with great frequency in a pumped storage plant. The gate may be operated under pulsating flow and flow concentrations, and can therefore be subject to dynamic stresses. Positive control in the standstill position, as well as during gate operation, is provided by the use of hydraulic cylinders. The same design consideration applies for trashracks and screens. Trashracks of many early pumped storage plants failed owing to flow-induced vibrations.

14. Project Arrangement. Project arrangement is the science or art of choosing the different elements in terms of type, size, and location, and fitting them together to arrive at an elegant solution for the entire project. The physical setting provides the basic rule under which to arrive at a technically sound and economical design concept.

In general, the upper and lower reservoirs are selected according to the required storage and the appropriate locations of the dams. Selection is controlled by the desire to obtain the maximum amount of storage and the minimum amount of dam volume. Two other objectives control. The first is to have the reservoirs as close as possible in the horizontal direction and as far apart as possible in the vertical direction—thus achieving a minimum ratio of the length of water conductors to operating head. The second is to have topographic and geologic factors that favor the placement of the powerhouse and the alignment of the water conductors. (See Fig. 10.)

Powerhouses can be of the underground, surface, and pit or shaft type. Water conductors can be entirely underground (tunnels and tunnel penstocks), entirely

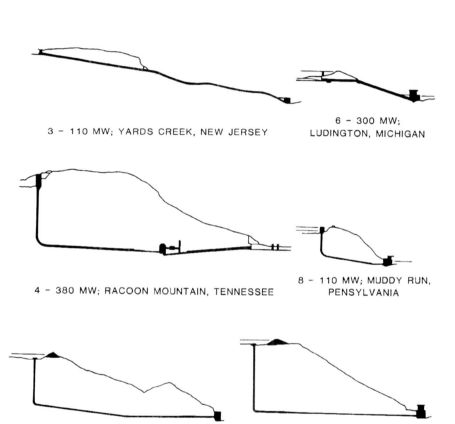

FIGURE 10 Typical U.S. pumped storage plants.

surface, or a combination. (See Figs. 11, 12, and 13 for typical designs.) Where rock conditions are favorable, an underground powerhouse and underground water conductors, which can rely on the rock mass to take up hydrostatic stresses, would produce the most economic layout. Recent trend has been toward more underground designs for two reasons. First, they are suited to the use of high-specific-speed pump-turbines with deep settings, and second, they provide better

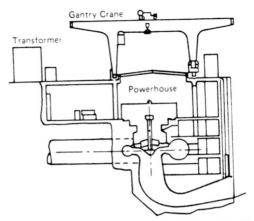

FIGURE 11a 512 MW, Fairfield, S.C., small tailwater fluctuation, outdoor powerhouse.

aesthetics. As an alternative to underground, the shaft or pit type also permits deep settings. This type has the advantage of shorter construction time and is especially suitable for small- to medium-sized installation.

15. Pumping and Generating Equipment. A modern pumped storage plant almost always utilizes reversible pump-turbines and generator-motors. The Francis-type design is most often used because of its wide operating range—from under 200 ft up to 2500 ft. For heads under 400 ft, the Deriaz type may be appropriate. At low and extremely low heads, the Kaplan or bulb type, respectively, can also be considered. However, pumped storage is rarely justified at such low heads for electric utility operation unless there are other controlling factors for their development. At very high heads, two-stage or multistage units may be used. The two-stage unit may still have the low-head (upper) stage regulable, but the second stage cannot be regulated.

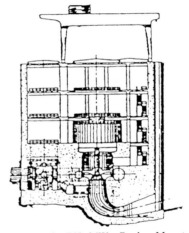

FIGURE 11b 765 MW, Rocky Mountain, Ga., large tailwater fluctuation, indoor powerhouse.

Large pumped-storage installations utilizing Francis-type pump-turbines employ vertical shafts to connect the generator-motor with the pump-turbine. This design is used to account for the very large upward hydraulic thrust that cannot be conveniently counteracted in a horizontal design. Also this design is appropriate for the very high submergence requirements for the pump-turbine. Setting of the pump-turbine is controlled by pumping operation which results in deep setting, 60 to 200 ft, depending on specific speed and head, and unit size. In a ver-

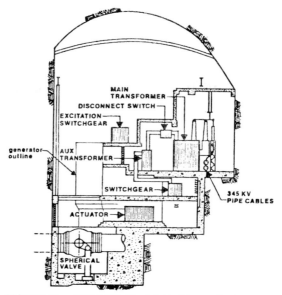

FIGURE 12a 1000 MW, Northfield, Mass., single cavern with transformers in main hall.

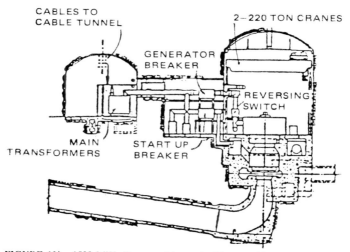

FIGURE 12b 1500 MW, Raccoon Mountain, Tenn., with separate transformer cavern.

tical setting, the generator-motor can be set above the pump-turbine, and access to most of the electrical control and protection equipment can be placed at a higher level.

Pump-Turbine. Pump-turbine design is controlled by the choice of specific speed. A high-specific-speed machine results in reducing the physical size of both

20.18   PUMPED STORAGE

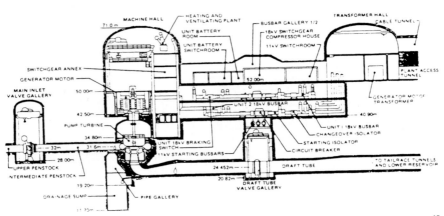

FIGURE 12c   1800 MW, Dinoring, Wales, with separate caverns for values and transformers.[10]

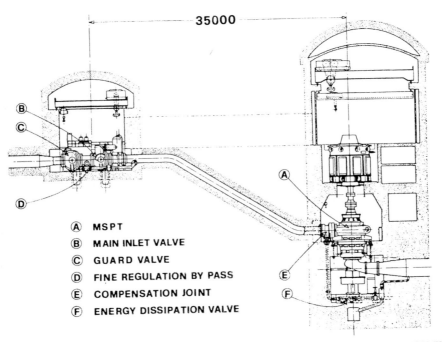

Ⓐ   MSPT
Ⓑ   MAIN INLET VALVE
Ⓒ   GUARD VALVE
Ⓓ   FINE REGULATION BY PASS
Ⓔ   COMPENSATION JOINT
Ⓕ   ENERGY DISSIPATION VALVE

FIGURE 12d   1000 MW, Edolo, Italy, with five-stage reversible pump-turbines each rated at 125 MW.

the pump-turbine and the generator-motor. However, such machines require a deep setting, lengthening the water conductors and tending to increase the cost of the civil features. The selected rotating speed must correspond to a synchronous speed, since the pump-turbine is directly connected to the generator-motor.

    The modern pump-turbine is usually a single stainless-steel runner-impeller.

ELEMENTS OF PUMPED STORAGE

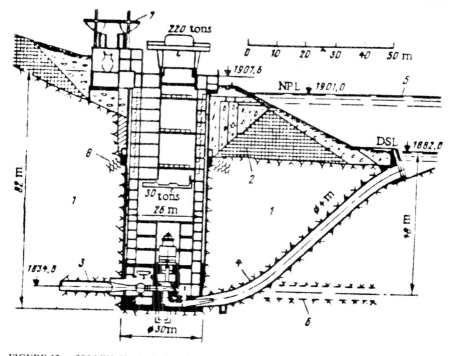

FIGURE 13a 286 MW, Kuehtai, Austria, circular shaft powerhouse.[11]

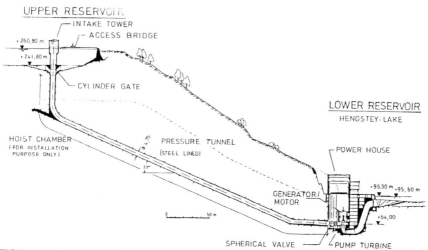

FIGURE 13b 153 MW, New Koepchenwork, Germany, with a shaft powerhouse and inclined penstock tunnel.[11]

This type of machine provides good resistance to cavitation and greater strength than a machine with mild steel. In addition, the design may also call for removable rotating wearing rings on the crown or on the runner, to permit convenient replacement. The use of wearing rings permits a close-clearance design, which improves the pump-turbine efficiency. Replacement capabilities enable maintenance of high efficiency throughout operating life.

Manufacturers have attained very high efficiencies in modern pump-turbines through three-dimensional finite-element design of the machine. The efficiency of the pump-turbine is now 92 percent in the pump mode and 91 percent in the turbine mode in the average operating head and gate range. See Sec. 21 for further discussion.

Generator-Motor. The generator-motor is sized to match the maximum output and input of the pump-turbine. In addition, it must provide the reactive capacity, which is controlled by the system characteristics defined by transmission-line voltages and line loadings under normal and contingency conditions. Since the predicted conditions might be exceeded, it is prudent to size the generator-motor at a slightly larger rating with a margin of 2 to 5 percent depending on the degree of accuracy of the estimate.

The generator-motor design has a strong influence on the powerhouse. The choice of the conventional unit or a close-coupled unit design is particularly important. In the close-coupled design, the thrust bearing of the machine is supported from the head cover of the pump-turbine, and the guide is supported from a bracket in the turbine. This results in a much shorter shaft, which improves the stiffness of the entire machine, thus enhancing stability and reducing vibration.

The choice of the close-coupled design eliminates possible consideration of an intermediate shaft, which facilitates maintenance of the pump-turbine. However, an intermediate shaft is not often used because of increased cost and other associated design problems for large machines.

16. Other Power-Plant Equipment. Other power-plant equipment includes starting equipment for pump operation, control and support equipment for power-reversal operation, and the accessory equipment normally required in a conventional hydroelectric power plant.

Starting Equipment. A pumped storage plant starts in the generating mode like a conventional power plant. In the pumping mode, starting is more difficult because of the large amount of power that will have to be input to the machine. The usual procedure calls for depressing tailwater with compressed air to a level below the runner. This reduces the initial power-input requirement. The machine is then accelerated by an appropriate means to synchronous speed and is connected to the system. Air is released and water is allowed to enter the spiral case, after which the wicket gates are opened to the desired setting. The table in Fig. 14 provides the alternative starting methods.

The full-voltage, reduced-voltage, and part-winding methods draw power directly from the system, resulting in a large power surge. The machine will have to operate like an induction motor, requiring a continuous amortisseur. The methods are applicable only for small (100 MW or less) units, because of the large demand of instantaneous power inflow.

The wound-rotor scheme involves placing a pony motor on top of the generator-motor to bring it to speed. It is used for units large and small. For flexibility of operation, every one of the units requires a pony motor. Thus, this scheme is less often used where there are more than two or three units. Helms project is the most recent installation relying on the pony motor.

The static converter system is often adopted for multiunit, relatively large in-

Starting method	Size applicable to	Approximate starting kVA from system	Approximate starting time	Per unit approximate relative cost (electric only)	Continuous amortisseur required
Full voltage	Moderate-sized units	100%	20–30 s	Lowest in applicable range	Yes
Reduced voltage	Moderate-sized units	100%	2–3 min	Lower in applicable range	Yes
Part winding	Moderate-sized units	100%	2–3 min	Low in applicable range	Yes
Wound-rotor motor	All ratings	5%	10 min	High on small units, low on large units	No
Synchronous	All ratings	0	1–5 min	High unless station has double bus	No
Converter	All ratings	5%	10 min	Low as number of units increases	No

FIGURE 14 Alternative starting methods.

stallations. In such an application, the system is reliable and energy-efficient. The Raccoon Mountain project relies on this scheme. The system consists of a rectifier, inverter, dc reactor, ac reactor, arrester, and circuit breaker. The static converter system can be used as a dynamic brake to stop a unit. This can be done at a high rotating speed, whereas the mechanical brakes can be applied only when the unit reaches a much lower speed. Since the units are stopped many times, this system offers a clear advantage over other systems.

In many cases, two starting schemes are built into the plant, with the second scheme involving back-to-back starting, where one of the units in the plant would be generating supply power to pump-start the other units. The hydraulic, mechanical, and electrical systems need to be specially designed for this purpose. A double-bus system will be needed. The back-to-back or synchronous system may permit starting of all but the last unit in the plant. As seen in the table, the back-to-back system provides a very fast startup sequence.

Other Electrical Equipment. In addition to the startup equipment mentioned above, a pumped storage plant requires control equipment for reversing the operation including the phase-reversal switches and the breakers associated with the switches. Some plants are designed and built to provide extensive redundancy and flexibility of operation. These additional features add to the complexity of operation and maintenance.

An important and essential feature involves providing for direct manual control at the plant even if the design is for unattended operation. Direct control is needed for startup when the plant is first commissioned as well for maintenance purposes.

The main power transformers may be located in or next to the powerhouse. In the case of an underground power station, the transformers may be located in the main cavern or in a separate cavern. High losses and difficulty of bringing low-voltage buses out to the surface usually induce the designer to place the transformers underground, the only exception being the Bad Creek project.

Mechanical Equipment. Turbine inlet valves are used to protect the turbine, prevent leakage when the plant is not operating, and pump-start when water has to be depressed below the runner. Spherical valves are most often used because they are reliable and leakproof and incur neither hydraulic losses nor vibration.

Butterfly valves, the alternative, are applicable for lower-head installations and where the valve diameter is large.

Pumped storage plants should be equipped with inlet valves except for the very low head units. In their absence, the intake gates must perform the functions described above. In that case, individual water conductors are needed for each unit.

Valve failure imposes a grave hazard to the power plant. To assure safety, upstream and downstream seals should be specified for the valve. Also, conservative design should be incorporated in such features as balancing lines and valves, including the use of stainless and corrosion-resistant steel.

Draft tube gates are also needed. In an underground plant, the bonneted-type lift gate or butterfly valve can be considered for draft tube gate purposes. In a surface plant, some plant operators would like to see draft tube gates provided for each unit, to permit servicing all units at the same time.

17. Design for Safety and Reliability. Experience with operating and maintaining existing pumped storage plants has shown that the units are subject to a great deal of abuse from cycling and frequent mode changing. Pumped storage plants have much higher outages—both maintenance and forced—than conventional plants. Design of future plants should therefore recognize the difference in expected operation and allow for much more conservatism. Suggested items include:

- Use of unit sizes and speeds within present manufacturer experience. The 250-MW unit seems to have the greatest experience.
- Use of lower-speed units (or adequate margin of submergence) to minimize cavitation
- Conservative design of the upper reservoir ring dike, recognizing that the geologic characteristics are, under most conditions, unsuited for water retention.
- Conservative design of the water-conductor system, recognizing the dynamic hydraulic and mechanical forces resulting from reversal of flows.

PART C. RATING, PERFORMANCE, AND OPERATION

18. Introduction. This part describes how a pumped storage plant is rated, what its performance characteristics are, how it is operated in an electric system, and how the existing plants operate. It ends with a discussion on how computer models are used to analyze pumped storage operation.

19. Rating. A pumped storage plant should be characterized by both a capacity rating and an energy storage rating. The capacity rating should give the rating in both the generating and the pumping mode, and the energy storage rating should be in both modes as well.

The plant capacity rating, in megawatts, can be given as (1) the generating capability at the minimum head, (2) the nameplete rating of the generator-motor, or (3) the average plant output in its drawdown cycle. While there is no uniformity in the industry, one can assume that the plant rating reflects the way the value of a plant may be assessed. The energy storage rating provides the capability or size of the reservoirs expressed in megawatthours, which reflects the conversion efficiency and operating head.

In addition to plant rating, reference is also often made to the ratings of the generator-motor and the pump-turbine. These ratings are associated with the design of the machines. The generator-motor rating refers to its nominal nameplate rating. Over the years, generator and motor ratings have changed in part because of the type of insulation used, and there are also differences between U.S. standards and foreign standards. To avoid confusion, a generator-motor rating should be given in MVA, power factor, and temperature rise.

The pump-turbine rating is given in both the turbine and pump modes. The turbine mode is given at a specified head, which can be the maximum head, some point near the average head, and the minimum head. The author prefers to use the average head, which would likely be near the rated head, and its full-gate output as the turbine rating, where the machine would be designed to give optimum performance. Turbine capability at the minimum head should also be given. This is the minimum output assured by the plant. Turbine capability at the maximum head is not important, since it would be rare that the machine would be operated at the full gate at its maximum head.

Rating in the pumping mode is given to correspond to the minimum head. This would give the maximum pump input throughout the operating head range. The pump rating is often given in cubic feet per second, but the writer prefers to see a kilowatt or horsepower rating as well.

20. Performance. The performance of a pumped storage plant cannot be totally described by its rating. In addition to the rating, the following characteristics of a pumped storage plant are important:

- Cycle efficiency—the ratio, expressed in percentage, of energy output to energy input, measured at the generator terminal voltage or the high-voltage side of the power transformer.
- Charge-discharge ratio—the ratio of power input to power output. Power input is the average pump load over the pumping cycle, and power output corresponds to the plant rated generating capacity. Generally, the larger the ratio, the larger the plant's ability to refill the upper reservoir.

A large modern pumped storage plant is expected to achieve a cycle efficiency of 75 to 80 percent, based on the following range of component efficiencies:

These efficiency values are based on the actual efficiencies of the plants completed since the late 1970s such as Raccoon Mountain, Helms, and Bath County where the efficiencies are between 78 and 80.0 percent. Efficiencies of the plants of the 1965–1975 vintage are much lower, in the 67 to 75 percent range, owing primarily to a low efficiency of the pump-turbine and degradation from rated values. Figure 15 shows how the power-plant input and output capabilities vary as a function of head.

Dynamic Performance. A pumped storage plant can be designed to provide excellent dynamic operating characteristics if required. The dynamic operating characteristics can be separated into two categories:

Expected Composition of Cycle Efficiencies

Generating:	
Water conductors	97.4–98.5
Pump-turbine	91.5–92.0
Generator-motor	98.5–99.0
Transformer	99.5–99.7
Subtotal	87.3–89.4
Pumping:	
Water conductors	97.6–98.5
Pump-turbine	91.6–92.5
Generator-motor	98.7–99.0
Transformer	99.5–99.8
Subtotal	87.7–99.8
Operational:	98.0–99.5
Total	75.0–80.0

1. Cycling and load following capability—associated with the ability of the plant to operate at part load efficiently, to be started and stopped quickly, and to be

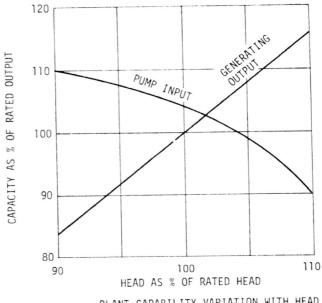

PLANT CAPABILITY VARIATION WITH HEAD

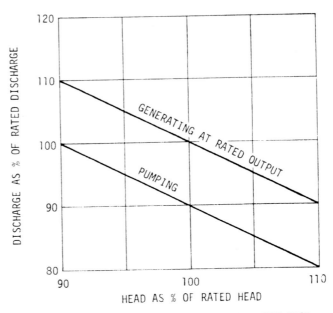

POWERPLANT FLOW VARIATION WITH HEAD

FIGURE 15 Power plant capability curves.

mode of operation to the other conveniently. A modern pumped storage plant is expected to achieve the following capabilities:

- From pumping to full-load generation—2 to 10 min
- From generation to pumping—5 to 20 min
- From shutdown to full-load generation—1 to 5 min
- From shutdown to pumping—3 to 15 min

These capabilities permit pumped storage to perform as peaking units and for cycling operation, as required by the system dispatcher.

2. System regulation—by maintaining frequency and voltage, performing spinning reserve duty, or by synchronous condenser operation.

These regulation characteristics permit dispatchers to operate pumped storage plants under automatic generation control and to stabilize the system under steady-state and contingency conditions and for system-restoration purposes. The regulation characteristics are measured in response times of seconds when synchronized to load.

Pumped storage is capable of delivering full power on a sustained basis within 10 to 30 s. These characteristics are functions of the hydraulic and mechanical characteristics. The hydraulic characteristics are related to the velocity and length of water conductors, which define the water-column inertia. A low value of water-column inertia allows the plant to respond quickly to load changes. Steam plants may have a short response time but are not capable of sustaining the power delivery as the boiler loses steam.

21. Dependable Capacity and Capacity Credit. Pumped storage is often assigned a dependable capacity and sometimes a capacity credit. The dependable capacity is determined as the capacity that the system can rely on the plant to provide under the peak demand of the year. The capacity credit is determined as what the plant can replace as alternative generation. In general, capacity credit recognizes that a plant is expected not only to operate under the load curve but also to provide operating reserve, whether quick reserve or spinning reserve. Furthermore, capacity credit may recognize the difference in outage rates of the different types of power plants.

Dependable Capacity. A pumped storage plant, like a conventional hydroelectric power plant, is assigned a dependable capacity that the Federal Energy Regulatory Commission defines as "its load-carrying capacity under adverse conditions for the time interval and period specified when related to the characteristics of the load, in conjunction with other plants in the system." This determination, for conventional hydropower, relates to the amount of firm energy that is available under a critically dry hydrologic cycle. For pumped storage, it relates to the following:

- The machine generating capability, which must exceed the designated dependable capacity
- The shape of the load curve in the peaking hours, which determines the amount of hours and the integrated energy which the plant must operate to bring the remaining load down to be met by the rest of the system generation
- The shape of the load curve in the off-peak period during which pumping can take place, and the availability of sufficient pumping energy to refill the upper reservoir
- The size of the energy storage to permit those generating and pumping operating cycles.

Experience with systems in the United States indicates that a pumped storage plant would have its dependable capacity equal to the machine capability, provided that the energy storage is at least 8 h and the plant capacity is less than 10 percent of total system demand.

Capacity Credit. In comparing alternative types of generation, the term capacity credit may be used to refer to the equivalent amount of capacity with which one type of generation may replace another type. In determining capacity credit, the factors to be taken into account include dependable capacity, operating reserve capacity, and outage rates. Today, electrical systems consist of a large number of plants, often predominantly thermal. System demand is met by dispatching the available generation and allowing for an operating reserve equal to 5 to 10 percent of the demand to meet contingency conditions. Each plant that is capable of providing such operating reserve is expected to operate at less than its full capacity. In the case of pumped storage, the generating capacity is typically greater than the dependable capacity, and the difference can easily be assigned as operating reserve. This cannot often be done for other generating plants such as combustion turbines because the entire plant is already assigned as reserve. Thus, the average generating capacity of pumped storage may be used for determining capacity credit. Finally, pumped storage has a higher availability and reliability factor than other types of generation that should be incorporated in the analysis.

Energy Generation. The energy generation of a pumped storage plant is naturally related to the plant and energy storage characteristics, to the load characteristics, to the availability of low-cost off-peak energy in relation to on-peak energy, and to the shape of the load curve. Assuming that the system characteristics are favorable, a pumped storage plant is expected to operate between 1000 and 2000 h at rated generating capacity, equivalent to total hours connected to load in the generating mode of 1500 to 3000 h. This also means that the plant will operate for an additional 1200 to 2400 h in the pumping mode.

22. Integrated Operation in Utility System. The operation of a group of power plants in a utility system depends on the characteristics of all the plants, as well as the system operating requirements. The combined capability of all the plants must exceed the expected load by an appropriate margin. In addition to meeting the load, the next most important criterion is to operate the system in the most economic manner. The next point is to meet the load changes and other contingencies. Thus, the following conditions must be met:

- Capacity—the total available capacity must be equal to the expected peak demand plus scheduled and operating reserve.
- Energy—the total energy demand must be met by the combined production of the plants in operation.
- Dynamics—the dispatched plants must be able to follow load fluctuations, maintaining frequency and voltage levels.

Not included but certainly appropriate to mention are such other requirements as transmission stability and area protection.

Operating Characteristics of Generating Plants. Generating plants have different operating characteristics as well as different energy production costs depending on their conversion efficiency and cost of fuel (see Fig. 16). Based on these characteristics, generating units are base-loaded, cycled, operated on-peak,

	Nuclear	Coal steam	Oil steam	Combustion turbine	Cogeneration	Pumped storage
Normal duty cycle	Base	Base	Intermediate	Peaking	Base	Peaking
Ramping and load-following capability	Poor	Poor	Moderate	No	No	Good
Operating reserve:						
Quick-start	No	No	No	Yes	No	Yes
Spinning reserve	Yes	Yes	Yes	No	No	Yes
Frequency regulation	No	Yes	Yes	No	No	Yes
Power system restoration	No	No	No	Yes	No	Yes

FIGURE 16 Operating characteristics and generating unit dispatch on load.

or relegated to a reserve status. In any case, some fraction of capacity of most of the operating units would be assigned to operating reserve.

Nuclear units operate on-base at their rated capacity since their energy cost is lowest and the units are not suitable for changes in loading. Coal-fired steam plants are normally operated on-base, this being applicable for mine-mouth units, scrubbed units, and units generally with low fuel cost and low heat rates. Some coal units are designed for cycling operation, which means that they can be backed down to 50 percent of load conveniently and in a relatively short period of time, say within an hour or two.

Oil- and gas-fired units normally operate above coal units on the load curve since their fuel costs and hence energy costs tend to be higher than those of coal generation. These steam units tend to be assigned to provide spinning reserve since they have a short response time by "fast valving" of their steam supply system. They are also used to follow load, although their load-following capability is generally limited.

Combustion turbines are ideal for peaking purposes. They can be turned on and shut off with load changes each day. The units do not cycle well since their part-load capability is limited. On the other hand, combustion turbines are relatively small in size and the units can be easily added or taken off the line.

Pumped storage plants are capable of operating to meet all dynamic needs—cycling, load following, and frequency regulation—provided that there is sufficient energy storage to meet the generating and pumping needs.

Renewable-energy plants are becoming more important especially as the world grapples with the acid rain problem and global warming caused by emission of carbon dioxide from fossil generation. Important renewable plants (excluding hydroelectric) are solar and wind power plants. Their generation is not dispatchable and does not match the pattern of peaks and troughs of the electrical demand. In addition, cogeneration plants—which are built to produce both steam and electricity—are being added at a rapid pace. These plants are not dispatchable—meaning they cannot be turned on or off in response to electrical load changes. The addition of these plants, as they become a more important part of the generation mix, requires that a greater proportion of conventional generation be allocated for operating reserve. This cannot be conveniently done by coal plants and nuclear plants since that would mean removing a good share of such capacity—built at a high investment cost—from base-load operation. The solution is more pumped storage capacity.

PART D. COSTS AND ECONOMICS

23. Introduction. In this part, the capital cost of pumped storage projects is first reviewed, followed by a discussion on how the economics of pumped storage is borne out in the electric system.

Pumped storage costs fall within a relatively narrow range when compared with those of conventional hydroelectric plants. For the latter, the cost in dollars per kilowatt varies over a factor of 10. For pumped storage, the range is likely to be within a factor of 2. Pumped storage costs can be estimated from limited information with reasonable accuracy. In the following sections the historical costs are first reviewed, followed by a presentation of simplified curves for estimating the costs for specific sites.

24. Historical Costs. The historical costs of pumped storage are available from files of the Federal Energy Regulatory Commission.[12] When indexed to the 1990 price level, the construction costs are as shown in Fig. 17. The costs are broken into the capacity component and the energy-storage component. The capacity component includes water conductors and powerhouse and equipment. The energy-storage component includes the dams and spillways, reservoirs, land and access, and water-supply system. The costs are separated and presented in this way because the capacity and energy-storage components can be sized independently, and the separation facilitates comparison. These costs exclude interest during construction and transmission. Historical costs serve as a useful starting point for estimating future costs, provided that one understands the background of these costs and applies them with appropriate adjustments.

With reference to Fig. 17, the plants of the early years generally have lower construction costs than the more recent plants for several reasons:

1. Earlier plant costs are subject to a large indexing to the current date, resulting in possible inaccuracy.
2. Earlier plants did not have as rugged machinery as the more recent plants.

Project	Initial operation	Capacity, MW	Energy storage, h	Gross head, ft	Length/head, L/H	Capacity component, $/kW	Energy storage, $/kWh	Total, $/kW
Taum Sauk	1963	350	7.7	809	8.7	340	19.5	504
Yards Creek	1965	330	8.7	723	5.1	286	8.0	362
Muddy Run	1967	855	14.3	386	3.3	286	4.2	351
Cabin Creek	1967	280	5.8	1159	3.7	287	24.1	440
Seneca	1969	380	11.2	736	3.4	396	13.8	550
Northfield	1972	1000	10.1	772	8.8	280	3.1	314
Bleneheim Gilboa	1973	1930	11.6	1099	4.0	239	8.7	350
Ludington	1973	1888	9.0	337	3.7	249	18.7	410
Jocassee	1973	628	93.5	310	5.5	279	1.8	460
Bear Swamp	1974	540	5.6	725	2.8	337	35.5	553
Raccoon Mountain	1978	1370	24.0	968	3.8	274	1.9	323
Fairfield	1978	512	8.1	163	13.0	421	24.9	639
Helms	1984	1200	118.0	1640	12.5	670	0.0	670
Bath County	1985	2100	11.3	1180	8.0	540	12.5	697

FIGURE 17 Historical construction costs January 1990, excluding interest during construction.

Some of the earlier plants have simpler control and protection schemes, and pump-start equipment.

3. Earlier plants were not subject to the same kind of elaborate licensing process as the recent plants, and had less extensive environmental mitigation provision.
4. Earlier plants have lower efficiencies designed in their mechanical and electrical components.

The owners of a few of the recent plants chose to stretch out or delay the construction of their plants to match their load growths, which were lower than earlier projections. These delays resulted in high indirect and administrative costs.

The experience gained from these projects can be applied to developing future pumped storage projects. Among the lessons learned are the need to design a more rugged and efficient plant as well as avoiding certain technical difficulties.

25. Parameters Affecting Costs. Pumped storage capital costs are influenced by several parameters which are easily quantified. The major ones include (1) project size, i.e., generating capacity and energy storage, and (2) site characteristics, specifically operating head and water conductor length.

The smallest pumped storage plant in the United States has a generating rating of 280 MW, indicating that sizes smaller than 300 MW are likely to be uneconomical owing to high unit costs with economy of scale. With economy of scale, unit cost tends to decrease with increase in plant size up to about 1000 MW. Beyond 1000 MW there is no measurable evidence in cost advantage. With a very large plant, the number of units tends to increase, and any cost advantage would be counterbalanced by the longer time required to complete the plant.

Site characteristics control project cost. Higher heads generally bring lower unit cost. A minimum head of 300 ft is preferred, with the desirable head ranging between 700 and 1200 ft. At heads above 1200 ft, there does not appear to be any cost advantage. Further, as operating head exceeds 1500 ft there may be an increase in unit cost. This is because the pump-turbines and water conductors are difficult to design for extremely high heads.

Water-conductor length affects the cost of the capacity component in two ways. First, the cost increases with increase in water-conductor length. In addition, longer water-conductor length causes an increase in hydraulic head loss, requiring an enlargement of the water conductor if the operating efficiency is to be maintained. Longer water-conductor length also increases the water-column inertia, resulting in a less responsive power plant. Water-conductor length is usually expressed in terms of its ratio with the operating head. An economic pumped storage plant should have a ratio of water-conductor length to head of at most 10, preferably much less.

26. Cost-Estimating Curves. Figure 18 presents a cost curve for the purpose of making ball-park estimates of pumped storage plant costs. The curve has been derived by updating and simplifying cost-estimating curves given in the report "Pumped Storage Planning and Evaluation Guide" published by Electric Power Research Institute. The energy-storage component is highly variable depending on the site characteristics, and yet it is a small component of the total cost. For the purpose of constructing the curves, it has been assumed to be 20 percent of the total project cost.

Estimating curves are to provide no more than a general guide on what a project might cost. And the estimates are subject to refinement and confirmation from subsequent studies. At best, the estimates from the cost curve represent a potential deviation of plus or minus 20 percent as a minimum. Cost-estimating

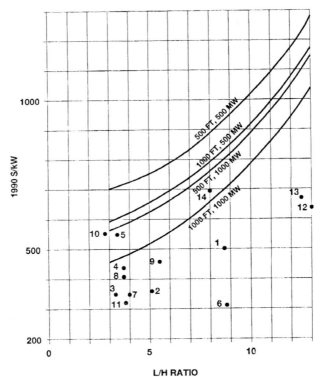

FIGURE 18 Pumped storage cost-estimating curves.

curves do not recognize nonquantifiable parameters which could significantly influence the cost:

1. Site characteristics which could add to or reduce the cost by up to 20 percent.
2. Efficiency and performance—cost can be reduced by up to 10 percent if a lower efficiency than 75 percent is acceptable.
3. Geologic factors which could add 20 percent to the costs or reduce them 10 percent.

27. Implementation and Construction Schedule. Implementing a pumped storage development can vary from 9 to 12 years from initial planning. The time period depends on the complexity in the planning and licensing process and the size of the project. Planning and licensing can take 5 to 7 years. Construction may take between 4 and 6 years, although some plants have taken only 3 years and some as long as 8 years. The short time schedules involved small plants with minimum underground construction. The long periods were associated with prolonged construction interruptions or difficulties. The following is a reasonable range:

	Underground power station	Surface power station
1000 MW and greater	5½ years	4½ years
500 MW and less	4½ years	3½ years

Figure 18 shows overnight costs; i.e., they exclude interest during construction, which can be estimated approximately using the above construction time. For a more rigorous analysis, the estimator may choose to factor in costs incurred prior to actual construction as well as to develop a construction schedule. Preconstruction costs are small relative to the total cost of the project, amounting to between 5 and 10 percent including planning, licensing, technical specification, and land acquisition.

28. Economics. Pumped storage provides a system with (1) capacity, (2) energy production, and (3) dynamic benefits through supply-side management. Pumped storage is economical if it can serve a system at a lower cost than any other type of new generation. Since pumped storages can be expected to represent only a small portion of the total system capacity, and since it does require other generation to provide the pumping energy, its economics is very much dependent on the system load profile and the generation mix.

System load profile determines the time that a pumped storage plant may operate, first in the generating mode and then in the pumping mode. Pumped storage would be very useful where the peak load period is not excessively long, say not more than 14 h, and where there is a period where pumping may take place, preferably not less than 8 h.

A favorable generation mix for pumped storage is one where the capacity of base-load plants is substantially greater than the minimum load and the rest of the capacity is in peaking plants burning fuel oil. There is then a significant difference in the costs of on-peak and off-peak generation which encourages pumped storage operation and enhances the economics of the project.

Figure 19 shows a typical chart often used to compare the relative attractiveness of alternative types of generation. The figure is useful for screening purposes, to establish which types of generation are worthy of further consideration.

A coal plant cannot be justified if it will operate for less than 3000 h a year. The same can be said about combustion turbines operating for more than 3000 h a year. The goal is therefore to find the right mix and number of plants to arrive at the least-cost system. Several conclusions can be drawn from Fig. 19:

1. A combustion turbine is competitive for reserve and peaking when operating for less than 1000 h a year.

2. Pumped storage is competitive for peaking and shoulder operation between 1000 and 2000 h a year.

3. The comparisons are only for the costs of individual plants. Since the operation of each plant affects the operation of the remaining system, a rigorous economic analysis must embrace a study of the whole system.

29. Economic Analysis. Economic analysis is conducted to determine whether it is preferable to build one type of power plant or another, when and what size. As noted, the analysis has to be done on an entire system basis. Fur-

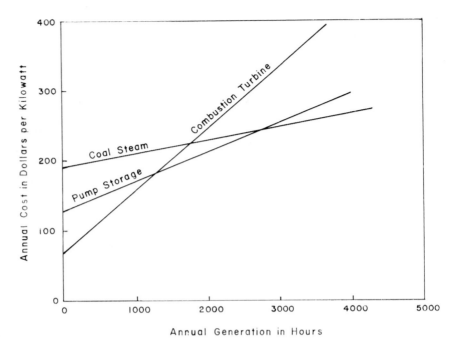

FIGURE 19 Comparative costs of alternative generation.

thermore, since the system changes over time, the analysis has to be done over a long period of years to cover the life cycle of a plant. Such an analysis is generally referred to as *least-cost planning*. If the planning also encompasses the demand side, the term *integrated resource planning* is often used. Regardless of the term, the economic analysis involves the development of year-by-year capital and operating costs of alternative system expansions over a long period of years. The projects that form the expansion with the lowest cost are then those most economic to develop. While the economic analysis involves evaluating the performance of the system over a long period of time, it must begin with how the plant performs in the system at a given point in time.

30. Computer Models for Economic Analysis. For evaluating the economics of pumped storage, it is necessary to utilize a model which accurately simulates operation of a generation system. Numerous models are now available. Called *production costing* models, they simulate operation and estimate fuel costs over a period of a week or a multiple of weeks to a full year or more. A typical model generally consists of the following features or submodels:

- *Load,* chronological or duration
- *Generating unit,* defining unit characteristics such as minimum and maximum loading, heat rates, ramp rates, outage rates, maintenance schedule, and startup cost
- *Forced outage,* whether probabilistic, Monte Carlo, or deterministic
- *Unit commitment*—which units are dispatched to serve the system load
- *Energy-storage scheduling*—how pumped storage is dispatched to meet demand, produce economy, and optimize the timing of charging and discharging

- *Transactions,* between utilities and power pools
- *Output,* presentation format of results

Older production costing models tend to utilize the load-duration method, while newer models tend to use the chronological approach. The use of load-duration curves causes the peaks and troughs within the time period to be lumped together. For evaluation of pumped storage, it is essential that a chronological production costing model be used. Models which simulate ramping constraints of alternative generation and allocate and optimize operating and spinning reserves are needed to best quantify the benefits of pumped storage.

Though they are useful, production costing models do not recognize how a generation system responds to minute changes in load for maintaining system frequency and voltage level, by placing certain units under *automatic generation control.* This type of analysis requires different models and analytical techniques but could be very important for recognition of the capability of pumped storage and its value in an electric system.

New models or new versions of old models are coming out rapidly with the advent of the personal computer. Some of them can be procured or accessed relatively easily. The reader is encouraged to select a model with the following characteristics:

Load

- Shortest time step of 1 h or less
- Chronological load shape
- One week, multiple of weeks, to a year or more in each run

Unit commitment

- Daily update of unit commitment
- Spinning reserve
- Loss of load probability to match supply with load

Thermal unit characteristics

- Forced outage by Monte Carlo or probabilistic methods
- Unit heat rates at three or more points, including minimum, maximum, and straight-line incremental
- Ramp rate constraint capability

Pumped storage model

- Modeling individual units and plants
- Unit efficiency with part-load generating and pumping
- Storage tracking

Pumped storage operation

- Dispatch to meet dependable capacity
- Capability of maximizing storage benefit, optimum timing of generation and pumping operation with daily and weekly look-ahead feature

Figure 20 presents the graphical output of a production costing model. The system presented contains three categories of generating plants: (1) base-load mine-mouth steam, which has the lowest fuel cost, (2) coal-fired and oil-fired steam with intermediate fuel cost, and (3) combustion turbines burning light fuel oil. The operation with the addition of either (1) combustion turbine or (2) pumped storage is shown in separate figures.

Figure. 20*a* shows the case with combustion turbines operating on-peak. It is

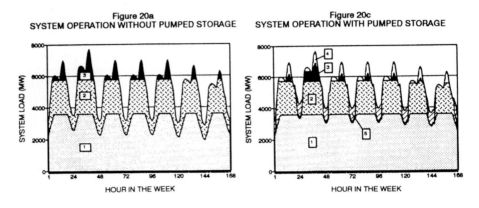

LEGEND:
1. COAL - STEAM
2. HEAVY FUEL OIL - STEAM
3. LIGHT FUEL OIL - COMBUSTION TURBINE
4. PUMPED STORAGE - GENERATION
5. PUMPED STORAGE - PUMPING

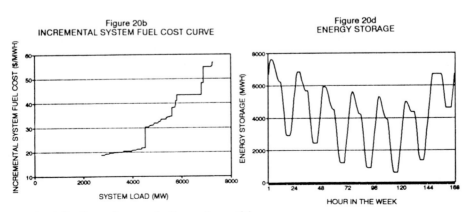

FIGURE 20 Output from production costing model.

necessary to keep the steam units operating in the off-peak period in order for it to be available the next day. Because of minimum load constraint on the steam units—usually 30 percent of rated capacity—a noticeable reduction in mine-mouth coal-plant output, which has very low fuel cost, is seen in the off-peak hours. Also, note the steepness of the curve, which shows the ramp rate of the plants, the maximum allowable rate of change in output from a power plant. The ramp rate is established to prevent undesirable effect due to rapid change in loading or (in the case of hydroelectric plants) discharge.

Figure 20b is the system fuel cost curve of one weekday. It shows the increasing cost of energy production as the load increases.

Figure 20c shows system operation with pumped storage. The plant would operate in the generating cycle in the peaking hours each day. Pumped storage would basically be operated to complement Fig. 20b, which suggests when to pump or generate in order to save system fuel costs. Refilling of the upper reservoir would take place the following night. However, the upper reservoir would continue to be drawn down further on each successive day, owing to insufficient pumping hours at night. During the weekend, extensive pumping returns the upper reservoir to full pond.

Referring to Fig. 20c, the minimum load is increased substantially, enabling the base-load plants to operate at a near constant level. Figure 20d shows the variation of the energy-storage content of the upper reservoir during the week.

In addition to production models, there are *generation expansion* models which optimize the generation additions in terms of types of plants, sizes, and timing. A generation expansion model is designed to evaluate, screen, and optimize the generation system expansion by selecting the timing, type, and size of new generation in the system. The usual overall period of study is 30 years, within which the expansion period could be 10 to 20 years. Generation expansion programs are complex since the system is not static but varies over time, and the costs include not only operation costs but also costs of new capital projects. In order for the model to be manageable, the programs must take shortcuts. One essential shortcut is to simplify the production costing feature by the use of load-duration curves. For this reason, generation expansion models are not sufficiently accurate to evaluate the pumped storage performance in the utility system. If a generation expansion model is to be used, it should be validated for specific years with a suitable production costing model. In this way annual fuel and operation costs can be benchmarked.

PART E. DEVELOPING PUMPED STORAGE

31. Introduction. Pumped storage tends to be large and capital-intensive. Its development requires a long lead time from initial concept through operation. Pumped storage is typically developed by electric and water utilities which are either public agencies or private companies that are regulated by the government. Pumped storage involves the construction of reservoirs. Consequently, laws and regulations governing water rights, water use, and environmental protection often govern the development process.

In the United States, the Federal Energy Regulatory Commission (FERC) is the lead agency for issuing licenses for the construction and operation of pumped storage. The agency consults and coordinates with other resource and regulatory agencies in processing the license application for pumped storage development.

The reviews address technical feasibility, economic soundness, and environmental acceptability. These are the same issues that the owner would evaluate for the financing agencies when they are requested to provide construction loans. Success in obtaining the approvals and financing depends on the attractiveness of the project, which the developer must demonstrate.

32. Feasibility Studies. Studies leading to the determination of project construction are usually called feasibility studies. They cover technical, economic and financial, and environmental issues. The document presenting the feasibility study serves as the vehicle for application of licenses and/or loans for project construction. Guidelines on the scope of the studies are usually available from regulatory agencies such as the FERC and international lending agencies such as the World Bank. The objectives and the range of the studies prescribed by the agencies are similar. They want a demonstration that the project is needed; that it is technically sound, economically viable, and environmentally acceptable; that laws and regulations are in place and adhered to; and further, that the studies have been performed to the level of accuracy so that there is some assurance most estimates and projections will not be out of line. Close adherence to the guidelines is essential to obtain the necessary approvals.

These studies tend to be extensive and time-consuming, especially if reviews by the regulatory agencies and lending institutions are included. If implemented strategically, the studies need not be overbearing or costly. When properly handled, the studies generally form four sequential phases and two planning categories outlined below:

Phases

1. Conceptual—identification and screening of options
2. Prefeasibility—selection of the preferred project
3. Feasibility—optimization and definition of proposed project
4. Funding and licensing—funding from lending agencies, execution of power sales contracts, and approvals from regulatory agencies

Study Areas

1. System planning—energy and electricity sector consideration, demand- and supply-side planning, generation options, system performance under alternative expansion scenarios, economic and financial analyses, with decisions to proceed to the next step.
2. Project planning—siting, project layout, conceptual design of features, capacity and energy-storage sizing, performance, and cost estimates. Office studies and field investigation covering hydrology, topography, geology, environment, and socioeconomic factors.

The phased approach is generally effective, where decisions can be made at the end of each phase to establish whether and how to proceed in the next step recognizing the risks involved under alternative scenarios. For each phase, an appropriate level of study can be designed for each study area.

The *conceptual study* determines from a system standpoint (1) if pumped storage is a potential option for the system in the next 10- to 20-year time frame, and (2) if reasonably attractive project sites exist. The system planning study could be handled manually or with a generation expansion planning model. The study may be approached using a breakeven analysis or data from (2). The study will determine if pumped storage is worthy of further consideration (and if so what size and time frame) and should be pursued. Under (2) results conceptual-level study for project siting is sometimes called reconnaissance study. The reconnaissance study is conducted with available topographic maps and information on geology, hydrology, and environment, supplemented by field inspection. If there are a number of pumped storage sites, the study should be directed to determining which of the sites is more attractive from cost, risk, and environmental standpoints. Three or more sites may be selected for further consideration. The program for the next phase can then be designed.

The *prefeasibility study* provides a more accurate assessment of the attractiveness of pumped storage in the system, as well as a more accurate estimate of developing a specific site potential. The system planning study should identify the desirable capacity and energy-storage component of pumped storage. A chronologic production costing model should be used. The project planning study should involve studies of each of the selected three or more sites. For each site, project-arrangement studies should be conducted considering appropriate locations of principal structures and their types and basic designs as a basis to develop cost estimates. Consideration should be given to environmental impact and mitigation opportunities. From this study the most attractive site should be selected from the short list of alternative sites. Permits for further studies should be applied for. A subsurface exploration program and data-gathering and -monitoring programs should be established and initiated.

The *feasibility study*, being the last phase in system and project planning, is directed toward providing all necessary information for seeking approvals, licenses, and financing for construction. The system planning study should establish how the project will operate in the system in the future, and the benefits therefrom in comparison with the least-cost alternative. Recovery of project investment will be demonstrated. The project planning studies optimize and define the project design, including the various components. A preliminary bill of material, construction cost estimates, and schedule and disbursement schedule will be established. Programs for subsequent project implementation will be proposed.

The *funding and licensing phase* involves those activities by the lending institutions and regulatory agencies in processing the applications. The agencies may raise questions and request additional information to which the applicant may have to respond. This phase could well have been started as early as the conceptual phase 1. In the application for an FERC license, a three-step process is involved which requires the applicant to undertake the planning study in close coordination with the resource and regulatory agencies. Thus there will be no surprises from the standpoint of either the applicant or the agencies.

REFERENCES

1. HARZA ENGINEERING COMPANY, *Pumped Storage Planning and Evaluation Guide*, EPRI Report GS-6669, Electric Power Research Institute, Palo Alto, Calif., January 1990.
2. ZAGARS, A., and J. M. HAGOOD, *Bath County, a 2100 MW Development in the USA*, Water Power & Dam Construction, 1977.
3. WATER POWER AND DAM CONSTRUCTION, "The World's Pumped-Storage Plants—Survey," *Water Power & Dam Construction*, vol. 42, no. 4, April 1990.
4. WATER POWER AND DAM CONSTRUCTION, "Pumped Storage Capacity by Country (graphics)," *Water Power & Dam Construction*, vol. 43, no. 2, February 1991.

5. CHEN, HENRY H., and IRWIN A. BERMAN, *Commonwealth Edison Company's Underground Pumped-Hydro Project,* AIAA/EPRI International Conference on Underground Pumped Hydro and Compressed Air Energy Storage, San Francisco, Calif., September 1982.
6. CHEN, HENRY H., and L. DOW NICHOL, *Sizing of Underground Pumped Storage Plant in a Major Utility System,* International Symposium and Workshop on the Dynamic Benefits of Energy Storage Plant Operation, Boston, Mass., May 1984.
7. CHACOUR, S. A., J. R. DEGNAN, D. M. LOSASSO, and D. R. WEBB, *Design Considerations for a 1000 Meter Head, 500 Megawatt Single and Double Stage Reversible Pump/Turbine.* The Joint ASME-CSME Applied Mechanics, Fluids Engineering Conference, Niagara Falls, N.Y., June 18–20, 1979, American Society of Mechanical Engineers.
8. COATES, M. S., *Subsurface Geologic Considerations in Siting a UPH Project Using the Illinois Deep Hole Project as Example of Geologic Exploration,* AIIA/EPRI International Conference on Underground Pumped Hydro and Compressed Air Energy Storage, San Francisco, Calif., September 1982.
9. KOKUSHO, T., "Pumped Storage Projects under Construction in Japan," *Water Power & Dam Construction,* February 1991.
10. CENTRAL ELECTRICITY GENERATING BOARD, *Dinorwig,* Central Electricity Generating Board, England, 1984.
11. INSTITUTE OF CIVIL ENGINEERS, Proceedings of the *Pumped Storage* Conference at Imperial College of Science, Technology and Medicine, London, England, Thomas Telford, London.
12. U.S. DEPARTMENT OF ENERGY, Energy Information Administration, *Historical Plant Cost and Annual Production Expenses for Selected Electric Plants* (DOE/EIA-0455), Washington, D.C., for years 1982, 1983, 1984, and 1985.

SECTION 21

HYDRAULIC MACHINERY

By James H. T. Sun[1]

TURBINES

1. Introduction. Hydraulic turbines involve in their action a continuous transformation of the potential energy of the fluid into kinetic energy (and in reaction turbines a subsequent reconversion) and a conversion of kinetic energy, or of both kinetic and potential energy, into useful work.

Hydraulic turbines are divided according to their hydraulic action into two main classes: impulse turbines and reaction turbines.

Impulse turbines are represented in modern practice by the Pelton waterwheel, illustrated in Fig. 1. In an impulse turbine, all the available head is converted into kinetic energy or velocity head in one or more constricting nozzles by which the water is formed into one or more free jets before acting upon the runner. Other impulse turbines are represented by the cross-flow and Turgo turbines.

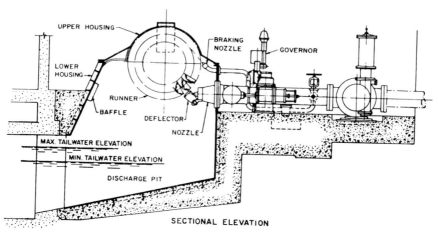

FIGURE 1 Impulse turbine (Pelton type) rated 12,000-hp, 2490-ft net head, 600 rpm. Typical horizontal-shaft single-jet arrangement with single overhung runner. Straight-flow nozzle with external needle servomotors. (*Voith Hydro Inc.*)

---

[1]Acknowledgment is made to Lewis F. Moody and Thaddeus Zowski for material in this section which appeared in the third edition (1969).

In reaction turbines, the entire flow from headwater to tailwater takes place in a closed conduit system. At the entrance to the runner, only a part of the available head is converted into kinetic energy. A substantial part of the available head remains in the form of pressure head which varies throughout the turbine water passage.

In modern practice, reaction turbines are represented by the following types: radial-flow Francis turbines (see Fig. 2); axial-flow Kaplan turbines with adjustable runner blades (see Fig. 3); axial-flow propeller turbines with fixed runner blades; diagonal-flow Deriaz turbines with adjustable runner blades. S-type (see Fig. 4), bulb (see Fig. 5), and Straflo units are other examples of axial-flow turbines. (The following abbreviations are used in this section: axial-flow turbines with adjustable runner blades—Kaplan turbines; axial-flow turbines with fixed runner blades—propeller turbines. The term axial-flow turbine refers to both Kaplan and propeller turbines.)

In the Francis turbine, the flow passes inwardly through a circular series of wicket gates, pivotally adjustable for regulation and forming contracting passages in which a part of the head is converted into velocity head. The streams issue from the wicket gates in a direction having both radial inward and tangential velocity components. The streams then merge within the space between the wicket

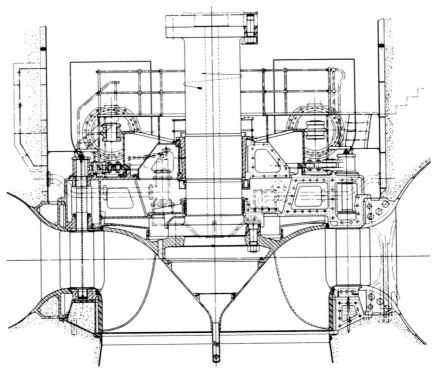

FIGURE 2 Sectional elevation of 210,000-hp turbine for the Robert Moses Niagara power plant. (*Seven turbines by Baldwin-Lima-Hamilton Corp., six turbines by Newport News Shipbuilding and Dry Dock Co.*)

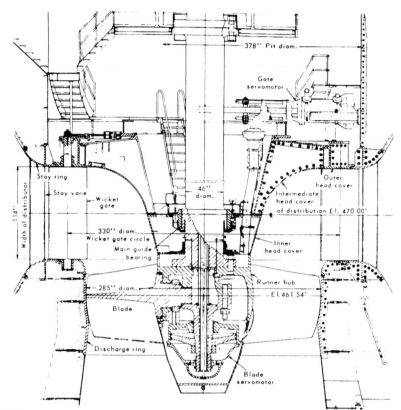

FIGURE 3 Sectional elevation of 120,000-hp, 80-ft net head, 85.7-rpm Kaplan-type turbine for Wanapum power plant. (*Ten turbines by Dominion Engineering Works Ltd.*)

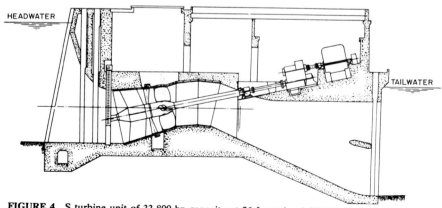

FIGURE 4 S turbine unit of 33,800-hp capacity at 26-ft net head (32.3-ft rated head). 60/514.3 rpm for the Ozark lock and dam powerhouse of the U.S. Army Corps of Engineers. Runner diameter approximately 315 in. (*Five units by Voith Hydro Inc.*)

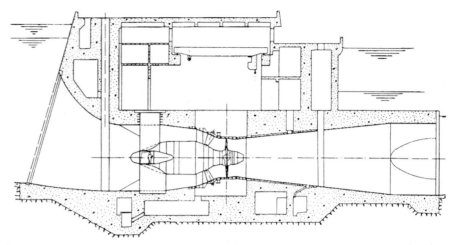

FIGURE 5 Sectional elevation of one of four bulb units for Pierre Benite power plant in France. Nominal capacity 20,000 kW each at 26-ft net head, 83.3 rpm. Runner diameter 240 in. (*Turbines by Neyrpic, generators by Alsthom.*)

gates and the runner to form a continuous ring of revolving and inwardly progressing water. The water then enters the runner passages in which the radial component of motion is gradually turned between the hub and outer band into the axial direction, while the tangential or whirl component is gradually deflected by the vanes until at discharge from the runner only a small whirl component remains. The flow then passes through a draft tube which by means of gradually increasing cross-sectional areas reduces the velocity, thus reconverting a large part of the residual kinetic energy.

In the propeller or Kaplan turbine, the flow through the runner is axial in the meridional plane as it passes through the annular throat section of the water passage between the runner hub and discharge ring. The runner has relatively few blades, usually between three and ten, with free outer ends revolving within the stationary discharge ring with as small running clearance as is practical. The relative velocity between the blades and water is high and undergoes comparatively little change as the water passes through the runner.

The Kaplan turbine has individual runner blades pivotally mounted in the hub so that their inclination may be adjusted during operation, by governor action, simultaneously with the adjustment of the wicket gates, to meet changing power demands and changes in head for better efficiency than the propeller turbine. The blades are adjusted by means of a mechanism which connects them with a servomotor. The servomotor is usually operated by oil pressure admitted from the governor through a blade control valve.

Reaction turbines have been applied commercially for heads up to 2205 ft. Impulse turbines of the Pelton type have been used for heads up to 5800 ft. Axial-flow turbines have been employed for heads up to nearly 290 ft.

Except for small installations under low heads, reaction turbines are equipped with spiral casings, these being usually circular and constructed of cast or plate steel. In low-head plants (less than about 100 ft), the spiral casings are usually of partly rectangular section, formed in the concrete substructure of the powerhouse. For small installations under very low head (about 30 ft or less), open-flume settings have been used, the runner and wicket gates being placed in an open rectangular or partly spiral flume into which the free surface of headwater extends.

Many high-speed Francis and axial-flow turbines are equipped with automatically controlled valves for admitting air to the water passages in or near the runner during part-gate or light-load operation. This admission of limited quantities of air improves the operation of Francis turbines under small loads, and the air is shut off progressively and automatically as the wicket gates are opened. In propeller turbines air admission is frequently advantageous nearly up to normal gate opening.

Impulse turbines are regulated by needle nozzles; one type is shown in Fig. 1. The straight-flow nozzle eliminates the disturbance of the jet caused by curvature of water passages in an elbow-type nozzle. Since the flow in the penstock cannot be suddenly altered for quick regulation because of its great inertia, the needle nozzle is usually supplemented by either a jet deflector or an auxiliary nozzle. Upon load rejection, the governor acts immediately to bypass a part of the jet by means of either the deflector or auxiliary nozzle so that a part of the flow is removed from acting on the runner, after which a slow-closing action controlled by a governing system permits the needle to close slowly while the deflector or auxiliary nozzle is simultaneously withdrawn from action.

In reaction turbines having long penstocks, as in high-head installations, pressure regulators or relief valves are used to temporarily allow part of the water to bypass the runner by flowing from the turbine spiral case directly to the tailrace during quick closure of the wicket gates. This is done in order to prevent excessive water hammer, which is the severe pressure rise that results from rapid deceleration of the penstock water column. Governor control permits the subsequent slow closure of the relief valve to minimize loss of water. Pressure regulators of several forms, such as the angle-needle-valve type with matching energy absorbers and the fixed-cone valve (Howell-Bunger valve), have been applied to this service.

For economical utilization of low heads up to about 50 ft, special axial-flow turbines of the tubular type have been developed. Conventional-type turbines with their spiral or semispiral cases and elbow-type draft tubes involve considerable costs in the structural features of the powerhouse. This is due to the relatively large water passages needed to pass sufficient flow through the machines to develop significant power outputs under the low heads. This results in powerhouse structures which have large dimensions relative to the available head. Tubular turbines, by straightening the flow through them from forebay to tailrace, confining it to a substantially axial direction and, placing the shaft axis in a horizontal or slightly inclined position, enable the use of a more compact and simplified powerhouse structure. Furthermore, the straight axial flow enhances turbine performance, permitting increased operating speed and output, thereby reducing the size and cost of the machinery for a given output.

Tubular turbines are a development of an idea originated and patented in America in the early 1920s by L. F. Harza and utilized subsequently in a number of European low-head installations. The original design approached the ideal axial-flow machine by passing the water through a rim-type generator, which had its rotor attached to the periphery of the runner blades. Initial mechanical difficulties due to water leakage and excessive frictional losses in the water seals were overcome in later designs. A more recent development is the bulb-type turbine in which the generator is placed in a bulb-shaped watertight steel housing located in the center of an enlarged water passage (Fig. 5). Many variations of the bulb-turbine design, with generator driven by the turbine either directly or through gears, have been built and put into successful service.

A type of tubular turbine developed in America and known as the S turbine avoids the use of a bulb in the S-shaped water passageway. It uses a conventional horizontal-type generator located outside the water passages where it is fully accessible. By inclining the turbine-generator shaft, the generator can be placed above the water passageway without requiring deep excavation for the draft tube (Fig. 4).

2. Net Head, Power, Efficiency. Figure 6 illustrates diagrammatically a reaction turbine supplied by a penstock and discharging through its draft tube into the tailrace. The turbine proper is, according to standard practice, taken to begin at the entrance to the turbine casing and to end at a section of the tailrace just beyond the physical end of the draft tube. The turbine comprises the casing, distributor, runner, and draft tube.

The net head under which the turbine operates corresponds to the difference between the total energy contained in the water immediately before its entrance into the turbine and immediately after discharge from the turbine draft tube.

The hydraulic turbine converts the potential energy into mechanical energy expressed in the following equation:

$$P = \frac{HQw\eta}{550} = \frac{HQ\eta}{8.81}$$

where P = turbine output, hp (also expressed in kilowatts where 1 hp = 0.7457 kW)
H = net head, ft
Q = turbine discharge, cfs
w = specific weight of water (standard conditions) = 62.4 pcf
η = turbine efficiency (losses in hydraulic resistances and eddies within the turbine, water leakage around the runner, frictional losses such as disk friction on a Francis runner crown and bearing and seal friction on the rotating shaft)

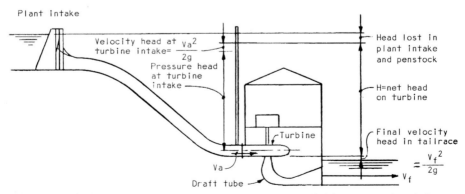

FIGURE 6 Diagrammatic arrangement of a reaction turbine supplied by a penstock and discharging through its draft tube into the tailrace.

TURBINES

3. Proportionality Laws for Homologous Turbines. Turbines that have geometrically similar water passages are homologous turbines. The proportionality laws for homologous turbines are shown in Table 1.

TABLE 1 Proportionality Laws

Constant head (H)	Constant runner diameter (D)	Variable runner diameter and head
$P \alpha D^2$	$P \alpha H^{3/2}$	$P \alpha D^2 H^{3/2}$
$N \alpha 1/D$	$N \alpha H^{1/2}$	$N \alpha \dfrac{H^{1/2}}{D}$
$Q \alpha D^2$	$Q \alpha H^{1/2}$	$Q \alpha D^2 H^{1/2}$

H = net head, ft
P = turbine output, hp
D = runner discharge diameter, ft
N = turbine rotating speed, rpm
Q = turbine discharge, cfs

4. Specific Speed. Specific speed may be defined as the speed of a homologous turbine producing one unit of power under one unit of head; it is calculated from the following equation:

$$N_s = N \frac{\sqrt{P}}{H^{5/4}}$$

where N_s = specific speed
N = turbine rotating speed, rpm
P = turbine output, hp
H = net head, ft

Since the specific speed remains constant for all turbines of the same design, the corresponding speed of another homologous turbine can be found by solving the above equation.

In a multiple-runner turbine, the specific speed is usually expressed as that of a single runner, and in a multiple-jet impulse wheel, the specific speed is normally based on the power of a single jet.

As might be expected, some forms and proportions in turbine design are particularly favorable to high efficiency with a corresponding range of specific-speed values. Specific speeds that are abnormally high or low compared with this normal range naturally entail impaired efficiencies.

Since for a given power capacity and head the actual speed is directly proportional to the specific speed, as shown by the specific-speed formula, and since high rpm results in reduced diameters and weights of the generating unit and also in reduced size and cost of the powerhouse structure, there is from economic considerations a strong attraction toward higher and higher specific speeds, even at the expense of some sacrifice in efficiency. It should be recognized, however, that in hydroelectric developments the major portion of the cost is usually in the dams, reservoirs, headworks pipelines, transmission lines, and similar fixed works and only a small portion in the powerhouse and machinery; so a saving in first cost of the latter elements must be weighed against the loss of earnings on the whole project entailed by a reduction in efficiency. In most cases a sizable increase in specific speed will be beneficial only when the impairment in efficiency is slight. These considerations are of special importance in low-head installations, where the actual speeds are low and the machinery large. The continual demand for the highest possible speeds has resulted in a wide range of specific speeds being available without significant sacrifice in efficiency. Over a wide range of specific speeds, peak efficiencies in the range of 93 to 96 percent may reasonably be expected under favorable conditions.

The curve of efficiency vs. power output for a given turbine operated at varying gate opening has different characteristic forms for various turbine types and specific speeds, as illustrated in Fig. 7. The Pelton impulse wheel, having very low specific speeds, gives a flat-topped efficiency curve with high part-load and overload efficiencies and only small impairment of efficiency over a wide range of nozzle openings and outputs. The low-specific-speed Francis turbine has similar characteristics. Both types are therefore suitable for plants serving to regulate a power system and take the load variations. High-specific-speed Francis turbines show poor part-load efficiencies and little overload capacity beyond the point of maximum efficiency; i.e., they have a sharply peaked efficiency curve and are therefore best suited to operation under block load or within a narrow range of outputs. This characteristic is further intensified in propeller turbines. In run-of-river plants, such as low-head developments with limited pondage, where under

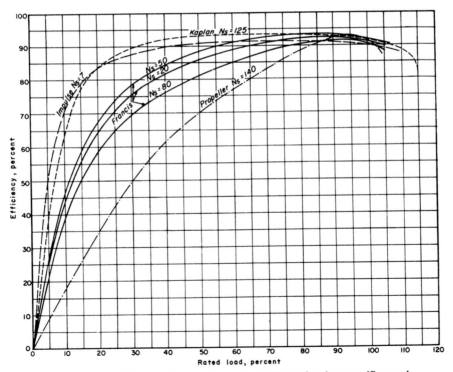

FIGURE 7 Turbine-efficiency variation with load for turbines of various specific speeds.

reduced demand the unused flow is wasted over the spillway, the low part-gate efficiencies are of little consequence.

When a high degree of regulation is of importance in low-head plants, high part-gate and over-gate efficiencies may be secured in high-speed axial-flow turbines by making the runner blades automatically adjustable in response to changes in the wicket gate openings, under governor control. This is the method of operation of the Kaplan turbine, which is capable of giving an efficiency-power curve of even more favorable form than low-speed Francis or Pelton units. Another advantageous characteristic of propeller and Kaplan units is their ability to operate under reduced heads while maintaining good power output, a fortunate property for low-head developments to which these types are suited, for such developments are usually subject to wide head variations.

In plants having high heads even a low specific speed in combination with the high head will usually give a reasonably high rpm. On the other hand, with low heads and large unit capacities, a low specific speed would result in extremely low rpm, very large diameters of turbine and generator, and large powerhouse structures. Consequently, high-specific-speed types are particularly desirable for low-head plants.

In selecting the specific speed suitable for a given head, cavitation requirements are usually the controlling consideration.

5. Cavitation. Suppose that both headwater and tailwater are progressively lowered by equal amounts. The net head H will remain constant, but the tailwater pressure will reduce so that the absolute pressure head at the point of minimum pressure will be progressively lowered. When the absolute-pressure head of the water is reduced to the vapor-pressure head, the water begins to boil and the passages become partly occupied by vapor cavities within the flowing stream. The formation of these vapor-filled cavities in the stream is called *cavitation.*

When the pressure at some point in the turbine reaches the vapor pressure, critical conditions ensue and the turbine becomes subject to the undesirable consequences of cavitation, namely, erosion (known as pitting), noise, and vibration of the machine and surrounding structures. When the extent of the cavitation increases, an impairment of power and efficiency results. This phenomenon is not limited to turbines but may also occur in pumps and in stationary conduits at points of low pressure and high velocity as, for example, in sluiceways through dams. Guarding against its occurrence is a vital consideration in the selection of type and specific speed of a turbine and in fixing the runner elevation in relation to tailwater.

The cavitation phenomenon can be explained as follows: Consider the flow through a turbine runner where the total draft head is excessive and where there is a failure of the vane (blade) contour to conform to the natural flow lines, because the curvature is too sharp for the pressure and velocity conditions. At such a point in the runner, usually on the back of the vanes near the discharge orifice and near the outer periphery, where the relative velocity is high, the flowing stream parts from the vane surface and leaves a void filled with eddies; and when the absolute pressure is reduced to the vapor pressure, this void or cavity becomes filled with water vapor, air, and other gases. As the flow continues downstream, the static pressure rises again and then exceeds the vapor pressure. Moreover the flow in large conduits at high velocities is never actually steady but is turbulent and subject to continual variations of velocity and pulsations of pressure. Hence, when a particle of the flowing liquid reaches a point where the local pressure just attains the vapor-pressure limit, at one instant the particle will be under this pressure and vaporize, forming a cavity filled with vapor; at the next instant, the pressure will rise above the vapor pressure and the vapor will suddenly condense and return to liquid, producing a collapse of the cavity and an explosion or, more strictly, an *implosion*. This action is not confined to the larger cavities but extends into the pores of the metal. The water rushing in to fill the collapsed cavity will also enter the vapor-filled pores until instantaneously stopped by the bottom surface of the pore, where water-hammer action takes place. This is capable of producing pressure intensities on the areas of the same order as the tensile strength of the metal, and under the continual repetition of the shocks, the metal fails locally under fatigue and small particles are irregularly broken away, giving the surface a peculiar spongy appearance. The pitting action is thus believed to be primarily mechanical, as just described; it was formerly thought to be mainly chemical, in the nature of rusting, or electrolytic; but it can be produced in wood, concrete, and even in glass, which points to a mechanical origin as outlined in the preceding theory.

It should be possible to prevent pitting by avoiding the occurrence anywhere in the turbine of a local pressure head so low as to approach the vapor pressure of the water. This method of avoiding cavitation and pitting has been found from experience to be effective.

The Thoma formula for the critical value of the dimensionless ratio known as the Thoma cavitation coefficient can be expressed as follows:

$$\sigma_c = \frac{H_a - H_v - H_s}{H}$$

where σ_c = cavitation coefficient (critical sigma)
H_a = atmospheric pressure, psi (see Table 2)
H_v = vapor pressure of water, ft (see Table 2)
H_s = distance from runner throat or blade centerline to minimum tailwater (see Fig. 8)
H = net head, ft

If the turbine in a given plant is installed at too high a setting of the runner in relation to tailwater, so that the actual plant sigma is lower than the critical sigma, σ_c, then portions of the water passages will be occupied by vapor instead of liquid and cavitation will occur.

TABLE 2 Atmospheric Pressure and Vapor Pressure of Water

	Atmospheric pressure				
Altitude, ft	H_a, psi	H_a, ft H$_2$O	Altitude, m	H_a, mmHg	H_a, m H$_2$O
0	14.696	33.959	0	760.00	10.351
1,000	14.17	32.75	500	715.99	9.751
2,000	13.66	31.57	1000	674.07	9.180
3,000	13.17	30.43	1500	634.16	8.637
4,000	12.69	29.33	2000	596.18	8.120
5,000	12.23	28.25	2500	560.07	7.628
6,000	11.78	27.21	3000	525.75	7.160
7,000	11.34	26.20	3500	493.15	6.716
8,000	10.91	25.22	4000	462.21	6.295
9,000	10.50	24.27			
10,000	10.10	23.35			

TABLE 2 (*Continued*)

Vapor pressure of water			
Temperature, °F	H_v, ft	Temperature, °C	H_v, m
40	0.28	5	0.089
50	0.41	10	0.125
60	0.59	15	0.174
70	0.84	20	0.239
80	1.17	25	0.324

Source: U.S. Bureau of Reclamation (USBR).

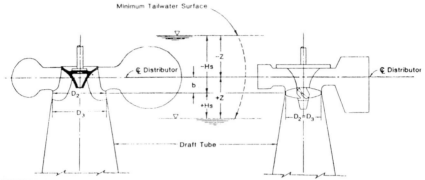

FIGURE 8 Vertical Francis and propeller or Kaplan turbine setting and submergence.

6. Setting and Speed of Turbine. Since the critical sigma and degree of resistance to cavitation of a particular turbine are greatly affected by its design, so that two turbines of the same specific speed may differ considerably in this respect, a cavitation test on a homologous model, supplemented, if possible, by actual field experience from a nonpitting homologous installation, is needed if a new installation is to be carried out safely without using a considerable margin in the value of the plant sigma, and therefore in the value of H_s.

Such information is of special importance in selecting Kaplan and propeller turbines for which the critical sigma is greatly dependent on the particular design, especially on the proportional blade area of the runner. Ample blade area is necessary to keep sigma within reasonable limits. In the absence of definite information, a fair degree of guidance for estimating the safe sigma may be obtained from curves based on pitting experience and available cavitation tests on representative turbines of various specific speeds.

The critical sigma of turbines is greatly dependent on their specific speed; therefore, values of critical sigma of turbines of normal or conventional design plotted as a function of N_s provide a basis for plotting curves which are useful for preliminary selection of turbines. Figure 9 shows the range of critical sigma values found in practice for reaction turbines. The curves shown in solid lines represent USBR and Harza experience curves for minimum values of plant sigma, for preliminary selection purposes. At these values of plant sigma, turbines of normal design, which are intended to develop their rated capacity at or near full

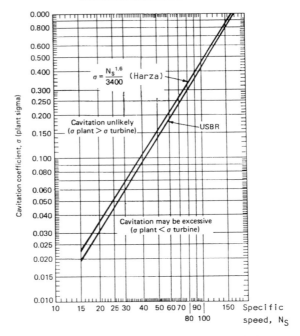

FIGURE 9 USBR experience curve with modified version used by author's firm for planning. (*Harza Engineering Company.*)

gate opening, can reasonably be expected to have a small margin of safety against detrimental pitting or objectionable noise due to cavitation.

Since the designer of a hydroelectric plant usually wishes to avoid undue depth of excavation for the powerhouse foundation, a fairly large value of H_s is usually desired. On the other hand, the engineer for the turbine manufacturer is faced with the vital necessity of limiting H_s in accordance with a safe value of sigma and at the same time providing as high a specific speed as is feasible, in the interest of economy in cost of turbine, generator, and powerhouse. The economy secured by increase of specific speed is so substantial and indeed necessary, in the case of low-head plants where the machinery is relatively large, that it becomes advantageous to adopt high-specific-speed turbines, and to secure the necessary plant sigma by setting the runner below the tailwater elevation, thus using a negative value of H_s.

Space does not permit covering selection of the type and speed of turbines. Turbine-selection methods are covered in Chapter 2.6 in *Standard Handbook of Powerplant Engineering*.[1]

7. Characteristic Proportions of Turbine Runners. Figure 10 shows typical forms of runner corresponding to various specific speeds. The three runners are plotted to the same scale, the dimensions being consistent with a constant output under a constant head, so that the figures show the great change in dimensions of a turbine of a given capacity involved in a change of specific speed. Since under a constant head the actual rpm is directly proportional to the specific speed, the values of N_s shown are a direct indication of the comparative speeds at which the

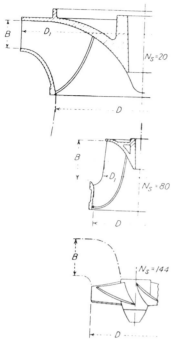

FIGURE 10 Characteristic proportions of turbine runners corresponding to various specific speeds.

various types would operate at a given power and head. Evidently, for example, a high-speed Francis turbine substituted for a propeller turbine would require a much heavier runner and larger turbine and would run only about one-half as fast, a generator of nearly twice the diameter of that needed for the propeller turbine being required. If a Pelton-type impulse wheel were substituted for the axial-flow turbine, it would require an enormous increase in dimensions and would run at a speed of the order of about one-fiftieth of that of the axial-flow unit.

The wide range of specific speeds available is obtained by altering the design proportions. Thus for a very low specific-speed Francis turbine, a large entrance diameter D_1 is used (see Fig. 10) and a small throat diameter D, the breadth B of the distributor being small. The disk friction on the runner hub and shroud ring and the leakage loss through the runner clearances are high, and the maximum efficiency is somewhat less than for a more normal specific speed; but the efficiency remains high for part loads and overloads, and the efficiency curve is flat and of favorable form for operation under widely varying load demands.

Increase in specific speed requires the reduction of D_1 to suit a higher rpm without going to a peripheral velocity out of proportion to the water velocities corresponding to the head. To accommodate the required quantity of flow, B must be increased to compensate for the reduced diameter and thus to maintain a proper entrance area. For a high-speed Francis turbine, D_1 becomes smaller than D, the runner vanes are large and are usually given a complex spoon formation, and the high relative velocity of the vanes through the water involves high surface friction. Still higher specific speeds have been made possible by the development of the propeller turbine in which the shroud-ring friction is eliminated; the runner vanes become nearly flat blades inclined at a small angle to the tangential direction and have a high relative velocity through the water.

At the lower end of the specific-speed range, reduction below the speeds of a low-speed Francis turbine is effected by going to partial admission; i.e., instead of admitting water all the way around the runner periphery, it is admitted at only one to six points; and to reduce the friction and turbulence of the entering flow, this admission takes place through circular contracting nozzles equipped with adjustable needles for regulation, forming the flow into undisturbed jets with little loss of head. The most important and valuable feature of the Pelton wheel is the use of free jets, surrounded by air, and all the space in the runner not occupied by the flow is filled with air. The pressure through the runner space is not appreciably reduced at any point below the atmospheric value, and the possibility of cavitation is greatly reduced.

8. Pelton Impulse Wheels. In considering the design proportions of Pelton impulse wheels, space does not permit the inclusion of the complex subject of bucket design, and the discussion is limited to the determination of the leading dimensions of the wheel or runner and the jet. The most important dimensions are the pitch diameter D of the runner and the diameter d of the jet after it has reached its vena contracta, after which its diameter remains practically uniform until it enters the buckets. Both diameters are expressed here in feet. The diameter ratio (D/d) is an important factor and is the principal one in fixing the specific speed. Consider the most common type of Pelton wheel, that with a single jet and a single runner. Naturally the effective speed of a unit having two single-nozzle runners or a single runner with two nozzles will be 1.41, or $\sqrt{2}$ times the specific speed of the single-nozzle runner, for the specific speed is proportional in the square root of the power. The pitch diameter of the runner is defined as the diameter of a circle tangent to the centerline of the nozzle and jet, and the characteristic ϕ is the ratio of the velocity of the runner at the pitch diameter to $\sqrt{2gH}$, H being the net head.

The net head under which the turbine operates corresponds to the difference between the total energy contained in the water immediately upstream from the nozzle and the elevation of the lowest point of the pitch circle of the runner buckets. The pitch circle is that circle which is tangent to the axis of the power jet. (See ASME Test Code for Hydraulic Prime Movers.)

This is the net head on the machine and is the amount properly chargeable against it; if the tailwater level were brought up to the point where it would just clear the buckets, then, by disregarding the negligible depth of buckets beyond the pitch circle, it would be the available head at the powerhouse. Actually, however, the normal tailwater level must be some distance below the wheel to provide sufficient clearance between the runner and tailwater surface under all operating conditions, including those of high tailwater occurring during periods of flood or surges in the tailrace. Hence there is always a free drop between runner and tailwater which is entirely lost. Operation of the discharge passage as a draft tube, creating partial vacuum in the runner discharge pit to regain the head lost between the runner and the tailwater, has not been found generally practical, because of the complications in regulating the free-water level and the relatively small amount of head thus saved in comparison with the high heads usually available at impulse-turbine installations. Furthermore, experiments have shown no perceptible gain attributable to draft-tube action through reduction of windage loss of the runner rotating in partial vacuum.

To allow development of the maximum head in impulse-turbine installations subject to considerable variation of the tailwater surface above normal level, it is sometimes economically justified to provide tailwater depression systems utilizing compressed air. In those cases, the impulse runners are set with the necessary minimum clearance above normal tailwater level and this clearance is maintained during periods of higher tailwater by admitting compressed air to the runner housing and discharge pit, which must be made airtight.

9. Scale Effect. In a series of homologous turbines, scale effect is the effect of size of turbine on its efficiency. Hence a given turbine should have a substantially constant efficiency when operated at its proper speed under various heads. It was also noted, however, that so-called *homologous* turbines of different actual dimensions, geometrically similar in design and proportions, are not completely homologous with respect to the surface roughness. The variation in degree of relative roughness with change of size of turbine will cause small, but appreciable, variation in the proportion of the effective head lost in hydraulic friction; therefore, the efficiency will change somewhat with change of size.

Of course, the hydraulic friction is not the entire source of loss in a turbine. There is also mechanical bearing friction, which, however, is very small in units of usual size, although appreciable in a laboratory model; and there are eddy or velocity head losses due to the final kinetic energy loss at the exit from the draft tube and at points of sudden enlargement in the turbine, as at the outlet from the wicket gates and runner where the vane thicknesses cause small eddy spaces at their ends. Such enlargement and outlet losses will tend to bear a fixed ratio to the velocity heads and therefore to the head on the turbine and consequently represent an unvarying element of proportional loss, not causing any change in efficiency.

The Moody formula has been found to be reasonably applicable to centrifugal pumps under certain conditions, but its general application to pumps is limited since the relation between small and large pumps is not usually the same as that between model and full-sized turbines.

The efficiency step-up by the Moody formula

$$\Delta \eta = (1 - \eta)\left[1 - \left(\frac{D}{D_1}\right)^{1/5}\right]$$

is based on the maximum efficiency point of the model (peak of the efficiency hill) and the efficiencies of the full-sized turbine at all points are customarily obtained by uniformly adding to the corresponding model efficiencies the same increment $\Delta \eta$ as for the point of maximum efficiency. The runner diameters of model and full-size turbines are D and D_1, respectively.

The Moody formula has been found to give somewhat excessive efficiency step-up between models and large modern turbines, especially for Kaplan and propeller turbines. Consequently, it has become common practice to apply only a part (about 0.6 to 0.75) of the Moody step-up to propeller turbines. Purchase specifications for Kaplan and propeller turbines in America now usually specify application of two-thirds of the Moody formula step-up in correcting the model efficiency to that which may be expected from the full-sized turbine. The trend in recent specifications for large Francis turbines, particularly those operating under relatively low to medium heads, is also toward application of two-thirds of the Moody step-up.

Although the efficiency of homologous reaction turbines changes rather consistently with change in size, their power output has not been found to change quite so consistently from proportional power, that is, proportional to the squares of the diameters and to the three-half powers of the heads. Therefore, it has become usual practice in America, when performance guarantees for full-sized turbines are to be verified by means of homologous model tests, to require that the model efficiency be corrected by the step-up formula, but that the principles of similarity be applied without correction to power.

Other forms of efficiency step-up formulas are used in various countries, and such formulas may be found in some of the national test codes for hydraulic turbines; space does not permit citing them here. However, it is of interest to note that the International Code for Model Acceptance Tests of Hydraulic Turbines, Publication 193, recommends the use of the Moody formula for Francis turbines and the Hutton formula for Kaplan and propeller turbines, unless some other formula has been specified or agreed upon in advance. The applicable Hutton formula is

$$\frac{1 - \eta_1}{1 - \eta} = 0.3 + 0.7\left(\frac{R}{R_1}\right)^{1/5}$$

$$= 0.3 + 0.7\left[\frac{D}{D_1}\frac{v_1}{v}\left(\frac{H}{H_1}\right)^{1/2}\right]^{1/5}$$

where η = turbine efficiency
 R = turbine Reynolds number = $VD/v = (2gH)^{1/2}\,D/v$
 D = turbine runner diameter, ft
 v = kinematic viscosity of the fluid, ft^2/s
 H = net head, ft
  Subscript 1 denotes values for the full-size turbine

PUMPS

10. General Classification. Pumps are machines that convert mechanical energy into hydraulic energy, a process which is the reverse of that of prime movers. Considered broadly, pumps comprise classes corresponding to those mentioned under prime movers: gravity lift, such as chain pumps, rarely used; displacement pumps, of the reciprocating type or piston pumps, or rotary as in lobed-wheel pumps and blowers or gear pumps, in which the fluid is moved without change in velocity in the spaces between intermeshing rotors or gear teeth from a point of low pressure to one of high pressure; and pumps that utilize the reversed reaction-turbine process, involving transformations between pressure head and velocity head and conversion of mechanical into hydraulic energy.

Only the last class will be considered here. There is no conventional name for this class as a whole, which includes centrifugal pumps, in which the flow through the runner or impeller is radially outward, and axial-flow propeller pumps (Figs. 11 and 12, respectively). Mixed-flow or diagonal-flow pumps, intermediate between these types, are included under centrifugal pumps as a subclassification.

Centrifugal pumps thus correspond to Francis turbines with reversed flow direction, and axial-flow pumps correspond to axial-flow turbines reversed. A turbine of normal design cannot be arbitrarily reversed in direction of rotation and employed as a pump; the changed conditions in operation require fundamental changes in the design. It is possible, however, to design a reversible machine with special design characteristics so that the same machine will operate as either a pump or a turbine (see Reversible Pump-Turbines, below).

A typical centrifugal pump is the simple combination of an axial inlet or suction pipe, a radial outward-flow impeller, and a volute or spiral casing. Usually no stationary guide vanes are used at either the inlet or the outlet. The impeller transforms mechanical energy into both pressure head and velocity head in the fluid, and the volute casing, often supplemented by a diverging conical discharge pipe, acts as a diffuser which gradually decelerates the flow velocity and converts most of the velocity head into pressure head at the discharge flange of the pump. Diffusion vanes are used sometimes in single-stage pumps and particularly in axial-flow pumps of high specific speed (Fig. 12).

Pumps of small and moderate sizes are commonly arranged with horizontal shafts, and a wide range of sizes may be secured from the pump manufacturers as standardized or stock products with known performance characteristics determined from shop tests. A common arrangement is with double-suction impellers, equivalent to two single-suction impellers placed back to back and made as a single casting; this arrangement gives balanced end thrust on the shaft (Fig. 11).

For low and moderate heads and large capacities, as in water supply, drainage

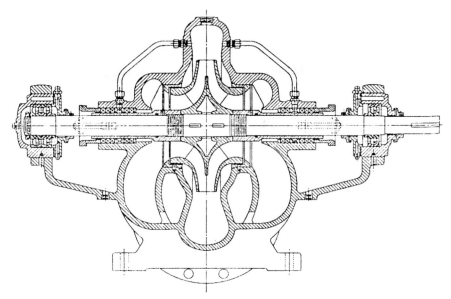

FIGURE 11 Double-suction single-stage horizontal-shaft split-casing pump. (*Peerless Pump, FMC Corp.*)

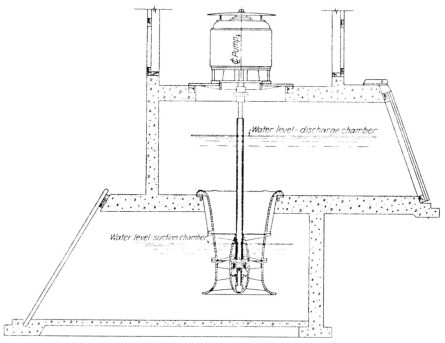

FIGURE 12 Axial-flow propeller pump with diffusion vanes. (*Byron Jackson Pumps, Inc.*)

and irrigation systems, and condenser pumps, mixed-flow and axial-flow pumps with high specific speeds are often used.

For high heads and small capacities requiring very low specific speeds, as in high-lift pumps such as mine drainage, deep-well pumps, and boiler-feed pumps, the method of placing a number of pumps in series is adopted, as many as 10 or 12 stages being employed. The successive impellers are often placed in a single casing containing the proper water passages and forming a single multistage unit.

11. Head, Power, Efficiency. The head (H) against which a pump operates is called the *total head* and is defined by the ASME Test Code for Centrifugal Pumps, PTC 8.2, as follows:

"Pump total head (H) is the energy imparted to the liquid by the pump, expressed in foot-pounds per pound of liquid. It is the algebraic difference between the discharge head and the inlet head: $H = h_d - h_s$."

The definition is the same in substance as that of the net head on a turbine.

The water horsepower is also defined exactly as for turbines, whp = $wQH/550$. The efficiency of a pump is the ratio of the water horsepower, which is the useful work delivered, to the energy supplied to the pump by the shaft of the driving motor, or the input horsepower, bhp:

$$\eta = \frac{\text{whp}}{\text{bhp}} = \frac{wQH}{550 \text{ bhp}}$$

No attempt will be made here to cover the detailed methods of pump design, because excellent designs are available for a wide range of specific speeds, and pumps of satisfactory performance are standard products of the leading pump manufacturers. Attention will therefore be directed to the selection and application of pumps for various conditions of use.

12. Specific Speed. The principles of dynamic similarity explained in the case of turbines are equally valid when applied to pumps; however, this is not so convenient a relation in the pump field as another, which will be used instead. In specifying or designing a pump, it is useful to remember that pumps are rated according to discharge capacity rather than power requirements. We therefore obtain a more convenient relation if we deal with Q instead of P. By applying the same principles of similarity as before, $Q \propto D^2 H^{1/2}$ when $N \propto H^{1/2}/D$. Eliminating D, we have

$$N \propto \frac{H^{1/2}}{\sqrt{Q/H^{1/2}}} \propto \frac{H^{3/4}}{\sqrt{Q}}$$

If this is expressed as an equation instead of a proportionality,

$$N = (\text{const}) \frac{H^{3/4}}{\sqrt{Q}}$$

By putting H equal to 1 ft and Q equal to 1 cfs, the constant is seen to be the speed, in rpm, of a homologous pump of such size that it would deliver 1 cfs against 1-ft head. This is taken as the specific speed of a pump and is commonly denoted N_s. When it is desired to apply both specific-speed functions in the same problem, this new specific speed, based on quantity, may be distinguished by using the notation N_{sq}. This specific speed is equally applicable to turbines, although not so convenient in most turbine problems as the specific speed based on P. By expressing turbine specific speeds as N_{sq}, a comprehensive chart can be drawn showing both turbine and pump characteristics on a single diagram. When not otherwise noted, N_s will be used in this discussion to mean specific speed based on quantity, when dealing with pumps.

As in the case of turbines, this new N_s (or N_{sq}) is a constant for all geometrically similar pumps of any size and head, each being operated at its proper speed corresponding to the head and size. The type of pump can therefore be determined by inserting its values of Q, H, and N in

$$N_s = \frac{N\sqrt{Q}}{H^{3/4}}$$

and finding its N_s.

We also use for pumps the specific speed for the point of best efficiency, which should be the point of normal operation. Figure 13 shows the efficiency

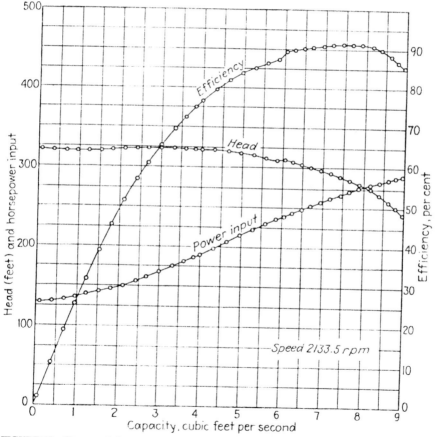

FIGURE 13 Characteristic curves of a Pelton-Byron Jackson model pump tested at California Institute of Technology for Metropolitan Water District of Southern California.

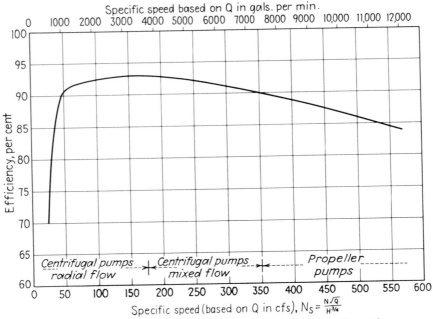

FIGURE 14a Typical maximum pump efficiencies obtained at various specific speeds.

and head characteristic curves for pumps of moderate specific speed and high efficiency.

As in the case of turbines, there are forms and proportions of pumps favorable to high efficiency, corresponding to a particular range of specific speeds; and abnormally high or low specific speeds require abnormal proportions and are accompanied by impaired efficiencies. Figure 14a indicates the trend of maximum efficiencies, obtained in nearly every case from tests of large-capacity pumps. Small-capacity pumps and those of the usual commercial sizes cannot be expected to approach these values, since just as in the case of turbines, efficiencies increase with an increase in dimensions. Figure 14b shows the effect of size approximately, in terms of capacity, on the efficiencies of commercial pumps.

For very high heads and small quantities to be pumped, as in boiler-feed pumps or mine pumps, the specific speed would work out far below the values for normal design and good efficiency, even when extremely high-speed driving motors are used. Instead of attempting to develop the head in a single impeller, the head is subdivided in a number of stages, and several pumps are placed in series, either as separate units or as a series incorporated in a single unit and with all the impellers on one shaft and within a common casing. For example, if s stages are used and the specific speed of the entire unit is $N_s = N\sqrt{Q}/H^{3/4}$, the specific speed of one stage corresponding to a single impeller will be increased to $N_{s1} = N[\sqrt{Q}/(H/s)^{3/4}] = s^{3/4} N_s$, which can be increased to a value that will fall within the range of good design and normal proportions.

Another reason for using a number of stages in a high-head pump is the desire not to exceed about 500 to 600 ft head per stage to avoid abrasion or wear of the

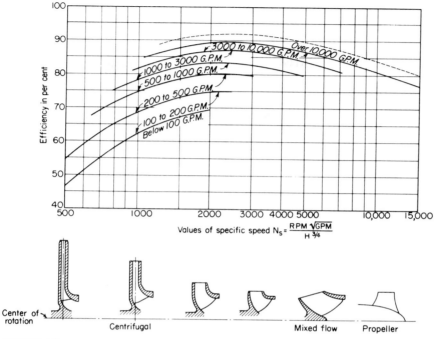

FIGURE 14b Effect of rated capacity on efficiency of good commercial pumps.

parts exposed to high velocity, and frequently the head per stage is limited by cavitation requirements.

The specific speed of a pump unit may also be increased by placing a number of impellers on the same shaft arranged to pump in parallel. If the number of impellers in parallel is n, the specific speed of the unit will be

$$N_s = N(\sqrt{nQ}/H^{3/4}) = N_{s1}\sqrt{n}$$

where N_{s1} is the specific speed of a single-suction impeller. Thus the specific speed of a double-suction pump, having a double impeller, will be $N_s = N(\sqrt{nQ}/H^{3/4}) = 1.41 N_{s1}$. The value of the specific speed to be used in determining the form and characteristics of the impeller is that corresponding to one stage and single suction, which is the value used in the curves shown here.

13. Cavitation. The same reasoning set forth in Art. 5 for turbines applies equally to pumps, including the Thoma formula for the cavitation coefficient sigma $\sigma = (H_a - H_v - H_s)/H$, the criterion of cavitation.

Values of sigma were plotted against specific speed, and the zones of safe and dangerous operation from the standpoint of cavitation were delineated. The charts, reproduced here in Figs. 15 and 16, show the relationship between the total pump head and the recommended upper limits of specific speed under various conditions of suction lift and suction head.

The charts are intended to apply to the normal rated operating conditions, on the assumption that the pump at rated conditions is operating at or near its point

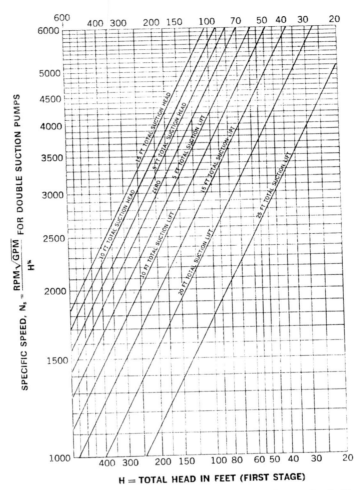

FIGURE 15 Hydraulic Institute, upper limits of specific speeds, double-suction pumps, handling clear water at 85°F at sea level. (*Reprinted from "Standards of the Hydraulic Institute," 1983 ed., copyright by the Hydraulic Institute, 9 Sylvan Way, Suite 180, Parsippany, NJ 07045.*)

of maximum efficiency. Since pumps are normally applied for rated conditions near their maximum efficiency point, their rated specific speed is usually sufficiently close to the specific speed at point of best efficiency. However, when a pump is expected to operate at greater than normal discharges, as in cases where the operating head is subject to considerable variation, lower specific-speed values should be used.

The values of suction lift and suction head used in the charts are measured by pressure gage at the pump suction flange, corrected for velocity head, and referred to the elevation of the shaft centerline for horizontal pumps, to the

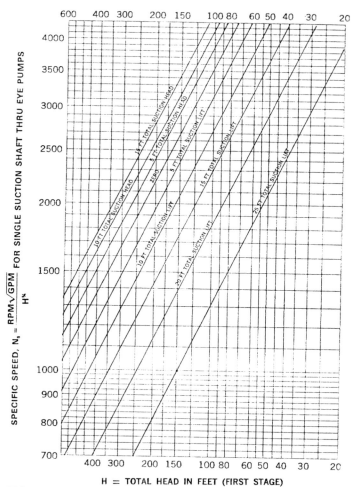

FIGURE 16 Hydraulic Institute, upper limits of specific speeds, single-suction overhung impeller pumps, handling clear water at 85°F at sea level. (*Reprinted from "Standards of the Hydraulic Institute," 1983 ed., copyright by the Hydraulic Institute, 9 Sylvan Way, Suite 180, Parsippany, NJ 07045.*)

centerline of the pump casing for vertical double-suction pumps and to the entrance eye of the first-stage impeller for single-suction vertical centrifugal and mixed-flow pumps. For vertical-shaft axial-flow pumps, the suction lift or suction head is measured to the midheight of the impeller.

With reference to Fig. 17, h_p is the height of water column in a gage glass connected to the pump suction pipe at a point z ft below the shaft centerline. h_p must be increased by the velocity head in the suction pipe at the section where the gage is connected. The top of the water column so corrected is the elevation of the equivalent or virtual sump elevation shown by the dashed line. The value of

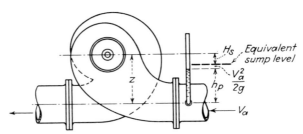

FIGURE 17 Pump setting and submergence.

H_s referred to the pump axis is therefore $H_s = z - (h_p + V_a^2/2g)$. When the pump operates under positive suction pressure, this is termed *suction head*, which is merely $-H_s$. In applying the specific speed-limit charts to large horizontal-shaft installations, it is advisable to figure the suction lift or suction head in the large pump to the upper side of the impeller entrance diameter rather than to the shaft centerline.

The Thoma cavitation formula $\sigma = (H_a - H_v - H_s)/H$ is usually expressed for pump application in the form $\sigma = H_{sv}/H$, in which H_{sv} is the excess of the absolute suction head above the vapor pressure, or the net positive suction head. The net positive suction head can also be introduced into the specific-speed relation. By using H_{sv} instead of H in the specific-speed formula, a useful parameter called the "suction specific speed" is obtained:

$$S = \frac{N\sqrt{Q \text{ gpm}}}{H_{sv}^{3/4}}$$

The suction specific speed expresses the combination of suction conditions which, if kept constant, will give similarity of flow and cavitation in pumps which are geometrically similar in the suction passageways and in the low-pressure portions of the impeller. The suction specific speed and the Thoma cavitation coefficient are related as follows:

$$\sigma = \frac{H_{sv}}{H} = \frac{(N\sqrt{Q \text{ gpm}}/S)^{4/3}}{H} = \left(\frac{N\sqrt{Q \text{ gpm}}}{SH^{3/4}}\right)^{4/3} = \left(\frac{N_{sg}}{S}\right)^{4/3}$$

or

$$S = \frac{N_{sg}}{\sigma^{3/4}}$$

The solid line of Fig. 18 corresponds to a value of S of 8990 or, in round numbers,

$$\sigma_c = \left(\frac{N_{sg}}{9000}\right)^{4/3}$$

Model Tests. In cases involving pumps of large size, model tests are of great utility. Even when it might be feasible to test the large pump in the shop, a model may often be tested with greater accuracy and thoroughness; and by adopting a standardized size of model for various pumps, properly comparable performances can be secured. Model testing in advance of final design and installation

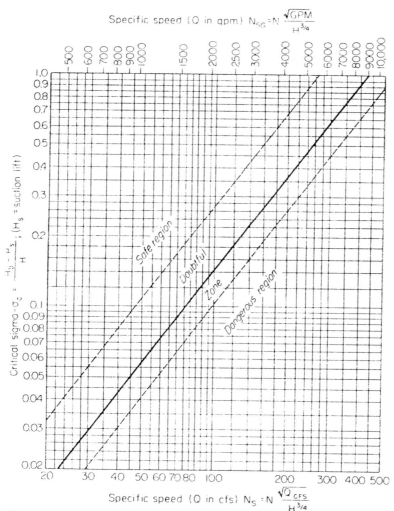

FIGURE 18 Approximate limits of cavitation-coefficient sigma at various specific speeds for single-stage single-suction pumps with overhung impellers.

of a large unit, as standard procedure, not only provides advance assurance of performance but makes alterations possible in time for incorporation in the prototype. It is not, in general, essential that the model test head be the same as that in the prototype. A model pump should be tested at such conditions that complete turbulent flow will be maintained in all flow passages at all times. In general, this means that the model head will be the same as that of the prototype.

The model should have complete geometric similarity with the prototype, not only in the pump but in the intake and discharge conduits as specified for tests on full-sized pumps. The model should be run at such speed under the test head that the specific speed remains the same as that of the installed unit; and if cavitation tests are not available, the suction head or lift should be such as to give the same sigma value as in the installation.

If corresponding diameters of model and prototype are D_1 and D, respectively, the model speed N_1 and capacity Q_1 under the tests head H_1 must agree with the relations

$$\frac{N_1}{N} = \frac{D}{D_1}\sqrt{\frac{H_1}{H}} \quad \text{and} \quad \frac{Q_1}{Q} = \left(\frac{D_1}{D}\right)^2 \sqrt{\frac{H_1}{H}}$$

In testing a model of reduced size, the above conditions being observed, complete hydraulic similarity will not be secured unless the relative roughness of the impeller and pump casing surfaces is the same. With the same surface texture in model and prototype, the model efficiency, η_1, will be lower than that of the larger unit; and greater relative clearances and shaft friction in the model will also reduce its efficiency.

The efficiency of a pump model can conveniently be stepped up to the prototype by applying a formula of the same general form as the Moody formula used for hydraulic turbines.

$$\frac{1 - \eta_1}{1 - \eta} = \left(\frac{D}{D_1}\right)^n$$

The exponent n should be determined for a given laboratory and given type of pump on the basis of an adequate number of comparisons of the efficiencies of models and prototypes, with consistent surface finish of the models and prototypes. The Standards of the Hydraulic Institute state that the values for the exponent n have been found to vary from zero when the surface roughness and clearances of the model and of the prototype are proportional to their size, to 0.26 when the absolute roughness is the same in both model and prototype. Experience data accumulated from continued research and testing by leading pump manufacturers have enabled them to narrow the range of n values applicable to their individual laboratories and methods of finishing the surfaces of models and prototypes. Some manufacturers have found from tests of centrifugal-pump models and prototypes that the use of an exponent of $1/5 = 0.20$, as used for hydraulic turbines, is reasonably justified; others have found that the average value of the exponent applicable to their test results is about $1/7.5 = 0.13$. Some European manufacturers consider an average exponent of $1/6.5$ reasonably justified for general pump application. In any case, it is apparent that, because the Moody step-up applies only to the friction losses, which are relatively small in pumps of modern design, reasonable differences in the value of the exponent have rather minor effect on the calculated efficiency of the prototype.

When model tests are to serve as acceptance tests, it is generally recommended that the efficiency guarantees be stated in terms of model performance, rather than in terms of calculated prototype performance. In the absence of such provision, the efficiency step-up formula and the numerical value of its exponent should be clearly specified, or agreed upon in advance of tests.

Not all pumps are well adapted for model testing. In installations in which free-water surface phenomena may affect the performance, complete hydraulic similarity prescribes constant Froude number, which requires a constant ratio of pumping head to the linear dimensions of the pump. To keep sigma the same, it is necessary to reduce the barometric pressure in the same ratio by enclosing the whole setting in a closed chamber. These conditions apply, for example, when a limited depth of free sump surface over the suction inlet may cause prerotation, or vortices drawing air into the pump. In pumps such as condensate pumps intended to operate under cavitating conditions, when vapor separation produces free surfaces within the pump, it is recommended that the pumps be tested in their full size at full head and speed.

REVERSIBLE PUMP-TURBINES

14. General Classification

Definition. A pump-turbine is a dual-purpose hydraulic machine that combines the functions of a pump and a turbine in a single machine; it is in a class distinct from both the pump and the turbine. Pump-turbines are represented in current practice mainly by the reversible-type machine which operates in one direction of rotation as a pump and in the opposite direction of rotation as a turbine. Although pump-turbines of the nonreversible type have also been developed, they have not yet gained wide acceptance, and relatively little operating experience is available thus far from such machines.

Classification. Pump-turbines may be divided into three principal types analogous to those of reaction turbines and pumps:

1. Radial-flow or Francis type
2. Mixed-flow or diagonal-flow type
3. Axial-flow type

The diagonal-flow and axial-flow types are subdivided into fixed-blade and adjustable-blade types. Pump-turbines are also classified according to the position of their main shaft as vertical-shaft and horizontal-shaft machines, and according to the number of stages as single-stage and multiple-stage units.

15. Basic Performance Relationships

Interrelation of Pumping and Turbine Performance. Since a reversible pump-turbine is both a pump and a turbine, a definite relationship exists between its pumping and generating capacities. Selection of a pump-turbine for a proposed installation ordinarily begins with the determination of the desired generating capacity at minimum net head. Establishment of the generating capacity for a machine of given type and specific speed substantially determines the pumping capacity also. Similarly, if a pumping capacity is established for a specified total head, the generating capacity for the corresponding turbine net head is also fixed within relatively narrow limits. The relationship of performances can be modified only slightly by design changes.

Cavitation, Setting, and Speed. The operating duties of a pump-turbine are inherently less favorable from the cavitation standpoint when pumping than when generating, because head losses associated with flow through the suction-draft tube in the pumping direction always act to decrease the available net positive suction head. Furthermore, pump-turbine impellers are generally more prone to cavitation than turbine runners designed for the same specific speed. To avoid cavitation when pumping, it is necessary to set a pump-turbine lower in relation to tailwater than would be required for turbine operation alone. As in the case of conventional turbines the selection of an appropriate specific speed for a pump-turbine depends not only on its operating head but also largely on the setting in relation to tailwater; essentially, it depends on the depth below tailwater to which it is practical to submerge the pump-turbine. To utilize relatively high specific speeds and consequently smaller, more economical machines with higher operating speeds, pump-turbines require considerably deeper settings than turbines.

The economics of pumped storage generally require that the pump-turbines have high efficiency both as a pump and as a turbine. It is also desirable, from the standpoint of design and cost of the generator-motor, that the pump-turbine have the same speed of rotation in both directions of operation. However, the hydraulic design of reversible pump-turbines with fixed blades is such that the point of best efficiency occurs at a higher value of peripheral-speed coefficient when pumping than when operating as a turbine. This means that, for operation of such machines at a single rotative speed, the head for pumping should preferably be lower than the head for operation as a turbine. Unfortunately, except in unusual installations, the effect of head losses in the penstock and related water-supply and discharge conduits is such that, for a given gross head, the total pump head will generally be higher than the net head during turbine operation by approximately twice the head losses for flow in one direction. However, through careful hydraulic design of reversible machines with fixed blades, the difference in heads required for optimum efficiency in both modes of operation at a single speed can be reduced appreciably. By use of adjustable blades, the best efficiency points for pumping and generating can be substantially brought together.

16. Francis-Type Pump-Turbines

General Design Characteristics. Radial-flow or Francis-type single-stage reversible pump-turbines are being applied for operating heads from about 75 to 1300 ft. The continuing development of their design may extend the upper limit to 1900 ft or more. It is evident that the overall range of heads for which Francis pump-turbines are applied approaches that for which conventional Francis turbines are used. The upper limits exceed the usual maximum for single-stage centrifugal pumps of conventional design.

Francis pump-turbines, though basically similar in construction to Francis turbines, differ from them significantly in the hydraulic design of their principal components. For a given head and power output, the reversible machine will be larger and will usually be set lower in relation to tailwater than a conventional Francis turbine. Since it has been found that a good centrifugal pump will perform well as a turbine in the reverse direction of rotation, but a good Francis turbine will not perform satisfactorily as a pump, the impeller runner of a reversible Francis pump-turbine is designed essentially as a pump impeller rather than as a turbine runner. It has fewer blades than a Francis turbine runner, and the blades usually wrap around from throat to tip diameter by more than 90°, sometimes up to about 180°. A conventional turbine runner, because of relatively short blades, is not well adapted to efficient deceleration of flow in its water passages or to the cavitation requirements, when it is operated in the reverse direction for pumping. The ratio of tip to throat diameter of a reversible impeller runner is substantially larger than for an equivalent turbine runner. The relatively large tip diameter, approximately equal to that of an equivalent pump impeller and approximately 1.4 times that of an equivalent Francis turbine runner, is necessary to obtain efficient diffusing passages and to reduce the velocity-head component of the total energy discharged from the impeller runner; it also provides a head-discharge characteristic favorable for starting of pump operation. The spiral case of a pump-turbine usually has its volute section on a larger base circle than the casing of a centrifugal pump for the same rating. Since the impeller runner is larger than an equivalent turbine runner and the casing of a pump-turbine is larger than that of a pump, the resulting overall dimensions of the reversible pump-turbine are larger than those of either the equivalent turbine or the equivalent pump.

Because of the comparatively large tip diameter of the impeller runner the Francis pump-turbine inherently has a lower runaway speed than a Francis turbine. A further significant characteristic inherent in the hydraulic design of medium- to high-head (low-specific-speed) Francis pump-turbines is their decrease in discharge with increase in speed above normal operating speed. This characteristic, in installations involving long water conduits, can cause substantial water-hammer effects when runaway speed occurs.

Single-Speed Operation. The majority of reversible Francis pump-turbines operate at a single speed. As previously indicated, to attain optimum efficiencies in both the pumping and generating cycle with single-speed operation, the head on a Francis pump-turbine should preferably be lower when pumping than when generating. The gap between the heads for peak efficiency is narrowed by careful hydraulic design but is difficult to eliminate entirely, as theoretical considerations indicate that the optimum efficiencies are not obtained with the same peripheral-speed coefficient in both pump and turbine operation. Nevertheless, good overall performance with efficiencies exceeding 90 percent is usually attainable when the ratio of maximum pumping head to minimum generating head does not exceed about 1.25. Since the difference in optimum efficiency heads diminishes somewhat with decreasing values of specific speed, high efficiency in both modes of operation is attained more readily by Francis pump-turbines of low specific speed which are used for high operating heads.

Two-Speed Operation. For those pump-turbines which must operate over a large range of heads, or where the pumping heads are considerably higher than the generating heads, two-speed operation to allow the maximum efficiency points for both cycles to be utilized may be justified. Specially designed two-speed synchronous generator-motors, which electrically change the effective number of poles for motor and generator operation, have been developed for pump-turbine application. Since a two-speed generator-motor is somewhat larger and its cost substantially greater than that of a single-speed generator-motor, a careful evaluation is usually necessary to determine whether the added cost of a two-speed machine is economically justified for a proposed installation.

Adjustable Wicket Gates. Reversible pump-turbines may be built with or without adjustable wicket gates. In the great majority of pumped storage applications, it is desirable to provide means for varying the power output of the pump-turbine. Adjustable wicket gates provide excellent regulation of power output and good control for starting and synchronizing the unit in turbine operation. In pumping operation, adjustable wicket gates are relatively ineffective for regulating discharge but enable operating at optimum efficiencies with varying heads. As the head increases with filling of the storage reservoir, the quantity of water pumped decreases; as the discharge decreases, the wicket-gate opening can be reduced to provide proper flow conditions for best efficiency. The use of adjustable wicket gates also facilitates starting and stopping in pump operation. Moreover, wicket gates provide a means of controlling the speed in the reverse direction after reversal of flow upon sudden loss of power to the generator-motor while pumping.

If adjustable wicket gates are not provided, the flow is usually controlled by means of the main shutoff valve. This valve usually opens and closes at a slower rate than adjustable wicket gates, because of its size and amount of travel as compared with wicket gates. Omission of adjustable wicket gates may be justified in some installations where the pump-turbine is used primarily for pumping service, and little adjustment of power and flow is required during turbine operation. Pump-turbines without wicket gates have the advantage of lower first cost, relative simplicity of operation, and reduced maintenance. They also are adaptable to slightly higher heads than those with wicket gates.

Starting Procedures. Starting a reversible pump-turbine for turbine operation does not present any special problems. If it is equipped with adjustable wicket gates, the machine can be started, brought up to speed, synchronized, and loaded in the same manner as a conventional turbine. If it does not have adjustable wicket gates, start-up is usually controlled by the main shutoff valve. However, starting the machine and bringing it up to speed in the reverse direction for pumping is associated with problems requiring special consideration.

Because of the water in the upper reservoir, the pump-turbine must obviously be started in the pumping direction with closed wicket gates, or with a closed shutoff valve or other device, if the machine does not have adjustable wicket gates. With the impeller runner submerged as necessary for pumping, the power input to start and accelerate the machine to synchronous speed and to develop shutoff head is quite high for large-capacity machines, despite the fact that the power for shutoff head on medium- and high-head pump-turbines is lower than at rated head and discharge. To reduce the power required for starting, compressed air is admitted to the space inside the wicket-gate circle to depress the water level below the bottom of the impeller runner so that it rotates in air. With the water level depressed, the machine is brought up to speed and synchronized. Means are usually provided to drain off the water that may leak from the spiral case through the wicket-gate clearances into the space between the outer periphery of the impeller runner and wicket gates, where it forms a rotating ring of water which increases the torque requirements. To minimize the starting torque due to bearing friction, the generator-motor thrust bearings for large pump-turbines are provided with high-pressure oil-starting equipment, which forces oil at high pressure between the stationary and rotating faces of the thrust bearing, thus lifting the rotating face free of metal-to-metal contact. Nevertheless, the power required for starting a large reversible Francis pump-turbine and its generator-motor from rest, and overcoming its inertia in accelerating it to synchronous speed, is considerable.

Several methods are in use for starting reversible pump-turbines for pumping operation; selection of the most suitable method depends largely on the size of the machine in relation to the capacity of the electrical-power system. The simplest method of starting consists of applying full voltage from the power system and letting the machine come up to speed as an induction motor, with the damper winding on the rotor acting as the squirrel-cage winding of an ordinary induction motor. However, this method will disturb the power system severely, unless the system and connecting transmission line are sufficiently larger in relation to the generator-motor capacity. In some installations, a similar method is used, except that the voltage is reduced to limit the starting current. In other cases, including several recent installations with pump-turbines of large capacity, a starting motor of the wound-rotor type is provided on the generator-motor shaft for accelerating the unit from rest to synchronous speed. Another method employs so-called back-to-back synchronous starting in which a nearby generator or generator-motor is connected electrically to the generator-motor to be started with both machines at rest, then started and accelerated by means of its turbine, thereby starting and bringing the pump-turbine up to speed. A hydraulic starting method utilizing a starting turbine mounted on the shaft of the reversible machine is also available but has not found broad application thus far to reversible pump-turbines, mainly because of design complications arising from problems of setting in relation to tailwater and properly accommodating the discharge from the starting turbine.

After the pump-turbine unit is synchronized to the power system while running in the pumping direction with the impeller runner rotating in air, the air must be released to admit water to the impeller runner. Proper procedures must be followed in releasing air and admitting the water, so as to control the rate of power increase up to the point where the unit is operating in water at shutoff head against closed wicket gates. The shutoff-power requirement of reversible Francis pump-turbines pumping against closed wicket gates can be from about 25 percent of full-load pumping power on high-head units to about 100 percent on low-head units. After full shutoff head is developed, the wicket gates are slowly opened. The increase in power demand is then controlled by the rate of opening the wicket gates to their normal operating position. The normal gate position for pumping should be kept at or near the best efficiency point; therefore, as the head on the pump-turbine increases during the pumping cycle, the wicket-gate opening should be gradually decreased.

Reversible pump-turbines are normally shut down at the end of the pumping cycle by slowly closing the wicket to zero opening and disconnecting the generator-motor from the power system. When the speed of the machine has dropped to below half speed, the brakes on the generator-motor are applied to bring it to a complete stop.

17. Diagonal-Flow Adjustable-Blade Pump-Turbines. Diagonal-flow, adjustable-blade-type, reversible pump-turbines, also called Deriaz-type pump-turbines, are being applied for heads from about 35 to 300 ft. These machines are the reversible version of the diagonal-flow adjustable-blade turbine and hydraulic design is based primarily on mixed-flow pump practice. They usually have between five and nine blades, somewhat fewer than a corresponding turbine. The axes of the adjustable blades are inclined to the main shaft axis at an angle of about 30 to 60°. The blades are of simpler shape than those of Kaplan machines and can be designed so that adjacent blades touch one another along their entire length in the closed position. This is unlike Kaplan runner blades which, when closed, may be in contact at the periphery while a large gap remains between blades near the hub.

Owing to their adjustable blades, the diagonal-flow pump-turbines have greater operating flexibility and flatter efficiency curves than reversible Francis-type pump-turbines. They are suitable for low- to medium-head pumped storage applications where the head and discharge vary considerably. Their adjustable blades provide effective regulation of pumping discharge and power, in contrast to the relatively ineffective regulation provided by adjustable wicket gates on Francis-type pump-turbines. By controlling simultaneously the blade angle and the wicket-gate opening of diagonal-flow machines, it is possible not only to vary the turbine output or maintain uniform output with varying head but also to vary the pump discharge while maintaining uniform head, without substantial decline in efficiency. Their ability to vary pump discharge while maintaining good efficiency permits these machines to follow system-load variations closely during pumping. Their high part-load efficiency in turbine operation, a feature which is always desirable, has special significance in connection with the present trend of selecting a small number of large units for pumped storage plants to gain economy in first cost and in annual cost of operation and maintenance. The high part-load efficiency may make it feasible in some cases to keep the units running as turbines at small load, in readiness for sudden increase in demand, thus providing highly valued instantaneous availability of power. These advantages should lead to increased application of adjustable-blade diagonal-flow pump-turbines in the range of low to medium heads, below the range of best application of Francis-type pump-turbines.

However, the cost of adjustable-blade diagonal-flow pump-turbines is substantially higher than that of Francis pump-turbines because of their greater complexity of design. Their cavitation characteristics are such that a deeper submergence is generally required than for Francis pump-turbines.

The adjustable-blade diagonal-flow machines have a relationship of optimum blade angle to wicket-gate opening, known as the cam relationship, resembling that of Kaplan turbines. Since the optimum relationship for turbine operation differs considerably from that for pump operation, an adjustable-blade diagonal-flow pump-turbine requires either two sets of cams or a somewhat more complex three-dimensional cam system suitable for both turbine and pump operation. However, modern digital microprocessor-controlled governors greatly facilitate obtaining the optimum settings. As in the case of Kaplan turbines, diagonal-flow machines have both an on-cam and an off-cam runaway speed, the latter being the higher of the two.

The shape of the adjustable blades of a diagonal-flow pump-turbine is such that, when closed, they form a relatively smooth cone; consequently, a comparatively small torque is required for rotating the impeller runner in water. As a result, it has not been found necessary to depress the water level below the blades for pump starting, as is done with Francis-type pump-turbines. When the machine has been brought up to speed, the blades and wicket gates are gradually opened. The discharge increases smoothly to full pumping capacity. This procedure permits making a changeover from generating to pumping in less time than usually required for Francis pump-turbines.

18. Axial-Flow Pump-Turbines. Axial-flow, reversible pump-turbines are utilized for operating heads from about 3 to 45 ft. The axial-flow pump-turbines have impeller runners resembling those of axial-flow turbines, and usually have a tubular-type arrangement. When provided with adjustable blades, they permit substantial variations in head and discharge with good efficiency. The adjustable blades also permit a considerable reduction in the required torque for starting in pumping operation.

Axial-flow pump-turbines with horizontal-shaft arrangement are well adapted to tidal-power applications. To increase the flexibility of operation in tidal plants or for special cases of inland installations, the machines can be designed to operate as a pump or as a turbine with either direction of flow, giving them four-way operating capability.

The extensive development work now under way in several countries on low-head pump-turbines for tidal-power application can be expected to result in further significant advances in the design of axial-flow pump-turbines.

TURBINE SPEED GOVERNORS

19. General. Turbines, depending on their types, are provided with wicket gates, adjustable blades, flow-control nozzles, deflectors, which are the devices used to adjust (throttle) or shut off the flow through the turbine and thereby provide a means of controlling the speed or the output of the turbine. The hardware for adjustment (positioning) of these devices is operated by high-pressure oil servomotors furnished with the turbine. The movements of the servomotors are controlled by the governor acting on input signals of speed, acceleration, and turbine output.

Modern governing systems are of the digital, microprocessor-controlled, three-term PID (with proportional, integral, and derivative functions), solid-state electric-hydraulic type. These governors are provided with speed and output power sensing and regulation, stabilizing, and diagnostic functions. All control and diagnostic functions are accomplished through a digital processor providing a control signal to an electrohydraulic transducer that controls positioning of the main oil distributing valve (actuator) directing pressure oil to the servomotors to position the turbine control devices. The turbine control devices are the runner blades and wicket gates for Kaplan, bulb, and Deriaz turbines; the wicket gates for Francis turbines; and the regulator needles and deflectors (not used for normal control) for Pelton units. In the case of double regulated turbines (such as Kaplan, bulb, and Deriaz, provided with double control devices), the governor also coordinates the optimum setting of both control devices. This function was provided on older, mechanical-type governors with a three-dimensional cam which was difficult to manufacture and to adjust at the site, whereas on the modern digital governors this function is provided in the software and can be easily adjusted as required.

Modern digital governors have a number of operating modes and functions and provide operating ease and flexibility which was not possible with the previous type of mechanical governor. Digital governors have proved to be very accurate and provide reliable service in an increasing number of installations, completely replacing mechanical governors for new installations.

20. Operating Modes. The basic operating modes of the digital governors are as follows:

Speed Control. Speed-control governing mode uses the speed droop function; that is, when the unit is producing power, the position of the turbine control devices is a function of the unit speed; as the speed drops (normally within a preset range of 5 percent), the control devices are opened to permit more flow through the turbine, producing more output, thereby compensating the increased demand that caused the drop in the frequency of the network (or the unit speed) in the first place. This mode is used to maintain synchronous speed and provide system stability. Depending on the size of the network (large, small, or isolated), the digital governors use different sets of control parameters to provide fast response when the unit is connected to a large network, or to provide the required stability when operating on an isolated or small network. For isolated or small networks this is the only operating mode for units producing power.

Speed Regulation or Power Control. Speed regulation is similar to speed control, except that in this case the output power remains at a preset constant value and the governor adjusts the droop in function of the network frequency deviations. This mode is normally used when the unit is operating on a large network. It is new for electric and digital electronic governors but is not available for mechanical governors.

Frequency Adaptation. The frequency adaptation mode is used while the unit is not loaded, to bring the unit up to synchronous speed and to match the unit and network frequencies for synchronization.

Gate Position Control. This mode is used to maintain the turbine control devices (wicket gates, runner blades, or control needles) in a preset position regardless of speed or load changes. This mode is used when the unit is connected to a large network, or for testing under manual control.

In addition to the above basic operating modes, the following operating modes and functions are available on modern digital governors:

- Water-level control
- Flow control
- Water-passage transient control
- Head-dependent opening limitation
- Joint control
- Automatic Pelton nozzle selection

The digital governor control logic also provides for automatic mode transfer when parameters for the selected operating mode are outside of a preset range, operating conditions change, and certain faults occur. The following operating-mode changes can be provided:

- Change from frequency adaptation to gate position or speed control upon closure of the unit circuit breaker.
- Change from any operating mode to frequency adaptation mode upon opening the unit circuit breaker.
- Change from any operating mode (except frequency adaptation mode) to speed control isolated mode if the turbine speed is outside a programmable speed deadband.
- Change to speed-control mode from speed-regulation mode in case of the fault of the active output power feedback signal.
- Change to gate-position control in case of the fault of the speed feedback signal.
- Unit emergency shutdown in case of loss of wicket gate, runner blade, or control needle position feedback.

All mode changes are bumpless, a condition that is achieved by appropriately controlling the wicket gates or other control devices. When a change is made to gate-position control, the gate-position setpoint is set at the position required by the previous mode of operation immediately before the change. If the change is made to water-level or flow control mode, the gate setpoint is first set at 100 percent opening and the gates start with a smooth movement to the position required by the new operating mode.

21. Operating Sequences

Automatic Start Sequence. When receiving the automatic start command the governor first verifies that all conditions for starting are met and then proceeds with the actual start-up sequence. In order to accelerate the unit as quickly as possible and to avoid operation at lower speeds, which could be damaging to the turbine bearings due to inadequate lubrication, the wicket gates are set at larger openings than for speed-no-load condition. When the unit reaches a preset speed, the governor starts moving the wicket gates back to the speed-no-load position and assumes the frequency-adaptation mode for synchronizing the unit. Once the unit is synchronized, the operating mode is automatically transferred to gate-opening or speed-control mode. Thereafter the operator may change the operating mode as required.

Normal Shutdown. When receiving a signal for normal shutdown, the governor will start the normal shutdown sequence by starting to close the wicket gates at a predetermined rate for gradual reduction of load on the unit, until the wicket gates are completely closed. When the wicket gates are fully closed, squeeze will be applied, and the wicket gate lock is automatically applied (if used). When the unit speed reaches a predetermined value (usually not more than 25 percent of normal speed), the governor will apply the brakes. The brakes are released after a few minutes of holding period after the unit indicators show complete stop.

Emergency Shutdown. The emergency shutdown sequence overrides the governor wicket-gate position control used for normal shutdown and the wicket gates close with maximum speed permitted by the governor hydraulic system. In case of Pelton turbines, the deflectors move with maximum speed and the regulator needles remain under governor control, and in the case of Kaplan and bulb turbines the runner blades also move to their highest tilt to reduce runaway speed. Application of the brakes and gate lock remains under the control of the governor.

Partial Shutdown. Partial shutdown is similar to normal shutdown, but the wicket gates close to the speed-no-load position, the unit maintains rated speed at no load, and operating mode switches to frequency adaptation.

22. Operating Parameters

Definitions. Speed droop, speed regulation, speed deadband, gains, various time constants, and other terms are defined in Section 2, "Glossary of Terms, Functions and Characteristics" of IEEE Standard 125.[2]

Speed Deadband. The speed deadband of digital governors is typically not more than 0.02 percent of the rated speed at any gate opening and at rated speed of the unit with the speed droop set at 5 percent.

In addition to the inherent speed deadband, digital governors are also provided with a software adjustable artificial deadband. The typical range for the artificial deadband is ±0.5 Hz, and it is used in certain operating modes to monitor if operating mode transfer is required to isolated speed-control mode. This artificial frequency deadband is not functional when the governor is in the isolated speed-control mode.

Integral Gain. Typical adjustment range of the integral gain is 0 to 100 s^{-1}.

Derivative Gain. Typical adjustment range of the derivative gain is 0 to 20 s.

Proportional Gain. Typical adjustment range of the proportional gain is 0 to 160 percent.

Damping Device Time Constant. Typical adjustment range of the damping device constant is 1 to 20 s.

Dead Time. Dead time is the time interval between the initiation of a specified change in steady-state speed and the first detectable movement of the turbine control servomotor. Typically the dead time between a speed change of 0.1 percent of the rated speed of the unit and the first detectable movement of the wicket-gate servomotor due to a sudden load change of more than 10 percent of the rate power output of the turbine is not more than 0.2 s for a digital governor.

Stability. With the water conductor–unit system inherently stable the governing system must be capable of controlling the unit with stability when operating at rated speed and no load or when operated at rated speed with isolated load at all turbine power outputs, including maximum output. The governing system must also be capable of controlling with stability the turbine power output between zero and maximum power output, inclusive, when the generator is operating on a large network. The following stability requirements can be expected from modern digital governors:

- The magnitude of the sustained speed oscillation caused by the governing system does not exceed ±0.15 percent of rated speed with the generator operating at rated speed and no load, or operating at rated speed and isolated sustained load with the permanent droop set at 2 percent or above.

- The magnitude of the sustained power-output oscillation caused by the governing system does not exceed ±0.4 percent of the rated power output of the turbine with the generator operating under sustained load demand operating on a network with the speed regulation set at 2 percent or above.

The dynamic performance of the governing system is achieved with proportional, integral, and derivative function blocks, each with independent continuously adjustable gains. On digital governors normally several sets of gains are provided and the governor automatically switches between the sets of gains as required for the mode of operation. Gains are set for quick response when the unit is on a large network (and the network provides the stability) and to provide high stability when the unit is operating on a small isolated network or is in the frequency-adaptation operating mode.

Dynamic Characteristics. The following dynamic characteristics can be expected from modern digital governors:

- After maximum load rejection, a peak speed rise in excess of 3 percent of the rated speed should not appear more than two times.
- The elapsed time between the first movement of the servomotors in the direction of the wicket-gate opening and the point at which the speed fluctuation exceeds +0.5 percent of the rated speed should not be more than 40 s.

The speed rise following a full-load rejection depends on the parameters of the installation including the water starting time, mechanical starting time, wicket-gate maximum closing speed with respect to pressure rise in the penstock, and corresponding pressure drop in the draft tube. Once the unit reaches its maximum speed and starts slowing down, the governor takes over to control the speed of the units (see Fig. 19). The following performance can be expected in general from modern digital governors for the phase following the peak speed rise.

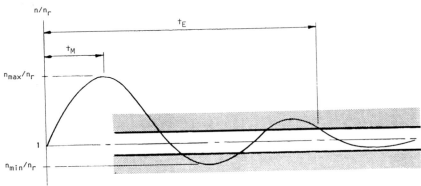

FIGURE 19 Load rejection: speed vs. time typical curve.

$$2.5 < t_E/t_M < 4 \quad \text{and} \quad 0.85 < n_{min}/n_r < 0.95$$

where t_E = time after which speed deviation from the rated speed remains below ±1 percent
n_{max} = maximum speed during load rejection

t_M = time between opening the main circuit breakers and reaching maximum speed, n_{max}
n_{min} = minimum speed after load rejection
n_r = rated speed

23. Physical Arrangement of Governor. The typical governing system consists of the following basic functional components: electronic control cabinet, electrical control cubicle, governor pumping unit, actuator assembly, oil pressure tank, speed-sensing elements, and devices for feedback. The electronic control cabinet usually contains the digital processor, input-output modules, speed monitoring and supervision system, and one set of indicating and operating instruments for manual local control. The electrical control cubicle usually contains the oil pump controls, motor starters, and protection devices. If the electrical cubicle and electronic cabinet are not located close to each other, a second set of indicating and operating instruments can be provided on the front panel of the electrical cubicle. The electrical cubicle, governor pumping unit, and actuator assembly can be included in a common enclosure or provided as separate items. All cabinets and enclosures should be provided with access doors for adjustment and maintenance of the equipment. The face of the governor electrical cubicle and electronic cabinet should include a section for mounting the instruments and controls.

REFERENCES

1. ESSER, CARLYLE, and JAMES H. T. SUN, Chapter 2.6, "Hydraulic Turbines," in T. C. ELLIOTT, *Standard Handbook of Powerplant Engineering*, McGraw-Hill, New York, 1989.
2. IEEE Standard 125, "Recommended Practice for Preparation of Equipment Specifications for Speed Governing of Hydraulic Turbines Intended to Drive Electric Generators," Institute of Electrical and Electronics Engineers, New York, 1988.

SECTION 22

HYDRAULIC TRANSIENTS

By James E. Borg

GENERAL

Hydraulic transients consist of pressure disturbances within closed-conduit systems when the system undergoes a change from one operational steady-state condition to another. The disturbances are initiated by the application of a definite action and are dissipated during the flow-transition period to the successive steady state by some form of damping within the system. In the case of hydroelectric or water-supply projects, hydraulic transients are usually initiated by an adjustment in the setting of a control valve or the change in operation of hydromachinery such as turbines or pumps. The damping of pressure transients during the transition to the second steady-state flow condition is achieved when system energy loss occurs in the form of conduit friction or minor losses.

Consideration of the unsteady-flow phenomenon in closed-conduit water-conveyance systems is essential for the determination of system design pressures and the establishment of operational procedures for achieving the desired project performance. Underestimation of the effects of hydraulic transients has resulted in system instabilities which either have placed severe operational constraints on the project or have resulted in system failure, causing excessive damage.

The hydraulic design engineer is mainly concerned with identifying how the unsteady-flow phenomenon is initiated and what governs the degree of the resulting pressure fluctuations. The engineer must be familiar with the theory of hydraulic transients and the development of basic water-hammer equations from the equations of continuity, motion, and wave speed. From these equations a method of analysis must be developed by which studies can be made to determine the response of the system to transient flow conditions. Hydraulic transients analyses are routinely made to establish operational procedures, identify pressure control device requirements, and determine system design pressures.

DEFINITIONS

The flow of water through a conduit is classified as steady-state if the velocity and pressure at any location in the system remain constant and do not vary with time. Conversely, in unsteady flow, the conduit flow conditions at any point may change over time. The term transient flow defines the unsteady-flow condition during which the conveyance system undergoes a change from one steady-state condition to another. The terms water hammer, hydraulic transients, and transient flow all refer to transient flow conditions within a system. Water-column separation refers to the situation where transient flow reduces the pressure at a point in the system sufficiently to cause the formation of a vapor cavity within the conduit. When identical unsteady-flow conditions recur in a fixed time interval, the flow is said to be steady-oscillatory, or periodic. The time in which the exact flow conditions are repeated is called the period.

System resonance refers to an oscillatory transient condition in which the amplitude of the unsteady oscillations increases over time until periodic flow with large pressure fluctuation results. Resonance in pipelines and penstocks frequently occurs at one of the natural periods of the system and often results in system failure.

BASIC WATER-HAMMER EQUATION

Consider the instantaneous closure of a valve at the downstream end of a frictionless pipeline (Fig. 1). At the instant of valve closure, the water at the downstream end of the pipeline comes to rest and the flow velocity changes from V_0 to zero by the pressure rise at the face of the valve. With time successive layers of water come to rest behind the high-pressure wave as it advances upstream at some sonic wave speed a.

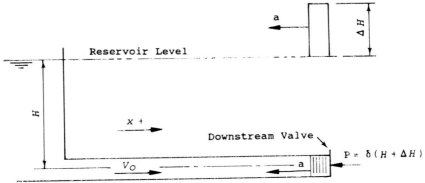

FIGURE 1 Flow condition in frictionless pipeline immediately following instantaneous closure of downstream valve.

Applying the momentum equation to the control volume shown in Fig. 2, it can be seen that

$$\gamma H A - \gamma(H + \Delta H)A = (\rho + \Delta\rho)(V_0 + a + \Delta V)^2 A - \rho(V_0 + a)^2 A \qquad (1)$$

and by further reduction by simplifying and eliminating terms of higher order:

$$-\gamma \Delta H = 2\rho V_0 \Delta v + 2\rho \Delta V a + \Delta\rho(V_0^2 + 2V_0 a + a^2) \qquad (2)$$

Applying the equation of conservation of mass to the control volume and assuming the equivalent steady-state condition:

$$0 = (\rho + \Delta\rho)(V_0 + a + \Delta V)A - \rho(V_0 + a)A \qquad (3)$$

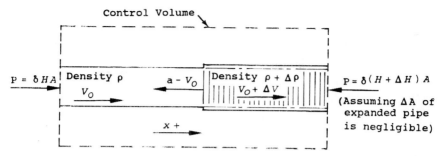

FIGURE 2 Equivalent steady-state condition assuming a stationary pressure wave front.

which when simplified becomes

$$\Delta \rho = - \frac{\rho \Delta V}{V_0 + a} \tag{4}$$

Since V_0 is usually quite small in comparison with the pressure wave speed a, Eq. (4) may be rewritten as

$$\Delta \rho = - \frac{\rho \Delta V}{a} \tag{5}$$

By substituting Eq. (5) into Eq. (2), simplifying, and neglecting terms of higher order, the relationship between pressure rise and velocity change for partial or incremental closure of the valve becomes

$$\gamma \Delta H = -\rho \Delta V a \tag{6}$$

or

$$\Delta H = \frac{-a \Delta V}{g} \tag{7}$$

This relationship between the change in velocity and pressure rise is known as the basic equation of water hammer. The equation is valid as long as the time associated with the valve movement is less than the time for the pressure wave to travel to the upstream end of the pipeline and return to the valve as a reflected wave. The time required for the pressure wave to travel twice the length of the pipeline is $2L/a$. The basic equation of water hammer can also be applied to valves at the upstream end of the pipeline by merely changing the sign of the wave speed to negative for pressure waves in the downstream direction.

As indicated in Fig. 1, the pressure wave resulting from the instantaneous closure of a downstream gate advances up the pipeline at a velocity of a. As the pressure wave moves upstream, the water velocity immediately behind the wave becomes zero, the density of the downstream water is increased, the pipe section is expanded, and the pressure is increased to a value in excess of that corresponding to the upstream reservoir level. Hence when the pressure wave arrives at the upstream reservoir at time L/a an unstable energy condition occurs at the entrance to the pipeline. Since the pressure within the pipeline exceeds the reservoir level, a flow reversal occurs producing a reflected pressure wave having a sonic velocity a. As the reflected wave travels downstream, the pressure behind the wave front is reduced to that corresponding to the reservoir level, the density of the water decreases, and the pipe contracts from its expanded cross-sectional area. At time $2L/a$ the reflected pressure wave reaches the gate. At this time the water in the pipeline is flowing toward the reservoir with a velocity of V_0. This upstream velocity creates a second reflected pressure wave at the face of the gate. Since the sign of the velocity change is now positive, the resulting pressure change, computed from Eq. (7), has a negative magnitude. As the wave travels upstream the second time, the water density is decreased and the pipeline section is reduced in its wake. When the wave reaches the pipeline entrance it is once again reflected to restore an unbalanced energy condition. When the wave returns to the gate at $4L/a$, the water is once again flowing downstream with a velocity of V_0 and the transient process will be repeated as long as system damping does not occur.

WAVE SPEED IN TUNNELS AND CONDUITS

The basic water-hammer equation indicates that the pressure change resulting from a change in pipeline velocity is directly proportional to the velocity of the pressure wave. Furthermore, the reflection time of pressure waves that govern the selection of gate-closure timings and design of pressure-relief devices is also a function of the pressure wave speed. It is therefore important to determine as accurately as possible the system wave speeds during the transient analyses of conduit conveyance systems. The detailed development of the equation for the computation of the pressure wave speed, or celerity, is presented by Parmakian,[1] and Wylie and Streeter.[2]

In the case of free-standing or buried penstocks and pipelines, the wave speed is a function of the diameter, the wall thickness, the modulus of elasticity, and the density and volume modulus of water. The wave speed is also affected by the degree of support, which may limit the longitudinal movement and axial expansion and contraction of the conduit. The resulting wave speed a is computed as

$$a = \sqrt{\frac{1}{\left(\dfrac{W}{g}\right)\left(\dfrac{1}{K} + \dfrac{DC_1}{Ee}\right)}} \tag{8}$$

where a = pressure wave speed, fps
 g = acceleration of gravity, ft/s^2
 D = diameter of pipeline or penstock, ft
 e = thickness of wall of pipeline or penstock, ft
 E = Young's modulus of elasticity of pipeline or penstock material, psf
 K = volume modulus of water, psf
 W = specific weight of water, pcf
 C_1 = coefficient indicating degree of freedom of lateral movement and circumferential expansion

The coefficient C_1 applies to relatively thin-walled conduits where the ratios of lateral to axial stresses are affected by the system of conduit support. Computations of the coefficient C_1 for various supporting arrangements are as follows:

$C_1 = 1 - \mu/2$ free-standing pipeline anchored at its upstream end only

$C_1 = \mu^2$ pipeline continuously anchored to prevent longitudinal movement throughout its length

$C_1 = 1$ pipeline continuously anchored throughout its length with expansion joints to permit longitudinal movement

where μ = Poisson's ratio defined as $-$lateral unit strain/axial unit strain

Table 1 shows the values of Young's modulus of elasticity and Poisson's ratio for various materials used for pressure conduits.

Reinforced-concrete pressure conduits rely mainly on the steel or prestressed wire reinforcement to withstand the internal water pressures. Additional strength is often provided by a thin steel liner inserted for leakage control. In order to compute the wave speed for a reinforced-concrete pipeline, the composite conduit wall of concrete, intermittent steel reinforcement, and steel liner must be converted to an equivalent steel wall thickness. This conversion is based on first determining the equivalent continuous thickness of the reinforcing steel and second converting the concrete thickness to an equivalent steel thickness by multiplying it by the ratios of steel to concrete modulus of elasticity. Both converted thicknesses of reinforcement steel and concrete are then added to each other, and to the steel liner, if any, to arrive at the total steel thickness equivalent.

A similar approach is followed in determining the equivalent wall thickness for computing wave speed in wood-stave pipe. The equivalent steel thickness of the wooden staves is added to the equivalent thickness for continuous area of the external bands to arrive at the equivalent steel wall thickness required for wave-speed calculations.

For lined and unlined tunnels the wave speed is a function of the modulus of rigidity of the material surrounding the tunnel. The wave speed in an unlined circular tunnel is computed as

TABLE 1 Elastic Properties of Pressure Conduit Materials

Material	E, Young's modulus of elasticity, 1,000,000 psi	Poisson's ratio
Cast steel	28.5	0.265
Cold-rolled steel	29.5	0.287
Stainless steel 18-8	27.6	0.305
All other steels, including high-carbon, heat-treated	28.6–30.0	0.283–0.292
Cast iron, tensile strength 23.5–50 ksi	13.5–21.0	0.211–0.299
Malleable iron	23.6	0.271
Copper	15.6	0.355
Brass, 70-30	15.9	0.331
Cast brass	14.5	0.357
Tobin bronze	13.8	0.359
Phosphor bronze	15.9	0.350
Aluminum alloys, various	9.9–10.3	0.330–0.334
Monel metal	25.0	0.315
Concrete	3.4	0.15–0.25

$$a = \sqrt{\frac{1}{\left(\frac{W}{g}\right)\left(\frac{1}{K} + \frac{1}{G}\right)}} \tag{9}$$

where G = modulus of rigidity, psf

Examples of modulus of rigidity for various tunneling media are presented in Table 2.

Wave speeds in steel- or reinforced-concrete-lined tunnels are slightly higher than those encountered in unlined tunnels. The composite lining material is first reduced to an equivalent steel thickness t in the manner previously prescribed for reinforced-concrete or wood-stave pipelines. The wave speed may then be computed as

$$a = \sqrt{\frac{1}{\left(\frac{W}{g}\right)\left(\frac{1}{k} + \frac{DC_2}{Ee}\right)}} \tag{10}$$

where $C_2 = Et/(GD + Et)$

Wave speeds in conduits having noncircular or rectangular cross sections can be determined if the relationship between the relative change in cross-sectional

TABLE 2 Modulus of Rigidity for Commonly Encountered Tunneling Media

Description of tunneling medium	G, modulus of rigidity, 1,000,000 psi
Mass concrete	1.40
Berea sandstone	1.01
Navajo sandstone	1.94
Tensleep sandstone	1.25
Hackensack siltstone	1.56
Monticello Dam s.s. (greywacke)	1.35
Solenhofen limestone	3.58
Bedford limestone	1.60
Tavernalle limestone	3.11
Oneata dolomite	2.37
Lockport dolomite	2.76
Flaming Gorge shale	0.32
Micaceous shale	0.63
Dworshak Dam gneiss 45° to foliation	2.90
Quartz mica schist 1 schistocyte	1.14
Baraboo quartzite	5.76
Taconic marble	2.48
Cherokee marble	3.23
Nevada test site granite	4.39
Pikes Peak granite	4.33
Cedar City tonalite	1.19
Palisades diabase	4.63
Nevada test site basalt	1.92
John Day basalt	4.71
Nevada test site tuff	0.21

*From R. E. Goodman, *Rock Mechanics*, John Wiley & Sons, New York.

area with change in pressure can be established. For a detailed discussion of the determination of wave speeds in noncircular conduits, the reader should consult the references by Wylie and Streeter[2] based on the work of Jenker.[3]

BASIC DIFFERENTIAL EQUATIONS FOR TRANSIENT FLOW

Basic differential equations of continuity and motion can be developed for the solution of hydraulic transient-flow problems using the method of characteristics. The detailed development of these two equations can be found in references by Chaudry,[4] Parmakian,[1] and Wylie and Streeter.[2]

1. Equation of Continuity. The continuity equation for elastic-water-column theory is based on the assumption that the rate of increase of mass in a control volume equals the net rate of mass inflow.

From Fig. 3 it can be seen that

$$\rho A V = \rho A V + \frac{\partial}{\partial x}(\rho A V)\delta x \tag{11}$$

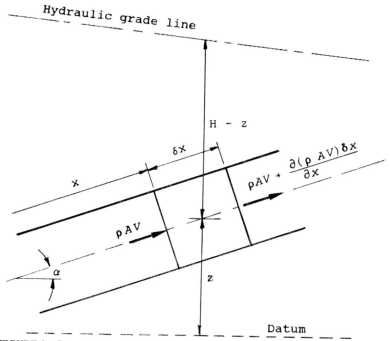

FIGURE 3 Control volume for the derivation of the equation of continuity.

and from expanding and rewriting:

$$\frac{1}{A}\frac{dA}{dt} + \frac{1}{\rho}\frac{d\rho}{dt} + \frac{\partial V}{\partial X} = 0 \tag{12}$$

In the first term, dA/dt is the rate of change in conduit area with time, which is a function of the rate of change in internal pressure and tensile stress in the conduit as well as the rate of radial expansion of the cross section.

From Fig. 4 it can be seen that

$$T = \frac{PD}{2} \tag{13}$$

and that the change of tensile force with time equals

$$\frac{dT}{dt} = \frac{D}{2}\frac{dP}{dt} \tag{14}$$

The conduit wall tensile strain σ can be determined by converting the wall tensile force to stress by dividing by the wall thickness and dividing by Young's modulus of elasticity.

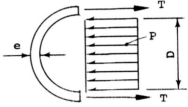

FIGURE 4 Relationship between internal pressure and tensile force.

$$\frac{d\sigma}{dt} = \left(\frac{D}{2eE}\right)\frac{dP}{dt} \tag{15}$$

In order to determine the rate of radial expansion, multiply by $D/2$:

$$\frac{dR_x}{dt} = \left(\frac{D}{2}\right)\left(\frac{D}{2eE}\right)\frac{dP}{dt} \tag{16}$$

Finally, the rate of area change is obtained by multiplying the rate of radial change by the conduit perimeter:

$$\frac{dA}{dt} = (\pi D)\left(\frac{D}{2}\right)\left(\frac{D}{2eE}\right)\frac{dP}{dt} = \left(\frac{\pi D^3}{4eE}\right)\frac{dP}{dt} \tag{17}$$

and

$$\frac{1}{A}\frac{dA}{dt} = \left(\frac{4}{\pi D^2}\right)\left(\frac{\pi D^3}{4eE}\right)\frac{dP}{dt} = \left(\frac{D}{eE}\right)\frac{dP}{dt} \tag{18}$$

The second term of Eq. (12) refers to the rate of change of fluid density. The bulk modulus of a fluid is by definition

$$K = \frac{dP}{(d\forall/\forall)} = \frac{dP}{(d\rho/\rho)} \tag{19}$$

and the second term becomes

$$\frac{1}{\rho}\frac{d\rho}{dt} = \frac{1}{K}\frac{dP}{dt} \tag{20}$$

Substituting Eqs. (18) and (20) into Eq. (12) yields

$$\frac{D}{eE}\frac{dP}{dt} + \frac{1}{K}\frac{dP}{dt} + \frac{\partial V}{\partial t} = 0 \tag{21}$$

Substituting Eq. (8) for wave speed in thin-walled conduits into the second term:

$$\frac{dP}{dt}\left(\frac{D}{eE} + \frac{1}{K}\right) + \frac{\partial V}{\partial t} = 0 \tag{22}$$

$$\frac{1}{\rho}\frac{dP}{dt}\left(1 + \frac{DK}{eE}\right)\frac{\rho}{K} + \frac{\partial V}{\partial t} = 0 \tag{23}$$

$$\frac{1}{\rho}\frac{dP}{dt} + \left[\frac{(K/\rho)}{\left(1 + \frac{DK}{eE}\right)}\right]\frac{\partial V}{\partial t} = 0 \tag{24}$$

$$\frac{1}{\rho}\frac{dP}{dt} + a^2\frac{\partial V}{\partial x} = 0 \tag{25}$$

And since $P = \gamma(H - Z)$,

$$\frac{1}{\rho}\frac{dP}{dt} = \frac{1}{\rho}\frac{\gamma(H-Z)}{dt} = g\left(\frac{\partial H}{\partial t} + \frac{V\partial H}{\partial x} - \frac{V\partial Z}{\partial x}\right) \tag{26}$$

Assuming that the last term in Eq. (26) is small compared with the other terms and can be omitted, Eq. (25) becomes

$$\frac{\partial H}{\partial t} + \frac{V\partial H}{\partial x} + \frac{a^2}{g}\frac{\partial V}{\partial z} = 0 \tag{27}$$

This form of the continuity equation shows V and H as dependent variables and x and t as the independent variables. The expression for the wave speed in thin-walled conduits addresses the elastic properties of the conduit walls and the properties of the fluid.

2. Equation of Motion. The equation of motion, or Newton's second law of motion, assumes that the summation of the forces on the fluid in the control volume equals the mass of the fluid times its acceleration.

From Fig. 5,

$$PA - PA - \frac{\partial(PA)}{\partial x}\delta x + P\frac{\partial A}{\partial x}\delta x + \frac{\partial P}{\partial x}\frac{\partial A}{\partial x}\frac{\delta x^2}{2} - \gamma A\delta x \sin\alpha - \tau_0 \pi D \delta x$$

$$= \rho A \delta x \frac{dV}{dt} \tag{28}$$

Assuming that δx^2 is small and can be omitted, and simplifying, yields

$$\frac{\partial P}{\partial x}A + \tau_0\pi D + \rho g A \sin\alpha + \rho A \frac{dV}{dt} = 0 \tag{29}$$

Assuming that the pipe flow is turbulent and the velocity is steady, the force resulting from the shear stress may be defined in terms of the Darcy-Weisbach expression for friction losses, that is,

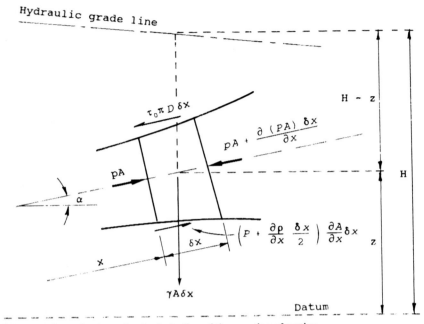

FIGURE 5 Control volume for the derivation of the equation of motion.

$$\tau_0 \pi D = \Delta P_f A + \rho g \Delta H_f A = \rho g \frac{f \delta x}{D} \frac{V^2}{2g} A \qquad (30)$$

Substituting Eq. (3) into (29) and dividing by the area yields

$$\frac{\partial P}{\partial x} + \rho g \frac{f \delta x}{D} \frac{V^2}{2g} + \rho g \sin \alpha + \rho \frac{\partial V}{\partial t} = 0 \qquad (31)$$

Knowing that $P = \rho g (H - Z)$,

$$\frac{\partial P}{\partial x} = \rho g \left(\frac{\partial H}{\partial x} - \frac{\partial Z}{\partial x} \right) = \rho g \left(\frac{\partial H}{\partial x} - \sin \alpha \right) \qquad (32)$$

The acceleration of the control mass can be expressed in terms of the partial derivative of the velocity with respect to t and x:

$$\frac{dV}{dt} = V \frac{\partial V}{\partial x} + \frac{\partial V}{\partial t} \qquad (33)$$

Substituting Eqs. (32) and (33) into (31), the resulting expression for the equation of motion becomes

$$g \frac{\partial H}{\partial x} + V \frac{\partial V}{\partial x} + \frac{\partial V}{\partial t} + \frac{fV^2}{2D} = 0 \qquad (34)$$

As with the equation of continuity, this form of the equation of motion shows V and H as dependent variables and x and t as independent variables.

CHARACTERISTICS METHOD OF ANALYSIS

The partial differential equations of continuity and motion are expressed in terms of two dependent variables of velocity and pressure, and two independent variables of time and position. These two equations can be developed into four ordinary differential equations by the method of characteristics. These four equations may then be converted into finite-difference equations which can be readily solved once pipeline locations and specified time intervals are introduced. The method of characteristics analysis of hydraulic transients can be easily applied to digital computers for numerical solutions.

During most engineering applications, the terms $V \partial H/\partial x$ and $V \partial V/\partial x$ are usually considered negligible in comparison with the remaining terms and can be omitted from the equations of continuity and motion. The resulting simplified forms of Eqs. (27) and (34), designated L_1 and L_2, become

$$L_1 = \frac{\partial H}{\partial t} - \frac{a^2}{g}\frac{\partial V}{\partial x} = 0 \qquad (35)$$

Development of the four ordinary differential equations involves multiplying the equation of motion (36) by an unknown multiplier λ and adding it to the equation of continuity (35):

$$L_2 = g\frac{\partial H}{\partial x} + \frac{\partial V}{\partial t} + \frac{fV^2}{2D} = 0 \qquad (36)$$

$$L = L_1 + \lambda L_2 = \left(\frac{\partial H}{\partial t} + \frac{a^2}{g}\frac{\partial V}{\partial t}\right) + \lambda\left(g\frac{\partial H}{\partial x} + \frac{\partial V}{\partial t} + \frac{fV^2}{2D}\right) = 0 \qquad (37)$$

while the further combining of terms yields

$$L = \lambda\left(\frac{\partial H}{\partial t} + \frac{g}{\lambda}\frac{\partial H}{\partial x}\right) + \left(\frac{\partial V}{\partial x}\frac{\lambda a^2}{g} + \frac{\partial V}{\partial t}\right) + \frac{fV^2}{2D} = 0 \qquad (38)$$

Since both V and H are functions of both t and x, their total derivatives may be expressed as follows:

$$\frac{dH}{dt} = \frac{\partial H}{\partial x}\frac{dx}{dt} + \frac{\partial H}{\partial t} \quad \text{and} \quad \frac{dV}{dt} = \frac{\partial V}{\partial x}\frac{dx}{dt} + \frac{\partial V}{\partial t} \qquad (39)$$

By inspection it can be seen that the first term of Eq. (38) can be set equal to the total derivative of H if the unknown multiplier is selected such that

$$\frac{dx}{dt} = \frac{\lambda a^2}{g} \qquad (40)$$

Likewise, the second term can be set equal to the total derivative of V if

$$\frac{dx}{dt} = \frac{g}{\lambda} \tag{41}$$

The solution of Eqs. (40) and (41) yields

$$\lambda = \pm \frac{a}{g} \tag{42}$$

and Eq. (38) then reduces to the ordinary differential equation

$$\lambda \frac{dH}{dt} + \frac{dV}{dt} + \frac{fV^2}{2D} = 0 \tag{43}$$

By definition, the change in position of the pressure wave with time is the wave speed a, that is,

$$\frac{dx}{dt} = \pm a \tag{44}$$

The sign of the wave speed must be consistent in both Eqs. (42) and (43). In addition, an absolute value of the velocity should be introduced to the friction-loss term of the equations to ensure that losses always occur in the direction of the flow. The resulting four ordinary differential equations can be grouped into pairs of equations identified below as the $C+$ and $C-$ equations:

$C+$ equations

$$+\frac{g}{a}\frac{dH}{dt} + \frac{dV}{dt} + \frac{fV|V|}{2D} = 0 \tag{45}$$

$$\frac{dx}{dt} = +a \tag{46}$$

$C-$ equations

$$-\frac{g}{a}\frac{dH}{dt} + \frac{dV}{dt} + \frac{fV|V|}{2D} = 0 \tag{47}$$

$$\frac{dx}{dt} = -a \tag{48}$$

A plot of the solution of the four equations in the independent variable, or xt, plane as shown in Fig. 6. The slopes of the $C+$ and $C-$ characteristics lines are equal to the positive and negative wave speed values. Equation (45) is valid only along the $C+$ line and Eq. (47) is valid only along its $C-$ line. For this reason, the latter equations are sometimes referred to as compatibility equations.

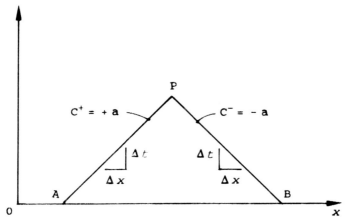

FIGURE 6 Characteristics lines in the xt plane.

FINITE-DIFFERENCE CHARACTERISTICS EQUATIONS

The simultaneous solution of Eqs. (45) and (47) involves multiplying the equations by dt and integrating them along their respective characteristics lines, that is,

$C+$:

$$\int_{H_A}^{H_P} dH + \frac{a}{g} \int_{V_A}^{V_D} dV + \frac{f}{2gD} \int_{x_a}^{x_p} V|V|dx + 0 \tag{49}$$

$C-$:

$$\int_{H_B}^{H_P} dH + \frac{a}{g} \int_{V_B}^{V_D} dV + \frac{f}{2gD} \int_{x_b}^{x_p} V|V|dx + 0 \tag{50}$$

The first two integrals can be solved directly. The integration of the friction-loss term cannot be evaluated since the variations of V and H along the characteristics lines are not yet known. However, it has been shown that a reasonable approach is to assume that the friction loss along each characteristics line is a function of its corresponding velocity at the previous time step. The resulting integrations become

$C+$:

$$H_P - H_A + \frac{a}{g}(V_P - V_A) + \frac{f\Delta x}{2gD} V_A|V_A| = 0 \tag{51}$$

$C-$:

$$H_P - H_B + \frac{a}{g}(V_P - V_B) + \frac{f\Delta x}{2gD} V_B|V_B| = 0 \qquad (52)$$

Modifying the characteristics equations by converting velocities to discharges and solving for H_P yields

$C+$:

$$H_P = H_A - \frac{a}{gA}(Q_P - Q_A) - \left(\frac{f\Delta x}{2gDA^2}\right) Q_A|Q_A| = 0 \qquad (53)$$

$C-$:

$$H_P = H_B + \frac{a}{gA}(Q_P - Q_B) + \left(\frac{f\Delta x}{2gDA^2}\right) Q_B|Q_B| = 0 \qquad (54)$$

By setting $B = a/gA$ and $R = f\Delta x/(2gDA^2)$ the compatibility equations simplify to

$C+$:

$$H_P = H_A - B(Q_P - Q_A) - RQ_A|Q_A| \qquad (55)$$

$C-$:

$$H_P = H_A - B(Q_P - Q_B) - RQ_B|Q_B| \qquad (56)$$

Rearranging the equations to solve for H_P and Q_P at an interior grid point i.

$C+$:

$$H_{Pi} = CP - BQ_{Pi} \qquad (57)$$

$C-$:

$$H_{Pi} = CM + BQ_{Pi} \qquad (58)$$

where

$$CP = H_{i-1} + BQ_{i-1} - RQ_{i-1}|Q_{i-1}| \qquad (59)$$

$$CM = H_{i+1} + BQ_{i+1} - RQ_{i+1}|Q_{i+1}| \qquad (60)$$

A pipeline can be divided into N equal segments with each reach defined as Δx. If a computational time step equal to the reach divided by the wave speed is selected, the characteristic equations (46) and (48) are immediately satisfied, and the simultaneous solution of the compatibility equations to determine the values of H_P and Q_P can proceed. Equations (57) and (58) can be used for the solution of transient conditions at the interior points of a conduit. As Fig. 7 indicates, only one of the characteristic equations holds true at each endpoint of the system. Ad-

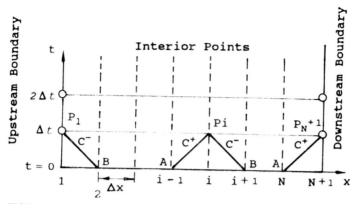

FIGURE 7 Example xt grid for the solution of interior and endpoint characteristics for a single pipeline problem.

ditional equations must therefore be developed to supply information about the behavior of the boundary to the conduit.

DEVELOPMENT OF BOUNDARY CONDITIONS FOR METHOD OF CHARACTERISTICS

A number of boundary conditions are commonly encountered in water-resource and hydroelectric developments. The development of these boundary conditions for application in the method of characteristics analysis of hydraulic transients is included in this section. These boundary conditions include:

1. A reservoir at either the upstream or downstream end of the system
2. A series connection between two conduits of different diameters, roughness, and wave speeds
3. A branch connection between a single connection and any number of additional conduits
4. A conduit dead end
5. An open surge tank
6. A pressurized air chamber
7. A control or pressure-relief valve located at the downstream end of a conduit
8. An "in-line" valve
9. Hydraulic machinery

1. Reservoir. The entrance and exit boundary conditions at the junction of a conduit with a constant-level reservoir are illustrated in Fig. 8. For flow entering the pipeline, the pressure at the boundary is the reservoir elevation minus the

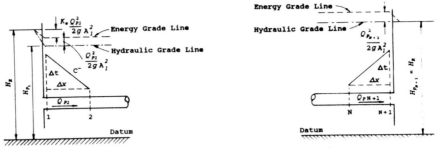

FIGURE 8 Boundary conditions at upstream and downstream reservoirs.

conduit velocity head and the entrance losses. Assuming that the reservoir level is H_R and the entrance loss coefficient is defined as K_e, the equation for the pressure at the upstream boundary becomes

$$H_{P_1} = H_R - (1 + K_e)\frac{Q_{P_1}^2}{2gA_1^2} \tag{61}$$

Similarly, assuming that the exit losses at the downstream boundary are equal to one velocity head, the equation for the pressure at the downstream end becomes

$$H_{P_{N-1}} = H_R \tag{62}$$

Once the pressure H_P has been established at the boundary, the boundary discharge Q_P can be computed directly from the corresponding $C+$ or $C-$ characteristics equations (57) or (58).

Examples of entrance loss coefficients for conventional hydroelectric intake structures are available in the U.S. Army Corps of Engineers "Hydraulic Design Criteria."[5] Minor loss coefficients for intake-outlet structures for pumped-storage projects can vary widely not only from project but also as the levels of the reservoirs fluctuate during the daily pump-generate cycle. Recent studies have been made at the Alden Research Laboratory[6] in determining some of the submergence requirements and minor loss coefficients for selected pumped-storage intake-outlet structures.

2. *Series Connection.* On many occasions water conduits change diameter, roughness, or wave speed. For example, many projects feature transitions from concrete-lined tunnels of one diameter to free-standing steel penstocks of a different diameter. This type of transition, shown in Fig. 9, is referred to as a series connection. At the junction, the $C+$ characteristics equation is available for the upstream pipe and the $C-$ is available for the downstream pipe. Assuming that the internal pressures are common at the ends of both conduits and that continuity of flow must be maintained:

$$H_{P_{1,N+1}} = H_{P_{2,1}} \quad \text{and} \quad Q_{P_{1,N+1}} = Q_{P_{2,1}} \tag{63}$$

The solution for the values of pressure and flow may then be carried out directly using the appropriate $C+$ and $C-$ characteristics equations (57) and (58), or

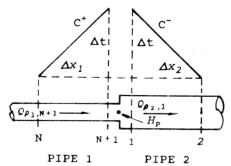

FIGURE 9 Series connection.

$$Q_{P_{2,1}} = \frac{CP_1 - CM_2}{B_1 + B_2} \tag{64}$$

It must be remembered that during the development of the series connection boundary equations the assumption is that the pressure, or hydraulic grade line, at just inside the ends of the two pipelines is common. This assumption is valid in systems where the flow velocities are relatively small compared with the overall head of the project, or when the change in velocity head at the connection is small. In systems such as modern pumped-storage projects, flow velocities as high as 100 fps or more are frequently encountered in penstocks. The corresponding increase in velocity head downstream of the transition from a pressure conduit having velocities as high as even 30 fps would be large enough to affect the results of an analysis based on the common pressure assumption at a connection. If large changes in velocity heads exist at a connection, the development of the boundary equation should follow the assumption that the energy grade line, rather than the hydraulic grade line, is common at the ends of the conduits.

3. *Branch Connection.* The assumptions made during the development of the boundary equations for a branch connection are identical to those of the series connection, i.e., that flow continuity is maintained at the connection and that a common pressure exists at the ends of the conduits. Based on the conduit designations defined in Fig. 10:

$$H_P = H_{P_{1,n+1}} = H_{P_{2,n+1}} = H_{P_{3,1}} = H_{P_{4,1}} \tag{65}$$

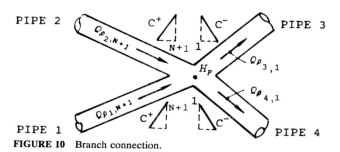

FIGURE 10 Branch connection.

$$Q_{P_{1,n+1}} = -\frac{H_P}{B_1} + \frac{CP_1}{B_1} \tag{66}$$

$$Q_{P_{2,N+1}} = -\frac{H_P}{B_2} + \frac{CP_2}{B_2} \tag{67}$$

$$-Q_{P_{3,1}} = -\frac{H_P}{B_3} + \frac{CM_3}{B_3} \tag{68}$$

$$-Q_{P_{4,1}} = -\frac{H_P}{B_4} + \frac{CM_4}{B_4} \tag{69}$$

$$\sum Q_P = 0 = -H_P \sum \left(\frac{1}{B}\right) + \frac{CP_1}{B_1} + \frac{CP_2}{B_2} + \frac{CM_3}{B_3} + \frac{CM_4}{B_4} \tag{70}$$

$$H_P = (CP_1|B_1 + CP_2|B_2 + CM_3|B_3 + CM_4|B_4)\left[\sum\left(\frac{1}{B}\right)\right] \tag{71}$$

Once the common end pressure has been determined, the appropriate compatibility equations may be used to compute the various conduit discharges.

As in the case with the series connection, attention should be given to the change in the velocity within the branch to determine if the approach assuming the common energy grade line at the connection is more appropriate than the use of a common hydraulic grade line. Additional consideration should be given to significant minor losses which can be encountered in manifolds of hydroelectric power plants. Since the lengths of power-plant manifolds are relatively short compared with the lengths of the power conduits and penstocks, the manifold losses can easily be lumped into the factor for friction losses, if the overall flow distributions within the manifold remain fairly constant throughout the analysis. Losses in manifolds for preliminary studies can be determined from publications by Idel'chik,[7] Miller,[8] and Levin.[9] However, for final design analyses of systems with unsymmetrical, high-velocity manifolds where a wide range of unit operating conditions are anticipated, the manifold losses can be quite large. For such projects, hydraulic model studies to closely determine manifold branch loss coefficients are economically justified.

4. *Dead End.* Dead ends exist in water-development projects in the form of blanked take-offs reserved for the future installation of hydroelectric units or pumps, or branch conduits leading to adjacent pressure relief, low-level outlet, emergency evacuation, or irrigation release valves which are assumed to be permanently closed during the transient analysis. When the dead end is located at the downstream end of a conduit, it is assumed that the flow at the downstream end is zero. Hence the equations for the determination of discharge and pressure at the dead end are

$$Q_{P_{N+1}} = O \quad \text{and} \quad H_{P_{N+1}} = CP \tag{72}$$

5. *Open Surge Tank.* Open surge tanks are vertical shaft storage reservoirs which are generally provided at a point or points in the system for three primary reasons: (1) As described earlier, open reservoirs reflect pressure waves by allowing a release of the excess energy stored in the form of conduit expansion and increased fluid density. An intermediate surge tank located near a power or pumping station will introduce an additional free reservoir surface to the system

which will reflect compensating pressure waves and reduce the magnitude of conduit pressure fluctuations. (2) Although the introduction of a surge tank will limit the amount of conduit which will experience the complete stoppage of flow, the water in the upstream power conduit will continue to flow at its initial velocity. This flowing water will be directed into the surge tank and stored until its kinetic energy is converted into potential energy equal to the risen water level in the tank. Likewise, during a full load acceptance at the downstream units, the tank can quickly release water to the system until the upstream conduit water column is accelerated. (3) The surge tank will improve the operational stability of the system during flow changes associated with load fluctuations by effectively reducing the length of the water column in which immediate flow accelerations or decelerations occur. Surge tank surface areas must be large enough to ensure that water-level fluctuations during load changes are minimal to allow the upstream conduit friction to damp the surge.

Open surge tanks are traditionally located as close to the power plant or pumping station as possible to have the greatest effect on transient suppression and operational stability. Since the top of the surge tank must be set no lower than the level of the upstream reservoir, the location of the surge tank is governed by site topography. Frequently upstream surge tanks are located at a vertical curve to reduce its overall height, or at the transition between a lined or unlined power tunnel and a high-pressure steel penstock. Tailrace tunnel surge tanks are located adjacent to power plants to protect hydromachinery from negative pressures and water-column separation during plant shutdown. Figure 11 shows the arrangement of a hydroelectric project featuring both upstream and downstream open surge tanks.

a. Operational Stability. The stability of a hydroelectric system and its ability to follow load changes is a function of the inertia of the water column and the mechanical inertia of the unit. The mechanical inertia, or flywheel effect, is defined in terms of the start-up time of the unit. The start-up time in seconds T_M required for torque to accelerate the rotating unit mass from zero to the rotational speed is

$$T_m = \frac{WR^2 n^2}{(1.6 \times 10^6)P} \tag{73}$$

where n = rotational speed of the unit, rpm
P = full-gate turbine capacity, hp
WR^2 = product of the revolving parts of the unit and the square of the radius of gyration (turbine runner, shaft, and generator rotor), lb-ft^2

For preliminary design studies in which the unit WR^2 is not known, the values may be estimated from the following U.S. Bureau of Reclamation[10] formulas:

$$\text{Turbine } WR^2 = 23{,}800 \, (P/n^{3/2})^{5/4}$$

$$\text{Generator } WR^2 = 356{,}000 \, (kVA/n^{3/2})^{5/4}$$

where 1 hp = 0.67 kVA

The water-column inertia can be expressed as the time it takes for the head to accelerate the conduit flow from zero to maximum velocity. The water start-up time T_W is defined as

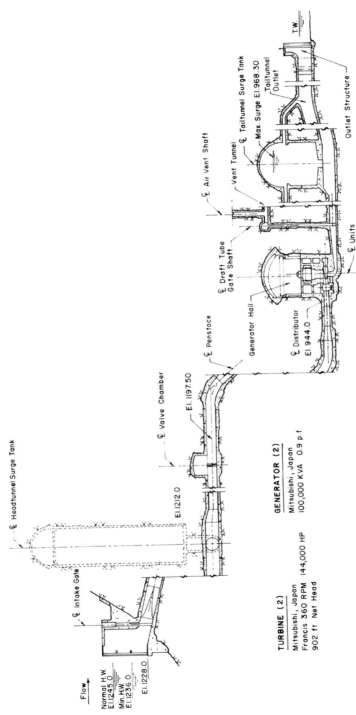

FIGURE 11 Lower Tachien Project.

$$T_W = \frac{\sum LV}{hg} \qquad (74)$$

The sum of the product of the lengths and corresponding maximum velocities is only for conduits connecting the unit forebay and draft tube to the nearest reservoir, open surge tank, or pressurized air chamber. The accelerating head h is defined as the combined regulating heads from the unit to its upstream and downstream free-surface reservoirs after final steady-state flow conditions have been established. The final parameter g is the gravitational constant.

The ratio T_m/T_W is referred to as the stability index of a unit and indicates the regulating characteristics of the unit and the ability of the unit governor to follow load changes. For units with mechanical governors, the stability index for synchronization and start-up should not be less than 2. The recent development of proportional-integral-derivative electronic governors has allowed successful plant operation with indexes as low as 1. This analysis for system hydraulic stability is applicable only when the plant units supply power to a large distribution grid and the total rotational inertia of the plant is less than 25 percent of the total rotational inertia of the electrical system. When plant inertia is more than 45 percent of the inertia of the entire network, the plant is considered to be isolated and complex analyses outside the scope of this presentation are required to determine plant stability.

It should be apparent that the location of an open surge tank has a profound effect on the operational stability of a hydroelectric project, as well as the selection of conduit diameters and unit inertia. However, sufficient surge tank cross-sectional area must be provided to ensure that level fluctuations during small load changes are minimal and damped out by the power-conduit friction. Insufficient area will result in large water-level fluctuations which could significantly overshoot the desired second steady-state tank water level. The resulting change in tank water level may be large enough to accentuate the change in the pressure and speed at the unit and cause the governor to once again readjust the gate position in an effort to maintain unit speed and load demand. The subsequent pressure wave initiated at the unit would travel to the tank and magnify the surge to such an extent that the friction damping effects of the upstream conduit would be insufficient to prevent periodic unsteady-flow conditions from occurring.

System parameters influencing the determination of surge tank stability of a hydroelectric project are shown in Fig. 12. The basic equation for determining the minimum tank surface area for stability during small load changes was developed by Dr. Dieter Thoma.[11] This classic equation is based on the assumption that during small load changes, the effect of riser and entrance port and orifice losses is negligible and they may be entirely removed from the analysis. The resulting tank area F, as defined by the following equation, is that which is just large enough to damp oscillations and return the flow to steady-state in a time just short of eternity. In order to ensure that damping of surges occurs in a reasonable amount of time, it is recommended that the area be increased a minimum of 50 percent for simple surge tanks and 25 percent for surge tanks featuring throttling orifices, ports, and risers.

$$F = \frac{AL}{2gcH_s} \qquad (75)$$

where A = power-conduit cross-sectional area, ft^2
L = power-conduit length, ft

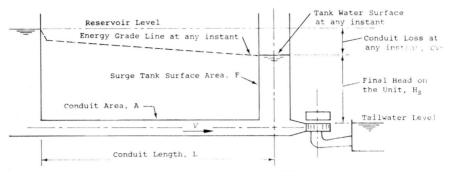

FIGURE 12 Parameters for determination of surge tank stability.

g = gravitational constant, 32.3 ft/s^2
c = conduit loss coefficient equal to the friction head loss divided by the square of the conduit velocity, s^2/ft
H_s = final steady-state head on the unit, taken from the nearest upstream and downstream open reservoirs, water surfaces, or air chambers

The above analysis for surge-tank stability is based on the assumption that the velocity head in the conduit beneath the tank is negligible compared with the total conduit loss. If the velocity head in the conduit beneath the tank is significant, as could be the case when the tank is located over a high-velocity penstock, the head-loss coefficient c must be adjusted to include the effects of the velocity head on the tank water surface, that is,

$$c = H_L/V^2 + (V_p^2/2g)/V^2 \qquad (76)$$

where V_p = velocity in high-velocity penstock beneath surge tank
H_L = friction loss divided by the square of the upstream conduit velocity
V = Velocity in low velocity conduit

b. Pressure Relief and Surge Control. Open surge tanks are designed to provide sufficient cross-sectional area and height to adequately contain upsurges and prevent overtopping. On rare occasions, site conditions allow the use of an overflow spillway to reduce the tank height. The bottom of the surge tank should always be low enough to prevent complete drainage and the introduction of air into the conduit during downsurges. Additional downsurge freeboard or flow-straightening vanes should be considered if exit velocities are high enough to induce air-entraining vortices. Some surge tank designs incorporated multi-sectioned geometries featuring additional upper and lower storage chambers for surge control connected by a smaller-diameter vertical shaft sized for operational stability. Entrance geometries can vary from the simple tank with a direct connection between the tank bottom and conduit, to that featuring a flow-throttling orifice. Differential surge tanks feature combinations of entrance orifice, risers, and port systems to throttle the flow entering and leaving the tank.

Entrance orifice openings can be designed as a simple orifice plate with similar entrance and exit loss coefficients, to flow nozzles with different loss coefficients for entrance and exit flow. Examples of exit and entrance loss coefficients for tank orifices can be found in the Corps of Engineers "Hydraulics Design Criteria,"[5] and in references by Parmakian,[1] Idel'chik,[7] Miller, Streeter,[12]

Rouse,[13] Pickford,[14] and Li.[15] A classic example of the riser and port throttling orifice design for the Appalatia differential surge tank is described by Rich.[11] The development of surge-tank designs featuring expanded upper and lower storage chambers is discussed in great detail by Jaeger.[16]

Since transient effects in penstocks usually have a much shorter period than that of the tank water-level fluctuations, the surge oscillation analysis can often be performed independently. In many instances where conduit lengths between the tank and downstream unit are relatively short, preliminary surge-tank requirements can be easily estimated assuming that the flow change at the unit is instantaneous. System parameters for the instantaneous stopping of flow with a throttled surge tank are shown in Fig. 13. Excellent design nomographs have been developed by R. S. Jacobson[17] to show the relationship between maximum change in water level in a tank having a known surface area and entrance throttling loss coefficient and the upstream conduit friction-energy loss for the instantaneous starting and stopping of flow in the downstream conduit. The nomographs for instantaneous starting and stopping of flow are shown in Figs. 14 and 15, respectively. The term "balanced design" refers to the surge-tank design in which the conduit pressure beneath the tank entrance resulting from the maximum water change is equal to that required to overcome the initial entrance throttling loss resulting from flow being diverted into or out of the tank. Surge-tank design analyses are performed assuming maximum upstream conduit friction and minimum headwater level for downsurge studies and minimum friction and maximum headwater for upsurge studies.

The development of the boundary equations for the throttled surge tank shown in Fig. 16 for the method of characteristics analysis of system transients, is based on the assumptions that the pressure at the junction of the upstream and downstream pipelines located beneath the entrance to the tank is common, and that there is continuity of flow at the junction:

$$H_{P_{1,N+1}} = H_{P_{2,1}} = H_{P_T} \tag{77}$$

$$\sum Q = Q_{P_{1,N+1}} + Q_{P_{2,1}} + Q_{P_T} = 0 \tag{78}$$

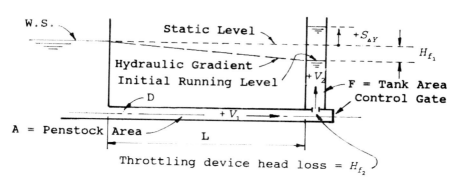

(H_{f_1}) max. $= C_1 (V_1$ max.$)^2$
(H_{f_2}) max. $= C_2 (V_2$ max.$)$
$Q_0 =$ intial flow

FIGURE 13 Surge-tank analysis for instantaneous stopping of downstream flow.

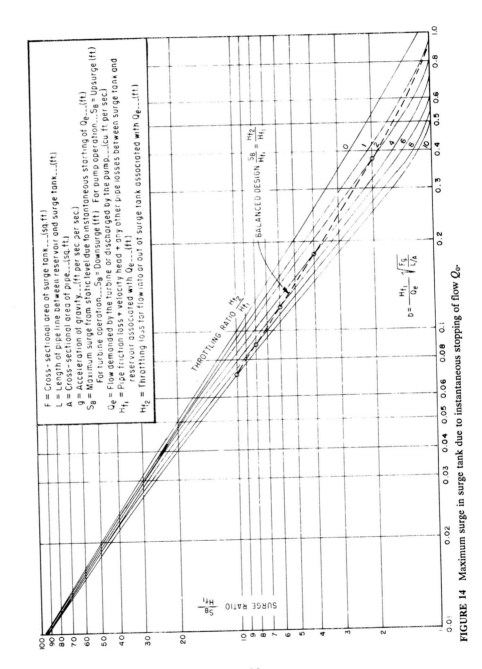

FIGURE 14 Maximum surge in surge tank due to instantaneous stopping of flow Q_0.

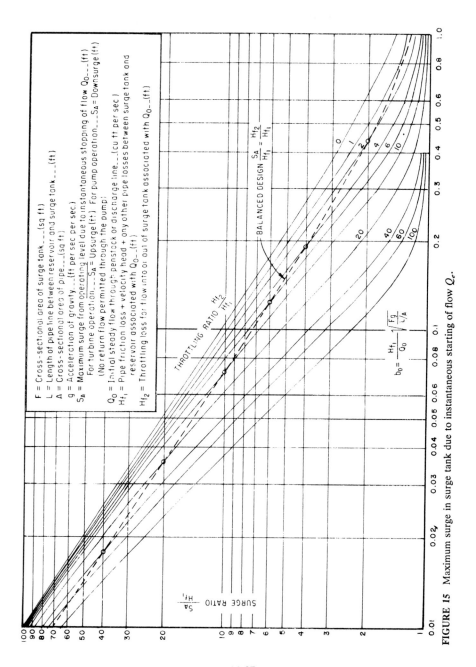

FIGURE 15 Maximum surge in surge tank due to instantaneous starting of flow Q_e.

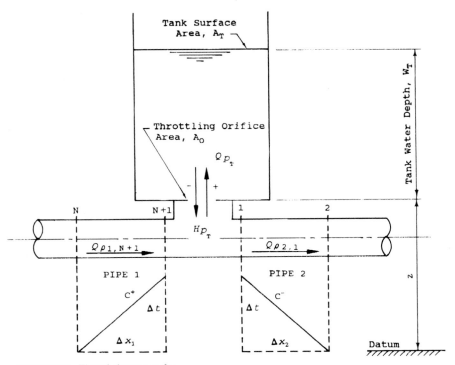

FIGURE 16 Throttled surge tanks.

The amount of flow entering or leaving the tank is a function of the pressure differential across the orifice, and the orifice area and discharge coefficient. When the pressure at the junction exceeds the water level in the tank, the discharge equation becomes

$$Q_{P_T} = C_d A_0 \sqrt{2g(H_{P_T} - W_T - Z)} \tag{79}$$

When the water level in the tank is greater than the conduit pressure, the release discharge is

$$Q_{P_T} = C_d A_0 \sqrt{2g(W_T + Z - H_{P_T})} \tag{80}$$

The discharge in water level in the tank with time is equal to the discharge entering or leaving the tank divided by the cross-sectional area of the tank:

$$\frac{dWT}{dt} = \frac{Q_{P_T}}{A_T} \tag{81}$$

In order to simplify the analysis, the discharge from the previous time increment is often used to compute the water-level change for the computation of the tank inflow or outflow for the current time step. Equations (77), (78), (81), the compatibility equations (57) and (58), and either (79) or (80) are used to solve for the variables H_{P_T}, Q_{P_T}, $Q_{P_{1,N+1}}$, Q_{P_1}, and W_T. For more complicated surge tanks with long connection risers having appreciable friction losses, surge tanks with small or varying cross-sectional areas, or overflow surge tanks, additional equations must be developed to establish orifice differential-discharge relationships and rate of water-surface-level change with time.

6. *Pressurized Air Chamber.* Pressurized surge tanks, or air chambers, can also be used to provide additional operational stability and surge protection. Air chambers are commonly used to protect pumps connected to long transmission pipelines from the vapor pressures and subsequent excessive pressure rise from the rejoining of the water column which can occur after power failure. Underground air chambers have also been used successfully as a cost-effective surge-protection alternative for high-head hydroelectric projects in Norway. Because of severe environmental and topographic constraints, Harza utilized an underground air chamber over a conventional surge tank for operational stability and surge control on the 12.2-MW Moose River project[19] in upstate New York (Fig. 18).

As with the open surge tank, a minimum water surface area is required to dampen system instabilities during small load changes in a time just short of infinity. The incipient stability criteria for pressurized air chambers developed by Svees[20] can be used to compute the required surface area F_a:

$$F_a = F\left[1 + n\left(\frac{H_0}{\gamma A_0}\right)\right] \qquad (82)$$

where n = polytropic exponent varying from 1 for isothermal and 1.4 for adiabatic thermodynamic processes
F = critical area for incipient stability in an open surge tank
H_0 = absolute pressure in the air chamber
γ = specific weight of water
A_0 = vertical distance from the air chamber roof to the water surface in a vertical-walled chamber

Figure 17 shows a section through a pressurized air chamber with a throttled entrance. The assumptions of continuity of flow, common pressure beneath the chamber entrance, losses through the throttling orifice, and the use of the previous time step discharge into the chamber to compute the change in chamber water level are the same as those used to develop the boundary equations for the open surge tank. The relationship between the water level and the pressure within the chamber, however, is governed by the gas law:

$$H_a \forall_C^n = C \qquad (83)$$

where H_a = absolute pressure of the air, ft of water
$\forall_C^n$ = volume of the air in the chamber, ft^3
n = polytropic exponent
C = gas constant established during the initial filling

At steady-state conditions, the relationship between the chamber pressure and the pressure in the conduit is

$$H_{P_C} = H_a + Z + W_C \qquad (84)$$

and assuming that the change in water level with time is

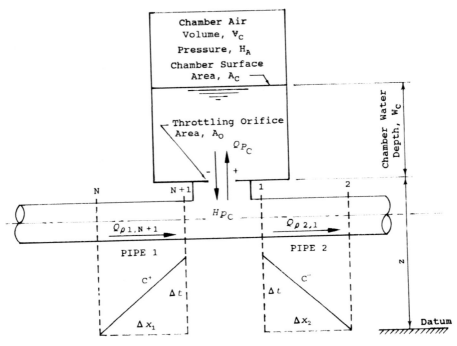

FIGURE 17 Throttled air chamber.

$$\frac{dW_c}{dt} = \frac{Q_{P_C}}{A_c} \quad (85)$$

If the conduit pressure exceeds the chamber pressure, the flow entering the chamber is

$$Q_{P_C} = C_d A_0 \sqrt{2g(H_{P_C} - H_a - Z - W_C)} \quad (86)$$

and the discharge equation for flow leaving the chamber becomes

$$Q_{P_C} = C_d A_0 \sqrt{2g(H_a + Z + W_C - H_{P_C})} \quad (87)$$

The gas law for flow entering the chamber becomes

$$[(H_{P_C} - Z - W_C - (Q_{P_C}|C_d A_0)^2/2g](\forall_C - Q_{P_C})^n = C \quad (88)$$

and for flow leaving the chamber:

$$[H_{P_C} - Z - W_C + (Q_{P_C}|C_d A_0)^2/2g](\forall_C + Q_{P_C})^n = C \quad (89)$$

The simultaneous solution of the equations of low continuity and common pressures at the junction, the appropriate throttling discharge and gas law equation, the compatibility equations (57) and (58), and Eq. (81) will yield the values for the variables $H_{P_C}, Q_{P_C}, Q_{P_{1,N+1}}, Q_{P_1}, W_C, H_a$, and $\forall_C$.

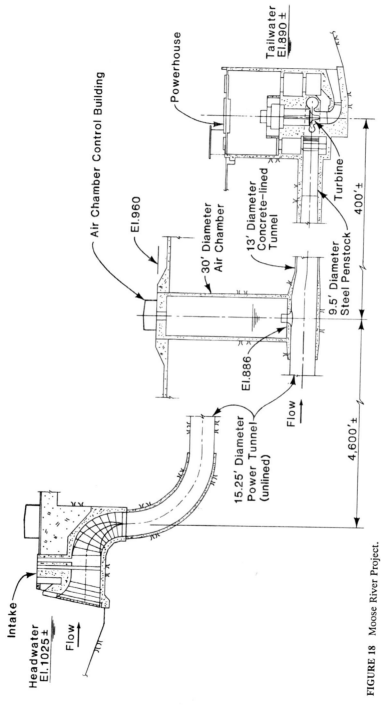

FIGURE 18 Moose River Project.

7. Downstream Valve. Water-supply projects frequently regulate discharges with downstream flow dispersion valves. Hydroelectric projects also use free discharge pressure-relief valves at powerhouses to protect penstocks and hydromachinery from pressure rise during unit shutdown after load rejection. Downstream valves for flow regulation and pressure relief include hollow-jet valves, hollow-cone valves, needle valves, jet-flow gates, and multiport sleeve valves. The relationship for steady-state flow through the fully opened downstream valve shown in Fig. 19 is based on the orifice equation

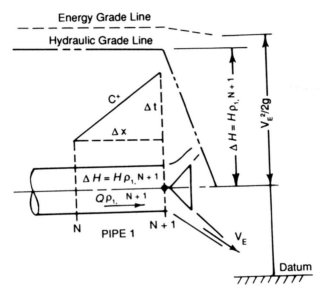

FIGURE 19 Downstream valve.

$$Q_{V0} = C_{V0} A_v \sqrt{2gH_{V0}} \tag{90}$$

where Q_{V0} is the discharge, C_{V0} is the discharge coefficient, A_v is the valve area, and H_{V0} is the drop in hydraulic grade line across the valve. For any other valve opening, the equation for discharge through the valve becomes

$$Q_P = C_V A_V \sqrt{2g\,\Delta H} \tag{91}$$

Setting τ equal to a dimensionless valve opening variable:

$$\tau = \frac{C_V A_V}{C_{V0} A_V} = \frac{C_V}{C_{V0}} \tag{92}$$

Assuming $H_{P_{1,N+1}} = \Delta H$, the boundary equation for a valve at the downstream end of a conduit becomes

$$Q_P = \tau Q_{V0} \sqrt{\Delta H/H_{V0}} = \tau Q_{V0} \sqrt{H_{P_{1,N+1}}/H_{V0}} \tag{93}$$

and the valve discharge can be determined from the simultaneous solution of the CM compatibility equation (57) and Eq. (58), or

$$Q_{P_{1,N+1}} = -B_1 C1 + \sqrt{(B_1 C1)^2 + 2C1 CP} \tag{94}$$

where

$$C1 = (Q_{V0}\tau)^2/2H_{V0} \tag{95}$$

Discharge coefficients for various dispersion valves can be readily obtained from valve manufacturers. However, excellent articles are available which define not only the discharge coefficients but maximum allowable pressure differentials, cavitation indexes, and required closure torques. Watson[21] describes the development and performance of multiport sleeve valves. The description of dispersion valves and limits of application is covered in a paper by Lewin.[22] Discharge coefficients for hollow-jet valves are available in the Corps of Engineers "Hydraulics Design Criteria."[5] The characteristics of cone valves are also described in detail in a paper by Elder and Dougherty.[23] The performance of jet-flow gates is documented in a Bureau of Reclamation Report HYD-569.[24] Discharge characteristics for needle valves are also available from the Bureau of Reclamation.[25]

8. *In-Line Valve.* In-line valves are commonly used as emergency shutoff of flow rather than for flow regulation or downstream flow dispersion and energy dissipation. Guard valves of hydroelectric units fall into this category. Valves designed for this application include butterfly valves and gate valves for low- and medium-head projects and ball valves and conical plug valves for developments operating under medium to high heads.

The development of the boundary equations for the in-line valve is also based on the orifice equation and the assumption that flow continuity is maintained at the valve. The application is complicated, however, by the possibility of pressure differential and flow reversal. For the condition of positive flow through the valve shown in Fig. 20 the orifice equation for discharge is

$$Q_{P_{1,N+1}} = Q_{P_{2,1}} = \left(\frac{Q_{V0}\tau}{\sqrt{H_{V0}}}\right)\sqrt{H_{P_{1,N+1}} - H_{P_{2,1}}} \tag{96}$$

And when combined with the compatibility equations (57) and (58) yields the boundary equation

$$Q_{P_{1,N+1}} = -C1(B_1 + B_2) + \sqrt{C1^2(B_1 + B_2)^2 + 2C1(CP_1 - CM_2)} \tag{97}$$

whereas with the downstream valve,

$$C1 = (Q_{V0}\tau)^2/2H_{V0} \tag{98}$$

For reverse flow through the valve the orifice equation becomes

$$Q_{P_{1,N+1}} = Q_{P_{2,1}} = -\left(\frac{Q_{V0}\tau}{\sqrt{H_{V0}}}\right)\sqrt{H_{P_{2,1}} - H_{P_{1,N+1}}} \tag{99}$$

and the boundary equation becomes

$$Q_{P_{1,N+1}} = -C1(B_1 + B_2) - \sqrt{C1^2(B_1 + B_2)^2 - 2C1(CP_1 - CM_2)} \tag{100}$$

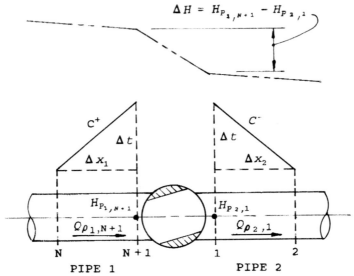

FIGURE 20 In-line valves.

The direction of the flow can be quickly determined by noting the relationship between *CP* and *CM* of the compatibility equations. If $CM > CP$, the flow is in the negative direction as defined by Fig. 19 and Eq. (97) should be used to solve for Q_P. If $CP \geq CM$, the flow is in the positive direction and the discharge can be determined by Eq. (100). The compatibility equations (57) and (58) are then used to solve for Q_P and H_P.

The discharge characteristics for shutoff valves can also be obtained from manufacturers. Characteristics and discussions on the uses of various shutoff valves are given in articles and papers by Strohmer,[26] Barp and Grein,[27] Guins,[28] and the Corps of Engineers.

9. *Hydraulic Machinery.* A hydraulic machine is a mechanical device which is used to transfer energy to or from a fluid medium. Hydraulic machinery commonly encountered in closed-conduit water-resources projects includes pumps and hydraulic turbines and pump-turbines. In pumping installations, electrical energy is converted by the motor into torque and conveyed to the fluid by the rotating pump impeller. Conversely, impulse, Francis, and propeller turbine runners convert the potential and kinetic energy of the fluid in the conduit to torque on a rotating shaft and ultimately the production of electrical energy at the generator.

Hydraulic transients are initiated within the closed-conduit system when flow conditions are altered by hydraulic machinery. Pressure changes occur in water supply or irrigation systems when the pump speed or discharge valve position is adjusted to follow the change in demand. Under normal operating conditions the changes in water demand are usually gradual and scheduled, and the corresponding pressure transients are small and well within design limits. However, since the system is dependent on an electrical power supply, it is vulnerable to abrupt changes in operation from unplanned power failure or operator error. During loss of power, referred to as pump trip-out, the reactive torque of the fluid initially reduces the rotational speed of the impeller and a drop in the pressure head in the pump discharge line results. The resulting negative pressure wave is transmitted down the conduit at the sonic wave speed a. In high-head pumping systems, a flow reversal through the pump will generally occur within a few seconds and cause a reversal in the rotation of the pump. As the pump rotational speed of the pump increases, the losses through the pump also increase, causing a reduction of the reverse flow. Eventually, the system will stabilize in a reverse steady-state discharge condition with the pump rotating at "runaway" speed in which zero torque is produced by the rotating shaft, impeller, and motor. The development of undesirable reverse steady-state flow conditions can be prevented with emergency check valves on the pump discharge line which automatically close when the flow reverses, or with vacuum relief valves to supply air at high points in the downstream conduit vulnerable to negative pressures and the formation of vapor cavities which could lead to conduit collapse.

In conventional hydroelectric projects with Francis or Kaplan turbines, governor-controlled wicket gates are used to change the turbine discharge and maintain the rotational speed of the unit during changes in power demand. During these controlled load changes, the rotational inertia of the power grid adds significantly to the hydraulic stability of the unit and helps maintain its rotational speed. Constant unit speed during load change greatly reduces the magnitude of the resulting system transients. During a load rejection, the unit becomes temporarily disconnected from the power grid and the resulting decrease in rotational inertia to that of the machine alone causes an increase in rotational speed. During normal shutdown, the wicket gates close to prevent a prolonged overspeed condition which can result in damage to the unit. The magnitude of the resulting pressure rise, discharge, and speed rise of the unit is dependent on the rate at which the wicket gates are closed.

Failure of the wicket or emergency shutdown gates to close will cause the system eventually to arrive at a steady-state flow condition in which the unit will rotate at a runaway speed with zero torque. The runaway flow condition for a Francis turbine will have less flow through the unit and more head than the corresponding flow condition prior to rejection. The opposite is true for a Kaplan turbine subsequent to load rejection.

Emergency shutoff valves are generally provided at high-head hydroelectric projects to provide additional system safety and allow routine maintenance of downstream turbine equipment. Emergency valve closure is predicated by the failure of the wicket gates to close. The resulting pressure and unit speed rise and change in unit discharge is again directly related to the closure rate of the valve.

a. Performance Characteristics. In order to develop boundary conditions for the method of characteristics analysis of systems with hydraulic machinery, the operational characteristics of the machine must be defined. The performance characteristics include the total head, the discharge, the rotational speed, and the shaft torque. The performance characteristics are generally determined from physical hydraulic model studies performed by the manufacturer on a geometrically similar design under steady-state flow conditions. The results of the model studies are converted using the relationships for homologous units to demonstrate the shaft torque and discharge at a given rotational speed for a geometrically similar machine having a unit throat diameter operating at a unit head. Figures 20 and 21 show the resulting torque and discharge, respectively, at various wicket-gate opening positions, for different rotational speeds of a model pump-turbine having a 1 ft throat diameter operating at a head of 1 ft.

The model performance characteristics for turbomachinery cover four zones,

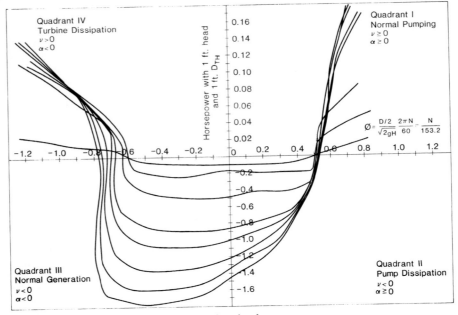

FIGURE 21 Model performance characteristics—head.

or quadrants, of operation which the unit may encounter. The quadrants shown in Figs. 21 and 22 are defined in terms of rotational speed and flow directions as follows:

Quadrant I—normal pumping zone
 A Flow is in the pumping direction ($Q > 0$).
 B Shaft rotation is in the pumping direction ($n > 0$).

Quadrant II—pump energy dissipation zone
 A Flow is in the turbine direction ($Q < 0$).
 B Shaft rotation is in the pumping direction ($n > 0$).

Quadrant III—normal generating zone
 A Flow is in the turbine direction ($Q < 0$).
 B Shaft rotation is in the generating direction ($n < 0$).

Quadrant IV—turbine energy-dissipation zone
 A Flow is in the pumping direction ($Q > 0$).
 B Shaft rotation is in the generating direction ($n < 0$).

The performance characteristics curves for a conventional pump would be similar to the pump-turbine curves for full wicket-gate opening. Performance curves for variable-blade-position Kaplan propeller units would feature a family of curves similar to those in Figs. 21 and 22 for each blade position. However, since wicket-gate operation is generally quicker than blade operation, only curves for full blade opening are generally required for controlled-closure transient studies. Likewise, only the blade-opening characteristics with full wicket-gate opening are required to study emergency closure of the blades during wicket-gate fail-

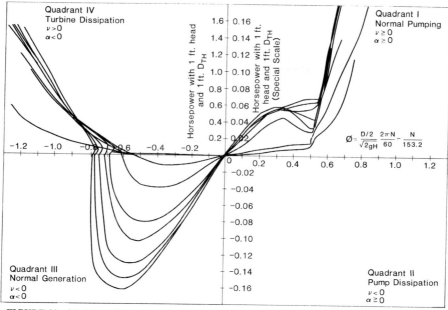

FIGURE 22 Model performances characteristics—torque.

ure cases. Since flow reversal is impossible through conventional hydroelectric turbines, only the third quadrant of the performance characteristics is required for transient studies.

b. Specific Speed. Turbomachines are considered to be geometrically similar, or homologous, if their specific speeds are equal. The specific speed of a pump is defined differently from that of a turbine. The specific speed of a series of homologous pumps is defined as the speed of a pump of a size required to pump a unit discharge at a unit head at the performance point of greatest efficiency. Since pump design criteria generally predominate during the selection of a pump-turbine, the pump specific speed is used to define the homologous series of these reversible-flow machines. The relationship for computing the specific speed of a pump N_{SP} is

$$N_{SP} = \frac{n\sqrt{Q}}{H^{3/4}} \tag{101}$$

where N_{SP} = specific speed, ft-gal/min (m-m³/s)
  n = rotational speed, rpm
  Q = pump discharge, gpm (m³/s)
  H = unit head of 1 ft (1 m)

The specific speed of a series of homologous turbine designs is defined as the speed of a turbine of a certain size required to produce a unit of power at a unit head at the best efficiency point. The formula for determining the specific speed of a turbine N_{St} is defined as

$$N_{St} = \frac{n\sqrt{P}}{H^{5/4}} \qquad (102)$$

where N_{St} = specific speed, ft-hp (m-hp)
n = rotational speed, rpm
P = power output, hp (hp)
H = unit head of 1 ft (1 m)

c. Normalization of Performance Characteristics. The relationships for determining the performance characteristics of one unit from those of a homologous unit of a different size and speed are

$$\frac{Q_2}{Q_1} = \frac{n_2 D_2^3}{n_1 D_1^3} \qquad \frac{H_2}{H_1} = \frac{n_2^2 D_2^2}{n_1^2 D_1^2} \qquad \frac{P_2}{P_1} = \frac{T_2}{T_1} = \frac{n_2^3 D_2^5}{n_1^3 D_1^5} \qquad (103)$$

By assuming that the equations hold true for any given machine, i.e., $D_1 = D_2$, the homologous equations may be used to establish ratios of operational characteristics such that

$$\frac{Q_1}{n_1} = \frac{Q_2}{n_2} \qquad \frac{H_1}{n_1^2} = \frac{H_2}{n_2^2} \qquad (104)$$

and

$$\frac{T_1 n_1}{Q_1 H_1} = \frac{T_2 n_2}{Q_2 H_2} \qquad (105)$$

which can be combined to define unit operation characteristics in terms of ratios of torque to speed squared, head to discharge squared, and torque to discharge squared, or

$$\frac{T_1}{n_1^2} = \frac{T_2}{n_2^2} \qquad \frac{H_1}{Q_1^2} = \frac{H_2}{Q_2^2} \qquad \frac{T_1}{Q_1^2} = \frac{T_2}{Q_2^2} \qquad (106)$$

Unit performance characteristics can be normalized to allow easy application to studies of numerous systems having similar units by dividing torque, discharge, head, and rotational speed by corresponding reference values. Reference values are generally selected as those related to the performance of the machine at best, or maximum, efficiency. The resulting dimensionless characteristics become

$$h = \frac{H}{H_R} \qquad v = \frac{Q}{Q_R} \qquad \alpha = \frac{n}{n_R} \qquad B = \frac{T}{T_R} \qquad (107)$$

where the subscripted denominations represent the reference values at best efficiency. Equations (107) and the performance characteristics from the model study curves can now be established in dimensionless form as follows:

$$\frac{h}{\alpha^2} \text{ vs. } \frac{v}{\alpha} \qquad \frac{\beta}{\alpha^2} \text{ vs. } \frac{v}{\alpha} \qquad \frac{h}{v^2} \text{ vs. } \frac{\alpha}{v} \qquad \frac{\beta}{v^2} \text{ vs. } \frac{\alpha}{v}$$

DEVELOPMENT OF BOUNDARY CONDITIONS

The inherent weakness of arranging the dimensionless relationships in this fashion is that zero values of α and v may be encountered during the transient analysis, prompting division by zero errors. DeFazio[29] solved this problem by developing a series of continuous performance curves for head and torque similar to those shown in Figs. 23 and 24, respectively, to prevent the division by zero values of α or v during the analysis. A method to prevent division by zero can thereby be developed by rearranging the dimensionless performance relationships as follows:

$$\frac{h}{\alpha^2 + v^2} \text{ vs. } \tan^{-1}\frac{v}{\alpha} \qquad \frac{\beta}{\alpha^2 + v^2} \text{ vs. } \tan^{-1}\frac{v}{\alpha}$$

Division by zero can occur only when v and α values are both zero at the same time, a condition which can occur only with the wicket gates closed. Figures 25 and 26 show the application of this unique approach to the normalization of the full and 10 percent gate performance characteristics similar to those shown in Figs. 22 and 23. The values of $h/(\alpha^2 + v^2)$ are defined as $FH(x)$ where x is the value $\pi + \tan^{-1}(v/\alpha)$. Similarly the values of $\beta/(\alpha^2 + v^2)$ are defined as $FT(x)$. The dimensionless homologous performance characteristics curves are presented in a continuous manner, which greatly simplifies the solution procedures.

d. Efficiency Step-up Relationships. The development of the homologous equations is based on the assumption that the losses and efficiencies of the homologous units are equal. Smaller units, however, have lower Reynolds numbers which result in higher friction factors and losses. Thus an adjustment in the losses and efficiencies must be computed when converting performance characteristics from known values of one unit to another. Procedures for adjustment of losses have been established by the International Electrotechnical Commission Publica-

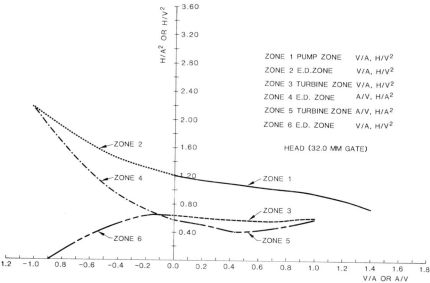

FIGURE 23 Dimensionless head curve.

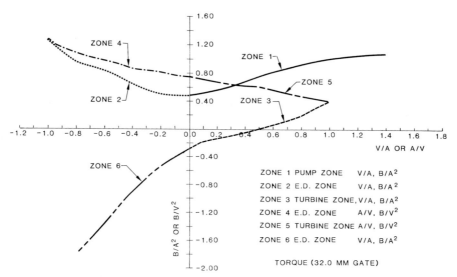

FIGURE 24 Dimensionless torque curve.

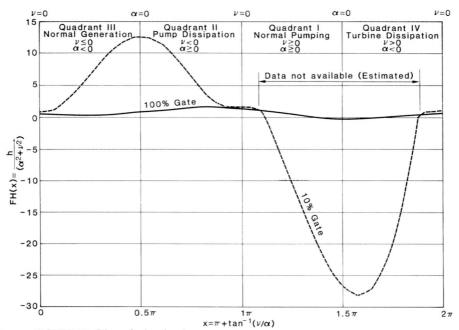

FIGURE 25 Dimensionless head curve.

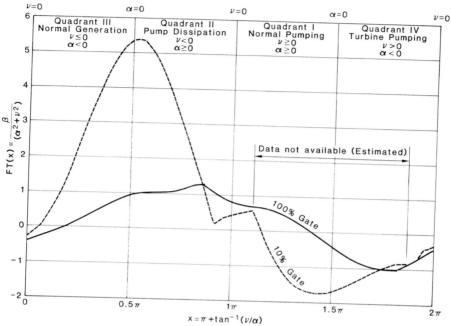

FIGURE 26 Dimensionless torque curve.

tion 193.[30] The formulas for correction for scale effects for Kaplan and propeller turbines are based on the Hutton formula:

$$\frac{\text{Prototype loss}}{\text{Model loss}} = 0.3 + 0.7\left(\frac{R_{e_m}}{R_{e_P}}\right)^{1/5} \quad (108)$$

where R_e = turbine Reynolds number = $\dfrac{D}{v}\sqrt{2gH_n}$
 D = turbine runner diameter, ft (m)
 v = kinematic viscosity of fluid, ft²/s (m²/s)
 H_n = head, ft (m)

The correction formula for losses in reaction turbines and pump-turbines is based on the Moody formula:

$$\frac{\text{Prototype loss}}{\text{Model loss}} = \left(\frac{D_m}{D_P}\right)^{1/5} \quad (109)$$

where D_m = model turbine runner diameter, ft (m)
 D_P = prototype turbine runner diameter, ft (m)

The use of model data of performance characteristics for transient studies of the prototype system is based on the assumptions that homologous relationships between model and prototype are valid and that steady-state model results can be used to simulate unsteady transients behavior of the unit. Transient conditions initiated by turbomachine operation require the development of two additional equations which, when solved simultaneously at each time step, will provide the necessary head and discharge values for the solution of the adjacent upstream and downstream characteristics equations. These equations used the dimensionless-homologous performance characteristics to solve for the head across the unit and the change in the unit speed. The equations are the head-balance equation and the speed-change equation.

e. Turbine Head-Balance Equation. Consider the steady-state generating operation of the underground pump-turbine shown in Fig. 27. Assuming negligible losses through the spherical guard valve and continuity of flow through the unit, the head across the unit as defined in terms of the dimensionless performance curves is

$$H_{unit} = hH_R = H_R(\alpha^2 + v^2) FH(x) \tag{110}$$

where

$$x = \pi + \tan^{-1}(v/\alpha) \tag{111}$$

The hydraulic grade line at the ends of the upstream and downstream power tunnels adjacent to the unit can be defined in terms of the positive and negative characteristics equations where

$$H_{P_U} = H_{P_{1,N+1}} = CP - B_1 Q_{P_{1,N+}} \tag{112}$$

$$H_{P_D} = H_{P_{2,1}} = CM + B_2 Q_{2,1} \tag{113}$$

The head across the unit then becomes

$$H_{unit} = H_{P_{1,N+1}} - H_{P_{2,1}} = CP - B_1 Q_{P_{1,N+1}} - CM - B_2 Q_{P_{2,1}} \tag{114}$$

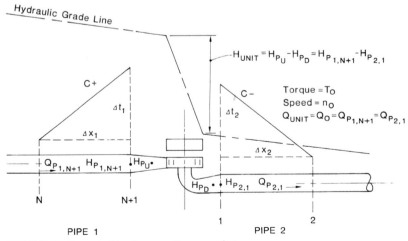

FIGURE 27 Pump-turbine in generation operation.

DEVELOPMENT OF BOUNDARY CONDITIONS

Assuming continuity of flow and setting the discharge equal to

$$Q_{P_{1,N+1}} = Q_{P_{2,1}} = vQ_R \tag{115}$$

The head-balance equation becomes

$$CP - CM - (B_1 + B_2)vQ_R = H_R(\alpha^2 + v^2)FH\left(\pi + \tan^{-1}\left(\frac{v}{\alpha}\right)\right) \tag{116}$$

The solution for $FH[\pi + \tan^{-1}(v/\alpha)]$ from the dimensionless-homologous curve can be simplified greatly by first dividing the curve into a series of I linear segments and solving interpolation once the general location of the solution is established from the solution of the previous time step. This simplified method of solution may introduce errors and instabilities into the analysis, particularly when the computational time step is relatively large, the number of curve segments is small, or the curve contains abrupt discontinuities. Problems associated with this simplified approach can be prevented by merely providing a large number of segments or providing additional segments in areas of local discontinuities.

The solution consists of converting the curved portion shown in Fig. 28 into an

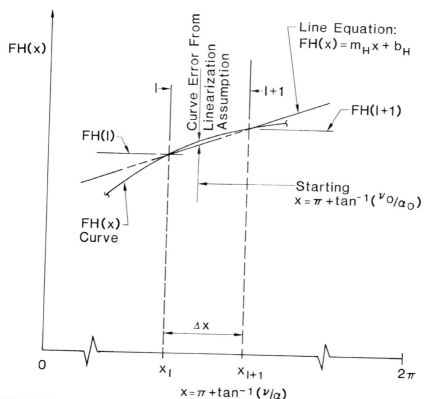

FIGURE 28 Local linearization of dimensionless head curve.

approximate equivalent linear equation using the values of FH at locations X_I and X_{I+1}:

$$FH(x) = M_h X + b_h \tag{117}$$

where

$$m_h = [FH(I+1) - FH(I)]/\Delta x \quad \text{and} \quad b_n = FH(I) \tag{118}$$

Substituting the linear equation into Eq. (116) and simplifying yields

$$HBE = DC - BUv + H_R(\alpha^2 + v^2)\left\{b_h + m_h\left[\pi + \tan^{-1}\left(\frac{v}{\alpha}\right)\right]\right\} \tag{119}$$

where $DC = CP - CM$
$BU = (B_1 + B_2)Q_R$

which is the head-balance equation in terms of v and α.

f. Turbine Speed Equation. Consider that a full-load rejection occurs on the pump-turbine shown in Fig. 27 and the wicket gates fail to close. The rotational speed of the unit will change from the unbalanced torque acting on it. The relationship between the unbalanced torque and the change in the angular velocity with time is

$$T_u = -\frac{WR^2}{g}\frac{dw}{dt} \tag{120}$$

The unbalanced torque is assumed to be the average of the torque at the beginning of the time step T_0 and that at the end of the time step T. Assuming a finite time step, the relationship then becomes

$$\frac{T + T_0}{2} = -\frac{WR^2}{g}\frac{dw}{\Delta t} = -\frac{WR^2}{g}\frac{\alpha_0 - \alpha}{\Delta t} \tag{121}$$

Knowing that

$$w = N_R \frac{2\pi}{60}\alpha \qquad \beta_0 = \frac{T_0}{T_R} \qquad \beta = \frac{T}{T_R} \tag{122}$$

The speed-change equation becomes

$$\frac{\beta_0 + \beta}{2} = -\frac{WR^2}{g}\frac{N_R}{T_R}\frac{\pi}{30}\frac{(\alpha_0 - \alpha)}{\Delta t} \tag{123}$$

or

$$\beta_o = -\frac{WR^2}{g}\frac{N_R}{T_R}\frac{\pi}{15}\frac{(\alpha_0 - \alpha)}{\Delta t} - \beta \tag{124}$$

By introducing an inertia constant:

$$AKL = +\frac{WR^2}{g}\frac{N_R}{T_R}\frac{\pi}{30}\frac{1}{\Delta t} \tag{125}$$

The speed-change equation is further reduced to

$$\beta + \beta_0 - AKL(\alpha_0 - \alpha) = 0 \tag{126}$$

By applying the simplified solution of linearizing the dimensionless torque curve in a manner similar to that used for the dimensionless head curve, the torque can be determined locally by the linear equation

$$\frac{\beta}{\alpha^2 + v^2} = FT(x) = b_t + m_t\left[\pi + \tan^{-1}\left(\frac{v}{\alpha}\right)\right] \qquad (127)$$

or

$$\beta = (\alpha^2 + v^2)\left\{b_t + m_t\left[\pi + \tan^{-1}\left(\frac{v}{\alpha}\right)\right]\right\} \qquad (128)$$

By substitution of Eq. (128) into Eq. (126), the speed-change equation in terms of v and α becomes

$$SC = (\alpha^2 + v^2)\left\{b_t + m_t\left[\pi + \tan^{-1}\left(\frac{v}{\alpha}\right)\right]\right\} + \beta_0 - AKL(\alpha_0 - \alpha) = 0 \qquad (129)$$

The simultaneous solution of the head-balance and speed-change equations will yield values of v and α which can be used to solve for h and β. Throughout the transient analysis for runaway of the unit, the solution of the two turbine equations will be found along the full-gate FH and FT curves. Cases involving the controlled closure of wicket gates will involve computing the gate position with time as defined by the operating governor tau curve and interpolating between head and torque curves for the gate position at the beginning and end of the time step to arrive at the simultaneous solution of the speed-change and head-balance equations. Similarly, if the system incorporated a guard valve with measurable losses or valve movement, the loss term must be included in the head-balance equation. The valve tau curve will be used to define the guard valve movement in the same manner as that of an in-line valve. Valve losses must take into consideration the direction of flow which is especially important in transient analysis of pump-storage projects.

REFERENCES

1. PARMAKIAN, JOHN, *Waterhammer Analysis,* Dover Publications, New York, 1963.
2. WYLIE, E. B., and V. L. STREETER, *Fluid Transients,* McGraw-Hill, New York, 1978.
3. JENKER, W. R., "Über die Druckstross-geschwindigkeit in Rohrleitungen mit quadratischen and rechteckigen Querschnitten," Schweiz. Bauzeitung, 89 Jahrgang, Heft, Feb. 4, 1971.
4. CHAUDHRY, M. H., J. A. ROBERSON, and J. J. CASSIDY, *Hydraulic Engineering,* Houghton Mifflin, Boston, 1988.
5. U.S. DEPARTMENT OF THE ARMY, Waterways Experiment Station, Corps of Engineers, "Hydraulics Design Criteria," Wicksburg, 1988.
6. HECKER, G. E., and B. J. PENNINO, "A Synthesis of Model Data for Pumped Storage Intakes," Alden Research Laboratory, Worcester, 1979.
7. IDELLCHIK, I. E., *Handbook of Hydraulic Resistance,* 2d ed., Hemisphere Publishing Corp., Washington. 1986.
8. MILLER, D. S., *Internal Flow—A Guide to Losses in Pipe and Duct Systems,* British Hydromechanics Research Association, Cranfield, England, 1971.

9. Levin, L., *Formulaire des Conduites Forcées Oleoducs et Conduits D'Aeration,* Dunod, Paris, 1968.
10. U.S. Department of the Interior, Bureau of Reclamation, "Selecting Hydraulic Reaction Turbines," Engineering Monograph no. 20, 1976.
11. Rich, G. R., *Hydraulic Transients,* 2d ed., Dover Publications, Inc., New York, 1963.
12. Streeter, V. L., *Fluid Mechanics,* 4th ed., McGraw-Hill, New York, 1966.
13. Rouse, H., *Elementary Mechanics of Fluids,* Wiley, New York, 1957.
14. Pickford, J. A., "Throttled Surge Tanks," *Water Power,* November 1965.
15. Li, Y. T., "Orifice Head Loss in the T-Section of a Throttled Surge Tank," *Water Power,* September 1972.
16. Jaeger, C., "Economics of Large Modern Surge Tanks," *Water Power,* May 1958.
17. Jacobson, R. S., "Charts for Analysis of Surge Tanks in Turbine or Pump Installations," U.S. Department of the Interior, Bureau of Reclamation, Report 104, Denver, February 1952.
18. Water Power '89, Proceedings of the International Conference on Hydropower, ed. A. J. Eberhardt, ASCE, 1989.
19. Stewart, E. H. III, and J. E. Borg, "Moose River Air Chamber Design and Performance," *Water Power '89.*
20. Svee, R., "Surge Chamber with an Enclosed Compressed Air-Cushion," International Conference on Pressure Surges, BHRA Fluids Engineering, 1972.
21. Watson, W. M., "The Evolution of the Multi-Jet Sleeve Valve," *Water Technology/Distribution Journal* AWWA, June 1977.
22. Lewin, J., "Valves in Reservoir Outlets," *Water Power,* September 1988.
23. Elder, R. A., and G. B. Dougherty, "Characteristics of Fixed-Dispersion Cone Valves," *ASCE Transactions,* Paper 2567.
24. U.S. Department of the Interior, Bureau of Reclamation, "Calibration of a 10-inch Jet Flow Gate," Report HYD-569, Denver, February 1969.
25. Peterka, A. J., "Hydraulic Design of Stilling Basins and Energy Dissipators," Engineering Monograph no. 25, U.S. Department of the Interior, Bureau of Reclamation, Washington, 1978.
26. Strohmer, F., "Investigating the Characteristics of Shut-off Valves by Model Tests," *Water Power,* July 1977.
27. Barp, B., and H. Grein, "The Biplane Butterfly Valve," *Water Power,* September 1968.
28. Guins, V. G., "Flow Characteristics of Butterfly and Spherical Valves," *ASCE Hydraulics Journal,* May 1968.
29. DeFazio, F. G., "Transient Analysis of Variable-Pitch Pump-Turbines," *Journal of Engineering for Power,* October 1967.
30. International Electrotechnical Commission, "International Code for Model Acceptance Tests of Hydraulic Turbines," Publication 193, Geneva, 1965.

SECTION 23

NAVIGATION LOCKS

By Istvan T. Laczo

1. General. Navigation locks are structures on waterways that permit navigation for vessels that have to overcome a sudden change in water-level elevation because of either natural or manmade obstacles, such as rapids or dams in the waterway. These structures consist of the lock chamber, upstream and downstream approach walls, approach channels, gates, and auxiliary equipment. Locks are usually provided with mooring facilities for waiting vessels and with breakwaters. The lock chamber is confined between two lateral walls and closed by gates on the upstream and downstream ends. Lock filling and emptying is usually by gravity flow from the upstream water pool into the lock chamber and from the lock chamber to the lower water pool. A variety of filling and emptying systems can be provided depending mainly on the lift and required operating time of the lock.

For a vessel to pass through the lock, it moves into the lock chamber with the lock chamber gate on the exit side and the filling and emptying gates closed. After the vessel is in the lock chamber, the inlet lock chamber gate is closed and the filling or emptying gates are opened. If a vessel is passing from the high level to the low level, the lock chamber is drained to the lower level. If passing from the lower level to the higher level, the chamber is filled from the upstream water pool. After the water level in the lock chamber reaches the pool level in the direction of the vessel's travel, the lock chamber gate on the exit side is opened, allowing the vessel to exit.

The examples of locks discussed in this section are for inland locks located on rivers and canals. Maritime locks are similar to inland locks, but they are normally larger because of the larger-sized vessels they have to handle, and maritime locks normally have to deal with the additional problem of preventing saltwater intrusion. Many of the examples cited in this chapter refer to the newly constructed (opened to traffic in 1990) navigation lock of the Yacyreta project on the Paraná River between Argentina and Paraguay. The Yacyreta lock and facilities were designed by Harza y Consorciados, Consultores Internacionales de Yacyreta.

LOCK LAYOUTS

2. General. The prime factors which must be considered when laying out a lock are navigable depth, location of existing navigation channels; unhindered ingress and egress, sufficient width for maneuvering vessels near the lock; freedom from hazardous currents, floating ice, and debris; good visibility; navigation during construction; foundations; and future expansion. Locks should be located in a relatively straight reach of the waterway, free from side currents, on one side of the waterway. If a spillway or powerhouse is constructed at the same site, the lock should be located away from the spillway and the powerhouse to minimize the tendency for side currents in the lock approaches.

The lock walls are usually concrete, gravity-type walls with inside vertical surfaces designed to avoid the variation of ship clearances. The ends of the lock walls, called the lock heads, are increased in thickness, to accommodate mechanical and hydraulic installations. The lock chamber gates are installed in the lock heads, and frequently, depending on the filling-emptying system used, the filling-emptying culvert gates are also installed in the lock heads. If the culvert-type filling-emptying system is used, the culverts are located in the lower part of the lock walls, and they discharge into the lock chamber through side ports located at the bottom of the lock walls or through discharge manifolds in the lock floor. The lock walls are normally protected against horizontal and vertical vessel movements inside the lock by either wooden, steel, or hard-rubber fenders.

When required for highway traffic, a bridge is installed across the lock. It is usually located on the downstream side of the lock, above the lower pool, where the most headroom can be provided for vessels. The lock control room is in a superstructure on one of the lock walls in a location with good visibility for the lock operator. Drainage systems, equipment galleries, and access shafts are located in the lock walls and floor.

Locks are usually provided with approach walls on both the upstream and downstream sides of the lock. The approach walls are the extensions of the lock walls, but their function and construction are different. Depending on their function, approach walls are referred to as guide walls or guard walls. Guide walls are long and are used to guide long tows into the lock and to provide mooring facilities for waiting vessels. Guard walls are short and are used in conjunction with guide walls to facilitate entrance to the lock and to act as barriers against vessel movement toward navigation hazards. Figures 1 and 2 show typical arrangements of guide and guard walls.

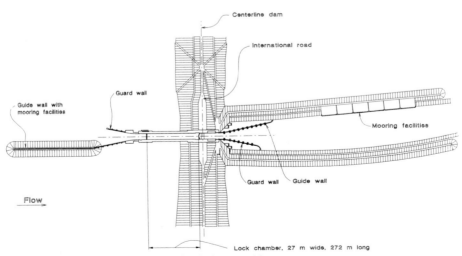

FIGURE 1 Yacyreta project navigation lock, general layout.

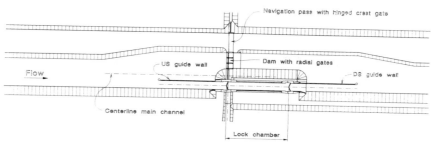

FIGURE 2 Calion lock layout.

Approach walls normally flare out to provide ample width for entrance. The slope of the flares is normally between 1:4 and 1:8, the gentler slope being preferred for locks where large, hard-to-steer tows are expected and the steeper slope for locks where traffic consists mostly of self-powered vessels. The mooring facilities on the guide walls are normally located beyond the flared part of the guide wall. This section of the guide wall is usually parallel to the lock centerline, and it is at least as long or longer than the lock chamber. The guide wall can be placed on either side of the lock. However, when the lock is located on the outside of a river bend, a landside guide wall is preferred.

The approach walls are normally not watertight. They are slotted buttress-type walls, or they are supported on piers or are floating walls (Ice Harbor Lock, United States, or Wijk bij Duurstede lock, Netherlands). Impact by vessels must be considered in the design of the approach walls. Protective, energy-absorbing fenders of wood or rubber are provided, and the wall supports may be designed for energy absorption.

3. Lock Dimensions. The size and number of navigation locks at the same site are based on several considerations, including expected maximum size of vessels or tows, expected frequency of traffic, and differential head. One lock sized to handle the maximum size tow is the most economical choice in most cases, though parallel twin locks are often considered when the frequency of smaller tows is expected to be high. High initial costs for the twin locks may be offset by the benefits of more flexible lock operation.

Navigation locks should have as high a differential head (lift) as possible to reduce the number of locks. The maximum single lift is limited to about 40 m for hydraulic considerations of the filling and emptying system. Two or more locks should be considered when the level difference is greater than the proved maximum head for the lock filling system.

Navigation lock dimensions should be kept to a minimum to reduce locking time, water losses, and construction costs. The size of the lock will be dictated by the maximum size of the vessels to be towed. As a general rule, a clearance of at least 10 percent of the vessel's beam should be provided on each side and 1 m between the ship's keel and the lock sill. Length of the lock will be based on the maximum length of tows with safety margins between the upstream and downstream lock gates. If safety barriers are also installed in the lock chamber, the chamber length will need to be increased to provide clearance for them. The dimensions of the Yacyreta lock, previously mentioned, demonstrate such considerations. The lock is 27 m wide, with a clear length between lock chamber gates of 272 and 238.7 m between the safety barriers within the lock chamber. These dimensions permit the locking of a tow of six barges 59.4 m long and 12.2 m wide and a 23.3-m-long and 5.5-m-wide tugboat between the safety barriers. Lowering the safety barriers, the lock could accept as a maximum eight such barges and the tugboat.

Minimum freeboard for lock walls and lock chamber gates is 1 to 1.5 m above the maximum upstream pool level for navigation. The freeboard should be at least 1.5 m for push tows. When the freeboard for navigation locks on rivers with run-of-the-river power plants is determined, surges due to powerhouse shutdown should also be considered.

On waterways where locks are in operation, the smallest lock will determine the size of the maximum tow for that waterway. Barges and ships are built to suit existing locks, and new locks on the same waterway usually have dimensions similar to existing locks. The original locks on the Ohio and Mississippi rivers in the United States were built with 110 by 600 ft (33.6 by 183 m) dimensions, based on the maximum tow sizes around the turn of the century. These lock dimensions resulted in standardization of the sizes of barges and vessels, and most locks constructed later followed this standardization. These dimensions were also used for the recently constructed locks on the Tennessee-Tombigbee waterway. Some longer locks have been constructed in the United States, for example, Lock and Dam No. 26 on the Mississippi River, T.J. O'Brien lock on the Illinois waterway, and at least one lock at each of the lock sites on the Ohio River, but the maximum width of 110 ft has been retained.

LOCK HYDRAULICS

4. General. The hydraulic design of a lock must find a balance between the opposing requirements of a fast locking operation and minimizing water turbulence to reduce the hydrodynamic forces acting on the vessels. Additionally, cavitation in the filling and emptying conduits must be avoided during filling and emptying, and air entrainment into the lock chamber must be minimized. Lock hydraulics must be carefully studied, and the lock should preferably be model tested.

The hydraulic problems in lock design are related to lock filling and emptying. Filling and emptying of locks can be accomplished by several means, basically depending on the lift. For low-head installations (up to about 10 m head) the following types of filling and emptying systems are customary:[1]

1. *Opening lock chamber gates.* The lock chamber gates for these installations must be of a type suitable for operation under unbalanced water head. These gates include the following types: vertical-lift wheel gate; double-leaf, hook-type vertical-lift wheel gate; radial Tainter gate; segment gates with vertical axis of rotation.
2. *Opening valves installed in the lock chamber gates.* In these arrangements normally more than one sluice gate or valve is incorporated in the miter or vertical-lift gates.
3. *Using short culverts in the upstream and downstream lock heads.*

For intermediate- and high-head installations (lifts more than 10 m), longitudinal culverts are normally used for filling and emptying. The most common designs include the following:

1. *Wall culverts with side ports.* Longitudinal culverts connect the upstream and downstream pools with filling and emptying gates or valves at the upstream and downstream ends of the lock chamber. The culverts are connected to the lock chamber by ports in the lock walls located immediately above the lock chamber floor. This arrangement is limited to moderate heads, up to about 18 m.

2. *Wall culverts with bottom laterals.* This arrangement is similar to the previous one, but instead of the side ports, transverse culverts are provided in the floor of the lock chamber, with discharge ports located on the sides of the transverse culverts. This filling and emptying system provides better flow distribution in the lock chamber than the side port system and should be used for higher heads. For an example of this filling and emptying system see Fig. 3 (Yacyreta project navigation lock).

3. *Wall culverts with bottom, longitudinal central culvert.* This system uses a longitudinal culvert in the floor of the lock chamber. Water from one of the wall culverts is delivered to the central culvert by a short transverse culvert located at the quarter point of the length of the central culvert, the other wall culvert is connected to the central culvert at the three-quarters point. Water flows from the central culvert laterally to stilling chambers and then passes vertically to the lock chamber.

4. *Dynamically balanced lock filling systems.* These filling systems were developed to reduce hawser loads during lock filling and are normally used for the higher-lift locks. There are several variations of this system, but the basic principle is the same. Water from the lateral wall culverts enters the transverse culverts at the midpoint of the wall culverts. The transverse culverts lead to a system of longitudinal culverts in the lock chamber which discharge

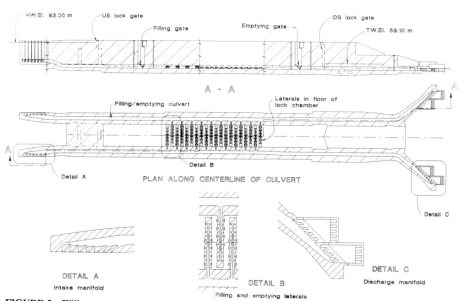

FIGURE 3 Filling system using longitudinal culverts and bottom laterals (Yacyreta).

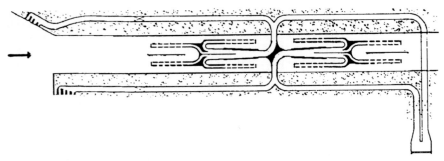

FIGURE 4 Balanced filling system (Lower Granite).[1]

the water symmetrically in the lock chamber. An example is the Lower Granite lock filling system, shown in Fig. 4.[1]

5. Lock Filling and Emptying. Locks should be designed so that the total filling or emptying time is approximately as determined from the following formula:[2]

$$T_t = K_1 \sqrt{\frac{h_s}{2g}} \tag{1}$$

where T_t = total time required to fill or empty the lock
h_s = differential water head at the start of filling or emptying
g = acceleration of gravity
K_1 = constant, between 720 (for bottom lateral filling system) and 960 (side port filling system). A value of 840, which is an average for all types of filling systems, can be used for preliminary calculations.

English or metric units can be used in the equations and formulas in this section as long as they are consistent.

For calculating lock filling and emptying times, the filling and emptying system of the lock can be considered as an orifice. The following additional notations are used:

A_l = lock chamber plan area

A_c = filling or emptying water passage (culvert) cross-section area

H_{gt} = gate opening height

H_c = height of culvert or gate opening

C_o = overall discharge coefficient

C_g = gate discharge coefficient

HW = headwater elevation

TW = tailwater elevation

v_g = gate opening velocity

t = time

T_i = lock filling and emptying with instantaneous gate opening

T_g = gate opening time

h = differential head (between HW and lock chamber water elevation or between lock chamber elevation and TW, positive value)

h_g = differential water head when gate opening is completed

Q = flow through the filling or emptying culvert

For the case of the gates fully open, the following equation can be written

$$-A_l dh = Q\, dt = C_o A_c \sqrt{2gh}\, dt \tag{2}$$

The time required for filling the lock, assuming instantaneous gate opening, can be calculated by

$$T_i = \frac{A_l}{C_o A_c} \int_{h_s}^{0} \frac{dh}{\sqrt{h}} = \frac{2 A_l \sqrt{h_s}}{C_o A_c \sqrt{2g}} \tag{3}$$

In practice the filling or emptying of the lock occurs in two phases (1) during gate operation and (2) after the gates have been completely opened.

For phase 1 of the filling operation, when the gate is opened at a constant velocity of v_g, the following equation can be written, assuming that C_o discharge coefficient (or A_c conduit cross section) changes linearly from the gate closed to gate open position:

$$-A_l dh = \frac{v_g t}{H_c} C_o A_c \sqrt{2gh}\, dt \tag{4}$$

and

$$\int_{t_1}^{t_2} t\, dt = -\frac{A_l H_c}{v_g C_o A_c \sqrt{2g}} \int_{h_1}^{h_2} \frac{dh}{\sqrt{h}} \tag{5}$$

resulting in

$$[t^2]_{t_1}^{t_2} = \frac{4 A_l H_c}{v_g C_o A_c \sqrt{2g}} [\sqrt{h}]_{h_2}^{h_1}$$

or

$$(t_2^2 - t_1^2) \frac{v_g C_o A_c \sqrt{2g}}{4 A_l H_c} = \sqrt{h_1} - \sqrt{h_2} \tag{6}$$

Using the substitution $Z_1 = v_g C_o A_c \sqrt{2g} \big/ 4 A_l H_c$, this can be written as

$$h_2 = [\sqrt{h_1} - (t_2^2 - t_1^2) Z_1]^2 \tag{7}$$

This equation is valid during the first phase of filling, while the gates are opening, for a given time and head if the initial or previous conditions are known. The head at the end of the gate-opening phase h_g can be obtained by substituting the initial conditions $t_1 = 0$, $t_2 = t_g$, and $h_1 = h_s$; then

$$h_g = (\sqrt{h_s} - t_g^2 Z_1)^2 \tag{8}$$

Equation (2) is applicable for the second phase of the filling operation when the gates are already open. Integrating Eq. (2) with respect to time, we get

$$[t]_{t_1}^{t_2} = -\frac{2A_l}{C_o A_c \sqrt{2g}} [\sqrt{h}]_{h_1}^{h_2} \tag{9}$$

The head can be calculated from this equation at any time; it is

$$\sqrt{h_2} = \sqrt{h_1} - (t_2 - t_1) \frac{C_o A_c \sqrt{2g}}{2A_l} \tag{10}$$

and substituting $Z_2 = C_o A_c \sqrt{2g} / 2A_l$ we get

$$h_2 = [\sqrt{h_1} - (t_2 - t_1) Z_2]^2 \tag{11}$$

and

$$t_2 = \frac{\sqrt{h_1} - \sqrt{h_2}}{Z_2} + t_1 \tag{12}$$

The total time of filling the lock can be calculated by making the following substitutions for the second phase of the filling process: $h_1 = h_g$, $h_2 = 0$, $t_1 = T_g$, and $t_2 = T_t$. T_t is then

$$T_t = \frac{\sqrt{h_g} - T_g^2 Z_1}{Z_2} + T_g = T_i + \frac{T_g}{2} \tag{13}$$

or the total filling time equals the filling time for instantaneous gate opening plus one-half of the gate operating time. These equations are also applicable for lock emptying using the appropriate boundary conditions.

The above equations can be used to calculate the head (or water level in the lock) for a given time. Once the head is known, flows can readily be calculated using the following equations:

For phase 1:

$$Q = C_o A_c \frac{v_g t}{H_c} \sqrt{2g\, h} = C_o A_c \frac{H_{gt}}{H_c} \sqrt{2g\, h} \tag{14}$$

where H_{gt} = gate opening height = $v_g t$

For phase 2:

$$Q = C_o A_c \sqrt{2g\, h} \tag{15}$$

These equations, although theoretical, show good agreement with model test results and can be used with confidence in estimating the time required for filling and emptying the lock. To use the equations, one needs only the basic parameters for the lock, such as lock chamber plan area, total cross section of the filling and emptying conduits at the gates or valves, opening time, and the overall discharge coefficient. With the exception of the overall discharge coefficient, all the

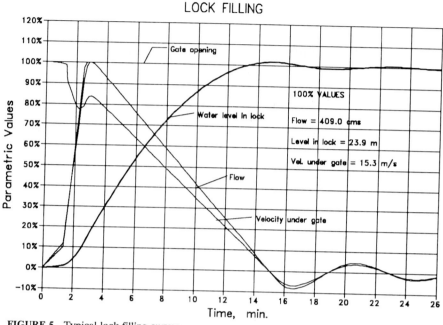

FIGURE 5 Typical lock-filling curves.

data are well defined. The overall discharge coefficient, if model test data are available, such as the filling curves shown in Fig. 5, can be readily determined from the phase 2 (gates are open) region of the filling or emptying curve, and it is

$$C_o = \frac{Q}{A_c\sqrt{2gh}} \qquad (16)$$

If no model test data are available, the discharge coefficient can be calculated as

$$C_o = \frac{1}{\sqrt{\Sigma \zeta}} \qquad (17)$$

where $\Sigma \zeta$ is the sum of all head-loss coefficients in the water passage. The head-loss coefficients can be defined as $\zeta = \Delta h 2g/v^2$, and with reference to the water passage cross section at the filling and emptying gate, the head-loss coefficients can be written as $\zeta = \Delta h 2g A_c^2/Q^2$. The discharge coefficient may not be the same for filling and emptying operations because of the possible differences in the water passages. For preliminary calculations the following discharge coefficients can be used:

Filling and emptying system	Discharge coefficient
Longitudinal culvert, lateral ports	0.75–0.85
Longitudinal culvert, bottom ports	0.60–0.80
Cracking lock gates, or valves in gates	0.80–0.95

These values represent conditions with the gates and valves fully open. Assuming a constant discharge coefficient during lock filling or emptying provides reasonably accurate estimates of filling time, considering the fact that the discharge coefficient is not constant during gate opening. When model test results for a particular installation indicate deviation from the calculated filling time, it signifies sensitivity to variation in discharge coefficient assuming that the overall discharge coefficient C_o has been correctly determined. In that case, a multiplier k should be used for the gate-opening time for subsequent calculations. Equation (13) would be modified by the k factor determined as

$$k = (T_t - T_i)\frac{2}{T_g} \tag{18}$$

If the gate-opening velocity is not constant for the entire gate operation, the above equations must be used, performing step-by-step calculations to include each instance of gate-velocity change, from the start to the end of a complete phase of gate operation.

For long culverts, the momentum effect of accelerating and decelerating the water in the water passages will have significant influence on the results of the calculations. Considering this effect, Eqs. (2) and (4) should be modified as follows:

$$-A_l dh = Q\, dt = C_o A_c \sqrt{2g\left(h - \frac{L}{g}\frac{dQ}{A_c dt}\right)}\, dt \tag{19}$$

and

$$-A_l dh = \frac{v_g t}{H_c} C_o A_c \sqrt{2g\left(h - \frac{L}{g}\frac{dQ}{A_c dt}\right)}\, dt \tag{20}$$

where L = conduit length.

The water momentum or inertia effect will appear as an "overtravel" of the water level in the lock chamber; that is, the water level in the lock chamber will overshoot the headwater level at the end of lock filling, or it will descend below tailwater level at the end of lock emptying; and it will then oscillate with decreasing amplitudes until the kinetic energy of the water column is dissipated by friction and local losses. This can occur even if the culvert gates are closed at the moment when the water levels in the lock chamber and the upstream or downstream pools are equalized. Depending on the prevailing conditions the action may produce reverse head on the culvert gates.

The approximate amount of overtravel d can be calculated by applying the expression determined by Pillsbury:[2]

$$d = \frac{KC_o^2 A_c^2}{A_l} \tag{21}$$

where $K = \Sigma[(Q_i/Q)^2(l_i/a_i)]$
Q = total flow in culvert
Q_i = flow in individual branch or section of culvert
l_i = length of individual branch or section of culvert
a_i = cross-section area of individual branch or section of culvert

Considering the overtravel, Eq. (3) should be modified as follows:

$$T_i = \frac{2A_l(\sqrt{h_s + d} - \sqrt{d})}{C_o A_c \sqrt{2g}} \tag{22}$$

This expression can be used for estimation of filling and emptying times. However, it has limited use for calculating intermediate values of flows and heads, especially during phase 1 (gate operation).

For calculating intermediate flow and head values, considering water inertia, a step-by-step calculation can be performed. A computer program or spreadsheet can be set up to perform the calculations based on Eqs. (19) and (20).

For the step-by-step calculation we can write the following equations:

$$Q_t = K_{1t}\left(h_t - K_2 \frac{\Delta Q_t}{\Delta t}\right)^{0.5} \tag{23}$$

and

$$h_t = h_{t-1} - \frac{1}{A_l}\left(Q_{t-1} + \frac{\Delta Q_t}{2}\right)\Delta t = h_{t-1} - \Delta h_t \tag{24}$$

where $K_1 = C_e A_c \sqrt{2g}$

$$K_2 = \frac{L_e}{gA_c}$$

L_e is an equivalent conduit length and can be calculated as

$$L_e = \frac{\sum L_i v_i}{v_c} = \frac{A_c}{Q_c} \sum L_i \frac{Q_i}{A_i}$$

L_e, and consequently K_2, are constant for the entire operation. C_e, or the equivalent discharge coefficient, includes the effect of partial gate openings, and it is variable for the gate-opening phase and constant when the gates are fully open. C_e can be calculated as

$$C_e = \frac{1}{\sqrt{\sum K}}$$

where

$$\sum K = \left(\frac{1}{C_{gt}^2} - 1\right)\left(\frac{H_c}{H_{gt}}\right)^2 + \frac{1}{C_o^2}$$

and C_{gt} can be approximated by $C_{gt} \cong 0.58 + 0.42(H_{gt}/H_c)^6$, H_{gt} being the gate opening at time t.

Substituting Eq. (24) into Eq. (23) and rearranging, we get the following second-degree equation for ΔQ_{tg}:

$$\Delta Q_t^2 + \Delta Q_t\left(2Q_{t-1} + \frac{\Delta t K_1^2}{2A_l} + \frac{K_1^2 K_2}{\Delta t}\right)$$
$$+ \left(Q_{t-1}^2 - K_1^2 h_{t-1} + \frac{Q_{t-1}\Delta t K_1^2}{A_l}\right) = 0 \tag{25}$$

or

$$\Delta Q_t^2 + B \, \Delta Q_t + C = 0 \qquad (26)$$

where

$$B = 2Q_{t-1} + \frac{\Delta t \, K_1^2}{2A_l} + \frac{K_1^2 K_2}{\Delta t}$$

and

$$C = Q_{t-1}[Q_{t-1}] - K_1^2 \, h_{t-1} + \frac{Q_{t-1} \Delta t \, K_1^2}{A_l}$$

(Note that Q_{t-1}^2 has been converted to $Q_{t-1}[Q_{t-1}]$ to account for flow reversal). The spreadsheet can be then set up with the following columns:

Column	Content
1	Time, $t_t = t_{t-1} + \Delta t$
2	Gate velocity, v_g
3	Gate opening, $H_{gt} = H_{gt-1} + v_{gt}(t_t - t_{t-1}) \leq H_c$
4	Gate discharge coefficient, C_g
5	Culvert discharge coefficient, C_o
6	Total loss coefficient, ΣK (see equation above)
7	Equivalent culvert discharge coefficient, C_e
8	K_1^2
9	B coefficient for second-degree equation
10	C coefficient for second-degree equation
11	$\Delta Q_t = -B + \sqrt{B^2 - 4C}\,/\,2$
12	$\Delta h_t = -(Q_{t-1} + .5\Delta Q_t)(t_t - t_{t-1})/A_l$
13	$h_t = h_{t-1} + \Delta h_t$
14	$Q_t = Q_{t-1} + \Delta Q_t$

This method does not provide an exact solution to the underlying problem, but the results do yield reasonably accurate estimates for general use and investigations. Using smaller time increments improves accuracy, and it is recommended that $1/1000 \, t_t$ increments be used for the first 10 time periods, while $1/100 \, t_t$ increments are sufficient thereafter. The time increments should be selected in such a way that significant events in gate operation, such as arriving at the fully open position or change in gate velocity, coincide with the end of a period.

Efficient lock design must combine fast filling and emptying rates with low hydraulic loads on the vessels during lock operation. These hydraulic loads are mostly caused by uneven water surface in the lock chamber during lock operation, but the impact of water jets discharged from the filling system, or air entrained in the water may also influence the loads.

As water is admitted or released from the lock chamber, the change in water level is not uniform over the entire chamber. The change will start at the location of water entrance or exit, creating an unbalance in the level and starting a flow in the chamber to equalize the water level. Instantaneous gate opening would produce a vertical-front wave of Z height traveling at C velocity on the surface and

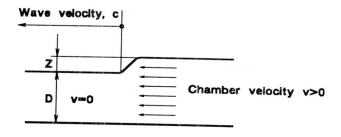

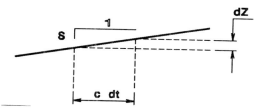

FIGURE 6 Wave propagation in lock chamber.

producing v water velocity in the chamber (see Fig. 6). When the wave arrives at the end of the lock chamber, the flow of water in the lock chamber is stopped at the wall and the wave is reflected in the opposite direction with an additional Z increment.

Applying the momentum equation and the principle of hydraulic continuity, the following equations can be obtained:[3]

$$c = \sqrt{gD} \qquad (27)$$

and

$$Z = \frac{Dv}{c} = \frac{Dv}{\sqrt{gD}} \qquad (28)$$

When the gate is opened gradually, the water surface will exhibit a relatively flat slope rocking back and forth between the ends of the lock rather than a vertical-front wave traveling across the lock chamber. The slope of the water surface S is

$$S = \frac{1}{g}\frac{dv}{dt} \qquad (29)$$

Considering that $Q = bDv$ and $dv = dQ/bD$, the slope can be written in terms of chamber flow as

$$S = \frac{1}{gbD}\frac{dQ}{dt} \qquad (30)$$

The unbalanced force acting on the vessel in the lock chamber is proportional to the slope of the water surface in the lock chamber and can be written as

$$F_u = KS = \frac{K}{g}\frac{dv}{dt} = \frac{K}{gbD}\frac{dQ}{dt} \qquad (31)$$

where F_u = unbalanced force on vessel
  K = force coefficient depending on size and shape of vessel
  S = slope of water surface
  b = width of lock chamber
  D = depth of water in lock chamber

Wave reflection and velocity of propagation will be affected by the presence of the vessel in the lock chamber. Therefore, these equations are not suitable for quantitative evaluation of the unbalanced forces during locking. In the case of a lock with high differential head, the impulse of the water emerging from the filling ports will have considerable influence on the magnitude of the unbalanced force. To determine the unbalanced forces on vessels during lockage with the accuracy required for finalizing the lock design, model tests are indispensable. These equations can provide preliminary basic information on the hydraulics of lock operation. The following conclusions can be made:

1. The rate of change in flow, rather than the rate of flow, is one of the major factors determining the magnitude of unbalanced forces on vessels during lockage.
2. The other important variable is the water depth in the lock chamber. With shallower depths the unbalanced forces acting on the vessel are higher.

The greatest value of dQ/dt occurs during the opening of the gates. Therefore, a slow rate of gate opening is recommended in order to keep the mooring-line loads at acceptable levels. Once the gates are open, dQ/dt will be almost constant, and its value will be lower than the maximum reached during gate opening. Because of this, and because the mooring-line loads are independent of the flow rate (assuming that no impulse forces act on the vessel), the filling and emptying water passages can be of large cross section to reduce filling and emptying times. It can also be concluded that for the same gate-opening velocity, the mooring-line loads will be higher during lock filling than for lock emptying because the depth of water at the beginning of the filling operation, when the gates are opened, is much shallower than at the beginning of the emptying operation. Conversely, for equal maximum mooring-line loads, the opening velocity of the emptying gates can be higher than that of the filling gates. These general conclusions have been proved by model test results.

The maximum rate of water-level change in the lock chamber normally adopted for lock designs is about 3.5 to 4 m/min. Higher maximum flow rates can be achieved with careful hydraulic design to assure that cavitation, surging, and turbulence in the lock chamber are within safe limits.

6. Air Entrainment and Cavitation. When the filling and emptying culvert gates are opened, the flow downstream of the gates contracts and accelerates, reaching its maximum velocity in the vena contracta at some distance from the gate before it slows farther downstream in the culvert. Because of the high local velocity, the velocity head increases, reducing the pressure. Depending on the submergence of the culvert and the velocity of the flow, subatmospheric pressures can develop at the top of the culvert during the gate-opening phase, reaching minimum values with the gates at 30 to 70 percent open. These subatmospheric pressures will allow air entrainment in the water, if air can reach the low-pressure zone, and can also produce cavitation. Both effects are undesirable and should be avoided by means as described below.

There are also other sources of air entrainment. Air could enter the culvert through the vortex formation at the culvert intake. Air entrainment may also occur if any part of the culvert between the filling and emptying gates is above the tailwater level allowing air to enter into the section during lock emptying or when the water level in the lock chamber is at tailwater level.

When air enters the culvert, it is carried by the flowing water until it can escape to the atmosphere. If the release of air occurs through the filling and emptying ports inside the lock chamber, the air will surface in large bursting pockets, creating very high turbulence and increasing hawser loads on the vessels. Entrained air not only produces undesirable effects in the lock chamber but may also reduce the capacity of the filling and emptying system. Air entrainment is normally more of a problem for the filling system than it is for the emptying system. The following design measures should be taken to minimize air entrainment in the flowing water:

1. Filling culvert intakes should have sufficient submergence and low approach velocities to prevent the formation of vortices which cause air cavities to extend into the culvert.
2. The roof of the culvert between the filling and emptying gates should be kept below the minimum tailwater level. If this is not practical, because of excavation costs or other reasons, vents should be provided at the roof of the culvert.
3. Filling gate design and location should be carefully selected. Downstream sealing gates with an upstream gate slot, such as reversed Tainter gates or downstream sealing vertical-lift gates, are preferred. Several early, low-head locks have been built with Tainter gates sealing on the upstream side. In the 1930s reverse Tainter gates were adopted. These seal on the downstream side to avoid air entrainment. The filling gate should have adequate submergence so that the pressure gradient during gate operations is always above the culvert roof. If maintenance bulkheads are provided downstream of the filling gate, their slots should be located at a distance from the filling gate, where the pressure gradient is safely above the culvert roof. Alternatively, if the bulkhead slot is close to the gate (to avoid excessive pumping when unwatering for maintenance), the bulkhead slot should be provided with an airtight cover safely anchored against uplift.

Cavitation occurs when the absolute pressure on the downstream side of the gate drops below the saturated vapor pressure of the water. Under these conditions, vapor-filled cavities develop in the low-pressure zone and are carried downstream in the flow where they collapse instantaneously upon reaching a higher-pressure zone. The instantaneous collapse of the cavities creates high-pressure shock waves in the water, similar to water-hammer effect. Surface irregularities or sharp corners in the low-pressure zone, enhance localized cavitation. Cavitation can cause gate vibrations and pitting damage in the culvert walls. Once pitting of the culvert walls is started, the rate of the damage will accelerate because of the surface irregularities caused by the initial pitting.

The tendency and extent of cavitation is best identified by model tests. The cavitation potential for a project can be determined and preliminary design parameters can be adjusted using the cavitation coefficient K_c. The cavitation coefficient can be expressed as

$$K_c = \frac{\Delta H_g - \sum h_{lus} + (1/\gamma_w)(p_a - p_v)}{v_c^2/2g} \qquad (32)$$

where ΔH_g = $HW.El$ − culvert roof elevation downstream of the gate
  h_{lus} = head losses upstream of the point in consideration, including head loss across the gate
  γ_w = specific weight of water
  p_a = atmospheric pressure
  p_v = saturated vapor pressure of the water at the prevailing temperature
  v_c = maximum flow velocity in the vena contracta

The gate head-loss coefficient for the range of gate openings of interest (between 30 and 70 percent) with reference to the flow velocity in the culvert is approximately

$$K_{gate} = 10^{[2.2-3.2(H_g/H_c)]} \quad \text{and} \quad h_{l\,gate} = K_{gate}\frac{1}{2g}\frac{Q^2}{A_c^2}$$

where the K_{gate} equation is from Ref. 4. The maximum flow velocity in the vena contracta v_c can be calculated as

$$v_c = \frac{1}{\mu}\frac{QH_c}{A_c H_g}$$

and μ is 0.65 to 0.75 in the range of 30 to 70 percent gate openings. The cavitation coefficient K_c should be at least 1.0 in order to prevent cavitation. U.S. Army Corps of Engineers tests on the Holt and John Day locks indicate that the minimum required cavitation coefficient can be reduced if the culvert cross section is gradually increased downstream of the gate by sloping the culvert roof upward. Their tests indicate that the required minimum cavitation coefficient K_c is a linear function of the culvert cross section at the vena contracta,[4] and it is

$$K_{cminr} = 1 - \frac{H_{gmax}}{0.75H_{cvc}} \tag{33}$$

where H_{gmax} = culvert height immediately downstream of the gate
  H_{cvc} = culvert height at the vena contracta

The reduced cavitation intensity, when the culvert roof is sloped to increase culvert cross section, can be explained by the presence of a reverse backflow of water into the low-pressure zone, resulting in increased pressure at the culvert roof.

In summary, cavitation damage and/or cavitation itself can be reduced or eliminated by the following measures:

1. Increasing submergence of the gate by lowering the elevation of the gate.
2. Reducing gate-opening velocities, which results in lower velocities in the vena contracta at the critical gate openings.
3. Increasing local pressure by admitting water or a controlled quantity of air to the low-pressure zone. Water can be admitted by sloping the roof of the culvert, or via a bypass connection from the upstream side of the gate. Air can be admitted through embedded pipes leading to the roof of the culvert and/or to the floor immediately downstream of the gate. Admission of large, uncontrolled amounts of air is detrimental to lock operation, whereas admission of

small amounts of air will reduce the tendency for cavitation but will also reduce hawser loads.

4. Reduction of surface irregularities at the zone of high water velocities. Reverse Tainter gates without lateral slots produce the least disturbance.

5. Providing a cavitation-resistant liner in the culvert immediately downstream of the gate. A stainless-steel liner in the zone where cavitation can be expected will resist pitting damage better than concrete. The liner will have a hydraulically smoother surface; therefore, cavitation effects would not be enhanced by surface irregularities, which are inherent in concrete.

7. Water Management. The available water supply is of major importance in the selection of the lock basic parameters. The water supply may be limited for locks on navigation canals, especially on canals crossing a high divide, such as part of the Rhine-Main-Danube canal in Germany. Water can be also valuable in power generation when the lock utilizes water that could otherwise be used for hydroelectric generation. In such cases, consideration should be given to water-saving features in the lock design. All water-saving lock arrangements, however, require additional expenditures, and the cost-benefit ratio must be carefully analyzed. The water used for locking on canals provided with several locks in series should be coordinated so that each lock is using about the same volume of water for locking.

The least complicated arrangement to save water for locking is by using an intermediate lock-chamber gate. For smaller tows only part of the lock is filled, using the intermediate gate as the upstream or downstream lock gate. This arrangement also reduces the time required for locking smaller vessels or tows and permits repairs on one gate at a time without closing the lock to traffic.

Water savings can also be achieved by building two locks in series with the total lift equally divided between the two locks, instead of one lock for the total lift. Since the lift of each of the two locks in series is one-half of the replaced lock, 50 percent water saving can be achieved assuming that lock-chamber lengths and widths are the same. Although the individual lifts are reduced, the costs will be higher than for a single lock, and there will also be some penalty in locking times because the vessels will have to pass two locks instead of one.

Side-by-side twin locks incorporating filling and emptying systems that permit filling one lock chamber with water from the other lock chamber can also be used for water saving. In this design the two locks are connected by a culvert with a sluice gate designed for loading from either side. When one lock is emptied, the water is directed to the other lock chamber until the water levels in the two lock chambers are almost equal, thus saving about 50 percent of the water. This arrangement is preferred for moderate lifts and when heavy waterway traffic is expected.

The amount of water saved with any of the above designs is limited to 50 percent. When higher water savings ratio is required, a single lock with one or more (normally up to four) water-saving basins can be used. During the emptying of the lock, water from the lock chamber is released into the water-saving basins until the water level in the lock chamber equalizes with the water level in the lowest savings basin. The water remaining in the lock chamber is then released to the tailwater. To fill the lock, filling starts from the lowest saving basin, progressing to the higher saving basins. The water level in the lock chamber is still below the headwater level when water from the highest saving basin has been used to fill the lock chamber. The remaining part of the lock chamber is then filled from the headwater. A schematic of the working principle is shown in Fig. 7.

The amount of water saved with this design is a function of the ratio of saving

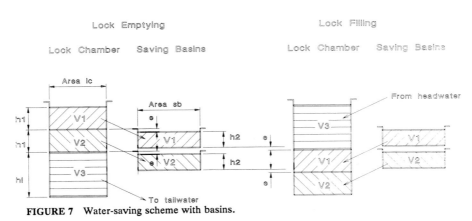

FIGURE 7 Water-saving scheme with basins.

basin and lock chamber surface areas m and the number of saving basins n. The higher m and n are, the more water is saved. However, the benefits of increasing m beyond 1 and n beyond 3 become increasingly small. Near optimum design can be achieved with $m = 1$ and $n = 3$ (Ref. 1). The water-saving basins can be constructed at the side of the lock chamber, attached or detached, or the saving basins can be incorporated in the lock walls.

EQUIPMENT

8. Gate Equipment. Several special types of gates have been developed for navigation locks. The most frequently used such gates are the miter, reverse Tainter, and large, counterballasted wheel gate.

Miter Gates. Miter gates consist of two gate leaves rotating about vertical axes. When the miter gate is open, its leaves are parallel to the lock-chamber centerline, retracted into recesses in the lock walls. When the gate is closed, the leaves meet to form a mitered joint and the predominant load in the gate leaves is compression in addition to bending and slight twisting of the gate structure. The compression load at the center of the gate is transferred from one gate leaf to the other through the miter bearings which are heavy stainless-steel plates. The miter bearings are located to coincide with the horizontal girders of the gate structure and the bearing plates on one of the gate leaves are provided with a large-radius curvature to compensate for mitering tolerances, whereas the matching miter bearing on the other gate leaf has a flat surface. Compression loads from the gate leaves are transmitted to the lock walls through similar bearings, flat bearing plates embedded in the concrete walls, and cylindrical bearings mounted on the gate leaves. Each gate leaf is supported by a pintle bearing installed at the center of rotation underneath the gate leaf and balanced at the top by the gudgeon pin with bushing anchored to the top of the concrete wall. A typical miter gate is shown in Fig. 8.

Large miter gates such as those shown in Fig. 8 are heavy; the gate is 30 m wide by 30 m high and each gate leaf weighs about 340 tons. Flotation chambers can be provided in the gate leaves to reduce the pintle bearing and gudgeon pin loads during normal operation. Diagonals are normally provided on the gate

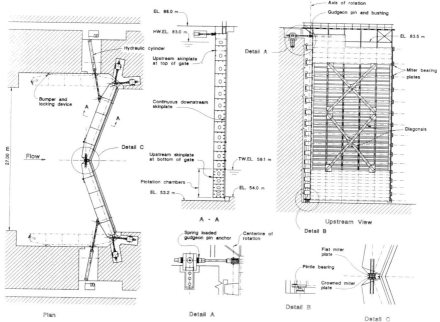

FIGURE 8 Large miter gate (Yacyreta project).

leaves. The diagonals increase the torsional rigidity of the leaf and, depending on the design, can also be used to adjust the flatness of the gate leaf. Miter gates are operated only in balanced head conditions. Hydraulic cylinders are used almost exclusively to operate miter gates, replacing the previously favored crank and lever, "Panama"-type operating machinery. The gate leaves are provided with bumpers for protection against hitting the lock walls and automatic, hydraulically operated latching devices to hold the leaves to the wall when open.

Reverse Tainter Gates. Reverse Tainter gates are the most common type of gate in the United States for lock filling and emptying, because of their reliable operation and simple structure. Reverse Tainter gates eliminate the danger of uncontrolled air entrainment, and the upstream gate shaft will act as a surge chamber in case of fast closure. A typical reverse Tainter gate installation is shown in Fig. 9.

The gate consists of a radial structure with vertical framing. The radial structure is connected to the gate support beam by means of gate arms. Trunnion bearings with self-lubricating bronze bushings are provided at the connection of the gate arms to the support beam. The support beam is a steel structure designed to be removable for maintenance and repairs.

The depth of the two horizontal beams on the radial structure, which are required for the connection of the gate arms, should be kept as small as possible. The bottom horizontal beam should be located as far as practicable from the bottom lip of the gate. This will reduce fluctuating loads on the gate due to turbulent-flow conditions in the gate shaft and on the upstream side of the gate. These flow conditions tend to increase hoist loads at small gate openings and to reverse hoist

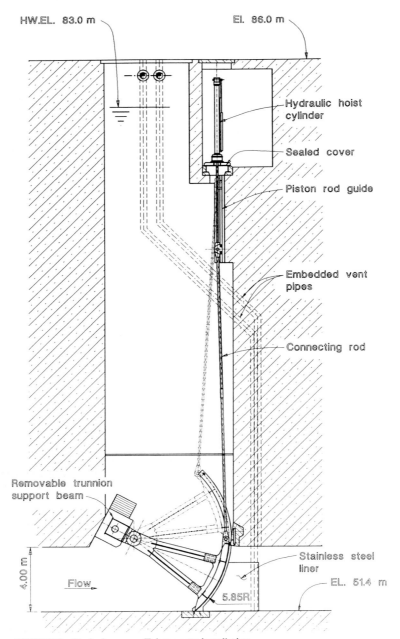

FIGURE 9 Typical reverse Tainter gate installation.

loads at large gate openings. These flow conditions can also induce vibrations because of their unstable nature. The gate arms should be slender in plan view, and protruding flanges should be covered with flow guide plates. Careful design and model tests are required to verify the design. The gate shown in Fig. 9 (Yacyreta project, 23.9-m lift) was model tested, and, with minor modifications on the lower horizontal beam, the vibrations and hoist load reversal were eliminated. Tainter valves for higher-head applications may require additional ballasts or compression-resisting hoist arrangements. Design criteria for the gates should include the possibility of surges in the culvert. Maximum differential head acting on the gate can be significantly higher than normal loading if the gate is closed rapidly. Load reversal on the gate should also be checked by the model. The reversal can occur when, after the initial pressure rise on the upstream side of the gate and the pressure drop on the downstream side, the transient pressures reverse owing to the momentum of water in the culvert.

Stainless-steel liners are normally provided on the downstream side of the gate to protect the culvert concrete walls against cavitation. Embedded piping, leading to the downstream side of the gates, may also be required to admit a controlled amount of atmospheric air to reduce the risk of cavitation damage.

Wheel Gates. Vertical-lift wheel gates are normally provided on the upstream end of the lock chamber, where the height of the lock chamber gate is smaller than the one on downstream end. A typical wheel gate is shown in Fig. 10.

When the lock is provided with separate filling gates or valves, the wheel gate can operate in balanced-head conditions. Thus, the wheels will be required to roll only under small loads when the gate is raised or lowered; they need not be operable when the hydrostatic load is acting on the gate. A wheel suspension detail of a gate using disk springs is shown in Fig. 11. The springs are designed to deflect under 0.5 m unbalanced head on the gate, and, when the springs are deflected, the gate is supported as a slide gate on the embedded track beam. This design reduces the cost of the gate (wheels, axles, gate structure endplates) and its embedded parts.

The gate is counterbalanced to reduce hoist capacity. The gate shown in Fig. 10 (Yacyreta project) weighs 143 tons. Two counterweights are used, each weighing 57 tons. Two 30-ton-capacity chain hoists are provided. The difference between the sums of the hoist capacities and the weights of the ballasts, and the gate weight accounts for the friction load plus an additional safety margin. For passage of vessels, the gate is lowered below the lock-chamber upstream sill. For maintenance the gate can be raised above the deck level, where it can be dogged. Each counterweight is connected to the gate by two steel cables that are provided with a load-compensating device and routed over large-sized return sheaves located at the top of the hoist building. The counterweights can be dogged in their shafts to permit replacement of the cables. The two hoists, which are located on opposite lock walls, are electrically synchronized.

9. Crane Equipment. Cranes are normally required for locks to handle the large and heavy bulkhead sections used on the upstream and downstream sides of the lock chamber to unwater the lock for inspection and maintenance. The most frequently used type of crane for this application is the stiff-leg derrick crane. Figure 12 shows an example of a stiff-leg derrick crane handling a bulkhead section.

The stiff-leg derrick crane consists of a latticed boom and a vertical mast mounted on a rotating platform. The top of the mast is connected to two stiff-legs using an arrangement permitting the rotation of the mast. The stiff-legs are installed at an approximate 45° angle in the elevation and at approximately 90° angle in plan relative to each other, permitting approximately 270° rotation of the mast and the boom about the vertical axis of the mast. The stiff-legs are designed for both tension and compression because of the load changes as the crane rotates. The boom can be lowered or raised to adjust the reach of the crane. While handling the load, the compression in the boom increases as it is lowered. Be-

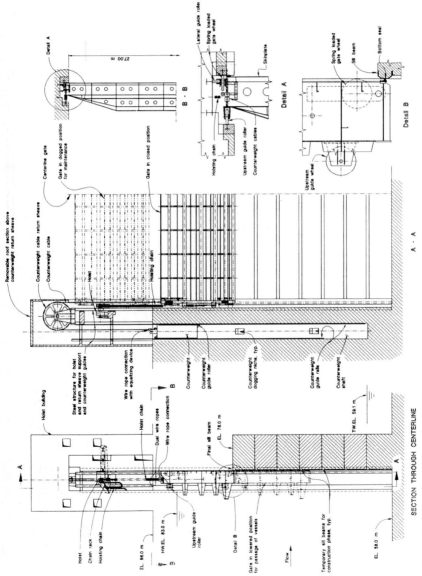

FIGURE 10 Lock-chamber wheel gate.

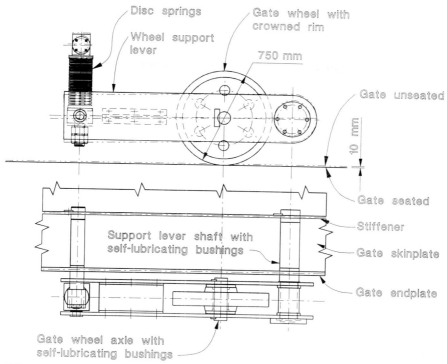

FIGURE 11 Spring-loaded wheel suspension.

cause of this, the capacity of the stiff-leg derrick crane must always be specified in relation to its corresponding reach.

Machinery and control equipment for modern derrick cranes are located on the rotating platform. Platform rotation is accomplished by a cogwheel drive. The pinion and the drive mechanism are installed on the platform, and the bullgear (a circular rail with round bars) is anchored to the concrete deck of the lock wall. For safety of operation the cranes are provided with a hoisting-rope slack switch, a directly activated load block upper-limit switch, and a load block lower-limit switch. Operation of one of these switches stops hoisting. Additional limit switches are provided for top and bottom boom positions and for the limits of crane rotation. Since the crane must rotate, the power-supply cable is usually routed from an outside source through one of the stiff-legs to descend through the mast to the rotating platform where the control and power equipment is located.

10. Safety Barriers. Safety barriers are normally provided on locks to protect the lock chamber gates against impact damage by the vessels. The safety barriers are energy-absorbing devices to stop a runaway vessel before it might hit a lock chamber gate. They are normally installed on both sides of the upstream lock-chamber gate and on the upstream (lock-chamber) side of the downstream gate. The required energy-absorbing capacity of the safety barriers, based on the kinetic energy of the approaching vessel, is

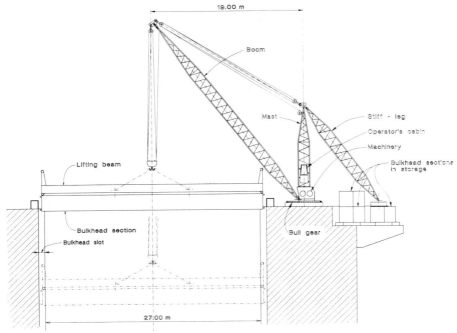

FIGURE 12 Stiff-leg derrick crane handling a bulkhead section.

$$U = \frac{mv^2}{2} \qquad (34)$$

where U = energy to be absorbed by the safety barrier
m = mass of the vessel
v = velocity of the vessel

Various kinds of safety barriers are used for locks. Sometimes these are beams across the lock chamber which can be lowered or raised to permit passage of the vessel and are provided with springs or hydraulic shock-absorbing devices to dissipate the kinetic energy of the vessel. Though shock-absorbing devices can be mounted on the lock-chamber gates themselves to absorb impact, the gates will still be subjected to the applied load. Another type of barrier utilizes a steel beam which swings across the lock chamber. A heavy braking cable is carried by the beam and is latched into an anchoring device on the opposite side of the lock when the beam is lowered into the horizontal position. After latching the braking cable in position the support beam may or may not be raised, depending on the design. If the design is such that the beam remains in a horizontal position, it will absorb part of the energy by bending. The remaining energy will be dissipated by an energy-absorbing device, such as a hydraulic cylinder, attached to the cable.

An improved arrangement of cable type of the aforementioned safety barrier is shown in Fig. 13. This design does not use a steel beam to move and remove the braking cable. The cable remains installed across the lock, but its elevation can be adjusted by raising or lowering carts on both sides of the lock. The elevation

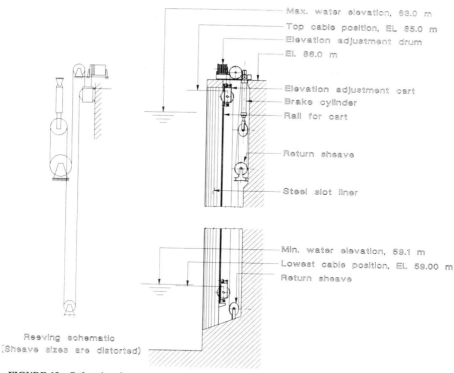

FIGURE 13 Safety barrier.

adjustment carts have spring-loaded wheels, and, when the tension is slightly released in the braking cable by lowering the pressure in the brake cylinder, the elevation of the carts can be adjusted as required for the water level and for the type of approaching vessel. Once the cable is positioned, the pressure in the brake cylinder is increased, the suspension of the elevation adjustment cart wheels deflects, and the carts are then supported by the cart frame resting on rails, in position to absorb the impact. To permit passage of vessels, the braking cable is lowered below the vessel. The safety barriers shown in Fig. 13 have been designed for 500 tons-meter capacity, that is, to stop a 1000-ton vessel moving at 1 m/s.

11. Mooring Equipment. Navigation locks must have sufficient mooring equipment within the lock chamber and also on the approach walls. Mooring equipment on the approach walls consists of stationary mooring bitts, provided mainly for vessels waiting for locking. When conditions require, tow haulage equipment, consisting of a hydraulic winch and traveling mooring bitts, may be also provided to handle split-barge tows. The traveling mooring bitts are installed on rails extending from the guide wall to the lock wall. Such tow haulage equipment is provided, for example, at Lock and Dam No. 1 (on the Mississippi River).

Inside the lock chamber stationary mooring bitts are mounted on top of the lock walls and in recesses in the sides of the walls. Recessed bitts are normally installed with a vertical spacing of approximately 1 m and require an access lad-

23.26 NAVIGATION LOCKS

der installed nearby in a vertical recess. As a rule, the ladder should be located not more than about 0.85 m from the bitts.

In addition to stationary mooring bitts, locks, especially the ones with higher lifts, may be equipped with floating mooring bitts. Figure 14 shows a typical floating mooring bitt. The guides of the floating mooring bitt are equipped with bumper stops at the top to prevent the floating mooring bitt from leaving the guides in case of extreme water levels or surges in the lock chamber. A check valve is provided on the floating mooring bitts flotation tank above the normal water level. The check valve is arranged to flood the flotation tank in order to protect it against collapse if the mooring bitt becomes jammed in the guides and water level in the lock chamber rises above the jammed bitt. Hooks are provided on the top of the bitt to facilitate lifting of the flooded bitt. Two pipes are installed in the flotation tank of the floating mooring bitt, one pipe leading to the bottom of

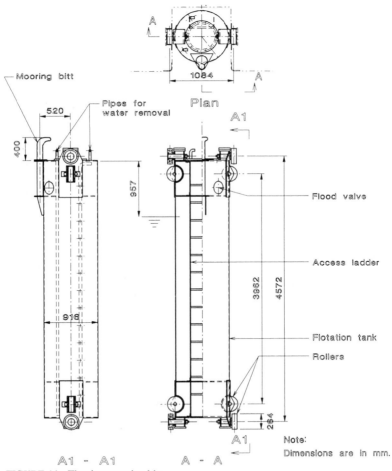

FIGURE 14 Floating mooring bitt.

the flotation chamber to remove water from the flotation chamber by using compressed air.

The mooring bitts must withstand the maximum mooring line, or hawser loads during lock operations. The maximum mooring-line load in most U.S. navigation locks is 5 tons, but a 10-ton mooring-line load is permitted for Great Lakes vessels and oceangoing vessels navigating on the St. Lawrence Seaway. In Germany and France the maximum mooring-line loads are limited to 1/600 of the total gross tonnage of the tow, and in the Netherlands they are limited to 1/1000 of the tow's tonnage or to 10 metric tons, whichever is smaller.

Floating fenders or mooring floats are used for smaller craft, permitting mooring of several vessels at a time. Floating fenders are hermetically sealed steel boxes, designed to float, provided with guide wheels on the sides which are similar to the wheels of floating mooring bitts, and several mooring buttons are provided on the top. Floating fenders are recessed into the lock walls so that the full width of the lock chamber is maintained. The floating fenders provided for Lock and Dam No. 1 on the Mississippi River are up to 26 ft wide, 11 ft high, and 1 ft deep. Each floating fender is provided with four 3000-lb-capacity mooring buttons.

12. Auxiliary Mechanical Equipment. Navigation locks normally require several special-purpose auxiliary mechanical equipment items or systems, including:

Fire-Protection Equipment. Although vessels are required to carry their own fire-extinguishing equipment, navigation locks should also be equipped with fire-protection equipment to extinguish fires within the lock chamber, to protect vessels in the lock chamber, and more importantly, to protect the lock gates in case of a fire. Typical modern lock fire-protection equipment includes, in addition to portable extinguishers for small fires, several water hydrants on top of the lock chamber and approach walls. Criteria adopted by the U.S. Army Corps of Engineers require one pump of 200 to 250 gpm capacity, with at least 100 psi pressure rating, to provide a flow of 50 gpm to each of four or five fire hoses simultaneously, with an effective pressure of at least 60 psi at the nozzles. Water spray systems are used to wash burning flammables away from the gates, to cool gates exposed to fires, and to assist in controlling fires on board vessels. Foam-type fire-extinguishing equipment with monitors on both sides of the lock chamber should also be provided, especially if vessels carrying flammable materials are regularly passing through the lock. The foam-type fire-extinguishing equipment consists of foaming agent storage tanks, piping, and pumps to deliver the foaming agent to mix with water at a ratio of about 4 to 6 percent of the foaming agent to water, depending on the type of agent, and high-pressure pumps and piping to deliver the mixture to the monitors. Modern monitors are motorized and can be remote-controlled from the navigation-lock control room.

Drainage Pumps. Drainage pumps are provided in sumps in the lower galleries of the lock to remove infiltrated water. The drainage pumps are normally of the self-contained, float-operated, submersible type, connected to the piping system discharging to the tailwater.

Unwatering Pumps. Locks are normally unwatered to the tailwater level by gravity, using the normal lock-emptying procedure. The water remaining in the lock chamber and culverts must then be pumped out to complete the unwatering of the lock for inspection and maintenance. Portable, submersible, construction-type pumps can be used for unwatering the lock. For larger locks, however, a permanently installed unwatering system can be considered, and it may prove more convenient than portable construction-type pumps. The permanently installed pumps should be located in a pump room at the bottom of the lock structure. The pumps should be connected to the lowest point in the lock chamber or in the filling and emptying culvert and should discharge through permanently installed piping to tailwater. Pump selection should be such that the unwatering of the lock can be completed in not more than 1 or 2 days.

Compressed-Air System. The compressed-air system in navigation locks may be required to provide air for a bubbler system in addition to station service needs. Bubbler systems are used in cold-climate locks to prevent ice formation in gate slots and in the front of miter gates.

Emergency Diesel Generator. Emergency diesel generators are required for navigation locks if the lock is at a remote location and/or the lock has an unreliable power supply. The emergency diesel generator should be sized to handle essential electrical loads which include (the largest power requirement for operating one gate at a time) signaling equipment, essential lighting, drainage pumps, and power for at least partial operation of the fire-protection equipment.

13. Control and Signaling Equipment. The large number of gates or valves and other equipment to be operated during a lockage and the precise sequence of operations to be followed require extensive control equipment permitting remote operation of the lock from the control room. The control room is normally located in a control building above the lock walls, preferably at the downstream lock-chamber gate, where best visibility is provided for the lock master.

Interlocks for sequencing gate operations typically include:

1. For filling culvert gate; opening is permitted only if:
    a. Emptying gate is closed.
    b. Downstream lock chamber gate is closed.
2. For emptying culvert gate, opening is permitted only if:
    a. Filling gate is closed.
    b. Upstream lock-chamber gate is closed.
3. For upstream lock-chamber gate, opening is permitted if:
    a. Downstream lock-chamber gate is closed.
    b. Emptying gate is closed.
    c. Differential water head acting on the gate is within the allowable range for gate operation.
4. For downstream lock-chamber gate, opening is permitted only if:
    a. Upstream lock-chamber gate is closed.
    b. Filling gate is closed.
    c. Differential water head acting on gate is within allowable range for gate operation.

When safety barriers are provided, their operation can also be included in the controlled sequence of operations. If safety barriers are provided of the type shown in Fig. 13, which can be adjusted for the water level in the lock chamber, controls can be added so that the safety barrier will automatically follow the water level in the lock chamber, with a preset level difference to compensate for the size and type of the vessel and whether it is loaded or empty.

Other equipment used in locks to facilitate control and operation of the lock includes closed-circuit television equipment for visual surveillance of the extremities of the lock and radio equipment for communication between lock personnel and with vessels.

Signaling devices are required in locks for directing traffic and as navigation aids. Navigation aids include navigation and nose lights at the ends of the approach walls and lock-chamber walls as well as buoys in the upstream and down-

TABLE 1 Some Recently Constructed Navigation Locks Listed in Order of Their Active Volume

Name	Country	Length, m	Width, m	Lift, m	Max. depth, m	Filling and emptying system	Filling and emptying gate	U.S. lock gate	DS lock gate
Volgograd*	Russia	290	30	27.5	5.7	Floor culverts	Slide gate	Miter	Miter
Zaporojie	Russia	290	18	39.2	5.5	Floor culverts	Rev. Tainter	Vert. lift	Miter
Votkinsk	Russia	289	29	23	4	Floor culverts	Slide	Vert. lift	Miter
John Day	United States	205.7	26.2	34.4	4.6	Wall culverts, floor diff.	Rev. Tainter	Vert. lift	Vert. lift
Pickwick	United States	304.8	33.5	18	4	Wall culverts, side disch.	Rev. Tainter	Miter	Miter
Barkley	United States	243.8	33.5	22.3	3.4	Wall culverts, floor diff.	Rev. Tainter	Miter	Miter
Yacyreta	Argentina/Paraguay	272	27	23.9	3.65	Wall culverts, floor diff.	Rev. Tainter	Vert. lift	Miter
Lower Granite	United States	205.7	26.2	32	4.6	Dynamically balanced	Rev. Tainter	Subm. Tainter	Miter
Lower Monumental	United States	205.7	26.2	31.4	4.6	Wall culverts, floor diff.	Rev. Tainter	Subm. Tainter	Vert. lift
Ice Harbor	United States	205.7	26.2	31.4	4.6	Wall culverts, floor diff.	Rev. Tainter	Subm. Tainter	Vert. Lift
Little Goose	United States	205.7	26.2	30.8	4.6	Wall culverts, floor diff.	Rev. Tainter	Subm. Tainter	Miter
Perm	Russia	240	30	21.8	5	In lock gates	Tainter	Horiz. slide	Horiz. slide
Bay Springs	United States	182.9	33.5	25.6	4.6	Wall culverts, bott. long.	Rev. Tainter	Miter	Miter
Balakovo	Russia	290	30	16.4	5	Opening lock gates	Wheel gate	Wheel gate	Miter
McAlpine	United States	365.8	33.5	11.3	3.7	Wall culverts, floor diff.	Rev. Tainter	Miter	Miter
Markland 1	United States	365.8	33.5	10.6	4.6	Wall culverts, floor diff.	Rev. Tainter	Miter	Miter
Bankhead	United States	182.9	33.5	21	4.1	Wall culverts, bott. long.	Rev. Tainter	Miter	Miter
Port Allen	United States	365.8	25.6	13.7	4.2	Wall culverts, side disch.	Rev. Tainter	Miter	Miter
Holt	United States	182.9	33.5	19.4	4	Wall culverts, floor diff.	Rev. Tainter	Miter	Miter
Bonneville	United States	205.7	26.2	21.3	6.1	Dynamically balanced	Rev. Tainter	Subm. Tainter	Miter
Meldahl 1	United States	365.8	33.5	9.1	4.6	Wall culverts, floor diff.	Rev. Tainter	Miter	Miter
Dardanelle	United States	182.9	33.5	16.5	4.3	Wall culverts, bott. long.	Rev. Tainter	Miter	Miter
Wheeler 1	United States	182.9	33.5	15.7	4	Wall culverts, side disch.	Rev. Tainter	Miter	Miter
Old River	United States	365.8	22.9	11.3	4.9	Wall culverts, side disch.	Rev. Tainter	Miter	Miter
W.F. George	United States	137.2	25.6	26.8	4	Wall culverts, floor diff.	Rev. Tainter	Miter	Miter
Altenwörth†	Austria	230	24	17.04	N/A	Short culverts	Wheel gate	Double-leaf hook-type gate	Miter
Cannelton 1	United States	365.8	33.5	7.6	4.6	Wall culverts, side disch.	Rev. Tainter	Miter	Miter

*An intermediate miter gate is also installed in the lock.
†Twin lock.

stream approaches of the lock marking the navigable channels for entering and leaving. Buoys should be lighted for navigation at night. The color, shape, and light characteristics of navigation lights are controlled in the United States by U.S. Coast Guard standards. Traffic-control signaling devices are provided in the form of traffic lights at the upstream and downstream ends of locks and signaling horns. Where required, a small-craft signaling system is provided to enable small boats to signal for lockage. The small-craft signal system consists of spring-loaded lever activated switches which can be operated by pulling on a cord. When the switch is activated, the system will alert lock personnel that small craft are requesting lockage.

LOCK EXAMPLES

The basic design parameters of some recently constructed navigation locks are indicated in Table 1.

All the navigation locks discussed in this section are conventional locks, which rely on flow by gravity to fill and empty lock chambers for lockage. In addition to these conventional locks, there are several different designs to permit passage of vessels. These include ship elevators of the vertical-lift type with flotation chambers (Henrichenburg in Germany, 93 m long, 12 m wide, with a lift of 14.7 m), ship elevators on inclined slopes (Ronquieres, Belgium, 90 m long, 12 m wide, lift about 3.3 m), dry slips, cranes, etc. Such designs are relatively uncommon and have not been discussed in this section.

REFERENCES

1. Permanent International Association of Navigation Congresses: Final Report of the International Commission for the Study of Locks (General Secretariat of PIANC, Residence Palace, Quartier Jordaens, rue de le Loi, 155, 1040 Brussels, Belgium).
2. U.S. Army Corps of Engineers, Engineering Manual EM 1110-2-1604, Part CXVI chap. 4, Hydraulic Design—Navigation Locks.
3. RICH, G. R., Navigation Locks, in Davis and Sorensen, *Handbook of Applied Hydraulics,* 3d ed., chap. 32.
4. U.S. Army Corps of Engineers, Engineering Manual EM 1110-2-1610, Hydraulic Design of Lock Culvert Gates.

SECTION 24

IRRIGATION

By David B. Palmer[1]

In determining the feasibility of a contemplated irrigation project, three fundamental questions must be carefully considered: (1) Are the project lands suited to agricultural use from topographic and soil-structure viewpoints? (2) Are available water supplies sufficient in quantity and suitable in quality to meet the irrigation requirements? (3) Will the cost of the necessary engineering works per acre of irrigable land be low enough to justify construction?

The greater part of this section is devoted to the physical aspects of irrigation work, such as water supply, water requirements, water losses, consumptive use, and irrigation methods. The principles governing the design and construction of irrigation works are discussed elsewhere.

LAND CLASSIFICATION

Land classification must be carefully considered in planning new projects. Such classifications must be made by agricultural and soil experts working in cooperation with engineers. Some lands, which may be easily and economically watered, may not merit inclusion in the irrigable acreage because of unproductive soil composition, undesirable soil texture, difficult drainage possibilities, presence of undesirable soluble salts, or danger of developing alkali surfaces under continued irrigation. Other lands, which may cost more to irrigate, may deserve inclusion because of ideal soil characteristics, easy drainage, and high prospects of developing into permanent and prosperous productive areas.

The methodology for land classification described in this section is adapted from that of the U.S. Bureau of Reclamation and assumes that no prior classification studies have been performed on project lands. Where it is found that prior studies have been made, the information therefrom should be used to the maximum extent possible. For example, taxonomic soil surveys, as made by the U.S. Soil Conservation Service in cooperation with state soil survey organizations, often furnish an abundance of data useful in the preparation of an irrigation suitability land classification. An experienced land-classification specialist is often able to develop a land classification from the soil survey report with a minimal amount of field checks.

1. Purposes of Classification. Land classification is undertaken to provide an inventory of the relative suitability of available lands for sustained production under irrigation. This primary purpose usually can be served by a survey of reconnaissance scope. Following the selection of the outlines of the project area on the

---
[1]Acknowledgment is made to Rolland F. Kaser for material in this section which appeared in the third edition (1969).

basis of the reconnaissance survey, a semidetailed survey is undertaken to provide the basis for selection of the lands to be served with irrigation water, for determination of the probable productivity of those lands and the crops for which they are suitable, and for determination of the criteria for design of irrigation and drainage works. The semidetailed survey also provides information on the need for soil reclamation to remove deleterious salts and on the characteristics and location of the lands requiring such treatment prior to successful cropping. In some cases, a detailed land-classification survey will be necessary in specific locations where critical problems such as soil-reclamation needs, farm-drainage requirements, or other considerations require more precise definition than that furnished by the semidetailed survey.

2. Physical Properties. Physical properties which are important to the irrigability and drainability of land include (1) topographic relief and slope, which influence the cost and type of water-distribution facilities and the labor requirements for water applications to the field; (2) texture, grading, and depth of the surface soil and subsoil, which determine the water-holding capacity and drainability of the soil; and (3) natural drainage characteristics, such as surface drainage channels, drainage outlets, and subsurface materials—gravel layers, hardpan formations, impervious soil or rock strata—which influence the drainability and reclaimability of the soil. The land classification must locate and identify changes in each of these characteristics with an accuracy commensurate with the scope of the survey needed.

Air drainage should be studied in classifying lands where appreciable areas may be adapted to fruit or winter vegetable culture. Experience has shown that crops of such nature, grown on low-lying lands where wind movements are obstructed by surrounding hills and consequently are not adequate to provide satisfactory air drainage, may suffer severe damage from frost at times when similar crops, grown on nearby sloping topography where wind movements provide ample air drainage, may suffer no damage at all. For example, orchards on the sloping hillsides of the Roza and Tieton divisions, Yakima project, Washington, frequently are undamaged by frost when similar orchards on low-lying lands near the town of Yakima are severely damaged. Many comparable examples occur in the citrus-fruit sections of Arizona and California, as well as in other sections of the West. Consequently, it may sometimes be desirable to include sloping hillside regions in the irrigable areas, even though the preparation of land and conveyance of water may cost considerably more per acre than in the low-lying, more level divisions of the project.

3. Chemical Properties. Chemical properties of the soil determine the need for reclamation treatment, the reactions to be expected when irrigation waters having certain chemical characteristics are applied, and the fertilizers and other treatments required to obtain optimum crop yields. In irrigation work, the term *alkali* is used to mean soluble salts that may be brought to the surface by capillary soil-moisture movements and precipitated as the moisture evaporates. The commonest alkali compounds are the sodium sulfates, chlorides, and carbonates. Other alkali compounds are the potassium and magnesium carbonates, chlorides, and sulfates, and the calcium chlorides and nitrates. Sodium carbonates, which cause the decomposition of organic matter and the formation of a dark-colored crust at the surface of the ground, are called *black alkali*. Sodium carbonate (salt soda), sodium bicarbonate (baking soda), and sodium chloride (table salt) are especially detrimental to plant growth, the first two being the more objectionable and more difficult to remove from the soil. Noncaustic compounds that do not deflocculate the soil, such as sodium chloride and sulfate, are often called *white alkali*.

The proportions of alkali salts that the soil may contain and still be irrigated profitably vary with the character of the soil, fertility, drainage, methods of irrigating, and crops grown, as well as with the kinds of salts present. The danger of nonalkali lands developing alkali surfaces under continued irrigation depends on the permeability of the soil, quantity of water applied, drainage facilities, and saline content of the irrigation water. Persons classifying alkali lands, or areas that may develop alkali surfaces, should consult agricultural references which present the findings of the latest research activities applied to similar problems. With a few exceptions, grasses, small grains, alfalfa, and root crops are more resistant to alkali than are corn, beans, peas, melons, and fruits.

The presence of noticeable quantities of alkali on the ground surface or considerable proportions of salts in the soil may not be serious if the ground is permeable and can be economically drained, and if sufficient water can be applied to maintain percolation toward drainage outlets. If the lands are properly drained, objectionable quantities of alkali sometimes can be leached out of the soil by flooding the surface during the winter months, by growing crops such as rice which require or can tolerate the application of large amounts of water, or by other special applications of water for the purpose of leaching out the salts to achieve reclamation.

4. Drainability. One of the principal purposes of a land-classification survey is to assess the properties of the land which influence surface and subsurface drainage. In addition to the physical properties listed above, chemical properties of the soil may also influence drainability. Excessive concentrations of sodium or sodium salts in the soil or in the irrigation water can cause reduction in permeability and, under extreme conditions, render the soil unproductive. Although physically feasible, reclamation of that soil may not be economically feasible.

Drainage is important in the suitability of land for permanent productivity under irrigation. Excessive and prolonged inundation resulting from heavy rainfall on the land can be destructive to growing crops and can adversely affect soil structure and permeability. High water tables adversely affect most irrigated crops by reducing the thickness of the root zone which the plants can utilize for essential nutrients (see Sec. 25).

Reclamation of saline and alkaline soils usually can be accomplished by the passage of large amounts of water through the soil profile, and this cannot be done efficiently if the soil permeability is low. A gradual buildup of salts will occur in the soils under continuing irrigation unless natural drainage processes or constructed drainage facilities are effective in removing from the soil, on an annual basis, a volume of salt equal to that brought in by the irrigation water. Accordingly, the soil permeability must be adequate for leaching, i.e., the application of sufficient irrigation water both to meet the evapotranspiration needs and to furnish the water required to convey excess salts to the water table. Subsurface drainage must be adequate to carry away enough groundwater to prevent the water table from entering the plant root zone with duration and frequency sufficient to cause uneconomic crop yields.

5. Sampling and Testing. Land-classification surveys include observation of the land and soil in the field, collection and laboratory analyses of samples of the soil and groundwater, field testing of the soil permeability, and examination of aerial photographs. The field observations and collection of soil samples are facilitated by the use of auger holes, test pits, and occasional deep drilling. The density of the sampling, i.e., number of auger holes and test pits of various depths per square mile, depends upon the purpose and scope of the survey and the variability of conditions within the area being surveyed. General criteria established by the Bureau of Reclamation for these surveys are summarized in Table 1.

Auger sampling of agricultural soils is usually done by manually operated augers, although modern equipment includes power-operated augers mounted on vehicles of various types. Vehicular-mounted augers take much less time per hole and permit a land classifier to cover more acres per day than would be possible with a manual auger. This advantage is somewhat offset by the fact that better identification of thin soil layers can be obtained with the manual auger. Test pits permit the classifier to inspect the various soil layers down to a considerable depth to identify the soil structure and to obtain undisturbed and other samples of the several layers. These test pits are usually dug manually, although truck-mounted large-diameter power augers and backhoes can be used to expedite the work.

TABLE 1 Minimum Requirements by Types of Classification*

			Detailed	
	Reconnaissance	Semidetailed	New lands	Fully developed or highly uniform new land areas
Land classes recognized†	1,2,3,6	1,2,3,6	1,2,3,4,5,6	1,2,3,4,5,6
Scale of base maps	1:24,000	1:12,000	1:4,800	1:12,000
Max distance between traverse, miles	1	0.5	0.25	0.5
Accuracy, %	75	90	97	97
Field progress per day (one land classifier and crew), mi²	3–5	1–3	0.25–1	1–3
Min area of class 6 to be segregated from larger arable areas, acres	4	0.5	0.2	0.2
Min area for change to lower class of arable land, acres	40	10	2	10
Min area for change to higher class of arable land, acres	40	20	10	20
Min soil and substrata examination:				
Borings or pits (5 ft deep) per mi²	1	4	16	4
Deep holes (10 ft or more) per township	1	2	4	2

*From "Irrigated Land Use," Bureau of Reclamation Manual, vol. 5.
†See Table 2 for a description of these land classes.

Field permeability tests are usually necessary where there are clay soils or other indications of low permeability. This is because experience has shown that laboratory tests of undisturbed soil samples often result in misleading permeability values and because of difficulties in collecting, transporting, and testing the "undisturbed" sample. Up-to-date reference sources should be consulted for details of equipment and procedures which may be employed under various field conditions and in field laboratories.

Soil samples collected for land-classification purposes must be properly tagged and identified by number, location of hole, depth below surface, and any other pertinent information. The samples should be tagged also to indicate the tests which should be made on the contents. It may be feasible on large projects to construct and equip a special laboratory for the purpose of testing the soil and water samples collected for that project; however, on most projects it probably will be desirable to have the samples tested by a commercial laboratory. Because many of the details of project formulation and design depend upon results of the land-classification survey—the conclusions of which are based on the field observations and laboratory analyses—it is essential that the survey be scheduled to begin early in the project investigation and that time be allowed for the collection and analysis of soil samples and interpretation of results.

6. Classification Standards. Land-classification surveys must be conducted and the results interpreted in accordance with standard specifications in order that data within and between areas and regions may be compared for purposes of resource appraisal and development. Generalized land-classification specifications used by the Bureau of Reclamation, and developed from the extensive experience of that agency in project development and operation, are summarized in Table 2.

The generalized specifications are broad in scope and apply to gravity irrigation in potential and existing project areas. These specifications must be reviewed and refined as necessary to meet the needs of each land-classification survey to be undertaken. The amount of refinement necessary will increase with the detail of the classification. The project specifications must result in significant differences in productivity and payment capacity between land classes and must assure that subclasses within land classes will have comparable payment capacity. In preparing for land-classification surveys, reference should be made to up-to-date bulletins and manuals presenting detailed descriptions of procedures and methods.

CROP EVAPOTRANSPIRATION

In the process of applying irrigation water to crops, water is returned to the atmosphere by evaporation and by transpiration. Water used for transpiration enters the roots of plants from the surrounding soil water and moves upward through the plants. Small fractions of this water are either used by chemical processes within the plant or incorporated into the cells of the growing plants. The major portion of the water which enters the root system is transpired through the stomata of the plant to the atmosphere.

The additional irrigation water which is evaporated is a function of several variables including temperature, relative humidity, wind, crop canopy, and method of irrigation being used. Evapotranspiration is the sum of the irrigation

TABLE 2 Land-classification Specifications—General*

Land characteristics	Class 1—Arable	Class 2—Arable	Class 3—Arable
Soils			
Texture	Sandy loam to friable clay loam	Loamy sand to very permeable clay	Loamy sand to permeable clay
Depth:			
To sand, gravel, or cobble	36 in plus—good free-working soil of fine sandy loam or finer; or 42 in of sandy loam	24 in plus—good free-working soil of fine sandy loam or finer; or 30–36 in of sandy loam to loamy sand	18 in plus—good free-working soil of fine sandy loam or finer; or 24–30 in of coarser-texture soil
To shale, raw soil from shale or similar material (6 in less in each instance to rock and similar material)	60 in plus; or 54 in with minimum of 6 in of gravel overlying impervious material or sandy loam throughout	48 in plus; or 42 in with minimum of 6 in of gravel overlying impervious material or sandy loam throughout	42 in plus; or 36 in with minimum of 6 in of gravel overlying impervious material or sandy loam throughout
To penetrable lime zone	18 in with 60 in penetrable	14 in with 48 in penetrable	10 in with 36 in penetrable
Alkalinity	pH less than 9.0 unless soil is calcareous, total salts are low, and evidence of black alkali is absent	pH 9.0 or less, unless soil is calcareous, total salts are low, and evidence of black alkali is absent	pH 9.0 or less, unless soil is calcareous, total salts are low, and evidence of black alkali is absent
Salinity	Total salts not to exceed 0.2%. May be higher in open permeable soils and under good drainage conditions	Totals salts not to exceed 0.5%. May be higher in open permeable soils and under good drainage conditions	Total salts not to exceed 0.5%. May be higher in open permeable soils and under good drainage conditions
Topography			
Slopes	Smooth slopes up to 4% in general gradient in reasonably large-sized bodies sloping in the same plane	Smooth slopes up to 8% in general gradient in reasonably large-sized bodies sloping in the same plane; or rougher slopes which are less than 4% in general gradient	Smooth slopes up to 12% in general gradient in reasonably large-sized bodies sloping in the same plane; or rougher slopes which are less than 8% in general gradient
Surface	Even enough to require only small amount of leveling and no heavy grading	Moderate grading required but in amounts found feasible at reasonable cost in comparable irrigated areas	Heavy and expensive grading required in spots but in amounts found feasible in comparable irrigated areas
Cover (loose rock and vegetation)	Insufficient to modify productivity or cultural practices, or clearing cost small	Sufficient to reduce productivity and interfere with cultural practices. Clearing required but at moderate cost	Present in sufficient amounts to require expensive but feasible clearing

TABLE 2 Land-classification Specifications—General* (*Continued*)

Soil and topography	Drainage		
	Soil and topographic conditions such that no specific farm-drainage requirement is anticipated	Soils and topographic conditions such that some farm drainage will probably be required but with reclamation by artificial means appearing feasible at reasonable cost	Soil and topographic conditions such that significant farm drainage will probably be required but with reclamation by artificial means appearing expensive but feasible

Class 4—Limited Arable

Includes lands having excessive deficiencies and restricted utility but which special economic and engineering studies have shown to be irrigable

Class 5—Nonarable

Includes lands which will require additional economic and engineering studies to determine their irrigability and lands classified as temporarily nonproductive pending construction of corrective works and reclamation

Class 6—Nonarable

Includes lands which do not meet the minimum requirements of the next higher class mapped in a particular survey and small areas of arable land lying within larger bodies of nonarable land

*From "Irrigated Land Use," Bureau of Reclamation Manual, vol. 5.

water which returns to the atmosphere through the transpiration and evaporation processes. In addition to these two processes the total irrigation water which is required must provide for inefficiencies and losses in water conveyance and application, such as seepage from canals, runoff from fields, and seepage through the root zone to depths not accessible to plant roots.

Several methods have been developed for estimating the evapotranspiration requirements of crops. The method which will yield the best estimate for a given situation depends upon the basic data available and the nature of the estimate required. Brief discussions of some of the methods now in use, and the circumstances under which each of them is usable, are given in the following paragraphs. For additional information on the selection and use of methods for estimating evapotranspiration, see (1) Evapotranspiration and Irrigation Water Requirements, American Society of Civil Engineers, Manuals and Reports on Engineering Practice no. 70, 1990, and (2) Crop Water Requirements, FAO Irrigation and Drainage Paper 24, Food and Agriculture Organization of the United Nations, revised 1977.

7. Penman-Monteith Method. This "combination" method, so named because it includes both energy balance and heat and mass transfer mechanisms, incorporates aerodynamic and surface resistance terms.

The Penman-Monteith combination method equations and relationships are:

$$E_{tr} = \frac{\Delta}{\Delta + \gamma^*}(R_n - G) + \frac{\gamma}{\Delta + \gamma^*} K_1 \frac{0.622 \lambda \rho}{P} \frac{1}{r_a}(e_z^0 - e_z)$$

where

$$r_a = \frac{\ln[(z_w - d)/z_{om}] \ln[(z_p - d)/z_{ov}]}{(0.41)^2 u_z}$$

and

$$\gamma^* = \gamma(1 + r_e/r_a)$$

where E_{tr} = rate at which water, if readily available, would be removed from soil and plant surfaces, expressed as a depth of water evaporated and transpired from a reference crop (alfalfa), mm day^{-1}
  Δ = slope of the saturation vapor pressure-temperature curve, kPa °C^{-1}
  γ = psychrometric constant, kPa °C^{-1}
  γ^* = psychrometric constant modified by the ratio of canopy resistance to atmospheric resistance, kPa °C^{-1}
  R_n = net radiation, MJ m^{-2} day^{-1}
  G = soil heat flux, MJ m^{-2} day^{-1}
  K_1 = dimension coefficient, 8.64 × 10^4
  λ = latent heat of vaporization, MJ kg^{-1}
  ρ = air density, kg m^{-3}
  P = atmospheric pressure, kPa
  r_a = diffusion resistance of air layer, m^{-2}
  r_e = crop canopy resistance, m^{-2}
  e_z^0 and e_z = saturation and actual vapor pressures at the Z level above the surface, kPa
  d = zero plane displacement of wind profile, cm
  z_w = height of wind speed measurement, cm
  z_{ov} = roughness length, heat, and water vapor, cm
  z_{om} = roughness length, momentum, cm
  z_p = height of humidity and temperature measurements, cm
  u_z = horizontal wind speed at height z_w, m s^{-1}

A modification of the Penman-Monteith method is presented in the FAO Irrigation and Drainage Paper 24. The FAO 24 presentation includes a format and representative calculations.

8. FAO-24 Blaney-Criddle Method. This "temperature" method is suggested for arid locations where available climatic data cover air temperature data only. Crop evapotranspiration E_t estimates are made in two steps. Initially, a grass-related reference crop evapotranspiration E_{to} is estimated and then E_{to} is applied to grass-related crop coefficients K_c (grass) to obtain crop E_t.

The empirical formula used is

$$E_{to} = a + bf = a + bp(0.46T + 8.13)$$

where E_{to} = grass reference E_t, mm day^{-1}
  T = mean air temperature, °C
  a,b = adjustment factors based on minimum relative humidity, sunshine hours, and daytime wind
  p = mean daily percent of annual daytime hours, monthly $p/$(days per month)
  a = 0.0043 RH$_{min}$ − n/N − 1.41
  b = 0.82 − 0.41 × 10^{-2} RH$_{min}$ + 1.07 (n/N) + 0.066U_d − 0.60 × 10^{-2} RH$_{min}$ (n/N) − 0.60 × 10^{-3} RH$_{min}$ U_d
  U_d = daytime wind at 2-m height, m s^{-1}

Estimates of E_{to} can be made using Table 3 and Fig. 1. If numerical values of relative humidity, sunshine duration, and daytime wind are not available, qualitative estimates may be made which will permit the use of Fig. 1. Figure 2 is a calculation format for the Blaney-Criddle method.

9. FAO-24 Radiation Method. This "radiation" method is suggested for areas where available climatic data include measured air temperature and sunshine, cloudiness or radiation, but not measured wind and humidity. Knowledge of general levels of humidity and wind is required, and these are to be estimated using published weather descriptions and extrapolation from nearby areas or from local sources.

The relationship is expressed as

$$E_{to} = c(W \times R_s)$$

where E_{to} = reference crop (grass) evapotranspiration, mm day^{-1}
 R_s = solar radiation in equivalent evaporation, mm day^{-1}
 W = weighting factor which depends on temperature and altitude
 c = adjustment factor which depends on mean humidity and daytime wind conditions

The amount of radiation received at the top of the atmosphere R_a is a function of latitude and time of year and is given in Table 4. The fraction of R_a which transits the cloud cover and reaches the earth's surface is the energy available for evapotranspiration. R_s can be measured directly but is frequently not available for the area of investigation. In this case, R_s can be obtained from measured sunshine duration records as follows:

$$R_s = (0.25 + 0.50 \, n/N) \times R_a$$

TABLE 3 Mean Daily Percentage (p) of Annual Daytime Hours for Different Latitudes

Latitude	North	Jan	Feb	Mar	Apr	May	June	July	Aug	Sept	Oct	Nov	Dec
	South	July	Aug	Sept	Oct	Nov	Dec	Jan	Feb	Mar	Apr	May	June
60°		.15	.20	.26	.32	.38	.41	.40	.34	.28	.22	.17	.13
58		.16	.21	.26	.32	.37	.40	.39	.34	.28	.23	.18	.15
56		.17	.21	.26	.32	.36	.39	.38	.33	.28	.23	.18	.16
54		.18	.22	.26	.31	.36	.38	.37	.33	.28	.23	.19	.17
52		.19	.22	.27	.31	.35	.37	.36	.33	.28	.24	.20	.17
50		.19	.23	.27	.31	.34	.36	.35	.32	.28	.24	.20	.18
48		.20	.23	.27	.31	.34	.36	.35	.32	.28	.24	.21	.19
46		.20	.23	.27	.30	.34	.35	.34	.32	.28	.24	.21	.20
44		.21	.24	.27	.30	.33	.35	.34	.31	.28	.25	.22	.20
42		.21	.24	.27	.30	.33	.34	.33	.31	.28	.25	.22	.21
40		.22	.24	.27	.30	.32	.34	.33	.31	.28	.25	.22	.21
35		.23	.25	.27	.29	.31	.32	.32	.30	.28	.25	.23	.22
30		.24	.25	.27	.29	.31	.32	.31*	.30	.28	.26	.24	.23
25		.24	.26	.27	.29	.30	.31	.31	.29	.28	.26	.25	.24
20		.25	.26	.27	.28	.29	.30	.30	.29	.28	.26	.25	.25
15		.26	.26	.27	.28	.29	.29	.29	.28	.28	.27	.26	.25
10		.26	.27	.27	.28	.28	.29	.29	.28	.28	.27	.26	.26
5		.27	.27	.27	.28	.28	.28	.28	.28	.28	.27	.27	.27
0		.27	.27	.27	.27	.27	.27	.27	.27	.27	.27	.27	.27

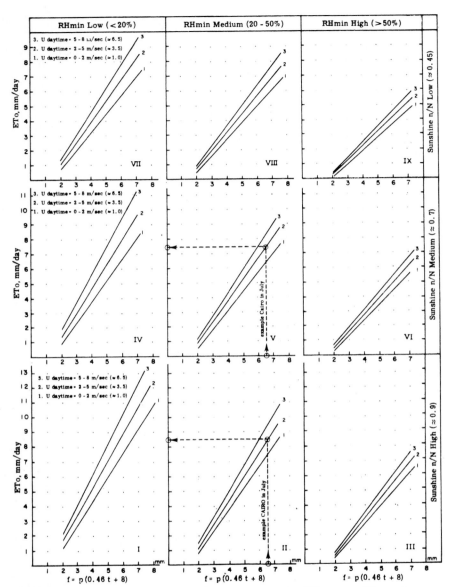

FIGURE 1 Prediction of E_{to} from Blaney-Criddle f factor for different conditions of minimum relative humidity, sunshine duration, and daytime wind.

where n/N = ratio between actual measured bright sunshine hours and maximum possible sunshine hours

Values of N for different months and latitudes are given in Table 5. Values of n, using the Campbell-Stokes sunshine recorder, may be available locally. If not, estimates can be made from cloudiness observation records. Such records may use "oktas" or "tenths." The n/N ratios are given in Table 6.

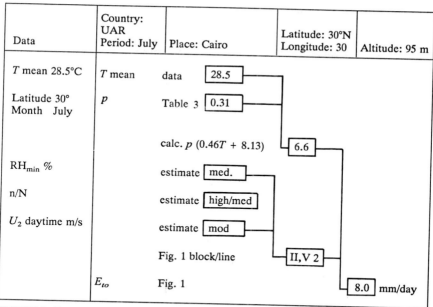

FIGURE 2 Format for calculation of Blaney-Criddle method.

The weighting factor W reflects the effect of temperature and altitude on the relationship between R_s and reference crop evapotranspiration E_{to}. Values of W as related to temperature and altitude are given in Table 7. Temperature reflects the mean air temperature in degrees Celsius for the period considered. Where temperature is given as T_{max} and T_{min}, the temperature $(T_{max} + T_{min})/2$ should be used.

The adjustment factor c depends greatly on general levels of mean relative humidity and daytime wind (see Fig. 3). Figure 4 is an example and format for calculations of radiation method.

10. FAO-24 Pan Evaporation Method. Evaporation pans provide a measurement of the integrated effect of radiation, wind, temperature, and humidity on evaporation from a specific open water surface. In a similar fashion the plant responds to the same climatic variables, but several major factors may produce significant differences in loss of water. These factors include (1) differences in reflection of solar radiation, (2) storage of heat within the pan, and (3) differences in turbulence, temperature, and humidity of the air immediately above the surfaces. Notwithstanding these deficiencies, the use of pans to predict crop water requirements for periods of 10 days or longer is still warranted.

Reference crop evapotranspiration E_{to} can be obtained from

$$E_{to} = K_p \times E_{pan}$$

where E_{pan} = pan evaporation, mm/day; represents the mean daily value of the period considered
K_p = pan coefficient

Values for K_p are given in Table 8 for the U.S. Class A pan for different hu-

TABLE 4 Extra Terrestrial Radiation (Ra) Expressed in Equivalent Evaporation in mm/day

Northern Hemisphere											
Jan	Feb	Mar	Apr	May	June	July	Aug	Sept	Oct	Nov	Dec
3.8	6.1	9.4	12.7	15.8	17.1	16.4	14.1	10.9	7.4	4.5	3.2
4.3	6.6	9.8	13.0	15.9	17.2	16.5	14.3	11.2	7.8	5.0	3.7
4.9	7.1	10.2	13.3	16.0	17.2	16.6	14.5	11.5	8.3	5.5	4.3
5.3	7.6	10.6	13.7	16.1	17.2	16.6	14.7	11.9	8.7	6.0	4.7
5.9	8.1	11.0	14.0	16.2	17.3	16.7	15.0	12.2	9.1	6.5	5.2
6.4	8.6	11.4	14.3	16.4	17.3	16.7	15.2	12.5	9.6	7.0	5.7
6.9	9.0	11.8	14.5	16.4	17.2	16.7	15.3	12.8	10.0	7.5	6.1
7.4	9.4	12.1	14.7	16.4	17.2	16.7	15.4	13.1	10.6	8.0	6.6
7.9	9.8	12.4	14.8	16.5	17.1	16.8	15.5	13.4	10.8	8.5	7.2
8.3	10.2	12.8	15.0	16.5	17.0	16.8	15.6	13.6	11.2	9.0	7.8
8.8	10.7	13.1	15.2	16.5	17.0	16.8*	15.7	13.9	11.6	9.5	8.3
9.3	11.1	13.4	15.3	16.5	16.8	16.7	15.7	14.1	12.0	9.9	8.8
9.8	11.5	13.7	15.3	16.4	16.7	16.6	15.7	14.3	12.3	10.3	9.3
10.2	11.9	13.9	15.4	16.4	16.6	16.5	15.8	14.5	12.6	10.7	9.7
10.7	12.3	14.2	15.5	16.3	16.4	16.4	15.8	14.6	13.0	11.1	10.2
11.2	12.7	14.4	15.6	16.3	16.4	16.3	15.9	14.8	13.3	11.6	10.7
11.6	13.0	14.6	15.6	16.1	16.1	16.1	15.8	14.9	13.6	12.0	11.1
12.0	13.3	14.7	15.6	16.0	15.9	15.9	15.7	15.0	13.9	12.4	11.6
12.4	13.6	14.9	15.7	15.8	15.7	15.7	15.7	15.1	14.1	12.8	12.0
12.8	13.9	15.1	15.7	15.7	15.5	15.5	15.6	15.2	14.4	13.3	12.5
13.2	14.2	15.3	15.7	15.5	15.3	15.3	15.5	15.3	14.7	13.6	12.9
13.6	14.5	15.3	15.6	15.3	15.0	15.1	15.4	15.3	14.8	13.9	13.3
13.9	14.8	15.4	15.4	15.1	14.7	14.9	15.2	15.3	15.0	14.2	13.7
14.3	15.0	15.5	15.5	14.9	14.4	14.6	15.1	15.3	15.1	14.5	14.1
14.7	15.3	15.6	15.3	14.6	14.2	14.3	14.9	15.3	15.3	14.8	14.4
15.0	15.5	15.7	15.3	14.4	13.9	14.1	14.8	15.3	15.4	15.1	14.8

midity and wind conditions and pan environments. The K_p values relate to pans located in an open field with no crops taller than 1 m within some 50 m of the pan. Immediate surroundings, within 10 m, are covered by a green, frequently mowed, grass cover or by bare soils. The pan station is placed in an agricultural area. The pan is unscreened.

In Table 8 a separation is made for pans located within cropped plots surrounded by or downwind from dry surface areas (case A) and for pans located within a dry fallow field but surrounded by irrigated or rained upwind cropped areas (case B). Where pans are placed in a small enclosure but surrounded by tall crops, for example, 2.5-m-high maize, the coefficients in the table will need to be increased by up to 30 percent for dry, windy climates, whereas only a 5 to 10 percent increase is required for calm, humid conditions.

| Lat | \multicolumn{12}{c}{Southern Hemisphere} |
	Jan	Feb	Mar	Apr	May	June	July	Aug	Sept	Oct	Nov	Dec
50°	17.5	14.7	10.9	7.0	4.2	3.1	3.5	5.5	8.9	12.9	16.5	18.2
48°	17.6	14.9	11.2	7.5	4.7	3.5	4.0	6.0	9.3	13.2	16.6	18.2
46°	17.7	15.1	11.5	7.9	5.2	4.0	4.4	6.5	9.7	13.4	16.7	18.3
44°	17.8	15.3	11.9	8.4	5.7	4.4	4.9	6.9	10.2	13.7	16.7	18.3
42°	17.8	15.5	12.2	8.8	6.1	4.9	5.4	7.4	10.6	14.0	16.8	18.3
40°	17.9	15.7	12.5	9.2	6.6	5.3	5.9	7.9	11.0	14.2	16.9	18.3
38°	17.9	15.8	12.8	9.6	7.1	5.8	6.3	8.3	11.4	14.4	17.0	18.3
36°	17.9	16.0	13.2	10.1	7.5	6.2	6.8	8.8	11.7	14.6	17.0	18.2
34°	17.8	16.1	13.5	10.5	8.0	6.8	7.2	9.2	12.0	14.9	17.1	18.2
32°	17.8	16.2	13.8	10.9	8.5	7.3	7.7	9.6	12.4	15.1	17.2	18.1
30°	17.8	16.4	14.0	11.3	8.9	7.8	8.1	10.1	12.7	15.3	17.3	18.1
28°	17.7	16.4	14.3	11.6	9.3	8.2	8.6	10.4	13.0	15.4	17.2	17.9
26°	17.6	16.4	14.4	12.0	9.7	8.7	9.1	10.9	13.2	15.5	17.2	17.8
24°	17.5	16.5	14.6	12.3	10.2	9.1	9.5	11.2	13.4	15.6	17.1	17.7
22°	17.4	16.5	14.8	12.6	10.6	9.6	10.0	11.6	13.7	15.7	17.0	17.5
20°	17.3	16.5	15.0	13.0	11.0	10.0	10.4	12.0	13.9	15.8	17.0	17.4
18°	17.1	16.5	15.1	13.2	11.4	10.4	10.8	12.3	14.1	15.8	16.8	17.1
16°	16.9	16.4	15.2	13.5	11.7	10.8	11.2	12.6	14.3	15.8	16.7	16.8
14°	16.7	16.4	15.3	13.7	12.1	11.2	11.6	12.9	14.5	15.8	16.5	16.6
12°	16.6	16.3	15.4	14.0	12.3	11.6	12.0	13.2	14.7	15.8	16.4	16.5
10°	16.4	16.3	15.5	14.2	12.8	12.0	12.4	13.5	14.8	15.9	16.2	16.2
8°	16.1	16.1	15.5	14.4	13.1	12.4	12.7	13.7	14.9	15.8	16.0	16.0
6°	15.8	16.0	15.6	14.7	13.4	12.8	13.1	14.0	14.0	15.7	15.8	15.7
4°	15.5	15.8	15.6	14.9	13.8	13.2	13.4	14.3	15.1	15.6	15.5	15.4
2°	15.3	15.7	15.7	15.1	14.1	13.5	13.7	14.5	15.2	15.5	15.3	15.1
0°	15.0	15.5	15.7	15.3	14.4	13.9	14.1	14.8	15.3	15.4	15.1	14.8

The level at which the water is maintained in the pan is very important; resulting errors may be up to 15 percent when water levels in Class A pans fall 10 cm below the accepted standard of between 5 and 7.5 cm below the rim. Screens mounted over pans will reduce E_{pan} by up to 10 percent.

11. Crop Coefficients. Coefficients which reflect the water-using characteristics of each crop are necessary for estimating the evapotranspiration requirements. Since the actual crop evapotranspiration is difficult to measure accurately under field conditions, and requires extensive and complex instrumentation, basic data for the determination of crop coefficients are available for only a few experimental sites. In the United States, most of these data result from the cooperative investigations of the Agricultural Research Service of the U.S. Department of Agriculture and universities in states in which irrigation is practiced ex-

TABLE 5 Mean Daily Duration of Maximum Possible Sunshine Hours (N) for Different Months and Latitudes

Northern Lats	Jan	Feb	Mar	Apr	May	June	July	Aug	Sept	Oct	Nov	Dec
Southern Lats	July	Aug	Sept	Oct	Nov	Dec	Jan	Feb	Mar	Apr	May	June
50	8.5	10.1	11.8	13.8	15.4	16.3	15.9	14.5	12.7	10.8	9.1	8.1
48	8.8	10.2	11.8	13.6	15.2	16.0	15.6	14.3	12.6	10.9	9.3	8.3
46	9.1	10.4	11.9	13.5	14.9	15.7	15.4	14.2	12.6	10.9	9.5	8.7
44	9.3	10.5	11.9	13.4	14.7	15.2	15.2	14.0	12.6	11.0	9.7	8.9
42	9.4	10.6	11.9	13.4	14.6	15.0	14.9	13.9	12.6	11.1	9.8	9.1
40	9.6	10.7	11.9	13.3	14.4	14.5	14.7	13.7	12.5	11.2	10.0	9.3
35	10.1	11.0	11.9	13.1	14.0	14.0	13.3	13.5	12.4	11.3	10.3	9.8
30	10.4	11.1	12.0	12.9	13.6	13.7	13.9*	13.2	12.4	11.5	10.6	10.2
25	10.7	11.3	12.0	12.7	13.3	13.3	13.5	13.0	12.3	11.6	10.9	10.6
20	11.0	11.5	12.0	12.6	13.1	13.0	12.2	12.8	12.3	11.7	11.2	10.9
15	11.3	11.6	12.0	12.5	12.8	12.7	12.9	12.6	12.2	11.8	11.4	11.2
10	11.6	11.8	12.0	12.3	12.6	12.4	12.6	12.4	12.1	11.8	11.6	11.5
5	11.8	11.9	12.0	12.2	12.3	.32	12.3	12.3	12.1	12.0	11.9	11.8
0	12.1	12.1	12.1	12.1	12.1	12.1	12.1	12.1	12.1	12.1	12.1	12.1

TABLE 6 Cloudiness Ratios

Cloudiness, oktas	n/N	Cloudiness, tenths	n/N
0	0.95	0	0.95
1	0.85	1	0.85
2	0.75	2	0.80
3	0.65	3	0.75
4	0.55	4	0.65
5	0.45	5	0.55
6	0.35	6	0.50
7	0.15	7	0.40
8	—	8	0.30
		9	0.15
		10	—

TABLE 7 Values of Weighting Factor W_0 for the Effect of Radiation on E_{to} at Different Temperatures and Altitudes

Temperature $e°C$	2	4	5	6	10	12	13	16	18	20	22	23	26	28	30	32	34	36	38	40
W at altitude m																				
0										0.68	0.71	0.73	0.75	0.77*	0.78	0.80	0.82	0.83	0.84	0.85
500										0.70	0.72	0.74	0.75	0.78	0.79	0.81	0.82	0.84	0.85	0.86
1000										0.71	0.73	0.75	0.74	0.79	0.80	0.82	0.83	0.85	0.86	0.87
2000										0.73	0.75	0.77	0.79	0.81	0.82	0.84	0.85	0.86	0.87	0.88
3000										0.75	0.77	0.79	0.81	0.82	0.82	0.85	0.86	0.88	0.88	0.89
4000										0.78	0.79	0.81	0.83	0.84	0.85	0.86	0.88	0.89	0.90	0.90

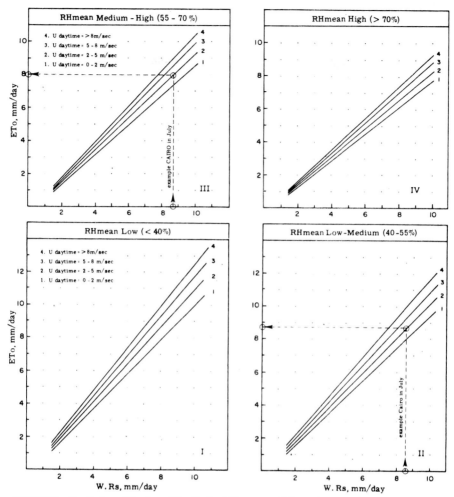

FIGURE 3 Prediction of E_{to} from $W \times R_s$ for different conditions of mean relative humidity and daytime wind.

tensively. Only a few are equipped with large-scale lysimeters capable of measuring evapotranspiration under field conditions. Consequently, it is necessary to extrapolate laboratory-type data for a relatively few locations to field conditions throughout the world where irrigation is practiced or may be beneficial.

The three FAO-24 methods described above predict the effect of climate on reference crop evapotranspiration E_{to}. To account for the effect of the crop characteristics on crop water requirements, crop coefficients K_c are presented to relate E_{to} to crop evapotranspiration E_{crop}. The K_c values relate to evapotranspiration of disease-free crops grown in large fields under optimum soil water and fertility conditions and achieving full production potential under the given growing environment. E_{crop} can be found by

EXAMPLE

Given:

Cairo; latitude 30°N; altitude 95 m; July. T_{mean} = 28.5°C; sunshine (n) mean = 11.5 h/day; wind daytime, U = moderate; RH_{mean} = medium.

Calculation:

From Table 4	R_A	=	16.8 mm/day
$(0.25 + 0.50\ n/N)R_a$	n	=	11.5 h/day
From Table 5	N	=	13.9 h/day
	n/N	=	0.83
	R_s	=	11.2 mm/day
From Table 7	W	=	0.77
	$W \times R_s$	=	8.6 mm/day
From Fig. 4, Blocks II and III, line 2,	E_{to}	=	8.4 mm/day

Yearly Data

Cairo, with solar radiation (R_s) given in mm/day

	J	F	M	A	M	J	J	A	S	O	N	D
T_{mean}, °C	14	15	17.5	21	25.5	27.5	28.5	28.5	26	24	20	15.5
R_s, mm/day	5.0	6.4	8.5	9.9	10.9	11.4	11.2	10.4	9.1	7.1	5.5	4.6
RH_{mean}	III	III	III	II	II	II	av. II, III	av. II, III	III	III	av. III, IV	av. III, IV
Wind daytime	av. 1, 2	av. 1, 2	av, 1, 2	av. 1, 2	2	2	2	av. 1, 2	av. 1, 2	av. 1, 2	av. 1, 2	av. 1, 2
W	0.61	0.62	0.65	0.70	0.74	0.76	0.77	0.77	0.75	0.73	0.68	0.63
$W \times R_s$	3.0	4.0	5.5	6.9	8.1	8.7	8.6	8.0	6.8	5.2	3.7	2.9
E_{to}, mm/day	2.5	3.4	4.8	6.7	8.2	8.8	7.4	6.0	4.5	3.0	3.0	2.2
mm/month	78	95	149	201	254	264	260	229	180	140	90	68

Format

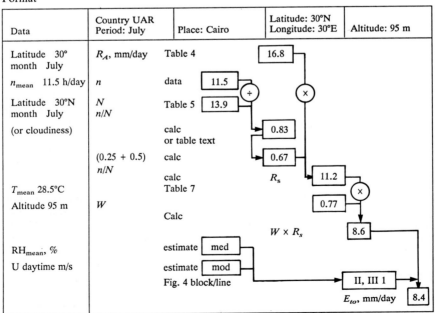

FIGURE 4 Example and format for calculation of radiation method. (R_s, as measured or obtained from regional or worldwide maps of solar radiation.)

TABLE 8 Pan Coefficient K_p for Class A Pan for Different Ground Cover and Levels of Mean Relative Humidity and 24 h Wind

Class A pan	Case A: pan placed in short green cropped area				Case B: pan placed in dry fallow area			
RH mean %		low <40	medium 40–70	high >70		low <40	medium 40–70	high >70
Wind, km/day	Windward side distance of green crop, m				Windward side distance of dry fallow, m			
Light <175	1	0.55	0.65	0.75	1	0.7	0.8	0.85
	10	0.65	0.75	0.85	10	0.6	0.7	0.8
	100	0.7	0.8	0.85	100	0.55	0.65	0.75
	1000	0.75	0.85	0.85	1000	0.5	0.6	0.7
Moderate 175–425	1	0.5	0.6	0.65	1	0.65	0.75	0.8
	10	0.6	0.7	0.75	10	0.55	0.65	0.7
	100	0.65	0.75	0.8	100	0.5	0.6	0.65
	1000	0.7	0.8	0.8	1000	0.45	0.55	0.6
Strong 425–700	1	0.45	0.5	0.6	1	0.6	0.65	0.7
	10	0.55	0.6	0.65	10	0.5	0.55	0.65
	100	0.6	0.65	0.7	100	0.45	0.5	0.6
	1000	0.65	0.7	0.75	1000	0.4	0.45	0.55
Very strong >700	1	0.4	0.45	0.5	1	0.5	0.6	0.65
	10	0.45	0.55	0.6	10	0.45	0.5	0.55
	100	0.5	0.6	0.65	100	0.4	0.45	0.5
	1000	0.55	0.6	0.65	1000	0.35	0.4	0.45

$$E_{crop} = K_c \times E_{to}$$

Each of the three methods predicts E_{to}, and only one set of crop coefficients is required. Procedures for selection of appropriate K_c values are given which take into account the crop characteristics, time of planting or sowing, and stages of crop development and general climatic conditions. These procedures are given for some of the more common crops. For other crops and for more detailed discussion of K_c selection, see "Crop Water Requirements," FAO Irrigation and Drainage Paper 24, revised 1977.

For general project planning it may be useful to refer to Table 9, which gives approximate ranges of seasonal E_{crop} for different crops. The magnitude shown will change according to the factors discussed, i.e., mainly climate, crop characteristics, length of growing season, and time of planting.

Crop coefficients K_c have been derived for different soil profile and surface moisture conditions and different reference crops, e.g., alfalfa and grass. Crop coefficients presented here are based on alfalfa and assume that root zone soil water is not limiting and that the soil surface is visually dry.

The crop planting or sowing date will affect the length of the growing season, the rate of crop development of full ground cover, and the onset of maturity. For instance, depending on climate, sugar beets can be sown in autumn, spring, and

TABLE 9 Approximate Range of Seasonal E_{crop}

Seasonal E_{crop}	mm	Seasonal E_{crop}	mm
Alfalfa	600–1500	Orange	600–950
Avocado	650–1000	Potatoes	350–625
Bananas	700–1700	Rice	500–950
Beans	250–500	Sisal	550–800
Cocoa	800–1200	Sorghum	300–650
Coffee	800–1200	Soybeans	450–825
Cotton	550–950	Sugar beets	450–850
Dates	900–1300	Sugarcane	1000–1500
Deciduous trees	700–1050	Sweet potatoes	400–675
Flax	450–900	Tobacco	300–500
Grains (small)	300–450	Tomatoes	300–600
Grapefruit	650–1000	Vegetables	250–500
Maize	400–750	Vineyards	450–900
Oil seeds	300–600	Walnuts	700–1000
Onions	350–600		

summer with a total growing season ranging from 230 to 160 days. For soybeans, the growing season ranges from 100 days in warm, low-altitude areas, to 190 days at 2500-m altitudes in Equatorial Africa and for maize 80 to 240 days, respectively. Crop development will also be at a different pace; as shown in Fig. 5 for sugar beets, the time needed to reach full development or maximum water demand varies from up to 60 percent of the total growing season for an autumn-sown crop to about 35 percent for an early-summer sowing. In selecting the appropriate K_c value for each period or month in the growing season for a given crop, the rate of crop development must be considered.

The mean crop coefficients shown in Table 10 divide the crop growth period into two parts, from planting to effective cover and days after effective cover. The planting to effective cover period can vary considerably for a given crop depending on temperature regime, latitude, and tillage practices. For this reason the crop coefficients throughout this period are given at percentage times. Dates of various crop growth stages for several crops are shown in Table 11. For other crops and locations, assistance from an agronomist may be needed to develop this information.

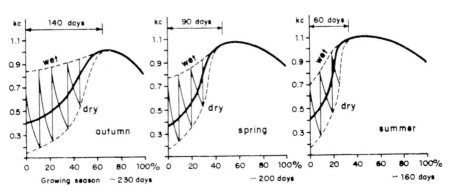

FIGURE 5 Duration and frequency of rainless periods.

TABLE 10 Mean E_t Crop Coefficients K_{cm} for Normal Irrigation and Precipitation Conditions, for Use with Alfalfa Reference $e_T E_{tR}$

	Mean E_T crop coefficients K_{cm}										
	P_{CT}, time from planting to effective cover, %										
Crop	0†	10	20	30	40	50	60	70	80	90	100
Spring grain*	0.20	0.20	0.21	0.26	0.39	0.55	0.66	0.78	0.92	1.00	1.00
Peas	0.20	0.20	0.21	0.26	0.36	0.43	0.51	0.62	0.73	0.85	0.93
Sugar beets	0.20	0.20	0.21	0.22	0.24	0.27	0.33	0.45	0.60	0.80	1.00
Potatoes	0.20	0.20	0.20	0.22	0.31	0.41	0.51	0.62	0.70	0.76	0.78
Corn	0.20	0.20	0.20	0.20	0.23	0.32	0.42	0.55	0.70	0.85	0.95
Beans	0.20	0.20	0.20	0.26	0.35	0.45	0.55	0.66	0.80	0.90	0.95
Winter wheat	0.20	0.30	0.30	0.50	0.75	0.90	0.98	1.00	1.00	1.00	1.00

	Dt. days after effective cover										
	0†	10	20	30	40	50	60	70	80	90	100
Spring grain*	1.00	1.00	1.00	1.00	0.90	0.50	0.30	0.15	0.10	—	—
Peas	0.93	0.93	0.70	0.53	0.35	0.20	0.12	0.10	—	—	—
Sugar beets	1.00	1.00	1.00	1.00	0.98	0.94	0.89	0.85	0.80	0.75	0.71
Potatoes	0.78	0.78	0.76	0.74	0.71	0.67	0.63	0.59	0.36	0.25	0.20
Field corn	0.95	0.96	0.95	0.97	0.90	0.85	0.79	0.74	0.35	0.25	—
Sweet corn	0.95	0.94	0.93	0.90	0.85	0.75	0.58	0.40	0.20	0.25	—
Beans	0.95	0.95	0.90	0.67	0.33	0.15	0.10	0.05	—	0.10	—
Winter wheat	1.00	1.00	1.00	1.00	0.95	0.55	0.25	0.15	0.10	—	—

	Time from new growth or harvest to harvest, %										
	Mean E_T crop coefficients K_{cm}										
	0†	10	20	30	40	50	60	70	80	90	100
Alfalfa (1st)‡	0.55	0.70	0.82	0.91	0.96	0.99	1.00	1.00	0.98	0.96	0.94
Intermediate	0.30	0.40	0.50	0.80	0.96	0.99	1.00	1.00	0.98	0.96	0.94
Last	0.30	0.40	0.50	0.60	0.65	0.63	0.61	0.59	0.57	0.55	0.50

	Total season (days from beginning of spring growth)										
	0†	20	40	60	80	100	120	140	160	180	200
Alfalfa	0.45	0.69	0.87	0.88	0.70	0.75	0.88	0.81	0.88	0.71	0.65
Seasonal	0.50	0.85	0.85	0.85	0.85	0.85	0.85	0.85	0.85	0.75	0.60
Overall seasonal mean						0.82					
Grass (perennial ryegrass) (8–15 cm)	0.60	0.70	0.78	0.78	0.78	0.78	0.78	0.78	0.78	0.76	0.75

Source: J. L. Wright, "New Evapotranspiration Crop Coefficients," *J. Irrig. and Drain. Div.,* ASCE, 108 (IR2):57–74, 1982. Minor changes from Wright (1982) reflect additional data for some crops (Wright, 1984, personal communication).

*Spring grain includes barley and wheat.

†The value 0.2 is appropriate for relatively dry surface soil conditions from planting until significant crop development. For moderately wet surface soil, as with preemergence irrigation(s) or some precipitation, use 0.35, and for very wet conditions use 0.50.

‡1st denotes first harvest; intermediate harvests may be 1 or more depending on length of season. The last harvest is when crops become dormant in cool weather. Cultivar used was Ranger.

TABLE 11 Dates of Various Crop Growth Stages Identifiable for Crops Studies at Kimberly, Idaho, 1968–1979

	Date of occurrence, month/day							Days	
Crop	Planting	Emergence	Rapid growth	Full cover	Heading or bloom	Ripening	Harvest	Planting to full cover	Full cover to harvest
Spring grain*	4/01	4/15	6/10	6/10	6/10	7/20	8/10	70	61
Peas	4/05	4/25	6/05	6/05	6/15	7/05	7/25	60	50
Sugar beets	4/15	5/10	6/01	7/10	—	—	10/15	85	95
Potatoes	4/15	5/25	6/10	7/10	7/01	9/20	10/10	75	90
Field corn	5/05	5/25	6/10	7/15	7/30	9/10	9/20	72	67
Sweet corn	5/05	5/25	6/10	7/15	7/20	—	8/15	72	30
Beans	5/22	6/05	6/15	7/15	7/05	8/15	8/30	55	45
Winter wheat*	(2/15)†	(3/01)	3/20	6/05	6/05	7/15	8/10	(110)	60
Alfalfa‡ (1st)	4/01		4/20				6/15		75
Alfalfa‡ (2d)	6/15		6/25				7/31		46
Alfalfa‡ (3d)	7/31		8/10				9/15		46
Alfalfa‡ (4th)	9/15		10/01				10/30		46

Source: J. L. Wright, "New Evapotranspiration Crop Coefficients," *J. Irrig. and Drain Div.,* ASCE, 108 (IR2):57–74, 1982. Minor changes from Wright (1982) reflect additional data for some crops (Wright, 1984, personal communication).
*Spring grain includes barley and wheat.
†Effective dates in parentheses. Crop was planted on 10/10 and emerged 10/25.
‡Effective planting date for established alfalfa is date growth begins in spring or harvest of preceding crop. Dates for these cuttings are indicated. Final harvest is date crop becomes dormant.

FARM-IRRIGATION REQUIREMENTS

Requirements for deliveries of irrigation water to farms depend upon numerous factors including (1) the acreage devoted to each of the several crops grown, (2) the evapotranspiration requirements of those crops under the climatic conditions prevailing in the area, (3) the effective rainfall in the area, (4) the need, if any, for irrigation prior to planting, (5) the farm-irrigation efficiency, and (6) other factors such as the quality of the irrigation water or the need for leaching of previous accumulations of salts from the soil. The farm-irrigation efficiency is the percentage of water delivered that is utilized in crop evapotranspiration, and it is influenced by the size of the farms because of the effect of conveyance losses between the point of delivery to the farm and the several fields. Farm conveyance losses are influenced by the same factors which affect losses from the larger conveyance facilities of an irrigation project, as discussed elsewhere under Conveyance Losses and Waste.

12. Cropping Pattern. Irrigated farms and projects usually raise a diverse pattern of crops. In order to assure that the distribution facilities for the irrigation water have sufficient capacity to meet the peak water requirements, it is necessary to identify representative sequences of crops and to determine their water requirements.

Cropping sequences for single farms are usually different from those for a large project, even when the particular farm is included in the project. This is because the growing seasons of crops grown on a single farm must be compatible, thereby avoiding interference in use of the fields and providing time for harvesting of one crop and for seedbed preparation before planting of the next crop. This is not required in cropping distributions for large project areas because of diversity among the cropping of the several farms. It is quite possible that no single farm would utilize the cropping distribution that would apply for the project.

The distribution of crops is influenced by many factors including (1) suitability of the soil for various crops, (2) suitability for the climatic characteristics for various crops, (3) economic conditions which influence the return the farmer may expect to receive for labor and investment from various crops, (4) governmental controls on acreage devoted to crops in which surplus production has been experienced, (5) limitations imposed by project regulations (generally to avoid having a large percentage of the project land in a single crop which would result in a high peak water requirement, possibly exceeding delivery capabilities or seasonal water supplies), and (6) the preferences and experience of the farmers in growing certain crops. In the United States it can be assumed that a high technical level of farm management will be employed in irrigation enterprises. This will assure the use of rotations of crops to improve or preserve the soil fertility and crop distributions which will make possible efficient employment of labor and equipment. Under such a level of management, it can be assumed that the project cropping will change over a period of years as market and economic conditions change. These assumptions may not be warranted or realistic in the planning of irrigation developments in many of the developing nations of the world. In any event, agricultural economic studies are required in order to establish crop distributions for projects which may be expected to be representative of future conditions with respect to water requirements, agricultural production, and farmers' incomes during the economic life of the project works.

13. Effective Rainfall. Part of the rainfall which occurs during the growing season of a crop is lost through surface runoff or deep percolation below the root zone of the soil profile. The portion of the growing-season rainfall which is assumed to be utilized in meeting evapotranspiration requirements of crops is termed "effective rainfall" in irrigation studies. In arid areas effective rainfall may be so small as to be of little consequence, while in humid areas it may provide a major portion of the evapotranspiration requirements for optimum growth of many crops. Growth of irrigation in humid areas in the United States in recent years is evidence of the economic significance of irrigation as an aspect of efficient farm management. This is because the time distribution of rainfall rarely suffices to maintain the soil moisture in the root zone within the range necessary for optimum growth during all parts of a growing season. Information in Fig. 6 is presented to illustrate this point. Curve 3 shows that at Corsicana, TX, there were 10 periods of 24 days or longer in which no rain occurred during the growing seasons of 10 consecutive years. This means an average of one such period each year. Since moisture stored in the soil root zone could supply crop needs for only about 10 days during the time of greatest water use, such a rainless period would result in some yield reduction, if not a complete failure. Curves 1 and 2 represent more favorable rainfall distributions for crop reduction.

In estimating irrigation-water requirements the engineer must estimate the effective rainfall from records of experienced rainfall in the area of the farm or project. In making these estimates, the average moisture level of the soil root zone at the beginning of rainfall is important. This moisture level is influenced by (1) the consumptive use rate of the crop, (2) the moisture-holding capacity of the soil root zone, and (3) the frequency and depth of applications of irrigation water

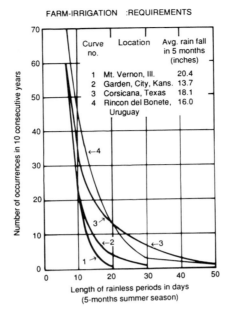

FIGURE 6 Duration and frequency of rainless periods.

or periods of rainfall. It may be assumed that through infiltration from rainfall or irrigation applications, the moisture level of the soil root zone will be maintained between field capacity and a level somewhat above that at which crop growth would be retarded. This means that even in arid climates some of the rainfall will occur on soil that is at or near field moisture capacity and will not be effective in supplying evapotranspiration. The chance of such a situation is reduced if the rate of evapotranspiration of the crop is high and if the soil root zone has a large range of usable moisture capacity.

Studies by the Soil Conservation Service for 50 years of record at 22 Weather Bureau stations throughout the continental United States have resulted in the relationship between monthly mean rainfall, monthly effective rainfall, and average monthly consumptive use (evapotranspiration) shown in Table 12. Values in that table may not be applicable to areas where soil intake rates are low or rainfall intensifies are consistently high. In such cases, the engineer must make adjustments to reflect the local conditions.

The engineer must consider, also, the frequency distribution of effective rainfall. While the crop evapotranspiration requirement will usually vary from year to year by a relatively small amount, there may be a large variation from year to year in rainfall. It may be desirable, for example, to plan for an irrigation supply that will be fully adequate for evapotranspiration requirements in 8 out of 10 years. In such a case it could be assumed that the ratio of the desired effective rainfall to the effective rainfall determined by use of Table 12 would be the same as the ratio of the growing season rainfall that would be equated or exceeded in 8 years out of 10 to the sum of the monthly mean rainfall values in the growing season. This will require the collection and analysis of rainfall records in the project area for the full period of available records. The adequacy of the irrigation

TABLE 12 Average Monthly Effective Rainfall,* as Related to Mean Monthly Rainfall and Average Monthly Consumptive Use

Monthly mean rainfall r_t, in	Average monthly consumptive use u, in									
	1.0	2.0	3.0	4.0	5.0	6.0	7.0	8.0	9.0	10.0
	Average monthly effective rainfall r_e, in									
0.5	0.20	0.25	0.30	0.30	0.30	0.35	0.40	0.45	0.50	0.50
1.0	0.55	0.60	0.65	0.70	0.70	0.75	0.80	0.85	0.95	1.00
2.0	1.00	1.25	1.35	1.55	1.55	1.55	1.60	1.70	1.85	2.00
3.0	1.00	1.85	1.95	2.10	2.20	2.30	2.40	2.55	2.70	2.90
4.0	1.00	2.00	2.55	2.70	2.90	2.90	3.15	3.30	3.50	3.80
5.0	1.00	2.00	3.00	3.25	3.50	3.60	3.85	4.05	4.30	4.60
6.0	1.00	2.00	3.00	3.80	4.10	4.25	4.50	4.80	5.10	5.40
7.0	1.00	2.00	3.00	4.00	4.60	4.80	5.05	5.40	5.70	6.05
8.0	1.00	2.00	3.00	4.00	5.00	5.30	5.60	5.90	6.20	
9.0	1.00	2.00	3.00	4.00	5.00	5.75	6.05	6.35		

Source: From U.S. Department of Agriculture Soil Conservation Service Technical Release 21.
*Based on 3-in net depth of application. For other net depths of application, multiply by the factors shown below.

Net depth of application	0.75	1.0	1.5	2.0	2.5	3.0	4.0	5.0	6.0	7.0
Factor	0.72	0.77	0.86	0.93	0.97	1.00	1.03	1.04	1.06	1.07

supply to be provided is a matter of economics, and it may be desirable for high-value crops to provide a water supply that will be fully adequate except in very rare circumstances.

14. Irrigation Applications. Irrigation involves the application of surface or groundwater to fields to supplement the soil moisture supply provided by natural rainfall as required for increased growth and production of crops. Accordingly, the frequency of irrigation-water applications will vary during the growing season of each crop depending upon (1) the evapotranspiration rate, (2) the distribution and amount of rainfall, (3) the capacity of the usable soil moisture reservoir, and (4) the design capacities and method of operation of the irrigation-distribution facilities. The depth of water applied at each irrigation varies from about 2 to 6 in, with frequent light applications being required on sandy soils or during the early part of the growing season and less frequent heavier applications being permissible later in the growing season or on soils with greater water-holding capacity.

15. Effect of Root Depth. At the beginning of the growing season the emerging roots can obtain moisture and plant nutrients from the zone of soil and the depth the seeds were planted. As the vegetative growth develops above the ground, the root system develops in the upper soil layers as required to furnish the moisture and plant foods needed to sustain that growth. Each crop has characteristic rooting habits which it will tend to follow if the soil is deep and uniform and equally moist throughout. The depth of rooting increases during the entire growing period. Crops which mature in 2 months usually penetrate only 2 to 3 ft, while crops requiring 6 months to mature may penetrate 6 to 10 ft or more.

Normally the greatest concentration of plant roots is in the upper layer of soil, and this concentration may be accentuated if the root zone is restricted by a high water table, shallow soil, or compacted layers (plow pan). When the upper portion of the soil is kept moist, plants will obtain most of their moisture supply from near the surface. As the moisture content of the upper layers decreases, the plants withdraw more water from the lower layers. While this may tend to encourage more root development in the lower levels, fewer roots exist in the lower portion of the root zone, and wilting may result, even though moisture is available, because of the inability of the root system to extract enough moisture from the lower levels. Irrigation practice in arid regions usually results in the extraction of 40, 30, 20, and 10 percent of the moisture supply, respectively, from succeeding quarters of the root zone (from the surface downward).

Average root-zone depths for many of the crops grown under irrigation are listed in Table 13. Generally, these depths are reached by the time the foliage of the plant has reached its maximum size. Root-zone depths are limited to the soil depth above the water table.

The moisture reservoir available to the plants is determined by the depth of the plant roots at the stage of growth under consideration and by the moisture-retention characteristics of the soil. The variable factor is the root depth, for a particular field, which has a major influence on the depth and frequency of irrigation-water applications and on the changes in those applications during the growing season.

16. Soil Moisture Capacity. The volume of soil moisture available to the plant is a function of root depth and the moisture-holding capacity of the soil between field moisture capacity and the minimum moisture content at which optimum plant growth can be sustained. Field moisture capacity is defined as the maximum moisture which can be retained in the soil against the forces of gravity. It does not include water which might be in the soil under saturated conditions and which, in time, will percolate downward through the soil profile. Usually, determinations of field moisture capacity are made 2 days after irrigation.

The moisture content of the soil which results in permanent wilting of plants is called the "wilting point." The wilting point varies with temperature and stage of

TABLE 13 Normal Root-Zone Depths of Mature Irrigated Crops Grown in a Deep, Permeable, Well-Drained Soil*

Crop	Ft	Crop	Ft	Crop	Ft
Alfalfa	5–10	Corn (sweet)	3	Peas	3–4
Artichokes	4	Corn (field)	3–5	Potatoes (Irish)	3–4
Asparagus	6–10	Cotton	4–6	Potatoes (sweet)	4–6
Beans	3–4	Cranberries	1–2	Pumpkins	6
Beets (sugar)	4–6	Deciduous orchards	6–8	Radishes	1
Beets (table)	2–3	Grain	4	Spinach	2
Broccoli	2	Grapes	4–6	Squash	3
Cabbage	2	Grass pasture	3–4	Strawberries	3–4
Cantaloupes	4–6	Hops	5–8	Tomatoes	6–10
Cane berries	3–4	Ladino clover	2	Turnips	3
Carrots	2–3	Lettuce	1–1½	Walnuts	12
Cauliflower	2	Mint	3–4	Watermelons	6
Celery	3	Onions	1		
Citrus	4–6	Parsnips	3		

*From *Engineering Handbook, Far Western States and Territories,* U.S. Department of Agriculture, Soil Conservation Service, sec. 15, part I, May 1957.

growth of the plant—higher values apply as rates of evapotranspiration increase. Generally, the wilting point is from 40 to 50 percent of field moisture capacity. The "available moisture capacity" of a soil is the amount of water per unit of soil depth between field moisture capacity and the wilting point. This amounts to 50 to 60 percent of the field moisture capacity and represents the amount of water that can be stored in the soil and used to sustain optimum plant growth. Consequently, it represents the maximum range of fluctuation of soil moisture content that can be tolerated between rains or irrigation-water applications without depression of crop yields.

The available moisture capacity of soils is primarily a function of soil texture. Coarse sands have the least moisture capacity and heavy clays the most. Common ranges of available moisture capacities for soils of different textures are shown in Table 14.

TABLE 14 Available Moisture vs. Soil Texture

Soil texture	In of water per ft of soil
Very coarse sands	0.40–0.75
Coarse to loamy sands	0.75–1.00
Sandy loams and fine sandy loams	1.00–1.50
Loams and silt loams	1.50–2.30
Sandy clay loams, silty clay loams, clay loams	1.75–2.50
Sandy clays, silty clays, and clays	1.60–2.50
Peat and mucks	2.00–3.00

17. Soil Moisture Reservoir. Irrigation studies which involve water conveyance and short-period delivery requirements (generally on a monthly basis) must consider the operation of the soil moisture reservoir. The usable capacity of that reservoir increases as the root system develops. The reservoir content diminishes as moisture is withdrawn by evapotranspiration and increases as moisture is added by rainfall or application of irrigation water.

In arid regions it is often necessary to irrigate the fields in advance of planting. This "preplanting" irrigation is required when the soil moisture reservoir is depleted and the soil is too dry for preparation of an efficient seedbed, for germination of the seeds, and for initial growth of the plants. The preplanting irrigation application is additional to the irrigation water necessary to supply the crop evapotranspiration requirements, since it makes up for deficient antecedent rainfall (during a period of months prior to the crop growing season under consideration). The amount of water to be applied depends upon the length of time following irrigation applications on the same land for the previous crop, the condition and covering of the soil surface, and the climatic conditions (especially rainfall). Generally, irrigation studies provide for sufficient preplanting irrigation to bring the soil to field capacity throughout the depth of soil to be used by the roots of the crop to be grown. This will supply moisture needed for seed germination and for growth of the young plants until they are able to withstand mechanical operations required for application of irrigation water and the irrigation water itself.

Irrigation-water applications will be relatively light and frequent during the early parts of the growing season—while the root zone is shallow. Later the applications can provide greater depths of water and the frequency will depend upon the rate of evapotranspiration by the crop. Often irrigation applications must be made at intervals of a week or less during the period of peak water use, particularly if the soil is sandy or shallow.

It is generally desirable to discontinue irrigation-water applications several weeks before the crop is harvested. This practice provides a dry soil surface for harvesting operations, reduces the moisture content of the vegetative growth, and conserves irrigation water. During this period of maturing of the crop, the moisture needed can be drawn from the soil moisture reservoir. The engineer making irrigation studies should recognize that the use of the soil moisture reservoir is an essential feature of irrigation agriculture and that it results in requirements for irrigation-water application which are phased earlier in time, and adjusted in amount, from the actual crop evapotranspiration.

18. Computation Procedure. In estimating requirements for application of irrigation water for a project area, the engineer must adapt the information given in the preceding paragraphs of this section to the local conditions expected to prevail at the time to be represented by the estimate. The cropping pattern should represent the distribution of crops to be grown in the entire project area, and the planting and harvesting seasons selected or those crops should reflect the fact that those operations take considerable time and that they are not done simultaneously on all the farms in the project. This can be handled in the computations by assuming, for example, that one-fourth of the area in a crop would be planted at the beginning and at the end of the planting season and that half the area would be planted on the median date of that season. Crop-use factors would then be weighted to account for the three planting dates to obtain weighted use factors for that crop which would be applicable to the entire project area.

The depth of water to be applied in preplanting irrigations must be estimated for each crop from an analysis of the factors influencing the probable average moisture condition in the soil and the time of seedbed preparation. Usually the engineer will select a depth of water in inches of depth over the area to be planted to the crop as the amount of soil moisture to be added by preplanting irrigations to bring the moisture level to field capacity. In some cases it may facilitate the computations to express that depth as a crop-use factor to be applied for the appropriate month to the evapotranspiration index used in the estimating method selected. Such factors can be added to the normal crop-use factors and included in the weighting computations to obtain project crop factors.

Another adjustment may be made to the crop-use factors to account for the withdrawal of moisture from the soil at the end of the growing season. The engineer must select an appropriate depth of water which can be withdrawn from the soil after the last irrigation to meet the needs of the maturing crop. The amount of water can be expressed as crop factors to be applied to the selected estimating index for the appropriate month or months. Incorporation of this adjustment in the project crop factors will involve reduction of the normal crop factors by the appropriate amounts and weighting the reduced factors for the selected planting dates to obtain the project factors.

The method of handling the estimated effective rainfall is important if that rainfall is sufficient to supply more than the needs of one of the crops in one or more months. In such cases it is necessary to consider the effective rainfall separately for each crop, rather than lumping it all together as an adjustment to the total project water requirement to determine the amount to be supplied by irrigation water. Otherwise, excess rainfall on one field would be assumed (erroneously) to be utilized on another field or crop requiring a greater amount of water.

19. Deep Percolation and Leaching. The principal factors which affect the field irrigation efficiency are those which influence the passage of water through the root zone. These include the method of applying the water (row irrigation, wild flooding, borders, or sprinkler systems); the texture and condition of the soil; the slope of the land and the care with which it has been leveled, ditched, or bordered; the rate of flow available to the irrigator in relation to the size of field to be irrigated; and the skill of the irrigator. Experience has shown that field efficiencies under management levels which prevail in the United States generally fall between 60 and 75 percent, with the higher value being achieved with sprinkler systems and/or a very high level of management. In general, the minimum allowance for deep percolation is about 20 percent if adequate irrigation is to be accomplished. It may be permissible to use a higher field efficiency for a project computation than would be appropriate for a single farm, if it can be assumed that the surface waste which would escape the farm would be captured and used elsewhere in the project.

If the irrigation water applied is saline, or if the soil is saline or alkaline and requires reclamation, it may be necessary to pass additional water through the root zone to achieve and maintain a permanent irrigation agriculture. Many areas of the world have gone out of agricultural production because inadequate attention was given to these requirements, and reclamation of those lands to restore their productivity may be uneconomical. The use of saline waters for irrigation supplies should be under the guidance of an experienced soil and water chemist, since the relationships are so complex as to require the services of a specialist. The considerations are (1) that the flow of water through the soil must be sufficient to keep the concentration of salts in the soil solution below levels harmful to the plants being grown and (2) that a salt-inflow–salt-outflow balance must be maintained, preferably with the level of salinity in the drainage water such that reuse of the water at downstream points will be possible.

Extra water applications made for the purpose of permitting the use of saline irrigation waters or of leaching harmful accumulations of chemicals from the soil root zone are designated "leaching requirements." Some indication of the magnitude of these requirements is given by data in Table 15, in which the leaching requirement (in percent of the applied water) is related to the quality of the irrigation and drainage waters. As shown in that table by interpolation, a normal deep percolation allowance of 20 percent—about the minimum to assure adequate irrigation in all parts of the fields—provides adequate leaching if the quality of the irrigation water does not exceed about 1300 ppm of dissolved salts and if the quality of the drainage water can be about 6000 ppm of dissolved salts. While maintenance of a salt balance is important for a permanent agriculture, it is not the sole criterion, since damaging concentrations in the soil solution could occur. The engineer should consult recent published literature on this subject and obtain the advice of a specialist if preliminary studies indicate water quality may be a problem.

20. Surface Waste and Farm Conveyance Losses. Conveyance losses on the farm vary from near zero to about 15 percent of the water delivered to the farm. The lower loss rates are associated with the use of pipelines or lined channels, while the higher values are associated with unlined ditches and coarse soil texture.

Surface waste from individual fields varies with the nature of the soil, slope of ground surface, method of preparing the land, and depth of irrigation. Naturally, larger quantities of water run off the fields during the greater depths of irrigation. Waste is more difficult to control on steep slopes than on flat slopes and more important on clay soils than on sandy soils, since either a flat slope or a sandy soil is more conducive to a higher rate of soil absorption. Waste from individual fields

TABLE 15 Leaching Requirements Related to Quality of Drainage and Irrigation Water for an Equilibrium Condition Ignoring Influence of Rainfall*

Quality of irrigation water				% leaching requirements for given qualities of drainage water			
Micromhos/ cm* EC × 10^6	ppm	meq/ liter	Tons/ acre-ft	5 millimhos/ cm* EC × 10^3	10 millimhos/ cm* EC × 10^3	15 millimhos/ cm* EC × 10^3	20 millimhos/ cm* EC × 10^3
100	63	1	0.09	2	1	0.7	0.5
200	125	2	0.17	4	2	1	1
400	250	4	0.34	8	4	3	2
800	600	8	0.82	16	8	5	4
1600	1000	16	1.36	32	16	11	8
3200	2000	32	2.72	64	32	21	16
6400	4000	64	5.44	128	64	43	32

Source: Orson W. Israelsen and Vaughn E. Hansen, *Irrigation Principles and Practices*, 3d ed., Wiley, New York, 1962.

*Electrical conductance, which is the reciprocal of resistance, is expressed in "reciprocal ohms" or "mhos," and electrical conductivity is expressed in mhos per centimeter of distance between contact points. Since most soil solutions have a conductivity of less than 1 mho/cm, smaller units have been designated as follows: 1 mho/cm = 1000 millimhos/cm = 1,000,000 micromhos/cm.

should be kept as low as possible, although some surface runoff is not serious if collected and used on lower areas. Surface waste can be practically eliminated, or effectively recovered, by constructing adequate levees, or drainage ditches, along the lower edges of fields. Charges for irrigation water based on actual deliveries and limitations on water supplies or canal deliveries encourage or force the use of practices which reduce wastage of irrigation water to a minimum.

Average quantities of water actually lost to beneficial use by surface waste on large irrigation projects are considerably smaller than the quantities lost on individual fields. Investigations in the Cache La Poudre valley, northern Colorado, showed an average surface runoff equal to 6 percent of the quantity applied. An allowance of 10 percent of the quantity delivered to the fields will usually be an ample provision for surface waste in considering a large irrigation project as a whole.

21. Effect of Irrigation Methods. The method of applying irrigation water to the field to obtain the most efficient water distribution and the best use of labor and equipment will vary according to the crops grown, the soil texture, the land slope, the levelness of the field surface, the frequency of required irrigations, and the rate of flow available to the irrigator. The adaptations and limitations of the common irrigation methods are summarized in Table 16. These methods are described in detail in standard texts and reference books.

Irrigation efficiencies vary with the methods of application used and with the experience of the irrigator. If the method is suited to the crops and local conditions, the irrigation structures are properly designed and constructed, and the irrigator is experienced and efficient, field efficiencies in the range of 50 to 70 percent can be achieved. The higher efficiencies are generally associated with sprinkler systems which generally have pipe conveyance systems, apply the water uniformly, and can be controlled readily to apply the exact amount of water desired. Lower efficiencies generally result from application by wild flooding, im-

TABLE 16 Adaptation and Limitations for Common Irrigation Methods*

Furrow	Light-, medium, and fine-textured soils; row crops; small stream	Slopes up to 3% in direction of irrigation; row crops; 10% cross slope
Corrugation	Light-, medium-, and fine-textured soils; close-growing field crops; small stream flows	Slopes up to 12% with semipermanent crops; 8% with annual crops; 5% cross slope; rough for equipment
Border	All soils; close-growing field crops; large streams	Slopes up to 3% for annual crops; slopes to 8% for sodded pastures; good leveling required; 0.3% cross slope; uniform grade; problem of starting crops in soils which puddle readily
Sprinklers	All slopes; soils; crops; and stream size	High initial equipment cost; lowered efficiency in windy and hot climate
Check (or ponding)	Light, medium, and heavy soils; large stream	Deep soils; high cost of land preparation; slopes less than 2%
Subirrigation	Free lateral movement of water in soils, rapid capillary rise, underlain by low-permeability layer, all crops; large quantities of water	Special soil and annual precipitation conditions; usually causes drainage problems elsewhere

*Prepared by Max Jensen and Claude H. Pair, "Water" Yearbook of Agriculture, 1955, 84th Congress, 1st Session, House Document, 32.

properly designed systems, inadequate land leveling, or overwatering because of inexperience or lack of restraints on use of water.

CONVEYANCE LOSSES AND WASTE

This section is concerned with that portion of the project water supply which is lost (so far as local use is concerned) between the point of diversion from a stream or reservoir and the points where deliveries are made to the farms. Part of the water lost finds its way to the groundwater and another part represents operational waste which is discharged into drainage channels or streams. Those waters may be recaptured and used downstream within the same project or elsewhere. The remainder of the water lost in conveyance is nonbeneficial evapotranspiration.

22. Evaporation Losses. A part of the water diverted into the canal system is lost by evaporation from the water surfaces of the flowing canals and laterals. Evaporation losses vary with the areas of the canal water surface and the prevailing rate of evaporation. They may be practically eliminated by constructing closed conduits. However, such losses usually are not great enough or of sufficient value to warrant the expense of such construction. For example, a canal surface 20 ft wide, evaporating water at a rate of ½ in/day, would lose about 2 acre-ft a day in a length of 20 miles, a quantity equivalent to a continuous flow of about 1 cfs—on the order of one-half of 1 percent of the canal flow.

23. Seepage Losses. Seepage losses from canals depend upon (1) the wetted area of the bed and banks, (2) the permeability of the canal bed and the underlying soil, and (3) the difference in level of the water in the canal and the adjacent groundwater table. Such losses are greatest where sandy or other permeable soils predominate and where the groundwater table is low. Sometimes the losses are materially reduced with time through sealing of the canal bed by the deposition of fine sediments brought into the canal in suspension with the diverted water during periods of high concentrations of suspended sediment in the streams. The most efficient method of reducing concentrations is the lining of the canal with concrete or other materials. An example of a canal having minimal conveyance losses is shown in Fig. 7. Losses are low in the East Ghor irrigation canal because of the large size of the canal and because it is concrete-lined. The U.S. Bureau of Reclamation has carried on extensive research in low-cost canal linings of various types.

Careful measurements, made on irrigation projects in Idaho, showed that small farm ditches, carrying less than 1 cfs, may lose half their flow in a length of 1 mile. The measurements made on canals carrying 10 to 3000 cfs through sections of different soil texture, showed seepage losses per mile varying from less than 0.1 to 10.8 percent of the flow. Measurements made on canals in the Salt

FIGURE 7 East Ghor Irrigation Canal, Jordan. (*Harza Engineering Company.*)

River valley, Arizona, showed an average seepage rate of approximately 0.34 acre-ft/acre of wetted area/day.

In making detailed studies of irrigation requirements and canal operations, it must be recognized that higher than average conveyance losses (in terms of percent of the diverted flow) occur when relatively small diversions are being made and the percent lost is least when the canal system is operating at capacity. The practice of rotating flows between the smaller laterals during periods of small diversions, with each of those laterals being dry during a considerable portion of those periods, tends to make the conveyance losses therein more of a function of the flow carried. On the other hand, check structures must be used to a greater extent during canal operations with small rates of flow in order to serve the several offtakes adequately, and the backwater thus created increases the wetted area of the canals and tends to increase the seepage losses.

24. Operational Wastes. Operational wastes of water are practically unavoidable if optimum service is to be given to all farms. These wastes result from more water arriving at critical points in the canal system than can be carried. This situation results from rainfall or other factors which produce unexpected canal flows or reduce the irrigation requirements below that which has been scheduled, and from unavoidable inabilities to dispatch and to measure diversions and deliveries to achieve a perfect balance between diversions, conveyance losses, channel storage, and irrigation delivery requirements.

Waste on U.S. Bureau of Reclamation projects varies from a minimum of 2 percent on the Boise project in southwestern Idaho to a maximum of 58 percent on the Yuma project in southeastern California and southwestern Arizona. The relatively high waste on the Yuma project is due partly to the use of water for power development and partly to the operation of the canals at high levels in order to reduce silt deposition. Wastage on some of the other projects is high because ample supplies are available and the excess diversions can be returned to the streams for rediversion to lower areas. Quantities given as waste do not include water used in sluicing sand and silt deposits at points of diversion.

REUSE OF DRAINAGE WATER

When fields are irrigated adequately to sustain optimum crop production and to prevent gradual soil salinization, it is inevitable that some deep percolation to drains or underlying aquifers will take place. Additional deep percolation results from seepage from canals and laterals. Drainage flows also include water wasted at the lower ends of fields and excess water discharges from canals through wasteways to avoid overflows or breaches of the canal banks. Such excess canal flows are normal occurrences in irrigation-project operation and result from errors in measuring or controlling diversions or deliveries, unanticipated reductions in farm requirements caused by rainfall after water orders had been placed, or smaller conveyance losses than had been provided for in the water-scheduling computations. Failure to use or to remove these drainage waters can impair land productivity through waterlogging and soil salinization. In many areas drainage waters provide a bonus value of additional water supplies which augment the original surface-water source and provide for the most efficient development and use of the water resource.

25. Integration of Ground- and Surface-Water Use. Many areas irrigated from a surface-water source are underlain by aquifers which are recharged by the irrigation operations or by lateral groundwater movement from external recharge areas. When the groundwater level is near the zone of soil penetrated by crop roots the ideal irrigation development involves coordinated and integrated use of groundwater and surface water to maintain a balance between groundwater pumping and recharge. The amount of water pumped must include the volume used for irrigation on the overlying lands and an amount to be exported from the area to carry away an annual volume of salts equal to the salts brought to the area in the imported surface water. Because the salt concentration in the groundwater is usually considered higher than that in the surface water, the amount of water to be exported for salt-balance purposes is usually comparatively small.

The discussion in the preceding paragraph is predicated on pumping of the groundwater, and that pumping is usually accomplished by the use of turbine pumps sized to comply with irrigation needs, aquifer characteristics, and the designs of the wells and screens. In many areas the aquifer is deep enough to provide a groundwater reservoir of large capacity which can be used to compensate for year-by-year variations in surface-water supplies. This will require deep-well pumps to be operable over a wide range of water-table fluctuations.

Coordinated use of ground and surface waters requires controlled use of those groundwaters which contain potentially damaging concentrations of salts, sodium, or bicarbonates. This may require that the groundwater be pumped into large-capacity canals so that it will be diluted prior to irrigation use. In other cases it may suffice to alternate irrigation applications between ground and surface supplies so that adequate dilution is accomplished in the soil. The practice to be followed in a particular case will depend upon the chemical qualities of the ground and surface water and upon the chemical properties of the soil and should be determined by a qualified, experienced soil and water chemist or soil scientist.

Irrigated areas that are not underlain by a usable aquifer, or which have an impermeable stratum near the crop root zone, must be drained by open or tile drains. The drainage water collected will usually be of adequate quality to be reused for irrigation. This reuse can be accomplished within the project area by discharging the drains into canals by gravity or by pumping, or the drainage waters can be delivered into the stream channels for diversion at a point some distance downstream.

26. Quality of Irrigation Water. The usability of water for irrigation of crops, or the conditions under which a particular water may be used, is determined by the chemical quality of the water, the sensitivity of the crops to salts and water-soluble elements, and the chemical characteristics of the soil to which the water will be applied. The most important quality consideration of irrigation water is the total salt content, which is usually expressed in terms of electrical conductivity—micromhos per centimeter—or in parts per million of total dissolved solids. A value of 1000 micromhos/cm is equivalent to about 640 ppm of total dissolved solids.

Irrigation waters are classified into four groups: low salinity, less than 250 micromhos/cm; medium salinity, 250 to 750; high salinity, 750 to 2250; and very high salinity, greater than 2250 micromhos/cm. About half the irrigation waters now in use in the Western United States fall in the medium-salinity classification.

The second most important factor in irrigation-water quality is the relationship of the cations of sodium to those of calcium and magnesium. The most satisfactory expression of that relationship is the sodium adsorption ratio (SAR), defined as

$$SAR = \frac{Na^{++}}{\sqrt{\frac{(Ca^{++} + Mg^{++})}{2}}}$$

Because of the highly significant relationship between the SAR of an irrigation water and the exchangeable sodium percentage of the soil irrigated with that water, it is possible to anticipate the effect of a water on the soil. Irrigation waters with SAR values of 8 or less are probably safe, those with values of 12 to 15 are marginal, and continued use of waters with values greater than 20 could lead to serious sodium problems. Sodium soils are relatively impermeable to air and water, and hard when dry, are difficult to till, and are plastic and sticky when wet. These conditions retard or prevent germination and are unfavorable for plant growth. The sodium ion is toxic to many plants.

Boron is an essential element to normal plant growth, but concentrations only slightly above optimum are toxic to many plants. Only a few surface waters contain harmful concentrations of boron but many groundwaters and many saline soils are contaminated. Permissible limits of boron in irrigation water are 0.1 to 0.3 ppm for sensitive crops, 1.0 to 2.0 ppm for semitolerant crops, and 2.0 to 4.0 ppm for tolerant crops. The sensitive crops are citrus fruits, nuts, and beans; the semitolerant include cereals, some vegetables, and cotton; and the tolerant group includes alfalfa, sugar beets, and asparagus.

The concentration of bicarbonate is another major consideration in the quality of irrigation water. The use of waters that are low in total salts but high in bicarbonate may aggravate the sodium problem if the amount of bicarbonate is considerably in excess of the calcium and magnesium present. This excess bicarbonate is referred to as residual sodium carbonate. When an irrigation water containing residual sodium carbonate evaporates in the soil, calcium and magnesium carbonates precipitate, and the sodium percentage of the soil solution increases. Then sodium replaces calcium on the soil particles, the exchangeable-sodium percentage of the soil increases, and the physical condition of the soil, especially the permeability, may be impaired. In addition, the pH may increase and organic matter may be dissolved, giving the dark color typical of a so-called black alkali soil.

Crops should be selected on the basis of their salt tolerance and the salt content of the irrigation water and the soil. The relative salt tolerance of the more important crops is shown in Table 17.

Water-quality considerations are most important when groundwaters or drainage flows are being used. This is because those waters characteristically contain concentrations of salts and other potentially harmful elements, since they result from the passage of water through the soil and the chemical concentrations are increased by the removal of pure water through evapotranspiration.

Brackish water may be used safely for irrigation of crops in some cases when such waters are available and supplies of more suitable water are temporarily deficient. The amount of brackish water that can be used depends on the salt concentration of the water, the number of irrigations between leaching rains or irrigation rains or irrigations with fresh water, the salt tolerance of the crop, and the salt content of the soil before irrigation.

RESULTS FROM IRRIGATION

Irrigation is a management practice which has been utilized for centuries for the purpose of increasing agricultural production and income. Feasibility studies of potential irrigation developments, whether for individual farms or for projects covering thousands of acres, must include determinations of the agricultural production volumes and values, the production costs, and the gross and net farm and project incomes for conditions with and without irrigation.

27. Crop-Yield Responses. The direct result which irrigation must achieve if the effort and expenses involved are to be justified is an increase in crop yields over those which can be obtained without irrigation. The contribution of irrigation ranges from 15 to 20 percent of yield without irrigation in the more humid regions to the entire yield achieved under irrigation in areas which have little or no rainfall. Yield response for a particular crop in a specific location can usually be estimated from yields realized in the area under irrigated and unirrigated conditions. Such evaluations should be made by experienced agriculturists who can interpret the effects of differences in management levels, use of fertilizers and plant-protection measures, and other inputs which influence crop yields.

TABLE 17 Crop Salt Tolerances

Salt-sensitive	
Avocado	Prune
Citrus	Apple
Strawberries	Pear
Peach	Beans
Apricot	Celery
Almond	Radish
Plum	Clover
Medium tolerance	
Grape	Olive
Cantaloupe	Fig
Cucumber	Pomegranate
Squash	Cauliflower
Peas	Cabbage
Onion	Broccoli
Carrot	Tomato
Peppers	Oats
Potato	Wheat
Sweet corn	Rye
Lettuce	Alfalfa
High tolerance	
Asparagus	Cotton
Garden beets	Barley
Sugar beets	

Production responses from irrigation include not only increases in yields of particular crops but those resulting from the situations where irrigation makes possible the growth of high-value or high-yielding crops which could not be justified without irrigation. Irrigation also may make it possible to obtain highly beneficial results from the application of fertilizers. In those cases part of the total yield response is attributable to the irrigation water alone and part is due to the integrated action of the water and the fertilizers, plant-protection measures, special seeds, and any other aspects of modern, efficient farm management made possible by a dependable supply of soil moisture. The services of a skilled agronomist are required to apportion the total production increase between water and other agricultural input factors if each of those factors is to be justified independently. Frequently it will suffice to make a collective evaluation of the entire irrigation enterprise, and in that case it is not necessary to identify the contribution of each factor to the total production increase.

28. Effect of Water Shortages. Since irrigation involves the application of developed surface or groundwater supplies to supplement those provided by natural rainfall, the engineering of the irrigation development requires the determination of the increase in the water supply which can be supported economically. Experience has shown that it is rarely justifiable to provide an irrigation water supply of a size that will support optimum plant growth under all possible conditions of natural rainfall. Ideally several alternative levels of irrigation-water supplies should be evaluated economically to determine the net farm or project income which could be realized. The crop reduction estimated in each case should reflect the yield depressions which might be expected to result from shortages in irrigation-water supplies over 50 years or other periods selected for analysis. The level of irrigation-water supply to be provided could then be selected as that offering the optimum benefits. As a general rule, one 100 percent shortage in irrigation-water supply might be tolerated in a 50-year period, and the maximum tolerable average annual shortage would be about 5 percent.

29. Economic Analysis. Decisions and support for the undertaking of irrigation developments or projects are based primarily on economic findings concerning the soundness of the enterprise, although frequently social and political aspects, which are not susceptible to expression in monetary terms, are also important considerations. The economic analyses must cover conditions with and without the irrigation development, thereby determining by differences the economic values which will result from the investment in initial and annual costs. The end results to be provided by these analyses include determinations of the production increases to be realized, the direct economic benefit on the investment expressed in terms of the ratio of benefits to costs and in percentage of annual return on the investment; the economic benefits to the farmers in terms of increased profit from their investments and returns for their labor; and the benefits to the community and the nation in terms of employment opportunities, business opportunities, reduction in welfare costs, foreign-exchange earnings or reductions in foreign purchases, and increased tax revenues.

The economic analysis usually begins with the preparation of farm budgets for typical or representative farms in the project area under conditions with and without the project. Those budgets include estimates of the gross income from the area devoted to each crop in terms of production sold and production used on the farm; production costs including land leveling, farm ditches and drains, seeds, seedbed preparation and tillage, fertilizers, plant-protection measures including insecticides and fungicides, harvesting, storage, and delivery to market, hired labor, and investment in work animals, breeding stock, mechanical equipment, and storage facilities; an appropriate allowance for family labor and return on investment; and the net amount available annually to the farmer for profit, payments for irrigation water delivered, and repayment of the capital and operation and maintenance costs of the irrigation project. In addition to providing an evaluation of the capability of the farmers to repay costs of building and operating the irrigation facilities, the farm budgets provide information which can be combined to develop the direct economic benefits from the project.

Project costs include the capital costs involved in formulating the project and bringing the physical works into being and the costs of operating those works and maintaining them or replacing them as required for continuing service. The project works include storage facilities, diversion dams, canals and laterals to deliver water to each farm, drainage facilities as needed to collect drainage waters from the on-farm works required such as desilting works, dredges, pumping plants, or wells. The capital costs include engineering services for appraisals and feasibility studies, designs and tender documents, and supervision of construction, construction costs; administrative costs by the owner or owning agency; and financing costs including interest during construction, water scheduling and control of deliveries, and the maintenance and repair of the project facilities.

The benefit-cost ratio and the percentage return on investment are two measures used in expressing the economic viability of an irrigation development. The benefit-cost ratio will vary with the rate of interest charged on borrowed capital; hence the analysis must cover the range of interest rates which apply to the available sources of financial assistance at the time the study is being made. A benefit-cost ratio may be determined by using primary benefits, and a value greater than unity on that basis will be recognized as indicating justifiability from an economic standpoint. Primary benefits in this case mean the increase in the net income of agricultural production resulting from the project. Usually these benefits are determined as the annual average over a 50-year period of analysis, discounted to allow for a reasonable development after completion of the works before realization of full production from the entire project area. Costs for comparison with average annual benefits would be annual values including the amount necessary to amortize the capital investment over the repayment period and the costs of operation, maintenance, and replacements. A benefit-cost ratio may be determined also from "present worth" values computed for both benefits and costs to the date when the decision to undertake the project is to be taken, using an appropriate discount rate. The discount rate could result in a benefit-cost ratio of unity when the benefits include all values expressed in monetary terms and is the percentage return on the investment. That percentage return is a value frequently used in comparing the relative desirability of undertaking or financing alternative schemes or projects. The percentage return or benefit-cost ratio will be considered along with intangible social and political factors in the legislative process involved in authorization of irrigation projects by governmental agencies.

Economic analyses of irrigation developments are complex undertakings which should be performed by experienced agricultural economists, particularly where the irrigation enterprise is a part of a multiple-purpose project for utilization of land and water resources. The information to be developed by the analyses and the methods of presentation will be influenced to some extent by the requirements of the lending agency or agencies which will be approached for financial assistance.

SECTION 25

DRAINAGE

By DAVID B. PALMER[1]

1. Scope of the Section. This section deals with principles and practices applicable to agricultural drainage. A detailed and complete discussion of this subject can be found in Engineering Practices of the American Society of Agricultural Engineers.[15] Applicable practices are:

- EP260.4 Design and Construction of Subsurface Drains in Humid Areas
- EP302.2 Design and Construction of Surface Drainage Systems on Farms in Humid Areas
- EP369 Design of Agricultural Drainage Pumping Plants
- EP407 Agricultural Drainage Outlets—Open Channels
- EP463 Design, Construction and Maintenance of Subsurface Drains in Arid and Semiarid Areas

Drainage investigations are treated in "Drainage Manual," U.S. Bureau of Reclamation, 2d Printing, 1984.

2. Drainage Needs. Drainage problems are both natural and man-made. For the most part, natural drainage problems occur in the humid areas of the world and man-made problems occur in the arid areas. As used here, "humid" refers to locations having about 30 in or more of precipitation per year. A UNESCO[1,*] study in 1961 revealed there are over 5.4 million mi^2 of arable land and nearly 10 million mi^2 of meadows and pastures in the world. Perhaps one-fourth of this land would benefit materially by the application of drainage engineering.

Man-made drainage problems usually develop as a consequence of irrigation. Historic evidence of this can be found on every continent of the world. A major contribution to the decline and disappearance of some ancient civilizations can be attributed to their failure to solve the drainage problem. A conservative estimate would be that 150 to 200 million acres of cropland are affected to some degree by man-made drainage problems around the world. This figure is obtained by taking the world acreage of irrigated land,[2] 300 million acres, and assuming that about one-half to two-thirds of this land has potential drainage problems.

3. Drainage Principles. A properly designed drainage system removes excess surface water and/or lowers the groundwater level to prevent waterlogging. Ordinary farm crops cannot grow in soil that is saturated with water. Where subsoil waters contain salts, these mineral elements migrate to the ground surface because of evaporation. This concentration in the root zone greatly retards plant growth and prevents seed germination.

---

[1]Acknowledgment is made to William W. Donnan for material in this section which appeared in the third edition (1969).

*Superscripts indicate items in the References at the end of this section.

Surface drainage is accomplished by the construction of lateral ditches and open drains. Subsurface drainage and lowering of the water-table are accomplished by a system of open drains or buried tile lines into which the gravity water seeps. Water collected in drains is conveyed to a suitable outlet. Subsurface drainage can also be accomplished by pumping from wells to lower the water table.

DRAINAGE SURVEYS AND INVESTIGATIONS

Drainage problems differ widely because of the varied nature of the physical land and hydrologic conditions. There are no fixed shortcut methods of investigation that are uniformly applicable.[3] Some problems are fairly simple and their solution is quickly apparent. Others require only limited investigation. Generally, however, the topography, hydrology, soils, and crop-management practices vary so greatly, both individually and in their total effect, that a complete and thorough evaluation is needed to determine the specific causes of undesirable drainage conditions and their correction. Holes must be bored, observation wells installed, soils examined, and hydrologic measurements taken. Every source of information relating to the problem must be explored and the information analyzed.

The basic information important in any drainage investigation deals with the following factors: (1) topography, (2) sources of water, (3) soils, (4) salinity, and (5) water tables.

4. Topography. Good topographic maps are necessary in order to plan an adequate drainage system. The topographic survey is the basis for all subsequent investigations since it is the framework upon which are built the soil survey, water level, drain location, and depth and outlet feasibility. A topographic survey should determine the surface configuration, including the surface slopes, the direction of natural drainage, and potential drainage outlets. This survey gives a clue to the type of drainage needed. It gives positive information upon which to base specific drainage plans.

If suitable topographic maps are not available, it is recommended that they be made by photogrammetric methods. Aerial coverage at an elevation compatible with the preparation of photogrammetric contour maps on a scale of 1:5000 with 1-m contour intervals and machine interpolated to ½-m contours is advisable.[4]

For example, analysis of the topography may reveal the lack of natural outlets for drainage water. Irrigation systems are usually developed on broad, flat expanses of land which are often devoid of natural drainage channels. Therefore, drainage systems in irrigated areas usually require a trunk outlet system. Broad, flat fields are ideal for tiling in a grid pattern, while benches and swales call for interceptor patterns. A basin type of topography often lends itself to pumping for drainage. The objective is to key the drainage system to the topography.

SOURCES OF WATER

The source of all waters coming into the area must be determined. The water-sources survey provides a key to the measures needed to remedy undesirable drainage conditions. More specifically, the water source often governs the type of drainage to be installed. Thus, if excess water is due to precipitation, the remedial measure would probably be better surface drainage; if due to canal seepage, an interception drain may be indicated; and if due to artesian pressure, pumped wells may provide the most practicable remedy. In some problem areas the source of water is obvious. In others the sources of water may be both numerous and complex, making specific origins difficult to discern. A consideration of all the pertinent information on geology, topography, soil strata, and water table is needed to determine the source of the water.

The common sources of water of major importance in drainage problems are precipitation, irrigation, seepage, and hydrostatic pressure.

5. Precipitation. (Refer to Sec. 1 for a detailed discussion on precipitation.) A positive correlation between the distribution of precipitation during the year and the fluctuations in water-table elevations may be evidence that seasonal precipitation is one of the chief sources of water. Lack of such correlation indicates that precipitation probably has little effect on the water table.

An attempt should be made to determine whether long-term cycles of precipitation are related to long-term hydrographs of water levels. For example, wet cycles may be followed by rising water tables. Deep seepage to artesian or other aquifers, though slow, may often be manifested by a rise in water levels, sometimes years after the peak of a precipitation cycle has passed.

Precipitation affects artesian wells, deep static wells, and shallow piezometer wells differently. The response of water-surface levels in these wells provides indications of the degree, mode, and duration of influence of precipitation on the water table.

6. Irrigation. (Refer to Sec. 24, Art. 4.) Many drainage problems in irrigated areas are caused by the application of too much irrigation water. Studies should be made of (1) the effect on the water table of single irrigations, (2) water-table fluctuations throughout the irrigation season, and (3) long-time changes in water-table elevation over a period of years subsequent to the beginning of irrigation.

Poor methods of water application and inefficient use of irrigation water are likely to result in the loss of large amounts of water by deep percolation and surface runoff. Methods of applying water vary widely, depending on such factors as soil, slope, crops, size of field, delivery schedule of water, and availability of water.

7. Seepage. Seepage is a major source of water in many drainage problem areas. A study should be made to determine the locations and amounts of excessive seepage from canals. A comparison should be made of water-table hydrographs when canals are full of water and when they are empty to show any correlative relationship.

8. Hydrostatic Pressure. Water originating from precipitation or seepage may cause drainage problems in areas far removed from the water source. The locations of springs, seeps, and abandoned wells may be important clues to the source of water.

SOILS

The soil-stratum survey, which gives the location, extent, and physical characteristics of the various underlying soil layers, is probably the most important single technical phase of the drainage investigation.[5]

No drainage system can be adequately designed without a knowledge of the soil profile and the characteristics of the subsurface strata. Points which should be considered are (1) kinds of soils, (2) thickness of the various strata, (3) continuity of strata, and (4) position of the various strata with respect to the ground surface and to each other.

Some soils drain easily; others are extremely difficult to drain. Generally speaking, the coarse-textured soils drain better than the fine-textured soils. In many irrigated areas the soils are formed into complex profile patterns. Stratified sands, silts, and clays are commonly found. Fine-textured clay layers are often underlain or overlain by coarse-textured sands. The sequence of permeable and impermeable soils and their ability to transmit water determine both the type of system that should be installed and the design. For example, open drains at intervals of 1 mi may be adequate for drainage of coarse-gravel subsoils, whereas a fine-textured clay soil to a depth of 10 or 12 ft might require mole drains spaced at 30 ft. Lack of drainable aquifers at the 3- to 5-ft level may make drainage by tile lines unfeasible. Deep underground sand and gravel are usually easy to drain with pumps.

9. Borings. The soil borings should be made in a grid over the project area. A minimum of one hole per square mile to a depth of 12 ft should be used to characterize the soils in the project. In addition to this reconnaissance grid there should be a series of detailed spot boring grids to determine the variability of soil parameters. These surveys should consist of a very detailed soils investigation of a 1000-acre parcel in each 20,000-acre block of the project. On these smaller parcels the borings would be spaced at about 400-ft intervals or at such spacing as to map carefully individual soil types and series.

10. Permeability. An estimate of the permeability of the strata underlying the soil surface is essential in developing sound techniques of land drainage. Water-transmission rates should be determined in quantitative terms to be of practical use in this connection.

Coefficient of permeability may be defined as the rate of flow of water through a unit cross-sectional area under a unit head during a unit period of time. For convenience in making comparisons, coefficient values are stated in terms of flow of water through saturated soil.

Methods of accurately determining the coefficient of permeability may be grouped in three broad classes, as follows:

1. Field Measurements.[6] (a) Direct measurement of the permeability of an entire soil profile, based on pumped-well data. A drawdown curve and data on quantity of water pumped are used to compute the coefficient. (b) Direct measurements of the permeability of individual strata by means of small tubes, piezometers, or auger holes.[7]
2. Laboratory measurements utilizing a permeameter device and either in-place undisturbed specimens, taken in the field by means of one of the various sampling devices, or samples of disturbed soil prepared for laboratory examination by drying the soil and packing it into the permeameter.
3. Indirect evaluations of permeability based on physical and chemical soil properties.

Each of these methods has its merits and drawbacks. The particular method selected will depend upon the requirements of the drainage survey, the availability of appropriate measuring devices, and the degree of accuracy desired.

11. Indirect Evaluations of Permeability. Some of the soil characteristics that control the movement of water through the soil are type of structure, arrangement of aggregates, grain size, texture, pore space, dispersion, swelling, and type of clay mineral. In many locations in the United States, visible soil characteristics have been correlated with measured percolation rates, and the soil permeability is graded in accordance with a classification which has been used extensively by the U.S. Soil Conservation Service[8] in describing mapping units of soil-conservation surveys. This classification follows:

Permeability class	Permeability	Percolation rate, in/h/, through saturated undisturbed cores under ½ in head of water
Very slow	1	Less than 0.05
Slow	2	0.05–0.2
Moderately slow	3	0.2–0.8
Moderate	4	0.8–2.5
Moderately rapid	5	2.5–5.0
Rapid	6	5.0–10.0
Very rapid	7	More than 10.0

12. Texture Index. One mappable characteristic which can be used to describe permeability is texture. The following arbitrary relationship between texture and permeability is used in the determination of soil drainability.

Textures	Average permeability		
	gal/day	ml/h	m/day
Coarse sand	2500	509	120
Sand	250	50	12
Fine sand	100	20	4.8
Very fine sand	50	10	2.4
Loamy sand	25	5	1.2
Sandy loam	5	1	0.24
Very fine sandy loam	2.5	0.5	0.12
Loam	1.0	0.2	0.048
Silty loam	0.5	0.1	0.024
Silty clay loam	0.25	0.05	0.012
Silty clay	0.05	0.01	0.0024
Clay	0.025	0.005	0.0012

SALINITY

The diagnosis and treatment of saline and alkali soils is a problem in soil chemistry, but because it is so frequently associated with areas needing drainage, especially in arid and semiarid regions, it is necessary for the drainage engineer to become familiar with this problem.

There must be a thorough and detailed study of the salinity problem of the drainage area. This should include a study of the salinity of the surface soils and of the groundwater.

Soil samples taken from boreholes made on the soil-survey grid should be analyzed for mineral content. A minimum of one sample for each square mile of the project area is suggested. In addition, spot checks should be made at intervals so as to be able to draw a map showing the location and degree of severity of saline and alkaline deposits.

Excessive quantities of mineral elements in the root zone of the soil inhibit plant growth. They must be leached downward and disposed of by the drain system.

Saline and alkaline soils have been separated into three groups, namely, (1) saline, (2) saline-alkali, and (3) nonsaline-alkali. The following discussion is taken largely from U.S. Department of Agriculture Handbook 60, "Diagnosis and Treatment of Saline and Alkali Soils," 1954.

13. Saline Soils. The term saline is used in connection with soils for which the conductivity of the saturation extract is more than 4 mmhos/cm at 25°C and the exchangeable-sodium percentage is less than 15. Ordinarily, the pH is less than 8.5.

Saline soils can usually be improved through leaching, as the soluble salts present will go into solution and be removed with the drain water. Leaching in areas of high precipitation is usually a natural process after subsurface drainage is established. In arid and semiarid regions it is necessary to supply irrigation water to accomplish this leaching. In summary, the reclamation of saline soils can usually be accomplished through some time of leaching without the addition of chemical amendments. Adequate subsurface drains are, of course, a prerequisite.

14. Saline-Alkali Soils. The term saline-alkali is applied to soils for which the conductivity of the saturation extract is greater than 4 mmhos/cm at 25°C and the exchangeable-sodium percentage is greater than 15. These soils form as a result of the combined process of salinization and alkalization. As long as excess salts are present, the appearance and properties of these soils are generally similar to those of saline soils. Under conditions of excess salts, the pH readings are seldom higher than 8.5 and the particles remain flocculated. If the excess soluble salts are leached downward, the properties of these soils may change markedly and become similar to those of nonsaline-alkali soils.

Drainage of these soils may require certain chemical amendments based on laboratory analysis of soil samples. Insofar as the engineer is concerned, the difficult problem with saline-alkali soils is their identification, as they may exhibit characteristics of both saline and nonsaline-alkali soils. Saline-alkali soils may be flocculated because of the presence of excess salts and may have a permeability equal to or higher than nonsaline soils. This is often misleading and may give the impression that the soils can be reclaimed through simple leaching. Actually, this may not be the case because leaching will remove the soluble salts, thereby causing the soils to become strongly alkaline, and the permeability will be materially reduced.

15. Nonsaline-Alkali Soils. The term nonsaline-alkali is applied to soils for which the exchangeable-sodium percentage is greater than 15 and the conductivity of the saturation extract is less than 4 mmhos/cm at 25°C. The pH readings usually range between 9.5 and 10. These soils frequently occur in semiarid and arid regions in small irregular areas, which are often referred to as "slick spots."

It is highly important that nonsaline-alkali soils be recognized as such before any attempt to establish subsurface drainage. This is important because these soils have lost some of their internal drainage characteristics and may not drain properly, regardless of the type of drainage system installed. Where it is economically feasible to reclaim these soils, chemical treatment may be necessary to flocculate the soil particles and restore soil permeability before leaching and drainage. Some of the chemical amendments commonly used are calcium chloride, gypsum (calcium sulfate), sulfur, and sulfuric acid. The kind and amount of amendment applied must be based on recommendations following an analysis of representative soil samples.

16. Water Quality. A knowledge of the salinity of the groundwater in the drainage problem area is important because of the adverse effect of saline water on the growth and production of beneficial plants. In areas of good-quality groundwater, the solution to the drainage problem is largely a matter of lowering the water table below the root zone of the plants to keep excess water from interfering with normal plant-growth processes. Some latitude in the fluctuation of the water-table elevations is permissible.

In areas of saline groundwater, however, the water table not only must be lowered to a point well below the feeding zone of the plant roots but must also be kept from rising above that level so that evaporation and capillary forces will not concentrate the salts in the root zone. The presence of large amounts of harmful mineral elements in the root zone of the soil intensifies the drainage problem since steps must then be taken to drain not only the excess water initially present but also any additional quantities that must be applied to leach the salts out of the soil. The danger of losing essential plant nutrients in the leaching process further complicates the problem of devising effective drainage systems and measures.

WATER TABLES

The water-table survey will indicate the height, movement, and cyclic trend of groundwater levels. These data can be further analyzed to pinpoint the sources of excess water, the quantity, and the direction of the underground flow. If observations indicate artesian pressure from deep aquifers, relief wells and pumpage are usually necessary. Where rainfall contributes to the rise in water table some type of surface drain may be indicated. If seepage from an adjacent canal or reservoir can be detected, but not controlled, an interceptor drain may solve the problem.

Records of water levels, maps of depths to water table, and other hydrologic data for past years are often available for many drainage problem areas. Such data, when correlated with current conditions, furnish clues to the cause of fluctuations in the water table. The plotting of water-table hydrographs in conjunction with data on precipitation, irrigation, runoff, effect of pumping, and other hydrologic phenomena provides a useful basis of analysis.

17. Observation Wells. A grid of observation wells should be established at the beginning of any project study. This grid should consist of cased holes to a depth of 15 to 20 ft spaced at about one per square mile. Water-table observations should be carried on periodically, at sufficient frequency to obtain an adequate seasonal and annual hydrograph. The records for any given project area should be related and compared with hydrographs from other observation wells having long-term hydrographs.

18. Piezometers. A useful drainage-investigation tool is the groundwater piezometer, an unperforated small-diameter pipe so designed and installed that after it is driven or jetted into the ground, water can enter only at the bottom end.[9] The piezometer registers the hydrostatic pressure of the underground water at whatever level it is terminated. Since underground water moves from a point of high pressure to one of low pressure, the piezometer opens up a wide range of possibilities for investigating groundwater movement. With sets of piezometers installed at different depths and spaced at intervals the hydrostatic pressures of an entire soil profile may be determined, and seepage movement detected.

SURFACE DRAINS

19. System Layout. All the data developed from the drainage surveys and investigations should be utilized in the design of both surface drains and subsurface drains. Surface drains are usually constructed as open ditches. The main outlet drain and its principal branches or laterals form the backbone of the drain system, and they are usually installed as open unlined channels.

The topography of the area is generally the controlling factor in the location of the main and lateral drains. Topography also influences the spacing of the laterals. For very flat areas, the laterals should be not more than ½ mile apart. In no case should the laterals be more than 1 mile apart. The open lateral thus provides an outlet for both surface and subsurface drain water. Property lines or government subdivision lines should be considered in determining the locations of the laterals. Drainage ditches in irrigated areas usually serve primarily to control surface and subsurface irrigation waste and seepage waters. Their location is often dictated by the location of the irrigation-canal system. Main outlet ditches may follow natural streams and provide channel capacity to handle flood runoff in addition to the drainage waters from irrigated lands.

20. Rate of Discharge. The rate of discharge is generally expressed in cubic feet per second per square mile of drainage area. Various empirical formulas have been developed for determining runoff into drainage channels (see Sec. 1, Art. 32).

21. Rational Formula. A formula developed by Ramser for estimating stormwater runoff from agricultural areas is as follows:

$$Q = CIA$$

where Q = maximum rate of runoff, cfs
C = runoff coefficient dependent on topography, vegetation, etc.
I = rainfall intensity, cfs/acre, which corresponds to intensity, in/h
A = watershed area, acres

General values of C runoff coefficient are as follows:

Kind of watershed	Slope %	Value of C
Cultivated gentle	0–5	0.40
Cultivated rolling	5–10	0.60
Cultivated hilly	10–30	0.72
Pasture gentle	0–5	0.25
Pasture rolling	5–10	0.36
Pasture hilly	10–30	0.42
Timber gentle	0–5	0.15
Timber rolling	5–10	0.18
Timber hilly	10–30	0.21

22. Southwestern Drainage Formula. A formula developed for conditions in the southwestern United States but which has been used in many other areas is as follows:

$$Q = CM^{5/6}$$

where Q = runoff, cfs
 C = coefficient dependent upon the topography, soils, and land use
 A = watershed area, mi^2

Some values of coefficients are as follows:

C = 125 for hill areas (maximum)

C = 75 for hill areas (minimum)

C = 45 for coastal area, cultivated land

C = 40 for delta area, cultivated land

C = 35 for western plains, cultivated land

C = 30 for improved pasture

C = 22 for rice land

C = 15 for range land

In irrigated areas the mains and laterals must have a capacity to remove rainfall runoff as well as excess irrigation water and inflow from tile drains. Where appreciable rainfall occurs, one of the above empirical formulas should be used to develop runoff design capacities. Where little or no appreciable rainfall occurs, capacities of 6-ft-depth open drains are seldom if ever exceeded. Studies made by the U.S. Bureau of Reclamation and others indicate that the overall yield of seepage into an open drain system is about 0.01 to 0.15 cfs per square mile of irrigated land.

HYDRAULIC DESIGN

The hydraulic design of open drains with respect to capacity, permissible velocity, roughness factors, grade, side slopes, and bottom width is all adequately covered in Secs. 7 and 17. For the most part, open drainage channels are designed in the same manner as open irrigation canals. Some additional pertinent factors are given below.

23. Grade. The longitudinal slope or grade of a ditch is determined almost entirely by local topographic and soil conditions. The grade of a ditch must follow more or less closely the slope of the ground surface along the centerline of the ditch, and its value is fixed within narrow limits by topographic conditions. Because of this fact, the grade is the first of the hydraulic elements entering into the design of a ditch which is considered.

It is impracticable to distinguish between the grades of mains, submains, and laterals, since these terms are relative, and the lateral ditches in a large drainage system may be larger than the main ditch of a small system. In any one system, however, the grade of the main ditch will usually be flatter than those of the laterals.

The elevation of the bed of the outlet stream gives a starting point for the grade line of the main ditch. A tentative grade is then assumed for the main which establishes tentative starting elevations for the submains. Tentative grades are then chosen for the submains, which in turn establish the elevations of the ends of the laterals. As the study progresses, the tentative grades must usually be modified somewhat in order to meet satisfactorily the requirements of the other ditches. Even after these changes have been made, further adjustments may be necessary later when the size of channels and velocity of flow are considered. The nature of the soil through which a ditch is to be constructed also has a bearing on the grade line, since it may affect the depth of the ditch.

24. Depth. Ditches for agricultural drainage are usually 6 to 12 ft in depth. The depth of each lateral is first determined. The criterion is that each lateral must be of sufficient depth to serve as an outlet for the tile underdrains that are to discharge into it. The flatter the topography and the greater the spacing of laterals, the deeper will the laterals have to be. Thus depth and spacing of laterals are interrelated. The laterals are the most important part of the drainage system. Each submain must be of sufficient depth to receive the discharges of the group of laterals that it serves. The depth of the submain, therefore, will be determined by either the longest lateral or the lateral from the flattest portion of the area.

Likewise, the depth of the main ditch is determined by the elevations of the bottoms of the submains at their ends. In very flat country, this may result in too low a grade for the main ditch. In this case, the best solution is to reduce the area served by a main ditch, i.e., divide the total area to be drained into two or more parts and design an independent system of drainage for each part.

In muck and peat soils, deeper drainage channels are necessary than in other soils to allow for the subsidence that occurs when the soil is drained and cultivated. In the virgin muck soils of the Florida Everglades, it has been found necessary to make the laterals 8 ft deep in order for a depth of 5½ ft to remain 3 or 4 years after construction.

25. Spacing. It will rarely be necessary in agricultural drainage to place the lateral ditches closer than ½ mi. With this spacing no portion of the area will be more than ¼ mi from an outlet, and even if the ground is very flat, a satisfactory underdrainage system can be designed. Where the natural surface slopes 5 to 10 ft/mi, a spacing of laterals of 1 mi is satisfactory. The mile has been used as the unit of spacing, since ditches are usually located on government section or quarter-section lines.

In situations where open lateral drains are designed to accomplish the entire dewatering of the problem area, the spacing criteria should follow that prescribed for tile spacing (see Arts. 28 to 30). In other words, one 6-ft-depth open drain is analogous to one 6-ft-depth tile line in drawdown of subsurface water.

Generally speaking, open drains are not designed for agricultural purposes except in coarse-textured soils, since their installation at close spacings removes a considerable portion of the crop land from cultivation and their maintenance becomes a problem.

DRAINAGE STRUCTURES

See Sec. 16 for design criteria on chutes, drops, transitions, inverted siphons, flume crossings, and/or other types of structures employed in the development of a drainage system.

SUBSURFACE DRAINS

Most subsurface horizontal drains are plastic tubing. The tubing is manufactured in lengths of 100 to 300 ft depending on the diameter. The tubing is perforated to permit entry of the drainage water.

26. Layout of Subsurface Drainage Systems. The herringbone system consists of parallel laterals that enter the main at an angle, usually from both sides. This system is adapted to fields where the main or submain lies in a slight depression. It may also be used where the main is located in the direction of greatest slope and steeper grades for laterals are obtained by angling the laterals upslope. The gridiron-tile system consists of parallel laterals located perpendicular to the main tile. It is used on flat, regularly shaped fields and in uniform soil.

The double-main system is a modification of the gridiron or herringbone system and is applicable when a depression, which is frequently a natural watercourse, divides the field where tile is to be installed. Placing a main on each side of the depression may serve to drain the waterway and provide an outlet for the laterals. Parallel mains are also used to reduce the size of the main.

A random system of tile is used where the topography is undulating or soils vary and fields contain isolated wet areas.

The interception system intercepts seepage moving down a slope. The interceptor usually should be placed at about the upper boundary of the wet area as determined by drainage investigations.

DESIGN CRITERIA

The design of a closed subsurface drainage system involves the determination of depth, spacing, and size of drain, together with an adequate outlet and appurtenant works. Depth and spacing are roughly proportional, depending on the permeability of subsurface materials. Generally, the greater the depth of drain the wider the spacing between drains. The choice of depth and spacing is often an economic consideration.

27. Depth. The depth of tile drains is often controlled by depth of the outlet system. Where this circumstance does not prevail, depth is determined by a combination of such factors as crop type, soils, drainable strata, and spacing. For best growth of ordinary crops in humid regions, the water table should be drained to at least 3 ft below ground surface. This would require a depth of drain of from 4 to 5 ft.

In arid areas of general high water table or where salts occur in the water and soil, drains should be relatively deep. It is necessary to maintain a water table at such a depth that water rising by capillarity from the water table will not reach the ground surface from which it can evaporate and deposit the dissolved salts. For medium-textured soils, the water table should be a minimum of about 4 ft, and for fine-textured soils about 5 ft. This means that for salt control, drains should be placed at least 6 ft in depth.

There is opportunity for crop diversification on many irrigated farms. Different crops have different rooting characteristics, with some rooting to shallow depths and some deeper, and ranging from 1 to 10 ft. Any curtailment of natural rooting will have an adverse effect on production. Since tile drains are relatively permanent, they should be installed as deep as possible to develop the deepest root area for any crop that may be grown. A drain cannot lower the water table below the depth to which the tile is laid. Tile laid above the water table will not intercept downward percolating water and no water will be collected until the water table has risen to the tile. Except in the immediate vicinity of the tile line the water table will always stand higher than in the tile.

The deeper the tile, the greater the tile spacing can be to obtain the same minimum depth of water table. In stratified soils it is advisable to place tile in the most permeable layer—provided, of course, this is below the depth to which the water table should be lowered and within a depth that can be economically reached. Where stratifications are undulating or discontinuous, it may not to possible to place the tile in the more permeable layer, since the tile must be on grade and continue to a point of discharge. In summary, it may be said that considering cost and soil conditions, tile should be placed as deep as possible.

28. Spacing. Spacing between the tile lines depends upon the texture of the soils and the depth of tile below the ground surface. Since water usually stands nearer the ground surface midway between drains, the depth at this point determines whether or not the drains are lowering the water table satisfactorily. Water usually moves through coarse-textured soils more rapidly than it does through fine-textured ones. Therefore, drains can be spaced greater distances apart in coarse-textured than in fine-textured soils. Soil texture is generally used as a guide but, under conditions where such things as the method of soil deposition may alter the conductivity, soil texture may be misleading. Differences may be detected by reliable investigations, as previously discussed.

In an irrigated area with a general high-water-table condition and with tile installed at 6- to 8-ft depths, drains may be placed 300 to 600 ft apart in sandy soils and 100 to 300 ft apart in clay soils.

29. Donnan Spacing Formula. Several theories covering the flow of water to tile drains have been proposed. These theories have modified the approach of the drainage engineer toward the solution of many tile drainage problems. Considerable progress has been made toward the rational design of drainage systems using such theories.

Among these formulas is the one commonly known as the Donnan formula,[10] which is typical. The formula was developed for relief drains and is based upon certain barrier conditions. This is illustrated by the following expression:

$$S = \frac{4P(b^2 - a^2)}{Q_d}$$

where S = spacing of drain lines, ft
P = hydraulic conductivity or coefficient of permeability, gal/ft^2/day
b = distance from the average tile depth to barrier stratum at the midpoint between the tile lines, ft
a = distance from the average tile depth to barrier stratum, ft
Q_d = quantity of water to be drained, gal/ft^2/day

Where no barrier stratum is present, a barrier should be assumed at a depth equal to twice the drain depth. Figure 1 is a sketch showing the above relationship.

The units for hydraulic conductivity or the coefficient of permeability P and Q_d can both be expressed in gallons per square foot per day or in cubic inches per square inch per hour in this formula without changing its validity.

The Donnan and other formulas require that the average effective permeability and barrier conditions be established by field investigations. The quantity of water to be drained Q_d is dependent upon a multitude of complex interrelationships of soil-water hydrology. All the various factors, such as rainfall, irrigation, slope, soil type, ponding time, waste disposal, crop, infiltration rate, evapotranspiration, and seepage in and out of the problem area, enter into the final determination of the amount of water which must be drained by the drainage facility. It is

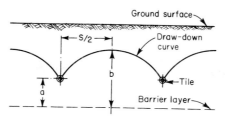

FIGURE 1 Development of the Donnan spacing formula.

essential that a reasonable determination of this amount be made since it is a factor in the design of the facility.

Where precise criteria are lacking, the designer can arbitrarily take a percentage of the irrigation water applied for estimating Q_d. In humid areas the "drainage coefficient" can be used to determine Q_d. *The drainage coefficient is the discharge of the underdrainage system, expressed in inches of depth of water which must be removed from the area in 24 h.* The drainage coefficient is a measure of the maximum rate at which the water will move through the soil to the laterals. The proper drainage coefficient for a specific section of the country can be stated in general terms only, since it varies not only with the rainfall intensity but also with the soil texture, topography, and depth and spacing of laterals. The following general values, based on annual rainfall, will aid in the choice: for 30 in of annual rainfall, a coefficient of ¼ in; for each additional 5 in of annual rainfall, an additional 1/16 in the coefficient.

Where salt is a problem, research has indicated that at least 8 to 10 percent of the water applied must drain down and out of the root zone in order to maintain a favorable salt balance in the soil. Therefore, under these conditions, not less than 10 percent of the water applied would be considered the amount to be drained.

30. DRAINMOD Water Management Model. The computer model DRAINMOD[11,12] was developed for design and evaluation of multicomponent water-management systems on shallow-water-table soils. It simulates subsurface drainage, surface drainage, subirrigation, controlled drainage, and surface irrigation. Input data include soil properties, crop parameters, drainage system parameters, and climatological and irrigation data. The model may be used to simulate the performance of a water-management system over a long period of climatological record. Approximate methods are used to simulate infiltration, drainage, surface runoff, evapotranspiration (ET), and seepage processes on an hour-by-hour, day-by-day basis. Water-table position and factors such as the ET deficit are calculated to quantify stresses due to excessive and deficient soil water conditions. Stress-day-index methods are used to predict relative yields as affected by excessive soil water conditions, deficit or drought conditions, and planting date delay.

31. Grade. Tile lines laid in most soils with little or no grade tend to silt up. On flat lands a minimum grade for tile lines should be established based on site conditions. Grades should be not less than the following limits, with the specific limitations on length:

Minimum grades of subsurface drains for clay-loam soils		
	%	Suggested max length for min. grade, ft
4-in drains	0.10	1500
5-in drains	0.07	2000
6-in drains	0.05	3000

Under exceptional conditions, grades less than those recommended have been justified where the soil was cohesive and where the quality of installation and local experience indicated that lesser grades would give satisfactory performance.

Under many site conditions where, as an example, sandy or other noncohesive soils are present, the minimum allowable tile grade of laterals should be 0.30 percent or greater. However, for such soils, the permissible grade of mains may need to be set at a lower value and sediment traps installed so that the system design is feasible.

Although the maximum permissible grade for subsurface drains varies with different soils, special precautions should be considered if the grades exceed 2 percent. Drains laid in sand and in sandy-loam soil on grades more than 1 percent may result in soil erosion which fills the drains and causes their misalignment.

32. Yield from Tile Drains. The average "runoff" from subsurface drainage systems will vary depending on spacing, soil type, irrigation practice, rainfall intensity, and many other interrelated factors. For humid regions the tile-drainage chart in Fig. 2 may be used to compute flow and required diameters for clay and concrete drain tile. The chart is a graphical solution of the continuity and Manning formulas.

For corrugated plastic drainage tubing, Fig. 3 may be used to compute flow and required diameters.

For irrigated regions, yield from subsurface drainage systems can be estimated from the following criteria:[13]

Size of area drained, hectares	Average yield from area, liters/s
0–16	11
16–32	20
32–400	20 plus 6 liters/s for each added 16 ha over the original 32
400–1200	140 plus 3 liters/s for each added 16 ha over the original 400

DRAIN SIZE

Four-inch tubing is the smallest diameter in general use in the United States. The problems of maintaining an accurate grade and the possibility of some settlement tend to limit the useful life of subsurface drainage systems. Accordingly, the usual practice calls for tubing size to be greater than that required

DRAIN SIZE

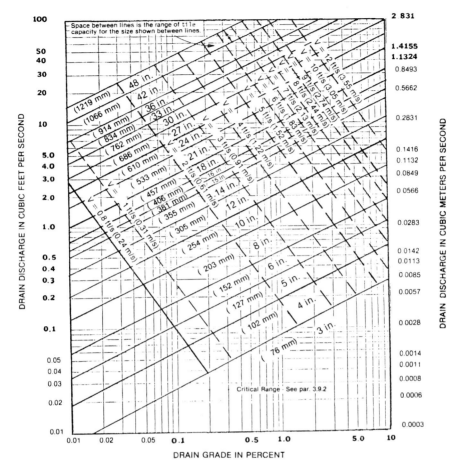

FIGURE 2 Guide for determining the required size of clay and concrete drain tile ($n = 0.013$).[16]

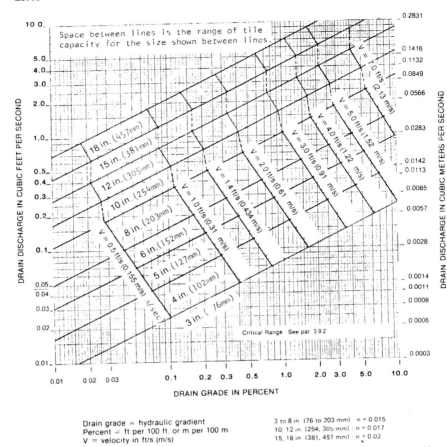

FIGURE 3 Guide for determining the required size of corrugated plastic drainage tubing.[16]

by capacity needs. High labor costs generally make it impractical to clean out tubing.

Given the above, in designing subsurface drainage systems for capacity, the size of the drain conduit is determined from its design flow, slope, and roughness. Manning's formula, $Q = 1.486 R^{2/3} S^{1/2} A/n$, is commonly used to determine required size. In so determining size, R and A are those values for the design flow Q in the conduit at a depth not greater than 70 percent of the inside diameter or height of the conduit. Recommended values of n for various conduit materials are:[14]

Conduit material	Minimum Manning's n	Recommended Manning's n
Clay and concrete tile	0.011	0.012–0.014
Perforated plastic tubing:		
75 to 200 mm diameter	0.015	0.015–0.016
200 to 380 mm diameter	0.017	0.017–0.018
Perforated corrugated-metal pipe	0.021	

MATERIALS AND INSTALLATION

33. Pipe Materials. Commonly used pipe materials are made from clay, concrete, and corrugated plastic. Other materials such as sewer pipe and corrugated metal are used in special situations, mainly where the grade line is so near to ground surface that there is danger that surface loads from vehicle traffic will deform or crush the pipe. Depending upon the length and location of these materials, it may be necessary that joints or perforations be used to provide for adequate water entry. The type of pipe material to be used on a project depends upon the availability of pipe manufactured from the various materials, the availability of suitable installation equipment, and cost.

34. Clay and Concrete Drain Pipe. Clay and concrete drain pipe should meet requirements specified by the American Society for Testing and Materials (ASTM). Quality classes for concrete pipe are designated in ASTM Standards C14 Specification for Concrete Sewer, Storm Drain, and Culvert Pipe; C118 Specification for Concrete Pipe for Irrigation or Drainage; C412 Specification for Concrete Drain Tile; and C444 Specification for Perforated Concrete Pipe. In addition, unreinforced-concrete drainage pipe should be Class C manufactured and tested according to U.S. Bureau of Reclamation Standard Specification for Unreinforced Concrete Drainage Pipe, Feb. 15, 1971. These specifications give criteria for pipe strength, frost resistance, acid and alkali resistance, exposure to sulfate soils, water absorption, and cement content and curing.

Quality classes for clay pipe are designated in ASTM Standards C4 Specifications for Clay Drain Tile; C498 Specifications for Perforated Clay Drain Tile, and C700 Specification for Vitrified Clay Pipe, Extra Strength, Standard Strength, and Perforated. An alternative to ASTM Standard C700 is Federal Specification SS-P-361E Pipe, Clay, Sewer (Vitrified Fittings and Perforated Pipe).

35. Corrugated Plastic Drain Tubing. The quality of corrugated plastic drain tubing is specified in ASTM Standards F405 Specification for Corrugated Polyethylene (PE) Tubing and Fittings; F667 Specification for 8, 10, 12, and 15-in Corrugated Polyethylene Tubing and F800 Specification for Corrugated Poly(vinyl chloride) Tubing and Compatible Fittings. Other applicable standards are USDA Soil Conservation Service (SCS) Specification 606 Subsurface Drain; SCS Specification 606R Manufacturing Corrugated Plastic Tubing Using Reprocessed Polyethylene Resin, and U.S. Bureau of Reclamation Standard Specification for Polyethylene or Polyvinyl-Chloride Corrugated Plastic Drainage Tubing.

36. Envelope Materials. Subsurface drains should be installed with suitable envelope or filter material around the conduit. The purposes of placing filter material around the drains are to prevent sediment from entering the drains and to allow water to flow freely into the drains. In most installations in the humid area of the United States, only the topsoil or selected permeable soils from the sides of the trench are placed around the drains. This is called "blinding." However, before "blinding" clay or concrete tile the joints are covered, if required, with tar-impregnated roofing paper, glass-fiber filter, plastic guards, burlap, or other materials. This joint covering is usually laid around the upper two-thirds of the tile to prevent sediment from entering the lines.

Envelope or filter materials are used in the construction of subsurface drain lines in unstable soil situations, such as very fine sand or other cohesion-less soil. Conditions where such installations are needed include (1) soils that easily fill a drain with sediment, (2) soils that do not provide a stable foundation, and (3) soils that tend to seal or clog drain openings and limit water entry into the drain. Envelope materials are vegetative, prefabricated, and mineral. Vegetative envelopes include straw, ground corncobs, and wood chips. Local experience should be considered before vegetative materials are used.

Prefabricated materials include fiberglass sheets and nylon fabric. Caution should be exercised in soils where iron or manganese oxide and other chemical deposits are probable because openings may become sealed, restricting water entry to the drain. Again, local experience should be considered.

Mineral materials include gravel and crushed stone. Design criteria for gravel envelopes have been developed and are published in Section 16, Drainage of Agricultural Land, U.S. Soil Conservation Service National Engineering Handbook. Mineral envelopes should be at least 3 in thick and completely surround the drainpipe. Pit-run sand and gravel may meet the criteria. Fines passing a No. 60 sieve should not exceed 5 percent of the envelope material.

The availability of filter material, its costs, and the extent of local successful experience usually determine the type used. In irrigated areas of the United States sand and gravel filters are generally used. The minimum thickness of filter material used around the conduit is 3 in. Design criteria are available for filter material based on research by the U.S. Bureau of Reclamation and the Corps of Engineers, U.S. Army.

37. Installation of Subsurface Drains. Digging of the trench for subsurface drains should start at the outlet end and proceed upgrade. The alignment of the trench should be such that the tile can be laid in straight lines or smooth curves. The trench width is often fixed by the width of available trenching machines. The trench width, measured at the top of the drain, may be equal to the outside diameter of the tile, plus about 0.5 ft. This clearance between the tile and the sides of the trench is necessary for proper bedding and blinding of the tile. It is very important that the bottom of the trench be cut accurately to grade and shape, and then the tile set to grade. If the trench is overcut, the depth should be sufficient to place gravel filter material under the tile. It should be backfilled either with graded gravel or with pulverized soil and tamped sufficiently to provide a firm foundation. The bottom of the trench is then recut to grade and shape.

The bottom of the trench should be rounded so that the drain will be embedded in undisturbed soil for the last 60° of its circumference. Some trenching machines shape the bottom of the trench as a part of trenching operations.

For corrugated plastic tubing, installation criteria are given in ASTM Standard F449 Recommended Practice for Subsurface Installation of Corrugated Thermoplastic Tubing for Agricultural Drainage or Water Table Control. The main concern is to assure that the shape of the bottom of the trench will furnish satisfactory alignment and resistance to deformation of the tubing.

38. Joint Spacing. Laying tile should begin at the lower end of the line and progress upgrade. A tight fit is required in noncohesive soils having a high percent of silt or sand. The gap between tile may be about 1/16 to 1/8 in for clay and clay-loam soils. However, local experience may indicate a wider spacing for peat and muck soils up to 1/4- to 3/8-in spacing. Where large gaps occur between tile, as on the outer side of a curve, the joints should be covered by broken tile, plastic, etc.

Perforated pipe should be laid with the perforations on the underside of the line.

39. Laying Tile in Unstable Soil. Special construction methods should be used when tile are laid through unstable pockets of soil such as saturated fine sand. One method is to place stable soil, coarse hay or straw, tough sod, crushed limestone, or gravel in the bottom of the trench before laying the tile. Another method is use of a broad cradle to support the pipe. A third method is use of a tightly sealed sewer pipe, continuous pipe, including corrugated-metal pipe, or nonperforated bituminized fiber or plastic pipe.

PUMPING FOR DRAINAGE

Most subsurface drains have gravity outlets. However, this is not always possible. Where the water surface of the receiving open ditch or natural stream is higher than the outlet end of the subsurface drain, pumps must be used. Pumping for drainage falls into two categories. Low-lift drainage pumps are used to pump surface water to lower the water level behind a levee or to lift water out of a shallow sump. High-lift drainage pumps are installed in deep wells in a grid pattern for the purpose of dewatering a large contiguous area to effect a general lowering of the water table.

40. Low-Lift Pumping. The factors that determine the location of a pumping plant for a large drained area are topography, nature of foundation, and proximity to towns. Generally, the ditch system is first designed in conformity with the topography and the pumping plant is located at the lower end of the main ditch. An unstable foundation for the building and pumping machinery at this point, however, may make a change of location necessary. The other factors may at times be the controlling ones.

41. Pumping Capacity. The rate at which the drainage water will have to be removed by the pumps depends upon the amount and distribution of rainfall, the size of the drained area, the nature of the soil, the quantity of water contributed by the higher lands outside the district, the storage capacity of the ditches, the amount of seepage under the levee, and the completeness of drainage desired.

The pumping capacity required is expressed in inches of depth of water over the entire drainage area which must be removed in 24 h. The capacities of pumping plants in successful operation vary from 0.25 to 1.50 in. The average capacity of a number of plants on the Illinois River is 0.36 in. The rate of removal in terms of inches of depth in 24 h is then converted into discharge in cubic feet per second. Most pump manufacturers rate their pumps for a discharge velocity of 10 fps, but as operated in practice this velocity is about 8 fps. With these data, the size of pump can be determined.

42. Types of Pumps. Centrifugal, rotary displacement, plunger, and sewer pumps have all been used, but the centrifugal pump is generally better adapted to drainage district pumping and is used almost exclusively at present.

Drainage pumping plants are operated only a small part of the time and rather intermittently. For a large part of the time, the pumps will operate at about one-third capacity. For this reason, it is desirable to have two or more small pumps rather than one large one. Where two pumps are used, one should have about twice the capacity of the other. Where three or more are required, they should be of the same size.

43. Inlet and Outlet Pipe. The suction lift of the pump should be as low as possible. Generally, the elevation of the floor of the pumping plant is about 1 ft above the highest elevation to which the water in the suction bay can rise during periods of excess rainfall with the pumps not operating.

Suction pipes are generally of riveted steel, although reinforced concrete has also been used. They must be airtight, smooth on the inside, and as straight as possible. The lower end should be expanded so that the entrance velocity will be not more than 2 fps.

The discharge pipe should be gradually enlarged immediately after leaving the pump to about twice the area of the pump discharge opening, so that the discharge velocity will be not more than 5 fps. The end of the discharge pipe should be submerged.

The discharge pipe may pass either through the levee or over its top. The latter arrangement is preferable, since it eliminates danger of seepage through the levee along the pipe and prevents the water in the river from flowing back through the pump when it is not operating.

44. Power. Steam engines, electric motors, and internal-combustion engines are all used to operate pumps. Steam engines are found in all the older plants, but the newer ones use either electric motors or diesel engines. Each type has its advantages and disadvantages.

The water, or theoretical, horsepower developed by a pump at maximum head and rate of discharge can be computed from the equation

$$\text{Water hp} = \frac{62.5\,hQ}{550} = 0.1136hQ$$

in which h is the dynamic head in feet and Q the maximum discharge in cubic feet per second.

The brake horsepower required will be greater than the water horsepower and depends upon the efficiency of the pump. With a pump efficiency of 70 percent, the brake horsepower will be 1.43 times the water horsepower.

45. Operation. Drainage pumping plants are run only a small part of the time, generally 60 to 90 days a year. During the fall months, after the crops are harvested, the pumps are not used. During the winter, the pumps are operated occasionally to remove the water that accumulates during winter storms. In the late winter and early spring, the pumps are operated continuously, not necessarily at full capacity, to remove the water from the saturated soil and to place the soil in condition for early plowing and planting. During the later part of this period, the pumps will be run more or less intermittently, and it is at this time that the advantage of gasoline engines or electric motors is most apparent. During the growing season, pumping will be required after each storm period to keep the groundwater at the desired elevation. At this time, the full capacity of the plant may be needed.

46. Pumping from Wells for Drainage. Drainage by pumping from wells is usually practiced in irrigated areas. Under some conditions, pumping groundwater to provide land drainage is an effective method of lowering a high groundwater table and reducing salinity hazards. In areas where the method has proved successful it has solved the drainage problems effectively and eliminated the need for open and closed subsurface drains to control the water table. In most instances the drain water is used to supplement the supply of irrigation water. However, careful engineering investigations, including pumping tests, should precede the installation of wells to ensure the success and economy of the project. One or more of the following conditions may contribute to the use of wells for drainage:

1. Large areas of flat lands with extensive high water table and salinity problems
2. Well-defined contiguous permeable aquifers underlying the wet areas
3. Aquifer deep enough to permit an efficient well
4. Groundwater under artesian pressure
5. Groundwater having good quality suitable for irrigation use with or without mixing with other irrigation waters
6. Electric power available at reasonable cost

In some areas it is possible to determine subsurface conditions from logs of existing wells. In other areas deep borings, often to depths of 100 to 200 ft, are necessary to explore underground aquifers. Nearly always, it is advisable to operate one or more test wells at capacity and make observations of the rates of pumping and the drawdown by lines of observation wells radiating from the test wells. If the well produces a significant drawdown at distances from 300 ft to ½ mi from the well, the method of drainage by pumping deserves additional study, especially if there is a use for the pumped water.

47. Spacing. The radius of influence of a drainage well will depend on the soil strata, rate of pumping, and other related factors. It is not unusual to find radii of 1000 to 4000 ft around a pumped well. The well spacing should be less than twice the radius of influence or drawdown cone of depression around the well. Drainage wells should be placed in a grid, and there should be a sufficient number of wells in the grid so as to command an extensive contiguous area. The spacing and location might be dictated by location of power supply or the location of disposal facilities for the pumped water. A grid of 1-mi spacing for 200-ft-depth gravel-packed wells having a 3-cfs pumping capacity has been used with success.

REFERENCES

1. Production Yearbook 1962, vol. 16, Food and Agricultural Organization of the United Nations, Rome.
2. "Worldwide View of Irrigation Developments," *Proceedings of the ASCE, Journal of Irrigation and Drainage* Div. 84 (IR3), 1958.
3. DONNAN, W. W., and G. B. BRADSHAW, "Drainage Investigation Methods for Irrigated Areas of Western United States," *USDA Tech. Bull.* 1065, 1952.

4. Drainage Surveys, Investigations, and Reports, *U.S. Department of Agriculture Soil Conservation Service National Engineering Handbook,* chap. 2, Sec. 16.
5. DONNAN, W. W., G. B. BRADSHAW, and H. F. BLANEY, Drainage Investigation in Imperial Valley, California, *U.S. Department of Agriculture Soil Conservation Service Tech. Pub.* 120, 1954.
6. DONNAN, W. W, "Field Experience in Measuring Hydraulic Conductivity for Drainage Design," *Journal of the American Society of Agricultural Engineers,* vol. 40, no. 5, pp. 270–273, 1959.
7. FREVERT, R. K., and D. KIRKHAM, "A Field Method for Measuring the Permeability of Soil Below the Water Table," *Proceedings of the Highway Research Board,* vol. 28, pp. 433, 442, 1948.
8. "Soil Survey Manual," USDA Soil Conservation Service Agricultural Handbook 18.
9. CHRISTIANSEN, J. R., "Groundwater Studies in Relation to Drainage," *Journal of the American Society of Agricultural Engineers,* vol. 24, pp. 339–342, 1943.
10. DONNAN, W. W., "Model Tests for Tile Spacing Formula," *Soil Science Society of America Proceedings,* vol. 11, pp. 131, 136, 1946.
11. SKAGGS, R. W., "A Water Management Model for Shallow Water Table Soils," *University of North Carolina Water Resources Research Institute Technical Report* 134, 1978.
12. CHANG, A. C., et.al., "Application of DRAINMOD on Irrigated Cropland," American Society of Agricultural Engineers Paper 81-2534, 1981.
13. WEEKS, L. O., "Drainage in the Coachella Valley of California," *Proceedings of the ASCE, Journal of Irrigation and Drainage Division* 85 (IR3), 1959.
14. "Design and Operation of Farm Irrigation Systems," American Society of Agricultural Engineers Monograph no. 3, revised printing, p. 258, 1983.
15. "Standards, Engineering Practices, and Data," American Society of Agricultural Engineers, *Standards* 1992, 39th ed.
16. American Society of Agricultural Engineers, *Standards* 1992, 39th ed., pp. 576–577.

BIBLIOGRAPHY

ROE, HARRY B., and Q. C. AYRES, *Engineering for Agricultural Drainage,* McGraw-Hill, New York, 1954.

ISRAELSON, O. W., and VAUGHN E. HANSON, *Irrigation Principles and Practices,* Wiley, New York, 1962.

WHITE, GILBERT F., editor, *The Future of Arid Lands,* American Association for the Advancement of Science, Publication 43, 1956.

LUTHIN, JAMES N., editor, *Drainage of Agricultural Lands,* American Society of Agronomy, 1957.

U.S. Soil Conservation Service, *Drainage of Agricultural Land,* National Engineering Handbook, sec. 16, 1971.

U.S. Bureau of Reclamation, *Drainage Manual,* 2d printing, 1984.

American Society of Agricultural Engineers Engineering Practices:
  EP260.4 Design and Construction of Subsurface Drains in Humid Areas, 1983.
  EP302.2 Design and Construction of Surface Drainage Systems on Farms in Humid Areas, 1985.
  EP369 Design of Agricultural Drainage Pumping Plants, 1982.
  EP407 Agricultural Drainage Outlets-Open Channels, 1982.
  EP463 Design, Construction and Maintenance of Subsurface Drains in Arid and Semiarid Areas, 1985.

U.S. Soil Conservation Service, *Soil Survey Manual,* Agriculture Handbook 18.

American Society of Agricultural Engineers, *Design and Operation of Farm Irrigation Systems,* Monograph 3, revised printing, chap. 7, Drainage Requirements and Systems, 1983.

SECTION 26

IRRIGATION STRUCTURES

By Nicholas M. Hernandez[1]

Structures on irrigation projects include dams, spillways, desilting works, intakes, pumping stations, canals, tunnels, pipelines, flumes, inverted siphons, chutes, drops, checks, turnouts, measuring devices, wasteways, culverts, overchutes, drainage systems, and other features peculiar to the project.

Dams vary from low diversion weirs to high storage structures. Distribution channels vary from small farm ditches, where the flow can be controlled by movable canvas barriers, to huge supply canals carrying several thousand cfs.

The Marala-Ravi canal in Pakistan has a maximum capacity of 22,000 cfs. Here in the United States, some sections of the All-American canal, built to carry water to Imperial Valley, southern California, have a bottom width of 160 ft, a permissible flow depth of 22 ft, and a capacity of 15,000 cfs. Other irrigation structures vary almost as widely.

Hydraulic features of reservoirs, dams, spillways, tunnels, canals, and other major structures are described in other sections. The following discussions are confined to irrigation structures not treated elsewhere and to such special considerations as may be necessary from the viewpoint of irrigation. Except as otherwise noted, designs included as illustrations are presented through the courtesy of the Bureau of Reclamation.

DIVERSION WEIRS

Irrigation projects supplied by gravity diversions from natural streams usually require the construction of weirs near the upper limits of irrigable lands, the purpose being to raise river surfaces high enough to permit controlled diversion of irrigation-water requirements. When two or more feasible sites are available, the weir location should be determined on the basis of economic considerations. Ordinarily, adoption of a site farther upstream permits a reduction in height of weir, since the canal can usually be built on a grade somewhat flatter than the river slope. However, savings resulting from a reduced height may be offset by costs of extending the canal upstream. When weir-height minimization is under consideration, care should also be taken to ensure that flood tailwater will not exceed the elevation of the canal system, thereby subjecting it to damage during flood flows.

1. Types of Diversion Weirs. Since canal flows are relatively small proportions of total stream discharges during flood periods, diversion weirs must be of

---

[1]Acknowledgment is made to Ivan E. Houk for the illustrations, tables, and other material in this section which also appear in Section 18, Irrigation Structures, in "Handbook of Applied Hydraulics," 2d ed., McGraw-Hill Book Company, 1952.

overflow or open-dam types or provided with bypass channels capable of carrying flood discharges. If bypass channels are not feasible, the dams must either be capable of carrying flood flows over their crests or be provided with enough gates or collapsible elements to pass flood discharges without damage. Weirs designed with gates to pass flood flows are sometimes called *barrages*. They are usually built in bays, separated by piers, and surmounted by operating bridges. Flow between piers may be controlled by collapsible shutters; by Stoney, radial, roller, or drum gates; or, in comparatively low weirs of inexpensive design, by horizontal removable flashboards.

Overflow weirs may vary from low temporary cobble and brush barriers, resembling beaver dams, to costly concrete arch or gravity structures. Intermediate designs may be built of logs, piles, cribs, rock, timber, steel, masonry, or reinforced concrete. Overflow weirs of permanent construction are usually provided with gate-controlled sluiceways, so that detritus may be flushed out periodically. Fish ladders or other provisions for passing fish over the dam may be required on streams where fishing is important, as in northwestern United States (see Sec. 18). Earth or earth and rockfill embankments extending to high ground are often built at the ends of weirs, in both overflow and open-dam types.

The most desirable type of diversion weir for a given site depends on height of weir, foundation conditions, streamflow, permissible upstream flooding, available construction materials, and amount of expense justified. Overflow types are desirable from the viewpoint of floating material, since they offer less obstruction to passage of ice, logs, brush, and miscellaneous debris. Gate installations are advantageous from the viewpoint of detritus problems, since deposits above the weir are scoured downstream during flood periods. Gate installations are also advantageous from the viewpoint of canal operation, since they permit some regulation of river water surfaces at intake structures.

2. Design of Diversion Weirs. Diversion weirs should be designed as dams. Concrete weirs on rock foundations should be made safe against failure by sliding, along any plane in the structure or its foundation, overturning, crushing, or rupture by tension. Analysis should include adequate allowances for wind, wave action, hydrostatic uplift along the foundation, horizontal and vertical water loads, and the structure dead load. Where conditions dictate, silt and ice, and dynamic loads due to the seismic acceleration of the structure and the mass being retained, that is, water, silt, or ice, are also considered. Weirs on sand, gravel, and other earth foundations should be made safe against the same forces and should also provide ample bearing resistance, percolation distance, and protection against erosion. An analysis of piping and percolation movements with the investigation of over 200 masonry dams on earth foundations has been published.[1,*] Detailed methods of design are discussed in Secs. 9 and 10.

Figure 1 shows some condensed cross sections of diversion weirs built on earth foundations. Figure 2 shows some cross sections of weirs built on rock foundations.

RIVER INTAKES

In rivers having stable beds and low-water discharges considerably greater than diversion requirements, intake structures may be built without constructing di-

---

*Superscripts indicate items in the References at the end of this section.

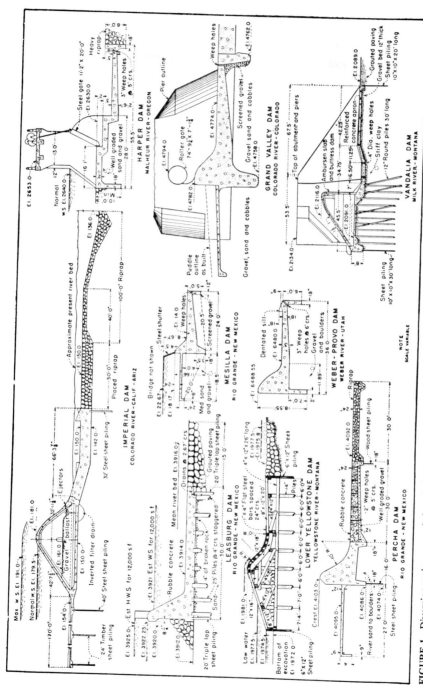

FIGURE 1 Diversion weirs on earth foundations.

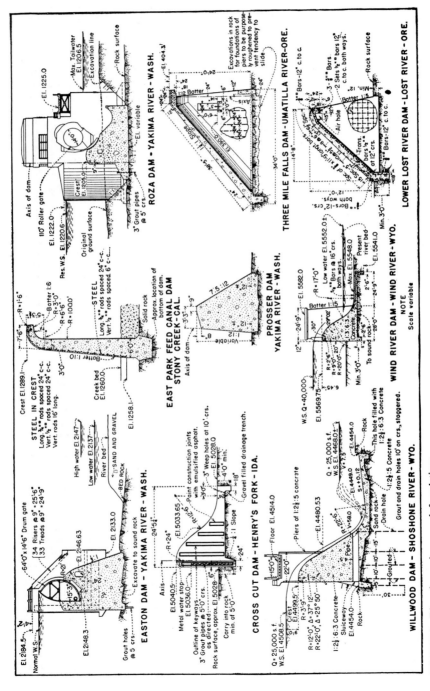

FIGURE 2 Diversion weirs on rock foundations.

version dams. However, such cases are rare. Ordinarily, intake structures are appurtenant parts of diversion weirs and are built at one or both ends of the weirs. Sometimes they are constructed short distances upstream, as when water is diverted into tunnel sections instead of open channels. When the intake is to be placed on one side of the river, it is advantageous from the standpoint of reducing the amount to detritus entering the intake to locate it just downstream of a bend in the river on the concave side. Figure 3[2] indicates results of model tests on bed-load distribution between the main course and branch watercourses. If intakes are to be on both sides of the river, the diversion should take place in a straight reach of the river. Figure 4 indicates a diversion scheme in a straight reach. Since obstruction of river flow causes detritus deposition above the weir, sluices often must be constructed. This is especially true when weirs are of the overflow type.

Depending upon the habits of the river's aquatic life, a means for excluding them from the distribution system may be required at the intake. A discussion of the various devices to effect this is covered in Sec. 18.

3. Design of Intakes. Intakes should be designed to control and regulate water drawn from river channels with minimum entrance losses and as little disturbance as possible. Temporary timber structures may be built on small projects, but permanent concrete or masonry construction usually is desirable. Ordinarily, a permanent structure consists of one or more bays, controlled by radial or vertical slide gates, the bays being separated by piers and surmounted by an operating platform. Vertical slide gates are suitable for small intakes; but radial gates are usually more economical and easier to operate. Stoney or roller gates may be used at large diversion structures. Overpour types have the advantage of drawing water from river surfaces, but undershot types are less likely to become clogged by sediment.

In diversions to concrete-lined conduits, where water is carried at relatively high velocities, total areas of gate openings are usually made about the same as conduit areas. In diversions to earth canals, where water is carried at relatively low velocities, total gate areas are usually based on velocities of about 5 fps. However, full gate openings can be utilized only during periods of low river flow and high irrigation demand. At other times, water is diverted through partial gate openings and gate velocities may be considerably higher than 5 fps.

A transition structure, consisting of a concrete floor and diverging curved or

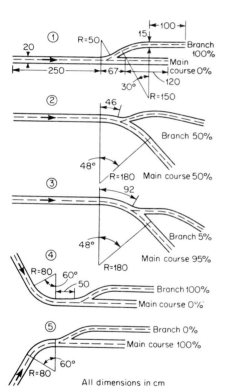

FIGURE 3 Typical shapes of diversion. Scale-model tests.

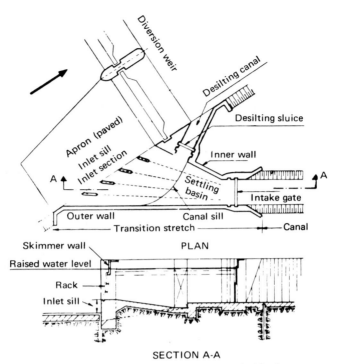

FIGURE 4 Canal intake on rivers carrying heavy bed load.

warped sidewalls, should connect the gate section with the head of the canal. The length of the transition should be great enough to quiet entrance disturbances and permit the gradual adjustment of flow to canal dimensions. Transition design will be discussed in Art. 12. The gate sill, with connecting transition floor, is usually placed at the same elevation as the canal bed. Portions of the abutment walls and piers upstream from the gates should be rounded to reduce entrance losses. Figure 5 shows an intake structure built on the Salt Lake Basin project, northern Utah.

4. Sluices. Sluices are usually located at ends of diversion weirs, just below or adjacent to intake structures (see Fig. 4), so as to draw water from the front of the intake gates. Sluiceway sills should be placed somewhat lower than intake sills, but above the riverbed. The size and number of sluice gates required depend on the detritus-carrying characteristics of the stream.

DISTRIBUTION SYSTEM

5. General Discussion. The distribution system includes conduits, conveyance structures, regulating works, measuring devices, protective elements, and miscellaneous features needed to deliver water to all parts of the irrigable area. On projects where irrigable lands are located on both sides of the river, or where

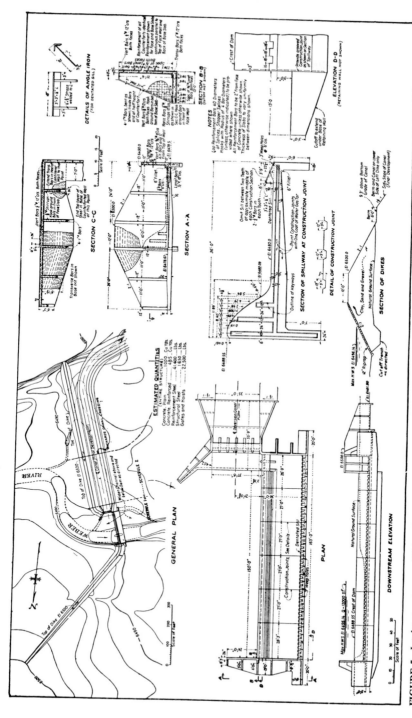

FIGURE 5 Intake structures at Weber-Provo Diversion Canal, Salt Lake Basin Project, Utah.

water is diverted at more than one point, the distribution system may include two or more separate units. When parts of the irrigable area cannot be supplied by gravity flow, the system may also include pumping units, consisting of pumping plants, pressure pipelines, and various incidental structures required by such methods of distribution.

On some gravity projects, as where water supplies are limited and where seepage and evaporation losses are to be minimized, distribution systems may consist principally of pipelines, or closed conduits, with their necessary appurtenant works. However, distribution systems on most gravity projects consist primarily of open canals and appurtenant canal structures.

Conduits on a gravity project usually include main canals, branch canals, laterals, sublaterals, such additional ditches as may be necessary to convey water to individual farm units, and such waste canals as may be needed to carry surplus delivered and undelivered water back to natural channels. Conveyance structures may include chutes, drops, transition sections, inverted siphons, lined or unlined tunnels, flumes, aqueducts, closed conduits, or other crossings at railways, highways, or drainage courses. Regulating works usually consist of canal checks and major gate structures such as division works, large lateral intakes, and wasteways. Protective elements include automatic spillways, tributary flood overchutes, settling basins, and sediment traps. Miscellaneous features include drainage inlets, drainage culverts, farm and highway bridges, measuring structures, and other incidental construction.

Naturally, features for two or more purposes are often combined in one structure. For instance, box culverts, checks, radial gate wasteways, and siphon spillways were combined in one structure at two locations on the Casper canal, Kendrick project, Wyoming. Furthermore, a single feature may serve dual purposes. Wasteways, although essentially of a regulating nature, may also constitute protective structures, since they may be used to sluice sand and silt deposits off canal beds, to divert canal flow during emergencies, or to discharge excess drainage waters which enter canal sections during storm periods. Various elements of the distribution system are discussed on the following pages.

CANALS

In general on gravity-flow irrigation projects, canals are used as the primary water-transporting medium. Canals in the majority of cases are excavated in earth or man-placed embankments. Occasionally rock may be encountered along the proposed alignment, and this may require the use of other types of conveyance structures to reduce costs related to excavation.

6. The Need for Canal Lining. The decision whether to have a lined canal or to have no lining is an economic one. The relevant factors in making this decision are (1) permeability of the native soil, (2) its resistance to erosion, (3) the capitalized cost of water diversion, and (4) the amount of water available for diversion.

When canals are excavated in permeable soils and it is evident that the seepage losses through the soil will not substantially change with aging of the canal, some type of lining or surface treatment should be used because (1) the area below the canal can become waterlogged and nonproductive, and (2) the diversion structure, canal section, and conveyance structures will require a greater capacity to compensate for the water lost. Canals excavated in erodible material will require lower operational velocities; consequently larger canal sections and larger maintenance costs can also be expected. The capitalized cost of water diversion is influenced by the size of the intake and its appurtenant features. If the available water for irrigation diversion is exceeded by the diversion demand, then a more impermeable canal section is required to maximize the area that can be served.

7. Canal Linings. The term canal lining implies a treatment applied to an excavated canal prism to increase the impermeability of the canal section. This treatment can be as simple as filling the canal section with silty water and allowing it to percolate through the foundation where the silt can fill the voids. Or it can be as sophisticated as precast-concrete elements placed on the foundation.

For the most part, linings can be classified as (1) exposed linings, (2) buried linings, (3) earth linings, and (4) soil sealants. Exposed linings are placed upon the prepared earth foundation. The commonest linings are constructed from portland-cement concrete (unreinforced and reinforced), pneumatically applied portland-cement mortar, asphaltic concrete, asphalt macadams, soil-cement mixtures, precast units of portland-cement concrete or asphaltic mixtures, and stone and brick masonry.

Buried linings are impervious membranes that are placed against the excavated earth prism and backfilled with a layer of earth. This layer of earth forms a protective cover for the membrane, since the membranes are fragile, have little structural value, and rely upon the foundation for full support. Sprayed asphalt, prefabricated asphalt, rubber sheet, polyvinyl and polyethylene film, butyl-coated fabrics, and montmorillonite clay blankets are examples of buried linings.

Earth linings are blankets of earth placed on the excavated earth prism. The blankets may be compacted to a true thickness of 18 to 24 in along the invert and 18 to 54 in along the sides for medium- to large-sized canals. Thin earth linings composed of a 6- to 12-in layer of compacted cohesive soil protected by a blanket of coarser soil also have been used. Loose blankets composed of selected fine-grain soils can be used where suitable equipment for compaction is not available. The latter two linings require the use of selected materials and will require excavation and haul from borrow areas.

Soil sealants for increasing the impermeability of the canal are a solution offering the least initial cost. Sealants used are silt, diversion water sediments, and chemical sealants. Three types of chemical sealants have been used by the U.S. Bureau of Reclamation,[3] (1) a resinous polymer, (2) a petroleum-based emulsion, and (3) a cationic asphalt emulsion. These and the earth sealants are applied to the canal by introducing the sealant either into the canal flow or into the stagnant water caused by damming the canal at intervals. The latter method has proved more effective, with reduction in seepage from 25 to 99 percent having been recorded.

The preceding discussion justifies a canal lining on the basis of reducing water losses. Exposed linings may be required where steep slopes require conveyance of water at relatively high velocities or in deep cuts where savings in excavation costs, resulting from carrying water at high velocity, overbalance lining costs. Lined canals are usually preferable from an operation and maintenance viewpoint, except that maintenance costs may be relatively high for concrete-lined canals in cold climates. For a more detailed discussion of canal linings see Sec. 7.

8. Canal Location. Main canals are usually located along higher edges of irrigable areas but may follow interior ridges, depending on local topographic conditions. Branch and lateral canals may follow interior ridges or may be located at approximately right angles to general ground slopes. Canal locations should follow roads or property lines whenever possible. On the Garland division, Shoshone project, northern Wyoming, the main canal parallels a railroad right-of-way which runs diagonally across the irrigable area, lateral canals branching off on each side at 1.5- to 3.0-mi intervals and running approximately parallel around the comparatively uniform basin slopes. The basic criterion for canal location is that maximum water surfaces in laterals and sublaterals must be high enough to permit deliveries to farm units.

In planning a canal system for a proposed irrigation project, approximate office locations of principal canals are made on topographic maps. Final locations are then made in the field, where the engineer can give more adequate consideration to the site conditions and requirements.

9. Canal Capacities. Capacities for which canals must be designed should be based upon the consumptive-use requirements of the crops anticipated after implementation of irrigation in the area.

This quantity of water should then be increased because of inefficiencies in the conveyance system and the application of the water to the land. Conveyance losses can be attributed to seepage, operational losses, and evaporation. The losses in this category represent losses in the distribution system up to farm delivery. Farm application efficiency is a measure of the ability to store in the root zone the crop's water demand. The shape and slope of the farm unit, method of application, and type of soil are all factors which affect this efficiency.

The period between irrigation and the selection of the period with the least rainfall are then used to determine the canal capacity.

$$Q = \frac{\sum(W_c A) - R}{12 E_c E_a}$$

where Q = ft^3 of water required during an irrigation period
W_c = consumptive use of each crop, in of water
A = area under each crop, ft^2
R = estimated effective rainfall, in, during dry period
E_c = conveyance efficiency
E_a = application efficiency

Capacities for laterals and sublaterals should be determined in a similar method but should be increased by 25 percent because of a greater sensitivity to changes in crops and water demands.

10. Canal Cross Sections. Unlined earth sections should ordinarily be provided with side slopes of 1½:1 to 2:1, depending on earth materials. Side slopes of 1:1 have been used. Side slopes of 1¾:1 and 2:1 have been used on the All-American canal. Rock sections and lined earth sections may be provided with steeper side slopes. Relatively large canals are often designed with berms between excavated sections and waste banks. When canal sides are in fill and the excavated earth is not uniformly satisfactory, core banks of selected fine materials should be specified along the inner slopes, extending from natural ground levels to a minimum height of 12 in above maximum canal water surfaces. In certain cases, it may be desirable to compact the core material by sprinkling and rolling. Freeboard provisions should vary with size of canal, nature of canal banks, possible variation in water surface during full operation, and extent of

TABLE 1 Values of Kutter's Roughness Factor for Use in Designing Irrigation Canals

Wetted perimeter	Canal description	Roughness factor, range
Concrete	Sections free from curvature	0.013–0.014
Concrete	Sections containing curvature	0.015–0.017
Concrete and gunite	Lined bottom with gunited rock side slopes	0.020–0.025
Concrete and rock	Retaining wall and lined bottom, one rock side	0.020–0.025
Concrete and rock	Retaining wall, unlined bottom and one rock side	0.025–0.030
Rock	Main, branch, and large lateral canals	0.030–0.035
Rock	Small canals, rough excavation	0.035–0.040
Earth	Main, branch, and large lateral canals	0.020–0.025
Earth	Small laterals and farm ditches	0.025–0.030

damage that may result from breaks in canal banks. Concrete linings usually should be carried 9 to 24 in above maximum canal water surfaces.

Canal cross sections may be designed by Kutter's formula, using roughness factors n selected from Table 1. All-American canal sections were designed for roughness factors of 0.020 to 0.025 in unlined earth sections, 0.014 in concrete-lined sections, and 0.035 in unlined rock sections. Proper factors for unlined or partly lined rock sections depend on size of canal, smoothness of excavation, and proportion of wetted perimeter lined with concrete. Sometimes flow in rock canals can be economically improved by lining the bottom with concrete and guniting the side slopes. Gunite may also be used in repairing damaged linings.[4] Many comprehensive tables for use in designing canal cross sections have been published.[5]

CONVEYANCE STRUCTURES

11. General. Conveyance structures, as the name implies, are the structures necessary to maintain the flow of water when the use of the canal section is no longer feasible. Examples where conveyance structures may be required are

1. Changes in grade caused in crossing existing highways or railways
2. Topography causing limitations on the breadth of the water conductor
3. Abrupt changes in grade due to topography
4. River- or drainage-course crossings
5. The need for maintaining the water surface over low-lying areas
6. Where topography greatly exceeds the canal water surface in elevation

Flumes, inverted siphons, and tunnels are examples of conveyance structures.

Conveyance-structure design differs from that of canals in that, although they rely upon the earth foundation for support, they are designed for either or both water loads and external earth pressures. Because of this, they are generally built of reinforced concrete and in some instances pneumatically applied mortar, steel, or timber. The use of timber for the water-carrying part of the conveyance structure should be limited to small structures and those which operate more or less continually. Cyclic operation may cause warping of the timber, and leaking.

12. Transitions. The change from a canal section to a conveyance structure requires the use of a transition. Transition sections are built to conserve head. Their purpose is to minimize losses where the velocity is increasing and to recover as much head as possible where the velocity is decreasing. They are used at inlets and outlets of flumes, inverted siphons, and closed conduits, as well as at places where the shape of the canal cross section suddenly changes. In open conduits, they are usually concrete sections with gradually converging or diverging sidewalls but may also include gradually changing bottom grades. In closed conduits, gradual transformations are made at tops and bottoms of sections as well as at sides. Sharp angles should be avoided in all major transition structures. Transitions in small lined ditches and at ends of small flumes and turnouts may include straight bottom grades and straight converging or diverging sidewalls; but all large transitions should be designed with curved or warped section transformations.

The proper length of a transition depends on the relative change in shape of section, initial velocity, and velocity change, the longer transitions being required for the higher velocities and greater velocity changes. Outlet transitions, where velocities are decreasing and flow conditions relatively unstable, should be 10 to 20 percent longer than inlet transitions, where velocities are increasing and flow conditions are relatively stable. Curves in alignment, either within or close to transition sections, have disturbing effects that tend to increase hydraulic losses. Laboratory measurements in small rectangular channels, containing 180° curves with inner radii equal to channel widths, have shown curve losses as great as two-tenths the velocity head.[6] However, irrigation conduits seldom, if ever, contain such pronounced curvature.

In carefully designed, warped transition structures, free from curves in alignment, hydraulic losses are probably less than 0.1 the difference in velocity head at the inlet and less than 0.2 the difference in velocity head at the outlet.

As a general rule, designs for transition structures should include the following allowances for hydraulic losses, including friction, the exact allowance to be made in a given case depending on the size of the transition and the care exercised in design.

1. Inlet transitions, free from curves in alignments, 0.1 to 0.3 the velocity head of the smaller cross section
2. Outlet transition, free from curves in alignments, 0.2 to 0.5 the velocity head of the smaller cross section
3. Inlet and outlet transitions, curved alignments, and 0.05 to 0.1 the head to values considered applicable for straight alignments

Careful hydraulic calculations should be made in planning all important transition structures, in order to secure efficient designs and avoid complications that may be experienced if flow occurs at or near critical depths.[7] Ample freeboard should be provided at the outlet, so that the water surface will not rise above any part of the structure in case the velocity head recovered should be greater than assumed.

Figure 6 shows well-designed, warped transitions constructed between lined earth and rock sections on the Kittitas main canal, Yakima project, Washington. Figure 14 shows warped inlet and outlet transitions at a monolithic reinforced-

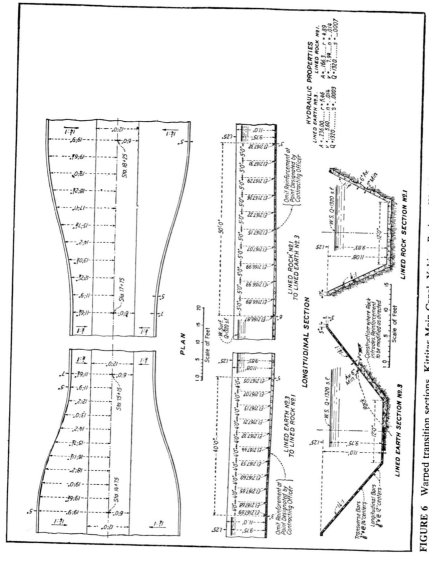

FIGURE 6 Warped transition sections, Kittitas Main Canal, Yakima Project, Washington.

concrete siphon built on the Heart Mountain canal, Shoshone project, Wyoming. Figure 7 shows the table and sample computations necessary for the design of the lined-earth to lined-rock transition shown in Fig. 6. The outlet transition-design method would be similar, except that a recovery of a 0.8 change in velocity head would determine the change in water surface (ΔWS).

13. Chutes. Chutes may be pipes, closed box sections, fumes, or open lined channels. A typical chute includes an inlet structure downstream of the transition, to control upstream water surfaces and regulate inflow; a relatively long inclined section in which the greater portion of the drop takes place; and an outlet structure designed to destroy the excessive energy developed in the inclined section. Chutes are often used on relatively steep slopes where a single drop or series of drops would be more expensive or less desirable from other viewpoints. Open, concrete-lined rectangular sections are preferable for larger discharges, though concrete trapezoidal sections have been used for smaller discharges.

Velocities in long inclined portions of open chutes are generally greater than critical; the drop through critical stage occurs in the transition if the inlet has no obstruction to flow such as a gate or stop planks.

Water-surface curves in inclined portions usually belong in the class where neutral depths are less than critical depths and actual depths are intermediate between critical depths and neutral depths, the neutral depth being the depth at which the water-surface slope parallels the bottom slope.[8] Water surfaces along inclined portions of chutes can be determined by backwater calculations, proper allowance being made for changes in velocity head as well as friction losses. Friction losses in concrete-lined chutes may be computed by Manning's or Kutter's formula, using roughness factors of 0.012 to 0.014, depending on smoothness. A sample backwater calculation for a chute follows.

Given: a rectangular chute 10 ft wide discharging 64 cfs per ft of width $n = 0.014$, slope of chute $S_0 = 0.10$.

$$\text{Critical depth } d_c = \sqrt[3]{\frac{Q^2}{g}} = \sqrt[3]{\frac{64.0^2}{32.2}}$$

$$= 5.02 \text{ ft}$$

$$V = \frac{640}{50.2} = 12.75 \text{ fps}$$

Assuming that the depth of flow at the inlet passes through critical, what will be the depth of flow 45 ft down the chute?

Check neutral depth d. Manning's equation

$$Q = \frac{1.486}{n} AR^{2/3} S^{1/2}$$

$$AR^{2/3} = \frac{Qn}{1.486} S^{1/2}$$

$$10d \left(\frac{10d}{10 + 2d}\right)^{2/3} = \frac{640(0.014)}{1.486} (0.10)^{1/2}$$

A trial-and-error solution gives a neutral depth of 0.38 ft.

For nonuniform flow the length L between two sections designated 1 and 2 can be determined by the following equation:

Rock lining	$V = 7.94$ fps	$h_v = 0.980$	El. W.S. at beginning of transition $= 2{,}177.44$, i.e., 77.44
Earth lining	$V = 5.60$ fps	$h_v = 0.489$	0.489 with El. $2{,}100$ as base. Entrance loss $= 1.1h_v$

Assume a water-surface profile of reversed parabolas.
$\Delta h_v + 1.1\Delta h_v = 0.541$ Transition length $= 40\text{'-}9''$

Station, ft

	0	4	8	12	16	20	24	28	32	36	40
ΔW.S. (neglects friction), drop in W.S.	0	0.011	0.043	0.094	0.173	0.27	0.367	0.446	0.497	0.529	0.541
$\Delta h_v = \Delta$W.S. $\div 1.1$		0.01	0.039	0.086	0.157	0.246	0.334	0.405	0.451	0.481	0.491
$h_v = 0.489 + \Delta h_v$	0.489	0.490	0.528	0.575	0.646	0.735	0.823	0.894	0.940	0.970	0.980
$V = \sqrt{2gh_v}$	5.60	5.62	5.83	6.08	6.44	6.88	7.28	7.59	7.78	7.90	7.94
Area $= Q \div V$	236.0	234.9	226.4	217.1	205.0	191.9	181.3	173.9	169.7	167.1	166.3
$\frac{1}{2}T = \frac{1}{2}$ width at W.S.	18.19	18.05	17.55	16.87	15.76	14.54	13.22	12.18	11.45	11.09	10.92
$\frac{1}{2}B = \frac{1}{2}$ bottom width	6.0	6.0	6.0	6.0	6.0	6.0	6.0	6.0	6.0	6.0	6.0
Average width	24.19	24.05	23.60	22.87	21.76	20.54	19.22	18.18	17.45	17.09	16.92
$d = $ area $\div$ avg width	9.75	9.76	9.60	9.49	9.43	9.34	9.43	9.57	9.73	9.79	9.83
$S_f = $ friction slope	0.00029	0.0012	0.00033	0.00035	0.00040	0.00047	0.00053	0.00060	0.00063	0.00066	0.00068
h_f (each station)		0.0012	0.0012	0.0014	0.0015	0.0017	0.0020	0.0023	0.0025	0.0026	0.0026
h_f (cumulative)		0.0012	0.0024	0.0038	0.0053	0.0070	0.0090	0.0113	0.0138	0.0164	0.0190
W.S. El. $= 77.04 - \Delta$W.S. $- h_f$	77.44	77.428	77.395	77.342	77.262	77.163	77.064	76.983	76.929	76.895	76.881
Grade $= $ W.S. El. $- d$	67.69	67.668	67.795	67.852	67.832	67.823	67.634	67.413	67.199	67.105	67.051
$\frac{1}{2}T - \frac{1}{2}B$	12.19	12.05	11.60	10.87	9.76	8.54	7.22	6.18	5.45	5.09	4.92
Side slopes	1.25	1.24	1.19	1.11	1.00	0.877	0.741	0.631	0.555	0.517	0.500
$H = $ top lining El. $-$ grade El.	11.00	10.966	10.783	10.670	10.614	10.587	10.720	10.885	11.043	11.081	11.079
Side slope $\times H = \frac{1}{2}W - \frac{1}{2}B$	13.75	13.58	12.83	11.83	10.614	9.27	7.95	6.86	6.13	5.73	5.54
$\frac{1}{2}W = \frac{1}{2}$ top lining width	19.75	19.58	18.83	17.83	16.614	15.27	13.95	12.86	12.13	11.73	11.54
$\frac{1}{2}W$ (to nearest $\frac{1}{2}$ in.)	19'-9''	19'-7''	18'-10''	17'-10''	16'-7½''	15'-3''	13'-11½''	12'-10½''	12'-1½''	11'-9''	11'-6½''

NOTES:
1. Values for $\frac{1}{2}T$ are scaled from a trial plan showing water intersection with transition walls.
2. A tangent to this water-surface boundary should have an included angle between it and the transition center line of not more than 22°30'.
3. The term $\frac{1}{2}B$ is also varied when the transition requires a change in base widths.
4. Differences between the above and dimensions shown on Fig. 6 can be attributed to differences in water-surface profile assumptions and numerical rounding off.

FIGURE 7 Comparative computations for inlet transition (Fig. 6).

$$L = \frac{(V_2^2/2g + d_2) - (V_1^2/2g + d_1)}{S_0 - (nV/1.486\,R^{2/3})^2}$$

where $V_{1,2}$ = velocity at the particular section
$d_{1,2}$ = depth at the particular section
S_0 = channel slope
V = mean velocity between sections
R = mean hydraulic radius between sections

Assume a change in depth of 1 ft; then $d_2 = 4.02$ ft; therefore,

$$V_2 = \frac{640}{4.02(10)} = 15.92$$

$$V = \frac{640}{4.52(10)} = 14.17 \quad \text{and} \quad R = \frac{4.52 \times 10}{10 + 2(4.52)} = 2.37$$

$$L = \frac{(15.92^2/64.4 + 4.02) - (12.75^2/64.4 + 5.02)}{0.10 - [0.014(14.17)/1.486(2.37)^{2/3}]^2}$$

$L = 43.8$ ft; therefore, the depth of flow at the required section will be approximately 4 ft.

Open chutes, carrying water at high velocities, often contain waves and agitated water surfaces, even though hydraulic jumps occur at the outlets. Consequently, ample freeboard must be provided in inclined sections. The U.S. Bureau of Reclamation[9] has recommended that the freeboard be not less than $0.4 d_c$ (critical depth) or the following:

Capacity Q, cfs	Freeboard, in
100 or less	12
101–500	15
501–1000	18
Over 1000	24

The above freeboard should be added to the depth of flow in the chute section computed using a roughness factor of 0.014. In long chutes the width of the chute section can be narrowed a short distance below the intake to effect an economy. A spreading transition will then be required between the narrow chute section and the stilling pool. The angle of flare on each side of the spreading transition should not exceed arctan $(gd/3V)$ where V = velocity at beginning of transition with $n = 0.010$, d = corresponding depth of flow.

The longitudinal profile of the spreading transition is usually to a curve defined by the chute slope (tan ϕ) and the desired slope of the transition at the intersection with the stilling pool (tan α). A slope of 2 horizontal to 1 vertical for the latter is preferred, with upper and lower limits being 1½ horizontal to 1 vertical and 3 horizontal to 1 vertical.

The curved profile of the transition reduces the hydrostatic pressures on the transition floor. To ensure positive pressures on the floor a K value of 0.5 or less is desirable:

$$K = \frac{(\tan \alpha - \tan \phi) \, 2h_v \cos^2 \phi}{L_T}$$

where L_T = horizontal length of the curved profile
h_v = velocity head at beginning of transition using $n = 0.010$

The curved profile is defined by the equation

$$Y = X \tan \phi + \frac{(\tan \alpha - \tan \phi) \, X^2}{2L_T}$$

where X = horizontal length from origin of curve
Y = vertical drop from origin of curve to point X on curve

The outlet structure is some form of stilling pool to dissipate energy to a level that will not cause damage by erosion, and also will allow a steady state of flow in the canal. The length of the stilling pool, excluding the outlet transition, should be four times the depth of flow after the hydraulic jump for structures in use for long duration and three times this depth for intermittent or short-duration use. The stilling-pool invert should have no longitudinal slope; thus, using the computed depth of flow at the end of the chute section d_1, the depth of flow d_2 after the hydraulic jump will be

$$d_2 = -\frac{d_1}{2} + \sqrt{\frac{2V_1^2 \, d_1}{g} + \frac{d_1^2}{4}}$$

Hydraulic-energy losses in the outlet transition are ignored; consequently, the difference between d_2 and the normal depth of flow plus velocity head in the canal below the outlet will determine the difference between canal and stilling-pool inverts. A chute embodying the design features discussed is shown in Fig. 8.

Model studies are often desirable in designing outlet sections for chutes carrying large discharges.[10]

14. Drops. Drops fulfill essentially the same purpose as chutes but are used when the required lowering of water surfaces is small. Drops can be either vertical or inclined; of the latter there are two types; the open channel, as in the chute, and the pipe drop.

The vertical drop, as the name implies, causes an abrupt change in elevation of the canal inverts. For a concrete drop structure in an earth canal, the change in water-surface elevation should not exceed 3 ft. A structure typical of this type is shown in Fig. 9. In concrete or hard-surfaced canals, the drop can be as high as 8 ft. The length L of the stilling pool can be determined by

$$L = \left[2.5 + 1.1 \frac{d_c}{h} + 0.7 \left(\frac{d_c}{h}\right)^3\right] \sqrt{hd_c}$$

where d_c = critical depth
h = drop in canal inverts

The pool invert should be a distance of $d_c/2$ below the invert of the downstream canal section. To effect relatively tranquil flow, the downstream water surface should be at least $0.4d_c$ below the upstream water surface. This allows some plunge to the jet and reduces waves. The use of vertical drops is dictated by

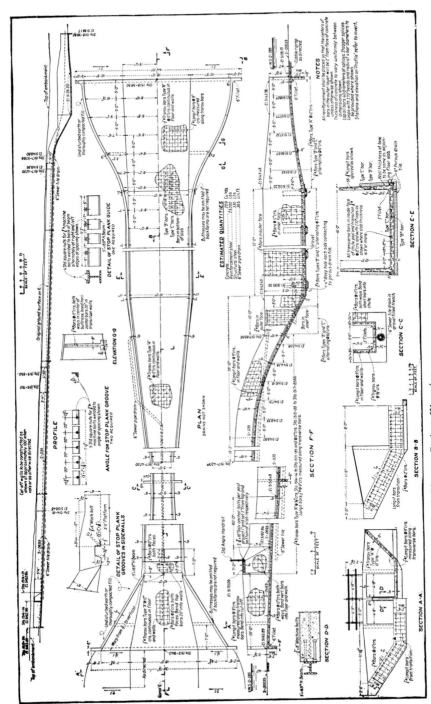

FIGURE 8 Open lined chute on Pilot Canal, Riverton Project, Wyoming.

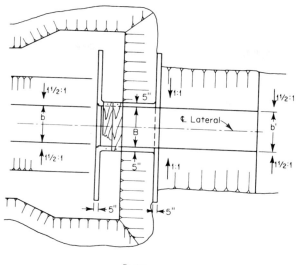

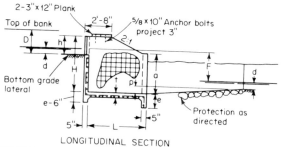

FIGURE 9 Vertical concrete drop.

an abrupt change in ground-surface elevation and the need to maintain tranquil flow in adjacent canal sections when water surfaces differ by several feet in elevation.

Inclined drops are used where the drop in water surface is greater than that allowed in a vertical drop and the distance between sections having tranquil flow is not limited. Figures 10 and 11 show typical rectangular and trapezoidal inclined drops. The hydraulic design of the inclined open-channel drop parallels that discussed in the section on chutes.

Pipe drops accomplish the desired change in water-surface elevations by introducing the canal flow into a steeply sloping closed conduit. The sloping conduit effects the elevation change, and an outlet structure dissipates the energy. The outlet structure may be of the stilling-pool type, but more often an impact stilling basin is used (Fig. 12, Table 2). This stilling basin relies upon the impact with the vertical baffle for energy dissipation. A downstream water surface is not required.[11]

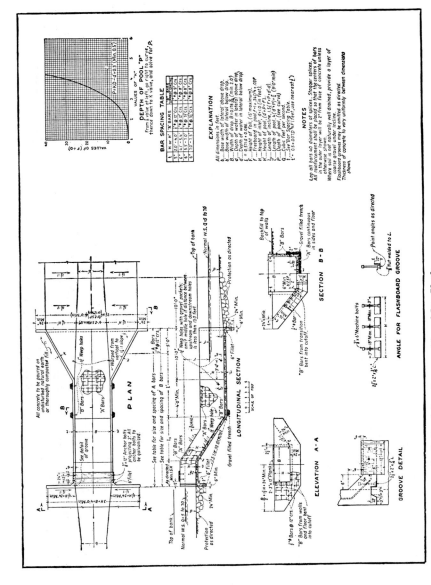

FIGURE 10 Rectangular inclined drop for capacities of 5 to 70 cfs.

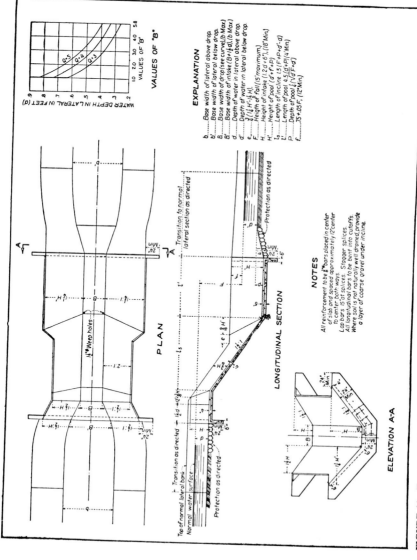

FIGURE 11 Trapezoidal inclined control—section drop, maximum capacity 5 cfs.

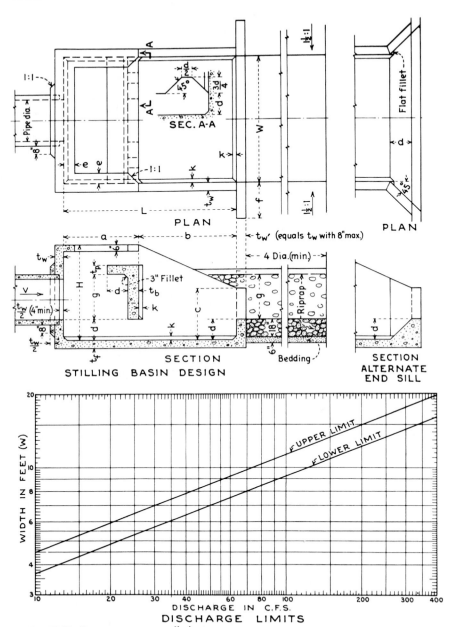

FIGURE 12 Impact-type energy dissipator.

TABLE 2 Stilling-Basin Dimensions. Impact-type Energy Dissipator

Suggested pipe size*						Feet and inches											Inches		
Diam, in (1)	Area, sq ft (2)	Max discharge (3)	W (4)	H (5)	L (6)	a (7)	b (8)	c (9)	d (10)	e (11)	f (12)	g (13)	t_w (14)	t_f (15)	t_b (16)	t_p (17)	K (18)	Suggested Riprap size (19)
18	1.77	21†	5-6	4-3	7-4	3-3	4-1	2-4	0-11	0-6	1-6	2-1	6	5½	6	6	3	4.0
24	3.14	38	6-9	5-3	9-0	3-11	5-1	2-10	1-2	0-6	2-0	2-6	6	6½	6	6	3	7.0
30	4.91	59	8-0	6-3	10-8	4-7	6-1	3-4	1-4	0-8	2-6	3-0	6	6½	7	7	3	8.5
36	7.07	85	9-3	7-3	12-4	5-3	7-1	3-10	1-7	0-8	3-0	3-6	7	7½	8	8	3	9.0
42	9.62	115	10-6	8-0	14-0	6-0	8-0	4-5	1-9	0-10	3-0	3-11	8	8½	9	8	3	9.5
48	12.57	151	11-9	9-0	15-8	6-9	8-11	4-11	2-0	0-10	3-0	4-5	9	9½	10	8	4	10.5
54	15.90	191	13-0	9-9	17-4	7-4	10-1	5-5	2-2	1-0	3-0	4-11	10	10½	10	8	4	12.0
60	19.63	236	14-3	10-9	19-0	8-0	11-0	5-11	2-5	1-0	3-0	5-4	11	11½	11	8	6	13.0
72	28.27	339	16-6	12-3	22-0	9-3	12-9	6-11	2-9	1-3	3-0	6-2	12	12½	12	8	6	14.0

*Suggested pipe will run full when velocity is 12 fps or half full when velocity is 24 fps. Size may be modified for other velocities by Q = AV, but relation between Q and basin dimensions shown must be maintained.
†For discharges less than 21 sec-ft, obtain basin width from curve of Fig. 12. Other dimensions proportional to W: H = 3W/4, L = 4W/3, d = W/6, etc.

The design of the inlet to the drop structure should ensure water-surface control. If the structure is adjacent to a farm delivery or a division in canal flows, the design canal water surface should be maintained in the canal section; otherwise the depth of flow at the control could be at critical depth or above. Methods by which the water surface may be controlled are weirs, flashboards, slide gates, or on larger structures, radial gates.

15. Inverted Siphons. Inverted siphons are used to carry canal discharges under highways, railways, and streams or across relatively long and deep drainage courses where construction of flumes or other aqueducts at or near canal grades would not be feasible. They usually consist of reinforced-concrete inlet and outlet transitions with connecting barrels of reinforced concrete, steel, or wood staves. Rectangular or square reinforced-concrete sections are used when the head is less than 30 ft. Reinforced-concrete precast and monolithic circular sections have been used for heads in excess of 100 ft. Prestressed-concrete pipe and steel plate can be used for higher heads. Steel plate and reinforced concrete were combined in certain sections of the Soap Lake siphon in the state of Washington.[12]

In some cases the siphon barrel may be lowered to an elevation where it may cross the depressed ground surface on a trestle. Conduit weight then becomes a factor in the design of the trestle, and wood staves in smaller siphons, or steel plate, are most generally used. The use of steel plate in water conduits has greatly advanced in recent years with the use of phenolic, vinyl, and more recently, coal-tar epoxy paints to deter corrosion.

The structural design of circular barrel siphons is based upon a hoop stress due to the hydrostatic head at the centerline of the section, a moment due to the weight of the water contained in the section, and the reaction of the foundation assumed in contact with the pipe.[13]

In reinforced-concrete siphons, the allowable stresses in the reinforcement may be reduced as the head increases.[14] Cognizance of the fact that Hooke's law is not appropriate in curved members is also necessary. Design for an external loading when the conduit is empty should also be considered.

Provisions for draining the siphon should be made either by connecting a valved pipe at the low point of the siphon or by pumping the water out. If the siphon is beneath a watercourse, it must be capable of resisting the buoyant forces exerted by the watercourse and the undermining of its foundation by scour.

Siphons should be inspected and cleaned occasionally. The use of sand and gravel traps upstream of the inlet will reduce the amount of cleaning required. Sand traps are discussed under Protective Structures.

The velocity of flow in the siphon is dependent upon the available head and economic considerations. Siphons with water containing abrasive suspended material should have a velocity less than 10 fps.

The hydraulic design of siphons should consider the hydraulic losses in the inlet transition, closed transition to pipe section (circular-section siphons only), friction and bend losses in the siphon barrel, and the hydraulic losses in the outlets of closed and open transitions. Friction losses in the transition are usually discounted, but a safety factor of 10 percent is added to the losses computed.

Figures 13 and 14, respectively, show a sample computation of siphon head losses, and the siphon on the Heart Mountain canal of the Shoshone project in Wyoming. Siphons designed to have a water surface subject to atmosphere within the closed sections of the siphon may entrain air with serious consequences.[15]

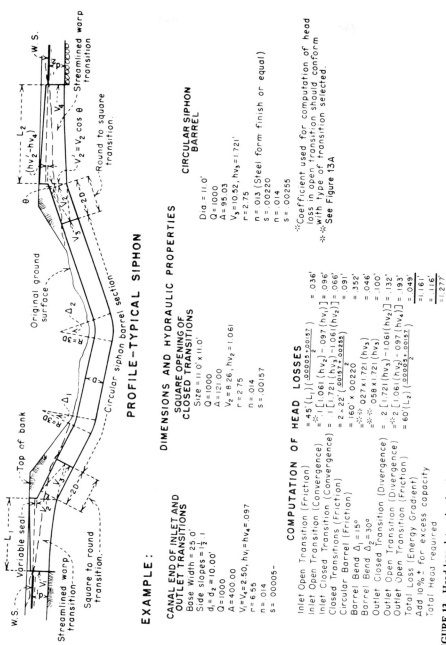

FIGURE 13 Head loss-determination for monolithic concrete siphons.

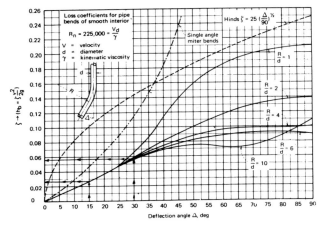

FIGURE 13 (*Continued*) Head loss-determination for monolithic concrete siphons.

16. Flume Crossings. Flume crossings at canal grades may include timber, wood-stave, metal, or reinforced-concrete aqueducts of either monolithic or precast construction. They may be supported on piles, pile bents, structural-steel bents on concrete pedestals, concrete piers, structural-steel trusses, or other construction. Timber-box flumes, supported on pile or pile bents, are often used to carry farm deliveries; metal flumes supported on pile or structural-steel bents are frequently used to carry lateral discharges; and large metal or reinforced-concrete flumes, supported on concrete piers, steel trusses, or other construction, are used to carry main-canal, branch-canal, and large lateral discharges.

Practically all flumes connecting canal sections should be provided with inlet and outlet structures (see discussion of transitions). In relatively small flumes, such as those required for farm deliveries, outlet losses need no special consideration, except that ample freeboard should be provided to prevent overlapping of the banks or outlets. Riprap or other protection against scour may be needed for short distances beyond outlet transitions, depending on flume velocities and composition of canal beds. The flume sections are designed in the same manner as canals, except that economic considerations dictate that the flume section be optimized consistent with allowable head losses. The freeboard on the flume should match in elevation that of the canal section or be greater if wave action losses from the flume are to be avoided.

Semicircular flumes, free from curves in alignment, are usually provided with freeboards varying from 6 to 10 percent of the diameter, the larger percentages being used for the smaller flumes, and no freeboards of less than 3 in should be used in any installation. Curved flumes require greater freeboards, especially along outer edges of curves.

Figure 15 shows a reinforced-concrete aqueduct, built to carry the Milner-Gooding canal over the Big Wood River, Minidoka project, southern Idaho.

The structural design of the box flume section considers the vertical walls as beams spanning between supports carrying half the flume and water weight.

When the supporting structure is founded on rock or other firm foundation, not subject to appreciate settlement, the flume can be monolithically continuous

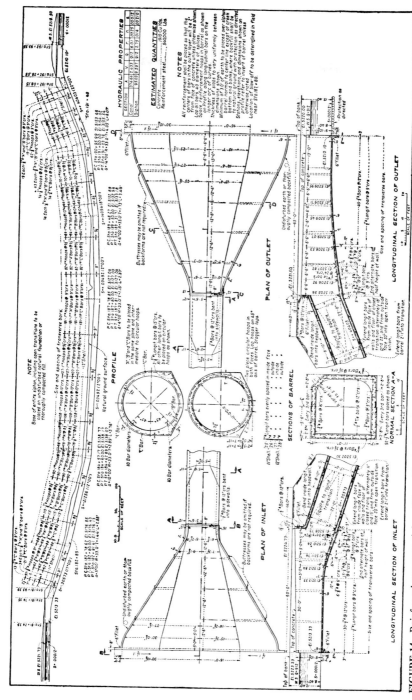

FIGURE 14 Reinforced-concrete siphon, Heart Mountain Canal, Shoshone Project, Wyoming.

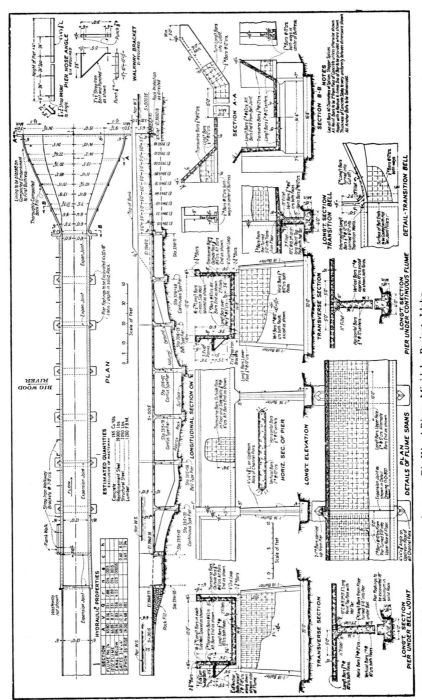

FIGURE 15 Reinforced concrete aqueduct over Big Wood River, Minidoka Project, Idaho.

over several supports; otherwise the flume spans should be simply supported. The flume walls should also be designed as vertical cantilevers fixed at the flume floor. In some instances struts across the top of the flume are used. These struts are either tension ties or compression struts, depending upon the proportion of flume width to water height. This scheme reduces the amount of cantilever reinforcement required and offers lateral support to the compression zone of the wall, but also reduces the effective freeboard. Another means of obtaining restraint at the top of the wall, without the above drawback, may be achieved by adding a corbel to the top of the wall and vertical buttresses on the outside of the wall at the flume supports. The corbel would act as a beam spanning between the vertical buttresses and subject to deflection in a horizontal plane. The flume floor should be designed for a moment due to the water weight minus the cantilever end moments. Axial thrust on the slab caused by hydrostatic pressure on the wall should also be considered.

REGULATING STRUCTURES

Regulating structures are built where canal water surfaces or discharges must be controlled. Checks are built to raise canal water surfaces so that deliveries can be made to relatively high lands. Major gate structures are built where main canals divide into branch canals, usually called *division works,* at large lateral intakes and at wasteways. Wasteways structures often include automatic spillway elements.

Temporary regulating works and minor control structures may be constructed of timber, but permanent works of major dimensions are usually built of reinforced concrete. Water surfaces and discharges are regulated by timber flashboards or various types of steel gates.

17. Checks. Most check structures on irrigation canals with depth of flows generally less than 5 ft are designed for flashboard regulation, although checks of this size and larger have been designed for a combination of flashboards and a radial gate. Check structures on laterals control water surface by use of either flashboards or slide-leaf gates.

Since a check structure has the possibility of having a design water surface on its upstream side, and being unwatered on the downstream side, it must be designed against sliding and overturning. In analyzing this, hydrostatic uplift on the structure's base should also be considered. Seepage calculations[16] along the base will give values for this force.

The velocity of flow through check structures with flashboards should not exceed 3.5 fps, because of the difficulty of placing and removing the flashboards. Checks with gates can tolerate velocities greater than 5 fps. Naturally, this must be tempered to meet acceptable head losses. The total head loss through a check structure can be estimated at 0.5 times the difference in the velocity heads of the upstream canal section and the check opening.

Some riprap protection may be required downstream on the check structure in earth canals because of the turbulence caused by partial flows over the flashboards, or the jet coming from underneath the gate. Figure 16 shows a concrete check constructed on the South canal, Succor Creek division, Owyhee project.

18. Division Works and Intakes. Division works and large lateral intakes are similar to river intakes but are subjected to less uncontrolled variation in water surfaces at upstream ends. They usually consist of one or more bays where flow

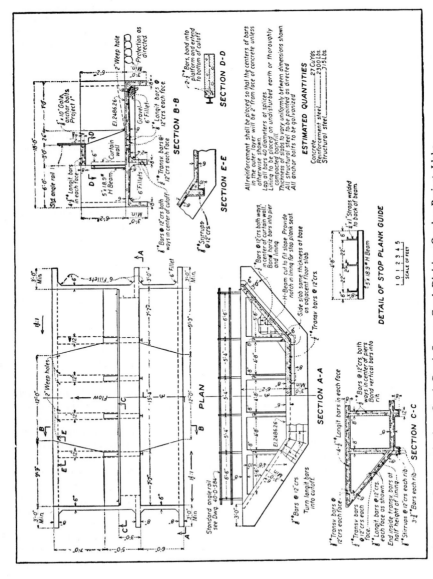

FIGURE 16 Concrete check on South Canal, Succor Creek Division, Owyhee Project, Idaho.

is controlled by gates. Radial or roller gates may be used in structures of unusual size; but radial or vertical slide gates are commonly installed in structures of ordinary dimensions.

In division structures, gates are usually installed at the head of each branch canal, so that the division of flow can be regulated at all times and so that flow in any branch can be shut off when desired. At lateral intakes, main canal water surfaces may be controlled by checks.

Total gate openings at lined canal intakes are approximately equivalent in area to the canal flow sections, since the canals are designed to carry water at relatively high velocities. At earth canal intakes, total gate openings are usually based on a velocity of about 5 fps. However, actual gate velocities are often higher, owing to operation at partial openings. Reinforced-concrete transition structures are provided below the gates, so that velocities can adjust to canal requirements before leaving transition sections. Transition sidewalls and floors may also be needed above the gates, depending on the design of canal sections and layout of gate structures.

Hydraulic losses through gate openings are seldom controlling factors in designing large structures. When gates are operated at full openings, entrance losses are simply transition losses. When operated at partial openings, available heads are not being fully utilized, so that increased losses due to gate contractions should not be important. Figure 17 shows the South Branch canal headworks, Kittitas main canal, Yakima project, Washington.

19. Wasteways. Wasteways serve two basic functions: (1) to empty the canal section and (2) to remove uncontrolled excess canal flows. The wasteways for these functions are respectively manually and automatically operated.

The manually operated wasteway is designed to discharge the canal design flow at a water surface equal to or lower than canal design water surface. The wasteway should be located so it may discharge into a natural watercourse capable of accommodating the wasteway discharge, plus the flow of the watercourse. For this type of wasteway to operate effectively, it must be located upstream of a check structure or division works, which must be closed to keep water from going farther down the canal.

Wasteway discharge is usually controlled by radial gates on larger canals. The use of flashboards on medium canals has the drawback of leakage. On smaller canals, wasteways may have slide-gate controls.

The hydraulic design of a wasteway is generally not concerned with conservation of head; therefore, velocities may be high.

The operation of a wasteway usually produces nonuniform flow conditions in the canal, with velocities approaching critical velocity, which results in increased scour. Earth canals should be paved for adequate distances adjacent to the wasteway, or a control section should be used in the wasteway to keep the canal velocities low.

If the difference in invert elevations of the canal and the wasteway discharge channel is great, chute and stilling-pool elements will be required, design of which is the same as discussed in Art. 13.

The automatic wasteway may be either a side-channel spillway or a siphon spillway. The former is a long weir, either forming a part of the canal lining or adjacent to the canal section, with its crest above the normal operating water surface. When excess canal flows raise the water surface in the canal, the water will flow over the weir into a collector which discharges into a drainage course. The efficiency of this weir is reduced by the fact that the weir is parallel to canal flow. Standard weir coefficients will have to be modified to reflect this condition. The

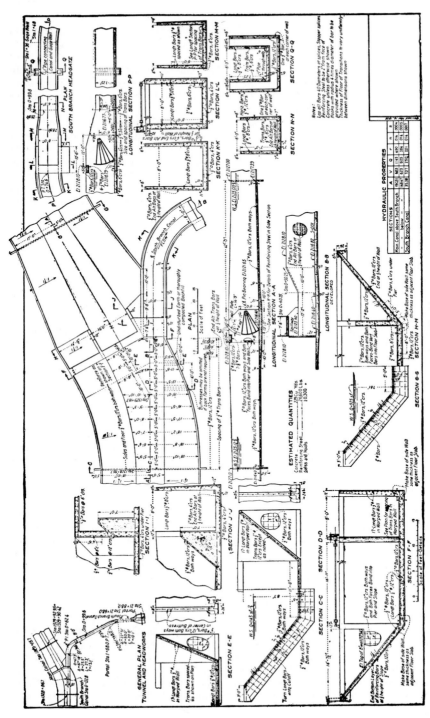

FIGURE 17 South Branch Canal Headworks, Kittitas Main Canal, Yakima Project, Washington.

velocity of flow in the main canal should be low for more efficient operation. Weir formulas are approximately correct for canal velocities less than 2½ fps. For canals with high velocities, the momentum of the water flowing parallel to the weir will cause the upstream end of the weir to be less efficient than the remainder of the weir because change of flow direction cannot be effected immediately. The use of side-channel spillways requires increased freeboard requirements unless small discharges over the weir, due to waves or higher water-surface elevations caused by operational errors, can be tolerated.

The siphon spillway represents a more sophisticated approach to an automatic wasteway. By developing a head approximately equal to the change in water surface between canal and wasted water, but not in excess of the atmospheric pressure in feet of water, the discharge obeys the orifice equation

$$Q = CA\sqrt{2GH}$$

Figures 18 and 19[17] offer a design guide for low-head siphon spillways with various outlet conditions.

Flexibility may be obtained by using an adjustable crest made of steel. The siphon breaker and seal portion of the upper leg then should also be adjustable.

Both types of spillways can be incorporated with other canal structures to effect an economy. It is not necessary that the adjoining structures be a wasteway since the requirements for the location of a manual-operated wasteway are not necessarily the same as those for an automatic wasteway. Culverts passing beneath the canal and overchutes are typical of other structures that may be used to accept wasted canal discharges.

PROTECTIVE STRUCTURES

Protective structures are built to prevent damage to the distribution system by flooding or silt deposition. Automatic spillways discussed under Regulating Structures may also be considered a protective structure. Overchutes are sometimes built to carry drainage course flows across canals at places where aqueducts, flume crossings, or inverted siphons are impracticable, and to keep flood flows from entering canal sections. Culverts are used to carry drainage-course flows beneath the canal prism. Settling basins are built to remove the heavier sand and silt loads in river diversion, which otherwise would settle to the canal beds and impair the efficiency of the system. Sand traps are built in locations where local sand troubles are pronounced.

20. Overchutes. An overchute is usually a type of flume running transversely over the canal. It is provided with inlet and outlet sections, consisting of erosion-protected floors and training walls of masonry or riprapped embankments, designed to prevent drainage-course flows from overtopping canal banks and discharging into the canal section. It may or may not be provided with a stilling basin at the outlet end, depending on local soil and topographic conditions. Overchutes are used where ordinary types of canal crossings are impractical and where heavy loads of sand, gravel, and debris carried by drainage courses would soon clog culverts under canal sections.

Adjacent to, and crossing the canal, the flow should have a high velocity to ensure that the drainage-course bed load is carried over the canal. The use of an

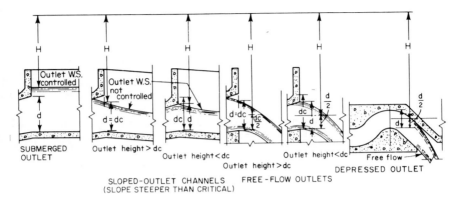

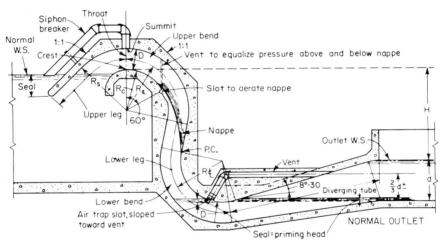

FIGURE 18 Typical low-head siphon spillway.

overchute requires the drainage-course bed to be above the canal water prism. This requirement usually places the canal in a relatively deep cut, and unless the drainage course has a steep slope, it will have to be deepened below the canal. Figure 20 shows an installation near Yuma, Ariz.

21. Culverts. Culverts are structures that carry drainage-course flow beneath canal sections. Principally they are composed of a closed conduit, or multiple conduits, which may or may not have inlet and outlet transitions. Figure 21 shows a single-barrel box culvert, built on the South Canal of the Succor Creek division, Owyhee project, Idaho.

The flow condition in the culvert can be categorized as being controlled by either the inlet or the outlet. If the former, the discharge will be dependent upon the headwater elevation, entrance shape, and conduit size. This condition occurs when the conduit has the capability of transporting the water faster than it can enter the conduit. Outlet control will exist when the discharge is limited by the

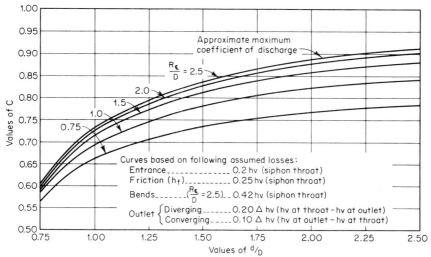

FIGURE 19 Siphon spillway coefficients of discharge.

capability of the conduit and not the inlet. The condition where the tailwater submerges the outlet is indicative of an outlet controlling discharge.

Referenced publications[18] and texts[19] cover culvert hydraulic design for the flow conditions mentioned.

22. Settling Basins. Settling basins, properly planned and constructed, are effective means of removing suspended sand loads and some of the heavier silt loads. However, only relatively small proportions of the finer silt particles carried in suspension can be removed by settling-basin operation. Such particles are usually carried through the basins and later deposited on canal beds or farmlands. Ordinarily, settling basins should be built just below diversion structures, where deposited materials can be sluiced back to river channels. When properly designed basins are built at such locations, major sand-removal structures are seldom required at other places along the canal system.

A typical settling basin usually has a skimming weir at the downstream end, over which canal flow is withdrawn; a gate-controlled sluiceway at one end of the weir, for sluicing deposited material off the floor of the basin; and a sluiceway channel, to carry the sluicing discharge back to the river at relatively high velocities. The skimming weir should be provided with gates or flashboards, so that canal flow can be shut off during sluicing operations. The sides and floor of the basin should be riprapped, or lined with concrete, to prevent excessive erosion during sluicing operations. In order to minimize interference with canal operation, it usually is desirable to design the sluiceway gates and wasteway channel so that velocities developed are capable of removing the deposited material from the basin.

The size of the settling basin is determined by the canal discharge and the maximum velocity that can be maintained through the basin with the sluice gates closed, without carrying sediment over the skimming weir. Since it is seldom practicable to remove all deposited material during a sluicing period, the pres-

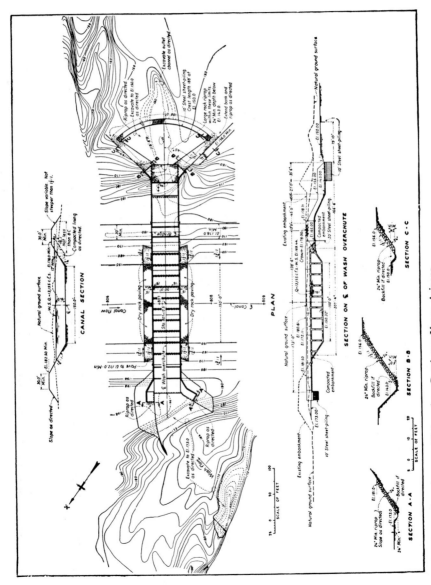

FIGURE 20 Overchute on All-American Canal near Yuma, Arizona.

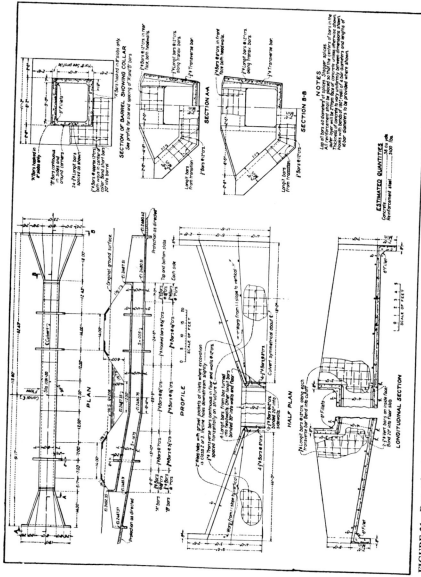

FIGURE 21 Box culvert on South Canal, Succor Creek Division, Owyhee Project, Idaho.

ence of some sediment on the basin floor must be considered in calculating basin velocities during normal canal operation. In most cases, it is probably desirable to proportion basin dimensions so that mean velocities do not exceed 1 fps during normal full canal discharge. However, effective results were secured at the Fort Laramie canal basin, North Platte project, Wyoming, where the sediment was mostly sand and the basin was designed for a maximum velocity of 1.25 fps.[20] Figure 22 shows the Fort Laramie installation.

23. Sand Traps. Sand traps are built to trap and remove bed load, moving along or suspended near the canal invert. Sand traps may be located in the headworks, preceding a settling basin, at siphons or drainage inlets, and in canal sections where the canal capacity is reduced by sediment deposition. For effective removal of sediment, the sand trap should be located near a drainage course so the sediment can be sluiced from the canal.

A sand trap may consist of a short depressed length of canal section provided with a sluice to remove sediment. In current favor is the vortex-tube sand trap, which had its beginning in the early 1930s.[21] It consists of a slot in the canal invert, usually circular in cross section, with the center located a distance less than the radius below the invert. The tube crosses the canal invert at an acute angle with the direction of flow existing in the conveyance channel. The outlet should be provided with a gate which can be adjusted to correspond to sediment concentrations. Sections other than circular have been tested[22] but presented no overall material advantage. The vortex-tube discharge varies from 5 to 15 percent of the conveyance-channel discharge. The spiral flow developed in the tube will remove small cobbles, and a properly designed tube will remove over 90 percent of material 1 mm in diameter and about 35 percent of material 0.3 mm in diameter.

Laboratory tests and a survey of literature[23] on vortex sand traps produced the following design criteria for effective operation.

1. The Froude number of the conveyance channel at the vortex tube should be approximately 0.8.
2. The width of the slot should be between 0.5 and 1.0 ft.
3. The ratio of tube length to opening width L/D should not exceed 20, with the maximum length being about 15 ft.
4. The tube angle should be about 45°.
5. The elevation of the upstream and downstream lips of the tube should be set at the same elevation; raising the downstream lip has little effect on operational efficiency.
6. Tubes of constant cross section operate as well as tubes with varying cross sections (tapered).
7. The required area of the tube can be approximated by the relationship $A_T = 0.06DL \sin \phi$, where D = slot width, L = tube length, and ϕ = tube angle with flow.

The discharge of the tube can be approximated by the equation for orifice discharge. The coefficient of discharge should be modified for the velocity of approach and the tube geometry.

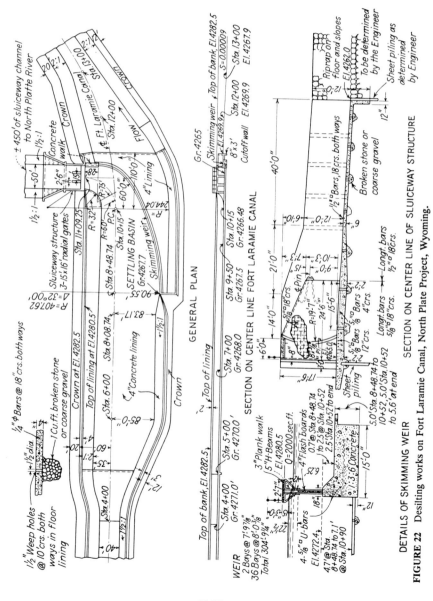

FIGURE 22 Desilting works on Fort Laramie Canal, North Plate Project, Wyoming.

DELIVERY STRUCTURES

24. General. Delivery structure, measuring device, or farm turnout, regardless of what it is called, performs the final function of the irrigation system. The design and construction skill that went into the diversion, canals, conveyance structures, regulating structures, and protective structures will be judged predominantly on the operation of the delivery structure. It is here that nature's life-giving substance is turned over to the irrigator. Any inadequacy in the delivery structure's operation will most likely produce condemnation of the entire irrigation system. Therefore, it cannot be overstressed that considerable thought should be given to this structure in the planning and design stages.

The perfect delivery structure should incorporate the following: (1) be capable of delivering the water an irrigator may need to meet the consumptive use of any feasible cropping pattern, (2) accurately deliver the rate prescribed (within 5 percent is acceptable), (3) have minimal loss in hydraulic head to effect the delivery, (4) have a discharge rate of water delivered that is oblivious to fluctuation in the canal water surface, (5) have a rate of water delivered that is also oblivious to fluctuations in the water surface of the irrigator's head ditch, (6) be of practical and economical construction, (7) have a structure that is easy to adjust for varied discharges, and (8) be tamperproof; that is to say, it should not be possible to adjust it illegally. Needless to say, no one delivery structure embodies all the above requirements.

The key to which of the above items may be dismissed is in the distribution system's method of operation. The distribution-system methods are (1) rotation method—a fixed quantity of water is delivered to each irrigator at predetermined intervals; (2) demand method—each irrigator specifies the quantity of water and time of each delivery; and (3) continuous method—the delivery structure is operating continually. Distribution may also be accomplished by a combination of the above methods.

The rotation method, with water being available at intervals for a specified length of time, requires a delivery structure with an adjustable discharge. As an example, if an irrigator who is delivered water for 12 h every 10 days decides to put in crops that require a 50 percent increase in delivered water, the only recourse with a fixed delivery time is to increase the delivery rate.

The demand method, by virtue of not having any limitations on the length of time the water is delivered, allows the use of a delivery structure with a constant or limited range of discharge.

The continuous method would require an adjustable delivery to allow for any increase or decrease in water demand.

25. Classification of Delivery Structures. Delivery structures can be categorized into three classifications. Structures in which fluctuations in the water surface of either the canal or the head ditch have relatively no effect on the discharge rate will be called *invariant delivery structures*.

Structures in which the rate of discharge remains relatively constant, with (1) fluctuations in the head ditch water surface, as long as the canal water surface remains constant and (2) fluctuations in the canal water surface, as long as the head ditch water surface remains constant, will be called *semivariant delivery structures*.

Structures where the rate of discharge is affected by changes in canal and head ditch water surface and which therefore require water-surface control structures in the canal will be called *variant delivery structures*.

26. Invariant Delivery Structures. A most interesting delivery structure of this type is the autoregulator of Italian design (see Fig. 23).

The device is a cylindrical sleeve with circumferential apertures which act as weirs. The cylinder is supported by floats or pontoons at the canal water-surface level. Thereby, regardless of fluctuations in the canal water surface, the weir crest remains at the original preset distance below the canal water surface. The originator of this device claims an accuracy of delivery of over 99 percent.

A delivery structure of somewhat less accuracy for a fluctuating canal water surface is the single-baffle distributor.[24]

This device consists of a battery of slide leaves, each of a different rated discharge. Altering discharge rates may be achieved by raising the leaves in combi-

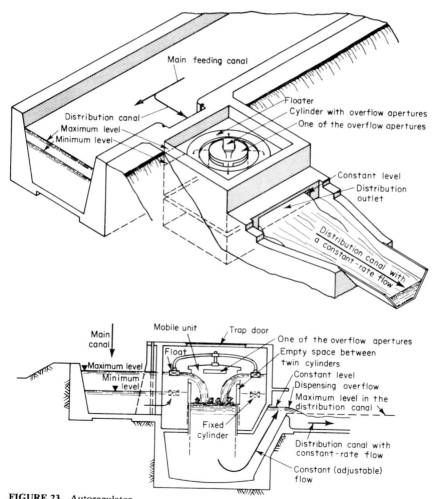

FIGURE 23 Autoregulator.

nation or singly. The distributor discharge remains unaffected by variations in the downstream level because it is designed to pass water through critical depth. So long as the downstream water level does not drown out the hydraulic jump, it will exert no influence on the discharge rate. This principle has been used on many delivery structures in the semivariant classification. The single-baffle distributor, being sensitive to changes in the canal water surface, has been improved upon by the advent of the double-baffle distributor. See Fig. 24 for operating curves for the two distributor types. The second baffle presents a smaller orifice for discharge with the increased head, and at the lower water surface, the contraction of the vena is sufficient to pass unrestricted through the orifice caused by the second baffle. The maximum change in water surface for comparable single- and double-baffle distributors to maintain discharge within ±5 percent of the nominal discharge would be 3⅛ and 9 in, respectively. See Thomas[25] for descriptions of other invariant delivery structures.

27. Semivariant Delivery Structures. Critical-flow flumes,[26] of which the commonest in the United States is the Parshall measuring flume,[27] form the major type under this classification.

Figure 25 shows a Parshall measuring flume with a 30-ft throat. Measurement of discharge is based on coefficients determined through experiments on prototype flumes.[28]

The discharge Q, in cubic feet per second, may be estimated as

$$Q = K_e b Y_1^n$$

where K_e = discharge coefficient ≈ 4.0
b = channel width, ft
Y_1 = depth at measuring section, ft
$n \approx 1.55$

The exact values of K_e and n for the particular flume sizes are given in the Parshall references cited above.

Weirs are a most common semivariant delivery structure. The discharge is quite sensitive to changes in water-surface heights above the weir crest because the discharge is proportional to the water height raised to a power. Criteria for design and construction and tables for discharge computations are available.[29]

28. Variant Delivery Structures. There are many delivery structures that come under this classification, but the two commonest in the United States will be covered here. The meter gate is a gated-orifice-type delivery (Fig. 26)[30] where the head loss through the gate is measured and, with the known gate opening, the discharge may be determined from tables.[31]

For accurate water measurement the following should be noted: (1) the meter gate should be submerged by one diameter, (2) the outlet should be sufficiently submerged so as to establish a pressure gradient on the crown of the pipe, and (3) the length of pipe downstream of the gate should be long enough to allow a uniform velocity distribution in the pipe at the outlet.

The constant-head-orifice delivery structure is a two-gated structure. The first gate is calibrated for discharges due to a head differential of 0.2 ft; the second gate is adjusted so that this differential is obtained. If calibration tables are not available, the discharge can be computed using the orifice equation and discharge coefficients of 0.70 and 0.75 for the first and second gates, respectively. Figure 27[32] lists the dimensional criteria for a constant-head-orifice delivery structure.

DELIVERY STRUCTURES

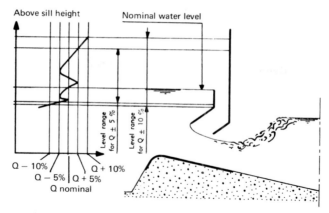

(a) Diagrammatic layout and operating curve for double-baffle distributors

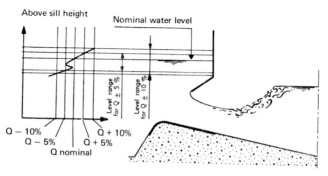

(b) Diagrammatic layout and operating curve for single-baffle distributors

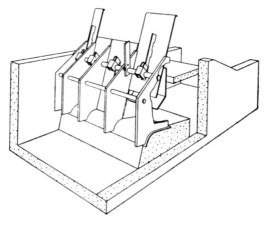

(c) Single-baffle distributor

FIGURE 24 Operation curves for the two distributor types.

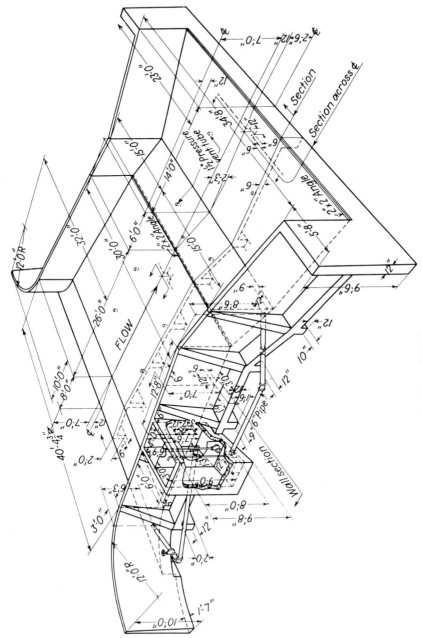

FIGURE 25 Reinforced-concrete parshall measuring flume, 30-ft throat.

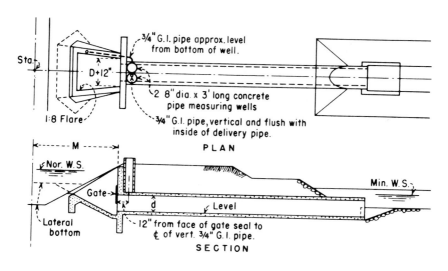

(a)–Typical field installation of metergate

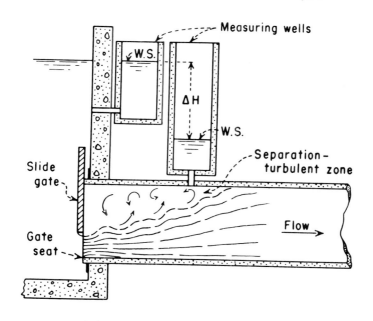

(b)–Metergate flow-measuring principle

FIGURE 26 Installation of meter gate and meter-gate flow-measuring principle.

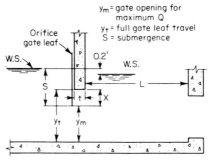

X must be equal to or greater than "t" for maximum Q.
S is equal to or greater than "y_m" for good accuracy.
For Q up to 10 cfs, L must be at least $2\frac{1}{4}$ y_m or $1\frac{3}{4}$ y_t, whichever is greater (3'-6" minimum).
For Q above 10 cfs, L = $2\frac{3}{4}$ y_m minimum.

FIGURE 27 Dimensions for a constant-head orifice.

MISCELLANEOUS STRUCTURES

Miscellaneous structures not discussed on the preceding pages may include drainage inlets, farms bridges, highway bridges, fish-control structures, and other construction incidental to irrigation developments. Drainage inlets are provided to supplement the canal flow and are located to intercept flows from springs or drainage courses. Farm and highway bridges are built where needed for crop and rural population transportation. Fish-control structures are required by law in many states on hydroelectric and irrigation developments. Whether the interests are commercial or sporting, considerable expenditures have been made for the protection of fish.

29. Drainage Inlets. Drainage inlets may be pipes or open channels running through canal banks. Riprap or concrete protection against erosion should be provided at both ends. Automatic flap gates should be provided at outlet ends when pipes are placed below canal water surfaces. However, the use of flap gates requires a water surface or head in the inlet greater than the water surface in the canal. If the bed of the drainage course is higher than the canal water surface, a gate is not required. In this case an open-channel inlet can be used.

Open-channel inlets can be lined with concrete, rubble masonry, or riprap, depending upon the inlet section and the structural requirements for retaining the embankment through which it passes. Figure 28 shows the drainage-inlet standards for earth canals used on the Deschutes project in Oregon; the inlet capacities are from 10 to 100 cfs. Inlets may terminate at the upper edge of the canal lining when discharging into lined canals but should be carried down the side slope to the bottom of the canal when discharging into earth canals.

Design discharge capacities for inlet structures should be based upon streamflow data if they are available, or determined by gathering maximum rain-

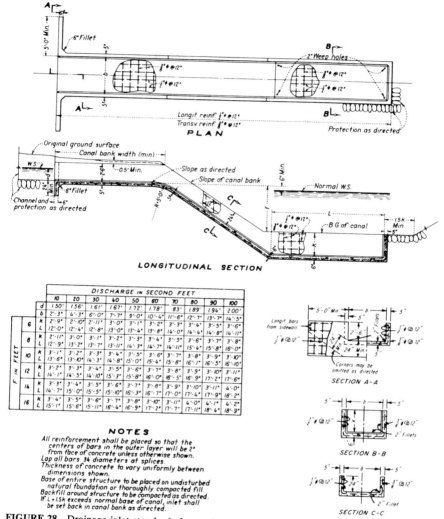

FIGURE 28 Drainage-inlet standards for earth canals. $Q = 10$ to 100 cfs.

fall and frequency data, watershed area, and runoff conditions, using methods described in Sec. 25. The flow intercepted should not carry appreciable sediment nor should the discharge exceed that which can safely be admitted into the canal. The U.S. Weather Bureau has available many technical papers and reports prepared in cooperation with other federal agencies relating precipitation rates to frequency of occurrence.

30. Farm and Highway Bridges. Farm bridges are usually timber-trestle structures. Highway bridges may be of steel or reinforced concrete. Bridges on state or interstate thoroughfares are designed in accordance with state or federal specifications. Bridge members spanning the canal should not encroach upon the prescribed canal freeboard. Bridge piers located in the canal flow section require no special consideration, although the canal may be widened to compensate for their obstruction. Piers in canals will cause the formation of eddies that may require the earth canal bed to be stabilized around them.

A thorough technical discussion of pumps is found in Sec. 21 and specific to irrigation in the references.[33-35]

31. Fish-Control Structures. Fish screens may be needed at canal intakes and fish ladders or other means of facilitating fish migrations may be needed at diversion weirs; the latter is covered in Sec. 18.

PUMPING INSTALLATIONS

32. General. Pumping installations are used in delivering water to irrigable areas that cannot be reached economically or otherwise by gravity systems. Large areas are usually supplied from control pumping plants, which are generally cooperatively installed and operated. Most large plants pump water from surface sources, i.e., from canals, rivers, or lakes. Farm plants are usually installed and operated by individual landowners. They generally pump from underground sources but sometimes draw water from surface supplies. In all cases, pumped water is delivered to the highest part of the irrigable area, so that further distributions can be made by gravity.

33. Pumps. In irrigation projects where water is pumped the types of pumps encountered fall in one of four broad categories.

1. Reciprocating pumps, where water is moved by the displacing action of pistons or plungers.
2. Propeller pumps develop head by the propelling or lifting action of the propeller in the water. The mixed-flow pump, which may be considered in this category, develops its head in part as above and in part by centrifugal force.
3. Centrifugal pumps develop head by the impeller accelerating the water from the center of rotation.
4. Turbine pumps develop head by a rotation in a radial plane between the impeller vanes as the impeller rotates, the pressure increasing in the water between each vane as it rotates from the suction inlet to the discharge outlet.

The reciprocating pump is used in small installation such as a single farm, because its capacity is limited. Turbine pumps are prone to wear if the water has a silt or sand content. The capacity is somewhat limited, but head over 2500 ft can be developed in the smaller pumps.

A thorough technical discussion of pumps is found in Sec. 21, and specific to irrigation in the references.[34,35]

REFERENCES

1. LANE, E. W., "Security from Under-seepage, Masonry Dams on Earth Foundations," *Transactions of the ASCE,* vol. 100, pp. 1235-1351, 1935.
2. From experiments conducted by F. Habermaas; figure from Emil Mosonyi, "Water Power Development," vol. 1, Hungarian Academy of Sciences, 1963.

REFERENCES

3. "Linings for Irrigation Canals," U.S. Bureau of Reclamation, 1963.
4. Reeves, A. B., "Concrete Rehabilitation Work on the Uncompahgre Project," *Journal of the American Concrete Institute,* January–February 1937, pp. 303–310.
5. "Hydraulic and Excavation Tables," 11th ed., U.S. Bureau of Reclamation, 1957.
6. Yarnell and Woodward, "Flow of Water around 180-degree Bends," *U.S. Department of Agriculture Technical Bulletin* 526, 1936.
7. Scobey, Fred C., "The Flow of Water in Flumes," *U.S. Department of Agriculture Technical Bulletin* 393, 1933.
8. Woodward, Sherman M., "Theory of the Hydraulic Jump and Backwater Curves," part III, *Technical Report,* The Miami Conservancy District, Dayton, Ohio, 1917.
9. "Canals and Related Structures," U.S. Bureau of Reclamation, Design Standards 3.
10. Elevatorski, Edward A., *Hydraulic Energy Dissipators,* McGraw-Hill, New York, 1959.
11. "Hydraulic Design of Stilling Basins and Energy Dissipators," U.S. Bureau of Reclamation, Engineering Monograph 25, July 1963.
12. "Soap Lake Siphon," U.S. Bureau of Reclamation, Engineering Monograph 5.
13. "Stress Analysis of Concrete Pipe," U.S. Bureau of Reclamation, Engineering Monograph 6.
14. "Soap Lake Siphon," U.S. Bureau of Reclamation, Engineering Monograph 5.
15. "Entrainment of Air in Flowing Water—Closed Conduits," *Transactions of the ASCE,* vol. 108, p. 1435, 1943.
16. "Design of Small Canal Structures," U.S. Bureau of Reclamation, 1974.
17. "Canals and Related Structures," U.S. Bureau of Reclamation, Design Standards 3.
18. Bossy, H. G., "Hydraulics of Conventional Highway Culverts," Division of Hydraulic Research, Bureau of Public Roads, Washington, D.C., "Culvert Design Aids: An Application of the U.S. Bureau of Public Roads Culvert Capacity Charts," Portland Cement Association, 1962.
19. *Design of Small Dams,* U.S. Bureau of Reclamation, 1987. *Handbook of Concrete Culvert Pipe Hydraulics,* Portland Cement Association, 1964.
20. Houk, Ivan E., "Sand Control Works at Fort Laramie Canal Intake," *Engineering News-Record,* vol. 100, pp. 922–926, 1928.
21. Parshall, R. L., "Control of Sand and Sediment in Irrigation, Power and Municipal Water Supplies," Water Works Association Annual Meeting, Denver, Colo., October 1933, p. 18.
22. Rowher, Carl, "Effect of Shape of Tube Efficiency of Vortex Tube Sand Traps for Various Sizes of Sand," Ft. Collins, Colo., 1935.
23. Robinson, A. R., "Vortex Tube and Sand Trap," *Proceedings of the ASCE, Journal of Irrigation Drainage Division,* December 1960.
24. "Irrigation," Laboratoire Dauphinois d'Hydraulique Neyrpic, July 1951.
25. Thomas, C. W., "World Practices in Water Measurement at Turnouts," *Transactions of the ASCE,* vol. 126, no. III, pp. 715–741, 1961.
26. Parshall, R. L., "The Improved Venturi Flume," *Transactions of the ASCE,* vol. 89, pp. 841–860, 1926.
27. Parshall, R. L., "Parshall Flumes of Large Size," *Colorado Agricultural Experiment Station Bulletin* 386, May 1932.
28. Parshall, R. L., "Parshall Measuring Flumes of Small Size," *Colorado Agricultural Experiment Station Bulletin* 423, March 1936.
29. "Water Measurement Manual," U.S. Bureau of Reclamation, May 1953.
30. Ball, J.W., "Limitations of Metergates," *Proceedings of the ASCE, Journal of the Irrigation Drainage Division,* vol. 88 (IR4), part I, December 1962.

31. "Water Measurement Tables for the Armco Metergate," Armco Metal Products, Denver, Colo., 1951.
32. "Canals and Related Structures," Chap. 6, Water Measurement Structures, U.S. Bureau of Reclamation.
33. "Turbines and Pumps," U.S. Bureau of Reclamation, Design Standards 6, 1960.
34. FINKEL, H. J. "Handbook of Irrigation Technology," vol. 1, CRC Press, 1982.
35. *SCS National Engineering Handbook,* sec. 15, Irrigation, Chap. 8, "Irrigation Pumping Plants," Soil Conservation Service.

SECTION 27

WATER DISTRIBUTION AND TREATMENT

By John P. Velon and Thomas J. Johnson[1]

INTRODUCTION

The previous sections of this handbook describe the hydraulics of natural or engineered systems commonly used to manage and utilize water resources for a variety of purposes including drainage, flood control, power generation, navigation, and irrigation. In this section, applications of hydraulic principles related to the treatment and distribution of potable water are presented. Specific topics addressed include:

- Potable-water requirements associated with residential, commercial, industrial, and other uses
- Hydraulics of water-treatment systems
- Hydraulics of water-distribution systems

POTABLE-WATER REQUIREMENTS

The critical first step in the planning, design, or analysis of potable-water supply facilities involves the definition of the water requirements for the area being studied. Determining the demands within a water system is comparable with determining the design loadings for a structure. The demands become the basis for evaluating the capacity of the system. For a water system, it is important to define both the quantity of water needed and how that need varies over the course of a day, a week, or a year.

1. Demand, Consumption, and Unaccounted-for Water. At the start of any discussion on estimating potable-water requirements, it is important to clearly define some of the terms used to describe water usage. Several terms merit special attention.

Water Demand. In this presentation, water demand is defined to be the total quantity of water which must be delivered to a water system to meet all of the system's needs, including water lost through physical leakage.

Consumption. Consumption is the actual quantity of water used by consumers for their specific needs. Consumption does not include water lost through physical leakage.

---

[1] Acknowledgment is made to Thomas R. Camp and Joseph C. Lawler for material in this section which appeared in the third edition (1969).

Unaccounted-for Water. In metered water systems, the difference between the total water usage and the sum of the metered and authorized unmetered usage is called the unaccounted-for water. This amount includes both water lost through physical leaks and water used but not accounted for through metering or an estimate.

The ratio of unaccounted-for water to total water production is a common indicator of the extent of leakage and incomplete or inaccurate metering. Results from a review of water use by nearly 200 water systems in the metropolitan Chicago area during 1987 found estimates of unaccounted-for flow ranging from less than 1 percent to greater than 30 percent.[1] For well-established systems, levels of unaccounted-for flow less than 8 percent of total production are generally considered acceptable.

2. Types of Water Users. In most urban water systems, users of water can be classified as falling under one of five major categories: residential, commercial, industrial, institutional, or public. Residential water use is typically associated with single or multifamily homes, apartments, or other dwellings and is estimated on the basis of per person or per capita consumption. Commercial water users range from small retail businesses to office buildings to hotels and restaurants. Water used for manufacturing or processing operations falls into the category of industrial water use. Institutional water users include schools, hospitals, and churches, while public users include municipal or governmental facilities which use water.

In areas where potable water is used for watering of lawns and/or gardens, special consideration must also be given to outdoor water use. Outdoor water use can significantly increase both average and peak water requirements depending on the climate of the area.

3. Typical Water Usage. Water usage varies widely from system to system as a result of the mixture of users among the categories given above, climate, and a number of other factors. A common indicator used to characterize the level of water use within a system is the average daily per capita consumption. Per capita consumption is the ratio of the average amount of water used in a system on a daily basis to the population served. Table 1 illustrates the range of per capita consumption for several communities in the Chicago area. Population and em-

TABLE 1 Per Capita Water Consumption in Selected Metropolitan Chicago Communities, 1985

City	Service population	Total employment	Manufacturing employment	Average per capita demand
Chicago	3,001,001	1,499,261	252,443	262
Evanston	71,822	41,368	3,757	158
Franklin Park	17,906	30,009	15,172	298
Lake Forest	16,003	5,809	238	196
Oak Brook	7,526	36,174	4,793	460
Naperville	60,265	32,841	2,795	137
Oak Lawn	58,435	19,424	631	120
Round Lake Beach	13,802	728	109	78
Schaumburg	58,734	59,660	4,340	140

Source: Ref. 1.

ployment data for each community are also given to illustrate the variation in per capita usage with local characteristics. Table 2 shows representative per capita usage rates for each of the major categories of water users. Actual usage rates will vary with local conditions, especially if a system serves large industrial or commercial users.

TABLE 2 Municipal Uses of Water and Representative Quantities in the United States

Use	Quantity, gpcpd	
	Range	Typical
Domestic	40–80	65
Commercial and industrial	10–75	40
Public uses	15–25	20
Loss and waste	15–25	20
Total	80–205	145

Adapted from Ref. 2.

4. Estimating Water Use. Estimates of water requirements for a given area can be developed using several different approaches. The approach selected is usually a function of the type of data available. In well-developed urban areas, extensive data pertaining to past, current, and projected population growth, land-use characteristics, commercial and industrial development, and water use may be readily available. Historic metering and billing records maintained by existing water utilities may also provide detailed information about water use. Where this type of existing data is available, projections of potable-water requirements can be readily developed with a high degree of confidence.

In rapidly developing areas, however, historic data may be inadequate to accurately represent the type and variety of potable-water users that may influence the demands of the area over the course of the planning period. Additionally, past records of use may be unavailable. In either case, it is necessary to project the future characteristics of the area to establish estimates of future water demands. Factors commonly considered in projecting future water use include population, housing, commercial and/or industrial area, institutional operations, employment characteristics, and land-use patterns. Once these factors have been used to characterize future conditions in the study area, typical water-usage rates may be assumed as the basis for estimating current and future water needs.

Commonly used water-usage rates for residential, commercial, and institutional users are summarized in Tables 3, 4, and 5.

Industrial water use can vary dramatically from site to site depending on the size of the plant and the way in which water is used. Facilities which use water in their manufacturing process will most often have large water requirements, while other plants at which water use is limited to sanitary and clean-up purposes may need relatively small amounts of flow. When developing estimates of water requirements for areas which include large industrial users, site-specific needs should be reviewed and usage patterns developed based on historical or expected practices.

In addition, it should be noted that all estimates of water requirements must include consideration of water loss and leakage in the system as discussed at the start of this section. Even in well-managed and -maintained systems, physical

TABLE 3 Typical Rates of Residential Water Use

Type of residence	Per capita water usage, gpcpd	
	Range	Typical value
Apartment/condominium	80–105	95
Average residence	50–160	110
Average residence (unmetered)	105–210	140
Luxury home	130–250	150
Summer home	30–50	40
Trailer park	60–90	80

Note: For users with metered connections to a public supply, except as noted. Adapted from Refs. 3, 4, and 5.

TABLE 4 Typical Rates of Commercial Water Use

Type of water	Water usage, gal/unit/day		
	Unit	Range	Typical value
Airport	Passenger	25–50	3.0
Automobile service station	Set of pumps	475–575	525
	Vehicle served	10–15	13
Hotel	Guest	40–60	50
	Employee	8–13	10
Laundry (self-service)	Machine	400–650	550
Motel	Person	25–40	35
Motel (with kitchen)	Person	50–60	55
Office	Employee	8–17	15
Restaurant (including toilet):			
Average	Meal	7–12	9
Short order	Meal	2.5–5.0	3.0
Bar and lounge	Customer	1.5–5.0	2
Department store	Restroom	420–635	525
	Employee	8–13	10
Shopping center	Parking space	0.5–2.0	1.0
	Employee	8–13	10
Theater (indoor)	Seat	2–4	3

Adapted from Refs. 3, 4, and 5.

leakage often accounts for 5 to 10 percent of the total water production. In poorly maintained systems, water lost through leakage may account for 30 percent or more of the total water produced.

In existing systems, estimates of water loss through leakage may be developed from the results of leak-detection studies, district measurements, or detailed water audits. However, where such information is unavailable or where estimates are to be developed for new or proposed water systems, an alternate procedure is required.

One such procedure is used by the Illinois Department of Transportation Division of Water Resources (IDOT–DOWR) in its program of regulating use of water from Lake Michigan.[6] The IDOT–DOWR procedure results in the computation of an acceptable allowance for "unavoidable leakage" in a water system based on the size and age of the distribution network as well as the pipe material

TABLE 5 Typical Rates of Water Use for Institutional Facilities

Facility	Unit	Water usage, gal/unit/day	
		Range	Typical
Hospital, medical	Bed	130–260	175
	Employee	5–15	10
Hospital, mental	Bed	80–145	105
	Employee	5–15	10
Prison	Inmate	80–160	120
	Employee	5–15	10
Rest home	Resident	50–120	90
	Employee	5–15	10
School, day:			
With cafeteria, gym, showers	Student	15–35	20
With cafeteria, but without gym and showers	Student	10–20	15
Without cafeteria, gym, or showers	Student	5–18	10
School, boarding	Student	50–105	75

Adapted from Refs. 3, 4, and 5.

used. Equation (1) gives the general form of the expression used to estimate the allowable leakage for a given system. Table 6 summarizes the leakage allowance factors used by IDOT–DOWR for various pipe materials and ages.

$$L = \sum_i \sum_j (P_{ij})(LR_{ij}) \quad (1)$$

where L = unavoidable leakage allowance, gal/day
P_{ij} = miles of pipe of material i, j years old
LR_{ij} = allowable leakage rate for pipe of material i, j years old, gal/day/miles of pipe

The allowance computed using the IDOT–DOWR formula represents the "unavoidable leakage" that would be considered acceptable for a given system. In areas where leakage is thought to be greater than the "acceptable" amount predicted by the IDOT–DOWR formula, an increased allowance expressed as a percentage of the system's total water requirements may have to be included in water-requirement estimates.

5. Variations in Water Use. Even within a single water-supply system, water demands fluctuate with time. Seasonal changes in climate affect water use in

TABLE 6 Unavoidable Leakage Factors for Water-Distribution Systems

Age of pipe	Unavoidable leakage allowance, gal/day/mile of pipe	
	Cast-iron pipe with lead joints	All other pipe and joint materials
> 60 years	3000	2500
40–60 years	2500	2000
20–40 years	2000	1500
20 years or less	1500	1000

Source: Ref. 6.

many parts of the United States. Outdoor water use for lawn watering or gardening frequently results in higher usage during summer months. In other systems, temporary increases in population associated with seasonal tourism or school operations may affect water-use patterns.

Significant fluctuations in water demand occur over the course of a day as well. In many water systems, the resulting diurnal pattern will include water-usage quantities which may range from as low as 25 to 40 percent of the average daily demand during the hours between midnight and 6:00 A.M. to 150 to 175 percent of the average demand during the morning or evening peak periods. As an example, Fig. 1 shows the range of daily and hourly variations over the course of a year in Valparaiso, Ind.[7]

Because of the fluctuations in water demand experienced in every water system, it is important to define and consider the range of conditions that may occur. Typically the range of conditions is defined in terms of demand factors such as the ratio of the maximum daily demand to the annual average daily demand, the ratio of the peak hourly demand to the annual average daily demand, and the ratio of the minimum hourly demand to the annual average daily demand. Other factors, such as seasonal demand factors, may also help to define the range of conditions which affect a water system. Table 7 gives representative ratios for several of the most commonly used demand factors.

TABLE 7 Typical Water-Demand Factors

	Ratio to average annual day	
Demand condition	Range	Typical value
Daily average in maximum month	1.10–1.50	1.20
Daily average in maximum week	1.20–1.60	1.40
Peak day	1.50–3.00	1.80
Peak hour	2.00–4.00	3.25
Minimum hour	0.20–0.60	0.30

Adapted from Refs. 2 and 8.

Like per capita usage rates, demand factors for individual water systems vary with local conditions. Where available, historical data should be reviewed to determine these factors. Consideration of local conditions is particularly important in the determination of the peak day and peak hour factors since these values define the critical demand conditions in many systems.

6. Fire-Fighting Demands. When evaluating water requirements for a water-supply system, fire-protection needs must also be considered. While water required to fight fires is generally needed only for a period of several hours, the high rates of flow required frequently result in the rapid use of extremely large volumes of water within a limited area. Consideration of these extreme demands is particularly important in the design or evaluation of small water systems where the rate of flow and total volume of water required to fight a fire may be significantly greater than the flow and volume required to meet normal peaks in domestic and commercial, industrial, and institutional water usage.

The rate of flow required for fire protection within the service area of a given water system was for many years estimated on the basis of population using an expression developed by the National Board of Fire Underwriters (now the American Insurance Association).[9] More recently, however, more site-specific

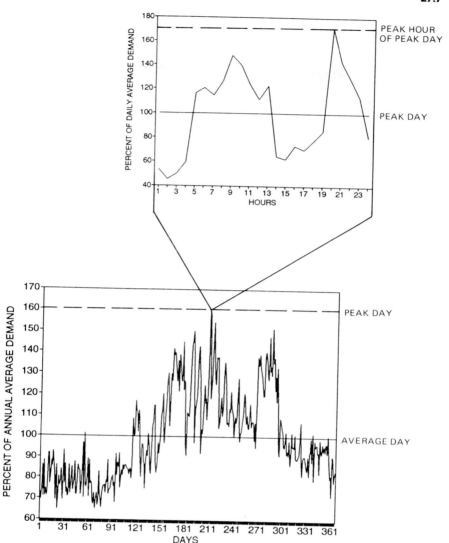

FIGURE 1 Daily and hourly variation in water consumption at Valparaiso, Ind.

methods for estimating fire-flow requirements have been formulated in response to the changing patterns of development affecting urban areas and improvements in fire-suppression capabilities. Three of these methods are described in the AWWA Manual of Practice M31, *Distribution System Requirements for Fire Protection*.[11]

The most detailed of these methods is the approach developed by the Insurance Services Office (ISO).[12] The ISO method computes the needed fire flow (NFF) for specific structures on the basis of four factors: construction (C_i), occupancy (O_i), exposure (X_i), and communication (P_i) using Eq. (2).

$$NFF_i = (C_i)(O_i)(X + P_i) \tag{2}$$

These factors take into consideration many features of the structure including its size, material and manner of construction, purpose and contents, and location relative to adjacent structures. Computed values of the needed fire flow range from 500 to 1500 gpm for one- and two-family dwellings less than or equal to two stories in height. For other habitable buildings, the NFF may range up to a maximum of 3500 gpm. Designers should refer to Ref. 12 for detailed information related to the computation of needed fire flow for specific structures.

Fire flows greater than 3500 gpm may be required for certain high-hazard or high-risk industrial facilities. However, such facilities are often constructed with on-site fire-suppression systems including sprinklers, fire pumps, and/or dedicated water storage. The availability of these on-site resources can greatly reduce the need for high fire flows from the local distribution system.

In planning for fire-protection needs in a water system, it is also important that the duration for which fire flows must be maintained is considered. Table 8 summarizes recommended fire-flow durations as published by the National Fire Protection Association.

TABLE 8 Recommended Fire-Flow Duration Requirements

Required fire flow, gpm	Recommended duration, h
Up to 2500	2
3000–3500	3
4000–12,000	4

Source: Ref. 10.

When defining the critical fire situation for a water system, the fire-flow requirements computed using the ISO or other methods most be added to the normal water needs of the system. Typically, to provide a conservative estimate of the total requirements, fire flows are superimposed on the maximum daily demand for a given system. Thus the fire situation demand for a system can be computed as

$$D_F = D_{PD} + NFF \tag{3}$$

where D_F = fire situation demand, gpm
D_{PD} = peak day demand, gpm
NFF = needed fire flow, gpm

Where a water-distribution system consists of two or more pressure zones, the critical fire situation demand must be computed separately for each zone, and the impact of each fire condition on the total system demand must be considered. The computed fire situation demands then define additional demand conditions which the water system must be capable of meeting.

7. Importance of Demand Projections. Because of the basic role of water-demand projections in the planning and design of water-supply facilities, it is important that time and effort be made to define the water needs of the system as accurately as possible. In addition, demand projections must be developed so as to properly represent the full range of current and future conditions including fire-fighting conditions which can be expected within the design life of the water-system facilities.

The proper demand condition to be used for planning or design of various facilities within the water-supply system must be defined. For example, because water- and -distribution mains must convey adequate flow throughout the service area under all conditions, these facilities are typically sized on the basis of the peak hour demand or the maximum fire-flow condition, whichever is more critical. Distribution system storage facilities are also sized on the basis of peak hour and fire-fighting conditions, since their primary purpose is to supplement flows into the system during these high-demand periods. Where treated water storage is provided, water-treatment plants are generally sized to meet only peak day demands. During high-demand periods, the required additional flow is supplied from distribution storage. Lastly, pumping stations may be sized for a variety of conditions ranging from average daily demand up to peak hour or fire-flow situations depending on their function in the system. In any case, estimates of demand must represent both current and projected conditions for all facilities through the end of the planning or design period.

HYDRAULICS OF WATER-TREATMENT SYSTEMS

Prior to use for consumption, most waters must be treated to achieve acceptable chemical, microbiological, and aesthetic standards. It is as a result of advances in the design of water-treatment facilities that the occurrence and transmittal of waterborne disease have now been almost entirely eliminated in the United States.

8. Water-Treatment-Plant Design. The design of a water-treatment plant typically involves two interrelated activities. These are:

- The design of the treatment process
- The design of the hydraulic devices and structures within which the treatment process takes place

The first of these two tasks, the process design, involves the selection and design of treatment techniques which will result in the production of a finished water that meets water-quality criteria set for the facility. Key factors in this stage of the design include the characterization of the raw water to be treated, the selection of the specific unit processes to be used, and the determination of operating criteria for those treatment processes.

In addition to the selection of the treatment processes, however, the hydraulic characteristics of the treatment system must be considered. The channels, pipes, and structures in the treatment system must be designed to provide the required flow conditions and to allow for the production of the required volume of water in the plant. Thus, the treatment capacity of a given water-treatment facility depends both on the capacity of the physical and chemical processes selected for treatment of the raw water and on the hydraulic capacity of the facilities in which the treatment is to take place.

The focus of this presentation is the hydraulic analysis and design of water-treatment systems. Detailed descriptions of procedures and criteria for the design of water-treatment processes are presented elsewhere.[13,14]

Also, many of the basic hydraulic principles commonly used in the analysis and design of water-treatment systems are presented in detail in Sec. 2. Descriptions of the basis of these principles will not be repeated here. Rather the balance of this section on hydraulics of water-treatment systems will focus on the determination of head losses in components of a treatment system and consideration of other important hydraulic conditions.

9. Water-Treatment Systems. Water-treatment systems generally consist of a number of physical and chemical unit processes connected in series to improve the quality of a given water. Given the wide range of treatment technologies currently in use, the number and type of process configurations possible are almost limitless. Figure 2 shows the schematic layout of two common water-treatment systems. The system shown in Fig. 2a is a conventional coagulation, sedimentation, and filtration system used widely in the United States for the treatment of surface waters. The system shown in Fig. 2b is a typical treatment facility designed to remove iron and manganese from well water.

While the process configurations found in modern water-treatment plants may vary widely, the basic hydraulic principles which govern flow through the plants are relatively consistent for all types of facilities. From a hydraulic standpoint, a water-treatment plant is simply a system of interconnected pipes, basins, and channels through which water moves. Within the treatment train, the flow rate and the level of the water at any given point are governed by the geometry of the system and the head loss occurring at control points such as overflow weirs or launders.

10. Hydraulic Design Criteria for Water-Treatment Plants. The primary factor in the hydraulic design of a water-treatment plant is its rated treatment capacity. For water-supply systems which include finished water storage

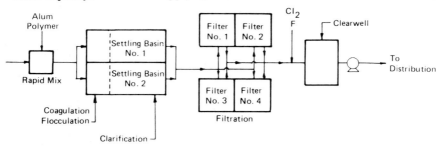

Fig. 2a. TYPICAL SURFACE WATER TREATMENT PLANT FLOWSHEET

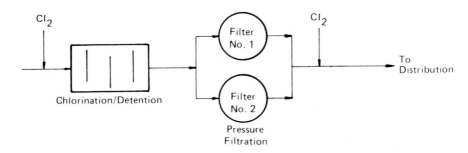

Fig. 2b. TYPICAL IRON/MANGANESE REMOVAL SYSTEM FLOWSHEET

FIGURE 2 Typical water-treatment plant flow sheets.

capable of supplementing water supplied to meet peak demands, water-treatment facilities must be capable of delivering the maximum daily demand for the planning period. In systems where distribution storage is limited or pumping rates greater than the maximum daily rate are required, treatment systems may have to be designed for higher production rates. In either case, treatment plants are commonly sized to provide hydraulic capacity equal to 110 to 125 percent of the nominal design capacity. This excess capacity provides both for minor variations in the hydraulic conditions within the plant and for flexibility in normal plant operations.

Where the potential exists for future increases in the process capacity of a plant through improvements to specific treatment units, the hydraulic design of the facility must also include consideration of the increased future capacity. This factor is particularly important since process upgrades frequently require only the installation or replacement of elements of a treatment unit while increases in plant hydraulic capacity may require significant piping improvements and expensive structural modifications.

For example, consider a treatment plant originally designed for filtration rates of 2 gpm per square foot. If, through media improvements, the filtration rate can be increased to 3 gpm per square foot, the treatment capacity of the process will be increased by 50 percent. Therefore, the original design of the treatment system should include sufficient hydraulic capacity to accommodate the potential increase in capacity without major structural improvements.

In treatment systems where residuals from specific unit processes are wasted (e.g., chemical sludges, filter backwash water), an allowance in the design capacity must be made to account for the difference between the raw water flow entering the plant and the finished water produced. In order to obtain the desired treated water capacity at the discharge from the plant, certain process units must be sized to handle more than the nominal design flow.

Also, in establishing hydraulic design criteria for specific unit processes, consideration must be given to units which are taken out of service frequently, such as rapid sand filters, or on a more periodic basis, such as sedimentation basins. In a filtration plant, the treatment system must have adequate capacity to maintain the maximum required production rate while filters are being backwashed. This is an important factor in the design since backwashing of filters is a routine operation which goes on regularly under all operating conditions.

Similar consideration must be given to plant operations with typically larger units out of service for routine maintenance. While it may not be practical to design every facility with sufficient capacity to meet maximum day requirements with a major unit out of service for maintenance or repair, these conditions must be considered in the development of the hydraulic design criteria for a system.

11. Hydraulic Profile. The hydraulic analysis of a water-treatment system is typically performed through the development of a series of hydraulic profiles. The hydraulic profile through a treatment plant is a plot of the critical water-surface elevations (for gravity systems) or hydraulic grade-line elevations (for pressurized systems) associated with the flow through the system.

The hydraulic profile is a design tool that is used to define head losses, water-surface elevations, and flow conditions in a treatment facility. The profile is developed by systematically computing and adding the head losses associated with individual components of the treatment system starting at the downstream water surface or hydraulic grade-line elevation. To verify that the hydraulic capacity of the system is adequate, profiles must consider the full range of current and future flow conditions expected at the plant. Attention must also be given to hydraulic conditions during periods of maintenance or repair, when a given treatment unit may be out of service.

The hydraulic profile through most treatment systems is a function of the head loss and water-surface elevations at a series of hydraulic control points. These points may be overflow weirs, orifice plates, pressure- or flow-regulating devices, or other hydraulic devices, depending on the type and configuration of facility. Specific conditions that must commonly be addressed in the development of a hydraulic profile include:[13]

1. The loss of energy through trash racks and/or fine screens
2. The loss of energy through flow-measuring devices
3. Entrance losses associated with the equal division of flow among parallel treatment trains
4. Losses in open and closed conduits and fittings between treatment processes
5. Losses through control valves and rate of flow devices associated with plant piping and treatment units
6. Set losses through filters and contactors, where the buildup of material in the process affects the unit head loss and rate of flow

By considering these factors in combination it is possible to define other critical hydraulic features of the treatment system. These include:

1. The water surface or hydraulic grade-line elevation needed at the head of the treatment plant to move water through the facility at the design rates
2. The total head loss through the treatment system under various current and future demand conditions

Given the widespread use of specialized equipment and proprietary processes, the development of a hydraulic profile for a modern water-treatment system is usually accomplished through a combination of basic hydraulic computations and close coordination with manufacturers. Losses through plant piping and fittings are readily determined using the Darcy-Weisbach, Hazen-Williams, or Manning friction formulas. Open-channel flow conditions can be analyzed in similar fashion using Manning's equation and hydraulic formulas for various flow conditions as well as expressions for specific devices including sharp-crested weirs and overflow launders. Minor loss relationships can be used to express entrance and exit losses, losses associated with fittings and valves, and other minor flow constraints, in terms of velocity.

However, losses associated with specialized treatment plant components such as screens and trash racks, flowmeters, and filtration systems are the result of complex hydraulic phenomena and cannot in most cases be readily computed from basic hydraulic formulas. For these items, designers should consult with manufacturers' literature or technical representatives who can provide data based on testing and experience with the specific items.

The balance of this presentation on water-treatment-plant hydraulics presents summaries of the basic hydraulic formula used in the analysis of pressure flow and minor losses in pipes, flow over sharp-crested weirs, and flow in effluent launders. The hydraulics of open-channel flow are mentioned briefly here but are presented in detail in Sec. 2. A brief discussion of other hydraulic considerations related to water-treatment-plant design is included at the end of this discussion.

12. Pressure Flow through Pipes. Head losses associated with pressure flow through pipes in a water-treatment plant are the result of friction losses in straight sections of pipe or channel combined with head losses at entrances, exits, bends, tees, and other valves and fittings. For most treatment-plant applications, these losses can be adequately described using either the Darcy-Weisbach, Hazen-Williams, or Manning equations listed below and minor loss coefficients.[4]

Darcy-Weisbach:

$$h_f = \frac{8fLQ^2}{\pi^2 g D^5} \qquad (4)$$

Hazen-Williams:

$$h_f = \frac{4.73 Q^{1.85} L}{C^{1.85} D^{4.87}} \qquad (5)$$

Manning:

$$h_f = 4.66 n^2 \frac{Q^2 L}{D^{16/3}} \qquad (6)$$

where h_f = head loss, ft
L = length of pipe, ft
Q = flow rate, cfs
D = pipe diameter, ft
g = acceleration due to gravity, 32.2 ft/s²
f = Darcy-Weisbach factor of friction
C = Hazen-Williams roughness coefficient
n = Manning's roughness coefficient

Friction or roughness coefficients are required to use each of these equations. The Darcy-Weisbach f is determined based on Reynolds number, pipe roughness, and pipe diameter using the Moody diagram and values of roughness given in Sec. 2. Commonly used values of the Hazen-Williams C factor and Manning's n for pipe materials found in water-treatment plants are given in Table 9.

While the equations above adequately describe head loss due to friction in straight sections of pipe, they do not account for losses associated with bends, elbows, partially closed valves, and other fittings. These minor losses frequently make up a large portion of the head loss which occurs in treatment-plant piping. A common method of representing minor losses in hydraulic computations involves the use of a minor loss coefficient K_L and the velocity head $V^2/2g$ as shown in Eq. (7).

$$H_L = K_L \frac{V^2}{2g} \qquad (7)$$

where H_L = minor head loss, ft
K_L = minor loss coefficient
g = acceleration due to gravity, ft/s²

TABLE 9 Typical Values of Pipe Roughness Coefficients

	Values of C		Values of n	
Pipe material	Range	Typical	Range	Typical
Cast iron:				
New, unlined	120–140	110	—	—
Old, unlined	40–100	80	—	—
Cement-lined and sea-coated	100–140	120	0.011–0.015	0.013
Ductile iron, cement-lined	100–140	120	0.011–0.015	0.013
Steel:				
Welded and seamless	80–150	120		
Riveted		110	0.012–0.018	0.015
Concrete pipe	100–140	110	0.011–0.015	0.015
Plastic pipe (smooth)	120–150	130	0.011–0.015	0.011

Adapted from Refs. 2 and 4.

Values of K_L range widely depending upon the type of loss being represented. For example, $K_L = 0.3$ is a typical value for a standard 90° bend. But a half-closed 8-in diameter gate valve gives a $K_L = 3.0$. Table 10 provides a summary of commonly used minor loss coefficients for standard entrances, exits, fittings, and valves.

13. Open-Channel Flow. Much flow through conventional water-treatment plants falls under the category of open-channel flow. In treatment plants, this category includes the flow of water through open channels and basins, over weirs and in troughs, and through partially filled pipes. In fact, open-channel flow includes all types of flow where a free surface is present.

Because of the presence of a free surface, the hydraulics of flow in open channels are more complex than those principles which apply to steady flow in pressurized pipes. Under open-channel conditions, the water surface and the hydraulic grade line are the same. However, depending on the geometry of the channel, the depth and velocity of the flow may vary. In particular, the water-surface elevation at any given point may depend on either upstream or downstream conditions. Figure 3 illustrates several ways in which channel characteristics and hydraulic controls may affect the water surface in a hypothetical channel, resulting in transitions from subcritical to supercritical flow. Detailed descriptions of the hydraulics of open channels are presented in previous sections. A detailed description of flow in partially full pipes is included in the discussion of sewers in hydraulics in Sec. 28.

14. Weirs. Weirs are commonly used to control the outlet elevation of flow from a basin or channel in water-treatment systems. In some instances they may also be used as a means of measuring the rate of flow at a point.

The depth of flow over a weir depends on the type and geometry of the weir, the weir length, and the flow rate across the weir. Flow across a weir may also be affected by downstream conditions, if the weir is located within the influence of backwater from a downstream control point or if the underside of the flow nappe downstream of the weir is not fully ventilated and at atmospheric pressure.

Figure 4 illustrates the basic characteristics of the sharp-crested weirs most commonly found in water-treatment systems. Expressions for flow over the weirs shown are given below.

The general equation for describing flow over a rectangular weir such as those shown in Fig. 4a and b is given in Eq. (8).

TABLE 10 Minor Loss Coefficients for Pipe Flow

Type of minor loss	Loss in terms of $V^2/2g$
Pipe fittings:	
90° elbow, regular	0.21–0.30
90° elbow, long radius	0.14–0.23
45° elbow, regular	0.2
Return bend, regular	0.4
Return bend, long radius	0.3
AWWA tee, flow through side outlet	0.5–1.80
AWWA tee, flow through run	0.1–0.6
AWWA tee, flow split side inlet to run	0.5–1.8
Valves:	
Butterfly valve ($\theta = 90°$ for closed valve)*	
$\theta = 0°$	0.3–1.3
$\theta = 10°$	0.46–0.52
$\theta = 20°$	1.38–1.54
$\theta = 30°$	3.6–3.9
$\theta = 40°$	10–11
$\theta = 50°$	31–33
$\theta = 60°$	90–120
Check valves (swing check) fully open	0.6–2.5
Gate valves (4 to 12 in) fully open	0.07–0.14
¼ closed	0.47–0.55
½ closed	2.2–2.6
¾ closed	12–16
Sluice gates	
As submerged port in 12 in wall	0.8
As contraction in conduit	0.5
Width equal to conduit width and without top submergence	0.2
Entrance and exit losses:	
Entrance, bellmouthed	0.04
Entrance, slightly taunted	0.23
Entrance, square edged	0.5
Entrance, projecting	1.0
Exit, bellmouthed	$0.1 = \left(\dfrac{V_1^2}{2g} - \dfrac{V_2^2}{2g}\right)$
Exit, submerged pipe to still water	1.0

*Loss coefficients for partially open conditions may vary widely. Individual manufacturers should be consulted for specific conditions.
Adapted from Refs. 4 and 19.

$$Q = 3.33(L - 0.1nh)[(h + h_v)^{3/2} - h_v^{3/2}] \qquad (8)$$

where Q = flow rate, cfs
L = length of weir crest, ft
n = number of end contractions
h = upstream head above weir crest, ft
h_v = velocity head associated with the flow approaching the weir, ft

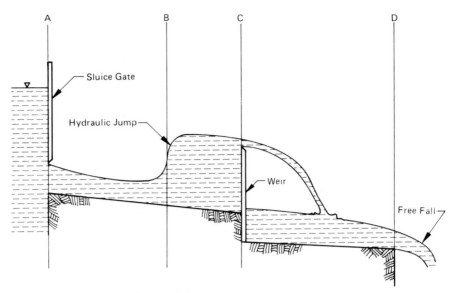

FIGURE 3 Open-channel flow conditions.

The term n in the equation defines the number of end contractions associated with a given weir. An end contraction occurs when the length of the weir crest is less than the full width of the approach channel upstream of the weir as shown in Fig. 4b. Under these conditions the streamlines upstream of the weir must contract to allow all of the flow to pass through the opening. When the weir length is equal to the channel width, as in Fig. 4a, the end contractions are said to be suppressed and $n = 0$.

The term h_v is included in Eq. (8) to account for the effect of the upstream approach velocity on the weir discharge, and can be written as

$$h_v = \frac{V^2}{2g} = \frac{Q^2}{A^2} \frac{1}{2g} \qquad (9)$$

where V = approach velocity in the channel upstream of the weir, fps
g = acceleration due to gravity, 32.2 ft/s^2
A = cross-sectional area of the flow upstream of the weir, ft^2
Q = discharge over weir, cfs

However, the incorporation of this term into the equation forces the use of a trial-and-error solution method since both sides of the equation now depend on Q.

The solution of this problem can be quickly achieved by first assuming that h_v is small in comparison with h. Rewriting Eq. (8) with this assumption yields

$$Q = 3.33(L - 0.1nh)h^{3/2} \qquad (10)$$

Solving Eq. (10) for Q, an estimate of the discharge can be obtained and used to estimate h_v in Eq. (9). If h_v is in fact small in comparison with h, then the dis-

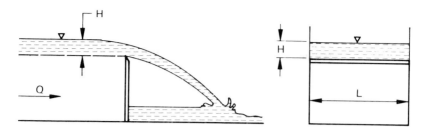

Fig. 4a. RECTANGULAR WEIR—NO END CONTRACTIONS

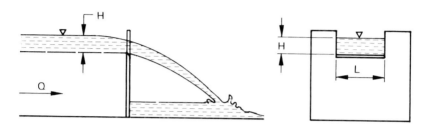

Fig. 4b. RECTANGULAR WEIR—WITH END CONTRACTIONS

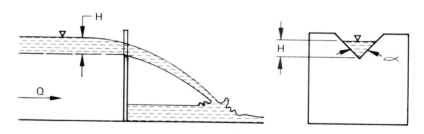

Fig. 4c. V-NOTCH WEIR

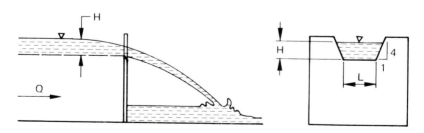

Fig. 4d. CIPOLLETTI WEIR

FIGURE 4 Sharp-crested weirs.

charge rate computed with Eq. (10) is correct. Otherwise, the computed value of h_v can be substituted into Eq. (8) to obtain a revised discharge.

When the velocity in the approach to a rectangular weir is small in relation to the head on the weir crest, and no end contractors exist, Eq. (10) can be reduced to:

$$Q = 3.33Lh^{3/2} \qquad (11)$$

Expressions for flow across triangular V-notch weirs and the trapezoidal Cipolletti weir shown in Fig. 4c and d are as follows:[15]

For V-notch weirs:

$$Q = Xh^{2.48} \qquad (12)$$

For Cipolletti weir:

$$Q = 3.367Lh^{3/2} \qquad (13)$$

where Q = flow rate, cfs
X = 2.49 for α = 90°
    = 1.443 for α = 60°
    = 1.035 for α = 60°
    = 0.497 for α = 22.5°
h = upstream head over vertex or crest of weir, ft
L = length of weir crest, ft

15. Launders. A special application of hydraulics commonly found in water-treatment plants is the use of launders or troughs for the collection of flow at the outlet of a basin. Launders are most commonly used at the outlets of settling basins or clarifiers where it is important to maintain relatively low and well-distributed flow velocities. Rectangular basins are usually constructed with straight launders while circular basins often have circular effluent launders installed around their perimeters. Washwater collection troughs in rapid sand filters also function as launders.

Hydraulically, a launder functions as an open channel with a uniform addition of water along its length. In designing water-treatment facilities, it is necessary to determine the maximum water-surface elevation at the upstream end of the launder in order to verify that the weir elevation and launder depth are sufficient for the design flow.

The procedure for determining the maximum water-surface elevation in a launder depends on the discharge condition. In most cases, it is desirable to maintain a free outlet at the discharge of the launder. However, where the energy associated with the free fall is too great, floc being carried by the flow may be damaged. As a result, some launders may be designed to operate with a submerged discharge. Figure 5 shows the configuration of typical launders under both free-fall and submerged conditions.

A detailed procedure for the determination of upstream flow depth in a launder was developed by Li.[16] This procedure utilizes a correction factor to account for frictional losses associated with flow along the launder. However, for most treatment plant design applications, simplified expressions as presented below can be used for this problem.

The simplified general expression for flow depth at the upstream end of a rectangular launder receiving uniformly distributed flow can be given as[15]

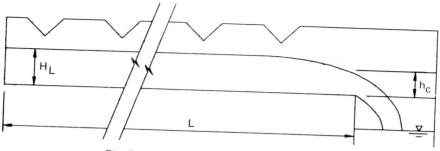

Fig. 5a. FREE FALL DISCHARGE, $h_o = h_c$

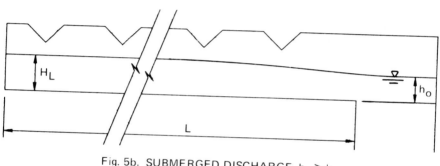

Fig. 5b. SUBMERGED DISCHARGE, $h_o > h_c$

FIGURE 5 Flow conditions in a launder.

$$H_L = \sqrt{(2h_c^3/h_o) + \left(h_o - \frac{1}{3}iL\right)^2} - \frac{2}{3}iL \tag{14}$$

where H_L = flow depth at upstream end of launder, ft
h_c = critical depth in launder, ft = $(Q^2/gb^2)^{1/3}$
L = length of channel
h_o = depth at discharge location, ft
i = launder slope, ft/ft
g = acceleration due to gravity, 32.2 ft/s²
b = channel width, ft
Q = total discharge from launder, cfs

For launders with zero slope, $i = 0$ and the expression can be reduced to

$$H_L = \sqrt{(2h_c^3/h_o) + h_o^2} \tag{15}$$

Lastly, for launders which have a free discharge with $h_o = h_c$,

$$H_L = \sqrt{3}\, h_c = 1.732\, h_c \tag{16}$$

These expressions are based on the integration of the general differential equation for a lateral spillway channel based on the momentum theory. The equations do not include consideration of the effect of friction along the channel. However, experimental studies have shown that friction losses generally account for approximately 10 percent of the water-surface drawdown in the channel. Therefore, a correction factor of 1.1 can be used in most design applications to adjust for frictional effects as shown in Eq. (17).

$$H'_L \cong C_F H_L \qquad (17)$$

where H'_L = flow depth at upstream end of launder, corrected for friction, ft
C_F = friction correction factor = 1.1
H_L = uncorrected flow depth at upstream end of launder, ft

Regardless of the method used to compute H_L, standard practice is to provide a minimum of 4 in of freeboard between the computed maximum water-surface elevation and the bottom of the weir crest along the side of the launder.

Upstream flow depths in troughs used for collection of backwash water may also be estimated using Eqs. (16) and (17).

16. Filtration. Filters are one of the most common treatment devices used in water-treatment plants. They can be constructed in a variety of configurations, using different types of media depending on their intended function in the treatment process.

From a hydraulic standpoint, the filtration process involves the flow of water through a filter medium such as sand or anthracite. Viscous flow of clean water through clean sand and other loose granular media, which has been studied by many investigators, appears to be represented very well in an equation developed rationally by Kozeny[17] from Poiseuille's law of flow through capillary tubes. The same equation was later developed by Fair and Hatch.[18] This equation states that the hydraulic slope through the medium is

$$i = \frac{dh}{dl} = \frac{\beta \mu}{g \rho} \frac{(1-p)^2}{p^3} \frac{v}{d^2} \qquad (18)$$

where dh = head loss through bed thickness dl, ft
p = porosity ratio
d = diameter of the grains, ft
v = rate of filtration as a velocity, fps
β = dimensional friction factor
μ = absolute viscosity of the fluid, lb-s/ft^2
ρ = density of the fluid, slugs/ft^3
g = acceleration due to gravity, 32.2 ft/s^2

During the hydraulic design of filtration facilities, consideration must be given to the range of head losses that will occur as particles removed from the water begin to build up in the filter media. In practice, for rapid sand, dual media, or mixed media gravity filters, the total head loss through the filter is typically 4 to 10 ft. Head losses through pressure filters are usually in a similar range but may vary with system design up to 25 ft.

Also, head loss is commonly used as one of the controls for filtration processes. Along with effluent turbidity, head loss is usually the factor that governs the duration of a filter run and the timing of backwashing. In gravity filters, flow-control valves on the discharge piping are frequently used to regulate the rate of filtration by adjusting to maintain a relatively constant level in the filters. Therefore, increasing head loss in the filter has limited effect on upstream hydraulic conditions. In systems which include pressure filters, however, head loss that builds up in the filter increases the hydraulic grade line at the inlet to the filter and at all points upstream. As a result, the full range of filter head loss must be considered in the hydraulic analysis of the treatment system. In cases where proprietary filter designs are used, manufacturers can provide the most accurate information related to filter head loss.

Filter influent, effluent, wash, and waste piping is generally sized on the basis of minimizing cost while limiting velocities and head loss. Final design criteria must consider site-specific conditions, but general design guidelines are as shown in Table 11.

TABLE 11 Typical Guidelines for Velocities in Filter Piping

Piping system	Maximum velocity at plant capacity or maximum wash rate, fps
Filter influent	2
Filter effluent	5
Backwash supply	12
Backwash drain	8
Filter to waste	15

17. Basin Hydraulics. Hydraulic conditions within the basins that make up a water-treatment plant are frequently very important to the performance of the treatment system. This is particularly true in settling basins and clarifiers where the performance of the unit is directly related to the proper hydraulic design of the tank.

However, even when uniform dispersion of and collection of flow are accomplished at the basin inlet and outlet, the velocity in a typical rectangular basin is not uniform over the cross section. Because of the drag on the walls and floor, the velocity at these boundaries is zero, and it is greater than the average at some points out from the boundaries. The velocity distribution over the cross section of most settling basins is not stable, moreover, owing to the disturbing influences of masses of water of varying density. This variation in density, which is due to temperature differences and differences in concentration of solids and entrained gases, though slight, is nevertheless sufficient to cause vertical movements of water masses, dead spaces, and reversals in the direction of flow. As a result of these disturbing influences, the probable time of passage of all the particles in a given volume of water will be less than the retention period, and the volume will disperse itself while passing through the basin so that the time interval between the passing of the first and last particles at the basin outlet will be much greater than at the inlet. This phenomenon is called *short-circuiting*. If two or more succeeding volumes of water passing through the basin at the same rate of flow require markedly different times for passage, the basin lacks stability of flow.

A particular type of short-circuiting common to many settling basis is caused by *density currents*. If the incoming suspension is heavier than the basin contents, and the basin velocity is insufficient to cause mixing, the heavy suspension will flow along the bottom as a density current. Similarly, light suspensions will flow along the surface. Since the incoming suspension contains more solids than the clarified water in the basin and is therefore likely to be heavier, it is better to introduce it near the bottom and to make the basin shallow.

Short-circuiting may be measured by inserting a charge of dye or other tracer into the basin influent and observing its concentration in the effluent after various time intervals as the slug of water containing the charge passes out of the basin. Short-circuiting studies are usually made on model tanks operating in accordance with Froude's law. Figure 6 shows the results of several such studies plotted in dimensionless terms; with the relative concentration c/c_0 as ordinates and the relative time t/T as abscissa, where c is the concentration of the tracer in the effluent after time t, C_O is the concentration at $t = 0$ which would result with instantaneous dispersion of the tracer throughout the tank, and T is the retention period equal to the volume of the basin divided by the flow rate. If all the tracer charge is accounted for in the effluent, the area under the curve will be 1.0. If the flow pattern is stable and there are no dead spaces, the value of $t/4$ to the center of gravity of the curve will also be 1.0. A stable flow pattern is indicated when the same curve can be reproduced with repeated runs under the same conditions.

In practice the relative time to the center of area t_A/T is usually less than 1.0. Smaller values indicate more severe short-circuiting. Since half the particles of water pass in less time, t_a is the *probable flowing through time*. The dispersion of the water is measured approximately by the time of initial appearance t_i of the tracer in the settling-tank effluent. Small values of t_i/T and t_A/T indicate serious short-circuiting; and small values of Froude's number F indicate unstable flow patterns. Table 12 shows the hydraulic characteristics of the curves of Fig. 6.

Curves B, C, D, and E of Fig. 6 are characteristics of progressively better types of settling tanks under stable flow conditions. The ideal dispersion tank, the baffled mixing chamber, and the ideal settling tank have been added for the purpose of comparison. A rough estimate of the effect of short-circuiting on removal may be developed if it is assumed that the suspension is subjected to varying settling times distributed as indicated by the dispersion curves.

Experiments indicate that, with well-designed inlets and outlets, both the inlet and the outlet zone in a basin will extend for a distance about equal to the basin depth. Since the floor area in these zones is ineffective for settling, it is obvious that the length of a settling basin should be great compared with its depth. Since radial-flow circular tanks and square tanks are quite inefficient for this reason, the design of the inlets and outlets for such tanks is a critical problem. In order to effect good distribution at the inlet of such a tank and good flow distribution over the cross section at the outlet end, with a minimum length for both inlet and outlet zones, it is essential to use orifice walls to the full depth and to have a relatively high velocity through the orifices. Unfortunately, such a wall is detrimental at the inlet end since the high velocity through the orifices may destroy floc.

TABLE 12 Hydraulic Characteristics of Typical Tanks

Curve (see Fig. 6)	Type of tank	B, ft	H, ft	L, ft	V, fpm	$F = \dfrac{V^2}{Rg}$	t_i/T	t_A/T
A	Ideal dispersion						0.0	0.694
B	Radial-flow circular		14.2		1.11	7.5×10^{-7}	0.14	0.831
C	Wide rectangular	135	18.5	330	3.12	5.8×10^{-6}	0.30	0.925
D	Narrow rectangular	16	14	273	5.17	4.53×10^{-6}	0.52	0.903
E	Baffled mixing	2	22.3	1,056	66.6	4×10^{-2}	0.74	0.988
F	Ideal settling						1.0	1.0

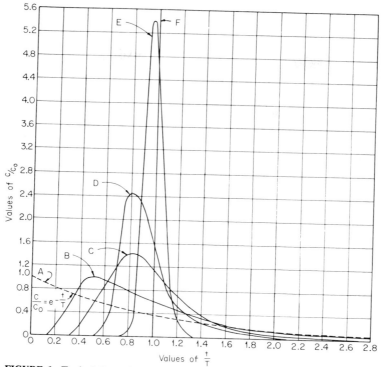

FIGURE 6 Typical dispersion curves for tanks.

The problem of inlet and outlet design is much simplified if the length of the basin is made so great as compared with its depth that the inlet and outlet zones are a negligible part of the gross length. For example, the effective settling area in a basin with a 20:1 ratio of length to depth is about 90 percent of the total floor area. Orifice walls are of no benefit in such a tank.

18. Inlet and Outlet Devices. The purpose of properly designed inlets and outlets is to distribute the water uniformly among basins, to distribute water uniformly over the cross section of each settling basin at the inlet end, and to collect the effluent uniformly at the outlet end. Properly designed inlets and outlets assist in the reduction of short-circuiting and are very important for short basins with low velocities. The velocity in the basin must be reduced to less than 1 percent of the velocity in the influent conduit in some cases.

The water may be distributed across the width of the plant by bringing it in through several pipes at intervals across the width, as shown in Fig. 7a, or by bringing it in through a single conduit to a transverse influent flume, as shown in Fig. 7b, which in turn distributes water across the width through orifices or sluice gates. An equal division of flow to all the basins is most readily approached if the water level is nearly the same in all; this may be accomplished by means of freely discharging effluent weirs all at the same level, or by means of an effluent equalizing channel with a substantially level water surface together with submerged effluent slots having the same capacity for all tanks.

A uniform distribution of flow through inlet pipes or orifices, all of the same size, may be approached by making the head loss at the inlet pipes or orifices large as compared with the maximum difference in energy head available at the inlets. The maximum difference in available energy head in Fig. 7 will be between inlets A and B owing to the change in energy head through the pipe CD or the flume. The head loss at one of the two inlets considered is $h_o = hq^2$ where q is the discharge through the inlet. If h_f is the difference in energy head available at A and B and m is the ratio of the rates of discharge at the two inlets, the head loss at the other inlet is $h_o - h_f = k(mq)^2$. Then the *distribution formula* is

$$\frac{h_o}{h_o - h_f} = \frac{1}{m^2} \tag{19}$$

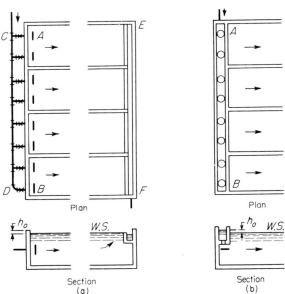

FIGURE 7 Flow distribution and collection in basins.

This equation may be used to compute the required inlet head loss h_o for any desired variation in discharge between the inlets. For example, if it is desired to limit the variation in flow between points A and B to not more than 5 percent, m must be equal to or greater than 0.95. Solving for h_o yields $h_o \geq 10 h_f$. Thus the head loss through the orifice at point A should be at least 10 times as great as the head loss occurring between points A and B.

The actual value of h_f may be estimated from friction losses and velocity head changes.[20]

The size of orifices or gates may be determined by introducing the required value of h_o into the orifice formula with the proper discharge coefficient. The leading edges of all ports should be rounded to approach 1.0 for the coefficient of contraction and thus reduce the required port area. The proper size of inlet pipes may be determined by making the velocity head in the pipes equal to the required value of h_o. The design should be based upon the peak flow and should be checked for minimum-flow conditions.

The effluent from settling tanks may be collected uniformly across the width of each basin, and the basin water levels may be kept nearly the same by means of freely discharging weirs at the same level across the width of all basins at the effluent end. Such weirs discharge into effluent flumes, such as EF (Fig. 7a), which hydraulically are lateral spillway channels. The hydraulics of these channels are described in the previous discussion of launders.

The *distribution formula* (19) has a variety of other uses, including the design of equalizing channels, effluent slots, and orifice walls, the design of filter underdrains for uniform distribution of washwater, and the design of manifold systems for feeding of chemical solutions, air, and carbon dioxide.

HYDRAULICS OF WATER-DISTRIBUTION SYSTEMS

The distribution of water to individual consumers is typically the final step in the process of water supply. For the purpose of this presentation, water distribution will be defined to include the conveyance of treated water from one or more sources of supply (wells, treatment plants, storage facilities) to a large number of users located throughout a given service area. The water-distribution system then is the network of physical facilities (pipes, pumps, reservoirs, etc.) which distribute this water to the users.

Within a water-distribution system, the performance of individual system components is based on basic principles of hydraulics as presented in previous sections. Movement of water through pipes in the system is governed by the principles of viscous fluid flow in closed conduits. Pressure is supplied, maintained, and regulated by pumping equipment, elevated storage facilities, and specialty valves. Thrust blocks and restraining equipment provide resistance to hydraulic forces, and flowmeters and pressure gages provide indications of flow and energy conditions within the system. A good knowledge of these basic principles of hydraulics is the key to understanding the operation of water-distribution systems.

This section focuses on the application of these principles to the operation and analysis of water-distribution facilities working together as a system.

19. Components of a Distribution System. The facilities in a typical water-distribution system include pipelines, pumping stations, storage facilities, valves, and meters which direct and control the flow of water throughout the service area. The primary purpose of the distribution system is to deliver the required supply of water to all parts of the system service area while maintaining acceptable levels of pressure. In addition, most water-distribution systems also serve as the primary supply of water for fire-protection purposes within their service area.

Because of the need for water-distribution systems to convey water to a large number of users spread out over a wide service area, these systems are frequently both large and complex. Many miles of pipe may be required to move water from treatment facilities to users throughout a system. At the same time, storage tanks and pumping stations are often required to provide flexibility within the system and to help move the required water over long distances or widely varying topography. In some cases, a water-distribution system may include several pressure zones where pumps, valves, and tanks maintain reasonable service

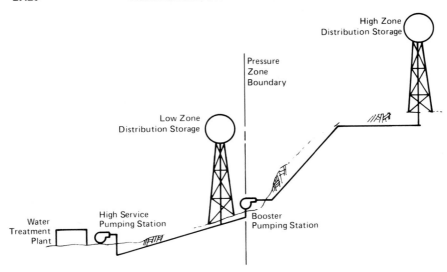

FIGURE 8 Typical two-pressure-zone system.

pressures for users distributed over wide ranges of ground-surface elevations. Figure 8 shows a schematic of a typical two-pressure-zone system in which a pumping station boosts water from the low-pressure zone into the higher-pressure service area.

20. The Piezometric Surface—Hydraulic Grade Line Elevations. While water-distribution systems consist of networks of discrete components including pipes, pumps, valves, and tanks, the hydraulics of such systems include the operation of all the components as a system. One common way to visualize this aspect of water-distribution-system hydraulics is in terms of a piezometric surface. An extension of the concept of the hydraulic grade line to three dimensions, this surface represents the level to which water would rise in piezometers connected to points throughout the distribution system.

By looking at the piezometric surface over a distribution system, it is possible to readily see the manner in which water is moving through the network since the elevation and slope of the surface vary in relation to the direction of flow and rate of head loss occurring. Rapid decreases in the elevation of the piezometric surface indicate areas of high head loss, while high and low spots correspond to the location of major inflows and withdrawals to the system. In addition, the difference between the piezometric surface and the ground-surface elevations across a system gives the pressure in the system at any given point.

The elevation of the piezometric surface across a system is a function of several factors. The maximum level of the surface is typically controlled by the discharge head of the supply pumps for the system, or by the level of a high elevation tank which feeds the balance of the network. As mentioned above, the changes in the surface are indicative of the direction of flow of water through the system, and the friction loss associated with that flow.

Figure 9 shows the profile of the piezometric line or hydraulic grade line (HGL) for a simple system under two conditions. During periods of low demand, water is being pumped from the source across the distribution system to fill the

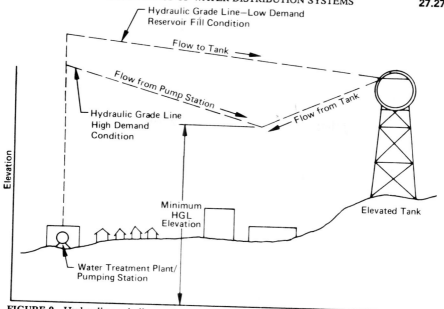

FIGURE 9 Hydraulic grade line under two demand conditions.

elevated tank. As a result, the HGL is highest at the source pumping station, and slopes downward to meet the free water surface at the tank. The slope of the HGL gives an indication of how much energy is required to drive the water across the system and into the tank. During a high demand period, heavy water use between the source and the tank requires water be supplied from both points with the HGL dropping to a minimum in the area of high demand.

In more complex systems, examination of three-dimensional or contour plots of HGL elevations is particularly helpful in defining the ways in which water moves through the distribution-system network.

By reviewing plots of pressure contours for the study area, it is possible to quickly identify areas subject to low pressures under various demand conditions. Plots of HGL elevations, reviewed in conjunction with the pressure plots, clearly show where friction losses in the distribution system are high, and where existing facilities limit the ability of the system to meet demand requirements. Together, these graphical tools provide a convenient means for reviewing and analyzing new or existing water-system operations.

21. Operating Criteria for Water Distribution on Networks. The primary functions of a typical water-distribution network can be summarized as follows:

1. The supply of adequate water to users while maintaining acceptable levels of pressure for service and protection against contamination
2. The supply of water as required for fire protection at specific locations within the system while maintaining acceptable levels of pressure for normal service throughout the remainder of the system
3. Provision of sufficient redundancy to provide a minimum level of reliable service during emergency conditions such as an extended loss of power or a major water-main failure

The tangible criterion upon which the performance of a water system is measured is service pressure. When service pressures are adequate, users get the water they require. If service pressures are not adequate, the distribution system may not be able to supply customers with the amount of water that they expect. Table 13 summarizes typical criteria for distribution system service pressures as recommended by the Great Lakes Upper Mississippi Board of State Public Health and Environmental Managers.[21] Actual criteria for specific systems may vary. However, these values provide an indication of common conditions in modern water-supply systems.

TABLE 13 Service Pressure Criteria for Distribution Systems

Demand condition	Range of acceptable service pressures, psi
Average annual day	35–60
Peak day	35–60
Peak hour	35–60
Fire situation	> 20
Emergency conditions	> 20

Source: Ref. 21.

As discussed previously in this section, the conditions under which a distribution system must meet these criteria also vary depending upon factors such as season, water-use characteristics, and time of day. For example, it may be more difficult to maintain the desired level of service within a system during the peak period of a hot summer day than it would be in the middle of a cool fall night. Therefore, it is important that design or evaluation of a system consider the full range of demand conditions that may be experienced.

22. Conditions for Assessing System Capacity. When evaluating which conditions may be most critical for a system, it is also important to consider how the distribution network responds hydraulically to the demand condition. While peak-hour or fire-fighting conditions may require the greatest short-term supply of water to a specific location, flow from storage facilities can usually be assumed to provide some of the supply necessary for those conditions. On the other hand, the system must be capable of meeting the maximum 24-h daily demand with no net withdrawal from storage. In other words, the system should provide for replenishment of the system's storage facilities. This means that the network must have the capacity to refill its storage facilities during the nighttime or slack demand period.

The AWWA Manual of Practice M32, *Distribution Network Analysis for Water Utilities,* summarizes general criteria for the design or analysis of distribution piping, storage, and pumping systems. Modified to include a condition for the combined capacity of pumping and storage systems, these criteria can be stated as follows:[8]

- The most limiting demand conditions for system piping are maximum-day demand plus fire-flow demand, maximum storage-replenishment rate, and peak-hour demand.

- The most limiting demand conditions for system storage are peak-hour demand, and maximum-day demand plus fire-flow demand

- The most limiting demand conditions for pumping systems alone are maximum-day demand, maximum-day demand plus fire-flow demand, and peak-hour demand.
- The most limiting demand conditions for pumping and storage systems in combination are maximum-day demand plus fire-flow and peak-hour demand.

23. Example of Distribution-System Operating Criteria. An example of the formulation of operating criteria for a hypothetical water system is presented in this section. Consider the water system shown in Fig. 10. The system includes two pressure zones served from a single treatment plant. In its existing condition, the low zone of the distribution system is served by the pumps at the treatment plant and a 2.0-million-gallon elevated water-storage tank. Water is supplied to the high elevation zone through a booster pumping station. A small elevated tank (0.2 million gallons) provides storage in the high zone. Table 14 summarizes the existing demand conditions in the system.

Based on these demands, critical conditions can be defined by applying the limiting condition criteria. The maximum-day and peak-hour requirements can be taken directly from Table 14.

	Limiting requirement		
	Low zone	High zone	Total system
Maximum day, mgd	7.0	0.6	7.6
Peak hour, mgd	14.0	1.2	15.2

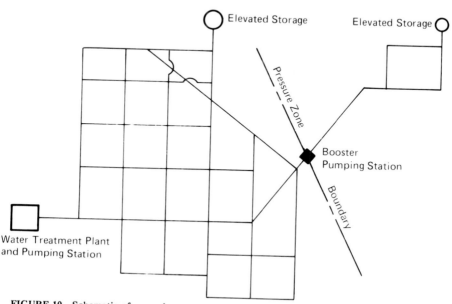

FIGURE 10 Schematic of example system.

TABLE 14 Water Demands for Example System

Demand condition	Low zone	High zone	Total system
Average day, mgd	4.0	0.4	4.4
Maximum day, mgd	7.0	0.6	7.6
Peak hour, mgd	14.0	1.2	15.2
Fire flow required, mgd	2.9	1.1	
Fire duration, h	4.0	2.0	

The limiting fire-flow conditions are determined by adding the required fire-flow rates to the maximum-day demands for each zone individually.

	Limiting requirements		
	Low zone	High zone	Total system
Fire situation—fire in low zone:			
Base demand, mgd (maximum day)	7.0	0.6	7.6
Fire flow, mgd	2.9	0.0	2.9
Total, mgd	9.9	0.6	10.5
Fire situation—fire in high zone:			
Base demand, mgd (maximum day)	7.0	0.6	7.6
Fire flow, mgd	0.0	1.1	1.1
Total, mgd	7.0	1.7	8.7

To determine the storage-replenish rate criteria, it is typically assumed that the system must be capable of replenishing 50 percent of the total system storage during a 6-h slack demand period on the maximum day. If the average demand during the slack period is assumed to be 70 percent of the peak day demand, the required replenishment criteria are as follows:

	Limiting requirement		
Storage replenishment	Low zone	High zone	Total system
Base demand, mgd*	4.9	0.4	5.3
Reservoir fill, mgd†	4.0	0.4	4.4
Total, mgd	8.9	0.8	9.7

*Base demand = 70 percent of maximum-day demand.
†Reservoir fill rate based on need to fill 50 percent of storage capacity in 6-h period.

Applying the criteria to the demands computed above yields the limiting conditions given in Table 15.

TABLE 15 Summary of Limiting Conditions

	Limiting condition		
System component and demand condition	Low zone	High zone	Total system
System piping:			
Maximum day plus fire flow	9.9	1.7	10.5
Storage replenishment rate	8.9	0.8	9.7
Peak-hour demand	14.0	1.2	15.2
Limiting condition	14.0	1.7	15.2
System storage:			
Peak-hour demand	14.0	1.2	15.2
Maximum day plus fire flow	9.9	1.7	10.5
Limiting condition	14.0	1.7	15.2
Pumping, maximum-day demand	7.0	0.6	7.6
Pumping plus storage:			
Peak-hour demand	14.0	1.2	15.2
Maximum day plus fire flow	9.9	1.7	10.5
Limiting condition	14.0	1.7	15.2

HYDRAULIC ANALYSIS OF WATER-DISTRIBUTION SYSTEMS

24. Objectives. Although the hydraulic principles which govern the performance of the individual components of a water-distribution system are relatively straightforward, distribution systems do not operate as a group of independent components. The elements of a distribution system depend directly on each other and affect the performance of one another. The purpose of analyzing a water-distribution system, therefore, is to determine how the system will perform under certain demand and operating conditions. The results of the analysis can be important in a number of situations including:

- The design of a new distribution network
- The design of an extension to an existing network
- Determination of the cause of a problem and definition of corrective actions required
- Evaluation of system reliability
- Preparation for planned outages (maintenance)
- Optimization of system performance and operations

For example, assume that a developer approaches the utility operating the example system described above with a plan for the development of a large office and industrial park complex to be located in the high-elevation section of town. How should the utility engineer assess the impact of the new development on the existing water system?

To effectively analyze the effect of the new development on the existing distribution system, the engineer needs a tool that will allow him or her to determine flow and pressure conditions in the network under a variety of operating conditions. That tool is a mathematical model of the distribution system.

Analysis of a water-distribution system requires as basic input information on the configuration and operating characteristics of the system's components and information on the quantity and distribution of water demands in the system. When completed, the analysis can provide information on flow and hydraulic grade line elevations throughout the system. From these results, other information can be developed including pressures, flow velocities, and indications of head loss in pipes.

25. Basic Pipe Network Equations. Formulation of a mathematical model representing a water-distribution system requires the development of a system of simultaneous equations that can be solved to determine flow and pressure conditions. The criteria that are the basis for this system of equations are as follows:

Energy Balance. The algebraic sum of the energy losses around any closed circuit in the network must be zero, and the total energy loss between any two points in the system must be equal to the sum of the energy losses in all the elements along any path between those two points.

Flow Balance. At each junction of pipes, the flow entering the junction must be equal to the flow leaving the junction.

Friction Loss. The appropriate relationship between flow rate and energy loss must be maintained for all the pipes in the system. Typically, this relationship is based on either the Hazen-Williams or the Darcy-Weisbach formula and takes the form[24]

$$h_l = k_p Q^x \qquad (20)$$

where h_l = energy lost in a pipe section, ft
k_p = a pipe coefficient which is a function of pipe diameter, pipe length, and pipe roughness or resistance to flow
Q = flow through the pipe section, cfs
x = exponential flow coefficient

For the Darcy-Weisbach equation, the terms k_p and x are defined as follows:

$$k_p = \frac{8fL}{\pi^2 g D^5} \qquad (21)$$

$$x = 2 \qquad (22)$$

Substituting into the general friction formula then gives the full Darcy-Weisbach equation

$$h_f = \frac{8fL}{\pi^2 g D^2} Q^2 \qquad (23)$$

where h_f = head loss, ft
f = friction factor
L = length of pipe, ft
Q = flow rate, cfs
D = pipe diameter, ft
g = acceleration due to gravity, 32.2 ft/s^2

For the Hazen-Williams expression,

$$k_p = \frac{4.73L}{C^{1.85}D^{4.87}} \qquad (24)$$

$$x = 1.85 \qquad (25)$$

Thus the full Hazen-Williams expression is given by

$$h_f = \frac{4.73L}{C^{1.85}D^{4.87}} Q^{1.85} \qquad (26)$$

where h_f = head loss, ft
L = length of pipe, ft
Q = flow rate, cfs
C = Hazen-Williams roughness coefficient
D = pipe diameter, ft

Extensive tables summarizing the relationships between the variables in the Hazen-Williams formula are presented in Ref. 22 for various values of C and commercially available pipe diameters ranging from 6 to 144 in. These tables are frequently useful in quickly evaluating simple pipe problems and determining trial values of pipe diameter.

In a simple network of pipes, where determination of the flow distribution and head loss in each pipe is the objective, these criteria can be readily used to develop the required equations for solution of the problem. As an illustration, consider the simple pipe networks shown in Fig. 11.

System A consists of two pipes in series. If the flow into one end of the system is known, it is possible to determine the flow and head loss in each of the segments in the system. (While this problem may seem trivial, it is included to demonstrate the basis for solution of more complex problems.) Let Q represent the rate of flow into the system at point A. The unknowns to be determined include q_1 and q_2, the flow rates in pipes 1 and 2, respectively; h_1 and h_2, the friction losses in pipes 1 and 2; and h, the total head loss between points A and C. Based on the energy balance, the total head loss between A and C must equal the sum of the head losses in pipes 1 and 2. Applying the flow balance criteria, the sum of flows at point A must equal zero, as must the sum of flows at point B. Lastly, the head losses in pipes 1 and 2 can be related to the flows in pipes 1 and 2 by one of the friction-loss equations of the form $h = kq^x$.

In this simple example, the five equations produced can be simplified and readily solved by substitution.

System B shows a slightly more complicated network that can be represented mathematically using the same approach. Given Q in this problem, the network criteria can again be used to define the required system of equations. Based on the energy equation, the head loss between points A and B must be equal in each of the pipes in the system. Second, the sum of the flows in the three segments must equal the total flow entering and leaving the system. Lastly, the friction loss equations written for each of the pipes in the network provide the additional required expressions. As with the series pipe example above, it is possible to solve this system using the appropriate substitutions.

System C presents a more complex problem which cannot be readily simplified. However, using the criteria described above, it is possible to readily define equations for the flow and head loss in each of the pipes as indicated. Application

27.34 WATER DISTRIBUTION AND TREATMENT

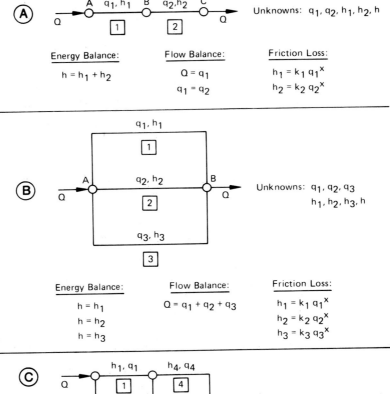

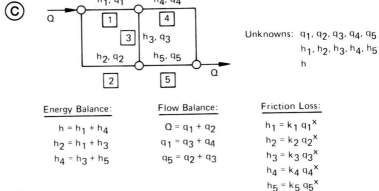

FIGURE 11 Simple pipe networks.

of the energy balance for each of the two independent loops and across the entire system yields three equations. Flow balances at points A, C, and D provide three more equations, and friction-loss expressions for each of the five pipes provide the remaining additional relationships necessary to determine the flow and energy loss in each pipe and the total energy loss across the system.

Unlike the equations formulated for the examples above, these equations require a two-step solution process. Through substitution, the problem can be reduced to six equations which can be solved simultaneously for the flow in each pipe and h, the total head loss across the system. Once the distribution of flow is determined, the computed values can be back-substituted into the friction-loss expressions to yield the energy loss in each pipe segment.

26. Series and Parallel Pipe Problems. An alternative approach to the solution of simple networks of pipes in series or parallel as shown in Fig. 11a and b involves the principle of equivalent pipes. Two pipes are said to be equivalent when, for the same head loss, both deliver the same rate of flow. Using the Hazen-Williams expression:

$$h_{f_1} = \frac{4.73 L_1 Q_1^{1.85}}{C_1^{1.85} D_1^{4.87}} = h_{f_1} = \frac{4.73 L_2 Q_2^{1.85}}{C_2^{1.85} D_2^{4.87}} \tag{27}$$

Since by the definition of equivalent pipes, $Q_1 = Q_2$ when $h_{f_1} = h_{f_2}$, Eq. (27) can be simplified to

$$\frac{L_1}{C_1^{1.85} D_1^{4.87}} = \frac{L_2}{C_2^{1.85} D_2^{4.87}} \tag{28}$$

This relationship is extremely useful in the simplification of pipe network problems. For example, the system of pipes in series shown in Fig. 11a can be reduced to a single pipe with an equivalent diameter of D_{EQ}, length $L = L_1 + L_2$, and roughness coefficient C_{EQ} using Eq. (29).

$$D_{EQ} = \left(\frac{L_1 + L_2}{C_{EQ}} \cdot \frac{1}{\dfrac{L_1}{C_1^{1.85} D_1^{4.87}} + \dfrac{L_2}{C_2^{1.85} D_2^{4.87}}} \right)^{0.205} \tag{29}$$

where D_{EQ} = diameter of equivalent pipe, ft
C_{EQ} = Hazen-Williams roughness coefficient
L_1 = length of pipe 1, ft
C_1 = Hazen-Williams roughness coefficient for pipe 1
D_1 = diameter of pipe 1, ft
L_2 = length of pipe 2, ft
C_2 = Hazen-Williams roughness coefficient for pipe 2
D_2 = diameter of pipe 2, ft

In a similar manner, the equivalent pipe method can be used to simplify systems of parallel pipes such as the one shown in Fig. 11b. Any set of parallel pipes can be reduced to a single pipe of equivalent diameter D_{EQ} with length L_{EQ} and C_{EQ} as shown in Eq. (30).

$$D_{EQ} = \left[\frac{C_1}{C_{EQ}} \left(\frac{L_{EQ}}{L_1} \right)^{0.54} D_1^{2.63} + \frac{C_2}{C_{EQ}} \left(\frac{L_{EQ}}{L_2} \right)^{0.54} D_2^{2.63} \right]^{0.38} \tag{30}$$

where D_{EQ} = diameter of equivalent pipe, ft
C_{EQ} = Hazen-Williams roughness coefficient
L_1 = length of pipe 1, ft
C_1 = Hazen-Williams roughness coefficient for pipe 1

D_1 = diameter of pipe 1, ft
L_2 = Length of pipe 2, ft
C_2 = Hazen-Williams roughness coefficient for pipe 2
D_2 = diameter of pipe 2, ft

Sets of more than two parallel pipes must be simplified two pipes at a time. However, the equation to be used does not change.

27. Formulation of equations for Distribution Networks. While the criteria illustrated above are applicable regardless of the complexity of a water-distribution network, analysis of actual systems is typically complicated by the presence of pumps, tanks, regulating valves, and other components. Thus the formulation of equations for analyzing such a system requires further consideration.

Three different approaches are commonly considered for formulating the mathematical relationships used to analyze complex distribution networks. These approaches as described by Walski[23] include the use of

- Flow equations
- Node equations
- Loop equations

Formulation of the network equations in terms of pipe flows involves defining equations in which flow rate is the unknown. These include one energy equation for each independent loop (including pseudo-loops between tanks) and one continuity equation for each node. The end result is a system of P nonlinear equations, where P is equal to the number of pipes in the network. The general form of the equations produced as presented by Walski is given in Eqs. (31) and (32).[23]

$$\sum_{i=1}^{m_l} h_{il} - \sum_{k=1}^{P_t} h_{pkl} = dh_l \quad (l = 1, 2, \ldots, L) \tag{31}$$

$$\sum_{i=1}^{n_q} Q_{iq} = U_q \quad (q = 1, 2, \ldots, N) \tag{32}$$

where h_{il} = head loss in ith pipe in lth loop, ft
h_{pkl} = head provided by kth pump in lth loop, ft
dh_l = change in head between constant head nodes in lth loop, ft
m_l = number of pipes in lth loop
P_t = number of pumps in lth loop
Q_{iq} = flow into qth node from ith pipe connected to the node, cfs
U_q = consumptive use at qth node, cfs
n_q = number of pipes into qth node

A second approach to the formulation of the network equations involves writing the energy and continuity equations for each pipe in terms of the heads at the endpoints of the pipe. When these expressions are combined, the result is a system of j nonlinear equations where j is equal to the number of junctions in the network. The form of these equations as presented by Walski is given in Eq. (33).

$$\sum_{k=1}^{m_j} \text{sgn}(H_k - H_i)\left(\frac{H_k - H_i}{K_{ki}}\right)^{1/n_{ki}} = U_i \tag{33}$$

where H_i = head at node i, ft
K_{ki} = head-loss coefficient for pipe between node k and node i
n_{ki} = exponent in head-loss equation for pipe between node k and node i
U_i = demand at node i, cfs

The third approach to the formulation of network equations as described by Walski depends on the development of a system of loop equations. In this approach, the energy is written around each independent loop in the system in terms of the flows in the various pipes in each loop. To reduce the number of unknowns, however, the actual flow in each pipe is written in terms of an assumed flow Qi_i and a correction factor ΔQ_l associated with each loop. Thus, for all the pipes in a specific loop,[23]

$$Q_i = Qi_i + \Delta Q_l \tag{34}$$

where Q_i = flow in pipe i
Qi_i = assumed initial flow in pipe i
ΔQ_l = flow correction for loop l

This approach results in a system of l nonlinear equations written in terms of l unknowns ΔQ_l. The form of these expressions as presented by Walski is given in Eq. (35).[23]

$$dh_l = \sum_{i=1}^{m_l} K_i [\text{sgn}(Qi_i + \Delta Q_l)]Qi_i + |\Delta Q_l|^n \quad \text{for } l = 1,2,3,\ldots,L \tag{35}$$

where dh_l = change in head between constant head nodes in lth loop
K_i = head-loss coefficient for pipe i
m_l = number of pipes in lth loop
L = number of loops

Because of their nonlinearity, none of the systems of equations produced by the approaches described above can be solved directly. In all three cases, an iterative numerical solution using a technique such as the Newton-Raphson or Hardy Cross methods is required.

Algorithms for the solution of complex network problems using most of these techniques have now been structured to run on computers, greatly simplifying the effort required to obtain solutions for a variety of conditions.

28. Algorithm for Solution of Network Problem using a Computer. The approach for solution of complex network problems developed by Wood at the University of Kentucky[24] is one of the algorithms most commonly used in modern network analysis programs. In this formulation, the distribution system is represented as a system of pipes and nodes. Each pipe is defined by two nodes, one at each end. Water-storage facilities are included as fixed-grade nodes at which the hydraulic grade line is fixed and known. Within the network, individual, independent closed loops of pipes are known as primary loops. For a given system, the relationship between the number of pipes, junctions, fixed grade nodes, and primary loops can be stated as follows:

$$p = j + l + f - 1 \tag{36}$$

where p = number of pipes
j = number of junctions

l = number of primary loops
f = number of fixed-grade nodes (tanks)

This relationship provides the basis for the formulation of a system of p loop equations as described above. These equations can then be used to determine the distribution of flow throughout a network.

First, considering the requirement for conservation of energy, the energy-balance equation can be written around each independent closed loop in the network to give l equations of the form[24]

$$\sum h_l = \sum E_p \quad (37)$$

where h_l = energy loss in each pipe (including minor losses)
E_p = energy input to the system by a pump

Where a given loop includes no pumps, $E_p = 0$ and the equation becomes

$$\sum h_l = 0 \quad (38)$$

To account for the presence of fixed-grade nodes in a system it is also necessary to write an energy-balance equation between each pair of fixed-grade points. This will result in $f - 1$ equations of the form

$$\Delta E = \sum h_l - \sum E_p \quad (39)$$

where ΔE is the difference between the two fixed-grade nodes.

Any path between each pair of fixed-grade nodes can be used to formulate this equation. However, the paths must not be redundant.

A set of j additional equations is produced by the formulation of flow-balance equations at each junction in the system. These equations take the form

$$\sum Q_{in} - \sum Q_{out} = Q_e \quad (40)$$

where Q_{in} = flow toward the junction
Q_{out} = flow away from the junction
Q_e = external demand at the junction

In combination, the loop equations, the fixed-grade node equations, and the junction equations provide a system of $j + l + f - 1$ equations written in terms of flow rate and head loss in each pipe segment.

In order to make this system solvable, it is necessary to substitute into the energy-balance equations expressions for energy loss in each pipe section and energy input from each pump written in terms of the flow in each pipe. The energy-loss equation for each pipe in the network is typically written in the form $h_l = k_p Q^x$ with the pipe coefficient as defined by either the Hazen-Williams or the Darcy-Weisbach friction-flow formula. For the Hazen-Williams formula, the relationship is[24]

$$h_l = k_p Q^{1.852} \quad (41)$$

where $k_p = 4.73 L / C^{1.852} D^{4.87}$
h_l = energy loss in the pipe section, ft
L = pipe length, ft
C = Hazen-Williams roughness coefficient

D = pipe diameter, ft
Q = flow rate, cfs

For the Darcy-Weisbach relationship, the expression is written

$$h_l = k_p Q^2 \qquad (42)$$

where $k_p = 8fL/gD^5\pi^2$
h_l = energy loss in the pipe section, ft
f = Moody friction factor
L = pipe length, ft
g = acceleration due to gravity, 32.2 ft/s²
D = pipe diameter, ft

Minor losses in system piping can also be introduced into the energy equations using the minor loss expression

$$h_{lm} = M(V^2/2g) \qquad (43)$$

where h_{lm} = minor energy loss, ft
M = minor loss coefficient
V = velocity in pipe, fps
g = acceleration due to gravity, 32.2 ft/s²

Substituting for velocity in terms of flow rate and pipe diameter, the minor loss expression becomes

$$h_{lm} = k_m Q^2 \qquad (44)$$

where $k_m = 8M/\pi^2 D^4$.

Flow-discharge relationships for pumps included in distribution networks can be incorporated into the model using a quadratic equation to represent the pump performance curve:

$$E_p = A + BQ + CQ^2 \qquad (45)$$

where E_p = energy input to the system by the pump, ft
Q = flow through the pump, cfs
A, B, C = curve-fitting coefficients which define the pump operating characteristics

Using the relationships for pipe friction loss, energy input by a pump, and minor losses, the final system of equations for the solution of the network problem can be formulated as follows:

$l + f - 1$ energy-balance equations:

$$\Delta E = \sum (k_p Q^x + k_m Q^2) - \sum (A + BQ + CQ^2) \qquad (46)$$

where ΔE = 0 in the equation for each primary loop with no fixed-grade nodes or
ΔE = difference in elevation between each pair of fixed-grade nodes for each pseudo-loop

j flow-balance equations:

$$\sum Q_{in} - \sum Q_{out} = Q_e \tag{47}$$

This approach yields a system of p nonlinear algebraic equations written in terms of p unknowns, that is, the flow rate in each pipe segment, that must be solved simultaneously. In the Kentucky formulation, the energy equations are linearized using iterative gradient methods to simplify the solution process. A detailed description of the solution technique used is given in Ref. 24.

29. Computer Models of Distribution-System Networks. With the widespread availability of computers, the use of detailed mathematical models for the analysis of water-distribution systems has become standard practice. Software tools now available give users the ability to quickly evaluate the performance of a distribution system for a wide range of operating conditions.

Several types of analysis can be performed using currently available distribution-system models. These include:

- Steady-state analysis of pressure and flow conditions
- Time-dependent analysis of pressure and flow conditions under changing system demands
- Time-dependent analysis of flow patterns and basic water quality

Steady-state analyses are generally adequate for the evaluation of the capacity of distribution-system facilities including piping, storage tanks, and pumping systems. The results of steady-state simulations performed for critical conditions such as peak hour, peak day plus fire flow, peak day, and storage replenishment provide a "snapshot" of the conditions in the networks at a given point in time.

Time-dependent simulations, sometimes called extended-period simulations, provide indications of system performance over time. These simulations typically consist of a series of steady-state runs in which the results of one simulation are extrapolated to define the starting conditions for the next run. These time-dependent simulations are most useful for evaluating the operation of facilities such as pumping stations or variable-level storage tanks. In addition, the results of this type of analysis provide the flow and velocity data necessary for modeling of water-mixing and/or water-quality variations in a distribution system over time.

The use of network models for the analysis of water-quality changes in distribution systems has been developed in response to increasing concerns about changes in water quality within distribution systems.

30. Application of Distribution-System Computer Models. Although computers and analysis software greatly increase the ability of engineers, operators, and managers to accurately analyze the performance of complex distribution systems, the process of developing and applying these tools still requires significant effort to achieve accurate and useful results. Key steps in this process are described briefly here. For more detailed information, readers are referred to the American Water Works Association Manual of Water Supply Practices M32 on *Distribution Network Analysis for Water Utilities.*[8] This manual presents a detailed guide to the selection, development, and application of distribution models.

Model Formulation. Prior to starting the development of the computer model, it is important that the objectives of the analysis, the general characteristics of the water system, the availability of accurate data, and the time and manpower available for the work be considered. These factors will affect the type of model to be used, the level of detail to be included, and the types of analysis to be performed.

Data Collection. In order to develop a useful model of a distribution system, accurate data on the type, size, location, capacity, and condition of the physical components of the system must be compiled. As indicated above, the amount of data required will depend to a great extent on the level of detail to be used in the model. For analysis of general system performance or design of major transmission and distribution facilities, it may be useful to "skeletonize" the system, including in the model only the major facilities required to describe the basic structure of the network. On the other hand, models which are to be used to analyze specific problems such as low pressures in a small service area or fire flows at a specific site, may require that all the distribution mains in the system be considered.

Definition of the Network. Once the required data have been collected, they must be used to define the system of pipes, nodes, tanks, pumps, and other facilities to be included in the model. It is this step of the process that defines the geometry of the system. Several modeling packages now available facilitate this step of the process by allowing graphical data entry using a digitizer or by organizing the information on the physical components of the system into a database that can also be used for inventory and maintenance purposes or as a link to intelligent mapping or facilities management systems.

Field Measurement and Testing. Accurate data on existing water-system facilities are not always readily available. Nor do records or drawings give accurate indications of the condition of the facilities. To address these problems, field measurement and testing of actual conditions in the system are usually required. Typical measurements that may be made include loss-of-head tests to define actual pipe roughness characteristics; fire-flow tests to define the system's ability to deliver large flows to specific locations; pump efficiency tests to define pump operating conditions, and pressure and level measurements used to determine normal operating conditions in water storage facilities. Procedures for performing these tests are presented in Ref. 23.

Definition and Allocation of Demands. The definition and allocation of demands throughout the distribution network is a critical step in the model development process. The demands imposed on the system are the driving force for the performance of the distribution system. Procedures for projecting demand estimates for the analysis of distribution systems are discussed in a previous part of this section. However, when allocating demands throughout the distribution network, additional factors must be considered. In particular, it is beneficial to have the ability to quickly modify usage data for a variety of demand conditions, both current and future. For a 1986 study of the Phoenix, Ariz., water distribution system, this ability was provided by an interactive demand generator system.[25] The system used past billing records, land-use characteristics, and projected demand factors organized in a computer-based database to enable operators to quickly prepare demand data sets for model simulations.

Model Calibration. Model calibration is the process by which the completed distribution system model is adjusted and refined to accurately reflect "real world" conditions. The adjustment is normally accomplished by comparing the results of field measurements made in the system (fire-flow tests, loss-of-head tests, etc.) with model results obtained by simulating the test conditions. Most commonly, calibration adjustments are made to the roughness coefficients input to the model or to the demand distribution, since these factors are subject to the greatest level of uncertainty. However, calibration testing in conjunction with field checks also frequently identifies unknown conditions in the system such as partially or completely closed valves or pipeline connections not noted in system records. The end result of the calibration process is a distribution-system model which accurately reflects the condition and capacity of the actual system.

Model Application. Once calibrated, the distribution-system model can be used to reliably analyze the system's performance and operation. It is important, however, to determine what simulations are necessary to effectively analyze the problem being considered. Although computers provide the power to run multiple simulations of all but the most complex distribution networks quickly, in practice most problems can be solved using the results of only a few well-planned simulations. This aspect of applying the model is particularly important when steady-state simulations are being used. It is useful to develop a plan of the simulations which are expected to be needed before starting to use the model. This plan should clearly define the purpose of each simulation and the conditions associated with it. Then, as the actual simulations are performed, the model and expected results should be compared and used to refine the plan and to define additional runs which are required.

31. Interpretation of Distribution-System Model Results. Results produced by the analysis of a water-distribution system usually include listings of pressure and/or hydraulic grade line elevation at all nodes; listings of flow, velocity, and head loss through all pipes; indications of operating characteristics for all pumps; and rates of flow into and out of storage facilities. To make use of all this information, it is important that the modeler know how to interpret the output. Pressures are not always the only indicator of system performance that need to be considered. At a minimum, pump operating rates and tank fill and draw rates must also be reviewed to provide a comprehensive understanding of network operations. Table 16 summarizes some of the critical indicators that can be used in the evaluation of model results.

TABLE 16 Distribution System Performance Criteria

Demand condition	System performance indicators
Peak day demand	Service pressures in acceptable range Flow into and out of tanks balanced
Peak hour demand	Service pressures in acceptable range Flow out of tanks in acceptable range
Peak day + fire flow	Service pressure in acceptable range Flows out of tanks in acceptable range
Storage replenishment	Service pressures in acceptable range Flow into tanks in acceptable range

WATER-DISTRIBUTION PUMPING SYSTEMS

Within most water-distribution systems, service pressures are established and maintained by pumping. Except in mountainous regions where supply reservoirs are located at elevations above the service area, pumps are used to lift water from sources of supply, convey it to treatment facilities, and distribute it throughout service zones that may be located at widely varying elevations.

32. Types of Pumping Installations. Pumps are used for a variety of purposes within water-supply and -distribution systems. The most common types of installations are:

- Low service pumping stations
- High service pumping stations
- Distribution booster pumping stations
- Fire service pumping stations

Low service pumps are typically low-head, high-rate units which convey water from a raw-water supply or storage facility to treatment facilities.

High service pumps are most commonly used to pump finished water from treatment facilities out into distribution networks at pressures suitable for normal service to customers.

Where portions of a distribution system are separated by long distances or large changes in elevations booster pumps are commonly used to add energy to the system and to maintain acceptable service pressures.

Fire-service pumps may be used to provide additional capacity for emergency fire protection to facilities which have great value, are particularly susceptible to damage by fire, or are located at the extreme edges of a distribution system.

33. Pumping-Station Hydraulics. The basic hydraulic principles which govern pumping systems are presented in detail elsewhere in this text. This section focuses on the hydraulic aspects of pumping systems most often used in the distribution of water.

The operation of a given pumping system is a function of both the characteristics of the pumping equipment (number of units, impeller size, motor horsepower) and the hydraulic characteristics of the distribution system. For a simple pumping system, the relationship between the pump performance and the system hydraulics can be represented by two curves as shown in Fig. 12.

The conditions under which the pump will operate are defined by the intersection of the system head curve and the pump performance curve. At this point the discharge from the pump is balanced by the combination of the static and friction heads against which the unit must operate. As conditions in the system change, so does the operating condition of the pump.

As a result, the operation of the pump will vary in direct response to fluctuations in the head against which it is pumping. Over a longer time frame pumping operations will vary as the roughness of the piping changes with age. Thus, in selecting a pump for a given system it is important to consider the full range of conditions under which the system must operate and the possible effects of time on system conditions and pump performance.

In a water-distribution network the relationship between the pump curve and the system head curve is further complicated by the wide range of flow-head conditions that may occur. The numerous flow patterns that can occur in a network system mean that the hydraulic characteristics of the networks are defined by a family of system head curves rather than a single curve. Therefore, the pumping facilities selected to serve a given system must be chosen to provide acceptable levels of service throughout the full range of potential operating conditions, rather than at a few selected points.

Figure 13 shows an example of the full range of operating conditions that can

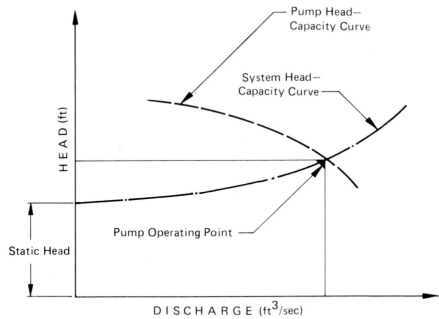

FIGURE 12 Typical pump performance and system head-capacity curve.

be associated with pumping in a distribution network. In most cases this range of conditions is too wide to be met by a single pump. Rather a multiple pump installation or variable-speed pumps may be required. Details of these types of pumping systems are presented in Sec. 28.

WATER-DISTRIBUTION STORAGE FACILITIES

34. Classification of Storage Facilities. Water-storage capacity is a key element in most water-distribution systems. Storage facilities provide operational flexibility that allows a system to more effectively respond to constantly changing demand conditions.

Distribution-system storage facilities are typically classified as either ground or elevated storage depending on their hydraulic function in the network. Ground storage facilities are generally constructed at about ground level and discharge water to the distribution system through a pumping station. Elevated storage facilities are constructed so that the potential energy associated with the elevation of the water in the tank is adequate to deliver water into the distribution system at the required pressure. Elevated storage may be provided by a tall tower or pedestal structure supporting the actual water tank. Or in hilly areas it may be provided by a shorter tank constructed on high ground at sufficient elevation to provide adequate pressure to the service area.

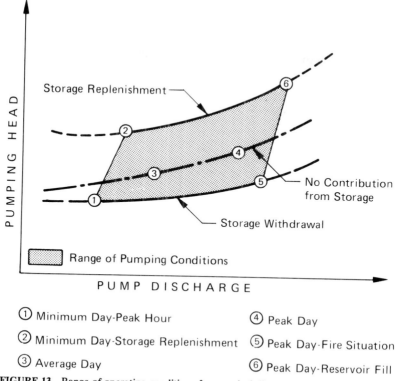

FIGURE 13 Range of operating conditions for a typical distribution pumping system.

The difference between ground and elevated storage can also be visualized in terms of the relationship between the water level in the storage tank and the hydraulic grade line (HGL) elevation in the adjacent distribution system. In ground storage facilities, the water level in the tank is generally much lower than the required distribution system. As a result, pumping is required to drive the water into the network. Elevated storage facilities, however, are usually connected directly to the distribution network so that the water level in the tank fluctuates with changes in the system HGL. Elevated storage facilities of this type are said to "float" on the system because of this direct relationship between tank level and system HGL.

35. Purpose of Storage. Distribution storage may be included in a specific water system for a variety of reasons. However, the primary functions commonly served by these storage facilities can be described as follows:

Equalization of Supply and Demand. By utilizing their supplemental supply and available storage capacity, distribution storage facilities can be used to simplify and/or manage the effective operation of treatment and pumping facilities. During periods of demand greater than the normal system supply rate, water can be drawn from storage to supplement the supply and meet the demand requirements. When demand drops to levels less than the normal supply rate, the excess water being delivered to the system replenishes the storage tanks. In this way, the rate at which water must be treated and pumped into the system may be held relatively constant, allowing for more efficient operation of the facilities.

Alternatively, distribution storage may be used to manage the operation of distribution facilities in other ways. For example, in areas where electric rates vary depending upon the time of day that power is used, it may be advantageous to provide and operate distribution storage facilities in such a way as to minimize pumping during the peak rate period. This can be accomplished by providing sufficient storage to meet all system needs during the peak period and pumping at an increased rate during off-peak hours to replenish the storage.

Provision of Supply for Fire Protection. Water-storage facilities are frequently used to provide large volumes of water for immediate availability in case of fire. Especially in industrial or high-hazard areas, water stored for fire fighting provides increased protection and reduces the need for extremely high capacity conveyance mains to bring all the needed water to the site from the system's source or point of supply.

Fire storage may be provided in either elevated or ground storage facilities. Elevated storage has the advantage of not requiring pumping since the height of the water in an elevated tank provides the energy needed to deliver the required flow. However, elevated storage is typically more costly per unit volume, because of the structure required to support an elevated system. Where ground storage is used for fire protection, pumping is required.

Provisions of Supply for Emergency Conditions. Stored water in a distribution system can provide a source of water for use during emergency conditions when the normal supply of water to an area is interrupted. In this situation, water from storage may be used to supplement a reduced supply or, in some cases, to serve as the sole source of supply.

As with the storage for fire protection, emergency storage is most reliable when provided in an elevated facility, since no pumping is required to distribute the water.

Equalization of Distribution-System Pressures. In addition to the features described above, elevated distribution storage facilities which float on the network also serve to regulate pressures in the system. By providing a location for supplying or storing water as needed, storage facilities function as a "bulge" in a distribution system. Fluctuations in flow into or out of the elevated tank result in only gradual changes in water level. Thus pressures in the area around the tank also see only minor fluctuations in HGL elevation and pressure.

36. Capacity of Storage. In designing new distribution storage facilities, the required capacity for the proposed tank is the first criterion which must be determined. The required capacity depends to a great extent on the function that the tank is intended to serve. For multipurpose tanks, the relative volume required for each function must be considered, and where appropriate, added together to determine the total required volume.

Distribution storage capacity for accommodating normal fluctuations in demand is determined based on the diurnal demand pattern and the desired level of pumping. Common design practice is to rely on pumping to meet daily average demands up to the maximum day. Distribution storage is then sized to provide the necessary additional flow during peak demand portions of the day. Where detailed demand data are not available, a commonly used guideline is that the storage available to supply peak demand periods should equal 20 to 25 percent of the total peak day demand volume. For example, in a typical system with a peak day demand of 10 mgd, approximately 2.0 to 2.5 million gallons of available storage would be needed to supplement pumping during peak periods. In systems where either greater or less dependence on supply pumping is desired, corresponding adjustments to the required storage volume are necessary.

Sizing of storage facilities for provision of fire protection is usually based on the critical fire situation for the zone being served. It is assumed that normal supply pumping will continue to keep up with system demands up to the maximum day rate. Therefore, the volume of additional water that must be supplied from storage is equal to the product of the critical fire-flow requirement and the corresponding flow-duration requirement. For example, in a zone serving an industrial area with a needed fire flow of 3500 gpm for a duration of 4 h the fire-flow storage volume required is given by

$$\frac{3500 \text{ gal}}{\text{min}} (4 \text{ h}) \left(\frac{60 \text{ min}}{1 \text{ h}}\right) = 840{,}000 \text{ gal} \qquad (48)$$

No firm guidelines exist for establishing emergency storage volumes in distribution systems. Rather the volume of distribution storage reserved for emergency conditions is usually determined based on a policy decision by the municipality or utility management. In theory, sufficient storage should be provided to meet normal demands for the duration of time that the system could be affected by a possible emergency. However, estimating the duration of outages associated with potential emergencies is difficult unless one can draw on prior experience. As a result guidelines for emergency storage volume for most municipal water-supply systems vary from 1 to 2 days of supply capacity at average usage rates. For example, the Recommended Standards for Water Works developed by the Great Lakes Upper Mississippi River Board of State Public Health and Environmental Managers suggest a minimum total system storage capacity equal to the average daily system demand. In contrast, some systems target total system storage capacity equal to twice their average daily usage.

In most public water-supply systems, storage facilities must provide capacity for all three of the conditions described above. In practice, storage facilities are generally sized using these conditions and the following criteria.

- Distribution storage facilities should have adequate volume to supply peak demands in excess of the maximum day average using no more than 50 percent of the available capacity.
- Distribution storage facilities should have adequate volume to supply the critical fire demand in addition to the volume required for meeting daily demand fluctuations.
- Total distribution storage capacity in a system should be adequate to supply the average demand of the system for the estimated duration of a possible emergency. The minimum storage capacity should be equal to approximately the average daily demand.

Consideration of these criteria provides a general basis for the sizing of multipurpose distribution storage facilities such as those shown in Fig. 14.

37. Height. In addition to the capacity to be provided by a given storage facility, the height of the structure is also an important factor to be considered in the planning and design of elevated storage. In these facilities the height of the structure limits the HGL range over which the storage volume of the tank is effective.

For most facilities providing elevated storage, the minimum acceptable water level in the tank can be determined by computing the minimum acceptable HGL

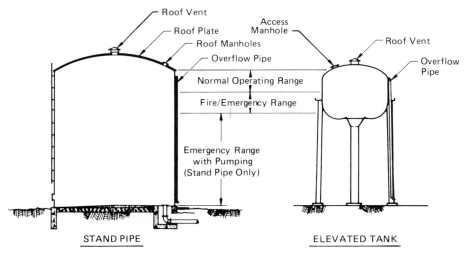

FIGURE 14 Distribution-system storage facilities.

in the service area and adding to that an estimate of the head loss between the critical service point and the storage tank under peak-day conditions. The overflow elevation for the tank can then be estimated by computing the target HGL in the critical section of the system and adding to that an allowance for the head loss between the critical point in the system and the tank at peak-hour conditions. The difference between these elevations defines the normal operating range for the tank. For most elevated storage tanks used to help regulate distribution-system pressures, the normal operating range is limited to 15 to 20 ft so that fluctuations in pressure are limited to 5 to 10 psi. Also, where possible, it is desirable to have the normal operating range in the upper half of the tank with the lower portion being reserved for fire-fighting or emergency conditions as shown in Fig. 14.

In systems where storage requirements for fire protection or emergency conditions are large, additional capacity may be economically provided in the lower portion of a standpipe. However, pumping facilities are required to make effective use of such storage.

Table 17 lists typical characteristics of several types of distribution facilities.

38. Additional Design Considerations. Although the determination of the type, size, and elevation are the primary steps in the planning of a water-

TABLE 17 General Characteristics of Distribution Storage Facilities

Facility type	Capacity range, millions of gallons	Typical head range,* ft
Ground storage reservoirs (steel or prestressed concrete)	0.05–10.0	20–40
Standpipes	0.05–5.0	30–100
Single-pedestal elevated tanks	0.025–3.0	15–45
Multiple-support elevated tanks	0.025–3.0	12.5–40

*Head range given for ground storage reservoirs and standpipes includes total usable water depth.

distribution storage facility, the detailed design of such a structure requires consideration of a number of other factors. These include:

- Structural and foundation design (including consideration of earthquake loadings as appropriate)
- Selection of material of construction
- Compliance with siting regulations related to proximity to groundwater table, potential sources of contamination, floodplains
- Design of inlet and outlet piping and valves
- Provisions of adequate venting and overflow piping to prevent pressurization or depressurization during tank fill and draw cycles
- Provision of security and screening devices to protect the quality of the stored water
- Provision of corrosion-control systems
- Provision of adequate safe access for inspection, maintenance, and repair

Detailed standards and guidelines pertaining to the factors are contained in the American Water Works Association Standards D100, D102, and D110 and in Ref. 22. However, brief mention of several piping-related design considerations is made here.

Inlet and Outlet Piping. Inlet and outlet piping configurations for water-distribution storage facilities vary widely depending upon the intended purpose of the structure. Ground storage tanks which serve as receiving tanks or fire or emergency storage commonly have separate inlet and outlet piping. The inlet piping serves as the supply for water to the facility while the outlet piping may be connected to a pumping station for distribution of water. Elevated storage facilities which float on the distribution system grade line commonly have only one pipe connection which serves as both inlet and outlet depending on the system conditions.

In both cases, the piping facilities should be sized to allow flow into and out of the tank in excess of the expected maximum future fill and draw rates with minimal head loss. This approach reduces the risk that inadequate piping may restrict tank operations under extreme conditions. Valving should be provided to allow the facility to be isolated from the system and drained as required for inspection or maintenance. Special valves or meters may also be provided in some cases to monitor or control the movement of water into or out of the tank.

Overflow Piping and Vents. All closed water-storage facilities should have overflow piping and vents as required to eliminate the potential for pressurization or depressurization of the tank during fill and draw cycles. Both the overflow and vent systems should be sized to accommodate fill or draw conditions in excess of the expected maximum rates. As with all openings to a treated water-storage facility, overflow and vent piping should be screened and located so as to minimize the potential for contamination of the stored water.

REFERENCES

1. "Regional Lake Michigan Water Demand Study," prepared for the Illinois Department of Transportation, Harza Environmental Services, April 1989.

2. LINSLEY, RAY K., and JOSEPH B. FRANZINI, *Water Resources Engineering,* McGraw-Hill, New York, 1979, p. 404.
3. METCALF AND EDDY, INC., *Wastewater Engineering: Treatment Disposal Reuse,* McGraw-Hill, New York, 1979.
4. METCALF AND EDDY, INC., *Wastewater Engineering: Collection and Pumping of Wastewater,* McGraw-Hill, New York, 1981.
5. TCHOBANOGLOUS, GEORGE, and EDWARD D. SCHROEDER, *Water Quality,* Addison-Wesley, Reading, Mass., 1985.
6. "Rules and Regulations for the Allocation of Water from Lake Michigan," Illinois Department of Transportation—Division of Water Resources. January 1990.
7. COOTE, PHILIP, General Manager, Department Water Works, Valparaiso, Ind., personal communication, 1991.
8. *Distribution Network Analysis for Water Utilities,* American Water Works Association Manual of Water Supply Practices M32, 1989.
9. "Standard Schedule for Grading Cities and Towns of the United States with Reference to Their Fire Defenses and Physical Conditions," National Board of Fire Underwriters, 1956.
10. COLE, A. E., and J. L. LINVILLE, eds., *Fire Protection Handbook,* National Fire Protection Association, Quincy, Mass. (16th ed., 1986).
11. *Distribution System Requirements for Fire Protection,* American Water Works Association Manual of Water Supply Practices M31, 1989.
12. *Grading Schedule for Municipal Fire Protection,* Insurance Services Office, New York, 1974.
13. *Water Treatment Plant Design,* American Society of Civil Engineers, American Water Works Association, McGraw-Hill, New York, 1990.
14. *Water Quality and Treatment,* American Water Works Association, McGraw-Hill, New York, 1991.
15. WILLIAMS, ROBERT B., and GORDON L. CULP, eds., *Handbook of Public Water Systems,* Van Nostrand Reinhold, New York, 1986.
16. LI, WEN-HSIUNG, "Open Channels with Non-Uniform Discharge," *Transactions of the ASCE,* vol. 120, 1955.
17. KOZENY, J., *Wasserkraft and Wasserwirtschaft,* vol. 22, p. 67, 1927.
18. FAIR and HATCH, *Journal of the American Water Works Association,* vol. 25, p. 1551, 1933.
19. "Hydraulics and Useful Information," Bulletin 9900, FMC Corporation, 1981.
20. CAMP, T. R., and S. D. GRABER, "Dispersion Conduits," *Journal of the Sanitary Engineering Division, ASCE,* 1968.
21. *Recommended Standards for Water Works, 1987,* Great Lakes Upper Mississippi River Board of State Public Health and Environmental Managers, Health Research, Inc., 1987.
22. WILLIAM, GARNER S., and ALLEN HAZEN, *Hydraulic Tables,* Wiley, New York, 1965.
23. WALSKI, THOMAS M., *Analysis of Water Distribution Systems,* Van Nostrand Reinhold, New York, 1984.
24. WOOD, DON J., *Computer Analysis of Flow in Pipe Networks Including Extended Period Simulations,* Office of Continuing Education and Extension of the College of Engineering of the University of Kentucky, 1980.
25. VELON, J. P., et al., *Distribution Modeling Tools: Making Them Flexible and Friendly,* Presented at the 1986 AWWA Annual Convention, Denver, June 1986.

SECTION 28

WASTEWATER CONVEYANCE AND TREATMENT

By John P. Velon and Richard J. Persaud[1]

INTRODUCTION

1. Introduction. This section covers the application of hydraulic principles to the design of facilities required for the conveyance and treatment of wastewaters including sanitary sewage, commercial and industrial wastewater, and stormwater.

Wastewaters generated in all communities must be collected and conveyed to points of treatment and disposal. In the United States, the final disposal of treated wastewater is governed by the provisions of the Clean Water Act. In practice, the relevant regulatory agency implements the act by granting a National Pollutant Discharge Elimination System (NPDES) permit to the responsible party. This permit identifies maximum contaminant concentrations and total pollutant loads in the wastewater which can be discharged to the receiving stream. These limits are established after careful consideration of the assimilative capacity of the stream and the anticipated contaminant types and concentrations in the wastewater.

Hydraulic principles are important to the design of both the wastewater collection and conveyance systems and treatment facilities because wastewater is more than 99 percent water and behaves as such. However, the final design of all facilities must consider not only hydraulic aspects but also other factors such as treatment process considerations, geotechnical and structural considerations, and costs.

This section presents information on the use of hydraulic principles for the analysis and design of wastewater systems. For convenience, the section is subdivided into two parts. Part A discusses the use of hydraulic principles in the design of wastewater collection and conveyance systems. Part B considers the application of these hydraulic principles to the design of wastewater treatment facilities.

PART A. WASTEWATER COLLECTION AND CONVEYANCE

2. General. Wastewater is usually collected and conveyed from its source to the point of treatment and disposal in a network of pipes or sewers called a sewer

---

[1] Acknowledgment is made to Samuel A. Greeley, William E. Stanley, Donald Newton, and Kenneth V. Hill for material in this section which appeared in the third edition (1969).

system. Sewer systems are classified according to the type of wastewater being conveyed. Three general types of systems are considered:

- Sanitary sewer systems
- Storm sewer systems
- Combined sewer systems

A *sanitary sewer system* is comprised exclusively of sanitary sewers which convey liquid wastes from residences, commercial buildings, industrial plants, and institutions. This system is designed to exclude, as far as possible, stormwater runoff and groundwater. However, the system will receive some storm runoff, surface water, and groundwater from inflow and infiltration.

A *storm sewer system* includes only storm sewers which convey stormwater runoff from buildings, streets, and other surfaces but excludes domestic and industrial and commercial wastewaters. Stormwater runoff is that portion of precipitation which flows over these types of surfaces during and after a storm.

A *combined sewer system* is comprised of a network of pipes designed to collect and convey both sanitary sewage and stormwater runoff.

In all sewer systems, the sewers are also classified according to their function as follows:

Lateral sewers serve individual streets and have no upstream sewers tributary to them.

Branch sewers receive flows from small areas served by laterals and discharge to a main or trunk sewer.

Main sewers are the principal sewers to which branch sewers discharge and are also called trunk sewers.

The laws of hydraulics are applied to sewer systems for three general purposes:

1. To determine the flow in an existing sewer as a means of estimating existing wastewater flow rates
2. To evaluate the capacity of an existing sewer system and to determine its ability to convey existing and anticipated future wastewater flow rates
3. To design new sewer systems to convey wastewater from currently unsewered areas or to design relief systems to convey excess wastewater flows from currently sewered areas

As will be noted in this section, a high degree of refinement in hydraulic computations for sewers is not considered necessary because of the practical limitations on the accuracy with which wastewater flow rates and future hydraulic characteristics of sewers can be estimated.

SEWAGE QUANTITIES

3. Overview. The term sewage or wastewater is used to describe the spent water supply of a community. Design or evaluation of a sanitary sewer system requires, as a first step, the estimation of the quantity of sewage that is or would be carried by the system. This estimate is usually made by multiplying the current and/or future population by the probable per capita sewage contribution. As a consequence, these two factors—tributary population and per capita sewage contribution—have the greatest influence on the estimate of flow rate. Other factors affecting the flow rate are the number, size, and type of industry; the extent of commercial activity; and the number and size of institutions.

Today, population projections are done by municipal planning agencies, and the engineer's role in this process is being reduced. Sewage flow rates are generally projected by the engineer using these population data and other information. However, the engineer should be aware of the general processes utilized in population projection and should understand the bases of development of the flow rates. As a consequence, the following sections give a general overview of the processes only, and the reader may consult other references[1,2,3] for more details.

4. Design Period. The length of time over which the capacity of the sewer system will be adequate is called the design period. This period could be limited in some instances by the economic or useful life of sections of the system. Thus, a design period of 50 years may not be considered appropriate for the mechanical components of a pump station which have an economic life of only 20 years.

Buried pipes are considered to have unlimited useful life. Therefore, for lateral and branch sewers which serve small areas with well-defined tributary populations, an indefinite design period is considered. These lateral and branch sewers must be designed to accommodate the peak flows from the tributary area, for the maximum development and saturation population density anticipated within the area.

For trunk or main sewers, a design period of 25 to 50 years may be considered. Final choice of design period in this case is therefore determined by other factors such as population growth rates and cost considerations. In rapidly growing areas, a long design period may be uneconomic. Also, each community must consider the price that the present population is willing to pay to provide excess system capacity for future populations.

5. Population Estimates. After the design period is chosen, the next step in determination of the design flow rate is to estimate the future population at the end of that period. The present population is also necessary as a basis for forecasts. In the United States, the U.S. Bureau of the Census publishes population data for all communities each decade. These data provide the best basis for most population forecasts. However, because of the 10-year gap in the census data, it is often necessary to estimate both present and future populations.

In many urban areas, the population estimates are worked out in great detail by specialists in this area such as state or local planning agencies and the results are made available in planning reports. Since such reports have the sanction of governing bodies, it is prudent to utilize this information when available. In smaller rural areas and urban areas without such data, the engineer may be called upon to make the estimate of future population.

Several methods which have been used to forecast future population from census data are as follows:

1. Arithmetical method—in which a constant increase in population is considered for each period (year or decade) under consideration
2. Uniform percentage rate of growth—in which population growth in any period is considered to be a uniform percentage of the population in the preceding period
3. Graphical method—in which the curve of past populations in the community is extended into the future
4. Graphical comparison method—in which a graphical comparison is made of

the growth of other similar but larger communities who had reached the present population of the community under consideration some time ago

5. Decreasing rate of increase method—in which a decreasing rate of increase is applied to each period under consideration in the future

In all cases where the engineer is developing population projections, it is important that good judgment be exercised and that input be obtained from local and state officials.

The design or evaluation of a sewer system depends not only on total flow rates from the entire community but also on flows tributary to all local sewers from subareas within the community. It is therefore useful to estimate population distributions within the community. This distribution can best be estimated by obtaining the number of houses in each subarea from aerial photographs, by field counts, or from other community tax records. Then by using other factors such as existing population densities, type of development, zoning regulations, and other data, future population distributions can be obtained.

Saturation population densities vary widely and range from 18 to 30 persons per acre for areas zoned for single-family residences to 25 to 60 persons per acre for areas zoned for two-family residences and may exceed 1000 persons per acre for multistory apartment buildings.

6. Per Capita Sewage Flow Rate. Sanitary sewage and industrial wastewater are derived largely from the water supply of the community. The actual percentage of the water supply reaching the sewer is related to the type of community and the water-use habits of the residents. In general, the entire public water supply does not reach the sewer because of losses due to lawn sprinkling, use in manufacturing processes, and exfiltration through leaking pipes. However, these losses are made up by discharges of sewage from industrial plants with private water-supply systems and from inflow and infiltration of clear ground water to the sewers. As a consequence, the average rate of sewage flow can be taken as the average rate of water consumption of the community.

The average per capita water consumption rate varies from 40 to 250 gallons per day (gpd). These variations are due to many factors such as size of city, number of industries, quality and cost of the water, climate, characteristics of the population, the extent of metering, and the efficiency of the system. Table 1 presents typical per capita wastewater flow rates from residential communities.[1]

TABLE 1 Average Wastewater Flows from Residential Sources*

Source	Unit	Flow, gal/unit/day Range	Typical
Apartment	Person	50–90	70
Hotel, resident	Resident	40–60	50
Individual dwelling:			
Average home	Person	50–90	75
Better home	Person	65–105	80
Luxury home	Person	80–145	100
Semimodern home	Person	30–65	50
Summer cottage	Person	30–60	50
Trailer park	Person	30–50	40

*Adapted from Ref. 1.

7. Industrial and Commercial Wastewater Flow Rates. The per capita sewage flow rate described above, which includes industrial and commercial wastewater contributions, may be adequate for evaluation or design of major components of a sewer system such as the trunk sewers. However, careful consideration must also be given to actual or anticipated quantities of industrial and commercial wastewater flow rates for the evaluation or design of lateral or branch sewers which receive flows directly from these concerns.

Industrial wastewater flow rates will vary significantly with the size and type of industry, the degree of water reuse, and the extent of on-site wastewater-treatment facilities. Estimates of actual and projected wastewater flow rates will therefore require special studies of the industry. For initial planning purposes, use can be made of general estimates. For industries without wet processes, a wastewater flow rate of 5000 gal/acre/day may be considered. For industries with wet processes and in which the nature of the process is known, wastewater flow rates may be developed from water-use data of the industry. In addition, consideration must be given to the sanitary sewage contributed by industrial concerns which may vary from 10 to 25 gal per capita per day (gpcd).

Commercial wastewater flow rates also vary significantly, and unit flow rates of between 4500 and 160,000 gpd/acre have been utilized in various areas.[2] Estimates of wastewater flow rates from specific commercial areas should therefore be based on a special study of the areas. This study may include review of wastewater contributions from similar areas and comparable data from other communities.

It should be noted that since commercial areas are generally well defined and the proportion of the city used for this purpose is small, the probable future expansion of the area should be taken into consideration when estimating wastewater flows. Also, generous allowances should be made for anticipated flows because of the high cost and disruption which may be caused by upsizing or modifying sewers in the future. Table 2 provides a summary of average wastewater flow rates from commercial areas.[2]

TABLE 2 Average Wastewater Flows from Commercial Sources*

Type of establishment	Avg. flow, gpcd
Stores, offices, and small businesses	12–25
Hotels	50–150
Motels	50–125
Drive-in theaters (3 persons per car)	8–10
Schools (no showers), 8-h period	8–35
Schools (with showers), 8-h period	17–25
Tourist and trailer camps	80–120
Recreational and summer camps	20–25

*Adapted from Ref. 2.

Consideration must also be given to the quantities of wastewater from institutions such as hospitals, schools, and prisons and from recreational sources. These flow rates may vary significantly and must be evaluated carefully. Typical wastewater flow rates from recreational areas are shown in Table 3.[1]

8. Infiltration. Infiltration is the water entering a sewer system through defective joints and structures and is unavoidable. The rate of infiltration into the system is a function of many factors such as the length of sewers, topography, soil types, quality of construction, type of pipe joints, and type of pipe.

TABLE 3 Wastewater Flows from Recreational Sources*

Source	Unit	Flow, gpd/unit	
		Range	Typical
Apartment, resort	Person	50–75	60
Cabin resort	Person	35–50	40
Cafeteria	Customer	1–3	2
	Employee	8–15	10
Campground (developed)	Person	21–40	30
Cocktail lounge	Seat	15–30	20
Coffee shop	Customer	4–8	5
	Employee	8–15	10
Country club	Member present	65–130	100
	Employee	10–15	13
Day camp (no meals)	Person	10–15	13
Dining hall	Meal served	4–10	8
Dormitory, bunkhouse	Person	20–45	40
Hotel, resort	Person	40–65	50
Laundromat	Machine	475–700	580
Store, resort	Customer	1–5	3
	Employee	8–15	10
Swimming pool	Customer	5–15	10
	Employee	8–15	10
Theater	Seat	3–4	4
Visitor center	Visitor	4–8	4

*Adapted from Ref. 1.

In many separate sewered areas in the United States, extensive work has been done to quantify and reduce the amount of infiltration to the systems. This work has included TV and other inspection of sewer lines to identify broken or damaged sections and extensive flow monitoring to identify variations in wastewater flow rates with storm events.

Flow monitoring in sewers is usually carried out by constructing a weir in the sewer and measuring the depth of flow over the weir. A rating curve is developed for the particular weir such that the actual flow rate can be determined from the depth of flow. Depths can be measured continuously and recorded, or at other specific times as required.

Flow monitoring in sewers can also be carried out using a data logger equipped with a probe or sensor that measures both depth and velocity of flow. Along with the diameter of the sewer, these measured values of depth and velocity are used to compute wastewater flows.

The results of flow monitoring can be used to establish base flow or average dry weather flow rates and peak flow rates, from which the quantity of infiltration can be estimated.

For new sewers, specifications are usually written to limit infiltration to between 200 and 500 gpd per inch-diameter per mile of sewer immediately after installation. Allowances for infiltration at the end of the design period range from 10,000 to 40,000 gpd/mi of sewer.

9. Flow Variations. The flow in any sanitary sewer system will vary throughout the day and also with the seasons. The flow rates will also generally increase with time as the areas develop and populations grow. Thus the absolute minimum flows will be experienced immediately after the system is installed and the maximum flows toward the end of the design period. Both of these flow rates are important to the design of a new system since the sewers must be able to convey peak flows without surcharge and also to convey minimum flows at adequate velocities to prevent deposition of solids.

Figure 1 shows the ratios of peak and minimum flows to average daily flows as a function of the population served. It will be noted that for populations less than 5000, the ratio of maximum to average flow is greater than 3 but for a population of 50,000, the ratio is approximately 2.5. Similarly, the ratio of minimum to the average flow varies from less than 0.3 for a population of 5000 to about 0.4 for a population of 50,000.

10. Regulatory Capacity Requirements. For design purposes, some state regulatory agencies specify certain minimum capacity requirements for sewers. For example, the Recommended Standards for Sewage Works[4] issued by the Great Lakes Upper Mississippi River Board of State Sanitary Engineers includes the following minimum requirements:

Design Capacity. In general sewer capacities should be designed for the estimated ultimate tributary population, except in considering parts of the systems that can be readily increased in capacity. Similarly, consideration should be given to the maximum anticipated capacity of institutions, industrial parks, etc. Where future relief sewers are programmed, economic analysis of alternatives should accompany initial permit applications.

In determining the required capacities of sanitary sewers, the following factors should be considered:

1. Maximum hourly domestic sewage flow
2. Additional maximum sewage or waste flow from industrial plants
3. Inflow and groundwater infiltration

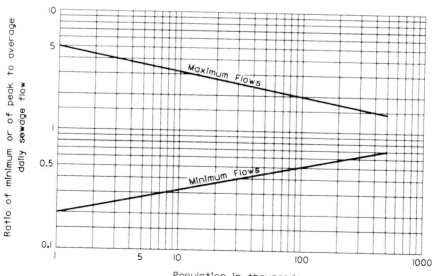

FIGURE 1 Ratio of extreme flows to average daily flow. (*Adapted from Ref. 3.*)

4. Topography of area
5. Location of sewage-treatment plant
6. Depth of excavation
7. Pumping requirements

The basis of design for all sewer projects should accompany the plan documents. More detailed computations may be required by the appropriate reviewing agency for critical projects.

Design Flows

- Per capita flow—new sewer systems should be designed on the basis of an average daily per capita flow of sewage of not less than 100 gpd (0.38 m^3/day). This figure is assumed to cover normal infiltration, but an additional allowance should be made where conditions are unfavorable.
- Peak design flow—sanitary sewers should be designed on peak design flow basis using one of the following methods:
  1. The ratio of peak to average daily flow as determined from curves similar to Fig. 1
  2. Values established from an infiltration and inflow study acceptable to the approving agency

Design of Sewers. Use of other values for peak design flow will be considered if justified on the basis of extensive documentation.

11. Summary. Development of estimates of sewage and wastewater flow rates to be conveyed by a sanitary or combined sewer system involves careful consideration of a number of factors including the following:

1. The design period during which the maximum flows will not exceed the capacity of the system. This is taken as 25 to 50 years for trunk sewers and indefinite for laterals serving small, well-defined areas.
2. The sanitary or domestic sewage contribution based on the future population and the anticipated per capita sewage contribution. The variations in flow rate and the population distribution within the community, which will affect the source of the wastewater flow, are also important.
3. Industrial and commercial area contributions based on careful consideration of the size of the tributary area and the type of industry or commercial activity. In the absence of actual flow data, a per acre allowance can be made for commercial properties based on flows from comparable areas. For industries, estimates of wastewater flow can be made based on the type of process and average water-use rates.
4. The estimated infiltration rate based on considerations of ground topography, soil type, water table, type of pipe, type of joint, and method of construction.
5. Regulatory requirements which may dictate minimum flow rates for different sections of the sewer system.

12. Design Example for Calculating Wastewater Flows. Determine the average and peak wastewater design flows for the downstream end of a sanitary sewer which is to serve an area of 340 acres. Current residential population is 1000, primarily in single-family homes, and the projected ultimate population in the future is 3500 with 3000 people in single-family homes and 500 people in apartments. Current commercial and institutional employment in the service area is 150 employees, primarily in office and non-water-intensive businesses. Future commercial and institutional employment is projected to be 900 in similar types of business. There is negligible existing industrial development, but an 80-acre industrial park is planned.

Solution. The solution proceeds stepwise through the parameters to be considered:

1. *Select the design period of the sewer.* For this sewer, the design condition is the ultimate future condition, since the area is relatively small and the ultimate population is known.

2. *Estimate the average unit rate of flow of domestic sewage from residential areas.* Based on water billing records of existing residential single-family accounts, average water consumption during winter months (November to January) is 80 gpcd and during summer months (June to August) is 110 gpcd.

Water use during the summer includes significant quantities for lawn sprinkling, car washing, and other outdoor uses. This water does not enter the sewer system. Winter water usage is therefore considered more indicative of wastewater flow rates in this small area. Also, no change in per capita water consumption is anticipated in the future. Apartments in adjacent communities have winter water unit consumption rates of 65 gpcd.

Current average wastewater flow from the residential population in single-family homes is therefore

$$80 \text{ (gpcd)} \times 1000 \text{ (people)} = 80{,}000 \text{ gpd}$$

For future conditions, the average wastewater flow includes a contribution from single-family homes plus the contribution from apartments and is

$$(80 \times 3000) + (65 \times 500) = 272{,}500 \text{ gpd}$$

3. *Estimate commercial and institutional wastewater flows.* Using flow rates from similar establishments in adjacent communities, it was found that an average per employee flow rate of 25 gpd was appropriate. Current flow from this sector is therefore

$$25 \text{ (gpcd)} \times 150 \text{ (people)} = 3750 \text{ gpd}$$

and future average flow is

$$25 \text{ (gpcd)} \times 900 \text{ (people)} = 22{,}500 \text{ gpd}$$

4. *Estimate the industrial wastewater contribution.* Current industrial flow is negligible. Projection of future flow requires a thorough analysis of the planned industrial park. It was determined in consultation with the developer, municipal officials, and the zoning board, that the ultimate future industrial park would have 800 employees and one significant water-intensive industry. The water-intensive industry will discharge process water to the sewer at the rates of 40,000 gpd (average) and 80,000 gpd (peak hour). The estimated average future sanitary wastewater flow from a projected 800 employees is estimated based on an allowance of 25 gpd per employee or

$$25 \text{ (gpcd)} \times 800 \text{ (people)} = 20{,}000 \text{ gpd}$$

5. Estimate an allowance for the peak quantity of future infiltration. Because the sewers are of new construction, the future allowance for infiltration will be 15,000 gpd per mile of sewer. Assume a sewer length of approximately 1 mile of sewer per 25 acres of service area. Estimated length of sewers is therefore

$$\frac{340}{25} \times 1 = 13.6 \text{ miles}$$

and the infiltration allowance is

$$13.6 \text{ miles} \times 15,000 \text{ (gpd/mile)} = 204,000 \text{ gpd}$$

6. Determine peaking factors. The peaking factor is a function of the total tributary population and is taken from Fig. 1. Where significant nonresidential sources are present, it is common to substitute a population equivalent for flows from commercial and industrial areas. The population equivalents are estimated by dividing the commercial and industrial employee wastewater contribution by the weighted per resident water-use rate. The weighted per resident water-use rate is

$$\frac{(3000 \times 80) + (500 \times 65)}{3500} = 77.8 \text{ say } 78.0 \text{ gpd}$$

The population equivalent for future commercial and industrial flow is therefore

$$\frac{22,500 + 20,000}{78} = 544.8 \text{ say } 545 \text{ people}$$

The total population equivalent is 3500 + 545 = 4045, and from Fig. 1 the peaking factor is 3.90.

7. Estimate peak design flows. The peak design flow is determined by addition of the peak sanitary flow, the peak industrial contribution, and the infiltration.

The peak sanitary flow is first determined by applying the peaking factor to the average flows:

Average future sanitary wastewater flow:

Residential	= 272,500 gpd
Commercial	= 22,500 gpd
Industrial	= 20,000 gpd
Total	= 315,000 gpd

Peak design flows:

Peak future sanitary flow	= 3.9 × 315,000 =	1,228,500 gpd
Peak industrial flow	=	80,000 gpd
Peak infiltration	=	204,000 gpd
Total peak wastewater flow	=	1,512,500 gpd

STORM-WATER RUNOFF FLOW RATES

13. Bases of Storm-Sewer Design. Storm and combined sewers are designed to collect and convey that portion of the precipitation which flows over the ground surface during and after the precipitation event and is termed runoff. The rate of storm-water runoff is extremely variable, and the value to be used in design is difficult to determine. Precipitation, which causes runoff, is itself variable. The amount of water lost by infiltration into the soil and the water left in surface depressions or on plant and other surfaces is also subject to extreme variations.

Under these circumstances, the design of storm or combined sewers involves a decision as to the quantity of storm-water runoff for which sewer capacity is to be provided. It is not economically feasible to construct sewers of sufficient size to convey runoff from extreme precipitation events which occur at infrequent intervals. Thus sewers will surcharge when the runoff rate exceeds the design capacity of the system. This surcharging of sewers may cause flooding of basements and streets, health hazards, and public nuisances.

Selection of the design runoff rate or the design quantity of storm water for which sewer capacity should be provided therefore involves selection of the level of protection to be provided against property damage, nuisances, and public inconvenience. Surcharging of combined sewers is more objectionable than surcharging of storm sewers because streets and basements are flooded by sewage rather than storm water. The level of protection may therefore be greater for combined sewers than for storm sewers and involves consideration of costs and benefits. Economic factors to be considered include the first cost of the system, the capitalized cost of potential damage to property, the hazards to health, and the public inconvenience.

The design runoff rate for which sewer system capacity will be provided is usually defined as the runoff rate from a precipitation event with a specific recurrence interval. Thus, for example, sewers may be designed to convey the runoff from a 2- or 5-year storm event of 1 or 2 h duration. The level of protection can therefore be defined as the period over which the capacity of the sewers will be adequate and no sewer surcharging will occur.

Many urban communities in the United States attempt to define a desired level of protection against sewer system surcharge and the problems associated with such surcharge. This level of protection is mainly driven by public opinion but also takes into consideration regulatory and other requirements. Typically, storm and combined sewers are designed to convey the runoff from the rain event that has a probability of recurrence of 3 to 10 years. Sanitary sewers are not designed to receive runoff flows and are required to convey the maximum design flow without surcharge.

Selection of the recurrence interval of the design storm is only the first step in defining the design runoff rate. As discussed above, the runoff rate is also a function of the intensity and duration of the design storm, the soil conditions, and ground characteristics. Rainfall-intensity-duration-frequency curves are usually developed by public agencies and made available for most areas. Soil and ground conditions can be established through field surveys.

Two basic approaches are currently used to compute runoff for the selected storm. In one method, the runoff rate is developed after subtracting losses due to infiltration, retention, and temporary storage from the rainfall rate. This method has been incorporated into public domain computer programs such as the Stormwater Management Model (SWMM).[6] In the second method, runoff is considered to be proportionate to the rainfall rate with the proportionality factor being related to soil and ground conditions. This second method is called the rational method.

Other methods used to develop peak storm-water flows are the hydrograph method and the unit hydrograph method, which are discussed in some detail in Sec. 1. As a consequence, only the rational method which is widely used by drainage engineers will be discussed in detail in this section. The abstraction method which is incorporated into public domain computer programs will also be presented.

14. Rational Method. The rate of storm-water runoff from a watershed is related to rainfall intensity by the formula

$$Q = CiA$$

where Q = storm-water runoff rate, cfs
C = a runoff coefficient representing the ratio of runoff to the rainfall rate, a function of the ground cover and soil characteristics
i = average rainfall intensity, in/h, for a period equal to the time of concentration t_c
A = tributary area, acres

Note that, for the equation to be dimensionally correct, the units for rainfall intensity should be cfs/acre. However, since 1.0 in/h of rainfall is equal to 1.008 cfs/acre, either unit may be used.

The rational method is based on several assumptions as follows:

1. The runoff to any point in the drainage area may increase as the rainfall continues until the entire area is contributing flow. The time required for the runoff to become established and flow from the most remote point to the point under consideration is called the time of concentration t_c.
2. The peak runoff rate at the design location is therefore a function of the average rainfall intensity or rate during the time of concentration.
3. The maximum rate of rainfall occurs during the time of concentration.
4. The frequency of the peak discharge at the point under consideration is equal to the frequency of the average rainfall intensity for the tributary area.

As a consequence, in using the rational method, the critical storm duration is considered equal to the time of concentration for the area and point under consideration. The greatest rainfall rate to be expected for this critical duration at the desired frequency is therefore the one that will produce the maximum runoff rate to be used for design purposes.

15. Area to Be Served. The area served or to be served by the sewer system is the only element in the rational method which can be determined precisely and must be measured as accurately as possible. The tributary area to the point under consideration will be established primarily by ground topography but will also be influenced by the arrangement of streets, the layout of the sewer system, and the location of street inlets.

Topographic maps prepared at scales of 1 in = 200 ft to 1 in = 500 ft by aerial topographic methods are most suitable for definition of tributary areas. The layout of the existing or proposed sewer system is superimposed on the topographic map and the points of inlet of runoff to the system identified. Boundaries of each drainage area or area tributary to specific inlets can then be identified and measured.

In order to compute the runoff rate to any point, the characteristics of the drainage area that are likely to effect runoff should also be identified. This information will include the percentage of the area that is impervious, the type of soil, the type and extent of grass cover, and the average ground slopes.

In estimating future rainfall runoff rates, it may also be necessary to estimate future development within the tributary areas. The present development patterns will serve as a guide in predicting future development but such factors as zoning ordinances and other city or area planning documents must be evaluated.

16. Rainfall Intensity. Rainfall intensities to be used in the rational formula are generally obtained from rainfall-intensity-duration-frequency curves. Typical curves as shown in Fig. 2 can be prepared from records of rainfall data at particular rainfall gaging stations.

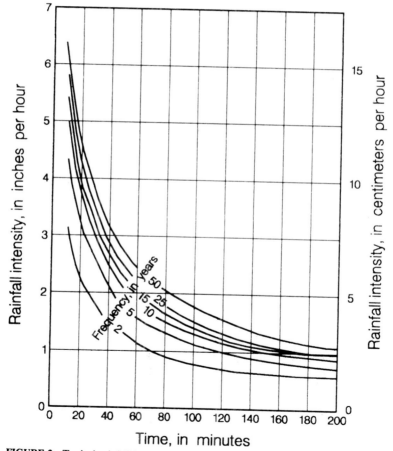

FIGURE 2 Typical rainfall-intensity duration curves.

Efforts have also been made to develop mathematical formulas in the form of

$$i = A(t + b)$$

where i is rainfall intensity in inches per hour, t is the storm duration in minutes and A and b are constants, to represent the time-intensity relation for rainfall with a particular frequency of occurrence.

Rainfall formulas have not found widespread use because they are generally applicable only to the area in which they are developed. However, they may be useful in communities where rainfall records are insufficient or have not been analyzed.

In order to establish the design rainfall intensity to be used in the rational method from rainfall intensity-duration curves or from equations, two other parameters must be known. These are the frequency and the duration of the design storm. The duration of the design storm is usually taken as the time of concentration for the area and point under consideration.

The design frequency is developed from considerations of the level of protection to be provided. As discussed earlier, the choice of level of protection is largely a matter of economics with typical ranges of values as follows:

1. For storm sewers in residential areas, 2 to 15 years, with a 5-year level of protection most commonly used. This means that sewers can convey the runoff from a storm with a 5-year return period without surcharge.
2. For storm sewers in commercial and high-value districts, a 10- to 50-year level of protection is chosen depending on economic factors.
3. For flood-protection works, a level of protection of 50 years or more is selected.

17. Time of Concentration. The time required for runoff to flow from the most remote location to the point under consideration consists of the overland flow time and the channel flow time. In urban areas, the overland flow time, referred to as the inlet time, is the time required for runoff to reach a defined channel such as a street gutter plus the gutter flow time to an inlet.

Channel flow time can be easily determined from the hydraulic properties of the conduit. However, inlet time is difficult to determine and will vary with such factors as the surface slope, surface roughness, flow distance, infiltration capacity, depression storage, and rainfall intensity. Because of these factors, values of design inlet flow times of 5 to 30 min are used in practice. In urban areas with closely spaced inlets, inlet times of 5 to 15 min are common. For similar areas with relatively flat slopes 10 to 15 min is used, and for very flat areas with widely spaced inlets 20 to 30 min has been used.

Attempts have also been made to determine the inlet flow time by breaking the flow path into its components such as grass flow, flow over concrete or asphalt, and so on and then computing the times of flow over each surface. The Illinois Department of Transportation[5] has developed a formula for estimating overland flow time as follows:

$$t = \frac{56 L^{0.6} n_0^{0.6}}{i^{0.4} S^{0.3}}$$

where t = overland flow time, s
 L = overland flow length, ft
 n_0 = roughness coefficient, similar to Manning's n factor

i = rainfall intensity, in/h
S = average slope of flow path, ft/ft

Solution of this equation for overland flow time requires a trial-and-error process using the following steps:

1. Assume a trial value of rainfall intensity i.
2. Find the overland travel time t from the equation.
3. Use t from step 2 to determine the actual rainfall intensity from rainfall curves.
4. Compare the trial and actual rainfall intensities; if they are not similar, select a new trial rainfall and repeat the process.

Values of n_0 to be used in the equation for overland flow time are shown on Table 4.

TABLE 4 Values of n_0 to Be Used in Computing Overland Flow Time from
$$t = \frac{56 L^{0.6} n_0^{0.6}}{i^{0.4} S^{0.3}}$$

Type of surface	n_0 Recommended value	Range of values
Concrete	0.011	0.01–0.013
Asphalt	0.012	0.01–0.015
Bare sand	0.010	0.010–0.016
Graveled surface	0.012	0.012–0.030
Bare clay-loam (eroded)	0.012	0.012–0.033
Fallow (no residue)	0.05	0.006–0.16
Chisel plow (<1/4–1 ton/acre residue)	0.07	0.006–0.16
Disk/harrow (<1/4 ton/acre residue)	0.08	0.008–0.41
No till (<1/4 ton/acre residue)	0.04	0.03–0.07
Plow (fall)	0.06	0.02–0.10
Range (natural)	0.13	0.01–0.32
Range (clipped)	0.08	0.02–0.24
Grass (bluegrass sod)	0.45	0.39–0.63
Short grass prairie	0.15	0.10–0.20
Dense grass	0.24	0.17–0.30
Bermuda grass	0.41	0.30–0.48
Woods	0.45	

*Adapted from Ref. 5.

18. Runoff Coefficient. The runoff coefficient C is the component of the rational formula that the engineer is least able to determine precisely, and it therefore requires the greatest judgment. The coefficient C as used in the rational formula suggests a fixed ratio of runoff to rainfall. However, in reality, the value is not fixed and may vary for a given area depending on such factors as the infiltration capacity of the soil, interception by vegetation, depression storage, and antecedent soil moisture conditions.

It should be noted, however, that for highly impervious areas, the coefficient approaches unity, and is less variable. This is true because for these impermeable surfaces, seasonal meteorological or antecedent conditions have less influence on the rate of runoff flow. Thus it is fair to say that the rational method is better suited for use in urban areas, where large percentages of the drainage areas are impervious.

In these urban areas, the common engineering practice is to use average values of the runoff coefficient for various surface types. Typical values of the runoff coefficients used by various cities for different surface types are shown in Table 5. These values are assumed to remain constant throughout the duration of the design storm.

19. Application of Rational Method. In order to apply the rational method for computing peak runoff rates the following procedure must be followed:

1. Select a level of protection which in turn defines the frequency of the design storm.
2. Identify the tributary area to each inlet point under consideration. Measure the area and determine its characteristics such as slope, percent impervious, and soil cover.
3. Estimate the time of concentration t_c using all available data.
4. Determine the design rainfall intensity from intensity-duration-frequency curves or empirical equations.
5. Establish the runoff coefficient C to be used for the specific area.
6. Compute the peak runoff rate by applying these values to the rational formula.

This procedure should be repeated for each point in the sewer system under consideration starting at the upper end of the system and moving downstream.

Based on experience, the current practice is to use the rational method for urban areas of less than 5 mi^2 even though the basic principles of the method are applicable to larger areas. This limitation on use of the rational method is considered prudent because of the unequal distribution of rainfall that can occur in larger areas, and in these cases, other runoff methods can be utilized.

20. Urban Runoff Models. Most urban runoff models such as the SWMM model[6] and the ILLUDAS model[7] attempt to determine the runoff from an area by deducting all losses from the total rainfall rate to arrive at an excess rainfall rate. The losses or abstractions that are generally recognized are:

1. Interception losses, which represent rainfall intercepted by vegetation. This water evaporates or reaches the ground after the storm.
2. Evaporation, which represents water lost due to evaporation from the ground surface.
3. Depression storage, which includes all rainfall that is stored in surface depressions that act as small reservoirs and release water (runoff) only after they are filled.
4. Infiltration, which includes all water that percolates into the receiving soil or rock.

Overland flow occurs only when the rainfall rate has satisfied the first three losses and exceeds the infiltration capacity of the soil.

Losses due to interception, evaporation, and depression tend to be minor compared with infiltration losses and are sometimes neglected. The rate at which a particular soil absorbs rainfall is a function of the rate of infiltration through the

TABLE 5 Runoff Coefficients for Various Cities

City	Type of area	Runoff coefficients
Baltimore County, Md.*	Roofs, pavements, and walls	0.95
	Pervious areas, varies with slope and soil type	
	Sparse vegetation	0.30–0.55‡
	Lawns	0.12–0.20‡
	Dense vegetation	0.08–0.12‡
Buffalo, N.Y.†	Residential	0.48–0.58
	Apartments	0.60–0.65
	Commercial	0.60–0.70
	Industrial	0.55–0.60
Cincinnati, Ohio	Suburban (large lots)	0.30
	Residential	0.35–0.40
	Apartments	0.50–0.65
	Retail business and downtown	0.70–0.90
Milwaukee County Metropolitan Dist., Wis.	Most dense community	0.75
	Adjoining densely built-up	0.65
	Residential, well built-up	0.53
	Adjoining built-up residential	0.45
	Suburban	0.30
Pittsburgh, Pa.	Varies with % impervious and slope	
	0–10% impervious	0.20–0.30
	50% impervious	0.50–0.57
	100% impervious	0.90
Rochester, N.Y.	Residential	0.25–0.40
	Commercial	0.50–0.90
	Industrial	0.60
St. Louis, Mo.	Varies with % impervious and storm duration	15 min 120 min
	0% impervious	0.30 0.60
	50% impervious	0.50 0.78
	100% impervious	0.70 0.95
San Francisco, Calif.	Industrial	0.60–0.90
	Commercial	0.70–0.80
	Apartments and flats	0.60–0.75
	Residential, attached houses	0.45–0.60
	Residential, detached houses	0.40–0.50
	Suburban	0.25–0.35
	Parks	0.10–0.20

*"Baltimore County Design Manual," Department of Public Works Baltimore County, Md., 1955.
†Stanley, W. E., and W. J. Kaufman, "Sewer System Capacity Design Practice," Boston Society of Civil Engineers, October 1953.
‡For average slopes, 2.7 to 7 percent, and sand to clay, extremes range from 0.03 to 0.70.

soil surface into the soil and the rate of percolation or movement within the soil. However, the combined effect is referred to as the infiltration rate.

Infiltration capacity is a function of many factors including soil type, moisture content, vegetative cover, and season of the year. Porous soils tend to have greater storage capacities and provide less resistance to flow within the soil. Thus infiltration tends to increase with porosity. The initial infiltration at the start of the storm also varies with the soil moisture content and is reduced in wet soils. Vegetal cover tends to increase infiltration when compared with barren soil because it retards surface flow, giving the water additional time to enter the soil. Vegetal cover also shields the soil from raindrop impact, which causes packing of the surface soil, and the vegetation root systems make the soil more pervious.

Most urban runoff modes like SWMM and ILLUDAS allow the user to input data on soil characteristics, vegetal cover, and antecedent moisture content which are applied to the area in determining runoff rates. In general these models utilize systems for classifying soils relative to their infiltration capacity similar to the U.S. Soil and Conservation Service system which is described in detail in Ref. 12.

HYDRAULICS OF SEWERS

21. General. Sanitary, storm, and combined sewers must convey their maximum design flows and transport suspended solids in an efficient manner to prevent the deposition of solids in the sewers and minimize odor and other nuisances. Thus design or evaluation of a sewer system must take into consideration the anticipated peak flows, the expected fluctuations in flow, and the nature of the suspended matter.

Except for the considerations of solids transport, the hydraulic properties of sewage or wastewater are considered identical to the properties of water in sewer design and evaluation. Hydraulically, the flow of wastewater in sewers may occur under open-channel or pressure-conduit (closed-conduit) conditions. In open-channel flow, which can occur in an open channel or a partially full closed channel, the wastewater surface is exposed to the atmosphere. In addition, the hydraulic grade line lies in or on the surface of the flowing sewage. In pressure-conduit flow, the liquid fills the conduit and the hydraulic grade line does not lie on the surface of the liquid, as shown in Fig. 3.

Sanitary sewers are usually designed to flow by gravity as open channels. This allows for ventilation of the sewers and provides a factor of safety in system capacity. Storm sewers, some deep sewers, and pump station discharge lines are designed for pressure-conduit flow. In sewer systems designed for pressure-conduit flow, the pipes may surcharge at peak design flow and wastewater will rise in manholes. This type of design permits sewers to convey greater flows than when flowing partially or just full. However, this method (closed-conduit flow) should not be used for sanitary sewers which have house service connections, since wastewater may back up into homes.

The general laws of hydraulics which apply to both open-channel and closed-conduit flow are discussed in detail in Sec. 2. The following paragraphs will be used to discuss in more detail those laws which find direct application to the design and evaluation of sewer systems.

22. Basic Hydraulic Considerations. When water flows through a pipe or channel at a constant rate, the flow is termed steady uniform flow. Steady flow occurs when the same volume of liquid flows past a given point per each unit of time. Uniform flow occurs when there are no changes in velocity or depth of flow along the course of the conduit or channel.

In sewers, flow is generally unsteady. However, in the usual sewer design or evaluation problem, the assumption of steady flow yields practical design results.

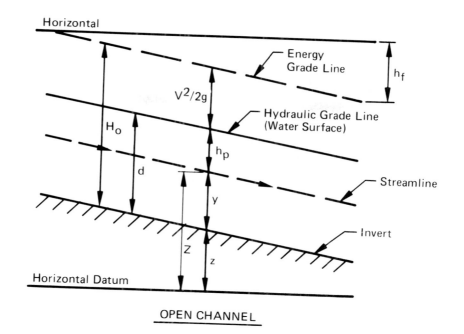
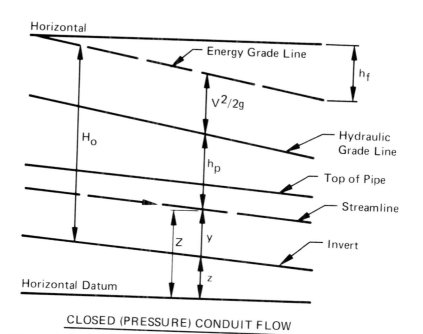

FIGURE 3 Comparison of uniform flow in an open channel and a pressure conduit.

Therefore, steady flow may be assumed and used to simplify the hydraulic analysis of most sewer systems. Uniform flow may be expected in straight sections of sewers but velocity changes may be expected at obstacles or changes in cross section and may have to be considered in some cases. Also, in some instances, the impacts of unsteady, nonuniform flow may be important, as in the design of force mains or large storm sewers where water hammer or surge waves may develop. The analysis of these conditions is beyond the scope of this section, and the reader should check other references for a discussion of these items.[8,9]

When water moves downstream through a pipe or channel, the available energy will be used up in overcoming frictional resistance to flow and in attaining kinetic energy or velocity head. The frictional resistance is related to the roughness of the pipe or channel, the area of the contact surface, the velocity of flow, and the density of the liquid. The general friction formulas discussed below have been developed to show the rate of energy dissipation for various conditions.

In addition, a quick review of other hydraulic principles which impact the solution of the general friction formulas and which are useful to the sanitary engineer involved in sewer design and evaluation is provided.

Continuity Principle. For steady flow, the continuity equation is expressed as

$$Q = A_1 V_1 = A_2 V_2$$

where Q is the discharge rate, A is the cross-sectional area, V is the mean velocity, and the subscripts designate different conduit or channel sections. Appropriate units of Q, A, and V can be used.

Energy Principle. The specific energy of a fluid is defined as the energy per unit weight at any point along a streamline in the conduit or channel, relative to the bottom of the channel, and is given by the equation

$$H_0 = y + h_p + v^2/2g$$

This is called the specific-energy equation, where H_0 is the specific energy, y is the height of the streamline above the bottom of the channel or conduit in feet, h_p is the pressure head, feet, and $v^2/2g$ is the kinetic energy per unit weight or the velocity head, feet, of the streamline as shown in Fig. 3. Note that v^2 is in units of ft^2/s^2 and $2g$ is equal to 64.4 ft/s^2.

The specific-energy equation includes the three basic forms of energy to be considered in connection with flow of liquids: the elevation or potential energy y, which is due to the elevation of the liquid above a certain datum; the pressure energy h_p, which is due to the pressure of the liquid; and the kinetic energy or velocity head $V^2/2g$, which is derived from the velocity of the liquid.

If an arbitrary datum plane is chosen instead of the bottom of the channel or conduit, the energy equation becomes

$$E = Z + h_p + v^2/2g$$

where Z is the height above (or below) the datum plane and is equal to $(z + y)$ as shown in Fig. 3.

It can also be shown that for liquids, $h_p = p/w$ where p is the pressure in psf and w is the weight per unit volume of the liquid in pcf. Substituting for these terms, the equation for total energy then becomes

$$E = Z + p/w + v^2/2g$$

This expression is referred to as the Bernoulli constant. In the equation, E represents the total energy (or head) at the point under consideration. The energy grade line (EGL) indicates the variation in total energy or head over the reach of the conduit under consideration as shown in Fig. 3.

The sum of the first two terms in the equation $(Z + p/w)$ is called the piezometric head and represents the height to which the liquid will rise if it is free to do so. The hydraulic grade line (HGL) shows how the piezometric head varies over the reach under consideration.

The *law of conservation of energy* is expressed by the Bernoulli equation, which can be written as

$$Z_1 + \frac{P_1}{w} + \frac{V_1^2}{2g} = Z_2 + \frac{P_2}{w} + \frac{V_2^2}{2g} + H_L$$

This equation states that for steady flow, the total energy at any point (point 1) is equal to the total energy at a downstream point (point 2) plus the loss of energy between the two points, H_L. The energy terms in the Bernoulli equation are usually expressed in terms of feet of water for convenience. The energy loss H_L can then be described as the loss of head in feet between these two points.

This loss of head is caused by pipe friction and by changes in direction or velocity of flow. Pipe friction is termed the major loss and is a continuous loss that is assumed to occur at a uniform rate along the length of the conduit or channel, provided that the size and other characteristics of the conduit do not change. The other losses are termed minor losses and will occur at sudden contractions or enlargements in the cross section of flow, at obstructions to flow, and at bends or curves in the conduit.

23. Flow Friction Formulas. Many formulas have been developed to express energy loss in conduits algebraically. However, in present design practice, some formulas have been more frequently used for sewer design and evaluation, and these will be presented and discussed. For open-channel flow, the Manning formula is generally preferred. For closed-conduit flow, the Manning and Hazen-Williams formulas tend to be used. The Chézy formula is also presented, as it was one of the first friction formulas developed and serves as a basis for the development of the other formulas.

The *Chézy formula* relates the velocity of flow to the hydraulic radius of the conduit and slope of the energy grade line and is expressed as

$$V = C\sqrt{(RS)}$$

where V = velocity of flow, fps
R = hydraulic radius, ft, equal to the area of flow divided by the wetted perimeter
S = slope of energy grade line, dimensionless
C = Chézy coefficient of flow

The *Manning equation*, one of the best-known friction formulas, can be used for both pipes and open channels (conduits) of all shapes flowing full or partly full. The Manning equation is given by

$$V = \frac{1.486}{n} R^{2/3} S^{1/2}$$

where V = velocity, fps
S = slope of the energy grade line (dimensionless)
R = hydraulic radius, ft
n = Manning's roughness coefficient

Typical values of n are shown in Table 6.[2]

Graphs or charts such as the alignment chart shown in Fig. 4 are also available for solution of the Manning equation for circular pipes flowing full.

The *Hazen-Williams formula* was developed for designing water-supply systems utilizing pressure conduit or pipe flow. However, it is also used for open-channel conditions. The Hazen-Williams equation is given on page 28.24.

TABLE 6 Values of Friction Formula Coefficients*

Conduit material	Manning's n	Hazen-Williams C
Closed conduits:		
Asbestos-cement pipe	0.011–0.015	100–140
Brick	0.013–0.017	
Cast-iron pipe		
New, unlined		120–140
Old, unlined		40–100
Cement-lined and seal-coated	0.011–0.015	100–140
Concrete (monolithic)		
Smooth forms	0.012–0.014	120–140
Rough forms	0.015–0.017	50–120
Concrete pipe	0.011–0.015	100–140
Corrugated-metal pipe (½- × 2⅔-in corrugations)		
Plain	0.022–0.026	
Paved invert	0.018–0.022	
Spun asphalt lined	0.011–0.015	100–140
Plastic pipe (smooth)	0.011–0.015	100–140
Vitrified clay		
Pipes	0.011–0.015	100–140
Liner plates	0.013–0.017	
Open channels:		
Lined channels		
Asphalt	0.013–0.017	
Brick	0.012–0.018	
Concrete	0.011–0.020	
Rubber or riprap	0.020–0.035	
Vegetal	0.030–0.40	
Excavated or dredged		
Earth, straight and uniform	0.020–0.030	
Earth, winding, fairly uniform	0.025–0.040	
Rock	0.030–0.045	
Unmaintained	0.050–0.14	
Natural channels (minor streams, top width at flood stage <100 ft)		
Fairly regular section	0.03–0.07	
Irregular section with pools	0.04–0.10	

*Adapted from Ref. 2.

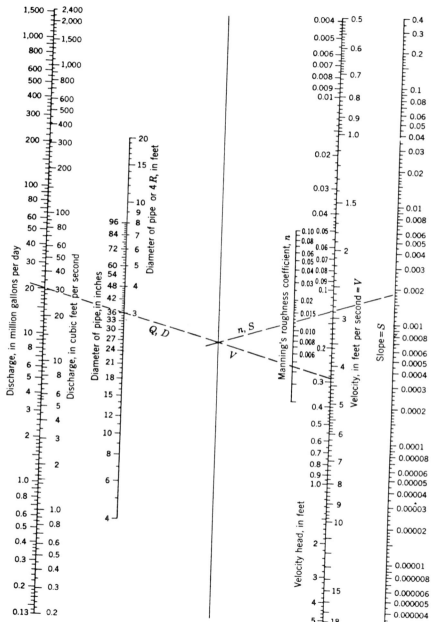

FIGURE 4 Alignment chart for solution of Manning's formula for full pipe flow. (*Adapted from Ref. 2.*)

$$V = 1.318\, CR^{0.63} S^{0.54}$$

where V = velocity of flow, fps
R = hydraulic radius, ft
S = slope of the energy grade line (dimensionless)
C = coefficient related to pipe or channel roughness

Typical values of C are given in Table 6.

Figure 5 is an alignment chart for solution of the Hazen-Williams formula for pipes flowing full.

24. Minor Losses. Minor energy losses occur at all transitions in a sewer system. These transitions connect conduits of different characteristics including differences in area, shape, grade, alignment, and conduit material. The energy losses in a transition are usually expressed as a function of the kinetic energy or velocity head ($V^2/2g$). Thus

$$h_m = K_m V^2 / 2g$$

where h_m = minor loss in energy
K_m = minor loss coefficient which varies with the nature and type of transition

Transitions may be streamlined carefully and be gradual or sudden.

25. Losses at Manholes. In gravity sewer systems, most transitions occur at manholes and such transitions could include any combination of changes in sewer characteristics. Losses through a manhole can be computed as the sum of the inlet loss caused by a sudden expansion and the contraction loss at the exit.

For a pipe flowing just full, through a manhole, the expansion loss H_e in feet is given by

$$H_e = k_e \frac{(V_1 - V_2)^2}{2g} = k_e \frac{V_1^2}{2g}\left[1 - \left(\frac{A_1}{A_2}\right)\right]^2$$

where K_e is a coefficient of expansion that varies from 0.2 for a well-designed transition to 1.0 for a sudden expansion.[2]

Similarly the contraction loss H_c in feet is given by

$$H_c = k_c \frac{V_3^2}{2g}\left[1 - \left(\frac{A_3}{A_2}\right)^2\right]^2$$

where k_c is a contraction coefficient that varies from 0.1 for a well-defined transition to 0.5 for a sudden contraction.[2] V is the velocity in fps and A is the area of flow in ft^2. The subscripts 1, 2, and 3 refer to the upstream pipe, the manhole, and the downstream pipe, respectively.

Consider a manhole in which the invert is fully developed with a semicircular bottom and with vertical sides from one-half depth up to the top of the pipe. In this case $A_1/A_2 \approx 0.88$. Also, if there is no change in pipe diameter through the manhole, $A_3/A_2 \approx 0.88$.

If the coefficient of expansion k_e is taken as 1.0, the expansion loss is

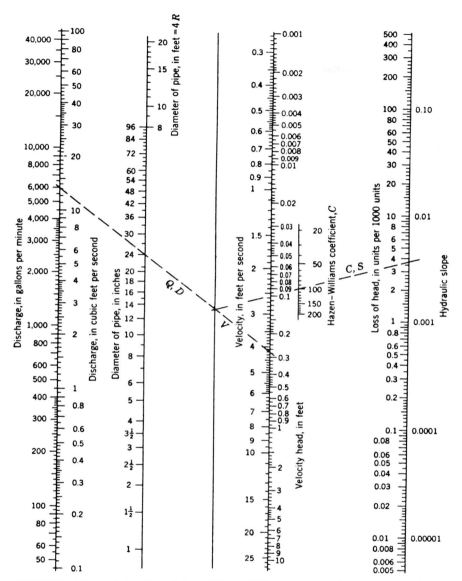

FIGURE 5 Alignment chart for solution of Hazen-Williams formula for full pipe flow. (*Adapted from Ref. 2.*)

$$H_e = 1.0 \frac{V_1^2}{2g}(1 - 0.88)^2 = 0.014 \frac{V_1^2}{2g}$$

If the downstream pipe has a bell, the coefficient of contraction k_c can be taken as 0.2 and the contraction loss is

$$H_c = 0.2 \frac{V_3^2}{2g}(1 - 0.88^2)^2 = 0.010 \frac{V_3^2}{2g}$$

Thus the total loss through the manhole H_M in feet is

$$H_M = 0.014 \frac{V_1^2}{2g} + 0.010 \frac{V_3^2}{2g}$$

This head loss is small, e.g., if $V_1 = V_3 = 6$ fps, $H_m = 0.024(36/64.4) = 0.013$ ft.

If the manhole invert is not fully developed and the manhole is large compared with the pipe, A_1/A_2 is small and the term $(1 - A_1/A_2)$ approaches unity. In this case the head loss through the manhole H_m in feet is the given by

$$H_m = 1.0 \frac{V_1^2}{2g}(1) + 0.5 \frac{V_3^2}{2g}(1)$$

where the coefficients k_e and k_c have been taken as 1.0 and 0.5 for a sudden expansion and contraction, respectively.

In this case, for $V_1 = V_3 = 6$ fps

$$H_m = 1.5 \left(\frac{36}{64.4}\right) = 0.84 \text{ ft}$$

The head loss through a manhole will therefore vary significantly and will be governed by the manner in which the manhole invert is developed. To minimize losses, the channel through the manhole should be a smooth continuation of the pipe as far as possible, with good hydraulic properties.

26. Losses at Bends. Minor losses in sewer systems can also occur at bends. The head loss at the bend in excess of that caused by an equivalent length of straight pipe is given by

$$h_b = \frac{k_b V^2}{2g}$$

and k_b has the following values for a 90° bend:

Radius to centerline of pipe	K_b
D (diameter of pipe)	0.50
2D to 8D	0.25

For bends less than 90°, the coefficient k_b may be reduced by the ratio $\theta/90$, where θ is the bend angle.

27. Hydraulic-Elements Graphs. Sewers which are designed to flow full will generally function as open channels during low-flow conditions. A hydraulic-elements graph can be used for the solution of open-channel flow problems in a

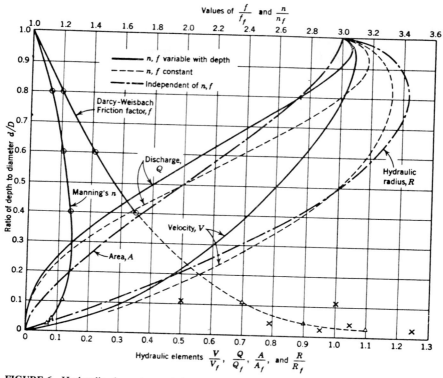

FIGURE 6 Hydraulic-elements graph for circular sewers. (*Adapted from Ref. 2.*)

closed conduit. A hydraulic-elements graph for circular sewers is shown in Fig. 6. As will be noted, the curves on this graph are plotted for both constant and variable n factors.

28. Self-Cleansing Velocities. An important consideration in the design of sewers is to maintain a minimum velocity of flow throughout the design period. Based on considerations of particle transport and experience, a minimum sewer velocity of 2 fps is normally recommended for sanitary sewers. For storm and combined sewers, where the particles being transported are mainly heavier sand and grit, a minimum velocity of 2.5 to 3.0 fps may be required.

The slope of a gravity sewer is generally chosen to provide the minimum required velocity when the sewer is flowing full. Also, regulations of many state agencies which govern the design of sewer systems, include minimum allowable slopes for various pipe sizes as shown below:

Sewer size	Minimum slope, ft/100 ft*
8 in (20 cm)	0.40
9 in (23 cm)	0.33
10 in (25 cm)	0.28

Sewer size	Minimum slope, ft/100 ft*
12 in (30 cm)	0.22
14 in (36 cm)	0.17
15 in (38 cm)	0.15
16 in (41 cm)	0.14
18 in (46 cm)	0.12
21 in (53 cm)	0.10
24 in (61 cm)	0.08
27 in (69 cm)	0.067
30 in (76 cm)	0.058
36 in (91 cm)	0.046

*Minimum slope required to give a minimum velocity of 2 fps for pipes with Manning's n = 0.013.

In very flat areas, where the minimum grade and velocity cannot be realized without excessive excavation and cost, it may be necessary to use flatter sewer grades. In this case, however, provisions should be made for flushing of the sewer to prevent stoppages or to remove deposits should they occur.

29. Maximum Velocities. Maximum velocities in sewers also need consideration because of the abrasive nature of the particles being transported and the potential damage to sewers due to erosion and abrasion of the sewer invert. To minimize these problems, a maximum velocity of between 8 and 10 fps should be considered in the design of sewers. However, larger velocities may be used if special pipe materials are considered or the pipe is lined.

30. Storm-Water Inlets. Inlets are drainage structures utilized to collect surface water (storm-water runoff) and convey it to storm or combined sewers. Inlets are generally either gutter or curb openings. Gutter inlets consist of an opening in the street gutter covered by a grate. Curb opening inlets consist of a vertical opening in the curb covered by a top slab. Sometimes, combination inlets which consist of both a curb opening and grate inlet acting as a unit are also utilized as shown in Fig. 7.

Inlets must be safe for travel by wheelchairs and cyclists and also be adequate to intercept the design flows with minimum head loss. Inlets may be located on a continuous grade or in a sag or depression in the street that is subject to flooding. For inlets on a continuous grade, the capacity varies with its geometry, the pavement cross slope and longitudinal slope, the total gutter flow, and depth of flow. Gutter inlets in a sag operate as weirs up to a certain depth of flow and as orifices at greater depths.

For gutter inlets on continuous grade, manufacturers' rating curves should be consulted to obtain inlet capacities. For gutter inlets in a sag, the capacity when acting as a weir can be computed from the typical equation[5] of flow over a weir which is given by

$$Q = CPd^{1.5}$$

where Q = flow, cfs
C = constant that varies from 3.0 to 3.3
P = perimeter of grate, ft, disregarding bars and the side against the curb
d = depth of water above grate, ft

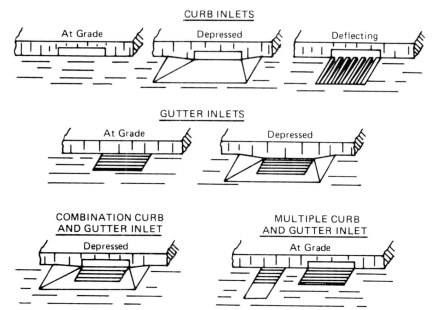

FIGURE 7 General arrangement of storm-water inlets.

When orifice-type flow occurs, the capacity in cfs can be estimated using the orifice equation,[5] which is given by

$$Q = CA(2gd)^{0.5}$$

where C = orifice coefficient and varies from 0.60 to 0.67
A = clear or open area of the grate, ft^2
d = depth of water over the grate, ft
g = 32.2 ft/s^2

In order to determine whether orifice or weir flow will be obtained, it is necessary to compute and compare the flows obtained using both equations, and to utilize the lower flow value at a particular depth of water.

SEWER-SYSTEM DESIGN

31. General Approach. Sewers are generally designed to collect wastewater and convey it to the point of disposal throughout their design life of more than 25 years. To function effectively, the sewers must be located deep enough to intercept tributary flows; must convey intercepted flows at minimum velocities to ensure self-cleansing of sewers while minimizing energy losses; must be resistant to corrosion and erosion of sewers; and must be strong enough to withstand external loads due to backfill, impact, and live loads. Other factors to be considered during the design of the system are operation and maintenance considerations and public safety.

To satisfy these requirements and to ensure long-term effectiveness of the sewer system, the following steps should be followed during design:

1. Preliminary investigations to determine the extent of the area to be served and to assist in selection of the type of sewer system to be utilized of the three general types:
   - Sanitary sewer system
   - Storm sewer system
   - Combined sewer system

2. A detailed survey of the service area including underground soil and utility surveys and surface surveys to identify topographic features and ground elevations.

3. The sewer design to include computation of flows and determination of pipe sizes, slopes, and elevations. This step will also include location of sewers, manholes, inlets, catch basins, and other appurtenant structures.

4. Preparation of plan and profile drawings to indicate the location and elevation of the proposed sewers and technical specifications.

5. The final design changes will be made during construction, and drawings must be corrected to reflect such changes.

32. Preliminary Investigations. The preliminary investigation is required to define the problems related to wastewater disposal in the area including the extent and adequacy of existing sewers if any. These investigations should provide initial information on population distributions and densities; location of industry and commercial centers and other data relevant to an estimation of wastewater sources and flow rates; location of potential treatment and disposal points; and general topographic data.

Data collected during the preliminary investigations will be used to help define existing and anticipated flow rates and to develop a preliminary alignment and cost of the proposed sewer system.

33. Detailed Survey. Detailed surveys are then necessary to confirm tributary areas and populations; to define the underground soil conditions which will impact the design of the proposed system; to confirm the location of existing utilities; and to provide detailed topographic and elevation information along the lines of proposed sewers.

Underground soil conditions will be obtained from boreholes suitably located along the proposed sewer alignment. Boreholes should extend at least one pipe diameter below the invert of the proposed sewer, and visual soil classifications should be confirmed by suitable laboratory tests on soil samples.

The detailed survey should provide adequate information to establish the street topography including spot elevations adequate for developing 1-ft contours; the types and locations of streets, alleys, and other paved areas; the presence and location of existing sewers, water mains, telephone and electricity lines, and any other utilities; the water levels in the stream or other water body that will receive the wastewater discharges; and the location and sizes of all buildings within 100 ft of the proposed sewers. A system of permanent bench marks should also be established.

Using these data, a map of the proposed system can be developed and the preliminary layout of the system checked. Note that several preliminary arrangements may be required to determine the most suitable and economical layout. After this layout is completed, manholes, inlets, and other appurtenances can be located. Manholes should be placed at changes in direction, grade, and size of sewers and at reasonable intervals of 300 to 600 ft along straight runs.

34. Sewer Design. After the layout of the sewer system is confirmed, the detailed design of the system can be carried out. The first step is to delineate the subareas tributary to each line using the topographic information obtained during the detailed surveys. The wastewater flow rates from each subarea can then be computed using methods described in previous paragraphs. For sanitary sewers, the design flows are computed by estimating the tributary population, the per capita wastewater flow rate, and the industrial and commercial wastewater contributions, making allowances for infiltration. Appropriate peaking factors are also applied. For storm sewers, the design flow can be determined by application of the rational method to the tributary area for the design storm selected. Finally, using the computed flows, the sewer sizes, inverts, and slopes can be determined.

In proceeding with the detailed design of the proposed sewer system, the engineer must also consider local, county, and/or state minimum requirements which may be applicable. In the United States typical design criteria as outlined below are published by such agencies as the Great Lakes Upper Mississippi River Board of State Sanitary Engineers.[4]

- *Minimum size for gravity sewers:* 8 in diameter.
- *Minimum depth:* Sufficient to receive sewage from basements and to prevent freezing.
- *Slope:* As required to produce velocities no less than 2.0 fps when flowing full at $n = 0.013$.
  Uniform slope between manholes.
  At velocities greater than 15 fps, sewers must be protected.
- *Alignment:* Sewers 24 in or less should be straight between manholes.
- *Changes in pipe size:* When a smaller sewer joins a larger pipe, the invert of the larger sewer should be lowered to maintain the energy gradient. This criterion is approximately met by matching the 0.8 depth point of both sewers, or by matching crowns in smaller pipes.
- *Manholes:* To be placed at the end of each line; at all changes in grade, size, or alignment; at all intersections of sewers and at intervals of not more than approximately 400 ft in smaller sewers and 600 ft in larger sewers.
- *Relation to water mains:* Sewers should be at least 10 ft horizontally from water mains, and where they cross, an 18-in minimum vertical separation should be maintained.
- *Design pressure:* Sewers convey wastewater under atmospheric pressure. In some instances, where the wastewater must be pumped, the pipe must be designed for maximum expected internal pressure.
- *Design temperature:* Wastewater and storm water are generally at ambient temperatures. Temperature of the storm water may vary significantly depending on the season, such as during winter and summer; however, no special consideration of temperature is warranted for the design of underground sewer pipes.

35. Organization of Computations. Sewer design computations are repetitious, and the same process needs to be carried out for each sewer line in the system. These computations can therefore be simplified by using tabular forms as shown in Tables 7 and 8 in Examples 2 and 3 below.

Because of the repetitious nature of sewer design problems, they can also be easily carried out using personal computers, and many specialized computer software programs are now available for this purpose. Some of these programs are still proprietary; others, such as the SWMM program[6] and the ILLUDAS program,[7] are available for use by the public.

The SWMM model simulates theoretical or real storm events on the basis of rainfall inputs and system characterization (area, imperviousness, slope, storage, and infiltration) to predict the flow rates over time or the inflow hydrograph at each node in the system, usually a manhole. The model then takes the hydrograph input and performs dynamic routing of the storm-water flows through the sewer system to the point of discharge. The system can model different types of channels and varying outfall conditions. The model determines the capacity of each sewer flowing full and the peak flows associated with the particular storm. Where peak flows exceed pipe capacities, the model also computes how high the water will rise in manholes and the volume of water flowing over the top of manholes that are fully surcharged. The ILLUDAS model performs similar tasks and can be used in both the evaluation and design modes.

As in any computer model, the importance of calibration of the model cannot be overstated. Calibration is the process of adjusting the model assumptions such that the computer-generated results match the "in-the-field" performance of the system.

36. Examples. **Example 1** *A 1000-ft-long, 36-in-diameter sewer is laid on a slope of 1.0 ft/1000 ft. Determine the following: (1) The full pipe capacity; (2) the water level in the upstream manhole when 50 cfs is conveyed through the pipe, and (3) the depth of flow in the sewer when 15 cfs is conveyed, assuming a free discharge at the downstream end of the sewer and n = 0.013.*

Solution. 1. Solve Manning's equation for full-flow conditions.

$$Q = \frac{1.486}{n} AR^{2/3} S^{1/2}$$

The terms in the equations are developed as follows:

$$n = 0.013; \text{ area of pipe, } A = \pi \times \left(\frac{36}{12}\right)^2 / 4 = 7.06 \text{ ft}^2$$

$$\text{Hydraulic radius } R = \frac{\pi D^2}{4} \times \frac{1}{\pi D} = \frac{D}{4} = \frac{36}{12} \times \frac{1}{4} = 0.75 \text{ ft}$$

$$\text{Slope of sewer } S = 1/1000 = 0.001$$

$$Q_{full} = \frac{1.486}{0.013} \times 7.06 \times 0.75^{2/3} \times 0.001^{1/2}$$

$$Q_{full} = 21.06 \text{ cfs}$$

2. Consider that the water level at the downstream end of the pipe is equal to the crown of the sewer and that flow through the pipe will be pressurized since

the flow (50 cfs) exceeds the full pipe capacity (21.06 cfs). In this case, the slope S in the Manning equation refers to the slope of the energy grade line and not the slope of the sewer. As before

$$Q = \frac{1.49}{n} AR^{2/3}S^{1/2}$$

Also

$$A = 7.06 \text{ ft}^2 \quad R = 0.75 \text{ ft} \quad \text{and} \quad n = 0.013$$

and

$$40 = \frac{1.49}{0.013}(7.06)0.75^{2/3}S^{1/2} \quad \text{which gives } S = 0.0056$$

For $S = 0.0056$, the head loss in 1000 ft of pipe is 5.6 ft (0.0056 × 1000). The change in pipe elevation is 1.0 ft in 1000 ft.
Therefore, depth of surcharge in upstream manhole is 5.6 − 1.0 = 4.6 ft.

3. Since the proposed flow (15 cfs) is less than the full pipe capacity, determine the depth of flow from a hydraulic-elements chart for circular sewers as shown in Fig. 6.

$$\text{Design flow } q = 15.00 \text{ cfs}$$
$$\text{Full pipe flow } Q = 21.06 \text{ cfs}$$
$$\therefore \frac{q}{Q} = \frac{15.00}{21.06} = 0.71$$

From Fig. 6, for $q/Q = 0.71$, the ratio of depth to diameter $d/D = 0.7$ for a variable n and 0.62 for constant n. Therefore, depth of flow = 0.7 × 36 = 25.2 in for a variable n and 0.62 × 36 = 22.3 in for a constant n.

Example 2 *Design of Sanitary Sewer System* Design a sanitary sewer system for the residential district shown in Fig. E-1. The district is two-thirds developed. It is estimated that the future average saturation population density will be 65 persons per acre. The maximum hourly rate of flow of sewage is estimated at 250 gpcd. The maximum rate of groundwater infiltration to the sewers, to be provided for, is 2000 gpd/acre.

The minimum size of sewer is to be 8 in. The minimum velocity of flow in the sewers when full is to be 2.0 fps. The capacity of the sewers will be determined using Manning's equation with a recommended n value of 0.013.

Since the homes in this area have basements, the minimum depth below the street surface to the top of the sewers will be 7.0 ft. (In areas where basements are not normally constructed, the depth of cover to the top of the sewer may be as little as 3.0 ft.) (Adapted from Ref. 13.)

Solution

1. Draw a line to represent each proposed sewer and the direction in which the wastewater is to flow. Except in special cases, the sewer should slope with the surface of the street.

   The lines representing the sewer system will often resemble a tree and its

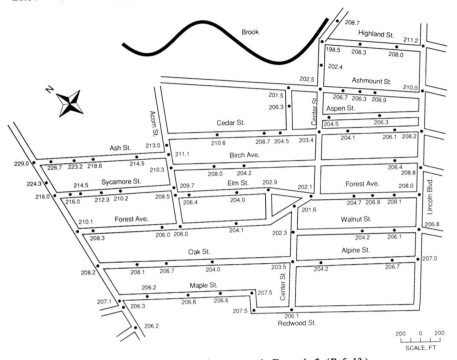

FIGURE E-1 Map used for design of sanitary sewer in Example 2. (*Ref. 13.*)

branches. In general, the laterals connect with the submains; and these, in turn, connect with the main or trunk sewer, which leads to the point of discharge.

2. Locate the manholes at all transitions and at minimum distances on straight sewers, giving each manhole an identification number.
3. Sketch the limits of the service areas for each sewer. The sewers should be extended far enough upstream such that the minimum size of sewer with the minimum slope will be adequate to serve the upstream area.

    Where the streets are laid out, the limits of the service areas may be assumed to occur midway between them. If the street layout is not shown on the plan, the limits of the different service areas cannot be determined easily and the area topography must be carefully evaluated.
4. Determine the size of the service areas using a planimeter or other means for measuring which will give similar results. At this point, the design may be represented as shown in the plan view in Fig. E-2.
5. Prepare a tabulation, such as that shown in Table 7, with columns for the different steps in the computation and a line for each section of sewer between manholes. This tabulation is a concise, timesaving method and shows both the data and the results in orderly sequence for subsequent use.

    Use col. 1 for line numbers starting with the manhole that is farthest from the point of discharge and moving downstream in order. On lines 2, 3, 4, and

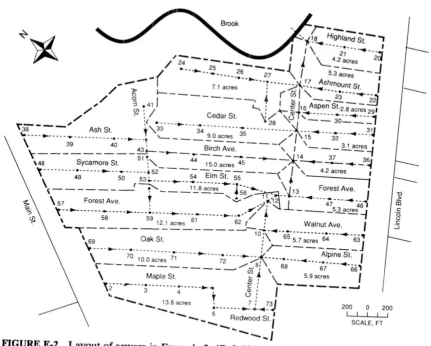

FIGURE E-2 Layout of sewers in Example 2. (*Ref. 13.*)

5 enter the upstream and downstream manhole numbers and the location and length of the sewer under consideration. The incremental area to be served by the sewer is entered in col. 6, and the total area in column 7. The total area is the sum of the area served by upstream sewers tributary to the one under consideration and the incremental area.

Enter in col. 8 the maximum rate of flow of sanitary sewage in the sewer. This rate is equal to the maximum per capita rate of sewage flow multiplied by the assumed future population density multiplied by the area shown in col. 7.

Enter in col. 9 the rate of allowance for groundwater infiltration, which is equal to the rate per acre to be provided for, multiplied by the area in col. 7. The total flow, which is the sum of the maximum sewage flow rate and the infiltration, is entered in col. 10 in mgd and in col. 11 in cfs, which is the more convenient way of expressing sewage flow rates.

The proposed sewer characteristics are shown in cols. 12 through 17, which are self-explanatory.

Example 3 *Design of Storm Sewer System* Design a storm-sewer system to provide a 5-year level of protection for the area shown in Fig. E-3. This sewer will discharge to an existing sewer system, and the invert elevation at the point of connection is known.

TABLE 7 Computations for a Sanitary Sewer

Line (1)	From manhole No. (2)	To manhole No. (3)	Location (4)	Length, ft. (5)	Area, acres Increment (6)	Area, acres Total (7)	Sewage, mgd* (8)	Ground water at 2,000 gpad (9)	Total maximum flow, sewage and ground water mgd (10)	Total maximum flow, sewage and ground water cfs (11)	Size of sewer, in. (12)	Slope, ft/ft (13)	Velocity, fps (14)	Capacity, cfs (15)	Surface elevation, upper end (16)	Invert elevation Upper end (17)	Invert elevation Lower end (18)
1	57	58	Forest Ave.	380	—	—	—	—	—	—	8	0.004	2.2	0.77	208.2	200.40	198.89
2	58	59	Forest Ave.	370	—	—	—	—	—	—	8	0.004	2.2	0.77	206.4	198.89	197.40
3	59	61	Forest Ave.	365	—	—	—	—	—	—	8	0.004	2.2	0.77	205.2	197.40	195.94
4	61	62	Forest Ave.	370	—	—	—	—	—	—	8	0.004	2.2	0.77	204.3	195.94	194.46
5	62	11	Forest Ave.	240	—	12.1	0.196	0.024	0.220	0.34	8	0.004	2.2	0.77	202.0	194.46	193.50
6	11	12	Forest Ave.	130	35.2	47.3	0.767	0.095	0.862	1.34	12	0.0023	2.2	1.70	201.6	189.30	189.00
7	12	13	Forest Ave.	82	11.8	59.1	0.960	0.118	1.078	1.67	12	0.0023	2.2	1.70	202.1	189.00	188.81
8	13	14	Center St.	280	5.3	64.4	1.046	0.129	1.175	1.82	15	0.0017	2.3	2.70	202.8	188.56	188.08
9	14	15	Center St.	275	19.2	83.6	1.360	0.167	1.527	2.37	15	0.0017	2.3	2.70	203.2	188.08	187.61
10	15	16	Center St.	113	12.1	95.7	1.555	0.191	1.746	2.70	15	0.0020	2.4	2.90	203.6	187.61	187.38
11	16	17	Center St.	245	2.8	98.5	1.600	0.197	1.797	2.78	15	0.0024	2.6	3.20	203.7	187.38	186.79
12	17	18	Center St.	375	12.4	110.9	1.800	0.222	2.022	3.13	15	0.0024	2.6	3.20	202.3	186.79	185.89
13	18	19	Right of way	130	4.2	115.1	1.871	0.230	2.101	3.26	15	0.0025	2.7	3.27	196.0	185.89	185.56

*Based upon a maximum rate of 250 gpcd and 65 persons per acre. Since the capacity of the minimum size of sewer with a minimum velocity of 2.0 fps is 0.7 cfs, equivalent to the maximum rate of discharge from 24.6 acres and since $[24.6(65 \times 250 + 2{,}000)1.55]/1{,}000{,}000 = 0.7$, laterals will be 8 in. in diameter (the minimum), as no lateral is to serve an area exceeding 24.6 acres.

Source: Adapted from Reference 13.

SEWER-SYSTEM DESIGN 28.37

FIGURE E-3 Map for storm sewer system design example. (*Ref. 13.*)

A careful study of local conditions indicates that 70 percent of the surfaces in the district are expected to be impervious. The inlet time has been assumed to be 20 min.

The rate of rainfall for the 5-year design storm is to be taken from the assumed formula $i = 20.4/t^{0.61}$, in which i is rainfall intensity in in/h and t is rainfall duration in minutes. The rainfall and runoff curves which are shown in Fig. E-4 are specific to the area under consideration and should not be used for other areas. Also shown in Fig. E-4 is a plot of i times A in the rational method equation $Q = CiA$. This term represents the runoff per acre.

In general, the storm sewers are to be designed with the crown at a depth of at least 5 ft below the surface of the street. The minimum size of sewer is to be 12 in. The assumed minimum velocity is 3.0 fps when flow is at full depth. Manning's n is to be taken as 0.013. (Adapted from Ref. 13.)

Solution

1. Draw a line to represent each storm sewer and indicate the direction of flow. The sewers should, in general, slope with the street surface.
2. Locate the manholes and give each an identification number.
3. Sketch the limits of the drainage areas tributary to each inlet. The assumed character of future development and the topography will determine the proper limits.

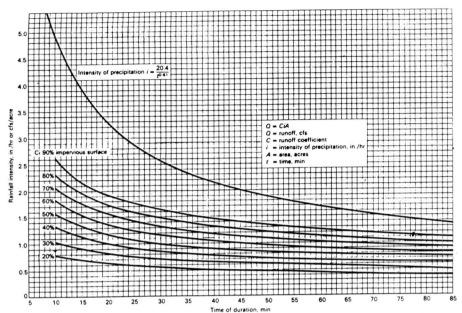

FIGURE E-4 Precipitation intensity curves for various degress of impervious. (*Adapted from Ref. 13.*)

4. Measure each individual area by planimeter or other methods that will give equally satisfactory results.
5. Prepare a tabulation as shown in Table 8 to record the data and steps in the computation of each section of sewer between manholes.

Computations are carried out as follows: Columns 1, 2, 3, 4, 5, and 6 are identified from the layout map of the system in Fig. E-3. Column 7 is the travel time to the upstream manhole considered. For line 1, 20.0 min is the time for overland flow to reach the farthest inlet. For line 2, the inlet time is 20.0 min plus the travel time (1.7 min) in the sewer from points 1 to 2 which is given in col. 8. Column 9 (runoff rate per acre) is read from Fig. E-4 using the 70 percent impervious curve for the time listed in col. 7. Column 10 (required capacity in cfs) is calculated by multiplying col. 9, the runoff rate per acre, by col. 6, the area in acres.

Columns 11, 12, 13, and 14 involve designing the pipe segment. A particular size and slope is determined from the Manning equation and the capacity and velocity at full flow are determined. Constraints are: (1) maintaining adequate cover of 5.0 ft over the crown of the pipe, (2) maintaining a minimum velocity at full flow capacity of 3.0 fps, and (3) avoiding excessive excavation.

Column 15 (surface elevation) is taken from street maps, survey information, or in preliminary design, from topographical maps. Column 16 (fall) is calculated by multiplying col. 12 (slope) by col. 4 (length). Column 17 (invert at upstream end) presents the invert at the upstream manhole. At the farthest manhole, the required depth of cover of 5.0 ft normally governs. For example, in line 1, the invert of 199.90 plus the pipe size of 1.30 ft (15-in pipe with allowance for pipe

TABLE 8 Computations for Storm Sewers

From (1)	To (2)	Location (3)	Length, ft (4)	Increment, acres (5)	Total acres (6)	To upper end, min (7)	In section, min (8)	Rate of run-off cfs per acre (9)	Required capacity (10)	Pipe size, in (11)	Slope, ft/ft (12)	Velocity, fps (13)	Capacity, cfs (14)	Surface elevation (15)	Fall, ft (16)	Invert Elevations Upper end (17)	Invert Elevations Lower end (18)
1	2	Maple St.	300	2.3	2.3	20.0	1.7	1.56	3.6	15	0.003	2.9	3.6	206.2	0.90	199.90	199.00
2	3	Maple St.	300	2.4	4.7	21.7	1.7	1.51	7.1	21	0.002	3.0	7.2	206.5	0.60	198.50	197.90
3	4	Maple St.	300	2.2	6.9	23.4	1.6	1.47	10.1	24	0.002	3.2	10.2	206.6	0.60	197.65	197.05
4	5	Maple St.	165	1.5	8.4	25.0	0.7	1.42	11.9	24	0.0027	3.8	11.9	207.1	0.45	197.05	196.60
5	6	Redwood St.	325	2.2	10.6	25.7	1.4	1.40	14.9	27	0.0023	3.8	15.0	207.4	0.75	196.35	195.60
6	7	Center St.	400	3.1	13.7	27.1	1.7	1.38	18.9	30	0.0021	3.9	18.8	206.1	0.84	195.35	194.51
7	8	Center St.	35	6.0	19.7	28.8	0.2	1.34	26.4	30	0.004	5.2	26.0	203.2	0.14	194.51	194.37
8	9	Center St.	230	10.2	29.9	29.0	1.0	1.34	40.0	42	0.0016	4.1	41.0	203.2	0.37	193.37	193.00
9	10	Center St.	240	5.7	35.6	30.0	0.9	1.32	47.0	42	0.0022	4.9	47.0	201.9	0.53	193.00	192.47
10	11	Forest Ave.	110	11.9	47.5	30.9	0.4	1.31	62.3	48	0.0018	4.9	62.0	201.6	0.20	191.97	191.77
11	12	Forest Ave.	95	11.1	58.6	31.3	0.3	1.30	76.3	54	0.0015	4.8	76.0	202.1	0.14	191.27	191.13
12	13	Center St.	295	5.6	64.2	31.6	0.9	1.30	83.3	54	0.0018	5.2	84.0	202.8	0.53	191.13	190.60
13	14	Center St.	260	17.2	81.4	32.5	0.8	1.28	104.0	60	0.0015	5.3	103.0	203.2	0.39	190.10	189.71
14	15	Center St.	145	13.7	95.1	33.3	0.5	1.27	121.0	66	0.0013	5.1	121.0	203.6	0.19	189.21	189.02
15	16	Center St.	225	2.9	98.0	33.8	0.7	1.26	124.0	66	0.0014	5.2	126.0	203.7	0.32	189.02	188.70
16	17	Center St.	380	13.2	111.2	34.5	1.1	1.25	139.0	66	0.0017	5.8	129.0	202.5	0.65	188.70	188.05
17	18	Private Land	165	4.1	115.3	35.6	—	1.24	143.0	48	0.0096	11.4	143.0	196.0	1.59	187.32	185.73

Source: Adapted from Reference 13.

thickness) gives an elevation of the top of the pipe as 201.20. This is 5.0 ft below the surface. Column 18 (invert elevation at downstream end) is calculated as col. 17 (upstream invert) minus col. 16 (fall). The lower or downstream elevation should be checked to ensure a minimum of 5.0 ft of cover is maintained. It should be noted that when pipe sizes change, the crowns of the sewers are generally matched. Therefore, col. 17 on line 2 is 0.50 ft lower than col. 18 for line 1. This is accounted for by the increase in pipe size from 15 to 21 in.

Finally, the outlet conditions at the downstream end of the system may be variable especially if the storm sewer discharges to a stream or river. In such cases, the designer must determine an outlet design condition (level of the river) to form the basis for the hydraulic computations. The outlet condition, which may be the 5-year recurrent stage level in the river, is determined in conjunction with policymakers. The outlet condition selected will, especially in flat areas, have a large impact on the design and its costs.

SEWAGE PUMPING STATIONS

37. Basic Concepts. Sewer systems are generally designed to flow by gravity. However, where the ground is very flat, the differences in elevation may be insufficient to permit gravity flow and external energy must be added to the system by pumps, to avoid use of very deep sewers. Also, since sewage-treatment facilities are usually located at the downstream end of the sewer system, where the sewers are deepest, pumps may be required to provide energy for flow through the wastewater treatment plant. Exhibit 1 shows the TARP mainstream pumping station in Chicago which is used to lift flows from the deep tunnels constructed in rock more than 200 ft below ground to the treatment plant at the ground surface. Sewage pumping stations may also be utilized for final discharge of wastewater when the elevation of the water in the receiving stream is higher than the invert elevation of the sewer.

Centrifugal pumps are the principal type used in sewer systems. These pumps may be submerged in the wastewater in the wet well or located in a dry well with suction pipes extended into the wet well. In either case, when the pumped liquid is released to atmosphere a short distance from the pumps, the station is referred to as a lift station. When the liquid is discharged into a pipe or closed conduit under pressure, the station is called a pumping station, even though the term pumping station can be applied to both situations. This section discusses the hydraulic factors critical to the design of a sewage pumping station such as system hydraulics including pump selection and wet well design. The design engineer must also consider general system requirements including pretreatment requirements.

38. System Hydraulics. The hydraulic design of a sewage pumping station requires the determination of the following major items:

1. The system capacity
2. The head against which the pumps must operate
3. The pump performance characteristics

The pump system capacity is generally estimated in the same manner as tributary flows to the sewer system described in previous sections. Estimates must

Exhibit 1 TARP mainstream pumping station, Chicago Metropolitan Sanitary District (Harza Engineering Company).

be made of the initial range of flows and also the anticipated future flows. System capacity will have to be adequate for anticipated future flows, and pumps must also operate efficiently during low flows in the initial period. As such, a decision must be made as to whether to install equipment capable of handling the full range of flows at the start, or whether to make provisions for increasing system capacity in the future. System capacity can be increased later by adding pumps, installing larger impellers in the same pumps, or replacing pumps by larger units.

39. System Head Curves. The head against which pumps must operate, for any given rate of pumping, is a function of the static head and the hydraulic losses in the system as shown in Fig. 8.

The static head is the difference in elevation between the free-water surface in the pump station wet well and the free-water surface at the point of discharge. The static head will therefore vary with changes in the water level in both the station wet well and the point of discharge.

The total head loss in the system includes friction losses in the pipeline and minor losses such as inlet and exit losses, losses in bends, and losses in appurtenances, as shown in Fig. 8. The friction losses in the pipeline can be computed by using the Manning's or Hazen-Williams formulas. Losses due to bends and other appurtenances can also be computed as described in this section.

Since both friction and minor losses vary with the capacity or discharge rate of the pumps, the total system head is also a function of capacity. This relationship between the total system head and capacity is best represented by a curve which is called a system-head curve as shown in Fig. 9. This curve can be developed by computing friction and minor losses as described above and

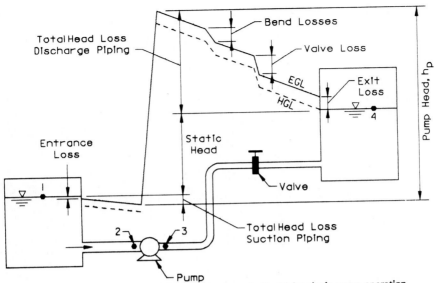

FIGURE 8 Sketch showing changes in total energy of a liquid for single-pump operation.

then plotting head vs flow. In selecting pumps it is usual to compute the energy losses for initial operation when the pipe is new and also for future conditions when the pipes will deteriorate and the head loss will increase. Both curves are shown in Fig. 9.

40. Pump-Head Curve. Each centrifugal pump has a unique relationship between the head developed and the pump discharge rate when the pump is operating at a constant speed. This relationship is represented by the pump performance curve, as shown in Fig. 9, which is generally prepared by the pump manufacturer. Curves showing the pump efficiency and brake horsepower developed over the operating range of the pump are also prepared by the manufacturer.

The pump operating point is determined by superimposing the system head curve on the pump performance curve. The point at which the two curves intersect represents the operating point of the pump for the particular conditions, as shown in Fig. 9.

The operating point obtained after consideration of one pump performance curve may not correspond to the most efficient operating condition for this pump, and the particular unit under consideration may not be the best choice for this application. In this case, other pump performance curves may be considered either for the same pump operating at different speeds or for a different unit.

41. Multiple-Pump Operation. In sewage pumping stations, it is not unusual to see two or more pumps operating in parallel and discharging into the same header. In order to select the pumps to be utilized and also to define the operating points for the pumps, the system-head curve must be compared with the combined pump performance curves. In this case, however, a modified approach as

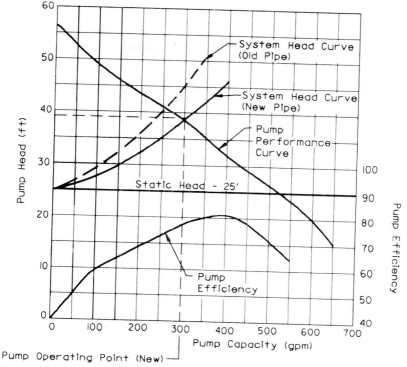

FIGURE 9 Selection of pump for single-pump operation.

described below must be used to develop both the system-head curve and the combined pump curve:

1. The system-head curve is developed only for that part of the system that is common to all pumps, and the losses in the suction and discharge piping of individual pumps are omitted.
2. The losses for the suction and discharge lines for individual pumps are subtracted from the head-capacity curves for the individual pumps, leading to development of modified pump curves. Each modified pump curve then represents the head capacity of the pump, together with its individual suction and discharge lines.
3. Develop the combined (modified) pump operating curve by adding the pump capacities at the same head for pumps in parallel. This gives the output capacity of the multiple pump operation with pumps working in parallel at any head.
4. Superimpose the system-head curve on the combined (modified) pump operating curve to determine the system-head capacity at the several points of operation.

Figure 10 shows graphically the development and comparison of system-head

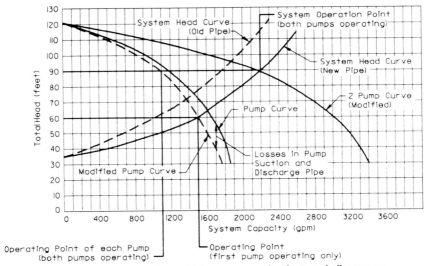

FIGURE 10 Selection of pumps for multiple-pump operation for two similar pumps.

and pump capacity curves for a typical station with two similar constant-speed pumps operating in parallel.

42. Wet Well Design. A typical layout and cross section of a sewage pumping station with the pump located in the wet well is shown in Fig. 11. The wet well receives influent flows and provides a storage area for excess flows when the inflow rate exceeds the pump capacity, or when the pumps are not working. This storage is necessary to ensure that total pump output always matches total influent flow rates whenever automatic controls and variable-speed motors are not provided.

In the usual case, with constant-speed motors, the inflow and discharge rates will rarely be equal and pumps will cycle on and off. Excessive cycling is undesirable and wet well volumes should be set such that the pump cycle time is not less than 20 min for large pumps and not less than 5 min for smaller units. However, the minimum cycle time will in most cases be dictated by the pump manufacturer.

The volume of the wet well between the start and stop elevations for a single pump or the incremental volume between the start and stop elevations for the next pump in multiple pump operation is given by

$$V = \frac{\sigma q}{4}$$

where V = required volume, gal
 σ = minimum time for one pumping cycle or time between successive starts, min
 q = pump capacity or increment in pump capacity, gpm, where one pump is already operating and a second pump is started

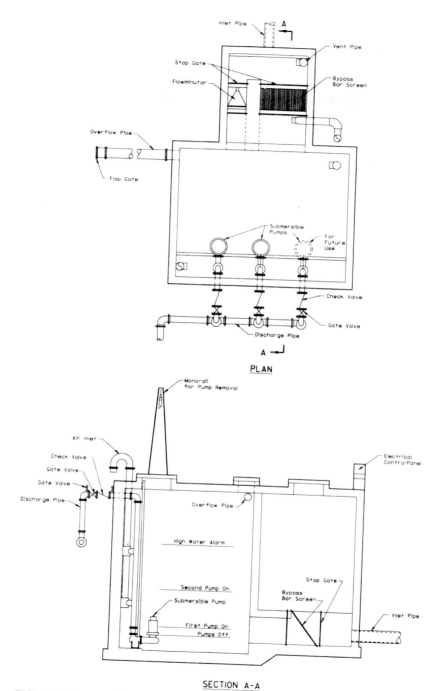

FIGURE 11 Layout of typical wet-pit pump station.

PART B. SEWAGE AND WASTEWATER TREATMENT

43. Objectives of Sewage and Wastewater Treatment. Sewage is the soiled water supply of a community and usually contains less than 0.1 percent of solid matter. About one-third of the solid matter is in suspension and the remaining two-thirds in solution. Suspended matter includes large floating material, heavy inert materials, and finer particles. Approximately half of all solid materials in wastewater is organic and the other half inorganic. Sewage also contains living organisms including many pathogens.

The discharge of industrial wastes to sewers has resulted in an increase in the concentration of radioactive materials, exotic and toxic chemicals, and other chemical compounds in the wastewater.

Sewage-treatment processes are designed to convert the contaminated wastewater into an acceptable final effluent and to dispose of the solids removed in the process. Sewage-treatment processes may be classified as primary, intermediate, secondary, and tertiary and may include physical, physiochemical, chemical, and biological processes. Typical methods of sewage treatment are listed in Table 9.

The type of treatment to be provided at a particular location is a function of the influent wastewater quality and the minimum effluent quality requirements established by regulatory agencies such as the U.S. Environmental Protection Agency. Effluent requirements are usually based on the ability of the receiving water to assimilate the wastes and the use to which the receiving waters are put.

Methods, structures, and equipment commonly used for wastewater treatment are briefly described in this section with special emphasis on the hydraulic design of the various treatment-plant elements.

44. Plant Hydraulics. The general arrangement of the various units in a wastewater-treatment plant will be dictated by the normal sequence of the units in the treatment process. The number and size of each unit will be governed by the total plant capacity, the requirements for redundancy in each unit process to permit ease of operation and maintenance, the range of flows anticipated over the design period, and plant hydraulics.

The hydraulic design establishes the minimum head required for flow through the various treatment units and dictates the vertical relationships between the units. If there is sufficient head for gravity flow through the units, the hydraulic requirements may also control plant layout. However, if intermediate pumping is utilized, a greater flexibility in plant layout is possible.

The hydraulic design is usually started after the number and size of units have been determined and the general arrangement of units established. Units are connected by pipes and/or channels as appropriate to permit flow between them.

The hydraulic design is then a trial-and-error process in which the size of the hydraulic components are assumed and the analysis of flow computed. The flow analysis is usually carried out for a particular flow condition and with a specific number of units in operation. Usually, several factors are evaluated including the water-surface elevations, the flow distributions, the energy losses in each unit, and the velocities of flow. If the results are unacceptable, the sizes of the hydraulic components are changed and the flow analysis is repeated.

TABLE 9 Typical Methods of Sewage Treatment and Disposal

Methods of sewage disposal
---
1. Dilution or disposal into water
2. Irrigation or disposal on land
    a. Application to surface
    b. Subsurface irrigation
3. Underground by deep-well injection

Methods of sewage treatment

I. Primary treatment
   A. Removal of floating solids and coarse suspended solids by
      1. Bar racks and screens
      2. Grit chambers
      3. Skimming tanks, with or without aeration
   B. Removal of fine suspended solids by
      1. Fine screens
      2. Sedimentation tanks
      3. Air flotation units
      4. Flocculation and chemical precipitation units
II. Secondary treatment
   A. Oxidation by
      1. Filters
         a. Intermittent sand filters
         b. Contact filters
         c. Trickling filters
      2. Aeration
         a. Activated sludge
         b. Contact aerators
III. Disinfection
   A. Chlorination
   B. Ozone
   C. Ultraviolet light
IV. Advanced wastewater treatment
   A. Physical processes
      1. Air stripping
      2. Filtration
      3. Microstainers
      4. Reverse osmosis
   B. Chemical processes
      1. Carbon absorption
      2. Chemical precipitation
      3. Ion exchange
   C. Biological processes
      1. Nitrification and denitrification

The water-surface elevations developed from the flow analysis can be summarized as the hydraulic profile for the plant for each scenario of flow and number of operating units considered. This hydraulic profile is useful to the engineer in determining whether adequate head is available for gravity flow through the plant or if intermediate pumping is required and also in determining whether flows will surcharge any units during peak flows or when individual units are taken out of service for maintenance.

The head losses through a treatment plant include losses in the pipes or conduits conveying flow between the treatment units; the losses in the units; and losses through equipment items such as bar screens. Losses associated with conveyance of flows between units will include friction losses in pipes and conduits; minor losses at bends, unit entrances, valves, and other appurtenances; head losses over weirs and other flow-control devices, including the head required for discharge and the free-fall allowance: and any head allowances for plant expansions.

45. Conduit Losses. Conduits between treatment units may flow as open channels or as pressure pipes (closed conduits). Head losses through conduits under these circumstances will include:

1. Losses due to friction in the conduits
2. Velocity head losses resulting from entrance disturbances, disturbances along the conduit, sudden enlargements, sudden contractions, gradual enlargements and contractions, obstructions, and bends

The head losses due to friction can be computed using the Chézy, Manning, and Hazen-Williams equations discussed in this section.

Entrance losses may be computed from the velocity-head formula

$$h = k \frac{V^2}{2g}$$

where h = head loss, ft
V = entrance velocity, fps
g = acceleration due to gravity, ft/s^2
k = constant for the particular condition, ft/s^2, with typical values of k as follows:

Type of entrance	Coefficient k
Inward-projecting, square-cornered	0.8–0.9
Square-cornered, flush with wall	0.50
Slightly rounded	0.20
Bellmouth	0.04

Losses due to enlargements and contractions can also be computed by use of the modified velocity-head formula[8] of $H = K(V_1^2/2g)$, where V_1 is the velocity in the smaller pipe and values of K for sudden and gradual enlargements are given in Tables 10 and 11, respectively. Values of K for sudden contractions are given in Table 12.

Head losses due to obstructions such as gate valves may also be computed using the velocity-head formula with values of K_g as given in Table 13.

46. Bar-Screen Losses. Bar screens are used to remove particles that are larger than the screen opening to protect the operation of mechanical equipment, to prevent clogging of flow channels, and to prevent fouling of the shoreline of receiving streams. Bar screens are usually located at the upstream end of a

TABLE 10 Values of K for Determining Loss of Head Due to Sudden Enlargements in Pipes, from the Formula $H = K(V_1^2/2g)$

$\dfrac{d_2}{d_1}$	Velocity V_1, fps												
	2	3	4	5	6	7	8	10	12	15	20	30	40
1.2	0.11	0.10	0.10	0.10	0.10	0.10	0.09	0.09	0.09	0.09	0.09	0.09	0.08
1.4	0.26	0.26	0.25	0.24	0.24	0.24	0.24	0.23	0.23	0.22	0.22	0.21	0.20
1.6	0.40	0.39	0.38	0.37	0.37	0.36	0.36	0.35	0.35	0.34	0.33	0.32	0.32
1.8	0.51	0.49	0.48	0.47	0.47	0.46	0.46	0.45	0.44	0.43	0.42	0.41	0.40
2.0	0.60	0.58	0.56	0.55	0.55	0.54	0.53	0.52	0.52	0.51	0.50	0.48	0.47
2.5	0.74	0.72	0.70	0.69	0.68	0.67	0.66	0.65	0.64	0.63	0.62	0.60	0.58
3.0	0.83	0.80	0.78	0.77	0.76	0.75	0.74	0.73	0.72	0.70	0.69	0.67	0.65
4.0	0.92	0.89	0.87	0.85	0.84	0.83	0.82	0.80	0.79	0.78	0.76	0.74	0.72
5.0	0.96	0.93	0.91	0.89	0.88	0.87	0.86	0.84	0.83	0.82	0.80	0.77	0.75
10.0	1.00	0.99	0.96	0.95	0.93	0.92	0.91	0.89	0.88	0.86	0.84	0.82	0.80
∞	1.00	1.00	0.98	0.96	0.95	0.94	0.93	0.91	0.90	0.88	0.86	0.83	0.81

d_2/d_1 = ratio of larger pipe

TABLE 11 Values of K for Determining Loss of Head Due to Gradual Enlargement in Pipes, from the Formula $H = K(V_1^2/2g)$

$\dfrac{d_2}{d_1}$	Angle of Cone												
	2°	4°	6°	8°	10°	15°	20°	25°	30°	35°	40°	45°	50°	60°
1.1	0.01	0.01	0.01	0.02	0.03	0.05	0.10	0.13	0.16	0.18	0.19	0.20	0.21	0.23
1.2	0.02	0.02	0.02	0.03	0.04	0.09	0.16	0.21	0.25	0.29	0.31	0.33	0.35	0.37
1.4	0.02	0.03	0.03	0.04	0.06	0.12	0.23	0.30	0.36	0.41	0.44	0.47	0.50	0.53
1.6	0.03	0.03	0.04	0.05	0.07	0.14	0.26	0.35	0.42	0.47	0.51	0.54	0.57	0.61
1.8	0.03	0.04	0.04	0.05	0.07	0.15	0.28	0.37	0.44	0.50	0.54	0.58	0.61	0.65
2.0	0.03	0.04	0.04	0.05	0.07	0.16	0.29	0.38	0.46	0.52	0.56	0.60	0.63	0.68
2.5	0.03	0.04	0.04	0.05	0.08	0.16	0.30	0.39	0.48	0.54	0.58	0.62	0.65	0.70
3.0	0.03	0.04	0.04	0.05	0.08	0.16	0.31	0.40	0.48	0.55	0.59	0.63	0.66	0.71
∞	0.03	0.04	0.05	0.06	0.08	0.16	0.31	0.40	0.49	0.56	0.60	0.64	0.67	0.72

d_2/d_1 = ratio of diameter of larger pipe to diameter of smaller pipe. Angle of cone is twice the angle between the axis of the cone and its side.

wastewater-treatment plant or immediately before an outfall used for discharging untreated wastes.

Bar screens are classified according to the size of openings and the method of cleaning which can be done manually or by mechanical means. Coarse bar screens may include bars 2 to 3 in wide and ⅓ to ⅜ in thick spaced ½ to 2 in apart. Common spacing is ½ to ¾ in for mechanically cleaned screens and 1 to 1½ in for manually cleaned units.

Fine screens are generally mechanically cleaned units equipped with a perforated plate, woven-wire cloth, or closely spaced bars with openings less than 3/16 in. Fine screens are not used extensively at municipal wastewater plants but are utilized as pretreatment units for industrial wastewater.

TABLE 12 Values of k_c for Determining Loss of Head Due to Sudden Contraction from the Formula $H = k_c(V_1^2/2g)$

$\dfrac{d_2}{d_1}$	Velocity V_1, fps											
	2	3	4	5	6	7	8	10	12	15	20	30	40
1.1	0.03	0.04	0.04	0.04	0.04	0.04	0.04	0.04	0.04	0.04	0.05	0.05	0.06
1.2	0.07	0.07	0.07	0.07	0.07	0.07	0.07	0.08	0.08	0.08	0.09	0.10	0.11
1.4	0.17	0.17	0.17	0.17	0.17	0.17	0.17	0.18	0.18	0.18	0.18	0.19	0.20
1.6	0.26	0.26	0.26	0.26	0.26	0.26	0.26	0.26	0.26	0.25	0.25	0.25	0.24
1.8	0.34	0.34	0.34	0.34	0.34	0.34	0.33	0.32	0.32	0.31	0.31	0.29	0.27
2.0	0.38	0.38	0.37	0.37	0.37	0.37	0.36	0.36	0.35	0.34	0.33	0.31	0.29
2.2	0.40	0.40	0.40	0.39	0.39	0.39	0.39	0.38	0.37	0.37	0.35	0.33	0.30
2.5	0.42	0.42	0.42	0.41	0.41	0.41	0.40	0.40	0.39	0.38	0.37	0.34	0.31
3.0	0.44	0.44	0.44	0.43	0.43	0.43	0.42	0.42	0.41	0.40	0.39	0.36	0.33
4.0	0.47	0.46	0.46	0.46	0.45	0.45	0.45	0.44	0.43	0.42	0.41	0.37	0.34
5.0	0.48	0.48	0.47	0.47	0.47	0.46	0.46	0.45	0.45	0.44	0.42	0.38	0.35
10.0	0.49	0.48	0.48	0.48	0.48	0.47	0.47	0.46	0.46	0.45	0.43	0.40	0.36
—	0.49	0.49	0.48	0.48	0.48	0.47	0.47	0.47	0.46	0.45	0.44	0.41	0.38

d_2/d_1 = ratio of larger to smaller diameter. V_1 = velocity in smaller pipe.

TABLE 13 Loss of Head Due to Gate Valves, Values of K_g in $H_g = K_g(V_2/2g)$

Nominal diam of valve, in	Ratio of height d of valve opening to diam D of full valve opening					
	1/8	1/4	3/8	1/2	3/4	1
1/2	450	60	22	11	2.2	1.0
3/4	310	40	12	5.5	1.1	0.28
1	230	32	9.0	4.2	0.90	0.23
1 1/2	170	23	7.2	3.3	0.75	0.18
2	140	20	6.5	3.0	0.68	0.16
4	91	16	5.6	2.6	0.55	0.14
6	74	14	5.3	2.4	0.49	0.12
8	66	13	5.2	2.3	0.47	0.10
12	56	12	5.1	2.2	0.47	0.07

The major hydraulic considerations in the design of bar screens are:

1. The velocity of flow in the approach and exit channels should be adequate to prevent deposition of solids in the channels but not so high as to dislodge screenings. The Ten State Standards[4] recommends that approach velocities should be not less than 1.25 fps and no greater than 3.0 fps at normal operating flow conditions.

2. The head loss through the screens should be minimized to prevent surcharging of the upstream influent sewers. The head loss through a bar screen will vary with the amount of material collected and the frequency of cleaning. The head loss through a clean screen can be computed from the formula.[10]

$$h_L = \frac{V^2 - v^2}{2g} \times \frac{1}{0.7}$$

where V = velocity of flow through screen, fps
v = approach velocity in channel, fps

Where different-shaped bars are utilized and the rack is not vertical, the head loss can be calculated from empirical formulas as developed by the screen manufacturer.

47. Grit Chambers. In wastewater treatment, the term grit is used to define small inorganic particles of sand or other mineral matter and other nonputrescible solids such as eggshells, bone chips, coffee grounds, and seeds which have higher specific gravities and subsiding velocities than organic solids. Grit removal from wastewater is essential to prevent abrasion of moving equipment and formation of grit deposits in sludge digestion tanks.

Grit-removal units can be horizontal-flow, velocity-controlled units or spiral-flow aerated units. In the horizontal-flow-type unit, the chamber is designed to maintain a flow velocity of 1 fps under all flow conditions. This velocity control is usually provided by a proportional weir or a Sutro weir at the downstream end of the unit. The proportional weir as shown in Fig. 12 maintains a constant velocity in the unit by varying the depth of flow with the flow rate. For proportional weirs to operate effectively they must have a free discharge.

In aerated grit tanks, spiral flow is induced in the tank by locating air diffusers along one side of the tank. The rate of air diffusion governs the velocity of roll in the tank and thereby the size of the particle to be removed. This air flow rate can be adjusted to control the roll velocity for varying conditions of flow, particle size, and specific gravity. The principal hydraulic consideration in the design of a

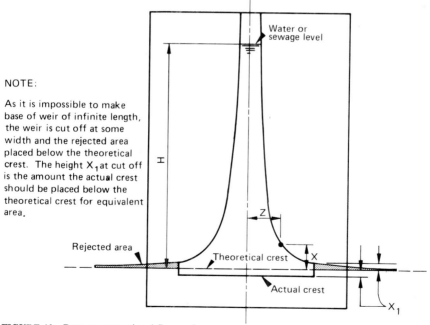

FIGURE 12 Rettger proportional-flow weir.

grit chamber is velocity control. The Sutro and the proportional weir serve to control the flow velocity by varying the depth and cross-sectional area of the flow in the channels in direct proportion to the rate of flow. The depth of flow through a proportional weir can be computed from the formula

$$Q = 7.82 \, b_r h$$

where Q = rate of flow, cfs
h = height of flow above weir crest, ft
b_r = weir constant

The constant b_r is computed after the height h is established for the maximum value of Q (the flow rate). The shape of the weir is computed from the formula

$$Z = \frac{b_r}{2\sqrt{X}}$$

where Z = distance from centerline to sides of openings
X = vertical distance above theoretical weir crest as shown in Fig. 12

Velocity control can also be achieved by use of a Parshall flume downstream of the grit tank. It can be shown that for good velocity control utilizing a Parshall flume, the channel cross section must approximate a parabola.

A Parshall flume consists of a converging section, a throat section, and a diverging section as shown in Fig. 13. This arrangement causes critical depth to occur near the beginning of the throat section and produces a backwater curve whose depth, measured at a point H_a, can be related to the discharge or flow rate. The relationship between the depth H_a and discharge for flumes of various throat sizes is given by the following equations:[11]

Throat width, in	Equation
3	$Q = 0.992 \, H_a^{1.547}$
6	$Q = 2.06 \, H_a^{1.58}$
9	$Q = 3.072 \, H_a^{1.53}$
12–96	$Q = 4W \, H_a^{(1.522W)^{0.026}}$

where Q = discharge rate, cfs
W = throat width, ft
H_a = upstream depth, ft

These equations provide accurate results provided that the flume is not submerged. The flume is considered to be submerged when the ratio of depth H_b near the end of the throat to the upstream depth H_a exceeds the following values:

W, throat size	H_b/H_a (maximum ratio for free discharge)
3–9 in	0.6
1–8 ft	0.7
10–50 ft	0.8

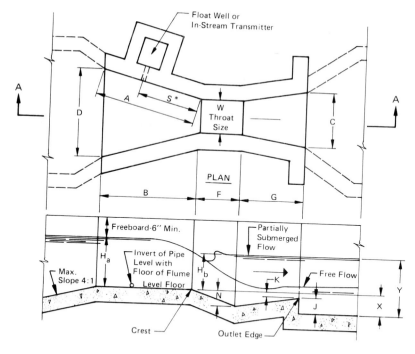

W	A	B	C	D	F	G	K	N	Free-flow capacity, cfs	
									Max	Min
0·3	1·6 3/8	1·6	0·7	0·10 3/16	0·6	1·0	0·1	0·2 1/4	1.1	0.03
0·6	2·0 7/16	2·0	1·3 1/2	1·3 5/8	1·0	2·0	0·3	0·4 1/2	3.9	0.05
0·9	2·10 5/8	2·10	1·3	1·10 5/8	1·0	1·6	0·3	0·4 1/2	8.8	0.09
1·0	4·6	4·4 7/8	2·0	2·9 1/4	2·0	3·0	0·3	0·9	16.1	0.35
1·6	4·9	4·7 7/8	2·6	3·4 3/8	2·0	3·0	0·3	0·9	24.6	0.51
2·0	5·0	4·10 7/8	3·0	3·11 1/2	2·0	3·0	0·3	0·9	33.1	0.66
3·0	5·6	5·4 3/4	4·0	5·1 7/8	2·0	3·0	0·3	0·9	50.4	0.97
4·0	6·0	5·10 5/8	5·0	6·4 1/4	2·0	3·0	0·3	0·9	67.9	1.26
6·0	7·0	6·10 3/8	7·0	8·9	2·0	3·0	0·3	0·9	103.5	2.63
8·0	8·0	7·10 1/8	9·0	11·1 3/4	2·0	3·0	0·3	0·9	139.5	4.62
10·0	14·3 1/4	10·0	12·0	15·7 1/4	3·0	6·0	0·6	1·1 1/2	200	6

* Dimension S = 2/3A except for W = 10′−0″ when S = 6′−0″

NOTE: All Dimensions given in Feet and Inches.

DIMENSIONS AND CAPACITIES

FIGURE 13 Parshall flume, general arrangement, dimensions, and capacities. (*U.S. Department of Agriculture,* Farmer's Bulletin 1683, *January 1932, revised October 1941.*)

Tables and charts such as Fig. 14 are available for estimating free-flow discharge at various heads H_a for the different sizes of flumes. Charts are also available to determine the discharge rate when the flume is submerged and the discharge rate is reduced. The reader should consult Ref. 11 for more details on submerged flume discharges and for charts and diagrams for computing submerged discharge rates.

Other hydraulic considerations in the design of a grit chamber include inlet and outlet losses and losses through flow-control devices such as gates, orifices, and/or weirs.

48. Sedimentation Tanks. Sedimentation is the process in which suspended particles that are heavier than water are separated from the liquid by the forces of gravity. In wastewater-treatment plants, sedimentation tanks are used in several ways as follows: (1) to remove readily settleable solids ahead of other treatment units as in grit tanks and primary settling tanks, (2) in secondary settling units, to remove biological solids produced by biological treatment processes such as the activated-sludge process, and (3) to thicken settled solids removed in primary and secondary tanks as in sludge thickeners.

In wastewater-treatment plants, particles will settle out of the suspension in one of four different ways depending on the solids concentration and the tendency of the particles to flocculate.[14]

Type I settling refers to the separation of nonflocculating discrete particles in a dilute suspension. In this case each particle settles as an individual unit and there is no interaction with neighboring particles. Settling is said to be unhindered and is a function of the fluid properties and characteristics of the particle only. Type I settling occurs in grit tanks.

Type II settling refers to the separation of flocculating particles from a dilute suspension. In this case, heavier solids with greater settling velocities overtake and coalesce with lighter particles to form larger floc particles with greater mass and settling velocities. As a result, removal of suspended solids depends not only on the clarification rate but also on the depth of the clarifier. This type of settling occurs in primary settling tanks and in the upper layers of secondary clarifiers.

There is no satisfactory formula for evaluating the effect of particle flocculation on sedimentation, and it is necessary to perform settling-column tests to determine the settling characteristics of the flocculent suspension. Based on these tests the minimum time required to obtain a desired removal rate at a certain depth and a minimum overflow rate can be established.

Type III settling refers to the separation of particles from a suspension of intermediate concentration in which particle forces are sufficient to hinder the settlement of neighboring particles. In this case, the particles form a zone in which all particles tend to remain in fixed positions relative to each other and settle collectively and at a reduced rate. This type of settlement occurs in secondary settling tanks used to remove biological solids after secondary treatment of wastewater. Settlement tests are also required to determine the settling characteristics of suspensions for type III settling.

Type IV settling refers to settling in which the particles form a structure and further settling occurs only by compression of the structure. Compression generally takes place because of the weight of the particles being added to the structure by settlement from the supernatant. This type of settlement occurs in the deep sections of clarifiers and in sludge-thickening facilities.

Sedimentation tanks are now generally designed on the basis of a surface loading rate (SLR) which is expressed in gallons per day per square foot and represents the value of the rate of flow divided by the surface area of the tank. Thus

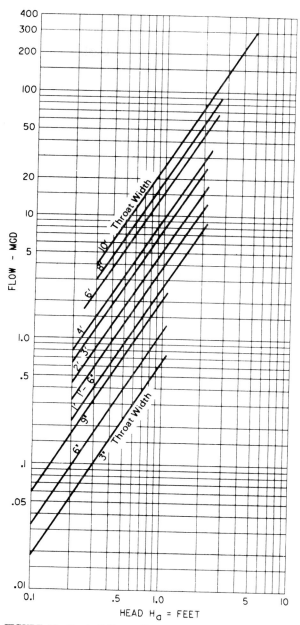

FIGURE 14 Parshall flume, flow rates and heads H_a. (*Chart plot from U.S. Department of Agriculture, Farmer's Bulletin 1683 and Colorado College Bulletin 426A.*)

$SLR = Q/A$, and it can be shown that for type I and type II unhindered settling, the surface settling rate should be less than the desired particle settling velocity to ensure that particles settle out. The surface settling rate is the principal hydraulic consideration in the design of sedimentation tanks, but other important hydraulic considerations are (1) detention time, (2) inlet design, and (3) outlet design and weir loading rates.

Detention Time. It can be shown that if all solids reaching a settling tank treating wastewater are discrete particles of uniform size and density, the efficiency of removal is a function of the surface area of the tank and is independent of the depth of the basin and the detention time.[10] However, the solids in most wastewater treatment units are not of such regular character, and the bulk of the particles are susceptible to flocculation. Since flocculation becomes more complete as time elapses, detention time is also a consideration in the design of settling tanks. Typical detention times for primary sedimentation tanks range from 1.5 to 2.5 h, with 2.0 h usually chosen. For secondary settling tanks, detention times are generally derived from consideration of the overflow rates and minimum tank depths.

Inlet Design. The inlet structure should be designed to provide horizontal and vertical distribution of influent flows while minimizing turbulence and short-circuiting of flows. The inlet structure must also dissipate the energy of the incoming flows and reduce the velocity into the inlet well to less than 1 fps at one-half the design flow.

When multiple settling tanks are used, the inlet structure also serves to subdivide the influent flows so that all tanks receive equal volumes and rates of flow. The hydraulics of inlet structures used for flow division between units is discussed in detail in Sec. 27. Typical design data for sedimentation tanks are given in Table 14.

Outlet Design. The outlet system includes the overflow weirs and the effluent launder or troughs with single or multiple outlets. The outlet system should

TABLE 14 Design Data for Sedimentation Tanks in Wastewater Treatment

Item	Value
Primary settling tanks	
Overflow rate	
At design average flow	1000 gpd/ft^2*
At peak hourly flow	1500 gpd/ft^2*
Detention time	1.5–2.5 h
Weir loading rate	10,000–15,000 gpd/ft
Secondary settling tanks†	
Overflow rate following	
Trickling filters	1200 gpd/ft^2 at peak hourly flow
Conventional activated sludge	1200 gpd/ft^2 at peak hourly flow
Extended aeration	1000 gpd/ft^2 at peak hourly flow
Nitrification	800 gpd/ft^2 at peak hourly flow
Solids loading rate	
Activated sludge	< 50 pfs/day at peak hourly flow
Weir loading rate	10,000–15,000 gpd/ft

*Ref. 4.
†This information should not be used for design unless settling column test data are unavailable.

be designed to minimize short-circuiting and to prevent velocity updrafts and surges in the tank which may cause settled solids to be drawn up and discharged with the effluent. These objectives are achieved by proper location of the weirs and prudent choice of weir loading rates.

For primary sedimentation tanks, the outlet weir is placed opposite the inlet structure, and a weir loading rate of between 10,000 and 15,000 gpd/ft is utilized. For secondary clarifiers, the outlet weirs are placed away from the upturn of the density current. For rectangular tanks this puts the outlet weir between half and two-thirds of the distance from the influent to the tank end wall. In circular tanks, the weirs are placed at two-thirds to three-fourths of the radial distance from the center of the tank. At these locations, a weir loading rate of between 15,000 and 20,000 gpd/ft can be utilized. The head required to obtain the desired flow over the weir is determined from weir flow equations presented in Secs. 2, 5, and 27.

Outlet launders or troughs act as lateral spillways and hydraulically are not unlike the washwater troughs serving rapid sand filters. Required dimensions of the trough can be developed in the same manner as for washwater troughs discussed in Sec. 27.

49. Biological Treatment Processes. The four common biological processes used for treating domestic wastewater are: (1) the activated sludge process, (2) the trickling filter, (3) the aerated lagoon, and (4) the oxidation pond. In these processes, finely divided and dissolved organic matter in the wastewater is converted into flocculent solids which can be removed by settling.

The oxidation pond is a shallow simple earthwork structure in which wastewater is treated by biological decomposition. Biological degradation in ponds can be carried out by aerobic, facultative, or anaerobic bacteria, but anaerobic systems are less widely used because of odor problems. In aerobic and facultative systems, the oxygen required for oxidation of organics is produced by green algae living in the pond and by oxygen transfer at the surface of the pond. The design of oxidation ponds is generally very imprecise but can be carried using empirical formulas.

The aerated lagoon is also a shallow basin in which wastewater is treated. However, in this case, oxygen is provided by mechanical surface aerators or diffused air instead of solely by algae. The design of the aerated lagoon is also generally based on empirical formulas.

50. Trickling Filters. The trickling filter is the most commonly used attached-growth biological wastewater-treatment system. In this process, the microorganisms responsible for conversion of the organic matter and other contaminants in the wastewater to settleable solids and gases are attached to the filter medium.

Filters currently used consist of a bed of rocks, 1 to 4 in in diameter, which is highly permeable and through which the wastewater is trickled. The depth of the medium varies from 3 to 8 ft with an average depth of 6 ft. Synthetic materials are also used as filter media.

In trickling filters, the wastewaters are not allowed to flood or drown the media but are forced to trickle in thin films over the biological growth covering the media stone. This trickling operation allows the pore spaces between the rocks to remain open and natural ventilation to occur. This ventilation helps to keep the surface layer of microorganisms around the rocks, called the slime layer, aerobic.

As organisms grow and the thickness of the slime layer increases, less oxygen and food penetrates to the microorganisms near the media face. These microorganisms then enter the endogenous phase of growth and lose their ability to cling to the rock. The slime layer then sloughs off and a new layer starts to form. The material sloughed off is usually intercepted in settling tanks downstream of the filter.

Recirculation of filter and clarifier effluent is sometimes undertaken, as this allows filters to treat higher rates of flow at the same efficiency as a standard filter with no recirculation.

The hydraulic analysis of trickling filters must include head-loss computations for the following components of the system:

1. The inlet piping system including control chambers
2. The distributor system including the piping to the distributor
3. The filter bed
4. The effluent system

The losses through the inlet piping system will include exit and entrance losses, friction losses in the connecting pipeline, and losses at bends, valves, and other appurtenances, which can be computed as outlined before in this section.

The head loss through the distributor will include the losses in the center column, the distribution arms, and the discharge orifices. The head losses in the center column and arms will be identified by the equipment manufacturer. The heads required for discharge through the orifices will be a function of the type of distributor used. If a hydraulic-reaction-driven distributor is used, the velocity head of discharge required to cause the reaction force of rotation must be considered.

The head loss in the filter bed will be the difference in elevation from the distributor to the lowest water surface in the underdrain system. This is the largest individual head loss in a trickling filter and cannot be reduced unless the bottom of the bed is submerged. This submergence is undesirable, since it will prevent ventilation of the bed.

Head losses in the effluent system include friction losses in the underdrainage system and in the piping between the filter and the settling tanks.

Typical overall head losses in a rotary trickling filter are as follows:

Filter element	Approximate head loss	
	Average	Range
Distributor	2.5	1.0–4.0
Filter depth	6.0	3.0–8.0
Outlet piping	2.5	2.0–3.0
Total	11.0	6.0–16.0

51. Activated-Sludge Process. In the activated-sludge process, a portion of the sludge produced in the sedimentation basins downstream of the activated-sludge aeration tank is returned to the aeration tank and mixed with the incoming sewage. The mixture is then aerated and agitated to encourage further microbial growth and enhance oxidation of finely divided and suspended solids in the wastewater.

The returned sludge serves two basic purposes. It serves as a vehicle for organisms which oxidize organic matter in the wastewater and also functions as nuclei for the formation of large floc particles by absorbing suspended and colloidal matter from the wastewater.

After aeration, the mixed sewage and sludge, which is usually called mixed liquor, passes to a final or secondary settling tank where the sludge settles out and a clear effluent is produced. A portion of the sludge produced is returned to the aeration tank and the remainder is wasted. The rate at which sludge is returned to the aeration tank is governed by the rate of wastewater flow; the influent suspended solids concentration, and the desired concentration of solids in the aeration tank.

In the aeration tank, the mixed liquor is aerated either by blowing air bubbles through it or by agitating the surface of the sewage while also stirring the tank contents. The processes used for aeration serve three purposes: (1) they provide the oxygen required for biodegradation of the organics under aerobic conditions, (2) they serve to keep the activated-sludge floc particles in suspension, and (3) they foster contact between the sludge particles and the finely divided and suspended particles in the wastewater.

Aeration tanks are usually designed as long channels with much transverse and some longitudinal mixing. The settling tank downstream of the aeration unit is designed as a rectangular or circular sedimentation basin as described in this section.

In the hydraulic design of the activated-sludge process, consideration must be given to head losses in the conduits connecting the units, inlet and outlet losses including losses across effluent weirs, losses in the tanks, and head losses in the air piping systems for diffused air systems.

Conduit, inlet, and outlet head losses can be computed as described in this section for sedimentation tanks. The head loss in the aeration tank is very small and can be neglected.

Air piping losses may be computed by the Fritzsche formula, which is written as

$$P = \frac{1.268(t + 460)Q^{1.852}L}{10^6(p + 14.7)d^{4.973}}$$

where P = drop in pressure, psi/ft of air piping
p = mean gage pressure, psi
t = mean air temperature, °F
d = pipe diameter, in
L = length of pipe, ft
Q = air flow rate, cfm, at 60°F and 760 mm pressure

Minor air pressure losses through venturi meters used for control of air flow rates are as follows:

Size of meter, in	Compressed-air factors		Pressure losses	
	Pressure, psia	Temperature, °F	% of meter differential head	psi per 1-ft differential
10 × 5	21.5	86	18	0.078
5 × 2	22.0	80	22	0.095
4 × 2	18.7	86	21	0.091
3 × 3/4	21.5	78	24	0.104

Air pressure losses through elbows and bends in air pipelines are computed as the loss through an equivalent length of pipe. Equivalent lengths for various diameter bends are as follows:

	Equivalent length of pipe, ft	
Nominal pipe diam, in	Elbow	Bend
1	1.5	0.23
2	4.9	0.74
3	9.4	1.41
4	14.5	2.2
6	25.9	3.9
8	38.0	5.7
10	50.7	7.6
12	63.7	9.6
14	76.7	11.5
16	90.1	13.5
18	104	15.5
20	117	17.5
24	144	21.6

Air pressure losses through globe valves, tees, and elbows in terms of equivalent lengths of straight pipe are given in Table 15.

TABLE 15 Equivalent Length of Straight Pipe for Air Piping Appurtenances

Nominal pipe diam, in	Additional length of straight pipe, ft	
	Globe valves	Tees and elbows
1	2	2
1½	4	3
2	7	5
2½	10	7
3	13	9
3½	16	11
4	20	13
5	28	19
6	36	24
7	44	30
8	53	35
10	70	47
12	88	59
15	115	77
18	143	96
20	162	108
22	181	120
24	200	134

52. Outfall. The final disposal of treated or untreated wastewater is generally accomplished by discharge into natural water bodies such as lakes, streams, rivers, and oceans. This discharge could be made through an exposed open end of a sewer with a concrete headwall on the bank, when the discharge is to a small stream. Sewers discharging into large water bodies are usually extended long distances beyond the bank and terminate with a multiport diffuser outfall section. Hydraulic losses through a diffuser outfall depend on the length of the outfall pipe, the size of the orifice, the depth of the diffuser section, the differential density of the effluent and the receiving water, and the variations in receiving water levels.

A typical multiport diffuser outfall is shown in Fig. 15. The head losses in this outfall include:

1. The entrance loss, which can be taken as $0.5\ V^2/2g$.
2. The pipe friction losses through the straight sections of pipeline, which can be computed by use of the Manning or Hazen-Williams formulas described in this section.
3. The head at the nozzle or diffuser required to force the flow through the opening. This head is computed from the formula for discharge through the port, which is given by

$$q = C_d A_p \sqrt{2gE}$$

where q = flow through the port
C_d = coefficient of discharge, a function of the velocity through the port
A_p = area of port under consideration
E = total energy of liquid immediately upstream of port

In calculating the energy at the port E, the relative difference in density of the effluent and receiving water must be taken into consideration. If the discharge is made to seawater with greater density than the effluent, the density head must be added to the energy of the water in the pipeline. This density head is given by

$$h_d = (\gamma_s/\gamma_f)h$$

where γ_s = specific weight of seawater, pcf
γ_f = specific weight of freshwater, pcf
h = depth of water at port, ft

Diffuser outfall design involves determination of a range of diffuser sizes that are required to discharge the effluent uniformly along the length of the diffuser section. The hydraulics of the diffuser section are complex and involve an iterative procedure which is beyond the scope of this section. The reader should consult other references[15] for more details on this subject.

53. Sludge Handling. Sludges from processing elements vary in their nature and solids concentration, which, together with temperature, determine flow behavior and hydraulic losses in the handling of sludge. Some processing elements produce sludge low in solids concentration and with characteristics close to the viscosity of water so that hydraulic losses may be computed for turbulent flow using common hydraulic formulas. Such hydraulic computations can be applied to systems for returning activated sludge and usually to sludge piping from intermediate and final sedimentation tanks used in conjunction with trickling filters. Other processing elements produce sludge of increasingly higher sludge concentration and of plastic or pseudoplastic nature with such shear resistance to flow that when overcome, the flow is laminar unless the pressure is increased to produce the velocity required for turbulent flow, resulting in greater hydraulic losses. Such process products are raw primary sludge, thickened sludge, concen-

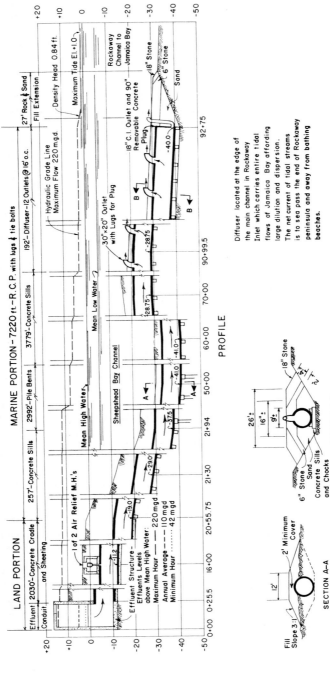

FIGURE 15 90-in R.C. Outfall, Coney Island, Sewage Treatment Plant, New York.

trated sludge, elutriated sludge, digested sludge, and scum. The characteristics of these sludges will vary further according to the quantity and type of solids added from other processing elements. Raw sludge has been found to give pressure drops greater than a digested sludge of the same solids concentration in the same pipeline.

Reference 16 contains references to many investigators whose papers with those of Behn give an understanding of the complexity of rheological sludge flow behavior and progress in quantitative development of reasonable predictions of pressure drops for laminar flow of non-Newtonian sludges in pipelines. Where long pipelines are to be used for transporting sludges of high concentration and plasticity, they must be designed for laminar flow to keep friction losses to the minimum.

At present, empirical methods with a large safety factor based on experience continue to be generally used. Head losses in piping are computed using the Hazen and Williams formula with a C factor range of 90 to 60, usually $C = 80$. Velocities range from 3 to 4 fps, increasing to the range from 5 to 8 fps for piping carrying heavy sludge and grease. Sludge piping has a minimum size of 6 or 8 in, with the length of suction piping to the pumps minimized, and, if possible, a positive head for self-pump priming. Provisions should be made for cleanouts, vent pipes, taps for steam cleaning, and access openings for mechanical cleaning.

Clogging of pipelines and pump stoppage have caused major maintenance costs and upset operation of sludge processing. Efficiency of sludge pumps should be subordinated to dependable and trouble-free operation.

Standby pumps are generally provided for scum and primary, secondary, and elutriated sludge systems. Standby provisions are not required where repair time is permissible in recirculation and sludge-transfer lines.

The handling, treatment, and disposal of sludges from wastewater comprise design problems comparable with the design of wastewater-treatment plants but with the selection of sludge-handling equipment more dependent on the peculiar characteristics of the sludges at the various stages of the sludge-processing system.

Applied hydraulics for sludge handling may reasonably include the foregoing described hydraulics for pipelines plus head losses for sludge-handling equipment based on experience and data obtainable from the manufacturer of the equipment anticipated to be selected. When other sludge equipment is used, in actual construction, some minor modifications in applied hydraulics may be necessary.

54. Summary. The foregoing indicates that some of the hydraulic losses in wastewater-treatment plants are readily computed, whereas others depend on special allowances made by the engineer to provide for satisfactory operation, future additions, and the like, based on judgment. Therefore, the computations should be related to and checked by actual plant-operating experience. The layout of a typical wastewater-treatment plant is shown in Exhibit 2.

The total head available, for a treatment plant, is the difference in elevation between the high level of the sewage in the inlet sewer and the high level of the water surface in the receiving waterway. Where this difference does not provide sufficient head for the operation of the treatment plant, pumping generally will be required.

Hydraulic profiles are computed to determine the head required for the operation of the treatment plant. The overall head requirements will vary for different types of sewage-treatment plants and for different plants of the same type, as illustrated by data in Table 16.

Sewage treatment by plain sedimentation requires the lowest overall head.

Exhibit 2 Wastewater collection and treatment plant at the Will County coal-fired station of Commonwealth Edison Company (Harza Engineering Company).

The greatest losses in this type of treatment occur through the preparatory elements including screens, grit chambers, and the conduits.

Chemical-treatment plants usually include a flocculation tank and some extra length of conduit and require some head in addition to that for plain sedimentation.

An effluent filter may be included and constructed either around the final sedimentation tanks or as a separate structure. Effluent filters may be operated on a 3-in normal and about 6-in maximum head. To provide a factor of safety, 8 to 9 in is sometimes allowed. A separate structure would require additional head for extra conduits.

Activated-sludge treatment plants will require additional head allowances for conduits, aeration tanks, and final sedimentation tanks.

Trickling-filter treatment plants usually require more head than other types, as head is required to distribute the sewage properly over the filter, to provide for the depth of the filter, and for the underdrainage system.

When a plant outfall is too short, a contact tank might be included for chlorination, in which case moderate additional losses will occur.

Hydraulic profiles illustrating head requirements are shown in Fig. 16 for a plain sedimentation plant and in Fig. 17 for a high-rate single-stage filter plant.

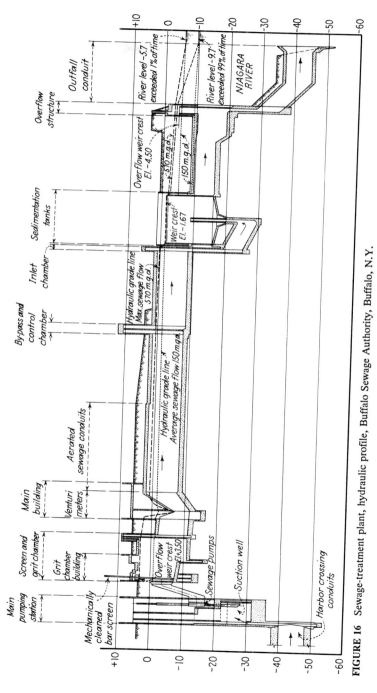

FIGURE 16 Sewage-treatment plant, hydraulic profile, Buffalo Sewage Authority, Buffalo, N.Y.

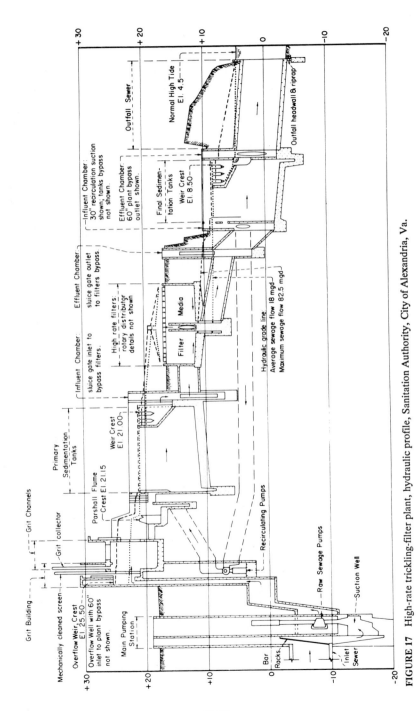

FIGURE 17 High-rate trickling-filter plant, hydraulic profile, Sanitation Authority, City of Alexandria, Va.

TABLE 16 Overall Head Allowances for Various Types of Wastewater-Treatment Plants

Treatment Process and Plant	Rated plant capacity, mgd		Loss of head through plant, at av. flow†
	Av flow	Max. flow	
Plain sedimentation:			
Yonkers joint plant, N.Y.	63.0	230.0	15.67
Richmond, Va.	70.0	200.0	7.92
Buffalo, N.Y.	150.0	570.0	4.34
Tampa, Fla.	36.0	86.4	4.92
Trickling filters:			
Alexandria, Va.	18.0	82.5	14.20
Austin, Minn.	9.3	12.2	58.29†
Knoxville, Tenn., Loves Creek	3.32	8.17	28.26†
Sioux Falls, S. Dak.*	4.9	8.9	46.43†
Rockford, Ill.*	31.0	39.0	17.63†
Tullahoma, Tenn.	1.5	4.2	10.53
Activated sludge:			
Appleton, Wis.	16.0	32.0	4.37
Sioux Falls, S. Dak.*	17.0	25.6	3.59
Rockford, Ill.*	39.0	56.0	7.43
Knoxville, Tenn.:			
Third Creek	31.4	60.0	13.88†
Fourth Creek	7.72	15.20	38.70†
North Shore Sanitary District, Lake County, Ill., Clavey Road	4.5	15.0	2.17
Nassau County, N.W., Sewage Disposal District No. 2	60.0	140.0	10.94†

*Integrated trickling filter and activated-sludge process.
†Includes head restored by pumping.

REFERENCES

1. METCALF and EDDY, INC., *Wastewater Engineering: Treatment/Disposal/Reuse*, 2d ed., McGraw-Hill, New York, 1979.
2. WPCF Manual of Practice 9 (ASCE Manuals and Reports on Engineering Practice 37), Design and Construction of Sanitary and Storm Sewers, 1969 Water Pollution Control Federation.
3. BABBITT, H. E., and E. R. BAUMANN, *Sewerage and Sewage Treatment*, Wiley, New York.
4. Recommended Standards for Sewage Works—A Report of Committee of the Great Lakes Upper Mississippi River Board of State Sanitary Engineers, 1978 ed.
5. "Drainage Manual," Illinois Department of Transportation, 1989.
6. "Storm Water Management Model," Environmental Protection Agency, vol I, 1971, METCALF and EDDY, INC., University of Florida, Gainesville, Fla., Water Resources Engineers, Inc.
7. TERSTRIEP, M. L., and J. B. STALL, "The Illinois Urban Drainage Area Simulator," ILLUDAS, Illinois State Water Survey, *Bulletin 58*, 1974.
8. BRATER, ERNEST F., and HORACE WILLIAMS KING, *Handbook of Hydraulics for the Solution of Hydraulic Engineering Problems*, 6th ed., McGraw-Hill, New York, 1976.

9. STREETER, VICTOR L., and E. BENJAMIN WYLIE, *Hydraulic Transients,* McGraw-Hill, New York, 1985.
10. MOP8, Wastewater Treatment Plant Design, 1977, "A Manual of Practice," Water Pollution Control Federation.
11. CHOW, VEN TE, *Open Channel Hydraulics,* McGraw-Hill, New York, 1959.
12. "Design of Small Dams," A Water Resources Technical Publication, 3d ed., U.S. Department of the Interior, Bureau of Reclamation, 1987.
13. METCALF and EDDY, INC., *Wastewater Engineering: Collection, Treatment and Disposal,* McGraw-Hill, New York, 1991.
14. CAMP, T. R., "Sedimentation and the Design of Settling Tanks," *Transactions of the ASCE,* vol. III, p. 895, 1946.
15. RAMON, A. M., F. R. BOWERMAN, and N. H. BROOKS, "Diffusers for Disposal of Sewage in Sea Water," *Transactions of the ASCE,* vol. 126, part III, p. 344, 1961.
16. BEHN, VAUGHN C., "Experimental Determination of Sludge Flow Parameters," *Proceedings of the ASCE,* vol. 88 (SA3), 1962.

INDEX

Accretion, in barrages and dams on permeable foundations, **12.**3
Adding powerhouses to gravity dams, **9.**27
Aeration devices, **3.**11
Aeration of downstream releases, **18.**21
Afflux, in barrages and dams on permeable foundations, **12.**3
Aging, effects of in closed conduits, **2.**12
Albinga Dam, Switzerland, **9.**20
All-American canal desilting works, **6.**14
Alluvial channels (*see* Channels in alluvium)
Anadromous fish, **18.**2
Analysis:
  of adopted plans for arch dams, **10.**14
  for hydraulic transients, **22.**11
  of preliminary plans for arch dams, **10.**10
  stability, criteria for, **9.**14
Anchorage and trunnion support for radial-gate, **17.**56
Anchors for prestressing and post-tensioning (*see* Cable and its anchors)
Antecedent or subsequent storms, **1.**39
Appalachia, tunnel, aging, Manning's *n*, **2.**12
Appalachia powerhouse, **19.**21
Arch dams, **10.**1
  analyses of preliminary plans, **10.**10
    full load on arches, **10.**10
      William Cain equations, **10.**10
    radial adjustment at crown, **10.**12
    radial adjustment at several sections, **10.**14
  analysis of adopted plans, **10.**14
  constant-radius, massive:
    Hoover Dam, **10.**39

Arch dams (*Cont.*):
  design, **10.**6
    allowable stresses, **10.**6
    constants needed in analyses, **10.**6
    foundations and abutments, **10.**9
    maximum stresses, **10.**6
    preliminary plans, **10.**9
  double-curvature:
    Karadj Dam, Iran, **10.**42
    Mossyrock Dam, Washington, **10.**42
    Strontia Springs Dam, Colorado, **10.**42
  finite element method, **10.**30
    criteria, **10.**31
    dynamic analysis, **10.**32
    current practice, **10.**32
    future research needs, **10.**33
    general, **10.**30
  general theory, **10.**2
    arch action only, **10.**2
    basic assumptions, **10.**3
    cantilever and arch action. **10.**3
  instrumentation, **10.**34
  loads, **10.**4
    ice pressure, **10.**4
    temperature, **10.**4
    uplift pressure, **10.**4
  model investigation, **10.**35
  stress distribution, **10.**5
    arch stresses, **10.**5
    cantilever stresses, **10.**5
    principal stresses, **10.**6
  trial-load method, **10.**14
    adjustments, **10.**29
      arch stresses, **10.**30
    arch analysis, **10.**18
      notation, **10.**18**

Arch dam, trial-load method (*Cont.*):
  signs, **10.**19
  cantilever analysis, **10.**16
    cantilever with parallel sides, **10.**16
  computer method, **10.**26
  manual method, **10.**20
    arch constants, **10.**24
    crown forces, **10.**22
    deflections, **10.**23
    load constants, **10.**26
    load formulas, **10.**25
    rock movements, **10.**15
  types, **10.**1
    design philosophy, **10.**1
    evolution of arch shapes, **10.**2
Area-capacity curves, reservoir, **4.**2
Area-reduction curves, **1.**17
Area-volume curves, reservoir, **4.**2
Areia concrete face rockfill dam (*see* Foz do Areia CFRD)
Arithmetic method, in flood routing, reservoir, **4.**16

Backwater curves, **2.**23, **5.**9
  basic data, **5.**10
  mathematical solutions, **5.**11
    Table 3, determination of, **5.**12
  problem and purpose, **5.**9
  theory and principles, **5.**9
    Manning formula, **5.**9
Baffle blocks, in barrages and dams on permeable foundations, **12.**7
Bank slopes, in canals and conduits, **7.**11
  Table 7, canal bank slopes, **7.**14
Banks, spoil, **7.**16
Banks, top width and thickness in canals and conduits, **7.**15 to **7.**16
Barrage and the river, in barrages and dams on permeable foundations **12.**1 to **12.**4
  accretion and retrogression, **12.**3
  definitions, **12.**3
  afflux, **12.**3
  introduction, **12.**2
    barrage, definition, **12.**1
  nomenclature, **12.**1
  orientation, **12.**2
  siting, **12.**1
Barrage appurtenant structures, barrages and dams on permeable foundations, **12.**20 to **12.**22
  flank walls, **12.**21
  gates and hoists, **12.**20

Barrage appurtenant structures, barrages and dams on permeable foundations (*Cont.*):
  guide banks, **12.**21
  marginal embankments or bunds, **12.**22
  regulators, **12.**21
Barrage-substructure design, barrages and dams on permeable foundations, **12.**10 to **12.**20
  exit gradient, **12.**20
    Khosla formula for, **12.**20
  sliding factor, **12.**20
  steel sheetpiling, **12.**10
    scour holes, **12.**11
  uplift, **12.**11
    Castle Rock Spillway, Wisconsin, **12.**18
    diagram, **12.**13
    fragments, method of, **12.**11
    independent variables, method of, **12.**11
    Khosla method, uplift and exit gradient determination, **12.**13 to **12.**18
    Marala Barrage, Pakistan, **12.**18
    Petenwell spillway, Wisconsin, **12.**18, **12.**22
    Santee Cooper spillway, South Carolina, **12.**18, **12.**22
Barrage-superstructure design, barrages and dams on permeable foundations, **12.**4 to **12.**10
  approach slab, **12.**8
    coefficients of friction, **12.**8
  baffle blocks, **12.**7
  barrage width, **12.**4
    design flood, **12.**4
    Lacey regime formula, **12.**4
    Lacey width, **12.**4
    looseness factor, **12.**4
  concrete block aprons and scour, **12.**8
    Lacey scour depth, **12.**9
  crest level, **12.**5
  design of weir section, glacis and downstream floor, **12.**8
    gravity design, **12.**8
    raft design, **12.**8
  downstream floor, **12.**6
    concentration factor, **12.**6
    Indus Basin project, Pakistan, **12.**6
    retrogression, **12.**6
  end sill, **12.**7

Barrage-superstructure design, barrages and dams on permeable foundations (*Cont.*):
  glacis, **12.5**
  piers, **12.5**
  stone aprons, **12.9**
  width of bays, **12.5**
Barrages and dams on permeable foundations, **12.1** to **12.23**
  barrage and the river, **12.1** to **12.4**
  barrage appurtenant structures, **12.19** to **12.21**
  barrage-substructure design, **12.10** to **12.19**
  barrage-superstructure design, **12.4** to **12.10**
  dams on permeable foundations, **12.21** to **12.23**
Basic concepts of pumped storage, **20.1**
Bath County pumped storage project, Virginia, **20.3**, **20.14**
  jet flow gates, **17.17**
Bear Valley, California, Arch Dam, built in 1888, **10.2**
Bend losses:
  in canals and conduits, **7.11** to **7.13**
  in closed conduits, **2.15**
Bernoulli equation, **2.2**
Bernoulli's theorem, **5.1**
Best hydraulic shape, in canals and conduits, **7.1** to **7.2**
Big Wood River, Minidoko Project, Idaho, **26.28**
Biological criteria, **18.6**
Bluestone Dam river diversion, **8.8**
Boulder Dam river diversion, **8.1**
Box Canyon Project, **16.7**
Boyd's Corner Dam, New York, spillway expansion, **11.12**, **11.21**
Bulkheads, **17.8**, **17.24**, **17.73**
Bunds, in barrages and dams on permeable foundations, **12.22**
Bureau of Reclamation pier-slot design studies, **16.12**
Butterfly valves, **17.43**

Cable and its anchorage, **11.6**
  anchor heads, **11.11**
  anchor stressing operations, **11.8**
  corrosion and corrosion protection, **11.7**
  epoxy coated, **11.7**
  types of anchors, **11.6**

Calion lock layout, **23.3**
Canals and conduits, **7.1** to **7.24**
  closed (*see* Closed conduits)
  conveyance losses, **7.7**
  design of canals, **7.13** to **7.16**
  flow resistance, **7.7** to **7.13**
  hydraulic computations, **7.17** to **7.19**
  hydraulic factors, **7.1** to **7.7**
  irrigation, **26.8**
  lining earth canals, **7.19** to **7.20**
  Portland-cement concrete lining, **7.20** to **7.23**
  regime (*see* Regime canals)
  seepage from canals, **7.23** to **7.24**
  typical sections, in canals and conduits, **7.13** to **7.14**
Carlson and Reading R values, **9.23**
Caruachi Dam river diversion, **8.3**, **8.9**, **8.11**
Catadromous species, **18.3**
Cavitation, incipient, and damage experience, **16.36**
Cethana concrete face rockfill dam, Tasmania, Australia, **14.17**
CFRD (*see* Concrete face rockfill dams)
Channels, natural (*see* Natural channels)
Channels in alluvium, **6.1** to **6.5**
  stable and regime channels, **6.1**
  tractive force, **6.2**
    channels with gravel beds, **6.4**
    channels with sand beds, **6.5**
    limiting tractive forces, **6.3**
    maximum shear stress, **6.2**
    slope formulas, **6.3**
    tractive force formula, **6.2**
Cheartas Dam, Algeria, post-tensioning, **11.12**
Chezy formula, **2.3**, **28.21**
Chicago, tunnel deposits, Manning's n, **2.12**
Chief Joseph Dam river diversion, **8.8**
Chopra, A. K., **9.21**
Churchill Falls powerhouse, **19.23**
Chute (or trough) spillways, **16.16**, **16.19**, **16.23**, **16.24**
Circular sections, properties of, **7.4**
Clay and concrete drain pipe, **25.17**
Closed conduits, **2.1** to **2.19**
  Bernoulli equation, **2.2**
  Chezy formula, **2.3**
  coefficient of roughness n, values of **2.6**
  coefficient, friction, evaluation of, **2.7** to **2.12**
  Colebrook and White, **2.10**

Closed conduits, coefficient, friction, evaluation of (*Cont.*):
    deposits and organic growth, **2.12**
    Moody Chart, **2.10**
    Nikuradse, **2.9**
    Reynolds, **2.7**
    von Karman and Prandtl, **2.10**
  conservation:
    of mass, **2.1**
    of momentum and energy, **2.1**
  corrosion in, **2.12**
  Darcy-Weisbach formula, **2.4**
  deposits and organic growth in, **2.12**
  friction losses, **2.3** to **2.7**
  Hazen and Williams formula, **2.7**
  Kutter formula, **2.4**
  losses:
    boundary **2.3**
    losses, form, **2.3**, **2.12** to **2.18**
    head, **2.3**
  Manning formula, **2.3**
  sudden enlargements in, **2.13**
Closed subsurface drainage system, design criteria, **25.11**
Coal-tar epoxy paint on gates, **17.64**
Coanda screens, **18.18**
Coefficients:
  of friction for approach slab barrages and dams on permeable foundations, **12.8**
  of resistance, f, **2.9**
  of roughness, n, **2.6**
Coefficients, discharge:
  C, values of, **2.19** to **2.20**
    gate opening, **2.20**
    orifices and tubes, **2.19** to **2.21**
    sharp-crested weirs, k, **2.30** to **2.33**
Cofferdams, **8.17**
  deflectors, **8.20**
Combined sewer systems, **28.2**
Compacted rockfill for concrete face rockfill dams, **14.8**
Concrete crack repair, typical methods, **11.20**
Concrete dams, **9.1**
  buttress dams, **9.1**
  criteria applicable to stability analysis results, **9.14**
    FERC criteria, **9.14**
  design considerations, **9.7**
    loading combinations, **9.9**
    loads, **9.8**

Concrete dams, design considerations (*Cont.*):
    dead, **9.8**
    earthquake, **9.9**
    ice, **9.9**
    silt, **9.9**
    thermal, **9.9**
    uplift, **9.9**
    water, **9.8**
  stresses in dams, **9.7**
  foundation stress, **9.12**
  sliding stability, **9.13**
  gravity dams, **9.1**
    examples:
      Daniel Johnson Dam (Quebec), **9.3**
      Dworshak Dam (Idaho), **9.1**
      Fontana Dam (Tennessee), **9.5**
      Grand Coulee Dam (Columbia River), **9.4**
      Grand Dixence Dam (Switzerland), **9.1**
      Guri Dam (Venezuela), **9.4**
      Hoover Dam (Colorado River), **9.3**
      Nurek Dam (USSR), **9.1**
      Three Gorges Project (China), **9.5**
    features, **9.26**
      adding powerhouses, **9.27**
      dam raising, **9.26**
      design for foundation movement, **9.29**
      post-tensioned anchors, **9.28**
      roller-compacted concrete, **9.28**
  method of analysis, **9.9**
    finite element analysis, **9.11**
    overturning stability, **9.10**
  seismic loading, **9.20**
    Chopra, A. K., **9.21**
    finite element method, **9.21**
    pseudo-static method, **9.20**
    seismic zone map, **9.20**
    University of California, **9.21**
  stability analysis criteria, **9.14**
  temperature and crack control, **9.22**
    Carlson and Reading R valves, **9.22**
    cause of cracking, **9.22**
    method of temperature control, **9.25**
    tensile stress equation, **9.23**
    USBR allowable temperature drop table, **9.23**
      research, uplift and drain holes, **9.17**
  uplift pressure, **9.15**
    assumptions, **9.16**

Concrete dams, uplift pressure (*Cont.*):
  and cracked base analysis, 9.17
  Fontana Dam diagram, 9.18
  Hiwasse Dam, 9.17
  uplift reduction and buttress dams, 9.18
  Albinga Dam (Switzerland), 9.20
Concrete drain pipe, clay and, 25.17
Concrete face rockfill dams (CFRD), 14.1
  compacted rockfill for, 14.8
    compaction, 14.8
    density, 14.9
    grading, 14.9
    modulus, 14.9
    placing, 14.8
    shear strength, 14.9
    use of water, 14.9
  consideration, general of, 14.1
    adaptability, 14.4, 14.5
    future raising, 14.5
    section and zoning, 14.2
    seismic considerations, 14.5
    sliding factor, 14.5
    spillways over CFRD, 14.5
    type and height, 14.1
  construction of, 14.14
    haul roads, 14.15
    quarry, 14.15
    specifications, 14.15
  design of, 14.10
    embankment zoning, 14.14
      plate vibrators, 14.14
      zone 2—face compaction, 14.14
    face slab, 14.12
      concrete, 14.12
      reinforcing, 14.12
      thickness, 14.12
      width and joints, 14.12
    foundation treatment, 14.10
      under embankment, 14.10, 14.15
      under toe slab, 14.10
    joints, 14.13
      details, 14.19, 14.26, 14.29
    parapet wall and camber, 14.13
    toe slab, 14.11
      dimensions, 14.11
      layout, 14.11
      reinforcing, joints and anchors, 14.11
  features of, 14.6
    construction, 14.7
    design, 14.6
    toe slab, (plinth), 14.9
    uplift, 14.7

Concrete face rockfill dams (CFRD) (*Cont.*):
  high, 14.33
    world's highest, 14.32
  performance of:
    leakage, 14.17
    settlement, 14,16
  typical designs, 14.15
    Cethana (Tasmania), 14.17
    Foz do Areia (Paraná, Brazil), 14.18, 14.20 to 14.22
    Khao Laem (Central West Thailand), 14,10, 14.24, 14.27
    Macagua (Puerto Ordaz, Venezuela), 14.23
    Salvajina, 14.10
    Segredo (Brazil), 14.28 to 14.30
Concrete-block aprons, in barrages and dams on permeable foundations, 12.8
Conduits (*see* Canals and conduits)
Conduits, closed (*see* Closed conduits)
Conical contraction, gradual, in closed conduits, 2.15
Conical expansion, gradual, in closed conduits, 2.15
Construction quality control, embankment dams, 13.59
Continuity, fundamental law of, 2.1
Contraction, in closed conduits, sudden, 2.14
Control, seepage, embankment dams, 13.18
Conveyance losses and waste, 24.29
  in canals and conduits, 7.7
    losses from canals, 7.7
    losses from concrete, metal and wood conduits, 7.7
Conveyance structures for irrigation, 26.11
Corps of Engineers (*see* U.S. Army Corps of Engineers)
Corrosion in closed conduits, 2.12
Corrugated plastic drain pipe, 25.17
Cost and economics of pumped storage, 20.28
Cranes, overhead bridge, gantry, and monorail, 17.73
Crest level, in barrages and dams on permeable foundations, 12.5
Critical flow:
  allowance for, in canals and conduits, 7.2
  in open channels, 2.19 to 2.20

Critical PMP duration, **1.**13
Critical velocity in open channels, **2.**21
Crop evapotranspiration (*see* Evapotranspiration, crop)
Cushma Dam (Conn.):
    No. 1: post-tensioned spillway piers, **11.**12
    No. 2: high radial gates, **17.**24
Cyclones, **1.**12

Dam raising, gravity dam, **9.**26 Dams:
    designed for foundation movement, **9.**29
    on permeable foundations, **12.**22 to **12.**23
        Pearl River spillway, **12.**23
        Petenwell spillway, **12.**23
        Santee Cooper spillway, **12.**23
Dams, rehabilitation (*see* Rehabilitation of dams)
Daniel Johnson Dam, Quebec, Canada, **9.**3
Darcy-Weisbach formula, **2.**4
Darcy-Weisbach head loss equation, **27.**13, **27.**39
Degree-day method, **1.**32
Delivery structures for irrigation, **26.**40
Deposits and organic growth in closed conduits, **2.**12
Depth-area-duration curves, **1.**16
    dimensionless, **1.**19
Derbendi Khan Dam spillway chute, **16.**21
Design considerations, concrete dams, **9.**1, **9.**7
Design considerations, equipment, gates, and valves, **17.**51
Design criteria for closed subsurface drainage system, **25.**11
Design of arch dams, **10.**6
Design of canals, in canals and conduits, **7.**13 to **7.**16
    bank slopes, **7.**14 to **7.**15
    canals, Table 7, **7.**14
    freeboard, **7.**15
    shape and size of waterway, **7.**16
    spoil banks and berms, **7.**16
    top width and thickness of banks, **7.**15 to **7.**16
Design of concrete face rockfill dams, **14.**10
Design of roller-compacted concrete dams, **15.**8 to **15.**19
Design of regime canals, **6.**15 to **6.**16
    procedures, **6.**15
    typical cross-section, **6.**15

Design storms, chronological sequence, **1.**14
    depth-duration frequency, **1.**14
    thunderstorm, **1.**14
Developing pumped storage, **20.**35
Differential equations for transient flow, **22.**7
Dimensionless depth-duration curves, **1.**19
Discharge capacity of spillways, **16.**1
Discharge coefficients (*see* Coefficients, discharge)
Discharge formulas, in hydraulics of natural streams, **5.**1
Discharge, determination of in natural channels, **5.**6
    area-velocity method, **5.**7
    current-meter measurements, **5.**7
    indirect determination, **5.**8
    methods, **5.**6
    slope-area method, **5.**7
    weir measurements, **5.**6
Distribution system for irrigation, **26.**6
Diversion:
    through concrete dams, **8.**8
    in deep channels, **8.**1
    over embankment dams, **8.**19
    in wide channels, **8.**3
    (*See also* River diversion)
Diversion discharge capacities, **8.**13
Diversion tunnels, **8.**15
Diversion weirs for irrigation, **26.**1
Donnan spacing formula for drainage systems, **25.**12
Double-mass curve, **1.**2
Downstream floor, length, in barrages and dams on permeable foundations, **12.**6
Drain pipe:
    clay and concrete, **25.**17
    corrugated plastic, **25.**17
Drain size, **25.**16
Drainage, **25.**1
    design criteria, **25.**11
    depth, **25.**11
    Donnan spacing formula, **25.**12
    DRAINMOD water management model, **25.**14
    grade, **25.**14
    spacing, **25.**12
    yield from drains, **25.**14
    drain size, **25.**16
    hydraulic design, **25.**9
        depth, **25.**10
        grade, **25.**9

Drainage, hydraulic design (*Cont.*):
  spacing, **25**.10
  materials and installation, **25**.17
    clay and concrete drain pipe, **25**.17
    corrugated plastic drain tubing, **25**.17
    envelope materials, **25**.18
    installation of subsurface drains, **25**.18
    join spacing, **25**.19
    laying tile in unstable soil, **25**.19
    pipe materials, **25**.17
  needs, **25**.1
  principles, **25**.1
  pumping for, **25**.19
    inlet and outlet pipe, **25**.20
    low-lift pumping, **25**.19
    operation, **25**.20
    power, **25**.20
    pumping capacity, **25**.19
    pumping from wells for drainage, **25**.21
    spacing, **25**.21
    types of pumps, **25**.20
  salinity, **25**.5
    nonsaline-alkali soils, **25**.6
    saline-alkali soils, **25**.6
    water quality, **25**.7
  soils, **25**.3
    borings, **25**.4
    indirect evaluations or permeability, **25**.4
    permeability, **25**.4
    texture index, **25**.5
  structures, **25**.10
  subsurface drains, **25**.10
    layout of, **25**.11
  surface drains, **25**.8
    rate of discharge, **25**.8
    rational formula, **25**.8
    southwestern drainage formula, **25**.8
    system layout **25**.8
  surveys and investigations, **25**.2
  water sources, **25**.2
    hydrostatic pressure, **25**.3
    irrigation, **25**.3
    precipitation, **25**.3
    seepage, **25**.3
  water tables, **25**.7
    observation wells, **25**.7
    piezometers, **25**.7
DRAINMOD Computer water management model, **25**.14
Droughts, definitions, **1**.8
  duration-frequency, **1**.9

Droughts, definitions (*Cont.*):
  low-flow sequence, **1**.9
Dworshak Dam (Idaho), **9**.1
  diversion tunnel, **8**.17
  river diversion, **8**.15, **8**.16

Earthquake considerations, embankment dams, **13**.50 to **13**.53
Economics of pumped storage, **20**.28 to **20**.32
Eicher screens, **18**.18, **18**.20
Elements of pumped storage, **20**.6 to **20**.21
Embankment dams, **13**.1
  basic requirements, **13**.1
  construction quality control, **13**.59
    general, **13**.59
    quality control, **13**.59
    records, **13**.60
  earthquake considerations, **13**.50
    defensive measures, **13**.51
    design criteria, **13**.53
    embankment dam performance, **13**.52
    general, **13**.50
    liquefaction analysis, **13**.53
  foundation treatment, **13**.6
    control of underseepage, **13**.10
    foundation treatment, **13**.8
    general considerations, **13**.6
    surface foundation treatment, **13**.9
  geology, **13**.2
    construction supervision, **13**.5
    design stage investigation, **13**.4
    early stage/prefeasibility investigation, **13**.2
    field supervision of investigations, **13**.5
    intermediate/feasibility stage investigations, **13**.2
    introduction, **13**.2
    tools and methods for subsoil investigation, **13**.3
  laboratory tests, **13**.17
    general considerations, **13**.17
    selected laboratory tests:
      ASTM Standards, **13**.19
      U.S. Army, **13**.20
      U.S. Bureau of Reclamation, **13**.20
    soil classification chart, **13**.14
    test procedures, **13**.17
  monitoring and performance evaluation, **13**.60
    general, **13**.60

Embankment dams, monitoring and performance evaluation (*Cont.*):
  instrumentation, **13.61**
  monitoring performance, **13.61**
  performance evaluation, **13.60**
 seepage analysis and control, **13.18**
  criteria for filters, **13.35**
  criteria for self-filtering, **13.35**
  Darcy's law for flow through soils, **13.22**
  discharge rate of seepage, **13.30**
  filters, **13.33**
   permeability criterion, **13.33**
   retention criterion, **13.33**
  flow conditions and forces acting on filters, **13.34**
  flow nets, **13.25**
  gradation limits to prevent segregation, **13.36**
  permeability and drainage characteristics of soils, **13.24**
  permeability defined, **13.20**
  piping, **13.32**
  seepage by relaxation methods, **13.27**
  seepage control within the dam, **13.36**
 settlement analysis, **13.54**
  general considerations, **13.54**
  post-construction settlement, **13.56**
  settlement from compression within embankment during construction, **13.55**
  settlement from consolidation of foundation, **13.55**
 slope protection and freeboard, **13.56**
  freeboard, **13.56**
  general considerations, **13.56**
  slope protection, **13.57**
   riprap layer thickness, **13.58**
   riprap sizes, **13.58**
   riprap weights, **13.58**
   types of slope protection, **13.57**
   wave height, **13.57**
 stability analysis, **13.38**
  basic concepts, **13.38**
  infinite-slope analysis, **13.43**
  shear strength parameters, **13.40**
   end of construction, **13.41**
   rapid drawdown, **13.41**
   steady state seepage, **13.41**
  slices method of stability analysis, **13.44**
   forces acting on a typical slice **13.46**
   use of electronic spreadsheet, **13.48**

Embankment dams, stability analysis (*Cont.*):
  sliding block method, **13.49**
  stability analysis procedures, **13.43**
  stability charts, **13.49**
 subsurface investigations, **13.9**
  field sampling and testing procedures U.S. Bureau of Reclamation, **13.12**
  general considerations, **13.9**
  in-situ permeability tests, **13.11**
  logging and test of soil, **13.10**
  logging and testing of rock, **13.15**
  pressure tests in rock, **13.16**
  selected field tests, ASTM, **13.11**
 types, **13.5**
  earth dams, **13.6**
  general considerations, **13.5**
  rockfill dams, **13.6**
  typical cross-sections, **13.7, 13.8**
End sill, in barrages and dams on permeable foundations, **12.7**
Energy, conservation of, in closed conduits, **2.1**
Enlargements, in closed conduits, sudden, **2.13**
Environmental aspects and fish facilities, **18.1**
 aeration of downstream releases, **18.21**
 biological criteria, **18.6**
  dissolved oxygen, **18.8**
  temperature, **18.7**
  velocity, **18.8**
 fish passage, **18.9**
  determination of biological criteria, **18.9**
  fish ladder components, **18.14**
  fish ladder design, **18.14**
  fish ladder hydraulics, **18.11**
  fish ladder types, **18.14**
   Denil and steep-pass fish ladders, **18.14**
   pool-and-weir/orifice fish ladders, **18.14**
   vertical-slot fish ladder, **18.14**
  fish locks and fish lifts, **18.15**
  preliminary design report, **18.12**
  quantification of hydrology, **18.10**
 fish protection, **18.16**
  behavioral systems, **18.16**
  diversion systems, **18.18**
  physical barriers, **18.17**

Environmental aspects and fish facilities (*Cont.*):
  general effects of impoundment, **18.**2
    altered flow regime, **18.**3
    physical barriers, **18.**2
    water quality effects, **18.**5
  multilevel outlet structures for temperature control, **18.**23
Environmental Protection Agency (EPA), **18.**1
Equipment design considerations, gates and valves, **17.**51
Equipment for navigation locks, **23.**18
Estimation of streamflow, **1.**39
Euler number, **16.**26
Evaporation, definitions, **1.**10
  evaporation pan, **1.**11
  geographic distribution, **1.**11
  measurement, **1.**10
  reservoirs, **1.**11
Evapotranspiration, crop, **24.**5 to **24.**20
  crop coefficients, **24.**13 to **24.**20
  estimating requirements:
    FAO-24 Blaney-Criddle Method, **24.**8 to **24.**9
    FAO-24 Pan Evaporation Method, **24.**11 to **24.**13
    FAO-24 Radiation Method, **24.**9 to **24.**11
    Penman-Monteith method, **24.**7 to **24.**8
Examples of prestressed/post-tensioned dams, **11.**12
  existing projects, **11.**12
  extensometers, **11.**14
Exit gradient, in barrages and dams on permeable foundations, **12.**18, **12.**20

FAO-24 Blaney-Criddle Method for estimating evapotranspiration, **24.**8
FAO-24 Pan Evaporation Method for estimating evapotranspiration, **24.**11
FAO-24 Radiation Method for estimating evapotranspiration, **24.**9
Farm irrigation requirements, **24.**20
Features of concrete face rockfill dams, **14.**6
Features of gravity dams, **9.**3
Federal Energy Regulatory Commission (FERC), **18.**1
  stability analysis criteria, **9.**14
Fetch, effective, **4.**12
Finite element analysis method for arch dams, **10.**30 to **10.**32

Finite-difference characteristic equations for hydraulic transients, **22.**13
Fish facilities (*see* Environmental aspects and fish facilities)
Fish ladder, **18.**11, **18.**14
Fish passage, **18.**9
Fish protection, **18.**16 to **18.**17
Flank walls, in barrages and dams on permeable foundations, **12.**21
Flip buckets, **3.**7
Flood frequency, confidence limits, **1.**38
  confidence interval, **1.**38
  expected probability, **1.**38
  goodness-of-fit, **1.**33
  plotting position, **1.**33
  probability distribution, **1.**33
  regional analyses, **1.**38
  reliability, **1.**38
  return period, **1.**36
Flood hydrographs, baseflow separation, **1.**30
  components, **1.**30
  continuous simulation models, **1.**33
  derivation of peak, **1.**39
  event simulation models, **1.**38
  multi-peaks, **1.**30
  return period selection, **1.**40
  single-peaks, **1.**30
Flood routing, reservoir, **4.**16
Flow:
  critical, **2.**19
  subcritical, **2.**19
  supercritical, **2.**19
Flow characteristics in natural channels, **5.**1
  critical flow, **5.**1
  subcritical flow, **5.**1
  supercritical flow, **5.**1
Flow formulas and frictional resistance, **2.**3 to **2.**7, **2.**23
Flow in pipes, computing, **2.**4 to **2.**6
Flow resistance, in canals and conduits, **7.**7 to **7.**13
  basic data, **7.**7
  bend losses, **7.**11 to **7.**13
  hard surface conduits, **7.**8
    Manning's n for hard-surface conduits (Table 5), **7.**10
  lined channels, **7.**9
    Manning's n for lined channels (Table 6), **7.**11
  pier losses, **7.**9 to **7.**10

Flow resistance, in canals and conduits (*Cont.*):
  unlined earth channels, **7.**8
    Manning's *n* for unlined channels (Table 4), **7.**8
Flow routing, in natural channels, **5.**11
  basic data requirements, **5.**13
  definition and methods, **5.**11
  mathematical solution, **5.**15
    Muskingum method, **5.**15
      routed hydrograph (Fig. 8), **5.**17
      routing computations (Table 5), **5.**18
  practical applications, **5.**11
  theory and principles, **5.**11
Flow sequence generation, long-term, **1.**29
Fontana Dam (Tennessee), **9.**5
  intake, **19.**7
Form losses, closed conduits, **2.**12 to **2.**18
  bend losses, **2.**15
  gradual conical contraction, **2.**15
  gradual conical expansion, **2.**15
  minor losses, **2.**16
  sudden contraction, **2.**14
  sudden enlargements, **2.**13
Fort Laramie Canal, North Platt Project, Wyoming, **26.**39
Fort Loudoun Dam river diversion, **8.**9, **8.**12
Foundation treatment, embankment dams, **13.**6
Foz do Areia CFRD (concrete face rockfill dam), Paraná, Brazil, **14.**18 to **14.**22
  characteristics, **14.**18
  construction, **14.**20 to **14.**22
  perimeter joint, **14.**20
  properties of rocks, **14.**19
  section and rockfill zoning, **14.**19
Free water surface evaporation, **1.**10
Freeboard:
  allowances, **4.**10
  in canals and conduits, **7.**15
  reservoir, definitions, **4.**2
Frequency analysis, rainfall, **1.**2
  floods, **1.**35 to **1.**38
Friction factor, f (Darcy-Weisbach), **2.**4, **2.**8 to **2.**11
Friction loss, **2.**2, **5.**10
  pipes flowing full, **2.**6
  tunnels and channels, **2.**5
Friction slope, **5.**9
Frontal-type storm, **1.**12
Froude number, **2.**21

Gantry cranes, **17.**73
Gated and orifice spillways, **16.**8
Gates and valves, **17.**1
  equipment design considerations, **17.**51
    corrosion protection, **17.**63
    cost estimates, **17.**71
    cranes, **17.**73
    gate wheels and embedded parts, **17.**53
    general, **17.**51
    hoists, **17.**57
    ice problems, **17.**64
    operation and maintenance, **17.**66
    radial-gate trunnion support and anchorage, **17.**56
    roller trains, **17.**55
    safety factors and friction coefficients, **17.**51
    seals, **17.**52
    weight estimates, **17.**67
  gates and bulkheads (or stoplogs) project lists, **13.**68
    bulkheads or stoplogs, **17.**73
    double-leaf fixed-wheel crest gates, **17.**72
    flap gates, **17.**75
    large fixed-wheel gates, **17.**70
    small fixed-wheel gates, **17.**72
    spillway tainter gates, **17.**68
    submerged tainter gates, **17.**69
  gates in barrages and dams on permeable foundations, **12.**20
  general, **17.**1
    definitions, **17.**2
      bulkhead gates, **17.**3
      gate, **17.**3
      guard gates or valves, **17.**3
      regulating gates and valves, **17.**3
      stop logs, **17.**3
      valve, **17.**3
    gates for orifice spillways, **17.**2
    outlet gates and valves, **17.**2
    scope, **17.**1
    spillway crest gates, **17.**1
  high head gates, **17.**8
    bulkheads, **17.**24
    cylinder gates, **17.**24
    general, **17.**8
    jet-flow gates, **17.**16
    radial gates, **17.**24
    ring-follower gates and ring-seal gates, **17.**16

INDEX

I.11

Gates and valves, high head gates (*Cont.*):
  slide gates, **17.**11
  wheel and roller-mounted gates, **17.**18
 hydraulic design factors, **17.**48
  gate bottom shapes, **17.**48
  gate slots, **17.**48
  general, **17.**48
  other hydraulic design factors, **17.**49
 spillway gates, **17.**3
  bulkheads, **17.**8
  flap gates, **17.**5
  flashboards, **17.**3
  inflatable gates, **17.**8
  radial gates, **17.**3
  vertical-lift gates, **17.**5
 valves, **17.**28
  butterfly valves, **17.**43
  fixed-cone (Howell-Bunger) valves, **17.**34
  general, **17.**28
  hollow-jet (Staats-Hornsby) valves, **17.**37
  needle valves, **17.**32
  sleeve-type valves, **17.**42
  sphere valves, **17.**45
  tube valves, **17.**32
General effects of impoundment, **18.**2
Generalized PMP, **1.**17
Geology, embankment dams, **13.**2
George B. Jarvis powerplant, **19.**26
Ghazi-Barotha (formerly Gariala) hydropower siphon spillway, **16.**45
Glacis, in barrages and dams on permeable foundations, **12.**5
Glazon and Pam studies free jet flow, **16.**33
Glen Canyon Dam:
 hollow jet valves, **17.**39
 spillway, **16.**26, **16.**44
Grand Coulee Dam on Columbia River, **9.**4
Grand Dixence Dam, in Switzerland, **9.**1
Gravity dams, adding powerhouses to, **9.**27
Gravity design, in barrages and dams on permeable foundations, **12.**8
Groundwater, aquifer, **1.**25
 artesian, **1.**25
 capillary fringe, **1.**25
 definitions, **1.**25
 depletion curve, **1.**26
 zone of aeration, **1.**25
Guide banks, in barrages and dams on permeable foundations, **12.**21

Gulliver Data, submergence and vortices, **16.**16
Guri Dam (Venezuela), **9.**4
 aeration device, **16.**38
 intakes, **19.**7
 uplift pressure diagram, **9.**18

Hartwell Project powerhouse, **19.**17, **19.**20
Harza method, in regime canals, **6.**7 to **6.**10
Hazen-Williams formula, **2.**7, **28.**22
Hazen-Williams head loss equation, **27.**13, **27.**38
Heart Mountain Canal, Shoshone Project, Wyoming, **26.**27
HGL (hydraulic grade line), **27.**28
High head gates, **17.**8
Hiwasse Dam, uplift pressures, **9.**17
Hoists, in barrages and dams on permeable foundations, **12.**20
Hoover Dam (Colorado River), **9.**3
 jet flow gates, **17.**17
 massive constant-radius arch dam, **10.**39
 spillway **16.**38, **16.**40
Horseshoe sections, properties of, **7.**5
Howell-Bunger valves, **18.**6
 fixed-cone, **17.**28, **17.**34
Hungry Horse Dam:
 hollow jet valves, **17.**39
 Morning-Glory spillway, **16.**41
Hydraulic analysis of water-distribution systems, **27.**31
Hydraulic computations, in canals and conduits, **7.**17 to **7.**19
 computation aids, **7.**17
 examples, **7.**17 to **7.**19
Hydraulic design:
 criteria, **16.**23
 factors:
  for drainage, **25.**9
  for gates, **17.**48
Hydraulic factors, in canals and conduits, **7.**1 to **7.**7
 allowance for critical flow, **7.**2
 best hydraulic shape, **7.**1 to **7.**2
  properties of circular sections (Table 1), **7.**4
  properties of horseshoe sections (Table 2), **7.**5
 effect of slope, **7.**1
 mean velocity, **7.**1
 permissible velocities, **7.**2 to **7.**7

Hydraulic factors, in canals and conduits, permissible velocities (*Cont.*):
  permissible canal velocities (Table 3), 7.6
Hydraulic grade line (HGL), 27.28
Hydraulic jump, 3.7
Hydraulic machinery, 21.1
  pumps, 21.16
    cavitation, 21.21
      model tests, 21.24
    general classification, 21.16
    head, power and efficiency, 21.18
    specific speed, 21.18
  reversible pump-turbines, 21.27
    axial-flow pump-turbines, 21.32
    basic performance relationships, 21.27
    diagonal-flow adjustable-blade pump-turbines, 21.31
    Francis-type pump-turbines, 21.28
      adjustable wicket gates, 21.29
      general design characteristics, 21.28
      single-speed operation, 21.29
      starting procedures, 21.30
      two-speed operation, 21.29
    general classification, 21.27
  turbine speed governors, 21.32
    general, 21.32
    operating modes, 21.33
      frequency adaptation, 21.33
      gate position control, 21.33
      speed control, 21.33
      speed regulation or power control, 21.33
    operating parameters, 21.35
      damping device time constant, 21.35
      dead time, 21.35
      definitions, 21.35
      derivative gain, 21.35
      dynamic characteristics, 21.36
      integral gain, 21.35
      proportional gain, 21.35
      speed deadband, 21.35
      stability, 21.35
    operation sequences, 21.34
      automatic start sequence, 21.34
      emergency shutdown, 21.35
      normal shutdown, 21.34
      partial shutdown, 21.35
    physical arrangement of governor, 21.37
  turbines, 21.1
    cavitation, 21.9

Hydraulic machinery, turbines (*Cont.*):
    characteristic proportions of turbine runners, 21.12
    introduction, 21.1
      axial flow turbines, 21.2
      Francis turbines, 21.2
      impulse turbines, 21.1
      Kaplan turbines, 21.2
    net head, power, efficiency, 21.6
    Pelton impulse wheel, 21.14
    proportionality laws for homologous turbines, 21.7
    scale effect 21.14
    setting and speed of turbine, 21.11
    specific speeds, 21.7
Hydraulic models, 3.1
  definition of a model, 3.1
  hydraulic modeling, principles of, 3.1
    dynamic similitude, 3.2
    geometric similitude, 3.1
    kinematic similitude, 3.1
  model scaling criteria, 3.3
    Euler number 3.3
    Froude number, 3.3
    Mach number, 3.3
    Reynolds number, 3.3
    Weber number, 3.3, 3.12
Hydraulic models, physical types, 3.4
  additional hydraulic models, 3.20
    harbor and tidal models, 3.20
    ice models, 3.21
    ship models, 3.20
    turbine models, 3.20
  models of hydraulic structures, 3.4
  river channel models, 3.14
    distorted models, 3.17
      fixed bed models, 3.17
      Knauss distortion factor, 3.19
      Manning's *n* scale, 3.20
      movable bed models, 3.19
    general consideration, 3.14
    undistorted models 3.15
      fixed bed models, 3.15
      movable bed models, 3.16
      Shield's diagram of sediment movement, 3.17
  spillway models, three dimensional, comprehensive, 3.5
    flip bucket, 3.8
    hydraulic jump, 3.8
    special spillway structures, 3.8
      fuseplug spillway, 3.10

Hydraulic models, physical types, spillway models three dimensional, comprehensive (*Cont.*):
    gated orifice spillway **3.9**
    labyrinth spillway **3.9**
    morning glory spillway **3.8**
    side channel spillway **3.8**
    syphon spillways, **3.9**
    spillway chute and spillway tunnel hydraulics, **3.6**
    spillway crests and discharge capacity, **3.5**
    spillway outlet structures, **3.6**
    submerged roller buckets, **3.8**
  spillway models two dimensional, sectional, **3.10**
    spillway with piers, **3.10**
    spillway without piers or gates, **3.10**
  structure models, miscellaneous, **3.10**
    aeration devices, **3.11**
    emergency gate loads, **3.10**
    fish conveyance facilities, **3.14**
    intake draft tube structures, **3.11**
    navigation locks, **3.13**
  structures in which friction is important **3.14**
    diversion tunnels, **3.14**
    lock flow conveyance system, **3.14**
    spillway tunnels, **3.14**
Hydraulic transients, **22.1**
  basic water-hammer equation, **22.2**
  characteristic method of analysis, **22.11**
  definitions, **22.1**
  development of boundary conditions for method of characteristics, **22.15**
    branch connection, **22.17**
    dead end, **22.18**
    downstream valve, **22.30**
    hydraulic machinery, **22.32**
      efficiency step-up relationships, **22.35**
      normalization of performance characteristics, **22.36**
      performance characteristics, **22.33**
      specific speed, **22.35**
      turbine head-balance equation, **22.40**
      turbine speed equation, **22.43**
    in-line valve, **22.31**
    open surge tank, **22.18**
      operational stability **22.19**
      pressure relief and surge control, **22.22**
    pressured air chamber, **22.27**

Hydraulic transients, development of boundary conditions for method of characteristics (*Cont.*):
    reservoir, **22.15**
    series connection, **22.16**
  differential equations for transient flow, **22.7**
  equation of motion, **22.9**
  equation on continuity, **22.7**
  finite-difference characteristics equations, **22.13**
  general, **22.1**
  wave speed in tunnels and conduits, **22.4**
Hydraulics:
  basic, **2.1** to **2.33**
  of natural streams, **5.1**
  navigation locks, **23.4**
  reservoir, **4.1** to **4.19**
  of sewers, **28.18**
  of water-distribution systems, **27.25**
  of water-treatment systems, **27.9**
Hydroelectric plants, **19.1** to **19.29**
  power from flowing water, **19.1**
    classification of power and energy, **19.5**
      average load, **19.6**
      capacity factor, **19.6**
      capacity of a powerplant, **19.6**
      dump energy, **19.6**
      firm power, **19.5**
      load factor, **19.6**
      peak load, **19.6**
      primary energy, **19.5**
      secondary energy, **19.6**
      surplus system capacity, **19.6**
      utilization factor, **19.7**
    efficiency, **19.4**
    energy and work, **19.1**
    energy line, **19.2**
    head, **19.3**
      gross head, **19.3**
      net or effective head, **19.3**
      operating head, **19.3**
    power formulas, **19.5**
    power, **19.1**
    the Bernoulli theorem, **19.3**
  powerhouse structures, **19.16**
    classification, **19.16**
    economic design, **19.29**
    integral powerhouse, **19.17**
    low-head plants, **19.24**
    powerhouse connected to intake by tunnels and penstocks, **19.21**

Hydroelectric plants, powerhouse structures (*Cont.*):
  separate powerhouse constructed at toe of dam, **19.**17
  underground powerhouse, **19.**22
  water conductor, **19.**7
    draft tubes, **19.**15
      energy relations, **19.**15
    forebay, **19.**7
    general, **19.**7
    intake, **19.**7
    penstocks, **19.**11
      mechanical starting time, **19.**14
      water starting time, **19.**13
    tailrace, **19.**16
    trashracks, **19.**7
    turbines, and pumps, **19.**15
Hydrograph recession, **1.**30
Hydrographs, flood (*see* Flood hydrographs)
Hydrologic analysis computer program, **1.**41
Hyetograph, **1.**13

Illinois Department of Transportation, division of water resources (IDOT DOWR), **27.**4
Illinois Urban Drainage Area Simulator (ILLUDAS), **28.**16, **28.**72
Impoundment, general effects of, **18.**2
Incipient cavitation and damage experience, **16.**36
Indus Basin project, in barrages and dams on permeable foundations, **12.**6
Infiltration losses, concept, **1.**33
  method of computations, **1.**35
Inflow design flood, antecedent and subsequent storms, **1.**39
  concept, **1.**38
  probable maximum flood, derivation, **1.**39
  standard project flood, **1.**40
Instrumentation of arch dams, **10.**34
Interpolation procedures for rainfall duration, **1.**12
Irrigation, **24.**1
  conveyance losses and waste, **24.**29
    evaporation losses, **24.**29
    operational wastes, **24.**31
    seepage losses, **24.**30
  conveyance structures for, **26.**11
  crop evapotranspiration, **24.**5
    FAO-24 Blaney-Criddle Method, **24.**8

Irrigation, crop evapotranspiration (*Cont.*):
    FAO-24 Pan Evaporation Method, **24.**11
    FAO-24 Radiation Method, **24.**9
    Penman-Monteith Method, **24.**5
  farm irrigation requirements, **24.**20
    computation procedures, **24.**26
    cropping pattern, **24.**20
    effect of irrigation methods, **24.**28
    effect of root depth, **24.**23
    effective rainfall, **24.**21
    irrigation applications, **24.**23
    soil moisture capacity, **24.**24
    soil moisture reservoir, **24.**25
    surface waste and farm conveyance losses, **24.**27
  land classification, **24.**1
    chemical properties, **24.**2
    classification standards, **24.**5
    drainability, **24.**3, **24.**3
    physical properties, **24.**2
    purpose of classification, **24.**1
    sampling and testing, **24.**2, **24.**3
  regulating structures for, **26.**29
  results from, **24.**33
    crop yield responses, **24.**34
    economic analysis, **24.**35
    effect of water shortages, **24.**34
  reuse of drainage water, **24.**31
    integration of ground and surface water use, **24.**32
    quality of water, **24.**32
Irrigation structures, **26.**1
  canals, **26.**8
    capacities, **26.**10
    cross sections, **26.**10
    linings, **26.**9
      need for, **26.**8
    location, **26.**9
  conveyance structures, **26.**22
    chutes, **26.**14
    drops, **26.**17
    flume crossings, **26.**26
    general, **26.**11
    invested siphons, **26.**24
    transitions, **26.**12
  delivery structures, **26.**40
    classification of delivery structures, **26.**40
    general, **26.**40
    invariant delivery structures, **26.**41
    semivariant delivery structures, **26.**42

Irrigation structures, delivery structures (*Cont.*):
  variant delivery structures, **26.42**
  distribution system, **26.6**
  diversion weirs, **26.1**
    design, **26.2**
    type, **26.1**
  miscellaneous structures, **26.46**
    drain inlets, **26.46**
    farm and highway bridges, **26.47**
  protective structures, **26.33**
    culverts, **26.34**
    overchutes, **26.33**
    sand traps, **26.38**
    settling basins, **26.35**
  pumping installations, **26.48**
    general, **26.48**
    pumps, **26.48**
  regulating structures, **26.29**
    checks, **26.29**
    division works and intakes, **26.29**
    wasteways, **26.31**
  river intakes, **26.2**
    design of intakes, **26.5**
    sluices, **26.6**

Kainji Project river diversion, **8.9**, **8.12**
Karadj Dam, Iran, double-curvature arch dam, **10.42**
  powerhouse, **19.17**
Kentucky Dam Power Station, **19.17**
  plant, **19.24**
Khao Laem (concrete face rockfill dam) (Central, West Thailand), **14,10**, **14.24**
Khosla method, uplift and exit gradient determination, **12.13** to **12.18**
Kinematic viscosity of water relative to temperature, **2.8**
King Talal Dam:
  powerhouse, **19.26**
  spillway, **16.32**
Kinzua pumped storage development, **20.8**, **20.13**
Kittitas Main Canal, Yakima Project, Washington, **26.13**, **26.33**
Koyna Gravity Dam (India), **9.21**
Kutter Formula, **2.4**

Laboratory tests, on materials for embankment dams, **13.17**
Labyrinth spillway, **3.10**
Lacey regime channel formulas, **6.6**

Lacey regime formula, for width, in barrages and dams on permeable foundations, **12.4**
Lacey silt factor, f_{VR}, in Harza method, regime canals, **6.7**
Lacey width, in barrages and dams on permeable foundations, **12.4**
Lake Brazos Project, Texas, **17.5**
Lake Chessman arch dam, Colorado, **10.1**
Land classification for irrigation, **24.1** to **24.5**
LG-2 Dam river diversion, **8.7**, **8.10**, **8.13** to **8.15**
Liang and Wang air concentration, **16.37**
Lining:
  joints and grooves in, in canals and conduits, **7.21** to **7.23**
  minimum reinforcement, in canals, and conduits, **7.20**
  thickness requirements, in canals and conduits, **7.20**
  suggested thicknesses (Table 8), **7.21**
Lining of earth canals, **7.19** to **7.20**
  history and progress of lining, **7.20**
  purposes of lining, **7.19** to **7.20**
Loads and arch dams, **10.4**
Lock and Dam 6 on Arkansas River, river diversion, **8.20**, **8.21**
Lock hydraulics, **23.4**
Locks:
  on Ohio and Mississippi Rivers, **23.4**
  recently constructed, table of, **23.29**
Long-term flow sequence generation, **1.29**
Looseness factor, in barrages and dams on permeable foundations, **12.4**
Losses:
  bend (*see* Bend losses)
  from canals, **7.7**
  in closed conduits bends, **2.15**
  from concrete, metal and wood conduits, **7.7**
  curvature, form and other, **5.4**
  minor, in valves and fittings, **2.16**, **2.18**
  pier (*see* Pier losses)

Macagua concrete face rockfill dam (Puerto Ordaz, Venezuela), **14.23**
Main Canal Headworks powerplant, **19.17**
Mangla Dam:
  intake and trashrack, **19.10**
  river diversion, **8.17**
  spillway, **16.14**, **16.16**

Manning equation, **28.**21
Manning formula, **2.**3 to **2.**4
   charts for solution of, **2.**5, **2.**6
   used for backwater curve, **5.**9
Manning head loss equation, **27.**13
Manning's n, in Harza method, regime canals, **6.**7
Marginal embankments, in barrages and dams on permeable foundations, **12.**22
Mass, conservation of, in closed conduits, **2.**1
Materials and installation of drains, **25.**17
Maximization and transposition of storms, **1.**18
Mayfield Dam river diversion, **8.**17
Melvin Price Locks and Dam river diversion, **8.**5
Method of fragments, in barrages and dams on permeable foundations, **12.**11
Method of independent variables, in barrages and dams on permeable foundations, **12.**11
Millers Ferry Lock and Dam river diversion, **8.**22
Miscellaneous structures for irrigation, **26.**46
Model investigation of arch dams, **10.**35
Model scaling criteria (*see* Hydraulic models)
Model testing, new types of gates and valves, **17.**48
Models, hydraulic (*see* Hydraulic models)
Models, structure (*see* Hydraulic models: physical types)
Momentum and energy, conservation of, in closed conduits, **2.**1
Monitoring and performance evaluation, embankment dams, **13.**60
Monorail cranes, **17.**73
Moody Chart, **2.**10, **2.**11
Moose River powerplant, **19.**27
Moritz formula for seepage losses, **6.**14
Morning glory spillway, **3.**8
   shaft and tunnel spillways, **16.**41
Mosel River hydroelectric stations, **19.**24, **19.**28
Moses-Saunders hydroelectric plant, **19.**17
Mossyrock Dam, Washington:
   double-curvature arch dam, **10.**42
   river diversion, **8.**1
   spillway plunge pool, **16.**46
Multilevel outlet structures for temperature control, **18.**23

n, values of for natural streams, **5.**3
Natural channels, **5.**1 to **5.**18
   backwater curves, **5.**9 to **5.**11
   determination of discharge, **5.**6 to **5.**8
   flow characteristics, **5.**1
   flow routing, **5.**11 to **5.**18
   hydraulics of natural streams, **5.**1 to **5.**6
   introduction, **5.**1
   streamflow records, **5.**8 to **5.**9
Natural streams, hydraulics of **5.**1
   Bernoulli's theorem, **5.**1
   curvature, form factors, and other losses, **5.**4
   discharge formulas, **5.**1
   roughness factor, **5.**2
      effective roughness, **5.**3
      values of n for, **5.**3
      Table 1, **5.**4
      Table 2, **5.**5
Navigation locks, **23.**1
   equipment, **23.**18
      auxiliary mechanical equipment, **23.**27
         compressed air system, **23.**28
         drainage pumps, **23.**27
         emergency diesel generator, **23.**28
         fire protection equipment, **23.**27
         unwatering pumps, **23.**27
      control and signaling equipment, **23.**28
      crane equipment, **23.**21
      gate equipment, **23.**18
         miter gates, **23.**18
         reverse tainter gates, **23.**19
         wheel gates, **23.**21
      mooring equipment, **23.**25
      safety barriers, **23.**23
   general, **23.**1
   lock examples, **23.**30
      recently constructed navigation locks, **23.**29
   lock hydraulics, **23.**4
      air entrainment and cavitation, **23.**14
      general, **23.**4
      lock filling and emptying, **23.**6
      equations for determining fill and emptying times, **23.**6 to **23.**14
      water management, **23.**17
   lock layouts, **23.**1
      lock dimensions, **23.**3
Needle valves, **17.**32
Net reservoir evaporation, **1.**11
Niagara River pumped storage plant, **20.**3
Nikuradse, criterion, **2.**9

Nurek Dam in USSR, **9.**1

Okinawa seawater pumped storage project, **20.**12
Open-channels, **2.**19 to **2.**33
  backwater curves, **2.**23
  critical flow, **2.**19
  critical velocity, **2.**21
  flow formulas and frictional resistance, **2.**23
  Froude number, **2.**21
  hydraulic jump:
    on horizontal floor, **2.**28
    on sloping apron, **2.**29
  negative surges, **2.**28
  sharp crested weirs, **2.**20
  solitary waves, **2.**26
  steady flow:
    nonuniform, **2.**23
    uniform, **2.**22
  surges in power channels, **2.**26
  typical wave velocities, **2.**26
  unsteady flow, **2.**23
  wave profiles and velocities, **2.**25
Operation and maintenance, gates and valves, **17.**66
Operation of pumped storage, **20.**22
Ord River Dam river diversion, **8.**10
Organic growth in closed conduits, **2.**12
Orientation, in barrages and dams on permeable foundations, **12.**2
Orifices, closed conduits, **2.**18 to **2.**19
  high head, **2.**18
  low head, **2.**19
Orographic rainfall, **1.**12
O'Shaughnessy Dam siphon spillway, **16.**45
Overchute on All-American Canal, Yuma, Arizona, **26.**39
Overfall spillways, **16.**2
Overhead cranes, **17.**73

Partial duration services, **1.**2
Pathfinder Dam, Wyoming, slide gates, **17.**2
Penman-Monteith method for estimating evapotranspiration, **24.**7
Performance evaluation, embankment dams, **13.**60
Performance of concrete face rockfill dams, **14.**15
Performance of pumped storage, **20.**22
Philadelphia powerplant, **19.**29

Physical properties of roller-compacted concrete, **15.**5
Pier losses, in canals and conduits, **7.**9 to **7.**10
Pier-slot design studies, Bureau of Reclamation, **16.**12
Piers, in barrages and dams on permeable foundations, **12.**5
Pilot Canal, Riverton Project, Wyoming, **26.**18
Piney Hydroelectric Project (Penn.), dam rehabilitation, **11.**15
Pipes, computing flow in, **2.**4
Point rainfall, **1.**13
Pontalto Arch Dam, Austria, built in 1611, **10.**1
Portland-cement concrete lining, **7.**20 to **7.**23
  joints and grooves in concrete and mortar lining, **7.**21 to **7.**23
  minimum reinforcement, **7.**20 to **7.**21
  thickness requirements, **7.**20
  suggested thicknesses (Table 8), **7.**21
Post-tensioned anchors, gravity dams, **9.**28
Post-tensioning (*see* Prestressing and post-tensioning)
Potable water requirements, **27.**1
Power from flowing water, **19.**1
Powerhouse structures, **19.**16
Precipitation, adjustment of record, **1.**2
  data sources, **1.**1
  definitions, **1.**1
  frequency analysis, **1.**2
  geographic distribution, **1.**3
  measurement, **1.**1
  missing records, **1.**2
  network design **1.**1
  seasonal variation, **1.**5
Prestressed/post-tensioned dams, examples **11.**12, **11.**14
Prestressing and post-tensioning, **11.**1
  analysis procedures and calculating loads, **11.**2
  basic mechanical methods, **11.**1
  bond length determination, **11.**3
  design criteria for dams and gravity structures, **11.**2
  foundation rock, **11.**5
  friction coefficient, **11.**3
  spacing, **11.**3
Prestressing/post-tensioning, cable and its anchorage (*see* Cable and its anchorage)

Priest Rapids, fish passing conduit, *n* values, **2.**12
Probability distributions, **1.**2
Probable maximum flood, derivations, **1.**39
Probable maximum precipitation, data sources, **1.**15
  definition, **1.**15
  generalized estimates, **1.**17
  maximization, **1.**15
  selection, **1.**15
  site-specific, **1.**18
  statistical procedures, **1.**20
  time distribution, **1.**20
  transportation, **1.**18
Protective structures for irrigation, **26.**33
Pump-turbines, reversible (*see* Hydraulic machinery: reversible pump-turbines)
Pumped storage **20.**1
  basic concepts, **20.**1
    current developments, **20.**3
    historical background, **20.**3
    pumped storage concept, **20.**1
    significant achievements, **20.**3
  costs and economics, **20.**28
    computer models for economics analysis, **20.**32
    cost-estimating curves, **20.**29
    economic analysis, **20.**31
    economics, **20.**31
    historical costs, **20.**28
    implementation and construction schedule, **20.**30
    introduction, **20.**28
    parameters affecting costs, **20.**29
  developing pumped storage, **20.**35
    feasibility studies, **20.**35
      phases, **20.**35
      study areas, **20.**35
    introduction, **20.**35
  elements of, **20.**6
    design for safety and reliability, **20.**22
    introduction, **20.**6
    lower reservoir, **20.**9
      seawater pumped storage, **20.**10
      underground pumped storage, **20.**10
    operating heads and cycling storage, **20.**6
    other power-plant equipment, **20.**20
      mechanical equipment, **20.**21
      other electrical equipment, **20.**21
      starting equipment, **20.**20
    physical setting, **20.**7

Pumped storage elements of (*Cont.*):
    power and energy equations, **20.**6
    principal physical elements, **20.**7
    project arrangement, **20.**14
    pumping and generating equipment, **20.**16
      generator-motor, **20.**20
      pump-turbine, **20.**17
    upper reservoir, **20.**8
      enclosed ring dikes, **20.**8
      low-level outlets, **20.**9
      spillways, **20.**9
    water-conductor system, **20.**10
      control of water-conductor system, **20.**14
      intakes, **20.**12
  rating, performance, and operation, **20.**22
    capacity and capacity credit, **20.**25
      capability credit, **20.**26
      dependable capacity, **20.**25
      energy generation, **20.**26
      integrated operation in utility system, **20.**26
      operating characteristics of generating plant, **20.**26
    introduction, **20.**22
    performance, **20.**23
      cycling and load following capability, **20.**23
      system regulation, **20.**25
    rating, **20.**22
Pumping for drainage, **25.**19
Pumping installations for irrigation, **26.**48
Pumps (*see* Hydraulic machinery: pumps)

Radial gates, **17.**3, **17.**24
  trunnion support and anchorage, **17.**56
Raft design, in barrages and dams on permeable foundations, **12.**8
Ramser stormwater runoff estimating formula, **25.**8
Rating, performance, and operating pumped storage, **20.**22
Regime canals, **6.**1 to **6.**17
  channels in alluvium, **6.**1 to **6.**5
  design of regime canals, **6.**15 to **6.**16
  effects of sediment and seepage, **6.**10 to **6.**15
  theory of regime channels, **6.**6 to **6.**10
Regime channels, **6.**1
  formulas, **6.**6
Regulating structures for irrigation, **26.**29

Regulators, in barrages and dams on permeable foundations, **12.21**
Rehabilitation of dams, **11.14**
  concrete dams, **11.14**
  embankment dams, **11.16**
  freeboard, **11.19, 11.21**
  laboratory test standards for investigation, **11.15**
  types of repairs, **11.16**
Rehabilitation of spillways, **11.16**
  embankment stability, **11.21**
  flip-bucket spillway, **11.21**
  fuse plugs, **11.23**
  inflow design flood, **11.21**
  National Dam Inspection Act, **11.21**
  PMF, **11.19, 11.21**
  PMP, **11.19, 11.21**
  roller-compacted concrete, **11.22**
  seepage analysis, **11.21**
  spillway capacity, **11.22**
  spillway expansion methods, **11.21**
  typical methods used in concrete crack repair, **11.20**
Reservoir flood routing (*see* Flood routing, reservoir)
Reservoir hydraulics, **4.1 to 4.19**
  analysis of water availability, **4.8**
  analysis of water requirements, **4.8**
  basic data for reservoir planning, **4.2**
  introduction, **4.1**
  reservoir evaporation, **4.8**
  reservoir flood routing, **4.16**
  reservoir selection withdrawal, **4.17**
  reservoir wave action, **4.10**
  sediment-storage requirements, **4.3**
Reservoir operation analysis, tabular, **4.9**
Reservoir planning, basic data, **4.2**
  climatological data, **4.2**
  reservoir area-volume curves, **4.2**
  sedimentation, **4.2**
  streamflow, **4.2**
Reservoir selective withdrawal, in reservoir hydraulics, **4.17**
  need for selective withdrawal, **4.17**
  reservoir thermal hydraulic and water quality modelling, **4.18**
Reservoir thermal hydraulic and water quality modelling, **4.18**
  HARZA/DYRESM dynamic reservoir simulation technique, **4.18**
Reservoir wave action, in reservoir hydraulics, **4.10**

Reservoir wave action, in reservoir hydraulics (*Cont.*):
  basic assumptions, **4.11**
  fetch, **4.12**
  wind, **4.11**
  freeboard allowances, **4.10**
  total allowance for wave action, **4.15**
  wave height and other characteristics, **4.12**
  wave run-up on slopes, **4.14**
  wind tide, **4.12**
Resistance (*see* Losses or flow formulas)
Results from irrigation, **24.33**
Retrogression, in barrages and dams on permeable foundations, **12.3**
Reuse of drainage water, **24.31**
Reversible pump-turbines (*see* Hydraulic machinery: reversible pump-turbines)
Reynolds number, **2.6 to 2.8**
Risk evaluation, **1.38**
River channel models (*see* Hydraulic models)
River diversion, **8.1**
  cofferdams, **8.17**
    Lock and Dam 6 on Arkansas River **8.20**
    Melvin price locks and dam river diversion, **8.19**
    Millers Ferry lock and dam river diversion, **8.22**
  diversion discharge capacities, **8.13**
  diversion in narrow and deep channels, **8.1**
    Boulder Dam river diversion, **8.1**
    Mossyrock Dam river diversion, **8.1**
    San Lorenzo Dam river diversion, **8.1**
    Wynoochee Dam river diversion, **8.2**
  diversion in wide stream channels, **8.3**
    Caruachi Dam river diversion, **8.3**
    LG2 Dam river diversion, **8.7**
    Melvin Price locks and dam, **8.4**
    rock material for diversion dikes, **8.7**
    steel sheetpile cellular cofferdams, **8.8**
    Yacyreta Dam river diversion, **8.4**
  diversion over and around embankment dams, **8.10**
    LG2 Dam river diversion, **8.10**
    Ord River Dam river diversion, **8.10**
  diversion through concrete dam powerhouses, **8.9**
    Fort Loudoun Dam river diversion, **8.9**

River diversion, diversion through concrete dam powerhouses (*Cont.*):
  Wanapum Dam river diversion, **8.9**
  diversion through concrete dams, spillways, **8.8**
    Bluestone Dam river diversion, **8.8**
    Caruachi Dam river diversion, **8.9**
    Chief Joseph Dam river diversion, **8.8**
  diversion tunnels, **8.15**
    Dworshak Dam, **8.15**, **8.17**
    Mangla Dam river diversion, **8.17**
    Mayfield Dam river diversion, **8.17**
River intakes for irrigation, **26.2**
Robert Moses Niagara powerplant, **19.22**
Rocky River pumped storage facility, **20.3**
Roller-compacted concrete (RCC) dams, **15.1**
  design of, **15.8**
    appurtenant structures, **15.18**
    design loads, **15.11**
    design philosophy, **15.8**
    factors of safety, **15.13**
    foundation properties, **15.9**
    galleries and adits, **15.18**
    loading combinations, **15.12**
    overturning stability (stress analysis), **15.13**
    seepage control, **15.14**
      extruded concrete curb, **15.16**
      precast panel with membrane liner, **15.15**
      reinforced concrete, **15.15**
      unreinforced concrete, **15.15**
    sliding stability, **15.13**
    structural stability, **15.10**
    summary, **15.19**
    thermal studies, **15.13**
  gravity dams, **9.28**
  physical properties of RCC, **15.5**
    durability, **15.7**
    permeability, **15.7**
    Poisson's ratio, **15.7**
    strength, **15.6**
    thermal properties, **15.8**
    unit weight, **15.8**
    volume changes, **15.8**
  types of RCC dams, **15.1**
    high-paste-content, **15.3**
    Japanese method type, **15.5**
    lean, **15.2**
Roller-mounted gates, **17.18**
Roosevelt Dam, Arizona, slide gates, **17.2**

Roseires Dam spillway, **16.14**
Rotary drum screens, **18.18**
Roughness factor, in hydraulics of natural streams, **5.2**
Rubber seals, **17.53**
Runoff, data sources, **1.24**
  data collection and transmission, **1.24**
  definition, **1.23**
  generation, **1.24**
  long-term variation **1.29**
  measurement, **1.24**
  missing record, **1.29**
  nature of, **1.23**
  river stages, **1.24**
  seasonal variation, **1.26**
  stage-discharge relationship, **1.26**
  units, **1.23**
  utilization studies, **1.29**

Salinity, **25.5**
Salvajina concrete face rockfill dam, **14.10**
San Lorenzo Dam river diversion, **8.1**
Sanitary sewer systems, **28.2**
Satellite imagery, **1.1**
Scour, in barrages and dams on permeable foundations, **12.8**
Scour holes, in barrages and dams on permeable foundations, **12.11**
Scour protection below overfall dams, **16.46**
Screens:
  Coanda, **18.18**
  Eicher, **18.18**, **18.20**
  rotary drum, **18.18**
  vertical traveling, **18.18**
Seals for gates and valves, **17.52**
Sediment and seepage, effects of, **6.10** to **6.15**
  sediment control, **6.13**
    desilting works, All-American Canal, **6.14**
    sediment excluders, **6.14**
  sediment transport, **6.10**
    fall velocity for quartz spheres, **6.12**
    sediment grade scale, **6.11**
    sediment size distribution, **6.12**
    transport capability of a channel, **6.13**
  seepage losses, **6.14**
    average observed losses in earth canals, **6.15**
    Moritz formula, **6.14**
Sediment control, in regime canals, **6.13**

Sediment excluders, in regime canals, **6.**14
Sediment rating curve, **4.**4
Sediment transport, in regime canals, **6.**10
Sedimentation, reservoir **4.**2
    rate, **4.**3
Sedimentation storage requirements, in reservoir hydraulics, **4.**3
    distribution of sediment deposits, **4.**8
    rate of sedimentation, **4.**3
        density of deposited sediments, **4.**6
        sedimentation inflow, **4.**3
        trap efficiency, **4.**5
Sediments, deposited, density, **4.**6 to **4.**8
Seepage analysis and control, embankment dams, **13.**18
    (*See also* Sediment and seepage, effects of)
Seepage from canals, **7.**23 to **7.**24
    lined canals, **7.**24
    table, **7.**23
    unlined canals, **7.**24
    losses from, **6.**14
    statistical losses (Table 9), **7.**22
Segredo concrete face rockfill dam, Brazil, **14.**28
Seismic loading on gravity dams, **9.**20
Seismic zone map, U.S. Army Corps of Engineers, **9.**20
Selective withdrawal, reservoir, **4.**17
    HARZA/DYRESM dynamic reservoir simulation technique, **4.**18
    need for, **4.**17
    reservoir thermal hydraulic and water quality monitoring, **4.**18
Settlement analysis, embankment dams, **13.**54
Sewage and wastewater treatment (*see* Wastewater collection and conveyance: sewage and wastewater treatment)
Sewage pumping stations (*see* Wastewater collection and conveyance: sewage pumping stations)
Sewage quantities (*see* Wastewater collection and conveyance: sewage quantities)
Sewer-system design (*see* Wastewater collection and conveyance: sewer-system design)
Shape, best hydraulic, in canals and conduits, **7.**1 to **7.**2
Shasta Dam, tube valves, **17.**2, **17.**17
Sheetpiling, steel, in barrages and dams on permeable foundations, **12.**10

Shepang Dam (Conn.), second-stage grouted type anchors, **11.**12
Side channel spillway, **3.**8
Side channel spillways, **16.**38
Siphon spillways, **16.**44
Site-specific PMP, **1.**18
Slide gates, **17.**11
Sliding factor, in barrages and dams on permeable foundations, **12.**20
Slope protection and freeboard, embankment dams, **13.**56
Slopes, bank (*see* Bank slopes)
Snow, data source, **1.**5
    definitions, **1.**5
    distribution, **1.**5
Snowmelt, estimation, **1.**5
    forecast, **1.**8
Soils, **25.**3
Sources of waters, **25.**2
South Canal, Succor Creek Division, Owymee Project, Idaho, **26.**30, **26.**37
Southwestern drainage formula, **25.**8
Sphere valves, **17.**45
Spillway and streamed protection works, **16.**1
    chute (or trough) spillways, **16.**16
        features, **16.**16
        floor slabs, **16.**24
        principles of design, **16.**19
        side walls, **16.**23
    discharge capacity, **16.**1
        approach flow conditions, **16.**1
        design flood, **16.**1
    gated and orifice spillway, **16.**9
        design of piers, **16.**16
        gate-controller ogee crests, **16.**9
        orifice spillway, **16.**13
    incipient cavitation and damage experience, **16.**26
    aerator spacing, **16.**36
    air supply system, **16.**33
    cavitation potential, **16.**26
    design procedure, **16.**37
    estimation of air entrainment capacity of the free jet, **16.**35
    examples of aeration device design, **16.**38
    finish requirement and surface treatment, **16.**29
    location of aerator devices, **16.**31
    need for aeration, **16.**30
    ramp design, **16.**31
    types of aerators, **16.**31

Spillway and streamed protection works (*Cont.*):
　morning-glory shaft and tunnel spillways, **16**.41
　　general, **16**.41
　　hydraulics, **16**.43
　　　with crest control, **16**.43
　　　with full pipe flow, **16**.43
　　　with tube or orifice flow, **16**.43
　　typical morning-glory spillway, **16**.41
　　typical tunnel spillway, **16**.44
　overfall spillways, **16**.2
　　overfall spillway, discharge coefficients, **16**.8
　　shape of crest, **16**.2
　scour protection below overfall dams, **16**.46
　　deflector buckets, **16**.47
　　plunge pools, **16**.46
　　stilling basins, **16**.47, **16**.49, **16**.50
　side channel spillways, **16**.38
　　flow characteristics, **16**.40
　　general, **16**.38
　siphon spillways, **16**.44
　　general, **16**.44
　spillway crests, **16**.2
　　types, **16**.2
　spillway design considerating cavitation and aeration, **16**.24
　　cavitation, **16**.25
　　introduction, **16**.24
Spillway crests, **16**.2
Spillway design considering cavitation and aeration, **16**.24
Spillway expansion methods, **11**.21
Spillway gates, **17**.3
Spillway models, three-dimensional (*see* Hydraulic models)
Spillway models, two-dimensional (*see* Hydraulic models)
Spillways, rehabilitation (*see* Rehabilitation of spillways)
Spoil banks and berms, in canals and conduits, **7**.16
Staats-Hornsby hollow-jet valves, **17**.28, **17**.37
Stability analysis:
　criteria applicable, **9**.14
　design of concrete dams, **9**.9
　embankment dams, **13**.38
Stable channels, **6**.1
Standard project flood, **1**.40

Statistical procedures for PMP, **1**.20
Steady nonuniform flow in open channels, **2**.23
Steady uniform flow in open channels, **2**.22
Stewart Mountain Dam, AZ, post-tensioning a thin-arch dam, **11**.16
Stone aprons, in barrages and dams on permeable foundations, **12**.9
Stoplongs (*see* Gates and valves: gates and bulkheads)
Storage, reservoir, definitions, **4**.1
Storm rainfall, causes, **1**.12
　basin average, **1**.12, **1**.13
　classification, **1**.12
　cyclones, **1**.12
　distribution, **1**.12
　storm type, **1**.12
　terrain influence, **1**.14
Storm sewer systems, **28**.2
Stormwater Management Model (SWMM), **28**.11, **28**.32
Stormwater runoff:
　quantities, **28**.11
　Ramser estimating formula, **25**.8
Streamflow models, **1**.33
Streamflow records, **5**.8
　gage height records and recorders, **5**.8
　stage-discharge relationships, **5**.8
Stress distribution in arch dams, **10**.5
Strontia Springs Arch Dam, Colorado, **10**.42
Structure models (*see* Hydraulic models)
Submerged roller buckets, **3**.8
Submergence and vortices, Gulliver Data, **16**.16
Subsurface drains, **25**.10
Subsurface investigations, embankment dams, **13**.9
Sudden contraction, in closed conduits, **2**.14
Sudden enlargements in closed conduits, **2**.13
Surface drains, **25**.8
Surges:
　in power channels, **2**.26, **2**.28
　negative, **2**.28
Sweetwater Arch Dam, California, built in 1888, **10**.1
Swimming speeds of some adult game fish, **18**.9

Taum Sauk pumped storage plant, **20**.3
Temperature and crack control, gravity dams, **9**.22

Tennessee Valley Authority spillways, **16.**9
Theory:
  of regime channels, **6.**6 to **6.**10
    formulas, **6.**6
      Harza method, **6.**7 to **6.**10
      Kennedy's formula, **6.**6
      Lacey's formulas, **6.**6
      Lindley's regime concept, **6.**6
    general, of arch dams, **10.**2
Thermal limits and preferred temperatures for selected North American game fish, **18.**8
Three Gorges Project, China, **9.**5
Time distribution of PMP, **1.**20
Tractive force, **6.**2
Transient flow, differential equation for, **22.**7
Trap efficiency, reservoir, **4.**5
Trial-load analysis method for arch dams, **10.**14
Trinity Dam, California, jet flow gate, **17.**17
Trough spillways (*see* Chute spillways)
Trunnion support for radial-gate, **17.**56
Turbine speed governors (*see* Hydraulic machinery: turbine speed governors)
Turbines (*see* Hydraulic machinery: turbines)
Typical fish passages layout, **18.**11

Underground powerhouses with caverns 21 meters or wider, **19.**27
Unit hydrograph, assumptions, **1.**30
  concept of application, **1.**30
  definition, **1.**30
  derivation methods, **1.**32 to **1.**33
University of California, investigate seismic behavior
University of Colorado, research, uplift and drain holes, **9.**17
Unsteady flow in open channels, **2.**23 to **2.**25
Uplift diagram, in barrages and dams on permeable foundations, **12.**13
Uplift pressure, gravity dam, conventional assumption, **9.**15
Uplift reduction and buttress dams, **9.**18
Uplift, in barrages and dams on permeable foundations, **12.**11
Upper Otay Arch Dam, California, built in 1900, **10.**1
Upper Stillwater roller-compacted concrete dam, **15.**1, **15.**16
Uribante-Doradas powerplant, **19.**22

U.S. Army Corps of Engineers:
  hydraulic design criteria, **16.**23
  seismic zone map, **9.**20
  Willow Creek roller-compacted concrete dam, **15.**1
U.S. Bureau of Reclamation, Upper Stillwater roller-compacted concrete dam, **15.**1
U.S. Fish and Wildlife Service, **18.**1, **18.**4, **18.**10, **18.**14
USBR allowable temperature drop table, **9.**23

Valves (*see* Gates and valves)
Velocities, permissible in canals and conduits, **7.**2 to **7.**7
  Table 3, **7.**6
Velocity:
  average, **2.**1
  Chezy formula, **2.**4
  critical, **2.**21
  distribution in laminar flow, **2.**9
  Hazen-Williams formula, **2.**7
  head, **2.**3
  Manning formula, **2.**4
  shear or friction, **2.**8
  temporal mean, **2.**1
  wave, **2.**25, **2.**26
Vertical traveling screens, **18.**18
von Karman and Prandtl resistance equation, **2.**10

Walls, flank, in barrages and dams on permeable foundations, **12.**21
Walski equations for analyzing complex distribution networks, **27.**36
Wanapum Dam:
  powerhouse, **19.**17
  river diversion, **8.**9
  spillway stilling basin, **16.**48
Wastewater collection and conveyance, **28.**1
  hydraulics of sewers, **28.**18
    basic hydraulic considerations, **28.**18
    flow friction formulas, **28.**21
    general, **28.**18
    hydraulic-elements graphs, **28.**26
    losses at bends, **28.**26
    losses at manholes, **28.**24
    minor losses, **28.**24
    self-cleansing velocities, **28.**27
    storm-water inlets, **28.**28

Wastewater collection and conveyance, (*Cont.*):
  sewage and wastewater treatment, 28.46
    activated-sludge process, 28.58
    biological treatment processes, 28.57
    conduit losses, 28.48
    grit chambers, 28.51
    objectives of sewage and wastewater treatment, 28.46
    outfall, 28.60
    plant hydraulics, 28.46
    sedimentation tanks, 28.54
    sludge handling, 28.61
    summary, 28.63
    trickling filters, 28.57
  sewage pumping stations, 28.40
    basic concepts, 28.40
    multiple-pump operating, 28.42
    pump head curves, 28.42
    system head curves, 28.41
    system hydraulics, 28.40
  sewage quantities, 28.2
    design example for calculating wastewater flows, 28.8
    design period, 28.3
    flow variations, 28.6
    industrial and commercial wastewater flow rates, 28.5
    infiltration, 28.5
    overview, 28.2
    per capita sewage flow rate, 28.4
    population estimates, 28.3
  sewer-system design, 28.29
    detailed survey, 28.30
    examples of computations, 28.32
    general approach, 28.29
    organization of computations, 28.32
    preliminary investigations, 28.30
    sewer design, 28.31
  storm-water runoff quantities, 28.11
    application of rational method, 28.16
    area to be served, 28.12
    bases of storm-sewer design, 28.11
    rainfall intensity, 28.13
    rational method, 28.12
    runoff coefficient, 28.15
    time of concentration, 28.14
    urban runoff models, 28.16
Wastewater flows:
  from commercial sources, 28.5
  from recreational sources, 28.6
  from residential sources, 28.4

Water availability, analysis of, in reservoir hydraulics, 4.8
  flow-duration analyses, 4.9
  mass-curve analysis, 4.9
  tabular reservoir-operation analysis, 4.9
Water conductors, 19.7
Water distribution and treatment, 27.1
  hydraulic analysis of water-distribution systems, 27.31
    algorithm for solution of network problem using a computer, 27.37
    application of distribution-system computer models, 27.40
    basic pipe network equations, 27.31
    computer models of distribution-system networks, 27.40
    formulation of equations for distribution networks, 27.36
    interpretation of distribution-system model results, 27.42
    objections, 27.31
    series and parallel pipe problems, 27.35
  hydraulics of water-distribution systems, 27.25
    components of a distribution systems, 27.25
    conditions for assessing systems capacity, 27.28
    examples of distribution-systems operating criteria, 27.29
    operating criteria for water distribution on networks, 27.28
    the piezometric surface-hydraulic grade line hydraulics, 27.26
  hydraulics of water-treatment systems, 27.9
    basin hydraulics, 27.21
    hydraulic design criteria for water treatment plants, 27.10
    hydraulic profile through a water treatment plant, 27.11
    inlet and outlet devices, 27.23
    launders, 27.18
    open-channel flow, 27.14
    pressure flow through pipes, 27.13
    water-treatment plant design, 27.9
    water-treatment systems, 27.10
    weirs, 27.14
  introduction, 27.1
  potable water requirements, 27.1
    demand, consumption and unaccounted for water, 27.1

Water distribution and treatment, potable water requirements (*Cont.*):
   estimating water use, **27.**3
   fire-fighting demands, **27.**6
   importance of demand projections, **27.**8
   types of water users, **27.**2
   typical water usage, **27.**2
   variations in water use, **27.**5
  water-distribution pumping systems, **27.**42
   pumping station hydraulics, **27.**43
   types of pumping installations, **27.**43
  water-distribution storage facilities, **27.**44
   additional design considerations, **27.**48
   capacity of storage, **27.**46
   classification of storage facilities, **27.**44
   height of storage facility, **27.**47
   purpose of storage, **27.**45
Water hammer equation, **22.**2
Water tables, **25.**7
Water-distribution pumping systems, **27.**42
Water-distribution storage facilities, **27.**44
WATSTORE, **1.**25
Wave:
  celerity, **2.**25, **2.**26
  height and other characteristics, **4.**12
  negative, **2.**25
  positive, **2.**25
  profiles, **2.**25

Wave (*Cont.*):
  run-up on slopes, **4.**14
  solitary, **2.**26
  velocities, **2.**25, **2.**26
Wave action:
  reservoir, **4.**10
  total allowance for **4.**15
Wave speed in tunnels and conduits, **22.**4
Weber-Provo diversion canal, Salt Lake Basin Project, Utah, **26.**7
Wei and DeFazio free jet flow, **16.**32
Wheel-mounted gates, **17.**18
Width, barrage, in barrages and dams on permeable foundations, **12.**4
Width, bays, in barrages and dams on permeable foundations, **12.**5
Willow Creek roller-compacted concrete dam, **15.**1
Wilson Dam spillway, **16.**12
Wind tide, **4.**12
Winfield Project, Corps of Engineers, radial gate hoisting system, **17.**61
World's highest concrete face rockfill dams, **14.**32
Wynoochee Dam river diversion, **8.**2

Xu and Gu free air jet problem, **16.**33

Yacyreta Dam river diversion, **8.**4
Yacyreta project navigation locks, **23.**2